FRONTIERS IN ELECTROMAGNETICS

IEEE Press
445 Hoes Lane, P.O. Box 1331
Piscataway, NJ 08855-1331

IEEE Antennas & Propagation Society, *Sponsor*
AP-S Liaison to IEEE Press, Robert Mailloux
IEEE Microwave Theory and Techniques Society, *Sponsor*
MTT-S Liaison to IEEE Press, Lawrence Dunleavy

Cover design: Caryl Silvers, *Silvers Design*

Technical Reviewers

Dr. Andreas Cangellaris, *University of Illinois*
Dr. Donald G. Dudley, *University of Arizona*
Dr. James W. Mintz, *North Carolina State University*
Dr. John Owens, *Auburn University, AL*

FRONTIERS IN ELECTROMAGNETICS

Edited by

Douglas H. Werner
Department of Electrical Engineering and
Applied Research Laboratory
The Pennsylvania State University

Raj Mittra
Department of Electrical Engineering
The Pennsylvania State University

IEEE Antennas & Propagation Society, *Sponsor*

IEEE Microwave Theory and Techniques Society, *Sponsor*

MTTS ®

IEEE
PRESS

IEEE Press Series on Microwave Technology and RF
Richard Booton and Roger Pollard, *Series Editors*

The Institute of Electrical and Electronics Engineers, Inc., New York

This book and other books may be purchased at a discount
from the publisher when ordered in bulk quantities. Contact:

IEEE Press Marketing
Attn: Special Sales
Piscataway, NJ 08855-1331
Fax: (732) 981-9334

For more information about IEEE Press products,
visit the IEEE Press Home Page: http://www.ieee.org/press

Printed in the United States of America

10 9 8 7 6 5 4 3 2 1

ISBN 0-7803-4701-3

IEEE Order No. PC5754

Library of Congress Cataloging-in-Publication Data

Frontiers in electromagnetics / edited by Douglas H. Werner, Raj Mittra.
p. cm. -- (IEEE Press series on Microwave Technology and RF)
"IEEE Microwave Theory and Techniques Society, sponsor. IEEE Antennas and Propagation Society, sponsor."
Includes bibliographical references and index.
ISBN 0-7803-4701-3
1. Electromagnetism. I. Werner, Douglas H., 1960– . II. Mittra, Raj.
QC760.25.F76 1999
537—dc21 99-34950
CIP

We dedicate this book to our friend and colleague,
Professor Anthony J. Ferraro of The Pennsylvania State University,
who has served as a mentor and an inspiration to many undergraduate as well as
graduate students during his long and distinguished career in academia.

CONTENTS

PART II OPTIMIZATION AND ESTIMATION

CHAPTER 15 FAST CALCULATION OF INTERCONNECT CAPACITANCES USING THE FINITE DIFFERENCE MODEL APPLIED IN CONJUNCTION WITH THE PERFECTLY MATCHED LAYER (PML) APPROACH FOR MESH TRUNCATION 644

Vladimir Veremey and Raj Mittra

CHAPTER 16 FINITE-DIFFERENCE TIME-DOMAIN METHODOLOGIES FOR ELECTROMAGNETIC WAVE PROPAGATION IN COMPLEX MEDIA 666

Jeffrey L. Young

PREFACE

The topics covered in this book are all relatively new and emerging areas of research in the field of electromagnetics. These topics were carefully selected not only because of their innovative nature, but also because they have the potential to make a significant impact on future directions in electromagnetics research. The chapters are designed to be as self-contained as possible with ample references provided for the benefit of interested readers. Many chapters also contain a brief tutorial intended to acquaint the unfamiliar reader with the mathematical foundations and fundamental concepts which form the basis for the more advanced material that follows.

The book contains 18 chapters that are organized into four sections. The first section (Chapters 1–7) addresses recent progress toward combining electromagnetic theory with concepts originating from several branches of mathematics including geometry, topology, and groups. State-of-the-art techniques in electromagnetic optimization and estimation are discussed in the second section of the book (Chapters 8–10). A variety of new developments in analytical and numerical methods for solving electromagnetics problems are considered in sections three (Chapters 11–13) and four (Chapters 14–18), respectively.

Fractal electrodynamics is the area of research that combines fractal geometric concepts with Maxwell's theory of electromagnetism in order to study a new class of radiation, scattering, and propagation problems. Recent advancements in fractal electrodynamics research are presented in Chapters 1–3. Chapter 1 starts out with an introduction to the properties of fractals, followed by an overview of research into the fundamental nature of electromagnetic wave interactions with fractal surfaces and superlattices. Applications of fractals to the design of antenna elements and arrays are discussed in Chapters 2 and 3, respectively. Chapters 4 and 5 deal with applications of group theory to the solution of electromagnetic problems that possess certain geometrical symmetries. The impact of reciprocity and geometrical symmetry of a target on the associated scattering dyadic is considered in Chapter 4. Chapter 5 introduces a generalized theory of self-complementary structures that is based on conformal and stereographic projections. The application of some topological results from knot theory to electromagnetics is addressed in Chapters 6 and 7. In Chapter 6, particular emphasis is placed on investigating the topological features of twisted or knotted field line configurations. The electromagnetic radiation and scattering properties of thin, knotted wires are discussed in Chapter 7.

Genetic algorithms are a group of powerful optimization methods that are based on the processes of procreation and natural evolution. Chapter 8 describes some novel approaches to antenna array beamforming based on genetic algorithms and neural networks. An approach for model-order reduction, known as *model-based parameter estimation*, has been successfully applied to expedite the solution of a wide variety of computational electromagnetics problems. Chapter 9 includes a brief background discussion of model-based parameter estimation techniques followed by several examples illustrating its many practical uses in computational electromagnetics. Wavelets have received a considerable amount of recent attention for the potential advantages they offer in the solution of many electromagnetics problems. A newly developed wavelet-based method for the adaptive decomposition of electromagnetic signals into a wide range of physically meaningful mechanisms is presented in Chapter 10.

A technique for finding analytical solutions to a special class of electromagnetics problems which relies on Lommel expansions is outlined in Chapter 11. Several examples are presented in Chapter 11 including the derivation of an exact representation for the cylindrical wire kernel, and of useful near-field expansions for the circular loop antenna. Fractional calculus is the branch of mathematics that deals with a generalization of the well-known operations of differentiation and integration to non-integer orders. Chapter 12 explores applications as well as physical interpretation of non-integer order differential and integral operators in electromagnetics. A vector spherical-multipole analysis technique is presented in Chapter 13 that may be used for deriving analytical solutions to a wide range of interesting scattering and diffraction problems.

The recently introduced concept of perfectly matched layers and their application to the general problem of mesh truncation in finite methods for computational electromagnetics analysis are discussed in Chapter 14. A new method for the rapid calculation of interconnect capacitances is introduced in Chapter 15 which combines an electrostatic finite differencing scheme with a perfectly matched layer approach for mesh truncation. Chapter 16 examines the most recent and popular advances in finite-difference time-domain algorithm development for analysis of wave propagation in complex media. A new discrete mechanics approach to computational electromagnetics is introduced in Chapter 17. This new computational method offers several advantages over conventional approaches for the simulation of the interaction between electromagnetic fields and physically realistic media. Chapter 18 begins with a background discussion on the classification of bianisotropic composites, which are formed by embedding miniature complex-shaped inclusions, such as helices or omegas, in a host medium. This is followed by a more in-depth coverage of both analytical and numerical methods for modeling the electromagnetic properties of individual bianisotropic inclusions as well as composites.

Douglas H. Werner
Department of Electrical Engineering and Applied Research Laboratory
The Pennsylvania State University

Raj Mittra
Department of Electrical Engineering
The Pennsylvania State University

LIST OF CONTRIBUTORS

Nicolaos G. Alexopoulos
U.C.I.—School of Engineering
305 Rockwell Engineering Center
Irvine, CA 92697-2700

Angel Cardama Aznar
Dept. Teoria del Senyal i Comunicacions
Universitat Politecnica de Catalunya (UPC)
Modul D3, Campus Nord, UPC
c/ Jordi Girona 1-3
08034, Barcelona, Spain

Carles Puente Baliarda
Dept. Teoria del Senyal i Comunicacions
Universitat Politecnica de Catalunya (UPC)
Modul D3, Campus Nord, UPC
c/ Jordi Girona 1-3
08034, Barcelona, Spain

Carl E. Baum
PL/WSQW
3550 Aberdeen Ave., SE
Kirtland AFB, NM 87117-5776

Siegfried Blume
Department of Electrical Engineering
Ruhr-Universität Bochum
Gebäude IC-FO 05/556
Universitätsstrasse 150
D-44780, Bochum, Germany

Joseph W. Burns
Research Engineer
Environmental Research Institute of Michigan
P.O. Box 134001
Ann Arbor, MI 48113-4001

Franco De Flaviis
U.C.I. – School of Engineering
305 Rockwell Engineering Center,
Irvine, CA 92697-2700

Rodolfo E. Diaz
Arizona State University
Mechanical and Aerospace Engineering
P.O. Box 876106
Tempe, AZ 85287-6106

Nader Engheta
Moore School of Electrical Engineering
University of Pennsylvania
200 South 33rd Street
Philadelphia, PA 19104-6390

Panyaiotis V. Frangos
National Technical University of Athens
Department of Electrical and Computer Engineering
Electroscience Division
9 Iroon Polytechniou Street
GR-157 73 Zografou
Athens, Greece

Randy L. Haupt
University of Nevada
Electrical Engineering Dept./260
College of Engineering
Reno, NV 89557-0153

Aaron D. Jaggard
Department of Mathematics
University of Pennsylvania,
Philadelphia, PA 19104-6395

Dwight L. Jaggard
113 Towne/6391
School of Engineering and Applied Science
University of Pennsylvania
Philadelphia, PA 19104-6391

Ludger Klinkenbusch
Computational Electromagnetics Group, University of Kiel,
Kaiserstrasse 2, D-24143 Kiel, Germany

Mustafa Kuzuoglu
Department of Electrical Engineering
Middle East Technical University
06531, Ankara, Turkey

Frédéric Mariotte
Laboratoire PIOM
ENSCPB
33402, Talence Cedex, France

Gerald E. Marsh
Argonne National Laboratory
9700 S. Cass Avenue, Building 900
Argonne, IL 60439

Edmund K. Miller
3225 Calle Celestial
Sante Fe, NM 87501-9613

Raj Mittra
The Pennsylvania State University
Department of Electrical Engineering
319 Electrical Engineering East
University Park, PA 16802

Massimo Noro
UCLA
Department of Chemistry
405 Hilgard Avenue
Los Angeles, CA 90095-1594

Teresa H. O'Donnell
Senior Scientist
ARCON Corporation
260 Bear Hill Rd
Waltham, MA 02451

Jordi Romeu Robert
Dept. Teoria del Senyal i Comunicacions
Universitat Politecnica de Catalunya (UPC)
Modul D3, Campus Nord, UPC
c/ Jordi Girona 1-3
08034, Barcelona, Spain

Tapan K. Sarkar
Department of Electrical Engineering and Computer Science
121 Link Hall
Syracuse University
Syracuse, NY 13244-1240

Bruno Sauviac
Laboratoire PIOM
ENSCPB
33402, Talence Cedex, France

Hugh L. Southall
RL/ERAA
31 Grenier Street
Hanscom AFB, MA 01824-3010

Nikola S. Subotic
Senior Scientist
Advanced Information Systems Group
Environmental Research Institute of Michigan
P.O. Box 134001
Ann Arbor, MI 48113-4001

Sergei Tretyakov
St. Petersburg State Technical University
Radiophysics Department
195251, St. Petersburg, Russia

Vladimir V. Veremey
The Pennsylvania State University
Department of Electrical Engineering
224 Electrical Engineering East
University Park, PA 16802

Douglas H. Werner
The Pennsylvania State University
Department of Electrical Engineering
211A Electrical Engineering East
University Park, PA 16802

Pingjuan L. Werner
The Pennsylvania State University
College of Engineering
DuBois, PA 15801

Jeffrey L. Young
Department of Electrical Engineering
University of Idaho
Moscow, ID 83844-1023

Chapter 1

FRACTAL ELECTRODYNAMICS: SURFACES AND SUPERLATTICES

Dwight L. Jaggard, Aaron D. Jaggard, and Panayiotis V. Frangos

ABSTRACT

We provide a selected review of *fractal electrodynamics* by investigating the nature of electromagnetic wave interactions with fractal surfaces and superlattices. Of particular interest are the ways in which these fractal objects imprint their characteristic geometry on scattered waves and the physical basis and explanation for these interactions. We review fractal geometry, sets, and descriptors with an emphasis on the roles of dimension and lacunarity.

Since multiscale rough surfaces are often modeled by fractals, we examine here the scattering from simple fractal surfaces as a function of roughness and scattering angle using both the Kirchhoff approximation and the T-matrix approach. We find a simple relation between fractal dimension and variations in scattering cross section for small or moderate roughness. For increasing roughness we observe the role of polarization.

Next, we formulate and examine the reflection characteristics of multiscale layered media and investigate the physical explanation for the distinctive and complex reflection characteristics from these superlattices as a function of both frequency and variation in fractal descriptors. Physically motivated models of interference provide a fundamental understanding of the reflection data as contained in *twist plots*. We review recent methods of extracting fractal descriptors from superlattice scattering data using both frequency-domain and time-scale techniques.

1.1 INTRODUCTION

1.1.1 Background

Fractal electrodynamics grew from the electromagnetics research of the past several decades that examined the blending of wave concepts and fractal geometry [1–4]. The goal of this work has been to study and understand the effect of canonical multiscale structures on the scattering, radiation, and guiding of electromagnetic waves in much the same way that the effect of simple Euclidean shapes on electromagnetic waves has been studied since the late nineteenth century. Several researchers examining scattering phenomena, circa 1980, found that in select cases the fractal dimension was encoded in the scattered wave in an easily decipherable way. The intriguing possibility arose that this phenomenon might be characteristic of a large class of fractal scatterers illuminated by electromagnetic waves. In the mid-1980s we set out to see if this was so and if this knowledge could be used in electromagnetic applications. More recently we have been looking to see if other fractal descriptors are similarly embedded in the scattering data and if these additional fractal attributes could aid in the synthesis of new devices and systems. To summarize some of these research results, we examine in this chapter two cases that are both straightforward to understand and physically illuminating—scattering from fractal surfaces and scattering from fractal superlattices.

Work in fractal electrodynamics builds on the initial literature on wave scattering and diffraction from fractal geometries starting with the pioneering work by Berry on *diffractals* [5], [6]; scattering from fractal surfaces and slopes by Jakeman [7–10]; initial studies of diffraction by fractal objects and apertures [11–21]; and X-ray, photon, and neutron scattering from fractal aggregates, colloids and porous media [22–33]. In addition, a number of books on fractals, several of which we note here [34–49], [78], provide a wealth of supporting material.

Although the primary goals of this chapter are to become familiar with fractals and waves, to analyze several canonical examples, and to understand their underlying physical principles, there are also a number of potential applications of this work. We envision these to include device synthesis for the microwave, millimeter wave, and optical regimes, as well as the use of fractals in remote sensing, characterization, and classification. Several examples of the synthesis problem are given in the following chapters where fractal antenna elements and arrays are considered.

1.1.2 Overview

We first introduce the fundamental concepts of fractal geometry, fractal sets and fractal descriptors. We examine the construction of statistical and more idealized fractals as well as the definition of the primary fractal descriptor, the *fractal dimension*. This is followed by an introduction to fractal texture through *lacunarity*, which provides a secondary fractal descriptor. We then briefly discuss the relationship between fractals and waves.

In the next section we investigate electromagnetic and optical wave scattering from fractally rough surfaces [7–10], [50–67]. Here we demonstrate the methods by which fractals imprint their distinctive geometry on interrogating waves to generate characteristic scattered waves. In all cases the spread of the scattered waves is related to surface roughness which can be quantified through fractal descriptors. We envision applications for this work that range from the radar and optical remote sensing of sea state to the characterization and classification of both natural and manufactured surfaces.

Finally, we investigate the reflection of electromagnetic and optical waves from fractal superlattices or multilayers [68–77]. We find that the superlattices also impress their fractal characteristics on scattered waves in quantifiable ways. We observe the symmetry of the superlattice structure in both the method of solution and the form of the scattering data. Of particular value is the use of *twist plots* to display frequency-domain results in an easily understandable way. Time-scale analysis using wavelets also provides insight into this scattering problem from an alternative and timely viewpoint. Potential applications of these results include the modeling and synthesis of resonant cavities, mirrors, multilayers and other devices for the microwave, millimeter wave, and optical wave regimes. In addition, our results have application to the analysis and modeling of wave propagation in finely divided layered media.

1.2 INTRODUCTION TO FRACTALS

A fractal is a shape made of parts similar to the whole in some way.
B. B. Mandelbrot [37]

Fractals allow us to describe the natural world according to *its* geometry. They appear irregular, yet have underlying order, and possess structure on all scales. It is not surprising, then, that the language of fractal geometry allows us to describe natural phenomena as diverse as fluid flow in porous materials and lightning strokes with greater ease than does Euclidean geometry.

Here we discuss some of the common characteristics of fractals, particularly their self-similarity and structure on all scales. It is these traits which make fractals, such as those in Fig. 1.1, useful for modeling naturally occurring structures, such as mountains. We also consider how one may describe a fractal using its fractal dimension, which need not be an integer, and its lacunarity, a more recently investigated means of quantifying the texture of a fractal set.

Figure 1.1 Fractal mountains including both a smoothly undulating (top) and a jagged and volume-filling (bottom) version. [*Figure taken from "Scattering from Fractally Corrugated Surfaces," D. L. Jaggard and X. Sun,* J. Opt. Soc. Am. A, **7**, 1131–1139 (1990). © *1990 Optical Society of America.*]

1.2.1 What Are Fractals?

1.2.1.1 Fractal Characteristics

Although there are no strict guidelines as to what constitutes a fractal, there are a number of qualities which frequently characterize fractals. The first is self-similarity and structure at all scales. A coastline with its various bays and peninsulas looks like a rough, jagged curve when viewed from hundreds of kilometers above the earth. When viewed by a person walking along its coves and points, this coastline again appears as a rough, jagged curve, and it retains this appearance even when considered from the perspective of an ant walking along individual grains of sand. This similar structure at many different scales is characteristic of fractals. In contrast, a beach ball on the same coastline looks like a point when seen from outer space, a sphere when viewed by a person walking along the shore, and a plane when considered by an ant on its surface. The substantially different appearance when viewed at different scales is more characteristic of non-fractal Euclidean structures.

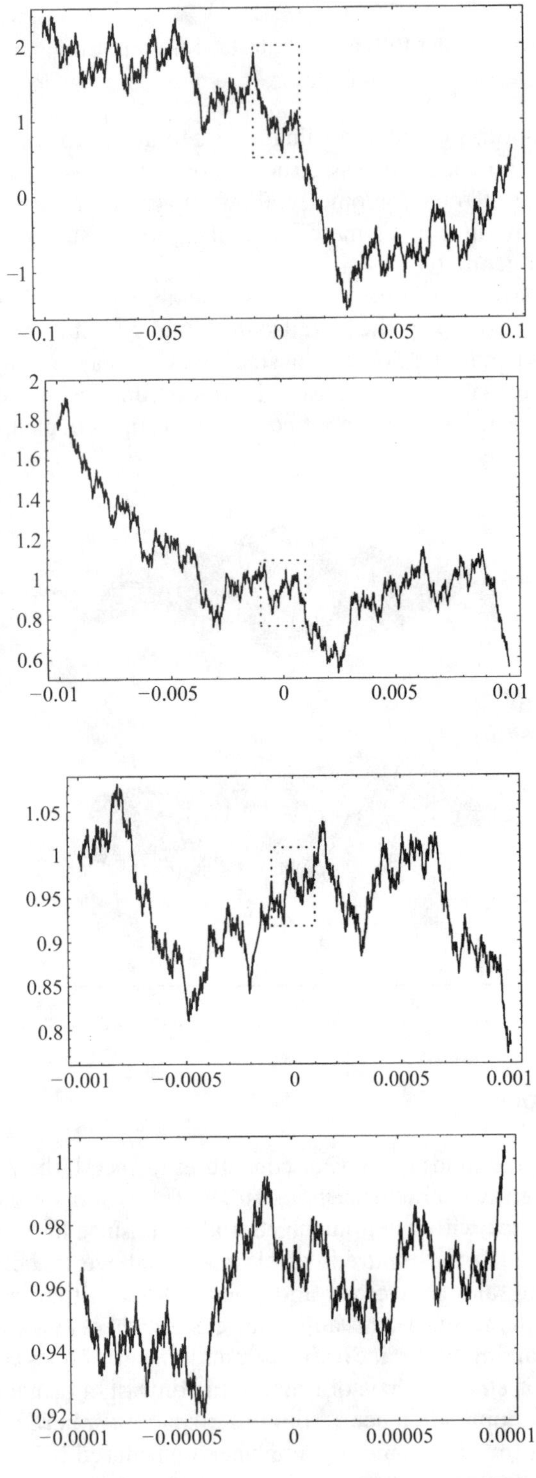

Figure 1.2 A bandlimited fractal Weierstrass function with dimension parameter $D = 1.5$, which has been shown to be an upper bound for the fractal dimension of the curve. As we zoom in on the curve, top to bottom with dashed lines indicating the magnified region, its general appearance does not change even though it does not possess strict self-similarity. This is considered to be self-similar in a statistical sense. [© *1999 Jaggard and Jaggard*.]

Self-similarity may occur in either a statistical or an exact sense. In the former case, seen in the curve in Fig. 1.2, zooming in on one portion of the set reveals a subset that has the same general appearance as the whole set, with similar roughness and irregularity, without being an exact copy of the whole set. Exact self-similarity is seen in the stylized fractal pictured in Fig. 1.3. In this case, zooming in on a subset of the whole reveals a scaled replica of the entire set.

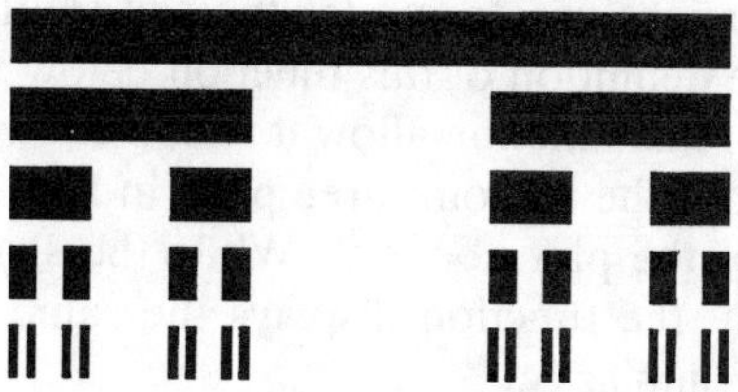

Figure 1.3 A triadic Cantor bar at stages of growth $S = 0, 1, 2, 3, 4$ (top to bottom), which has dimension $D_s = \ln 2/\ln 3 \approx 0.631$. This structure is strictly self-similar for discrete variations in magnification as each stage of growth contains two scaled copies of the set at the previous stage of growth.

Whether a fractal set has statistical or exact self-similarity, its similar appearance on all scales suggests that the set might be best constructed through recursive or iterative means. This stands in contrast to Euclidean structures, which may be more easily defined using formulae. We will also see that fractals may often be described with a non-integral dimension, in contrast to the integral dimension n assigned to Euclidean space $\mathbb{R}^n$. We summarize these and other qualitative differences between fractal and Euclidean geometry in Table 1.1.

TABLE 1.1 Heuristic Summary of Fractal and Traditional or Euclidean Geometrical Attributes. (Table after [1].)

Fractal Geometry	Euclidean Geometry
Often defined by iterative rule	Often defined by formula
Structure on many scales	Structure on one or few scales
Dilation symmetry (self-similarity)	No self-similarity
Fractional dimension possible	Integer dimension
Long-range correlation	Variable correlation
Described as ramified, variegated, spiky	Described as regular
Rough on most scales	Smooth on most scales

1.2.1.2 Bandlimited Fractals and Prefractals

While mathematically defined fractals exhibit an infinite range of scales, physical objects modeled by fractals, such as coastlines, trees, or naturally formed aggregates, cannot. Thus, as we try to model physical phenomena it is also useful to consider structures which display fractal characteristics over multiple scales but not the infinity of scales associated with fractals. In the case of statistical fractals, such structures are termed *bandlimited fractals*. These are constructed using the same methods as for true statistical fractals, but terminating construction when the additional structure to be added is smaller than a specified size, with the initiator providing the upper bound for the structure scales. For strictly self-similar fractals, the analogous concept is that of *prefractals*, which may be constructed by halting the fractal growth process after a finite number of steps. Bandlimited fractals and prefractals display self-similarity over a limited range of scales, looking like Euclidean space at sufficiently small scales.

While in a strict sense these are not fractals, we shall use the term *fractal* to describe fractals, bandlimited fractals, and prefractals.

1.2.1.3 Bandlimited Weierstrass Function

The fractal curve in Fig. 1.2 is the graph of a bandlimited Weierstrass function, which one might use to model the self-similar coastline described above. We will consider the definition of this function below, but at present we observe that the wild variations of this function allow it to somehow cover more of the plane than a smooth curve. Each of the bottom three plots in Fig. 1.2 is an enlarged picture of the middle one-tenth of the plot above it. While the shape of the graph is different with each enlargement, the function displays the same irregularity and roughness and thus a general self-similarity.

1.2.1.4 Triadic Cantor Set

The first fractal construction we will detail is the triadic Cantor set, one of the classic examples of a fractal set. As we broaden our ability to describe fractal sets, we will consider generalizations of this set that make use of our expanded means of characterizing fractals. The triadic Cantor set is formed by starting with the interval $[0, 1]$, the *initiator* of the fractal, and iteratively applying an operation which excises the middle third of each remaining closed interval, the *generator* of the fractal. In general, the initiator of an exact fractal is some structure taken as a starting point, and the generator is some operation—excision, addition, or replacement—which allows iterative modification of the initiator through repeated rescaling and application. Figure 1.3 shows stages of growth (= number of applications of the excising operation) $S = 0, 1, 2, 3$, and 4 of the triadic Cantor set.

Alternatively, we may construct the triadic Cantor bar by defining two functions $c_1: [0, 1] \to [0, 1/3]$ and $c_2: [0, 1] \to [2/3, 1]$ by

$$c_1(x) = \frac{x}{3} \tag{1.1}$$

$$c_2(x) = \frac{x + 2}{3} \tag{1.2}$$

These map the interval onto two disjoint subsets of itself, each shrinking the interval by a scaling factor $\gamma = 1/3$. In this method of construction, the Cantor bar at stage of growth $S + 1$ is formed by applying c_1 and c_2 to the Cantor bar at stage of growth S and taking the union of the two images, so that denoting by C_S the bar at stage of growth S, we have

$$C_{S+1} = c_1(C_S) \cup c_2(C_S) \tag{1.3}$$

with $C_0 = [0, 1]$. The triadic Cantor set C is the limit of this process as $S \to \infty$ and is seen to be the subset of $[0, 1]$ invariant under this mapping, that is,

$$C = c_1(C) \cup c_2(C) \tag{1.4}$$

Equation (1.4) highlights the self-similarity of C, which consists of two scaled copies of itself.

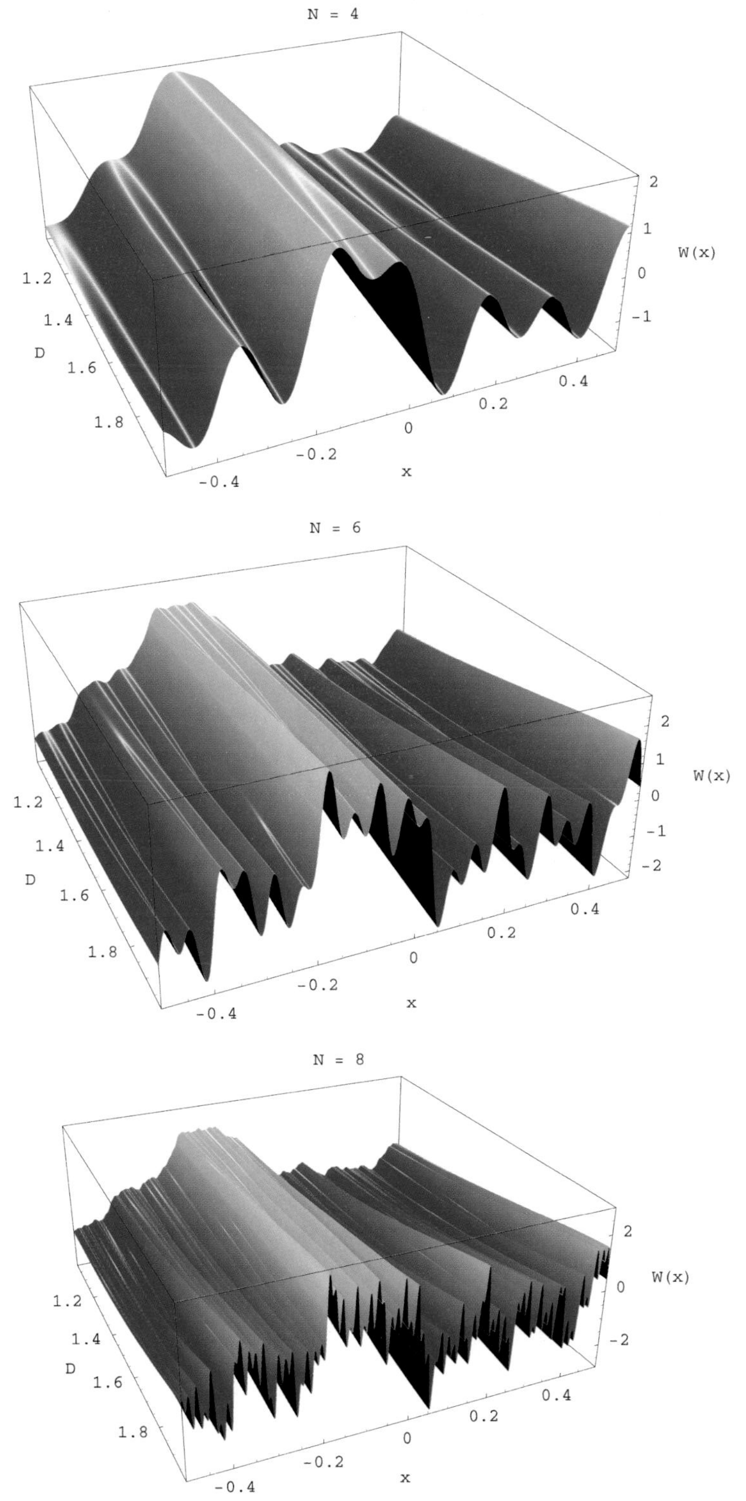

Figure 1.4 A bandlimited fractal Weierstrass function $W(x)$ as a function of coordinate x and dimension parameter $1 < D < 2$, $N = 4$ tones (top), $N = 6$ tones (middle), and $N = 12$ tones (bottom), with frequency parameter $b = 2e/3$. As $D \to 2$, the curve becomes area filling while as $D \to 1$ the curve becomes much smoother.

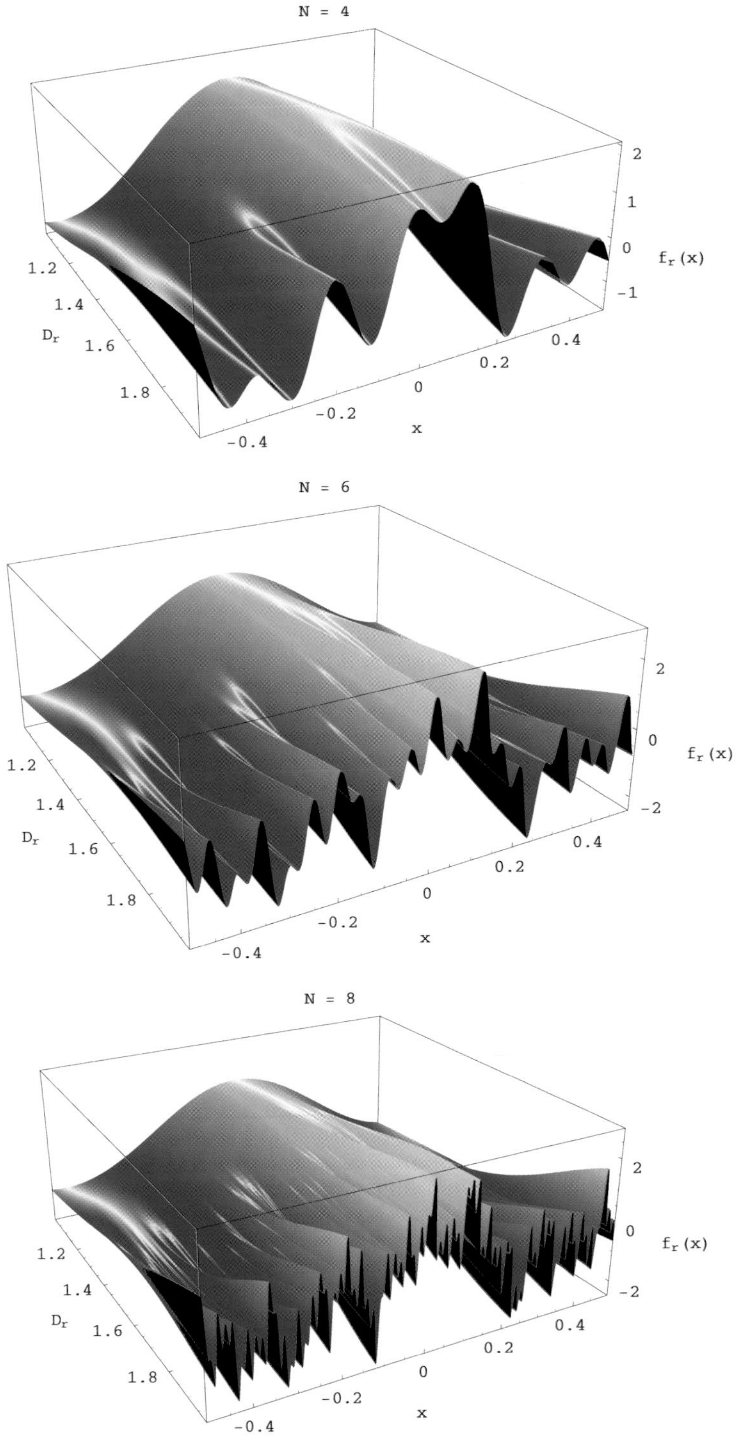

Figure 1.11 Plot of the fractal function $f_r(x)$ defined in relation (1.24) as a function of coordinate x for $1 < D_r < 2$ for $b = 2e/3$ and $N = 4$ tones (top), $N = 6$ tones (middle), and $N = 12$ tones (bottom). This self-similar function varies from smoothly undulating ($D_r \to 1$) to area-filling ($D_r \to 2$).

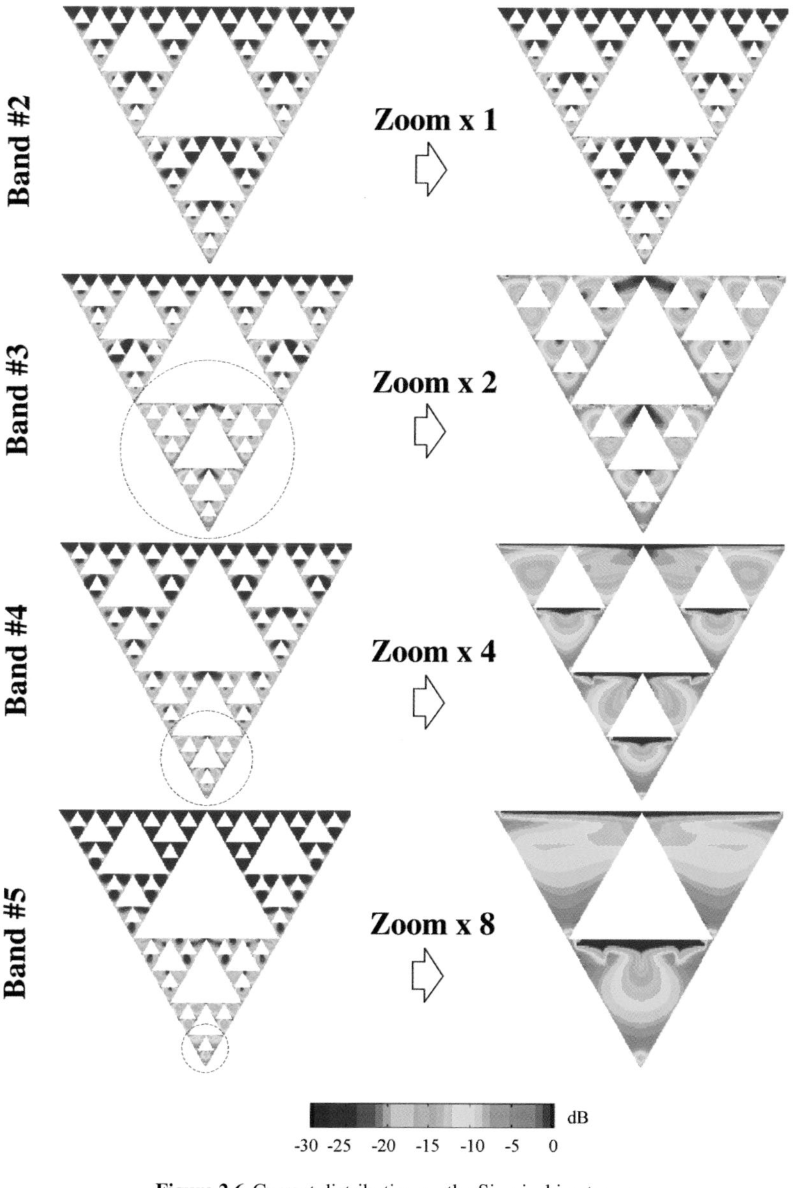

Figure 2.6 Current distribution on the Sierpinski antenna.

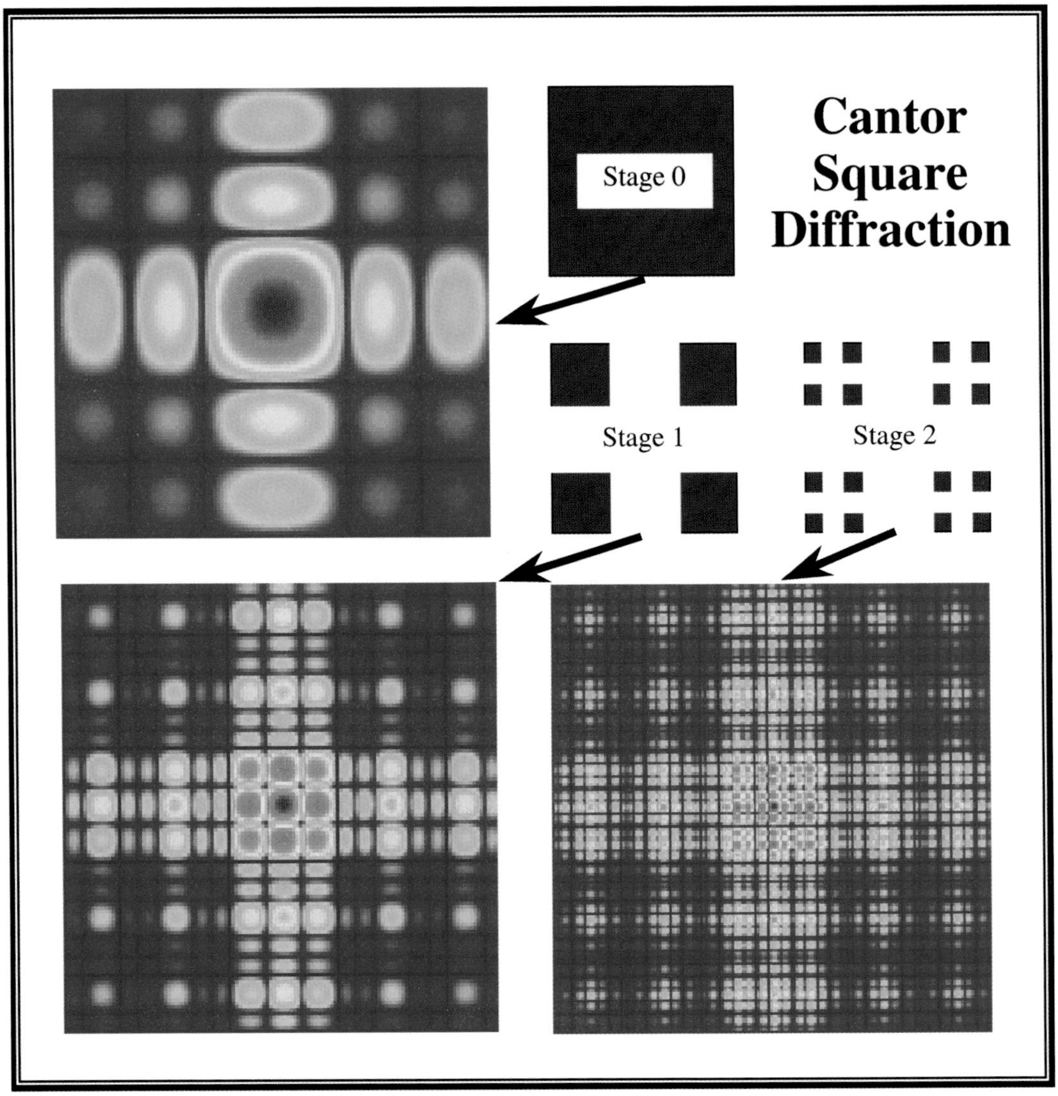

Figure 3.8 Aperture (top right) and diffracted field $|\Psi(x, y, z)/\Psi(0, 0, z)|$ in dB scale as a function of normalized spatial frequencies $f_x L$ and $f_y L$ for the three stages of growth $S = 0, 1, 2$ of a Cantor square. The axes are re-scaled by the factor $\gamma = 1/3$ for each successive stage of growth. The largest values are shown in dark red and the nulls in dark blue. (©1997 Jaggard and Jaggard [32].)

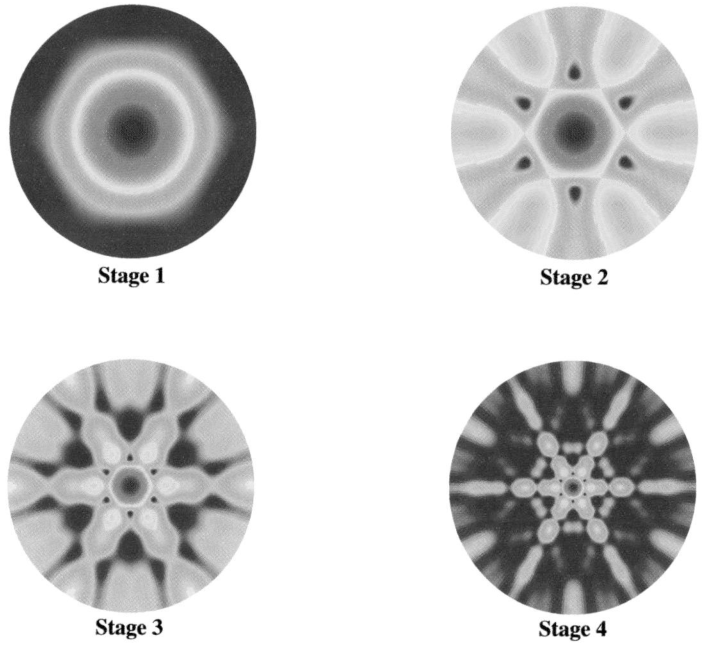

Figure 3.70 Color contour plots of the far-field radiation patterns produced by the four triangular arrays shown in Fig. 3.67. The angle ϕ varies azimuthally from 0° to 360°, and the angle θ varies radially from 0° to 90°.

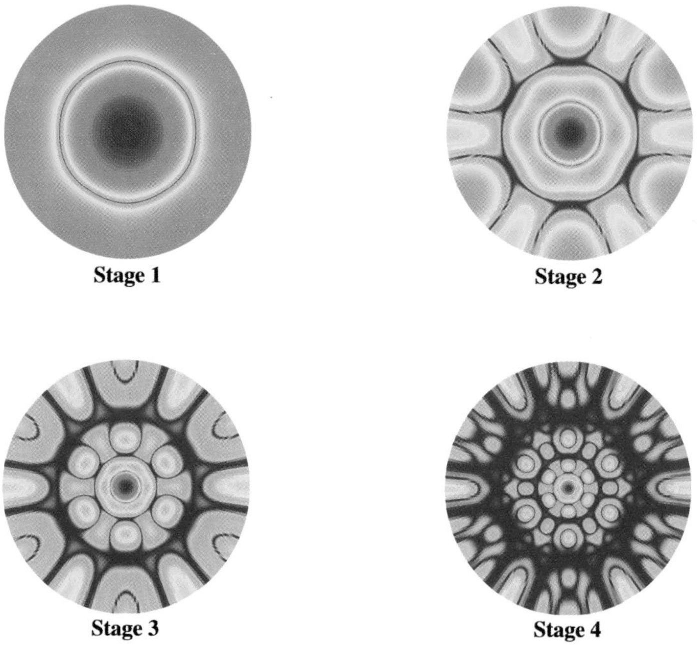

Figure 3.75 Color contour plots of the far-field radiation patterns produced by the four hexagonal arrays shown in Fig. 3.72 and Fig. 3.73. The angle ϕ varies azimuthally from 0° to 360°, and the angle θ varies radially from 0° to 90°.

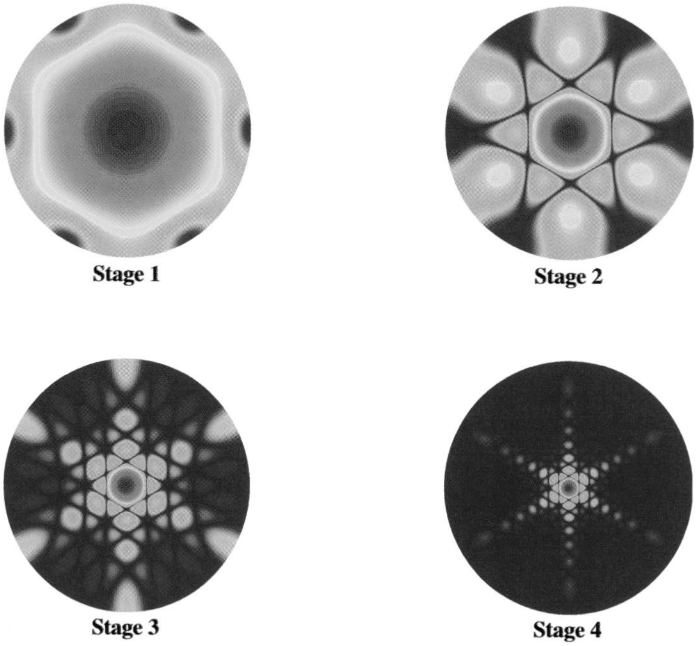

Figure 3.78 Color contour plots of the far-field radiation patterns produced by a series of four ($P = 1, 2, 3$ and 4) fully-populated hexagonal arrays generated with an expansion factor $\delta = 2$. The angle ϕ varies azimuthally from 0° to 360°, and the angle θ varies radially from 0° to 90°.

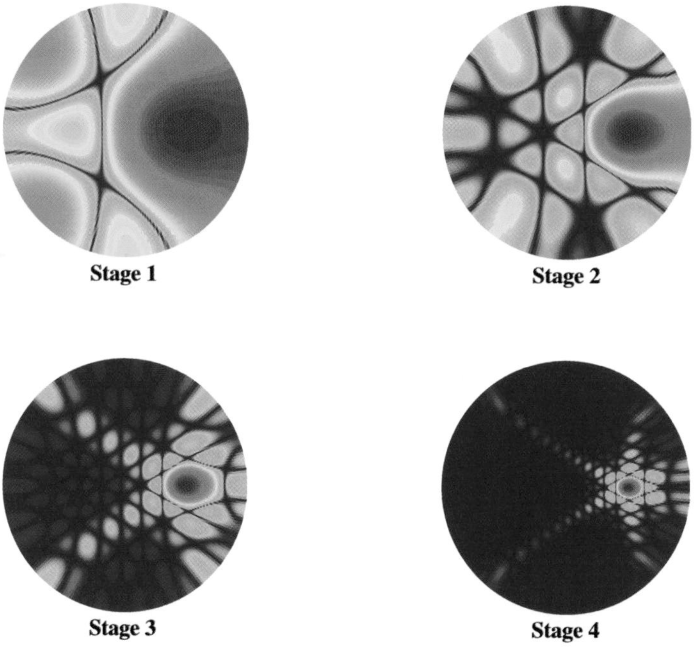

Figure 3.79 Color contour plots of the far-field radiation patterns produced by a series of four ($P = 1, 2, 3$ and 4) fully-populated hexagonal arrays generated with an expansion factor $\delta = 2$. In this case, the phasing of the generating subarray was chosen such that the maximum radiation intensity would occur at $\theta_0 = 45°$ and $\phi_0 = 90°$.

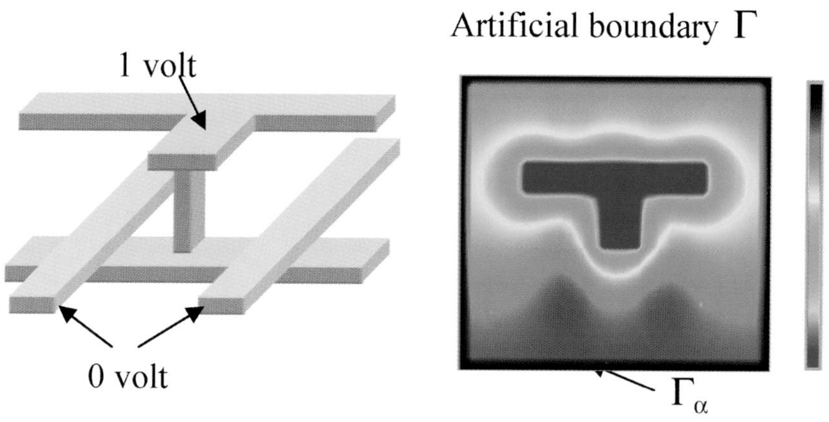

Figure 15.4 Potential distribution across the cross section that is parallel to the ground plane and located at the level of the upper T-shape conductor of the via.

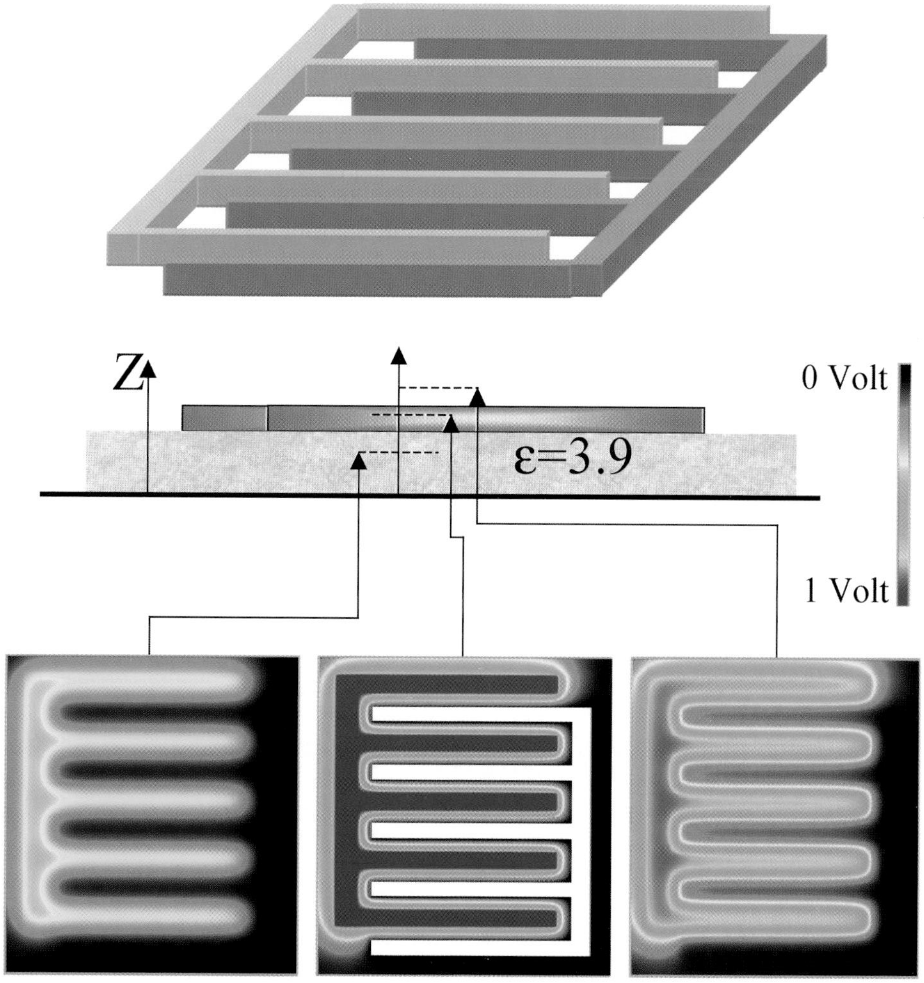

Figure 15.14 Two-comb structure. Potential distributions across the cross sections that are parallel to the ground plane.

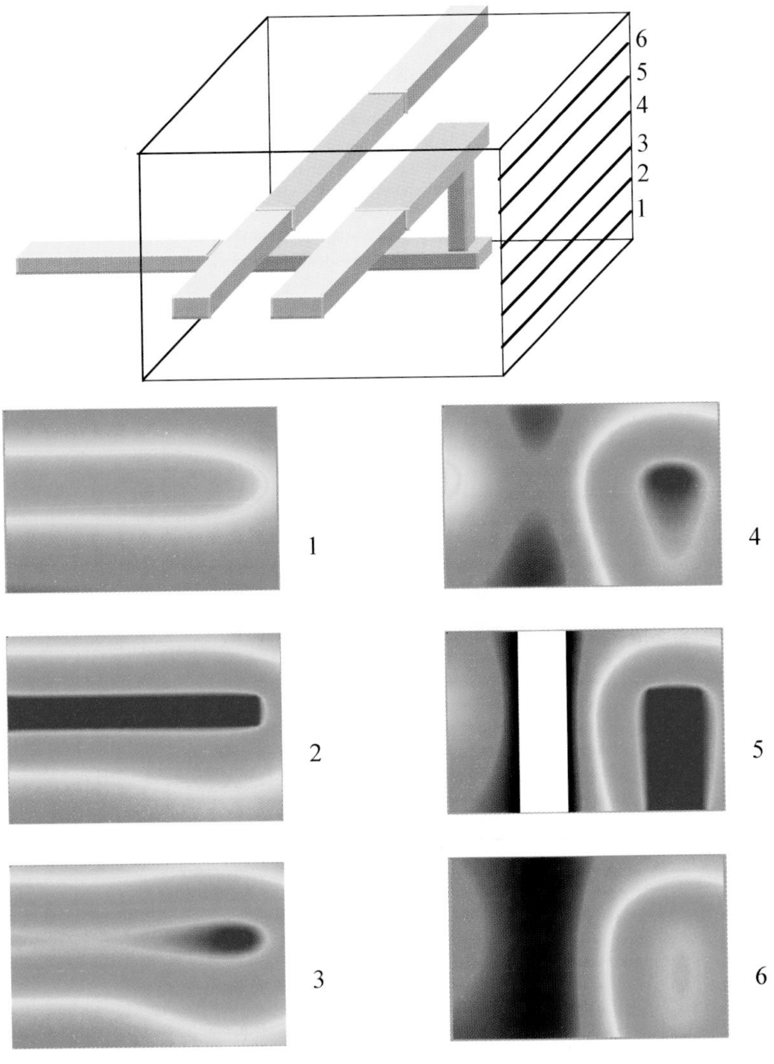

Figure 15.18 Potential distributions across the cross sections that are parallel to the x–z plane in subdomain 3 at different levels with respect to the ground plane.

1.2.2 Fractal Dimension

1.2.2.1 Motivation

Having seen two examples of fractal sets, we observe that fractals often seem to fill different amounts of space than objects described by Euclidean geometry. A fractal curve in the plane, like that in Fig. 1.2, with its wiggles on all scales, is able to cover more of the plane than a smooth curve, but still does not fill the entire plane. The limit of the Cantor set in Fig 1.3 contains uncountably many points, but contains no intervals and so seems to fill qualitatively less space than the line. Finally, because of the infinite range of scales in a fractal, we can never consider a small enough region that the fractal looks like an n-dimensional Euclidean object. These considerations suggest that fractals might be viewed as having dimensions different than those of traditional geometric objects, and cause us to revisit the concept of dimension.

The following argument may be made to extend the notion of dimension from Euclidean objects to structures that we would like to be of non-integral dimension. If we wish to measure the length of the unit interval $[0, 1]$, we may count the number of intervals of length δ needed to cover the interval. Using our standard ideas about length, we need

$$N_{[0,1]}(\delta) = \left(\frac{1}{\delta}\right)^1 \tag{1.5}$$

such intervals to cover $[0, 1]$. To measure longer or shorter intervals, we may multiply $N_{[0,1]}$ by an appropriate constant (the length of the interval), but the δ^{-1} dependence will remain. Similarly, if we wish to measure the area of the unit square $[0, 1] \times [0, 1]$ we may count the number of squares of side δ needed to cover it. This is easily seen to be

$$N_{[0,1]\times[0,1]}(\delta) = \left(\frac{1}{\delta}\right)^2 \tag{1.6}$$

If we measure larger or smaller squares, the result will be the same up to a constant factor, in particular retaining the δ^{-2} dependence of $N_{[0,1]\times[0,1]}$. In general, if we wish to measure the size of the n-dimensional unit cube $[0, 1]^n$, we may cover it by

$$N_{[0,1]^n}(\delta) = \left(\frac{1}{\delta}\right)^n \tag{1.7}$$

n-cubes of side δ. The number of n-cubes of side δ needed to cover a general n-cube is proportional to δ^{-n}.

1.2.2.2 Definition

This suggests that we may view the exponent in the δ dependence of $N_{[0,1]^n}(\delta)$ as the dimension of the set $[0, 1]^n$ being covered, and that dimension might be extended to fractal sets in this manner. We count the number $N_E(\delta)$ of sets needed to cover a fractal set $E \subset \mathbb{R}^n$ with n-cubes, or multiple copies of another n-dimensional set. If

$$N_E(\delta) = C\delta^{-D}, \tag{1.8}$$

we have

$$D = \frac{\ln C - \ln N_E(\delta)}{\ln \delta} \tag{1.9}$$

For bounded fractal sets and sufficiently large δ, N_E has no δ dependence. Additionally, our intuition suggests that it is the *local* structure of fractals which differentiates them from Euclidean space. In this spirit we consider small δ, taking the limit of (1.9) as $\delta \to 0$. The $\ln C$ term then becomes insignificant, and we may define the *disk covering dimension*

$$D_d = \lim_{\delta \to 0} \frac{\ln N_E(\delta)}{-\ln \delta} \tag{1.10}$$

Just as covering a square with lines or a cube with squares is of minimal utility in this discussion, so is covering a fractal set E by sets of dimension less than that of E. To avoid this, we use n-dimensional sets to cover a set $E \subset \mathbb{R}^n$.

Consider a strictly self-similar fractal E which contains N copies of itself, each scaled by a factor γ. The Sth stage of growth defines a cover of E by N^S sets of size $C\gamma^S$, where C is a constant depending on the size of the initiator (0th stage). Forming an expression similar to Eq. (10), we have

$$\frac{\ln N^S}{-\ln C\gamma^S} = \frac{S \ln N}{-(\ln C + S \ln \gamma)} \tag{1.11}$$

As $S \to \infty$ and the size of the covering sets $C\gamma^S \to 0$, the $\ln C$ term vanishes. This leads to the *similarity dimension* for strictly self-similar sets, defined as

$$D_s = \frac{\ln N}{-\ln \gamma} \tag{1.12}$$

for sets consisting of N non-overlapping copies of the whole, each scaled by a factor γ.

The triadic Cantor bar in Fig. 1.3 consists of two copies of itself scaled by 1/3. Thus, it has similarity dimension $\ln 2/\ln 3 \approx 0.631$. This agrees with our expectations for the dimension of this set, since it seems to be qualitatively more than a (zero-dimensional) point without completely filling the (one-dimensional) interval $[0, 1]$, and in fact not filling *any* interval $[a, b]$.

1.2.2.3 Extensions

The computation of the disk-covering dimension is suitable for adaptation to automated systems. These might use a fixed grid with cells of size δ, counting the number of cells intersecting the set E in question to find $N_E(\delta)$. This approach has applications to the computation of dimension for naturally occurring aggregates, electrical discharges, and digitized images. When implemented with a fixed grid, this is usually referred to as the *box-counting dimension*.

There are a number of other fractal dimensions used in describing fractal sets. Further discussion can be found in [36–37], [41], with a more formal approach given in [34].

Bandlimited fractals and prefractals have integral dimension, for at sufficiently small scales they look like Euclidean objects. However, we will still refer to the "fractal dimension" of these types of sets. For bandlimited fractals, we will often take as the dimension the value of D_d suggested by the disk-covering procedure using disks whose size is within the range of scales present in the structure of the prefractal. In the case of exact prefractals, we define the dimension as that of the fractal for which the prefractal is some finite stage of growth. The fractal dimension of these structures is useful in investigating their properties on scales at which they are indistinguishable from a true fractal, for example electromagnetic interrogation of a prefractal when the frequencies used are too low to resolve the omitted fine structure.

1.2.3 Fractals and Their Construction

Since one common characteristic of fractals is self-similarity, we look to the construction of self-similar sets as one means of obtaining additional examples of fractals. This will provide a few illustrations of the concept and computation of fractal dimension. We first give the construction of the Weierstrass function and then give that of the Sierpiński gasket, a strictly self-similar fractal in the plane $\mathbb{R}^2$. We conclude with a generalization of the triadic Cantor bar.

1.2.3.1 Bandlimited Weierstrass Function

We now construct the bandlimited Weierstrass function $W(x)$ shown in Fig. 1.2. The graph of this function is not strictly self-similar, as the Cantor set is, but does display self-similarity in the sense of having a similar appearance at different scales. This self-similarity is not surprising if we consider the definition of the function

$$W(x) = C \sum_{n=N_1}^{N_2} b^{n(D-2)} \cos(2\pi s b^n x + \theta_n) \tag{1.13}$$

where D is the dimension parameter, sb^{N_1} and sb^{N_2} $(b > 1)$ are the lowest and highest spatial frequencies, the θ_n are random phases in $[0, 2\pi]$, and C is an appropriate constant [1]. The plots in Fig. 1.2 are for $N_1 = 1$, $N_2 = 40$, $b = 2e/3$, $s = 1$, and $D = 1.5$. As N_2 increases, $W(x)$ contains increasingly fine structure due to the addition of high-frequency sinusoids. For a finite number of tones $N = N_2 - N_1 + 1$, $W(x)$ is a bandlimited fractal, displaying fractal characteristics on scales between $(sb^{N_2})^{-1}$ and $(sb^{N_1})^{-1}$.

As $D \to 1$ the lowest frequency sinusoid in the summation defining $W(x)$ becomes dominant and the function approaches a smoothly undulating curve. As $D \to 2$, the sinusoids in the summation have almost equal contributions and the curve becomes very jagged and starts to fill up the plane. This "dimension parameter" thus behaves much as we would expect a measure of fractal dimension to behave, and has in fact been shown to be equal to the box counting dimension in certain cases [41]. Functionally defined fractals such as $W(x)$ often have similar dimension parameters which control the apparent roughness of the function but which have not necessarily been proven to be equal to any of the measures of fractal dimension for the particular function. A generalization of the Weierstrass function is used to form the fractal mountains in Fig. 1.1.

In Fig. 1.4 (see full-color insert) is shown the Weierstrass function in relation (1.13) as a function of the continuous variable D and coordinate x for number of tones $N = 4$ and 12.

1.2.3.2 Sierpiński Gasket

The Sierpiński gasket is shown in Fig. 1.5 for the first six stages of growth. We construct this fractal by taking a filled equilateral triangle (top left) as the initiator and an operation which excises an inverted equilateral triangle which is the initiator inverted and scaled by one-half as the generator (top right). Application of the generator leaves three smaller filled triangles, to which we may apply a scaled copy of the generator. We iterate this process to obtain higher stages of growth.

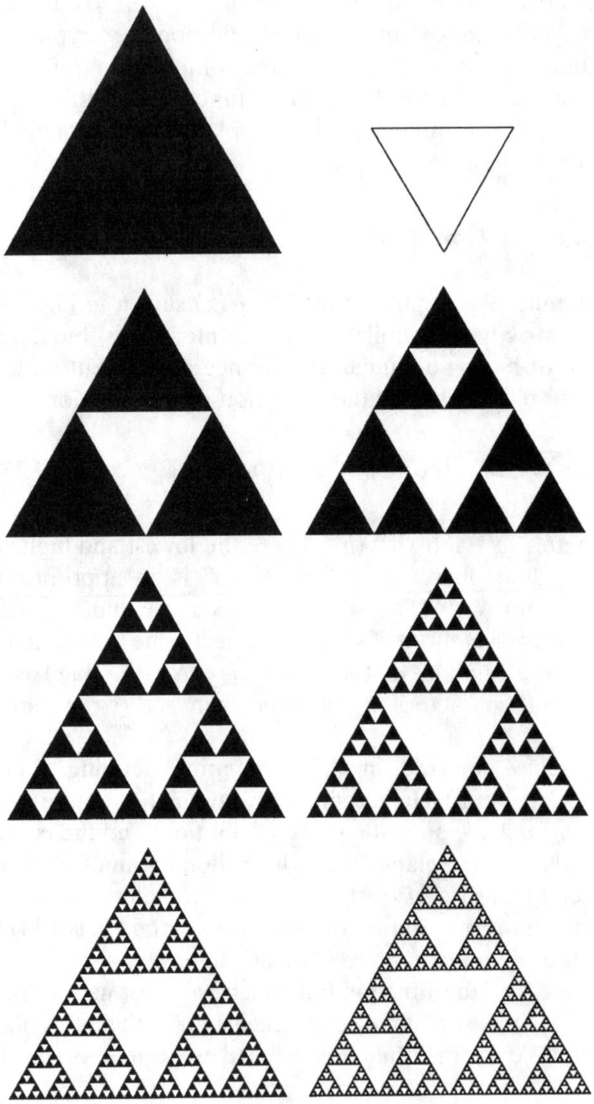

Figure 1.5 The construction of the Sierpiński gasket, which has dimension $D_s = \ln 3/\ln 2 \approx 1.58496$, with initiator (top left) and generator (top right) and stage of growth $S = 1, 2, 3, 4, 5, 6$. This dimension corresponds to our intuitive feeling for the dimension of the set, which contains straight lines, but is unable to "fill up" the plane as $S \to \infty$.

From the perspective of mappings from $\mathbb{R}^2 \to \mathbb{R}^2$, we may define

$$s_1(x,y) = \left(\frac{x}{2}, \frac{y}{2}\right) \tag{1.14}$$

$$s_2(x,y) = \left(\frac{x+1}{2}, \frac{y}{2}\right) \tag{1.15}$$

$$s_3(x,y) = \left(\frac{2x+1}{4}, \frac{2y+\sqrt{3}}{4}\right) \tag{1.16}$$

If the initiator (zeroth stage) of the Sierpiński gasket has side length 1 with its lower left vertex at the origin and the bottom edge coincident with the x-axis, the maps s_1, s_2, and s_3 take this triangle to the lower left, lower right, and top triangles seen in the first stage of construction. As with the triadic Cantor set, the $(S+1)^{\text{st}}$ stage of growth is the union of the images of the Sth stage of growth under the (in this case three) defining maps. The Sierpiński gasket is the limit of this process as $S \to \infty$ and is invariant under this operation.

1.2.3.3 Polyadic Cantor Bars—Minimal Lacunarity

The construction of the Cantor bar discussed above is easily generalized to give different values of D_s. The first generalization is a variation in γ, so that an amount other than the middle third is excised from each interval. We change the generator to be the operation which excises the middle $1-2\gamma$, $\gamma \in (0, \frac{1}{2})$, of each interval. Equivalently, these Cantor bars contain two copies of the whole set, each scaled by γ and thus have similarity dimension $D_s = \ln 2/\ln(1/\gamma)$. With the appropriate choice of γ, we may construct a "single-gap" Cantor bar with any value of D_s between 0 and 1 as shown in Fig. 1.6 for $D_s = 1/2$ and 3/4.

As D_s is a function of both γ and the number N of scaled copies of the whole, we might also try changing N. We extend the Cantor bar construction using a generator that excises $n_{gaps} > 1$ intervals. Equivalently we may define $N = n_{gaps} + 1 > 2$ maps $\{c_i\}_{i=1}^{N}$ of the unit interval into subintervals, analogous to the maps c_1 and c_2 that gave

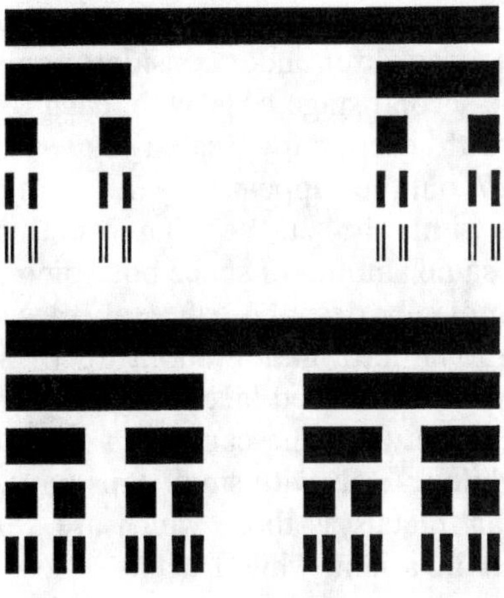

Figure 1.6 Stages of growth $S = 0, 1, 2, 3, 4$ for single-gap Cantor bars with $D_s = 1/2$ (top) and 3/4 (bottom).

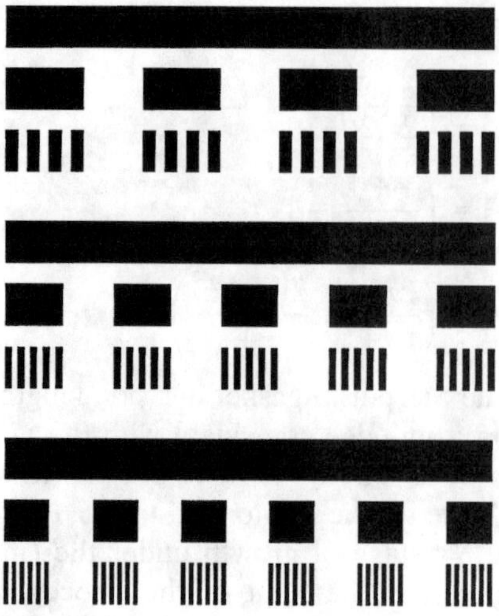

Figure 1.7 Stages of growth $S = 0, 1, 2$ for three-, four-, and five-gap Cantor bars (top to bottom) with the scaled copies of the set uniformly distributed within the unit interval. In each case $D_s = 3/4$.

the triadic Cantor bar. We will restrict our attention to Cantor bars in which the sizes of the intervals remaining after each excising operation (i.e., the scaling ratios of the N maps from the interval to subintervals) are the same value γ. In cases where these values are different ($= \{\gamma_i\}_{i=1}^{N}$) the fractal dimension is the value of D_s which satisfies $\Sigma_{i=1}^{N} \gamma_i^{D_s} = 1$ [37]. For the moment we also require the subintervals to which the interval is mapped to be uniformly spaced throughout the interval. Examples of such Cantor bars are pictured in Fig. 1.7, with $n_{gaps} = 3$, 4, and 5 (top to bottom). We will examine some of the effects of relaxing the uniform spacing condition below.

1.2.4 Lacunarity

1.2.4.1 Concept

Fractal dimension is the primary descriptor of fractal sets and structures. However, just as knowing the Euclidean dimension of an object does not determine the object (e.g., a curve of dimension one may be a straight line or a smoothly varying curve such as a sinusoid), knowing the fractal dimension D_s does not uniquely determine the fractal set under consideration. Figure 1.8 shows two square Sierpiński carpets at the second stage of growth, each of which consists of $N = 40$ copies of the whole set, with each copy scaled by $\gamma = 1/7$. Both have similarity dimension $D_s = \ln 40/\ln 7$, but they appear very different due to the different arrangements of the scaled copies within the set. We might also think of these two carpets as in some sense covering the same amount of space but doing so in different ways. Because dimension does not completely describe a fractal set, we look for other fractal descriptors to provide additional information about fractal structures.

Mandelbrot introduced lacunarity [78], from the Latin *lacuna* (gap), to describe the differences between the carpets in Fig. 1.8. Fractals with large gaps have large lacunarity, while fractals with small gaps have small lacunarity. We may also think of highly lacunar fractals as those which are very inhomogeneous and far from being translationally invariant, while fractals of low lacunarity are more homogeneous and

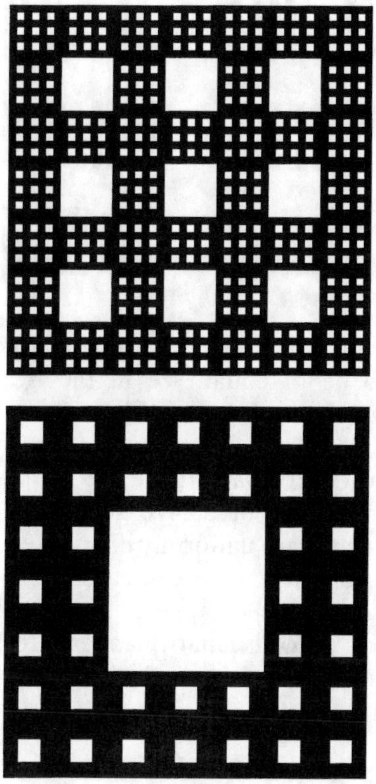

Figure 1.8 Two Sierpiński carpets of dimension $D_s = \ln 40/\ln 7$. Their different appearances suggest the need for additional fractal descriptors. The same fractal dimension for these sets means there is the same amount of black ink used in printing them, while the different lacunarities indicate that the ink is distributed differently. [*Figure after* [*78*].]

approach translational invariance. While the fractal dimension gives a measure of how much space is filled by a set, lacunarity gives an idea of the way in which it is filled or the texture of the set. The Sierpiński carpet on the top of Fig. 1.8, in which the black area is more homogeneously distributed throughout the square, has low lacunarity while the one on the bottom, with its large gap in the center and highly inhomogeneous appearance, has high lacunarity.

1.2.4.2 Examples—Polyadic Cantor Bars with Variable Lacunarity

Having seen one pair of fractals with differing lacunarity, we consider other examples to illustrate this measure of texture. Following the example of the Sierpiński carpets in Fig. 1.8, we take a previously defined self-similar fractal and rearrange the subcopies of the whole while continuing to keep them non-overlapping. One family of fractals we have seen which is immediately amenable to this generalization is the family of polyadic Cantor bars discussed in Section 1.2.3.3.

Because of physical considerations from the problem of reflection from a Cantor superlattice in Section 1.4, we continue to impose some restrictions on the positioning of the subintervals within the original interval. The Cantor bars we consider must be symmetric about their midpoint, that is, for a Cantor bar C constructed in $[0, L]$, $x \in C \Leftrightarrow L - x \in C$. Additionally, we require the outer $\lfloor (n_{gaps} - 1)/2 \rfloor$ gaps between subintervals be of equal size εL. For n_{gaps} odd or even, we have a central gap or bar,

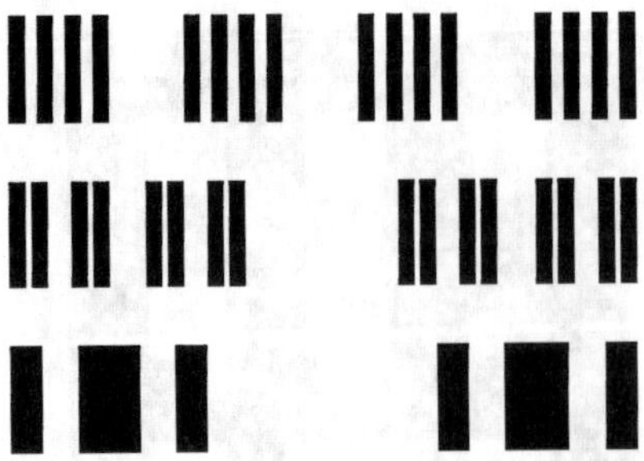

Figure 1.9 The second stage of growth for three-gap Cantor bars of low, medium, and high (top to bottom) lacunarity. The corresponding values of ε are $\varepsilon_{max} \approx 0.123$, $\varepsilon_{max}/2 \approx 0.0617$, and 0. Each bar has dimension $D_s = 3/4$.

respectively, with the midpoint symmetry forcing the two gaps on either side of the central bar to be of equal size in the n_{gaps} even case. All gaps other than the centermost gap(s) are of size εL. The value of ε determines the size of the centermost gap(s).

Three-gap Cantor bars at the second stage of growth are shown in Fig. 1.9 for three different values of ε. We see that for $\varepsilon = \varepsilon_{max} = (1 - N\gamma)/n_{gaps}$ (top), the subcopies of the set are uniformly distributed and the set appears fairly homogeneous and of low lacunarity. As ε decreases to 0 (middle and bottom), the central gap becomes larger and the set appears more "gappy" and inhomogeneous. As discussed below, over this range lacunarity appears to be a monotonically decreasing function of ε for polyadic Cantor bars, and we use ε to characterize the lacunarity of this family of Cantor bars.

1.2.4.3 Definition

As one potential measure of lacunarity, we adapt the *gliding box method* proposed by Allain and Cloitre [79]. The formulation given here is for linear fractals such as Cantor sets, but generalizations would allow measurement of the lacunarity of the planar carpets in Fig. 1.8. In this approach, we overlay a box of radius R on the set in question, and let $Q(M,R)$ be the probability that it covers mass M. If we define the qth moment of Q by

$$Z_Q^{(q)} = \sum_M M^q Q(M,R) \tag{1.17}$$

then a measure of lacunarity may be defined as

$$\Lambda(R) = \frac{Z_Q^{(2)}(R)}{[Z_Q^{(1)}(R)]^2} \tag{1.18}$$

If we let M_R be the mass covered by a box of radius R and denote by $\langle\,\rangle$ average value, we see that

$$\Lambda(R) = \frac{\langle M_R^2\rangle}{\langle M_R\rangle^2} \tag{1.19}$$

For a generalized Cantor bar $C_{(\varepsilon)}$ with outer gap width ε in which we are interested, we may define

$$K_\varepsilon^R(\tau) = \int_{\tau-R}^{\tau+R} \chi_{C_{(\varepsilon)}}(x)\,\mathrm{d}x = \int_{-\infty}^{\infty} \chi_{C_{(\varepsilon)}}(x)\,h_R(x-\tau)\,\mathrm{d}x \tag{1.20}$$

where

$$\chi_E(x) = \begin{cases} 1, & x \in E \\ 0, & x \notin E \end{cases} \tag{1.21}$$

is the characteristic function of the set E and we have defined the window function $h_R(x) = \chi_{[-R,R]}(x)$ to give K_ε^R the form of a cross-correlation. Analogously to (1.19) we integrate K_ε^R and $(K_\varepsilon^R)^2$ over all allowable values of τ, obtaining the expression

$$\Lambda_\varepsilon(R) = \frac{\displaystyle\int_{\tau_{min}}^{\tau_{max}} K_\varepsilon^R(\tau)^2 \, d\tau}{\left[\displaystyle\int_{\tau_{min}}^{\tau_{max}} K_\varepsilon^R(\tau) \, d\tau\right]^2} \tag{1.22}$$

for lacunarity in the setting of this family of Cantor bars. For a Cantor set in $[0, L]$, we take $[\tau_{min}, \tau_{max}]$ to be $[R, L - R]$ so that the support of the window function h_R is always contained in $[0, L]$. This expression for lacunarity is dependent upon the window radius R, but we may define $\tilde{R} = (\gamma + \varepsilon_{max})L/2$ and

$$\tilde{\Lambda}(\varepsilon) = \Lambda_\varepsilon(\tilde{R}) \tag{1.23}$$

As shown in [77], $\tilde{\Lambda}$ appears to be monotonic with ε over $[0, \varepsilon_{max}]$, suggesting that ε is a reasonable descriptor of lacunarity for this family of Cantor sets. Because ε has useful interpretations in fractal electrodynamics problems, we will use it as our primary means of quantifying lacunarity.

1.2.5 Fractals and Waves

We have seen that the common characteristics of fractals include self-similarity and structure at many different scales. These traits suggest that fractals may fall "in between" Euclidean spaces and possess non-integer dimension. Using the many-scale structure of fractals, we are able to define reasonable measures of dimension and texture for fractals.

Our discussion of fractals has made use of variable length yardsticks, both in describing the scales on which a fractal has structure and in computing a fractal's dimension. If we are investigating fractal electrodynamics problems, we might look for *electromagnetic yardsticks* of variable size. The wavelength and pulse width of interrogating signals give us variable scale tools with which to investigate fractal structures in the frequency-domain and time-domain, respectively. We now turn our attention to the use of these tools in the problems of wave scattering from fractal surfaces and superlattices.

1.3 SCATTERING FROM FRACTAL SURFACES

For decades, scattering of optical, electromagnetic and acoustical waves from rough surfaces has been a key theoretical and experimental problem in science and engineering. Often the goal has been the remote characterization of microscopically rough interfaces, sea surfaces, ocean bottoms, and rough terrain. Other applications include surface imaging, the classification or characterization of naturally occurring and manufactured objects, and non-destructive evaluation and testing. In these past studies, deterministic periodic functions, random functions or distributions, and random iterations have been used in the mathematical modeling of rough surfaces. Concepts developed in fractal electrodynamics have provided a useful tool to describe electromagnetic interactions with multiscale rough structures. Naturally occurring rough surfaces are often formed through repetitive actions of nature that yield bandlimited fractals, carefully balancing ordered and chaotic processes.

Here we examine the scattering of electromagnetic waves from rough surfaces first using the Kirchhoff or physical optics approximation and then using the exact transition matrix (T-matrix) approach or extended boundary condition method (EBCM). The approximate solution provides physical insight into the scattering, yields a simple relation between the scattering data and the fractal dimension, and in the appropriate limit, provides a touchstone for the subsequent numerical results. This approximate method involves a simplified scalar approach that does not take polarization into account. The T-matrix approach validates the approximate method in the regime of appropriately shallow and smooth corrugations for near-normal incidence illumination. The exact method also allows an investigation of the effect of increased roughness and an examination of the effect of polarization, both of which are inaccessible to most approximate techniques.

1.3.1 Problem Geometry

The geometry used here is given in Fig. 1.10 in which an incident plane wave illuminates a fractal surface patch of length $2L$ at an incident angle θ_i with respect to the surface normal of the reference ($z = 0$) plane. The incident and scattered wavenumbers are $\mathbf{k_i}$ and $\mathbf{k_s}$, respectively, and have magnitude equal to the free space wavenumber k. In the smooth surface or low-frequency limit, the scattered field has a specular component whose mainlobe width is inversely proportional to both the patch size and the wavenumber of the illuminating wave. As the roughness of the

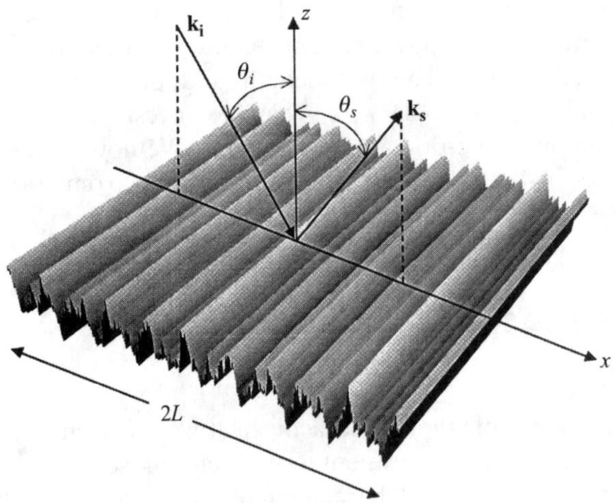

Figure 1.10 Geometry of rough surface scattering problem in which an incident plane wave illuminates a fractal surface patch of size $2L$ at an angle θ_i with respect to the reference plane ($z = 0$) normal. The scattering cross section is calculated for various scattering angles θ_s. The fractal function $f_r(x)$ defines the rough surface and $\mathbf{k_i}$ and $\mathbf{k_s}$ are the incident and scattered wavevectors, respectively. [© *1999 Jaggard and Jaggard.*]

surface increases, or as the frequency increases, the specular component of reflection is reduced as nonspecular components play an increasingly dominant role. As a result, larger amounts of the scattered power are redistributed over an increasing number of directions due to diffraction.

The surface patch roughness for both the approximate and exact cases considered in this chapter is described by the one-dimensional bandlimited *fractal function* that is a mathematical relative to the bandlimited Weierstrass function developed for other problems in fractal electrodynamics [1], [18], [19], [80] and defined in (1.13). This fractal function is given by

$$f_r(x) = \sigma C \sum_{n=0}^{N-1} (D_r - 1)^n \sin(K_0 b^n x + \phi_n) \tag{1.24}$$

where $D_r \in (1,2)$ is the *roughness fractal dimension*,[1] $K_0 = 2\pi/\Lambda_0$ is the fundamental spatial wavenumber of the surface, Λ_0 is the corresponding fundamental spatial frequency, $b > 1$ is the spatial frequency scaling parameter, ϕ_n are specified or random phases, and N is the number of tones or spatial frequencies describing the surface. The amplitude factor is given by

$$C = \sqrt{\frac{2D_r(2 - D_r)}{[1 - (D_r - 1)^{2N}]}} \tag{1.25}$$

and is chosen so that the function has a standard deviation or *rms* height σ. Note that $f_r(x) \approx (D_r - 1) f_r(bx)$ so that the function is discretely self-similar as the number of tones becomes sufficiently large. The *rms* slope of the surface given by (1.24) is found to be

$$\sigma_s = K_0 \sigma \sqrt{\frac{[1 - (D_r - 1)^2][1 - b^{2N}(D_r - 1)^{2N}]}{[1 - (D_r - 1)^{2N}][1 - b^2(D_r - 1)^2]}} \tag{1.26}$$

Plots of the fractal function (1.24) for $N = 4$ and 12 are shown in Fig. 1.11 (see full-color insert) as a function of continuous independent variable x and roughness fractal dimension D_r. As D_r increases the roughness increases due to the enhancement of the higher spatial frequencies in (1.24). Its value ranges from $D_r \rightarrow 1$ (smooth periodic curve) to $D_r \rightarrow 2$ (rough, area-filling curve). Likewise, as the number of tones N increase, small scale roughness becomes apparent for moderate and larger values of dimension, $D_r \gtrsim 1.5$.

1.3.2 Approximate Scattering Solution

1.3.2.1 Formulation of Approximate Surface Scattering Solution

Under the Kirchhoff, physical optics, or undisturbed field approximation, the scattered field is given by an integration over the illuminated surface patch of the

[1]From the plots it appears that D_r acts qualitatively as a fractal dimension and has the expected effect in the limiting cases $D_r \rightarrow 1$ and $D_r \rightarrow 2$. However, it has not been proven that D_r is equal to one of the common fractal dimensions.

incident field. The phase delay imprinted on the scattered field provides its fractal signature. The Kirchhoff approximation treats the rough surface as locally flat with the assumption that the wavelength of the incident wave is small compared to the radius of curvature of the surface corrugation. Incident angles close to normal incidence for the smooth surface will be used to avoid shadowing. The scattering results shown here satisfy all of these requirements. The geometry is shown in Fig. 1.10, and the analysis follows the formulation of Beckmann and Spizzichino [81].

The scattered field at the observation distance r from the origin is given by

$$\Psi_{sc}(\theta_i, \theta_s) = \frac{ikLe^{ikr}}{4\pi r} \iint_{S_{fractal}} [pf_r'(x') - q] e^{ik_x x' + ik_z f_r(x')} dx' dy' \tag{1.27}$$

with

$$\begin{aligned} p &= (1 - R)\sin(\theta_i) + (1 + R)\sin(\theta_s) \\ q &= (1 + R)\cos(\theta_s) - (1 - R)\cos(\theta_i) \\ k_x &= +k[\sin(\theta_i) - \sin(\theta_s)] \\ k_z &= -k[\cos(\theta_i) + \cos(\theta_s)] \end{aligned} \tag{1.28}$$

where R is the appropriate Fresnel reflection coefficient of the tangent plane at the point of interest and $f_r'(x)$ is the derivative with respect to its argument, and the integrals are evaluated over the illuminated patch $S_{fractal}$. We consider here the case of scattering from a perfectly conducting rough surface where the magnitude of the Fresnel reflection coefficient is unity.

Evaluating the integral in (1.27) we find that after integrating by parts and assuming that the patch size is much larger than a wavelength, the scattering cross section or scattering coefficient

$$\frac{d\sigma(\theta_i, \theta_s)}{d\theta_s} = \left| \frac{\Psi_{sc}(\theta_i, \theta_s)}{\Psi_{max}} \right|^2$$

is found to be [50]

$$\frac{d\sigma(\theta_i, \theta_s)}{d\theta_s} = \left| \sec(\theta_i) \frac{1 + \cos(\theta_i + \theta_s)}{\cos(\theta_i) + \cos(\theta_s)} \right|^2 \times$$

$$\left| \sum_{m_0, m_1, \ldots, m_{N-1} = -\infty}^{+\infty} e^{[i \sum_{n=0}^{N-1} m_n \phi_n]} \prod_{n=0}^{N-1} J_{m_n}[C(D_r - 1)^n k_z \sigma] \operatorname{sinc}[(m_n b^n K_0 + k_x) L/\pi] \right|^2 \tag{1.29}$$

where Ψ_{max} is the specularly scattered field when the surface is smooth and we use the notation $\operatorname{sinc}(x) \equiv \sin(\pi x)/(\pi x)$.

It can be shown from (1.29) that the decrease in the specular scattering is proportional to the variance of the surface height as given by the relation

$$\left| \frac{\Psi_{max}}{\Psi_{max}\big|_{k\sigma \to 0}} \right| \approx 1 - 2[k\sigma \cos(\theta_i)]^2 \tag{1.30}$$

in the limit $K_0 L \gg 1$ and for $k\sigma < 1$ [50]. This result is consistent with physical intuition and with an analogous result derived for the average scattering coefficient of random surfaces. It also demonstrates that under appropriate approximations the relative *rms* height of the surface determines the scattering intensity in the specular direction for both fractal and random surfaces.

1.3.2.2 Scattering Cross Sections for the Approximate Case

The next three figures display the scattering cross-section results from the Kirchhoff scattering approximation for fractally rough surfaces. In each case the scattering cross section (1.29) is plotted on a logarithmic scale as a function of scattering angle θ_s in the scattering plane formed by $\mathbf{k_i}$ and $\mathbf{k_s}$. Figure 1.12 displays the fundamental result of this section. Here the angular scattering is shown for the roughness fractal dimension $D_r = 1.05, 1.30, 1.50, 1.70$, respectively, from left to right and top to bottom. The frequency scaling parameter is $b = 2e/3 \approx 1.81$ and the number of tones is $N = 6$. The illuminating plane wave is incident at an angle $\theta_i = 30°$ with respect to the reference plane as indicated by the arrow on the left. The *rms* height of the surface is $\sigma = 0.05\lambda$ and the patch size is $2L = 40\lambda$. Each plot is the average result of ten members of the ensemble, each with a different set of randomly chosen phases ϕ_n in (1.24) and (1.29).

As expected, the scattered energy is increasingly dispersed in angle as the roughness or fractal dimension increases. Therefore, we anticipate that this dispersion will provide a method to distinguish and classify such surfaces.

The physical basis for the spreading of scattered energy away from the specular

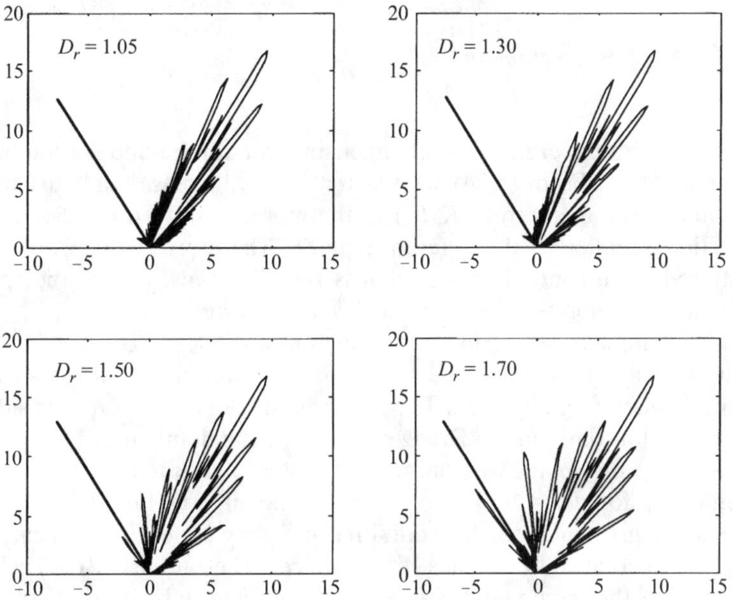

Figure 1.12 The angular variation of the scattering cross-section $d\sigma(\theta_i, \theta_s)/d\theta_s$ as a function of scattering angle θ_s for fractally corrugated surfaces with roughness fractal dimension $D_r = 1.05, 1.30, 1.50$, and 1.70, respectively, left to right, top to bottom. The incident wave is at $\theta_i = 30°$ (indicated by arrow) to the reference plane normal, and the patch size is $2L = 40\lambda$. These are polar plots with a dB scale shown for reference. [*Figure adapted from "Scattering from Fractally Corrugated Surfaces," D. L. Jaggard and X. Sun,* J. Opt. Soc. Am. A, **7**, 1131–1139 (1990). © *1990 Optical Society of America.*]

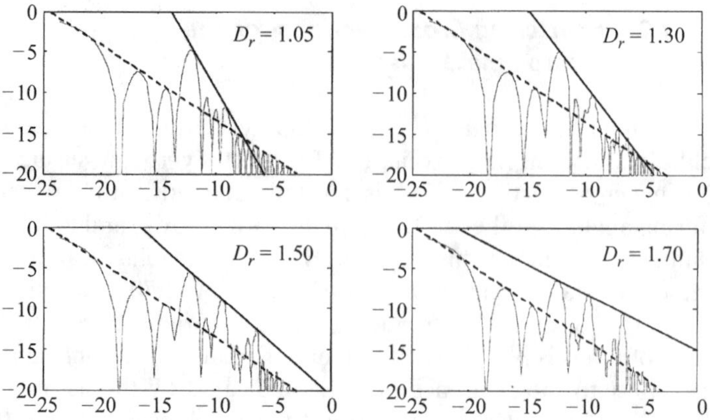

Figure 1.13 The scattering cross section $d\sigma(\theta_i, \theta_s)/d\theta_s$ as a function of $\sin[(\theta_s - 30°)/2]$, both in dB scales, for roughness fractal dimension D_r = 1.05, 1.30, 1.50, and 1.70, respectively, left to right, top to bottom. The incident wave is at $\theta_i = 30°$ to the reference plane normal, and the patch size is $2L = 40\lambda$. These are polar plots with a dB scale shown for reference. The envelope slopes of coupling sidelobes (solid lines) vary monotonically with the fractal dimension while the background slope (dashed lines) is constant for varying D_r. [*Figure adapted from "Scattering from Fractally Corrugated Surfaces," D. L. Jaggard and X. Sun,* J. Opt. Soc. Am. A, **7**, 1131–1139 (1990). © *1990 Optical Society of America.*]

direction with increasing fractal dimension can also be understood from conservation of momentum. From (1.24) we see that as D_r increases the surface will add larger amounts of momentum ($\sim K_0 b^n$) to the momentum of the incident wave ($\sim k$), in a direction parallel to the reference plane. Therefore, to conserve momentum, the scattered wave contains power that is redirected away from the specular direction. This phenomenon is often termed *Bragg coupling*.

The angular scattering cross section for $D_r = 1.05$ in Fig. 1.12, where the roughness is minimal and the surface is near sinusoidal, shows a main beam in the specular scattering direction. The two large sidelobes, one on each side of the specular scattering lobe, are due to Bragg coupling by the dominant ($n = 0$) sinusoid in (1.24). These *coupling lobes* are characteristic of any diffraction grating. With increasing roughness, for $D_r = 1.30$ and 1.50, additional coupling lobes emerge and their intensities grow due to the additional coupling by each significant sinusoid in the fractal function that defines the surface. The intensity of each coupling lobe approaches the same value as $D_r \rightarrow 2$. This effect can be partially observed in the plot for $D_r = 1.70$. Nulls in these plots are evident in the ensemble average and so must be due to the zeros of the sinc functions of (1.29).

In Fig. 1.13 we replot the scattering cross-section data of Fig. 1.12 for the region $30° < \theta_s < 90°$ as a function of $\sin[(\theta_s - 30°)/2]$, both quantities shown on logarithmic scales. Two envelopes are evident in these plots. The background envelope (dashed line with slope ≈ -1) is due to the finite patch size and consists of the specularly reflected main beam and its sidelobes. The coupling lobe envelope (solid line with variable slope) is due to surface coupling and consists of the first-order coupling of the

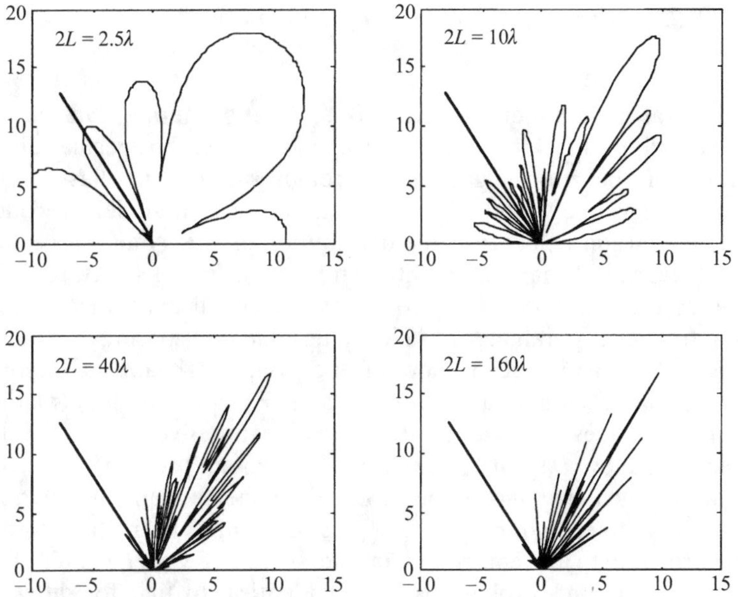

Figure 1.14 Polar plot of the angular variation of the scattering cross section $d\sigma(\theta_i, \theta_s)/d\theta_s$ as a function of scattering angle θ_s for fractally corrugated surfaces with dimension $D_r = 1.5$ and patch sizes $2L = 2.5\lambda$, 10λ, 40λ, and 160λ, from left to right, top to bottom. Here $\lambda/\Lambda_0 = 0.1$ and $\theta_i = 30°$. [*Figure adapted from "Scattering from Fractally Corrugated Surfaces," D. L. Jaggard and X. Sun,* J. Opt. Soc. Am. A, **7,** 1131–1139 (1990). © *1990 Optical Society of America.*]

surface harmonics. This latter slope is indicative of surface roughness and is given approximately as $-[2.7(2 - D_r)]$ for the cases examined here. This result is not dissimilar to that of scattering from fractal aggregates and porous material and suggests that this slope provides a means for quantifying fractal dimension and surface roughness. An alternative measure of fractal dimension may be obtained through the integration of the scattered power in successive bands of sidelobes. Examples of this for other diffraction problems are given elsewhere [12], [15], [82–86].

The variation of scattering cross section with patch size is of interest in remote sensing and surface characterization. This information is displayed in Fig. 1.14 where polar plots of the scattering cross section are given for $D_r = 1.50$ and $2L = 2.5\lambda$, 10λ, 40λ, and 160λ. The specular scattering mainbeam width is inversely proportional to the normalized patch size. As expected, the angular scattering cross section provides less information about surface roughness as the patch size $2L$ decreases below the critical value Λ_0 since there is not enough of the fractal surface to accurately imprint its fractal signature on the illuminating wave. Likewise, as seen from Fig. 1.14 the patch must contain at least 5 to 10 wavelengths of the incident wave to accurately provide information regarding its multiscale nature. Therefore, we suggest that the relation $2L > \Lambda_0 \gtrsim 5\lambda$ should hold to extract fractal information from multiscale surfaces.

1.3.2.3 Observations on the Approximate Case

Several observations on fractal surface scattering can be noted from the approximate results. First, the scattering intensity in the specular direction is a strong function of the *rms* height of the rough surface and has a similar functional dependence as one would expect for random surface models. Second, the roughness fractal dimension D_r controls the distribution of energy among scattering lobes. This result is typical of many wave interactions with fractal structures. As the dimension increases there are more opportunities for significant Bragg coupling. Third, the spatial frequency parameter b, which determines the spacing of the surface harmonics, controls the angular separation of the coupling beams in accordance with the conservation of momentum (Bragg's law). Fourth, the slope of the coupling lobe envelope provides both a qualitative and a quantitative measure of surface roughness as expressed by D_r. Fifth, for accurate remote probing of fractal surfaces, the scattering patch size should be larger than the fundamental period of the fractal surface and larger than five wavelengths for significant fractal information contained in the surface to be embedded in the scattered wave. Finally, we note that the formulation for this problem is almost identical to that for diffraction by a single fractal phase screen [80]. In each case, an incident field has impressed on its phase front information about a fractal object. The wave then propagates in accordance with diffraction theory but contains encoded fractal information which can be extracted from the scattered data. Alternative viewpoints of approximate scattering using the Rayleigh approximation [51], [52] and using a perturbational approach [54] can be found elsewhere. Applications to sea and related surface scattering are also available [56], [57], [63–67], [87–96].

We turn now to an exact method that yields quantitatively similar results but can be applied to surfaces which violate some of the restrictions (e.g., moderate or large radii of curvature, relatively small surface slopes, and relatively small heights of roughness) inherent in most approximate methods including the one used here. The exact method also takes into account electromagnetic wave polarization which was neglected in the scalar Kirchhoff approximation.

1.3.3 Exact Scattering Solution

We apply the T-matrix approach or EBCM of Waterman [97–99] to the problem of the previous section—scattering of electromagnetic waves from perfectly conducting fractal surfaces. The method formulation is exact, at least in principle, when applied to periodic surfaces and has been previously applied to the case of sinusoidal surfaces [100] and random rough surfaces [101].

The T-matrix approach involves the extinction theorem in which the surface field produces a contribution inside the object that exactly cancels the incident field throughout the interior of the object. The resulting transition-matrix or T-matrix $[T]$ relates incident and scattered fields. Details of the original method are readily available in several texts [102], [103].

Here we follow our prior work [55], [60] and expand the T-matrix method to a class of fractal surfaces characterized by the fractal function (1.24) which is subset of the class of almost-periodic functions. Using concepts developed for waves in almost-periodic media [104], we expand the surface fields in terms of generalized Floquet modes. This allows us to find closed-form expressions for the scattering amplitudes in the the TE and TM cases. Results for the angular scattering cross section are found numerically using infinite matrices truncated to the nth order.

1.3.3.1 Formulation of Exact Surface Scattering Solution

The T-matrix approach starts with Green's theorem below, assuming $e^{-i\omega t}$ time-harmonic variation,

$$\psi_i(\mathbf{x}) + \iint_{S_{fractal}} [\psi(\mathbf{x}')\hat{\mathbf{n}}\cdot\nabla_S' G(\mathbf{x},\mathbf{x}') - G(\mathbf{x},\mathbf{x}')\hat{\mathbf{n}}\cdot\nabla_S'\psi(\mathbf{x}')]\,dS' = \begin{cases} \psi(\mathbf{x}), & z > f_r(x) \\ 0, & z < f_r(x) \end{cases} \tag{1.31}$$

where $\psi(\mathbf{x}')$ is the electric (TE or perpendicular polarization) or magnetic (TM or parallel polarization) surface field transverse to the scattering plane formed by $\mathbf{k_i}$ and $\mathbf{k_s}$. Here $\psi(\mathbf{x})$ and $\psi_i(\mathbf{x})$ are the total and incident transverse fields, respectively. As before, $f_r(x)$ is the fractal function defined in (1.24). The upper inequality in (1.31) provides the total field at an arbitrary point $\mathbf{x}$ above the conducting surface as the sum of incident and scattered fields, with the scattered fields represented by the surface integral which extends over the fractal surface $S_{fractal}$. The lower inequality corresponds to the extended boundary condition in which the surface field and its normal derivative give rise to a wave which cancels the incident field in the region below the surface.

The Green's function in (1.31) is the two-dimensional free-space form given by

$$\begin{aligned} G(\mathbf{x},\mathbf{x}') &= \frac{i}{4}H_0^{(1)}(k|\mathbf{x}-\mathbf{x}'|) \\ &= \frac{i}{4\pi}\int_{-\infty}^{\infty}\frac{1}{k_z}e^{[ik_x(x-x')+ik_z|z-z'|]}\,dk_x \end{aligned} \tag{1.32}$$

where $H_0^{(1)}$ is the Hankel function of the first kind and zeroth order and represents an outwardly propagating cylindrical wave.

Next the surface field and the field above the surface are both expanded into generalized Floquet series. This allows the surface field coefficients in the column vector $[\alpha]$ to be written in terms of the incident field coefficients of the column vector $[a]$. The relation is

$$[a] = [Q^-][\alpha] \tag{1.33}$$

where the square matrix $[Q^-]$ has elements containing exponentials and Bessel functions related to the surface geometry. Likewise it can be shown that the scattered field coefficients in the column vector $[b]$ can also be written in terms of the column vector $[\alpha]$ as

$$[b] = [Q^+][\alpha] \tag{1.34}$$

where $[Q^+]$ is a second square matrix containing information regarding surface geometry. The formulations are slightly different for the TE and TM polarizations and are carried out explicitly elsewhere [55], [60] for the quantities $[\alpha]$, $[a]$, $[b]$, $[Q^-]$ and $[Q^+]$. The result in both cases is a T-matrix $[T]$ of the form

$$[T] = [Q^+][Q^-]^{-1} \tag{1.35}$$

Using (1.33)–(1.35) we find

$$[b] = [T][a] \tag{1.36}$$

which relates incident and scattered field coefficients. The column vector $[b]$ contains the information from which the scattering cross sections $d\sigma(\theta_i, \theta_s)/d\theta_s$ can be calculated.

One of the major contributions of this work is the extension of the T-matrix approach from periodic surface scattering to multiscale, almost-periodic surface scattering. We turn now to these results.

1.3.3.2 Scattering Cross Sections for the Exact Case

The T-matrix numerical results of the next four figures show the scattering cross section $d\sigma(\theta_i, \theta_s)/d\theta_s$ as a function of scattering angle θ_s for a specified incident angle θ_i. These results appear to be numerically accurate in that they are self-consistent with respect to the energy balance between incident and scattered wave quantities and take into account interactions through the fifth order with little variation in the scattering results with respect to change in order. Typically we find the energy balance parameter to differ from unity by less than 0.1. Finally, in the appropriate limit, the T-matrix results approach results from both the Kirchhoff and Rayleigh approximations which provides additional validation of these results.

We first examine the comparison between the T-matrix results and those of the Kirchhoff approximation. This is followed by an examination of T-matrix results for the case where the surface profiles no longer satisfy the Kirchhoff assumptions because of increasing surface height or increasing illumination frequency. For these latter cases in which an exact formulation is needed, we also examine the effect of polarization.

The angular scattering cross sections for TE polarization shown in Fig. 1.15 demonstrate rather good agreement between the T-matrix method (solid line) and the Kirchhoff method (solid line with boxes) for the case of relatively small wavelength and *rms* surface height variations ($\lambda/\Lambda_0 = 0.2$ and $k\sigma = 0.5$). The results show that for modest fractal dimension the scattered field displays the specular peak and two large peaks due to the dominant sinusoid of the surface. The specular peaks predicted by the T-matrix and Kirchhoff methods are graphically indistinguishable. Higher spatial frequency sinusoids of the surface and the finite patch size contribute auxiliary sidelobes which appear on either side of the specular mainlobe. The Kirchhoff approximation accurately mimics the three major scattering lobes (mainlobe and two major sidelobes) and most of the smaller sidelobes found in the T-matrix calculation.

As the surface roughness and wavelength increases, the Kirchhoff assumptions are violated and this approximation technique gracefully degrades until it provides neither

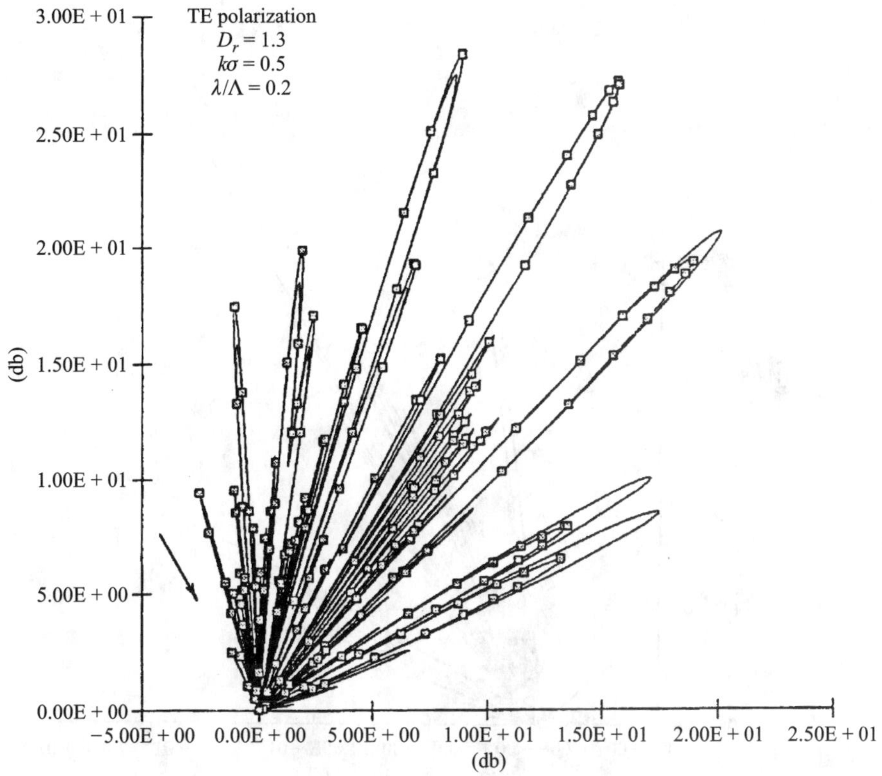

Figure 1.15 Polar plot of the angular variation of the scattering cross section $d\sigma(\theta_i, \theta_s)/d\theta_s$ as a function of scattering angle θ_s for fractally corrugated surfaces with dimension $D_r = 1.3$, spatial frequency scaling $b = 2e/3$, patch size $2L = 40\lambda$, *rms* height $k\sigma = 0.5$ and normalized incident wavelength $\lambda/\Lambda = \lambda/\Lambda_0 = 0.2$ for TE polarization. The incident angle $\theta_i = 30°$ is indicated by the arrow at left. Kirchhoff results (solid lines with boxes) and T-matrix results (solid lines) differ by only small amounts since the Kirchhoff approximations are valid. [*Figure adapted from "Scattering from Fractally Corrugated Surfaces: An Exact Approach," S. Savaidis, P. Frangos, D. L. Jaggard and K. Hizanidis,* Opt. Lett., **20,** 2357–2359 (1995). © *1995 Optical Society of America.*]

a quantitatively accurate nor a qualitatively accurate picture of the TE scattering as demonstrated in Fig. 1.16. Here the Kirchhoff results (solid line with boxes) differ greatly from the exact T-matrix results (solid line) and offer no reasonable approximation either in the mainlobe region or in the far sidelobe region. In particular the Kirchhoff results do not correctly identify the mainlobe in the specular direction and place large sidelobes where none exists. This is an example of the eventual failure of an approximation method for rough surface scattering.

The next figure displays results for *rms* surface height and illuminating wave frequency that again significantly violate the Kirchhoff approximation. The angular

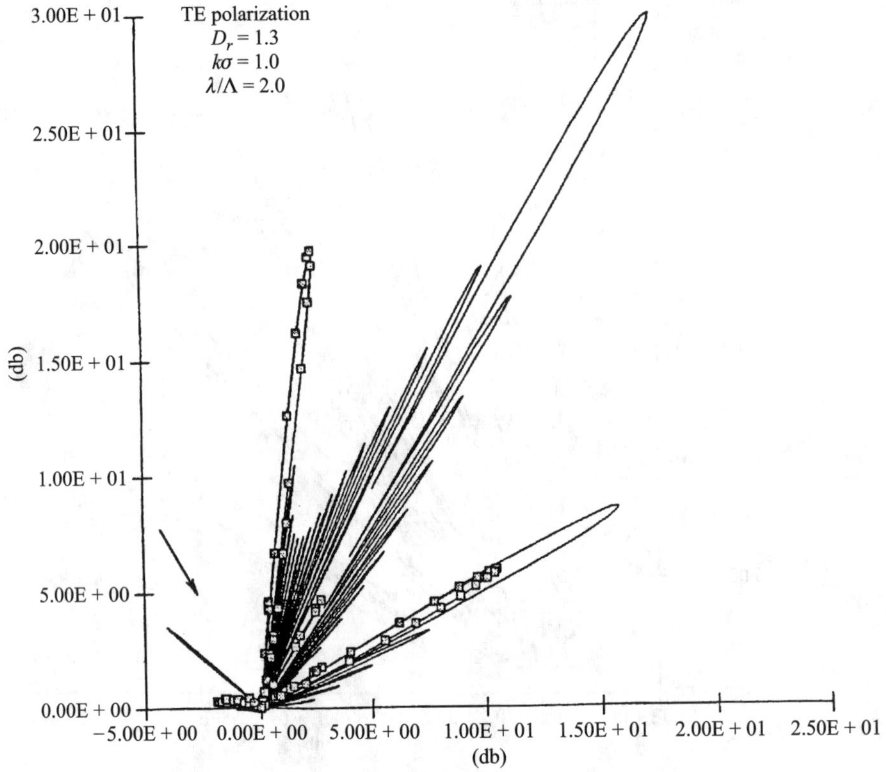

Figure 1.16 Polar plot of the angular variation of the scattering cross section $d\sigma(\theta_i, \theta_s)/d\theta_s$ as a function of scattering angle θ_s for fractally corrugated surfaces with dimension $D_r = 1.3$, spatial frequency scaling $b = 2e/3$, patch size $2L = 40\lambda$, *rms* height $k\sigma = 1.0$, and normalized incident wavelength $\lambda/\Lambda = \lambda/\Lambda_0 = 2.0$ for TE polarization. The incident angle $\theta_i = 30°$ is indicated by the arrow at left. Kirchhoff results (solid lines with boxes) and T-matrix results (solid lines) differ significantly since the Kirchhoff approximations are violated. Note that the Kirchhoff results do not provide even a qualitative picture of the T-matrix results. [*Figure adapted from "Scattering from Fractally Corrugated Surfaces: An Exact Approach," S. Savaidis, P. Frangos, D. L. Jaggard and K. Hizanidis,* Opt. Lett., **20,** 2357–2359 (1995). © *1995 Optical Society of America.*]

scattering cross-section for variations in normalized *rms* surface height $k\sigma$ is shown in Fig. 1.17 for both TE (left side) and TM (right side) polarizations as a function of *rms* surface height $k\sigma$ with constant roughness fractal dimension $D_r = 1.5$. It is clear that as the *rms* surface height increases (bottom of figure), the role of polarization becomes more important while for small surface variations (top of figure) polarization is less important. However, the effect of polarization is subtle for all examples in this figure.

Similar results appear in Fig. 1.18 for plots of angular scattering cross-section with increasing roughness as indicated by the fractal dimension D_r and constant *rms* surface height $k\sigma = 1.0$. In most cases the role of polarization is modest but greater than that displayed in the previous figure. As might be anticipated from the Kirchhoff results, the

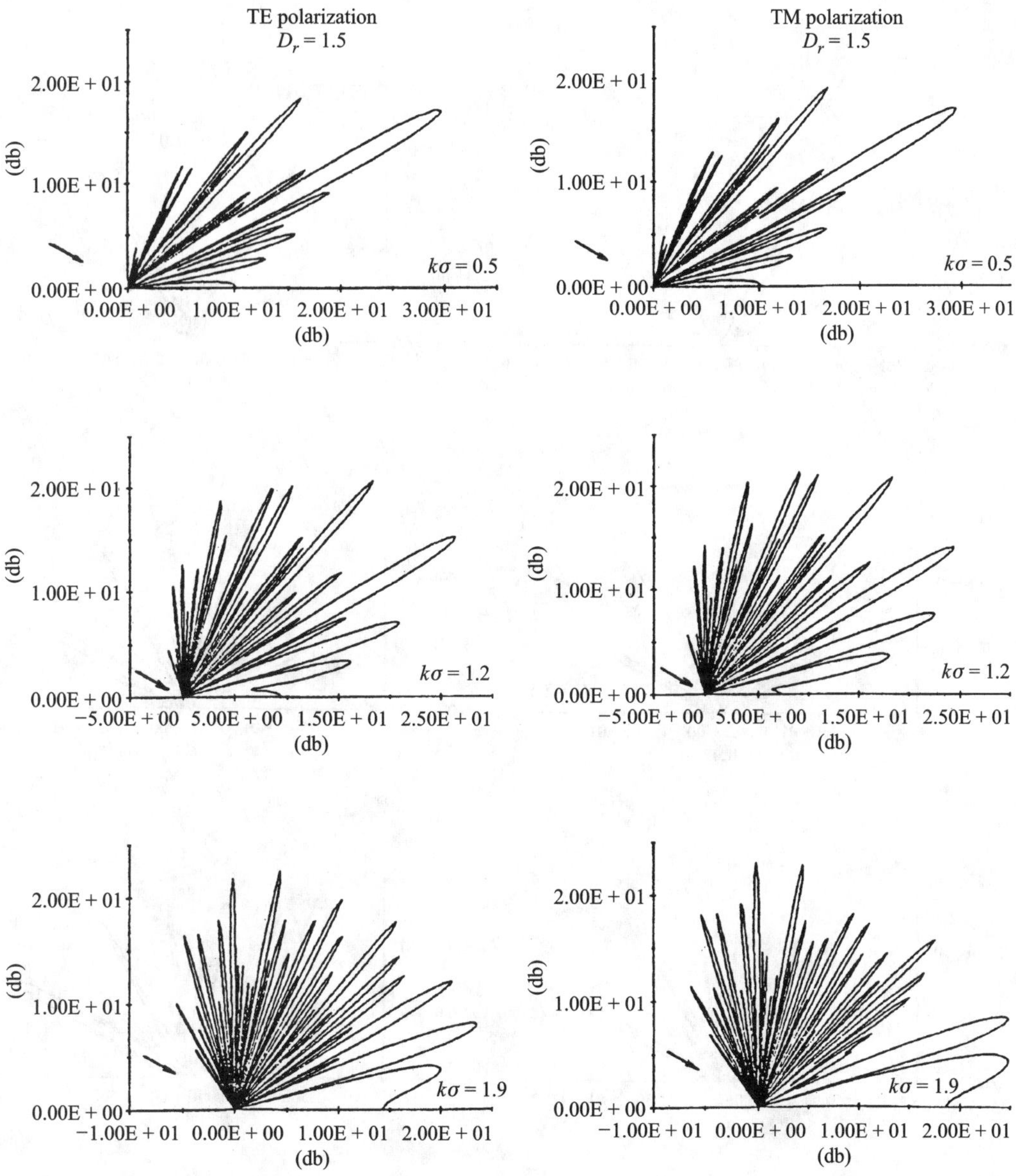

Figure 1.17 Polar plots of the angular variation of the scattering cross section $d\sigma(\theta_i, \theta_s)/d\theta_s$ as a function of scattering angle θ_s for TE (left) and TM (right) polarizations and *rms* surface height $k\sigma = 0.5$, 1.2, 1.9 (top to bottom) using the T-matrix approach. The fractally corrugated surfaces are characterized by dimension $D_r = 1.5$, spatial frequency scaling $b = 2e/3$, patch size $2L = 40\lambda$, normalized incident wavelength $\lambda/\Lambda_0 = 0.2$ with $N = 2$. The incident angle $\theta_i = 60°$ is indicated by the arrow at left. [*Figure adapted from "Scattering from Fractally Corrugated Surfaces with Use of the Extended Boundary Condition Method," S. Savaidis, P. Frangos, D. L. Jaggard and K. Hizanidis,* J. Opt. Soc. Am. A, **14**, 475–485 (1997). © *1997 Optical Society of America.*]

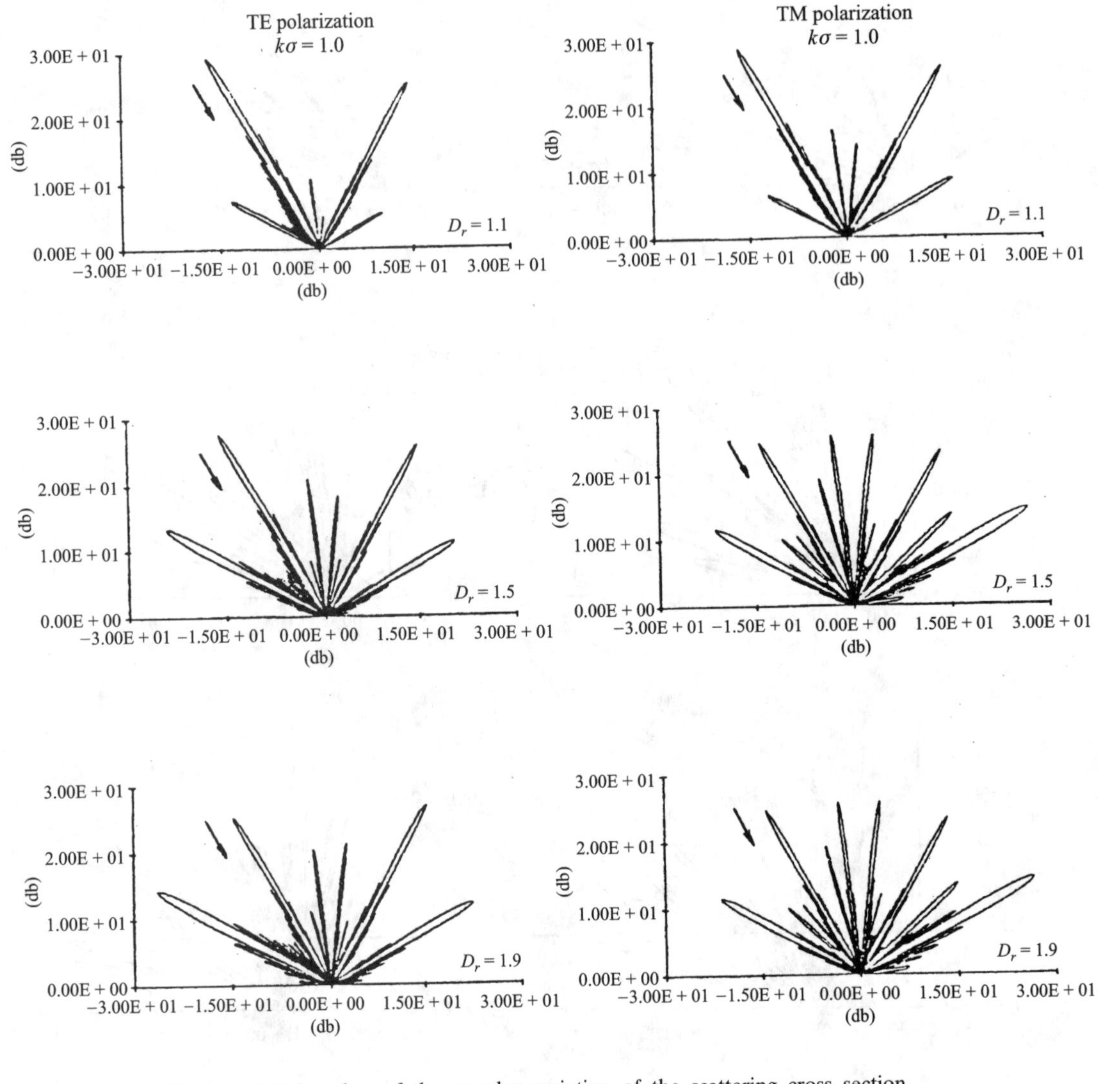

Figure 1.18 Polar plots of the angular variation of the scattering cross section $d\sigma(\theta_i, \theta_s)/d\theta_s$ as a function of scattering angle θ_s for TE (left) and TM (right) polarizations and dimension $D_r = 1.1, 1.5, 1.9$ (top to bottom) using the T-matrix approach. The fractally corrugated surfaces are characterized by *rms* surface height $k\sigma = 1.0$, spatial frequency scaling $b = 2e/3$, patch size $2L = 40\lambda$, normalized incident wavelength $\lambda/\Lambda_0 = 1.0$ with $N = 2$. The incident angle $\theta_i = 30°$ is indicated by the arrow at left. [*Figure adapted from "Scattering from Fractally Corrugated Surfaces with Use of the Extended Boundary Condition Method," S. Savaidis, P. Frangos, D. L. Jaggard and K. Hizanidis,* J. Opt. Soc. Am. A, **14**, 475–485 (1997). © *1997 Optical Society of America.*]

redistribution of energy away from the specular direction also increases with dimension for these T-matrix results.

1.3.3.3 Observations on the Exact Case

The T-matrix results summarized in this chapter have extended prior results for sinusoidal and other periodic surfaces to the case of almost-periodic and fractal surfaces. This analytic extension of previous results is significant since it enlarges the family of surfaces that can be considered using T-matrix techniques. The T-matrix numerical results have been validated by comparing them to the solution for a sinusoidal surface, by comparing them to both Kirchhoff and Rayleigh results in the appropriate limit, by confirming that energy is conserved, and by examining the stability of these results with respect to variations in the number of interactions that are included in the calculation. From all of these considerations it appears the T-matrix cross sections for the surfaces examined here are accurate and consistent with known results.

These considerations suggest that for sufficiently large surface variations or sufficiently large fractal dimension the polarization of the illuminating wave may play an important role. In particular there are more rapid variations in the scattering cross section for the TM case when compared to the TE case. This is consistent with the polarization effects for scattering from a variety of non-fractal objects. However, for none of the examples considered here are the scattering cross sections radically different for TE and TM polarizations. Instead the effects tend to be modest and in some cases rather subtle. Larger effects will occur as the roughness, number of tones in the surface function, frequency of illumination, and *rms* surface height increase. One of the research problems that remains to be solved is to find the general relation between the dispersion of power away from the specular direction and the fractal dimension D_r for arbitrary values of corrugation depth, surface slope, and frequency.

1.4 REFLECTION FROM CANTOR SUPERLATTICES

We now consider the reflection and transmission properties of fractal superlattices (multilayers) characterized by generalized Cantor bars. As in wave interactions with fractally rough surfaces and other structures based on fractal geometry, we see that an electromagnetic wave reflected from one of these superlattices is distinctively imprinted with the fractal characteristics of the superlattice. Our interest is in how this occurs and what information about the superlattice can be retrieved from the reflected wave.

We are able to exploit the fractal nature of these superlattices so that the structure of the solution parallels that of the fractal. Additionally, we can explain the main features of frequency-domain reflection data using physically motivated first-order arguments. This aids us in extracting from the reflection data information about the stage of growth S, number of gaps n_{gaps}, and fractal dimension D_s describing the superlattice. We also briefly discuss time-scale methods that use wavelets to extract fractal descriptors from time-domain reflection data.

1.4.1 Problem Geometry

We consider an electromagnetic wave incident upon a superlattice derived from a polyadic (multigap) Cantor bar C constructed as described in Section 1.2.4.2. A superlattice infinite in y and z may be constructed from C by defining the refractive index

$$n(x) = \begin{cases} n_1, & x \in C \\ n_0, & x \notin C \end{cases} \tag{1.37}$$

We take C to be a finite stage of growth Cantor bar, since as S increases the number of interfaces becomes large and the reflection coefficient approaches unity. As above, we will still refer to the "fractal dimension" of these prefractals and use the similarity dimension D_s of the limiting set to quantify this descriptor.

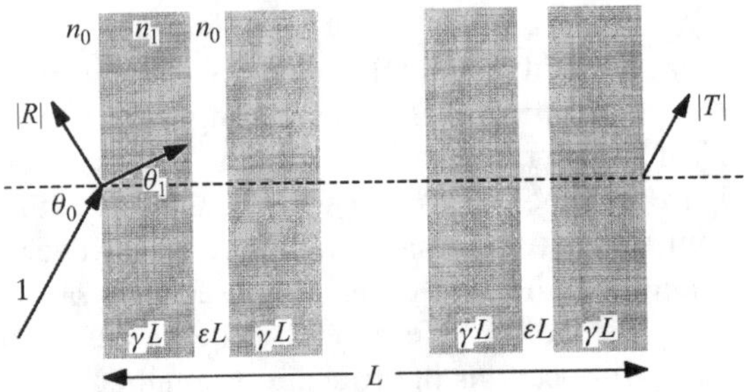

Figure 1.19 Electromagnetic wave obliquely incident from the left on a three-gap Cantor superlattice of length L at the first stage of growth. The angle of incidence and propagation in regions of refractive index (white) is θ_0, while the angle of propagation through layers of index n_1 (shaded) is θ_1 given by Snell's law. The individual refractive layers are of length γL and the outer gaps are of length εL. [*Figure after* [77].]

The problem geometry is shown in Fig. 1.19. An electromagnetic wave is incident from the left on the superlattice at an angle θ_0 from normal. The angle θ_1 from normal of propagation within the regions of index n_1 is given by Snell's law. We derive expressions for the magnitudes $|R|$ and $|T|$ of the reflected and transmitted fields using a recursive method that incorporates the symmetry of the superlattice structure.

1.4.2 Doubly Recursive Solution

The reflection and transmission coefficients of general superlattices may be computed using the chain-matrix method. However, for the family of superlattices defined above, we look for more efficient methods which might exploit the symmetry of these structures. We have developed a doubly recursive method of solution in which both recursions make use of the fractal structure to decrease the computational burden of calculating reflection and transmission coefficients [75], [77].

The first recursion is used to obtain the form for the coefficients of a collection of N identical objects arranged as the scaled copies of C are arranged within $[0, L]$. Following Knittl [105], we temporarily replace each of the N sub-multilayers with a single interface whose reflection and transmission coefficients x and y represent those of the entire sub-multilayer. We then separate the front interface from the back $N-1$ interfaces and express the coefficients of the N interfaces in terms of the coefficients of the 1st interface and the group of $N-1$ interfaces. The second step in the recursion gives the coefficients of the back $N-1$ interfaces in terms of the second interface and the remaining $N-2$ interfaces. This is repeated until only a single interface, with coefficients x and y, remains.

Formally, the reflection and transmission coefficients of $N = n_{gaps} + 1$ interfaces are of the forms

$$gen_r(n_{gaps}) = \frac{x + (y^2 - x^2)\,gen_r(n_{gaps} - 1)\exp(2in_0 k\alpha_i L\cos\theta_0)}{1 - x\,gen_r(n_{gaps} - 1)\exp(2in_0 k\alpha_i L\cos\theta_0)} \tag{1.38}$$

$$gen_t(n_{gaps}) = \frac{y\,gen_t(n_{gaps} - 1)\exp(in_0 k\alpha_i L\cos\theta_0)}{1 - x\,gen_r(n_{gaps} - 1)\exp(2in_0 k\alpha_i L\cos\theta_0)} \tag{1.39}$$

respectively, for $n_{gaps} \geq 1$ and k the free space wavenumber of the illuminating wave. Here the α_i ($i = 1, 2, \ldots, n_{gaps}$) denote the gap sizes normalized to the total length L and θ_0 is the angle of propagation with respect to the interface normal in the gaps. Relations (1.38–1.39) are subject to the initial conditions

$$gen_r(0) = x \tag{1.40}$$

and

$$gen_t(0) = y \tag{1.41}$$

corresponding to a single sub-multilayer.

Equations (1.38–1.41) are used in the first recursion to generate the reflection and transmission functions

$$g_r[x, y, L] = gen_r(n_{gaps}) \tag{1.42}$$

and

$$g_t[x, y, L] = gen_t(n_{gaps}) \tag{1.43}$$

for an n_{gaps}-gap Ca7ntor superlattice whose sub-multilayers have reflection and transmission coefficients x and y.

The dummy coefficients x and y in Eqs. (1.42–1.43) must now be replaced by the actual reflection and transmission coefficients of the sub-multilayers. We do this in the second recursion, with the fractal structure of the superlattice allowing us to write x and y as the coefficients of the superlattice at the previous stage of growth and of thickness γL. We thus have the reflection coefficient of the superlattice of thickness L generated by the stage S Cantor bar given by

$$R(S, L) = g_r[R(S - 1, \gamma L), T(S - 1, \gamma L), L] \tag{1.44}$$

with the corresponding transmission coefficient given by

$$T(S, L) = g_t[R(S - 1, \gamma L), T(S - 1, \gamma L), L] \tag{1.45}$$

When $S = 1$, the coefficients of the sub-multilayers no longer have the same form as that of the whole multilayer; they are instead simply the coefficients of a single dielectric slab of index n_1 embedded in media of index n_0. This leads to the initial conditions

$$R(0, d) = \frac{r_{01} + (t_{01}t_{10} - r_{01}r_{10})r_{10}\exp(2in_1 kd\cos\theta_1)}{[1 - r_{10}r_{10}\exp(2in_1 kd\cos\theta_1)]} \tag{1.46}$$

and

$$T(0, d) = \frac{t_{01}t_{10}\exp(in_1 kd\cos\theta_1)}{1 - r_{10}r_{10}\exp(2in_1 kd\cos\theta_1)} \tag{1.47}$$

where

$$r_{ij} = -\frac{\hat{n}_i - \hat{n}_j}{\hat{n}_i + \hat{n}_j} \tag{1.48}$$

and

$$t_{ij} = \frac{2\hat{n}_i}{\hat{n}_i + \hat{n}_j}\Theta_{ij} \tag{1.49}$$

are the Fresnel reflection and transmission coefficients, respectively, for an interface between media of refractive indices n_i and n_j. Here

$$\Theta_{ij} = \begin{cases} \cos\theta_i/\cos\theta_j & \text{for } \parallel \text{ or TM polarization} \\ 1 & \text{for } \perp \text{ or TE polarization} \end{cases} \tag{1.50}$$

$$\hat{n}_i = \begin{cases} n_i/\cos\theta_i & \text{for } \parallel \text{ or TM polarization} \\ -n_i\cos\theta_i & \text{for } \perp \text{ or TE polarization} \end{cases} \tag{1.51}$$

In all of the work reported here, we take $n_0 = 1$ and $n_1 = 1.5$.

These relations (1.38–1.51) give the reflection and transmission coefficients for the superlattice generated by any Cantor bar in this family (D_s, S, n_{gaps}, and ε arbitrary), with any values for the refractive indices n_0 and n_1, either polarization, and any angle of incidence θ_0. For triadic, single-gap, and other three-gap Cantor bars, these relations imply previously obtained results [68], [70], [75].

1.4.3 Results

1.4.3.1 Twist Plots

Using the expressions derived above, we plot the magnitude $|R|$ of the reflection coefficient for a Cantor superlattice against the normalized frequency kL of the incident wave in the left of Fig. 1.20. The curves shown are for $D_s = 3/4$, $n_{gaps} = 3$, and $\varepsilon = 0.1, 0.06, 0.02$ (top to bottom).

The differences between the plots in the left of Fig. 1.20 indicate that changes in ε have an effect on the reflection data, but these plots do not give a physical understanding of these changes. As one of our primary interests is the effect of lacunarity on the reflection coefficient, we use the *twist plots* introduced in [75] to investigate the relationship between this fractal descriptor and the scattering properties of the structure. In these plots, we show $|R|$ as a function of both normalized frequency kL and lacunarity parameter ε, with dark values representing minimal reflection and light values large reflection. The twist plot for $D_s = 3/4$ and $n_{gaps} = 3$ and $S = 1$ is shown in the right of Fig. 1.20, with dashed lines indicating the ε values corresponding to the plots on the left. Here, as in other twist plots, the bottom edge of the twist plot is the maximum lacunarity case $\varepsilon = 0$.

One of the most prominent features of the twist plot in Fig. 1.20 and of twist plots in general is the clearly visible null structure which gives the plots their twisting appearance. These dark lines, representing vanishing reflection, may be classified in

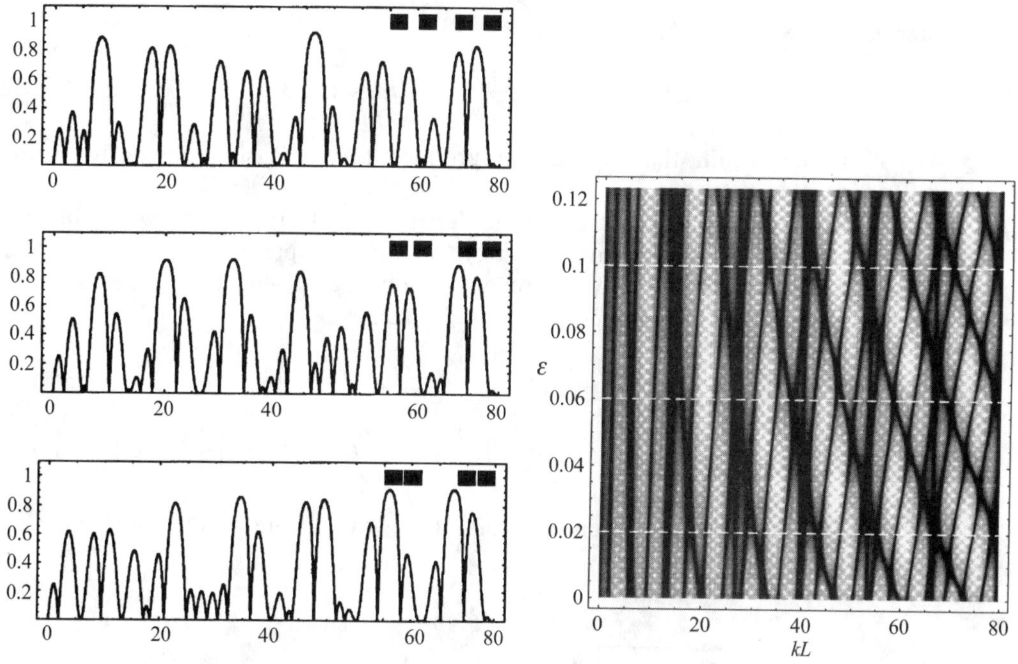

Figure 1.20 Plots of $|R|$ as a function of kL (left) with $\varepsilon = 0.1, 0.06, 0.02$ (top to bottom) for a three-gap Cantor bar with $D_s = 3/4$ and $S = 1$, and the twist plot (right) for this Cantor bar, with dashed lines indicating the values of ε for the plots at left. [*Figure from "Fractal Superlattices: A Frequency Domain Approach," A. D. Jaggard and D. L. Jaggard, presented at the* PIERS (Progress in Electromagnetics Research Symposium) 1998 Meeting, *Nantes, France (July 13–17, 1998). © 1998 Jaggard and Jaggard.*]

three families of non-intersecting nulls. These are the *vertical nulls*, running from the top to bottom of the twist plots, the *arc nulls*, running from the upper left to lower right, and the *striation nulls*, running from the lower left to upper right. We will first explain these nulls through first-order arguments, linking them to ε, and then use them as landmarks in assessing the effects of changes in stage of growth, fractal dimension, and number of gaps.

1.4.3.2 Nulls and Their Structure

We now consider the first-order interactions which cause these three families of nulls, and derive equations predicting the nulls from these interactions. The first family of nulls is the vertical nulls, which are independent of the lacunarity parameter ε. These nulls are caused by destructive interference between the front and back faces of individual slabs of index n_1 as shown in the top of Fig. 1.21. The distance γL between these interfaces is independent of ε. The reflection from the front and back face will be out of phase (noting the sign change in the reflection off the back interface) for values of kL given by

$$kL|_{\text{vertical nulls}} = \frac{m\pi}{\tilde{n}_1 \gamma} \qquad (m = 1, 2, 3, \ldots) \tag{1.52}$$

All N^S layers within the superlattice produce this null simultaneously, further reinforcing it.

The arc nulls arise from a collective destructive interference between the front faces of the individual layers within the front half of the superlattice. This occurs when the sum of the phases of the waves reflected from the sub-bar front faces is zero as shown in the middle of Fig. 1.21 or when kL satisfies

$$kL|_{\text{arc nulls}} = \frac{\left(2m + \frac{2l}{\left\lfloor \frac{N}{2} \right\rfloor}\right)}{2\tilde{n}_1 \gamma + 2\varepsilon}\pi \qquad \left(m = 1, 2, 3, \ldots; \quad l = 1, 2, \ldots, \left\lfloor \frac{N}{2} \right\rfloor - 1\right) \tag{1.53}$$

We observe that this interference occurs at the same time as the same collective

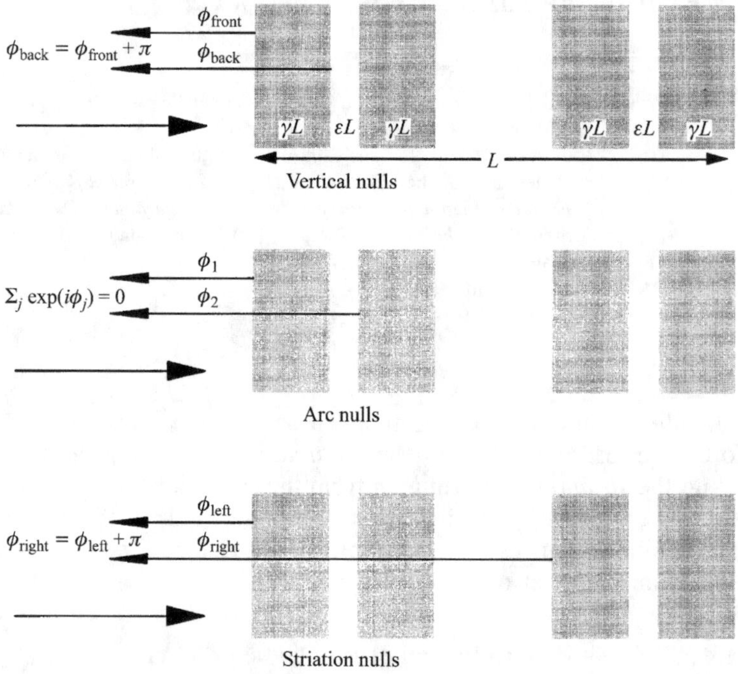

Figure 1.21 Diagram of the three first-order interferences which generate the vertical nulls, arc nulls, and striation nulls (top to bottom). Here, the ϕ's are the phases of waves reflected from the sub-bar surfaces. The vertical nulls are caused by the destructive interference between the front and back surfaces of individual sub-bars. The arc nulls are due to the collective destructive interference from front, or back, surfaces of neighboring sub-bars. Striation nulls arise from the destructive interference from the front, or back, surface of corresponding sub-bars on either side of the Cantor bar center. [© *1999 Jaggard and Jaggard.*]

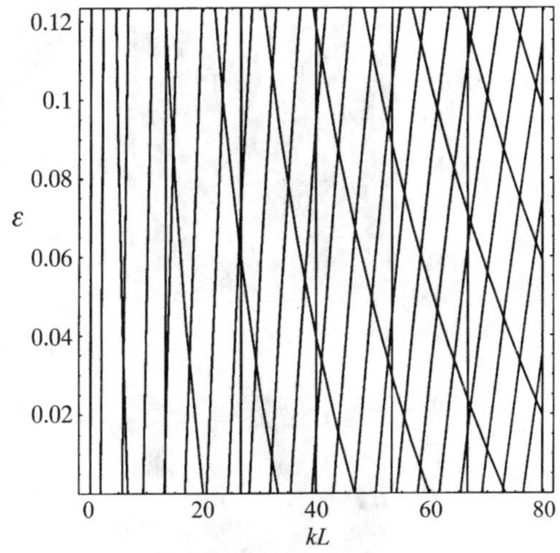

Figure 1.22 Null structure predicted by first-order analysis for the three-gap Cantor bar with $D_s = 3/4$ at the first stage of growth. Compare with the twist plot for this Cantor bar shown in Fig. 1.20. [*Figure taken from "Scattering from Fractal Superlattices with Variable Lacunarity," by A. D. Jaggard and D. L. Jaggard,* J. Opt Soc. Am A, **15**, 1626–1635 (1998). © *1998 Optical Society of America.*]

destructive interference within the back faces in the front half of the superlattice, the front faces in the back half, and the back faces in the back half.

The striation nulls are caused by destructive interference between the front (back) face of the ith slab in the front half of the superlattice and the front (back) face of the ith slab in the back half of the superlattice as pictured in the bottom of Fig. 1.21. These nulls are described by the equations

$$kL\big|_{\text{striation nulls}} = \frac{(2m+1)\,\pi}{\left(N - \left\lfloor \frac{N}{2} \right\rfloor\right)\tilde{n}_1\gamma + 1 - N\gamma - \left(\left\lfloor \frac{N}{2} \right\rfloor - 1\right)\varepsilon} \qquad (m = 1, 2, 3, \ldots) \qquad (1.54)$$

The interference occurs simultaneously for all such face pairs, reinforcing the striation nulls.

The null structure predicted by (1.52–1.54) is shown in Fig. 1.22 for a three-gap Cantor set with dimension $D_s = 3/4$ at stage of growth $S = 1$. We observe the close agreement between the nulls predicted by the first-order analysis and those shown in the corresponding twist plot in Fig. 1.20.

1.4.3.3 Polarization

Equations (1.50–1.51) allow the relations developed above for the reflection and transmission coefficients to be used for oblique angles of incidence, and we now consider the effect of polarization on the reflection data, again using twist plots to display the magnitude of the reflection coefficients.

Comparing twist plots for the parallel (TM) and perpendicular (TE) polarizations as shown in Fig. 1.23, we see that the parallel polarization reflection coefficient (left) is smaller than that for the perpendicular case (right). This is expected from the results for a half-space or a single refractive slab. As θ_0 nears the Brewster angle [$= \arctan(n_1/n_0) \approx 56°$] the reflection coefficient vanishes for parallel polarization, while for grazing incidence the magnitude approaches unity.

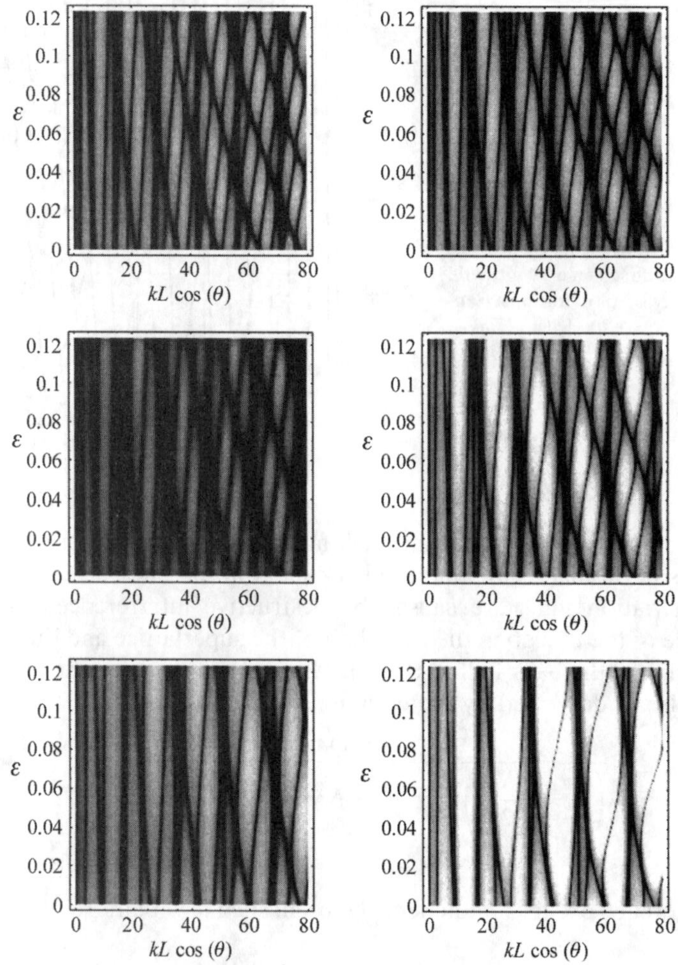

Figure 1.23 Twist plots of the absolute value of the reflection $|R|$ for the three-gap, first-stage Cantor bar with $D_s = 3/4$. These twist plots are for oblique incidence of $\theta_0 = 30°, 50°$, and $70°$ (top to bottom) for parallel (left) and perpendicular (right) polarizations. Note the decrease in reflection near the Brewster angle ($\theta_0 \approx 56°$) in the middle plot in the left column. [*Figure taken from "Scattering from Fractal Superlattices with Variable Lacunarity," A. D. Jaggard and D. L. Jaggard,* J. Opt. Soc. Am. A, **15**, 1626–1635 (1998). © *1998 Optical Society of America.*]

Although the reflection coefficient may differ by large factors between the two polarizations, we observe that the null structures are essentially the same for the perpendicular and parallel cases. This suggests that the null structure is a useful tool for analyzing the reflection coefficient, regardless of polarization. The null structure does, however, change with angle of incidence, as an increase in θ_0 represents an increased path length in each layer of the superlattice, effectively stretching the null structure along the kL axis. Other examples are given in the literature [77].

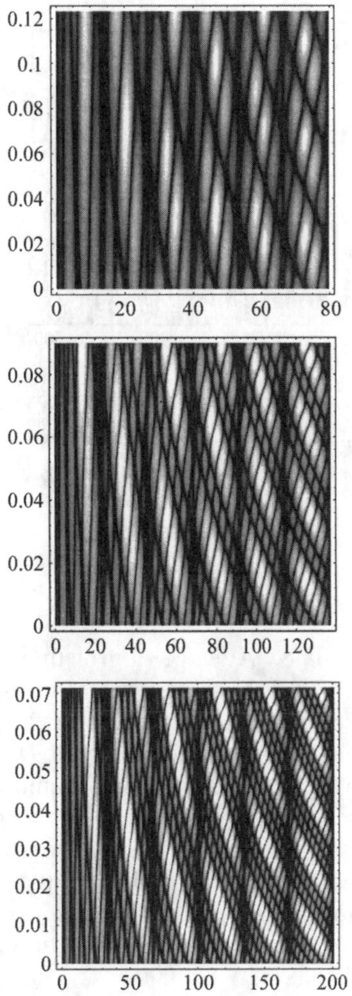

Figure 1.24 Twist plots for $n_{gaps} = 3$, 5, and 7 (top to bottom). Here $D_s = 3/4$ and $S = 1$. Note the grouping of the arc nulls as the number of gaps increases. [*Figure adapted from "Scattering from Fractal Superlattices with Variable Lacunarity," A. D. Jaggard and D. L. Jaggard,* J. Opt Soc. Am A, **15**, 1626–1635 (1998). © *1998 Optical Society of America.*]

1.4.4 Fractal Descriptors: Imprinting and Extraction

1.4.4.1 Frequency-Domain Approach

Number of Gaps n_{gaps}. We are able to retrieve the value of n_{gaps} through visual inspection of the twist plots. The arc nulls are grouped in sets of $\lfloor (N-1)/2 \rfloor$, with these groups separated by regions without arc nulls. This is illustrated in Fig. 1.24 where twist plots are shown for $n_{gaps} = 3$, 5, and 7 (top to bottom); these contain arc nulls in sets of 1, 2, and 3, respectively. We may explain this grouping from first principles and (1.52–1.54) [106]. For Cantor bars with n_{gaps} even and thus a central bar instead of a central gap, we see evidence of this geometrical change in the null structure [77]. We may use this to determine the parity of n_{gaps}, and along with the grouping of the arc nulls determine the exact value of n_{gaps}.

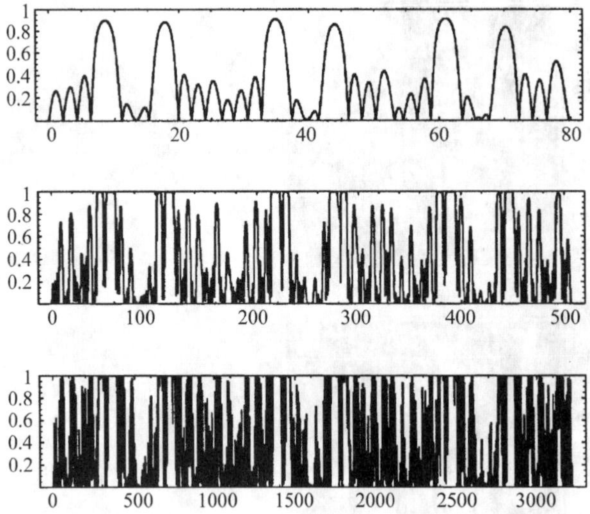

Figure 1.25 Plots of $|R|$ as a function of kL for $S = 1, 2, 3$ (top to bottom). Each stage reflection is rescaled by $1/\gamma$ in kL. Here $D_s = 3/4$, $n_{gaps} = 3$, and $\varepsilon = \varepsilon_{max}$ for minimum lacunarity. Note the similar overall structure, overlaid with increasingly rich detail as S increases.

Stage of Growth S. At stage of growth S, the intervals in the Cantor bar are of length $\gamma^S L$. Thus with increasing stage of growth, we see increasingly fine structure in the superlattice and richer structure in the twist plots. In Fig. 1.25 we plot the reflection coefficient for superlattices with $D_s = 3/4$, $n_{gaps} = 3$ for stages of growth $S = 1, 2, 3$ (top to bottom), and with minimum lacunarity $\varepsilon = \varepsilon_{max}$. Note that the kL axis has been rescaled by $1/\gamma$ between each stage; this corresponds to the scaling by γ of the smallest structure present in the superlattice. Here the reflection plot at stage of growth S exhibits structure similar to the (rescaled) reflection plot at stage of growth $S+1$. Knowing the length L and n_{gaps} of the superlattice it appears the stage of growth can be extracted from the reflection data.

Fractal Dimension D_s. From (1.52–1.54), we see that the locations of all three families of nulls are functions of γ and, in the case of the arc and striation nulls, N. Since these quantities determine D_s, we expect that variations in fractal dimension would have some effect on the null structure of twist plots. As demonstrated in [77], changes in D_s are seen in the corresponding twist plots, indicating that the fractal dimension of the superlattice is imprinted on the reflected wave. This is also evident from effective medium theory, which suggests that low-frequency reflection data will contain information about the average index of the superlattice since low-frequency waves in the superlattice "see" the refractive index averaged over the entire superlattice. With knowledge of the number of gaps, stage of growth, and values of n_0 and n_1, we may compute D_s from the average index of the superlattice.

We may consider the behavior at $kL = 0$ of the expression obtained above for the reflection coefficient to retrieve the average index of the superlattice and hence D_s [106], [107]. Preliminary results using this method are shown in Table 1.2 for a three-gap Cantor bar with $D_s = 1/2$, 3/4, 9/10 and $S = 1$, 2, 3. We are able to extract the value of D_s from the reflection data independent of the value of the lacunarity parameter ε.

TABLE 1.2 Values of Fractal Dimension Retrieved from Three-Gap Superlattices for Stages of Growth $S = 1, 2, 3$ and $D_s = 1/2, 3/4, 9/10$

D_s	Stage 1	Stage 2	Stage 3
1/2	0.529	0.519	0.513
3/4	0.778	0.775	0.771
9/10	0.915	0.914	0.913

1.4.4.2 Time-Scale Approach

Here, we briefly consider an alternative approach to the frequency-domain discussion above by using the impulse response of a Cantor superlattice and a time-scale analysis. We expect that the finer structure present in Cantor superlattices at high stages of growth will also be seen in their time-domain response and this is indeed true for continuous Weierstrass profiles [108]. The time-scale method we consider has recently been developed using wavelet analysis to extract the stage of growth and fractal dimension from time-domain data of Cantor superlattices [109], [110], [113]. This work has made use of the natural connection between the multiple scales present in fractal structures, and thus in their electromagnetic signatures, and the multiscale analysis for which wavelets and wavelet transforms are ideally suited.

We show that when the hierarchical structure of a wavelet-transform modulus maxima is revealed in time-scale data, this structure displays the construction rule of the Cantor set that governs the geometry of the superlattices. As an example we consider the wavelet analysis of a triadic Cantor superlattice at stage of growth $S = 6$. As displayed in Fig. 1.26 its time-domain response to a narrow width Gaussian impulse consists of a series of several large peaks followed by a series of smaller peaks which eventually decay. The peaks are due to the numerous multiple reflections inherent in this superlattice. For this example we choose an interrogating impulse

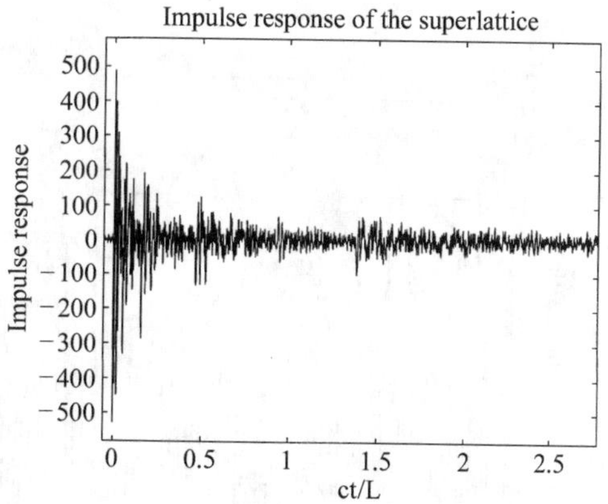

Figure 1.26 Impulse response of a triadic Cantor superlattice for $S = 6$. Here $L/\tau c = 900$ where c is the vacuum speed of light and τ is the temporal standard deviation of the incident Gaussian pulse. The pulse width is chosen to be on the order of the inverse of the highest spatial frequency of the superlattice. [*Figure adapted from "Fractal Superlattices and Thier Wavelet Analyses," H. Aubert and D. L. Jaggard,* Optics Comm., **149**, 207–212 (1998). © *1998 Elsevier Science B.V.*]

width close to the inverse of the highest spatial frequency of the superlattice. The incident wave can then access its finest structure. The continuous wavelet transform [111] of the impulse response is given in Fig. 1.27, showing both the modulus of the wavelet transform (top) and the maxima "skeleton" (bottom).

The wavelet skeleton displays the existence of a hierarchical distribution of singularities which in turn indicates both the stage of growth and fractal dimension. The stage of growth can be found from the number of hierarchical structures present in the skeleton. This skeleton "grows" from the bottom of the plot as the fractal is grown from one stage to the next and additional branching of the wavelet skeleton takes place. Thus the number of hierarchical levels in the wavelet plot reveals the stage of growth S if there is sufficient resolution in both the incident impulse and the wavelet analysis. This result appears to be valid for a variety of values of S and D_s as demonstrated in a series of examples examined recently [110] (not shown here).

Likewise, we note that a wavelet dimension D_w can be defined from the continuous wavelet transform and its associated skeleton through the scaling apparent in the hierarchical structure [109], [110], [113] (see Fig. 1.27). For refractive indices $n_0 = 1$ and $n_1 < 3.0$, the wavelet dimension is found to be almost identical to the similarity dimension D_s for a particular family of Cantor superlattices formed by

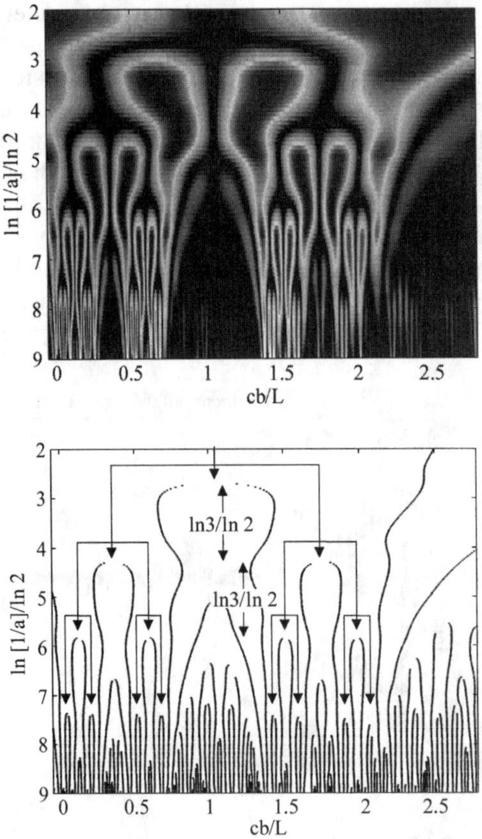

Figure 1.27 Modulus (top) of the continuous wavelet transform of the impulse response shown in Fig. 1.26 (black is zero). Here a is the scale parameter and b is the time parameter. The mother wavelet is the second derivative of the Gaussian function. Maxima (bottom) or skeleton of the wavelet transform modulus shown in the top plot. Each point indicates the location in time-scale domain of a maximum. A hierarchical structure of some wavelet maxima emerges clearly and reveals the construction rule of the associated Cantor set. The arrows indicate the branchings of such a structure (note that the scale factor between two successive branchings is constant). Here $D_s = \ln 2/\ln 3$ which is apparent from the skeleton. [*Figure taken from "Fractal Superlattices and Thier Wavelet Analyses," H. Aubert and D. L. Jaggard,* Optics Comm., **149**, 207–212 (1998). © *1998 Elsevier Science B.V.*]

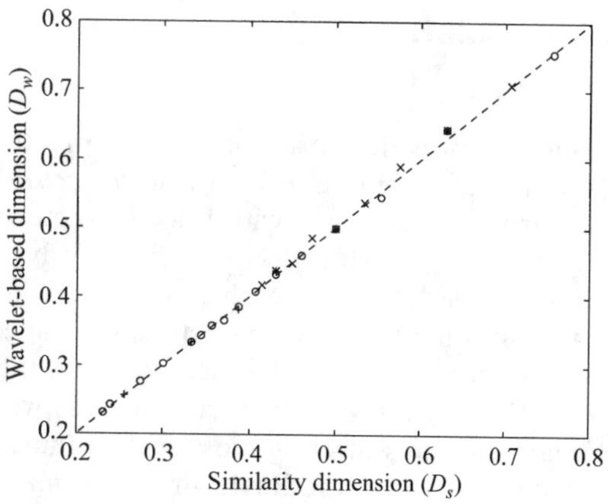

Figure 1.28 Wavelet-based dimension D_w as a function of the similarity dimension D_s for a class of Cantor superlattices: (+) $M - P = 2$, (○) $M - P = 4$ and (×) $M - P = 6$. Here each Cantor superlattice is constructed from a medium of index n_1 that is divided into M segments with the middle P segments removed and replaced with index n_0. [*Figure taken from "Fractal Superlattices and Thier Wavelet Analyses," H. Aubert and D. L. Jaggard,* Optics Comm., **149**, 207–212 (1998). *© 1998 Elsevier Science B.V.*]

taking a refractive bar of index n_1, dividing it into M segments and removing the middle P segments and replacing these segments with index n_0. A plot of the wavelet dimension D_w found from the continuous wavelet transform as a function of the similarity fractal dimension D_s used in the superlattice construction for some 25 examples of these superlattices is shown in Fig. 1.28. For these cases it appears the time-scale method is very useful for analyzing a variety of Cantor superlattices.

We see that the time-scale approach is a powerful method for examining fractal structures and appears to be inherently suited to extracting fractal descriptors from impulse response data. We also note its use in the analysis of slightly rough surfaces [112].

1.4.5 Observations on Superlattice Scattering

There are several results that should be noted. First, analytic expressions for the reflection and transmission coefficients of a family of Cantor superlattices are given as functions of fractal dimension, lacunarity, number of gaps, stage of growth, angle of incidence, and polarization. The scattering results are developed from a doubly recursive technique that efficiently yields numerical results. This method exploits the dilation symmetry of the structures so that the symmetry of the structure is mirrored in the symmetry of the calculations. Second, the concept of lacunarity is introduced to form increasingly diverse families of midpoint-symmetric polyadic Cantor sets. Lacunarity is connected to the geometric spacing of the associated Cantor set by a relation whose behavior matches our expectations for this descriptor. Third, the scattering results appear to be best summarized by a series of twist plots, particularly their null structure which is simply related to the geometry through interference arguments. Fourth, frequency-domain and time-scale approaches both appear to offer excellent ways to extract fractal descriptor information from the scattering data. For example, we find that fractal dimension, stage of growth and number of gaps can be found from reflection data using these methods. Both of these approaches are topics of continuing research.

1.5 CONCLUSION

Here we provide a brief introduction to fractal geometry, sets, and descriptors and summarize several aspects of *fractal electrodynamics*—the study of the interaction of electromagnetic waves with fractal structures. In this chapter our emphasis has been on the scattering of electromagnetic waves from fractally corrugated surfaces and fractal superlattices. In each case we have investigated methods through which fractal structures imprint their characteristic geometry on scattered waves, and techniques by which these waves can be made to yield the fractal descriptors of the scatterer.

For scattering from fractal surfaces we find that the fractal dimension is simply related to the scattering cross-section under the Kirchhoff approximation. A similar result holds under the Rayleigh approximation. The same results are evident for exact calculations of scattering cross-section in the case where the corrugation height is small or modest and the fractal dimension not too large. Likewise, similar trends appear in the exact case when approximate methods fail. However, simple connections between fractal dimension and cross section have not been found in the case of arbitrarily large corrugations or large fractal dimension, although it is quite possible they exist. An added complexity and opportunity for remote sensing is provided by the increasing role of polarization for high frequency, large dimension, or large corrugation.

For superlattices we find that many fractal descriptors are embedded in the scattering data in recognizable ways. We pay particular attention to the effect of lacunarity, "gappiness," or texture on scattering parameters. From frequency-domain data we find methods of extracting not only the fractal dimension but also the number of gaps and stage of growth for these Cantor superlattices. In an alternative approach, we review methods by which time-scale analyses of the superlattice impulse response may also be used to investigate the properties of such multilayers. We expect the wavelet methods will be a powerful tool for fractal scattering analysis, the idea being to use multiscale analytical tools to examine multiscale structures. From our initial results the concept appears to be most promising.

Although relatively young, fractal electrodynamics has already yielded useful and intriguing results in a number of areas covering the microwave, millimeter wave and optical regimes. We anticipate additional applications to the synthesis of new devices and in reaching a deeper understanding of wave interactions with multiscale structures. The optimal use of lacunarity and perhaps other descriptors in these problems remains an open question but appears to offer a useful additional degree of freedom in the design of fractal devices. As one particularly promising case, antenna elements and arrays which follow in subsequent chapters provide several examples of applications of fractals to radiation problems. For the case of fractal arrays we again find the concept of lacunarity to be useful.

REFERENCES

[1] D. L. Jaggard, "On Fractal Electrodynamics," in *Recent Advances in Electromagnetic Theory*, H. N. Kritikos and D. L. Jaggard, eds., Springer-Verlag, New York, 1990, pp. 183–224.

[2] D. L. Jaggard, "Fractal Electrodynamics and Modeling," in *Directions in Electromagnetic*

Wave Modeling, H. L. Bertoni and L. B. Felsen, eds., Plenum Publishing Co., New York, 1991, pp. 435–446.

[3] D. L. Jaggard, "Fractal Electrodynamics: Wave Interactions with Discretely Self-Similar Structures," in *Electromagnetic Symmetry*, C. Baum and H. Kritikos, eds., Taylor & Francis, Washington, D.C., 1995, pp. 231–281.

[4] D. L. Jaggard, "Fractal Electrodynamics: From Super Antennas to Superlattices," in *Fractals in Engineering: 1997*, J. L. Vèhel, E. Lutton, and C. Tricot, eds., Springer-Verlag, Berlin, 1997, pp. 204–221.

[5] M. V. Berry, "Diffractals," *J. Phys. A: Math. Gen.*, **12**, 781–797 (1979).

[6] M. V. Berry and T. M. Blackwell, "Diffractal Echoes," *J. Phys. A: Math. Gen.*, **14**, 3101–3110 (1981).

[7] E. Jakeman, "Scattering by a Corrugated Random Surface with Fractal Slope," *J. Phys. A*, **15**, L55–L59 (1982).

[8] E. Jakeman, "Fresnel Scattering by a Corrugated Random Surface with Fractal Slope," *J. Opt. Soc. Am.*, **72**, 1034–1041 (1982).

[9] E. Jakeman, "Fraunhofer Scattering by a Sub-Fractal Diffuser," *Optica Acta*, **30**, 1207–1212 (1983).

[10] E. Jakeman, "Scattering by Fractals," in *Fractals in Physics*, L. Pietronero and E. Tosatti, eds., Elsevier Science Publishers B.V., New York, 1986, pp. 55–60. See Ref. [35].

[11] C. Allain and M. Cloitre, "Optical Fourier Transforms of Fractals," in *Fractals in Physics*, L. Pietronero and E. Tosatti, eds., Elsevier Science Publishers B.V., New York, 1986, pp. 61–64. See Ref. [35].

[12] C. Allain and M. Cloitre, "Optical Diffraction on Fractals," *Phys. Rev. B*, **33**, 3566–3569 (1986).

[13] C. Bourrely, P. Chiappetta, and B. Torresani, "Light Scattering by Particles of Arbitrary Shape: A Fractal Approach," *J. Opt. Soc. Am. A*, **3**, 250–255 (1986).

[14] C. Bourrely and B. Torresani, "Scattering of an Electromagnetic Wave by an Irregularly Shaped Object," *Optics Comm.*, **58**, 365–368 (1986).

[15] C. Allain and M. Cloitre, "Spatial Spectrum of a General Family of Self-Similar Arrays," *Phys. Rev. A*, **36**, 5751–5757 (1987).

[16] D. L. Jaggard and X. Sun, "Scattering from Bandlimited Fractal Fibers," *IEEE Trans. Ant. and Propagat.*, **37**, 1591–1597 (1989).

[17] M. M. Beal and N. George, "Features in the Optical Transforms of Serrated Apertures and Disks," *J. Opt. Soc. Am. A*, **6**, 1815–1826 (1989).

[18] X. Sun and D. L. Jaggard, "Scattering from Fractally Fluted Cylinders," *J. Electromagnetic Wave Appl.*, **4**, 599–611 (1990).

[19] Y. Kim, H. Grebel, and D. L. Jaggard, "Diffraction by Fractally Serrated Apertures," *J. Opt. Soc. A*, **8**, 20–26 (1991).

[20] D. L. Jaggard, T. Spielman, and X. Sun, "Diffraction by Cantor Targets," in *Conference Proceedings of the 1991 AP-S/URSI Meeting*, London, Ontario, 1991.

[21] T. Spielman and D. L. Jaggard, "Diffraction by Cantor Targets: Theory and Experiment," in *Conference Proceedings of the 1992 AP-S/URSI Meeting*, Chicago, Illinois, 1992.

[22] H. D. Bale and P. W. Schmidt, "Small-Angle X-Ray-Scattering Investigation of Submicroscopic Porosity with Fractal Properties" *Phys. Rev. Lett.*, **53**, 596–599 (1984).

[23] D. W. Schaefer, J. E. Martin, P. Wiltzius, and D. S. Cannell, "Fractal Geometry of Colloidal Aggregates," *Phys. Rev. Lett.*, **52**, 2371–2374 (1984).

[24] B. H. Kaye, J. E. L. Blanc, and P. Abbott, "Fractal Description of the Structure Fresh and Eroded Aluminum Shot Fineparticles," *Part. Charact.*, **2**, 56–61 (1985).

[25] P. W. Schmidt and X. Dacai, "Calculation of the Small-Angle X-Ray and Neutron Scattering From Nonrandom (Regular) Fractals," *Phys. Rev. A*, **33**, 560–566 (1986).

[26] D. Rojanski, D. Huppert, H. D. Bale, X. Dacai, P. W. Schmidt, D. Farin, A. Seri-Levy, and D. Avnir, "Integrated Fractal Analysis of Silica: Adsorption, Electronic Energy Transfer, and Small-Angle X-Ray Scattering," *Phys. Rev. Lett.*, **56**, 2505–2508 (1986).

[27] J. Teixeira, "Experimental Methods for Studying Fractal Aggregates," in *On Growth and Form: Fractal and Non-Fractal Patterns in Physics*, H. E. Stanley and N. Ostrowsky, eds., Martinus Nijhoff Publishers, Boston, MA, 1986, pp. 145–162. See Ref. [36].

[28] J. G. Rarity and P. N. Pusey, "Light Scattering from Aggregating Systems: Static, Dynamic (QELS) and Number Fluctuations," in *On Growth and Form: Fractal and Non-Fractal Patterns in Physics*, H. E. Stanley and N. Ostrowsky, eds., Martinus Nijhoff Publishers, Boston, MA, 1986, pp. 218–221. See Ref. [36].

[29] B. H. Kaye, "Fractal Dimension and Signature Waveform Characterization of Fine Particle Shape," *American Laboratory*, pp. 55–63 (1986).

[30] Z. Chen, P. Sheng, D. A. Weitz, H. M. Lindsay, M. Y. Lin, and P. Meakin, "Optical Properties of Aggregate Clusters," *Phys. Rev. B*, **37**, 5232–5235 (1988).

[31] J. G. Rarity, R. N. Seabrook, and R. J. G. Carr, "Light-Scattering Studies of Aggregation," in *Fractals in the Natural Sciences*, M. Fleischmann, D. J. Tildesley, and R. C. Ball, eds., Princeton University Press, Princeton, NJ, 1989, pp. 89–102. See Ref. [40].

[32] M. Y. Lin, H. M. Lindsay, D. A. Weitz, R. C. Ball, R. Klein, and P. Meakin, "Universality of Fractal Aggregates as Probed by Light Scattering," in *Fractals in the Natural Sciences*, M. Fleischmann, D. J. Tildesley, and R. C. Ball, eds., Princeton University Press, Princeton, NJ, 1989, pp. 71–87. See Ref. [40].

[33] S. K. Sinha, "Scattering from Fractal Structures," *Physica D*, **38**, 310–314 (1989).

[34] K. J. Falconer, *The Geometry of Fractal Sets*, Cambridge University Press, Cambridge, 1985.

[35] *Fractals in Physics*, L. Pietronero and E. Tosatti, eds., Elsevier Science Publishers B.V., New York, 1986.

[36] *On Growth and Form: Fractal and Non-Fractal Patterns in Physics*, H. E. Stanley and N. Ostrowsky, eds., Martinus Nijhoff Publishers, Boston, MA, 1986.

[37] J. Feder, *Fractals*, Plenum Press, New York, 1988.

[38] M. Barnsley, *Fractals Everywhere*, Academic Press, New York, 1988.

[39] *The Science of Fractal Images*, H.-O. Peitgen and D. Saupe, eds., Springer-Verlag, New York, 1988.

[40] *Fractals in the Natural Sciences*, M. Fleischmann, D. J. Tildesley, and R. C. Ball, eds., Princeton University Press, Princeton, NJ, 1989.

[41] K. J. Falconer, *Fractal Geometry: Mathematical Foundations and Applications*, John Wiley and Sons, Chichester, 1990.

[42] A. L. Méhauté, *Fractal Geometries: Theory and Applications*, CRC Press Inc., Boca Raton, 1991, originally appeared in French as *Les Geometries Fractales: L'espace-temps brisé* in 1990 by Editions Hermes, Paris.

[43] M. Schroeder, *Fractals, Chaos, Power Laws: Minutes from an Infinite Paradise*, W. H. Freeman and Company, New York, 1991.

[44] *Fractals and Disordered Systems*, A. Bunde and S. Havlin, eds., Springer-Verlag, Berlin, 1991.

[45] *Fractals: Non-Integral Dimensions and Applications*, G. Cherbit, ed., John Wiley and Sons, Chichester, 1991.

[46] G. A. Edgar, *Measure, Topology, and Fractal Geometry*, Springer-Verlag, New York, 1990.

[47] D. L. Turcotte, *Fractals and Chaos in Geology and Geophysics*, Cambridge University Press, Cambridge, 1992.

[48] H.-O. Peitgen, H. Jürgens, and D. Saupe, *Chaos and Fractals: New Frontiers of Science*, Springer-Verlag, New York, 1992.

[49] T. Vicsek, *Fractal Growth Phenomena*, 2 edn, World Scientific, Singapore, 1992.

[50] D. L. Jaggard and X. Sun, "Scattering by Fractally Corrugated Surfaces," *J. Opt. Soc. A*, **7**, 1131–1139 (1990).

[51] X. Sun and D. L. Jaggard, "Wave Scattering from Non-Random Fractal Surfaces," *Opt. Comm.*, **78**, 20–24 (1990).

[52] D. L. Jaggard and X. Sun, "Rough Surface Scattering: A Generalized Rayleigh Solution," *J. Appl. Phys.*, **68**, 5456–5462 (1990).

[53] A. Dogariu, J. Uozumi, and T. Asakura, "Angular Power Spectra of Fractal Structures," *J. Mod. Optics*, **41**, 729–738 (1994).

[54] P. McSharry and P. J. Cullen, "Wave Scattering by a One-Dimensional Band-Limited Fractal Surface Based on a Perturbation of the Green's Function," *J. Appl. Phys.*, **78**, 6940–6948 (1995).

[55] S. Savaidis, P. Frangos, D. L. Jaggard, and K. Hizanidis, "Scattering from Fractally Corrugated Surfaces: An Exact Approach," *Opt. Lett.*, **20**, 2357–2359 (1995).

[56] J. Chen, T. K. Y. Lo, H. Leung, and J. Litva, "The Use of Fractals for Modeling EM Waves Scattering from Rough Sea Surface," *IEEE Trans. Geo. Remote Sensing*, **34**, 966–972 (1996).

[57] G. Franceschetti, M. Migliaccio, and D. Riccio, "An Electromagnetic Fractal-Based Model for the Study of the Fading," *Radio Sci.*, **31**, 1749–1759 (1996).

[58] K. Ivanova and O. I. Yordanov, "Bistatic Properties of Kirchhoff Diffractals," *Radio Sci.*, **31**, 1901–1906 (1996).

[59] C. A. Guérin, M. Holschneider, and M. Saillard, "Electromagnetic Scattering from Multi-Scale Rough Surfaces," *Waves in Random Media*, **7**, 331–349 (1997).

[60] S. Savaidis, P. Frangos, D. L. Jaggard, and K. Hizanidis, "Scattering from Fractally Corrugated Surfaces Using the Extended Boundary Condition Method," *J. Opt. Soc. Am. A*, **14**, 475–485 (1997).

[61] S. Rouvier and P. Borderies, "Ultra Wide Band Electromagnetic Scattering of a Fractal Profile," *Radio Science*, **32**, 285–293 (1997).

[62] J. A. Shaw and J. H. Churnside, "Fractal Laser Glints from the Ocean Surface," *J. Opt. Soc. A*, **14**, 1144–1150 (1997).

[63] F. Berizzi and E. D. Mese, "Fractal Analysis of the Signal Scattered from the Sea Surface," *IEEE Trans. Ant. and Propagat.*, **47**, to appear (1999).

[64] F. Berizzi, E. D. Mese, and G. Pinelli, "A One-Dimensional Fractal Model of the Sea Surface," *Sonar and Navigation*, **146**, 55–64 (1999).

[65] F. Berizzi and E. D. Mese, "A Fractal Theoretical Approach for the Sea Wave Spectrum Evaluation," in preparation (1998–1999).

[66] G. Franceschetti, A. Iodice, M. Migliaccio, and D. Riccio, "Scattering from Natural Rough Surfaces Modelled by Fractional Brownian Motion Two-Dimensional Processes," *IEEE Trans. Ant. and Propagat.*, **47**, to appear (1999).

[67] A. Collaro, G. Franceschetti, M. Migliaccio, and D. Riccio, "Gaussian Rough Surfaces and Kirchhoff Approximation," *IEEE Trans. Ant. and Propagat.*, **47**, to appear (1999).

[68] D. L. Jaggard and X. Sun, "Reflection from Fractal Layers," *Opt. Lett.*, **15**, 1428–1430 (1990).

[69] V. V. Konotop, O. I. Yordanov, and I. V. Yurkevich, "Wave Transmission Through a

One-Dimensional Cantor-Like Fractal Medium," *Europhys. Lett.*, **12**, 481–485 (1990).

[70] X. Sun and D. L. Jaggard, "Wave Interactions with Generalized Cantor Bar Fractal Multi-Layers," *J. Appl. Phys.*, **70**, 2500–2507 (1991).

[71] T. Megademini, B. Pardo, and R. Jullien, "Fourier Transform and Theory of Fractal Multilayer Mirrors," *Opt. Comm.*, **80**, 312–316 (1991).

[72] M. Bertolotti, P. Masciulli, and C. Sibilia, "Spectral Transmission Properties of a Self-Similar Optical Fabry-Perot Resonator," *Opt. Lett.*, **19**, 777–779 (1994).

[73] S. De Nicola, "Reflection and Transmission of Cantor Fractal Layers," *Opt. Comm.*, **111**, 11–17 (1994).

[74] S. A. Bulgakov, V. V. Konotop, and L. Vázquez, "Wave Interaction with a Random Fat Fractal: Dimension of the Reflection Coefficient," *Waves in Random Media*, **5**, 9–18 (1995).

[75] D. L. Jaggard and A. D. Jaggard, "Polyadic Cantor Superlattices with Variable Lacunarity," *Opt. Lett.*, **22**, 145–147 (1997).

[76] A. D. Jaggard and D. L. Jaggard, "Fractal Superlattices and Scattering: Lacunarity, Fractal Dimension, and Stage of Growth," in *Conference Proceedings of the 1997 AP-S/URSI Meeting*, Montréal, Canada, 1997.

[77] A. D. Jaggard and D. L. Jaggard, "Scattering from Fractal Superlattices with Variable Lacunarity," *J. Opt. Soc. Am. A*, **15**, 1626–1635 (1998).

[78] B. B. Mandelbrot, *The Fractal Geometry of Nature*, Freeman, San Francisco, 1983.

[79] C. Allain and M. Cloitre, "Characterizing the Lacunarity of Random and Deterministic Fractal Sets," *Phys. Rev. A*, **44**, 3552–3558 (1991).

[80] D. L. Jaggard and Y. Kim, "Diffraction by Bandlimited Fractal Screens," *J. Opt. Soc. Am. A*, **4**, 1055–1062 (1987).

[81] P. Beckmann and A. Spizzichino, *The Scattering of Electromagnetic Waves from Rough Surfaces*, Pergamon, New York, 1963.

[82] J. Uozumi, H. Kimura, and T. Asakura, "Fraunhofer Diffraction by Koch Fractals: The Dimensionality," *J. Mod. Optics*, **38**, 1335–1347 (1991).

[83] D. L. Jaggard and A. D. Jaggard, "Fractal Apertures: The Effect of Lacunarity," in *Conference Proceedings of the 1997 AP-S/URSI Meeting*, Montréal, Canada, 1997.

[84] D. L. Jaggard and A. D. Jaggard, "Cantor Ring Arrays," in *Conference Proceedings of the 1998 AP-S/URSI Meeting*, Atlanta, GA, 1998.

[85] D. L. Jaggard and A. D. Jaggard, "Fractal Arrays and Lacunarity," in *Proceedings of PIERS'98*, Nantes, France, 1998.

[86] D. L. Jaggard and A. D. Jaggard, "Fractal Ring Arrays," invited paper submitted to *Wave Motion* (1999).

[87] G. Franceschetti, M. Migliaccio, and D. Ricci, "Scattering from Natural Surfaces via a Two-scale Fractal Model," in *Proceedings of the URSI International Symposium on Electromagnetic Theory*, pp. 685–687, St. Petersburg, Russia, 1995.

[88] G. Franceschetti, M. Migliaccio, and D. Ricci, "Fractal Modelling of Microscopic Roughness Scattering," in *Proceedings of the URSI International Symposium on Wave Propagation and Remote Sensing*, pp. 160–163, Ahmedabad, India, 1995.

[89] G. Franceschetti, M. Migliaccio, and D. Ricci, "A Fractal-Based Approach to Electromagnetic Modelling," in *Proceedings of IGARSS'94*, pp. 1641–1643, Pasadena, CA, 1996.

[90] F. Berizzi and E. D. Mese, "Fractal Theory of Sea Fractal Scattering," in *Proceedings of 1996 CIE International Conference of Radar*, pp. 661–665, Beijing, China, 1996.

[91] G. Andreoli, F. Berizzi, E. D. Mese, and G. Pinelli, "Fractal Model of the Sea Surface and Statistical Analysis of the Sea-Scattered Signal," in *Proceedings of the Fractals in*

Engineering Conference, Arcachon, France, 1997.

[92] G. Andreoli, F. Berizzi, E. D. Mese, and G. Pinelli, "A Two-Dimensional Fractal Model of the Sea Surface and Sea Spectrum Evaluation," in *Proceedings of the IEE International Radar Conference*, pp. 189–193, Edinburgh, Scotland, 1997.

[93] G. Franceschetti, M. Migliaccio, and D. Ricci, "Fractal Models of Scattering from Natural Bodies," in *Proceedings of PIERS'97*, p. 74, Hong Kong, 1997.

[94] G. Franceschetti, A. Iodice, M. Migliaccio, and D. Ricci, "Backscattering from an fBm Surface," in *Proceedings of the URSI International Symposium on Electromagnetic Theory*, pp. 692–694, Thessaloniki, Greece, 1998.

[95] G. Franceschetti, A. Iodice, M. Migliaccio, and D. Ricci, "Scattering from Natural Rough Surfaces Described by the fBm Fractal Model," in *Proceedings of PIERS'98*, p. 1153, Nantes, France, 1998.

[96] G. Andreoli, F. Berizzi, E. D. Mese, and G. Pinelli, "A Two-Dimensional Fractal Model of the Sea Surface and Sea Spectrum Evaluation," in *Proceedings of IGARSS '98*, Seattle, Washington, 1998.

[97] P. C. Waterman, "Matrix Formulation of Electromagnetic Scattering," *Proc. IEEE*, **53**, 805–812 (1965).

[98] P. C. Waterman, "Symmetry, Unitarity, and Geometry in Electromagnetic Scattering," *Phys. Rev., D*, **3**, 825–839 (1971).

[99] P. C. Waterman, "Scattering by Periodic Surfaces," *J. Acoust. Soc. Am.*, **57**, 791–802 (1975).

[100] S. L. Chuang and J. A. Kong, "Scattering of Waves from Periodic Surfaces," *Proc. IEEE*, **69**, 1132–1144 (1981).

[101] J. A. Sanchez and M. Nieto-Vesperinas, "Light Scattering from Random Rough Dielectric Surfaces," *J. Opt. Soc. Am. A*, **8**, 1270–1286 (1991).

[102] J. A. Kong, *Electromagnetic Wave Theory*, John Wiley & Sons, New York, 1986.

[103] A. Ishimaru, *Electromagnetic Wave Propagation, Radiation, and Scattering*, Prentice Hall, Englewood Cliffs, NJ, 1991.

[104] A. R. Mickelson and D. L. Jaggard, "Electromagnetic Wave Propagation in Almost-Periodic Media," *IEEE Trans. Ant. and Propagat.*, **AP-27**, 34–40 (1979).

[105] Z. Knittl, *Optics of Thin Films (An Optical Multilayer Theory)*, John Wiley & Sons, London, 1976.

[106] A. D. Jaggard and D. L. Jaggard, "Cantor Superlattices: Extracting Fractal Descriptors," in preparation for publication (1999).

[107] A. D. Jaggard and D. L. Jaggard, "Fractal Superlattices: A Frequency Domain Approach," in *Proceedings of PIERS'98*, Nantes, France, 1998.

[108] Y. Kim and D. L. Jaggard, "Wave Interactions with Continuous Fractal Layers," in *Proceedings of the 1991 SPIE Meeting vol. 1558*, pp. 113–119, John Wiley & Sons, San Diego, 1991.

[109] H. Aubert and D. L. Jaggard, "Fractal Superlattices and Their Wavelet Analyses," *Opt. Comm.*, **149**, 207–212 (1998).

[110] H. Aubert and D. L. Jaggard, "Wavelet Analysis of Transients in Fractal Superlattices," to appear in *IEEE Trans. Ant. and Propagat.*, (1999–2000).

[111] I. Daubechies, *Ten Lectures on Wavelets*, S.I.A.M., Philadelphia, 1992.

[112] A. Dogariu, J. Uozumi, and T. Asakura, "Wavelet Transform Analysis of Slightly Rough Surfaces," *Optics Comm.*, **107**, 1–5 (1994).

[113] H. Aubert and D. L. Jaggard, "Continuous Wavelet Transform Analysis of Fractal Superlattices," in *Fractals: Theory and Applications in Engineering*, M. Dekking, J. Lévy Véhel, E. Lutton, and C. Tricot, eds., Springer, London, 1999, pp. 245–259.

Chapter 2

FRACTAL-SHAPED ANTENNAS

Carles Puente, Jordi Romeu, and Angel Cardama

2.1 INTRODUCTION

The strong relationship between the behavior of an antenna and its size relative to the operating wavelength has imposed, for decades, a tight constraint to the antenna designer. Antennas are usually designed to operate at a relatively narrow range of frequencies, typically on the order of 10~40% around a center wavelength, which imposes the size of the antenna (generally a quarter or a half wavelength in size). For long, such a constraint was believed to be an intrinsic restriction on the behavior of any antenna until the invention, in 1961, of frequency independent antennas [1–5]. Being aware that the size of the antenna was the main limiting parameter that imposed the operating frequency, people designed antennas that had no characteristic size, at least, ideally. The logarithmic spiral together with some other angle-defined structures were successfully used to design antennas with an almost frequency independent behavior [1], [2], [4]. The log-periodic dipole array was another example of the few geometrical shapes that were used for that purpose [1], [3], [5]. The fact is that, by that time, it was not trivial to imagine an antenna geometry with a scale-independent shape (and thus a wavelength independent performance). However, the work by B. Mandelbrot et al. in the last two decades drove the attention of the scientific community to some weird, complex, convoluted objects that were often left aside by mathematicians as 'pathological' counterexamples and defied the rules of the classic Euclidean geometry. The term *fractal* was coined by B. Mandelbrot to name those strange objects, which in fact are not that strange but actually among the most common forms in nature [6], [7]. Among the astonishing properties of fractal objects, one would observe that most of them are self-similar. That is, roughly speaking fractals are composed by many copies of themselves at different scales; the global fractal form is repeated at different sizes as many times as desired within the object structure such that the global object and its parts become identical. It is often said that fractals have no characteristic size [7], and it is precisely due to such a particular property that they appear specially attractive to defy, once again, the classical antenna's performance constraint on its size to wavelength ratio.

In 1993, [8], [9] fractals were suggested as suitable candidates to design multifrequency antenna arrays, i.e., arrays that were able to keep the same behavior at many frequencies. Some years before, D. L. Jaggard et al. had already shown how electromagnetic waves could acquire some special properties when interacting with fractal bodies [10–18], so it was actually not unreasonable to assume that the scale independent geometry of fractals might be appropriate to design wavelength independent antennas as well.

Regarding the development process, first some theoretical characteristic properties of self-similar fractal arrays and their multifrequency (multiband) behavior were described [9], [19], [20]. Later on, the Sierpiński antenna [21–26] was shown as the first reported practical example of fractal antenna which was able to keep the same basic performance at several (up to five) bands. That invention opened up the doors for a wide range of new possibilities for antenna designs and applications, which led the Universitat Politècnica de Catalunya (UPC) to apply for a patent on fractal and multifractal antennas [23].

But the interest on the research and development of fractal antennas currently extends beyond the field of multifrequency antennas. The same kind of constraint (the antenna size to wavelength ratio) that makes it difficult to design multifrequency antennas, is also responsible for the limitations on the design of small antennas and again, owing to their special geometrical features, fractals appear as magnificent candidates to become also efficient small antennas. A low radiation resistance, low efficiency and small bandwidth are some of the typical characteristics of an antenna when it is made much smaller than the operating wavelength. But once more, fractals can help to face that problem, taking into account that the common notion of size is broken into pieces when dealing with fractal geometry: Is a 10 m long but a 1 cm tall fractal antenna smaller or longer than a 10 cm wavelength? One cannot take for granted the same common geometrical assumptions that apply to classical Euclidean geometry and that is precisely what the designer of small fractal antennas tries to take advantage of. In fact, N. Cohen showed in 1995 some numerical results on large perimeter fractal loops [27–30] which evinced that such small fractal antennas might feature a low resonant frequency together with a large driving resistance. Recently, C. Puente et al. showed that by fractally shaping small monopoles, an improvement on the bandwidth, radiation resistance and reactance with respect to other classical Euclidean antennas could be achieved [26], [31].

Nowadays, the design of multifrequency and small antennas is still of major importance as an engineering topic. The growth of the telecommunication sector, and in particular the tremendous expansion of mobile telephony and personal communications systems are driving the engineering efforts to develop multiservice (multiband) and compact (portable) systems which require such kind of antennas. In the case of cellular systems, not only the handset antenna is important, but also those on base stations. The demand's growth has led to the introduction of new frequency bands and thus, to an increase on the number of base stations and antennas. Multifrequency operation through a single, multisystem antenna appears especially attractive there to minimize both the cost and visual impact of such base stations. It is in this framework where fractal technology appears potentially as a powerful tool to meet the telecommunication operator requirements. As a matter of fact, it is in this field where fractal-shaped antennas have found the first commercial applications; the *Fractus®-MSPK* omnidirectional antenna from the company FRACTUS S.A. (developed in cooperation with the Universitat Politècnica de Catalunya, UPC), is the first example of multiband antenna based on fractal technology that fulfills the requirements of GSM 900 MHz and DCS 1800 MHz systems simultaneously [32].

The present chapter summarizes most of the work carried out by the Electromagnetics and Photonics Engineering (EEF) group at the UPC in the field of fractal-shaped antennas. From theory to experiment, the following sections illustrate the basic features of multifrequency and small fractal-shaped antennas.

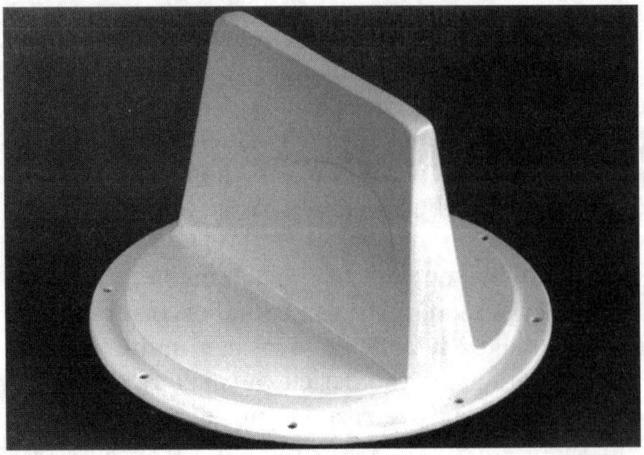

Figure 2.1 The Fractus®-MSPK antenna is a dualband antenna based on fractal technology that simultaneously meets the specifications for the GSM and DCS systems in a microcell environment. In the picture, the fractal element is under the protecting radome.

Section 2.2 is devoted to the basic theoretical aspects of fractal antennas; the basic fractal geometry properties concerning the antenna design, namely fractal self-similarity and fractal dimension, are outlined there together with an extensive discussion on the potentiality of fractals in this field. The next section deals with the description of multifrequency fractal-shaped antennas. It is basically focused on the performance of the Sierpiński antenna and it also includes some results on random fractal tree-like antennas. Finally, based on the example of the Koch monopole, Section 4 is dedicated to discuss the potential of fractals on the design of small antennas.

2.2 FRACTALS, ANTENNAS, AND FRACTAL ANTENNAS

2.2.1 Main Fractal Properties

2.2.1.1 Fractal Self-Similarity

Self-similarity is a rather intuitive concept that can be readily understood by taking a look to the geometrical features of the best known fractal shapes. When an object is composed by smaller copies of itself reduced to a smaller scale it is said to be *self-similar*. A self-similar object can be decomposed in a set of clusters that are identical to the whole object. Since each cluster is identical to the whole object, it will have to be composed itself by clusters, which will be composed by clusters as well, and so on. Hence, an infinite number of small copies of the whole object will be found in some part of the whole structure.

Sometimes, self-similarity is too tight a concept to describe the similarities found in several parts of an object. Often, the smaller copies of the whole structure are not identical to the whole object but are rather distorted, i.e., they might appear skewed or compressed by a different scale factor with respect to a different axis. In this case, the object is said to be *self-affine* instead of self-similar.

A more precise definition of self-similarity can be given based on the Hutchinson's geometrical transformation algorithms [7], that generate many fractal shapes by means of an iterative procedure. A Hutchinson operator $W[A]$ is a transformation over a subset A of a plane (for the sake of simplicity we will consider initially a two dimensional space) as follows:

$$W[A] = w_1[A] \cup w_2[A] \cup \cdots \cup w_N[A] \tag{2.1}$$

where $w_i[A]$ is an affine transformation [7]. An affine linear transformation is a composition of a linear transformation $\boldsymbol{F}$ of the kind

$$\boldsymbol{F} = \begin{pmatrix} r\cos\phi & -s\sin\psi \\ r\sin\phi & s\cos\psi \end{pmatrix} \tag{2.2}$$

plus a translation to a point (x_o, y_o). That is, if (x, y) is a point of the subset A, the point is transformed to (x', y') in the following way:

$$\begin{pmatrix} x' \\ y' \end{pmatrix} = \begin{pmatrix} r\cos\phi & -s\sin\psi \\ r\sin\phi & s\cos\psi \end{pmatrix} \begin{pmatrix} x \\ y \end{pmatrix} + \begin{pmatrix} x_o \\ y_o \end{pmatrix} \tag{2.3}$$

A linear transformation is a contraction if $0 < r < 1$ and $0 < s < 1$. An affine linear transformation is a similarity transformation if it is only composed by contractions or expansions, rotations, and translations, i.e., if $r = s$ and $\phi = \psi$. One can iteratively apply a Hutchinson operator (i.e., a composition of affine linear transformations) to a subset A in the following way:

$$\begin{aligned} A_1 &= W[A] \\ A_2 &= W[W[A]] \\ &\vdots \\ A_n &= W[A_{n-1}] \end{aligned} \tag{2.4}$$

This iterative process is called an *Iterated Function System* (IFS) and can be understood as a feedback process. It can be proved [7] that if the Hutchinson operator of an IFS is composed only by contractions, translations and rotations, the sequence $A, A_1 \ldots A_n$ converges to a subset A_∞ when iterated a large number of times ($n \to \infty$). This set A_∞ is called the *attractor* of the IFS and is independent of the initial subset A applied to the IFS. Hence, the attractor (the fractal shape) can be seen as a characteristic solution of the IFS. The convergence criteria applied here is the Hausdorff distance, i.e., a sequence of subsets A_n converges to a subset A_∞ if the Hausdorff distance between both subsets tends to zero [7].

An important property that follows directly from the convergence of the recursive system is that the attractor holds the invariance property with respect to the Hutchinson operator of the IFS, that is,

$$A_\infty = W[A_\infty] \tag{2.5}$$

An example of an IFS is the one that generates the fractal Sierpiński gasket. It is composed by three similarity transformations

$$\begin{aligned} w_1(x,y) &= \begin{pmatrix} 1/2 & 0 \\ 0 & 1/2 \end{pmatrix} \begin{pmatrix} x \\ y \end{pmatrix} + \begin{pmatrix} 1/4 \\ 0 \end{pmatrix} \\ w_2(x,y) &= \begin{pmatrix} 1/2 & 0 \\ 0 & 1/2 \end{pmatrix} \begin{pmatrix} x \\ y \end{pmatrix} + \begin{pmatrix} -1/4 \\ 0 \end{pmatrix} \\ w_3(x,y) &= \begin{pmatrix} 1/2 & 0 \\ 0 & 1/2 \end{pmatrix} \begin{pmatrix} x \\ y \end{pmatrix} + \begin{pmatrix} 0 \\ 1/2 \end{pmatrix} \\ W_{Sierpiński}[A] &= w_1(A) \cup w_2(A) \cup w_3(A) \end{aligned} \tag{2.6}$$

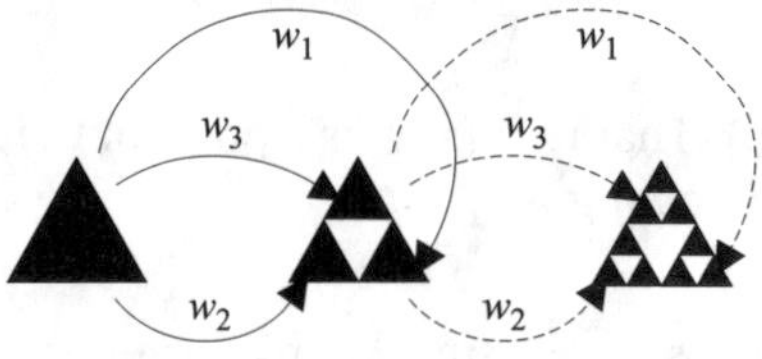

Figure 2.2 First stage of the IFS that generates the Sierpiński gasket.

Figure 2.2 illustrates the first two stages of the IFS, and Fig. 2.3 shows how the Sierpiński attractor is obtained independently of the initial subset A.

The IFSs are often named *Multiple Reduction Copy Machines* (MRCM) in the literature. The initial subset A is also named *generator* and the Hutchinson operator is the *iterator*. IFSs are most common algorithms for fractal generation. Other more sophisticated schemes include networks and cascading of IFSs to generate some complex forms such as the Barnsley or Sierpiński ferns [6], [7].

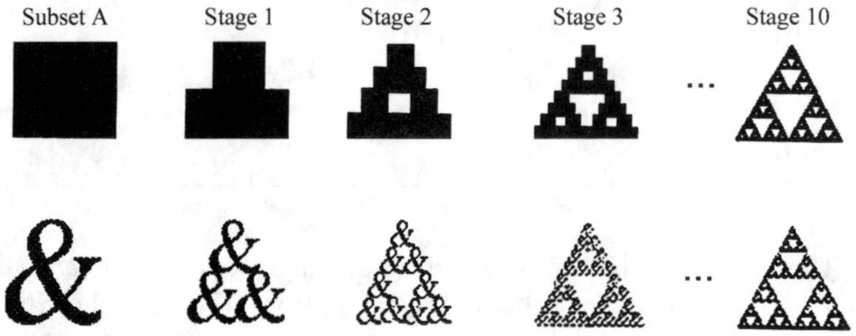

Figure 2.3 The IFS converges to the same solution despite the initial subset.

An attractor of an Iterated Function System (IFS) is a self-similar object if the corresponding generating Hutchinson operator is only composed by similarity linear transformations, i.e., translations, rotations, contractions or expansions. It is a self-affine object when the Hutchinson operator is composed by any set of affine-linear transformations as described by (2.3) which includes, for instance, reflecting and skewing transformations as well.

Again, these definitions of self-similarity and self-affinity are enough to characterize most of the antenna geometries described in this chapter, they are still too tight to describe some kinds of similarities commonly found in many fractal objects. For instance, the Julia and Mandelbrot sets [7] are obtained through a composition of non-linear transformations, leading to objects that have an extremely rich set of similarities within themselves but that can not be described by any composition of linear transformations. But even in those cases the words self-similar and self-affine are usually applied in a loose sense also to such kind of objects, restricting the attributes of *strict self-similarity* and *strict self-affinity* to describe the properties of the attractors of an IFS.

There are other kinds of similarity qualities that should also be taken into account. Mainly, these are self-similarity to a point and random self-similarity. The first applies, for instance, to the logarithmic spiral. The spiral is self-similar only in the sense that the central part of its shape is a scaled (and rotated) version of the whole spiral. However, if we randomly take a portion of the curve, it does not look as a whole spiral. Therefore, strictly speaking only the central point of the ideal spiral holds the property of similarity, and thus the name of self-similarity at a point. Usually, this kind of object, which by the way has been broadly used in the design of frequency independent antennas, is called a self-scalable object.

On the other hand, the *random* or *statistical self-similarity* property comes from the generalization of the fractal generating schemes to random processes. This can be carried out in many different ways. For instance, a random deviation to the translation or contraction transformations of a Hutchinson operator could be added at each step of an IFS. But randomness can be introduced through many different ways to generate a whole new family of fractal objects. The most common is the midpoint displacement scheme used to generate the fractional brownian motion random process [7], which is the basis for modeling fractal landscapes with random mountains, clouds, and coastlines [6]. Although the final object obtained by this method is not strictly self-similar, the details in the structure look indeed similar to the whole shape, in the sense that the same degree of detail is obtained at any scale, regardless of the precise shape of such details.

We should end this discussion of self-similarity by stressing that this property only applies strictly to ideal mathematical objects. Ideal fractal shapes are obtained through an infinite set of transformations which lead to objects with an infinite set of sizes and scales. Of course, such an idealization can only be approached, in the real world, by objects that hold some similarity properties up to a certain scale. The resulting truncated versions of the ideal fractal shapes are often named *band-limited fractal shapes* [11].

2.2.1.2 The Fractal Dimension

The common notion of dimension is rather intuitive and self-explanatory: a straight line is a one-dimensional object, a square is two-dimensional while a cube is three-dimensional. We also think of a point as a zero-dimensional object since it does not occupy any space in a straight line which has dimension 1, the same way an infinite number of segments can be fitted in a square, or an infinite number of squares fitted in a cube. In a slightly more mathematical approach, one would describe the dimension of an object as the number of parameters, i.e., coordinates, which are necessary to describe it. Nevertheless, such an intuitive way of understanding dimension is found in trouble when trying to describe fractal shapes. There are curves that can be fitted in a finite area but have an infinite length (Koch curves, Julia set contours and natural coastlines ...) and dusts of non-connected points that have a one-to-one correspondence with all the points in the unit segment (the Cantor set). Also, some curves are able to fill higher order two-dimensional objects (Peano and Hilbert curves), and hence these curves can be used to describe such two-dimensional objects with a single parameter instead of the usual two. Therefore, it appears necessary to carefully review the notion of dimension to properly characterize any kind of geometrical shape, including fractals.

The fact is that there can be roughly found about ten different definitions of dimension [7], all of them with some common features but each one of them with its own differences and particularities. Here, we will take the box-counting dimension (the most common simplification of the Hausdorff dimension [7]) as the definition of fractal dimension. Although a thorough discussion of the fractal dimension is beyond the scope of this chapter (the authors suggest [12], [26] for a deeper treatment of the topic), from the fractal antenna theory point of view the reader should just keep in mind that the fractal dimension can be interpreted as a measure of the space-filling properties and complexity of a fractal shape. For instance, the Sierpiński gasket described in this chapter has a fractal dimension of $D = 1.585$, i.e., a non entire dimension between one and two, which means that the fractal is halfway between a one-dimensional body (a curve) and a two-dimensional one (a surface). The ideal Sierpiński gasket (the attractor) has no surface, so it is not a two-dimensional object; however, somehow the shape seems to fill a plane in a better way than any standard, non-fractal Euclidean curve, so intuitively it should have a dimension larger than unity. Actually, the dimension of $D = 1.585$ (obtained, for instance, by applying the definition in the chapter on Fractal Antenna Arrays in this book) nicely fits the intuitive view of such a surprising space-filling property.

2.2.2 Why Fractal-Shaped Antennas?

It is well known that the existence of characteristic lengths in an antenna sets a limit on its operating bandwidth. Rumsey [4] established that an antenna with shapes only defined by angles would be frequency independent, since it would not have any characteristic size to be scaled with wavelength; spiral antennas were designed attending to this criterion. Self-scalable shapes are also the basis for log-periodic antennas.

Once the main fractal characteristics have been introduced, the reason for investigating fractal antennas should become apparent. Actually, such a reason is twofold. First, fractals have no characteristic size. They have an infinite range of scales within their structure, being composed of an infinite number of clusters which are equal to the whole shape but at a smaller scale. It has just been stated that characteristic sizes have to be avoided to design a frequency-independent antenna or at least, the antenna structure has to include many characteristic sizes to allow a multiwavelength operation. Thus, it becomes presumable that fractal shapes, having a geometrical form which is self-similar and including multiple copies of themselves within their structure at several scales, are good candidates for becoming frequency independent (or at least multiwavelength) antennas.

Second, fractals are highly convoluted, irregular shapes. As a matter of fact, they are the most irregular shapes described up to now. It is well known that sharp edges, corners, and discontinuities enhance radiation from electric systems. Thus, it seems logical to expect that curves and surfaces which have discontinuities everywhere, that have a characteristic highly irregular shape (which is fundamentally different from that of classical Euclidean shapes) should become efficient radiators. That point can be particularly relevant when dealing with the design of small antennas (i.e., antennas which are much smaller than the operating wavelength) which are usually poor radiators. These issues are addressed in the following discussion.

2.2.2.1 Multifrequency Fractal Antennas

The self-scaling principle for frequency independent antennas can be directly derived from Maxwell equations [4] under a coordinate transformation

$$x' = x/\lambda \qquad y' = y/\lambda \qquad z' = z/\lambda \tag{2.7}$$

where λ is the wavelength. Basically, under the scaling transformation (2.7) Maxwell Equations become independent of the wavelength. As is well known, this implies that the frequency behavior of an antenna is intimately related to its size, and that the same behavior at a particular wavelength can be obtained at a different wavelength, provided that the antenna is properly rescaled. And it also means that an antenna which is able to keep exactly the same shape under the scaling transformation (2.7), will feature a frequency-independent behavior, i.e., the same behavior at any frequency.

The scaling transformation in (2.7) can be seen as a particular case of the affine linear transformations used to formally state the IFSs, where the wavelength is used as the scale factor. Therefore, it appears natural to look for a similar reasoning to explore the frequency-dependent properties of IFS fractal attractors.

An IFS process can be described in an alternative way by means of delta functions and convolution operators. First, one has to translate the set theory notation to the function theory notation. That is, a subset of the plane A can be assigned to an index function $G(x, y)$ as follows:

$$G(x, y) = \begin{cases} 1 & \text{if} \quad (x, y) \in A \\ 0 & \text{if} \quad (x, y) \notin A \end{cases} \tag{2.8}$$

Then, a translation of A to a point (x_0, y_0) can be described as a convolution transformation in the following way:

$$G(x - x_0, y - y_0) = G(x, y) * \delta(x - x_0, y - y_0) \tag{2.9}$$

A Hutchinson operator composed only by similarity transformations will be noted as

$$W[G(x, y)] = \sum_{i=1}^{N} G(r_i x, r_i y) * \delta(x - x_i, y - y_i) \tag{2.10}$$

where N is the number of similarity transformations of the operator and r_i is the reduction factor of each transformation. Once the Hutchinson operator has been defined as a linear combination of convolution operations, an IFS can be readily understood as a sequential procedure where the operator in (2.10) is applied iteratively to the function $G(x, y)$.

Let's take an IFS where all the contractions reduce the input subset by the same scale factor r. Then one can separate the contraction and translation parts of (2.10) such that

$$W[G(x,y)] = G(rx,ry) * \sum_{i=1}^{N} \delta(x - x_i, y - y_i) \tag{2.11}$$

and write the Hutchinson operator in a more compact way,

$$W[G(x,y)] = G(rx,ry) * IF(x,y) \tag{2.12}$$

where the function $IF(x,y)$ has been properly defined as

$$IF(x,y) \equiv \sum_{i=1}^{N} \delta(x - x_i, y - y_i) \tag{2.13}$$

The function $IF(x,y)$ would be the equivalent of the iterator in fractal terminology, and $G(x,y)$ would be the equivalent of the generator. Now, the IFS can be described in a more compact way by a series of convolutions as

$$\begin{aligned} G_n(x,y) &= G(r^n x, r^n y) * IF(r^{n-1}x, r^{n-1}y) * \cdots * IF(rx,ry) \\ &= G(r^n x, r^n y) \mathop{X}_{m=0}^{n-1} IF(r^m x, r^m y) \end{aligned} \tag{2.14}$$

where X indicates the convolution operator. If n is a large enough number, $G(r^n x, r^n y)$ tends to collapse into a single point if the original subset A is finite. Since the attractor of the IFS $G_\infty(x,y)$ is the same irrespective of the shape of the generator $G(x,y)$ due to the invariance property stated in (2.5), one can take a delta function as a generator (i.e., a point-like generator) and simply describe the IFS as

$$G_n(x,y) = \mathop{X}_{m=0}^{n-1} IF(r^m x, r^m y) \tag{2.15}$$

Then, the attractor function will be described by

$$G_\infty(x,y) = \mathop{X}_{m=0}^{\infty} IF(r^m x, r^m y) \tag{2.16}$$

which is not strictly speaking a self-scalable function since

$$G_\infty(sx,sy) = \mathop{X}_{m=0}^{\infty} IF(sr^m x, sr^m y) \neq G_\infty(x,y) \tag{2.17}$$

due to the finite size of the attractor. However, we can generalize the concept of IFS by defining a Generalized Iterated Function System (GIFS) in such a way that

$$G_n(x,y) = \mathop{X}_{m=-(n-1)}^{n-1} IF(r^m x, r^m y) \tag{2.18}$$

That is, not only smaller copies of the whole shape are added at each iteration, but also a larger copy which is the union of N objects obtained from the previous iteration is constructed. Then, the attractor will have an infinite size and will be defined by the equation

$$G_\infty(x,y) = \sum_{m=-\infty}^{\infty} IF(r^m x, r^m y) \tag{2.19}$$

This attractor of the newly defined GIFS is self-scalable by its characteristic scale factor r and any power of it r^p, that is,

$$G_\infty(r^p x, r^p y) = \sum_{m=-\infty}^{\infty} IF(r^{m+p} x, r^{m+p} y) = \{_{q=m+p}\} = \sum_{q=-\infty}^{\infty} IF(r^q x, r^q y) = G_\infty(x,y) \tag{2.20}$$

Although such a derivation has been applied, for the sake of clarity, to a scalar function $G_\infty(x,y)$, the extension to a vector function $\vec{G}_\infty(x,y)$ is straightforward. Hence, one could freely extend such a result to a function describing, for instance, an electromagnetic field or a current density distribution over a fractal antenna structure A_∞. Now, the key point is noticing that (2.20) implies that a fractal distribution of current or fields over an ideal self-similar fractal-shaped antenna (constructed by an infinite GIFS algorithm) is also self-scalable. Therefore, owing to the scaling principle of Maxwell equations, if one considers a set of log-periodically spaced wavelengths $\lambda_n = r^n \cdot \lambda_o$, one would conclude that such a fractal antenna would have exactly the same behavior at such wavelengths (a multifrequency or multiband behavior) with the bands log-periodically spaced by a factor r.

Of course, an infinite structure is assumed here, which is not feasible for a real antenna. However, such a limitation is also found in shapes only defined by angles as for instance the spiral antenna, but a frequency independent behavior is still supported by those antennas because they hold the truncation principle. Similarly, it can be stated that a practical (finite-size) fractal antenna will have a multifrequency (multiband) behavior, provided that the truncation principle is observed.

2.2.2.2 Small Fractal Antennas

When the size of an antenna is made much smaller than the operating wavelength, it becomes highly inefficient. Its radiation resistance decreases, while proportionally the reactive energy stored in the antenna neighborhood increases rapidly. Both phenomena make small antennas difficult to match to the feeding circuit, and when matched they display a high Q, i.e., a very narrow bandwidth.

Many fractal curves such as the Koch, Minkowski, Peano, and Hilbert curves [7] can be fitted in a finite area although they have an infinite length. Such a striking property should lead one to question many common assumptions regarding Euclidean antennas. For instance, might such long structures support arbitrarily large wavelengths and couple them efficiently to free space? Also, fractal curves are highly irregular curves, their fractal dimension being a good parameter for their roughness characterization. Since it is well known that sharp shapes, sudden bends and discontinuities tend to enhance radiation [33], might highly convoluted fractal shapes improve the radiation properties of common antennas? Could fractal shapes increase the low radiation resistance of a small dipole or loop? Before trying to answer such questions, let us first discuss some of the common assumptions on common Euclidean antennas that can be challenged in the case of fractal geometry.

For instance, the radiation resistance of a small Euclidean circular loop can be consistently written either in terms of its area (A) or its perimeter (C) as [33]

$$R_r = \eta\left(\frac{2\pi}{3}\right)\left(\frac{kA}{\lambda}\right)^2 = 20\pi^2\left(\frac{C}{\lambda}\right)^4 \tag{2.21}$$

Obviously, such a dual description is not correct for fractal loops, since area and perimeter are not necessarily linked to each other as they are in Euclidean loops (a finite area fractal loop can have an infinite perimeter), and the two approaches would lead to completely different conclusions.

Another feature which is broadly assumed for an antenna much smaller than the operating wavelength is that its input impedance has a very large reactive component. When operated below resonance, the input reactance of a small antenna increases rapidly, which together with the fast decrease of the radiation resistance yields a poor Q factor and thus a narrow bandwidth. Again, some fractal loops might have surprisingly small resonant frequencies due to its large perimeters, which together with their larger radiation resistance might lead to a broader bandwidth than common antennas. As a matter of fact, the potentiality of fractal shapes for improving the radiation of common small loops was suggested by N. Cohen [27–30], who observed on some numerical results that the fractal Minkowski loop presents an unusual large input resistance together with a surprising small resonance frequency.

Regarding the small bandwidth (high Q) of small antennas, L. J. Chu [34] established in 1948 a theoretical fundamental limit [34–40] based on the assumption that any field distribution around the antenna (in particular any current over the antenna) can be written as a linear combination of orthogonal spherical modes. For a small antenna, the larger modes are said to be negligible because their argument ka (a being the largest distance from any part of the antenna to the origin) is very small and their real part becomes highly reduced. Consequently, only fundamental TE_{01} and TM_{01} modes are considered when deriving the fundamental limit. It has been shown that the higher order modes within a sphere of radius a become evanescent when $ka < 1$, then for a very small linearly polarized antenna its Q has the only contribution of the lowest TM mode which reduces to [34], [35], [39], [41]

$$Q = \frac{1}{k^3a^3} + \frac{1}{ka} \approx \frac{1}{k^3a^3} \tag{2.22}$$

where the approximate limit $1/(ka)^3$ was independently derived by H. Wheeler as well [38], assuming a net inductive or capacitive behavior for the input reactance of a small antenna. Equation (2.22) is important because it establishes an upper bound to the fractional bandwidth at resonance ($\Delta f/f_0 = 1/Q$) of a small antenna no matter what its shape. Whether a small antenna reaches such a limit or not depends on how efficiently it utilizes the available volume within the sphere [36]. In practice, common antennas such as small dipoles or loops are far away from such a limit; only some special cases such as the Goubau antenna [37] are about 65% away from it.

Again, such an assumption should be reexamined when dealing with 'pathological', nowhere differentiable fractal curves. The linear combination of modes does not need to converge, i.e., although high-order modes had a small argument, they could be weighted by large amplitude factors which would make their contribution non-negligible, resulting thus in an unbounded series of terms. These would suggest that the fundamental limit could not hold for a fractal curve and an arbitrarily short but electrically long antenna could be made from a fractal shape.

In order to gain some insight on the behavior of small fractal antennas, an experimental investigation on the Koch antenna was done [26], [31]. Its main results and their discussion are presented in Section 2.4.

2.3 MULTIFREQUENCY FRACTAL-SHAPED ANTENNAS

2.3.1 The Equilateral Sierpiński Antenna

2.3.1.1 The Sierpiński Gasket

A classical fractal shape is the Sierpiński gasket or Sierpiński triangle, named after the Polish mathematician Waclaw Sierpiński (1882–1969) who introduced it. Rather than the curve-like appearance of some well known fractals such as Koch, Peano, and Hilbert curves, the Sierpiński gasket has a surface-like look, although it is not and it actually has a fractal dimension $D = 1.585$ as discussed in the previous section. Figure 2.4 shows the implementation of the Sierpiński gasket used to explore its behavior as an antenna. The ideal structure has three exact copies of the whole shape scaled down by a factor of 2, 3^2 copies scaled down by a factor of 2^2 and so on. Thus, it is said that the Sierpiński gasket is strictly a self-similar object.

For several reasons the Sierpiński gasket is a good candidate to explore the multiband properties of fractal antennas. First, its behavior can be readily compared to the well-known [42] triangular (bow-tie) antenna. Second, its overall triangular shape provides a convenient way of feeding the antenna structure through one apex. Given the self-similarity property of its shape, one could expect electromagnetic waves traveling from the apex to the tips becoming radiated by the smaller subgaskets when the wavelength properly matches any gasket size. One could assume that for a particular wavelength, smallest details (i.e., the small fractal iterations) could not be resolved by such waves and thus do not significantly contribute to the overall performance of the antenna. Analogously, if the traveling waves become efficiently radiated in some part of the antenna, the rest of it could become effectively disconnected such that the larger scale would not contribute much to the radiation process either. This way, only a finite range of scales would intervene in the antenna behavior. Since such a range is available at several sizes within the antenna structure, a similar performance among the wavelengths matching such sizes should be obtained, yielding this way, a multifrequency operation.

2.3.1.2 Input Impedance and Return-Loss

Several prototypes of Sierpiński monopoles were constructed and tested [21], [22], [26] the first of them being a four-iteration structure of thin triangular copper plates properly welded together to conform to the antenna shape [26] (Fig. 2.4).

The printed structure was mounted over an 80×80 cm aluminum ground plane

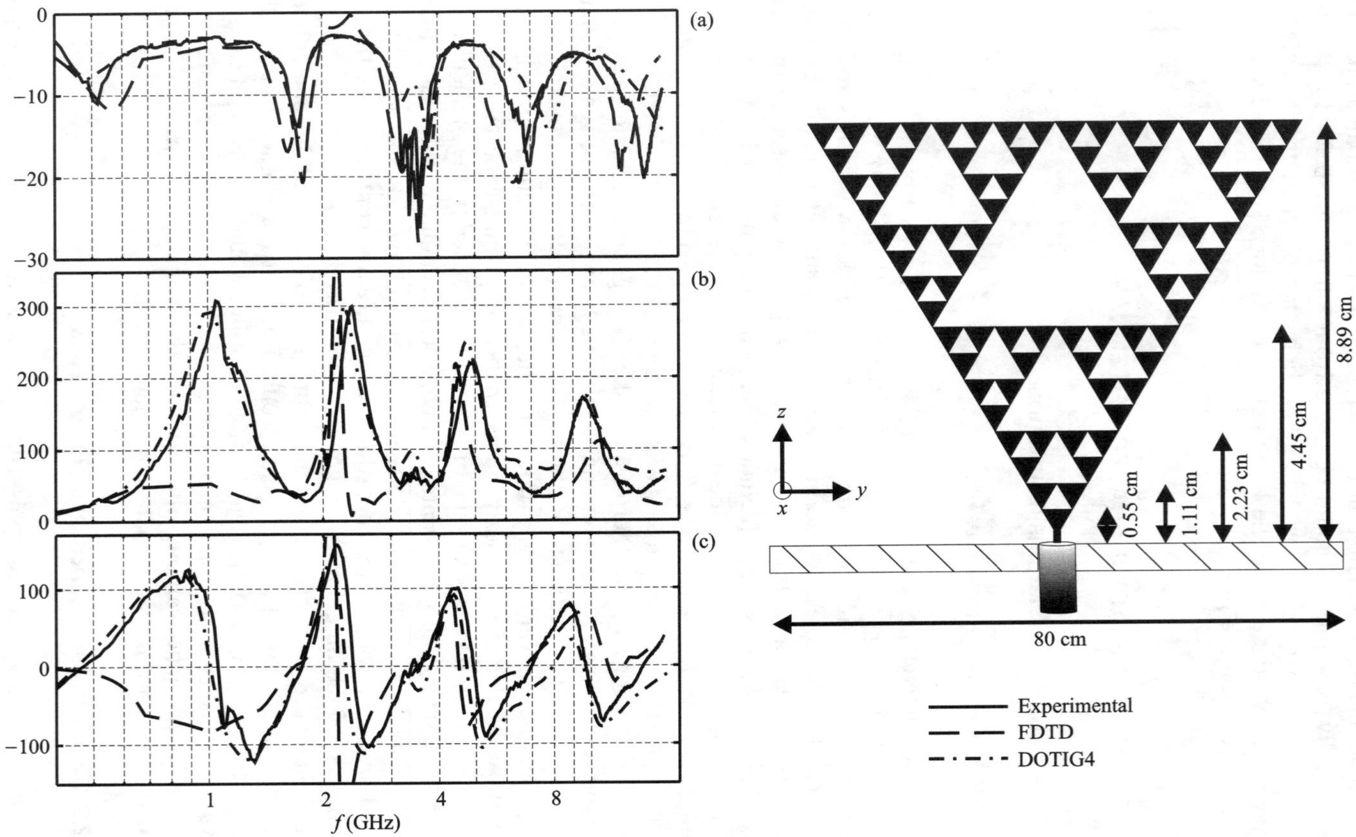

Figure 2.4 Input reflection coefficient, Γ_{in}, relative to 50 Ω (a), input resistance R_{in} (b) and input reactance X_{in} (c) of a five-iteration Sierpiński monopole. The monopole configuration is shown on the right.

and fed through a 50 Ω coaxial probe. The monopole configuration was chosen because it provides a simple way of feeding the antenna and avoids using a balun transformer. Since the main goal was investigating the input impedance behavior through a very large frequency range, a broadband balun would have been necessary for a dipole configuration.

The 8.89 cm tall fractal monopole was measured in an HP-8510B network analyzer in the 0.05 GHz–16 GHz frequency range. Also the antenna was numerically analyzed both using an FDTD algorithm [26], [45], [46]) running on a Connection Machine (a CM-200 at the CEPBA[1] and a CM-5 at the CNCPST[2]), and the DOTIG4[3] software package [47] running on a Sun Sparc workstation. The antenna structure, its input resistance R_{in} and reactance X_{in}, together with the magnitude of the input reflection coefficient Γ_{in} relative to 50 Ω (also named the input return loss L_r) are plotted versus a logarithmic frequency scale in Fig. 2.4.

The multiband behavior of the fractal antenna is precisely stated in this plot. Five log-periodically spaced bands are clearly distinguished in all three parameters. Experimental and numerical results are in good agreement and both yield to such a multiband behavior.

The Sierpiński monopole is well matched (VSWR < 2, L_r < 9.5 dB) at five bands because the input resistance approaches 50 Ω where the input reactance becomes negligible (odd resonances). The main features with regard the band locations, width and spacing are described in Table 2.1.

TABLE 2.1 Main Parameters of the Measured Sierpiński Monopole

n (band n^0)	f_n (GHz)	BW (%)	L_r (dB)	f_{n+1}/f_n	h_n/λ_n
1	0.52	7.15	10	3.50	0.153
2	1.74	9.04	14	2.02	0.258
3	3.51	20.5	24	1.98	0.261
4	6.95	22	19	2.00	0.257
5	13.89	25	20	—	0.255

The five band frequencies (f_n) are picked up at the lowest input return-loss points (band number in column 1, frequencies in column 2). The third column describes the relative bandwidth at each band for VSWR < 2; the fourth one the input return-loss; the fifth one represents the frequency ratio between adjacent bands, and the sixth one the ratio between the height (h_n) of the five subgaskets (circled in dashed lines in Fig. 2.4) and the corresponding band wavelength (λ_n).

It is interesting to notice that the bands are log-periodically spaced by a factor of $\delta = 2$, which is exactly the characteristic scale factor that relates the several gasket sizes within the fractal shape. The number of bands is directly associated with the number of fractal iteration stages, which indicates that one could freely design the

[1]CEPBA is *European Center for Parallelism of Barcelona.*

[2]CNCPST is *Centre National de Calcul Parallele en Sciences de la Terre* in Paris.

[3]The DOTIG4 simulation is a courtesy of Dr. Rafael Gomez from the University of Granada.

number of operating bands by properly choosing such a construction number. The antenna performance is kept similar through the bands, with a moderate bandwidth (~21%) at each one. From the band location and the height to wavelength ratio numbers, an empirical design equation for the Sierpiński monopole can be derived

$$f_n \approx 0.26 \frac{c}{h} \delta^n \tag{2.23}$$

where c is the vacuum speed of light, h is the height of the largest gasket (i.e., the whole antenna), and δ is the log-period ($\delta \approx 2$).

A significant deviation from the log-periodic behavior can be found at the first band. Neither the bandwidth nor the return-loss and the spacing with respect to the next band is kept similar to the remaining bands. Such a phenomenon can be mostly related to the truncation effect; the largest gasket (the overall structure) lacks larger fractal iterations breaking in this way the symmetry with respect to the central bands.

2.3.1.3 Radiation Patterns

A truly multifrequency (multiband) performance cannot be solely stated from the antenna input impedance behavior; the agreement in the radiation patterns among bands has to be proved too. While in many antennas usually the bandwidth constraints come from the input impedance rather than from the radiation patterns, here the whole frequency range is very large (1:8) and the agreement among patterns cannot be assumed. The antenna is about 4 wavelengths long at the highest frequency and one could expect many lobes to appear, similarly to what is found in common antennas. The full radiation patterns of the Sierpiński monopole, measured at the minimum VSWR frequency points, are shown in Fig. 2.5. It becomes apparent that their main features are kept similar through the bands: a two lobe structure with some tendency to enhance radiation in the x direction. Such a propensity was expectable if one takes into account that the antenna is planar and displaying the larger extension in the yz plane, orthogonal to the maximum direction. The antenna is basically linearly polarized in the θ direction, with an average axial ratio around 20 dB in the maximum direction. The valley between the main two lobes is kept approximately at 30° at all wavelengths. The patterns also display a minimum along the longitudinal direction (z axis), which could have been anticipated from the antenna geometry as well.

Beyond the overall similarities the patterns exhibit some disagreements among bands too. Mainly, the lobe closest to the zenith tends to display a higher ripple for increasing frequencies. One could link such behavior to the non-multiband performance of the finite ground plane. Since the squared plane is not self-scalable, its size relative to wavelength is larger at the upper bands, resulting in an interference pattern with faster variations which might explain the pattern ripple at the upper bands.

2.3.1.4 Current Density Distribution

In order to get a deeper physical insight on the behavior of the Sierpiński antenna, a Finite-Difference in the Time Domain algorithm (FDTD) was developed to calculate the current density distribution over the fractal surface [45], [46].

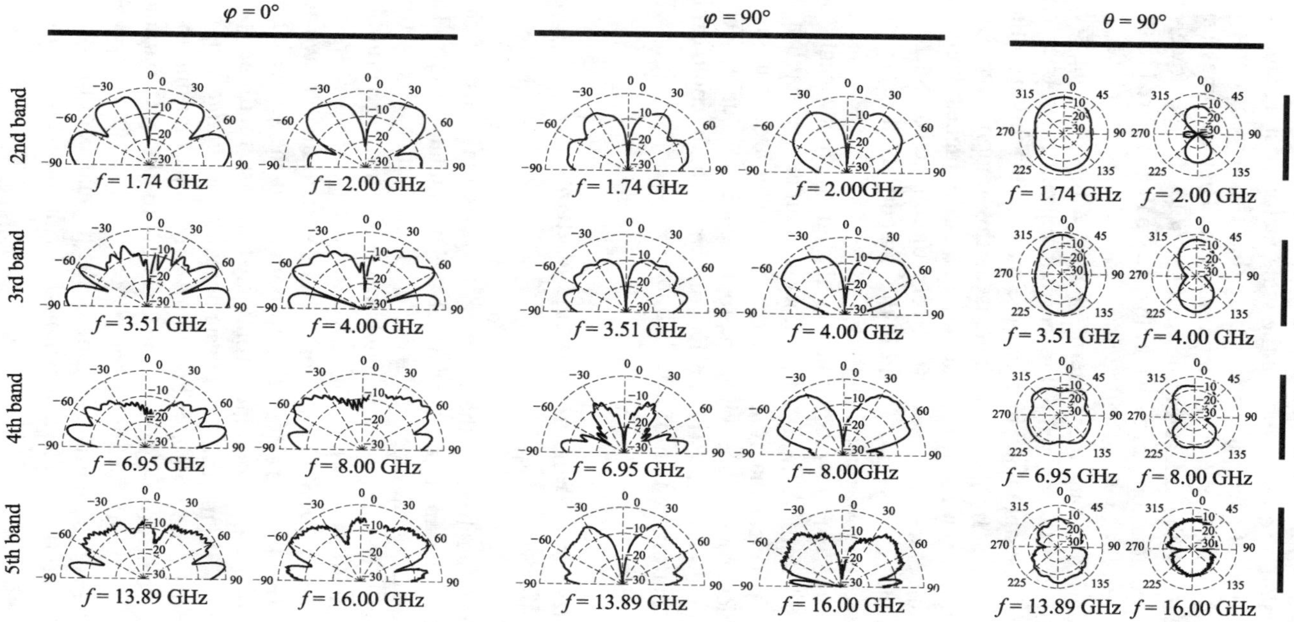

Figure 2.5 Radiation patterns (E_θ component) of the Sierpiński monopole at the four upper bands. The patterns have been measured at VSWR minima.

Figure 2.6 (see full-color insert) displays the magnitude $|\vec{J}|$ of the overall current density vector, (the magnitude plots are normalized with respect to its maximum and covering a 30 dB dynamic range). The right column of the plots presents an expanded view of the region where most of the current is concentrated at each frequency. All of them are scaled by a factor of two, the scale factor existing among bands. The most interesting feature of such plots is noticing that the current density distributions on the right column are very similar among them. That becomes especially apparent when one neglects the effect of the smaller holes at the lower bands. This is a reasonable approximation since such holes are small compared to the wavelength and the current density is lower at the regions where holes are placed.

The similarity among the patterns shown in the previous section can now be readily explained: at each band the current concentrates over a properly scaled substructure on the antenna which has the main contribution to the overall antenna performance. When frequency is increased, this region becomes smaller and the current does not reach the top of the antenna at the highest bands. This way, a large area of the structure becomes effectively disconnected. Such a phenomenon is equivalent to the active region found in the log-periodic arrays described in the early sixties by Carrel and Mayes [1].

Likewise, the current reconstruction also gives some insight on the truncation effect outlined before. Although it is difficult to precisely establish the bounds of the active region, it is clear that it actually covers a wider area than the encircled clusters of Fig. 2.6 which, roughly speaking, are about $\lambda/4$ tall. Actually, such a large extent of the current in the z direction (larger than one wavelength) could explain the similarity of the $\varphi = 0°$ pattern to that of a 1.5λ dipole.

2.3.1.5 Iterative Transmission Line Network Model

The FDTD analysis of the antenna suggests that most of the current density concentrates at the apex and borders of the triangular clusters. Therefore, it is not senseless to assume that most of the current flowing through any triangle comes from the direct ohmic contact from its neighbors. This way, a general Sierpiński structure could be analyzed as shown in Fig. 2.7.

That is, such a Sierpiński network can be seen as an iterative nesting structure of three-port networks. It is interesting to notice that the whole information upon the network behavior can be obtained from only two basic relationships [48]. On one side, the characteristic [S] matrix of the basic network (the *initiator* in fractal terminology), and on the other, the *generator* constitutive relations that link the [S] parameters of a particular stage of the fractal construction to the [S′] parameters of the next stage.

Regardless of the particular value of the [S] matrix of the triangular initiator, we know that due to the equilateral symmetry of the structure and its reciprocity, it has to be of the form

$$[S] = \begin{bmatrix} \alpha & \beta & \beta \\ \beta & \alpha & \beta \\ \beta & \beta & \alpha \end{bmatrix} \tag{2.24}$$

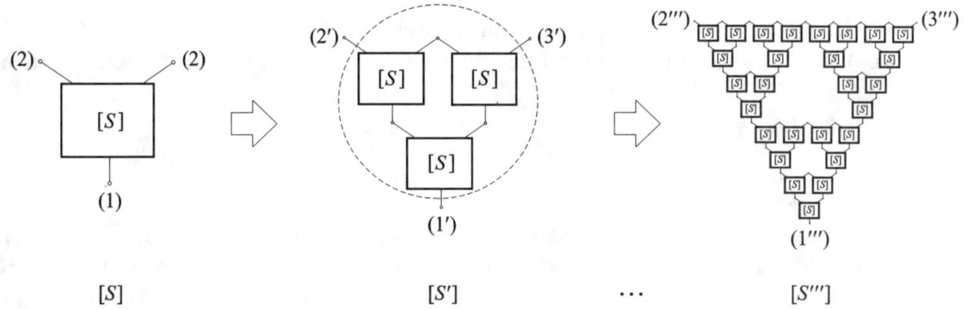

Figure 2.7 Iterative model for a Sierpiński Network.

After some algebraic manipulation, it can be shown that the generator relates the [S′] parameters of the following stage with the [S] of the previous one as

$$\alpha' = \alpha + \frac{2\beta^2\left(\alpha + \dfrac{\beta^2}{1-\alpha}\right)}{1-(\alpha+\beta)\left(\alpha + \dfrac{\beta^2}{1-\alpha}\right)}$$

$$\beta' = \frac{\beta^2\left(\alpha + \dfrac{\beta}{1-\alpha}\right)}{1-(\alpha+\beta)\left(\alpha + \dfrac{\beta^2}{1-\alpha}\right)} \tag{2.25}$$

One of the most interesting features of the recursive relations (2.25) is that they allow us to predict efficiently the antenna (network) behavior at any iteration stage without an increase of the model complexity.

In general, any symmetrical reciprocal three-port network could be used to model the triangular initiator. However, a transmission line network appears a natural approach as discussed before. In particular, the three models shown in Fig. 2.8 have been tested.

The length l of the triangular side has been chosen to match the side of the measured Sierpiński antenna, that is, $l = 6.4$ mm, which gives a length a of the Y-branch model of $a = 3.7$ mm.

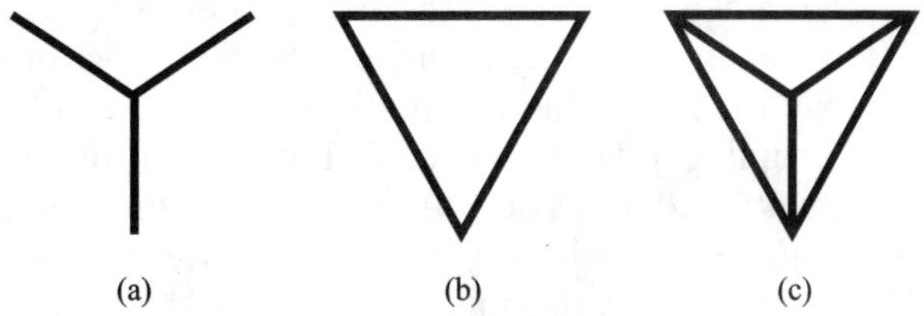

Figure 2.8 Three simple approximate models for the triangular initiator.

In order to take into account the losses due to radiation, a finite Q has been accounted for the transmission lines by introducing an attenuation factor α such as

$$Q = \frac{\beta}{2\alpha} \tag{2.26}$$

which has been empirically adjusted to $Q = 10$. Also the transmission line characteristic impedance has also been empirically adjusted to $5Z_o$; Z_o being the reference impedance used to compute the input reflection coefficient. Such coefficient for the whole three-port fractal structure can be calculated as

$$\Gamma_{in} = \alpha + \beta^2 \frac{\Gamma_2 + \Gamma_3 + 2(\beta - \alpha)\Gamma_2\Gamma_3}{(1 - \alpha\Gamma_2)(1 - \alpha\Gamma_3) - \beta^2\Gamma_2\Gamma_3} \tag{2.27}$$

where Γ_2 and Γ_3 are the normalized reflection coefficients of any couple of loads connected to ports 2 and 3, respectively. When modeling the one-port antenna network, such ports can approximately be considered to be open-circuited such that (2.27) becomes

$$\Gamma_{in} = \frac{\alpha^2 + \alpha(\beta - 1) - 2\beta^2}{\alpha + \beta - 1} \tag{2.28}$$

The three network models yield to a log-periodic behavior similar to that of the Sierpiński antenna. In Fig. 2.9 the input return loss and the input impedance for a five iteration structure is shown for the triangular model of Fig. 2.8(c). A better agreement is obtained when the structure is resistively terminated to take into account the additional radiation at the lowest band (see dashed reflection coefficient in Fig. 2.9). In particular, the Sierpiński network of Fig. 2.8 has been loaded with a frequency-dependent resistance R_L such as

$$R_L \propto \left(\frac{h}{c} f\right)^2 \tag{2.29}$$

h being the whole fractal monopole size, to take into account the parabolic characteristic radiation resistance of a small monopole at low frequencies. It is interesting noticing that such an extra load does not significantly affect the performance of the higher bands since it becomes large enough to be considered an open circuit.

Some relevant conclusions can be extracted from the iterative model just introduced. First, the log-periodic behavior is directly related to the antenna fractal geometry; no matter the details of the initiator, the multiband behavior is always obtained from the recursive model. Second, one can conclude that the overall behavior is mostly influenced by the direct ohmic contact of the triangular clusters with their neighbors. That is, it has not been assumed anywhere any kind of mutual coupling effects between clusters, yet the model still reproduces the expected multiband behavior. Also, the transmission line based model suggests that the antenna input impedance performance is mostly influenced by electromagnetic waves being propagated and reflected from the several parts of the metallic fractal structure,

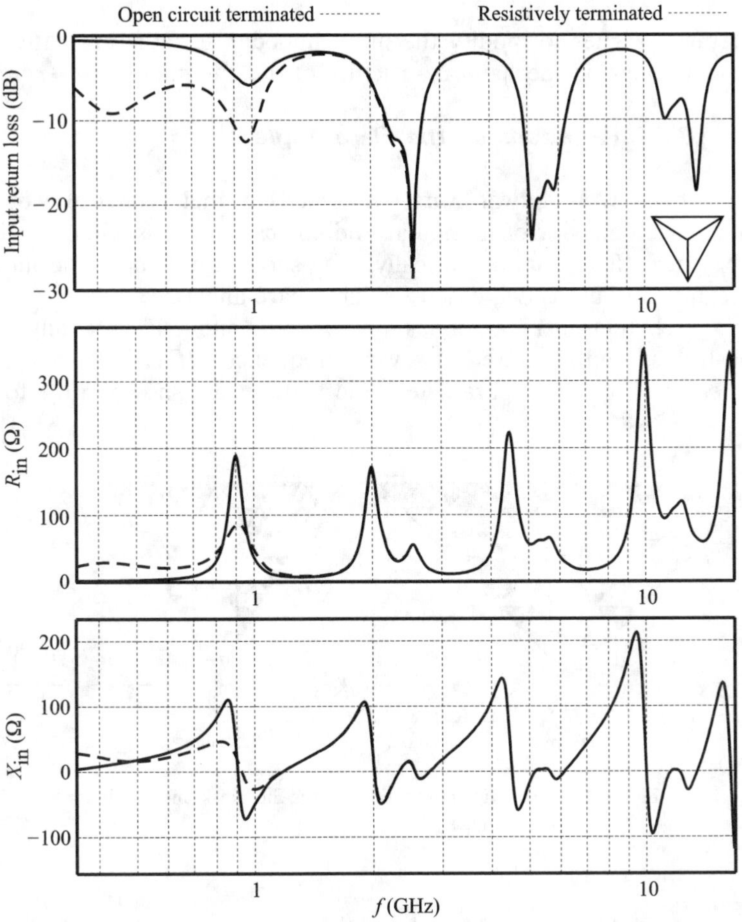

Figure 2.9 The Sierpiński network input impedance and return loss. The solid line displays the open circuit loaded network, while the dashed line displays the behavior of the resistively loaded network.

which must be linked to its geometrical shape. The algorithm even predicts the double resonant behavior found within each band.

2.3.2 Variations on the Sierpiński Antenna

Once the multiband performance of the fractal Sierpiński antenna has been demonstrated, the next logical step is to explore whether the shape of the antenna can be modified to tailor the antenna performance for several application requirements. Also, it is most interesting to investigate which are the limitations on distorting the antenna shape and yet not perturbing its multiband behavior. This section is devoted to the description of the antenna response when some variations are introduced upon its geometrical structure; mainly, by adjusting the flare angle, or changing the scale factor. The ultimate goal would be to explore the changes that would allow the antenna designer to modify the input impedance, radiation pattern and directivity, band spacing and mechanical structure of the antenna.

2.3.2.1 Variations on the Flare Angle

Brown & Woodward first described in their classical work [42] the input impedance behavior of triangular and conical antennas when modifying the feeding apex angle (flare angle). Basically, they showed how both the input resistance and reactance variations decreased for large flare angles.

Three Sierpiński antennas of 90°, 60°, and 30° flare angles (α) have been constructed and measured at several frequencies (Fig. 2.10), all of them having the same size. Their main parameters and features are shown in the following sections.

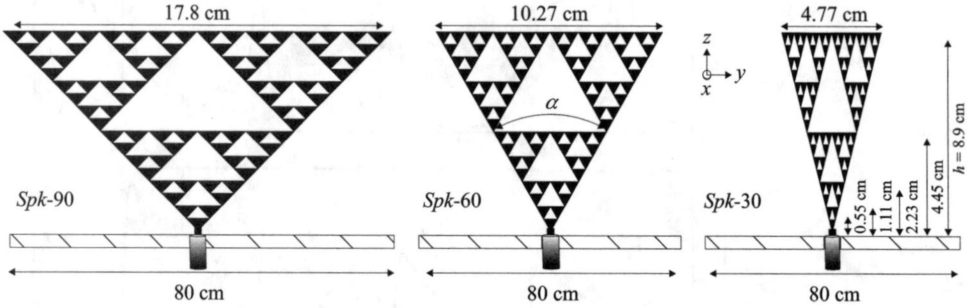

Figure 2.10 Several configurations of the Sierpiński monopole for 90°, 60°, and 30° flare angles.

The plots in Fig. 2.11 show the input parameters of the three Sierpiński monopoles, together with the same parameters for the analogous bow-tie antennas (results from [42]), in the low frequency region. The parameters are plotted with respect to the antenna electrical length ϕ, i.e.,

$$\phi = 360\frac{f.h}{c} \tag{2.30}$$

Both results, the measured Sierpiński parameters and the bow-tie ones exhibit a similar behavior at the first band (first odd and even resonances). The larger the flare angle, the smaller the input resistance and reactance variations. Also, resonant frequencies are consistently shifted toward the origin due to an overall increase of the antenna size. In the lower frequency region ($\phi < 60°$) the similar performance of the two antenna types is clearer, which was expectable since the contribution of the etched wholes on the Sierpiński antenna should be negligible for large enough wavelengths. Roughly speaking, the measured Sierpiński antenna works similarly to a slightly longer bow-tie antenna. Such an increase in the effective length when comparing first resonances must be related to the dielectric substrate supporting the fractal structure.

The performance of the two families of antennas starts to deviate significantly around the first even resonance. The input resistance and reactance of the Sierpiński

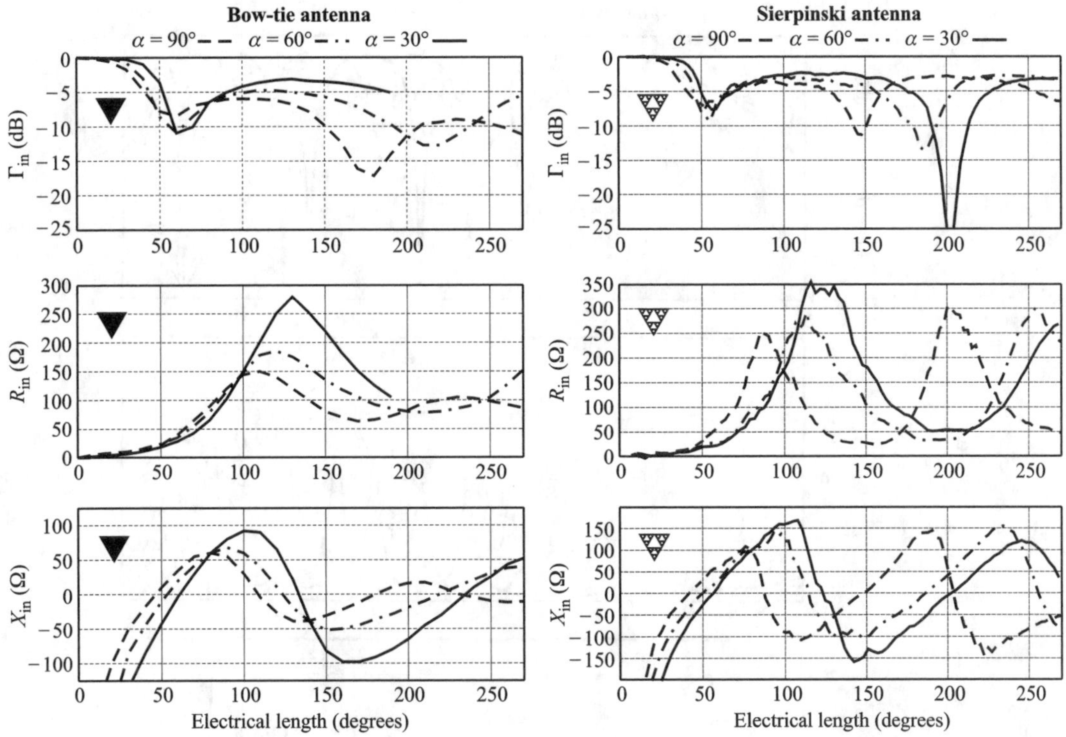

Figure 2.11 Comparison of the input parameters of a bow-tie [42] and a Sierpiński antenna, for several flare angles ($\alpha = 90°$, $\alpha = 60°$, and $\alpha = 30°$).

antenna suffer stronger variations at the third and fourth resonances. Actually, this phenomenon is on account of the first Sierpiński mode operation, which becomes significant at the second band and beyond. The shift of the resonant frequencies to longer wavelengths could be compared to that of a transmission line which has its distributed self-inductance increased (the holes on the Sierpiński structure would provide such an additional self-inductance); the additional self-inductance would reduce the phase propagation speed along the transmission line making the whole antenna appear slightly longer.

It is interesting noting that although the bow-tie and Sierpiński antennas work similarly at low frequencies, the fractal antenna deviates from the classical bow-tie operation mode at larger ones. In fact, the fractal antennas perform in a multifrequency, log-periodic fashion [26], [43], [44], while the Euclidean one keeps an almost periodic, harmonic behavior. The input resistance is not kept lower for wider apex angles, nor do the antennas maintain the same matching conditions. The antennas are best matched to 50 Ω at odd resonances, where the input resistances attain their minimum values. Such minima being reduced for larger flare angles, the input return-loss decreases below 10 dB in that case.

2.3.2.1.1 The Iterative Network Model. In Section 2.3.1.5 an iterative transmission line network model for the Sierpiński antenna was introduced. Being extremely simple and computing efficient, the algorithm is basically able to predict the

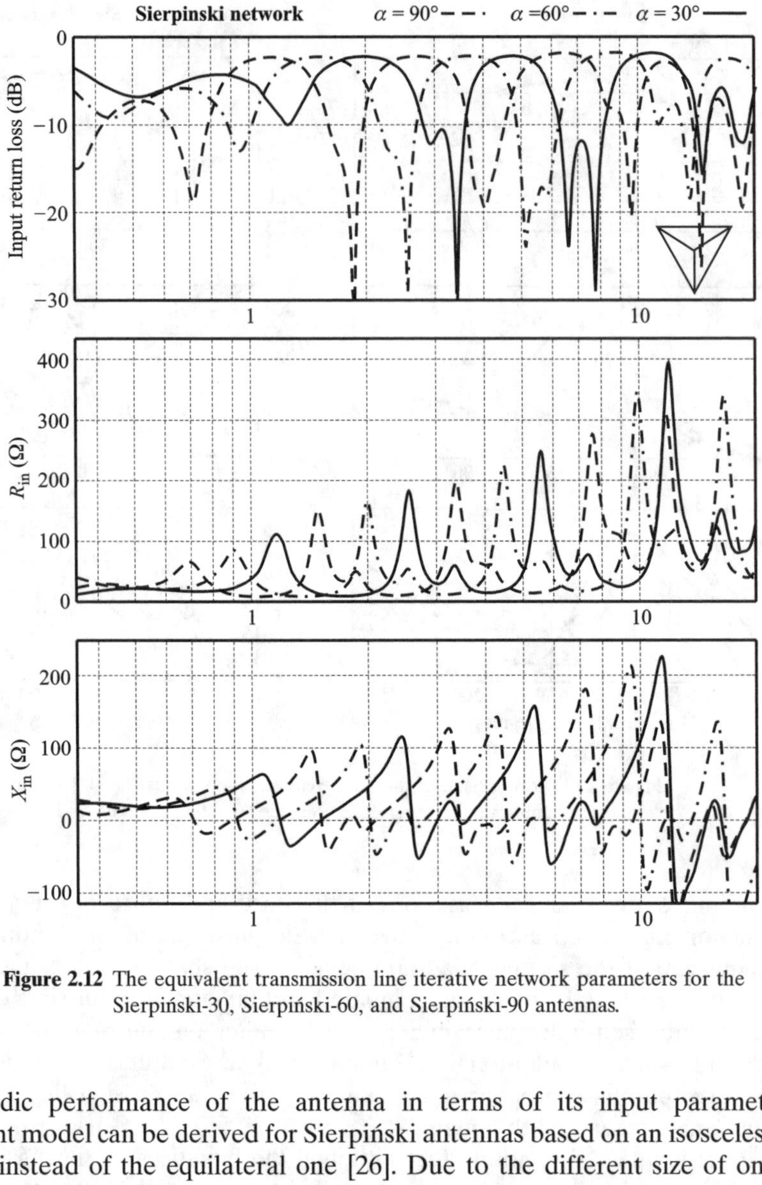

Figure 2.12 The equivalent transmission line iterative network parameters for the Sierpiński-30, Sierpiński-60, and Sierpiński-90 antennas.

log-periodic performance of the antenna in terms of its input parameters. An equivalent model can be derived for Sierpiński antennas based on an isosceles triangle initiator instead of the equilateral one [26]. Due to the different size of one of the triangle sides, the equilateral symmetry of the network is broken, which must now be characterized by the following [S] matrix:

$$[S] = \begin{bmatrix} \alpha & \beta & \beta \\ \beta & \gamma & \delta \\ \beta & \delta & \gamma \end{bmatrix} \tag{2.31}$$

Similar relations to those in (2.25) have been used to compute the input return-loss of three transmission line Sierpiński networks with 30°, 60°, and 90° flare angles (the same characteristic impedance $5Z_o$, quality factor $Q = 10$, and resistive loading as those of the previous chapter have been used). In general, the results shown in Fig. 2.12 are consistent with the experimental data in Fig 2.11 and in [26], [44]: large flare angles shift the bands toward lower frequencies, while narrower angles tend to present a double matching characteristic within each band. The two input resistance maxima at each band become more separated for the Sierpiński-30° equivalent network, leading to such an enhanced double resonant behavior. Also, the resistance minima where the networks are best matched are higher for the narrower angle case.

2.3.2.1.2 The Multiperiodic Traveling Wave Vee Model. It appears most interesting to investigate an approximate model that would predict the antenna response to flare angle variations. A traveling wave model looks attractive for several reasons: we know that the bow-tie antenna performs basically as a traveling wave vee antenna at large enough frequencies, and we assume that the Sierpiński antenna operates through a propagating wave that becomes attenuated at smaller gaskets at large enough frequencies.

The Sierpiński antenna can be modeled as a vee dipole with currents propagating on the edges of the structure, as shown in Fig. 2.13. A classical analysis of the radiated field would lead to the following expression for the radiation vector in the principal plane $\varphi = 0°$,

$$N_{zo} = 2Ih \cdot e^{-j\beta_\alpha h/2} \cdot \left(e^{jk_z h/2} \frac{\sin((k_z - \beta_\alpha)h/2)}{(k_z - \beta_\alpha)h/2} + e^{-jk_z h/2} \frac{\sin((k_z + \beta_\alpha)h/2)}{(k_z + \beta_\alpha)h/2} \right) \quad (2.32)$$

where the subscript 'zo' has been added to the radiation vector to denote the waves propagating outwards from the origin.

The traveling wave model just presented can be accurate when the equivalent vee arms are terminated with a matched load that completely absorbs the waves reaching the antenna tips. Although this model can be accurate enough if current becomes properly attenuated before reaching the antenna ends, it can be further improved by adding the contribution of waves being reflected at the extremes. By introducing a current reflection coefficient ρ_c, the radiation vector of such equivalent reflected waves N_{zp} would be

$$N_{zp} = 2Ih \cdot \rho_c e^{-j\beta_\alpha h} \cdot e^{-j\beta_\alpha h/2} \cdot \left(e^{jk_z h/2} \frac{\sin((k_z - \beta_\alpha)h/2)}{(k_z + \beta_\alpha)h/2} + e^{-jk_z h/2} \frac{\sin((k_z - \beta_\alpha)h/2)}{(k_z - \beta_\alpha)h/2} \right) \quad (2.33)$$

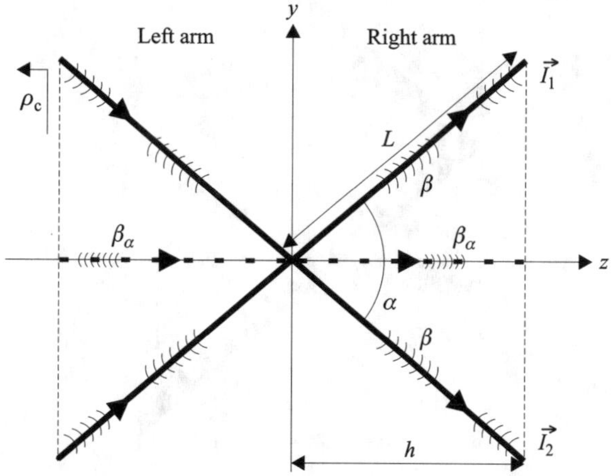

Figure 2.13 The balanced V model.

where the extra phase term $e^{-j\beta_\alpha h}$ can be interpreted as the average phase delay of the waves traveling from the average $|z| = h/2$ point to the antenna tips and back toward the start up point. The overall radiation vector N_z would reduce for such a full model to

$$N_z = N_{zo} + N_{zp} \tag{2.34}$$

Naturally, this equivalent model yields the same result as the well known sinusoidal standing wave pattern of a linear dipole when $\alpha = 90°$, $\rho = 1$ and $\beta = k$.

2.3.2.1.3 Propagation through Multiperiodic Media. Before illustrating our traveling wave vee model with some examples, some considerations regarding the phase propagation constant must be taken into account. Up to now, the model reduces the Sierpiński antenna to a double vee antenna model that supports edge current traveling waves with a phase propagation constant β. Such a propagation constant should be $\beta = k \,.\, n_{eff}$ to take into account the dielectric antenna coating (the effective refraction index was empirically adjusted to $n_{eff} \approx 1.1$ by comparing experimental and computed results). Although this model should be enough for a continuous shape, the fractal structure appears quite distinct. Again, if propagation along the antenna edges is primarily considered, the fractal shape comes into view basically as a periodic structure (see Fig. 2.14).

Propagation through a periodic structure is a rather particular phenomenon. The solution of the one-dimensional wave equation in periodic media yields an infinite superposition of n propagating modes,

$$E(l) = \sum_{n=-\infty}^{\infty} A_n e^{-j\beta_n l} \tag{2.35}$$

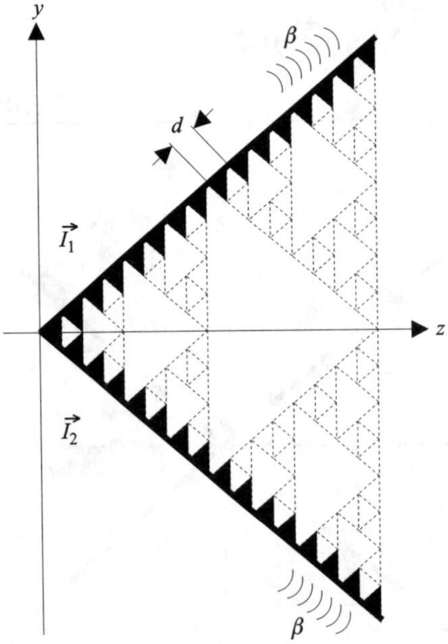

Figure 2.14 The edge of the Sierpiński antenna is a periodic structure.

with each mode characterized by a propagation constant β_n such that

$$\beta_n = \beta_o + \frac{2\pi}{d} n \tag{2.36}$$

where d is the characteristic structure period. Often, such modes are called *Hartree space harmonics* and the term corresponding to $n = 0$ is called the *fundamental harmonic* or the *fundamental mode*. One of the most interesting aspects of this phenomenon is the fact that a slow fundamental propagation constant β_o can yield a fast wave (therefore radiation) when the proper Hartree harmonic with a negative mode number n is considered. Also, depending on the sign of the resulting propagation constant β_n, such radiation can be given either in the forward direction (the same direction as the fundamental mode propagation) or backward direction (radiation opposite to propagation). In particular, it can be seen that forward radiation is given by modes with negative n numbers such that

$$\frac{d}{\lambda}(n_{eff} - 1) < |n| < \frac{d}{\lambda} n_{eff} \qquad \text{(forward)} \tag{2.37}$$

while backward radiation is supported by modes with

$$\frac{d}{\lambda} n_{eff} < |n| < \frac{d}{\lambda}(n_{eff} + 1) \qquad \text{(backward)} \tag{2.38}$$

n_{eff} being the fundamental mode's effective index of refraction. Again, some interesting conclusions can be derived from these relations. First, given a fixed period length d, there is a minimum wavelength that yields fast waves. That is,

$$\lambda < d(n_{eff} + 1) \tag{2.39}$$

This can be thought of as the cutoff wavelength of a high-pass system which, however, does not imply an abrupt cease of radiation when considering finite sources as discussed previously. As a rule of thumb and neglecting coating effects, such a cutoff wavelength is $\lambda/2 = d$. Second, backward radiation is allowed for lower frequencies than in the forward radiation case. Forward radiation can be thought of as having a smaller cutoff wavelength $\lambda = d \cdot n_{eff}$ (roughly speaking $\lambda = d$).

With these basic ideas in mind, some striking conclusions regarding our fractal Sierpiński antenna should be reached. One would usually think that in such case the characteristic period d is the length of the smallest triangle side. This would lead to the stunning conclusion that increasing the number of fractal iterations should convey to higher and higher cutoff frequencies that, in the ideal fractal structure, would forbid propagation of Hartree harmonics beyond the fundamental one. However, a careful view to the fractal shape evinces that it is in fact a multiperiodic structure! (See Fig. 2.15.)

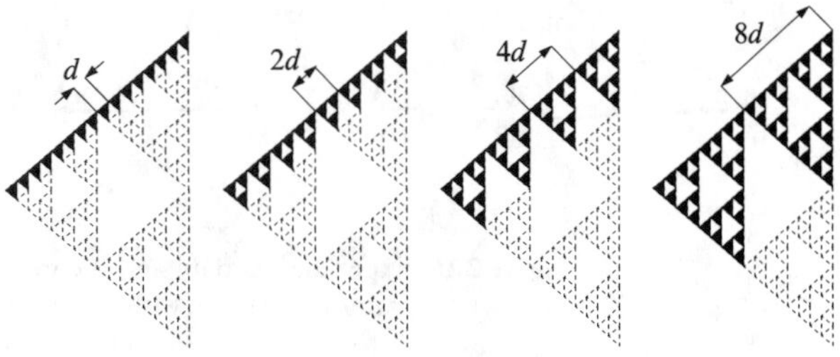

Figure 2.15 The fractal Sierpiński gasket seen as a multiperiodic structure.

With this new picture in mind, the periods $d_o, 2d_o, 4d_o \ldots, 2^m d_o$ become apparent, which means that wavelengths $\lambda_o, 2\lambda_o, 4\lambda_o \ldots, 2^m\lambda_o$ and frequencies $f_o, f_o/2, f_o/4, \ldots, f_o/2^m$ are actually supported. Of course, there must be a longest cutoff wavelength unless the structure is made infinitely large. In fact, the finite extent of the antenna only implies that most of the contribution to radiation comes from the fundamental mode.

2.3.2.1.4 Comparison with Experimental Results. The Sierpiński antenna is schematically described by means of a balanced two-vee model, which supports outward traveling waves. Such a model is completed by taking into account the proper reflected waves at the antenna tips. In the principal $\varphi = 0°$ plane, such a model is equivalent to a straight wire of the same size as the triangle height, aligned along the antenna axis direction (z axis). Waves propagate through this equivalent wire with a slower propagation constant β_α, which depends on the flare angle (2.36). A multiperiodic media approach suggests that not only a phase propagation constant along the antenna edges should be considered, but also several Hartree harmonics must be taken into account. The overall propagation constant can now be any within the set

$$\beta_{n\alpha} = \left(\beta_o + \frac{2\pi}{d}n\right)\cdot\frac{1}{\cos(\alpha/2)} \tag{2.40}$$

depending on which space harmonic (or harmonics) is considered.

In general it is not trivial determining how many Hartree harmonics must be included in the sum (2.35), or their relative amplitudes A_n. However, we can test the model at low frequencies where not many modes are expected to contribute to the overall sum. The plot in Fig. 2.16 displays the result of applying the current analysis to 30°, 60°, 90° flare angles. All antennas had the same physical parameters ($h = 8.9$ cm, $n_{eff} = 1.1$) except for the reflection coefficient at the antenna tips, which was chosen to be $\rho_c = -0.8$ for the two narrower antennas and $\rho_c = -0.2$ for the 90° one. That lower reflection coefficient for the 90° case has been empirically found to better match the nonideal null characteristic on the broadside direction of the experimental setup. All three antennas were tested at 2 GHz and a period equal to half of the antenna edge length was taken. It can be stated that both the theoretical model (solid line) and the experimental data (dashed) are in very good agreement.

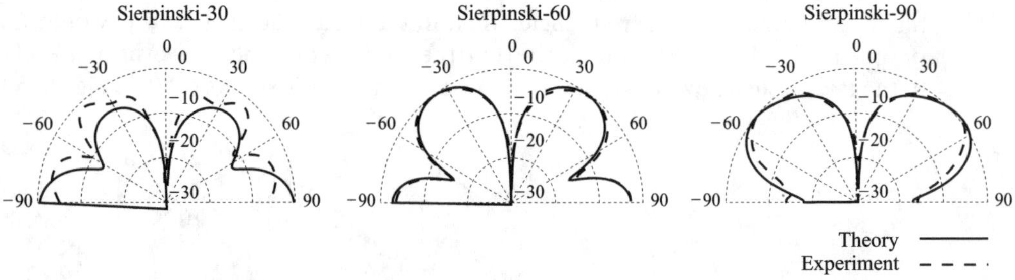

Figure 2.16 Experimental data and theoretical analysis for the Sierpiński antenna pattern variation with changing flare angle. The three models were tested at 2 GHz. Measured data are shown in dashed lines.

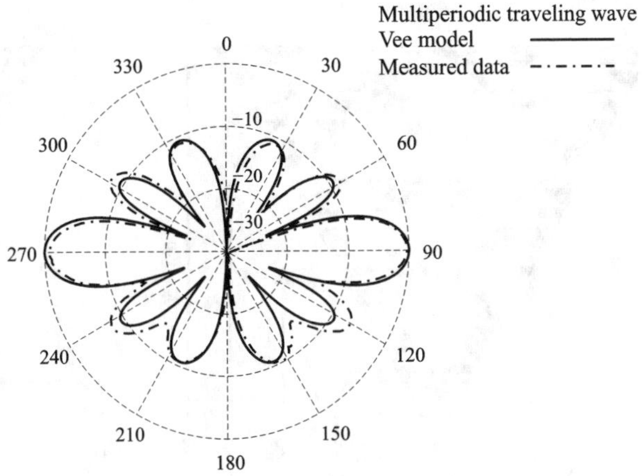

Figure 2.17 Theoretical and experimental patterns of the Sierpiński-60° antenna at 3.6 GHz (third band). The experimental pattern is slightly asymmetrical due to an imperfect balance of the balun.

The result just presented validates the theoretical approach at low frequencies, where the fundamental mode is basically dominant. At higher frequencies one must properly take into account other additional phenomena such as the reduction of the active region (which can be taken into account by reducing the effective antenna height h), the attenuation of the traveling wave due to radiation, and the multiperiodic nature of the system. To illustrate how the contribution of several space harmonics can modify the pattern that would predict the non-periodic model, another example is introduced. The Sierpiński-60 antenna at 3.6 GHz is now considered and space harmonics of first and second order are included.

When all modes are added with a proper amplitude factor (a relative weight amplitude $A_o = -2.43$ for the fundamental mode was the only required adjustment), the resulting pattern is the one in Fig. 2.17. Again, theoretical and experimental data are found to be in good agreement except for a nonideal behavior of the tapered balun used in the experimental setup.

It is very interesting to stress that the resulting superposition of modes yields a rather different result than the fundamental mode ($n = 0$) alone. That is, the traveling wave vee model itself is not enough to predict the fractal antenna behavior at large frequencies, but requires the periodic structure approach to give an accurate result.

2.3.2.2 Shifting the Operating Bands

It appears clearly from the results presented up to this point, that the multiband behavior of fractal antennas is a direct consequence of the geometric shape of the antenna body. It has been shown that the self-similarity scale factor is clearly linked to the band spacing in the log-periodic domain. For the classical Sierpiński gasket form, such a log-period was $\delta = 2$. Now, one should question whether such similarity relations would hold for other fractal shapes having different scale factors, and if one could properly modify a given fractal shape to allocate the operating bands where necessary.

The two novel designs are plotted in Fig. 2.18 together with the original Sierpiński antenna. Both have the overall form replicated at five different scales at the base of

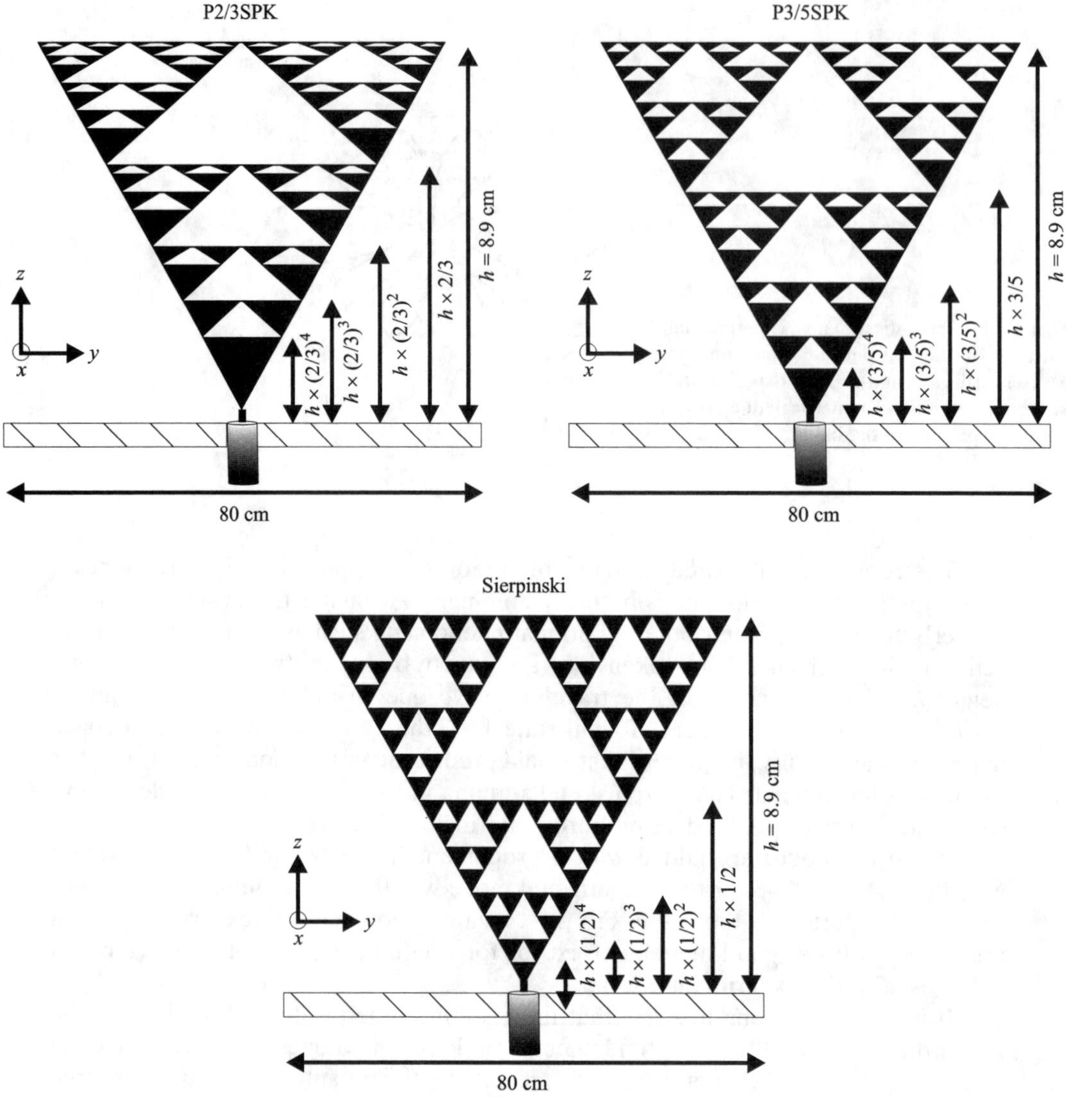

Figure 2.18 Evolution of the original Sierpiński-60° antenna towards the P2/3SPK and P3/5SPK models.

the antenna, the smaller copy being a single triangle. The antennas are designed by means of an iterative algorithm that basically consists in subtracting a scaled triangle from the original triangular form as described in [12], [24], which is a rather different approach than the IFS scheme. At each iteration, a reduction factor of 3/5 is found on the lower triangular cluster of one antenna (hereafter P3/5SPK) and a reduction factor of 2/3 is found on the other (hereafter P2/3SPK). It is interesting to notice that the two upper triangular clusters that remain after each subtraction are not proportional copies of the overall shape but are rather distorted. An affine transformation rather than a similarity one should be applied to go from the overall structure to these

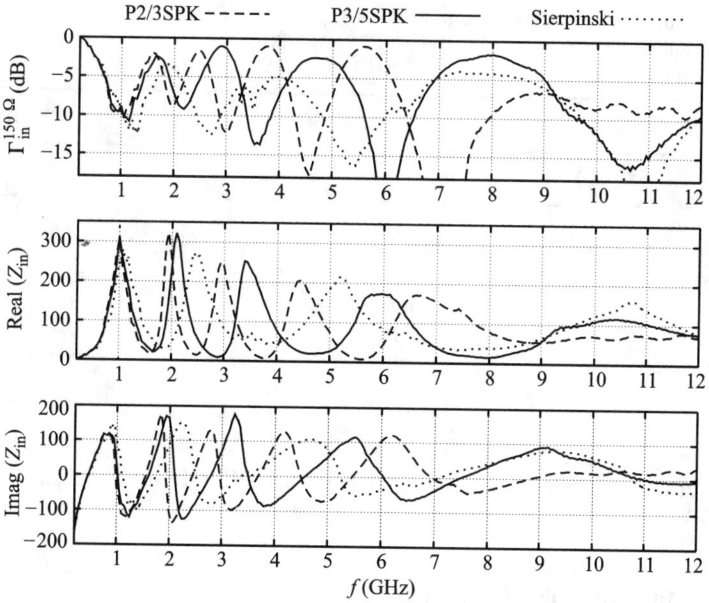

Figure 2.19 Input reflection coefficient relative to 150 Ω ($\Gamma_{in}^{150\Omega}$), input resistance and reactance of the three antennas. [*Figure from* [*24*].]

skewed gaskets, and hence the form is said to be self-affine [7] instead of self-similar. Nevertheless, a similarity relationship holds for the lower cluster where the active region is expected to concentrate, and hence a self-similar electromagnetic behavior can be expected for both structures.

Both antennas were constructed following exactly the same technique as for the Sierpiński antenna. The plots in Fig. 2.19 show the input parameters of the two novel antennas together with those of the Sierpiński-60. The input reflection coefficient has been normalized to a 150 Ω reference impedance which provides a better matching condition than the standard 50 Ω impedance. The input parameters clearly show a band compression toward the origin on the novel designs with respect to the original one. Such a compression effect is more remarkable in the P2/3SPK case, i.e., the design displaying the most compressed geometrical structure.

The link of the frequency response with the geometrical features of the antenna is clearly evinced in Table 2.2. It lists the first five resonant frequencies (odd resonances) together with the adjacent bands frequency ratio for the three fractal antennas. Clearly, the novel designs display a characteristic log-period equal to the characteristic scale factor of the fractal shape, i.e., $\delta = 1.66$ ($1/\delta = 3/5$) for the P3/5SPK antenna, and $\delta = 1.5$ ($1/\delta = 2/3$) for the P2/3SPK one. It is interesting to notice that once more, similarity is broken at the first band where a frequency ratio much larger ($\delta \approx 3.5$) is found between the first two odd resonances. Again, such an effect must be related to the truncation of the infinite ideal fractal structure which leads to a bow-tie mode operation at the longest wavelengths.

As a concluding remark, we should state that the Sierpiński gasket can provide a useful multiband antenna start-point design, which can be modified to match the specific needs of different applications. The perturbation of the scaling relation among

TABLE 2.2 Band Frequencies (odd resonances, GHz) and Scale Factor Between Adjacent Bands for the P3/5SPK, P2/3SPK and Sierpiński Antennas

P3/5SPK		P2/3SPK		Sierpiński	
f_n	f_n/f_{n-1}	f_n	f_n/f_{n-1}	f_n	f_n/f_{n-1}
0.45	—	0.44	—	0.46	—
1.61	3.57	1.54	3.46	1.75	3.77
2.76	1.71	2.43	1.57	3.51	2.00
4.54	1.65	3.69	1.51	7.01	2.00
7.52	1.66	5.43	1.47	13.89	1.98

fractal iterations appears a powerful tool to fulfill band location design requirements.

2.3.3 Fractal Tree-Like Antennas

Previous sections have been devoted to show that fractals can become multifrequency radiating systems, and this has been demonstrated both theoretically and experimentally. The Sierpiński antenna has been shown to operate similarly at a discrete set of frequencies and thus, it has been identified as a multiband antenna. Such feature has been shown to be clearly linked to the antenna fractal geometry, being the spacing among bands equal to the characteristic scale factor of the several fractal iterations.

One would naturally wonder what happened if such a characteristic scale factor was made arbitrarily close to unity. One might think that the several bands would become arbitrarily close and couple together into a single wide band, obtaining this way, a frequency-independent antenna. Actually, a frequency-independent antenna such as the log-periodic dipole array can be thought as a multiscale structure, smoothly varying by means of a small scale factor from small to large sizes. Therefore, one could reasonably expect that proper fractal shapes with small scale factors would operate similarly too.

However, it is not easy to find deterministic, regular self-similar fractal shapes with scale factors close to unity. Deterministic self-similar fractals such as the Cantor set, the Sierpiński gasket and Carpet, or the Dürer pentagon [7], are usually highly lacunar objects presenting large gaps within substructures and abrupt scale changes among iterations. In many cases, self-similarity is broken when the scale factor δ is made $\delta < 2$ and structures become at most self-affine. Among the fractal object families that allow small scale factors, the most common are random fractal shapes, and some relatives of Julia and Mandelbrot sets obtained by iterative transformations on the complex plane [6], [7]. Random fractal forms are among those that more clearly evince a continuous set of scales within their shape. Among the several families of random fractal objects such as Brownian processes, random IFS systems, etc. maybe the most appealing are the tree-like random ones. Many natural objects, such as corals, trees, leaves, weeds, and ferns have distinct fractal shapes [6], [7] with a rich scale distribution that look specially attractive to design antennas. Actually, one of the most popular fractal shapes (which however is not random), the Barnsley fern, is clearly similar to the log-periodic dipole array. Being extremely simple to feed, fractal tree-like monopoles have been chosen to investigate the feasibility of fractal wideband or frequency-independent antennas.

There are several ways of generating fractal random branching structures. Among them, the most common is the Diffusion Limited Aggregation (DLA) process. Such process can be either carried out through an experimental chemical process or by means of a computer simulated one. Both approaches lead to similar results.

The fractal shape was obtained by means of a quasi-two-dimensional electrodeposition experiment [49]. The fractal deposit obtained this way being a good conductor structure, the idea of using it directly as an antenna looked particularly attractive, since the generation process was quite simple, inexpensive, and fast (the deposit grew in a few minutes). However, the resulting structure is often too weak to be kept standing up once the glass plates are separated, and too thin to handle a significant amount of power. Therefore, the chemical deposit had to be recorded using a CCD camera, and the resulting image processed and printed over a Cuclad 250 dielectric substrate ($\varepsilon_r = 2.5$, $h = 1.588$ mm) using standard printed circuit techniques. Finally, the tree was mounted up in a monopole configuration over the same 80×80 cm aluminum ground plane used in previous fractal antenna designs (see Fig. 2.20).

The fractal dimension of this structure (D_f) was computed by means of the MFRAC software [49], leading to a value of $D_f = 1.71$. It is interesting to notice the statistical self-similarity [7] of the tree as shown in the blow-up in Fig. 2.20. One can expand an inner region around the feeding point and the resulting form still keeps a similar shape to the overall tree. An expansion by a factor of 7 still holds the main features of the structure and if one neglects the contribution of the smaller parts of the body to the radiation process at longer wavelengths, a similar electromagnetic behavior might be expected at a frequency up to seven times larger than the smaller one.

The antenna input reflection coefficient (Γ_{in}) relative to 50 Ω was measured from 900 MHz to 8 GHz. The resistance and reactance of the input impedance along with Γ_{in} are shown in plots (b), (c), and (d) of Fig. 2.20. The same parameters corresponding to a monopole of the same size (7.8 cm) are also plotted in dash-dotted lines for comparison. The fractal tree clearly displays a rich spectral structure with many resonant frequencies and is well matched (VSWR < 2) to 50 Ω at as many as 15 bands within the measured frequency range. This behavior is visibly distinct to that of the

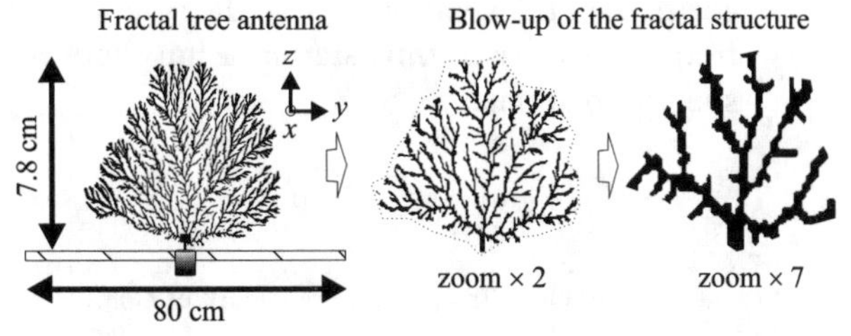

Figure 2.20 Fractal tree antenna and blow-ups around the feeding point. [*Figure from* [25].]

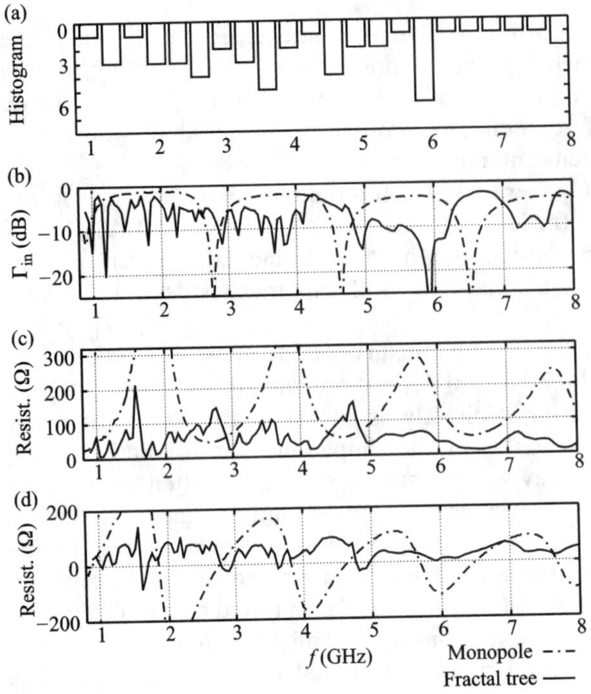

Figure 2.21 Fractal tree antenna input parameters (b), (c), (d), and equivalent distribution of lengths (a). [*Figure from* [25].]

previously tested deterministic fractal shapes, since the distribution of bands is much denser, as could be anticipated from the antenna geometry. Also, the difference with the harmonic behavior of the Euclidean linear dipole is manifest.

On the other hand, frequency-independent behavior cannot be stated for the fractal tree because the antenna does not consistently keep a uniform matching characteristic in the whole frequency range. Nevertheless, the tree improves the spectral behavior of the linear monopole because both the reactance and the resistance of the fractal keep a smaller deviation from 0 and 50 Ω respectively.

In order to link the spectral response of the antenna to its fractal geometry, the distribution of lengths on the structure was investigated as well. First, the recorded bitmap image of the tree had to be vectorized using a standard image segmentation software.[4] The several branches of the structure were converted this way to straight segments from which one could compute a length distribution histogram.

A histogram of the corresponding frequency distribution was computed and the result is shown in Fig. 2.21(a). The plot basically seems to evince that a certain correlation between the length distribution on the antenna structure and the matched frequencies exists, thus stating a link between the fractal antenna geometry and its spectral response.

[4] Corel OCR-Trace™ is a trademark of Corel Corporation.

2.4 SMALL FRACTAL ANTENNAS

2.4.1 Some Theoretical Considerations

Some fractal bodies have highly uneven shapes which might prove especially suitable for enhancing radiation. The Koch curve, for instance, is an example of an especially irregular one (Fig. 2.22). If one could make an antenna with its shape (for instance, a dipole or a monopole) the antenna would display an infinite set of sudden bends and sharp corners. The velocity vector associated to the current flowing along the curve would experience abrupt direction changes in such bends, which means that the electrical charges would be highly accelerated. Actually, the curve has no defined derivative anywhere, which means that such a sudden acceleration would be held everywhere within the curve. As discussed in [33], the acceleration of electric charges over an antenna surface is understood as the main mechanism to produce radiation, so fractal shapes such as the Koch curve seem magnificent candidates to become antennas.

2.4.1.1 About the Koch Curve

The Koch curve is named after the Swedish mathematician Helge von Koch who first introduced it in 1904 [50]. The curve can be constructed iteratively (Fig. 2.22)

Figure 2.22 Iterative construction of the Koch curve.

replacing the central segment of the unit interval by two segments of length 1/3, both forming the upper part of an equilateral triangle. The same procedure can be iterated an infinite number of times, each time replacing the central part of the remaining segments by a triangular cap, to obtain the ideal fractal Koch shape.

Some special features characterize this fractal form. The Koch curve is a self-similar object; many exact copies of the curve scaled down by a factor of three are found within the structure. Also, the curve is nowhere differentiable, i.e., no point of the ideal curve has a defined tangent [50]. Actually, the curve was first introduced as another example for the discovery of Karl Weierstrass who first described, in 1872, a non-differentiable curve.

Other striking properties regard the length and area of the curve. Although its shape looks basically the same when the number of iterations is large enough, the curve increases its length at each iteration up to an infinite length for the ideal fractal shape. That is, the length l_n^{Koch} of the nth iteration of the Koch curve is given by

$$l_n^{Koch} = a \cdot \left(\frac{4}{3}\right)^n \qquad l_n^{Koch} \underset{n \to \infty}{\longrightarrow} \infty \tag{2.41}$$

where a is the length of the initial segment if, in a more general case, the interval $[0, a]$ is taken instead of the unit interval as the fractal generator.

2.4.1.2 Theoretical Hypothesis

Several problems arise when trying to analyze, from the classical point of view, a fractal antenna based on a curve such as the Koch one. Assuming an ideal wire antenna which follows the shape of the curve, the vector potential equation would be calculated as

$$\vec{A}(\vec{r}, \vec{k}) = \frac{\mu}{4\pi} \int_C \vec{I}(l') \frac{e^{-jkR}}{R} dl' \tag{2.42}$$

which is a line integral with respect to the arc length, where C is the curve path defining the wire antenna shape and $I(l)$ is the linear current flowing along it. If C is the Koch or a similar fractal curve, can it be assumed that the integral along a finite size but infinite length curve is going to be bounded? Actually, line integrals with respect to the arc length are defined for rectifiable (finite length) curves [51] and the Koch curve is non-rectifiable. It could be argued that the exponential phasor term is also weighting the integral and might introduce some cancellations within the integral in case of a sign reversal due to proper phase retardation. In fact, the radiation mechanism is also understood as a process where a system, which is comparable in size to wavelength, introduces a sufficient phase retardation within its several parts to allow an in-phase radiation contribution from them. Therefore, an arbitrarily long (but not tall) Koch curve could fit any long wavelength which, contrasting to less convoluted non-fractal shapes, could provide the necessary phase retardation in an otherwise reduced space to constructively contribute to the radiation mechanism.

Another usual way to interpret (2.42) when evaluated at large distances from the source (i.e., in the far field), is as a Fourier transform of the current density distribution over the antenna structure. Again, the Fourier transform has been proved to converge for piecewise continuous functions with a piecewise continuous derivative

[51]. And the fractal Koch curve is continuous but does not have a piecewise continuous derivative, and consequently, one has to conclude it has not been proved that the implicit Fourier integral in (2.42) converges for this particular case. A similar argument can be applied to question the validity of Chu's theory, which is based on an expansion of the fields around the antenna in a summation of spherical harmonics that might not converge for a fractal case.

One could argue that an ideal fractal shape is unfeasible precisely due to its infinitely detailed structure and, in case of wired structures, due to the infinite length involved in their construction. Certainly, this would establish a technological practical limit; nevertheless a technological limit is a rather different concept than a fundamental limitation. In particular, the ideal fractal shape can be understood as a limit object obtained from a series of non-ideal objects, which progressively approach the ideal fractal shape at each iteration of a recursive generating algorithm (as shown in Fig. 2.22). That is, the ideal fractal shape can be approached to an arbitrary degree of similarity (i.e., with an arbitrarily small Hausdorff distance from the ideal fractal set [7]) with a non-ideal, non-infinite (and maybe practical and technologically feasible) shape. If the ideal fractal antenna had no limit on its Q factor, there might be a non-ideal and maybe technologically feasible shape which would go beyond the classical limit established for Euclidean shape antennas. In other words, its Q factor might be made arbitrarily small by choosing the suitable truncated non-ideal object to properly design the antenna shape.

2.4.2 The Small but Long Koch Monopole

2.4.2.1 Antenna Description

To test some of the hypotheses upon the feasibility of fractal-shaped small antennas, the first five iterations of the Koch curve were constructed. When analyzing the fractal antenna behavior, it looked especially interesting to compare it with that of the closest Euclidean version, i.e., a straight monopole. Hereafter we name such a straight monopole as $K0$ (the zeroth iteration of the fractal construction), while the remaining objects of the iterations will be referred as $K1, K2 \ldots KN$. Although the fractal shape might look too convoluted to be of practical application, it can be easily printed over a dielectric substrate using standard printed circuit techniques. The fabrication complexity in this case is exactly the same for the Euclidean antennas and for almost all fractal ones. Since one is interested in examining the low frequency behavior, a high performance (low-loss) microwave dielectric substrate is no longer required.

Up to five iterations of the fractal succession were constructed, all of them of the same overall height (but of course not the same length) and using the same substrate. The six antennas ($K0 \ldots K5$) where mounted over an 80×80 cm ground plane and measured over a 0.1 to 2 GHz frequency range. The antennas were also numerically analyzed using the frequency domain method of moments algorithm, written in a MATLAB™ code and run on an HP workstation. Figure 2.23 shows the fifth iteration version of the Koch monopole (i.e. $K5$), which has an overall height $h = 6$ cm, but a whole length of $l = h \cdot (4/3)^5 = 25.3$ cm. The results from the characterization of the six antennas are described and discussed in the following sections.

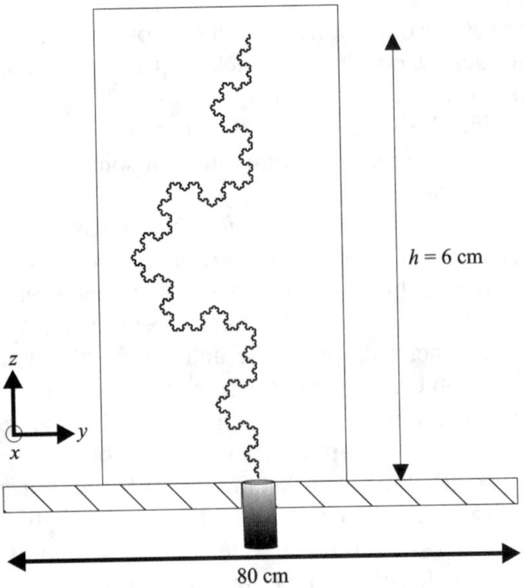

Figure 2.23 The five-iteration fractal Koch monopole ($K5$).

2.4.2.2 Input Parameters

The input resistance and reactance of the five measured Koch monopoles, together with the same parameters for a linear monopole are shown in Fig. 2.24. All antennas had the same height $h = 6$ cm, but of course a different length l. The method of moments data corresponding to the Euclidean antenna ($K0$) and the first three iterations of the fractal one ($K1, K2, K3$) are also shown for comparison. The copper etched wire had approximately a rectangular cross-section of width $w = 200\ \mu$m and thickness $t = 35\ \mu$m, except for the $K5$ model, where the wire width was reduced to $w = 150\ \mu$m. To get the best match between experimental and numerical data, an equivalent radius of $a = 120\ \mu$m was considered, following the equivalent radius approach outlined in [33]. Also, an equivalent input resistance due to ohmic losses was computed at each frequency by integrating the square of the current distribution over the whole antenna length. The skin effect was also taken into account there by assuming an equivalent cylindrical wire cross-section as described in [33] too. It is readily observed that a good match between numerical and experimental data is obtained.

Some interesting conclusions can be derived from the input parameters plot. First of all, in the low frequency region the input resistance increases with the number of iterations when comparing the six characteristics at a given frequency. While the linear monopole input resistance becomes very small below the first resonance (0.9 GHz), the $K5$ model is about its input resistance maximum value.

Analogously, the input reactance plot evinces that resonant frequencies are consistently shifted toward the lower frequency region at each fractal iteration. In particular, even all these particular Koch models can be considered small antennas ($kh < 1$) below $f \approx 0.8$ GHz, they are self-resonant, i.e., they have a vanishing input reactance without the need of an external compensating reactive element. The longest antenna ($K5$) reaches its second resonance about the same frequency where the linear dipole just has its first one.

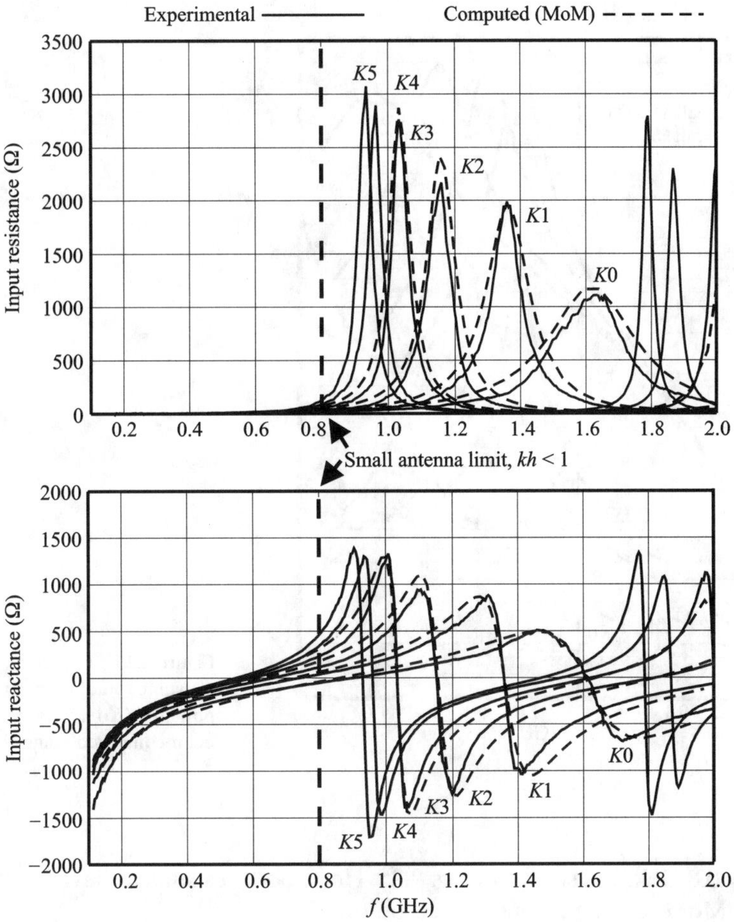

Figure 2.24 The Koch monopole input parameter evolution for several fractal iterations. [*Figure from* [*31*].]

The peculiar behavior of the Koch monopole must be linked to its geometrical shape. It is apparent that even though all models have the same size, they perform as longer antennas than would be predicted from their height alone. It seems that the electrical current propagates along the whole wire length despite its shape such that, the longer the whole wire, the lower the resonant frequencies regardless of the antenna size (height). This supports the idea that a sinusoidal-like current distribution along the whole wire should be considered. Such a distribution would introduce some phase shifts among the antenna that would enhance radiation with respect to the uniform current case where the contributions from some parts of the antenna might cancel.

It is important to stress that the input resistance increase is not due to an increase in the input loss-resistance with the wire length. Although both the resistance due to radiation and that due to ohmic losses behave similarly with frequency, the latter is consistently below the radiation resistance even in the small antenna frequency region

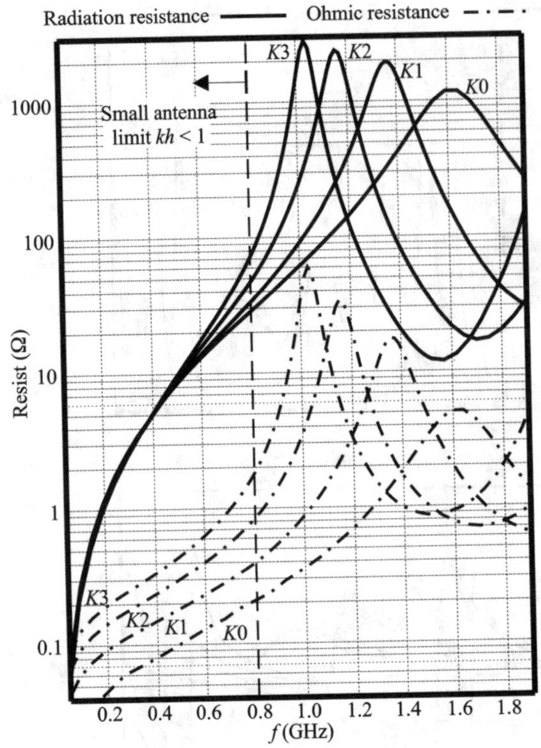

Figure 2.25 Frequency evolution of drive resistance radiation and ohmic loss components ($K0 \ldots K3$). Both features have been computed using the MoM technique.

($f < 0.8$ GHz), as shown in Fig. 2.25. Here, both resistances have been computed using the MoM technique as outlined before.

2.4.2.3 The Quality Factor

It has been shown that the fractal Koch antenna improves some features of a classical linear monopole when operating as a small antenna. Namely, resonant frequencies are shifted toward the longer wavelengths at each fractal growth iteration, making the antenna resonant even below the small antenna limit. Such a frequency shift makes the input resistance appear consistently larger in the fractal case than in the linear monopole one. However, one must take into account that not only the input resistance is raised, but also that the input reactance is increased. Actually, a figure of merit of the small antenna is its Q factor, which can be loosely estimated as the input reactance (X_{in}) to radiation resistance ratio (R_r).

The Koch antennas Q factors have been computed from both experimental and numerical data. The Q factor, rather than the Wheeler's power factor (PF) [35] has been chosen to check the antenna performance because the latter might lead to some wrong conclusions, as is readily shown. Both parameters are equivalent at the very low frequency range when the antenna is operated far from resonance. At such a low frequency region, the contribution to the input impedance of either the equivalent input inductor or capacitor almost vanishes and either PF or Q can be computed from the input impedance data as,

$$Q \approx \frac{1}{PF} \approx \frac{X_{in}}{R_r} \tag{2.43}$$

In such a very low frequency range, the Wheeler cap method [35], [38] can be used to estimate the antenna ohmic resistance and subtract it from the input resistance to evaluate the true antenna radiation resistance. However, even though the measured Koch antennas can be considered small, neither can the Wheeler cap method be applied, nor is its PF computation straightforward. Since the fractal monopoles are resonant, the inductive component contribution to the input impedance extends well beyond the small antenna limit. If one directly applied (2.43), one would get the wrong conclusion that a zero Q (and an infinite PF) were achieved at resonance. Also, the Wheeler cap method might lead to a wrong radiation resistance measurement since the metallic cap would slightly shift the antenna resonant frequencies where fast variations on the input parameters are obtained.

To properly evaluate the antenna Q, one must find a proper way to estimate either the average stored electric or magnetic energy and apply the following definitions [52]

$$\begin{aligned} Q &= \omega\frac{2W_e}{P_r} \qquad W_e > W_m \\ Q &= \omega\frac{2W_m}{P_r} \qquad W_m > W_e \end{aligned} \tag{2.44}$$

where W_e and W_m are the stored electric and magnetic energies respectively and P_r is the average radiated power. Of course, both definitions are equivalent if the antenna is self-resonant, which is not usual for small antennas. Harrington showed in [41] that the average stored electric and magnetic energies of a loss-less one-terminal microwave network can be related to the input reactance (X_{in}) and susceptance (B_{in}) as

$$\begin{aligned} W_e &= \frac{|I|^2}{8}\left(\frac{dX_{in}}{d\omega} - \frac{X_{in}}{\omega}\right) = \frac{|V|^2}{8}\left(\frac{dB_{in}}{d\omega} + \frac{B_{in}}{\omega}\right) \\ W_m &= \frac{|I|^2}{8}\left(\frac{dX_{in}}{d\omega} + \frac{X_{in}}{\omega}\right) = \frac{|V|^2}{8}\left(\frac{dB_{in}}{d\omega} - \frac{B_{in}}{\omega}\right) \end{aligned} \tag{2.45}$$

where I and V are the input terminals current and voltage, respectively. Since the power dissipated by the antenna is

$$P_L = \frac{1}{2}|I|^2 R_{in} = \frac{1}{2}|V|^2 G_{in} \tag{2.46}$$

the Q factor as defined in (2.44) can be computed as

$$\begin{aligned} Q &= \frac{\omega}{2R_{in}}\left(\frac{dX_{in}}{d\omega} - \frac{X_{in}}{\omega}\right) \qquad W_e > W_m \\ Q &= \frac{\omega}{2R_{in}}\left(\frac{dX_{in}}{d\omega} + \frac{X_{in}}{\omega}\right) \qquad W_m > W_e \end{aligned} \tag{2.47}$$

or equivalently as

$$Q = \frac{\omega}{2R_{in}}\left(\frac{dX_{in}}{d\omega} + \left|\frac{X_{in}}{\omega}\right|\right) \tag{2.48}$$

One must recall that (2.45) only strictly apply to one-port loss-less networks, but they become a good approximation for low loss, high Q networks, which is the case of small antennas. Actually, the definition in (2.47) is the same Chu used in his work [34] to derive the antenna Q fundamental limit.

Equation (2.48) has been applied to compute the Q factor over the low frequency range for the Euclidean and Koch fractal antennas (Fig. 2.26). Both experimental and numerical data have been used. In the later case, ideal loss-less Koch antennas have been considered to evaluate the antenna Q, that is, only considering power dissipation due to radiation. This has been done because the experimental data include the ohmic resistance that lowers the overall Q, which might lead us to the wrong conclusion that the Q reduction were only due to an increase of the ohmic losses.

The plot in Fig. 2.26 clearly evidences that the fractal antenna not only presents a lower resonant frequency and a larger radiation resistance, but it also improves the Q feature of the linear monopole. In a loose sense, such a Q can be interpreted as the inverse of the fractional bandwidth, which means that the fractal antenna features a broader bandwidth than the Euclidean one. Up to ~1.6 bandwidth improvement

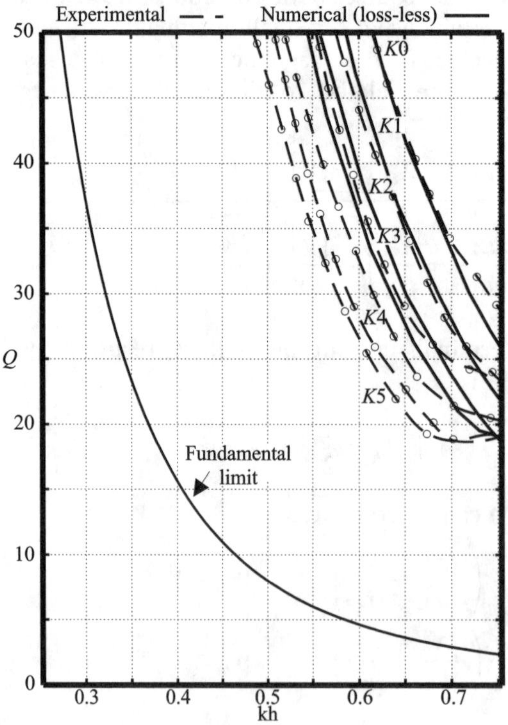

Figure 2.26 Q factor of the linear monopole and Koch antennas computed from numerical (solid lines) and experimental data (dashed lines). The Chu fundamental limit curve is shown for comparison. [*Figure from [31]*.]

is obtained when comparing the ideal loss-less monopole and the $K3$ antenna, while up to a ~2.25 bandwidth enhancement is obtained when comparing the monopole with the $K5$ antenna, which however has the contribution of larger ohmic losses. Anyway, it is clear that the Q factor is reduced at each fractal growth iteration and that, the larger the number of iterations, the closer the Q to the fundamental limit.

One can apply an argument based on geometrical considerations to explain such a result. It is commonly understood that an antenna Q factor depends on how efficiently it uses the available volume inside the imaginary radiansphere surrounding the antenna. As stated by Hansen in [36], the linear dipole being a one-dimensional object ($D = 1$), it inefficiently exploits such a radiansphere volume. Actually, as outlined in Section 2.2, fractal dimension is commonly loosely interpreted as a measure of the space-filling properties of the fractal object, therefore, it is not surprising that a fractal curve, featuring a fractal dimension $D > 1$ ($D \approx 1.262$ for the Koch monopole), is a better small antenna. Anyway, as a main conclusion one should state again that there exists a strong relation between the fractal geometric properties and the electromagnetic behavior of the antenna, and that such properties can be readily employed to design antennas with some useful properties.

2.4.2.4 Current Distributions

The method of moments analysis allows the reconstruction of the current distribution over the antenna. Figure 2.27 shows such a current distribution for the linear monopole and the $K3$ antenna. The fractal structure has been unfolded such that the current is shown over the whole antenna wire length. The wire length of both antennas has been normalized to the shorter one such that the further point on the longer current distribution yields the relative physical length, that is, $l^{K3} = (4/3)^3 l^{K0} = 2.37 \cdot l^{K0}$.

Both currents have been computed at four different frequencies. The first one corresponds to the first fractal resonance; the second one coincides with the small antenna limit; the third one is computed at the first linear monopole resonance, while the fourth one is about where the radiation pattern of the Koch antenna starts to become asymmetrical [26]. It is clear that the current distribution over the Koch antenna is very similar to that of a linear antenna with a length 2.37 times its height. Boundary conditions at the antenna tip force current to vanish and the wave propagating along the antenna wire becomes reflected toward the feeding point. Roughly speaking, a typical standing wave sinusoidal pattern is formed, which makes the input impedance oscillate more or less as it does in a typical transmission line. For instance, it is apparent that while the linear monopole still features an almost triangular current distribution, the sinusoidal pattern is already formed in the fractal case. At the first monopole's resonance, the current over the $K3$ antenna is about to reach its minimum value, i.e., the maximum impedance point. In the last case, the current amplitude over $K3$ starts to rise above the minimum point, and a π radian phase shift on the current distribution is obtained near to the feeding point. These results corroborate the idea introduced before that phase retardation can take place even in very small regions, and that such a retardation can enhance the radiation process.

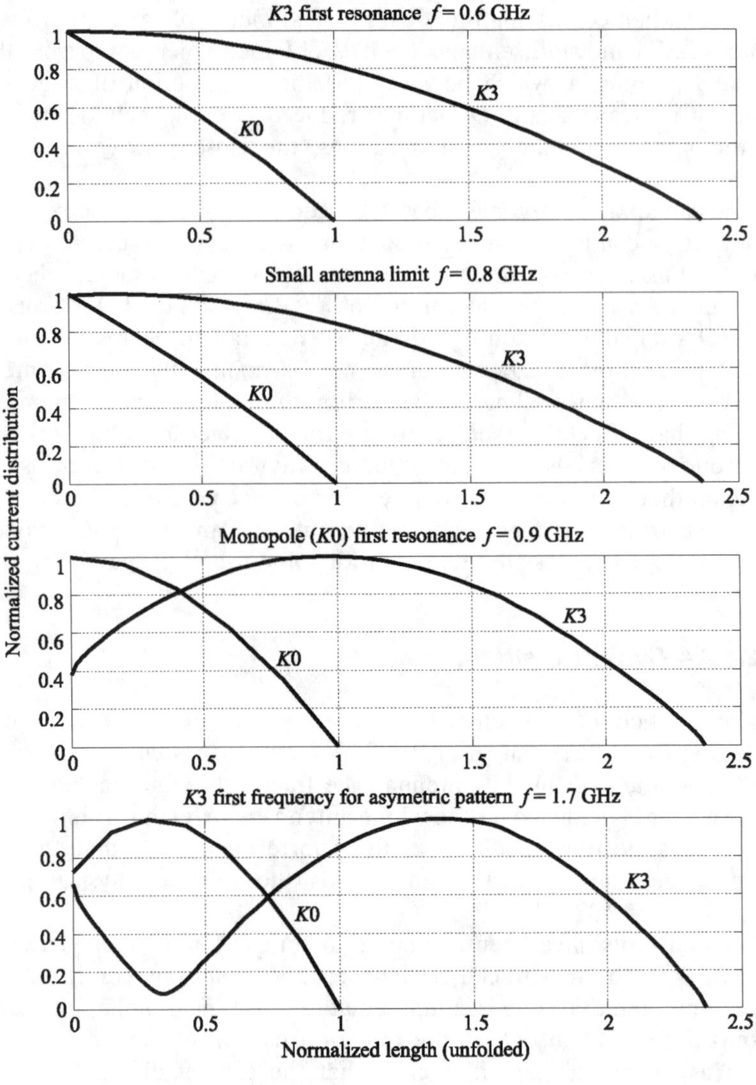

Figure 2.27 Current distributions over the monopole and Koch antenna ($K3$). From top to bottom, the currents have been computed at the first $K3$ resonance, the small antenna limit, the first monopole resonance, and the pattern asymmetry boundary. Length and current amplitude are normalized.

2.4.3 Conclusion

The potentiality of fractal antennas to become low-Q resonant small antennas has been proved through experimental and numerical results. The Koch monopole improves bandwidth, radiation resistance and reactance compared to those of a linear monopole of the same size. Equivalently, it can be thought that the size of a common monopole can be reduced by fractally shaping its linear wire. The antenna performance enhancement can be linked to the irregularity of the fractal shape, as well to its plane filling characteristic. Featuring a fractal dimension larger than unity, it can be thought that the fractal antenna better fills the available volume inside the radian-sphere that satellites the small antenna limit size. It has been shown too that it cannot be readily assumed that the Chu-Wheeler theoretical fundamental limit on small antennas also applies for ideal fractal antennas. One might think that the arbitrarily long length of the latter might yield an arbitrarily large radiation resistance even for arbitrarily small antennas. However, experimental and numerical results show that it is rather difficult to achieve a performance improvement beyond certain limits, and that both the electrical length, compression factor and radiation resistance enhancement characteristics seem to saturate. For fractal shapes, it seems that such limit should be related to the fractal dimension D since it describes the space-filling properties of the fractal object, but not a straightforward relation between D and such a theoretical limit (if existed) has been derived yet.

ACKNOWLEDGMENTS

The authors would like to thank P. Mayes from the University of Illinois and B. Mandelbrot from Yale University for encouraging this work. They also acknowledge R. Pous, J. Claret, F. Sagués, X. Garcia, F. Benítez, Ll. Milà, X. Fernández, A. Seguí, A. Hijazo, R. Bartolomé, A. Medina, J. Ramis, C. Borja, M. Navarro, J. Anguera, and J. Soler for several contributions to this work. This work has been partially supported by the Spanish Commission of Science and Technology (CICYT) and the European Commission through grant 2FD97-0135, and by the companies SISTEMAS RADIANTES F.MOYANO, FRACTUS, and AIRTEL.

REFERENCES

[1] P. E. Mayes, "Frequency-Independent Antennas and Broad-Band Derivatives Thereof," *Proc. of the IEEE*, **80** (1), January 1992.

[2] J. D. Dyson, "The Equiangular Spiral Antenna," *IRE Trans. on Antennas and Propagation*, **AP-7,** pp. 181–187, October 1959.

[3] D. E. Isbell, "Log-Periodic Dipole Arrays," *IRE Trans. on Antennas and Propagation,* **8,** pp. 260–267, May 1960.

[4] V. H. Rumsey, *Frequency-Independent Antennas*, Academic Press, New York, 1966.

[5] R. L. Carrel, "Analysis and Design of the log-periodic dipole antenna," Doctoral Thesis at the Dept. of Electrical Engineering of the University of Illinois at Urbana-Champaign, 1961.

[6] B. B. Mandelbrot, *The Fractal Geometry of Nature*, W. H. Freeman and Company, New York, 1983.

[7] H. O. Peitgen, H. Jürgens, and D. Saupe, *Chaos and Fractals, New Frontiers of Science*, New York, Springer-Verlag, 1990.

[8] C. Puente, "Fractal Design of Multiband Antenna Arrays," Electrical Engineering Dept. of the University of Illinois, Urbana-Champaign, ECE 477 term project, December 1993.

[9] C. Puente and R. Pous, "Fractal Design of Multiband and Low Side-Lobe Arrays," *IEEE Trans. Antennas and Propagation*, **44** (5), pp. 1–10, May 1996.

[10] Y. Kim and D. L. Jaggard, "The Fractal Random Array," *Proc. of the IEEE*, **74** (9), pp. 1278–1280, September 1986.

[11] D. L. Jaggard, "On Fractal Electrodynamics," in *Recent Advances in Electromagnetic Theory*, H. N. Kritikos and D. L. Jaggard, eds., Springer-Verlag, New York, 1990, pp. 183–224.

[12] D. L. Jaggard, "Prolog to special section on Fractals in Electrical Engineering," *Proc. of the IEEE*, **81** (10), pp. 1423–1427, October 1993.

[14] Y. Kim and D. L. Jaggard, "Optical Beam Propagation in a Bandlimited Fractal Medium," *J. Opt. Soc. Am.*, **A5**, pp. 1419–1426, 1988.

[15] D. L. Jaggard and X. Sun, "Scattering by Fractally Corrugated Surfaces," *J. Opt. Soc.*, **A7**, pp. 1131–1139, 1990.

[16] D. L. Jaggard and X. Sun, "Rough Surface Scattering: A Generalized Rayleigh Solution," *J. Appl. Phys.*, **68**, pp. 5456–5462, December 1990.

[17] D. L. Jaggard and X. Sun, "Reflection from Fractal-Layers," *Optics Let.*, **15**, pp. 1428–1430, 1990.

[18] D. L. Jaggard and T. Spielman, "Triadic Cantor Target Diffraction," *Microwave and Optical Technology Letters*, **5** (9), pp. 460–466, August 1992.

[19] C. Puente and R. Pous, "Diseño Fractal de Agrupaciones de Antenas," *IX Simposium Nacional de la URSI*, **1**, Las Palmas de Gran Canaria, September 1994 (*in Spanish*).

[20] D. H. Werner and P. L. Werner, "Frequency-independent Features of Self-similar Fractal Antennas," *Radio Science*, **31** (6), pp. 1331–1343, November–December 1996.

[21] C. Puente, J. Romeu, R. Pous, X. Garcia, and F. Benitez, "Fractal multiband antenna based on the Sierpiński gasket," *IEE Electronics Letters*, **32** (1), pp. 1–2, January 1996.

[22] C. Puente, J. Romeu, R. Pous, and A. Cardama, "On the Behavior of the Sierpiński multiband fractal antenna," *IEEE Trans. on Antennas and Propagation*, **46** (4), pp. 517–524, April 1998.

[23] C. Puente, R. Pous, J. Romeu, and X. García, "Antenas Fractales o Multifractales," Invention Patent, n°:P-9501019. Presented at the Oficina Española de Patentes y Marcas. Owner: Universitat Politècnica de Catalunya, 1995.

[24] C. Puente, J. Romeu, R. Bartolomé, and R. Pous, "Perturbation of the Sierpiński antenna to allocate the operating bands," *IEE Electronics Letters*, **32** (24), pp. 2186–2188, November 1996.

[25] C. Puente, J. Claret, F. Sagués, J. Romeu, M. Q. López-Salvans, and R. Pous, "Multiband properties of a fractal tree antenna generated by electrochemical deposition," *IEE Electronics Letters*, **32** (25), pp. 2298–2299, December 1996.

[26] C. Puente, "Fractal Antennas," Ph. D. Dissertation at the Dept. of Signal Theory and Communications, Universitat Politècnica de Catalunya (UPC), June 1997.

[27] N. Cohen, "Fractal Antennas: Part 1," *Communications Quarterly*, pp. 7–22, Summer 1995.

[28] N. Cohen, R. G. Hohlfeld, "Fractal Loops and The Small Loop Approximation," *Communications Quarterly*, pp. 77–81, Winter 1996.

[29] N. Cohen, "Fractal and Shaped Dipoles," *Communications Quarterly*, pp. 25–36, Spring 1996.

[30] N. Cohen, "Fractal Antennas: Part 2," *Communications Quarterly*, pp. 53–66, Summer 1996.

[31] C. Puente, J. Romeu, R. Pous, J. Ramis, and A. Hijazo, "Small but long Koch fractal monopole," *IEE Electronics Letters*, **34** (1), pp. 9–10, January 1998.

[32] Electromagnetics & Photonics Engineering group WWW site, http://www-tsc.upc.

es/eef/research_lines/antennas/fractals/fractal_antennas.htm

[33] C. A. Balanis, *Antenna Theory, Analysis and Design*, John Wiley & Sons, New York, 1982.

[34] L. J. Chu, "Physical Limitations on omni-directional antennas," *J. Appl. Phys.*, **19**, pp. 1163–1175, December 1948.

[35] H. A. Wheeler, "Fundamental Limitations of Small Antennas," *Proc. IRE*, pp. 1479–1488, December, 1947.

[36] R. C. Hansen, "Fundamental Limitations in Antennas," *Proc. of the IEEE*, **69** (2), February 1981.

[37] G. Goubau, "Multi-Element Monopole Antennas," *Proc. Workshop on Electrically Small Antennas* ECOM, Ft. Mammouth, N.J., pp. 63–67, May 1976.

[38] H. A. Wheeler, "Small Antennas," *IEEE Trans. on Antennas and Propagation*, **23** (4), pp. 462–469, July 1975.

[39] J. S. McLean, "A Re-Examination of the Fundamental Limits on the Radiation Q of Electrically Small Antennas," *IEEE Trans. on Antennas and Propagation*, **44** (5), May 1996.

[40] K. Fujimoto, A. Henderson, K. Hirasawa, and J. R. James, *Small Antennas*, Research Studies Press Ltd., West Sussex, February 1988.

[41] R. F. Harrington, *Time-Harmonic Electromagnetic Fields*, McGraw-Hill, New York, 1961.

[42] G. H. Brown and O. M. Woodward, "Experimentally Determined Radiation Characteristics of Conical and Triangular Antennas," *RCA Review*, pp. 425–452, December 1952.

[43] C. Puente, J. Romeu, R. Pous, and A. Cardama, "Multiband Fractal Antennas and Arrays," *Fractals in Engineering*, Springer-Verlag, New York, June 1997.

[44] C. Puente, M. Navarro, J. Romeu, and R. Pous, "Variations on the Fractal Sierpiński Antenna Flare Angle," *IEEE Antennas & Propagation* – URSI Symposium Meeting, Atlanta, June 1998.

[45] K. S. Yee, "Numerical Solution of Initial Boundary Value Problems Involving Maxwell's Equations in Isotropic Media," *IEEE Trans. on Antennas and Propagation*, **AP-11**, pp. 302–307, May 1966.

[46] G. Mur, "Absorbing Boundary Conditions for the Finite-Difference Approximation of the Time-Domain Electromagnetic-Field Equations," *IEEE Trans. on Electromagnetic Compatibility*, **EMC-23**, pp. 377–382, November 1981.

[47] A. Rubio, A. Salinas, R. Gómez, and I. Sánchez, "Time Domain Analysis of Dielectric-Coated Wire Antennas and Scatterers," *IEEE Trans. on Antennas and Propagation*, **42** (6), June 1994.

[48] C. Borja, C. Puente, and A. Mesa, "Iterative Network Model to Predict the Behavior of a Sierpiński Fractal Network," *IEE Electronics Letters*, **34** (15), pp. 1443–1445, July 1998.

[49] L. López-Tomàs, J. Claret, and F. Sagués, "Quasi Two-Dimensional Electrodeposition under Forced Fluid Flow," *Physical Review Letters*, **71** (26), pp. 4373–4376, December 1993.

[50] H. Koch, "Sur une courbe continue sans tangente, obtenue par une construction géometrique élémentaire," *Arkiv för Matematik*, **1**, pp. 681–704, 1904.

[51] T. M. Apostol, *Calculus, Multi-Variable Calculus and Linear Algebra, with Application to Differential Equations and Probability*, Blaisdell Publishing Company, Waltham, MA, 1967.

[52] R. E. Collin and S. Rothschild, "Evaluation of Antenna Q," *IEEE Transactions on Antennas and Propagation*, **12** (1), pp. 23–27, January 1964.

Chapter 3

THE THEORY AND DESIGN OF FRACTAL ANTENNA ARRAYS

Douglas H. Werner, Pingjuan L. Werner, Dwight L. Jaggard
Aaron D. Jaggard, Carles Puente, and Randy L. Haupt

ABSTRACT

One of the most fruitful areas of fractal electrodynamics research to date concerns the application of fractal geometry to antenna engineering. In this chapter we provide a comprehensive overview of recent developments in the field of fractal antenna engineering, with particular emphasis placed on the theory and design of fractal arrays. Several important properties of fractal arrays will be presented and discussed in this chapter. These include the frequency-independent multiband characteristics of fractal arrays, schemes for realizing low sidelobe designs, systematic approaches to thinning, and the ability to develop rapid beamforming algorithms by exploiting the recursive nature of fractals.

3.1 INTRODUCTION

The term *fractal*, which means broken or irregular fragments, was originally coined by Mandelbrot [1] to describe a family of complex shapes that possess an inherent self-similarity in their geometrical structure. Since the pioneering work of Mandelbrot and others, a wide variety of applications for fractals have been found in many branches of science and engineering. One such area is *fractal electrodynamics* [2–6] in which fractal geometry is combined with electromagnetic theory for the purpose of investigating a new class of radiation, propagation, and scattering problems. A brief introduction to the subject of fractal electrodynamics has been provided in Chapter 1. One of the most promising areas of fractal electrodynamics research is in its application to antenna theory and design.

Traditional approaches to the analysis and design of antenna systems have their foundation in Euclidean geometry. There has been a considerable amount of recent interest, however, in the possibility of developing new types of antennas that employ fractal rather than Euclidean geometric concepts in their design. We refer to this new and rapidly growing field of research as *fractal antenna engineering*. There are primarily two active areas of research in fractal antenna engineering, which include the study of fractal-shaped antenna elements as well as the use of fractals in antenna arrays. A comprehensive treatment of the subject of fractal antenna elements can be found in Chapter 2. The purpose of this chapter, on the other hand, is to provide an overview of recent developments in the theory and design of fractal antenna arrays.

The first application of fractals to the field of antenna theory was reported by Kim and Jaggard [7]. They introduced a methodology for designing low sidelobe arrays that is based on the theory of random fractals. The subject of time-harmonic and time-dependent radiation by bifractal dipole arrays was addressed in [8]. It was shown that, whereas the time-harmonic far-field response of a bifractal array of Hertzian dipoles is also a bifractal, its time-dependent far-field response is a unifractal. Lakhtakia et al. [9] demonstrated that the diffracted field of a self-similar fractal screen also exhibits self-similarity. This finding was based on results obtained using a particular example of a fractal screen constructed from a Sierpiński carpet. Diffraction from Sierpiński carpet apertures has also been considered in [6], [10], [11], [15], and [17]. The related problems of diffraction by fractally serrated apertures and Cantor targets have been investigated in [12–17].

The fact that self-scaling arrays can produce fractal radiation patterns was first established in [18]. This was accomplished by studying the properties of a special type of nonuniform linear array, called a Weierstrass array, which has self-scaling element spacings and current distributions. It was later shown in [19] how a synthesis technique could be developed for Weierstrass arrays which would yield radiation patterns having a certain desired fractal dimension. This work was later extended to the case of concentric-ring arrays by Liang et al. [20]. Application of fractal concepts to the design of multiband Koch arrays as well as low sidelobe Cantor arrays are discussed in [21]. A more general fractal geometric interpretation of classical frequency-independent antenna theory has been offered in [22]. Also introduced in [22] is a design methodology for multiband Weierstrass fractal arrays. Other types of fractal array configurations that have been considered include planar Sierpiński carpets [23–25] and concentric-ring Cantor arrays [26], [30], [36].

The distinction between deterministic and random fractals is discussed in Section 3.2. The properties of random fractals are then exploited to develop a new design methodology for sparse arrays that combines the attractive features of both periodic and random arrays. In Section 3.3, the radiation characteristics of fractal aperture antenna arrays are investigated as a function of stage-of-growth, fractal dimension, and lacunarity. Several techniques for the synthesis of fractal radiation patterns are proposed in Section 3.4. These techniques are based on a special class of self-scalable arrays of discrete elements that include Weierstrass linear arrays, Fourier-Weierstrass linear arrays, and Weierstrass concentric-ring planar arrays. Also discussed in Section 3.4 is a Fourier-Weierstrass fractal radiation pattern synthesis technique for continuous line sources. Section 3.5 considers fractal array factors and their role in the design of multiband arrays. In particular, two types of multiband array designs are introduced in Section 3.5, one which has a Weierstrass-type fractal array factor and the other which has a Koch-type fractal array factor. The theoretical foundation for the study of deterministic fractal arrays is developed in Section 3.6. Various types of fractal array configuration are considered including Cantor linear arrays, Sierpiński carpet planar arrays, and Cantor concentric-ring planar arrays. Finally, a more general and systematic approach to the design of deterministic fractal arrays is outlined in Section 3.7. This generalized approach is then used to show that a wide variety of practical array designs may be recursively constructed using a concentric-ring circular subarray generator.

3.2 THE FRACTAL RANDOM ARRAY

3.2.1 Background and Motivation

Apparently the first fractal antenna array was the *random fractal array* [7]. The idea was to synthesize a sparse or thinned array with low sidelobes that was robust with respect to element failure and variations in element location and current excitation. It was clear that periodic arrays have the advantage of simplicity and relatively low sidelobes. However, their periodicity is destroyed when an element fails and because of grating lobes these arrays are particularly ill suited for the design of sparse or thinned arrays. Periodic arrays also tend to be sensitive to element spacing and the current amplitude and phase in each element. Likewise, random arrays—which can be formed in several ways—are less easy to specify but are robust with respect to element failure, placement and current excitation. Their performance tends to degrade gracefully as many of these parameters vary slightly. However, random arrays typically have high sidelobes that are the remnants of the grating lobes that appear in the periodic case. These characteristics are summarized below in Table 3.1.

TABLE 3.1 Heuristic Comparison of Ordered and Disordered Arrays

Ordered (periodic)	Disordered (random)
√ Easy to specify	√ Harder to specify
√ Few degrees of freedom	√ Many degrees of freedom
√ Relatively low sidelobes	√ Relatively high sidelobes
√ Not robust to element failure	√ Graceful degradation with element failure
√ Sensitive to exact element placement/excitation	√ Relatively insensitive to exact element placement/excitation
√ Not suitable for thinned arrays	√ Suitable for thinned arrays

Here we bring together some of the advantages of both periodic and random arrays by using random fractals whose geometry is a careful blend of short-range disorder and long-range order as indicated in Fig. 3.1. In a traditional sense we can think of the fractal random array as being a *quasi-random* array in which fractal geometry provides the lead in determining the appropriate relative amount of order and disorder. This geometry ensures long-range order or correlation not typical of completely random arrays.

We exemplify the method of fractal array construction by first examining the underlying fractal structures. Consider the initiator and generator pictured in the left

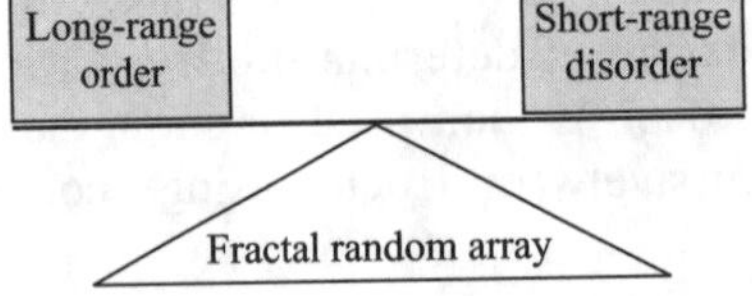

Figure 3.1 Illustration of the concept of using a balance of long-range order typical of fractals and short-range disorder typical of random arrays to create fractal random arrays.

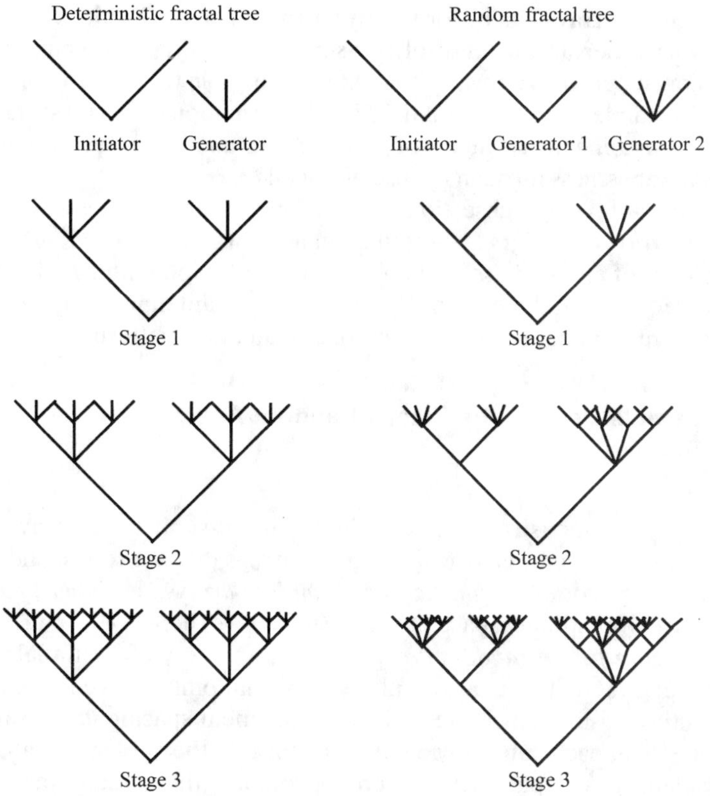

Figure 3.2 Construction of a deterministic (left) and random (right) fractal tree using initiators and generators for the first three stages of growth. In each case the initiator forms the general shape of the fractal and scaled forms of the generator are applied at each stage of growth to provide the necessary self-similarity. We choose randomly between the two generators to grow branches and leaves of the random fractal tree.

side of Fig. 3.2. By repeatedly applying a properly scaled version of the generator to the tips of branches of the initiator, a self-similar deterministic fractal tree begins to emerge. This structure is somewhat reminiscent of naturally occurring plants such as broccoli or cauliflower. However, we notice that this deterministic fractal tree lacks realism since it appears to be too regular when compared to common trees or plants. To reduce the regularity and increase the realism, we construct a random fractal tree as displayed in the right side of Fig. 3.2. Here we use a pair of generators with the same initiator. As the tree grows from one stage to the next, one of these generators is randomly chosen for each existing branch or leaf and applied to that branch or leaf. The self-similarity of the random fractal tree becomes evident if we consider average statistical properties of such trees.

The self-similarity characteristic of fractals is embodied by the phrase "generators of generators" which describes the construction of fractals when the initiator and generator are identical. This phrase provides a clue to the connection between fractal patterns and linear antenna array theory. These arrays, as noted by Schelkunoff [27] for amplitude-tapered linear arrays, can be formed by "arrays of arrays" which clearly implies both an underlying order and a sense of self-similarity or dilation symmetry. Based on this discussion, it becomes plausible that one may be able to design a useful fractal antenna array in a manner similar to that used for the formation of fractal structures. The central idea is to combine the strengths of periodic and random arrays with respect to sidelobe level and robustness through the use of fractals.

In the following subsection, random arrays are synthesized by imposing on a random array (initiator) a periodic subarray (generator) in order to control the distribution of energy in the sidelobes of the radiation pattern. In addition, we define a modified fractal dimension $\langle D \rangle$ suitable for antenna array theory in order to characterize the energy distribution in the radiation pattern.

3.2.2 Sample Design of a Fractal Random Array and Discussion

Here we demonstrate a method in which self-similar sub-arrays can be placed on randomly spaced centers to achieve sparse arrays that are robust and display relatively low sidelobes. For *deterministic fractal random arrays*, periodically spaced sub-arrays are placed on randomly spaced centers (not shown here) [7]. An attractive alternative is the *random fractal random array* in which a variety of irrational spacings are used for sub-arrays that are again placed on randomly spaced centers. Many other possibilities exist in which the variables of element spacing in the sub-arrays, number of elements in each sub-array, method of spacing the sub-arrays, and stage of growth are available design parameters. Here we compare the random fractal random array—here denoted simply as the *fractal random array*—with its traditional random array counterpart using the same number of elements and average element spacing.

The radiation pattern of a sparse random array is shown in the array factor $|AF|$ of Fig. 3.3 (top) (see the full-color insert) as a function of $\cos\theta$ where θ is the angle between the array axis and the observer. In this case 180 elements are distributed over 360 wavelengths (λ) so that the average element spacing is $\langle d \rangle = 2\lambda$. However, the peak sidelobe level has become relatively large at -5.7 dB, as predicted earlier, and its location is unknown a priori.

As an example of a random fractal array, six different periodic sub-arrays are used in a first stage random fractal tree. The location of the top leaves or branches defines the element positions along a line. The ratio of the element spacings of the various periodic sub-arrays are irrational numbers so that their radiation patterns do not share the same location of the extrema (nulls and secondary maxima). By doing so, the sidelobe radiation pattern is smeared and the large sidelobes typical of random arrays are avoided. This is observed in Fig. 3.3 (bottom) (see the full-color insert) where the maximum sidelobe level is -12.5 dB.

Following our prior work [7] we introduce a modified fractal dimension $\langle D \rangle^{\dagger}$ that describes the relative distribution of energy in the radiation pattern and explicitly

† We refer to $\langle D \rangle$ as a "modified fractal dimension" since when this definition is applied to the nodes of closed initiators, the Hausdorff-Besicovitch fractal dimension described by Mandelbrot and others is reproduced providing that common nodes are not double counted. Here a of (3.1) and (3.3) is analogous to a scaling ratio.

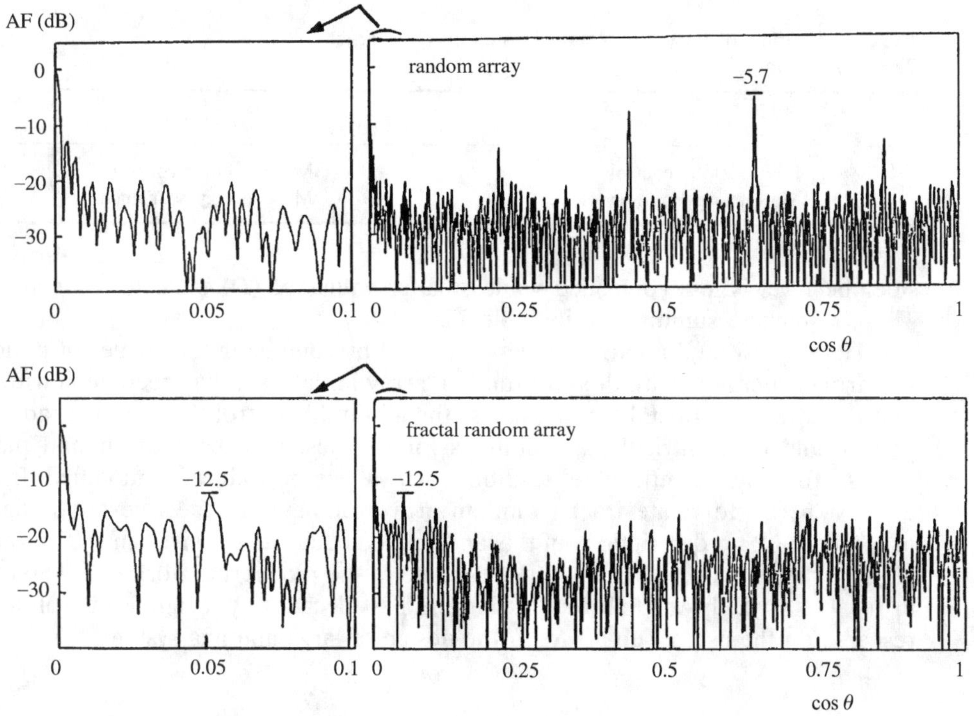

Figure 3.3 Array factor $|AF|$ in dB for a random array (top) and a fractal random array (bottom) as a function of $\cos\theta$, θ being the angle from the array axis to the observer. An inset (left of each plot) shows the details of the array factor near the mainbeam. These thinned arrays are composed of 180 elements spread over 360 wavelengths. The random array and fractal random array have sidelobe levels of -5.7 dB and -12.5 dB, respectively. These arrays have modified fractal dimension $\langle D\rangle = 1$ (random array) and $\langle D\rangle = 0.8$ (fractal random array). [*Figure adapted from "The Fractal Random Array," Y. Kim and D. L. Jaggard,* Proc. IEEE, **74**, 1278–1280 (1986). © *1986 IEEE.*].

shows the link between previous fractal theory and the fractal random arrays synthesized here. Let

$$\langle D\rangle = \frac{\ln(N)}{\ln(a)} \tag{3.1}$$

where

$$N = [\text{number of elements in the sub-array}] \tag{3.2}$$

$$a = \frac{[\text{average element spacing of initiator}]}{[\text{average element spacing of generator}]} \tag{3.3}$$

with N being the same for each sub-array. Using relations (3.1)–(3.3), the modified fractal dimension for the random array is $\langle D\rangle = 1$ while for the random fractal arrays it is $\langle D\rangle = 0.8$. From our work with these arrays, it appears that larger values of $\langle D\rangle$ indicate that the radiated energy is less concentrated near the main lobe and that the

TABLE 3.2 Effect of Varying Fractal Dimension $\langle D \rangle$ of Fractal Random Arrays

$\langle D \rangle \to 1$	$\langle D \rangle \to 0$
√ Less sidelobe control	√ More sidelobe control
√ More energy in far sidelobes	√ More energy near mainlobe

sidelobe level is less controlled while smaller values of $\langle D \rangle$ indicate the converse. These results are summarized in Table 3.2.

Here, quasi-random arrays are synthesized by combining the virtues of periodic sub-array generators with those of random array initiators. It has been demonstrated that these random fractal arrays possess the ability to control the sidelobe radiation pattern and offer intriguing possibilities for the design of both linear and planar arrays. To help quantify the resulting arrays, we introduce a modified fractal dimension $\langle D \rangle$ to relate fractal random array geometry to the relevant radiation properties. Other descriptions of this and related concepts are given in the literature [2], [6]. The precursors of this work originate in the pioneering work of Schelkunoff in the 1940s which was followed during the past decades by a significant volume of research on the use of sub-arrays in the design of large antenna systems.

3.3 APERTURE ARRAYS OR DIFFRACTALS

In this section we investigate the diffraction of electromagnetic waves by fractal objects or *diffractals*, a term originally coined by Berry [28]. In many of the cases studied here, the diffractals are apertures constructed from two-dimensional generalizations of the Cantor set described in a previous chapter (see Chapter 1). As is the case for many areas of *fractal electrodynamics* [2], [3], [4], [6], we are interested here both in ways that fractals imprint their characteristic signature on illuminating waves and in ways that they can be used in the design of new electromagnetic structures and devices, specifically antenna and aperture arrays.

We start with an examination of Cartesian diffractals and move to their polar counterparts. In the first case we consider a set of three apertures—the Cantor square, the Purina square or Vicsek fractal, and the Sierpiński square. The two families of polar diffractals considered in the second case are derived from the rotation of a Cantor set around either its midpoint or one of its endpoints. For all of these apertures, the physical optics approximation provides the desired far-zone diffraction pattern in terms of the Fourier transform of the aperture field. The self-similarity of the arrays is immediately evident in both the original diffractal and in its diffracted field. In particular the field at each stage of growth provides an envelope for a re-scaled version of the field at the following stage of growth. The diffracted fields can often be calculated using iterative techniques that parallel the growth of the fractals under investigation. These approaches are analogous to the iterative techniques used in calculating the scattering data for fractal superlattices examined in a previous chapter (see Chapter 1). A second self-similarity appears as various portions of the diffracted sidelobe pattern, when rescaled, mimic other portions. In particular, for a diffractal at stage of growth S, there are S similar bands in the sidelobe structure.

From our analysis we find that for the diffractals examined, the diffracted field of each has three characteristic regimes, one of which carries fractal information. The three regions of interest are the *subfractal*, *fractal*, and *superfractal* regimes. The subfractal regime reflects the gross geometry of the diffractal, but not its fractal structure, and includes the mainlobe. This is followed by the fractal regime in which the structure of the diffractal field contains information about the fractal geometry of the diffractal and the near sidelobes are controlled by the fractal geometry. The superfractal regime appears in the distant sidelobes where the fractal structure of the diffractal can no longer affect or control the sidelobes. Here the sidelobes are reminiscent of random array or random aperture results, although they have some structure in the form of "harmonic echoes" of the fractal region.

In one of the polar fractal examples we pay particular attention to the role of lacunarity as defined in Chapter 1. This parameter provides an added degree of freedom for designing fractal arrays and is used in Section 3.6.3 of this chapter to describe a design for a sparse low-sidelobe fractal array.

3.3.1 Calculation of Radiation Patterns

For an infinite planar conducting screen with aperture Ap, the diffracted or radiated electric field $\boldsymbol{E}(\boldsymbol{x})$ at the observation point $\boldsymbol{x}$ is given from diffraction theory [29] by

$$\boldsymbol{E}(\boldsymbol{x}) = 2\nabla \times \iint_{Ap} [\hat{\boldsymbol{n}} \times \boldsymbol{E}_{\mathrm{Ap}}(\boldsymbol{x}')]\,\mathrm{G}(\boldsymbol{x},\boldsymbol{x}')\,dS' \tag{3.4}$$

where $\boldsymbol{E}_{\mathrm{Ap}}(\boldsymbol{x}')$ is the exact aperture electric field, $\mathrm{G}(\boldsymbol{x},\boldsymbol{x}')$ is the free-space Green's function, $\hat{\boldsymbol{n}}$ is a unit vector along the position vector $\boldsymbol{x}$ of the observer, and the integration is taken over the aperture with $dS' = dx'\,dy'$. Here we invoke the Kirchhoff, high-frequency, or physical optics approximation so that the aperture field $\boldsymbol{E}_{\mathrm{Ap}}(\boldsymbol{x}')$ is replaced by the undisturbed illuminating electric field $\boldsymbol{E}_0(\boldsymbol{x}')$.

Ignoring polarization and a slowly varying obliquity factor, and scalarizing relation (3.4), we find the scalar diffracted field $\Psi(\boldsymbol{x})$ associated with $\boldsymbol{E}(\boldsymbol{x})$ is

$$\Psi(\boldsymbol{x}) \approx 2ik \iint_{Ap} [\Psi_0(\boldsymbol{x}')]\,\mathrm{G}(\boldsymbol{x},\boldsymbol{x}')\,dS' \tag{3.5}$$

where $\Psi_0(\boldsymbol{x}')$ is the scalar undisturbed illuminating field, λ is its wavelength, and $k = 2\pi/\lambda$ is the wavenumber. We have assumed an $\exp(-i\omega t)$ time-harmonic excitation. Using the far-zone approximation for the Green's function and taking into account the polarization and obliquity factors ignored in (3.5) we symbolically write the expression for the scalar diffracted field as

$$\Psi(\mathbf{x}) = \mathcal{F}\,[\Psi_0(\mathbf{x}')] \times [\text{Polarization Factor}] \times [\text{Obliquity Factor}] \tag{3.6}$$

where $\mathcal{F}$ indicates the Fourier transform. For Cartesian coordinates, the first term in the product is given explicitly as

$$\mathcal{F}\,[\Psi_0(\boldsymbol{x}')] = \mathcal{F}\,[\Psi_0(x',y')] = \iint_{Ap} \Psi_0(x',y',0) \cdot \exp[i(f_x x' + f_y y')]\,dx'\,dy' \Big|_{f_x = x/\lambda r,\, f_y = y/\lambda r} \tag{3.7}$$

where $r = \sqrt{x^2 + y^2 + z^2} = \sqrt{\rho^2 + z^2}$, and (x', y') and (f_x, f_y) form the appropriate Fourier conjugate pairs. Here f_x and f_y are the spatial frequencies along the x and y directions, respectively. The polarization and obliquity factors are slowly varying functions of the diffraction angle and are unimportant for the high frequency and small scattering angle case examined here. Therefore, in this section these terms in (3.6) will be approximated by a constant.

Relations (3.5)–(3.7) are the starting-point for calculations shown in the following discussions of radiation patterns and characteristics of diffractals. Many of the characteristics true of these diffractals also hold for the discrete case involving array factors and point sources since the former case and the latter case are related by multiplication of the array factor by the Fourier transform of the single aperture field distribution. Fractal array factors are treated explicitly in Sections 3.4 through 3.7.

3.3.2 Symmetry Relations

If aperture functions are self-similar, then so are their Fourier transforms which represent their radiated fields. As an example, consider a function $\mathrm{f}(x, y)$ which is self-similar, at least for discrete magnification. In this case we can write

$$\mathrm{f}(x, y) = \gamma^{2\alpha}\,\mathrm{f}(\gamma x, \gamma y) \tag{3.8}$$

Taking the Fourier transform we find that if

$$\mathcal{F}\{\mathrm{f}(x, y)\} = \mathrm{F}(f_x, f_y) \tag{3.9}$$

then

$$\mathrm{F}(f_x, f_y) = \gamma'^{2\alpha'}\,\mathrm{F}(\gamma' f_x, \gamma' f_y) \tag{3.10}$$

where $\alpha' = 1 - \alpha$ and $\gamma' = 1/\gamma$. Comparing (3.8) and (3.10) we find the transform of $\mathrm{f}(x, y)$ is self-similar in the same way the function itself is self-similar. This result is consistent with the literature [10], [21] and along with relations (3.5)–(3.7) demonstrates that self-similar aperture distributions yield self-similar diffracted or radiated field patterns. Since the arrays we consider are *prefractal* structures for which relation (3.8) holds for a limited range of scales, the symmetry relation (3.10) likewise holds over a limited range of scales. In particular, for the Sth stage of growth we can find S self-similar structures in both the aperture distribution and in its diffracted field. This is anticipated from the Scaling Theorem of Fourier transform theory and the fractal concept of stage of growth.

A second symmetry is present for a given stage of growth in which the diffraction pattern, when plotted on a log-log plot, repeats itself for a fractal aperture. The argument follows that of Puente [21]. Let us consider a fractal aperture that is composed of an infinite set of convolutions of suitable functions

$$\mathrm{f}(x, y) = \lim_{P\to\infty} \mathop{\ast}_{n=-P/2}^{P/2} \mathrm{f}(x\gamma^n, y\gamma^n) \tag{3.11}$$

where $\mathop{\ast}\limits_{n=-P/2}^{P/2}$ is a P-fold or $(P-1)$-fold convolution. We then find for the Fourier transform that

$$\mathrm{F}(f_x, f_y) = \lim_{P\to\infty} \prod_{n=-P/2}^{P/2} \mathrm{F}(f_x/\gamma^n, f_y/\gamma^n) \tag{3.12}$$

This implies that for fractal apertures characterized by (3.11), their array factor is log-periodic in spatial frequency and that on a log-log scale the radiation pattern repeats in each of an infinite number of bands. Again, since we are working only with *prefractals* (S finite), we find that these fractal array factors contain S replicas when plotted on a log-log scale. We will demonstrate this scaling for both the Cartesian and polar formatted arrays we examine next.

3.3.3 Cartesian Diffractals

Figure 3.4 shows a Cartesian fractal aperture Ap in the x'-y' plane ($z = 0$) illuminated from the left ($z < 0$) by a plane wave scalar field $\Psi_0(x', y', z')$. The observer at the point $x = (x, y, z) = (r, \theta, \phi)$ measures the diffracted field $\Psi(x, y, z)$. Using expression (3.7) we find the scalar diffracted field as

$$\Psi(x, y, z) = C\mathcal{F}\{\Psi_0(x', y', 0)\}|_{f_x=x/\lambda r, f_y=y/\lambda r} \tag{3.13}$$

where C is an unimportant constant.

The three Cartesian diffractals—Cantor square, Purina square or Vicsek fractal, and Sierpiński square—considered in this section are shown in Fig. 3.5 at the first several stages of fractal growth. In addition to stage of growth S, these fractals are described by their side length L and the scaling ratio γ used in their construction. From γ, the similarity fractal dimension $D^{\dagger}$ introduced in Chapter 1 is easily calculated for these regular fractals as noted below. The diffraction pattern is found by realizing the diffractal can be formed by the convolution of the array factor (a summation of delta functions and itself expressible as a convolution) with the shape of each individual opening (a square of side $L\gamma^S$). This provides an efficient method for

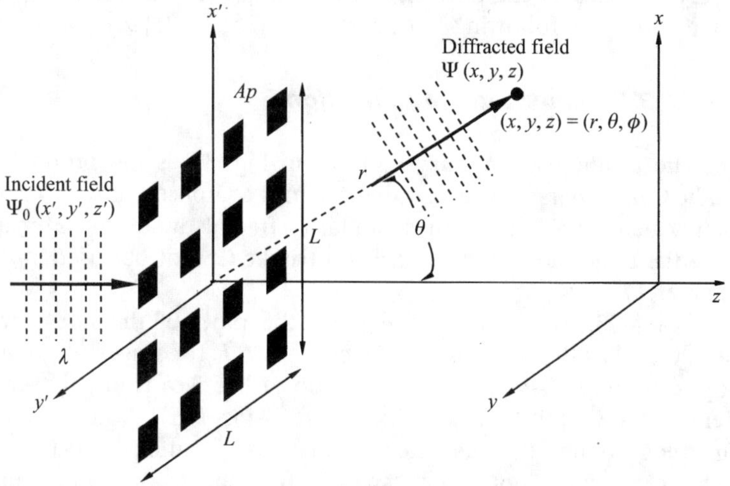

Figure 3.4 The problem geometry under consideration. A time harmonic wave $\Psi_0(x', y', z')$ of unit amplitude illuminates the Cartesian diffractal Ap in the plane $(x', y', 0)$ which generates the far-zone diffracted field $\Psi(x, y, z)$.

†Except in the case of the Weierstrass arrays, D will be used in this chapter to denote the similarity dimension D_s defined in Chapter 1.

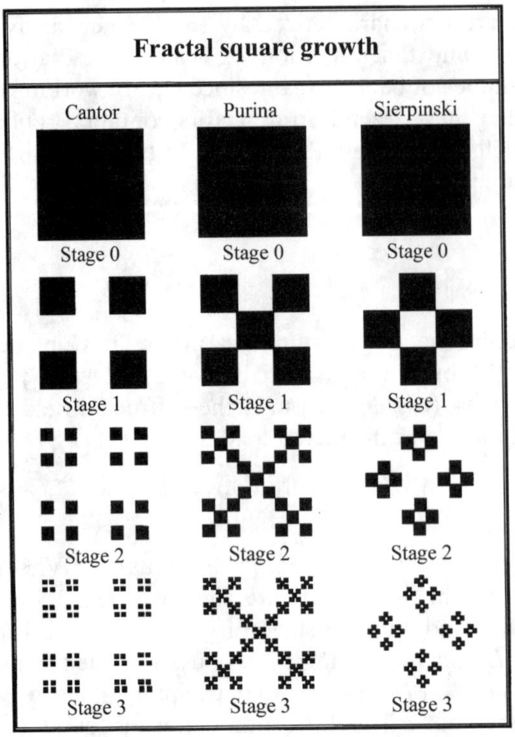

Figure 3.5 Stages of growth $S = 0, 1, 2, 3$ of the Cantor, Purina (Vicsek) and Sierpiński squares. The similarity fractal dimension of the Cantor and Sierpiński fractals is $D = \ln(4)/\ln(3) = 1.261$ while the Purina fractal has dimension $D = \ln(5)/\ln(3) = 1.465$. [*Figure adapted from* [15], [17].]

determining Ψ through (3.13) by multiplication in the spatial frequency domain that we use for each of the Cartesian diffractals. Alternative methods are used below for polar diffractals, following our prior work [15], [17], [26].

3.3.3.1 Cantor Square Diffraction

The triadic Cantor square shown in Fig. 3.5 is the product of two orthogonal triadic Cantor bars. Given that the fractal is composed of $N = 4$ copies of the original, each scaled by $\gamma = 1/3$, it has similarity fractal dimension $D = \ln(4)/\ln(3)$ which is twice the dimension of the associated triadic Cantor bar used in its construction (see Chapter 1).

Using (3.13) we show in Fig. 3.6 the plots of the normalized diffracted field $|\Psi(x, y, z)/\Psi(0, 0, z)|$ in dB as a function of $f_x L$ for stages of growth $S = 0, 1, 2, 3, 4, 5$ (top to bottom). The $S = 0$ pattern (top of Fig. 3.6) is just that of a square aperture. Here $|\Psi(x, y, z)/\Psi(0, 0, z)| = L^2|\text{sinc}(f_x L)\text{sinc}(f_y L)|$ where $\text{sinc}(x) \equiv \sin(\pi x)/(\pi x)$. This plot exhibits the characteristic decrease in intensity ($\sim[x^2 + y^2]^{-1} \sim [f_x^2 + f_y^2]^{-1}$) with increasing coordinate, spatial frequency, or polar angle $\theta = \cos^{-1}(z/r)$ characteristic of square apertures. The $S = 1$ pattern is re-scaled by the fractal scaling factor $\gamma = 1/3$ when compared to the $S = 0$ pattern. Here the $S = 0$ pattern provides an envelope for the $S = 1$ pattern and similarly for the other stages of growth. In general, the pattern for the Sth stage of growth provides the envelope for the $(S + 1)^{\text{st}}$ stage of growth when the latter is rescaled by γ. This follows directly from the self-similarity of the diffractal and the resulting self-similarity of its Fourier transform as given by (3.10) and from the multiplicative calculation of the diffraction pattern.

If we remove the effect of the size of individual apertures we get the diffraction pattern analogous to that of Fig. 3.6 which is shown in Fig. 3.7. The first figure is derived from the second by multiplying by the functions $\text{sinc}(f_x L)\,\text{sinc}(f_y L)$. Alternatively Fig. 3.7 can be thought of as the array factor for point sources located at the center of each aperture. More on these array factors will be noted in Section 3.6 of this chapter.

Another view of the scaling typical of these diffractals is given in the color plate of Fig. 3.8 (see full-color insert) in which the intensity in dB is shown on a color scale for $S = 0, 1, 2$ as a linear function of $f_x L$ and $f_y L$. Here the diffraction pattern is again rescaled in spatial frequency for different stages of growth so the scaling of each stage can be viewed as before.

A second scaling is evident by comparing different regions of sidelobes on log-log plots. In particular, the sidelobes contain S bands of similar structure, which form the fractal regime. We show this in Fig. 3.9 in which each of these $S = 5$ regions is shown and compared with the superfractal regime at high spatial frequency.

3.3.3.2 Purina Square Diffraction

The Purina square—so named because of its resemblance to a well-known corporate logo—is also known as the Vicsek fractal. The Purina square contains $N = 5$ copies of itself, each scaled by $\gamma = 1/3$, making its similarity fractal dimension $D = \ln(5)/\ln(3)$. As before, using (3.13) we display in Fig. 3.10 the plots of the normalized diffracted field $|\Psi(x, y, z)/\Psi(0, 0, z)|$ in dB as a function of $f_x L$ for stages of growth $S = 0, 1, 2, 3, 4, 5$ (top to bottom). The plots for the corresponding array factor are given in Fig. 3.11. In both cases we again see how the Sth stage diffraction pattern provides the envelope for the $(S + 1)^{\text{st}}$ stage, when the latter is rescaled by the scaling factor $\gamma = 1/3$. The diffraction pattern exhibits the log-log self-similarity across the sidelobes (not shown here) as demonstrated in Fig. 3.9 for the Cantor square.

3.3.3.3 Sierpiński Square Diffraction

The final example of Cartesian diffractals is the Sierpiński square diffractal. This fractal has the same fractal dimension $D = \ln(4)/\ln(3)$ as the Cantor square since it is also made from $N = 4$ copies, each scaled by $\gamma = 1/3$. The diffraction pattern for this diffractal is given in Fig. 3.12 and its associated array factor in Fig. 3.13. By comparing Fig. 3.12 with the Cantor square diffraction pattern in Fig. 3.6 we see that, as expected, the patterns differ. Therefore, the fractal dimension does not uniquely determine the radiation pattern of a diffractal. However, the characteristics of scaling across stage of growth S and scaling across sidelobes (not shown here), analogous to Fig. 3.9 for the Cantor square, do hold as expected.

3.3.3.4 Discussion

The three Cartesian diffractals examined here share a number of common characteristics. They are constructed using an iterative technique and this leads to the self-similarity of the prefractals investigated in this section. In turn, this self-similarity in the diffractal leads to self-similarity in the diffraction pattern with respect to stage of growth and log-periodic behavior with spatial frequency that leads to self-similarity across appropriately scaled sidelobes.

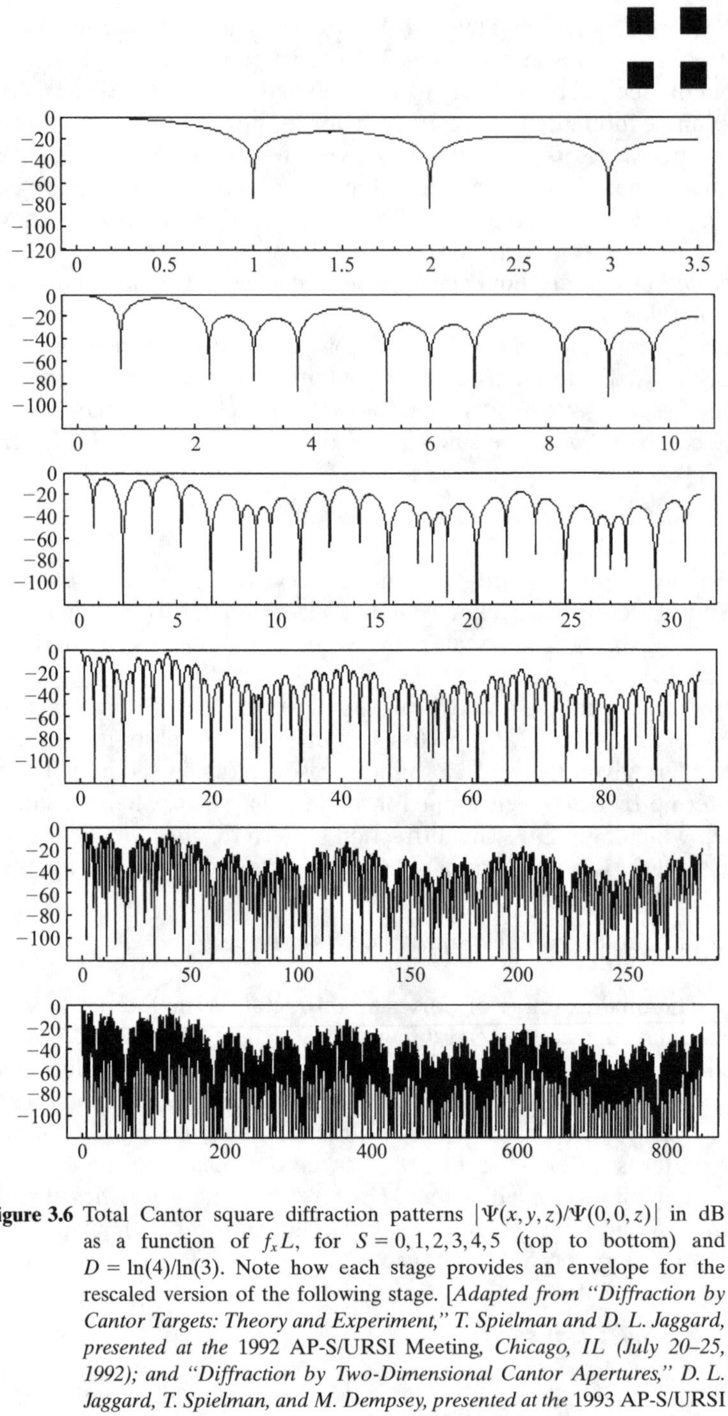

Figure 3.6 Total Cantor square diffraction patterns $|\Psi(x, y, z)/\Psi(0, 0, z)|$ in dB as a function of $f_x L$, for $S = 0, 1, 2, 3, 4, 5$ (top to bottom) and $D = \ln(4)/\ln(3)$. Note how each stage provides an envelope for the rescaled version of the following stage. [*Adapted from "Diffraction by Cantor Targets: Theory and Experiment," T. Spielman and D. L. Jaggard, presented at the* 1992 AP-S/URSI Meeting, *Chicago, IL (July 20–25, 1992); and "Diffraction by Two-Dimensional Cantor Apertures," D. L. Jaggard, T. Spielman, and M. Dempsey, presented at the* 1993 AP-S/URSI Meeting, *Ann Arbor, MI (June 28–July 2, 1993).*]

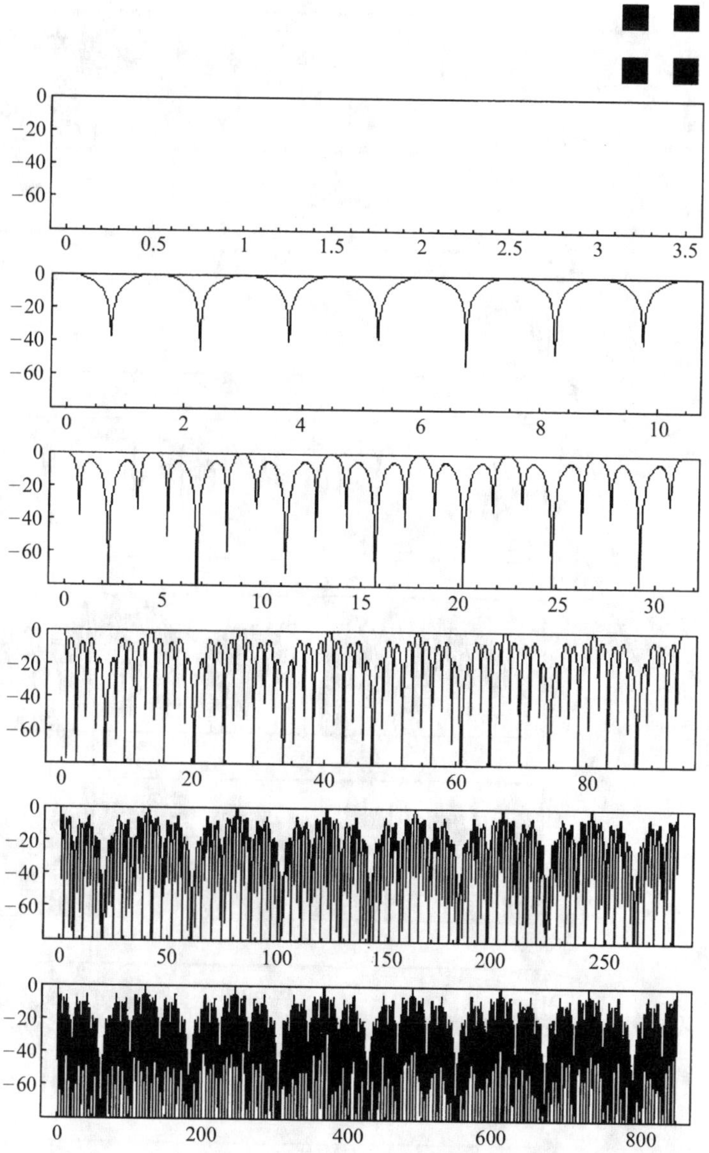

Figure 3.7 Cantor square diffraction patterns $|\Psi(x,y,z)/\Psi(0,0,z)|$ in dB as a function of $f_x L$, for $S = 0,1,2,3,4,5$ (top to bottom) and $D = \ln(4)/\ln(3)$ without effect of aperture. Note how each stage provides an envelope for the rescaled version of the following stage. [*Adapted from "Diffraction by Cantor Targets: Theory and Experiment," T. Spielman and D. L. Jaggard, presented at the* 1992 AP-S/URSI Meeting, *Chicago, IL (July 20–25, 1992); and "Diffraction by Two-Dimensional Cantor Apertures," D. L. Jaggard, T. Spielman, and M. Dempsey, presented at the* 1993 AP-S/URSI Meeting, *Ann Arbor, MI (June 28–July 2, 1993).*]

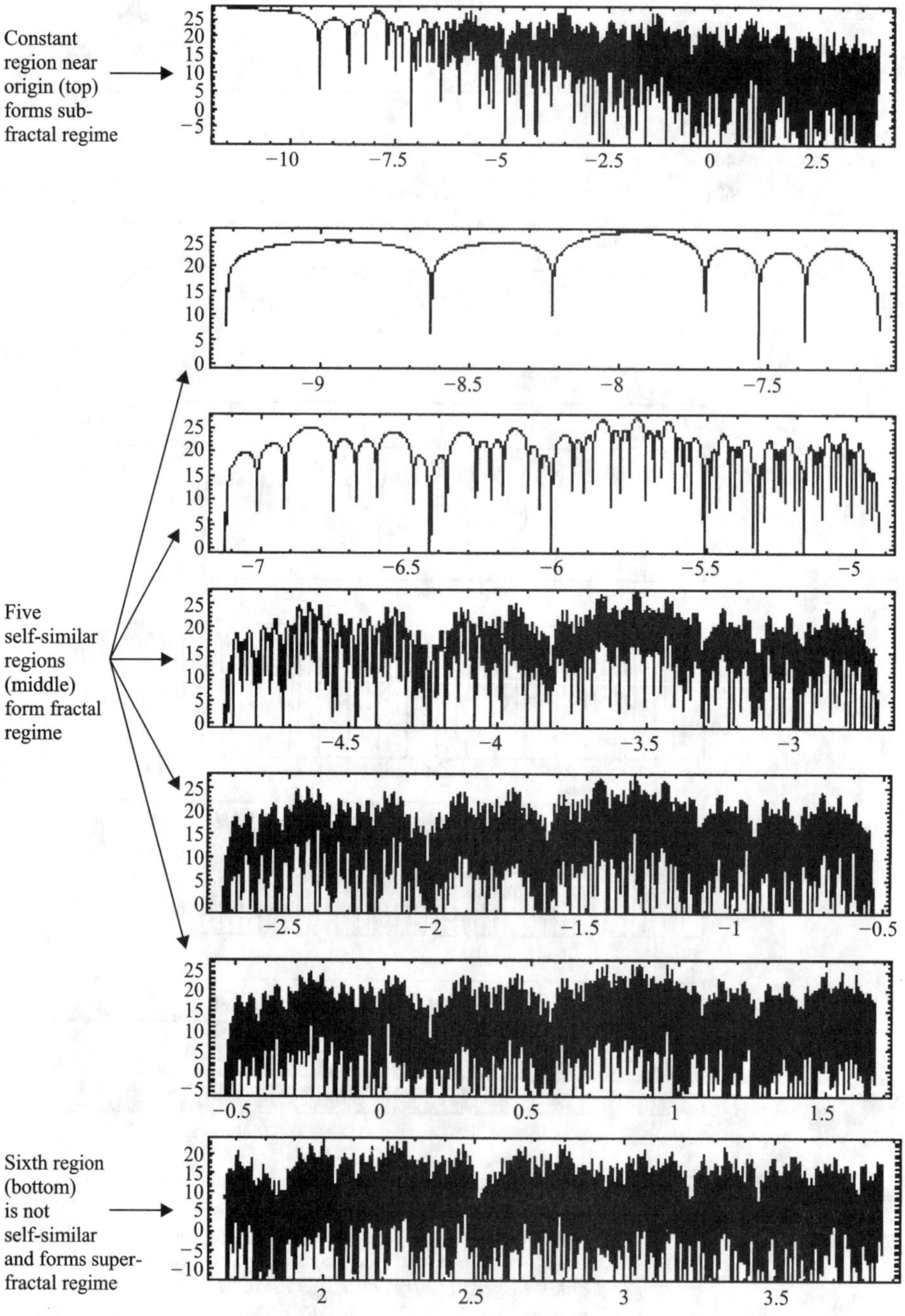

Figure 3.9 Normalized diffraction pattern $|\Psi(x, y, z)|$ in dB along the $f_x L$ axis, in dB, for the triadic Cantor square diffractal on a log-log scale with $S = 5$. Here the subfractal regime includes the mainlobe. This is followed by five similar bands that define the fractal regime. Finally, for higher spatial frequencies, the superfractal regime takes over in which the fractal nature of the diffraction pattern is no longer evident. [*Figure adapted from* [*11*] *© 1997 Jaggard and Jaggard.*]

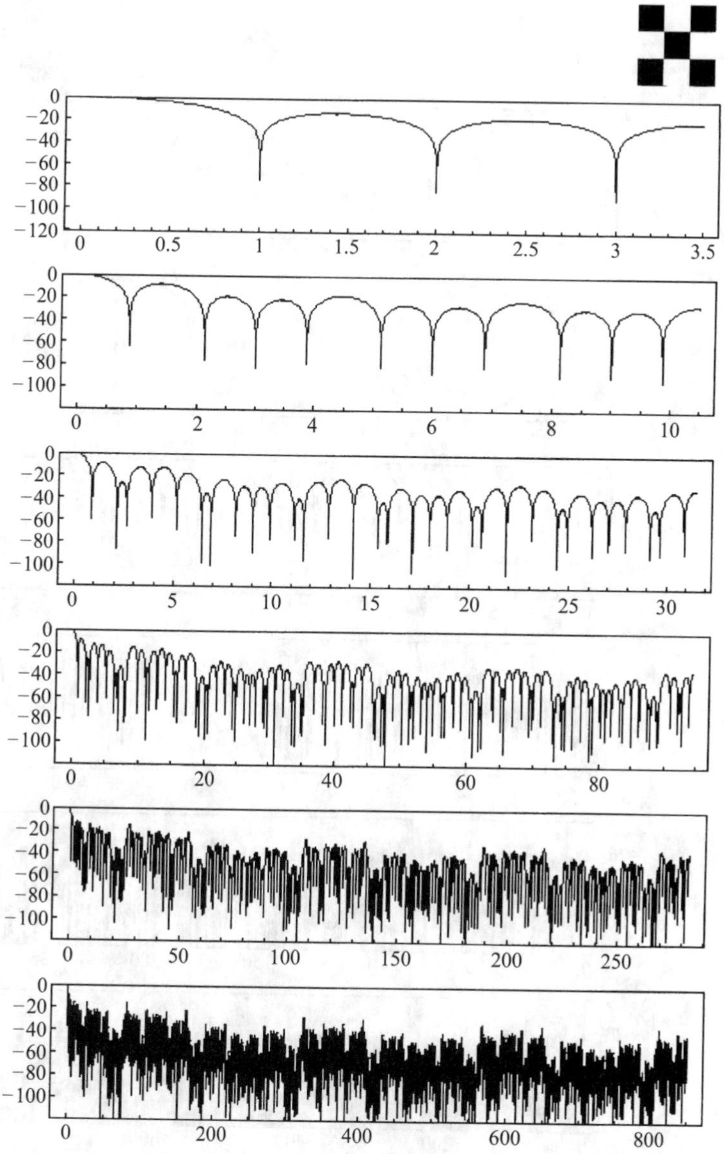

Figure 3.10 Total Purina square diffraction patterns $|\Psi(x, y, z)/\Psi(0, 0, z)|$ in dB as a function of $f_x L$, for $S = 0, 1, 2, 3, 4, 5$ (top to bottom) and $D = \ln(5)/\ln(3)$. Note how each stage provides an envelope for the re-scaled version of the following stage. [*Figure adapted from "Diffraction by Cantor Targets: Theory and Experiment," T. Spielman and D. L. Jaggard, presented at the* 1992 AP-S/URSI Meeting, *Chicago, IL (July 20–25, 1992); and "Diffraction by Two-Dimensional Cantor Apertures," D. L. Jaggard, T. Spielman, and M. Dempsey, presented at the* 1993 AP-S/URSI Meeting, *Ann Arbor, MI (June 28–July 2, 1993).*]

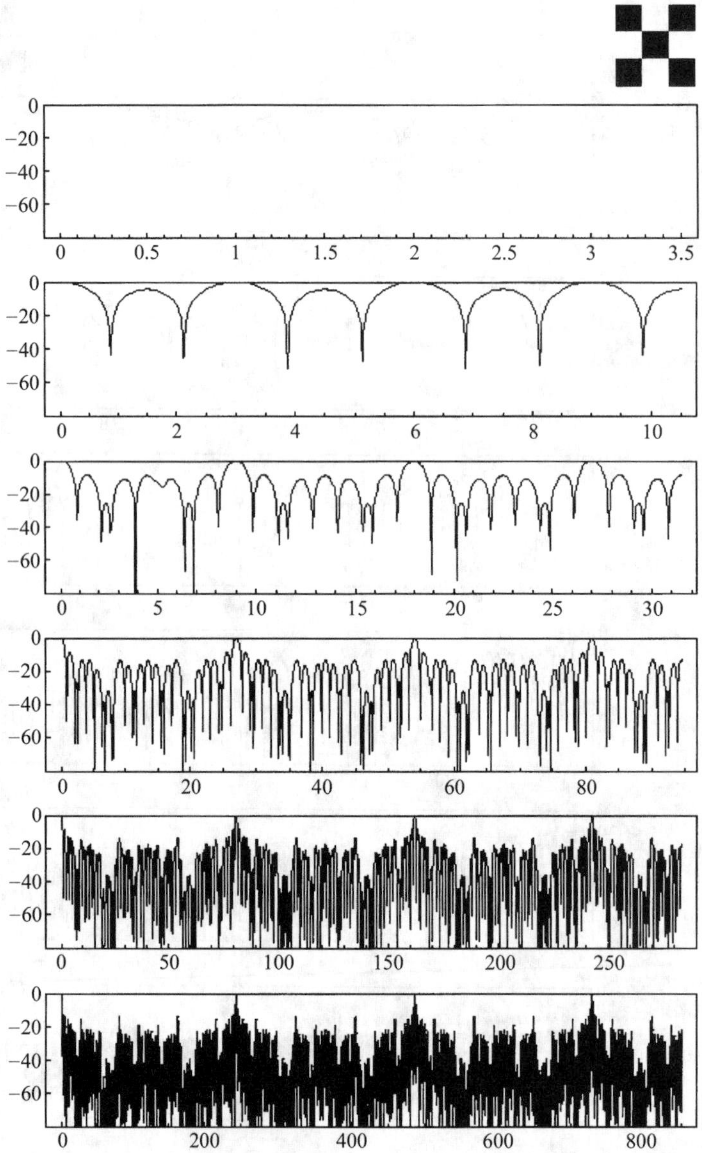

Figure 3.11 Purina square diffraction patterns $|\Psi(x, y, z)/\Psi(0, 0, z)|$ in dB as a function of $f_x L$, for $S = 0, 1, 2, 3, 4, 5$ (top to bottom) and $D = \ln(5)/\ln(3)$ without effect of aperture. Note how each stage provides an envelope for the re-scaled version of the following stage. [*Figure adapted from "Diffraction by Cantor Targets: Theory and Experiment," T. Spielman and D. L. Jaggard, presented at the* 1992 AP-S/URSI Meeting, *Chicago, IL (July 20–25, 1992); and "Diffraction by Two-Dimensional Cantor Apertures," D. L. Jaggard, T. Spielman, and M. Dempsey, presented at the* 1993 AP-S/URSI Meeting, *Ann Arbor, MI (June 28–July 2, 1993).*]

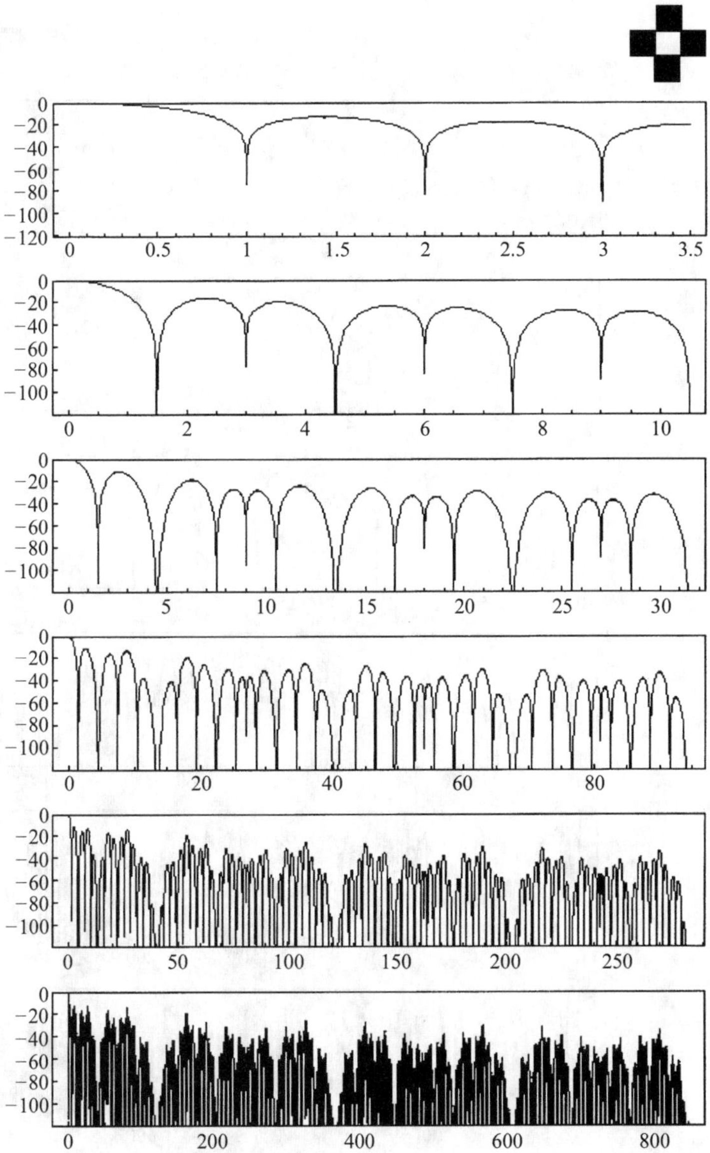

Figure 3.12 Total Sierpiński square diffraction patterns $|\Psi(x, y, z)/\Psi(0, 0, z)|$ in dB as a function of $f_x L$, for $S = 0, 1, 2, 3, 4, 5$ (top to bottom) and $D = \ln(4)/\ln(3)$. Note how each stage provides an envelope for the re-scaled version of the following stage. [*Figure adapted from "Diffraction by Cantor Targets: Theory and Experiment," T. Spielman and D. L. Jaggard, presented at the* 1992 AP-S/URSI Meeting, *Chicago, IL (July 20–25, 1992); and "Diffraction by Two-Dimensional Cantor Apertures," D. L. Jaggard, T. Spielman, and M. Dempsey, presented at the* 1993 AP-S/URSI Meeting, *Ann Arbor, MI (June 28–July 2, 1993).*]

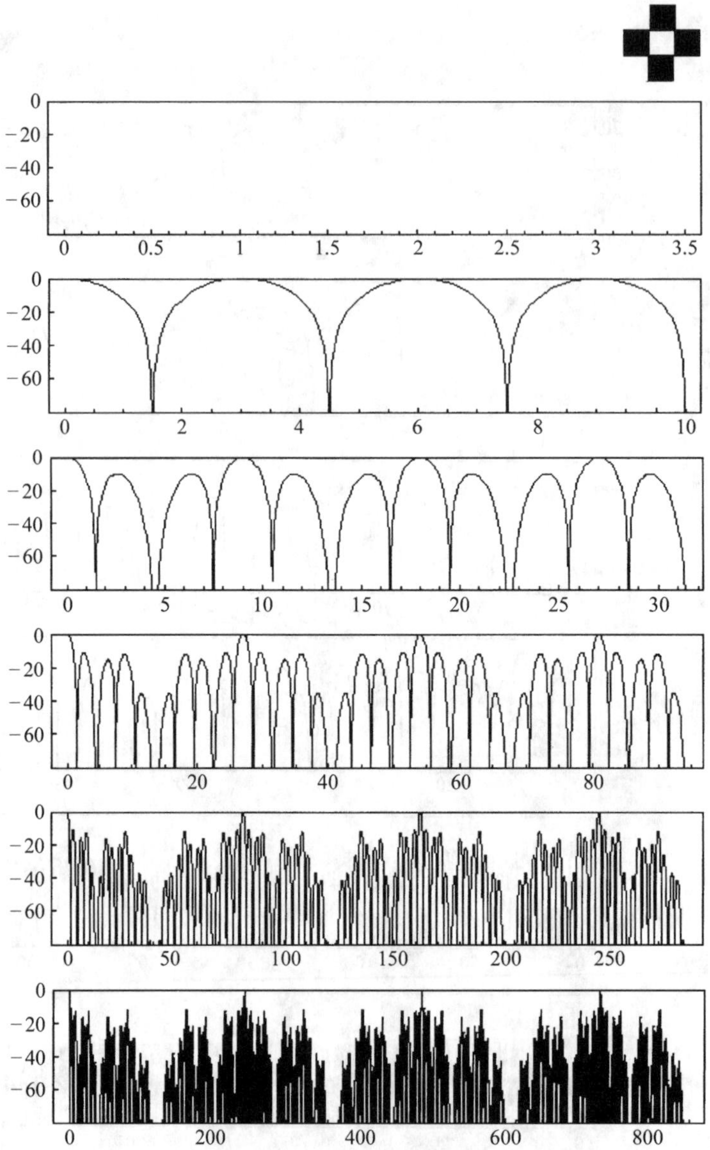

Figure 3.13 Sierpiński square diffraction patterns $|\Psi(x, y, z)/\Psi(0, 0, z)|$ in dB as a function of $f_x L$ for $S = 0, 1, 2, 3, 4, 5$ (top to bottom) and $D = \ln(4)/\ln(3)$ without effect of aperture. Note how each stage provides an envelope for the re-scaled version of the following stage. [*Figure adapted from "Diffraction by Cantor Targets: Theory and Experiment," by T. Spielman and D. L. Jaggard, presented at the* 1992 AP-S/URSI Meeting, *Chicago, IL (July 20–25, 1992); and "Diffraction by Two-Dimensional Cantor Apertures," D. L. Jaggard, T. Spielman, and M. Dempsey, presented at the* 1993 AP-S/URSI Meeting, *Ann Arbor, MI (June 28–July 2, 1993).*]

Although we have primarily examined the diffraction parallel to the axes of symmetry of the aperture, analogous results hold in any radial direction. For example, plots of the diffraction pattern in other radial directions yield results similar to those displayed with the exception of scaling—the features are stretched—and the more rapid decay characteristic of diffraction patterns of symmetric apertures when not viewed along an axis of symmetry.

3.3.4 Cantor Ring Diffractals

In looking at the Cartesian diffractals, we observed the preferred axes in the diffracted fields associated with such apertures. As there may be some advantage to structures in which the diffracted energy is more uniformly distributed, we now consider diffraction by circularly symmetric fractal apertures. Our particular focus is on apertures generated using triadic and generalized Cantor sets. We first review early work in the area [13], [15], [16] and then consider more recent generalizations including the use of polyadic (multigap) Cantor bars as well as variations in lacunarity [26], [30], [31], [66]. A discrete version of the Cantor ring diffractal—particularly useful in the design of sparse fractal arrays—is described in Section 3.6.3 of this chapter. Here we find the multiscale structure of the apertures to be evident in the radial structure of the diffracted field. We also observe the subfractal, fractal, and superfractal regimes in the diffraction patterns. These regimes are directly related to the geometry of the aperture.

In the Cantor ring diffractal problem we now consider, as in other fractal electrodynamics problems, we see several scaling properties connecting the geometrical structure of the fractal with its electromagnetic signature. These are the same relations observed in the Cartesian examples.

For Cantor ring diffractals, we consider a plane wave with time-harmonic dependence $\exp(-i\omega t)$ normally incident upon a rotationally symmetric aperture Ap constructed from a Cantor bar at finite stage of growth, as shown in Fig. 3.14. We investigate the properties of Cantor ring diffractals constructed using different methods and several different generating Cantor sets. The methods of aperture construction and the differences between them will be detailed in the following subsections. The geometric construction of Cantor bars is discussed in detail in Chapter 1, including variations in the number of gaps and in lacunarity. We will use the same notation here as in Chapter 1, denoting the similarity fractal dimension by D. For Cantor bars consisting of N copies of the whole set, each scaled by $\gamma < 1$, $D = [\ln N]/[-\ln \gamma]$. The diffractals studied here are constructed by rotating a Cantor set about either its endpoint or midpoint. The dimension of the aperture Ap so formed is $D_{Ap} = 1 + D$ although we shall generally use D_s to characterize the aperture.

We again use (3.5)–(3.7) to find the field diffracted by each of these apertures. Since we now take the aperture Ap to be independent of ϕ, its aperture field Ψ_0 is also azimuthally symmetric and (3.7) yields the Fourier-Bessel transform

$$\mathcal{F}\,[\Psi_0(\mathbf{x}')] = \mathcal{F}\,[\Psi_0(\rho', \phi')] = \mathcal{F}\,[\Psi_0(\rho')]$$

$$= 2\pi \int_{Ap} \Psi_0(\rho')\,\mathrm{J}_0(2\pi f_\rho \rho')\,\rho'\,d\rho' \bigg|_{f_\rho = \rho/\lambda r} \qquad (3.14)$$

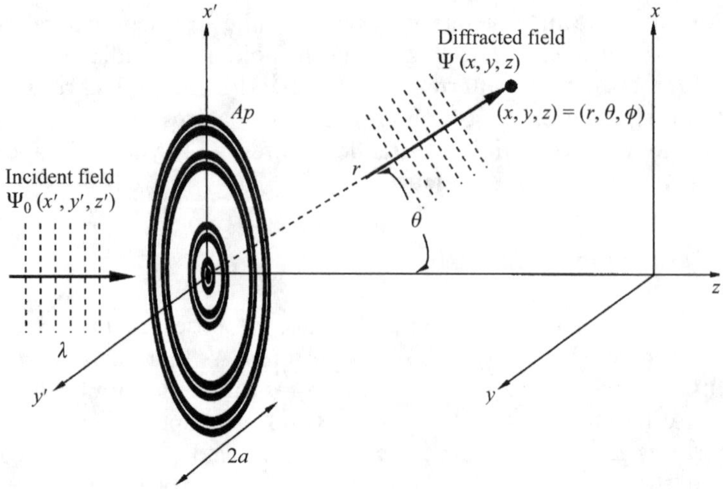

Figure 3.14 The problem geometry under consideration. A time harmonic wave $\Psi_0(x', y', z')$ of unit amplitude illuminates the Cantor ring diffractal *Ap* in the plane $(x', y', 0)$ which generates the far-zone diffracted field $\Psi(x, y, z)$. [*Figure adapted from "Diffraction from Triadic Cantor Target Diffraction," D. L. Jaggard and T. Spielman,* Microwave and Optical Technology Letters, **5**, 460–466 (1992). © *1992 John Wiley & Sons, Inc.*]

which gives

$$\Psi(x, y, z) = \Psi(\rho, z) = C \int_{Ap} \Psi_0(\rho')\, J_0(2\pi f_\rho \rho')\, \rho'\, d\rho' \Big|_{f_\rho = \rho/\lambda r} \tag{3.15}$$

as the diffracted field with C again being an unimportant constant. To calculate this diffracted field Ψ, it is convenient to find a method to represent the aperture field Ψ_0 in the aperture plane in terms that take into account fractal characteristics such as fractal dimension, lacunarity, dilation symmetry and stage of growth. We include here two different approaches to the problem, which are discussed separately below.

3.3.4.1 Triadic Cantor Ring Diffractal

Cantor ring diffractals were first considered as generated by triadic Cantor sets. These are constructed by removing the middle third of an interval, and then iteratively removing the middle third of each remaining interval. The finite stage of growth triadic Cantor sets formed in this way are then rotated about their endpoints to construct a fractal aperture *Ap* with circular symmetry.

Following [13], [16] we approach the problem of finding an expression for the aperture field Ψ_0 using multiple convolutions. In this method, the calculation has a structure that mimics the structure of the Cantor ring diffractal. To accomplish this, we define the aperture construction function

$$K_L(d) = \sum_{n=0}^{\infty} (-1)^n u(d - nL/2) \tag{3.16}$$

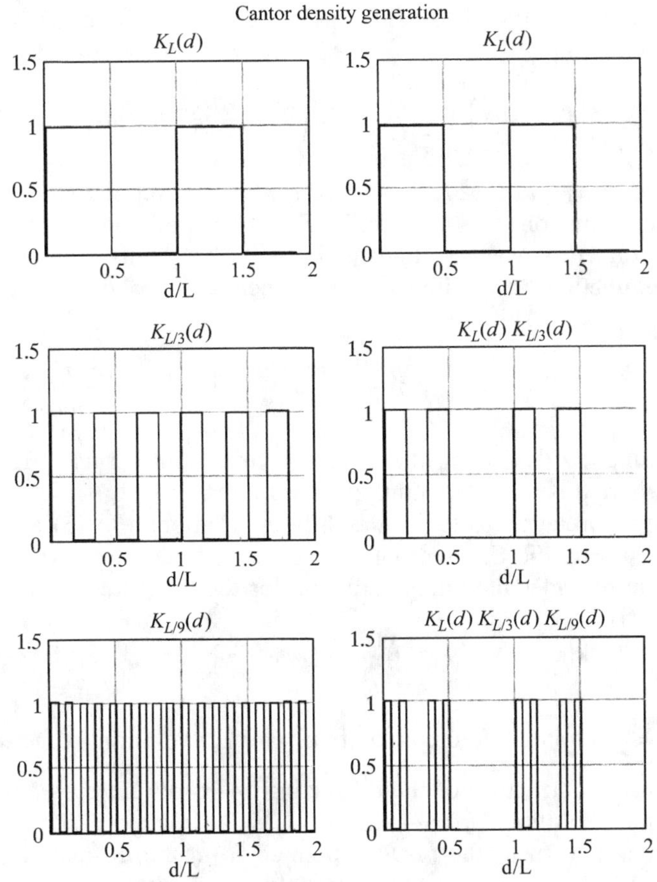

Figure 3.15 Incremental cumulative products of the aperture construction function $K_{L/3^i}(d)$. The non-zero windows which do not correspond to sub-bars in the finite stage Cantor set are filtered out by the multiplication. [*Figure adapted from "Triadic Cantor Target Diffraction," D. L. Jaggard and T. Spielman,* Microwave and Optical Technology Letters, **5**, 460–466 (1992). © *1992 John Wiley & Sons, Inc.*]

where $u(d)$ is the Heaviside step function. As seen in the left column of Fig. 3.15 this produces a "multiple window" function whose constituent windows include those which correspond to the intervals included in the Cantor set at a particular stage of growth. Multiplication of these functions, shown in the right column of Fig. 3.15 leaves the windows that correspond to the intervals in a finite stage triadic Cantor bar, although this pattern repeats and gives additional copies of the bar. Adding a window function corresponding to the length of the bar, we remove all windows but those corresponding to the first copy of the Cantor bar. We then define a Cantor density function in terms of the aperture construction function as

$$C_S(d) = \left[\prod_{m=0}^{S} K_{L/3^{m-1}}(d)\right][u(d) - u(d-3L/2)]. \qquad (S \geqslant 0) \qquad (3.17)$$

Combining relations (3.15) and (3.17) we find the alternative expression for the density function

$$C_S(d) = \left[\prod_{m=0}^{S}\sum_{n=0}^{\infty}(-1)^n u\left(d - \frac{nL}{2\cdot 3^{m-1}}\right)\right][u(d) - u(d - 3L/2)] \tag{3.18}$$

The term $[u(d) - u(d - 3L/2)]$ above may be removed if we restrict the summation to those step functions within $[0, 3L/2]$. This gives us a one-dimensional distribution for the Cantor set. As we are interested in Cantor ring diffractals in the plane, we replace the step functions $u(d)$ with their two-dimensional polar analog, $\mathrm{circ}(\rho) \equiv 1$ for $\rho < 1$, and 0 elsewhere. This leads to

$$C_S(\rho') = \left[\prod_{m=0}^{S}\sum_{n=0}^{3^m}(-1)^{n-1}\,\mathrm{circ}\left(\frac{3^m\rho'}{na}\right)\right] \tag{3.19}$$

as the function describing the desired triadic Cantor ring diffractal in the aperture plane where a is the maximum possible radius.

We use relation (3.19) in the diffraction integral (3.15) to obtain the diffracted field from a triadic Cantor ring diffractal. At stage of growth S, the Fourier-Bessel transform provides the final expression for the diffracted field Ψ,

$$\Psi(x, y, z) = \Psi(r, \theta) = 2\pi C \mathop{\ast}_{m=0}^{S} \left[\sum_{n=1}^{3^m}(-1)^{n-1}\mathrm{J}_1\left(2\pi\rho\frac{na}{3^m}\right)\frac{na}{3^m\rho}\right] \tag{3.20}$$

where $\mathop{\ast}_{i=0}^{S}$ denotes S-fold convolution. This convolution in the field solution reflects the geometric structure of the diffracting aperture and provides us with the desired closed-form solution for the field.

We now turn to the results obtained from the diffracted field calculated from relation (3.20). Since the diffracted field is azimuthally symmetric, all information about the diffraction patterns is found in any single radial cut of the diffraction pattern. Results for the triadic Cantor bar are shown in Fig. 3.16 for stages of growth $S = 0, 1, 2, 3$ and $D_S = \ln(2)/\ln(3)$. The diffracted field $|\Psi(x, y, z)/\Psi(0, 0, z)|$ is plotted in dB as a function of the radial Fourier variable or normalized spatial frequency $f_\rho a = \sqrt{f_x^2 + f_y^2}\,a$. At stage $S = 0$, the ring diffractal is an open disk and the diffraction calculation produces the typical Airy disk pattern.

The general scaling suggested by relation (3.9) and found in the Cartesian diffractals again becomes apparent for the triadic Cantor fractal as shown in the diffraction patterns of Fig. 3.16 for $S = 0, 1, 2, 3$. As expected, each pattern at stage S of growth forms an envelope for the $S + 1$ stage of growth diffraction when the latter is scaled by $\gamma = 1/3$ in spatial frequency.

3.3.4.2 Polyadic Cantor Diffractal

Having seen that triadic Cantor ring diffractals imprint some of their fractal structure on the diffracted field, we look to more general Cantor ring diffractals [30], [66] to see if they give similar results. In particular, we now vary the number of gaps and the fractal dimension. With three or more gaps in a generalized Cantor set, we may also vary the lacunarity as described in Chapter 1. As with the Cantor

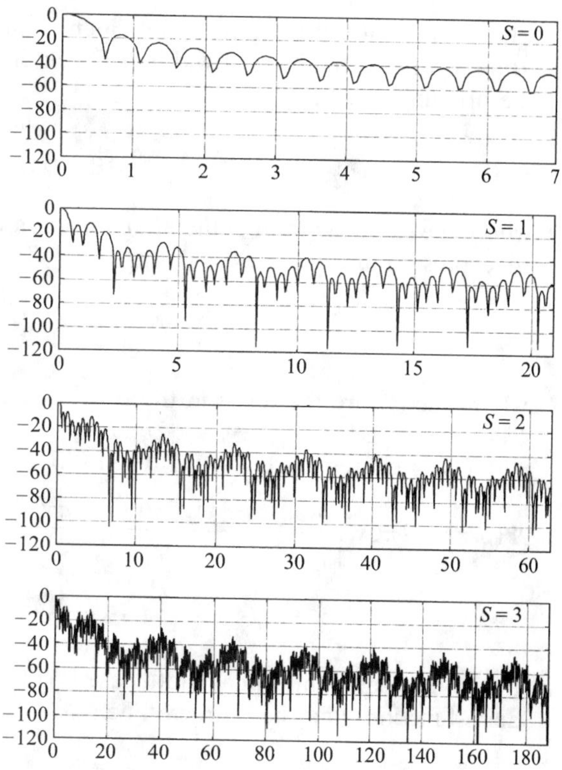

Figure 3.16 Linear-log plot of Cantor target diffraction pattern intensity $|\Psi(x,y,z)/\Psi(0,0,z)|$ in dB as a function of the normalized radial spatial frequency $f_\rho a = \sqrt{f_x^2 + f_y^2}\,a$, for increasing stages of growth $S = 0, 1, 2, 3$ (top to bottom). Each successive stage has been scaled by $\gamma = 1/3$ to exhibit the inherent self-similarity of the far-field diffraction pattern. [*Figure taken from "Triadic Cantor Target Diffraction," D. L. Jaggard and T. Spielman,* Microwave and Optical Technology Letters, **5**, 460–466 (1992). © *1992 John Wiley & Sons, Inc.*]

superlattices previously discussed, we are able to use twist plots to examine the effects of lacunarity on the electromagnetic wave interaction with these Cantor structures. These allow us to find ranges of lacunarity values for which Cantor ring diffractals have desirable properties when considered as aperture arrays. These techniques are also adaptable to Cantor ring diffractals in which the open annuli have infinitesimal width.

In addition to the fractal dimension D_s, we characterize these more general Cantor ring diffractals by the number of gaps n_{gaps} and the lacunarity parameter ε of the symmetric polyadic Cantor bar used to generate the Cantor ring diffractal. As described elsewhere [30], [32], [33], and Chapter 1, ε is the normalized outer gap width of the Cantor bar and is used to characterize the lacunarity of the set. We give examples for $n_{gaps} = 3$, although the methods described here may be generalized to sets with more gaps. Also, for the remainder of this discussion we construct the ring diffractals by rotating the generating Cantor set about its midpoint, rather than one of its endpoints.

For these more general polyadic Cantor ring diffractals, we take a somewhat different approach to the calculation than that used in the triadic case. As before, we look for a closed-form expression for the aperture field Ψ_0. Because of the azimuthal symmetry of the aperture, we may express the field as a function of ρ', the radial coordinate in the aperture plane. If the total length of the Cantor bar used to generate the target is $L = 2a$, the bar widths at stage of growth S will be $2a\gamma^S$, which is then also the width of each open annulus in the aperture. If we define $W(\rho')$ to be a window function of width $2a\gamma^S$ centered at the origin, we may then express the aperture field as

$$\Psi_0(x', y', 0) = \Psi_0(\rho') = \sum_{i=0}^{M-1} W(\rho') * \delta(\rho' - a_i) \tag{3.21}$$

where the a_i are the radii of the M annulus centers in the diffractal. These M radii may be expressed as

$$a_i = a \sum_{j=1}^{S} \gamma^{j-1} b_{c_j^{(i)}} \qquad (i = 0, \ldots, 2^{2S-1} - 1) \tag{3.22}$$

where the $c_j^{(i)}$ are the digits in the base 4 expansion of $i + 2^{2S-1}$ $[= (c_1^{(i)} c_2^{(i)} \cdots c_S^{(i)})_4]$, $b_0 = -(1 - \gamma)$, $b_1 = -(1 - 3\gamma - 2\varepsilon)$, $b_2 = (1 - 3\gamma - 2\varepsilon)$, and $b_3 = (1 - \gamma)$.

Using expression (3.21) for Ψ_0 in (3.15), the diffracted field is found to be

$$\Psi(\rho, z) = 2\pi C \sum_{i=0}^{M-1} \frac{(a_i + a\gamma^S) J_1[2\pi f_\rho(a_i + a\gamma^S)]}{f_\rho} - \left.\frac{(a_i - a\gamma^S) J_1[2\pi f_\rho(a_i - a\gamma^S)]}{f_\rho}\right|_{f_\rho = \rho/\lambda r} \tag{3.23}$$

In the limiting infinitesimal width case $[W(\rho') = \delta(\rho')]$, Ψ_0 is expressible as a sum of delta functions and the diffracted field is

$$\Psi(\rho, z) = 2\pi C \sum_{i=0}^{M-1} a_i J_0(2\pi f_\rho a_i)\Big|_{f_\rho = \rho/\lambda r} \tag{3.24}$$

The M summands in each of (3.23) and (3.24) correspond to the M annuli in the diffractal.

As for the triadic ring diffractals, all of the information about the diffracted field is contained in a radial cut of the diffraction pattern. The right of Fig. 3.17 shows three of these cuts, with the intensity of the diffracted field as a function of $f_\rho a$, and with the lacunarity parameter $\varepsilon = 60/1000$, $44/1000$, $20/1000$ (top to bottom). In this case, $n_{gaps} = 3$, $S = 1$, and $D = 9/10$. We might also vary ε continuously and plot the field intensity as a function of both ε and $f_\rho a$. This gives a twist plot that is similar to those developed for Cantor superlattices (see Chapter 1) and is shown in the left of Fig. 3.17. The twist plots for Cantor ring diffractals exhibit an envelope of $[f_\rho a]^{-2}$, corresponding to the diffraction envelope provided by the bounding disk, and seen both in the radial cuts and in the generally darker appearance of the twist plots as $f_\rho a$ increases.

The most prominent feature of the twist plots is their null structures, the dark lines and curves which give these plots their twisting appearance. As for the twist plots for Cantor superlattices, we are able to predict these nulls using first-order calculations. To do this, we separate the nulls into three families. The first is the *vertical nulls*. These become more pronounced with increasing spatial frequency since the Bessel function zero crossings become more nearly periodic as spatial frequency increases. The vertical nulls are caused by interference between the inner and outer edges of each annulus and are thus independent of lacunarity as is clear from the plots. Since these nulls arise from interference between the two edges of the same annular ring,

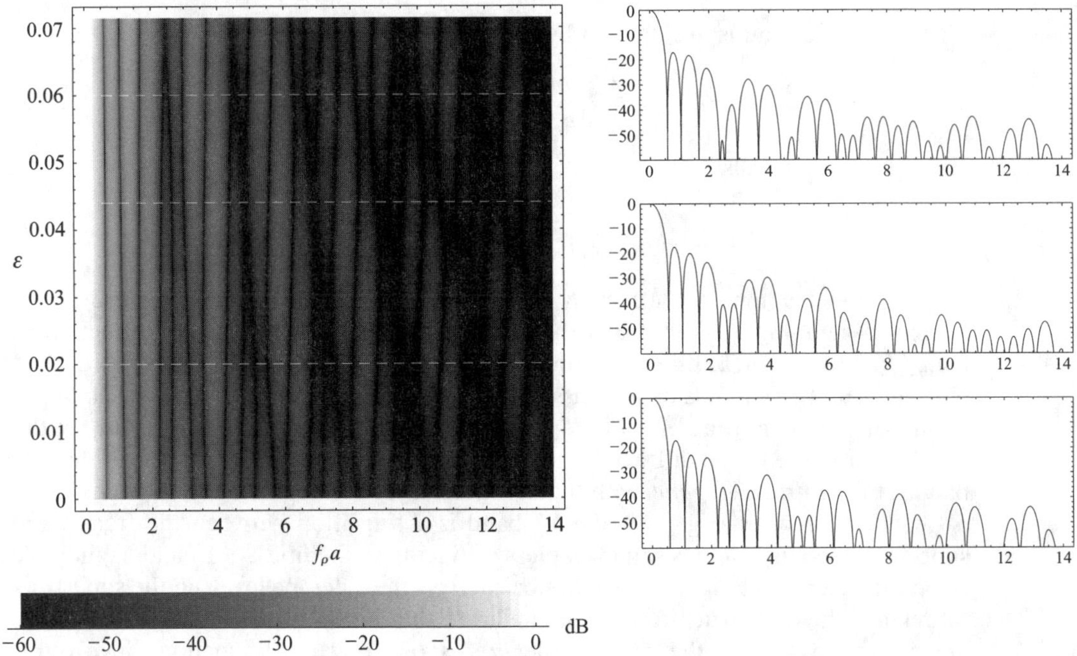

Figure 3.17 This figure displays a *twist plot* (left), along with three of the constituent radial cuts of the diffracted field intensity (right) for $\varepsilon = 60/1000, 44/1000$, and $20/1000$ (top to bottom) corresponding to the dashed lines on the twist plot. This three-gap Cantor ring diffractal has dimension $D = 9/10$, $D_{Ap} = 1 + 9/10$ and stage $S = 1$. In each case the diffracted intensity in dB is plotted as a function of normalized spatial frequency $f_\rho a$. [*Figure adapted from "Cantor Ring Diffractals," A. D. Jaggard and D. L. Jaggard,* Optics Communications **158**, 141–148 (1998). *© 1998 Elsevier Science B. V.*]

they are not seen in the limiting case of infinitesimal width rings [26], [31]. The nulls in the second family, which we denote *arc nulls*, are concave upward and curve from the upper left of the twist plots to the lower right. These nulls are caused by collective destructive interference between the edges in the set $\{(a_i + a\gamma^S, \theta)\}_{i=0}^{M-1}$. Simultaneously, there is collective destructive interference between the edges in each of the three sets $\{(a_i - a\gamma^S, \theta)\}_{i=0}^{M-1}$, $\{(a_i + a\gamma^S, \theta + \pi)\}_{i=0}^{M-1}$, and $\{(a_i - a\gamma^S, \theta + \pi)\}_{i=0}^{M-1}$. The third family of nulls, which we denote *striation nulls*, gives the finer structure of the twist plots with curves that are concave downward, running from the lower left to the upper right of the figure. These are caused by destructive interference between the edges at $(a_{M-1-i} \pm a\gamma^S, \theta)$ and $(a_i \mp a\gamma^S, \theta + \pi)$. Both arc nulls and striation nulls are clearly functions of ε since they involve aperture edges separated by one or more gaps.

For apertures generated by a first-stage three-gap Cantor bar, we find the vertical nulls described above to be given by

$$f_\rho a\big|_{\text{vertical nulls}} = \frac{m}{2\gamma^S} \qquad (m = 1, 2, 3, \ldots) \tag{3.25}$$

Likewise, the arc nulls are defined by

$$f_\rho a\big|_{\text{arc nulls}} = \frac{2m-1}{4[\varepsilon + 2\gamma^S]} \qquad (m = 1, 2, 3, \ldots) \tag{3.26}$$

and the striation nulls by

$$f_\rho a\big|_{\text{striation nulls}} = \frac{2m-1}{4[1-\varepsilon - 2\gamma^S]} \qquad (m = 1, 2, 3, \ldots) \tag{3.27}$$

We now examine the effect of fractal stage of growth on the structure of the diffracted field. Figure 3.18 shows the radial diffracted field plotted as a function of $\log_{10}(f_\rho a)$ for $S = 1, 2, 3, 4$ (top to bottom), $D_s = 9/10$, and $\varepsilon = 44/1000$. The dashed lines indicate the S bands of similar structure which are the *fractal regime*. The upper bound of this region increases by a factor of $1/\gamma$ with each stage of growth, as the smallest structure in the diffractal is scaled by γ between stages. To the left of the first band is the *subfractal regime*, with the *superfractal regime* beginning to the right of the Sth band. As seen in Fig. 3.18, the Sth band is a transition between the fractal and superfractal regimes and is somewhat less distinct than the first $S-1$ bands. While the fractal structure no longer determines or controls the sidelobe levels in the superfractal regime, the null structure there is similar to that in the fractal regime. These nulls appear to arise from the higher-order zeros of the Bessel functions involved—essentially "harmonic echoes" of the fractal regime. The three regimes are labeled on the $S = 2$ plot of Fig. 3.18.

As suggested by Fig. 3.16, we might examine twist plots rescaled by factors of γ^S. Such a comparison [26], [30] reveals a similar scaling as that seen in the triadic case, in which the diffraction pattern for one stage forms an envelope for the rescaled pattern for a higher stage of growth. The presence of this scaling in twist plots indicates that it is independent of lacunarity, the second axis in a twist plot.

Finally, as suggested by Allain and Cloitre [10], [34], [35] and demonstrated for other fractal structures, we may recover the fractal dimension of the diffracting aperture by averaging over the sidelobe bands in the fractal regime in the diffracted field. This holds for both the finite width and infinitesimal width ring diffractals, and appears to be independent of lacunarity as demonstrated elsewhere [26], [31].

3.3.4.3 Discussion

Here we examine both Cartesian and polar diffractals. In each case we observe three characteristic forms of scaling. In the first, there is scaling with stage of growth in which each stage of growth forms an envelope for the re-scaled plot of the diffraction pattern at the next stage of growth. This scaling has its origin in the fact that self-similar structures have self-similar Fourier transforms. In the second we find that the iterative construction of the fractal leads directly to a similarity in the sidelobe structure or to log-periodic behavior of the diffraction pattern. Of particular interest are the *subfractal*, *fractal*, and *superfractal* regimes of diffraction. In the first or subfractal regime, the sidelobe behavior is dominated by the mainlobe and so is dependent upon the total diameter of the diffractal rather than any of its fractal characteristics. In the second or fractal regime, the behavior is controlled by the fractal geometry and depends directly on fractal dimension, lacunarity, stage of growth, and

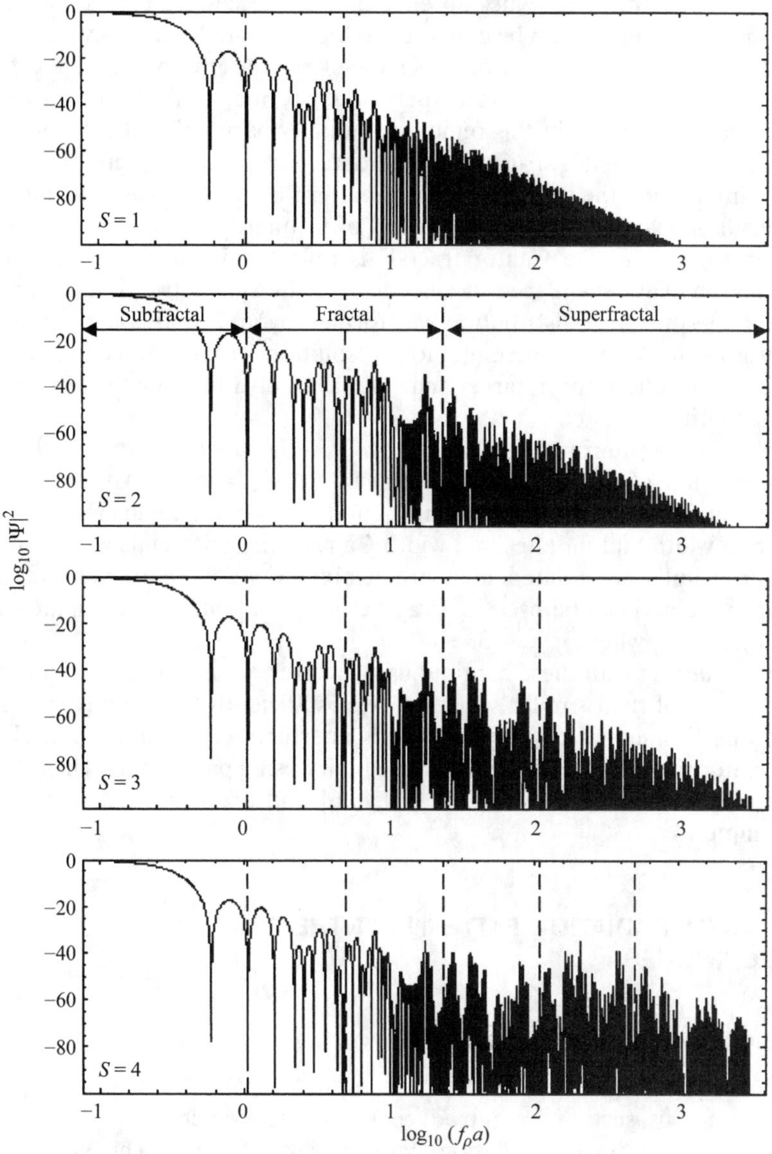

Figure 3.18 Radial cuts of the diffracted field in dB as a function of $\log_{10}(f_\rho a)$ for stages $S = 1, 2, 3, 4$ (top to bottom). In each case $D = 9/10$, $D_{Ap} = 1 + 9/10$ and $\varepsilon = 44/1000$. Dashed lines indicate similar bands seen in the diffraction pattern, with S bands visible at stage of growth S. The subfractal, fractal, and superfractal regimes are labeled on the $S = 2$ plot. [*Figure taken from "Cantor Ring Diffractals," A. D. Jaggard and D. L. Jaggard,* Optics Communications, **158**, 141–148 (1998). *© 1998 Elsevier Science B. V.*]

number of gaps. Log-log plots of the Sth stage radial diffraction pattern display S bands; each band provides an envelope for successive bands and has width $-\log \gamma$. This fractal regime can be extended to higher spatial frequency by increasing the stage of growth. For the third or superfractal regime, the order of the fractal no longer controls the sidebands since there are no fractal structures corresponding to the spatial frequencies in this regime, although "harmonic echoes" of the fractal regime are seen in the null structure of the data. The fractal dimension D_s can be retrieved by integrating the radial diffraction pattern over each band of similar sidelobes. This result appears to be valid regardless of lacunarity. Finally, we note that the iterative construction of the Cantor fractals is reflected in the iterative construction of the solution. This was placed in evidence for the symmetric triadic Cantor target where the aperture field distribution was formed by an S-fold convolution for the Sth stage of growth. An alternative method of solution was introduced for the generalizations of the triadic Cantor target that yielded polyadic Cantor diffractals with variable lacunarity.

For Cantor ring diffractals with $n_{gaps} = 3$, we found that for relatively large fractal dimension ($D \approx 9/10$) and small lacunarity ($\varepsilon \approx 44/1000$) there exists a window of operation where the sidelobe diffraction is relatively small. The null structure of both finite width and infinitesimal width Cantor ring diffractals is captured by twist plots. These nulls are related in a predictable way to the fractal characteristics of the aperture and can be used in array design [31], [36]. We anticipate similar results for other cases where $n_{gaps} > 3$.

Starting with these continuous structures, we may discretize them to form arrays. Because of their similar structure, over sufficiently low frequencies such arrays have properties similar to the continuous apertures considered here, allowing us to use Cantor ring diffractals to investigate their design parameters. These Cantor ring arrays are fully described elsewhere [31], [36] and are discussed in Section 3.6.3 of this chapter.

3.4 FRACTAL RADIATION PATTERN SYNTHESIS TECHNIQUES

3.4.1 Background

In this section we introduce techniques which have been developed for the synthesis of fractal radiation patterns [18–20], [37]. The goal of these synthesis techniques is to determine the required antenna current distributions and geometrical structures that would result in the realization of radiation patterns which exhibit certain desired fractal features. The first fractal radiation pattern synthesis technique to be discussed in this section is based on the self-scaling properties of a special class of non-uniformly but symmetrically spaced linear arrays, which we refer to as Weierstrass arrays. It will be demonstrated that the fractal dimension of the radiation patterns produced by these arrays may be controlled via the array element excitation currents. In addition to this, it is shown that the scale size of the fractal structure in the radiation patterns is governed by the number of elements contained in the array. The synthesis technique developed for Weierstrass arrays will then be generalized to include not only the freedom to choose a desired fractal dimension, but also the ability to select a suitable generating function. This more general class of self-scalable arrays has been called Fourier-Weierstrass arrays. A particular example will be presented which illustrates how a Fourier-Weierstrass array may be decomposed into a sequence of self-scaled uniformly spaced linear subarrays. Finally, a method for extending the synthesis technique developed for Weierstrass linear arrays to include concentric-ring planar arrays is briefly summarized.

In addition to linear and planar arrays of discrete elements, a fractal radiation pattern synthesis technique is also developed which is applicable to continuous line sources. It is shown that in the case of the line source, the desired fractal radiation pattern and the corresponding current distribution are related through a Fourier transform pair. The properties of infinite fractal arrays and line sources are first investigated, with further consideration given to the effects of truncation on the synthesis procedure.

3.4.2 Weierstrass Linear Arrays

Fractals are typically quantified and compared by using a measure known as the fractal dimension. There are several definitions of fractal dimension in common use [38], [39]. However, the box-counting definition is usually used for computational or empirical determination of fractal dimensions. For a given fractal F, the box-counting fractal dimension, denoted by $dim_B(F)$, is defined as [39]

$$\dim_B(F) = \lim_{\delta \to 0} \frac{\ln N_\delta(F)}{\ln(1/\delta)} \tag{3.28}$$

where N_δ is the smallest number of sets of diameter at most δ required to cover the fractal F. Voss [40] and Jaggard [2] have discussed the correlation between the box-counting definition of fractal dimension and the intuitive Euclidean concept of dimension.

The class of functions known as generalized Weierstrass functions may be represented in the form

$$f(x) = \sum_{n=1}^{\infty} \eta^{(D-2)n} g(\eta^n x) \tag{3.29}$$

where $1 < D < 2$, $\eta > 1$, and $g(.)$ is a suitable bounded periodic function [39], [41]. These generalized Weierstrass functions have the property that they are everywhere continuous but nowhere differentiable and exhibit fractal behavior at all scales. Suppose that $F = \text{graph}\,(f)$; then D represents the box-counting fractal dimension of F, that is, $dim_B\,(F) = D$, provided η is sufficiently large [39]. The fractal dimension D in this case is a fractional dimension that lies between the integer (Euclidean) dimensions of one and two. Of course, we expect the fractal dimension of F to be at least one because F is the graph of a continuous function with one-dimensional domain. On the other hand, we expect the fractal dimension of F to be no more than two because F is contained in the plane, a two-dimensional space. In the remainder of this section, a relationship between Weierstrass functions and classical antenna

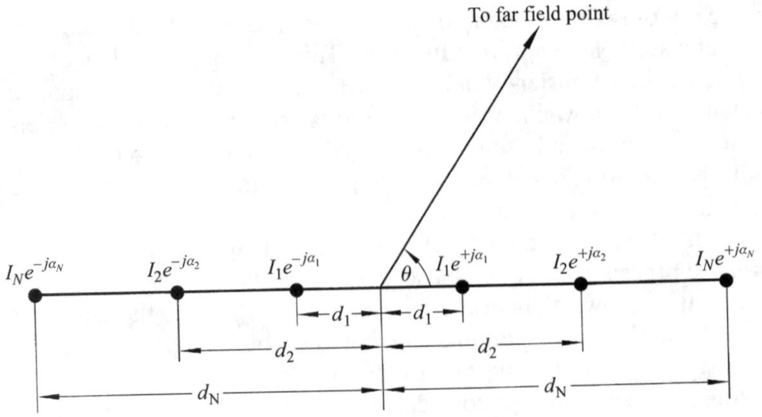

Figure 3.19 A non-uniformly but symmetrically spaced linear array of 2N elements. [*D. H. Werner and P. L. Werner,* Radio Science, **30**, pp. 29–45, January–February 1995. (*© American Geophysical Union.*)]

theory is established and used to investigate the geometrical properties as well as radiation characteristics of a new class of arrays, which we call Weierstrass arrays.

The subject of non-uniformly spaced antenna arrays has received considerable attention over the years and continues to be a topic of interest. This is primarily because the element spacings provide a third variable, in addition to the amplitude and phase of the array excitation current, with which to control the radiation pattern. The array factor for the nonuniformly but symmetrically spaced linear array of $2N$ elements illustrated in Fig. 3.19 may be expressed in the form [42]

$$f(\theta) = 2 \sum_{n=1}^{N} I_n \cos(kd_n \cos\theta + \alpha_n) \tag{3.30}$$

where

$$k = \frac{2\pi}{\lambda} \tag{3.31}$$

is the free-space wavenumber and λ is the corresponding free-space wavelength. The parameters I_n and α_n represent the amplitude and phase of the array excitation currents, respectively, while d_n represents the array element locations. Now, suppose the following Weierstrass function is considered

$$f(u) = \sum_{n=1}^{\infty} \eta^{(D-2)n} \cos(a\eta^n u) \tag{3.32}$$

where $1 < D < 2$, $\eta > 1$, a is an arbitrary constant, and $g(.)$ is chosen to be the cosine function. We note here that the properties of (3.32) are preserved when the series is multiplied by a constant or arbitrary phases are added to each term in the series. Therefore,

$$f(u) = 2 \sum_{n=1}^{\infty} \eta^{(D-2)n} \cos(a\eta^n u + \alpha_n) \tag{3.33}$$

also represents a Weierstrass function with a box-counting fractal dimension of D. This particular Weierstrass function may be interpreted as representing the array factor for a certain non-uniformly but symmetrically spaced linear array consisting of infinitely many elements [19]. This is easily verified by comparing (3.33) with (3.30) and recognizing that the required current amplitudes and element spacings for this array are

$$I_n = \eta^{(D-2)n} \tag{3.34}$$

$$kd_n = a\eta^n \tag{3.35}$$

with $u = \cos\theta$, $\eta > 1$, and $1 < D < 2$. A fractal radiation pattern resulting from (3.33) would possess structure over an infinite range of scales. It is instructive to study the properties of infinite arrays of elements from the theoretical point of view. However, physical arrays necessarily consist of a finite number of elements. The infinite series in (3.33) may be truncated to yield

$$f_N(u) = 2\sum_{n=1}^{N} \eta^{(D-2)n} \cos(a\eta^n u + \alpha_n) \tag{3.36}$$

This Weierstrass partial sum represents the array factor for a non-uniform linear array of $2N$ elements with current amplitudes and spacings given by (3.34) and (3.35), respectively. The Weierstrass partial sum of (3.36) may be classified as bandlimited since the resulting radiation pattern only exhibits fractal behavior over a finite range of scales. It should be pointed out that the box-counting dimension of bandlimited Weierstrass functions, such as (3.36), no longer necessarily yields the number D. However, it has been demonstrated through numerical experiment, that it approaches D for certain values of η [43]. The lower bound on the scale size for which the radiation pattern remains fractal is $2\pi/a\eta^N$. This suggests that the range of scales may be controlled by the number of elements in the array. That is, the addition of two or more elements to the array has the effect of enhancing the fine structure in the radiation pattern. In fact, the structure of the radiation pattern becomes finer and more detailed as the number of array elements is increased.

In order to ensure that the maximum value of (3.36) occurs at a certain specified angle θ_0, the excitation current phases must be chosen in the following way:

$$\alpha_n = -a\eta^n u_0 \tag{3.37}$$

where

$$u_0 = \cos\theta_0 \tag{3.38}$$

Hence, by substituting (3.37) into (3.36) and evaluating at $u = u_0$ we find that

$$f_N(u_0) = 2\sum_{n=1}^{N} \eta^{(D-2)n} = 2\eta^{(D-2)} \left[\frac{1-\eta^{(D-2)N}}{1-\eta^{(D-2)}}\right] \tag{3.39}$$

Dividing (3.36) by its maximum value (3.39) yields a normalized form of the Weierstrass array factor that is given by

$$g_N(u) = \left[\frac{1-\eta^{(D-2)}}{1-\eta^{(D-2)N}}\right] \sum_{n=1}^{N} i_n \cos(a\eta^n u + \alpha_n) \tag{3.40}$$

where

$$i_n = \eta^{(D-2)(n-1)} \tag{3.41}$$

represent the normalized excitation current amplitudes. Equations (3.34) and (3.41) suggest that the fractal dimension of the radiation pattern can be controlled by the array element current distribution.

Equation (3.35) may be used to show that the separation between any two consecutive array elements is given by

$$d_{n+1} - d_n = \left[\frac{a(\eta - 1)\,\eta^n}{2\pi}\right]\lambda \qquad n = 1, 2, \ldots, N-1 \tag{3.42}$$

Since $\eta > 1$, it follows that $\eta^n > \eta$ for $n > 1$. This inequality can be used to prove that

$$d_{n+1} - d_n > d_2 - d_1 \qquad n = 2, 3, \ldots, N-1 \tag{3.43}$$

Let h be a constraint that is imposed on the minimum separation between any two consecutive elements in the array. There are two possible cases in which this minimum spacing constraint may be satisfied. These cases are as follows:

$$\begin{aligned} &\text{Case 1} \qquad d_2 - d_1 = h \quad \text{and} \quad d_1 \geq h/2 \\ &\text{Case 2} \qquad d_1 = h/2 \quad \text{and} \quad d_2 - d_1 \geq h \end{aligned} \tag{3.44}$$

Note that if either condition in (3.44) is satisfied, then the spacing between all other pairs of consecutive array elements will automatically meet the minimum separation criterion h. This property is a direct consequence of inequality (3.43). An expression for a as a function of h and η can be derived by using (3.35) in conjunction with the conditions given in (3.44). The result is

$$a = \begin{cases} \dfrac{kh}{\eta(\eta - 1)} & \text{for} \quad 1 < \eta \leq 3 \\[2ex] \dfrac{kh}{2\eta} & \text{for} \quad \eta \geq 3 \end{cases} \tag{3.45}$$

It should be noted here that the parameter η governs the convergence of the Weierstrass array factor. The closer η is to one, the slower the array factor will converge and the more elements will be required in the array. At the same time, however, the spacing between consecutive array elements is decreasing as η approaches unity. Another interesting property of the Weierstrass array factor is that the array excitation current amplitudes (3.34), phases (3.37), and spacings (3.35) may all be obtained recursively. That is,

$$I_n = \eta^{(D-2)} I_{n-1} \qquad I_1 = \eta^{(D-2)} \tag{3.46}$$

$$\alpha_n = \eta \alpha_{n-1} \qquad \alpha_1 = -a\eta u_0 \tag{3.47}$$

$$d_n = \eta d_{n-1} \qquad d_1 = a\eta/k \tag{3.48}$$

where $n = 2, 3, \ldots, N$ in all three cases. From this we can conclude that Weierstrass arrays are examples of completely self-scalable arrays since their element spacings and current distributions, amplitudes as well as phases, obey the power law relationships given in

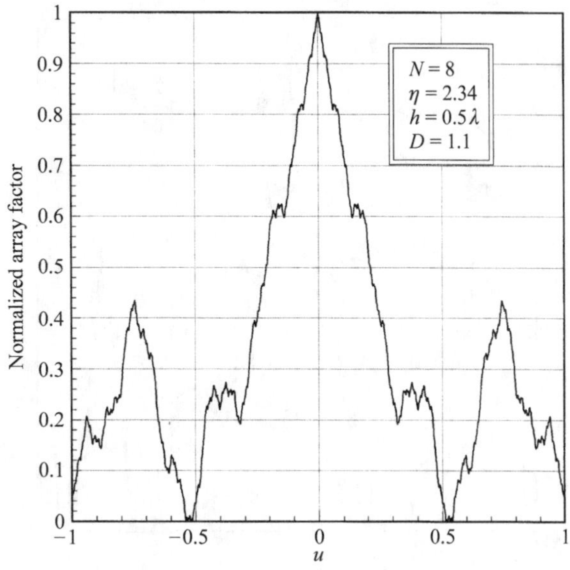

Figure 3.20 The normalized Weierstrass array factor for a 16-element array with $u_0 = 0$, $h = \lambda/2$, $\eta = 2.34$, and $D = 1.1$ [*D. H. Werner and P. L. Werner,* Radio Science, **30**, pp. 29–45, January–Feburary 1996. *(© American Geophysical Union.)*]

(3.46)–(3.48). Hence, we find that the self-scaling properties of these arrays are fundamentally linked to their ability to produce fractal radiation patterns.

Suppose we consider a Weierstrass array in which the minimum separation h between any two adjacent array elements is restricted to be a half-wavelength (i.e., $h = \lambda/2$). Further suppose that we choose a value of $\eta = 2.34$ so that, according to (3.45), the array design parameter a will be unity. These values of $\eta = 2.34$ and $a = 1.0$ may then be used in conjunction with (3.35) or (3.48) to determine the required element spacings for this array. Figure 3.20 shows the radiation pattern that is produced by a Weierstrass array consisting of 16 elements with excitation currents that follow a $D = 1.1$ distribution. By properly adjusting the current distribution on this same 16-element array, radiation patterns can be synthesized which have any desired fractal dimension D between the integer bounds of one and two. For example, plots of the radiation patterns that would result by specifying a fractal dimension of $D = 1.5$ and $D = 1.9$ are shown in Fig. 3.21 and Fig. 3.22, respectively. Table 3.3 contains a listing of element locations for a 16-element Weierstrass array whose geometry is prescribed by the parameters $h = \lambda/2$, $\eta = 2.34$, and $a = 1.0$.

The maximum directive gain or directivity associated with a nonuniform linear Weierstrass array of isotropic sources may be determined by using the bandlimited Weierstrass array factor (3.36) and its maximum value (3.39). In this case, the directivity may be expressed as

$$G(u_0) = \frac{2 f_N^2(u_0)}{\displaystyle\int_{-1}^{1} f_N^2(u)\, du} \tag{3.49}$$

where

$$f_N^2(u) = 4 \sum_{m=1}^{N} \sum_{n=1}^{N} \eta^{(D-2)(m+n)} \cos(a\eta^m u + \alpha_m) \cos(a\eta^n u + \alpha_n) \tag{3.50}$$

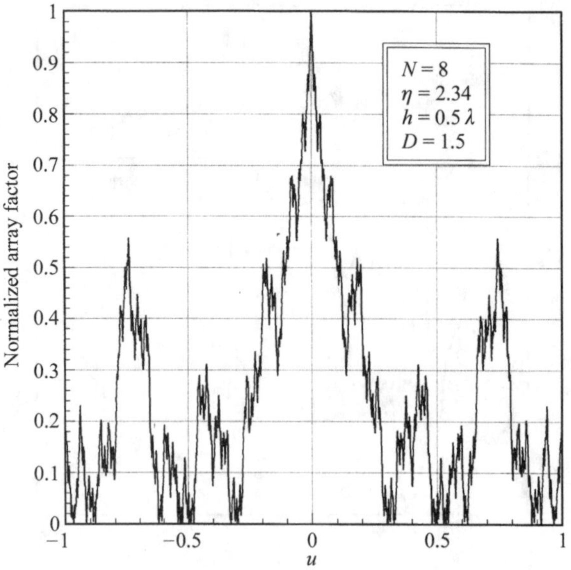

Figure 3.21 The normalized Weierstrass array factor for a 16-element array with $u_0 = 0, h = \lambda/2, \eta = 2.34$, and $D = 1.5$. [*D. H. Werner and P. L. Werner,* Radio Science, **30**, pp. 29–45, January–February 1995. *(© American Geophysical Union.)*]

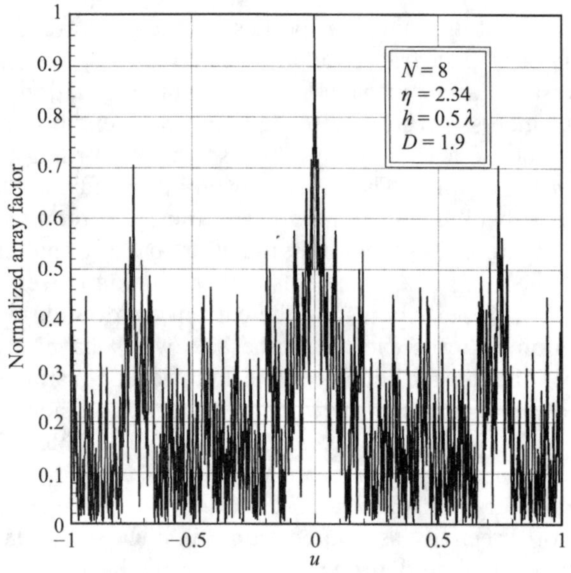

Figure 3.22 The normalized Weierstrass array factor for a 16-element array with $u_0 = 0, h = \lambda/2, \eta = 2.34$, and $D = 1.9$. [*D. H. Werner and P. L. Werner,* Radio Science, **30**, pp. 29–45, January–February 1995. *(© American Geophysical Union.)*]

and

$$f_N^2(u_0) = 4\eta^{2(D-2)}\left[\frac{1-\eta^{(D-2)N}}{1-\eta^{(D-2)}}\right]^2 \tag{3.51}$$

Integration of (3.50) yields

$$\int_{-1}^{1} f_N^2(u)\,du = 4\sum_{m=1}^{N}\sum_{n=1}^{N}\eta^{(D-2)(m+n)}\,\Lambda_{mn}(a,\eta,u_0) \tag{3.52}$$

TABLE 3.3 Element Geometry for a Symmetric 16-Element Weierstrass Array

Element, n	Element Location, d_n/λ
1	0.372
2	0.872
3	2.039
4	4.772
5	11.166
6	26.129
7	61.141
8	143.069

$h = 0.5\lambda$, $\eta = 2.34$, and $a = 1.0$

where

$$\Lambda_{mn}(a,\eta,u_0) = \int_{-1}^{1} \cos[a\eta^m(u-u_0)]\cos[a\eta^n(u-u_0)]\,du$$
$$= \cos[a(\eta^m+\eta^n)u_0)]\,\mathrm{sinc}[a(\eta^m+\eta^n)]$$
$$+ \cos[a(\eta^m-\eta^n)u_0]\,\mathrm{sinc}[a(\eta^m-\eta^n)] \tag{3.53}$$

and

$$\mathrm{sinc}(x) = \frac{\sin(x)}{x} \tag{3.54}$$

Finally, substituting (3.51) and (3.52) into (3.49) results in a closed-form expression for the directivity given by

$$G(u_0) = \left\{\frac{1}{2}\left[\frac{1-\eta^{(D-2)}}{1-\eta^{(D-2)N}}\right]^2 \sum_{m=1}^{N}\sum_{n=1}^{N} \eta^{(D-2)(n+m-2)}\Lambda_{mn}(a,\eta,u_0)\right\}^{-1} \tag{3.55}$$

Note that the special case when $D = 2$ corresponds to an array with uniform excitation. The expression for directivity in this case becomes

$$G(u_0) = \frac{1}{\dfrac{1}{2N^2}\displaystyle\sum_{m=1}^{N}\sum_{n=1}^{N}\Lambda_{mn}(a,\eta,u_0)} \tag{3.56}$$

It can be shown that (3.56) reduces to the well-known result

$$G(0) = 2N \tag{3.57}$$

for a half-wavelength uniformly spaced broadside linear array when η^n is replaced by $n\pi/a$ and $u_0 = 0$.

The directivity as a function of fractal dimension D is plotted in Fig. 3.23 for several different array sizes. A uniform current distribution is obtained on a Weierstrass array when $D = 2$, which represents an upper bound on the fractal

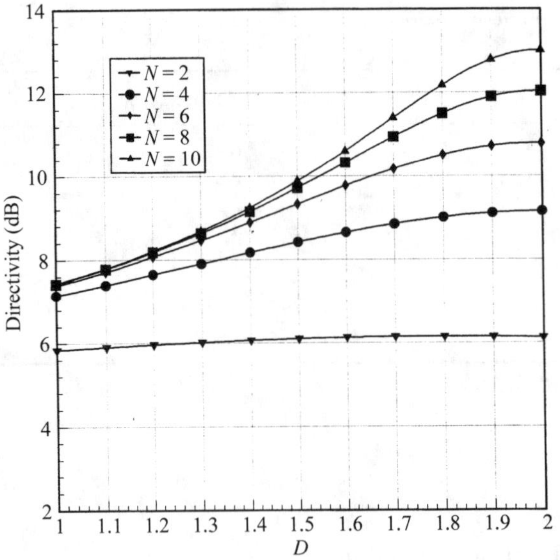

Figure 3.23 The maximum directivity versus fractal dimension for different-sized linear Weierstrass arrays of isotropic radiators. [*D. H. Werner and P. L. Werner,* Radio Science, **30**, pp. 29–45, January–February 1995. *(© American Geophysical Union.)*]

dimension. Figure 3.23 demonstrates that as the fractal dimension of the radiation pattern is decreased, the main beam broadens and the corresponding directivity also decreases.

3.4.3 Fourier-Weierstrass Line Sources

In this section we investigate the use of long, straight current-carrying antennas, known as line sources, for the synthesis of fractal radiation patterns [19]. The geometry for a line source of length L centered symmetrically about the origin of the z axis with a current distribution of $I(z)$ is shown in Fig. 3.24. For a line source of infinite length, the radiation pattern $F(u)$ and the current distribution $I(s)$ are related by the following Fourier transform pair [44]:

$$F(u) = \int_{-\infty}^{\infty} I(s)e^{j2\pi us}\,ds \tag{3.58}$$

$$I(s) = \int_{-\infty}^{\infty} F(u)e^{-j2\pi su}\,du \tag{3.59}$$

where

$$u = \cos\theta \tag{3.60}$$

$$s = z/\lambda \tag{3.61}$$

Any fractal can be constructed through the recursive application of an appropriate generating function. For instance, suppose that the radiation pattern of an

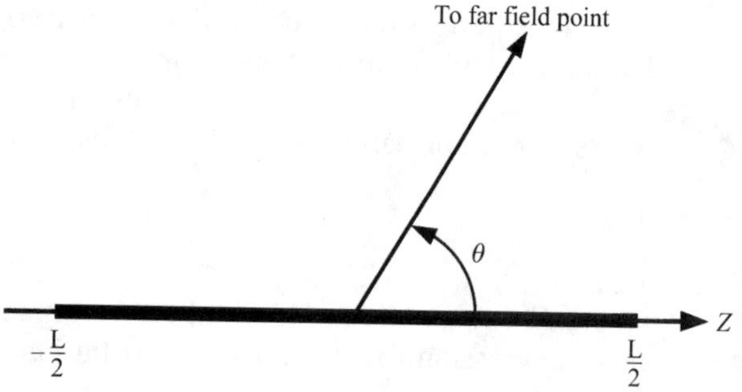

Figure 3.24 Geometry of a continuous line source of length L oriented along the z axis. [*D. H. Werner and P. L. Werner,* Radio Science, **30**, pp. 29–45, January–February 1995. (*© American Geophysical Union.*)]

infinite line source may be represented as a bandlimited generalized Weierstrass function of the form

$$F(u) = \sum_{n=0}^{N-1} \eta^{(D-2)n} g(\eta^n u) \tag{3.62}$$

where D represents the fractal dimension and $g(u)$ is an arbitrary generating function. Here we assume that the generating function $g(u)$ is periodic and even, i.e., $g(u+2) = g(u)$ and $g(-u) = g(u)$. Therefore, $g(u)$ may be expanded in a Fourier cosine series as

$$g(u) = \frac{a_0}{2} + \sum_{m=1}^{\infty} a_m \cos(m\pi u) \tag{3.63}$$

where the coefficients of (3.63) are determined from

$$a_m = 2\int_0^1 g(u)\cos(m\pi u)\,du \tag{3.64}$$

Substituting (3.63) into (3.62) leads to an expression for the radiation pattern given by

$$F(u) = \frac{a_0}{2}\frac{\eta^{(D-2)N} - 1}{\eta^{(D-2)} - 1} + \sum_{m=1}^{\infty}\sum_{n=0}^{N-1} a_m \eta^{(D-2)n} \cos(m\pi\eta^n u) \tag{3.65}$$

Without loss of generality we replace u by $u+1$ in (3.62), effectively mapping the interval $[-1, 1]$ to the interval $[0, 2]$. This leads to the result

$$F(u) = \frac{a_0}{2}\frac{\eta^{(D-2)N} - 1}{\eta^{(D-2)} - 1} + \sum_{m=1}^{\infty} a_m \left\{ \sum_{n=0}^{N-1} \eta^{(D-2)n} \cos[m\pi\eta^n(u+1)] \right\} \tag{3.66}$$

where

$$a_m = 2\int_0^1 g(u-1)\cos(m\pi u)\,du \tag{3.67}$$

with the requirements that $\eta > 1$ and $1 < D < 2$. Equation (3.66) represents a Fourier decomposition of the fractal radiation pattern $F(u)$ in which the basis functions are bandlimited Weierstrass cosine functions.

The line source current distribution required in order to produce the desired fractal radiation patterns may be obtained by evaluating the Fourier integral (3.59)

$$I(s) = \int_{-1}^{1} F(u)e^{-j2\pi su}\,du \tag{3.68}$$

where $F(u)$ is defined in (3.66). Performing the necessary integration of (3.68) results in an expression for the current distribution given by

$$\begin{aligned} I(s) = a_0 \frac{\eta^{(D-2)N}-1}{\eta^{(D-2)}-1}\operatorname{sinc}(2\pi s) + \sum_{m=1}^{\infty}\sum_{n=0}^{N-1} a_m\,\eta^{(D-2)n} \\ \cdot \{e^{jm\pi\eta^n}\operatorname{sinc}(2\pi s - m\pi\eta^n) + e^{-jm\pi\eta^n}\operatorname{sinc}(2\pi s + m\pi\eta^n)\} \end{aligned} \tag{3.69}$$

This equation represents the required current distribution for a line source of infinite extent. However, an approximation may be obtained for the current distribution on a finite length line source by truncating (3.69) in the following way:

$$\tilde{I}(s) = \begin{cases} I(s), & |s| \le \dfrac{L}{2\lambda} \\ 0, & |s| > \dfrac{L}{2\lambda} \end{cases} \tag{3.70}$$

The corresponding expression for the synthesized fractal radiation pattern associated with a finite line source of length L may be found by using (3.70) in conjunction with (3.58). Following this procedure yields

$$\begin{aligned} \tilde{F}(u) = \frac{a_0}{2\pi}\frac{\eta^{(D-2)N}-1}{\eta^{(D-2)}-1}\{Si[\pi(L/\lambda)(1+u)] + Si[\pi(L/\lambda)(1-u)]\} \\ + \frac{1}{2\pi}\sum_{m=1}^{\infty}\sum_{n=0}^{N-1} a_m\,\eta^{(D-2)n}\{\cos[m\pi\eta^n(u+1)]\,S_{m,n}(u) \\ + \sin[m\pi\eta^n(u+1)]\,C_{m,n}(u)\} \end{aligned} \tag{3.71}$$

where

$$\begin{aligned} S_{m,n}(u) = Si[\pi(L/\lambda - m\eta^n)(1+u)] + Si[\pi(L/\lambda + m\eta^n)(1+u)] \\ + Si[\pi(L/\lambda - m\eta^n)(1-u)] + Si[\pi(L/\lambda + m\eta^n)(1-u)] \\ \text{for} \quad -1 \le u \le 1 \end{aligned} \tag{3.72}$$

$$\begin{aligned} C_{m,n}(u) = Ci[\pi|L/\lambda - m\eta^n|(1+u)] - Ci[\pi(L/\lambda + m\eta^n)(1+u)] \\ - Ci[\pi|L/\lambda - m\eta^n|(1-u)] + Ci[\pi(L/\lambda + m\eta^n)(1-u)] \\ \text{for} \quad -1 < u < 1 \quad \text{and} \quad m\eta^n \ne L/\lambda \end{aligned} \tag{3.73}$$

$$\begin{aligned} C_{m,n}(u) = \ln\left[\frac{1+u}{1-u}\right] - Ci[2\pi(L/\lambda)(1+u)] + Ci[2\pi(L/\lambda)(1-u)] \\ \text{for} \quad -1 < u < 1 \quad \text{and} \quad m\eta^n = L/\lambda \end{aligned} \tag{3.74}$$

$$C_{m,n}(\pm 1) = \pm\{Ci[2\pi|L/\lambda - m\eta^n|] - Ci[2\pi(L/\lambda + m\eta^n)] - \ln|L/\lambda - m\eta^n| + \ln(L/\lambda + m\eta^n)\} \quad \text{for} \quad m\eta^n \neq L/\lambda \tag{3.75}$$

$$C_{m,n}(\pm 1) = \pm\{\gamma + \ln[4\pi(L/\lambda)] - Ci[4\pi(L/\lambda)]\} \quad \text{for} \quad m\eta^n = L/\lambda \tag{3.76}$$

in which

$$Si(x) = \int_0^x \frac{\sin t}{t} dt \tag{3.77}$$

$$Ci(x) = -\int_x^\infty \frac{\cos t}{t} dt = \gamma + \ln(x) + \int_0^x \frac{\cos t - 1}{t} dt \tag{3.78}$$

are the sine and cosine integrals, respectively, and the parameter $\gamma = 0.57721 \ldots$ is Euler's constant [45]. Finally, it can easily be shown that

$$\lim_{L\to\infty} \tilde{F}(u) = F(u) \tag{3.79}$$

since

$$Si(\infty) = \frac{\pi}{2} \tag{3.80}$$

$$Ci(\infty) = 0 \tag{3.81}$$

$$\lim_{L\to\infty} S_{m,n}(u) = 2\pi \tag{3.82}$$

$$\lim_{L\to\infty} C_{m,n}(u) = 0 \tag{3.83}$$

In order to illustrate the line source synthesis procedure as well as investigate the properties of the resulting fractal radiation patterns and associated current distributions, we consider the following example. Suppose we choose a generating function of the form

$$g(\theta) = 1 - |\cos\theta| \quad \text{for} \quad 0 \leq \theta \leq \pi \tag{3.84}$$

Transforming (3.84) into the u domain results in a triangular generating function given by

$$g(u) = 1 - |u| \quad \text{for} \quad |u| \leq 1 \tag{3.85}$$

The Fourier coefficients a_m that are associated with this triangular generating function can be found from (3.67). That is,

$$a_m = 2\int_0^1 u\cos(m\pi u)\,du = \begin{cases} 1, & m = 0 \\ -\left(\dfrac{2}{m\pi}\right)^2, & m = 1,3,5,\ldots \\ 0, & m = 2,4,6,\ldots \end{cases} \tag{3.86}$$

Substituting the Fourier coefficients (3.86) into (3.66) and choosing a value of $\eta = 2$ leads to the result

$$F(u) = \frac{1}{2}\left[\frac{2^{(D-2)N} + 2^{(D-2)} - 2}{2^{(D-2)} - 1}\right] - |u| - \left(\frac{2}{\pi}\right)^2 \sum_{m=1}^{\infty}\sum_{n=1}^{N-1} \frac{2^{(D-2)n}}{(2m-1)^2}\cos[(2m-1)2^n\pi u] \tag{3.87}$$

which represents the desired line source fractal radiation pattern, with fractal dimension D, formed by the recursive application of a triangular generating function. The fact that $F(\pm 1) = 0$ and $F(0) = 1$ may be easily verified using (3.87). The line source current distribution required in order to produce the fractal radiation patterns of (3.87) is given by

$$I(s) = \left[\frac{2^{(D-2)N} - 2^{(D-2)}}{2^{(D-2)} - 1}\right]\operatorname{sinc}(2\pi s) + \operatorname{sinc}^2(\pi s) - \left(\frac{2}{\pi}\right)^2 \sum_{m=1}^{\infty}\sum_{n=1}^{N-1} \frac{2^{(D-2)n}}{(2m-1)^2}\{\operatorname{sinc}[2\pi s - (2m-1)2^n\pi] + \operatorname{sinc}[2\pi s + (2m-1)2^n\pi]\} \tag{3.88}$$

This expression for the current distribution may be normalized by its peak value

$$I(0) = \frac{2^{(D-2)N} - 1}{2^{(D-2)} - 1} \tag{3.89}$$

which yields

$$i(s) = \left[\frac{2^{(D-2)N} - 2^{(D-2)}}{2^{(D-2)N} - 1}\right]\operatorname{sinc}(2\pi s) + \left[\frac{2^{(D-2)} - 1}{2^{(D-2)N} - 1}\right]\operatorname{sinc}^2(\pi s) - \left(\frac{2}{\pi}\right)^2\left[\frac{2^{(D-2)} - 1}{2^{(D-2)N} - 1}\right] \sum_{m=1}^{\infty}\sum_{n=1}^{N-1} \frac{2^{(D-2)n}}{(2m-1)^2}\{\operatorname{sinc}[2\pi s - (2m-1)2^n\pi] + \operatorname{sinc}[2\pi s + (2m-1)2^n\pi]\} \tag{3.90}$$

The line source current distributions of (3.88) or (3.90) may be used to synthesize radiation patterns with any desired fractal dimension, which are based on the

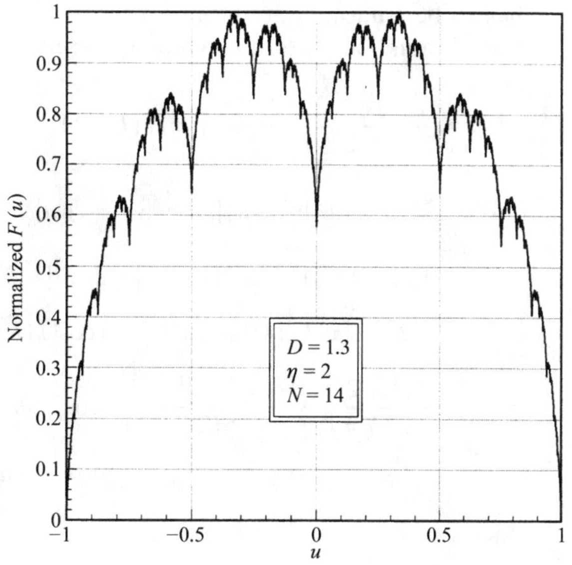

Figure 3.25 Synthesized fractal radiation pattern (normalized) for a line source with $D = 1.3$, $N = 14$, and a triangular generating function. [*D. H. Werner and P. L. Werner,* Radio Science, **30**, pp. 29–45, January–February 1995. *(© American Geophysical Union.)*]

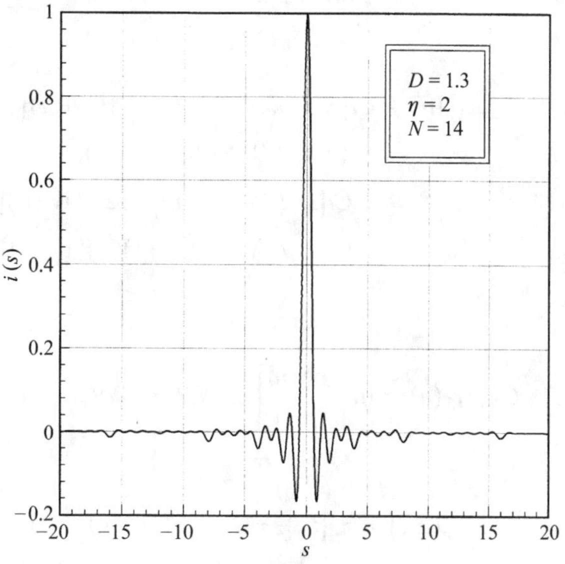

Figure 3.26 The normalized line source current distribution required in order to produce the fractal radiation pattern shown in Fig. 3.25. [*D. H. Werner and P. L. Werner,* Radio Science, **30**, pp. 29–45, January–February 1995 *(© American Geophysical Union.)*]

triangular generating function of (3.85). For example, the synthesized radiation pattern for a line source with a desired fractal dimension of $D = 1.3$ is shown in Fig. 3.25. Figure 3.26 illustrates the line source current distribution required in order to produce the fractal radiation pattern shown in Fig. 3.25.

If we consider a line source of finite length L, then the current distribution (3.88) has to be truncated according to (3.70) and, in this case, the resulting synthesized fractal radiation pattern may be expressed as

$$\begin{aligned}\tilde{F}(u) = &\frac{1}{2\pi}\left[\frac{2^{(D-2)N}-2^{(D-2)}}{2^{(D-2)}-1}\right]\{Si[\pi(L/\lambda)(1+u)] \\ &+ Si[\pi(L/\lambda)(1-u)]\} + \frac{1}{\pi}\{(1+u)\,Si[\pi(L/\lambda)(1+u)] \\ &+ (1-u)\,Si[\pi(L/\lambda)(1-u)] - 2u\,Si[\pi(L/\lambda)u] \\ &- 2\,\mathrm{sinc}[\pi(L/2\lambda)]\sin[\pi(L/2\lambda)]\cos[\pi(L/\lambda)u]\} \\ &- \frac{1}{2\pi}\left(\frac{2}{\pi}\right)^2 \sum_{m=1}^{\infty}\sum_{n=1}^{N-1}\frac{2^{(D-2)n}}{(2m-1)^2}\{\cos[(2m-1)2^n\pi u]\,S_{2m-1,n}(u) \\ &+ \sin[(2m-1)2^n\pi u]\,C_{2m-1,n}(u)\}\end{aligned} \tag{3.91}$$

where

$$\begin{aligned}S_{2m-1,n}(u) = &\,Si[\pi(L/\lambda-(2m-1)2^n)(1+u)] \\ &+ Si[\pi(L/\lambda+(2m-1)2^n)(1+u)] \\ &+ Si[\pi(L/\lambda-(2m-1)2^n)(1-u)] \\ &+ Si[\pi(L/\lambda+(2m-1)2^n)(1-u)] \qquad \text{for} \quad -1 \le u \le 1\end{aligned} \tag{3.92}$$

$$\begin{aligned}C_{2m-1,n}(u) = &\,Ci[\pi|L/\lambda-(2m-1)2^n|(1+u)] \\ &- Ci[\pi(L/\lambda+(2m-1)2^n)(1+u)] \\ &- Ci[\pi|L/\lambda-(2m-1)2^n|(1-u)] \\ &+ Ci[\pi(L/\lambda+(2m-1)2^n)(1-u)] \\ &\text{for} \quad -1<u<1 \quad \text{and} \quad (2m-1)2^n \neq L/\lambda\end{aligned} \tag{3.93}$$

$$\begin{aligned}C_{2m-1,n}(u) = \ln\left[\frac{1+u}{1-u}\right] - Ci[2\pi(L/\lambda)(1+u)] + Ci[2\pi(L/\lambda)(1-u)] \\ \text{for} \quad -1<u<1 \quad \text{and} \quad (2m-1)2^n = L/\lambda\end{aligned} \tag{3.94}$$

$$\begin{aligned}C_{2m-1,n}(\pm1) = \pm\{&Ci[2\pi|L/\lambda-(2m-1)2^n|] - Ci[2\pi(L/\lambda+(2m-1)2^n)] \\ &- \ln|L/\lambda-(2m-1)2^n| + \ln(L/\lambda+(2m-1)2^n)\} \\ &\text{for} \quad (2m-1)2^n \neq L/\lambda\end{aligned} \tag{3.95}$$

$$\begin{aligned}C_{2m-1,n}(\pm1) = \pm\{\gamma + \ln[4\pi(L/\lambda)] - Ci[4\pi(L/\lambda)]\} \\ \text{for} \quad (2m-1)2^n = L/\lambda\end{aligned} \tag{3.96}$$

Note that the term in (3.91) which contains $\sin[\pi(L/2\lambda)]$ vanishes if the half-length of the line source $L/2$ is an integer multiple of the wavelength λ.

As our next example, we consider an initial pattern of the form

$$g(\theta) = \sin^2 \theta \qquad \text{for} \qquad 0 \leq \theta \leq \pi \tag{3.97}$$

which is equivalent to the quadratic generating function

$$g(u) = 1 - u^2 \qquad \text{for} \qquad -1 \leq u \leq 1 \tag{3.98}$$

The required coefficients for the Fourier-Weierstrass expansion can be found by making use of (3.67) and (3.98), which results in

$$a_m = \begin{cases} \dfrac{4}{3}, & m = 0 \\ -\left(\dfrac{2}{m\pi}\right)^2, & m \neq 0 \end{cases} \tag{3.99}$$

Figure 3.27 illustrates the first eight stages (N = 1–8) in the construction of a fractal radiation pattern based on the quadratic generating function (3.98) with a desired dimension of $D = 1.1$ and a value of $\eta = 2$.

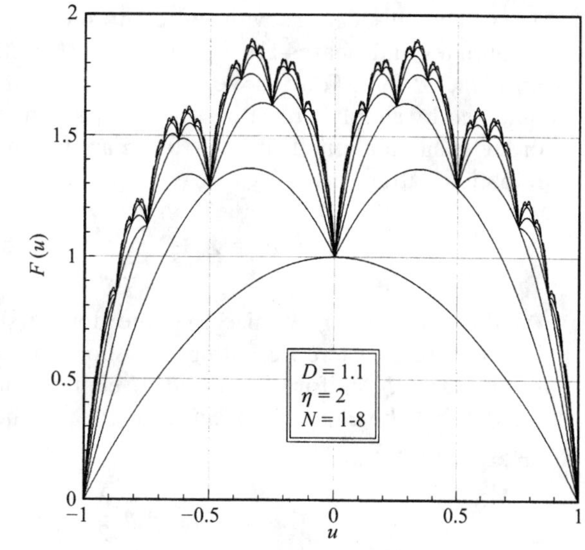

Figure 3.27 The first eight stages (N = 1–8) in the construction of a fractal radiation pattern based on a quadratic generation function with $D = 1.1$ and $\eta = 2$. [*D. H. Werner and P. L. Werner,* Radio Science, **30**, pp. 29–45, January–February 1995. *(© American Geophysical Union.)*]

3.4.4 Fourier-Weierstrass Linear Arrays

The technique introduced in this section is a generalization of the method of fractal radiation pattern synthesis using Weierstrass arrays which was presented in Section 3.4.2. This generalization involves introducing a Fourier-Weierstrass expansion of the type discussed in Section 3.4.3, which provides the additional flexibility of choosing a suitable generating function. In other words, while both synthesis techniques allow for the specification of a desired fractal dimension, the method based on Fourier-Weierstrass arrays provides for additional control of the overall shape as well as the underlying geometrical structure of the radiation pattern.

We begin our study of bandlimited Fourier-Weierstrass fractal arrays by expressing (3.66) in the following convenient form [19]:

$$F(u) = I_0 + 2\sum_{m=1}^{\infty}\sum_{n=0}^{N-1} I_{mn}\cos(kd_{mn}u + \alpha_{mn}) \tag{3.100}$$

where

$$I_0 = \frac{a_0}{2}\frac{1-\eta^{(D-2)N}}{1-\eta^{(D-2)}} \tag{3.101}$$

$$I_{mn} = \frac{a_m}{2}\eta^{(D-2)n} \tag{3.102}$$

$$kd_{mn} = m\pi\eta^n \tag{3.103}$$

$$\alpha_{mn} = m\pi\eta^n \tag{3.104}$$

and the Fourier coefficients a_m corresponding to a particular generating function may be obtained through the use of (3.67). Equation (3.100) represents the Fourier-Weierstrass array factor for a discrete array containing infinitely many elements. The array element locations with respect to the origin are d_{mn}, while the element current amplitudes and phases are I_{mn} and α_{mn}, respectively. A useful representation of the array factor for a Fourier-Weierstrass array with a finite number of elements may be obtained by simply truncating the outer summation in (3.100) and interchanging the order of summation. This leads to an approximate expression for the desired fractal radiation pattern given by

$$\tilde{F}(u) = I_0 + 2\sum_{n=0}^{N-1}\sum_{m=1}^{M} I_{mn}\cos(kd_{mn}u + \alpha_{mn}) \tag{3.105}$$

At this point, we recognize that the double summation appearing in (3.105) may be interpreted as representing the superposition of the radiation produced by a sequence of N uniformly spaced M-element linear arrays. It follows directly from (3.103) that the element spacings for each of the M-element uniformly spaced linear subarrays are

$$\Delta_n = d_{m+1n} - d_{mn} = \eta^n\frac{\lambda}{2} \quad \text{for} \quad n = 0, 1, 2, \ldots, N-1 \tag{3.106}$$

which may be used to derive the recurrence relation

$$\Delta_{n+1} = \eta\Delta_n \quad \text{where} \quad \Delta_0 = \lambda/2 \tag{3.107}$$

Hence, this property suggests that Fourier-Weierstrass arrays may be viewed as being composed of a sequence of self-scalable uniformly spaced linear sub-arrays. Recurrence relations for the excitation current amplitudes and phases may be found in a similar way using (3.102) and (3.104), respectively. These recurrence relations are

$$I_{mn+1} = \eta^{(D-2)}I_{mn} \quad \text{where} \quad I_{m0} = \frac{a_m}{2} \tag{3.108}$$

$$\alpha_{mn+1} = \eta\alpha_{mn} \quad \text{where} \quad \alpha_{m0} = m\pi \tag{3.109}$$

Finally, an expression for the normalized array current excitation amplitudes may be obtained by dividing (3.102) by (3.101), which yields

$$i_0 = 1 \tag{3.110}$$

$$i_{mn} = \left(\frac{a_m}{a_0}\right)\left(\frac{1-\eta^{(D-2)}}{1-\eta^{(D-2)N}}\right)\eta^{(D-2)n} \tag{3.111}$$

Suppose we wish to use a Fourier-Weierstrass array to synthesize a fractal radiation pattern based on the triangular generating function (3.85) with $\eta = 2$. The Fourier coefficients a_m which are associated with the triangular generating function were found in (3.86) and may be used to show that

$$I_0 = \frac{1}{2}\frac{1-2^{(D-2)N}}{1-2^{(D-2)}} \tag{3.112}$$

$$I_{mn} = -\frac{1}{2}\left(\frac{2}{\pi}\right)^2 \frac{2^{(D-2)n}}{(2m-1)^2} \tag{3.113}$$

$$\alpha_{mn} = (2m-1)2^n\pi \tag{3.114}$$

$$d_{mn} = (2m-1)2^{n-1}\lambda \tag{3.115}$$

Various stages in the construction of the Fourier-Weierstrass array corresponding to a triangular generating function are illustrated in Fig. 3.28. In this case, since $\eta = 2$, (3.107) becomes

$$\Delta_{n+1} = 2\Delta_n \qquad \text{where} \qquad \Delta_0 = \lambda \tag{3.116}$$

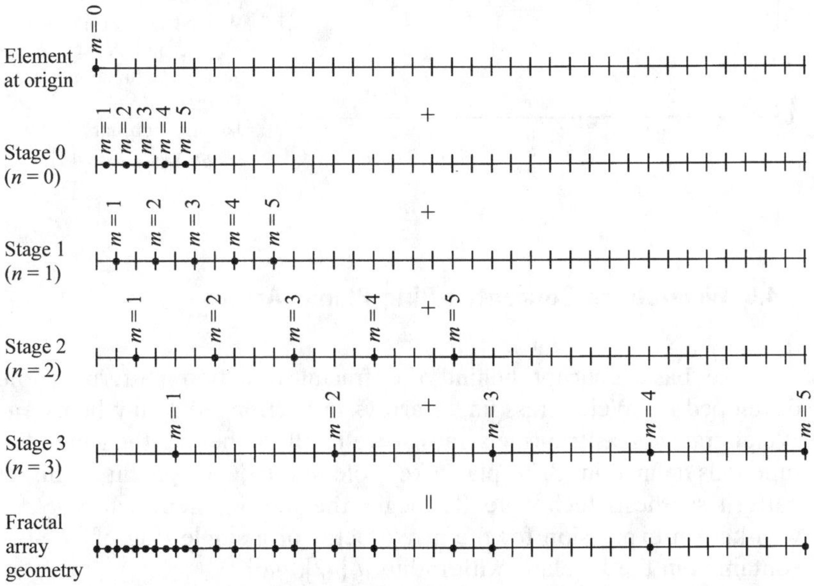

Figure 3.28 The construction of a Fourier-Weierstrass fractal array from a sequence of four ($N = 4$) self-scalable uniformly spaced five-element ($M = 5$) linear arrays. The subarray element spacings are scaled according to those associated with the triangular generating function where $\eta = 2$, i.e., $\Delta_{n+1} = 2\Delta_n$, and $\Delta_0 = \lambda$. [*D. H. Werner and P. L. Werner,* Radio Science, **30**, pp. 29–45, January–February 1995. *(© American Geophysical Union.)*]

which suggests that the element spacings for each consecutive sub-array may be obtained by doubling the spacing of the previous sub-array. This self-scaling property of the subarrays is clearly illustrated in Fig. 3.28.

Figure 3.29 shows a synthesized radiation pattern formed by a triangular generating function with $\eta = 2$ and a desired fractal dimension of $D = 1.1$. A Fourier-Weierstrass array with $M = 4$ and $N = 8$ was used to synthesize this radiation pattern. Table 3.4 contains a listing of the Fourier-Weierstrass array element spacings and excitation currents required in order to produce the fractal radiation pattern shown in Fig. 3.29.

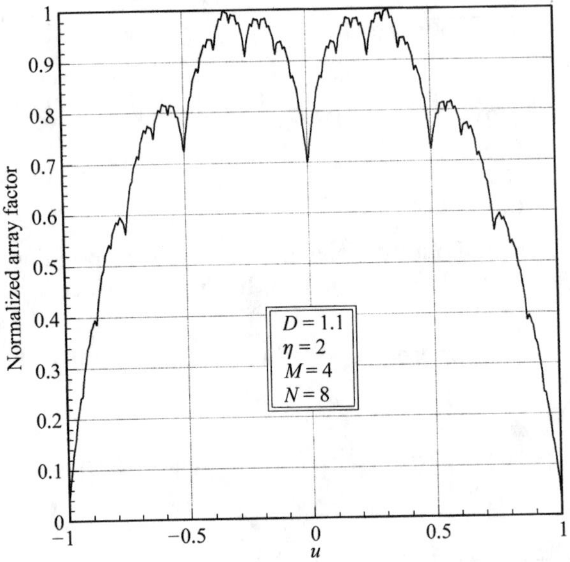

Figure 3.29 Synthesized fractal radiation pattern (normalized) for a Fourier-Weierstrass array with $D = 1.1$, $\eta = 2$, $M = 4$, and $N = 8$, and with a triangular generating function. [*D. H. Werner and P. L. Werner,* Radio Science, **30**, pp. 29–45, January–February 1995. (*© American Geophysical Union.)*]

3.4.5 Weierstrass Concentric-Ring Planar Arrays

The basic concept behind the fractal radiation pattern synthesis technique developed for Weierstrass linear arrays in Section 3.4.2 may be extended to include Weierstrass concentric-ring arrays as well [20]. As before, the generalized Weierstrass functions defined in (3.29) play a key role in the development of this fractal radiation pattern synthesis technique. To begin the development, suppose we consider the well-known expression for the array factor of a single ring of M isotropic elements contained in the xy plane with radius a [42], [46]

$$F(\theta, \phi) = \sum_{m=1}^{M} I_m e^{j[ka \sin\theta \cos(\phi - \phi_m) + \alpha_m]} \tag{3.117}$$

where I_m and α_m represent the magnitudes and phases of the excitation current for the mth element located at $\phi = \phi_m$ where $\phi_m = 2\pi m/M$ and $m = 1, 2, \ldots, M$. The desired

TABLE 3.4 Fourier-Weierstrass Array Element Spacings and Excitation Currents

p	d_p/λ	i_p	α_p (deg)
0	0.000	1.000E + 00	0
1	0.500	−2.026E − 01	180
2	1.000	−1.086E − 01	0
3	1.500	−2.250E − 02	180
4	2.000	−5.820E − 02	0
5	2.500	−8.100E − 03	180
6	3.000	−1.210E − 02	0
7	3.500	−4.100E − 03	180
8	4.000	−3.120E − 02	0
9	5.000	−4.300E − 03	0
10	6.000	−6.500E − 03	0
11	7.000	−2.200E − 03	0
12	8.000	−1.670E − 02	0
13	10.000	−2.300E − 03	0
14	12.000	−3.500E − 03	0
15	14.000	−1.200E − 03	0
16	16.000	−9.000E − 03	0
17	20.000	−1.200E − 03	0
18	24.000	−1.900E − 03	0
19	28.000	−6.364E − 04	0
20	32.000	−4.800E − 03	0
21	40.000	−6.685E − 04	0
22	48.000	−9.951E − 04	0
23	56.000	−3.411E − 04	0
24	64.000	−2.600E − 03	0
25	80.000	−3.582E − 04	0
26	96.000	−5.332E − 04	0
27	112.000	−1.828E − 04	0
28	160.000	−1.920E − 04	0
29	192.000	−2.858E − 04	0
30	224.000	−9.794E − 05	0
31	320.000	−1.029E − 04	0
32	448.000	−5.249E − 05	0

main beam maximum may be positioned at $\theta = \theta_0$ and $\phi = \phi_0$ by choosing the excitation current phases according to

$$\alpha_m = -ka \sin\theta_0 \cos(\phi_0 - \phi_m) \tag{3.118}$$

Suppose that the array contains a large number of elements M equally spaced around the circumference of the ring and that the main beam is pointing in the z direction (i.e., $\theta_0 = 0$). Further suppose that the array is uniformly excited such that $I_m = I_0$ for all values of m, then it can be shown that, under these conditions, (3.117) will reduce to [42], [46]

$$F(\theta) = I_t J_0(ka \sin\theta) \tag{3.119}$$

in which $I_t = MI_0$ represents the total current on the ring. For concentric-ring arrays, an expression for the array factor may be obtained by simply adding up the contributions due to each individual ring array. That is,

$$F(u) = \sum_{n=1}^{N} I_n J_0(ka_n u) \tag{3.120}$$

where $u = \sin\theta$, N is the total number of concentric rings, and a_n denotes the radius of the nth ring array. Next, if we let the total current on each ring be

$$I_n = \eta^{(D-2)n} \tag{3.121}$$

and choose the corresponding radii such that

$$ka_n = \eta^n \tag{3.122}$$

then (3.120) represents a generalized Weierstrass function of the form defined in (3.29), provided $\eta > 1$ and $1 < D < 2$. Finally, we note that even though $g(x) = J_0(x)$ is not a strictly periodic function in this case (but is a quasi-periodic function for large argument x), the resulting radiation patterns will still be fractal [20].

3.5 FRACTAL ARRAY FACTORS AND THEIR ROLE IN THE DESIGN OF MULTIBAND ARRAYS

3.5.1 Background

It has long been recognized that one of the fundamental properties of frequency-independent antennas is their ability to retain the same shape under certain scaling transformations. Examples of such antennas include designs based on angles, cones, and spirals [47–51]. The standard approach for categorizing frequency-independent antennas is to consider them as being constructed from a multiplicity of adjoining cells. Each cell is identical to the previous cell except for a scaling factor τ such that [52]

$$\tau = \Omega_n/\Omega_{n+1} \tag{3.123}$$

where Ω_n represents the dimension of the nth cell and Ω_{n+1} represents the dimension of the next adjoining cell. For instance, the logarithmic spiral is defined in polar coordinates (ρ, ϕ) by the simple relation [52], [53]

$$\rho = \rho_0 e^{\beta_0 \phi} \tag{3.124}$$

where

$$\rho_0 = b > 0 \tag{3.125}$$

$$\beta_0 = \frac{1}{\rho}\frac{d\rho}{d\phi} = \cot\alpha \tag{3.126}$$

The geometry for this logarithmic spiral is illustrated in Fig. 3.30, where b is the starting radius of the spiral and α represents the angle between a tangent to the curve at any point and a line drawn to the origin at that point. Scaling the spiral defined in (3.124) by a factor of τ yields the same spiral rotated by a constant angle $\ln(\tau)/\beta_0$. This suggests that the logarithmic spiral is self-scalable, with a scale factor $\tau = e^{2\pi|\beta_0|}$.

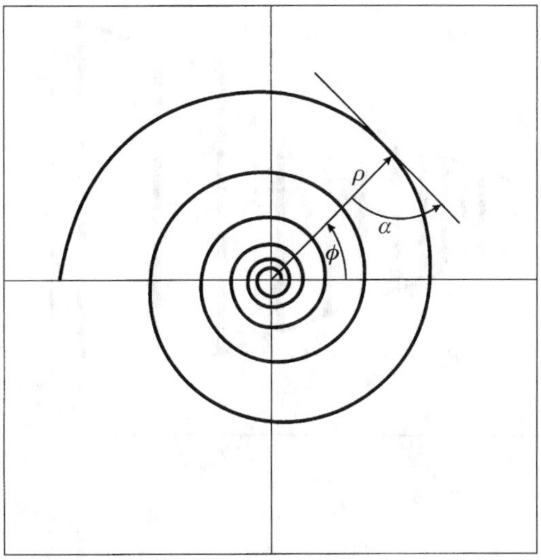

Figure 3.30 Logarithmic spiral antenna geometry. [*D. H. Werner and P. L. Werner,* Radio Science, **31.** pp. 1331–1343, November–December 1996. (© *American Geophysical Union.)*]

However, if rotations are disregarded, then the logarithmic spiral can be considered self-scalable for any real scaling factor τ.

Another example of a frequency-independent antenna in common use is the log-periodic dipole antenna [44], [54]. Log-periodic dipole antennas exhibit self-scalability in their geometric structure at discrete frequencies, as illustrated by Fig. 3.31. The lengths and spacings of adjacent dipole elements are both scaled by the same factor of τ. Figure 3.31 also demonstrates how the log-periodic dipole antenna may be constructed through the use of an initiator and an associated generator. The construction begins with an initial dipole of length L_1 located a distance R_1 from the origin as shown in Fig. 3.31. This initial dipole serves as an initiator for the recursive construction of the log-periodic dipole antenna. The generator is formed by first scaling and then translating the initiator by the factor of τ. In other words, the corresponding length and position of the generator dipole are determined by the relations $L_2 = \tau L_1$ and $R_2 = \tau R_1$, respectively. This process is then repeated until the entire log-periodic dipole antenna is constructed. Other concepts for broadband arrays of log-periodics which exploit this self-scaling concept are discussed by DuHamel and Berry [55], Mei et al. [56], Johnson and Jasik [57], and Breakall [58].

The recent introduction of fractal geometry has provided the framework for the study of a new and diverse class of self-similar structures which are said to possess no characteristic size [1], [59]. This has led to the notion that fractal geometric principles be used to provide a natural extension of the traditional approaches for classification, analysis, and design of frequency-independent antennas [21], [22]. In this section we introduce a special family of self-scalable radiating structures (antenna arrays) which produce fractal radiation patterns in the limit of infinite array size. This property is shown to play a central role in the development of a design methodology for multifrequency linear arrays that have radiation characteristics which are a log-periodic function of frequency. Two types of multifrequency arrays will be considered, one which generates Weierstrass fractal radiation patterns and the other which

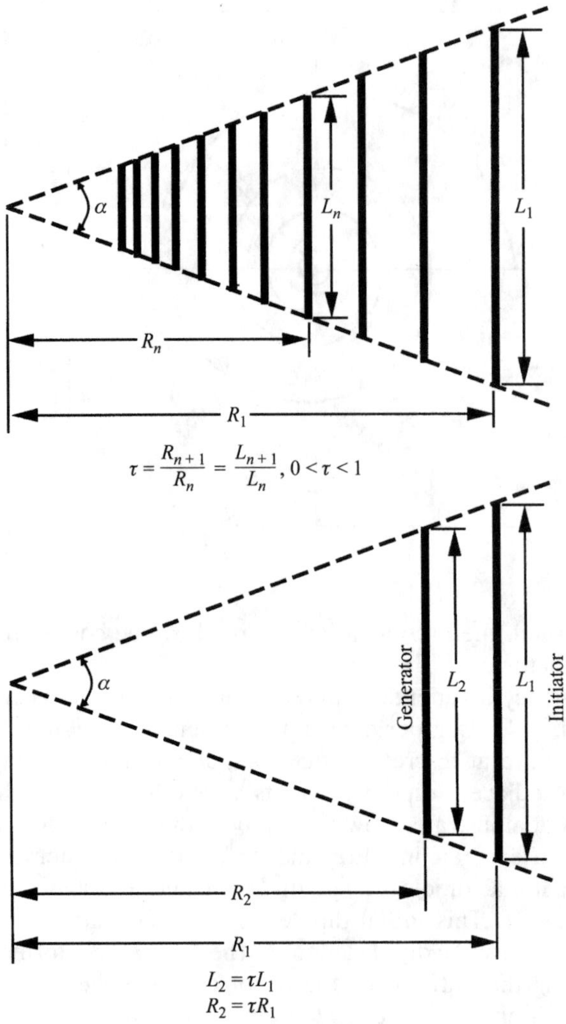

Figure 3.31 Geometric description of a log-periodic dipole antenna in terms of an initiator and generator. [*D. H. Werner and P. L. Werner,* Radio Science, **31**, pp. 1331–1343, November–December 1996. (© *American Geophysical Union.*)]

generates Koch fractal radiation patterns. These arrays are also shown to have power-law current distributions which present some interesting scaling properties as well.

3.5.2 Weierstrass Fractal Array Factors

The ability of Weierstrass arrays to produce fractal radiation patterns which are self-similar at a single frequency has already been demonstrated in Section 3.4.2. In this section, however, we will turn our attention to an investigation of the scaling properties of Weierstrass arrays with frequency. To begin this investigation, we first consider the geometry for a typical Weierstrass array depicted in Fig. 3.32. The locations of any two consecutive array elements with respect to the origin are scaled

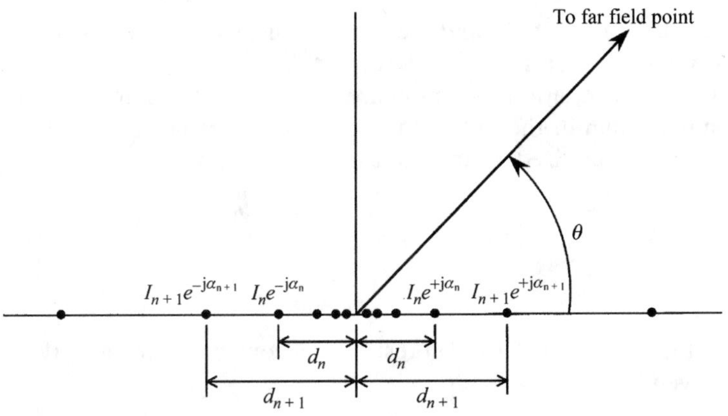

Figure 3.32 Self-scalable Weierstrass fractal array geometry. Array element spacings obey the scaling relation $d_n/d_{n+1} = \tau$. [*D. H. Werner and P. L. Werner,* Radio Science, **31**, pp. 1331–1343, November–December 1996. (© *American Geophysical Union.*)]

by a factor of τ where $\tau = 1/\eta$. In other words, the array spacings satisfy the self-scaling relation

$$d_n/d_{n+1} = \tau \tag{3.127}$$

Suppose we choose the array excitation current magnitudes I_n and element spacings d_n to be

$$I_n = \tau^{(2-D)n} \tag{3.128}$$

$$kd_n = a\tau^{-n} \tag{3.129}$$

where $k = 2\pi/\lambda$ is the wavenumber, λ is the wavelength, a is an arbitrary constant, $0 < \tau < 1$ and $1 < D < 2$. It can be shown that this choice of parameters leads to an expression for the normalized array factor given by

$$f_N(u) = \left[\frac{1-\tau^{(2-D)}}{\tau^{(2-D)N_L} - \tau^{(2-D)(N_U+1)}}\right] \sum_{n=N_L}^{N_U} \tau^{(2-D)n} \cos(a\tau^{-n}u + \alpha_n) \tag{3.130}$$

where $u = \cos\theta$, α_n are the array excitation current phases, and $N = 2(N_U + N_L + 1)$ is the total number of elements in the array. In the limit as the number of array elements approaches infinity, i.e., $N_L \to -\infty$ and $N_U \to \infty$, we define the array factor to be

$$f(u) = \lim_{N\to\infty} f_N(u) \tag{3.131}$$

It is easily verified that the element spacings defined in (3.129) obey the scaling relation (3.127), while the array current magnitudes of (3.128) scale according to

$$I_{n+1}/I_n = \tau^{(2-D)} \tag{3.132}$$

This may be classified as a Weierstrass array because of its unique scaling properties [19], [22]. Hence, the radiation pattern corresponding to this infinite array is fractal, provided the scale factor τ and the fractal dimension D are restricted to lie within the ranges of $0<\tau<1$ and $1<D<2$, respectively.

The array excitation current phases α_n can be chosen in such a way as to produce a beam maximum in the radiation pattern at some specified angle θ_0. The excitation current phases required to accomplish this are

$$\alpha_n = -a\tau^{-n}u_0 \tag{3.133}$$

where

$$u_0 = \cos\theta_0 \tag{3.134}$$

Substituting (3.133) into (3.131) results in a convenient form of the normalized array factor given by

$$f(v) = \lim_{N\to\infty} f_N(v) \tag{3.135}$$

where

$$f_N(v) = \left[\frac{1-\tau^{(2-D)}}{\tau^{(2-D)N_L} - \tau^{(2-D)(N_U+1)}}\right]\sum_{n=N_L}^{N_U} \tau^{(2-D)n}\cos(a\tau^{-n}v) \tag{3.136}$$

and $v = u - u_0$. At this point we recognize that (3.135) satisfies the scaling or self-similarity relation

$$f(\gamma v) = \gamma^{(2-D)} f(v) \tag{3.137}$$

when

$$\gamma = \tau^p \qquad \text{for} \qquad p = 0, \pm 1, \pm 2, \ldots \tag{3.138}$$

This property implies that the radiation pattern produced by this Weierstrass array has a fractal dimension D and is self-similar with a similarity factor τ.

The self-similar radiation patterns produced by Weierstrass arrays suggest that they may be used as multiband arrays which maintain the same radiation characteristics at an infinite number of frequencies. In other words, suppose we consider an array which is phasically sized to fit the requirement (3.129) at some frequency f_0 with corresponding wavelength λ_0. Under these conditions, (3.129) may be written as

$$k_0 d_n = 2\pi s_n = a\tau^{-n} \tag{3.139}$$

where

$$k_0 = 2\pi/\lambda_0 \tag{3.140}$$

$$d_n = s_n\lambda_0 \tag{3.141}$$

If the operating frequency of the array is now changed from f_0 to f, then it follows directly from (3.129) and (3.139) that

$$kd_n = \left(\frac{2\pi}{\lambda}\right)s_n\lambda_0 = (k_0 d_n)\gamma = (a\tau^{-n})\gamma \tag{3.142}$$

where

$$\gamma = \frac{\lambda_0}{\lambda} = \frac{f}{f_0} \tag{3.143}$$

Finally, according to (3.137), (3.138), and (3.143), multiband performance may be achieved by selecting a sequence of discrete frequencies which satisfy the relationship

$$f_p = \tau^p f_0 \qquad \text{where} \qquad p = 0, \pm 1, \pm 2, \ldots \tag{3.144}$$

The directivity for a non-uniform linear Weierstrass array of isotropic sources may be expressed in the form

$$G(u) = \frac{|f(u)|^2}{\frac{1}{2}\int_{-1}^{1} |f(u)|^2 du} \tag{3.145}$$

where $f(u)$ represents the fractal radiation pattern of the Weierstrass array as defined in (3.131). However, if the array excitation current phases are chosen according to (3.133), the directivity expression of (3.145) may be written as

$$G(v) = \frac{|f(v)|^2}{\frac{1}{2}\int_{-1-u_0}^{1-u_0} |f(v)|^2 dv} \tag{3.146}$$

where the expression for the radiation pattern $f(v)$ is given in (3.135). We next recognize that by making use of the multifrequency properties of (3.137), (3.138), and (3.144) in conjunction with (3.146), the extraordinary fact that [22]

$$G(\gamma v) = \frac{|f(\gamma v)|^2}{\frac{1}{2}\int_{-1-u_0}^{1-u_0} |f(\gamma v)|^2 dv} = \frac{|f(v)|^2}{\frac{1}{2}\int_{-1-u_0}^{1-u_0} |f(v)|^2 dv} = G(v) \tag{3.147}$$

may be easily demonstrated. This suggests that the directivity of an infinite Weierstrass array is a log-periodic function of frequency with a log period of τ, that is,

$$G(w) = G(w + pT) \qquad \text{for} \qquad p = 0, \pm 1, \pm 2, \ldots \tag{3.148}$$

where

$$w = \log v \tag{3.149}$$

$$T = \log \tau \tag{3.150}$$

The geometry for the doubly infinite Weierstrass array may be generated by the application of the recursive procedure illustrated in Fig. 3.32 to an infinitely small as well as infinitely large scale. This recursive process may be truncated to yield bandlimited Weierstrass arrays that contain a finite number of elements. An expression for the bandlimited array factor was given in (3.136) and may be obtained by truncating (3.135). Suppose we let $N_L = 0$ and $N_U = (N_0/2) - 1$, where N_0 is the total number of elements in the array; the unnormalized form of (3.136) may then be reduced to

$$f^0_{N_0}(v) = \sum_{n=0}^{N_U} \tau^{(2-D)n} \cos(a\tau^{-n} v) \tag{3.151}$$

Further suppose that $\gamma = \tau^{-p}$, where $p = 0, 1, 2, \ldots$; it can then be shown using (3.151) that

$$f^0_{N_0}(\gamma\nu) = \gamma^{(2-D)} f^0_{N_0}(\nu) - \gamma^{(2-D)} \sum_{n=0}^{p-1} \tau^{(2-D)n} \cos(a\tau^{-n}\nu) + \gamma^{(2-D)} \sum_{n=N_U+1}^{N_U+p} \tau^{(2-D)n} \cos(a\tau^{-n}\nu) \tag{3.152}$$

At this point we recognize that the contribution from the last series in (3.152) may be considered negligible, provided N_0 is sufficiently large. Therefore the expression for $f^0_{N_0}(\gamma\nu)$ may be approximated by

$$f^0_{N_0}(\gamma\nu) \approx \gamma^{(2-D)} f^0_{N_0}(\nu) - \gamma^{(2-D)} \sum_{n=0}^{p-1} \tau^{(2-D)n} \cos(a\tau^{-n}\nu) \tag{3.153}$$

The presence of the additional series in (3.153) suggests that as a consequence of truncation, the array factor is no longer strictly self-similar. However, a more careful examination of (3.153) reveals that this self-similarity property may be restored by simply including one additional pair of elements in the array for each desired frequency band of operation above the initial design frequency f_0 (i.e., for each $f_p = \tau^{-p} f_0$ such that $p > 0$).

This modified form of the array factor may be conveniently represented by introducing the following compact notation:

$$f^p_{N_p}(\nu) = \sum_{n=-P}^{N_U} \varepsilon_{pn} \tau^{(2-D)n} \cos(a\tau^{-n}\nu) \qquad \text{for} \qquad p = 0, 1, \ldots, P \tag{3.154}$$

where

$$\varepsilon_{pn} = \begin{cases} 1, & p+n \geq 0 \\ 0, & p+n < 0 \end{cases} \tag{3.155}$$

P is the maximum number of frequency bands desired above the initial band of $p = 0$, N_0 is the total number of active elements required for the lowest band of $p = 0$, and $N_P = 2P + N_0$ is the total number of active elements required for the highest band of P. The coefficients of the array factor (3.154) may be expressed as

$$I^p_n = \varepsilon_{pn} \tau^{(2-D)n} \tag{3.156}$$

from which the required current amplitude distribution on the array may be determined for each band p. Likewise, an expression may be found from (3.154) which relates the array element spacings to the wavelength associated with the initial (lowest) design band of $p = 0$. This formula is given by

$$d_n = \left(\frac{a}{2\pi}\right) \tau^{-n} \lambda_0 \tag{3.157}$$

The modified array design discussed above is now capable of producing radiation patterns which are roughly self-similar for each frequency band $p = 0, 1, \ldots, P$. This important property may be easily demonstrated by first using the array factor (3.154) to show that

$$f^p_{N_P}(\gamma\nu) = \gamma^{(2-D)} \sum_{n=0}^{N_U+p} \tau^{(2-D)n} \cos(a\tau^{-n}\nu) \tag{3.158}$$

Next, use is made of the fact that (3.158) may be written as

$$f_{N_P}^{p}(\gamma v) = \gamma^{(2-D)} f_{N_0}^{0}(v) + \gamma^{(2-D)} \sum_{n=N_U+1}^{N_U+p} \tau^{(2-D)n} \cos(a\tau^{-n} v) \tag{3.159}$$

in order to arrive at the conclusion

$$f_{N_P}^{p}(\gamma v) \approx \gamma^{(2-D)} f_{N_0}^{0}(v) \tag{3.160}$$

for a sufficiently large value of N_0.

An expression for the directive gain of a bandlimited Weierstrass array of isotropic radiators with $p = 0, 1, \ldots, P$ may be derived using (3.158) together with

$$G_{N_P}^{p}(\gamma v) = \frac{|f_{N_P}^{p}(\gamma v)|^2}{\frac{1}{2}\displaystyle\int_{-1-u_0}^{1-u_0} |f_{N_P}^{p}(\gamma v)|^2 \, dv} \tag{3.161}$$

which yields

$$G_{N_P}^{p}(\gamma v) = \frac{\displaystyle\sum_{m=0}^{N_U+p} \sum_{n=0}^{N_U+p} \tau^{(2-D)(m+n)} \cos(a\tau^{-m} v) \cos(a\tau^{-n} v)}{\frac{1}{2}\displaystyle\sum_{m=0}^{N_U+p} \sum_{n=0}^{N_U+p} \tau^{(2-D)(m+n)} \Lambda_{mn}(a, \tau, u_0)} \tag{3.162}$$

where

$$\begin{aligned}\Lambda_{mn}(a, \tau, u_0) &= \int_{-1-u_0}^{1-u_0} \cos(a\tau^{-m} v) \cos(a\tau^{-n} v) \, dv \\ &= \cos[a(\tau^{-m} + \tau^{-n}) u_0] \operatorname{sinc}[a(\tau^{-m} + \tau^{-n})] \\ &\qquad + \cos[a(\tau^{-m} - \tau^{-n}) u_0] \operatorname{sinc}[a(\tau^{-m} - \tau^{-n})]\end{aligned} \tag{3.163}$$

This result suggests that except for end effects, the directive gain is self-similar for frequency shifts up to order P. In other words,

$$G_{N_P}^{p}(\gamma v) \approx G_{N_0}^{0}(v) \qquad p = 0, 1, \ldots, P \tag{3.164}$$

which also follows directly from (3.160). Finally, an expression for the maximum directive gain as a function of p may be found from (3.162) by setting $v = 0$, which leads to

$$G_{N_P}^{p}(0) = \frac{2[1 - \tau^{(2-D)[(N_0/2)+p]}]^2}{[1 - \tau^{(2-D)}]^2 \displaystyle\sum_{m=0}^{N_U+p} \sum_{n=0}^{N_U+p} \tau^{(2-D)(m+n)} \Lambda_{mn}(a, \tau, u_0)} \tag{3.165}$$

Next, we will demonstrate the design procedure for a bandlimited Weierstrass array by considering a particular example. Suppose we want to design an array which yields the same directivity pattern at three distinct bands (i.e., $p = 0$, $p = 1$, and $p = 2$). If we choose the array design parameters $a = \pi$ and $\tau = 1/2$, then the expression for the array factor (3.154) becomes

$$f_{N_2}^{2}(v) = \sum_{n=-2}^{N_U} \varepsilon_{pn} (1/2)^{(2-D)n} \cos(\pi 2^n v) \tag{3.166}$$

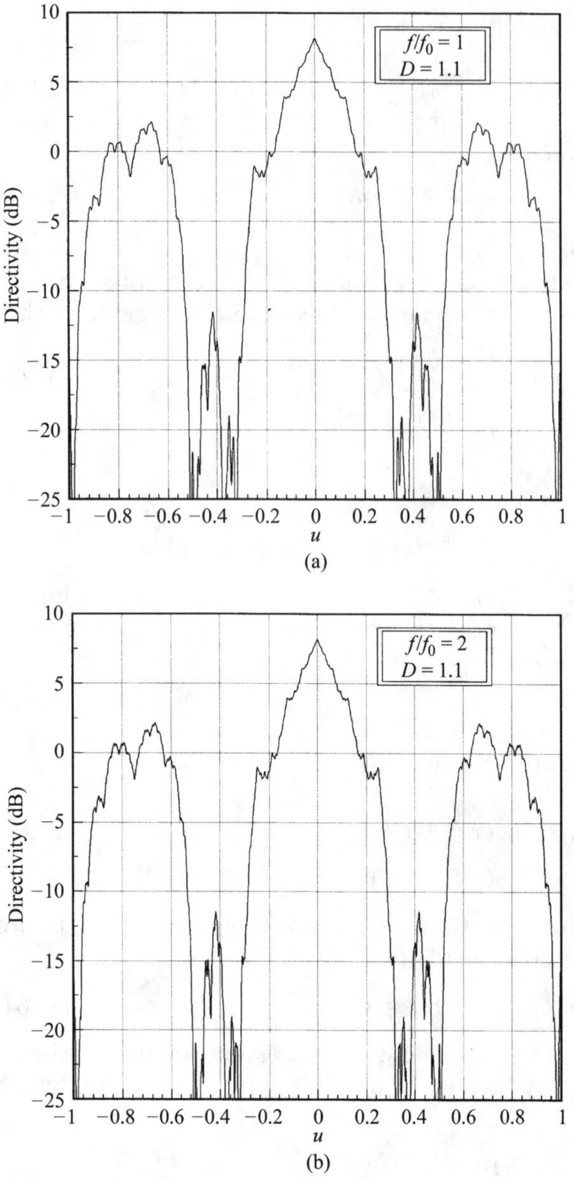

Figure 3.33 Directivity plots for a triband 20-element Weierstrass array where $a = \pi$, $\tau = 1/2$, and $D = 1.1$. (a) Lowband with $f/f_0 = 1$, (b) midband with $f/f_0 = 2$, and (c) highband with $f/f_0 = 4$. [*D. H. Werner and P. L. Werner,* Radio Science, **31**, pp. 1331–1343, November–December 1996. *(© American Geophysical Union.)*]

which is valid for bands $f_p = 2^p f_0$ where $p = 0$, 1, and 2. The element current amplitudes and spacings for this array may be obtained directly from (3.156) and (3.157), respectively, which yield

$$I_n^p = \varepsilon_{pn}(1/2)^{(2-D)n} \tag{3.167}$$

$$d_n = 2^{n-1}\lambda_0 \tag{3.168}$$

The next step in the process is to select the number of array elements to be used in the design. This is accomplished by first choosing the number of array elements desired for the lowest frequency band of operation (N_0) and then adding two more

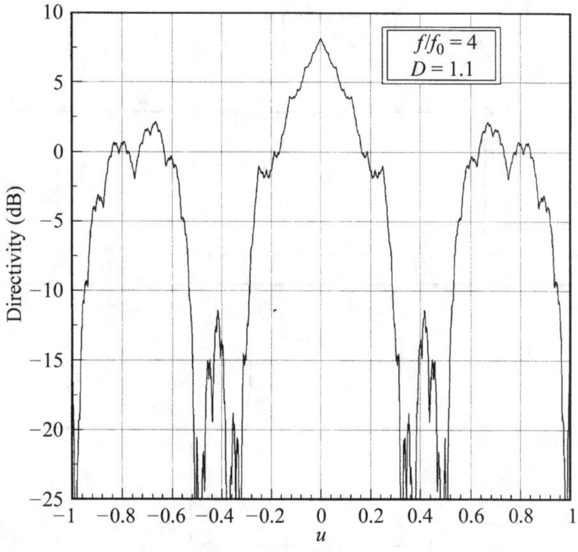

Figure 3.33 (continued)

elements to the array for each additional band desired ($N_P = 2P + N_0$). Hence, for the particular design being considered here, if we choose $N_0 = 16$ and since $P = 2$, then the array would require a total of twenty elements ($N_2 = 20$).

A plot of the directivity corresponding to the lowest design band ($f/f_0 = 1$) for this array is shown in Fig. 3.33a. Figure 3.33b contains a plot of the directivity which would be produced at the midband design frequency ($f/f_0 = 2$). Finally, the directivity pattern which would result from operation at the highest design band ($f/f_0 = 4$) is shown in Fig. 3.33c. A desired fractal dimension of $D = 1.1$ was chosen in order to generate the radiation patterns shown in Fig. 3.33. The multiband property of this array is clearly demonstrated by its ability to produce nearly identical directivity patterns at three different design frequencies.

The array current amplitude distributions required for each of the three bands are given in Table 3.5. For the lowest band of operation ($p = 0$), there are two pairs

TABLE 3.5 Current Amplitude Distributions Required for a Triband 20-Element Weierstrass Array with $\tau = 1/2$ and $D = 1.1$

	f_p/f_0		
I_n^p	1 ($p = 0$)	2 ($p = 1$)	4 ($p = 2$)
I_{-2}^p	0.0000	0.0000	3.4822
I_{-1}^p	0.0000	1.8660	1.8660
I_0^p	1.0000	1.0000	1.0000
I_1^p	0.5359	0.5359	0.5359
I_2^p	0.2872	0.2872	0.2872
I_3^p	0.1539	0.1539	0.1539
I_4^p	0.0825	0.0825	0.0825
I_5^p	0.0442	0.0442	0.0442
I_6^p	0.0237	0.0237	0.0237
I_7^p	0.0127	0.0127	0.0127

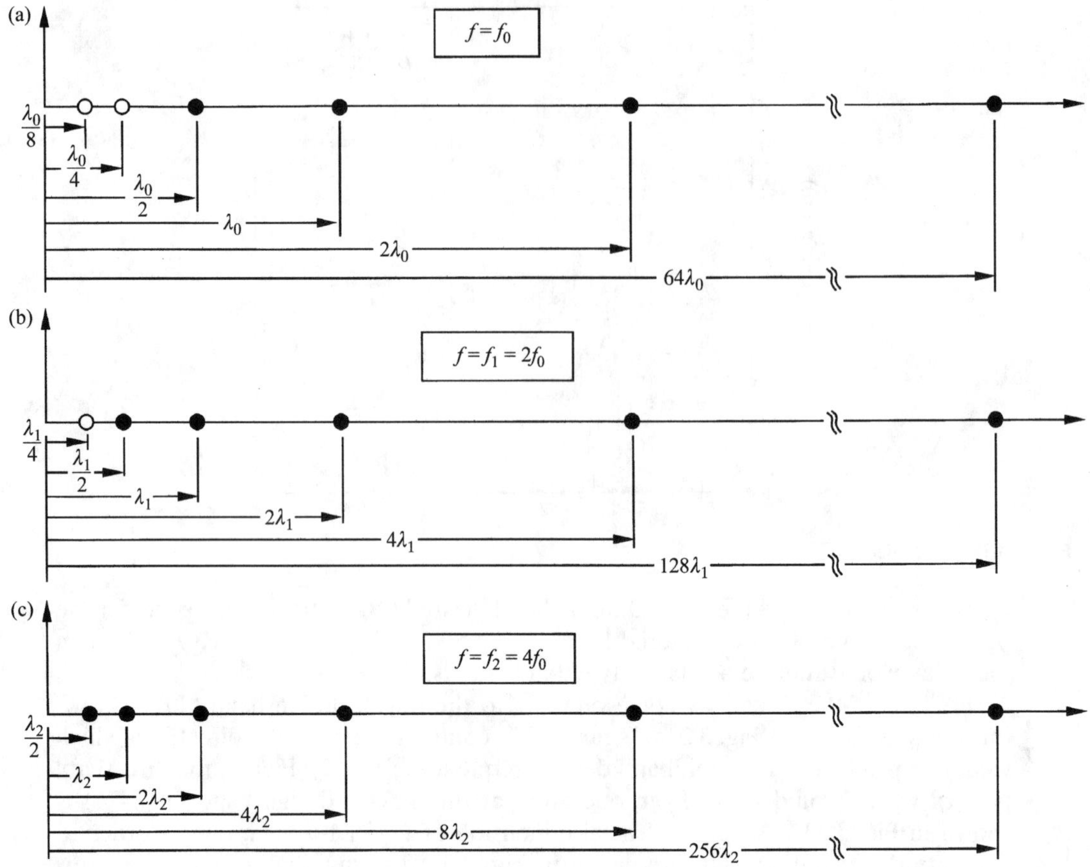

Figure 3.34 Element spacings in terms of wavelengths for a triband 20-element Weierstrass array. (a) Lowband with $f/f_0 = 1$, (b) midband with $f/f_0 = 2$, and (c) highband with $f/f_0 = 4$. Note that due to symmetry, only the geometry for one half of the array is shown for each band. [*D. H. Werner and P. L. Werner,* Radio Science, **31**, pp. 1331–1343, November–December 1996. *(© American Geophysical Union.)*]

of elements in the array which do not contribute to the overall radiation pattern and are therefore open-circuited. Table 3.5 indicates that operation at midband ($p = 1$) can be achieved by maintaining essentially the same current distribution on the array, with the exception of activating (switching on) an additional pair of previously open-circuited elements. A similar procedure is followed when switching to the highest of the three bands ($p = 2$). In this case, all 20 of the array elements are being excited.

The array geometry for the design under consideration is illustrated in Fig. 3.34. The element spacings in terms wavelengths for each of the three bands corresponding to $f = f_0$, $f = 2f_0$, and $f = 4f_0$ are shown in Figs. 3.34a, 3.34b, and 3.34c, respectively. The solid circles indicate the active elements for each band, while the hollow circles identify elements of the array which are open-circuited.

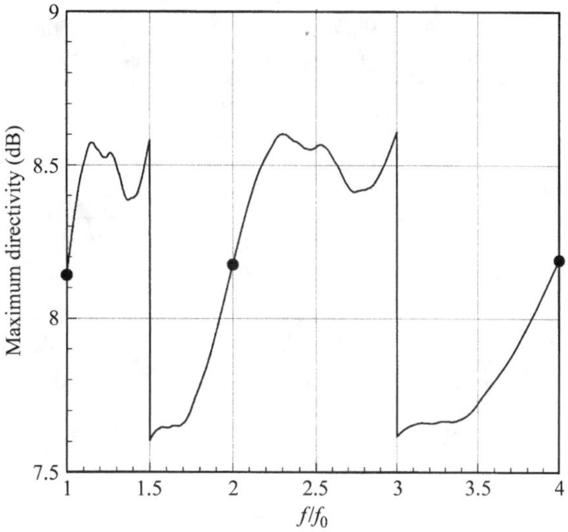

Figure 3.35 Plot of the maximum directivity versus f/f_0 for a triband 20-element Weierstrass array. [*D. H. Werner and P. L. Werner,* Radio Science, **31**, pp. 1331–1343, November–December 1996. *(© American Geophysical Union.)*]

An important issue to be addressed concerns the expected bandwidth around each of the three operating frequencies associated with the array. A plot of the maximum directivity for the array as a function of frequency over the range $1 \leq f/f_0 \leq 4$ is shown in Fig. 3.35. As expected, the value of directivity is the same at the three design frequencies of $f/f_0 = 1$, $f/f_0 = 2$, and $f/f_0 = 4$. There are two small discontinuities in the directivity which are observed as the frequency is increased from f_0: The first one occurs at $f/f_0 = 1.5$, and the second one occurs at $f/f_0 = 3$. These discontinuities are caused by the "switching on" of an additional pair of previously open-circuited array elements. Figure 3.35 demonstrates that there is only a slight variation in the array directivity of at most ± 0.5 dB over the entire frequency range of interest. This demonstrates that successful operation of the array can be achieved, not only at the three original bands specified in the design but also at any frequency which lies in between these bands. This also suggests that some type of frequency-independent antenna, such as a log-periodic, might be a suitable candidate for use as the elemental radiator in the array design.

The array current distributions given in Table 3.5 result in radiation patterns at the three design bands which are approximately the same, as illustrated by Fig. 3.33. However, by introducing a slight complication into the design, it is possible to modify these current distributions in such a way that the radiation patterns at each band will be identical. The current distributions required to accomplish this are listed in Table 3.6. Note that each time a pair of array elements is "switched on," another pair of elements is simultaneously "switched off."

3.5.3 Koch Fractal Array Factors

One of the most appealing characteristics of fractals is the richness and tremendous variety of shapes that can be obtained by applying a few basic design rules (see, for instance, the Hutchinson operators described in the previous chapter). For antenna design purposes, such diversity should translate into a high degree of

TABLE 3.6 Modified Current Amplitude Distributions Required for a Triband 20-Element Weierstrass Array with $\tau = 1/2$ and $D = 1.1$

	f_p/f_0		
I_n^p	1 ($p = 0$)	2 ($p = 1$)	4 ($p = 2$)
I_{-2}^p	0.0000	0.0000	1.0000
I_{-1}^p	0.0000	1.0000	0.5359
I_0^p	1.0000	0.5359	0.2872
I_1^p	0.5359	0.2872	0.1539
I_2^p	0.2872	0.1539	0.0825
I_3^p	0.1539	0.0825	0.0442
I_4^p	0.0825	0.0442	0.0237
I_5^p	0.0442	0.0237	0.0127
I_6^p	0.0237	0.0127	0.0000
I_7^p	0.0127	0.0000	0.0000

freedom when looking for desired antenna performance characteristics. When dealing with the design of antenna arrays, one is often bounded by constraints regarding, for instance, the main-lobe width, the sidelobe ratio, and the null positions. Classical signal processing techniques (window tapering, Fourier synthesis, etc.) are often applied to control some of these pattern parameters, and one might desire to extend such techniques to the fractal synthesis of multifrequency arrays. For instance, a high degree of control on the pattern shape is frequently desired. With this idea in mind, an alternative technique based on an iterative procedure for fractally shaping the array factor is explored in this section. An arbitrary shape window function is taken as the fractal generator, in order to construct array factors with some multifrequency properties.

The patterns designed in this section are based on a family of self-similar curves known as Koch curves [21], [60], [62]. The pattern construction algorithm is quite similar to that of the Koch fractal curves, but they are modified to provide a functional form. The shape and the scaling properties of this class of array factors (hereafter referred to as Koch array factors or simply Koch arrays) are summarized in the example shown in Fig. 3.36.

The main feature of this family of patterns is that each lobe of the curve is a scaled-down copy of the whole pattern. As described previously, the array factor can be interpreted as the Fourier transform of the discrete distribution of elements of the array; in particular, when a uniform spacing d and a phase shift β between elements is assumed, i.e., $d_n = nd$ and $\alpha_n = n\beta$, then a common term Ψ in the argument of the cosine function in (3.30) can be defined as

$$\Psi = kd\cos\theta + \beta \tag{3.169}$$

The argument Ψ is bounded between $-kd + \beta < \Psi < kd + \beta$ which is known as the visible range of the array factor. A change in frequency implies a contraction or expansion of the visible range around the central point β in the Ψ domain. Now, by designing an array with an array factor as shown in Fig. 3.36, and by reducing the visible range around one of the secondary lobes, the resulting visible pattern becomes

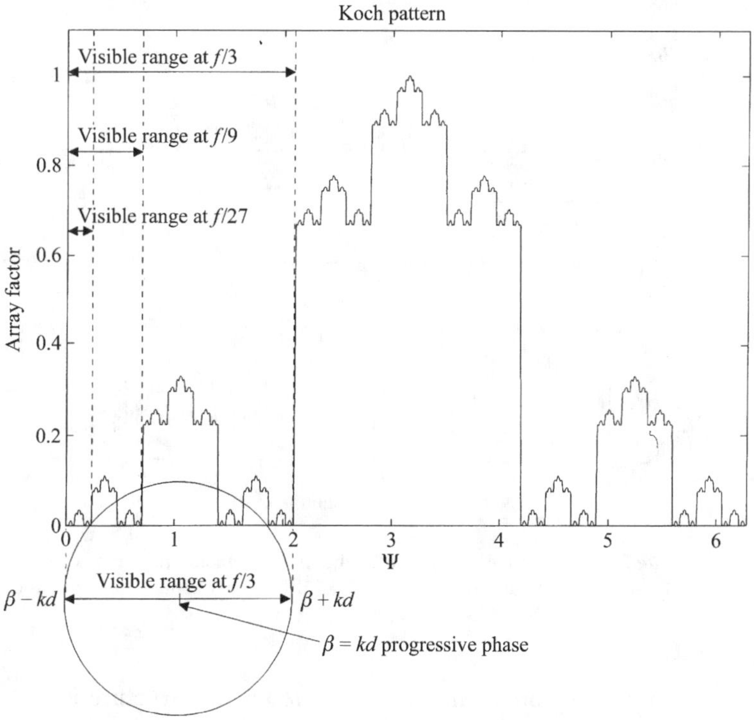

Figure 3.36 The Koch array factor. The curve keeps its similarity properties at six different scales (six iterations, $M = 6$). By adding a progressive phase $\beta = kd$ the visible range is always centered at a secondary lobe that has the same shape as the total pattern. The frequency change by a factor $\tau = 1/3$ reduces the visible range around this similar subpattern.

the same as the original one except for a scale factor. In particular, by choosing the phase shift factor β, as

$$\beta = kd \tag{3.170}$$

the visible range will cover the interval $[0, 2kd]$ at any frequency. For the specific case in Fig. 3.36, a frequency-reduction factor $\tau = 1/3^n$ is required in order to reduce the visible range around a lobe equal to the whole pattern. In other words, the array keeps the same pattern (and hence the same directivity, sidelobe ratio, null positions and so on) at frequencies f, $f/3$, $f/9$, ..., $f/3^n$. In that sense, the same kind of similarity as that stated in (3.137) is obtained. It can be seen that in general, the number of equal bands depends on the number of iterations taken in the pattern construction analogously to the fractal-shaped antennas described in the previous chapter ($M = 6$ iterations were taken in the example of Fig. 3.36 which corresponds to six bands).

Once the fractal array factor has been defined to achieve a certain multifrequency behavior, its relative current distribution among elements is readily derived by applying the Inverse Fourier Transform to the Koch pattern. The result, shown in Fig. 3.37 plotted on a semilogarithmic scale, indicates that the magnitude of the relative current distribution has a power-law (i.e., $1/|z|$) shape, which is a self-scalable (self-affine) function [60].

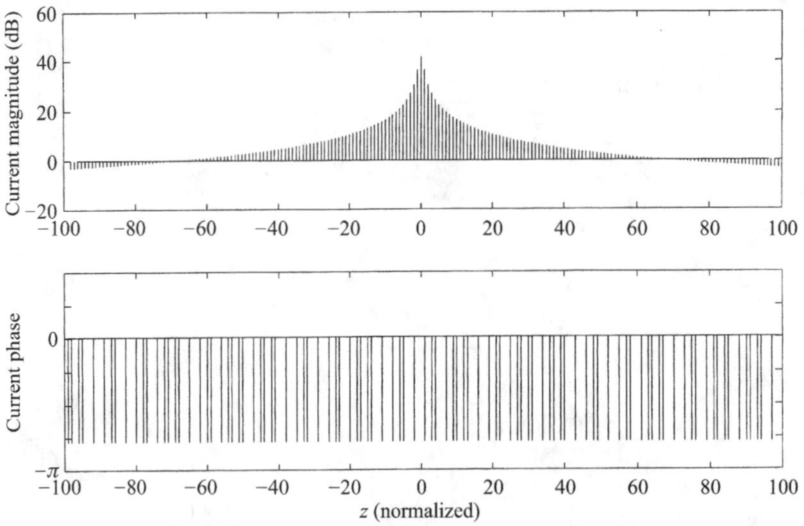

Figure 3.37 Magnitude and phase of the array elements required to produce the Koch pattern shown in Fig. 3.36. The current magnitude in this case has a hyperbolic-like shape.

A relevant issue concerning the computed array structure is that, although only a finite number of elements is required to synthesize the pattern, the array becomes defined by $3^6 = 729$ points (elements). This quantity comes from the design of the Koch curve itself, which requires a minimum of N points to achieve the desired degree of detail. This parameter N is defined to be

$$N = (\text{log-period})^{iterations} = \tau^M \tag{3.171}$$

where the number M of iterations used to construct the curve determines the number of times the curve will look similar under a τ factor scaling transformation. In other words, M is the number of bands or log-periods in which the array will have a similar pattern. This suggests that there is a trade-off between the size of the array and the number of operating bands. Hence, this would appear to be an intrinsic design limitation of these arrays, since the number of elements grows exponentially with the number of log-periodic bands. However, it will be demonstrated in the following sections that this problem can be overcome by application of a suitable truncation technique.

3.5.3.1 Reducing the Number of Elements: Array Truncation

The studies performed on this kind of pattern reveal that, for the vast majority of them, most of the current is concentrated around the central element (see, for instance, Fig. 3.40 and [21], [60]). Also, all relative current distributions have been shown to decay near the ends of the array. Hence, one might easily come to the conclusion that the number of elements could be reduced by merely truncating the array at its ends and retaining only the central part. Figure 3.38 shows the result of

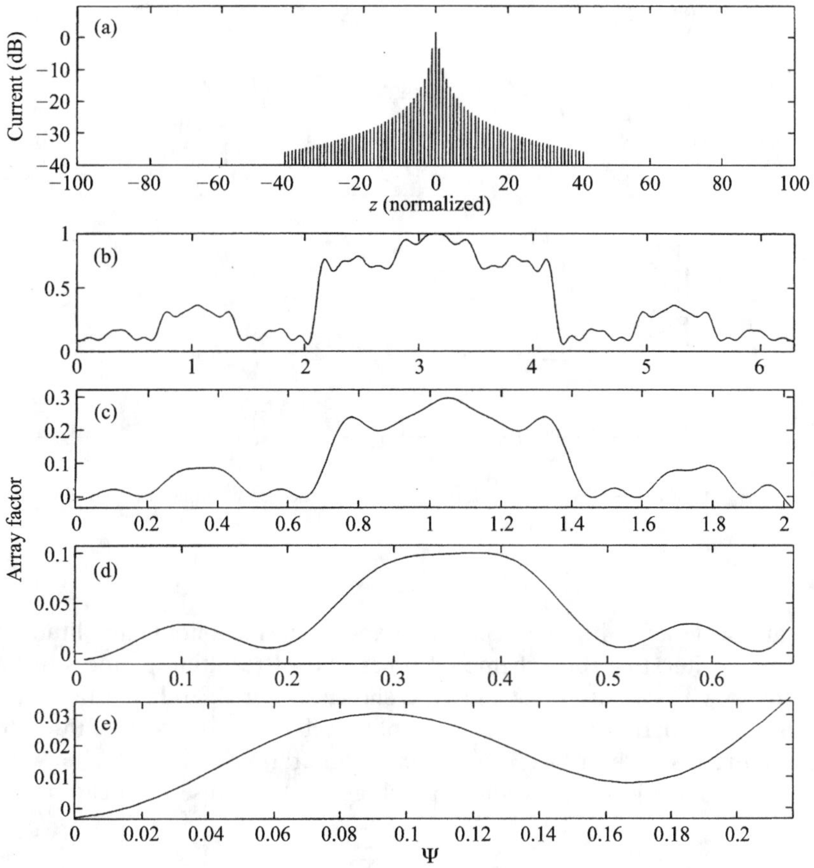

Figure 3.38 Truncation of the Koch array to the central 84 elements (a). The resulting patterns only keep similarity properties up to four bands at wavelengths $\lambda_0 = d/2$ (b); $\lambda_1 = 3\lambda_0$ (c); $\lambda_2 = 9\lambda_0$ (d); and $\lambda_3 = 27\lambda_0$ (e).

truncating the array depicted in Fig. 3.37 from 729 elements to only the first 84 centered around the maximum.

It becomes clear upon examination of Fig. 3.38 that the truncation in the spatial domain has low-pass filtered the patterns reducing their degree of detail. As a result, the pattern looks only similar at four different scales (four different bands), which is understandable if one takes into account (3.171) which implies that the curve for only four bands is defined by $3^4 = 81$ elements. Clearly, a deeper analysis and understanding of the Koch array structure is required in order to develop suitable techniques for reducing the number of required array elements.

3.5.3.2 Koch-Pattern Construction Algorithm

The proper interpretation of the array current distribution derived from fractal patterns requires a detailed understanding of the Koch-pattern construction algorithm (see Fig. 3.39). Suppose we take a periodic pulse train in the spatial-frequency domain

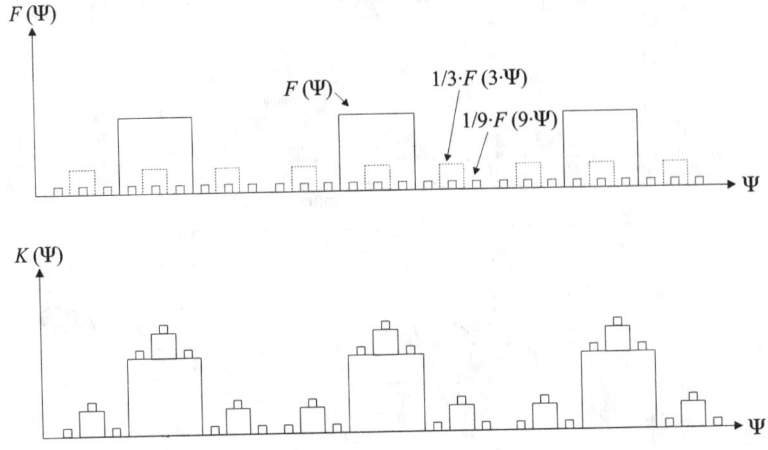

Figure 3.39 The Koch patterns can be constructed by adding several pulse trains at M different scales.

Ψ and scale its width by a factor of τ and its corresponding amplitude by a factor of $\alpha\tau$. After iterating this scheme M times, the M resulting patterns are added, which leads to a Koch pattern of the type shown in Fig. 3.36. In particular, a rectangular pulse, a log-period $\tau = 3$, and an amplitude factor $\alpha = 1$ were chosen for generating the pattern shown in Fig. 3.36 with $M = 6$ iterations.

The analytical expression for each generating pulse train can be written as

$$F(\Psi) * \sum_{n=-\infty}^{\infty} \delta(\Psi - n\Psi_T) \tag{3.172}$$

where $F(\Psi)$ is the single pulse function which, in general, can be taken to have any arbitrary window shape, and Ψ_T is the period of the pulse train. From (3.172), the analytical expression for the Koch pattern $K(\Psi)$ after adding the M scaled pulse trains is

$$K(\Psi) = \sum_{p=0}^{M-1} F(\tau^p \Psi) * \left\{ \frac{1}{\alpha^p \tau^p} \sum_{n=-\infty}^{\infty} \delta\left(\Psi - n\frac{\Psi_T}{\tau^p} \right) \right\} \tag{3.173}$$

Again, for the Koch pattern in Fig. 3.36, the reduction factor is chosen to be $\alpha = 1$, i.e., no extra reduction or magnification is applied. The resulting array patterns for several combinations of α, τ, and M are shown in Fig. 3.40, where a rectangular generating pulse is chosen for the first three examples, and a Blackman window for the last one. Once an expression for the Koch pattern has been derived, a corresponding expression for the Koch array element distribution $k(z)$ can be easily found by taking the Inverse Fourier Transform of (3.173):

$$k(z) = \frac{1}{\Psi_T} \sum_{p=0}^{M-1} \frac{1}{\tau^p} f\left(\frac{z}{\tau^p} \right) \left\{ \frac{1}{\alpha^p} \sum_{n=-\infty}^{\infty} \delta(z - nd\tau^p) \right\} \tag{3.174}$$

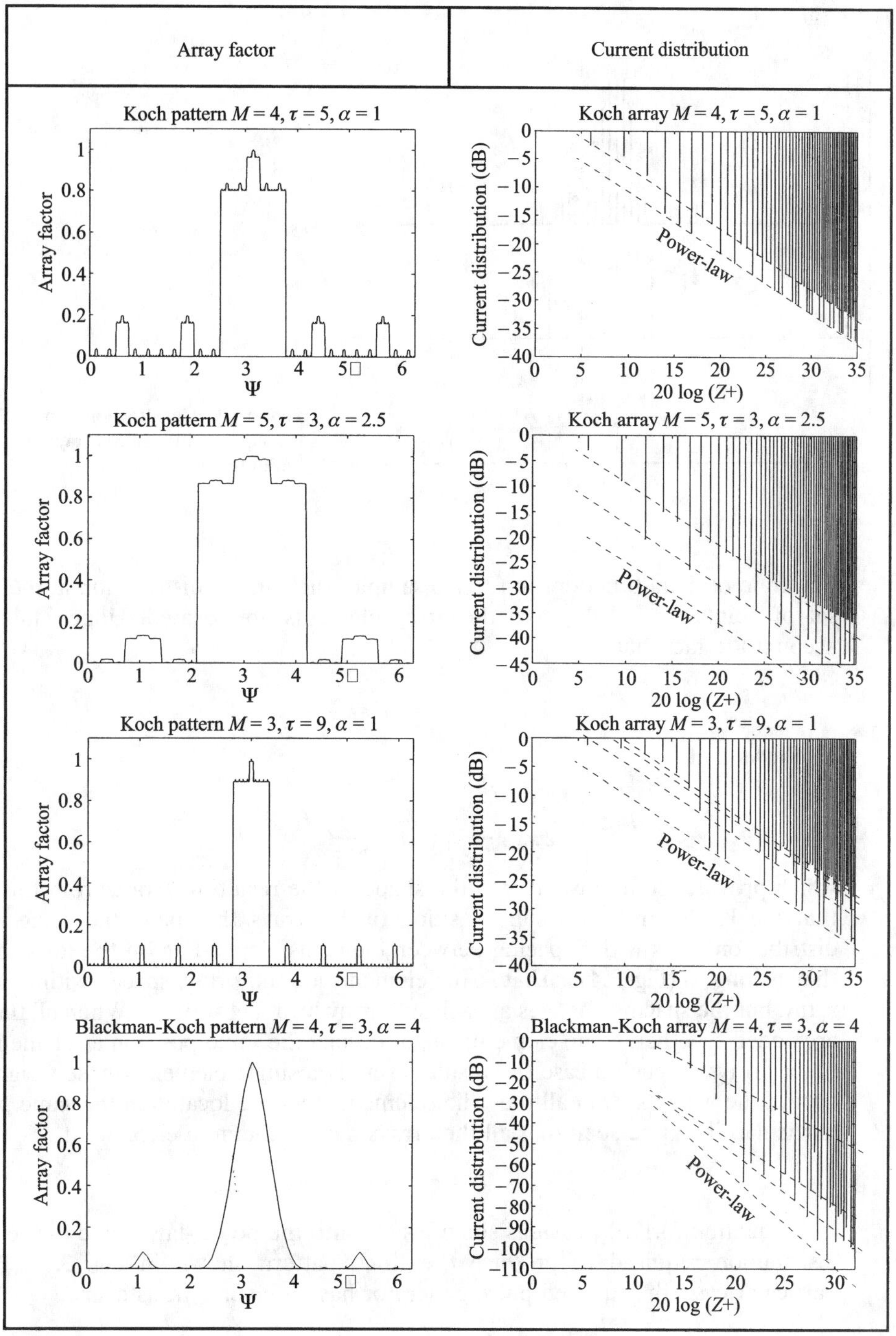

Figure 3.40 Koch patterns resulting from arrays constructed by interleaving hyperbolic distributions.

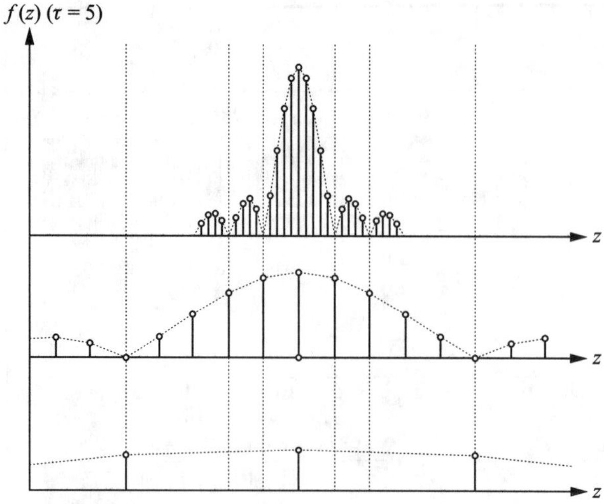

Figure 3.41 Construction of a Koch array as the superposition of M arrays of equal shape but different scales.

The train of delta functions in (3.174) samples the current distribution at the discrete set of points $z = nd\tau^p$ where the array elements are located. Hence, taking into account the fact that

$$\Psi_T = kd = \frac{2\pi d}{\lambda} \tag{3.175}$$

it follows that

$$k(z) = \frac{\lambda}{2\pi d} \sum_{p=0}^{M-1} \left\{ \left(\frac{1}{\alpha\tau} \right)^p \sum_{n=-\infty}^{\infty} f(nd)\, \delta(z - nd\tau^p) \right\} \tag{3.176}$$

which provides some insight into the shape of the resulting array. From this we see that the Koch array is a superposition of M arrays that have the same element distribution but a wider spacing between elements, depending on the iteration stage they belong to (Fig. 3.41). That is, the elements are uniformly spaced within the same array, but the spacing changes at each sub-array by a factor of τ^p. When all the arrays are added together, some elements might occupy the same position as elements from other arrays. In such a case the result is simply a single element whose weight is the sum of the weights from all the other elements that are located in the same place. In particular, it can be seen that all the arrays have a common element at

$$z = nd\tau^{M-1} \tag{3.177}$$

Equation (3.176) provides some insight into the power-law shape of the current distribution required to produce the Koch pattern shown in Fig. 3.36. For this particular case, the squared pulse generator has an inverse transform

$$f(z) = \frac{\sin\left(\dfrac{\pi}{d\tau} z \right)}{\pi z} \tag{3.178}$$

which corresponds to the shape of all the M sub-arrays that make up the Koch array.

The weight of each element can be easily found by sampling (3.178) at $z = md$. It is readily seen that for the example under study ($\tau = 3$) the magnitude of (3.178) reduces to

$$|f(md)| = \left| \frac{\sin\left(\frac{\pi}{3}m\right)}{m\pi d} \right| = \begin{cases} \left|\frac{\sqrt{3}}{2\pi d}\right| \cdot \frac{1}{|m|}, & m \neq 3^q \\ 0, & m = 3^q \end{cases} \tag{3.179}$$

which is, in absolute value, a power-law (hyperbolic) function of the index element m. Another important property associated with (3.179) is the fact that nulls will occur for values of m such that

$$m = \tau^q \tag{3.180}$$

where q is an integer. From this property, we are able to conclude that all of the sub-arrays contribute to the weight of the central element, but that they do not overlap at any other point since the nulls of each array are filled by an element corresponding to the sub-array in the next iteration. Therefore, the global array configuration obtained after M iterations is actually formed by interleaving the elements of M equal sub-arrays at M different scales. The result for the particular case $\tau = 3$ and the squared pulse generator is an equally spaced array with a hyperbolic distribution of the element-current magnitudes.

Two significant conclusions can be reached from the analysis of the generalized Koch array $k(z)$ described above. First, the shape of the M-superimposed sub-arrays depends on the shape of the pulse generator. Second, the superposition of the sub-arrays might result in the confluence of many elements in a single location or might result in an interleaving of the elements. As will be demonstrated in the following subsection, both facts contribute to the reduction of the number of elements in the Koch array.

3.5.3.3 The Blackman-Koch Array Factor

Since the array current distribution is basically a superposition of the inverse transforms of the pulse generator, one should reasonably choose a pulse generator (window) with low side-lobes in the transformed domain to allow a better truncation of the Koch array factors. One possible candidate for this pulse generator is the Blackman window, which is characterized as having a low side-lobe spectrum [61]. Therefore, it appears to be a suitable candidate for replacing the rectangular window in the Koch pattern generation. The result of applying such a technique is shown in Fig. 3.42 through Fig. 3.44.

Figure 3.42 demonstrates that the Blackman-Koch pattern results in a much smoother shape which maintains the same similarity properties as the Koch array factor of Fig. 3.36. The main advantage of this pattern is that the array relative current distribution has lower side-lobes and a better confinement around the central elements (Fig. 3.43a). This way, the array simplification becomes much easier. The logarithmic plot of the current distribution reveals some important isolated current peaks well beyond the center of the array. One should expect a critical contribution of these isolated elements to the global-pattern characteristics. Thus, instead of just truncating the ends of the array, a threshold level can be set to discern which elements

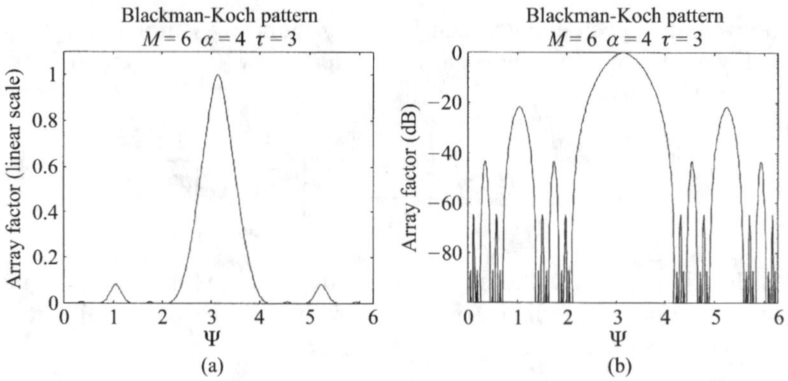

Figure 3.42 The Blackman–Koch pattern. The main construction parameters are $M = 6$, $\tau = 3$, and $\alpha = 4$. The reduction factor $\alpha = 4$ has been chosen here to improve the sidelobe ratio (SLR) with respect to the previous case shown in Fig. 3.36.

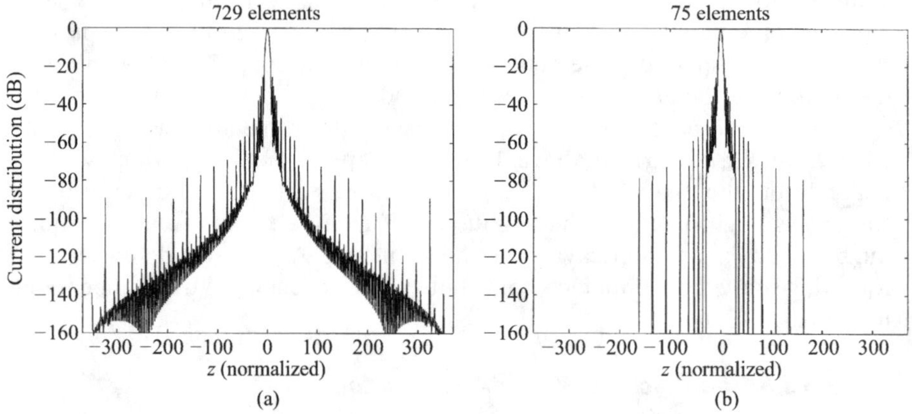

Figure 3.43 The Blackman–Koch array: current distribution, logarithmic scale (a). Current distribution after truncation to 75 elements (b).

are important in the pattern synthesis and which are not. The result is the simplified array shown in Fig. 3.43b. It can be seen (Fig. 3.44) that, although the array has been simplified to only 75 elements (as opposed to 729), it still keeps its self-similar behavior at five bands through a whole 81:1 frequency range, with a lower degree of pattern distortion with respect to the rectangular pulse case.

The resulting array is no longer a uniformly spaced array since the main 75 current elements are not placed together near the midpoint of the array. Thus, some elements are placed further from the origin than in the truncation scheme which means that faster variations will appear in the dual domain (the pattern domain). This explains why this method better keeps the multiband behavior for a larger number of bands than does the method based on the truncation scheme. Hence, the further the elements are placed from the origin, the finer will be the resulting lobe structure, which allows the pattern to maintain the same overall shape even with a further reduction of the visible range.

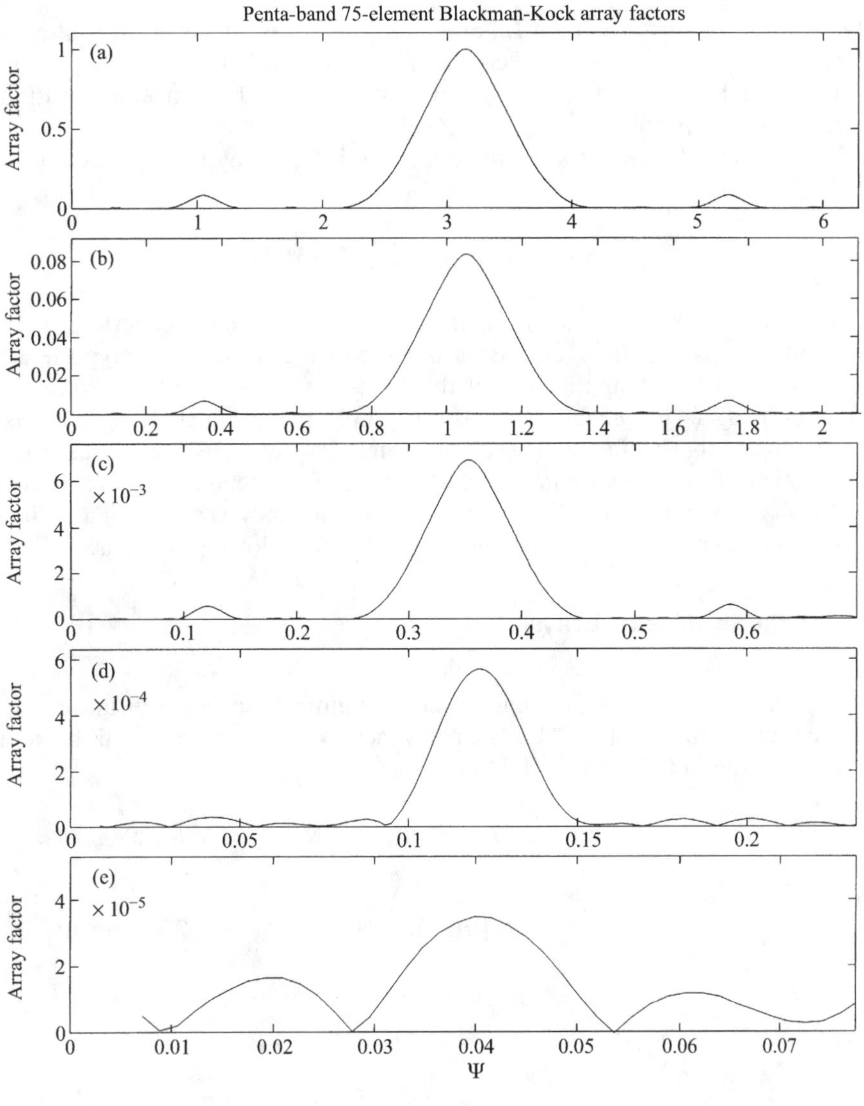

Figure 3.44 Resulting Blackman-Koch patterns after truncation to only 75 elements. The array factor is plotted for $\lambda_0 = d/2$ (a); $\lambda_1 = 3\lambda_0$ (b); $\lambda_2 = 9\lambda_0$ (c); $\lambda_3 = 27\lambda_0$ (d); and $\lambda_4 = 81\lambda_0$ (e). Notice the low degree of pattern distortion (five bands) as opposed to the patterns shown in Fig. 3.37 (four bands).

3.6 DETERMINISTIC FRACTAL ARRAYS

A rich class of fractal arrays exist which can be formed recursively through the repetitive application of a generating sub-array. A generating sub-array is a small array at scale one ($P = 1$) used to construct larger arrays at higher scales (i.e., $P > 1$). In many cases, the generating subarray has elements that are turned on and off in a certain pattern. A set formula for copying, scaling, and translating of the generating sub-array is then followed in order to produce the fractal array. Hence, fractal arrays which are created in this manner will be composed of a sequence of self-similar sub-arrays. In other words, they may be conveniently thought of as arrays of arrays [6].

The array factor for a fractal array of this type may be expressed in the general form [23–25]

$$AF_P(\psi) = \prod_{p=1}^{P} GA(\delta^{p-1}\psi) \tag{3.181}$$

where $GA(\psi)$ represents the array factor associated with the generating subarray. The parameter δ is a scale or expansion factor which governs how large the array grows with each recursive application of the generating sub-array. The expression for the fractal array factor given in (3.181) is simply the product of scaled versions of a generating subarray factor. Therefore, we may regard (3.181) as representing a formal statement of the *pattern multiplication theorem* for fractal arrays. Applications of this specialized pattern multiplication theorem to the analysis and design of linear as well as planar fractal arrays will be considered in the following sections.

3.6.1 Cantor Linear Arrays

A linear array of isotropic elements uniformly spaced a distance d apart along the z axis is shown in Fig. 3.45. The array factor corresponding to this linear array may be expressed in the form [44], [46]

$$AF(\psi) = \begin{cases} I_0 + 2\sum_{n=1}^{N} I_n \cos[n\psi], & \text{for} \quad 2N+1 \text{ elements} \\ 2\sum_{n=1}^{N} I_n \cos[(n-1/2)\psi], & \text{for} \quad 2N \text{ elements} \end{cases} \tag{3.182}$$

where

$$\psi = kd[\cos\theta - \cos\theta_0] \tag{3.183}$$

$$k = 2\pi/\lambda \tag{3.184}$$

These arrays become fractal-like when appropriate elements are turned off or removed such that

$$I_n = \begin{cases} 1, & \text{if element } n \text{ is turned on} \\ 0, & \text{if element } n \text{ is turned off} \end{cases} \tag{3.185}$$

Hence, fractal arrays produced by following this procedure belong to a special category of thinned arrays.

One of the simplest schemes for constructing a fractal linear array follows the recipe for the Cantor set [62]. Cantor linear arrays were first proposed and studied in [21] for their potential use in the design of low sidelobe arrays. Some other aspects of Cantor arrays have been investigated more recently in [23–25].

The basic triadic Cantor array may be created by starting with a three-element generating sub-array, and then applying it repeatedly over P scales of growth. The

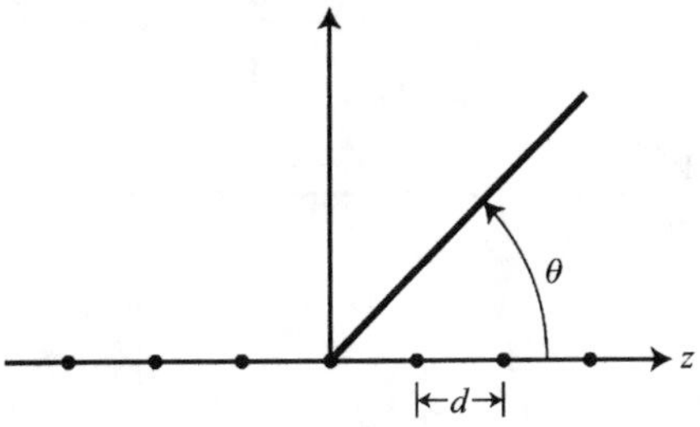

Figure 3.45 Geometry for a linear array of uniformly spaced isotropic sources.

generating sub-array in this case has three uniformly spaced elements with the center element turned off or removed, i.e., 101. The triadic Cantor array is generated recursively by replacing 1 by 101 and 0 by 000 at each stage of the construction. For example, at the second stage of construction ($P = 2$) the array pattern would look like

1 0 1 0 0 0 1 0 1

and at the third stage ($P = 3$) we would have

1 0 1 0 0 0 1 0 1 0 0 0 0 0 0 0 0 0 1 0 1 0 0 0 1 0 1

The array factor of the three-element generating sub-array with the representation 101 is

$$GA(\psi) = 2\cos(\psi) \tag{3.186}$$

which may be derived from (3.182) by setting $N = 1$, $I_0 = 0$ and $I_1 = 1$. Substituting (3.186) into (3.181) and choosing an expansion factor of three (i.e., $\delta = 3$) results in an expression for the Cantor array factor given by

$$\widehat{AF}_P(\psi) = \prod_{p=1}^{P} \widehat{GA}(3^{p-1}\psi) = \prod_{p=1}^{P} \cos(3^{p-1}\psi) \tag{3.187}$$

where the *hat* notation indicates that the quantities have been normalized. Fig. 3.46 contains plots of (3.187) for the first four stages in the growth of a Cantor array.

Suppose that the spacing between array elements is a quarter-wavelength (i.e., $d = \lambda/4$) and that $\theta_0 = 90°$, then an expression for the directivity of a Cantor array of isotropic point sources may be derived from

$$D_P(u) = \frac{\widehat{AF}_P^2(u)}{\frac{1}{2}\int_{-1}^{1} \widehat{AF}_P^2(u)\,du} \tag{3.188}$$

where $u = \cos\theta$ and

$$\widehat{AF}_P(u) = \prod_{p=1}^{P} \cos\left(\frac{3^{p-1}}{2}\pi u\right) \tag{3.189}$$

Substituting (3.189) into (3.188) and using the fact that

$$\frac{1}{2}\int_{-1}^{1} \prod_{p=1}^{P} [1 + \cos(3^{p-1}\pi u)]\,du = 1 \tag{3.190}$$

Scale	Array factor	Array factor plot
1	$\cos(\psi)$	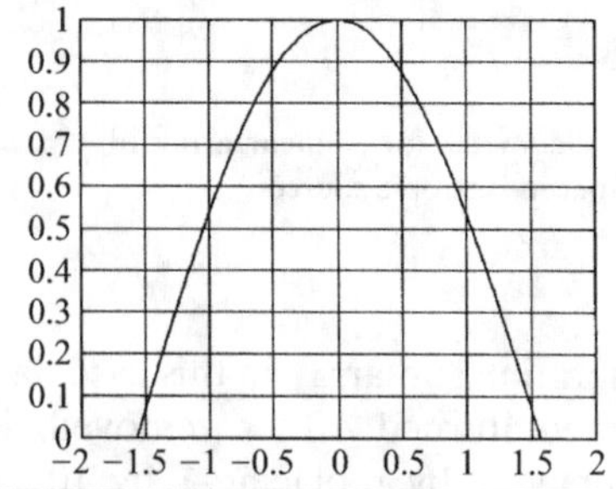
2	$\cos(\psi)\cos(3\psi)$	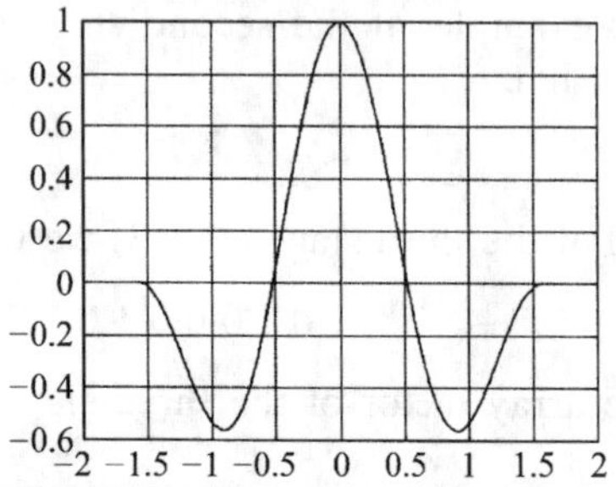
3	$\cos(\psi)\cos(3\psi)\cos(9\psi)$	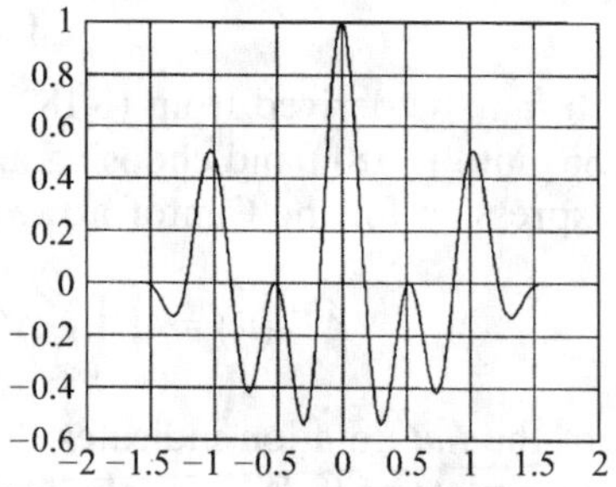
4	$\cos(\psi)\cos(3\psi)\cos(9\psi)\cos(27\psi)$	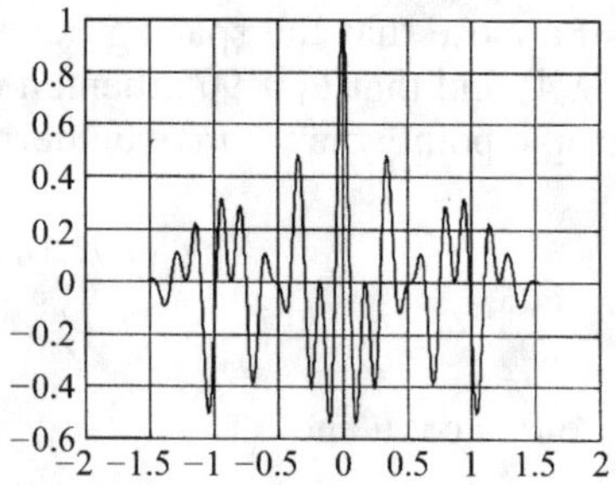

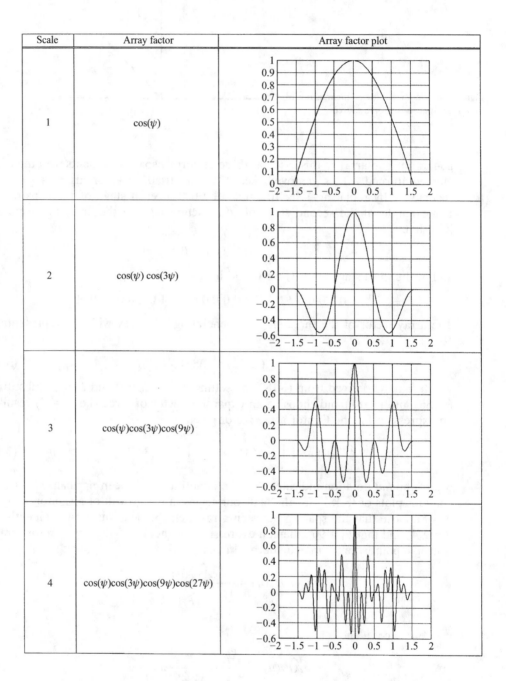

Figure 3.46 Plots of the triadic Cantor fractal array factor for the first four stages of growth. (a) $P = 1$, (b) $P = 2$, (c) $P = 3$, and (d) $P = 4$.

leads to the following convenient representation for the directivity

$$D_P(u) = \prod_{p=1}^{P} [1 + \cos(3^{p-1}\pi u)] = 2^P \prod_{p=1}^{P} \cos^2\left(\frac{3^{p-1}}{2}\pi u\right) \tag{3.191}$$

Finally, it is easily demonstrated from (3.191) that the maximum value of directivity for the Cantor array is

$$D_P = D_P(0) = 2^P \qquad \text{where} \qquad P = 1, 2, \ldots \tag{3.192}$$

or

$$D_P(\text{dB}) = 3.01P \qquad \text{where} \qquad P = 1, 2, \ldots \tag{3.193}$$

Locations of nulls in the radiation pattern are easy to compute from the product form of the array factor (3.181). For instance, at a given scale P, the nulls in the radiation pattern of (3.189) occur when

$$\cos\left(\frac{3^{P-1}}{2}\pi u\right) = 0. \tag{3.194}$$

Solving (3.194) for u yields

$$u_k^P = \pm(2k-1)(1/3)^{P-1} \qquad \text{where} \qquad k = 1, 2, \ldots, (3^{P-1}+1)/2 \tag{3.195}$$

Hence, from this we may easily conclude that the radiation patterns produced by triadic Cantor arrays will have a total of $3^{P-1}+1$ nulls.

The generating sub-array for the triadic Cantor array discussed above is actually a special case of a more general family of uniform Cantor arrays. The generating sub-array factor for this general class of uniform Cantor arrays may be expressed in the form

$$\widehat{GA}(\psi) = \left(\frac{2}{\delta+1}\right)\frac{\sin\left[\left(\frac{\delta+1}{2}\right)\psi\right]}{\sin[\psi]} \tag{3.196}$$

where

$$\delta = 2n+1 \qquad \text{and} \qquad n = 1, 2, \ldots \tag{3.197}$$

Hence, by substituting (3.196) into (3.181), it follows that these uniform Cantor arrays have fractal array factor representations given by [21], [23–25]

$$\widehat{AF}_P(\psi) = \left(\frac{2}{\delta+1}\right)^P \prod_{p=1}^{P} \frac{\sin\left[\left(\frac{\delta+1}{2}\right)\delta^{p-1}\psi\right]}{\sin[\delta^{p-1}\psi]} \tag{3.198}$$

For $n = 1$ ($\delta = 3$) the generating sub-array has a pattern 101 such that (3.198) reduces to the result for the standard triadic Cantor array found in (3.187). The generating sub-array pattern for the next case in which $n = 2$ ($\delta = 5$) is 10101 and, likewise, when

$n = 3$ ($\delta = 7$) the array pattern is 1010101. The fractal dimension D of these uniform Cantor arrays can be calculated as [21]

$$D = \frac{\log\left(\frac{\delta+1}{2}\right)}{\log(\delta)} \tag{3.199}$$

which suggests that $D = 0.6309$ for $n = 1$, $D = 0.6826$ for $n = 2$, and $D = 0.7124$ for $n = 3$.

As before, if it is assumed that $d = \lambda/4$ and $\theta_0 = 90°$, then the directivity for these uniform Cantor arrays may be expressed as [23–25]

$$D_P(u) = \left(\frac{2}{\delta+1}\right)^P \prod_{p=1}^{P} \frac{\sin^2\left[\frac{\pi}{4}(\delta+1)\,\delta^{p-1}u\right]}{\sin^2\left[\frac{\pi}{2}\delta^{p-1}u\right]} \tag{3.200}$$

where use has been made of the fact that

$$\frac{1}{2}\int_{-1}^{1}\prod_{p=1}^{P} \frac{\sin^2\left[\frac{\pi}{4}(\delta+1)\,\delta^{p-1}u\right]}{\sin^2\left[\frac{\pi}{2}\delta^{p-1}u\right]}\,du = \left(\frac{\delta+1}{2}\right)^P \tag{3.201}$$

The corresponding expression for maximum directivity is

$$D_P = D_P(0) = \left(\frac{\delta+1}{2}\right)^P \quad \text{where} \quad P = 1, 2, \ldots \tag{3.202}$$

or

$$D_P(\text{dB}) = 10P\log\left(\frac{\delta+1}{2}\right) \quad \text{where} \quad P = 1, 2, \ldots \tag{3.203}$$

A plot of the directivity for a uniform Cantor array with $d = \lambda/4$, $\theta_0 = 90°$, $\delta = 7$, and $P = 1$ is shown in Fig. 3.47. Figures 3.48 and 3.49 show the directivity plots which correspond to stages of growth for this array of $P = 2$ and $P = 3$, respectively.

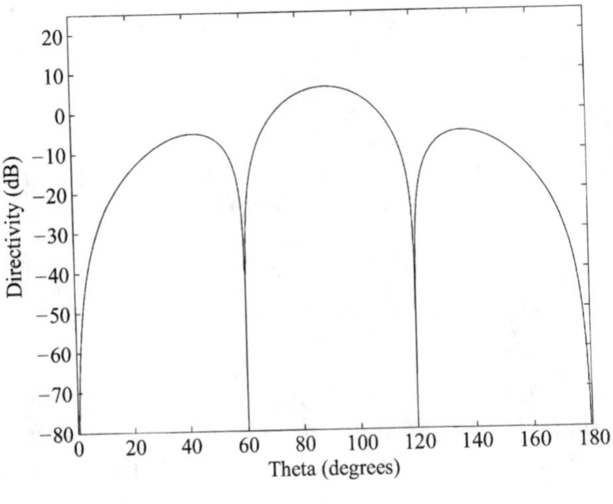

Figure 3.47 Directivity plot for a uniform Cantor fractal array with $P = 1$ and $\delta = 7$. The spacing between elements of the array is $d = \lambda/4$. The maximum directivity for stage 1 is $D_1 = 6.02$ dB.

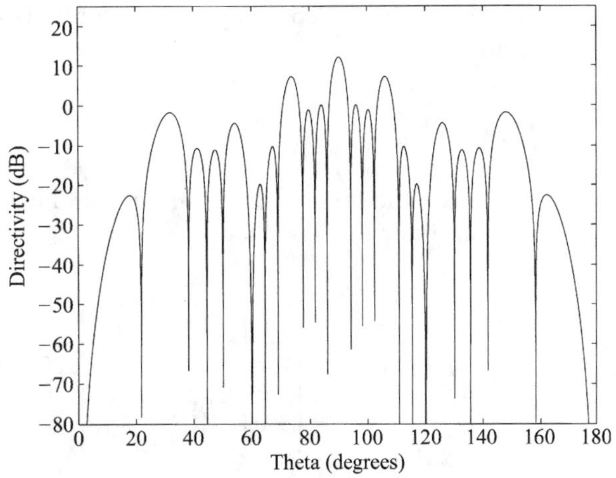

Figure 3.48 Directivity plot for a uniform Cantor fractal array with $P = 2$ and $\delta = 7$. The spacing between elements of the array is $d = \lambda/4$. The maximum directivity for stage 2 is $D_2 = 12.04$ dB.

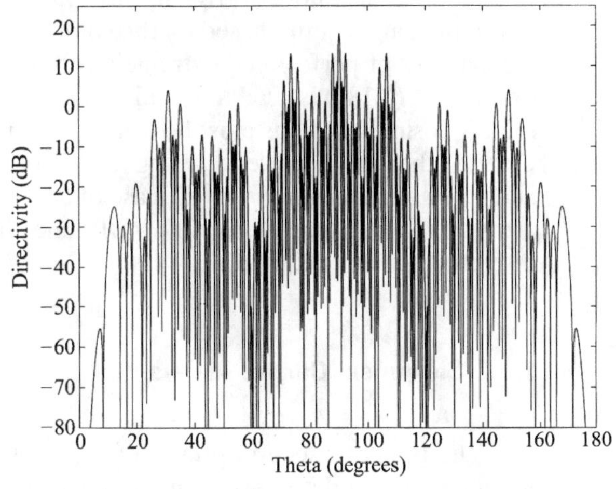

Figure 3.49 Directivity plot for a uniform Cantor fractal array with $P = 3$ and $\delta = 7$. The spacing between elements of the array is $d = \lambda/4$. The maximum directivity for stage 3 is $D_3 = 18.06$ dB.

The multiband characteristics of an infinite fractal array may be demonstrated by following a similar procedure to that outlined in [21] and [22]. In other words, suppose that we consider the array factor for a doubly infinite linear fractal (uniform Cantor) array which may be defined as

$$AF(\psi) = \prod_{p=-\infty}^{\infty} GA(\delta^{p-1}\psi) \tag{3.204}$$

Next, by making use of (3.204), we find that

$$AF(\delta^{\pm q}\psi) = \prod_{p=-\infty}^{\infty} GA(\delta^{p\pm q-1}\psi) = \prod_{n=-\infty}^{\infty} GA(\delta^{n-1}\psi) = AF(\psi) \tag{3.205}$$

which is valid provided q is a positive integer (i.e., $q = 0, 1, 2, \ldots$). The parameter $\delta^{\pm q}$

introduced in (3.205) may be interpreted as a frequency shift which obeys the relation

$$f_{\pm q} = \delta^{\pm q} f_0 \qquad \text{for} \qquad q = 0, 1, \ldots \tag{3.206}$$

where f_0 is the original design frequency. Finally, for a finite size array, (3.181) may be used in order to show that

$$AF_P(\delta^q \psi) = \left[\frac{\prod_{l=P+1}^{P+q} GA(\delta^{l-1}\psi)}{\prod_{l=1}^{q} GA(\delta^{l-1}\psi)} \right] AF_P(\psi) \qquad \text{for} \qquad q = 1, 2, \ldots \tag{3.207}$$

where the bracketed term in (3.207) clearly represents the "end-effects" introduced by truncation.

One of the more intriguing attributes of fractal arrays is the possibility for developing algorithms, based on the compact product representation of (3.181), which are capable of performing extremely rapid radiation pattern computations [23]. For example, if (3.182) is used to calculate the array factor for an odd number of elements, then N cosine functions must be evaluated and N additions performed for each angle. On the other hand, however, using (3.187) only requires P cosine function evaluations and $P - 1$ multiplications. In the case of an 81-element triadic Cantor array, the fractal array factor is at least $N/P = 40/4 = 10$ times faster to calculate than the conventional discrete Fourier transform.

3.6.2 Sierpiński Carpet Arrays

The previous section presented an application of fractal geometric concepts to the analysis and design of thinned linear arrays. In this section, these techniques are extended to include the more general case of fractal planar arrays. A symmetric planar array of isotropic sources with elements uniformly spaced a distance d_x and d_y apart in the x and y directions, respectively, is shown in Fig. 3.50. It is well known that the array factor for this type of planar array configuration may be expressed in the following way [46]:

$$AF(\psi_x, \psi_y) = \begin{cases} I_{11} + 2\sum_{m=2}^{M} \{I_{m1}\cos[m\psi_x] + I_{1m}\cos[m\psi_y]\} \\ \quad + 4\sum_{n=2}^{M}\sum_{m=2}^{M} I_{mn}\cos[m\psi_x]\cos[n\psi_y], & \text{for } (2M-1)^2 \text{ elements} \\ 4\sum_{n=1}^{M}\sum_{m=1}^{M} I_{mn}\cos[(m-1/2)\psi_x]\cos[(n-1/2)\psi_y], & \text{for } (2M)^2 \text{ elements} \end{cases} \tag{3.208}$$

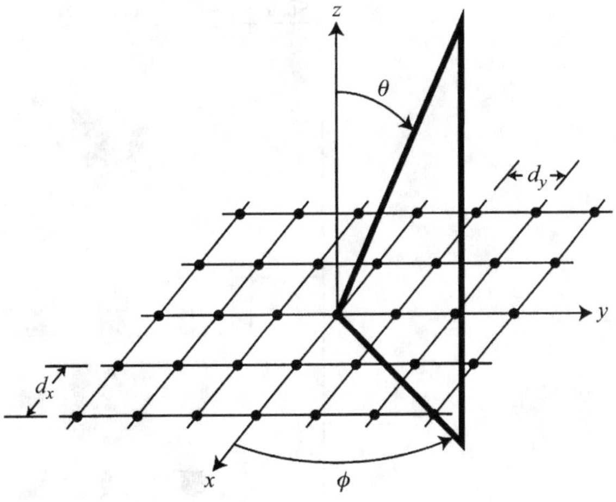

Figure 3.50 Geometry for a symmetric planar array of isotropic sources with elements uniformly spaced a distance d_x and d_y apart in the x and y directions, respectively.

where

$$\psi_x = kd_x[\sin\theta\cos\phi - \sin\theta_0\cos\phi_0] \tag{3.209}$$

$$\psi_y = kd_y[\sin\theta\sin\phi - \sin\theta_0\sin\phi_0] \tag{3.210}$$

As before, these arrays can be made fractal-like by following a systematic thinning procedure where

$$I_{mn} = \begin{cases} 1, & \text{if element } (m,n) \text{ is turned on} \\ 0, & \text{if element } (m,n) \text{ is turned off} \end{cases} \tag{3.211}$$

A Sierpiński carpet is a two-dimensional version of the Cantor set [62] and can similarly be applied to thinning planar arrays. Consider, for example, the simple generating sub-array

$$\begin{matrix} 1 & 1 & 1 \\ 1 & 0 & 1 \\ 1 & 1 & 1 \end{matrix}$$

The normalized array factor associated with this generating sub-array for $d_x = d_y = \lambda/2$ is given by

$$\widehat{GA}(u_x, u_y) = \tfrac{1}{4}\,[\cos(\pi u_x) + \cos(\pi u_y) + 2\cos(\pi u_x)\cos(\pi u_y)] \tag{3.212}$$

where

$$u_x = \sin\theta\cos\phi - \sin\theta_0\cos\phi_0 \tag{3.213}$$

$$u_y = \sin\theta\sin\phi - \sin\theta_0\sin\phi_0 \tag{3.214}$$

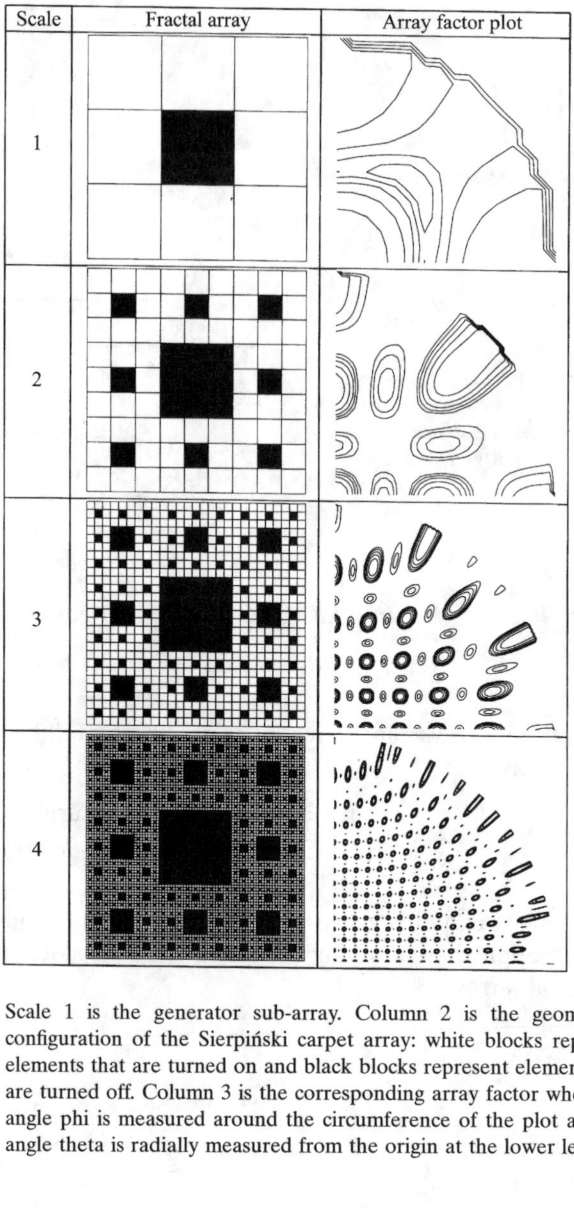

Figure 3.51 Scale 1 is the generator sub-array. Column 2 is the geometrical configuration of the Sierpiński carpet array: white blocks represent elements that are turned on and black blocks represent elements that are turned off. Column 3 is the corresponding array factor where the angle phi is measured around the circumference of the plot and the angle theta is radially measured from the origin at the lower left.

Substituting (3.212) into (3.181) with an expansion factor of $\delta = 3$ yields the following expression for the fractal array factor at stage P:

$$\widehat{AF}_P(u_x, u_y) = \frac{1}{4^P} \prod_{p=1}^{P} [\cos(3^{p-1}\pi u_x) + \cos(3^{p-1}\pi u_y) + 2\cos(3^{p-1}\pi u_x)\cos(3^{p-1}\pi u_y)] \qquad (3.215)$$

The geometry for this Sierpiński carpet fractal array at various stages of growth is illustrated in Fig. 3.51 along with a plot of the corresponding array factor. A comparison of the array factors for the first four stages of construction shown in Fig. 3.51 reveals the self-similar nature of the radiation patterns.

An expression for the directivity of the Sierpiński carpet array for the case in which $\theta_0 = 0°$ may be obtained from

$$D_P(\theta, \phi) = \frac{\widehat{AF}_P^2(\theta, \phi)}{\frac{1}{4\pi}\int_0^{2\pi}\int_0^{\pi} \widehat{AF}_P^2(\theta, \phi) \sin\theta \, d\theta \, d\phi}$$

$$= \frac{\widehat{AF}_P^2(\theta, \phi)}{\frac{1}{2\pi}\int_0^{\pi}\int_0^{\pi} \widehat{AF}_P^2(\theta, \phi) \sin\theta \, d\theta \, d\phi} \tag{3.216}$$

where

$$\widehat{AF}_P^2(\theta, \phi) = \frac{1}{16^P}\prod_{p=1}^{P} [\cos(3^{p-1}\pi \sin\theta \cos\phi) + \cos(3^{p-1}\pi \sin\theta \sin\phi) + 2\cos(3^{p-1}\pi \sin\theta \cos\phi)\cos(3^{p-1}\pi \sin\theta \sin\phi)]^2 \tag{3.217}$$

which follows directly from (3.215). The double integral which appears in the denominator of (3.216) does not have a closed form solution in this case and, therefore, must be evaluated numerically. However, this technique for evaluating the directivity is much more computationally efficient than the alternative approach which involves making use of the Fourier series representation for the Sierpinski carpet array factor given by (3.208) and (3.211) with

$$\psi_x = \pi \sin\theta \cos\phi \tag{3.218}$$

$$\psi_y = \pi \sin\theta \sin\phi \tag{3.219}$$

$$M = \frac{3^P + 1}{2} \tag{3.220}$$

The geometry for a stage 4 ($P = 4$) Sierpiński carpet array and its complement are shown in Figs. 3.52 and 3.53, respectively. These figures demonstrate that the full 81×81 element planar array may be decomposed into two sub-arrays, namely, a Sierpiński carpet array and its complement. This relationship may be represented formally by the equation [24], [65]

$$AF_P(u_x, u_y) + \overline{AF_P}(u_x, u_y) = AF(u_x, u_y), \tag{3.221}$$

where

$$AF_P(u_x, u_y) = 2^P \prod_{p=1}^{P} [\cos(3^{p-1}\pi u_x) + \cos(3^{p-1}\pi u_y) + 2\cos(3^{p-1}\pi u_x)\cos(3^{p-1}\pi u_y)] \tag{3.222}$$

is the Sierpiński carpet array factor, $\overline{AF_P}$ is the array factor associated with its complement, and AF denotes the array factor of the full planar array. It can be shown that the array factor for a uniformly excited square planar array (i.e., $I_{mn} = 1$ for all values of m and n) may be expressed in the form

$$AF(u_x, u_y) = [1 - 2f_P(u_x)][1 - 2f_P(u_y)] \tag{3.223}$$

where

$$f_P(x) = \frac{\cos\left[\frac{\pi}{4}(3^P - 1)x\right]\sin\left[\frac{\pi}{4}(3^P + 1)x\right]}{\sin\left[\frac{\pi}{2}x\right]} \tag{3.224}$$

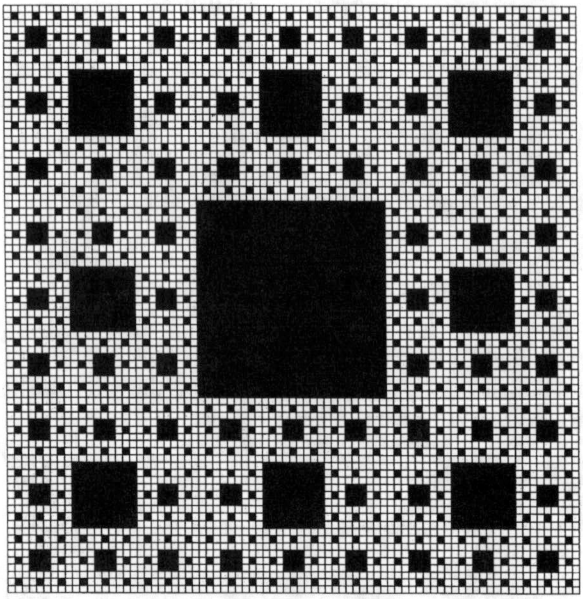

Figure 3.52 Sierpiński carpet array for $P = 4$.

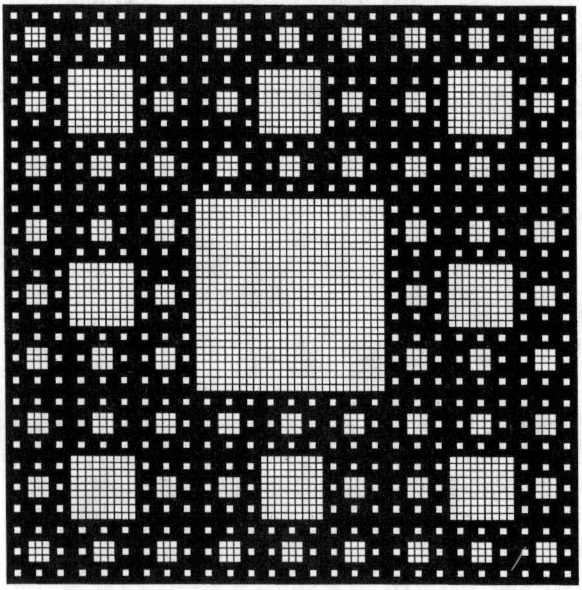

Figure 3.53 The complement of the Sierpiński carpet array for $P = 4$.

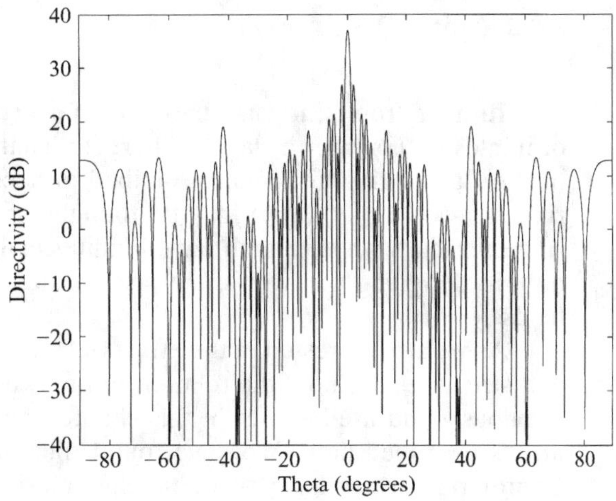

Figure 3.54 Directivity plot for a Sierpiński carpet array with $P = 4$. The spacing between elements of the array is $d = \lambda/2$ with $\phi = 0^0$. The maximum directivity for stage 4 is $D_4 = 37.01$ dB.

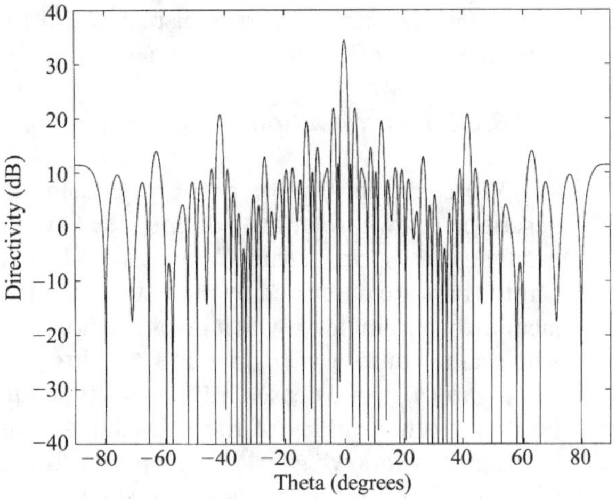

Figure 3.55 Directivity plot for the complement to the Sierpiński carpet array with $P = 4$. The spacing between elements of the array is $d = \lambda/2$ with $\phi = 0^0$. The maximum directivity for stage 4 is $D_4 = 34.41$ dB.

Finally, by using (3.223) together with (3.222), an expression for the complementary array factor $\overline{AF_P}$ may be obtained directly from (3.221). Plots of the directivity for a $P = 4$ Sierpiński carpet array and its complement are shown in Figs. 3.54 and 3.55, respectively.

The multiband nature of the planar Sierpiński carpet arrays may be easily demonstrated by generalizing the argument presented in the previous section for linear Cantor arrays. Hence, for doubly infinite carpets we have

$$AF(\psi_x, \psi_y) = \prod_{p=-\infty}^{\infty} GA(\delta^{p-1}\psi_x, \delta^{p-1}\psi_y) \tag{3.225}$$

from which we conclude that

$$AF(\delta^{\pm q}\psi_x, \delta^{\pm q}\psi_y) = AF(\psi_x, \psi_y) \qquad \text{for} \qquad q = 0, 1, \ldots \tag{3.226}$$

3.6.3 Cantor Ring Arrays

Turning from Cartesian-based fractal arrays to arrays described in polar coordinates we consider a class of discrete planar arrays whose structure is reminiscent of the Cantor ring diffractals described in Section 3.3 of this chapter. These Cantor ring arrays [26], [36], [66] may be thought of as discretized versions of the Cantor ring diffractals whose annuli have infinitesimal width [30]. This analogy is useful not only in designing the arrays, but also in interpreting the behavior of their radiated fields.

As we are interested in the effect of lacunarity on fractal electromagnetic devices and structures, we note the relationship between the lacunarity parameter ε, defined previously and used earlier in this chapter, and the performance of the Cantor ring arrays described here. We have found that there are ranges of ε values for which Cantor ring arrays have desirable characteristics and compare favorably with their thinned random counterparts under suitable conditions. Also, we have found that in addition to the three radiation regimes detailed in a previous section on Cantor ring diffractals, Cantor ring arrays display a fourth regime characteristic of their discrete nature. We describe these and other results below.

3.6.3.1 Formulation

We construct Cantor ring arrays using a polyadic Cantor bar from the family described elsewhere in this volume (see Chapter 1) in a manner similar to the way we used these bars to construct Cantor ring diffractals in Section 3.3.4 of this chapter. These Cantor bars are described by their similarity fractal dimension D, number of gaps n_{gaps}, lacunarity parameter ε which gives the normalized outer gap width, stage of growth S, and scaling ratio $\gamma = (n_{gaps} + 1)^{-1/D}$. Throughout this section we will be using the three-gap ($n_{gaps} = 3$) case with $\varepsilon = 44/1000$ and $D = 9/10$ although a similar analysis holds for other values of these fractal descriptors. These values correspond to the "lacunarity window" $\varepsilon \approx 44/1000$ previously found for three-gap Cantor ring diffractals with $D \approx 9/10$. In this window, the sidelobe levels of the Cantor ring diffractal were relatively low. We make use of that information here as we examine its discrete counterpart, the Cantor ring array, particularly for the case where the array is thinned. Since we are interested in discrete arrays, we use the distances to the centers of the annuli in the diffractal as a discrete set of radii at which to place array elements. As the Cantor bars used to generate these arrays are symmetric about their midpoints, rotating a stage S bar about its midpoint generates $M = (1/2)4^S$ distinct radii from the $4^S = (n_{gaps} + 1)^S$ sub-bars. We then place N_k elements uniformly spaced on the kth ring, which is at radius a_k from the origin in the $z = 0$ plane, where the a_k are given as

$$a_k = a \sum_{j=1}^{S} \gamma^{j-1} b_{c_j^{(k)}} \qquad (k = 0, \ldots, 2^{2S-1} - 1) \tag{3.227}$$

and the $c_j^{(k)}$ are the digits in the base 4 expansion of $k + 2^{2S-1} = (c_1^{(k)} c_2^{(k)} c_3^{(k)} \ldots c_S^{(k)})_4$, $b_0 = -(1-\gamma)$, $b_1 = -(1-3\gamma-2\varepsilon)$, $b_2 = (1-3\gamma-2\varepsilon)$, and $b_3 = (1-\gamma)$ as given in Section 3.3.4. The first element in the kth ring is placed at a random angle ϕ_k. For these arrays, we use the same linear density of elements in each ring, so that for each k, N_k is a constant times a_k to within rounding error. Given the discrete nature of these arrays, we have the additional variable of the linear density of the elements to describe each array.

If the elements are excited by identical currents with time-harmonic excitation $e^{-i\omega t}$, we find the array factor AF of this array to be [26], [36]

$$AF(x,y,z) = \sum_{k=0}^{M-1} a_k J_0(2\pi f_\rho a_k) + \sum_{k=0}^{M-1} R_k \tag{3.228}$$

where J_0 is the zeroth-order Bessel function, $f_\rho = \sqrt{f_x^2 + f_y^2} = \sqrt{x^2+y^2}/\lambda r$ is the radial spatial frequency, λ is the wavelength, and $r = \sqrt{x^2+y^2+z^2}$ is the observation distance from the origin. We observe that the first term in relation (3.228) is of the same form as the diffracted field of an infinitesimal width Cantor ring diffractal, found earlier in this chapter. This term represents the field radiated by the rings on which the elements lie if the elements are spaced an infinitesimal distance apart. The residuals R_k in the second term of (3.228) represent the azimuthal variations arising from the discrete nature of the array. These are given by

$$R_k = \sum_{l=1}^{\infty} a_k J_{lN_k}(2\pi f_\rho a_k)\{\exp[-ilN_k(\pi/2 - \phi + \phi_k)] + (-1)^{lN_k}\exp[ilN_k(\pi/2 - \phi + \phi_k)]\} \tag{3.229}$$

where J_{lN_k} is a Bessel function of order lN_k, and $\phi = \tan^{-1}(y/x)$ is the azimuthal angle of the observer with respect to the x axis.

3.6.3.2 Results and Discussion

In Fig. 3.56, we plot the array factor $|AF|$ in dB as a function of $\log_{10}[f_\rho a]$, fixing $\phi = 0$, for stages of growth $S = 1, 2$, and 3. The corresponding numbers of rings M and total numbers of elements N_{tot} are (2, 1960), (8, 7840), and (32, 31,360). While there is a small azimuthal variation for large spatial frequency due to the discrete nature of these arrays, these cuts along a fixed azimuthal angle are typical of the radiated field at an arbitrary azimuthal angle. These plots permit analysis of the array factors of Cantor ring arrays, in particular allowing us to note two significant scalings present in the radiated field.

The first scaling is seen in four spatial frequency regimes as shown in Fig. 3.56. First, the *subfractal* regime is the region of the mainbeam and first sidelobe and has width inversely proportional to the diameter of the array. Here the behavior of $|AF|$ is determined by the array diameter, and the fractal nature of the array is not evident. Next, the *fractal* regime extends from $f_\rho a \approx 1.12$ near the edge of the first sidelobe through the region containing repeated scaled versions of the same general pattern. This scaling depends on the stage of growth of the fractal. In particular, S bands of width $-\log_{10}[\gamma]$ exist, although the last band may be less distinct than the previous bands. This is consistent with prior results displayed for one-dimensional fractal arrays [21], [22], [35]. This regime extends to the *harmonic superfractal* regime that follows the S similar bands and begins at $f_{\rho harmonic} a \approx f_{\rho fractal} a/\gamma^S$. As for the continuous ring case, this region contains "harmonic echoes" of the fractal region in which the sidelobe levels no longer give repeated bands but the almost periodic zeros of the Bessel functions generate the same null structure as in the fractal regime. For discrete arrays with finite N_{tot}, either the fractal or harmonic superfractal region is truncated by the

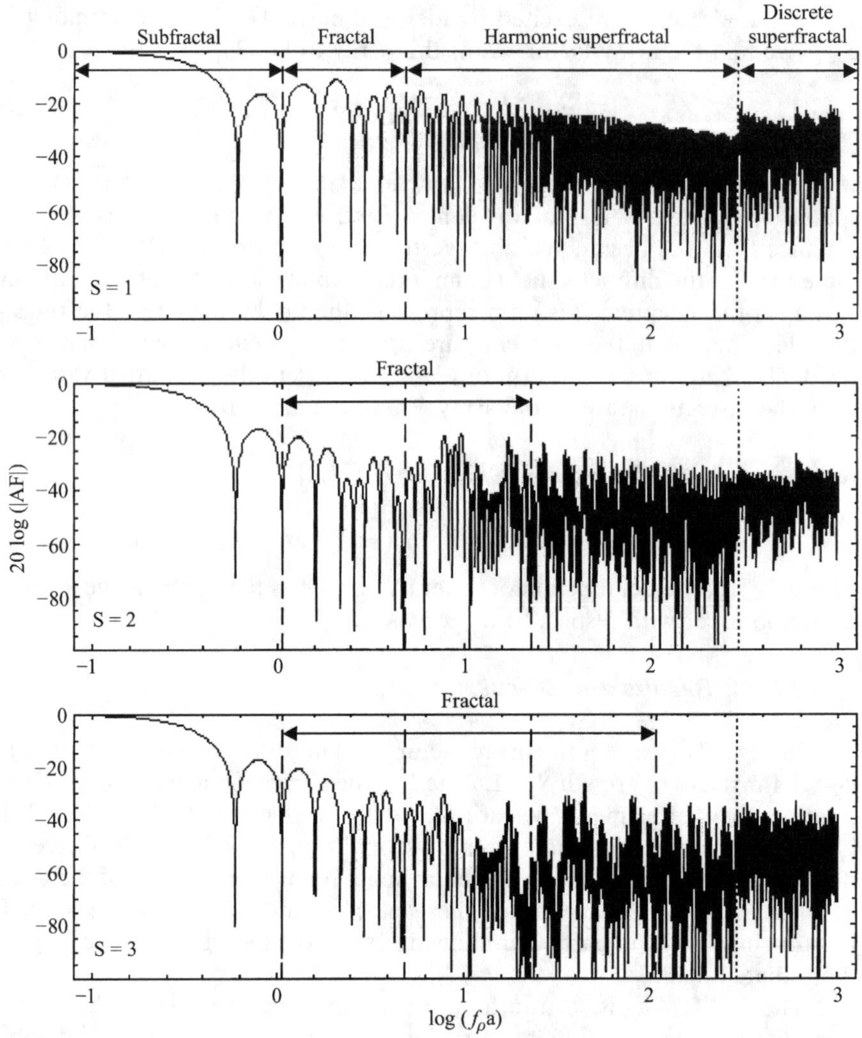

Figure 3.56 The array factor $|AF|$ as a function of normalized spatial frequency $f_\rho a$ on a log-log plot for the polyadic Cantor ring array with $D = 0.9$ and $\varepsilon = 0.044$. Three stages of growth $S = 1, 2, 3$ are displayed (top to bottom) with total number of elements $N = 1960, 7840, 31{,}360$, respectively. Note the similarity of the re-scaled patterns for each stage of growth and the self-affinity of the S bands of sidelobes (indicated by vertical dashed lines). *Subfractal*, *fractal*, and two *superfractal* regimes are indicated. The dotted line indicates $f_\rho a = f_{\rho_{discrete}} a$, beyond which the discrete nature of the array becomes the dominating factor. [*Figure from "Cantor Ring Arrays," D. L. Jaggard and A. D. Jaggard,* Microwave Opt. Technol. Lett., **19**, 121–124 (1998). © *1998 John Wiley & Sons, Inc.*]

discrete superfractal regime that arises from the discrete nature of the ring array. This truncation occurs when the closest elements in the array are a half-wavelength apart or, equivalently, when the residual terms in (3.229) become comparable to the first term in (3.228). Both calculations yield the same value $f_{\rho_{discrete}}a = N_k a/(2\pi a_k)$ which is independent of k due to the constant linear density of elements from ring to ring. Therefore, the fractal region consists of those values of $f_\rho a$ satisfying $f_{\rho_{fractal}}a \leq f_\rho a \leq \min\{f_{\rho_{harmonic}}a, f_{\rho_{discrete}}a\}$, with the upper boundary giving the lower boundary of the superfractal region. Note that in the discrete superfractal region, the sidelobes are no longer controlled by the fractal geometry of the array and they appear chaotic.

The four regions of operation are most easily identified in the bottom plot of Fig. 3.56, where for $S = 3$ we see the characteristic mainlobe and first sidelobe (subfractal regime), two to three bands of scaled log-periodic sidelobes (fractal regime), a region characterized by sidelobes with deep linear-periodic nulls or "harmonic echoes" (harmonic superfractal regime), and finally a region characterized by chaotic high-frequency variations and relatively high sidelobes (discrete superfractal regime). As the element spacing approaches zero and the array approaches the continuous ring case, the lower boundary $f_{\rho_{discrete}}a$ of the discrete regime migrates to higher spatial frequencies and disappears in the limit. Thus, the discrete case approximately tracks the continuous case for $f_\rho a \leq f_{\rho_{discrete}}a$. This is evident in looking at relation (3.228), since $f_\rho a \leq f_{\rho_{discrete}}a$ implies that the residual terms R_k are negligible and the array factor is approximately that of a Cantor ring diffractal with infinitesimal width annuli.

In the second scaling, the array factor of the array generated by a Cantor bar at stage S, rescaled in $f_\rho a$ by $1/\gamma$, forms an envelope for the array factor of the array generated by stage $S+1$ of the same bar. The scaling by $1/\gamma$ in the array factor corresponds to the scaling by γ of the smallest structure present in the generating Cantor bar. This scaling is present throughout both the fractal and harmonic superfractal regimes and has been seen often in other cases of radiation from fractal structures [6], [15], [16], [20], [21].

The distribution of elements (top) and the array factor (bottom) for the Cantor ring array are shown in Fig. 3.57 for the first (left) and second (right) stages of growth with numbers of elements $N_{tot} = 34$ and $N_{tot} = 548$, respectively. As for the arrays in Fig. 3.56, each array has fractal dimension $D = 0.9$ and lacunarity parameter $\varepsilon = 0.044$. We optimize the array by choosing the number of elements N_{tot} to satisfy approximately the relation $f_{\rho_{harmonic}}a \approx f_{\rho_{discrete}}a$ so that the array contains the fewest elements needed to prevent truncation of the fractal regime. The top of Fig. 3.57 shows the increase in element density required to have all S bands in the fractal regime as we increase S.

The first-stage case shown in Fig. 3.57 (left) is a relatively sparse array. In the limit of a continuous ring, the largest sidelobe will be -11 dB. For a -6 dB threshold, we find the visible range of the array extends to $f_\rho a = 26$ in the discrete case with a corresponding average element spacing of $\langle d \rangle = 7.9\lambda$. This represents an improvement over the corresponding random array results [26], [31], [36].

The second-stage case of Fig. 3.57 (right) is less sparse. Here the largest sidelobe in the limit of the continuous ring is -17 dB. For a -17 dB threshold, we find the visible range extends to $f_\rho a = 21$ with a corresponding average element spacing of $\langle d \rangle = 1.6\lambda$. We find that these less sparse Cantor ring arrays also merit investigation as alternatives to random arrays [26], [31], [36].

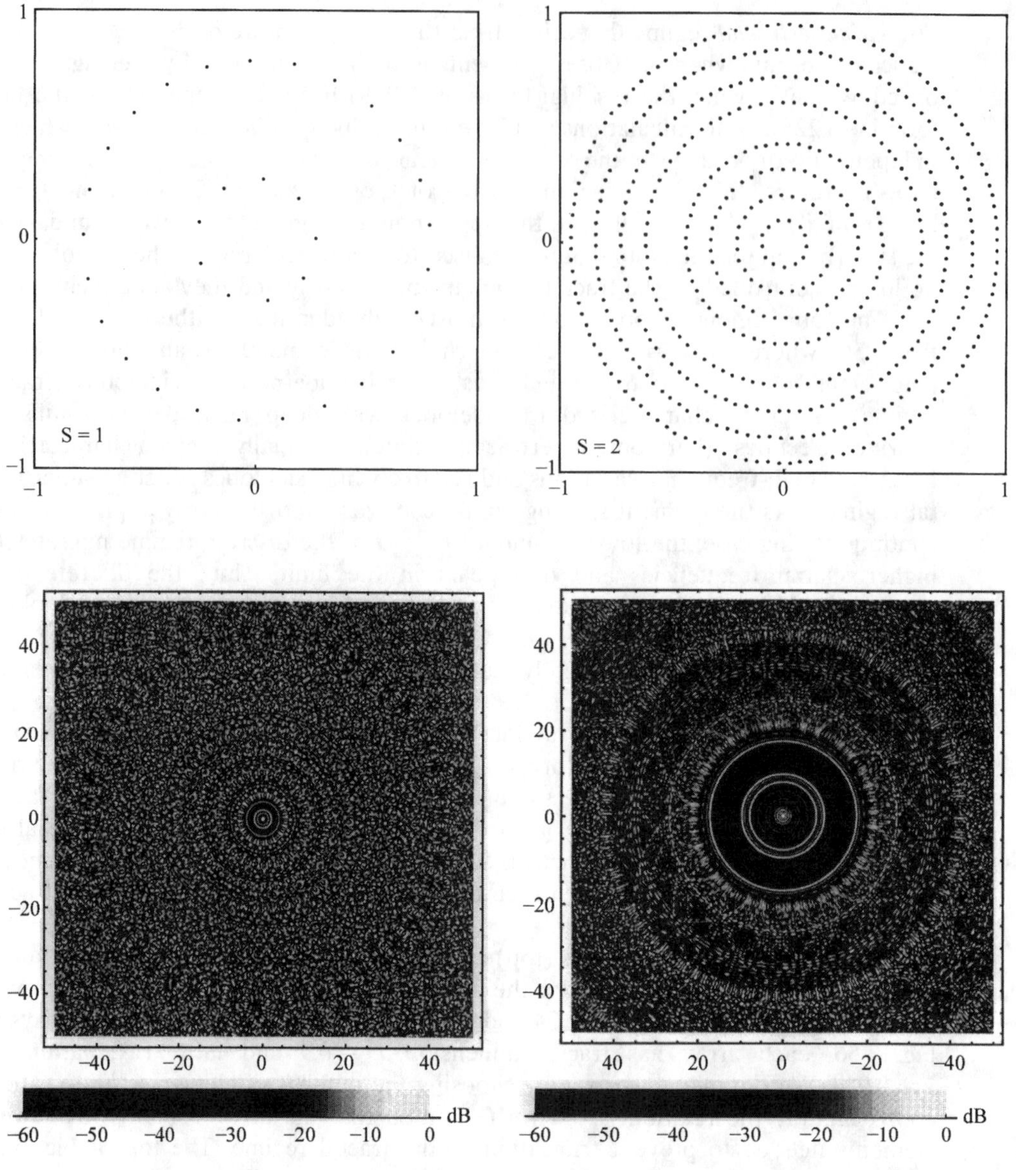

Figure 3.57 Element distributions (top), and array factors $|AF|$, in dB, (bottom) as a function of normalized spatial frequency $f_\rho a$ for a Cantor ring array with $S = 1$ (left) and $S = 2$ (right) where $D = 0.9$ and $\varepsilon = 0.044$. For $S = 1$ the number of rings is $M = 2$ and the number of elements in these rings is $(N_1, N_2) = (9, 25)$ for a total $N = 34$. For the $S = 2$ case, $M = 8$ and $(N_1, N_2, N_3, N_4, N_5, N_6, N_7, N_8) = (13, 27, 42, 57, 80, 95, 110, 124)$ for a total of $N = 548$. [*Figure from "Cantor Ring Arrays," D. L. Jaggard and A. D. Jaggard,* Microwave Opt. Technol. Lett., **19**, 121–124 (1998). *©1998 John Wiley & Sons, Inc.*]

We see from Figs. 3.56–3.57 that near the mainbeam, the fractal geometry of the array controls the distribution of power in the sidelobes which is relatively independent of azimuthal angle. This region is defined by the concentric dark circles in the figure and is the fractal regime noted previously. However, as the spatial frequency increases beyond the fractal regime, the higher-order correction terms in (3.229) eventually become dominant and the discrete superfractal regime is entered. Alternatively, as seen in Fig. 3.56 the harmonic superfractal regime is first seen, in which the fractal geometry has less control over the sidelobes due to the limited stage of growth of the fractal.

In review, we see that the previous construction of Cantor ring diffractals and the results of Section 3.3.4 suggested that we investigate the large dimension and small lacunarity case of the discrete array studied in this subsection. From the results of our research we found that for $D \approx 9/10$ and $N_{tot} < 100$, there is a lacunarity window $\varepsilon \approx 44/1000$ in which Cantor ring arrays provide lower sidelobes and enhanced quality of the mainbeam when compared to their random array counterparts. It would not be surprising if sparse arrays up to several hundred elements compare favorably to random arrays, perhaps for values of N_{tot} as high as 500, as long as the array is optimized so that $f_{\rho_{harmonic}} a \approx f_{\rho_{discrete}} a$.

Finally, we note that the fractal dimension can be retrieved from the radiation pattern, regardless of lacunarity, when $f_{\rho_{fractal}} a\gamma^{-3} \leq \min\{f_{\rho_{harmonic}} a, f_{\rho_{discrete}} a\}$ (not shown here [26], [36]). Therefore, as has been found with the diffraction of electromagnetic waves from a variety of fractal objects, information regarding the fractal dimension—and perhaps other fractal descriptors—is embedded in the scattering data and can be extracted from it. This is a major problem of research interest in the area of inverse problems associated with fractal electrodynamics.

3.7 THE CONCENTRIC CIRCULAR RING SUB-ARRAY GENERATOR

3.7.1 Theory

There has been considerable recent interest in the radiation characteristics of self-scalable and self-similar planar arrays. For instance, array configurations based on Sierpinski carpets have been considered in [23–25]. The properties of self-scalable concentric circular Weierstrass arrays and self-similar concentric circular Cantor arrays have also been investigated in [20] and [26], respectively. An alternative design methodology for the mathematical construction of self-scalable and self-similar (i.e., fractal) arrays will be introduced in this section [63]. This technique is very general and provides a high degree of flexibility in the design of fractal and related arrays. This is primarily due to the fact that the generator in this case is based on a concentric circular ring array.

The generating array factor for the concentric circular ring array may be expressed in the form [42]

$$GA(\theta, \phi) = \sum_{m=1}^{M} \sum_{n=1}^{N_m} I_{mn} e^{j\psi_{mn}(\theta,\phi)} \tag{3.230}$$

where

$$\psi_{mn}(\theta, \phi) = kr_m \sin\theta \cos(\phi - \phi_{mn}) + \alpha_{mn} \tag{3.231}$$

$$k = 2\pi/\lambda \tag{3.232}$$

$$M = \text{Total number of concentric rings} \tag{3.233}$$

$$N_m = \text{Total number of elements on the } m\text{th ring} \tag{3.234}$$

$$r_m = \text{Radius of the } m\text{th ring} \tag{3.235}$$

$$I_{mn} = \text{Excitation current amplitude of the } n\text{th element on the } m\text{th ring located at } \phi = \phi_{mn} \tag{3.236}$$

$$\alpha_{mn} = \text{Excitation current phase of the } n\text{th element on the } m\text{th ring located at } \phi = \phi_{mn} \tag{3.237}$$

A wide variety of interesting as well as practical fractal array designs may be constructed using a generating sub-array of the form given in (3.230). The fractal array factor for a particular stage of growth P may be derived directly from (3.181) by following a similar procedure to that outlined in the previous section. The resulting expression for the array factor was found to be

$$AF_P(\theta, \phi) = \prod_{p=1}^{P} \left\{ \sum_{m=1}^{M} \sum_{n=1}^{N_m} I_{mn} e^{j\delta^{p-1}\psi_{mn}(\theta,\phi)} \right\} \tag{3.238}$$

where δ represents the scaling or expansion factor associated with the fractal array.

A graphical procedure can be used to conveniently illustrate the construction process for fractal arrays which is embodied in (3.238). For example, suppose we consider the simple four-element circular array of radius r shown in Fig. 3.58a. If we regard this as the generator (stage 1) for a fractal array, then the next stage of growth (stage 2) for the array would have a geometrical configuration of the form shown in Fig. 3.58b. Hence the first step in the construction process, as depicted in Fig. 3.58, is to expand the four-element generator array by a factor of δ. This is followed by replacing each of the elements of the expanded array by an exact copy of the original unscaled four-element circular sub-array generator. The entire process is then repeated in a recursive fashion until the desired stage of growth for the fractal array is reached.

It is convenient for analysis purposes to express the fractal array factor (3.238) in the following normalized form:

$$\widehat{AF}_P(\theta, \phi) = \prod_{p=1}^{P} \left\{ \frac{\sum_{m=1}^{M} \sum_{n=1}^{N_m} I_{mn} e^{j\delta^{p-1}\psi_{mn}(\theta,\phi)}}{\sum_{m=1}^{M} \sum_{n=1}^{N_m} I_{mn}} \right\} \tag{3.239}$$

Taking the magnitude of both sides of (3.239) leads to

$$|\widehat{AF}_P(\theta, \phi)| = \prod_{p=1}^{P} \left| \frac{\sum_{m=1}^{M} \sum_{n=1}^{N_m} I_{mn} e^{j\delta^{p-1}\psi_{mn}(\theta,\phi)}}{\sum_{m=1}^{M} \sum_{n=1}^{N_m} I_{mn}} \right| \tag{3.240}$$

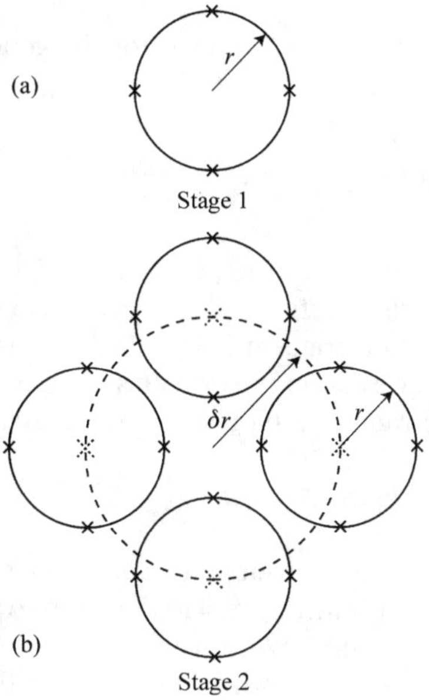

Figure 3.58 An illustration of the graphical construction procedure for fractal arrays based on a four-element circular sub-array generator. (a) Stage 1 ($P = 1$) and (b) stage 2 ($P = 2$).

which has a corresponding representation in terms of decibels given by

$$|\widehat{AF}_P(\theta, \phi)|_{\text{dB}} = 20 \sum_{p=1}^{P} \log \left| \frac{\sum_{m=1}^{M} \sum_{n=1}^{N_m} I_{mn} e^{j\delta^{p-1}\psi_{mn}(\theta,\phi)}}{\sum_{m=1}^{M} \sum_{n=1}^{N_m} I_{mn}} \right| \tag{3.241}$$

For the special case when $\delta = 1$, we note that (3.240) and (3.241) reduce to

$$|\widehat{AF}_P(\theta, \phi)| = \left| \frac{\sum_{m=1}^{M} \sum_{n=1}^{N_m} I_{mn} e^{j\psi(\theta,\phi)}}{\sum_{m=1}^{M} \sum_{n=1}^{N_m} I_{mn}} \right|^{P} \tag{3.242}$$

$$|\widehat{AF}_P(\theta, \phi)|_{\text{dB}} = 20P \log \left| \frac{\sum_{m=1}^{M} \sum_{n=1}^{N_m} I_{mn} e^{j\psi_{mn}(\theta,\phi)}}{\sum_{m=1}^{M} \sum_{n=1}^{N_n} I_{mn}} \right| \tag{3.243}$$

Another unique property of (3.239) is the fact that the conventional cophasal excitation [42]

$$\alpha_{mn} = -kr_m \sin\theta_0 \cos(\phi_0 - \phi_{mn}) \tag{3.244}$$

where θ_0 and ϕ_0 are the desired main-beam steering angles, can be applied at the generating sub-array level. To see this we recognize that the position of the main beam produced by (3.239) is independent of the stage of growth P since it corresponds to a value of $\psi_{mn} = 0$ (i.e., when $\theta = \theta_0$ and $\phi = \phi_0$). In other words, once the position of the main beam is determined for the generating sub-array, it will remain invariant at all higher stages of growth.

3.7.2 Examples

Several different examples of recursively generated arrays will be presented and discussed in this section. These arrays have in common the fact that they may be constructed via a concentric circular ring sub-array generator of the type considered in [63] and reviewed in the previous section. Hence, the mathematical expressions that describe the radiation patterns of these arrays are all special cases of (3.239).

3.7.2.1 Linear Arrays

Various configurations of linear arrays may be constructed using a degenerate form of the concentric circular ring sub-array generator introduced in Section 3.7.1. For instance, suppose we consider the two-element circular sub-array generator with a radius of $r = \lambda/4$ shown in Fig. 3.59. If the excitation current amplitudes for this two-element generating sub-array are assumed to be unity, then the general fractal array factor expression given in (3.239) will reduce to the form

$$\widehat{AF}_P(\phi) = \frac{1}{2^P} \prod_{p=1}^{P} \sum_{n=1}^{2} e^{j\delta^{p-1}[(\pi/2)\cos(\phi-\phi_n)+\alpha_n]} \tag{3.245}$$

where, without loss of generality, we have set $\theta = 90°$ and

$$\phi_n = (n-1)\,\pi \tag{3.246}$$

$$\alpha_n = -\frac{\pi}{2}\cos(\phi_0 - \phi_n) \tag{3.247}$$

Substituting (3.246) and (3.247) into (3.245) results in a simplified expression for the fractal array factor given by

$$\widehat{AF}_P(\phi) = \prod_{p=1}^{P} \cos\left[\delta^{p-1}\frac{\pi}{2}(\cos\phi - \cos\phi_0)\right] \tag{3.248}$$

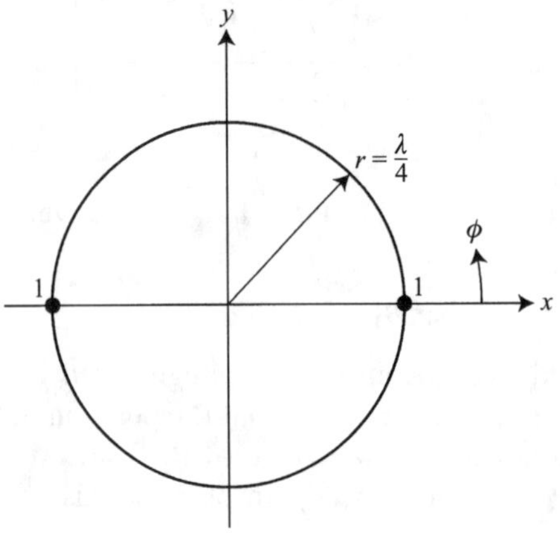

Figure 3.59 The geometry for a two-element circular sub-array generator with a radius of $r = \lambda/4$.

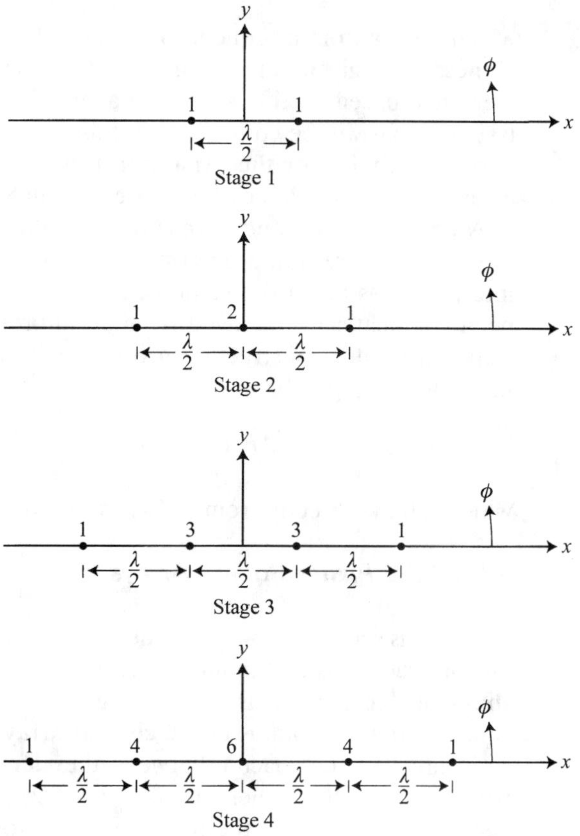

Figure 3.60 The first four stages in the construction process of a uniformly spaced binomial array.

If we choose an expansion factor equal to one (i.e., $\delta = 1$) then (3.248) may be written as

$$\widehat{AF}_P(\phi) = \cos^P\left[\frac{\pi}{2}(\cos\phi - \cos\phi_0)\right] \tag{3.249}$$

This represents the array factor for a uniformly spaced ($d = \lambda/2$) linear array with a binomial current distribution, where that total number of elements N_P for a given stage of growth P is $N_P = P + 1$. The first four stages in the construction process of these binomial arrays are illustrated in Fig. 3.60. The general rule in the case of overlapping array elements is to replace each of those elements by a single element which has a total excitation current amplitude equal to the sum of all the individual excitation current amplitudes. For example, in going from stage 1 to stage 2 in the binomial array construction process illustrated in Fig. 3.60, we find that two elements will share a common location at the center of the resulting three-element array. Since each of these two array elements has one unit of current, they may be replaced by a single equivalent element which is excited by two units of current.

Next we will consider the family of arrays which are generated when $\delta = 2$. The expression for the array factor in this case is

$$\widehat{AF}_P(\phi) = \prod_{p=1}^{P} \cos\left[2^{p-1}\frac{\pi}{2}(\cos\phi - \cos\phi_0)\right] \tag{3.250}$$

which results from a sequence of uniformly excited, equally spaced ($d = \lambda/2$) arrays. Hence, for a given stage of growth P, these arrays will contain a total of $N_P = 2^P$ elements spaced a half-wavelength apart with uniform current excitations. Finally, the last case that will be considered in this section corresponds to a choice of $\delta = 3$. This particular choice for the expansion factor gives rise to the family of triadic Cantor arrays which have already been discussed in Section 3.6.1. These arrays contain a total of $N_P = 2^P$ elements and have current excitations which follow a uniform distribution. However, the resulting arrays in this case are nonuniformly spaced. This can be interpreted as being the result of a thinning process in which certain elements have been systematically removed from a uniformly spaced array in accordance with the standard Cantor construction procedure. The Cantor array factor may be expressed in the form

$$\widehat{AF}_P(\phi) = \prod_{p=1}^{P} \cos\left[3^{p-1}\frac{\pi}{2}(\cos\phi - \cos\phi_0)\right] \tag{3.251}$$

which follows directly from (3.248) when $\delta = 3$.

3.7.2.2 Planar Square Arrays

In this section we will consider three examples of planar square arrays which can be constructed using the uniformly excited four-element circular sub-array generator shown in Fig. 3.61. This sub-array generator can also be viewed as a four-element square array. The radius of the circular array was chosen to be $r = \lambda/(2\sqrt{2})$, in order to insure that the spacing between the elements of the circumscribed square array would be a half-wavelength (i.e., $d = \lambda/2$). In this case, it can be shown that the general expression for the fractal array factor given in (3.239) will reduce to

$$\widehat{AF}_P(\theta, \phi) = \frac{1}{4^P}\prod_{p=1}^{P}\sum_{n=1}^{4} e^{j\delta^{p-1}[(\pi/\sqrt{2})\sin\theta\cos(\phi-\phi_n)+\alpha_n]} \tag{3.252}$$

where

$$\phi_n = (n-1)\frac{\pi}{2} \tag{3.253}$$

$$\alpha_n = -\frac{\pi}{\sqrt{2}}\sin\theta_0\cos(\phi_0 - \phi_n) \tag{3.254}$$

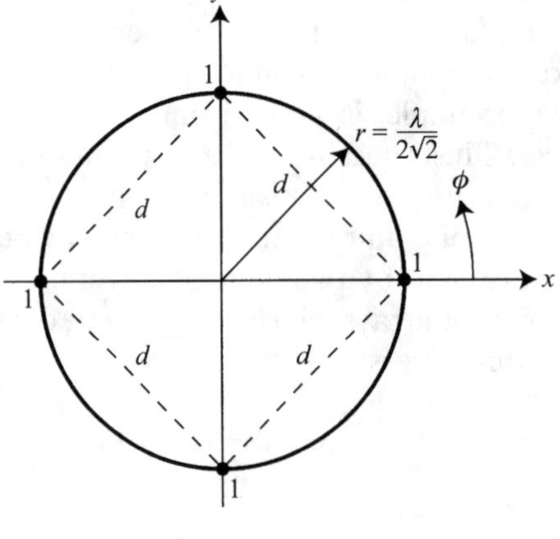

Figure 3.61 The geometry for a four-element circular sub-array generator with a radius of $r = \lambda/(2\sqrt{2})$.

If we define

$$\psi_n(\theta, \phi) = \frac{\pi}{\sqrt{2}}[\sin\theta\cos(\phi - \phi_n) - \sin\theta_0\cos(\phi_0 - \phi_n)] \tag{3.255}$$

then (3.252) may be written in the convenient form

$$\widehat{AF}_P(\theta, \phi) = \frac{1}{4^P}\prod_{p=1}^{P}\sum_{n=1}^{4} e^{j\delta^{p-1}\psi_n(\theta,\phi)} \tag{3.256}$$

where $\psi_n(\theta_0, \phi_0) = 0$.

Now suppose we consider the case where the expansion factor $\delta = 1$. Substituting this value of δ into (3.256) leads to

$$\widehat{AF}_P(\theta, \phi) = \left[\frac{1}{4}\sum_{n=1}^{4} e^{j\psi_n(\theta,\phi)}\right]^P \tag{3.257}$$

The first four stages of growth for this array are illustrated in Fig. 3.62. The pattern which emerges clearly shows that this construction process yields a family of square arrays with uniformly spaced elements ($d = \lambda/2$) and binomially distributed currents. For a given stage of growth P, the corresponding array will have a total of $N_P = (P+1)^2$ elements. Figure 3.63 contains plots of the far-field radiation patterns which are produced by the four arrays shown in Fig. 3.62. These plots were generated

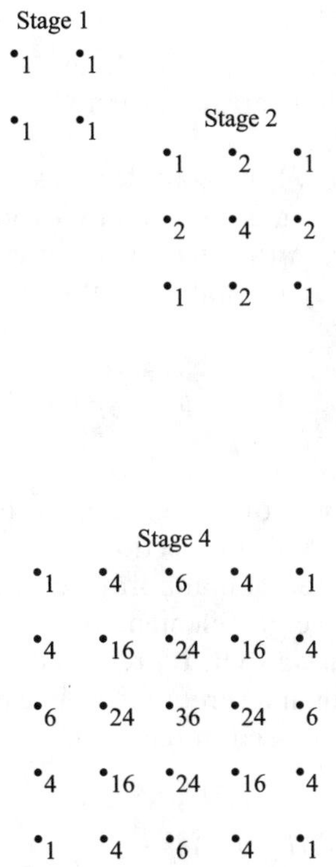

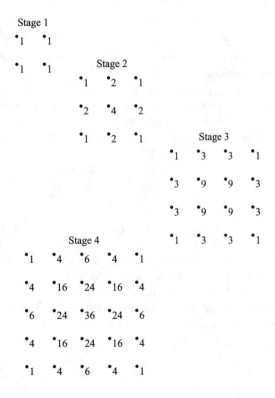

Figure 3.62 The first four stages of growth for a family of square arrays with uniformly spaced elements and binomially distributed currents.

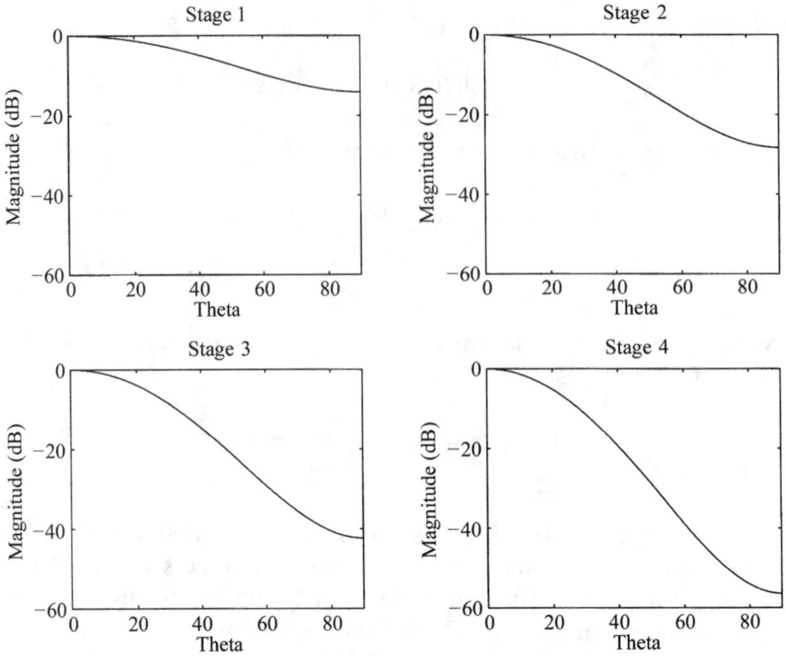

Figure 3.63 Plots of the far-field radiation patterns produced by the four arrays shown in Fig. 3.62.

using (3.257) for values of $P = 1$, 2, 3, and 4. It is evident from Fig. 3.63 that the radiation patterns for these arrays have no sidelobes, which is a feature characteristic of binomial arrays [44].

The next case that will be considered is when the expansion factor $\delta = 2$. This particular choice of δ results in a family of uniformly excited and equally spaced ($d = \lambda/2$) planar square arrays which increases in size according to $N_P = 2^{2P}$. The recursive array factor representation in this case is given by

$$\widehat{AF}_P(\theta, \phi) = \frac{1}{4^P} \prod_{p=1}^{P} \sum_{n=1}^{4} e^{j2^{p-1}\psi_n(\theta,\phi)} \tag{3.258}$$

For our final example of this section we return to the square Sierpiński carpet array previously discussed in Section 3.6.2. The generating sub-array for this Sierpiński carpet consisted of a uniformly excited and equally spaced ($d = \lambda/2$) 3×3 planar array with the center element removed. However, we note here that this generating sub-array may also be represented by two concentric four-element circular arrays. By adopting this interpretation it is easily shown that the Sierpinski carpet array factor may be expressed in the form

$$\widehat{AF}_P(\theta, \phi) = \frac{1}{8^P} \prod_{p=1}^{P} \sum_{m=1}^{2} \sum_{n=1}^{4} e^{j3^{p-1}\psi_{mn}(\theta,\phi)} \tag{3.259}$$

where

$$\psi_{mn}(\theta,\phi) = \sqrt{m}\,\pi[\sin\theta\cos(\phi-\phi_{mn}) - \sin\theta_0\cos(\phi_0-\phi_{mn})] \tag{3.260}$$

$$\phi_{mn} = \left(\frac{mn-1}{m}\right)\frac{\pi}{2} \tag{3.261}$$

$$N_P = 2^{3P} \tag{3.262}$$

3.7.2.3 Planar Triangular Arrays

A class of planar arrays will be introduced in this section which has the property that their elements are arranged in some type of recursively generated triangular lattice. The first category of triangular arrays that will be studied are those which can be constructed from the uniformly excited three-element circular sub-array generator shown in Fig. 3.64. This three-element circular array of radius $r = \lambda/(2\sqrt{3})$ can also be interpreted as an equilateral triangular array with half-wavelength spacing on a side (i.e., $d = \lambda/2$). The fractal array factor associated with this triangular generating sub-array is

$$\hat{AF}_P(\theta,\phi) = \frac{1}{3^P}\prod_{p=1}^{P}\sum_{n=1}^{3} e^{j\delta^{p-1}[(\pi/\sqrt{3})\sin\theta\cos(\phi-\phi_n)+\alpha_n]} \tag{3.263}$$

where

$$\phi_n = (n-1)\frac{2\pi}{3} \tag{3.264}$$

$$\alpha_n = -\frac{\pi}{\sqrt{3}}\sin\theta_0\cos(\phi_0-\phi_n) \tag{3.265}$$

The compact form of (3.263) is then

$$\hat{AF}_P(\theta,\phi) = \frac{1}{3^P}\prod_{p=1}^{P}\sum_{n=1}^{3} e^{j\delta^{p-1}\psi_n(\theta,\phi)} \tag{3.266}$$

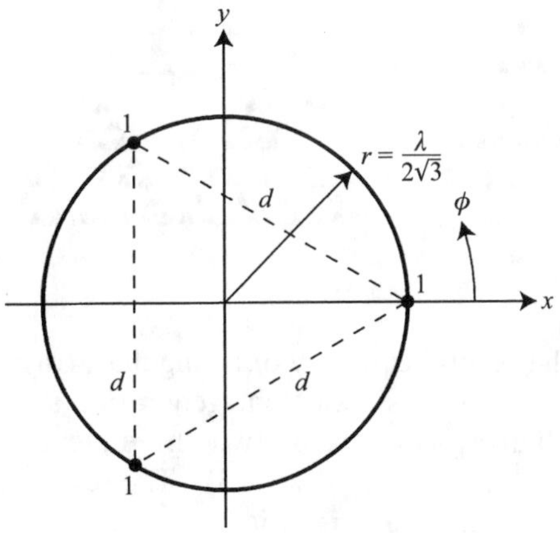

Figure 3.64 The geometry of a three-element circular sub-array generator with a radius of $r = \lambda/(2\sqrt{3})$.

where

$$\psi_n(\theta, \phi) = \frac{\pi}{\sqrt{3}}[\sin\theta\cos(\phi - \phi_n) - \sin\theta_0\cos(\phi_0 - \phi_n)] \tag{3.267}$$

If it is assumed that $\delta = 1$, then we find that (3.266) can be expressed in the simplified form

$$\widehat{AF}_P(\theta, \phi) = \left[\frac{1}{3}\sum_{n=1}^{3} e^{j\psi_n(\theta,\phi)}\right]^P \tag{3.268}$$

This represents the array factor for a stage P binomial triangular array. The total number of elements contained in this array may be determined from the formula

$$N_P = \sum_{p=1}^{P+1} p = \frac{(P+1)(P+2)}{2} \tag{3.269}$$

which has been derived by counting overlapping elements only once. On the other hand, if we choose $\delta = 2$, then (3.266) becomes

$$\widehat{AF}_P(\theta, \phi) = \frac{1}{3^P}\prod_{p=1}^{P}\sum_{n=1}^{3} e^{j2^{p-1}\psi_n(\theta,\phi)} \tag{3.270}$$

This array factor corresponds to uniformly excited Sierpiński gasket arrays of the type shown in Fig. 3.65. The construction process illustrated in Fig. 3.65 assumes that the array elements are located at the center of the shaded triangles. Hence, these arrays have a growth rate which is characterized by $N_P = 3^P$.

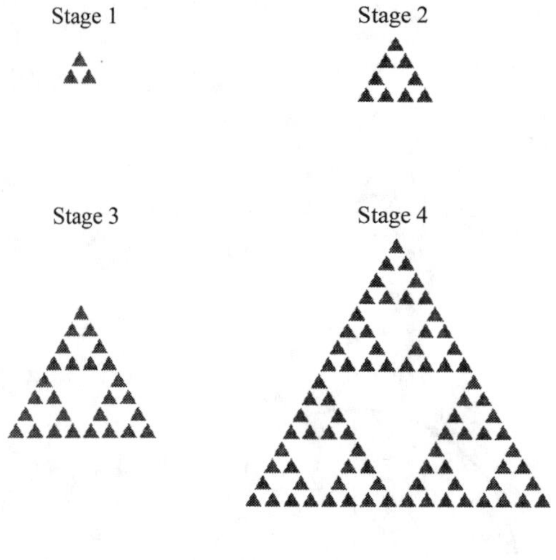

Figure 3.65 The geometry for a uniformly excited Sierpiński gasket array. The elements for this array are assumed to correspond to the centers of the shaded triangles.

The second category of triangular arrays that will be explored in this section are produced by the six-element generating sub-array shown in Fig. 3.66. This generating sub-array consists of two three-element concentric circular arrays with radii $r_1 = \lambda/(2\sqrt{3})$ and $r_2 = \lambda/(\sqrt{3})$. The excitation current amplitudes on the inner three-element array are twice as large as those on the outer three-element array. The

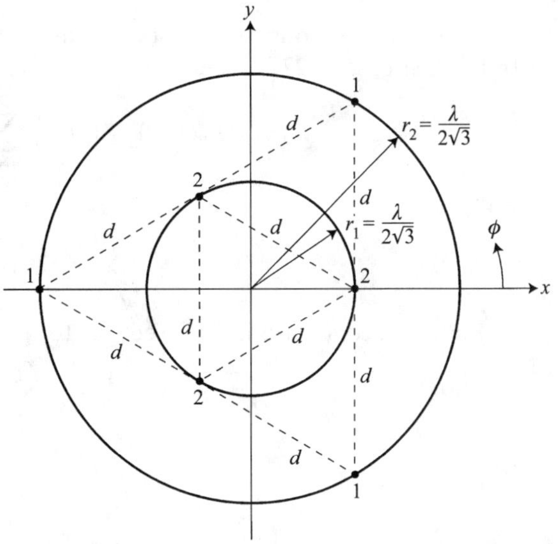

Figure 3.66 The geometry for a six-element generating sub-array which consists of two three-element concentric circular arrays with radii $r_1 = \lambda/(2\sqrt{3})$ and $r_2 = \lambda/(\sqrt{3})$.

dimensions of this generating sub-array were chosen in such a way that it forms a non-uniformly excited six-element triangular array with half-wavelength spacing between its elements (i.e., $d = \lambda/2$). If we treat the generating sub-array as a pair of three-element concentric circular ring arrays, then it follows from (3.239) that the fractal array factor in this case may be expressed as

$$\widehat{AF}_P(\theta, \phi) = \frac{1}{9^P} \prod_{p=1}^{P} \sum_{m=1}^{2} \sum_{n=1}^{3} I_{mn} e^{j\delta^{p-1}[kr_m \sin\theta \cos(\phi - \phi_{mn}) + \alpha_{mn}]} \tag{3.271}$$

where

$$I_{mn} = (2/m) \tag{3.272}$$

$$kr_m = \frac{m\pi}{\sqrt{3}} \tag{3.273}$$

$$\phi_{mn} = (2n + m - 3)\frac{\pi}{3} \tag{3.274}$$

$$\alpha_{mn} = -kr_m \sin\theta_0 \cos(\phi_0 - \phi_{mn}) \tag{3.275}$$

If we define

$$\psi_{mn}(\theta, \phi) = \frac{m\pi}{\sqrt{3}}[\sin\theta \cos(\phi - \phi_{mn}) - \sin\theta_0 \cos(\phi_0 - \phi_{mn})] \tag{3.276}$$

such that $\psi_{mn}(\theta_0, \phi_0) = 0$, then (3.271) may be written in the form

$$\widehat{AF}_P(\theta, \phi) = \frac{1}{9^P} \prod_{p=1}^{P} \sum_{m=1}^{2} \sum_{n=1}^{3} I_{mn} e^{j\delta^{p-1}\psi_{mn}(\theta, \phi)} \tag{3.277}$$

We will next consider two special cases of (3.277), namely when $\delta = 1$ and $\delta = 2$. In the first case (3.277) reduces to

$$\widehat{AF}_P(\theta, \phi) = \left[\frac{1}{9}\sum_{m=1}^{2}\sum_{n=1}^{3} I_{mn} e^{j\psi_{mn}(\theta,\phi)}\right]^P \tag{3.278}$$

where

$$N_P = \sum_{p=1}^{2P+1} p = (P+1)(2P+1) \tag{3.279}$$

and in the second case (3.277) reduces to

$$\widehat{AF}_P(\theta, \phi) = \frac{1}{9^P}\prod_{p=1}^{P}\sum_{m=1}^{2}\sum_{n=1}^{3} I_{mn} e^{j2^{p-1}\psi_{mn}(\theta,\phi)} \tag{3.280}$$

where

$$N_P = \sum_{p=1}^{2^{P+1}-1} p = 2^P(2^{P+1} - 1) \tag{3.281}$$

These are both examples of fully populated, uniformly spaced ($d = \lambda/2$) and nonuniformly excited triangular arrays. Hence, this type of construction scheme can be exploited in order to realize low-sidelobe array designs. To further exemplify this important property, we will focus here on the special case where $\delta = 2$. The first four stages in the construction process of this triangular array are illustrated in Fig. 3.67.

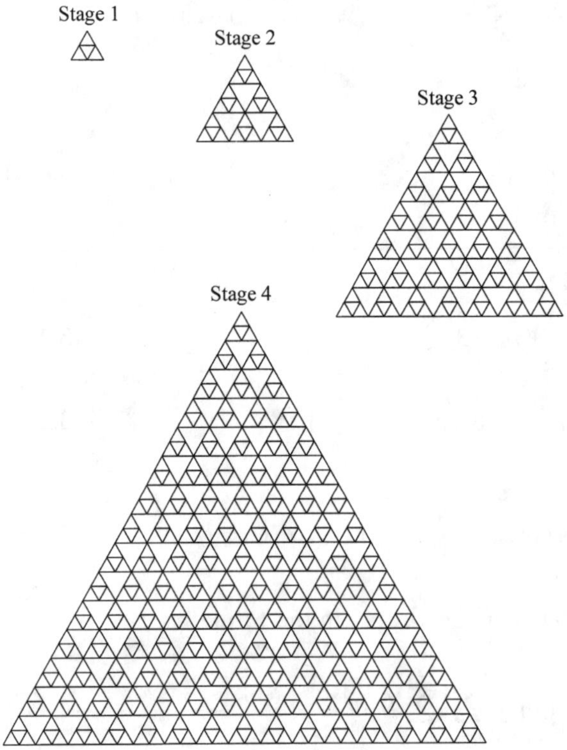

Figure 3.67 The first four stages in the construction process of a triangular array via the generating subarray illustrated in Fig. 3.66 with an expansion factor of $\delta = 2$. The element locations correspond to the vertices of the triangles.

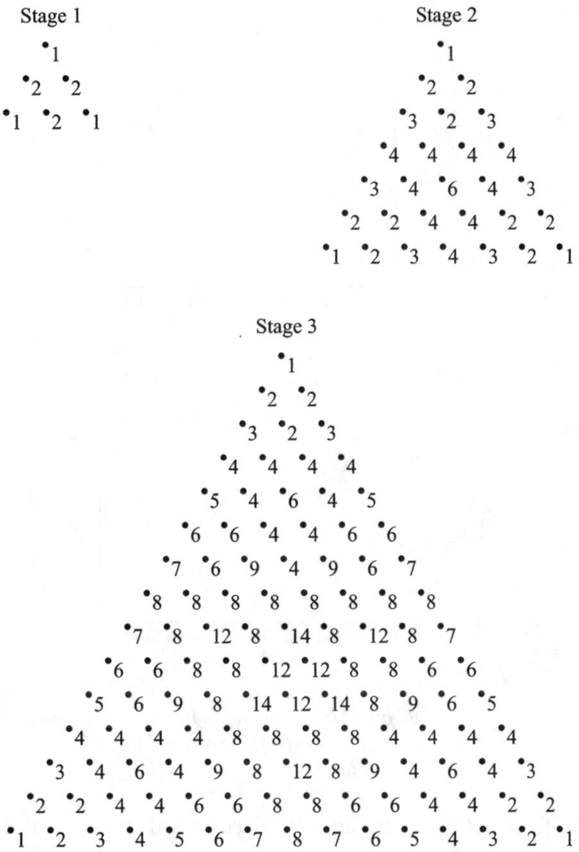

Figure 3.68 Current distributions for the first three triangular arrays shown in Fig. 3.67.

The triangular arrays shown in Fig. 3.67 are made up of many small triangular subarrays that have elements located at each of their vertices. Some of these vertices overlap and, consequently, produce a non-uniform current distribution across the array as shown in Fig. 3.68. Figure 3.69 contains a sequence of far-field radiation pattern plots, calculated using (3.280) with $\phi = 0°$, that correspond to the four triangular arrays depicted in Fig. 3.67. It is evident from the plots shown in Fig. 3.69 that a sidelobe level of at least -20 dB can be achieved by higher-order versions of these arrays. Color contour plots of these radiation patterns are also shown in Fig. 3.70 (see full-color insert). This sequence of contour plots clearly reveals the underlying self-scalability which is manifested in the radiation patterns of these triangular arrays.

3.7.2.4 Hexagonal Arrays

Another type of planar array configuration in common use is the hexagonal array. These arrays are becoming increasingly popular, especially for their applications in the area of wireless communications. The standard hexagonal arrays are formed by placing elements in an equilateral triangular grid with spacings d (see, for example, Fig. 2.8 of [64]). These arrays can also be viewed as consisting of a single element

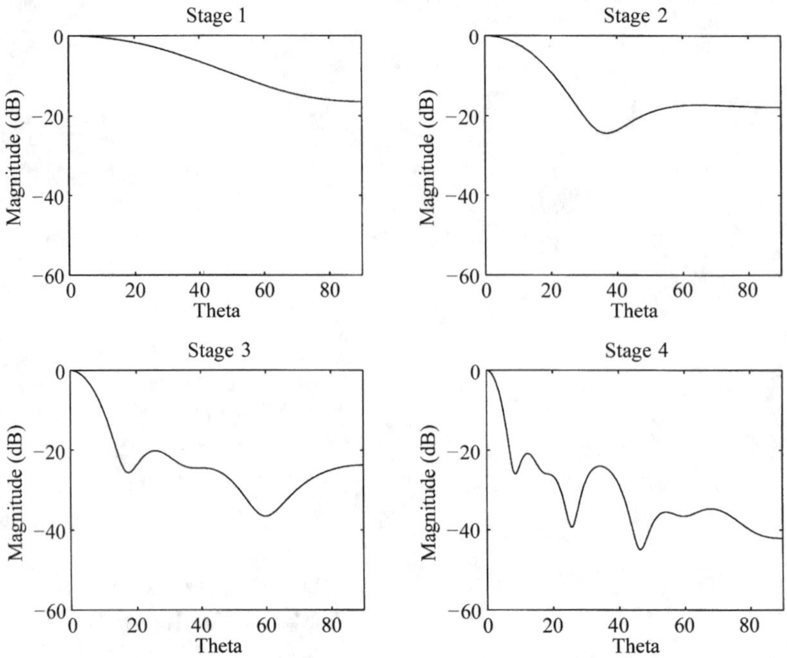

Figure 3.69 Plots of the far-field radiation patterns produced by the four triangular arrays shown in Fig. 3.67.

located at the center surrounded by several concentric six-element circular arrays of different radii. This property has been used to derive an expression for the hexagonal array factor [64]. The resulting expression, in normalized form, is given by

$$\widehat{AF}_P(\theta, \phi) = \frac{I_0 + \sum_{p=1}^{P} \sum_{m=1}^{p} \sum_{n=0}^{5} I_{pmn} e^{j[kr_{pm} \sin\theta \cos(\phi - \phi_{pmn}) + \alpha_{pmn}]}}{I_0 + \sum_{p=1}^{P} \sum_{m=1}^{p} \sum_{n=0}^{5} I_{pmn}} \tag{3.282}$$

where

$$r_{pm} = d\sqrt{p^2 + (m-1)^2 - p(m-1)} \tag{3.283}$$

$$\phi_{pmn} = \cos^{-1}\left[\frac{r_{pm}^2 + d^2p^2 - d^2(m-1)^2}{2r_{pm}dp}\right] + \frac{n\pi}{3} \tag{3.284}$$

$$\alpha_{pmn} = -kr_{pm} \sin\theta_0 \cos(\phi_0 - \phi_{pmn}) \tag{3.285}$$

and P is the number of concentric hexagons in the array. Hence, the total number of elements contained in an array with P hexagons is

$$N_P = 3P(P+1) + 1 \tag{3.286}$$

At this point we investigate the possibility that useful designs for hexagonal arrays may be realized via a construction process based on the recursive application of a generating sub-array. To demonstrate this, suppose we consider the uniformly excited six-element circular generating sub-array of radius $r = \lambda/2$ shown in Fig. 3.71.

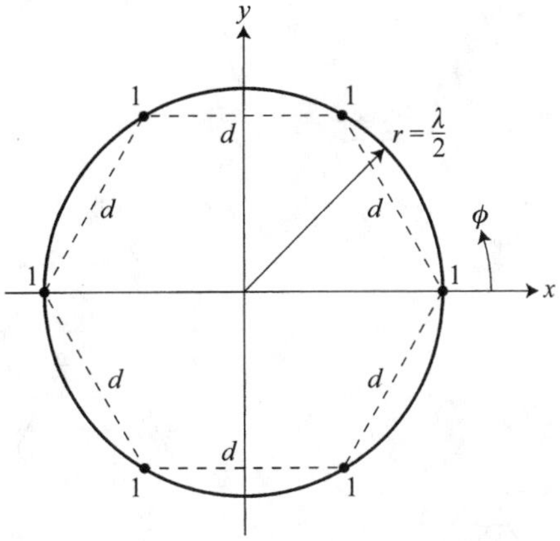

Figure 3.71 The geometry for a uniformly excited six-element circular sub-array generator of radius $r = \lambda/2$.

This particular value of radius was chosen so that the six elements in the array correspond to the vertices of a hexagon with half-wavelength sides (i.e., $d = \lambda/2$). Consequently, the array factor associated with this six-element generating sub-array may be shown to have the following representation:

$$\widehat{AF}_P(\theta, \phi) = \frac{1}{6^P} \prod_{p=1}^{P} \sum_{n=1}^{6} e^{j\delta^{p-1}[\pi \sin\theta \cos(\phi - \phi_n) + \alpha_n]} \tag{3.287}$$

where

$$\phi_n = (n-1)\frac{\pi}{3} \tag{3.288}$$

$$\alpha_n = -\pi \sin\theta_0 \cos(\phi_0 - \phi_n) \tag{3.289}$$

The array factor expression given in (3.287) may also be written in the form

$$\widehat{AF}_P(\theta, \phi) = \frac{1}{6^P} \prod_{p=1}^{P} \sum_{n=1}^{6} e^{j\delta^{p-1}\psi_n(\theta,\phi)} \tag{3.290}$$

where

$$\psi_n(\theta, \phi) = \pi[\sin\theta \cos(\phi - \phi_n) - \sin\theta_0 \cos(\phi_0 - \phi_n)] \tag{3.291}$$

We will first examine the special case where the expansion factor of the recursive hexagonal array is assumed to be unity (i.e., $\delta = 1$). Under these circumstances, (3.290) will reduce to

$$\widehat{AF}_P(\theta, \phi) = \left[\frac{1}{6}\sum_{n=1}^{6} e^{j\psi_n(\theta,\phi)}\right]^P \tag{3.292}$$

These arrays increase in size at a rate that obeys the relationship

$$N_P = 3P(P+1) + (1 - \delta_{P1}) \tag{3.293}$$

where δ_{P1} represents the Kronecker delta function defined by

$$\delta_{P1} = \begin{cases} 1, & P = 1 \\ 0, & P \neq 1 \end{cases} \tag{3.294}$$

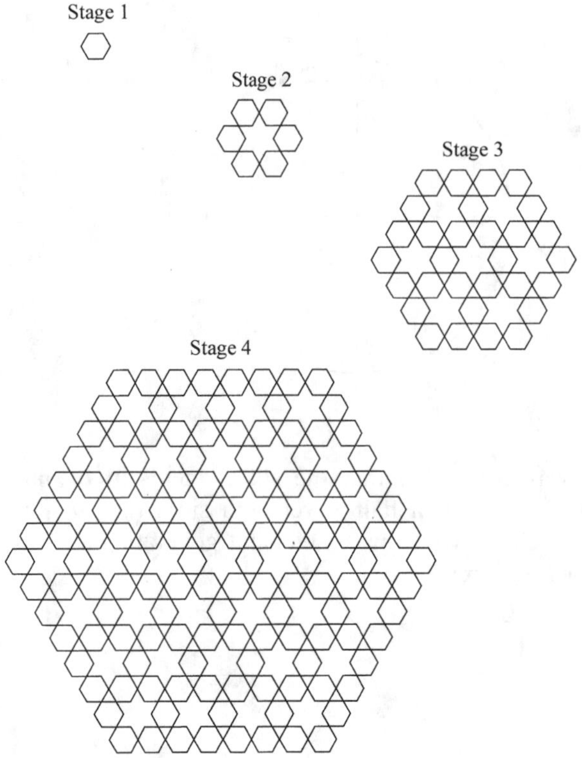

Figure 3.72 A schematic representation of the first four stages in the construction of a hexagonal array. The element locations correspond to the vertices of the hexagons.

In other words, every time this array evolves from one stage to the next, the number of concentric hexagonal sub-arrays contained in it increases by one.

The second special case of interest to be considered in this section results when a choice of $\delta = 2$ is made. Substituting this value of δ into (3.290) yields an expression for the recursive hexagonal array factor given by

$$\widehat{AF}_P(\theta, \phi) = \frac{1}{6^P} \prod_{p=1}^{P} \sum_{n=1}^{6} e^{j2^{p-1}\psi_n(\theta,\phi)} \tag{3.295}$$

where

$$N_P = 3[2^P(2^P - 1) - 2^{P-1}(2^{P-1} - 1)] \tag{3.296}$$

Clearly, by comparing (3.296) with (3.293), we conclude that these recursive arrays will grow at a much faster rate than those generated by a choice of $\delta = 1$. Schematic representations of the first four stages in the construction process of these arrays are illustrated in Fig. 3.72, where the element locations correspond to the vertices of the hexagons. Figure 3.73 shows the element locations and associated current distributions for each of the four hexagonal arrays depicted in Fig. 3.72. Figures 3.72 and 3.73 indicate that the hexagonal arrays which result from the recursive construction process with $\delta = 2$ have some elements missing, i.e., they are thinned. This is a potential advantage of these arrays from the design point of view since they may be realized with fewer elements. Another advantage of these arrays is

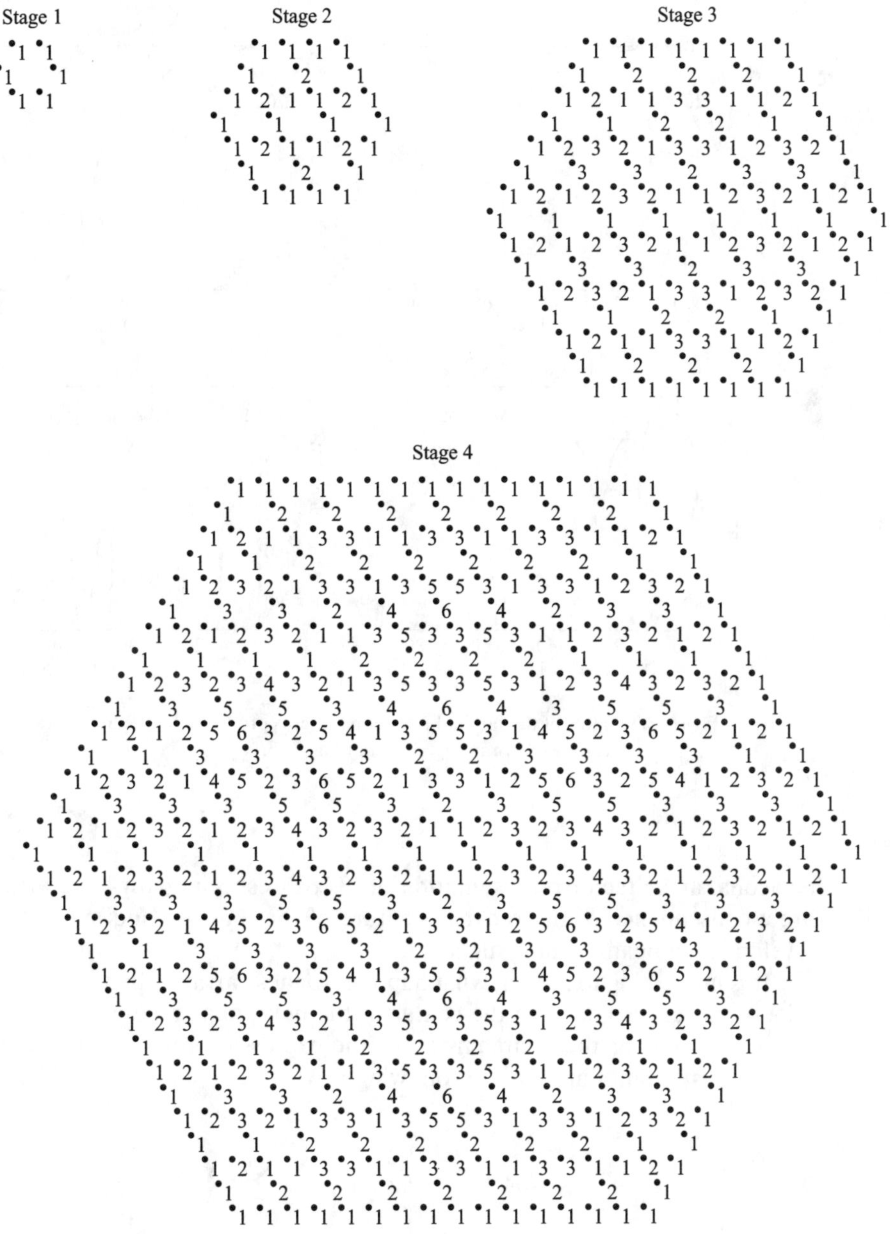

Figure 3.73 The element locations and associated current distributions for each of the four hexagonal arrays shown in Fig. 3.72.

that they possess low sidelobe levels, as indicated by the set of radiation pattern slices for $\phi = 90°$ shown in Fig. 3.74. Full-color contour plots of these radiation patterns have also been included in Fig. 3.75 (see full-color insert). Finally, we note that the compact product form of the array factor given in (3.295) offers a significant advantage in terms of computational efficiency when compared to the conventional

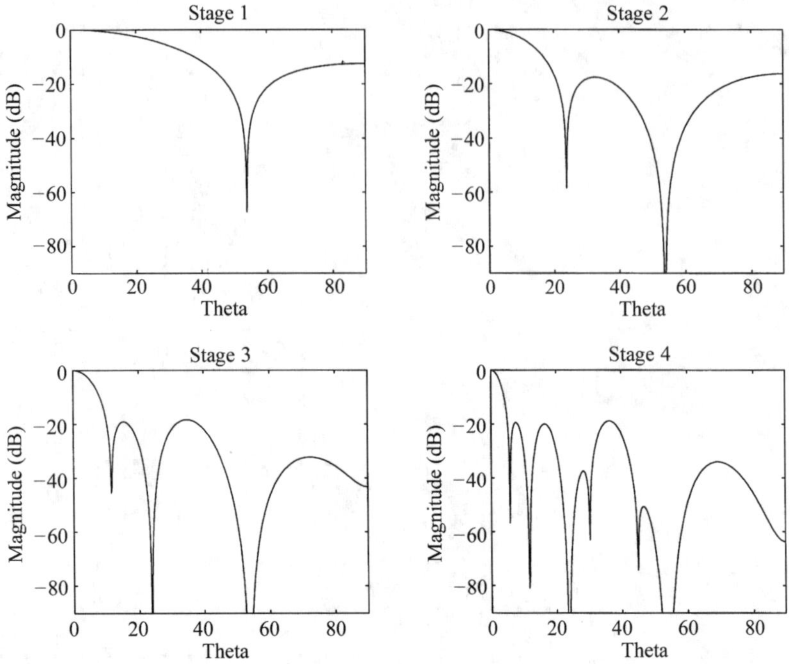

Figure 3.74 Plots of the far-field radiation patterns produced by the four hexagonal arrays shown in Fig. 3.72 and Fig. 3.73.

hexagonal array factor representation of (3.282), especially for large arrays. This is a direct consequence of the recursive nature of these arrays and may be exploited to develop rapid beamforming algorithms.

It is interesting to look at what happens to these arrays when an element with two units of current is added to the center of the hexagonal generating sub-array shown in Fig. 3.71. Under these circumstances the expression for the array factor given in (3.290) must be modified in the following way:

$$\widehat{AF}_P(\theta, \phi) = \frac{1}{8^P} \prod_{p=1}^{P} \left\{ 2 + \sum_{n=1}^{6} e^{j\delta^{p-1}\psi_n(\theta,\phi)} \right\} \tag{3.297}$$

Plots of several radiation patterns calculated from (3.297) with $\delta = 1$ and $\delta = 2$ are shown in Figs. 3.76 and 3.77, respectively. These plots indicate that a further reduction in sidelobe levels may be achieved by including a central element in the generating sub-array of Fig. 3.71. Figure 3.78 (see full-color insert) contains a series of color contour plots that show how the radiation pattern intensity evolves for a choice of $\delta = 2$. Finally, a series of color contour plots of the radiation intensity for this array is shown in Fig. 3.79 (see full-color insert), where the phasing of the generating sub-array has been chosen so as to produce a mainbeam maximum at $\theta_0 = 45°$ and $\phi_0 = 90°$.

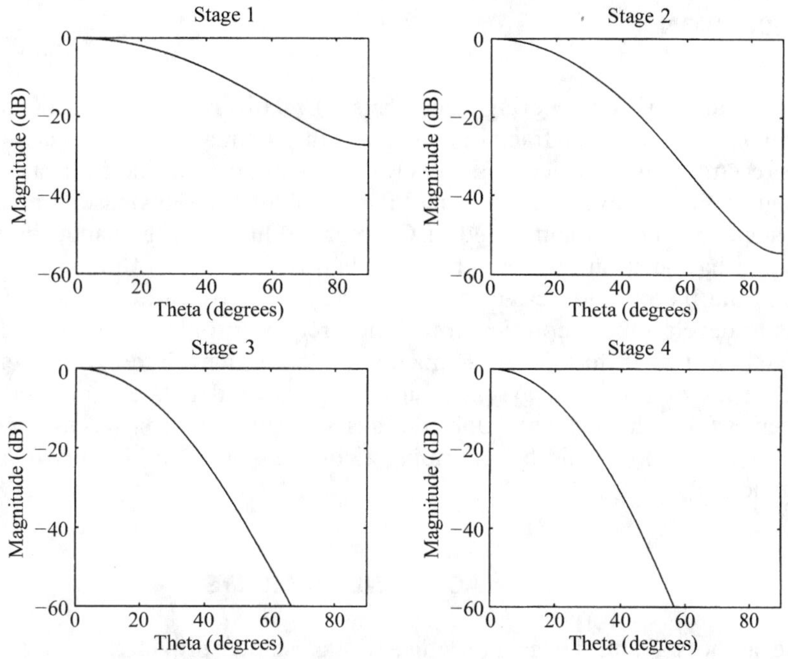

Figure 3.76 Plots of the far-field radiation patterns produced by a series of four ($P = 1, 2, 3,$ and 4) fully populated hexagonal arrays generated with an expansion factor of $\delta = 1$.

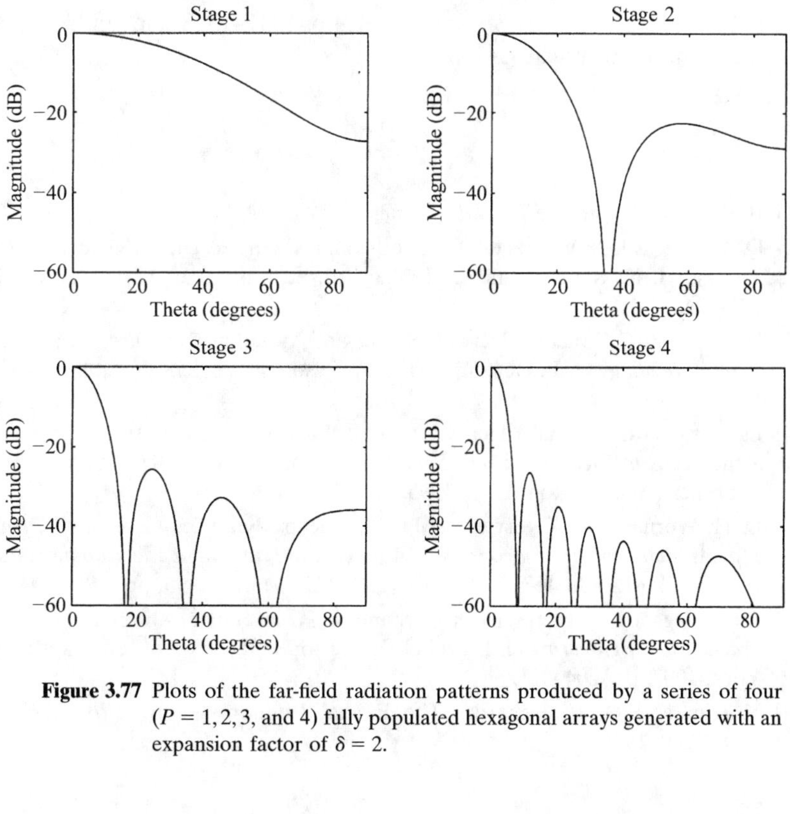

Figure 3.77 Plots of the far-field radiation patterns produced by a series of four ($P = 1, 2, 3,$ and 4) fully populated hexagonal arrays generated with an expansion factor of $\delta = 2$.

3.8 CONCLUSION

Fractal antenna engineering represents a relatively new field of research that combines attributes of fractal geometry with antenna theory. Research in this area has recently yielded a rich class of new designs for antenna elements as well as arrays. Progress in the development of multiband antenna designs based on fractal-shaped elements has been summarized in Chapter 2. On the other hand, the focus of this chapter has been on providing a comprehensive overview of the present state of the art in the theory and design of fractal arrays. The overall objective of this chapter was to develop the theoretical foundation required for the analysis and synthesis of fractal arrays. It has been demonstrated here that there are several desirable properties of fractal arrays including frequency-independent multiband behavior, schemes for realizing low sidelobe designs, systematic approaches to thinning, and the ability to develop rapid beamforming algorithms by exploiting the recursive nature of fractals.

ACKNOWLEDGMENTS

The authors gratefully acknowledge the assistance provided by Waroth Kuhirun, David Jones and Rodney Martin. Portions of the text in Section 3.4 have been reproduced with permission from D. H. Werner and P. L. Werner, *Radio Science*, Vol. 30, pp. 29–45, January–February 1995, ©American Geophysical Union, and portions of the text in Section 3.5 have been reproduced with permission from D. H. Werner and P. L. Werner, *Radio Science*, Vol. 31, pp. 1331–1343, November–December 1996 © American Geophysical Union.

REFERENCES

[1] B. B. Mandelbrot, *The Fractal Geometry of Nature*, W. H. Freeman, New York, 1983.

[2] D. L. Jaggard, "On Fractal Electrodynamics," in *Recent Advances in Electromagnetic Theory*, H. N. Kritikos and D. L. Jaggard, eds., Springer-Verlag, New York, 1990, pp. 183–224.

[3] D. L. Jaggard, "Fractal Electrodynamics and Modeling," in *Directions in Electromagnetic Wave Modeling*, H. L. Bertoni and L. B. Felson, eds., Plenum Publishing Co., New York, 1991, pp. 435–446.

[4] D. L. Jaggard, "Fractal Electrodynamics: Wave Interactions With Discretely Self-Similar Structures," in *Electromagnetic Symmetry*, C. Baum and H. Kritikos, eds., Taylor & Francis Publishers, Washington, D.C., 1995, pp. 231–281.

[5] D. H. Werner, "An Overview of Fractal Electrodynamics Research," *Proceedings of the 11th Annual Review of Progress in Applied Computational Electromagnetics (ACES)*, Vol. II, Naval Postgraduate School, Monterey, CA, March 1995, pp. 964–969.

[6] D. L. Jaggard, "Fractal Electrodynamics: From Super Antennas to Superlattices," in *Fractals in Engineering*, J. L. Vehel, E. Lutton and C. Tricot, eds., Springer-Verlag, New York, 1997, pp. 204–221.

[7] Y. Kim and D. L. Jaggard, "The Fractal Random Array," *Proc. IEEE*, **74** (9), pp. 1278–1280, 1986.

[8] A. Lakhtakia, V. K. Varadan, and V. V. Varadan, "Time-harmonic and Time-dependent Radiation by Bifractal Dipole Arrays," *Int. J. Electronics*, **63** (6), pp. 819–824, 1987.

[9] A. Lakhtakia, N. S. Holter, and V. K. Varadan, "Self-similarity in Diffraction by a Self-similar Fractal Screen," *IEEE Trans. Antennas Propagat.*, **AP-35** (2), pp. 236–239, February 1987.

[10] C. Allain and M. Cloitre, "Spatial Spectrum of a General Family of Self-similar Arrays," *Phys. Rev. A*, **36**, pp. 5751–5757, 1987.

[11] D. L. Jaggard and A. D. Jaggard, "Fractal Apertures: The Effect of Lacunarity," *Proceedings of the URSI Radio Science Meeting*, p. 728, Montréal, Canada, July 1997.

[12] M. M. Beal and N. George, "Features in the Optical Transform of Serrated Apertures and Disks," *J. Opt. Soc. Am.*, **A6**, pp. 1815–1826, 1989.

[13] D. L. Jaggard, T. Spielman, and X. Sun, "Fractal Electrodynamics and Diffraction by Cantor Targets," *Proceedings of the URSI Radio Science Meeting*, p. 333, London, Ontario, Canada, June 1991.

[14] Y. Kim, H. Grebel, and D. L. Jaggard, "Diffraction by Fractally Serrated Apertures," *J. Opt. Soc. Am.*, **A8**, pp. 20–26, 1991.

[15] T. Spielman and D. L. Jaggard, "Diffraction by Cantor Targets: Theory and Experiments," *Proceedings of the URSI Radio Science Meeting*, p. 225, Chicago, IL, July 1992.

[16] D. L. Jaggard and T. Spielman, "Diffraction From Triadic Cantor Targets," *Microwave and Optical Technology Letters*, **5**, pp. 460–466, 1992.

[17] D. L. Jaggard, T. Spielman, and M. Dempsey, "Diffraction by Two-dimensional Cantor Apertures," *Proceedings of the URSI Radio Science Meeting*, p. 314, Ann Arbor, MI, June/July 1993.

[18] D. H. Werner and P. L. Werner, "Fractal Radiation Pattern Synthesis," *Proceedings of the URSI National Radio Science Meeting*, Boulder, Colorado, January 1992, p. 66.

[19] D. H. Werner and P. L. Werner, "On the Synthesis of Fractal Radiation Patterns," *Radio Science*, **30** (1), pp. 29–45, 1995.

[20] X. Liang, W. Zhensen, and W. Wenbing, "Synthesis of Fractal Patterns From Concentric-Ring Arrays," *IEE Electronics Letters*, **32** (21), pp. 1940–1941, October 1996.

[21] C. Puente Baliarda and R. Pous, "Fractal Design of Multiband and Low Side-lobe Arrays," *IEEE Trans. Antennas Propagat.*, **44** (5), pp. 730–739, May 1996.

[22] D. H. Werner and P. L. Werner, "Frequency-independent Features of Self-similar Fractal Antennas," *Radio Science*, **31** (6), pp. 1331–1343, 1996.

[23] R. L. Haupt and D. H. Werner, "Fast Array Factor Calculations for Fractal Arrays," *Proceedings of the 13th Annual Review of Progress in Applied Computational Electromagnetics (ACES)*, Vol. I, Naval Postgraduate School, Monterey, CA, March 1997, pp. 291–296.

[24] D. H. Werner and R. L. Haupt, "Fractal Constructions of Linear and Planar Arrays," *Proceedings of the IEEE Antennas and Propagation Society International Symposium*, **3**, pp. 1968–1971, Montréal, Canada, July 1997.

[25] R. L. Haupt and D. H. Werner, "Fractal Constructions and Decompositions of Linear and Planar Arrays," submitted to *IEEE Trans. Antennas Propagat.*

[26] D. L. Jaggard and A. D. Jaggard, "Cantor Ring Arrays," *Proceedings of the IEEE Antennas and Propagation Society International Symposium*, **2**, pp. 866–869, Atlanta, GA, June 1998.

[27] S. A. Schelkunoff, "A Mathematical Theory of Linear Arrays," *Bell Syst. Tech. J.*, **22**, pp. 80–107, 1943.

[28] M. V. Berry, "Diffractals," *J. Phys. A: Math. Gen.*, **12**, pp. 781–797, 1979.

[29] J. D. Jackson, *Classical Electrodynamics*, John Wiley & Sons, New York, 1975.

[30] A. D. Jaggard and D. L. Jaggard, "Cantor Ring Diffractals," *Optics Communications*, **158**, pp. 141–148, 1998.

[31] D. L. Jaggard and A. D. Jaggard, "Fractal Ring Arrays," invited paper submitted to *Wave Motion*, 1999.

[32] D. L. Jaggard and A. D. Jaggard, "Polyadic Cantor Superlattices with Variable Lacunarity," *Opt. Lett.*, **22**, pp. 145–147, 1997.

[33] A. D. Jaggard and D. L. Jaggard, "Scattering from Fractal Superlattices with Variable Lacunarity," *J. Opt. Soc. Am. A*, **15**, pp. 1626–1635, 1998.

[34] C. Allain and M. Cloitre, "Optical Fourier Transforms of Fractals," in *Fractals in Physics*, L. Pietronero and E. Tosatti, eds., Elsevier Science, New York, pp. 61–64, 1986.

[35] C. Allain and M. Cloitre, "Optical Diffraction on Fractals," *Phys. Rev. B*, **33**, pp. 3566–3569, 1986.

[36] D. L. Jaggard and A. D. Jaggard, "Cantor Ring Arrays," *Microwave and Optical Technology Letters*, **19**, pp. 121–125, 1998.

[37] P. L. Werner, D. H. Werner, and A. J. Ferraro, "Fractal Arrays and Fractal Radiation Patterns," *Proceedings of the 11th Annual Review of Progress in Applied Computational Electromagnetics (ACES)*, **II**, Naval Postgraduate School, Monterey, CA, March 1995, pp. 970–978.

[38] M. F. Barnsley, *Fractals Everywhere*, Academic Press, New York, 1988.

[39] K. Falconer, *Fractal Geometry*, John Wiley & Sons, New York, 1990.

[40] R. F. Voss, "Fractals in Nature: From Characterization to Simulation," in *The Science of Fractal Images*, H.-O. Peitgen and D. Saupe, eds., Springer-Verlag, New York, 1988, pp. 21–70.

[41] M. V. Berry and Z. V. Lewis, "On the Weierstrass–Mandelbrot Fractal Function," *Proc. R. Soc. London A*, **370**, pp. 459–484, 1980.

[42] M. T. Ma, *Theory and Application of Antenna Arrays*, John Wiley & Sons, New York, 1974.

[43] D. L. Jaggard and X. Sun, "Scattering From Bandlimited Fractal Fibers," *IEEE Trans. Antennas Propagat.*, **37** (12), pp. 1591–1597, 1989.

[44] W. L. Stutzman and G. A. Thiele, *Antenna Theory and Design*, John Wiley & Sons, New York, 1981.

[45] L. C. Andrews, *Special Functions for Engineers and Applied Mathematicians*, Macmillan, New York, 1985.

[46] C. A. Balanis, *Antenna Theory: Analysis and Design*, John Wiley & Sons, New York, 1997.

[47] P. E. Mayes, G. A. Deschamps, and W. T. Patton, "Backward-wave Radiation From Periodic Structures and Application to the Design of Frequency-independent Antennas," *Proc. IRE*, **49**, pp. 962–963, May 1961.

[48] P. E. Mayes, "Frequency-independent Antennas and Broadband Derivatives Thereof," *Proc. IEEE*, **80** (1), January 1992.

[49] J. D. Dyson, "The Equiangular-spiral Antenna," *IRE Trans. Antennas Propagat.*, **AP-7**, pp. 181–187, 1959.

[50] G. A. Deschamps and J. D. Dyson, "The Logarithmic Spiral in a Single-aperture Multimode Antenna System," *IEEE Trans. Antennas Propagat.*, **AP-19**, pp. 90–96, January 1971.

[51] V. H. Rumsey, *Frequency Independent Antennas*, Academic Press, New York, 1966.

[52] Y. T. Lo and S. W. Lee, *Antenna Handbook*, Van Nostrand Reinhold, New York, 1993.

[53] M. Schroeder, *Fractals, Chaos, Power Laws: Minutes from an Infinite Paradise*, W. H. Freeman, New York, 1991.

[54] R. Carrel, "Log Periodic Dipole Arrays," *IRE International Convention Record*, Part 1, pp. 61–75, New York, 1961.

[55] R. H. DuHamel and D. G. Berry, "Logarithmically Periodic Antenna Arrays," *Wescon Convention Record*, Part 1, pp. 161–174, 1958.

[56] K. K. Mei, M. W. Moberg, V. H. Rumsey, and Y. S. Yeh, "Directive Frequency Independent Arrays," *IEEE Trans. Antennas and Propagat.*, **44** (50), pp. 730–739, 1996.

[57] R. C. Johnson and H. Jasik, *Antenna Engineering Handbook*, McGraw-Hill, New York, Chapter 14-4, pp. 14.32–14.37, 1984.

[58] J. K. Breakall, "Introduction to the Three-dimensional Frequency-independent Phased-array (3D-FIPA): A New Class of Phased Array Design," *Proceedings of the IEEE Antennas and Propagation Society International Symposium*, **III**, pp. 1414–1417, Chicago, IL, July 1992.

[59] M. F. Barnsley, R. L. Devaney, B. B. Mandelbrot, H. O. Peitgen, D. Saupe, R. F. Voss, Y. Fisher, and M. McGuire, *The Science of Fractal Images*, Springer-Verlag, New York, 1988.

[60] C. Puente, "Fractal Antennas," Ph.D. Dissertation, Department of Signal Theory and Communications, Universitat Politècnica de Catalunya, June 1997.

[61] A. Papoulis, *Signal Analysis*, McGraw-Hill, New York, 1977.

[62] H.-O. Peitgen, H. Jurgens, and D. Saupe, *Chaos and Fractals: New Frontiers of Science*, Springer-Verlag, New York, 1992.

[63] D. H. Werner, R. L. Haupt, and P. L. Werner, "Fractal Antenna Engineering: The Theory and Design of Fractal Antenna Arrays," Feature article to appear in the *IEEE Antennas and Propagation Magazine*, October 1999.

[64] J. Litva and T. K. Y. Lo, *Digital Beamforming in Wireless Communications*, Artech House, Boston, MA, 1996.

[65] D. H. Werner, K. C. Anushko, and P. L. Werner, "The Generation of Sum and Difference Patterns Using Fractal Subarrays," *Microwave and Optical Technology Letters*, **22**(1), pp. 54–57, July 1999.

[66] A. D. Jaggard and D. L. Jaggard, "Cantor Ring Arrays: Continuous and Discrete Cases," *Proceedings of the Fractals in Engineering Conference*, pp. 370–373, Delft, The Netherlands, June 1999.

Chapter 4

TARGET SYMMETRY AND THE SCATTERING DYADIC

Carl E. Baum

ABSTRACT

This chapter considers the impact of target symmetries including reciprocity, geometrical symmetry, and self-duality on the associated scattering dyadic. By orienting various rotation axes and reflection planes of the target in special ways with respect to both incidence and scattering directions, various symmetry-associated simplifications in the scattering dyadic can be made to occur. For special cases of back- and forward-scattering the various point symmetry groups are considered with preferred axes and planes now aligned according to the common axis defined by incidence and scattering directions, giving a rich structure to the scattering dyadic. For low frequencies (electrically small target) further simplifications occur, leading to further symmetries (including invariance to reversal of incidence and scattering directions) in the scattering dyadic. An additional symmetry one can impose on a target is that of self-duality, i.e., invariance to interchange of electric and magnetic parameters. With this symmetry the scattering dyadic describing the scattering of electromagnetic waves attains special properties. These include rotation of the scattered field by the same angle the incident field is rotated, and zero backscattering for a threefold or higher rotation axis aligned along the direction of incidence. Various point symmetries of the target combined with self-duality also introduce simplifications (symmetries) in the scattering dyadic.

4.1 INTRODUCTION

In electromagnetic scattering one takes an incident plane wave

$$\tilde{\vec{E}}^{(inc)}(\vec{r},s) = E_0\tilde{f}(s)\vec{1}_p e^{-\tilde{\gamma}\vec{1}_i\cdot\vec{r}}, \qquad \vec{E}^{(inc)}(\vec{r},t) = E_0\vec{1}_p f\left(t-\frac{\vec{1}_i\cdot\vec{r}}{c}\right)$$

$$\tilde{\vec{H}}^{(inc)}(\vec{r},s) = \frac{E_0}{Z_0}\tilde{f}(s)\vec{1}_i\times\vec{1}_p e^{-\tilde{\gamma}\vec{1}_i\cdot\vec{r}}, \qquad \vec{H}^{(inc)}(\vec{r},t) = \frac{E_0}{Z_0}\vec{1}_i\times\vec{1}_p f\left(t-\frac{\vec{1}_i\cdot\vec{r}}{c}\right)$$

$$\sim \;\equiv \text{Laplace transform (two-sided) over time} \tag{4.1}$$

$$s \equiv \Omega + j\omega \equiv \text{Laplace-transform variable or complex frequency}$$

$$c \equiv [\mu_0\varepsilon_0]^{-1/2}, \qquad Z_0 \equiv \left[\frac{\mu_0}{\varepsilon_0}\right]^{1/2}, \qquad \tilde{\gamma} = \frac{s}{c}$$

$$\vec{1}_i \equiv \text{direction of incidence}, \qquad \vec{1}_p \equiv \text{polarization}, \qquad \vec{1}_i\cdot\vec{1}_p = 0$$

and scatters it from a target (scatterer) contained in some volume V (here taken of finite linear dimensions) surrounded by a closed surface S as indicated in Fig. 4.1.

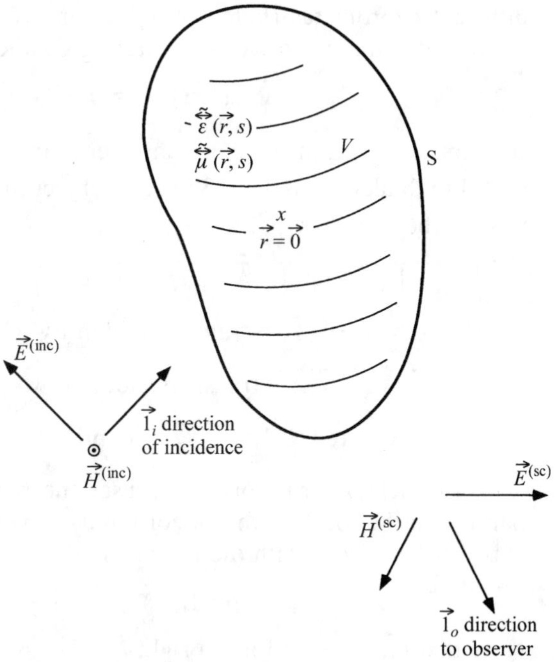

Figure 4.1 Scattering of incident plane wave.

For convenience the coordinate center $\vec{r} = \vec{0}$ is appropriately centered near or inside the target. The scattered field (superscript *sc*) is given in the far-field limit (subscript *f*) by

$$\tilde{\vec{E}}_f(\vec{r}, s) = \frac{e^{-\gamma r}}{4\pi r} \tilde{\overleftrightarrow{\Lambda}}(\vec{1}_o, \vec{1}_i; s) \cdot \tilde{\vec{E}}^{(inc)}(\vec{0}, s),$$

$$\vec{E}_f(\vec{r}, t) = \frac{1}{4\pi r} \overleftrightarrow{\Lambda}(\vec{1}_o, \vec{1}_i; t) \overset{\circ}{\cdot} \vec{E}^{(inc)}\left(\vec{0}, t - \frac{r}{c}\right)$$

$$\vec{H}_f(\vec{r}, t) = \frac{1}{Z_0} \vec{1}_o \times \vec{E}_f(\vec{r}, t), \qquad \vec{E}_f(\vec{r}, t) \cdot \vec{1}_o = \vec{0} \tag{4.2}$$

$$\tilde{\overleftrightarrow{\Lambda}}(\vec{1}_o, \vec{1}_i; s) \equiv \text{scattering dyadic},$$

$$\overleftrightarrow{\Lambda}(\vec{1}_o, \vec{1}_i; t)\circ \equiv \text{scattering dyadic operator}$$

$$\circ \equiv \text{convolution with respect to time},$$

$$\vec{r} = r\vec{1}_o, \qquad \vec{1}_o \equiv \text{direction to observer}$$

It is the properties of the scattering dyadic (usually written in frequency domain, but sometimes taken as a temporal operator) that are of interest. This gives us all our information concerning the target (by hypothesis) if one is using this scattering to identify the target (as a type of aircraft, etc.).

Note that $\vec{E}^{(inc)}$ and $\vec{E}_f$ are perpendicular to $\vec{1}_i$ and $\vec{1}_o$, respectively, and thereby have only two components each with which to be concerned. However, except in the special cases of forward- and back-scattering these components are referred to different coordinates (being contained in different planes). As such there are four important components of the scattering dyadic due to the constraints

$$\tilde{\Lambda}(\vec{1}_o, \vec{1}_i; s) \cdot \vec{1}_i = \vec{0}, \qquad \vec{1}_o \cdot \tilde{\Lambda}(\vec{1}_o, \vec{1}_i; s) = \vec{0} \tag{4.3}$$

In pairs, however, these four components are referred to two different planes. So it is useful to think of $\tilde{\Lambda}$ as a 3×3 dyadic subject to the above constraints. In another form this can be stated as

$$\begin{aligned} &\tilde{\Lambda}(\vec{1}_o, \vec{1}_i; s) = \overleftrightarrow{1}_o \cdot \tilde{\Lambda}(\vec{1}_o, \vec{1}_i; s) \cdot \overleftrightarrow{1}_i \\ &\overleftrightarrow{1}_o \equiv \overleftrightarrow{1} - \vec{1}_o \vec{1}_o \equiv \text{transverse identity with respect to } \vec{1}_o \\ &\overleftrightarrow{1}_i \equiv \overleftrightarrow{1} - \vec{1}_i \vec{1}_i \equiv \text{transverse identity with respect to } \vec{1}_i \\ &\overleftrightarrow{1} \equiv \vec{1}_x \vec{1}_x + \vec{1}_y \vec{1}_y + \vec{1}_z \vec{1}_z \equiv \text{identity (three-dimensional)} \end{aligned} \tag{4.4}$$

For backscattering one often uses the radar coordinates $\vec{1}_h$ (horizontal, usually parallel to the local earth horizon) and $\vec{1}_v$ (vertical, except not strictly so for targets above the horizon) with the relations

$$\vec{1}_v \times \vec{1}_h = \vec{1}_i, \qquad \vec{1}_i \times \vec{1}_v = \vec{1}_h, \qquad \vec{1}_h \times \vec{1}_i = \vec{1}_v \tag{4.5}$$

Note that $(\vec{1}_h, \vec{1}_v, -\vec{1}_i)$ form a right-handed system. In backscattering

$$\vec{1}_o = -\vec{1}_i \tag{4.6}$$

and $\vec{1}_h$ and $\vec{1}_v$ can be used for both transmitted and received (scattered) waves. In forward-scattering

$$\vec{1}_o = \vec{1}_i \tag{4.7}$$

and $\vec{1}_h$ and $\vec{1}_v$ can still be used for both waves.

For $\vec{1}_o$ not so simply related to $\vec{1}_i$, one can establish coordinates using what are called the scattering plane P_s and bisectrix plane P_b [19]. As indicated in Fig. 4.2, the scattering plane P_s is parallel to both $\vec{1}_i$ and $\vec{1}_o$. The bisectrix unit vector (direction) $\vec{1}_b$ lies between $-\vec{1}_i$ and $\vec{1}_o$ with equal angles ψ_b with

$$\vec{1}_b \cdot \vec{1}_o = -\vec{1}_b \cdot \vec{1}_i, \qquad \vec{1}_b \times \vec{1}_o = \vec{1}_b \times \vec{1}_i \tag{4.8}$$

The bisectrix plane P_b is parallel to $\vec{1}_b$ and perpendicular to P_s. For convenience construct unit vectors perpendicular to the two planes as

$$\vec{1}_s \perp P_s, \qquad \vec{1}_B \perp P_b, \qquad \vec{1}_b \times \vec{1}_B = \vec{1}_s, \qquad \vec{1}_B \times \vec{1}_s = \vec{1}_b, \qquad \vec{1}_s \times \vec{1}_b = \vec{1}_B \tag{4.9}$$

Note the relations

$$\vec{1}_i \cdot \vec{1}_s = 0 = \vec{1}_o \cdot \vec{1}_s, \qquad \vec{1}_i \cdot \vec{1}_B = \vec{1}_o \cdot \vec{1}_B$$

With these definitions one could choose two different "horizontal" unit vectors $\vec{1}_h^{(i)}$ and $\vec{1}_h^{(o)}$ as parallel to P_s, and "vertical" unit vectors $\vec{1}_v^{(i)}$ and $\vec{1}_v^{(o)}$ as perpendicular to P_s (and parallel to P_b) with

$$\begin{aligned} &\vec{1}_v^{(o)} \times \vec{1}_h^{(o)} = -\vec{1}_o, \qquad -\vec{1}_o \times \vec{1}_v^{(o)} = \vec{1}_h^{(o)}, \qquad -\vec{1}_h^{(o)} \times \vec{1}_o = \vec{1}_v^{(o)} \\ &\vec{1}_h^{(o)} \cdot \vec{1}_b = -\vec{1}_h^{(i)} \cdot \vec{1}_b, \qquad \vec{1}_h^{(o)} \times \vec{1}_b = -\vec{1}_h^{(i)} \times \vec{1}_b, \qquad \vec{1}_v^{(o)} = \vec{1}_v^{(i)} \end{aligned} \tag{4.10}$$

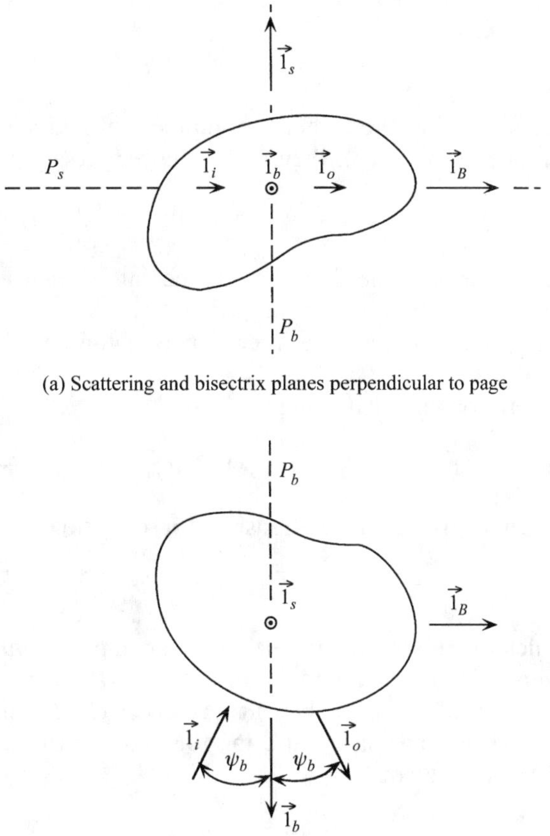
$\vec{1}_s$
P_s
$\vec{1}_i$
$\vec{1}_b$
$\vec{1}_o$
$\vec{1}_B$
P_b
(a) Scattering and bisectrix planes perpendicular to page
P_b
$\vec{1}_s$
$\vec{1}_B$
$\vec{1}_i$
$\vec{1}_o$
ψ_b
ψ_b
$\vec{1}_b$
(b) Scattering plane parallel to page and bisectrix plane perpendicular to page

Figure 4.2 Coordinates for scattering.

where the i-superscript vectors are as in (4.5). Note the minus sign used with $\vec{1}_o$ in (4.10) to make $\vec{1}_h^{(o)}$ the same as $\vec{1}_h^{(i)}$ in backscattering, the usual radar convention.

In the case of backscattering

$$\vec{1}_o = -\vec{1}_i = \vec{1}_b, \qquad P_s \perp P_b \tag{4.11}$$

but there are an infinite number of possible pairs of P_s and P_b given by rotation about their common axis parallel to $\vec{1}_i$. In forward-scattering

$$\vec{1}_o = \vec{1}_i, \qquad \vec{1}_b \cdot \vec{1}_i = 0, \qquad P_s \perp P_b \tag{4.12}$$

and while P_b is oriented perpendicular to $\vec{1}_i$, P_s can be arbitrarily rotated about an axis parallel to $\vec{1}_i$. These two special cases will reappear with special symmetries.

4.2 RECIPROCITY

Assuming that the target is comprised of reciprocal media (symmetric constitutive-parameter matrices), then scattering reciprocity gives

$$\tilde{\overleftrightarrow{\Lambda}}(\vec{1}_o, \vec{1}_i; s) = \tilde{\overleftrightarrow{\Lambda}}^T(-\vec{1}_i, -\vec{1}_o; s) \tag{4.13}$$

This expresses the fact that if one interchanges the role of transmit and receive antennas in the scattering experiment one gets the same result. Note that the transpose expresses the interchange of the coordinates for transmission and reception.

In forward-scattering we have

$$\vec{1}_o = \vec{1}_i, \qquad \tilde{\overleftrightarrow{\Lambda}}_f(\vec{1}_i, s) \equiv \tilde{\overleftrightarrow{\Lambda}}(\vec{1}_i, \vec{1}_i; s) = \tilde{\overleftrightarrow{\Lambda}}^T(-\vec{1}_i\, -\vec{1}_i; s) = \tilde{\overleftrightarrow{\Lambda}}_f^T(-\vec{1}_i, s) \tag{4.14}$$

which expresses a relationship for direction reversal. In backscattering we have

$$\vec{1}_o = -\vec{1}_i, \qquad \tilde{\overleftrightarrow{\Lambda}}_b(\vec{1}_i, s) \equiv \tilde{\overleftrightarrow{\Lambda}}(-\vec{1}_i, \vec{1}_i; s) = \tilde{\overleftrightarrow{\Lambda}}^T(-\vec{1}_i, \vec{1}_i; s) = \tilde{\overleftrightarrow{\Lambda}}_b^T(\vec{1}_i, s) \tag{4.15}$$

which expresses the fact that $\tilde{\overleftrightarrow{\Lambda}}_b$ is a complex symmetric dyadic and that $\overleftrightarrow{\Lambda}_b \circ$ is a real symmetric dyadic operator. In both forward- and back-scattering one can regard the dyadic as of size 2×2, being transverse to $\vec{1}_o$. In this form there are only four elements to consider. For backscattering this reduces to three elements since the off-diagonal elements are equal.

4.3 SYMMETRY GROUPS FOR TARGET

Let the target symmetry be expressed by a group G under which the target is invariant, given by

$$\begin{aligned}
&G = \{(G)_\ell \,|\, \ell = 1, 2, \ldots, \ell_0\}, \qquad (G)_\ell \equiv \text{group element} \\
&\ell_0 \equiv \text{group order (number of elements)}, \qquad (G)_\ell^{-1} \in G,\ (1) \equiv \text{identity} \in G \\
&(G)_{\ell_1}(G)_{\ell_2} \in G \text{ for all ordered pair of elements}
\end{aligned} \tag{4.16}$$

Let there be a set of 3×3 dyadics which form a representation of the group with

$$(G)_\ell \rightarrow \overleftrightarrow{G}_\ell, \qquad (1) \rightarrow \overleftrightarrow{1} \tag{4.17}$$

and group multiplication becoming the usual dot multiplication. For the point symmetry groups (rotations and reflections) with real coordinate transformations these dyadics are real and orthogonal with

$$\overleftrightarrow{G}_\ell^{-1} = \overleftrightarrow{G}_\ell^T, \quad \det\left(\overleftrightarrow{G}_\ell\right) = \begin{cases} +1 \Rightarrow \text{proper rotation} \\ -1 \Rightarrow \text{improper rotation (includes a reflection)} \end{cases} \tag{4.18}$$

An even number of reflections gives a proper rotation.

The order of a group element, or its dyadic representation, is n_ℓ, the smallest integer (≥ 1) such that

$$\overleftrightarrow{G}_\ell^{n_\ell} = \overleftrightarrow{1}, \qquad \frac{\ell_0}{n_\ell} = \text{positive integer} \tag{4.19}$$

The eigenvalues have the property

$$\lambda_\beta^{n_\ell}(\overleftrightarrow{G}_\ell) = 1, \qquad \beta = 1, 2, 3 \tag{4.20}$$

Combining with (4.18) gives

$$\det(\overleftrightarrow{G}_\ell) = -1 \Rightarrow n_\ell = \text{even} \tag{4.21}$$

so improper rotations all have even periods. The group (cyclic) of order n_ℓ

$$\begin{aligned} G_\ell &= \left\{(G)_\ell^n \,|\, n = 1, 2, \ldots, n_\ell\right\} \\ &= \text{subgroup of } G \text{ of smallest order containing } (G)_\ell \end{aligned} \tag{4.22}$$

is called the period of $(G)_\ell$.

A special case of interest for a group is one of order 2, called an involution [13], [16]. If $(G)_\ell$ is an element of order 2 then its period is an involution as

$$\text{period of } (G)_\ell = \left\{(1), (G)_\ell\right\} \tag{4.23}$$

In terms of the 3×3 dyadic representation then such a dyadic has the property

$$\overleftrightarrow{G}_\ell = \overleftrightarrow{G}_\ell^{-1} = \overleftrightarrow{G}_\ell^{T} \tag{4.24}$$

i.e., it is not only real and orthogonal, but also symmetric. Now any real, symmetric dyadic can be written as [4]

$$\begin{aligned} &\overleftrightarrow{G}_\ell = \sum_{\beta=1}^{3} g_{\ell,\beta} \vec{1}_{\ell,\beta} \vec{1}_{\ell,\beta}, \qquad g_{\ell,\beta} \equiv \text{real eigenvalues} \\ &\vec{1}_{\ell,\beta} \equiv \text{real eigenvectors} = \text{real unit vectors giving spatial directions} \\ &\vec{1}_{\ell,\beta_1} \cdot \vec{1}_{\ell,\beta_2} = 1_{\beta_1,\beta_2} \text{ (orthonormal)} \end{aligned} \tag{4.25}$$

Furthermore, we have

$$\overleftrightarrow{G}_\ell^2 = \overleftrightarrow{1} = \sum_{\beta=1}^{3} g_{\ell,\beta}^2 \vec{1}_{\ell,\beta} \vec{1}_{\ell,\beta}, \qquad g_{\ell,\beta}^2 = 1, \qquad g_{\ell,\beta} = \pm 1 \tag{4.26}$$

Rotate the Cartesian coordinates (x, y, z) to coincide with the $\vec{1}_\beta$ (all orthogonal, as required for such a correspondence). Then this involution dyadic gives combinations of reflections through the three coordinate planes ($x = 0$, $y = 0$, and $z = 0$) as the *only* possibilities. These can be categorized as

$$\begin{aligned} \text{all } g_{\ell,\beta} &= 1 \Rightarrow \text{identity } \overleftrightarrow{1} \text{ giving group order 1 (not an involution)} \\ \text{one } g_{\ell,\beta} &= -1 \Rightarrow \text{reflection } \overleftrightarrow{G} \text{ with respect to any plane (reflection} \\ &\qquad \text{symmetry } R\text{, improper rotation)} \\ \text{two } g_{\ell,\beta} &= -1 \Rightarrow \text{twofold rotation axis } \vec{1}_\beta \text{ corresponding to the} \\ &\qquad \text{one } g_{\ell,\beta} = 1 \text{ (rotation symmetry } C_2\text{, proper rotation)} \\ \text{all } g_{\ell,\beta} &= -1 \Rightarrow \text{inversion } -\overleftrightarrow{1} \text{ (inversion symmetry I, improper} \\ &\qquad \text{rotation)} \end{aligned} \tag{4.27}$$

Note that combinations of the above operations give groups with more than two elements.

4.4 TARGET SYMMETRY

By a symmetric target is meant one that is invariant under a coordinate transformation via the group elements discussed in Section 4.3. For each group element we have

$$\vec{r} = x\vec{1}_x + y\vec{1}_y + z\vec{1}_z, \qquad \vec{r}^{(2)} = \overleftrightarrow{G}_\ell \cdot \vec{r}^{(1)} \tag{4.28}$$

Whatever of the target is at $\vec{r}^{(1)}$ is also at $\vec{r}^{(2)}$, applying for all the elements of the symmetry-group representation. As in [10] transform the fields as

$$\vec{E}^{(2)}(\vec{r}^{(2)}, t) = \overleftrightarrow{G}_\ell \cdot \vec{E}^{(1)}(\vec{r}^{(1)}, t), \qquad \vec{H}^{(2)}(\vec{r}^{(2)}, t) = \pm\overleftrightarrow{G}_\ell \cdot \vec{H}^{(1)}(\vec{r}^{(1)}, t) \tag{4.29}$$

with + for proper rotations and − for improper rotations when dealing with magnetic parameters. This applies to all the fields and sources in the Maxwell equations and constitutive relations.

Applying this to the permittivity of the target gives

$$\begin{aligned}\tilde{\vec{D}}^{(2)}(\vec{r}^{(2)}, s) &= \overleftrightarrow{G}_\ell \cdot \tilde{\vec{D}}^{(1)}(\vec{r}^{(1)}, s) = \overleftrightarrow{G}_\ell \cdot \tilde{\overleftrightarrow{\varepsilon}}(\vec{r}^{(1)}, s) \cdot \overleftrightarrow{G}_\ell^T \cdot \overleftrightarrow{G}_\ell \cdot \tilde{\vec{E}}^{(1)}(\vec{r}^{(1)}, s) \\ &= \overleftrightarrow{G}_\ell \cdot \tilde{\overleftrightarrow{\varepsilon}}(\vec{r}^{(1)}, s) \cdot \overleftrightarrow{G}_\ell^T \cdot \tilde{\vec{E}}^{(2)}(\vec{r}, s) = \tilde{\overleftrightarrow{\varepsilon}}(\vec{r}^{(2)}, s) \cdot \tilde{\vec{E}}^{(2)}(\vec{r}^{(2)}, s)\end{aligned} \tag{4.30}$$

implying

$$\tilde{\overleftrightarrow{\varepsilon}}(\vec{r}^{(2)}, s) = \overleftrightarrow{G}_\ell \cdot \tilde{\overleftrightarrow{\varepsilon}}(\vec{r}^{(1)}, s) \cdot \overleftrightarrow{G}_\ell^T \tag{4.31}$$

This applies similarly to the conductivity (which can be absorbed into the permittivity) and permeability. So in a more general form we have

$$\tilde{\overleftrightarrow{\mu}}(\overleftrightarrow{G}_\ell \cdot \vec{r}, s) = \overleftrightarrow{G}_\ell \cdot \tilde{\overleftrightarrow{\mu}}(\vec{r}, s) \cdot \overleftrightarrow{G}_\ell^T, \qquad \tilde{\overleftrightarrow{\varepsilon}}(\overleftrightarrow{G}_\ell \cdot \vec{r}, s) = \overleftrightarrow{G}_\ell \cdot \tilde{\overleftrightarrow{\varepsilon}}(\vec{r}, s) \cdot \overleftrightarrow{G}_\ell^T \tag{4.32}$$

as the symmetry conditions of the target, noting that the above applies for all the $\overleftrightarrow{G}_\ell$ in the representation of the group G.

Applying (4.2) to the incident electric field gives

$$\vec{1}_i^{(2)} = \overleftrightarrow{G}_\ell \cdot \vec{1}_i^{(1)}, \qquad \vec{1}_p^{(2)} = \overleftrightarrow{G}_\ell \cdot \vec{1}_p^{(1)} \tag{4.33}$$

Similarly application to the far scattered field gives

$$\vec{1}_o^{(2)} = \overleftrightarrow{G}_\ell \cdot \vec{1}_o^{(1)} \tag{4.34}$$

Then transforming the scattering equation gives

$$\begin{aligned}\tilde{\vec{E}}_f^{(2)}(\vec{r}^{(2)}, s) &= \overleftrightarrow{G}_\ell \cdot \tilde{\vec{E}}_f^{(1)}(\vec{r}^{(1)}, s) = \frac{e^{-\gamma r}}{4\pi r} \overleftrightarrow{G}_\ell \cdot \tilde{\overleftrightarrow{\Lambda}}(\vec{1}_o^{(1)}, \vec{1}_i^{(1)}; s) \cdot \overleftrightarrow{G}_\ell^T \cdot \overleftrightarrow{G}_\ell \cdot \tilde{\vec{E}}_f^{(inc,1)}(\vec{0}, s) \\ &= \frac{e^{-\gamma r}}{4\pi r} \overleftrightarrow{G}_\ell \cdot \tilde{\overleftrightarrow{\Lambda}}(\vec{1}_o^{(1)}, \vec{1}_i^{(1)}; s) \cdot \overleftrightarrow{G}_\ell^T \cdot \tilde{\vec{E}}^{(inc,2)}(\vec{0}, s) \\ &= \frac{e^{-\gamma r}}{4\pi r} \tilde{\overleftrightarrow{\Lambda}}(\vec{1}_o^{(2)}, \vec{1}_i^{(2)}; s) \cdot \tilde{\vec{E}}^{(inc,2)}(\vec{0}, s)\end{aligned} \tag{4.35}$$

implying

$$\tilde{\overleftrightarrow{\Lambda}}(\overrightarrow{1}_o^{(2)}, \overrightarrow{1}_i^{(2)}; s) = \overleftrightarrow{G}_\ell \cdot \tilde{\overleftrightarrow{\Lambda}}(\overrightarrow{1}_o^{(1)}, \overrightarrow{1}_i^{(1)}; s) \cdot \overleftrightarrow{G}_\ell^T \tag{4.36}$$

So in a more general form we have

$$\tilde{\overleftrightarrow{\Lambda}}(\overleftrightarrow{G}_\ell \cdot \overrightarrow{1}_o, \overleftrightarrow{G}_\ell \cdot \overrightarrow{1}_i; s) = \overleftrightarrow{G}_\ell \cdot \tilde{\overleftrightarrow{\Lambda}}(\overrightarrow{1}_o, \overrightarrow{1}_i; s) \cdot \overleftrightarrow{G}_\ell^T \tag{4.37}$$

as the symmetry conditions for the scattering dyadic.

4.5 SYMMETRY IN GENERAL BISTATIC SCATTERING

As discussed in [19] there are various transformations of the target which give a scattering dyadic simply related to that for the original target. These are

(a) Rotation of the target by π about the bisectrix $\overrightarrow{1}_b$
(b) Reflection of the target with respect to the scattering plane P_s
(c) Reflection of the target with respect to the bisectrix plane P_b

The target need not have any of these symmetries; they still appear in the scattering dyadic.

Instead of transforming the target the above operations can be interpreted as transforming the coordinates, in particular $\overrightarrow{1}_i$ and $\overrightarrow{1}_o$. Define dyadics for the three operations above as

$$\begin{aligned} \overleftrightarrow{C}_2^{(b)} &\equiv -\overleftrightarrow{1}_b + \overrightarrow{1}_i\overrightarrow{1}_i = -\overleftrightarrow{1} + 2\overrightarrow{1}_b\overrightarrow{1}_b \equiv \text{rotation by } \pi \text{ (twofold axis) about } \overrightarrow{1}_b \\ &= \text{two-dimensional inversion transverse to } \overrightarrow{1}_b \\ \overleftrightarrow{R}_s &\equiv \overleftrightarrow{1} - 2\overrightarrow{1}_s\overrightarrow{1}_s \equiv \text{reflection through } P_s \\ \overleftrightarrow{R}_B &\equiv \overleftrightarrow{1} - 2\overrightarrow{1}_B\overrightarrow{1}_B \equiv \text{reflection through } P_b \end{aligned} \tag{4.38}$$

Each of these forms with the identity $\overleftrightarrow{1}$ an involution group. Noting that

$$\overleftrightarrow{R}_s \cdot \overleftrightarrow{R}_B \equiv \overleftrightarrow{R}_B \cdot \overleftrightarrow{R}_s = \overleftrightarrow{1} - 2\overrightarrow{1}_s\overrightarrow{1}_s - 2\overrightarrow{1}_B\overrightarrow{1}_B = -\overleftrightarrow{1} + 2\overrightarrow{1}_B\overrightarrow{1}_B = \overleftrightarrow{C}_2^{(b)} \tag{4.39}$$

then the three dyads in (4.38) with the identity $\overleftrightarrow{1}$ form a representation of the $C_{2a}^{(b)}$ group ($C_2^{(b)}$ twofold rotation axis $\overrightarrow{1}_b$, plus two axial symmetry planes P_s and P_b), a commutative group given by

$$C_{2a}^{(b)} = \left\{(1), (R_s), (R_B), (C_2^{(b)})_1\right\} = R_s \otimes R_B \tag{4.40}$$

formed from the two involution groups

$$R_s = \left\{(1), (R_s)\right\}, \qquad R_b = \left\{(1), (R_b)\right\} \tag{4.41}$$

Apply these symmetry operations to the scattering dyadic. For (a) we have

$$\begin{aligned} \overleftrightarrow{C}_2^{(b)} \cdot \overrightarrow{1}_i = -\overrightarrow{1}_o, \qquad \overleftrightarrow{C}_2^{(b)} \cdot \left(-\overrightarrow{1}_o\right) = \overrightarrow{1}_i \\ \tilde{\overleftrightarrow{\Lambda}}\left(-\overrightarrow{1}_i, -\overrightarrow{1}_o; s\right) = \tilde{\overleftrightarrow{\Lambda}}^{\mathrm{T}}\left(\overrightarrow{1}_o, \overrightarrow{1}_i; s\right) = \tilde{\overleftrightarrow{\Lambda}}\left(\overleftrightarrow{C}_2^{(b)} \cdot \overrightarrow{1}_o, \overleftrightarrow{C}_2^{(b)} \cdot \overrightarrow{1}_i; s\right) \end{aligned} \tag{4.42}$$

where the reciprocity relation (4.13) has shown that this rotation of the coordinates has not reproduced the scattering dyadic but rather its transpose. Clearly a second application of this rotation gives back the original scattering dyadic (identity operation). For (b) we have

$$\overleftrightarrow{R}_s \cdot \vec{1}_o = \vec{1}_o, \qquad \overleftrightarrow{R}_s \cdot \vec{1}_i = \vec{1}_i, \qquad \tilde{\overleftrightarrow{\Lambda}}\left(\vec{1}_o, \vec{1}_o; s\right) = \tilde{\overleftrightarrow{\Lambda}}\left(\overleftrightarrow{R}_s \cdot \vec{1}_o, \overleftrightarrow{R}_s \cdot \vec{1}_o; s\right) \tag{4.43}$$

For (c) we have

$$\begin{gathered} \overleftrightarrow{R}_B \cdot \vec{1}_i = -\vec{1}_o, \qquad \overleftrightarrow{R}_B \cdot \left(-\vec{1}_o\right) = \vec{1}_i \\ \tilde{\overleftrightarrow{\Lambda}}\left(-\vec{1}_i, -\vec{1}_o; s\right) = \tilde{\overleftrightarrow{\Lambda}}^T\left(\vec{1}_o, \vec{1}_i; s\right) = \tilde{\overleftrightarrow{\Lambda}}\left(\overleftrightarrow{R}_B \cdot \vec{1}_o, \overleftrightarrow{R}_B \cdot \vec{1}_i; s\right) \end{gathered} \tag{4.44}$$

with the same form as (4.42). Note that these are not the same results in general as in (4.37) for geometrical symmetries in the target since they come from the reciprocity symmetry.

Going a step further let each of these symmetry dyadics individually be symmetry dyadics $\overleftrightarrow{G}_\ell$ for the target. Then use (4.37) in combination with the above results for each of these three involution dyadics. For (a) we have

$$\tilde{\overleftrightarrow{\Lambda}}\left(\overleftrightarrow{C}_2^{(b)} \cdot \vec{1}_o, \overleftrightarrow{C}_2^{(b)} \cdot \vec{1}_i; s\right) = \overleftrightarrow{C}_2^{(b)} \cdot \tilde{\overleftrightarrow{\Lambda}}\left(\vec{1}_o, \vec{1}_i; s\right) \cdot \overleftrightarrow{C}_2^{(b)} = \tilde{\overleftrightarrow{\Lambda}}^T\left(\vec{1}_o, \vec{1}_i; s\right) \tag{4.45}$$

For (b) we have

$$\tilde{\overleftrightarrow{\Lambda}}\left(\overleftrightarrow{R}_s \cdot \vec{1}_o, \overleftrightarrow{R}_s \cdot \vec{1}_i; s\right) = \overleftrightarrow{R}_s \cdot \tilde{\overleftrightarrow{\Lambda}}\left(\vec{1}_o, \vec{1}_i; s\right) \cdot \overleftrightarrow{R}_s = \tilde{\overleftrightarrow{\Lambda}}\left(\vec{1}_o, \vec{1}_i; s\right) \tag{4.46}$$

For (c) we have

$$\tilde{\overleftrightarrow{\Lambda}}\left(\overleftrightarrow{R}_B \cdot \vec{1}_o, \overleftrightarrow{R}_B \cdot \vec{1}_i; s\right) = \overleftrightarrow{R}_B \cdot \tilde{\overleftrightarrow{\Lambda}}\left(\vec{1}_o, \vec{1}_i; s\right) \cdot \overleftrightarrow{R}_B = \tilde{\overleftrightarrow{\Lambda}}^T\left(\vec{1}_o, \vec{1}_i; s\right) \tag{4.47}$$

This last case of a bisectrix symmetry plane (c) is treated in some detail in [1], [17] in terms of symmetric and antisymmetric parts of the fields and associated natural modes. So, as the above results indicate, the particular geometrical symmetries combine with reciprocity to give more symmetry in the scattering dyadic.

Use of the h,v coordinates discussed in Section 4.1 is a natural choice for simplifying the representation of the scattering dyadic. In particular for (b) (target symmetric with respect to the scattering plane) the incident field is a symmetric field when polarized parallel to $\vec{1}_h^{(i)}$ giving a far scattered field parallel to $\vec{1}_h^{(o)}$. When the incident field is polarized parallel to $\vec{1}_v^{(i)}$ the far scattered field is polarized parallel to $\vec{1}_v^{(o)} = \vec{1}_v^{(i)}$. This can be regarded as some kind of generalized diagonal form.

4.6 SYMMETRY IN BACKSCATTERING

Specializing to the important case of backscattering, recall from (4.15)

$$\vec{1}_o = -\vec{1}_i, \qquad \tilde{\overleftrightarrow{\Lambda}}_b\left(\vec{1}_i, s\right) \equiv \tilde{\overleftrightarrow{\Lambda}}\left(-\vec{1}_i, \vec{1}_i; s\right) = \tilde{\overleftrightarrow{\Lambda}}_b^T\left(\vec{1}_i, s\right) \tag{4.48}$$

The scattering dyadic can be regarded as a symmetric 2×2 dyadic or matrix by considering only coordinates (h, v) transverse to $\vec{1}_i$. Also, as discussed in Section 4.1, the scattering and bisectrix planes are not uniquely specified, but are any pair of orthogonal planes with $\vec{1}_i$ parallel to their common axis of intersection.

Consider an orthogonal (real) dyadic (matrix) of the form

$$(T_{n,m}) = \begin{pmatrix} T_{h,h} & T_{h,v} & 0 \\ T_{v,h} & T_{v,v} & 0 \\ 0 & 0 & 1 \end{pmatrix}, \qquad (T_{n,m})^{-1} = (T_{n,m})^T \tag{4.49}$$

where the coordinates are arranged in a right-handed $(h, v, 0)$ coordinate system. Note that when operating on longitudinal vectors, no change is induced; i.e.,

$$\left(T_{n,m}\right)\cdot\vec{1}_i = \vec{1}_i, \qquad \left(T_{n,m}\right)\cdot\vec{1}_o = \vec{1}_o, \qquad \vec{1}_z = \vec{1}_o = -\vec{1}_i \tag{4.50}$$

Then for backscattering (4.37) becomes

$$\tilde{\overleftrightarrow{\Lambda}}_b\left(\vec{1}_i, s\right) = \left(T_{n,m}\right)\cdot\tilde{\overleftrightarrow{\Lambda}}_b\left(\vec{1}_i, s\right)\cdot\left(T_{n,m}\right)^T \tag{4.51}$$

provided $(T_{n,m})$ is a symmetry of the target for the $\vec{1}_i$ of interest. Note that in the Cartesian coordinates appropriate to (4.49) the z coordinate has been chosen as a symmetry axis or parallel to a symmetry plane of the target.

Now apply the general symmetries noted in Section 4.5. Since the bisectrix $\vec{1}_b = \vec{1}_o = -\vec{1}_i$, then for item (a) we have a twofold rotation axis as $\vec{1}_o$ or $-\vec{1}_i$. In two-dimensional form

$$C_2 = \{(1), (C_2)_1\}, \qquad (1)\rightarrow\overleftrightarrow{1}_i = \overleftrightarrow{1}_o \tag{4.52}$$

$$(C_2)_1\rightarrow\left(C_{n,m}(\pi)\right) = \begin{pmatrix} \cos(\pi) & -\sin(\pi) \\ \sin(\pi) & \cos(\pi) \end{pmatrix} = -\begin{pmatrix} 1 & 0 \\ 0 & 1 \end{pmatrix} = -\overleftrightarrow{1}_i = -\overleftrightarrow{1}_o$$

In three-dimensional form rotations about $-\vec{1}_i$ are generalized as

$$\left(C_{n,m}(\phi)\right) = \begin{pmatrix} \cos(\phi) & -\sin(\phi) & 0 \\ \sin(\phi) & \cos(\phi) & 0 \\ 0 & 0 & 1 \end{pmatrix} = \overleftrightarrow{C}(\phi) \tag{4.53}$$

but for present purposes the two-dimensional form is adequate. Applying this to the scattering dyadic gives

$$\tilde{\overleftrightarrow{\Lambda}}_b\left(\vec{1}_i, s\right) = \left(C_{n,m}(\pi)\right)\cdot\tilde{\overleftrightarrow{\Lambda}}_b\left(\vec{1}_i, s\right)\cdot\left(C_{n,m}(\pi)\right)^T \tag{4.54}$$

so the backscattering dyadic has C_2 symmetry, regardless of any such symmetry in the target with respect to the $\vec{1}_i$ axis.

Concerning items (b) and (c) dealing with reflection through P_s and P_b, these become any axial planes (containing $\vec{1}_i$). Let $\overleftrightarrow{R}_a$ represent any such reflection. Noting that the h, v coordinates can be arbitrarily rotated so that this reflection reverses only the h coordinate we have

$$\overleftrightarrow{R}_h \cdot \tilde{\overleftrightarrow{\Lambda}}_b\left(\vec{1}_i, s\right) \cdot \overleftrightarrow{R}_h = \begin{pmatrix} -1 & 0 \\ 0 & 1 \end{pmatrix} \cdot \begin{pmatrix} \tilde{\Lambda}_{h,h}\left(\vec{1}_i, s\right) & \tilde{\Lambda}_{h,v}\left(\vec{1}_i, s\right) \\ \tilde{\Lambda}_{v,h}\left(\vec{1}_i, s\right) & \tilde{\Lambda}_{v,v}\left(\vec{1}_i, s\right) \end{pmatrix} \cdot \begin{pmatrix} -1 & 0 \\ 0 & 1 \end{pmatrix}$$

$$= \begin{pmatrix} \tilde{\Lambda}_{h,h}\left(\vec{1}_i, s\right) & -\tilde{\Lambda}_{h,v}\left(\vec{1}_i, s\right) \\ -\tilde{\Lambda}_{v,h}\left(\vec{1}_i, s\right) & \tilde{\Lambda}_{v,v}\left(\vec{1}_i, s\right) \end{pmatrix} \tag{4.55}$$

$$\tilde{\Lambda}_{h,v}\left(\vec{1}_i, s\right) = \tilde{\Lambda}_{v,h}\left(\vec{1}_i, s\right) \text{ (reciprocity)}$$

which is a rather simple form for the transformation of the dyadic, but not an invariance. If the target has an axial symmetry plane and $\overleftrightarrow{R}_a$ is oriented to reflect through this plane, then we have

$$\tilde{\overleftrightarrow{\Lambda}}_b\left(\vec{1}_i, s\right) = \overleftrightarrow{R}_a \cdot \tilde{\overleftrightarrow{\Lambda}}_b\left(\vec{1}_i, s\right) \cdot \overleftrightarrow{R}_a = \left(-\overleftrightarrow{R}_a\right) \cdot \tilde{\overleftrightarrow{\Lambda}}_b\left(\vec{1}_i, s\right) \cdot \left(-\overleftrightarrow{R}_a\right) \tag{4.56}$$

as symmetries of the backscattering operator. To see the significance of this choose the (h, v) coordinates as in (4.55) to give

$$\overleftrightarrow{R}_h = \begin{pmatrix} -1 & 0 \\ 0 & 1 \end{pmatrix} = \text{reflection of } h \text{ coordinate}$$
$$\overleftrightarrow{R}_v = \begin{pmatrix} 1 & 0 \\ 0 & -1 \end{pmatrix} = -\overleftrightarrow{R}_h \equiv \text{reflection of } v \text{ coordinate} \tag{4.57}$$

So one axial symmetry plane in the target ($\overleftrightarrow{R}_a$ reflection) gives two symmetry planes ($\overleftrightarrow{R}_a$ reflection and $-\overleftrightarrow{R}_a$ reflection) in the backscattering dyadic. Viewed another way, the C_2 symmetry in $\tilde{\overleftrightarrow{\Lambda}}_b$ plus an axial symmetry plane (R_a symmetry) in the target give C_{2a} symmetry in $\tilde{\overleftrightarrow{\Lambda}}_b$, a twofold rotation axis with two axial symmetry planes. With this kind of symmetry (4.55) implies that with the orientation of one of the radar coordinates (either h or v) along this symmetry plane then $\tilde{\overleftrightarrow{\Lambda}}_b$ is a diagonal dyadic (matrix), making this a convenient choice of coordinate axes.

For $\vec{1}_i$ as a higher-order (N-fold) rotation axis we have C_N symmetry

$$C_N = \left\{(C_N)_n \middle| n = 1, 2, \ldots, N\right\},$$
$$(C_N)_n \equiv \text{rotation by } \frac{2\pi n}{N} \text{ (positive } \phi \text{ direction)}$$
$$(C_N) = (C_N)_1^n, \qquad (C_N)_N = (1) \tag{4.58}$$

$$(C_N)_n \rightarrow \left(C_{n,m}\left(\frac{2\pi n}{N}\right)\right) = \begin{pmatrix} \cos\left(\frac{2\pi n}{N}\right) & -\sin\left(\frac{2\pi n}{N}\right) \\ \sin\left(\frac{2\pi n}{N}\right) & \cos\left(\frac{2\pi n}{N}\right) \end{pmatrix} = \left(C_{n,m}\left(\frac{2\pi n}{N}\right)\right)^n$$

$$(C_N)_N \rightarrow \left(C_{n,m}(2\pi)\right) = \left(C_{n,m}(0)\right) = \begin{pmatrix} 1 & 0 \\ 0 & 1 \end{pmatrix} = \vec{1}_h\vec{1}_h + \vec{1}_v\vec{1}_v$$

This gives a special case of (4.51):

$$\tilde{\overleftrightarrow{\Lambda}}_b(\vec{1}_i, s) = \left(C_{n,m}\left(\frac{2\pi n}{N}\right)\right) \cdot \tilde{\overleftrightarrow{\Lambda}}_b(\vec{1}_i, s) \cdot \left(C_{n,m}\left(\frac{2\pi n}{N}\right)\right)^T \tag{4.59}$$

This is the same equation as derived from a different procedure in [2], [11], [17], where it is shown that with reciprocity as in (4.48) the only solution for $N \geq 3$ is

$$\tilde{\overleftrightarrow{\Lambda}}_b(\vec{1}_i, s) = \tilde{\Lambda}_b(\vec{1}_i, s)\overleftrightarrow{1}_i \tag{4.60}$$

This form of scattering dyadic, proportional to the transverse identity, is invariant to all transverse rotations and reflections. Such a scattering dyadic then has $C_{\infty a}$ symmetry which is also labeled as O_2.

These point-symmetry results are summarized in Table 4.1.

There are other kinds of target symmetries one can consider [9], [12]. One of interest here is continuous dilation symmetry (generalized conical symmetry [6], [7], [20], including cones, wedges, and half spaces. For a limited amount of time (the time of validity giving a partial symmetry) based on truncation of the target (to give a finite

TABLE 4.1 Point Symmetry Groups (rotation and reflection) for Backscattering Dyadic for Reciprocal Target

Symmetry in Target (transverse to $\vec{1}_i$)	Form of $\tilde{\overleftrightarrow{\Lambda}}_b$	Symmetry in $\tilde{\overleftrightarrow{\Lambda}}_b$
C_1 (no symmetry)	$\tilde{\overleftrightarrow{\Lambda}}_b = \tilde{\overleftrightarrow{\Lambda}}_b^T$	C_2 (twofold axis $\vec{1}_i$)
C_2 (twofold axis $\vec{1}_i$)	$\tilde{\overleftrightarrow{\Lambda}}_b = \tilde{\overleftrightarrow{\Lambda}}_b^T$	C_2 (twofold axis $\vec{1}_i$)
R_a (single-axial symmetry plane)	$\overleftrightarrow{\Lambda}_b$ diagonal when referred to axial symmetry plane or perpendicular axial plane	C_{2a} (twofold axis $\vec{1}_i$ with two axial symmetry planes)
C_N for $N \geq 3$ (N-fold axis $\vec{1}_i$)	$\tilde{\Lambda}_b \overleftrightarrow{1}_i$	$C_{\infty a} = O_2$ (continuous rotation axis $\vec{1}_i$ with all axial planes as symmetry planes)

TABLE 4.2 Generalized Conical Symmetry for Target and Backscattering Dyadic

Symmetry in Target	Form of $\overleftrightarrow{\Lambda}_{b^\circ}$	Symmetry in $\overleftrightarrow{\Lambda}_{b^\circ}$
Continuous dilation (partial symmetry due to truncations)	$\overleftrightarrow{K}_n I_t^n$ $\overleftrightarrow{K}_n = \overleftrightarrow{K}_n^T$ (real) (number of values of n: simple cone: 1 finite-length wedge: 2 finite-dimensioned half space: 3)	C_{2a} for each term (twofold axis $\vec{1}_i$ with two axial symmetry planes)

size) and arrival of multiple scattering at the observer (between tips for finite-length wedges, and tips and edges for finite-dimensioned half spaces) the scattering dyadic operator takes the form of one, two, or three kinds of terms. Each term contains a real symmetric 2×2 dyadic $\overleftrightarrow{K}_n(\vec{1}_i)$ times some order of temporal integration I_t^n (or power of $1/s$ in frequency domain). Each term can begin at different times, depending on the relative distances of tips and edges from the observer. As a real symmetric dyadic each of these $\overleftrightarrow{K}$ coefficient dyadics can always be diagonalized with real eigenvalues and real eigenvectors giving real and orthogonal spatial directions. The two eigenvectors specify two symmetry planes (except in the special case of equal eigenvalues). This is summarized in Table 4.2. So here is a case of no point symmetry in the target, yet still giving additional point symmetry in the dyadic scattering operator, here expressed in time domain recognizing the limited time of validity (partial symmetry).

4.7 SYMMETRY IN FORWARD-SCATTERING

For forward-scattering we have

$$\vec{1}_o = \vec{1}_i, \qquad \tilde{\overleftrightarrow{\Lambda}}_f(\vec{1}_i, s) = \tilde{\overleftrightarrow{\Lambda}}_f^T(-\vec{1}_i, s) \tag{4.61}$$

but even with reciprocity in the target the scattering dyadic is not necessarily symmetric. Now the bisectrix plane P_b is perpendicular to $\vec{1}_i$, but the scattering plane P_s is any plane parallel to $\vec{1}_i$.

Consider first two-dimensional rotations and reflections in the transverse (bisectrix) plane as was done in Section 4.6 for backscattering. As in (4.54), C_2 symmetry applies in forward-scattering as well, due to the fact that the rotation dyadic is just $-\overleftrightarrow{1}_i$. For reflections, as in (4.55), let $\overleftrightarrow{R}_a$ represent reflection through some axial symmetry plane P_a (a scattering plane). Orient the h, v coordinates so that this reflection reverses only the h coordinate. Then the incident electric field can be chosen as symmetric, i.e., parallel to $\vec{1}_h$, producing a symmetric forward-scattered electric field also parallel to $\vec{1}_h$ [2], [17], [18], [22], [23]. Similarly, considering antisymmetric fields an electric field parallel to $\vec{1}_v$ forward scatters with this same polarization. So for this choice of coordinates the forward-scattering dyadic is *diagonal*. Applying (4.37) for reflection symmetry gives

$$
\begin{aligned}
\tilde{\overleftrightarrow{\Lambda}}_f(\vec{1}_i, s) &= \overleftrightarrow{R}_h \cdot \tilde{\overleftrightarrow{\Lambda}}_f(\vec{1}_i, s) \cdot \overleftrightarrow{R}_h = (-\overleftrightarrow{R}_h) \cdot \tilde{\overleftrightarrow{\Lambda}}_f(\vec{1}_i, s) \cdot (-\overleftrightarrow{R}_h) \\
&= \begin{pmatrix} -1 & 0 \\ 0 & 1 \end{pmatrix} \cdot \begin{pmatrix} \tilde{\Lambda}_{f_{h,h}}(\vec{1}_i, s) & \tilde{\Lambda}_{f_{h,v}}(\vec{1}_i, s) \\ \tilde{\Lambda}_{f_{v,h}}(\vec{1}_i, s) & \tilde{\Lambda}_{f_{v,v}}(\vec{1}_i, s) \end{pmatrix} \cdot \begin{pmatrix} -1 & 0 \\ 0 & 1 \end{pmatrix} \\
&= \begin{pmatrix} \tilde{\Lambda}_{f_{h,h}}(\vec{1}_i, s) & -\tilde{\Lambda}_{f_{h,v}}(\vec{1}_i, s) \\ -\tilde{\Lambda}_{f_{v,h}}(\vec{1}_i, s) & \tilde{\Lambda}_{f_{v,v}}(\vec{1}_i, s) \end{pmatrix} \\
\tilde{\Lambda}_{f_{h,v}}(\vec{1}_i, s) &= 0 = \tilde{\Lambda}_{f_{v,h}}(\vec{1}_i, s)
\end{aligned}
\tag{4.62}
$$

confirming the diagonal nature of the dyadic as well as showing that the dyadic has two symmetry planes, the second one characterized by $\overleftrightarrow{R}_v$ which reflects the v coordinate. Now rotate the dyadic through any angle ϕ (about $\vec{1}_i$) as in (4.53) and take its transpose giving

$$
\begin{aligned}
\left[(C_{n,m}(\phi)) \cdot \tilde{\overleftrightarrow{\Lambda}}_f(\vec{1}_i, s) \cdot (C_{n,m}(\phi))^T \right]^T &= (C_{n,m}(\phi)) \cdot \tilde{\overleftrightarrow{\Lambda}}_f^T(\vec{1}_i, s) \cdot (C_{n,m}(\phi))^T \\
&= (C_{n,m}(\phi)) \cdot \tilde{\overleftrightarrow{\Lambda}}_f(\vec{1}_i, s) \cdot (C_{n,m}(\phi))^T
\end{aligned}
\tag{4.63}
$$

since a diagonal dyadic is a special case of a symmetric dyadic. So we have in general for any choice of the h, v coordinates

$$
\tilde{\overleftrightarrow{\Lambda}}_f(\vec{1}_i, s) = \tilde{\overleftrightarrow{\Lambda}}_f^T(\vec{1}_i, s) \tag{4.64}
$$

if the target has an axial symmetry plane.

For $\vec{1}_i$ as a higher order (N-fold) rotation axis given C_N symmetry in (4.58) and (4.59), now apply in forward-scattering. However, with reciprocity not implying a symmetric $\tilde{\overleftrightarrow{\Lambda}}$, then for $N \geq 3$ and following the procedure in [2], [17] we have

$$
\begin{aligned}
\tilde{\overleftrightarrow{\Lambda}}_f(\vec{1}_i, s) &= \left(C_{n,m}\left(\frac{2\pi n}{N} \right) \right) \cdot \tilde{\overleftrightarrow{\Lambda}}_f(\vec{1}_i, s) \cdot \left(C_{n,m}\left(\frac{2\pi n}{N} \right) \right)^T \\
\tilde{\Lambda}_{f_{1,2}}(\vec{1}_i, s) &= \tilde{\Lambda}_{f_{2,2}}(\vec{1}_i, s), \qquad \tilde{\Lambda}_{f_{1,2}}(\vec{1}_i, s) = -\tilde{\Lambda}_{f_{2,1}}(\vec{1}_i, s) \\
\tilde{\overleftrightarrow{\Lambda}}_f(\vec{1}_i, s) &= \tilde{\Lambda}_{f_{1,1}}(\vec{1}_i, s) \begin{pmatrix} 1 & 0 \\ 0 & 1 \end{pmatrix} + \tilde{\Lambda}_{f_{1,2}}(\vec{1}_i, s) \begin{pmatrix} 0 & 1 \\ -1 & 0 \end{pmatrix}
\end{aligned}
\tag{4.65}
$$

which is a transverse-identity part plus a rotation ($\pm\pi/2$) part. It is important that $N \geq 3$ since this assures that there is at least one n from $n = 1, 2, 3, \ldots, N$ such that $\sin(2\pi n/N) \neq 0$. This form of dyadic is invariant to any two-dimensional rotation so that it is characterized by the group C_∞ or O_2^+. Note that it is not invariant to axial reflections which can be seen by applying $\overleftrightarrow{R}_a$ to the above. If now the target has an axial symmetry plane (with C_N implying N such planes) then the target has C_{Na} symmetry and the forward-scattering dyadic has $C_{\infty a}$ or O_2 symmetry with

$$\tilde{\overleftrightarrow{\Lambda}}_f(\overrightarrow{1}_i, s) = \tilde{\Lambda}_f(\overrightarrow{1}_i, s)\begin{pmatrix} 1 & 0 \\ 0 & 1 \end{pmatrix} \tag{4.66}$$

Going on to symmetries that involve full 3×3 dyadics, let the target have inversion symmetry with group structure (an involution)

$$I = \{(1), (I)\}, \qquad (I)^2 = (1) \tag{4.67}$$

and dyadic representation

$$(1) \rightarrow \overleftrightarrow{1}, \qquad (I) \rightarrow -\overleftrightarrow{1} \tag{4.68}$$

Then (4.10) gives the general form

$$\tilde{\overleftrightarrow{\Lambda}}(-\overrightarrow{1}_o, -\overrightarrow{1}_i; s) = \tilde{\overleftrightarrow{\Lambda}}(\overrightarrow{1}_o, \overrightarrow{1}_i; s) \tag{4.69}$$

In forward-scattering we have

$$\overrightarrow{1}_o = \overrightarrow{1}_i, \qquad \tilde{\overleftrightarrow{\Lambda}}_f(-\overrightarrow{1}_i; s) = \tilde{\overleftrightarrow{\Lambda}}_f(\overrightarrow{1}_i; s) \tag{4.70}$$

Combining the above with reciprocity in (2.1) gives

$$\tilde{\overleftrightarrow{\Lambda}}_f(\overrightarrow{1}_i; s) = \tilde{\overleftrightarrow{\Lambda}}_f^T(\overrightarrow{1}_i; s) \tag{4.71}$$

Thus inversion symmetry is sufficient to give a symmetric scattering dyadic [24] for forward-scattering for *all* $\overrightarrow{1}_i$, a result which applies in backscattering in (4.15) for targets with no particular geometric symmetry.

In forward-scattering one can regard the scattering dyadic as a 2×2 matrix since incidence and scattering directions have the same axis. Since in (4.71) it is symmetric then only three of the four non-zero elements need to be considered. Inversion symmetry is rather general since it does not imply any rotation axes or symmetry planes in the target. There are many common shapes with inversion symmetry, e.g., rectangular parallelepipeds (bricks), finite-length straight wires, rectangular disks, circular loops, etc. Less common are strange shapes without symmetry planes and/or rotation axes.

If the target has one or more symmetry planes, and we restrict $\overrightarrow{1}_i$ to be perpendicular to one of these, then the result in (4.71) still holds for these particular forward-scattering directions. Describe this symmetry by the reflection group (transverse reflection plane, an involution)

$$R_t = \{(1), (R_t)\}, \qquad (R_t)^2 = (1) \tag{4.72}$$

and a dyadic representation

$$(1) \rightarrow \overleftrightarrow{1}, \qquad (R_t) \rightarrow \overleftrightarrow{R}_t = \overleftrightarrow{1} - 2\overrightarrow{1}_i\overrightarrow{1}_i \tag{4.73}$$

$\equiv$ reflection through a transverse (to $\overrightarrow{1}_i$) symmetry plane

This reflection is the same as $\overleftrightarrow{R}_B$ in (4.38) for the case of forward-scattering, in which case P_b is perpendicular to $\overrightarrow{1}_i$. Reflection through a transverse plane is to be distinguished from reflection $\overleftrightarrow{R}_a$ through an axial plane.

Note that

$$\overleftrightarrow{R}_t \cdot \overrightarrow{1}_i = -\overrightarrow{1}_i, \qquad \overleftrightarrow{R}_t \cdot \tilde{\overleftrightarrow{\Lambda}}_f(\overrightarrow{1}_i; s) \cdot \overleftrightarrow{R}_t = \tilde{\overleftrightarrow{\Lambda}}_f(\overrightarrow{1}_i; s) \tag{4.74}$$

since $\tilde{\overleftrightarrow{\Lambda}}$ now has no $\overrightarrow{1}_i$ components. Then (4.37) gives the same result as in (4.70), leading to the symmetric scattering dyadic as in (4.71), except that it applies only for particular $\overrightarrow{1}_i$.

Having a symmetric forward-scattering dyadic the results of Section 4.6 for the symmetric backscattering dyadic can now be directly applied. Adjoining the various symmetries transverse to $\overrightarrow{1}_i$ (two-dimensional) to the inversion symmetry I and reflection symmetry R_t discussed in this section, the same results are obtained as in Table 4.1. The dilation symmetry in Table 4.2 does *not* apply here since the scattering to the observer does not arrive first from cone tips and/or wedge edges, but rather from the entire scatterer (including truncations), all at the same time.

An additional point symmetry to be considered is rotation-reflection S_N. Keeping $\overrightarrow{1}_i$ as the rotation-reflection axis we have

$$\begin{gathered} S_N = \{(S_N)_\ell \,|\, \ell = 1, 2, \ldots, N\}, \qquad (S_N)_1 = (C_N)_1(R_t) = (R_t)(C_N)_1 \\ (S_N)_\ell = (S_N)_1^\ell, \qquad (S_N)_1^N = (1) \text{ for } N \text{ even} \end{gathered} \tag{4.75}$$

Note that N even is the only interesting case since N odd requires that $(C_N)_1$ and (R_t) be separate symmetry elements giving C_{Nt} symmetry [14], [18]. Defining for N even

$$N' \equiv \frac{N}{2} \tag{4.76}$$

we have

$$\begin{gathered} \overleftrightarrow{C}_N = \left(C_{n,m}\left(\frac{2\pi}{N}\right)\right) = \overleftrightarrow{C}\left(\frac{2\pi}{N}\right), \qquad \overleftrightarrow{C}_{N'} = \left(C_{n,m}\left(\frac{2\pi}{N'}\right)\right) = \overleftrightarrow{C}\left(\frac{2\pi}{N'}\right) = \overleftrightarrow{C}_N^2 \\ \overleftrightarrow{S}_N = \overleftrightarrow{C}_N \cdot \overleftrightarrow{R}_t = \overleftrightarrow{R}_t \cdot \overleftrightarrow{C}_N, \qquad \overleftrightarrow{S}_N^2 = \overleftrightarrow{C}_N^2 = \overleftrightarrow{C}_{N'} \end{gathered} \tag{4.77}$$

So $C_{N'}$ is a subgroup of S_N.

Applying $\overleftrightarrow{S}_N$ we have

$$\begin{gathered} \overleftrightarrow{S}_N \cdot \overrightarrow{1}_i = -\overrightarrow{1}_i \\ \tilde{\overleftrightarrow{\Lambda}}_f(-\overrightarrow{1}_i, s) = \overleftrightarrow{S}_N \cdot \tilde{\overleftrightarrow{\Lambda}}_f(\overrightarrow{1}_i, s) \cdot \overleftrightarrow{S}_N^T = \overleftrightarrow{C}_N \cdot \tilde{\overleftrightarrow{\Lambda}}_f(\overrightarrow{1}_i, s) \cdot \overleftrightarrow{C}_N^T = \tilde{\overleftrightarrow{\Lambda}}_f(\overrightarrow{1}_i, s) \text{ (reciprocity)} \end{gathered} \tag{4.78}$$

where the $\overleftrightarrow{R}_t$ part reverses $\overrightarrow{1}_i$ but does not operate on $\tilde{\overleftrightarrow{\Lambda}}$ which has only transverse components. This operation can be carried out an odd number of times as

$$\begin{gathered} N'' \equiv \text{odd integer}, \qquad \overleftrightarrow{S}_N^{N''} \cdot \overrightarrow{1}_i = -\overrightarrow{1}_i \\ \tilde{\overleftrightarrow{\Lambda}}_f(\overrightarrow{1}_i, s) = \overleftrightarrow{S}_N^{N''} \cdot \tilde{\overleftrightarrow{\Lambda}}_f(\overrightarrow{1}_i, s) \cdot \left(\overleftrightarrow{S}_N^{N''}\right)^T = \overleftrightarrow{C}_N^{N''} \cdot \tilde{\overleftrightarrow{\Lambda}}_f(\overrightarrow{1}_i, s) \cdot \left(\overleftrightarrow{C}_N^{N''}\right)^T = \tilde{\overleftrightarrow{\Lambda}}_f(\overrightarrow{1}_i, s) \end{gathered} \tag{4.79}$$

Now N is even, so consider the case that

$$N'' = N' = \frac{N}{2}, \qquad N' = \text{odd}, \qquad \overleftrightarrow{C}_N^{N'} = -\overleftrightarrow{1}_i, \qquad \tilde{\overleftrightarrow{\Lambda}}_f(\overrightarrow{1}_i, s) = \tilde{\overleftrightarrow{\Lambda}}_f^T(\overrightarrow{1}_i, s) \tag{4.80}$$

giving a symmetric scattering dyadic, as with some cases previously discussed. Note that for $N' = 1$ we have

$$S_2 = I \text{ (inversion)}, \qquad \overleftrightarrow{S}_2 = \overleftrightarrow{C}_2 \cdot \overleftrightarrow{R}_t = -\overleftrightarrow{1} \tag{4.81}$$

which is the case considered initially and does not have any particular rotation axis (i.e., is not related to $\overrightarrow{1}_i$). Other cases included in the above are S_6, S_{10}, ..., i.e., $N = 2N'$ with N' odd.

The case of N' even (N a multiple of 4) is more complicated since, analogous to (4.73), we have

$$\begin{aligned} N'' &= \text{even integer} = 2N'', \qquad \overleftrightarrow{S}_N^{N''} \cdot \overrightarrow{1}_i = \overrightarrow{1}_i \\ \tilde{\overleftrightarrow{\Lambda}}_f(\overrightarrow{1}_i, s) &= \overleftrightarrow{S}_N^{N''} \cdot \tilde{\overleftrightarrow{\Lambda}}_f(\overrightarrow{1}_i, s) \cdot \left(\overleftrightarrow{S}_N^{N''}\right)^T = \overleftrightarrow{C}_N^{N''} \cdot \tilde{\overleftrightarrow{\Lambda}}_f(\overrightarrow{1}_i, s) \cdot \left(\overleftrightarrow{C}_N^{N''}\right)^T \\ &= \overleftrightarrow{C}_{N'}^{N''} \cdot \tilde{\overleftrightarrow{\Lambda}}_f(\overrightarrow{1}_i, s) \cdot \left(\overleftrightarrow{C}_{N'}^{N''}\right)^T \end{aligned} \tag{4.82}$$

but without the reversal of $\overrightarrow{1}_i$ to give the relation to the transpose scattering dyadic. This does exhibit, however, C_N and $C_{N'}$ as subgroups for the symmetry of the scattering dyadic. At this point let us note that for $N \geq 4$ (since N is even) the result in (4.65) with only two distinct elements for the forward-scattering dyadic applies.

Extending this rotation-reflection symmetry let us adjoin axial symmetry planes. Let any one of these be characterized by, say, $\overleftrightarrow{R}_h$ as in (4.57), and align the h coordinate with this plane. Then letting the incident wave be symmetric or antisymmetric with respect to this plane (h and v polarization, respectively), the scattered fields including the forward-scattered fields will have the same symmetry properties. Hence, in this coordinate system the forward-scattering dyadic is diagonal (a special case of symmetric). Rotation about $\overrightarrow{1}_i$ (to any other choice of the transverse coordinates) of a symmetric dyadic gives another symmetric dyadic. Hence, adjunction of one or more axial symmetry planes makes

$$\tilde{\overleftrightarrow{\Lambda}}_f(\overrightarrow{1}_i, s) = \tilde{\overleftrightarrow{\Lambda}}_f^T(\overrightarrow{1}_i, s) \tag{4.83}$$

With a symmetric dyadic, then the result in (4.65) reduces to the transverse identity in (4.60) for backscattering as

$$\tilde{\overleftrightarrow{\Lambda}}_f(\overrightarrow{1}_i, s) = \tilde{\Lambda}_f(\overrightarrow{1}_i, s)\overleftrightarrow{1}_i \qquad \text{for} \qquad N \geq 4 \tag{4.84}$$

Note that adjunction of axial symmetry planes to S_N, giving S_{Na}, has $C_{N'a}$ as a subgroup and there are N' such axial symmetry planes. While S_N has N elements, S_{Na} has $2N$ elements and $C_{N'a}$ has N elements. These N' axial symmetry planes introduce N' secondary twofold rotation axes between these planes (diagonal) on the transverse rotation-reflection plane. This is dihedral symmetry $D_{N'}$ (N elements) with diagonal symmetry planes given $D_{N'd}$ symmetry ($2N$ elements) [14], [18]. So we have

$$S_{Na} = D_{N'd} \tag{4.85}$$

as two equivalent labels for the same group, giving a symmetric dyadic as in (4.83) for which case the various results for backscattering also apply for forward-scattering.

These point-symmetry results are summarized in Table 4.3. Note that due to the relevance of three-dimensional rotation-reflections in the target symmetry the list is considerably longer than for backscattering in Table 4.1.

TABLE 4.3 Point Symmetry Groups (rotation and reflection) for Forward-Scattering Dyadic for Reciprocal Target

Symmetry in Target ($\vec{1}_i$ as reference axis)	Form of $\overleftrightarrow{\tilde{\Lambda}}_f$	Symmetry in $\overleftrightarrow{\tilde{\Lambda}}_f$
C_1 (no symmetry)	General 2×2	C_2 (twofold axis $\vec{1}_i$)
C_2 (twofold axis $\vec{1}_i$)	General 2×2	C_2 (twofold axis $\vec{1}_i$)
R_a (single axial symmetry plane)	$\overleftrightarrow{\tilde{\Lambda}}_f = \overleftrightarrow{\tilde{\Lambda}}_f^T$ and diagonal when referred to axial symmetry plane or perpendicular axial plane	C_{2a} (twofold axis $\vec{1}_i$ with two axial symmetry planes)
C_N for $N \geq 3$ (N-fold axis $\vec{1}_i$)	$\tilde{\Lambda}_{f1,1}\begin{pmatrix}1 & 0\\ 0 & 1\end{pmatrix} + \tilde{\Lambda}_{f1,2}\begin{pmatrix}0 & 1\\ -1 & 0\end{pmatrix}$	$C_\infty = O_2^+$ (continuous rotation axis $\vec{1}_i$)
C_{Na} for $N \geq 3$ (N-fold axis $\vec{1}_i$ with N axial symmetry planes)	$\tilde{\Lambda}_f \overleftrightarrow{1}_i$	$C_{\infty a} = O_2$ (continuous rotation axis $\vec{1}_i$ with all axial planes as symmetry planes)
I (inversion, no special axes or planes) R_t (symmetry plane $\perp \vec{1}_i$)	$\overleftrightarrow{\tilde{\Lambda}}_f = \overleftrightarrow{\tilde{\Lambda}}_f^T$	C_2 (twofold axis $\vec{1}_i$)
C_{2t} C_{Nt} for $N \geq 3$	$\overleftrightarrow{\tilde{\Lambda}}_f = \overleftrightarrow{\tilde{\Lambda}}_f^T$ $\tilde{\Lambda}_f \overleftrightarrow{1}_i$	C_2 (twofold axis $\vec{1}_i$) $C_{\infty a} = O_2$ (continuous rotation axis $\vec{1}_i$ with all axial planes as symmetry planes)
S_N (rotation-reflection, N (even) $= 2N'$)		
N' odd	$\overleftrightarrow{\tilde{\Lambda}}_f = \overleftrightarrow{\tilde{\Lambda}}_f^T$	C_2 (twofold axis $\vec{1}_i$)
odd $N' \geq 3$	$\tilde{\Lambda}_f \overleftrightarrow{1}_i$	$C_{\infty a} = O_2$ (continuous rotation axis 1_i with all axial planes as symmetry planes)
even $N' \geq 2$	$\tilde{\Lambda}_{f1,1}\begin{pmatrix}1 & 0\\ 0 & 1\end{pmatrix} + \tilde{\Lambda}_{f1,2}\begin{pmatrix}0 & 1\\ -1 & 0\end{pmatrix}$	$C_\infty = O_2^+$ (continuous rotation axis $\vec{1}_i$)
$S_{4a} = D_{2d}$ (dihedral with two diagonal (axial) symmetry planes)	$\overleftrightarrow{\tilde{\Lambda}}_f = \overleftrightarrow{\tilde{\Lambda}}_f^T$	C_{2a} (twofold axis $\vec{1}_i$)
$D_{N'd}$ for $N' \geq 3$	$\tilde{\Lambda}_f \overleftrightarrow{1}_i$	$C_{\infty a} = O_2$ (continuous rotation axis $\vec{1}_i$ with all axial planes as symmetry planes)

4.8 SYMMETRY IN LOW-FREQUENCY SCATTERING

At low frequencies for which the target is electrically small ($\lambda \gg$ target dimensions), the scattering is dominated by the induced electric and magnetic dipole moments. The scattering dyadic then takes the form [5], [8]

$$\begin{aligned}
\tilde{\overleftrightarrow{\Lambda}}(\overrightarrow{1}_o,\overrightarrow{1}_i;s) &= \tilde{\overleftrightarrow{\Lambda}}^T(-\overrightarrow{1}_i,-\overrightarrow{1}_o;s)\ \text{(reciprocity)}\\
&= \tilde{\gamma}^2\left[-\overleftrightarrow{1}_o\cdot\tilde{\overleftrightarrow{P}}(s)\cdot\overleftrightarrow{1}_o+\overrightarrow{1}_i\times\tilde{\overleftrightarrow{M}}(s)\times\overrightarrow{1}_i\right]\ \text{as } s\to 0\\
\tilde{\overleftrightarrow{P}}(s) &= \tilde{\overleftrightarrow{P}}^T(s)\equiv \text{electric-polarizability dyadic}\\
\tilde{\overleftrightarrow{M}}(s) &= \tilde{\overleftrightarrow{M}}^T(s)\equiv \text{magnetic-polarizability dyadic}
\end{aligned}\tag{4.86}$$

These polarizabilities have dimensions of volume (m^3) and, with the incident electric and magnetic fields, give the induced electric and magnetic dipole moments. Reciprocity makes these dyadics symmetric. For perfectly conducting targets they are also frequency independent.

For convenience express the cross-products in dyadic/dot-product form [3]. Consider an arbitrary real unit vector $\overrightarrow{1}_a$ and form

$$\overleftrightarrow{\tau}_a\equiv\overrightarrow{1}_a\times\overleftrightarrow{1}=\overrightarrow{1}_a\times\overleftrightarrow{1}_a,\qquad \overleftrightarrow{1}_a\equiv\overleftrightarrow{1}-\overrightarrow{1}_a\overrightarrow{1}_a\equiv\text{identity transverse to }\overrightarrow{1}_a \tag{4.87}$$

This is the most general skew symmetric dyadic [15] with the relations

$$\overleftrightarrow{1}\times\overrightarrow{1}_a=\overleftrightarrow{\tau}_a=-\overleftrightarrow{\tau}_{-a}=\overleftrightarrow{\tau}_a^T,\qquad \overleftrightarrow{\tau}_a^2=-\overleftrightarrow{1}_i \tag{4.88}$$

where the subscript $-a$ refers to replacing $\overrightarrow{1}_a$ by $-\overrightarrow{1}_a$. This dyadic can be described as a rotation by $\pi/2$ of the coordinates transverse to $\overrightarrow{1}_a$ around $\overrightarrow{1}_a$ in a right-handed sense. The longitudinal coordinate (in the direction of $\overrightarrow{1}_a$) is multiplied by zero. To illustrate the form of this dyadic consider the right-handed coordinates based on $\overrightarrow{1}_h, \overrightarrow{1}_v, -\overrightarrow{1}_i$ satisfying (4.5). In this case

$$\overleftrightarrow{\tau}_{-i}=-\overrightarrow{1}_i\times\overleftrightarrow{1}_i=\begin{pmatrix}0&-1&0\\1&0&0\\0&0&0\end{pmatrix}=-\overleftrightarrow{\tau}_i=\overleftrightarrow{\tau}_i^T \tag{4.89}$$

gives a positive sense of rotation in the h, v plane in the usual radar coordinates. This is equivalent to $\left(C_{n,m}(\pi/2)\right)$ if only transverse coordinates are used. However, in three dimensions this is not an orthogonal dyadic, unless the $(3,3)$ position is changed from 0 to 1 as in (4.53) for $\overleftrightarrow{C}(\pi/2)$. Similarly for $\overrightarrow{1}_o$ we have

$$\overleftrightarrow{\tau}_o=\overrightarrow{1}_o\times\overleftrightarrow{1}_o=-\overleftrightarrow{\tau}_{-o}=-\overleftrightarrow{\tau}_o^T \tag{4.90}$$

Now rewrite (4.86) as

$$\tilde{\overleftrightarrow{\Lambda}}(\overrightarrow{1}_o,\overrightarrow{1}_i;s)=\tilde{\overleftrightarrow{\Lambda}}^T(-\overrightarrow{1}_i,-\overrightarrow{1}_o;s)=\tilde{\gamma}^2\left[-\overleftrightarrow{1}_o\cdot\tilde{\overleftrightarrow{P}}(s)\cdot\overleftrightarrow{1}_i-\overleftrightarrow{\tau}_o\cdot\tilde{\overleftrightarrow{M}}(s)\cdot\overleftrightarrow{\tau}_i\right]\ \text{as } s\to 0 \tag{4.91}$$

Interchanging $\vec{1}_o$ and $\vec{1}_i$ we have

$$
\begin{aligned}
\tilde{\overleftrightarrow{\Lambda}}(\vec{1}_i,\vec{1}_o;s) &= \tilde{\gamma}^2\left[-\vec{1}_i\cdot\tilde{\overleftrightarrow{P}}(s)\cdot\vec{1}_o - \overleftrightarrow{\tau}_i\cdot\tilde{\overleftrightarrow{M}}(s)\cdot\overleftrightarrow{\tau}_o\right] \text{ as } s\to 0 \\
\tilde{\overleftrightarrow{\Lambda}}^T(\vec{1}_i,\vec{1}_o;s) &= \tilde{\gamma}^2\left[-\vec{1}_i\cdot\tilde{\overleftrightarrow{P}}^T(s)\cdot\vec{1}_i - \overleftrightarrow{\tau}_o^T\cdot\tilde{\overleftrightarrow{M}}^T(s)\cdot\overleftrightarrow{\tau}_i^T\right] \text{ as } s\to 0 \\
&= \tilde{\gamma}^2\left[-\vec{1}_o\cdot\tilde{\overleftrightarrow{P}}(s)\cdot\vec{1}_i - \overleftrightarrow{\tau}_o\cdot\tilde{\overleftrightarrow{M}}(s)\cdot\overleftrightarrow{\tau}_i\right] \text{ as } s\to 0 \\
&= \tilde{\overleftrightarrow{\Lambda}}^T(-\vec{1}_i,-\vec{1}_o;s) \text{ as } s\to 0
\end{aligned}
\tag{4.92}
$$

so that we have the additional invariance (symmetry)

$$
\tilde{\overleftrightarrow{\Lambda}}(\vec{1}_i,\vec{1}_o;s) = \tilde{\overleftrightarrow{\Lambda}}(-\vec{1}_i,-\vec{1}_o;s) \text{ as } s\to 0 \tag{4.93}
$$

Note that this result, using the leading term (dipoles) for low frequencies, has only been shown to be valid in a low-frequency asymptotic sense, the error being thus far associated with the next higher-order (quadrupole) terms.

Applying this result to back- and forward-scattering gives

$$
\begin{aligned}
&\tilde{\overleftrightarrow{\Lambda}}_b(\vec{1}_i,s) = \tilde{\overleftrightarrow{\Lambda}}_b(-\vec{1}_i,s) = \tilde{\overleftrightarrow{\Lambda}}_b^T(\vec{1}_i,s), \\
&\qquad\qquad \tilde{\overleftrightarrow{\Lambda}}_f(\vec{1}_i,s) = \tilde{\overleftrightarrow{\Lambda}}_f(-\vec{1}_i,s) = \tilde{\overleftrightarrow{\Lambda}}_f^T(\vec{1}_i,s) \text{ as } s\to 0
\end{aligned}
\tag{4.94}
$$

For low frequencies the back- and forward-scattering dyadics are both symmetric and invariant to direction reversal.

With both $\tilde{\overleftrightarrow{\Lambda}}_b$ and $\tilde{\overleftrightarrow{\Lambda}}_f$ symmetric, then the results of Section 4.6 for point symmetries as in Table 4.1 apply to both. In addition, the fact that the target is electrically small means that the incident electric and magnetic fields are uniform (quasistatic) in the vicinity of the target, so the front and back of the target are seen "simultaneously." This is also exhibited in (4.91) where only the dipole terms are retained.

Writing $\tilde{\overleftrightarrow{\Lambda}}_i(\vec{1}_i,s)$ for both $\tilde{\overleftrightarrow{\Lambda}}_b(\vec{1}_i,s)$ and $\tilde{\overleftrightarrow{\Lambda}}_f(\vec{1}_i,s)$ we have, in the electrically small regime,

$$
\tilde{\overleftrightarrow{\Lambda}}_i(\vec{1}_i,s) = \tilde{\overleftrightarrow{\Lambda}}_i(-\vec{1}_i,s) = \tilde{\overleftrightarrow{\Lambda}}^T(\vec{1}_i,s) \tag{4.95}
$$

where we can take the dipole terms in (4.86) and (4.91) to *define* this case and give exact equality in (4.95). From (4.91) we have

$$
\tilde{\overleftrightarrow{\Lambda}}_i(\vec{1}_i,s) = \gamma^2\left[-\vec{1}_i\cdot\tilde{\overleftrightarrow{P}}(s)\cdot\vec{1}_i \pm \overleftrightarrow{\tau}_i\cdot\tilde{\overleftrightarrow{M}}(s)\cdot\overleftrightarrow{\tau}_i\right], \qquad \pm\Rightarrow\begin{cases} + \text{ for backscattering} \\ - \text{ for forward-scattering}\end{cases} \tag{4.96}
$$

Now the scattering dyadic is in a form in which the results of [10] can be used. In that paper the symmetries of the magnetic polarizability dyadic are analyzed. These also apply to the electric polarizability dyadic by appropriate interchange of electric and magnetic parameters. In accordance with (4.32) the same symmetries are taken for both electric and magnetic parameters so that $\tilde{\overleftrightarrow{P}}(s)$ and $\tilde{\overleftrightarrow{M}}(s)$ have the same symmetries.

TABLE 4.4 Decomposition of Polarizability Dyadic According to Target Point Symmetries

Form of $\tilde{\overleftrightarrow{D}}(s)$	Symmetry Type (groups)
$\tilde{D}_z(s)\vec{1}_z\vec{1}_z + \tilde{\overleftrightarrow{D}}_t(s)$ $\left(\tilde{\overleftrightarrow{D}}_t(s)\cdot\vec{1}_z = \vec{0}\right)$	R_z (single symmetry plane) C_2 (twofold rotation axis)
$\tilde{D}_z(s)\vec{1}_z\vec{1}_z + \tilde{D}_x(s)\vec{1}_x\vec{1}_x + \tilde{D}_y(s)\vec{1}_y\vec{1}_y$	$C_{2a} = R_x \otimes R_y$ D_2 (three twofold rotation axes)
$\tilde{D}_z(s)\vec{1}_z\vec{1}_z + \tilde{D}_t(s)\overleftrightarrow{1}_z$ $\left(\overleftrightarrow{1}_z = \overleftrightarrow{1} - \vec{1}_z\vec{1}_z \Rightarrow \text{double degeneracy}\right)$	C_N for $N \geq 3$ (N-fold rotation axis) S_N for N even and $N \geq 4$ (N-fold rotation-reflection axis) D_{2d} (three twofold rotation axes plus diagonal symmetry planes)
$\tilde{D}(s)\overleftrightarrow{1}$ $\left(\overleftrightarrow{1} \Rightarrow \text{triple degeneracy}\right)$	O_3 (generalized sphere) T, O, Y (regular polyhedral)

Write $\tilde{\overleftrightarrow{D}}(s)$ as a general dipole polarizability dyadic, applying to both electric and magnetic polarizabilities. Then Table 9.1 of [10] can be written as Table 4.4 here. Note that the axes (z, etc.) here are aligned according to the target and are not in general $\vec{1}_i$ unless so selected for a particular case.

Concerning rotations and axial reflections with respect to $\vec{1}_i$ as the symmetry axis (two-dimensional transformations), Table 4.1 is applicable. Including transverse reflections $\overleftrightarrow{R}_t$ changes the results from the previous results for back- and forward-scattering somewhat. Both inversion I and transverse reflection R_t symmetry do not add to the symmetries in $\overleftrightarrow{\Lambda}_d$ as they did for $\overleftrightarrow{\Lambda}_f$ in Section 4.7 because $\overleftrightarrow{\Lambda}_d$ is already symmetric. The discussion can then jump ahead to rotation-reflection symmetry in (4.75) through (4.77).

For our symmetric dyadic we now have

$$\begin{gathered}\tilde{\overleftrightarrow{\Lambda}}_i(\vec{1}_i, s) = \tilde{\overleftrightarrow{\Lambda}}_i(-\vec{1}_i, s) = \overleftrightarrow{S}_N^n \cdot \tilde{\overleftrightarrow{\Lambda}}_i(\vec{1}_i, s) \cdot \overleftrightarrow{S}_N^n = \overleftrightarrow{C}_N^n \cdot \tilde{\overleftrightarrow{\Lambda}}_i(\vec{1}_i, s) \cdot \overleftrightarrow{C}_N^n = \tilde{\overleftrightarrow{\Lambda}}^T(\vec{1}_i, s) \\ \overleftrightarrow{S}_N^n \cdot \vec{1}_i = (-1)^n \vec{1}_i, \qquad n = 1, 2, \ldots, N\end{gathered} \tag{4.97}$$

The invariance to $\vec{1}_i$ reversal avoids the effect of $\overleftrightarrow{S}_N$ on $\vec{1}_i$. The symmetry of the dyadic then makes

$$\tilde{\overleftrightarrow{\Lambda}}_i(\vec{1}_i, s) = \tilde{\Lambda}_i(\vec{1}_i, s)\overleftrightarrow{1}_i \qquad \text{for} \qquad N \geq 4 \ (N \text{ being even}) \tag{4.98}$$

which is $C_{\infty a}$ or O_2 symmetry. This result can also be found using Table 4.4 noting that for $\vec{1}_z = \vec{1}_i$ the transverse part of the polarizability dyadics have the form $\tilde{D}_t(s)\overleftrightarrow{1}_i$. So for low frequencies the results for rotation-reflection symmetry are considerably simpler than those in Section 4.7 for general forward scattering.

Referring to Table 4.4 we also have dihedral symmetry to consider. Here we see that D_2 with three twofold rotation axes (one taken as $-\vec{1}_i$ and the other two

TABLE 4.5 Point-Symmetry Groups (rotation and reflection) for Low-Frequency Back- and Forward-Scattering Dyadics

Symmetry in Target ($\overrightarrow{1}_i$ as reference axis)	Form of $\tilde{\overleftrightarrow{\Lambda}}_i$	Symmetry in $\tilde{\overleftrightarrow{\Lambda}}_i$
C_1 (no symmetry)	$\tilde{\overleftrightarrow{\Lambda}}_i = \tilde{\overleftrightarrow{\Lambda}}_i^T$	C_2 (twofold axis $\overrightarrow{1}_i$)
C_2 (twofold axis $\overrightarrow{1}_i$)	$\tilde{\overleftrightarrow{\Lambda}}_i = \tilde{\overleftrightarrow{\Lambda}}_i^T$	C_2 (twofold axis $\overrightarrow{1}_i$)
R_a (single axial symmetry plane)	$\tilde{\overleftrightarrow{\Lambda}}_i$ diagonal when referred to axial symmetry plane or perpendicular axial plane	C_{2a} (twofold axis $\overrightarrow{1}_i$ with two axial symmetry planes)
C_N for $N \geq 3$	$\tilde{\Lambda}_i \overleftrightarrow{1}_i$	$C_{\infty a} = O_2$ (continuous rotation axis $\overrightarrow{1}_i$ with all axial planes as symmetry planes)
S_N for $N \geq 4$ (rotation-reflection, N even)	$\tilde{\Lambda}_i \overleftrightarrow{1}_i$	$C_{\infty a} = O_2$ (continuous rotation axis $\overrightarrow{1}_i$ with all axial planes as symmetry planes)
D_2 (dihedral, three rotation axes at right-angles)	$\tilde{\overleftrightarrow{\Lambda}}_i$ diagonal when referred to transverse rotation axes	C_{2a} (twofold axis $\overrightarrow{1}_i$ with two axial symmetry planes)
D_{2d} (dihedral, two transverse rotation axes and two axial diagonal symmetry planes)	$\tilde{\Lambda}_i \overleftrightarrow{1}_i$	$C_{\infty a} = O_2$ (continuous rotation axis $\overrightarrow{1}_i$ with all axial planes as symmetry planes)
O_3 (all rotations and reflections) T (tetrahedral) O (octahedral) Y (icosahedral) (orientation of rotation axes arbitrary with respect to $\overrightarrow{1}_i$)	$\tilde{\Lambda}_i \overleftrightarrow{1}_i$ independent of $\overrightarrow{1}_i$	O_3 (all rotations and reflections of both polarization and $\overrightarrow{1}_i$)

perpendicular to $-\overrightarrow{1}_i$ and each other) is sufficient to give two axial symmetry planes in the polarizabilities. Note in (4.96) that $\overleftrightarrow{\tau}_i$ is a $-\pi/2$ rotation, merely rotating the axial symmetry planes into each other. So D_2 results in C_{2a} symmetry in $\tilde{\overleftrightarrow{\Lambda}}_i$. Going a step further D_{2d} symmetry makes the transverse part of the polarizabilities proportional to $\overleftrightarrow{1}_z$ so that (4.96) takes the simple form

$$\tilde{\overleftrightarrow{\Lambda}}_i(\overrightarrow{1}_i, s) = \gamma^2\left[-\tilde{P}_t(s) \mp \tilde{M}_t(s)\right]\overleftrightarrow{1}_i = \tilde{\Lambda}_i(s)\overleftrightarrow{1}_i \tag{4.99}$$

which is $C_{\infty a}$ or O_2 symmetry.

Finally, there are the symmetry groups which give $\tilde{D}(s)\overleftrightarrow{1}$, namely O_3 (invariance to all rotations and reflections), T (tetrahedral), O (octahedral), and Y (icosahedral) symmetries. This also gives $C_{\infty a}$ or O_2 as in (4.99) with the additional invariance of being independent of $\overrightarrow{1}_i$. One can think of this as O_3 symmetry. Remember that T, O, and Y give this result for only low frequencies for which the induced dipole moments dominate the scattering.

These point-symmetry results are summarized in Table 4.5. These low-frequency results can be compared to Tables 4.1 and 4.3 for general frequencies.

4.9 PRELIMINARIES FOR SELF-DUAL TARGETS

As indicated in Fig. 4.1, the scatterer is characterized by constitutive parameter dyadics

$$\tilde{\overleftrightarrow{\varepsilon}}(\vec{r},s) \equiv \text{permittivity}, \qquad \tilde{\overleftrightarrow{\mu}}(\vec{r},s) \equiv \text{permeability} \tag{4.100}$$

One can also have conductivity, but for present purposes, this can be included in the permittivity.

For later use, define some additional dyadics associated with incidence and scattering directions as

$$\overleftrightarrow{\tau}_i \equiv \vec{1}_i \times \overleftrightarrow{1}_i = \overleftrightarrow{1}_i \times \vec{1}_i = -\overleftrightarrow{\tau}_i^T = \begin{pmatrix} 0 & 1 & 0 \\ -1 & 0 & 0 \\ 0 & 0 & 0 \end{pmatrix} \text{ in } \vec{1}_h,\ \vec{1}_v,\ -\vec{1}_i \text{ coordinates}$$

$$= -\pi/2 \text{ rotation in } h, v \text{ plane (observer looking in } \vec{1}_i \text{ direction)}$$

$$\overleftrightarrow{\tau}_o \equiv \vec{1}_o \times \overleftrightarrow{1}_o = \overleftrightarrow{1}_o \times \vec{1}_o = -\overleftrightarrow{\tau}_o^T \tag{4.101}$$

with similar interpretation with respect to $\vec{1}_o$ direction

$$\overleftrightarrow{\tau}_i^2 = -\overleftrightarrow{1}_i, \qquad \overleftrightarrow{\tau}_o^2 = -\overleftrightarrow{1}_o$$

$$\overleftrightarrow{\tau}_i \cdot \overleftrightarrow{\tau}_i^T = \overleftrightarrow{1}_i, \qquad \overleftrightarrow{\tau}_o \cdot \overleftrightarrow{\tau}_o^T = \overleftrightarrow{1}_o \text{ (two-dimensional inverses)}$$

In terms of the standard h, v radar coordinates we have (with $\vec{1}_n, \vec{1}_v, -\vec{1}_i$ as a right-handed system)

$$\overleftrightarrow{1}_i = \vec{1}_h \vec{1}_h + \vec{1}_v \vec{1}_v, \qquad \overleftrightarrow{\tau}_i = \vec{1}_h \vec{1}_v - \vec{1}_v \vec{1}_h \tag{4.102}$$

Consider an angle ψ for positive rotation in the usual mathematical convention (counterclockwise) in the h, v plane

$$\begin{aligned} \overleftrightarrow{C}_i(\psi) &= \overleftrightarrow{1}_i \cos(\psi) - \overleftrightarrow{\tau}_i \sin(\psi) + \vec{1}_i \vec{1}_i \\ \overleftrightarrow{C}_i(-\psi) &= \overleftrightarrow{C}_i^T(\psi) = \overleftrightarrow{C}_i^{-1}(\psi), \qquad \overleftrightarrow{C}_i(\psi) \cdot \overleftrightarrow{C}_i(-\psi) = \overleftrightarrow{1} \text{ (identity)} \end{aligned} \tag{4.103}$$

Similarly for the scattering direction we have

$$\overleftrightarrow{C}_o(\psi) = \overleftrightarrow{1}_o \cos(\psi) - \overleftrightarrow{\tau}_o \sin(\psi) + \vec{1}_o \vec{1}_o \tag{4.104}$$

with similar properties. Then in the $\vec{1}_h, \vec{1}_v, -\vec{1}_i$ coordinates we have

$$\overleftrightarrow{C}_i(\psi) = \begin{pmatrix} \cos(\psi) & -\sin(\psi) & 0 \\ \sin(\psi) & \cos(\psi) & 0 \\ 0 & 0 & 1 \end{pmatrix} \tag{4.105}$$

as an explicit form of the rotation. A similar form can be written for the scattering. Note that for backscattering we have

$$\vec{1}_o = -\vec{1}_i, \qquad \overleftrightarrow{1}_o = \overleftrightarrow{1}_i, \qquad \overleftrightarrow{\tau}_o = -\overleftrightarrow{\tau}_i, \qquad \overleftrightarrow{C}_i(\psi) = \overleftrightarrow{C}_o(-\psi) \tag{4.106}$$

so that in backscattering both conventions merge with attention to signs.

4.10 DUALITY

By duality is meant the symmetry in the Maxwell equations on the interchange of electric and magnetic parameters [18]. Using the combined field and current density

$$\vec{E}_q(\vec{r},t) = \vec{E}(\vec{r},t) + jqZ_o\vec{H}(\vec{r},t), \qquad \vec{J}_q(\vec{r},t) = \vec{J}_e(\vec{r},t) + j\frac{q}{Z_o}\vec{J}_h(\vec{r},t) \tag{4.107}$$

$$q = \pm 1 \equiv \text{separation index}$$

the combined Maxwell equation is

$$\left[\nabla\times -j\frac{q}{c}\frac{\partial}{\partial t}\right]\vec{E}_q(\vec{r},t) = jqZ_o\tilde{\vec{J}}_q(\vec{r},t) \tag{4.108}$$

The electric and magnetic current densities can here be interpreted as

$$\tilde{\vec{J}}_e(\vec{r},s) = s\left[\tilde{\overleftrightarrow{\varepsilon}}(\vec{r},s) - \varepsilon_o\overleftrightarrow{1}\right]\cdot\tilde{\vec{E}}(\vec{r},s),$$
$$\tilde{\vec{J}}_h(\vec{r},s) = s\left[\tilde{\overleftrightarrow{\mu}}(\vec{r},s) - \mu_o\overleftrightarrow{1}\right]\cdot\tilde{\vec{H}}(\vec{r},s) \tag{4.109}$$

Source currents can also be included in (4.109) which in this form include electric and magnetic polarization currents, conductivity having been lumped in with the permittivity.

The duality transformation is just

$$\vec{E}_q^{(d)}(\vec{r},t) = -qj\vec{E}_q(\vec{r},t)$$
$$\vec{E}^{(d)}(\vec{r},t) = Z_o\vec{H}(\vec{r},t), \qquad \vec{H}^{(d)}(\vec{r},t) = -\frac{1}{Z_o}\vec{E}(\vec{r},t) \tag{4.110}$$

Applying this twice gives the negative of the original fields. The identity is recovered by four applications which makes this like C_4 symmetry in the complex plane [18]. Applying this to the current densities gives

$$\vec{J}_q^{(d)}(\vec{r},t) = -qj\vec{J}_q(\vec{r},t), \qquad \vec{J}_e^{(d)}(\vec{r},t) = \frac{1}{Z_o}\vec{J}_h(\vec{r},t),$$
$$\vec{J}_h^{(d)}(\vec{r},t) = -Z_o\vec{J}_e(\vec{r},t) \tag{4.111}$$

where we have

$$\tilde{\vec{J}}_e^{(d)}(\vec{r},s) = s\left[\tilde{\overleftrightarrow{\varepsilon}}^{(d)}(\vec{r},s) - \varepsilon_o\overleftrightarrow{1}\right]\cdot\tilde{\vec{E}}^{(d)}(\vec{r},s)$$
$$\tilde{\vec{J}}_h^{(d)}(\vec{r},s) = s\left[\tilde{\overleftrightarrow{\mu}}^{(d)}(\vec{r},s) - \mu_o\overleftrightarrow{1}\right]\cdot\tilde{\vec{H}}^{(d)}(\vec{r},s) \tag{4.112}$$

Comparing the dual to the original current densities gives

$$\frac{\tilde{\overleftrightarrow{\mu}}^{(d)}(\overrightarrow{r},s)}{\mu_o} = \frac{\tilde{\overleftrightarrow{\varepsilon}}(\overrightarrow{r},s)}{\varepsilon_o}, \qquad \frac{\tilde{\overleftrightarrow{\varepsilon}}^{(d)}}{\varepsilon_o} = \frac{\tilde{\overleftrightarrow{\mu}}(\overrightarrow{r},s)}{\mu_o} \tag{4.113}$$

as the relation of the dual target to the original one. From this we conclude that for a target to be self-dual we need

$$\frac{\tilde{\overleftrightarrow{\mu}}(\overrightarrow{r},s)}{\mu_o} = \frac{\tilde{\overleftrightarrow{\varepsilon}}(\overrightarrow{r},s)}{\varepsilon_o} \tag{4.114}$$

Scalar permeability and permittivity are merely a special case of this. Note that reciprocity (symmetric dyadics) has not been assumed for this result.

4.11 SCATTERING BY SELF-DUAL TARGET

From (4.2) scattering from the dual target is given by

$$\begin{aligned} \tilde{\overrightarrow{E}}_f^{(d)}(\overrightarrow{r},s) &= \frac{e^{-\gamma r}}{4\pi r}\tilde{\overleftrightarrow{\Lambda}}^{(d)}(\overrightarrow{1}_o,\overrightarrow{1}_i;s)\cdot\tilde{\overrightarrow{E}}^{(inc,d)}(\overrightarrow{0},s) \\ \tilde{\overrightarrow{H}}^{(inc,d)}(\overrightarrow{r},t) &= \frac{1}{Z_o}\overrightarrow{1}_i\times\overrightarrow{E}^{(inc,d)}(\overrightarrow{r},t) = \frac{1}{Z_o}\overleftrightarrow{\tau}_i\cdot\overrightarrow{E}^{(inc,d)}(\overrightarrow{r},t) \\ \tilde{\overrightarrow{H}}_f^{(d)}(\overrightarrow{r},t) &= \frac{1}{Z_o}\overrightarrow{1}_o\times\overrightarrow{E}_f^{(d)}(\overrightarrow{r},t) = \frac{1}{Z_o}\overleftrightarrow{\tau}_o\cdot\overrightarrow{E}_f(\overrightarrow{r},t) \end{aligned} \tag{4.115}$$

Relating the dual to the original fields as

$$\begin{aligned} \tilde{\overrightarrow{E}}^{(inc,d)}(\overrightarrow{0},t) &= Z_o\overrightarrow{H}^{(inc)}(\overrightarrow{r},t) = \overrightarrow{1}_i\times\overrightarrow{E}^{(inc)}(\overrightarrow{r},t) = \overleftrightarrow{\tau}_i\cdot\overrightarrow{E}^{(inc)}(\overrightarrow{r},t) \\ \tilde{\overrightarrow{E}}_t^{(d)}(\overrightarrow{0},t) &= Z_o\overrightarrow{H}_f(\overrightarrow{r},t) = \overrightarrow{1}_o\times\overrightarrow{E}_f(\overrightarrow{r},t) = \overleftrightarrow{\tau}_o\cdot\overrightarrow{E}_f(\overrightarrow{r},t) \end{aligned} \tag{4.116}$$

and substituting into the dual scattering equation we have

$$\tilde{\overrightarrow{E}}_f(\overrightarrow{r},s) = -\frac{e^{-\gamma r}}{4\pi r}\overleftrightarrow{\tau}_o\cdot\tilde{\overleftrightarrow{\Lambda}}^{(d)}(\overrightarrow{1}_o,\overrightarrow{1}_i;s)\cdot\overleftrightarrow{\tau}_i\cdot\tilde{\overrightarrow{E}}^{(inc)}(\overrightarrow{0},s) \tag{4.117}$$

Comparing this to (4.2) we have

$$\begin{aligned} \tilde{\overleftrightarrow{\Lambda}}(\overrightarrow{1}_o,\overrightarrow{1}_i;s) &= -\overleftrightarrow{\tau}_o\cdot\tilde{\overleftrightarrow{\Lambda}}^{(d)}(\overrightarrow{1}_o,\overrightarrow{1}_i;s)\cdot\overleftrightarrow{\tau}_i = -\overrightarrow{1}_o\times\tilde{\overleftrightarrow{\Lambda}}^{(d)}(\overrightarrow{1}_o,\overrightarrow{1}_i;s)\times\overrightarrow{1}_i \\ \tilde{\overleftrightarrow{\Lambda}}^{(d)}(\overrightarrow{1}_o,\overrightarrow{1}_i;s) &= -\overleftrightarrow{\tau}_o\cdot\tilde{\overleftrightarrow{\Lambda}}(\overrightarrow{1}_o,\overrightarrow{1}_i;s)\cdot\overleftrightarrow{\tau}_i = -\overrightarrow{1}_o\times\tilde{\overleftrightarrow{\Lambda}}(\overrightarrow{1}_o,\overrightarrow{1}_i;s)\times\overrightarrow{1}_i \end{aligned} \tag{4.118}$$

For a self-dual scatterer (as in (4.114)) we then have a restricted form of the scattering dyadic obeying

$$\tilde{\overleftrightarrow{\Lambda}}(\overrightarrow{1}_o,\overrightarrow{1}_i;s) = -\overleftrightarrow{\tau}_o\cdot\tilde{\overleftrightarrow{\Lambda}}(\overrightarrow{1}_o,\overrightarrow{1}_i;s)\cdot\overleftrightarrow{\tau}_i \tag{4.119}$$

This is an interesting kind of symmetry which makes the scattering invariant to rotation of the fields about the directions of incidence and scattering. From the basic scattering equation (4.2) we have

$$\begin{aligned}
\overleftrightarrow{\tau}_o \cdot \tilde{\vec{E}}_f(\vec{r},s) &= \frac{e^{-\tilde{\gamma}r}}{4\pi r}\overleftrightarrow{\tau}_o \cdot \tilde{\overleftrightarrow{\Lambda}}(\vec{1}_o,\vec{1}_i;s)\cdot \tilde{\vec{E}}^{(inc)}(\vec{0},s) \\
&= -\frac{e^{-\tilde{\gamma}r}}{4\pi r}\overleftrightarrow{\tau}_o \cdot \tilde{\overleftrightarrow{\Lambda}}(\vec{1}_o,\vec{1}_i;s)\cdot \overleftrightarrow{\tau}_i^{2} \cdot \tilde{\vec{E}}^{(inc)}(\vec{0},s) \\
&= \frac{e^{-\tilde{\gamma}r}}{4\pi r}\tilde{\overleftrightarrow{\Lambda}}(\vec{1}_o,\vec{1}_i;s)\cdot \overleftrightarrow{\tau}_i \cdot \tilde{\vec{E}}^{(inc)}(\vec{0},s)
\end{aligned} \tag{4.120}$$

Weighting this by $-\sin(\psi)$ and adding to (4.2) weighted by $\cos(\psi)$ gives

$$\overleftrightarrow{C}_o(\psi)\cdot \tilde{\vec{E}}_f(\vec{r},s) = \frac{e^{-\tilde{\gamma}r}}{4\pi r}\tilde{\overleftrightarrow{\Lambda}}(\vec{1}_o,\vec{1}_i;s)\cdot \overleftrightarrow{C}_i(\psi)\cdot \tilde{\vec{E}}^{(inc)}(\vec{0},s) \tag{4.121}$$

where the rotation-by-angle-ψ dyadics are in (4.103) and (4.104). So rotating the incident field by ψ (about $\vec{1}_i$) results in rotation of the far scattered field by the same angle ψ (about $\vec{1}_o$). The waveforms (incident and scattered) are invariant to this transformation. Applying this to the scattering dyadic itself, dot multiply (4.121) by $\overleftrightarrow{C}_o(-\psi)$ giving

$$\begin{aligned}
\tilde{\vec{E}}_f(\vec{r},s) &= \frac{e^{-\tilde{\gamma}r}}{4\pi r}\overleftrightarrow{C}_o(-\psi)\cdot \tilde{\overleftrightarrow{\Lambda}}(\vec{1}_o,\vec{1}_i;s)\cdot \overleftrightarrow{C}_i(\psi)\cdot \tilde{\vec{E}}^{(inc)}(\vec{0},s) \\
\tilde{\overleftrightarrow{\Lambda}}(\vec{1}_o,\vec{1}_i;s) &= \overleftrightarrow{C}_o(-\psi)\cdot \tilde{\overleftrightarrow{\Lambda}}(\vec{1}_o,\vec{1}_i;s)\cdot \overleftrightarrow{C}_i(\psi)
\end{aligned} \tag{4.122}$$

This is a generalization of (4.119) to arbitrary rotation angles. Note that

$$\overleftrightarrow{C}_i\left(-\frac{\pi}{2}\right) = \overleftrightarrow{\tau}_i + \vec{1}_i\vec{1}_i, \qquad \overleftrightarrow{C}_o\left(\frac{\pi}{2}\right) = -\overleftrightarrow{\tau}_o + \vec{1}_o\vec{1}_o \tag{4.123}$$

giving (4.119) as a special case.

4.12 BACKSCATTERING BY SELF-DUAL TARGET

Applying (4.119) to backscattering for a self-dual scatterer we have

$$\begin{aligned}
&\vec{1}_o = -\vec{1}_i, \qquad \overleftrightarrow{1}_o = \overleftrightarrow{1}_i, \qquad \overleftrightarrow{\tau}_o = -\overleftrightarrow{\tau}_i \\
&\tilde{\overleftrightarrow{\Lambda}}_b(\vec{1}_i,s) \equiv \tilde{\overleftrightarrow{\Lambda}}(-\vec{1}_i,\vec{1}_i;s) = \overleftrightarrow{\tau}_i\cdot \tilde{\overleftrightarrow{\Lambda}}_b(\vec{1}_i,s)\cdot \overleftrightarrow{\tau}_i
\end{aligned} \tag{4.124}$$

In terms of the more general rotation in (4.122) we have

$$\overleftrightarrow{C}_o(-\psi) = \overleftrightarrow{C}_i(\psi), \qquad \tilde{\overleftrightarrow{\Lambda}}_b(\vec{1}_i,s) = \overleftrightarrow{C}_i(\psi)\cdot \tilde{\overleftrightarrow{\Lambda}}_b(\vec{1}_o,\vec{1}_i;s)\cdot \overleftrightarrow{C}_i(\psi) \tag{4.125}$$

Note that in backscattering rotation about $\vec{1}_o$ is reversed when referred to $\vec{1}_i$. Also note in (4.125) that reciprocity has not yet been assumed.

Writing out the components of the backscattering dyadic we have, referred to the $\vec{1}_h, \vec{1}_v, -\vec{1}_i$ coordinate directions,

$$\begin{aligned}
\tilde{\overleftrightarrow{\Lambda}}_b(\vec{1}_i, s) &= \tilde{\Lambda}_{b_{h,h}} \vec{1}_h \vec{1}_h + \tilde{\Lambda}_{b_{h,v}} \vec{1}_h \vec{1}_v + \tilde{\Lambda}_{b_{v,h}} \vec{1}_v \vec{1}_h + \tilde{\Lambda}_{b_{v,v}} \vec{1}_v \vec{1}_v \\
&= \vec{1}_i \times \tilde{\overleftrightarrow{\Lambda}}_b(\vec{1}_i, s) \times \vec{1}_i \\
&= -\tilde{\Lambda}_{b_{h,h}} \vec{1}_v \vec{1}_v + \tilde{\Lambda}_{b_{h,v}} \vec{1}_v \vec{1}_h + \tilde{\Lambda}_{b_{v,h}} \vec{1}_h \vec{1}_v - \tilde{\Lambda}_{b_{v,v}} \vec{1}_h \vec{1}_h \\
\tilde{\Lambda}_{b_{h,h}} &= -\tilde{\Lambda}_{b_{v,v}}, \qquad \tilde{\Lambda}_{b_{h,v}} = \tilde{\Lambda}_{b_{v,h}} \\
\tilde{\overleftrightarrow{\Lambda}}_b(\vec{1}_i, s) &= \tilde{\Lambda}_{b_{h,h}}(\vec{1}_i, s)\left[\vec{1}_h \vec{1}_h - \vec{1}_v \vec{1}_v\right] + \tilde{\Lambda}_{b_{h,v}}(\vec{1}_i, s)\left[\vec{1}_h \vec{1}_v + \vec{1}_v \vec{1}_h\right] \\
&= \begin{pmatrix} \tilde{\Lambda}_{b_{h,h}}(\vec{1}_i, s) & \tilde{\Lambda}_{b_{h,v}}(\vec{1}_i, s) & 0 \\ \tilde{\Lambda}_{b_{h,v}}(\vec{1}_i, s) & -\tilde{\Lambda}_{b_{h,h}}(\vec{1}_i, s) & 0 \\ 0 & 0 & 0 \end{pmatrix} \\
\tilde{\overleftrightarrow{\Lambda}}(\vec{1}_i, s) &= \tilde{\overleftrightarrow{\Lambda}}_b^T(\vec{1}_i, s) \text{ (symmetric)}
\end{aligned} \tag{4.126}$$

Note that the backscattering dyadic is symmetric even though reciprocity has not been assumed. Furthermore, this dyadic has only two independent components.

Let us now consider some implications of geometrical symmetries combined with duality. Consider the case of an axial symmetry plane (containing $\vec{1}_i$) which we take without loss of generality as containing $\vec{1}_h$. Such reflection is described by the group and dyadic representation

$$R_h = \{(1), (R_h)\}, \quad (1) \rightarrow \overleftrightarrow{1}, \quad (R_h) \rightarrow \overleftrightarrow{R}_h = \overleftrightarrow{1} - \vec{1}_h \vec{1}_h = \begin{pmatrix} -1 & 0 & 0 \\ 0 & 1 & 0 \\ 0 & 0 & 1 \end{pmatrix} \tag{4.127}$$

Applying this to the backscattering dyadic as

$$\tilde{\overleftrightarrow{\Lambda}}_b(\vec{1}_i, s) = \overleftrightarrow{R}_h \cdot \tilde{\overleftrightarrow{\Lambda}}_b(\vec{1}_i, s) \cdot \overleftrightarrow{R}_h \tag{4.128}$$

gives

$$\tilde{\Lambda}_{b_{h,v}}(\vec{1}_i, s) = \tilde{\Lambda}_{b_{v,h}}(\vec{1}_i, s) = 0$$

$$\tilde{\overleftrightarrow{\Lambda}}_b(\vec{1}_i, s) = \tilde{\Lambda}_{b_{h,h}}(\vec{1}_i, s) = \begin{pmatrix} \tilde{\Lambda}_{b_{h,h}}(\vec{1}_i, s) & 0 & 0 \\ 0 & -\tilde{\Lambda}_{b_{h,h}}(\vec{1}_i, s) & 0 \\ 0 & 0 & 0 \end{pmatrix} \tag{4.129}$$

with now only one independent component. However, noting the minus sign in the v, v location, this is *not* proportional to $\overleftrightarrow{1}_i$. Note that this dyadic is also invariant with respect to another axial plane containing $\vec{1}_h$ as

$$R_v = \{(1), \ (R_h)\}, \quad \overleftrightarrow{R}_v = \overleftrightarrow{1} - \vec{1}_v \vec{1}_v = \begin{pmatrix} 1 & 0 & 0 \\ 0 & -1 & 0 \\ 0 & 0 & 1 \end{pmatrix} \tag{4.130}$$

$$\tilde{\overleftrightarrow{\Lambda}}_b(\vec{1}_i, s) = \overleftrightarrow{R}_v \cdot \tilde{\overleftrightarrow{\Lambda}}_b(\vec{1}_i, s) \cdot \overleftrightarrow{R}_v$$

This backscattering dyadic is then invariant to transformation by the direct product group

$$C_{2a} = R_h \otimes R_v \tag{4.131}$$

which is also described as a twofold rotation axis with two axial symmetry planes.

Now consider rotation about the $-\vec{1}_i$ axis. Such an N-fold rotation is described by

$$\begin{gathered} C_N = \left\{(C_N)_\ell \,|\, \ell = 1, 2, \ldots, N\right\}, \quad (C_N) = (C_N)_1^\ell \equiv \text{rotation by } \phi_\ell = \frac{2\pi\ell}{N} \\ (C_N)_1^N = (C_N)_N = (1) \end{gathered} \tag{4.132}$$

This has a dyadic (matrix) representation

$$\overleftrightarrow{C}_i(\phi_\ell) = \overleftrightarrow{1}_i \cos(\phi_\ell) - \overleftrightarrow{\tau}_i \sin(\phi_\ell) + \vec{1}_i\vec{1}_i = \begin{pmatrix} \cos(\phi_\ell) & -\sin(\phi_\ell) & 0 \\ \sin(\phi_\ell) & \cos(\phi_\ell) & 0 \\ 0 & 0 & 1 \end{pmatrix} \tag{4.133}$$

$$\overleftrightarrow{C}_i(\phi_\ell) = \overleftrightarrow{C}_i^\ell(\phi_1), \quad \overleftrightarrow{C}_i^N(\phi_1) = \overleftrightarrow{1}, \quad \overleftrightarrow{C}_i^{-1}(\phi_\ell) = \overleftrightarrow{C}_i(-\phi_\ell)$$

As discussed in [2], [17], [18] backscattering for targets with such symmetry has the form

$$\begin{aligned} \tilde{\overleftrightarrow{\Lambda}}_b(\vec{1}_i, s) &= \tilde{\Lambda}_{b_{1,1}}(\vec{1}_i, s)\overleftrightarrow{1}_i + \tilde{\Lambda}_{b_{1,2}}(\vec{1}_i, s)\overleftrightarrow{\tau}_i \\ &= \begin{pmatrix} \tilde{\Lambda}_{b_{1,1}}(\vec{1}_i, s) & \tilde{\Lambda}_{b_{1,2}}(\vec{1}_i, s) & 0 \\ -\tilde{\Lambda}_{b_{1,2}}(\vec{1}_i, s) & \tilde{\Lambda}_{b_{1,1}}(\vec{1}_i, s) & 0 \\ 0 & 0 & 0 \end{pmatrix} \quad \text{for} \quad N \geq 3 \end{aligned} \tag{4.134}$$

where reciprocity has not yet been assumed. (With reciprocity the coefficient of $\overleftrightarrow{\tau}_i$ is zero.) For $N = 2$ the above result is not obtained and a general transverse $\tilde{\overleftrightarrow{\Lambda}}_b$ (four non-zero components) is invariant to C_2 transformation which is equivalent to a sign reversal of the incident and scattered field, a consequence of linearity.

Comparing (4.134) to (4.126) the unique solution is

$$\tilde{\overleftrightarrow{\Lambda}}_b(\vec{1}_i, s) = \overleftrightarrow{0} \qquad \text{for} \qquad N \geq 3 \tag{4.135}$$

i.e., zero backscattering. This is a generalization of the result in [21] where zero backscattering is derived for the case of $N = 4$, invariance to rotation by $\pi/4$. Now the result is also generalized from scalar to dyadic permittivity and permeability.

4.13 FORWARD-SCATTERING BY SELF-DUAL TARGET

Applying (4.119) to forward-scattering from a self-dual target we have

$$\begin{gathered} \vec{1}_o = \vec{1}_i, \qquad \overleftrightarrow{1}_o = \overleftrightarrow{1}_i, \qquad \overleftrightarrow{\tau}_o = \overleftrightarrow{\tau}_i \\ \tilde{\overleftrightarrow{\Lambda}}_f(\vec{1}_i, s) \equiv \tilde{\overleftrightarrow{\Lambda}}(\vec{1}_i, \vec{1}_i; s) = -\overleftrightarrow{\tau}_i \cdot \tilde{\overleftrightarrow{\Lambda}}_f(\vec{1}_i, s) \cdot \overleftrightarrow{\tau}_i \end{gathered} \tag{4.136}$$

From (4.122) we have the more general rotation

$$\tilde{\overleftrightarrow{\Lambda}}_f(\vec{1}_i, s) = \overleftrightarrow{C}_i(-\psi) \cdot \tilde{\overleftrightarrow{\Lambda}}_f(\vec{1}_i, s) \cdot \overleftrightarrow{C}_i(\psi) = \overleftrightarrow{C}_i^T(\psi) \cdot \tilde{\overleftrightarrow{\Lambda}}_f(\vec{1}_i, s) \cdot \overleftrightarrow{C}_i(\psi) \tag{4.137}$$

so that the forward-scattering dyadic is invariant to rotation about $\vec{1}_i$ (independent of any symmetries in the target other than self-duality). Writing out the components as in (4.126) we have

$$\begin{aligned}
\tilde{\overleftrightarrow{\Lambda}}_f(\vec{1}_i, s) &= \tilde{\Lambda}_{f_{h,h}} \vec{1}_h \vec{1}_h + \tilde{\Lambda}_{f_{h,v}} \vec{1}_h \vec{1}_v + \tilde{\Lambda}_{f_{v,h}} \vec{1}_v \vec{1}_h + \tilde{\Lambda}_{f_{v,v}} \vec{1}_v \vec{1}_v \\
&= -\vec{1}_i \times \tilde{\overleftrightarrow{\Lambda}}_f(\vec{1}_i, s) \times \vec{1}_i \\
&= \tilde{\Lambda}_{f_{h,h}} \vec{1}_v \vec{1}_v - \tilde{\Lambda}_{f_{h,v}} \vec{1}_v \vec{1}_h - \tilde{\Lambda}_{f_{v,h}} \vec{1}_h \vec{1}_v + \tilde{\Lambda}_{f_{v,v}} \vec{1}_h \vec{1}_h \\
\tilde{\Lambda}_{f_{h,h}} &= \tilde{\Lambda}_{f_{v,v}}, \qquad \tilde{\Lambda}_{f_{h,v}} = -\tilde{\Lambda}_{f_{v,h}}
\end{aligned} \tag{4.138}$$

$$\tilde{\overleftrightarrow{\Lambda}}_f(\vec{1}_i, s) = \tilde{\Lambda}_{f_{h,h}}(\vec{1}_i, s)\overleftrightarrow{1}_i + \tilde{\Lambda}_{f_{h,v}}(\vec{1}_i, s)\overleftrightarrow{\tau}_i = \begin{pmatrix} \tilde{\Lambda}_{f_{h,h}}(\vec{1}_i, s) & \tilde{\Lambda}_{f_{h,v}}(\vec{1}_i, s) & 0 \\ -\tilde{\Lambda}_{f_{h,v}}(\vec{1}_i, s) & \tilde{\Lambda}_{f_{h,h}}(\vec{1}_i, s) & 0 \\ 0 & 0 & 0 \end{pmatrix}$$

which is a combination of the transverse identity and a rotation. The forward-scattering dyadic has only two independent components.

If the target has an axial symmetry plane parallel to $\vec{1}_h$, then the reflection described by (4.127) applied to the forward-scattering dyadic gives

$$\tilde{\Lambda}_{f_{h,v}}(\vec{1}_i, s) = \tilde{\Lambda}_{f_{v,h}}(\vec{1}_i, s) = 0, \qquad \tilde{\overleftrightarrow{\Lambda}}_f(\vec{1}_i, s) = \tilde{\Lambda}_f(\vec{1}_i, s)\overleftrightarrow{1}_i \tag{4.139}$$

which is symmetric even though reciprocity has not been invoked. Furthermore, there is no change in the polarization from incident to scattered field. This forward-scattering dyadic has $C_{\infty a}$ (or O_2) symmetry (rotation by any angle with infinitely many axial symmetry planes).

Considering $\vec{1}_i$ as an N-fold rotation axis, one can apply $\overleftrightarrow{C}_i(\phi_\ell)$ as in (4.133). However, note that this is just a special case of the general property of the forward-scattering dyadic in (4.137). So C_N symmetry here adds no new property to the forward-scattering dyadic.

Inversion symmetry is described by

$$I = \{(1), \ (I)\}, \qquad (1) \rightarrow \overleftrightarrow{1}, \ \ (I) \rightarrow -\overleftrightarrow{1} \tag{4.140}$$

Applying this to the forward-scattering dyadic for a target with this symmetry gives

$$-\overleftrightarrow{1} \cdot \vec{1}_i = -\vec{1}_i, \qquad \tilde{\overleftrightarrow{\Lambda}}_f(\vec{1}_i, s) = \tilde{\overleftrightarrow{\Lambda}}_f(-\vec{1}_i, s) \tag{4.141}$$

which with (4.138) gives

$$\tilde{\Lambda}_{f_{h,h}}(\vec{1}_i, s) = \tilde{\Lambda}_{f_{v,v}}(-\vec{1}_i, s), \qquad \tilde{\Lambda}_{f_{h,v}}(\vec{1}_i, s) = \tilde{\Lambda}_{f_{h,v}}(-\vec{1}_i, s) \tag{4.142}$$

Remember, however, that on inversion we have

$$(-\overleftrightarrow{1}) \cdot \vec{1}_i = -\vec{1}_i, \qquad \overleftrightarrow{\tau}_{-i} = -\overleftrightarrow{\tau}_i \tag{4.143}$$

and one of the h and v directions reverses. While inversion produces the above form for any $\vec{1}_i$, various other symmetries which reverse $\vec{1}_i$ (including reflection R_t through a transverse symmetry plane, as well as rotation, reflection and dihedral D_N symmetry with this principal axis) produce similar results.

Consider now the implication of reciprocity. From (4.13) we have

$$\begin{aligned}\tilde{\overleftrightarrow{\Lambda}}_f(-\overrightarrow{1}_i, s) &= \tilde{\overleftrightarrow{\Lambda}}_f^T(\overrightarrow{1}_i, s) = \tilde{\Lambda}_{b_{h,h}}(\overrightarrow{1}_i, s)\overleftrightarrow{1}_i - \tilde{\Lambda}_{b_{h,v}}(\overrightarrow{1}_i, s)\overleftrightarrow{\tau}_i \\ &= \tilde{\Lambda}_{b_{h,h}}(-\overrightarrow{1}_i, s)\overleftrightarrow{1}_i - \tilde{\Lambda}_{b_{h,v}}(-\overrightarrow{1}_i, s)\overleftrightarrow{\tau}_i \\ \tilde{\Lambda}_{b_{h,h}}(-\overrightarrow{1}_i, s) &= \tilde{\Lambda}_{b_{h,h}}(\overrightarrow{1}_i, s), \quad \tilde{\Lambda}_{b_{h,v}}(-\overrightarrow{1}_i, s) = -\tilde{\Lambda}_{b_{h,v}}(\overrightarrow{1}_i, s)\end{aligned} \tag{4.144}$$

which is a kind of invariance to direction reversal. Note the change of sign on $\overleftrightarrow{\tau}_i$ on direction reversal, and the change of the h direction or v direction.

Now combine inversion symmetry from (4.141) with reciprocity to give

$$\tilde{\overleftrightarrow{\Lambda}}_f(\overrightarrow{1}_i, s) = \tilde{\overleftrightarrow{\Lambda}}_f^T(\overrightarrow{1}_i, s) = \tilde{\Lambda}_b(\overrightarrow{1}_i, s)\overleftrightarrow{1}_i = \tilde{\overleftrightarrow{\Lambda}}_f(-\overrightarrow{1}_i, s) = \tilde{\Lambda}_b(-\overrightarrow{1}_i, s)\overleftrightarrow{1}_i \tag{4.145}$$

The coefficient of $\overleftrightarrow{\tau}_i$ vanishes and the scattering dyadic is proportional to the transverse dyadic with a coefficient which is invariant to reversal of $\overrightarrow{1}_i$. Similar results hold for other symmetries which reverse $\overrightarrow{1}_i$ for particular choices of $\overrightarrow{1}_i$.

4.14 LOW-FREQUENCY SCATTERING BY SELF-DUAL TARGET

For low frequencies the scattering is dominated by the induced electric and magnetic dipole moments via the respective polarizabilities [5], [8] as

$$\tilde{\overleftrightarrow{\Lambda}}(\overrightarrow{1}_o, \overrightarrow{1}_i; s) = \tilde{\gamma}^2\left[-\overleftrightarrow{1}_o \cdot \tilde{\overleftrightarrow{P}}(s) \cdot \overleftrightarrow{1}_i + \overleftrightarrow{\tau}_i \cdot \tilde{\overleftrightarrow{M}}(s) \cdot \overleftrightarrow{\tau}_i\right] \text{ as } s \to 0 \tag{4.146}$$

From (4.19) the dual scattering dyadic is

$$\begin{aligned}\tilde{\overleftrightarrow{\Lambda}}^{(d)}(\overrightarrow{1}_o, \overrightarrow{1}_i; s) &= \overleftrightarrow{\tau}_o \cdot \tilde{\overleftrightarrow{\Lambda}}(\overrightarrow{1}_o, \overrightarrow{1}_i; s) \cdot \overleftrightarrow{\tau}_i \\ &= \tilde{\gamma}^2\left[-\overleftrightarrow{1}_o \cdot \tilde{\overleftrightarrow{P}}^{(d)}(s) \cdot \overleftrightarrow{1}_i + \overleftrightarrow{\tau}_o \cdot \tilde{\overleftrightarrow{M}}^{(d)}(s) \cdot \overleftrightarrow{\tau}_i\right] \text{ as } s \to 0\end{aligned} \tag{4.147}$$

from which we identify

$$\tilde{\overleftrightarrow{P}}^{(d)}(s) = \tilde{\overleftrightarrow{M}}(s), \qquad \tilde{\overleftrightarrow{M}}^{(d)}(s) = \tilde{\overleftrightarrow{P}}(s) \tag{4.148}$$

This should be apparent from the interchanged roles of $\overleftrightarrow{\varepsilon}$ and $\overleftrightarrow{\mu}$ in (4.113) together with the interchange of electric and magnetic fields. For a self-dual target then evidently

$$\begin{aligned}\tilde{\overleftrightarrow{P}}(s) &= \tilde{\overleftrightarrow{M}}(s) \\ \tilde{\overleftrightarrow{\Lambda}}(\overrightarrow{1}_o, \overrightarrow{1}_i; s) &= \tilde{\gamma}^2\left[-\overleftrightarrow{1}_o \cdot \tilde{\overleftrightarrow{P}}(s) \cdot \overleftrightarrow{1}_i + \overleftrightarrow{\tau}_o \cdot \tilde{\overleftrightarrow{P}}(s) \cdot \overleftrightarrow{\tau}_i\right] \text{ as } s \to 0 \\ &\equiv \tilde{\gamma}^2 \tilde{\overleftrightarrow{X}}(\overrightarrow{1}_o, \overrightarrow{1}_i; s) \text{ as } s \to 0\end{aligned} \tag{4.149}$$

Note a general symmetry on reversing both directions of incidence and scattering, giving

$$\tilde{\overleftrightarrow{X}}(-\overrightarrow{1}_o, -\overrightarrow{1}_i; s) = \tilde{\overleftrightarrow{X}}(\overrightarrow{1}_o, \overrightarrow{1}_i; s) \tag{4.150}$$

For backscattering this becomes

$$\begin{aligned}\tilde{\overleftrightarrow{X}}_b(\overrightarrow{1}_i, s) &\equiv \tilde{\overleftrightarrow{X}}(-\overrightarrow{1}_i, \overrightarrow{1}_i; s) = -\overleftrightarrow{1}_i \cdot \tilde{\overleftrightarrow{P}}(s) \cdot \overleftrightarrow{1}_i - \overleftrightarrow{\tau}_i \cdot \tilde{\overleftrightarrow{P}}(s) \cdot \overleftrightarrow{\tau}_i \\ &= \left[\tilde{P}_{v,v}(s) - \tilde{P}_{h,h}(s)\right]\left[\overrightarrow{1}_h \overrightarrow{1}_h - \overrightarrow{1}_v \overrightarrow{1}_v\right] - \left[\tilde{P}_{h,v}(s) + \tilde{P}_{v,h}(s)\right]\left[\overrightarrow{1}_h \overrightarrow{1}_v + \overrightarrow{1}_v \overrightarrow{1}_h\right]\end{aligned} \tag{4.151}$$

This exhibits the fact that the backscattering dyadic is symmetric without invoking reciprocity.

In forward-scattering we have

$$\begin{aligned}\tilde{\overleftrightarrow{X}}_f(\overrightarrow{1}_i, s) &\equiv \tilde{\overleftrightarrow{X}}(\overrightarrow{1}_i, \overrightarrow{1}_i; s) = -\overleftrightarrow{1}_i \cdot \tilde{\overleftrightarrow{P}}(s) \cdot \overleftrightarrow{1}_i + \overleftrightarrow{\tau}_i \cdot \tilde{\overleftrightarrow{P}}(s) \cdot \overleftrightarrow{\tau}_i \\ &= -\left[\tilde{P}_{h,h}(s) + \tilde{P}_{v,v}(s)\right]\overleftrightarrow{1}_i - \left[\tilde{P}_{h,v}(s) - \tilde{P}_{v,h}(s)\right]\overleftrightarrow{\tau}_i\end{aligned} \tag{4.152}$$

consistent with the general form in (4.138). Invoking reciprocity gives

$$\begin{gathered}\overleftrightarrow{P}(s) = \overleftrightarrow{P}^T(s), \qquad \overleftrightarrow{M}(s) = \overleftrightarrow{M}^T(s) \\ \tilde{\overleftrightarrow{X}}_f(\overrightarrow{1}_i, s) = \tilde{\overleftrightarrow{X}}_f^T(\overrightarrow{1}_i, s) = -\left[\tilde{P}_{h,h}(s) + \tilde{P}_{v,v}(s)\right]\overleftrightarrow{1}_i = \tilde{\overleftrightarrow{X}}_f(-\overrightarrow{1}_i, s)\end{gathered} \tag{4.153}$$

Sections 4.13 and 4.14 consider the influence of geometrical symmetries on backscattering and forward-scattering, respectively. For low frequencies these results also apply. Of course, it is the symmetries in the polarizabilities, whether or not they come from geometrical symmetries that are important. Various geometrical symmetries (with reciprocity) induce simplifications (i.e., symmetries) in the forms of the polarizability dyadics as tabulated in Table 4.4. A special case is that of a generalized spherical (O_3), tetrahedral (T), octahedral (O), or icosahedral (Y) symmetry (including reciprocity), each of which gives

$$\begin{gathered}\tilde{\overleftrightarrow{P}}(s) = \tilde{P}(s)\overleftrightarrow{1}, \qquad \tilde{\overleftrightarrow{X}}(\overrightarrow{1}_o, \overrightarrow{1}_i; s) = \tilde{P}(s)\left[-\overleftrightarrow{1}_o \cdot \overleftrightarrow{1}_i + \overleftrightarrow{\tau}_o \cdot \overleftrightarrow{\tau}_i\right] \\ \tilde{\overleftrightarrow{X}}_b(\overrightarrow{1}_i, s) = \overleftrightarrow{0}, \qquad \tilde{\overleftrightarrow{X}}_f(\overrightarrow{1}_i, s) = -2\tilde{P}(s)\overleftrightarrow{1}_i\end{gathered} \tag{4.154}$$

Note that the forward scattering is polarization independent and independent of the orientation of the target. For a spherical target, this is immediately apparent, but for the regular polyhedral this relies on the fact that the multiple rotation axes make the polarizabilities (related to dipole moments, not including the higher-order multipoles) invariant to all rotations and reflections.

4.15 CONCLUSION

As shown by the various cases considered here, reciprocity is fundamental to the symmetry properties of the scattering dyadic, including both back- and forward-scattering. When combined with geometrical symmetries in the target this gives yet higher-order symmetries in the scattering dyadic as summarized in the tables. There are also symmetries in general bistatic scattering if the target has a symmetry plane aligned with the scattering plane or bisectrix plane. For low-frequency scattering (electrically small target) there is also the general bistatic symmetry of invariance to direction reversal for both incidence and scattering directions.

Another related topic concerns the impact of symmetry on the eigenmodes and natural modes. As discussed in [1], [2], [11], [17] there is an interesting case of twofold degeneracy for a body of revolution with axial symmetry planes ($C_{\infty a} = O_2$ symmetry). For backscattering, the rotation axis and the observer ($\vec{1}_i$) determine a symmetry plane. The symmetry plane makes the various residue vectors for the poles align parallel (symmetric) or perpendicular (antisymmetric) to the plane, thereby determining the target orientation. The twofold degeneracy for modes with Fourier-series terms $\cos(m\phi')$ and $\sin(m\phi')$ and $m \geqslant 1$ (ϕ' around rotation axis) give rise to two natural modes and residue vectors (one parallel and the other perpendicular to the plane) for each natural frequency s_α. Other types of symmetry (such as C_N for $N \geq 3$) can also lead to such twofold modal degeneracy. Such degeneracy is treated in detail for the magnetic polarizability dyadic (and by simple substitution the electric polarizability dyadic) in [10]. There various point symmetries gave rise to two- and three-fold degeneracies. Noting that only dipole terms are treated there, and that the scattering at higher frequencies (target not electrically small) involves in general all multipoles, then there can be various kinds of modal degeneracies associated with the various point symmetries. These considerations are beyond the scope of the present discussion.

Self-duality gives some special properties to target scattering. For general angles of incidence and scattering, rotation of the incident field produces an equal rotation of the scattered field. For backscattering, this means that the scattered field is rotated with sense opposite to the incident field when both are referred to the same direction, i.e., the direction of incidence. If, in addition, the target has a threefold or higher rotation axis parallel to the direction of incidence, the backscattering is zero. When combined with various geometrical symmetries and/or reciprocity in the target, the scattering dyadic attains various simplifications (symmetries) as well.

REFERENCES

[1] C. E. Baum, Scattering, Reciprocity, Symmetry, EEM, and SEM, Interaction Note 475, May 1989.

[2] C. E. Baum, SEM Backscattering, Interaction Note 476, July 1989.

[3] C. E. Baum, The Self-Complementary Rotation Group, Interaction Note 479, December 1989.

[4] J. Nitsch, C. E. Baum, and R. Sturm, Commutative Tubes in Multiconductor-Transmission-Line Theory, Interaction Note 480, January 1990.

[5] C. E. Baum, The SEM Representation of Scattering From Perfectly Conducting Targets in Simple Lossy Media, Interaction Note 492, April 1993.

[6] C. E. Baum, Scattering From Cones, Wedges, and Half Spaces, Interaction Note 493, May 1993.

[7] C. E. Baum, Scattering From Finite Wedges and Half Spaces, Interaction Note 495, May 1993.

[8] C. E. Baum, Low-Frequency Near-Field Magnetic Scattering From Highly, But Not Perfectly, Conducting Bodies, Interaction Note 499, November 1993. Also Chapter 6, pp. 163–218, in C. E. Baum, ed., *Detection and Identification of Visually Obscured Targets*, Taylor and Francis, Washington, D.C., 1998.

[9] C. E. Baum, Signature-Based Target Identification, Interaction Note 500, January 1994.

[10] C. E. Baum, The Magnetic Polarizability Dyadic and Point Symmetry, Interaction Note

502, May 1994. Also Chapter 7, pp. 219–242, in C. E. Baum, ed., *Detection and Identification of Visually Obscured Targets*, Taylor and Francis, Washington, D.C., 1998.

[11] C. E. Baum, E. J. Rothwell, K.-M. Chen, and D. P. Nyquist, The Singularity Expansion Method and Its Application to Target Identification, *Proceedings of IEEE*, 1991, pp. 1481–1492.

[12] C. E. Baum, Signature-Based Target Identification and Pattern Recognition, *IEEE Antennas and Propagation Magazine*, June 1994, pp. 44–51.

[13] E. P. Wigner, *Group Theory*, Academic Press, New York, 1959.

[14] M. Hamermesh, *Group Theory and Its Application to Physical Problems*, Addison-Wesley, Reading, MA, 1962.

[15] J. Van Bladel, *Electromagnetic Fields*, Hemisphere Publishing Corp, Taylor and Francis, Washington, D.C., 1985.

[16] P. B. Yale, *Geometry and Symmetry*, Dover, 1988.

[17] C. E. Baum, SEM and EEM Scattering Matrices, and Time-Domain Scatterer Polarization in the Scattering Residue Matrix, pp. 427–486, in W.-M. Boerner et al., eds., *Direct and Inverse Methods in Radar Polarimetry*, Kluwer Academic Publishers, Dordrecht, 1992.

[18] C. E. Baum and H. N. Kritikos, Symmetry in Electromagnetics, Chapter. 1, pp. 1–90, in C. E. Baum and H. N. Kritikos, eds., *Electromagnetic Symmetry*, Taylor and Francis, Washington, D.C., 1995.

[19] S. R. Cloude, Lie Groups in Electromagnetic Wave Propagation and Scattering, Chapter 2, pp. 91–142, in C. E. Baum and H. N. Kritikos, eds., *Electromagnetic Symmetry*, Taylor and Francis, Washington, D.C., 1995.

[20] C. E. Baum, Continuous Dilation Symmetry in Electromagnetic Scattering, Chapter 3, pp. 143–183, in C. E. Baum and H. N. Kritikos, eds., *Electromagnetic Symmetry*, Taylor and Francis, Washington, D.C., 1995.

[21] V. H. Weston, Theory of Absorbers in Scattering, *IEEE Trans. Antennas and Propagation*, 1963, pp. 578–584.

[22] G. E. Heath, Bistatic Scattering Reflection Asymmetry, Polarization Reversal Symmetry, and Polarization Reversal Reflection Symmetry, *IEEE Trans. Antennas and Propagation*, 1981, pp. 429–434.

[23] G. E. Heath, Properties of the Linear Polarization Bistatic Scattering Matrix, *IEEE Trans. Antennas and Propagation*, 1981, pp. 523–525.

[24] M. Davidovitz and W. M. Boerner, Reduction of Bistatic Scattering Matrix Measurements for Inversely Symmetric Radar Targets, *IEEE Trans. Antennas and Propagation*, 1983, pp. 237–242.

Chapter 5

COMPLEMENTARY STRUCTURES IN TWO DIMENSIONS

Carl E. Baum

ABSTRACT

Self-complementary antennas are based on the Babinet principle in which electric and magnetic fields are interchanged (duality) with the structure invariant to this transformation. This paper develops self-complementary structures from the electric and magnetic parts of a complex potential as used in conformal transformation. This leads to various geometries of electrically small impedance sheets with relatively simply calculable admittance properties. These are also related to the impedance properties of some TEM transmission-line structures (cylindrical and conical). The use of conformal and stereographic transformations allows one to generalize self-complementary structures on planes and spheres to structures having the self-complementary admittance properties without the restricted geometrical symmetries.

5.1 INTRODUCTION

A classical concept in antenna theory is the complementary antenna derived from the Babinet principle [9]. The complement is formed by replacing an aperture in a perfectly conducting planar sheet by a perfectly conducting disc of the same shape in free space. Then with an appropriate introduction of antenna terminals the input impedance of the two antennas has the relationship

$$\tilde{Z}(s)\,\tilde{Z}^{(c)}(s) = \frac{Z_0^2}{4}$$
$$Z_0 = \left[\frac{\mu_0}{\varepsilon_0}\right]^{1/2} \simeq 376.73\,\Omega \qquad \text{(wave impedance of free space)} \tag{5.1}$$

where the superscript c denotes complement. Which of the two antennas is the original one, and which is the complement, is arbitrary. The above definition of complement can be generalized to the case of portions of the plane of interest consisting of impedance sheets of scalar or even dyadic (2×2) form [4], [5], [14]. The complementary sheet impedance (which may vary with position) satisfies a formula like (5.1) (with a rotation in the case of a dyadic sheet impedance). If the original and the complement are the same except for a rotation and/or reflection, the antenna is said to be self-complementary and (5.1) directly gives

$$\tilde{Z}(s) = \tilde{Z}^{(c)}(s) = \frac{Z_0}{2} \simeq 188.4\,\Omega \tag{5.2}$$

which is now frequency independent. As discussed in various papers [10], [12] this relationship can be generalized to multiterminal cases.

As observed in [10] this self-complementary relationship can also be applied to the TEM modes on structures consisting of perfectly conducting cylindrical and conical sheets lying on a circular cylinder and circular cone, respectively, and having an N-fold rotation axis with alternating spaces of equal angular width to the conducting sheets. (This is described by C_{Nc} symmetry discussed in [5], [14].) Note that by the stereograph transformation [13] the conical case can be described by an equivalent cylindrical case, so that our present discussion can be carried out in terms of two dimensions on a plane.

As observed in [5], [14] the self-complementary relationship for impedance sheets, when considered for low frequencies (quasi-static), implies that such structures can be used to give lumped impedances which have the same value (ohms) as the sheet impedance (ohms per square). A more classical way to calculate the impedance of such structures is via the two-dimensional Laplace equation and the common technique of conformal transformation for solving this equation.

This suggests that one might approach the concept of complementarity from the point of view of complex variables such as are used in conformal transformation. As we shall see, there is a certain duality or complementarity between the electric and magnetic potentials making up the complex potential. This can be used to define complementary and, in turn, self-complementary two-dimensional structures.

5.2 QUASI-STATIC BOUNDARY VALUE PROBLEMS IN TWO DIMENSIONS

Consider, for example, a uniform impedance sheet illustrated in Fig. 5.1a. Here we write this as R_s, a simple resistance sheet, but later, this can be given frequency dependence as desired. This sheet is terminated in two separate perfectly conducting terminals forming a port with (from some source) a voltage V and current I. There are also two truncations of the resistive material as indicated so that the current is limited to some domain D_{in}. The electric field $\vec{E}$ and surface current density $\vec{J}_s$ are perpendicular to the terminal conductors (electric boundaries) and parallel to the other two truncations (magnetic boundaries). For the present discussion the surface current density is limited to the interior domain D_{in}. One can also consider the exterior domain D_{ex} as having an impedance sheet with similar results. Later we shall consider the union of these.

In our two-dimensional x, y coordinates we have

$$\begin{gathered} \vec{E}(x,y) = R_s\vec{J}_s(x,y) = -\nabla_s\Phi_e(x,y), \qquad \Phi_e(x,y) \equiv \text{electric potential} \\ \Phi_{e+} - \Phi_{e-} = V, \qquad \nabla_s = \vec{1}_x\frac{\partial}{\partial x} + \vec{1}_y\frac{\partial}{\partial y} \end{gathered} \tag{5.3}$$

With

$$\nabla_s \cdot \vec{J}_s(x,y) = 0 \text{ (equation of continuity, static)} \tag{5.4}$$

we have the Laplace equation

$$\nabla_s^2\Phi_e(x,y) = -\nabla_s\cdot\vec{E}(x,y) = -\nabla_s\cdot(R_s\vec{J}_s(x,y)) = -R_s\nabla_s\cdot\vec{J}_s(x,y) = 0 \tag{5.5}$$

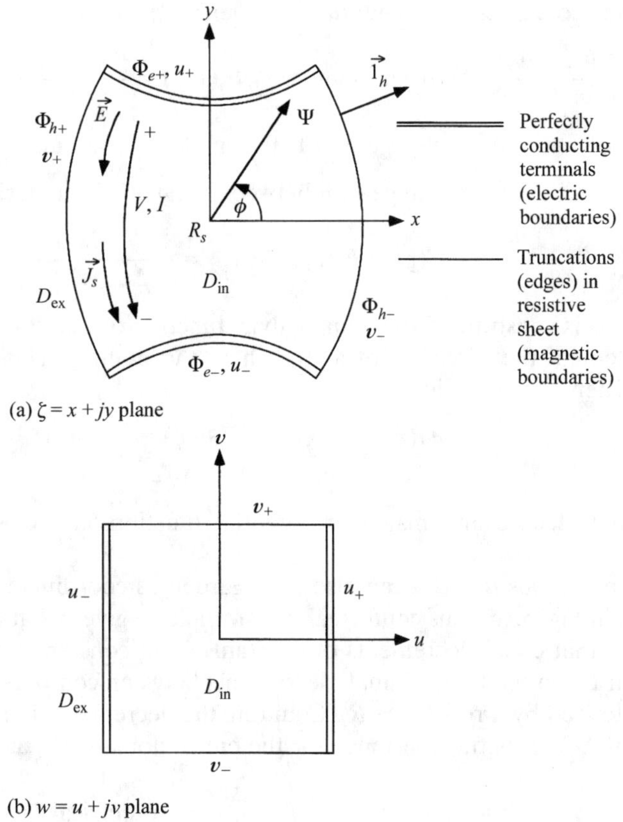

Figure 5.1 Uniform impedance sheet of finite dimensions with two terminals (single port).

since R_s is not a function of position. This two-dimensional potential function is to be solved subject the boundary conditions

$$\begin{aligned}
&\Phi_e(x, y) = \Phi_{e+}, \Phi_{e-} \text{ on electric boundaries} \\
&\vec{1}_h \cdot \nabla_s \Phi_e(x, y) = 0 \text{ on magnetic boundaries} \qquad (5.6) \\
&\vec{1}_h \equiv \text{unit normal (in } x, y \text{ plane) to magnetic boundaries}
\end{aligned}$$

For various irregular boundary shapes one may wish to solve this Laplace equation numerically. For present purposes of obtaining analytic results we introduce the usual complex coordinate and potential as

$$\zeta = x + jy, \qquad w(\zeta) = u(\zeta) + jv(\zeta) \tag{5.7}$$

where $w(\zeta)$ is also referred to as a conformal transformation. This complex potential can be used to express the potentials and fields [2], [15] as

$$\Phi_e(x,y) = \frac{\Phi_{e+} - \Phi_{e-}}{\Delta u} u(\zeta) + \text{constant}, \qquad \Phi_h(x,y) = \frac{\Phi_{h+} - \Phi_{h-}}{\Delta v} v(\zeta) + \text{constant}$$

$$\begin{aligned} \Delta u &= u_+ - u_- \equiv \text{change in } u \text{ between electric boundaries (terminals)} \\ \Delta v &= v_+ - v_- \equiv \text{change in } v \text{ between magnetic boundaries (edges)} \end{aligned} \tag{5.8}$$

$$E(\zeta) = E_x(\zeta) - jE_y(\zeta) = -\frac{V}{\Delta u}\frac{dw(\zeta)}{d\zeta}$$

Note that w is constrained to be an analytic function of ζ so that the derivative exists independent of the direction of approach to the point ζ. This is expressed by the Cauchy–Riemann conditions

$$\frac{\partial u(\zeta)}{\partial x} = \frac{\partial v(\zeta)}{\partial y}, \qquad \frac{\partial u(\zeta)}{\partial y} = -\frac{\partial v(\zeta)}{\partial x} \tag{5.9}$$

so that the electric and magnetic potential functions are closely related to each other.

The potentials u and v can also be regarded as coordinates in a (u, v) plane as indicated in Fig. 5.1b. This conformal transformation gives what are called curvilinear squares in that equal decrements of constant u and constant v contours give square patches in D_{in} in the limit of small decrements between contours. Such a small square is characterized by a resistance R_s. Counting the decrements in u (to total Δu) and in v (to total Δv) gives the resistance for the entire domain D_{in} as

$$R = \frac{V}{I} = f_g R_s, \qquad f_g = \frac{\Delta u}{\Delta v} \equiv \text{geometrical factor} \tag{5.10}$$

It is this geometrical factor which can be used to scale any sheet impedance, the wave impedance in the case of a cylindrical transmission line, and the inductance and capacitance of related two-dimensional structures (together with the length).

5.3 TWO-DIMENSIONAL COMPLEMENTARY STRUCTURES

Let us now define a complementary structure and boundary-value problem by interchanging the roles of u and v, i.e.,

$$w^{(c)}(\zeta) = u^{(c)}(\zeta) + jv^{(c)}(\zeta) = \pm v(\zeta) \pm ju(\zeta) + \text{constant} \tag{5.11}$$

As indicated, there are various choices of sign that one can use, and being potential functions, an arbitrary constant can be added. It is the change in the functions (taken positive) between the appropriate boundaries that is relevant. If desired, one can choose the complementary potential function as an analytic function of ζ, e.g.,

$$w^{(c)}(\zeta) = -jw(\zeta) = v(\zeta) - ju(\zeta) \tag{5.12}$$

which is a rotation in the u, v plane.

For our example problem in Fig. 5.1, this corresponds to interchanging the roles of the electric and magnetic boundaries. The perfectly conducting terminals are placed along the magnetic boundaries denoted by magnetic potentials v_+ and v_-, and removed from the electric boundaries denoted by electric potentials u_+ and u_-. This gives a complementary resistance

$$R^{(c)} = f_g^{(c)} R, \qquad f_g^{(c)} = \frac{\Delta u^{(c)}}{\Delta v^{(c)}} \equiv \frac{\Delta v}{\Delta u}, \qquad f_g f_g^{(c)} = 1, \qquad RR^{(c)} = R_s^2 \tag{5.13}$$

Comparing this to (5.1) we have the same form of complementary relationship with $Z_0/2$ replaced by R_s.

One way to define self-complementarity is then

$$f_g \equiv f_g^{(c)} \equiv 1, \qquad \Delta u \equiv \Delta v, \qquad R \equiv R^{(c)} \equiv R_s \tag{5.14}$$

This is rather general, but in the example in Fig. 5.1a, this may still allow rather unsymmetrical shapes. Note, however, in Fig. 5.1b, that in the w plane the domain D_{in} is a square. Keeping the distinction between electric and magnetic boundaries, this has a twofold rotation axis (C_2 symmetry, equivalent to inversion in two dimensions) with two axial symmetry planes (C_{2a} symmetry) and self-complementarity (C_{2ac} symmetry).

Comparing the highly symmetric form in Fig. 5.1b to the generally unsymmetric form in Fig. 5.1a, remember that they are related by the conformal transformation $w(\zeta)$ and have the same impedance properties. So we can take a symmetrical shape, including self-complementarity, and transform it to a variety of other shapes which can be analyzed as though they had the symmetries of the original symmetrical shape.

5.4 LOWEST-ORDER SELF-COMPLEMENTARY ROTATION GROUP: C_{2c} SYMMETRY

In the simplest form, C_{2c} symmetry, we have a twofold rotation axis (C_2 symmetry) and the operation of self-complement on rotation by $\pi/2$. By self-complement we take the interchange of the roles of electric and magnetic potentials as in Section 3, now on rotation by $\pi/2$, as

$$\begin{aligned} \zeta^{(c)} &= j\zeta = e^{j(\pi/2)}\zeta \ (\pi/2 \text{ rotation of coordinates}) \\ &= x^{(c)} + jy^{(c)} = -y + jx \\ w^{(c)}(\zeta^{(c)}) &= u^{(c)}(\zeta^{(c)}) + jv^{(c)}(\zeta^{(c)}) = w(\zeta) = u(\zeta) + jv(\zeta) \\ w^{(c)}(\zeta) &= -jw(\zeta), \qquad w(\zeta^{(c)}) = jw(\zeta) \end{aligned} \tag{5.15}$$

Note that taking the complement also involves the interchange of the role of electric and magnetic boundaries. So the complementary structure looks just like the original structure, except rotated by $\pi/2$, and it is therefore self-complementary. In terms of shape the structure has C_4 symmetry (fourfold rotation axis) except that there are two each of alternating electric and magnetic boundaries. Figure 5.1b gives a simple

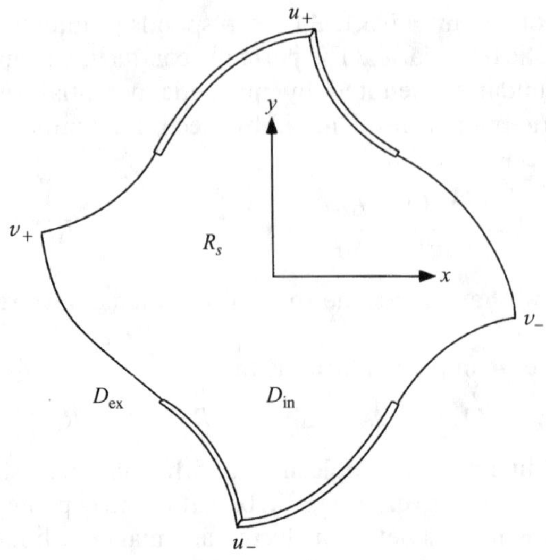

Figure 5.2 C_{2c} impedance sheet.

example of such a self-complementary two-terminal device with a square shape. Figure 5.2 gives a more general example. Other examples are given in [5] and [14]. There are other ways to define complement, such as by interchanging u and v as $\pm v$ and $\pm u$ with various combinations of signs. Which one chooses is but a matter of convenience.

Carrying the development further, the transformation in (5.15) can be carried full circle as

$$\begin{gathered} \zeta^{(n)} \equiv x^{(n)} + jy^{(n)} \equiv e^{j(n\pi/2)}\zeta^{(0)}, \qquad \zeta^{(4)} = \zeta^{(0)} \equiv \zeta, \qquad \zeta^{(1)} \equiv \zeta^{(c)} \\ w^{(n)}(\zeta^{(n)}) = w^{(0)}(\zeta^{(0)}) \equiv w(\zeta), \qquad w^{(c)}(\zeta) = w^{(1)}(\zeta) \end{gathered} \tag{5.16}$$

So C_{2c} symmetry has four group elements as special kinds of successive $\pi/2$ rotations. Note that

$$\begin{gathered} \zeta^{(2)} = -\zeta^{(0)}, \qquad \zeta^{(3)} = -\zeta^{(1)} \\ w^{(2)}(\zeta^{(0)}) = -w^{(0)}(\zeta^{(0)}) = w^{(0)}(-\zeta^{(0)}) \\ w^{(3)}(\zeta^{(0)}) = -w^{(1)}(\zeta^{(0)}) = w^{(1)}(-\zeta^{(0)}) \end{gathered} \tag{5.17}$$

This illustrates that rotation by π merely changes signs, the structure (including respective positions of electric and magnetic boundaries) being invariant to this rotation. This gives C_2 symmetry (which is also two-dimensional inversion) as a subgroup.

A consequence of this symmetry is that $w(\zeta)$ can be taken as an odd function of ζ, i.e.,

$$w(-\zeta) = -w(\zeta) \tag{5.18}$$

which specifies the arbitrary additive constant for potentials with now

$$w(0) = 0 \tag{5.19}$$

suggesting that our potential reference be at $\zeta = 0$. The boundary conditions as in Fig. 5.2 now take the form

$$u_+ = -u_- = v_+ = -v_- \tag{5.20}$$

Referring back to Section 2, this suggests that one express the electric and magnetic potentials in the symmetrical form

$$\Phi_{e+} + \Phi_{e-} = 0, \qquad \Phi_{h+} + \Phi_{h-} = 0 \tag{5.21}$$

so that voltages at and currents into the two terminals (electric boundaries) are equal and opposite for a completely differential system.

Here we have illustrated the case where D_{in} contains the sheet resistance. Moving R_s to D_{ex} also produces a two-terminal device with C_{2c} symmetry and resistance R_s. In this case the reference potential of zero can be taken at $\zeta = \infty$. Note that in the conformal transformations one encounters branch cuts as discontinuities in $v(\zeta)$, but these can be placed for convenience outside of D_{in} or outside of D_{ex}.

5.5 *N*-FOLD ROTATION AXIS: C_N SYMMETRY

As a prelude to C_{Nc} symmetry, consider C_N symmetry (N-fold rotation axis) which is a subgroup. We have the cylic group with N elements

$$\begin{aligned} C_N &= \{(C_N)_\ell \,|\, \ell = 1, 2, \ldots, N\} \\ (C_N)_\ell &= (C_N)_1^\ell \equiv \text{rotation by} \frac{2\pi\ell}{N} \\ (C_N)_1^N &= (C_N)_N = (1) \equiv \text{identity} \end{aligned} \tag{5.22}$$

This has a scalar representation

$$(C_N)_\ell \rightarrow C(\phi_\ell) = e^{j\phi_\ell}, \qquad \phi_\ell \equiv \frac{2\pi\ell}{N}, \qquad \ell = 1, 2, \ldots, N \tag{5.23}$$

Being a scalar representation, the group is clearly commutative. One can also have a commonly used matrix representation [14].

Compared to the previous section, for $N \geq 3$, there is a new complication in that the boundary conditions are not neatly equal and opposite for both electric and magnetic potentials as in (5.20). With three or more terminals there are $N-1 \geq 2$ potential differences to consider. We can generalize (5.20) and (5.21) by constraining

$$\sum_{n=1}^{N} u_n = 0, \qquad \sum_{n=1}^{N} v_n = 0 \tag{5.24}$$

where the u_n and v_n are the boundary conditions assumed on the successive electric and magnetic boundaries. This is illustrated for the case of $N = 5$ in Fig. 5.3 (taken as self-complementary). On a physical basis, the $N-1$ potential differences can be chosen independently, implying that $w(\zeta)$ is not unique. One can similarly interpret this in terms of $N-1$ currents that can be independently chosen, subject to the Kirchhoff condition that the sum of the N currents be zero. Note that the subscript

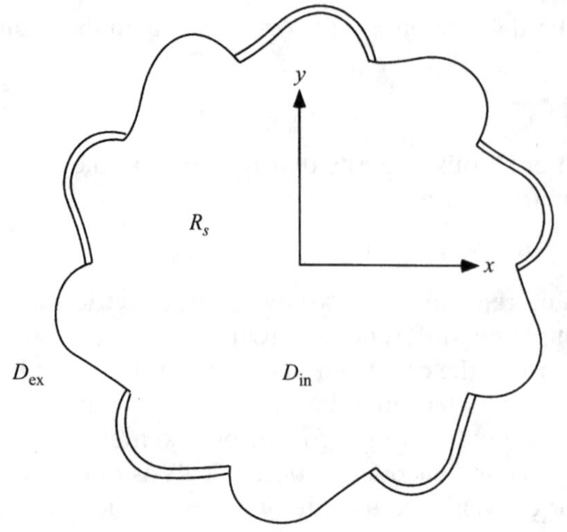

Figure 5.3 C_{5c} impedance sheet.

index n runs from 1 to N since only the alternate boundaries (the electric (N) or the magnetic (N), but not both) are counted.

For rotating the potential functions we can define a coordinate rotation by

$$\zeta^{(2\ell)} \equiv e^{j\phi_\ell}\zeta^{(0)}, \qquad \zeta^{(2N)} \equiv \zeta^{(0)} \equiv \zeta \tag{5.25}$$

and rotated potentials by

$$w^{(2\ell)}(\zeta^{(2\ell)}) \equiv w^{(0)}(\zeta^{(0)}) \equiv w(\zeta) \tag{5.26}$$

The factor of two in the superscripts is to allow for later inclusion of odd indices for self-complementary rotation.

Given a particular $w(\zeta)$ subject to (5.24) then a set of $w^{(2\ell)}(\zeta^{(2\ell)})$ can be generated as per (5.25) and (5.26). Considering first the electric potentials $u^{(2\ell)}(\zeta^{(2\ell)})$, form the sum

$$u'(\zeta) = \sum_{\ell=1}^{N} u^{(2\ell)}(\zeta) \tag{5.27}$$

which is also a solution of the Laplace equation for D_{in}. The potential at the electric boundaries is the sum of the electric boundary potentials for the $u^{(2n)}$. Define

$$\begin{aligned} \zeta_1 &\equiv \zeta_1^{(0)} \equiv \text{point on 1st electric boundary} \\ \zeta_\ell &\equiv \zeta_\ell^{(0)} \equiv e^{j\phi_{\ell-1}}\zeta_1 = e^{j(2\pi/N)(\ell-1)}\zeta_1 = e^{j\phi_\ell}\zeta_0 = e^{j\phi_\ell}\zeta_N \\ &\equiv \text{corresponding point on } \ell\text{th electric boundary} \end{aligned} \tag{5.28}$$

We then have

$$\begin{aligned} u_\ell &\equiv u_\ell^{(0)} = u^{(0)}(\zeta_\ell) \\ u'_\ell &\equiv u'(\zeta_\ell) = \sum_{n=1}^{N} u^{(2n)}(\zeta_\ell) = \sum_{n=1}^{N} u^{(2N)}(\zeta_\ell^{(2(N-n))}) \\ &= \sum_{n=1}^{N} u^{(0)}(e^{j\phi_1(N-h)}\zeta_\ell^{(0)}) = \sum_{n=1}^{N} u(\zeta_{\ell+N-n}^{(0)}) = \sum_{n'=1}^{N} u(\zeta_{n'}) \\ &= \sum_{n=1}^{N} u'_n = 0 \end{aligned} \tag{5.28}$$

Thus we have all the electric boundaries (all ℓ) with zero potential for computing $u'(\zeta)$. The only solution is

$$u'(\zeta) = 0 = \sum_{n=1}^{N} u^{(2n)}(\zeta) \tag{5.29}$$

Similarly, considering the magnetic potential and magnetic boundaries, the same steps as above lead to the same result for $v'(\zeta)$, giving

$$w'(\zeta) = 0 = \sum_{n=1}^{N} w^{(2n)}(\zeta) = \sum_{n=1}^{N} w^{(0)}(\zeta^{(2(N-n))}) = \sum_{n'=1}^{N} w^{(0)}(\zeta^{(2n')}) \tag{5.30}$$

Applying this result to $\zeta = 0$ we have

$$\zeta = 0 \Rightarrow \zeta^{(2m)} = 0 \text{ for all } m, \qquad 0 = \sum_{n=1}^{N} w^{(0)}(0) = Nw(0), \qquad w(0) = 0 \tag{5.31}$$

Thus C_N symmetry with the sum of the terminal potentials equal to zero [as in (5.24)] implies that the complex potential is zero at the coordinate origin. Considering a sheet impedance in D_{ex} instead of D_{in}, the same arguments lead to a zero of the potential at ∞. Note that only C_N (N-fold rotation axis) instead of the more restrictive C_{Nc} has been assumed for obtaining this result.

The N terminals have voltages V_n with

$$V_n = V_o\, u_n \tag{5.32}$$

and N currents with

$$I_n = I_o[v_n - v_{n-1}], \qquad v_0 = v_N \tag{5.33}$$

Since the ratio of voltage drop across an elementary curvilinear square (equal decrements of u and v) to the current through this curvilinear square is just R_s we have the normalizing condition

$$V_0 = I_0 R_s \tag{5.34}$$

Note that the V_n are referenced to $\zeta = 0$ which is not a terminal, and the voltages from an experimental viewpoint are voltage differences between terminals. In some problems (such as multiconductor transmission lines [6], [7]) there is another conductor (such as a cable shield or ground plane) which is taken as zero voltage, giving a total of $N+1$ conductors. If we define an impedance matrix via

$$(V_n) = (Z_{n,m}) \cdot (I_n), \qquad (Z_{n,m})^T = (Z_{n,m}) \tag{5.35}$$

we have the consequence

$$Z_{n,n} = \infty \tag{5.36}$$

i.e., a single current into the nth terminal is forced to zero by the constraint in (4.15). Defining an admittance matrix we have

$$(I_n) = (Y_{n,m}) \cdot (V_n), \quad (Y_{n,m})^T = (Y_{n,m}), \quad \sum_{n=1}^{N} Y_{n,m} = 0 \text{ for } m = 1, 2, \ldots, N \tag{5.37}$$

The zero row- and column-sum constraint comes from setting all voltages to zero except one and requiring the sum of the currents to be zero. Within these limitations, one can regard these two matrices as mutually inverse in a generalized sense.

As discussed in various contexts [6], [7], [10] the C_N symmetry makes $(Y_{n,m})$ *circulant*, i.e.,

$$Y_{n,m} = Y_{m-n+1} \tag{5.38a}$$

so that the admittance between two terminals is only a function of how far apart the two terminals are around a circle. Furthermore, due to reciprocity, the matrix is symmetric and it does not matter which direction one goes around the circle. The matrix is then *symmetric circulant* or *bicirculant* as

$$(Y_{n,m}) = \begin{pmatrix} Y_1 & Y_2 & Y_3 & \cdots & Y_N \\ Y_2 & Y_1 & Y_2 & \cdots & \vdots \\ Y_3 & Y_2 & Y_1 & \cdots & \vdots \\ \vdots & \vdots & \vdots & & \vdots \\ Y_N & \cdots & \cdots & \cdots & Y_1 \end{pmatrix}$$

$$Y_{N+2-\ell} = Y_\ell, \qquad \ell = 1, 2, \ldots, N, \qquad \sum_{\ell=1}^{N} Y_\ell = 0 \tag{5.38b}$$

with a similar form for the impedance matrix. Note that there are not N distinct entries, but only $N/2 + 1$ for N even and $(N + 1)/2$ for N odd. This can be diagonalized by the Fourier matrix (unitary) [6], [7] or other related forms.

The admittance matrix has the dyadic form [7]

$$\begin{aligned}
(Y_{n,m}) &= \sum_{\beta=1}^{N} y_\beta (x_n)_\beta (x_n)_\beta^* \qquad \text{(a purely real sum)} \\
y_\beta &= \sum_{\ell=1}^{N} Y_{\ell+1} e^{j2\pi(\beta/N)} \qquad Y_{N+1} = Y_1 \\
y_\beta &= y_{N-\beta} \qquad \text{(all real and } \geq 0\text{)} \\
(x_n)_\beta &= N^{-1/2}(e^{j2\pi(\beta/N)}, e^{j2\pi(2\beta/N)}, e^{j2\pi(3\beta/N)}, \ldots, e^{j2\pi\beta}) \\
(x_n)_\beta^* &= (x_n)_{N-\beta}
\end{aligned} \tag{5.39}$$

with the special case for $\beta = N$

$$y_N = \sum_{\ell=1}^{N} Y_\ell = y_0 = 0, \qquad (x_n)_N = N^{-1/2}(1, 1, 1, \ldots, 1) + (x_n)_0 \tag{5.40}$$

For more general cases of C_N symmetry, involving a reference conductor $y_N > 0$, the diagonalization still applies. There is an alternate form of the above which can simplify matters. For an even N we have $N/2 + 1$ eigenvalues from (5.39) with

$$\begin{aligned}
y_\beta &= Y_1 + 2\left[\sum_{\ell=1}^{(N/2)-1} Y_{\ell+1} \cos\left(2\pi \frac{\ell\beta}{N}\right)\right] + (-1)^\beta \frac{Y_N}{2} + 1 \\
\beta &= 0, 1, 2, \ldots, N/2 - 1
\end{aligned} \tag{5.41}$$

For an odd N, we have $(N+1)/2$ eigenvalues with

$$y_\beta = Y_1 + 2\left[\sum_{\ell=1}^{(N-1)/2} Y_{\ell+1}\cos\left(2\pi\frac{\ell\beta}{N}\right)\right]$$
$$\beta = 0, 1, 2, \ldots, (N-1)/2 \tag{5.42}$$

The eigenvectors can be put in a purely real form by taking linear combinations of the two in (5.39) associated with equal (non-distinct) eigenvalues to give two real eigenvectors associated with each distinct eigenvalue as

$$(x_n)_{\beta,e} = \left[\frac{2}{N}\right]^{1/2}\left(\cos\left(2\pi\frac{\beta}{N}\right), \cos\left(2\pi\frac{2\beta}{N}\right), \ldots, \cos\left(2\pi\frac{(N-1)\beta}{N}\right), 1\right)$$
$$(x_n)_{\beta,o} = \left[\frac{2}{N}\right]^{1/2}\left(\sin\left(2\pi\frac{\beta}{N}\right), \sin\left(2\pi\frac{2\beta}{N}\right), \ldots, \sin\left(2\pi\frac{(N-1)\beta}{N}\right), 1\right) \tag{5.43}$$
$$\beta = \begin{cases} 1, 2, \ldots, N/2-1 \text{ for } N \text{ even} \\ 1, 2, \ldots, (N-1)/2-1 \text{ for } N \text{ odd} \end{cases}$$

with the special case for $\beta = 0$ in (5.40) (only one eigenvector). The subscripts e (even) and o (odd) distinguish the two eigenvectors associated with a particular β. This is an example of twofold degeneracy of eigenmodes as discussed in [8].

Now any acceptable $w(\zeta)$ can be constructed in terms of the eigenvectors above. For each eigenvector $(x_n)_{\beta,{}^e_o}$, there is a complex (analytic) mode function $w_{\beta,{}^e_o}(\zeta)$ satisfying the electric boundary conditions (at ζ_n) given by the eigenvectors. A linear combination of these will give the desired electric boundary values subject to (5.24). For this purpose we have the identity

$$(1_{n,m}) = \sum_{\beta=1}^{N}(x_n)^*_\beta(x_n)_\beta = \sum_{\beta,{}^e_o}(x_n)_{\beta,{}^e_o}(x_n)_{\beta,{}^e_o} \tag{5.44}$$

for calculating the coefficients of each of the eigenmodes via

$$(u_n) = \sum_{\beta=1}^{N}(x_n)^*_\beta\left[(x_n)_\beta\cdot(u_n)\right] = \sum_{\beta,{}^e_o}(x_n)_{\beta,{}^e_o}\left[(x_n)_{\beta,{}^e_o}\cdot(u_n)\right] \tag{5.45}$$

The coefficient of the N (or zero) mode is zero due to (5.24).

5.6 SELF-COMPLEMENTARY ROTATION GROUP: C_{Nc} SYMMETRY

For constructing the self-complementary rotation group C_{Nc}, begin with the self-complementary angle

$$\phi_c \equiv \frac{\phi_1}{2} = \frac{\pi}{N} \tag{5.46}$$

so that twice application of this rotation gives rotation by ϕ_1, the elementary rotation for C_N. Consistent with this, define rotated coordinates

$$\zeta^{(n)} \equiv e^{jn\phi_c}\zeta^{(0)}, \qquad \zeta^{(2N)} = \zeta^{(0)} \equiv \zeta, \qquad \zeta^{(N)} = -\zeta \tag{5.47}$$

Taking the complement as in (5.15) we have the first application of complement

$$w^{(1)}(\zeta^{(1)}) = w^{(0)}(\zeta^{(0)}) \tag{5.48}$$

which interchanges the electric and magnetic boundaries. The second application of complement

$$w^{(2)}(\zeta^{(2)}) = w^{(1)}(\zeta^{(1)}) = w^{(0)}(\zeta^{(0)}) \tag{5.49}$$

returns the boundaries to their original type, but with the potentials rotated by $2\phi_c = \phi_1$. Generalizing this we have

$$w^{(n)}(\zeta^{(n)}) = w^{(0)}(\zeta^{(0)}), \qquad n = 1, 2, \ldots, 2N, \qquad w^{(2N)}(\zeta) = w^{(0)}(\zeta) \tag{5.50}$$

The group structure of C_{Nc} is the same as S_{2N} rotation reflection symmetry where interchange of electric and magnetic boundaries is replaced by reflection through a transverse plane; this is appropriate to a three-dimensional structure where rotation by ϕ_c is accompanied by reflection ($z \to -z$) through the x, y symmetry plane. These two groups are said to be *isomorphic*, and C_N is a subgroup of both.

Concentrating on the admittance form (to avoid unbounded elements) we have the original problem

$$\begin{aligned} &(I_n) = (Y_{n,m}) \cdot (V_n), \qquad (V_n) = V_0(u_n) \\ &(I_n) = I_0(v_1 - v_N, v_2 - v_1, \ldots, v_N - v_{N-1}), \qquad V_0 = I_0 R_s \end{aligned} \tag{5.51}$$

The complementary problem is found by interpreting the v_ns for voltage and u_ns for current (thus interchanging the roles of electric and magnetic boundaries) as

$$\begin{aligned} &(I_n^{(c)}) = (Y_{n,m}^{(c)}) \cdot (V_n^{(c)}), \qquad (V_n^{(c)}) = V_0(v_n), \qquad V_0 = I_0 R_s \\ &(I_n^{(c)}) = I_0(u_2 - u_1, u_3 - u_2, \ldots, u_N - u_{N-1}, u_1 - u_N) \end{aligned} \tag{5.52}$$

Imposing self-complementarity implies [from (5.34)]

$$(Y_{n,m}^{(c)}) = (Y_{n,m}) = \sum_{\beta=1}^{N} y_\beta (x_n)_\beta (x_n)_\beta^* \tag{5.53}$$

Here we have used the complex form of the eigenvectors in (5.39) for the reason that

$$\begin{aligned} (x_n)_\beta &= N^{-1/2}(e^{j2\pi(\beta/N)}, e^{j2\pi(2\beta/N)}, \ldots, e^{j2\pi(N-1)\beta/N}, e^{j2\pi(N\beta/N)}) \\ &= N^{-1/2} e^{j2\pi(\beta/N)}(e^{j2\pi(N\beta/N)}, e^{j2\pi(\beta/N)}, \ldots, e^{j2\pi(N-2)\beta/N}, e^{j2\pi(N-1)\beta/N}) \\ &= N^{-1/2} e^{-j2\pi(\beta/N)}(e^{j2\pi(2\beta/N)}, e^{j2\pi(3\beta/N)}, \ldots, e^{j2\pi(N\beta/N)}, e^{j2\pi(\beta/N)}) \end{aligned} \tag{5.54}$$

So, for these eigenvectors, shifting an index merely multiplies the eigenvector by a complex constant of magnitude one.

Considering the voltages and currents in (5.51), let us specify the (u_n) as proportional to one of the $(x_n)_\beta$ and obtain

$$(V_n) = V_0\left[(x_n)_\beta^* \cdot (u_n)\right](x_n)_\beta, \qquad (I_n) = V_0\left[(x_n)_\beta^* \cdot (u_n)\right] y_\beta (x_n)_\beta \tag{5.55}$$

so that the current is proportional to this same eigenvector. Then from the shifting property in (5.54) the v_ns in (5.51) can be expressed as

$$\begin{aligned} (v_n) \equiv (v_1, v_2, \ldots, v_N) &= \left[(x_n)_\beta^* \cdot (v_n)\right](x_n)_\beta \\ (v_N, v_1, \ldots, v_{N-1}) &= e^{-j2\pi(\beta/N)}(v_n) = e^{-j2\pi(\beta/N)}\left[(x_n)_\beta^* \cdot (v_n)\right](x_n)_\beta \\ (I_n) &= I_0\left[(x_n)_\beta^* \cdot (v_n)\right][1 - e^{j2\pi(\beta/N)}](x_n)_\beta \end{aligned} \tag{5.56}$$

Equating the two forms for the current vector gives

$$I_0\left[(x_n)_\beta^* \cdot (v_n)\right][1 - e^{-j2\pi(\beta/N)}] = V_0\left[(x_n)_\beta^* \cdot (u_n)\right] y_\beta \tag{5.57}$$

with the eigenvector removed as a common factor. Turning to (5.52) expand the (v_n) as proportional to one of the $(x_n)_\beta$ and obtain

$$(V_n^{(c)}) = V_0\left[(x_n)_\beta^* \cdot (v_n)\right](x_n)_\beta, \qquad (I_n^{(c)}) = V_0\left[(x_n)_\beta^* \cdot (v_n)\right] y_\beta (x_n)_\beta \tag{5.58}$$

Again using the shifting property the u_ns in (5.52) can be expressed as

$$\begin{aligned} (u_n) \equiv (u_1, u_2, \ldots, u_N) &= \left[(x_n)_\beta^* \cdot (u_n)\right](x_n)_\beta \\ (u_2, u_3, \ldots, u_N, u_1) &= e^{j2\pi(\beta/N)}(u_n) = e^{j2\pi(\beta/N)}\left[(x_n)_\beta^* \cdot (u_n)\right](x_n)_\beta \\ (I_n^{(c)}) &= I_0\left[(x_n)_\beta^* \cdot (u_n)\right][e^{j2\pi(\beta/N)} - 1](x_n)_\beta \end{aligned} \tag{5.59}$$

Equating the two forms for the complementary current gives

$$I_0\left[(x_n)_\beta^* \cdot (u_n)\right][e^{j2\pi(\beta/N)} - 1] = V_0\left[(x_n)_\beta^* \cdot (v_n)\right] y_\beta (x_n)_\beta \tag{5.60}$$

Combining this with (5.57) gives

$$\begin{aligned} \frac{V_0^2}{I_0^2} y_\beta^2 &= R_s^2 y_\beta^2 = [e^{j2\pi(\beta/N)} - 1][1 - e^{-j2\pi(\beta/N)}] = 4\sin^2\left(\pi\frac{\beta}{N}\right) \\ R_s y_\beta &= 2\sin\left(\pi\frac{\beta}{N}\right) \end{aligned} \tag{5.61}$$

where the positive square root has been chosen since the eigenvalues (admittances) must be real and non-negative. This is in agreement with [10] noting the replacement of $Z_0/2$ by R_s. Note that

$$y_N = y_0 = 0 \tag{5.62}$$

as required in (5.40). By the symmetry in the sine function about $\pi/2$ the eigenvalues pair as in (5.41) and (5.42) reducing the number of distinct ones to roughly $N/2$ as previously discussed.

Having the eigenvalues of the admittance matrix, as well as the eigenvectors, the elements of this bicirculant matrix may be obtained as

$$
\begin{aligned}
R_s Y_\ell &= R_s Y_{1,\ell} = R_s \sum_{\beta=1}^{N} y_\beta x_{1;\beta} x^*_{\ell;\beta} \\
&= -jN^{-1} \sum_{\beta=1}^{N} [e^{j\pi(\beta/N)} - e^{-j\pi(\beta/N)}] e^{j2\pi(\beta/N)} - e^{-j2\pi\ell(\beta/N)} \\
&= -jN^{-1} \sum_{\beta=1}^{N} [e^{j\pi[-2\ell+3](\beta/N)} - e^{j\pi[-2\ell+1](\beta/N)}] \\
&= -jN^{-1} \left\{ e^{j[-2\ell+3](\pi/N)} \frac{1 - e^{j[-2\ell+3]\pi}}{1 - e^{j[-2\ell+3](\pi/N)}} - e^{j[-2\ell+1](\pi/N)} \frac{1 - e^{j[-2\ell+1]\pi}}{1 - e^{j[-2\ell+1](\pi/N)}} \right\} \\
&= N^{-1} \left\{ -\frac{e^{j[-2\ell+3](\pi/2N)}}{\sin\left([2\ell-3]\dfrac{\pi}{2N}\right)} + \frac{e^{j[-2\ell+1](\pi/2N)}}{\sin\left([2\ell-1]\dfrac{\pi}{2N}\right)} \right\} \\
&= -N^{-1} \frac{\sin\left(\dfrac{\pi}{N}\right)}{\sin\left([2\ell-3]\dfrac{\pi}{2N}\right)\sin\left([2\ell-1]\dfrac{\pi}{2N}\right)} \\
&= \frac{2}{N} \frac{\sin\left(\dfrac{\phi_1}{2}\right)}{\cos\left([\ell-1]\dfrac{2\pi}{N}\right) - \cos\left(\dfrac{\pi}{N}\right)} \\
&= \frac{2}{N} \frac{\sin\left(\dfrac{\phi_1}{2}\right)}{\cos([\ell-1]\,\phi_1) - \cos\left(\dfrac{\phi_1}{2}\right)}
\end{aligned}
\tag{5.63}
$$

From these matrix elements various combinations of connections of the N terminals may be connected to give a single port with various impedances of the form of a constant times R_s. Examples of the same are included in [10] where R_s is replaced by $Z_0/2$.

5.7 RECIPROCATION OF TWO-DIMENSIONAL STRUCTURES

The self-complementary impedance sheet has been considered for the interior domain D_{in}. However, it applies equally well to the exterior domain D_{ex} (e.g., in Figs. 5.2 and 5.3). One way to see this is by the operation of reciprocation [15]. For this it is convenient to introduce complex cylindrical coordinates as

$$\zeta = x + jy = \Psi e^{j\phi} \tag{5.64}$$

Then introduce the analytic transform

$$\zeta_1 = x_1 + jy_1 = \Psi_1 e^{j\phi_1} = b^2/\zeta, \qquad \Psi_1 = b^2/\Psi, \qquad \phi_1 = \phi \tag{5.65}$$

which is a conformal transformation. The operation of reciprocation (non-analytic) is given by

$$\zeta_2 = x_2 + jy_2 = \Psi_2 e^{j\phi_2} = b^2/\zeta^* = \zeta_1^*, \qquad \Psi_2 = b^2/\Psi, \qquad \phi_1 = \phi \tag{5.66}$$

These formulas describe the simple transformation by which the angle ϕ is kept the same, but the radius Ψ is replaced by its reciprocal, appropriately scaled by some radius b.

As illustrated in Fig. 5.4a, consider some impedance sheet (here illustrated as C_{3c} self-complementary), where the sheet occupies D_{in} as previously discussed. Ac-

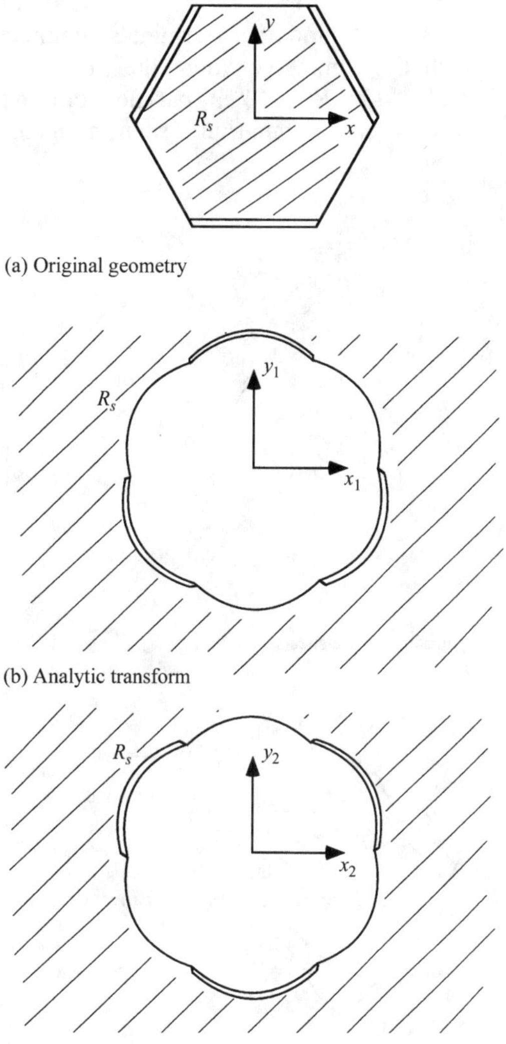

(a) Original geometry

(b) Analytic transform

Figure 5.4 Reciprocation of C_{3c} impedance sheet. (c) Reciprocal geometry complishing the analytic transform as in (5.65) produces the configuration in Fig. 5.4b. Since this is an analytic transformation ($\zeta \to \zeta_1$), then the values of u on the electric boundaries and v on the magnetic boundaries are unchanged and the admittance calculations as before are still applicable. Hence Fig. 5.4 also has C_{3c} symmetry and the previous results apply. Going to Fig. 5.4c the reciprocation transformation ($\zeta \to \zeta_2$) is just a reflection through the x axis. This is equivalent to looking at the figure from the back and the admittance results are the same. So both transformations of a self-complementary structure with C_{Nc} symmetry produce other C_{Nc} self-complementary structures and the same admittance results apply.

Going a step further, as in Fig. 5.5a, combine two self-complementary structures with the same C_{Nc} symmetry. Let one have $R_s^{(\text{in})}$ on the original D_{in}. Let the second have $R_s^{(\text{out})}$ occupying a subset of D_{ex} (by making the dimensions of the interior boundaries sufficiently large so as not to intersect D_{in}). Note that the outer structure need not be related to the inner one (or even have the same value for N). With the same values for N the inner and outer structures can have their corresponding electrical boundaries (terminals) connected together to give a combined structure with C_{Nc} symmetry where the previous admittance formulas can be used with R_s replaced by $R_s^{(\text{in})} // R_s^{(\text{out})}$ (parallel combination), or by $R_s/2$ if $R_s^{(\text{in})} = R_s^{(\text{out})}$. A limiting case of this combination is seen in Fig. 5.5b where the two sheets are made to occupy

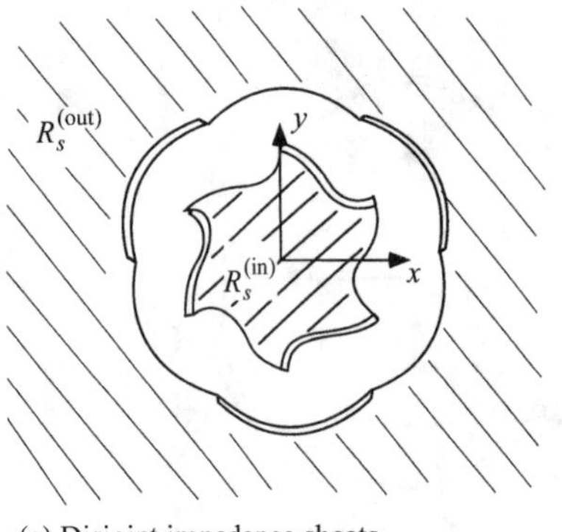

(a) Disjoint impedance sheets

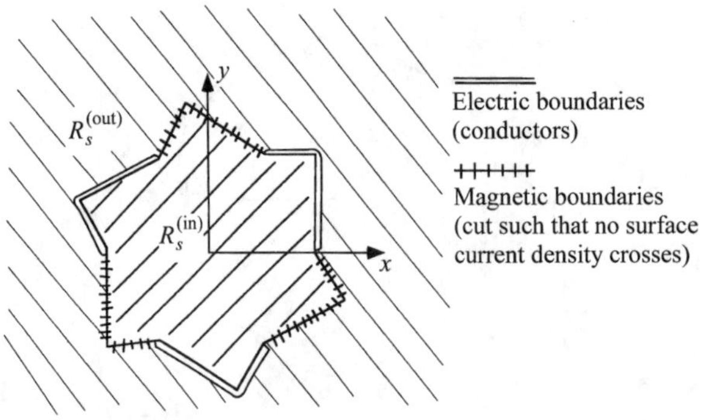

(b) Special case: $D_{\text{in}} \cup D_{\text{ex}}$

Figure 5.5 Self-complementary combination of internal and external impedance sheets with C_{3c} symmetry.

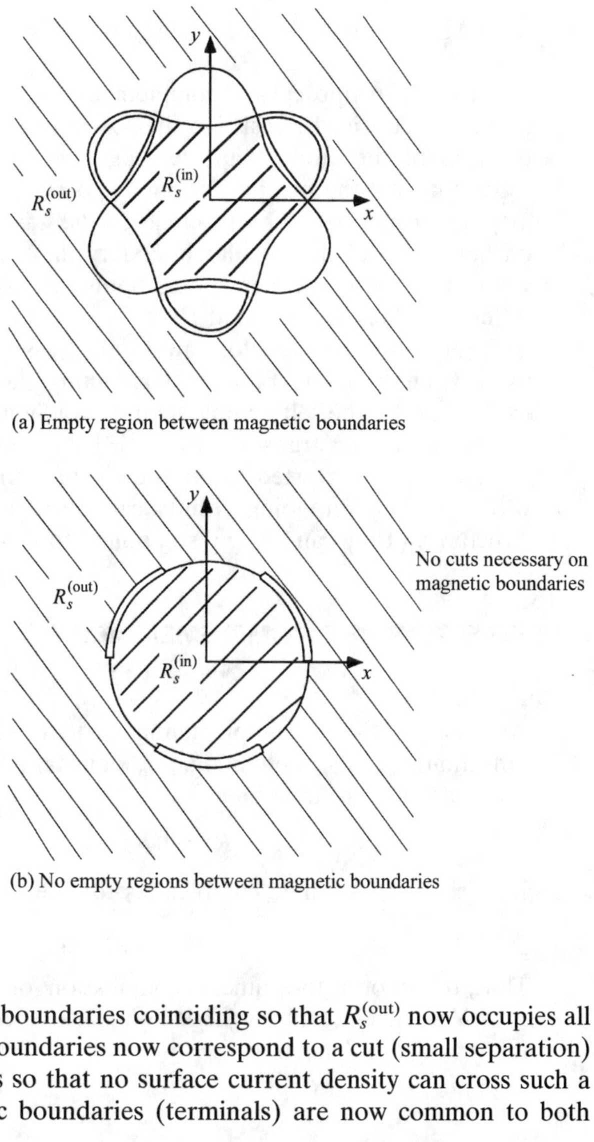

Figure 5.6 Self-reciprocal self-complementary structures.

the entire plane with respective boundaries coinciding so that $R_s^{(\text{out})}$ now occupies all of D_{ex}. Note that the magnetic boundaries now correspond to a cut (small separation) between the two resistive sheets so that no surface current density can cross such a magnetic boundary. The electric boundaries (terminals) are now common to both resistive sheets.

As discussed in [15], it is possible to have a self-reciprocal structure, i.e., one which maps into itself in the transformation (5.65). As shown in Fig. 5.6a, this could involve two resistive sheets with at least partly disjoint boundaries. The portion of the plane between the electric boundaries can be enclosed by a closed conducting boundary as in Fig. 5.6a. In the case of a TEM transmission line such closed electric boundaries correspond to a (perfectly) conducting tube. However, the corresponding portion of the plane between corresponding magnetic boundaries has no such simple analog in two-dimensional TEM transmission lines due to the difficulty in realizing magnetic boundaries in such a case. If we are to have the magnetic boundaries one and the same for the two resistive sheets under reciprocation, then the magnetic boundaries become arcs of a circle of radius b as in Fig. 5.6b. Self-complementarity then requires that the electric boundaries also lie on the same circle. Note that applied to TEM transmission-line configurations this last type of structure is that discussed in [10].

The self-reciprocal self-complementary configuration in Fig. 5.6b is a special case in that no cuts in the resistive sheets are needed to stop the current flow across the portions of the circle that are magnetic boundaries. The reciprocation principle assures us that the mapping from D_{in} to D_{ex} preserves electric potential (voltage) in mapping from ζ to ζ_2. On the circle $\zeta = be^{j\phi} = \zeta_2$ and the electric potential is the same on both sides of the circular boundary implying no current flow through the circle where there is to be a magnetic boundary. Note, however, that one can still have different values for $R_s^{(\text{in})}$ and $R_s^{(\text{out})}$.

The conformal transformation for C_{2a} symmetry (twofold rotation axis with two axial symmetry planes) for two conducting plates on the circular boundary is known explicitly [1]. (The self-complementary configuration C_{2ac} is a special case of this.) Via the stereographic transformation [13] a cylindrical TEM transmission line (plane wave) can be converted to an equivalent conical TEM transmission line (spherical wave), so the foregoing results can also be applied to conical-transmission-line structures. (They can also be so applied to sheets on a spherical surface.)

5.8 REFLECTION SELF-COMPLEMENTARITY

Another type of self-complementary structure is one which is its own complement on reflection. As illustrated in Fig. 5.7, let us take the reflection with respect to the x axis. The reflection group is then

$$R_y = \{(R_y), (1)\}, \qquad (R_y)^2 = (1) \tag{5.67}$$

If ζ_3 is the transformed complex coordinate, we have

$$\zeta_3 = x_3 + jy_3 = \zeta^* = x - jy \tag{5.68}$$

Thus, reflection is the same as conjugation, or reversing the y coordinate.

With electric and magnetic boundaries around D_{in} as in Section 5.2, we have the interchange on reflection

$$u_+^{(c)} = v_+, \qquad u_-^{(c)} = v_-, \qquad v_+^{(c)} = u_+, \qquad v_-^{(c)} = u_- \tag{5.69}$$

So we have

$$w^{(c)}(\zeta) = u^{(c)}(\zeta) + jv^{(c)}(\zeta) = v(\zeta) + ju(\zeta) = jw^*(\zeta) \tag{5.70}$$

This is one of the choices in (5.11), one which is appropriate to reflection (an improper rotation). Except for the coefficient, then conjugation of the complex coordinate results in conjugation of the complex potential. Furthermore from (5.69) we have

$$\Delta u = \Delta v, \qquad f_g = 1 \tag{5.71}$$

as discussed in Section 5.3.

In addition, the reflection symmetry gives

$$w(\zeta^*) = u(\zeta^*) + jv(\zeta^*) = v(\zeta) + ju(\zeta) = jw^*(\zeta), \qquad w^{(c)}(\zeta) = w(\zeta^*) \tag{5.72}$$

As previously discussed, this type of self-complementary structure can also be

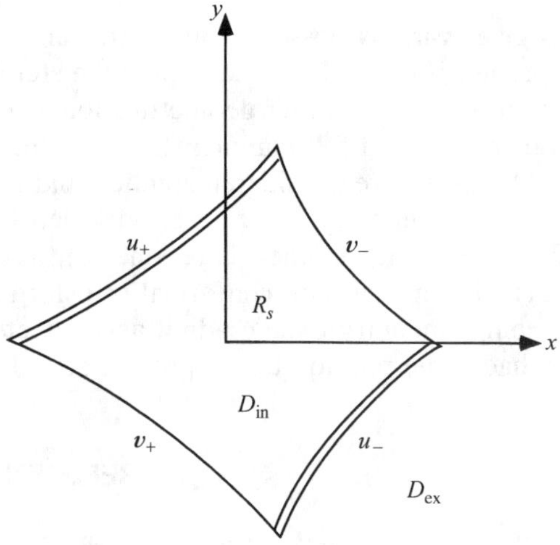

Figure 5.7 Reflection self-complementarity R_c.

transformed by conformal, reciprocal, and/or stereographic transformations to give other structures with the same impedance properties.

5.9 CONCLUSION

Complex potentials and the associated conformal transformations are then another way to define and analyze self-complementary structures. The impedance sheets with uniform R_s (or even with frequency dependence) are appropriate to electrically small structures which can be analyzed via the two-dimensional Laplace equation. However, the results are the same as those derived from the Babinet principle for more general electromagnetic scattering and antennas involving planar structures (not in general electrically small).

This interchange of electric and magnetic potential functions for defining self-complementarity leads to some generalization of the concept of self-complementary structures. Starting from some self-complementary structure involving geometrical symmetry (e.g., rotation, reflection, etc.), the original geometry can be changed to a generally non-symmetrical one via a conformal transformation. The values of the electric and magnetic potentials are unchanged by the transformation, thereby leaving the terminal impedance properties unchanged. So one can think of any single-port device (two terminals) with $f_g = 1$ (or $\Delta u = \Delta v$) as self-complementary, and similarly for N-terminal devices with an admittance matrix described as in Section 5.6.

By including both $R_s^{(in)}$ and $R_s^{(out)}$ to fill the entire plane, and invoking reciprocation symmetry so as to make the electric and magnetic boundaries all lie on a common circle, one has a self-complementary geometry applicable to a special cylindrical TEM transmission line (plane wave). The N terminals occupy equal arcs, equally spaced around the circle. By a stereographic transformation (plane to sphere) a conical TEM transmission line (spherical wave) is defined. However, there are various possible stereographic projections depending on the selected radius of the sphere, and point of tangency of the sphere to the plane (i.e., not necessarily in the center of the circle).

This gives various possible conical transmission lines, all with the same admittance properties. If desired, one can apply the stereographic projection back to a different plane to obtain yet other geometries for a cylindrical transmission line. All of these can be considered self-complementary in the generalized sense discussed above.

Here we have considered rotation and reflection symmetry, but various kinds of translation symmetry can also be included [3], [11], [14]. In terms of the complex-variable approach studied here, one can also approach periodic arrays of resistive sheets with appropriate conformal transformations. In [4], [5], [14] special kinds of self-complementary dyadic admittance sheets are found. It would be interesting if they had some relation to the present complex-variable approach.

REFERENCES

[1] T. K. Liu, Impedances and Field Distributions of Curved Parallel-Plate Transmission-Line Simulators, Sensor and Simulation Note* 170, February 1973.

[2] C. E. Baum, D. V. Giri, and R. D. Gonzales, Electromagnetic Field Distribution of the TEM Mode in a Symmetrical Two-Parallel-Plate Transmission Line, Sensor and Simulation Note 219, April 1976.

[3] C. E. Baum, Self-Complementary Array Antennas, Sensor and Simulation Note 374, October 1994.

[4] C. E. Baum and B. K. Singaraju, Generalization of Babinet's Principle in Terms of the Combined Field to Include Impedance Loaded Aperture Antennas and Scatterers, Interaction Note 217, September 1974.

[5] C. E. Baum, The Self-Complementary Rotation Group, Interaction Note 479, December 1989.

[6] J. Nitsch, C. E. Baum, and R. Sturm, Commutative Tubes in Multiconductor-Transmission-Line Theory, Interaction Note 480, January 1990, and, Analytical Treatment of Uniform Multiconductor Transmission Lines, *IEEE Trans. EMC*, 1993, pp. 285–294.

[7] J. Nitsch, C. E. Baum, and R. Sturm, The Treatment of Commuting Nonuniform Tubes in Multiconductor-Transmission-Line Theory, Interaction Note 481, May 1990, and, Analytical Treatment of Circulant Nonuniform Multiconductor Transmission Lines, *IEEE Trans. EMC*, 1992, pp. 28–38.

[8] C. E. Baum, The Magnetic Polarizability Dyadic and Point Symmetry, Interaction Note 502, May 1994; also Chapter 7, pp. 219–242. in C. E. Baum (ed.), *Detection and Identification of Visually Obscured Targets*, Taylor & Francis, Washington, D.C., 1998.

[9] H. G. Booker, Slot Aerials and Their Relation to Complementary Wire Aerials (Babinet's Principle), *J. IEE*, pt. IIIA, **93**, 1946, pp. 620–626.

[10] G. A. Deschamps, Impedance Properties of Complementary Multiterminal Planar Structures, *IRE Trans. Antennas and Propagation*, 1950, pp. S371–S378.

[11] N. Inagaki, Y. Isogai, and Y. Mushiake, Self-Complementary Antenna with Periodic Feeding Points, *Trans. IECE*, **62-B** (4), 1979, pp. 388–395, in Japanese. Translation NAIC-ID(RS) T-0506-95, December 11, 1995, National Air Intelligence Center.

*Listings of the "Notes" can be found in the *IEEE Antennas and Propagation Magazine*. Copies can be requested from the author, the Defense Documentation Center, Cameron Station, Alexandria, VA 22314, or from the editor C. E. Baum. These are also available at many universities and companies involved in electromagnetics.

[12] Y. Mushiake, Self-Complementary Antennas, *IEEE Antennas and Propagation Mag.*, **34** (6), December 1992, pp. 23–29.

[13] W. R. Smythe, *Static and Dynamic Electricity*, Hemisphere, Taylor & Francis, Washington, D.C., 1989.

[14] C. E. Baum and H. N. Kritikos, Symmetry in Electromagnetics, Chapter 1, pp. 1–90, in C. E. Baum and H. N. Kritikos (eds.), *Electromagnetic Symmetry*, Taylor & Francis, Washington, D.C., 1995.

[15] E. G. Farr and C. E. Baum, Radiation from Self-Reciprocal Apertures, Chapter 6, pp. 281–308, in C. E. Baum and H. N. Kritikos (eds.), *Electromagnetic Symmetry*, Taylor & Francis, Washington, D.C., 1995.

Chapter 6

TOPOLOGY IN ELECTROMAGNETICS

Gerald E. Marsh

ABSTRACT

Topological concepts have increasingly found application in a number of areas of physics including hydrodynamics, electromagnetics, and particularly the magnetic configurations used to model eruptive solar flares. An understanding of energy storage in such fields, which are generally force-free in the sense that the Lorentz force density vanishes, requires an understanding of the concepts of helicity and magnetic energy in multiply connected domains. Because a discussion of the physics of solar prominences can provide such a good introduction to the topological aspects of magnetic energy and helicity, it will be used here to introduce the concepts involved.

Conceptually, helicity can be interpreted as a measure of the linkage of the magnetic field lines. Fields having non-vanishing helicity have the common feature of being twisted or knotted, motivating the introduction of some topological results from knot and braid theory. The concepts of linkage and linking number directly lead to the introduction of the Hopf invariant. For magnetic configurations having field lines that do not close, the formalism associated with dynamical systems may be used to introduce the idea of an asymptotic linking number and subsequently the asymptotic Hopf invariant. Unlike the classical Hopf invariant, which is restricted to integer values, the asymptotic Hopf invariant can have any real value. Magnetic helicity can be identified with the average value of the asymptotic linking number of a pair of field lines and is equivalent to the asymptotic Hopf invariant.

6.1 INTRODUCTION

Topological concepts have increasingly been recognized as playing an important role in a variety of areas of physics and engineering. One of the first, and perhaps the most dramatic example of the effect of a non-trivial topology in physics, was the experiment proposed in 1959 by Aharonov and Bohm [1] who showed that, although a magnetic field may vanish in a given region, measurable effects may nevertheless exist because of the non-vanishing of the vector potential. In their proposed experiment, interference between the components of a split, coherent electron beam, each of which passes on either side of a region confining a magnetic field, is shown to depend on the magnitude of the field. Although the field vanishes in the region through which the electron beams pass, the vector potential cannot vanish if the confined magnetic field is to be non-zero. The phase of the electron wave functions, and hence the interference pattern, is affected by the value of the vector potential. The Aharonov and Bohm effect, while it is quantum mechanical in origin, nonetheless challenges the classical perception that measurable effects in electromagnetism are due to local Maxwell fields and not the potentials.

The experiment was first successfully performed by Chambers [2] and forms the experimental basis for classifying electromagnetism as a gauge field. Note that, as far as the phase of the electron is concerned, the confined magnetic field in this experiment effectively makes the space doubly connected. It is not necessary to physically excise a portion of the space in order to produce a non-trivial topology.

There are magnetic field configurations for which the computation of the field energy explicitly requires that the non-trivial nature of the topology be taken into account. Since such configurations often occur in nature, they should not be viewed as being in any sense exotic. Indeed, one of the most important applications of the concepts discussed in this chapter is to solar prominence models, and for that reason the physics of solar prominences will be used to introduce the topological aspects of magnetic energy and helicity. A more extensive discussion of many of these topics is given in Marsh [3].

Solar prominences are arch-like configurations of plasma that emerge from the solar surface. Viewed on the limb of the sun, they often have the appearance of being twisted; viewed from above, they typically have backward "S" shapes in the north solar hemisphere and forward "S" shapes in the southern hemisphere. Prominences may remain quiescent for many days, but can erupt on a time scale of minutes, ejecting plasma and magnetic field into interplanetary space. Such eruptions often occur when the twisting of the prominence exceeds a threshold value. Magnetic storms are generated when these disturbances intercept the Earth's magnetosphere.

The magnetic field associated with a solar prominence is often modeled as a twisted flux tube, and these are usually approximated as being force-free in the sense that the Lorentz force density vanishes. The first discussion of force-free magnetic fields in an astrophysical context was apparently by Lüst and Schlüter [4]. To see how they arise, consider the condition for pressure equilibrium in a plasma (the origin of which will be discussed shortly),

$$(\nabla \times \boldsymbol{H}) \times \boldsymbol{B} = \nabla p \tag{6.1}$$

where it is assumed that $\boldsymbol{B} = \mu_0 \boldsymbol{H}$, and either $\boldsymbol{B}$ or $\boldsymbol{H}$ will be used for convenience or to maintain consistency with the literature.

When the gradient of the gas pressure, p, is small enough—as it is for many astrophysical applications—the right-hand side of (6.1) may be set equal to zero. This means the field is force-free and consequently satisfies the relation

$$\nabla \times \boldsymbol{H} = \alpha \boldsymbol{H} \tag{6.2}$$

where α may be a function of position. When a plasma has finite conductivity, force-free configurations having α constant in both space and time have the unique [5] property of allowing resistive decay of the field without introducing magnetic stresses that act on the plasma [6].

Given the scale of solar prominences, it is important to realize that the currents are determined by $\nabla \times \boldsymbol{H}$ rather than the conductivity, which merely affects the very small electric field needed to produce the currents. This can be understood by observing that the change in the magnetic field is given by

$$\frac{\partial \boldsymbol{B}}{\partial t} = \nabla \times (\boldsymbol{v} \times \boldsymbol{B}) + \eta \nabla^2 \boldsymbol{B} \tag{6.3}$$

where the magnetic diffusivity is $\eta = (\mu\sigma)^{-1}$. If the material is at rest so that the velocity satisfies $\boldsymbol{v} = 0$, this expression reduces to the diffusion equation, while if the last term vanishes, the field is frozen into the plasma and moves with it. The latter will be the case when the magnetic Reynolds number, $R_m = LV/\eta$, where L and V are the scale length and velocity, is large compared to unity. In the corona of the sun, the relevant $\boldsymbol{v}$ is the Alfvén velocity $[v_a^2 = B^2(\mu\rho)^{-1}]$, and even though η is similar to that for copper, it is nevertheless true that $R_m \gg 1$ for values of L appropriate for a prominence.

The condition for pressure equilibrium in a plasma, (6.1), comes from the equation of motion for a plasma under the influence of a pressure gradient, a magnetic field, and gravity. If the viscous force per unit volume can be neglected, as is the case here, the equation of motion is

$$\rho \frac{d\boldsymbol{v}}{dt} = -\nabla p + \boldsymbol{j} \times \boldsymbol{B} + \rho \boldsymbol{g} \tag{6.4}$$

When the flow speed $\boldsymbol{v}$ is much smaller than both the isothermal velocity of sound and the velocity of Alfvén waves, (6.4) reduces to the equation of magnetohydrostatics

$$\nabla p = \boldsymbol{j} \times \boldsymbol{B} + \rho \boldsymbol{g} \tag{6.5}$$

which in principle can be used to describe the static, preflare configuration of a solar flare [7]. If it is furthermore the case that the ratio of the plasma pressure to magnetic field pressure, known as the plasma β, is negligible compared to unity, and the height of the preflare configuration is much smaller than the scale height so that the gravity term can be neglected, the field is approximately force-free satisfying (6.2). With regard to terminology in the astrophysical literature, it is common to consider potential fields, where the current vanishes, to be force-free with vanishing α; linear force-free fields to be those with α equal to a constant; and non-linear force-free fields to be those having α a function of position.

There are a number of models used to describe the creation of solar prominences. Generally, one begins with a potential field and develops a mechanism for twisting the field. Consider a flux tube emerging from the sun's photosphere into the chromosphere to form an arch. In the photosphere or below, material motions determine the magnetic field structure, and the configuration will be assumed to be unknown; in the chromosphere, and above in the corona, the gas pressure is small, the magnetic pressure is greater than that of the plasma, and the field must consequently be approximately force-free (a potential field, as defined above, is force-free with vanishing α). Nevertheless, one should keep in mind that in order to confine the flux tube the material pressure within the tube must be less than that outside the tube. The force-free nature of the magnetic field configuration will allow the forces needed to balance the outward pressure of the electromagnetic field to be reduced in magnitude, by spreading them out over the bounding surface, but they cannot vanish. In the case of a flux tube, the magnetic energy density at the bounding surface of the tube is

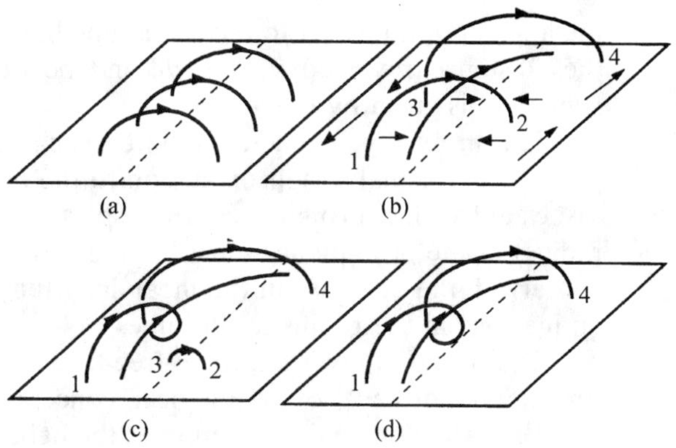

Figure 6.1 Formation of a twisted magnetic field by shear and flux cancellation. (a) shows the original potential field; (b) the sheared field with photospheric shear and convergence flows; (c) the reconnected field; and (d) the resulting twisted field after submergence of the high curvature flux loop. [*Figure taken from* Force-Free Magnetic Fields: Solutions, Topology and Applications, *Gerald E. Marsh*, p. 105 (1996). *© 1996 World Scientific Publishing Co. Pte. Ltd.*]

comparable to the thermal energy of the surrounding chromospheric material. In addition to this pressure differential, there must be a surface current flowing in a boundary sheath, where the field is not force-free, in order to satisfy the magnetic boundary conditions at the surface of the flux tube.

The flux tube can become twisted by either photospheric motions (due to Coriolis forces) at the locations where the base of the arch enters the photosphere [8], or by flux cancellation; i.e., by shear and convergence along the neutral line separating two regions of opposite magnetic polarity, followed by reconnection of the field lines [9]. In either case, currents are introduced that flow along the field lines in the force-free region. While the twisting of a flux tube is easily visualized, the second mechanism is not and is illustrated in Fig. 6.1(a)–(d). Here the original potential field is shown bridging the neutral line in (a). Subsequently, magnetic shear is produced by photospheric flows both along the neutral line as well as toward the neutral line as shown in (b). This model assumes that the small, high curvature loop formed during the reconnection of the magnetic field lines [between 2 and 3 in (c)] submerges into the photosphere leaving the twisted field, (d), which has a larger radius of curvature. The latter is sustained above the photosphere by magnetic buoyancy forces. If relatively cool plasma resides in the magnetic configuration, the structure forms a prominence.

The type of twisted field that ultimately results from either twisting mechanism is relevant to the discussion of both helicity and magnetic energy. Parker [10] has made the point that one cannot expect to find in nature flux ropes having a complexity greater than a more or less uniform twist across the radius of the tube. A good example of such a uniformly twisted flux tube, having the $\hat{z}$ and $\hat{\phi}$ components of the field decreasing monotonically with increasing radius, is given by [11]

$$\boldsymbol{B} = \left\{0, \frac{B_0}{(1 + r^2/a^2)} \frac{r}{a}, \frac{B_0}{(1 + r^2/a^2)}\right\} \tag{6.6}$$

where a is a constant that determines the twist. As can be readily verified, this field satisfies the force-free field equations, (6.2), with α being given by

$$\alpha = \frac{2}{a(1 + r^2/a^2)} \tag{6.7}$$

That the field is uniformly twisted can be seen from the fact that $B_\phi/B_z = r/a$. Defining the angle θ with respect to the z axis, one has that $\tan\theta = B_\phi/B_z = r\,d\phi/dz$. Therefore the z distance traversed is $z = a\int d\phi$, independent of r. Note that the case of a constant axial field is given by $a \to \infty$.

For an axially symmetric solution such as that given by (6.6) and (6.7) to be a physically reasonable approximation to the twisted flux tube used to model the field associated with a prominence, the aspect ratio (the ratio of the principal to minor radius) of the prominence must be large. Since the flux along the axis of the tube is conserved, the decrease in the mean longitudinal field as the tube is twisted leads to an increase in the radius of the tube.

The uniformly twisted, force-free field of (6.6) and (6.7) can serve as a paradigm for the importance of topological concepts in electromagnetism. This becomes apparent when one naively computes the helicity, as will be done in Section 6.3, or the magnetic energy contained in the field, as will be done in Section 6.6.

6.2 MAGNETIC FIELD HELICITY

The concept of the helicity of a magnetic field is due to Woltjer [12]. The term "helicity" was first introduced in a hydrodynamical context by Moffatt [13], who also discussed its topological interpretation. The helicity of a magnetic field can be expressed as

$$\mathcal{H} = \int_V \boldsymbol{A}\cdot\boldsymbol{B}\,dV \tag{6.8}$$

It is a constant of the motion of the equations of non-dissipative magneto-hydrodynamics, and is gauge invariant in simply connected domains. Helicity provides a measure of the linkage of the field lines. This can be demonstrated in a simply connected domain by choosing to work in the Coulomb gauge where the vector potential satisfies $\nabla\cdot\boldsymbol{A} = 0$. Then each rectangular component of the vector potential will satisfy Poisson's equation, and the solution for $\boldsymbol{A}$ can be written in integral form as

$$\boldsymbol{A}(\boldsymbol{r}) = \frac{\mu_0}{4\pi}\int \frac{\boldsymbol{J}(\boldsymbol{r}')}{|\boldsymbol{r}-\boldsymbol{r}'|}\,dV' \tag{6.9}$$

Since $\boldsymbol{J}(\boldsymbol{r}') = \nabla'\times\boldsymbol{H}(\boldsymbol{r}')$, this expression for $\boldsymbol{A}$ may be written as

$$\boldsymbol{A}(\boldsymbol{r}) = \frac{\mu_0}{4\pi}\int \nabla'\times\frac{\boldsymbol{H}(\boldsymbol{r}')}{|\boldsymbol{r}-\boldsymbol{r}'|}\,dV' + \frac{\mu_0}{4\pi}\int \boldsymbol{H}(\boldsymbol{r}')\times\nabla'\left(\frac{1}{|\boldsymbol{r}-\boldsymbol{r}'|}\right)dV' \tag{6.10}$$

where the vector relation

$$\nabla'\times\frac{\boldsymbol{H}(\boldsymbol{r}')}{|\boldsymbol{r}-\boldsymbol{r}'|} = \frac{1}{|\boldsymbol{r}-\boldsymbol{r}'|}\nabla'\times\boldsymbol{H}(\boldsymbol{r}') + \nabla'\left(\frac{1}{|\boldsymbol{r}-\boldsymbol{r}'|}\right)\times\boldsymbol{H}(\boldsymbol{r}') \tag{6.11}$$

has been used. The first integral in (6.10) may be expressed as a surface integral, and for a bounded current distribution, will vanish far outside the distribution. Performing the indicated gradient operation in the second integral, the vector potential can be written as

$$\boldsymbol{A}(\boldsymbol{r}) = \frac{1}{4\pi}\int \boldsymbol{B}(\boldsymbol{r}') \times \frac{\boldsymbol{r}-\boldsymbol{r}'}{|\boldsymbol{r}-\boldsymbol{r}'|^3} dV' \tag{6.12}$$

When this expression for the vector potential is substituted into (6.8), the helicity becomes

$$\begin{aligned} \mathcal{H} &= \frac{1}{4\pi}\iint_{V,V'} \boldsymbol{B}(\boldsymbol{r}) \cdot \frac{\boldsymbol{B}(\boldsymbol{r}') \times (\boldsymbol{r}-\boldsymbol{r}')}{|\boldsymbol{r}-\boldsymbol{r}'|^3} dV'\, dV \\ &= \frac{1}{4\pi}\iint_{V,V'} \frac{[\boldsymbol{B}(\boldsymbol{r}) \times \boldsymbol{B}(\boldsymbol{r}')] \cdot (\boldsymbol{r}-\boldsymbol{r}')}{|\boldsymbol{r}-\boldsymbol{r}'|^3} dV'\, dV \end{aligned} \tag{6.13}$$

Now, by letting the domains of integration be two untwisted flux filaments Γ and Γ', so that for each domain the field and volume element may be combined to give $\boldsymbol{B}(\boldsymbol{r})\, dV = \boldsymbol{B}(\boldsymbol{r}) \cdot d\boldsymbol{S}\, d\boldsymbol{l} = \delta\Phi\, d\boldsymbol{l}$ and $\boldsymbol{B}(\boldsymbol{r}')\, dV' = \boldsymbol{B}(\boldsymbol{r}') \cdot d\boldsymbol{S}'\, d\boldsymbol{l}' = \delta\Phi'\, d\boldsymbol{l}'$, the helicity—as given by (6.13)—may be expressed in terms of the Gauss linking number, n, as

$$\mathcal{H} = \frac{\Phi\Phi'}{2\pi}\iint_{\Gamma,\Gamma'} \frac{[d\boldsymbol{l} \times d\boldsymbol{l}'] \cdot (\boldsymbol{l}-\boldsymbol{l}')}{|\boldsymbol{l}-\boldsymbol{l}'|^3} = 2n\Phi\Phi' \tag{6.14}$$

While the absence of linkage implies that the linking number vanishes, the converse is not true, as illustrated by the Whitehead link [3]. Clerk Maxwell was apparently the first to realize that the vanishing of the linking number did not necessarily imply that two curves were not linked. The topological interpretation of helicity in terms of the Gauss linking number and its limiting form, the Călugăreanu invariant, has been discussed by Moffatt and Ricca [14].

Helicity and magnetic field energy are closely related, as is easily demonstrated for the case of force-free fields. The energy for a general force-free field is

$$E = \frac{1}{2}\int_V \boldsymbol{J} \cdot \boldsymbol{A}\, dV = \frac{1}{2\mu_0}\int_V \boldsymbol{A} \cdot (\nabla \times \boldsymbol{B})\, dV = \frac{1}{2\mu_0}\int_V \alpha \boldsymbol{A} \cdot \boldsymbol{B}\, dV \tag{6.15}$$

When α is a constant, it may be taken outside the integral and the energy can be written as the product of α and the total magnetic helicity.

Helicity can also be identified with the asymptotic Hopf invariant introduced by Arnold [15]. It is defined as the mean value of the asymptotic linking number of a pair of field lines. The asymptotic Hopf invariant is applicable to cases where the magnetic field may have a complicated topology and the field lines may not close. The definition of the asymptotic Hopf invariant requires the introduction of "short curves" that do indeed close the field lines. However, Arnold was able to show that the asymptotic linking number is independent of this family of short curves. The asymptotic Hopf invariant will be discussed in Section 6.5.

6.3 SOLAR PROMINENCE HELICITY

The magnetic field configuration of a solar prominence within and below the photosphere of the sun is unobservable and therefore unknown. Consequently, only relative helicities, in the sense of Berger and Field [16], where the magnetic field below the chromosphere is assumed to be the same for different field configurations in the force-free region (within the chromosphere and above in the corona), can be computed. From a physical point of view, this is not a particularly serious limitation in that the helicity associated with the field below the chromosphere is not readily transferred to the field above because the field below is "frozen" into the plasma.

Consider again the uniformly twisted field given by (6.6) and (6.7). In order to compute the helicity of this field, an expression for the vector potential is needed. By symmetry, $\boldsymbol{A} \neq f(\phi, z)$, and $A_r = 0$. The relation $\boldsymbol{B} = \nabla \times \boldsymbol{A}$ then allows the non-zero components of $\boldsymbol{A}$ to be written in integral form as

$$A_\phi = \frac{B_0}{r} \int \frac{r}{\left(1 + \frac{r^2}{a^2}\right)} dr$$

$$A_z = -\frac{B_0}{a} \int \frac{r}{\left(1 + \frac{r^2}{a^2}\right)} dr \tag{6.16}$$

Thus, $A_\phi/A_z = -a/r$, and since (6.6) implies that $B_\phi/B_z = r/a$, one has that $\boldsymbol{A} \cdot \boldsymbol{B} = \left(0, -A_z(a/r), A_z\right) \cdot \left(0, B_z(r/a), B_z\right) = 0$. This means that the helicity vanishes! Yet, the field lines are clearly linked, and (6.14) implies that the helicity cannot vanish.

The resolution to this apparent paradox is topological in nature: first, the vector potential computed in (6.16) is determined only up to the gradient of a scalar function $\nabla\chi$ which leads to an additional helicity term $\int \nabla\chi \cdot \boldsymbol{B}\, dV$; and second, for force-free fields satisfying $\nabla \times \boldsymbol{H} = \alpha \boldsymbol{H}$, the field lines lie on the surfaces $\alpha =$ constant [provided α is not globally constant, which is the case here since α is given by (6.7)]. That the second condition must hold can be seen by taking the divergence of $\nabla \times \boldsymbol{H} = \alpha \boldsymbol{H}$ to obtain $\nabla\alpha \cdot \boldsymbol{B} = 0$. Cowling's theorem [17] then tells us that the magnetic surfaces corresponding to $\alpha =$ constant cannot be simply connected.

There is also a theorem by Hopf [18] which states that the torus and the Klein bottle are the only smooth, compact, connected surfaces without boundary that can have a nonsingular (nowhere vanishing) vector field. The Klein bottle not being a real possibility, nonsingular magnetic surfaces can be expected to have the topology of nested tori. If α is globally a constant, the field can have far more complicated topologies with some field lines being semi-ergodic so that they fill a region of 3-dimensional space [19].

The solution given by (6.6) to the force-free field equation can be given the topology of a torus by identifying $z = \pm z_0$ which, in the case of a solar prominence, can be thought of as the two regions where the prominence enters the photosphere. The helicity $\Delta\mathcal{H}$ introduced by the additional term $\int \nabla\chi \cdot \boldsymbol{B}\, dV$, when the volume of integration is a torus, is [20]

$$\Delta\mathcal{H} = \int_D \nabla\chi \cdot \boldsymbol{B}\, dV = \int_D \nabla \cdot (\chi\boldsymbol{B})\, dV = \int_{\partial D} (\chi\boldsymbol{B}) \cdot d\boldsymbol{S} + \int_{\Sigma_1} [\chi]\boldsymbol{B} \cdot d\Sigma_1 \tag{6.17}$$

where $[\chi]$ is the jump across Σ_1, the cut introduced to make χ single valued [see Fig. 6.2]. Since the bounding surface of the torus is a magnetic surface, the first term on the far right-hand side vanishes. It is the second term, and the doubly connected

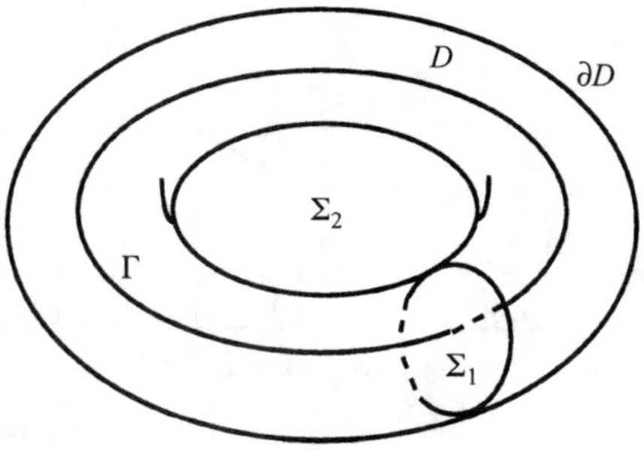

Figure 6.2 Σ_1 is the cut need to make the scalar function χ single valued. The jump $[\chi]$ in the value of χ across the surface Σ_1 is constant on Σ_1, so that Γ may be taken as bounding Σ_2.

nature of the torus, that is responsible for the helicity not vanishing. The jump $[\chi]$ is constant on Σ_1, and may therefore be taken outside the integral [see the discussion following (6.54)]. Thus, Γ may be chosen to bound Σ_2.

Now for any two points p_1 and p_2, the difference in the value of the scalar function χ is $\chi_{p2} - \chi_{p1} = \int_{p_1}^{p_2} \nabla\chi \cdot d\boldsymbol{l}$. If the path of integration is closed, but does not cross the cut needed to make χ single valued, the integral will vanish by Stokes's theorem. When the path of integration Γ does cross the cut Σ_1, the value of the jump $[\chi]$ is $\int_\Gamma \nabla\chi \cdot d\boldsymbol{l} = \int_{\Sigma_2} \boldsymbol{B} \cdot d\Sigma_2 = \Phi_{\Sigma_2}$. Thus,

$$\Delta\mathcal{H} = \int_{\Sigma_1} [\chi]\boldsymbol{B} \cdot d\Sigma_1 = \int_\Gamma \nabla\chi \cdot d\boldsymbol{l} \int_{\Sigma_1} \boldsymbol{B} \cdot d\Sigma_1 = \Phi_{\Sigma_1}\Phi_{\Sigma_2} \tag{6.18}$$

The helicity explicitly depends on the topology. Without the gauge term, and the multiply connected nature of the space, the helicity would vanish. It is also an explicit physical realization of the exchange of twist helicity for link helicity, as discussed in the following section.

6.4 TWIST, KINK, AND LINK HELICITY

Coronal magnetic structure may be topologically far more complex than a simple twisted flux tube. While magnetic flux emerging from the photosphere tends to be localized into discrete flux tubes, in the chromosphere and corona these may be braided due to photospheric motions. Since the field is unknown within and below the photosphere, for the purpose of computing relative helicities one need only consider the braided field above the photosphere. The study of knotted or linked flux tubes thus arises in a natural way since every knot and link is equivalent to the closure of a braid.

A simple example is given in Fig. 6.3 for the closure of the two-stranded braid. $K_{2,q}$ alternates between links and a subset of the torus knots, with knots occurring when 2 and q are relatively prime [21] and links when they are not (the general torus knot, $K_{p,q}$, has p and q relatively prime). Another way of looking at this is to consider the two strands of the braid to be the edges of a band, and determine the twist number of the closure of the braid: links occur when the band is twisted by an integral multiple of 2π, and knots when the number of π twists is odd.

If, instead of two strands, n-strands are twisted by 2π, the closure results in a

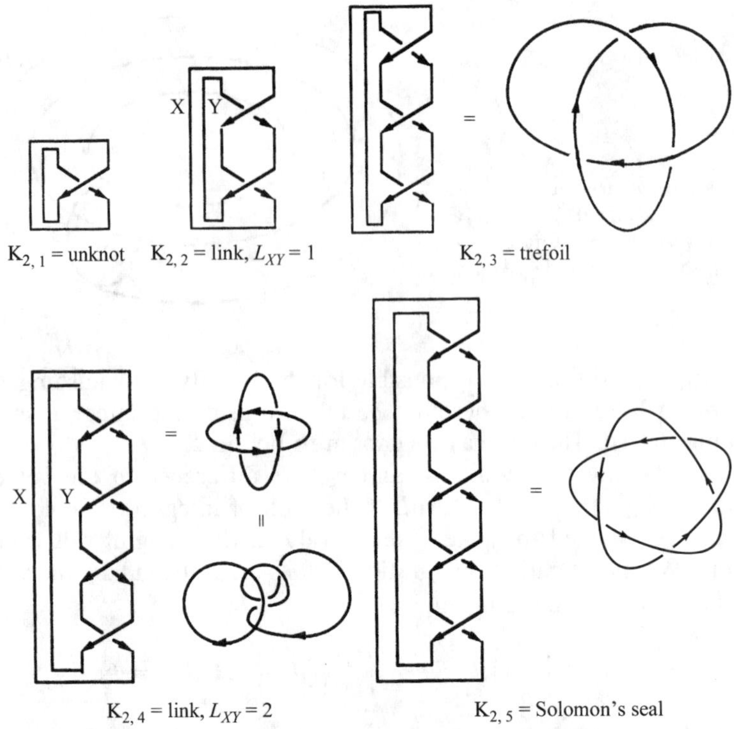

Figure 6.3 The closure of the two-stranded braid gives $K_{2,q}$, a subset of the torus knots, which alternates between links and knots depending upon whether or not 2 and q are relatively prime.

configuration where each pair of strands is linked once; adding an extra twist of π results in a torus knot of type $K_{n,n\pm1}$, depending on whether the π-twist is in the same or opposite direction as the 2π twist.

The closure of a braid may then be thought of as a flux tube which is either knotted or linked with other flux tubes.

In Section 6.2 it was seen that helicity is a measure of the linkage of the field lines, while the example of Section 6.3 showed that helicity could be non-local and gave an explicit physical example of the exchange of twist and link helicity. This type of helicity exchange is related to the work of Berger and Field [22] who used a geometric invariant of a space curve, the writhing number introduced by Călugăreanu [23], to decompose the helicity of a magnetic field into the sum of "twist" and "kink" helicities and, based on the work of Fuller [24], defined the helicity of open field structures. They made use of a theorem [25] from knot theory [26] that states that the linking number of two curves X and Y without common points can be written as the sum of the twist number, T_W, and the writhing number, W_R,

$$L_{XY} = T_W + W_R \tag{6.19}$$

Moffatt and Ricca have called this relation the Călugăreanu theorem.

Consider a flux tube in the form of a torus with circular cross section and a uniform constant flux Φ. If the tube is cut, twisted by $2\pi L_{XY}$ and reattached, each pair of field lines will be linked L_{XY} times. The link helicity generated by such twisting is thus

$$\mathcal{H}_L = 2L_{XY}\Phi^2 \tag{6.20}$$

since two singly linked flux tubes carrying a flux Φ have a helicity [27] of $2\Phi^2$. Now allow the flux tube to be knotted, and define the twist, kink, and link helicity of the tube to be

$$\mathcal{H}_T = T_W\Phi^2, \qquad \mathcal{H}_K = W_R\Phi^2, \qquad \mathcal{H}_L = 2L_{XY}\Phi^2 \tag{6.21}$$

A twist of 2π corresponds to a twist helicity of Φ^2; the writhing number W_R is the sum of the positive and negative $[\sigma(p) = \pm 1]$ crossovers—as determined by the right-hand rule—of the flux tube when projected onto the plane; and

$$L_{XY} = \frac{1}{2}\sum_{p\in X\cap Y} \sigma(p) \tag{6.22}$$

Here, $X \cap Y$ denotes the set of crossings at the points p of the link X with the link Y.

Moffat [28] has pointed out that knotted flux tubes may always be decomposed into two or more linked tubes. Thus a flux tube twisted by 4π is equivalent to two linked but untwisted flux tubes. This can be seen as follows: a twist of 4π means that $T_W = 2$ so that $\mathcal{H}_T = 2\Phi^2$. Let the total conserved helicity be given by the sum of the twist helicity $\mathcal{H}_T$, the kink helicity $\mathcal{H}_K$, and the link helicity $\mathcal{H}_L$. Following Berger and Field, the "twist" helicity can be shared with a "kink" helicity [29] [Fig. 6.4(b)]. In Fig. 6.4(c) the crossing is switched by introducing a link helicity; and in Fig. 6.4(d) the kink is eliminated by straightening out the figure eight, leaving only a link helicity. One can convince oneself of the validity of these exchanges by playing with a piece of ribbon.

Another example is given by setting $T_W = 1/2$ so that a twist of π is introduced into the flux tube. The twist helicity is then $\mathcal{H}_T = \Phi^2/2$ and $\mathcal{H}_K = \mathcal{H}_L = 0$. Such a configuration is essentially a Möbius strip as can be seen by considering any pair of

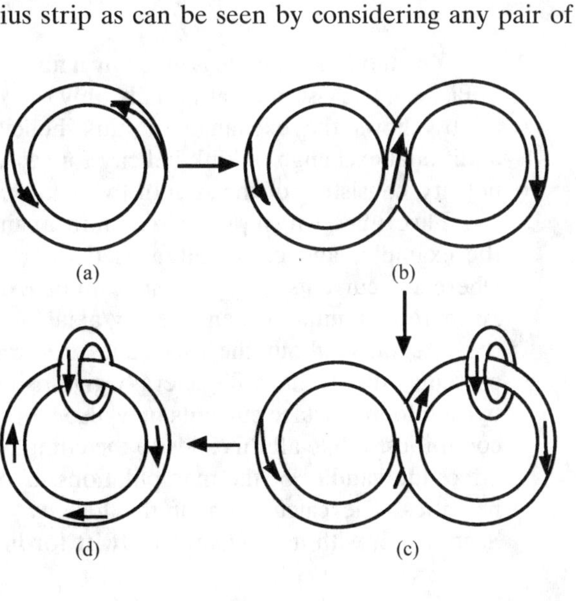

Figure 6.4 The sequence of manipulations that exchange a twist helicity of $2\Phi^2$ for a link helicity. (a) $\mathcal{H}_T = 2\Phi^2$, $\mathcal{H}_K = 0$, $\mathcal{H}_L = 0$; (b) $\mathcal{H}_T = \Phi^2$, $\mathcal{H}_K = \Phi^2$, $\mathcal{H}_L = 0$; (c) $\mathcal{H}_T = \Phi^2$, $\mathcal{H}_K = -\Phi^2$, $\mathcal{H}_L = 2\Phi^2$; (d) $\mathcal{H}_T = 0$, $\mathcal{H}_K = 0$, $\mathcal{H}_L = 2\Phi^2$.

field lines as the edges of a band. If the tube is now cut along the plane defined by a pair of field lines constituting such a band; i.e., along the length of the tube following the twist, the flux is divided in half and one obtains a tube with a twist of 4π ($T_W = 2$) carrying a flux of $\Phi/2$. The twist helicity is still $\mathcal{H}_T = \Phi^2/2$. By defining $\Phi/\sqrt{2}$ as a new flux, this is the same as the starting-point for Fig. 6.4. It is more interesting, however, to again cut the flux tube along a pair of field lines as before. The result is two linked flux tubes each carrying a flux of $\Phi/4$. The complete set of cuts is shown in Fig. 6.5.

Note that the twist helicity and total helicity are conserved in going from (a) to (b), while only the total helicity is conserved from (b) to (c). The total helicity in (c) has the components

$$\mathcal{H} = \mathcal{H}_T^1 + \mathcal{H}_T^2 + \mathcal{H}_K^1 + \mathcal{H}_L = \left(\frac{\Phi}{4}\right)^2 + 2\left(\frac{\Phi}{4}\right)^2 + \left(\frac{\Phi}{4}\right)^2 + 4\left(\frac{\Phi}{4}\right)^2 = \frac{\Phi^2}{2} \tag{6.23}$$

The loops in (c) can be exchanged without affecting the result.

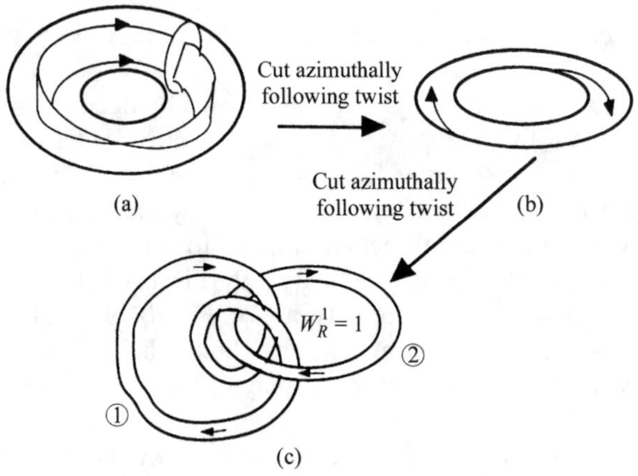

Figure 6.5 A series of azimuthal cuts following the twist of a flux tube: (a) $T_W = 1/2$, $\mathcal{H}_T = \Phi^2/2$, $\mathcal{H}_K = \mathcal{H}_L = 0$; (b) $T_W = 2$, $\mathcal{H}_T = 2(\Phi/2)^2 = \Phi^2/2$, $\mathcal{H}_K = \mathcal{H}_L = 0$; (c) $T_W^1 = 1$, $T_W^2 = 2$, $L_{12} = 2$, and the helicity of the components is given in (6.23).

Yet another example is given by a flux tube in the form of a trefoil knot, as shown in Fig. 6.6(a). Assume that initially only the kink helicity is different from zero. Figure 6.6(b) shows the exchange of kink helicity for link helicity, and Fig. 6.6(c) an additional exchange of kink helicity for twist helicity. Note that at each step the total helicity, consisting of the sum of twist, kink, and link helicities, is again conserved.

Flux tubes, which preserve a uniform and constant flux during deformation, as in the examples above, are often used for heuristic purposes. But twisted flux tubes, where no currents are present, cannot exist in nature. This is a consequence of Ampère's circuital law: an interior, axial current *must* be present since $\oint \boldsymbol{B} \cdot \mathrm{d}\boldsymbol{l}$, where the integral is about the tube, does not vanish. Thus, twisted flux loops will have currents present that will generally give rise to additional magnetic fields that link the twisted loop. Surface currents may also be present so as to match particular boundary conditions. While all this adds to the complexity of the field configuration, it does not affect the validity of the manipulations involved in exchanging twist, kink, and link helicities. The calculation of the helicity of the solar prominence model used in Section 6.3, with its exchange of twist for link helicity, is an example of this.

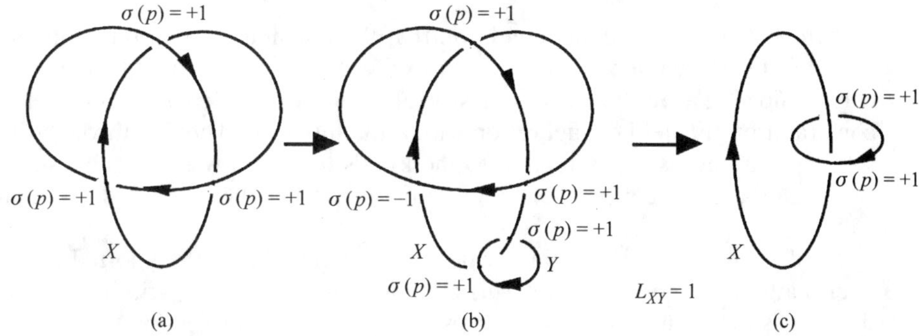

Figure 6.6 A trefoil knot: (a) $\mathcal{H}_T = 0$, $\mathcal{H}_K = 3\Phi^2$, $\mathcal{H}_L = 0$; (b) shows an exchange of kink helicity for link helicity and $\mathcal{H}_T = 0$, $\mathcal{H}_K = \Phi^2$, $\mathcal{H}_L = 2\Phi^2$; (c) shows a further exchange of kink helicity for twist helicity and $\mathcal{H}_T = \Phi^2$, $\mathcal{H}_K = 0$, $\mathcal{H}_L = 2\Phi^2$.

The existence of currents due to twisted flux tubes becomes important when discussing magnetic energy in multiply connected domains and the flux through Seifert surfaces, which have been useful for studying non-dissipative magnetic configurations. Akhmet'ev and Ruzmaikin [30] have used them to describe a fourth-order topological invariant [31], and Chui and Moffatt [32] to investigate minimum energy magnetostatic equilibrium states.

It is a perhaps somewhat counterintuitive fact that every knot or link can be spanned by an orientable surface. Such a surface, when constructed by an algorithm due to Seifert [33], is called a Seifert surface. In 3- dimensional space, a Seifert surface for a knot or link is a connected, orientable, and compact surface whose boundary is the knot or link.

Because the boundary of any compact surface is a disjoint union of circles, by adding disks to the surface, one for each boundary component, an associated *closed* 2-manifold may be constructed. This resulting closed, orientable, and connected 2-manifold is homeomorphic to one of the following: a sphere, a torus T^2, or a connected sum of tori $T^2 \# \ldots \# T^2$, for some n (a sphere with n handles). The integer n is called the genus of the surface. Every Seifert surface has an associated closed manifold homeomorphic to a connected sum of tori. It is an interesting and suggestive fact that any knot can be placed, without any crossings (embedded), on a surface of some genus n; the figure eight knot, for example, can be embedded in $T^2 \# T^2$, a closed surface of genus 2.

Seifert surfaces are often considered in combination with a "framing" of a knotted flux tube, the magnetic axis of which is the boundary of the surface. A "framing" is a frame of reference defined on the flux tube by choosing a longitude and meridian. The longitude may be chosen on the surface of the tube such that it is the intersection of the knot's Seifert surface and the boundary of the tubular neighborhood of the knot corresponding to the flux tube. The linking number of this longitude with the knot itself (the magnetic axis of the flux tube) will vanish since the knot does not intersect the portion of the Seifert surface lying between the longitude and the axis. Such a choice corresponds to a "zero-framing." Since the linking number vanishes, (6.19) implies that the twist of the chosen longitude with the axis will always compensate the writhe of the flux tube.

Imposing the additional condition that the magnetic flux through the portion of the Seifert surface outside the flux tube itself vanishes means that the net longitudinal current flowing around the knot must vanish; this, in turn, imposes a rather stringent constraint on the field configuration within the flux tube. The flux through the Seifert surface of a knot is easily related to the knot's twist and writhe. Doing so, however, requires that the concept of the writhing number be extended to non-integer values.

All of the examples of this section have assumed that the knotted flux tube has been flattened onto the plane. This assumption allowed the writhing number to be defined as the number of positive crossings of the knot projection minus the number of negative crossings (as determined by the right-hand rule). But, consider the simplest case of the un-knot, a circular flux tube lying in the plane. The linking number of the flux within the tube vanishes, and the Seifert surface is the disk bounded by the axis of the flux tube. The intersection of the Seifert surface with the boundary of the flux tube is a circle concentric with the magnetic axis of the tube. Since the linking number vanishes, if the flux tube is smoothly deformed out of the plane, the writhe should continuously vary so as to compensate for the resulting twist.

The general expression for the writhe, $\mathscr{W}$, bears a formal similarity to the Gauss linking number given in (6.14):

$$\mathscr{W}=\frac{1}{4\pi}\iint_{C,C}\frac{[d\boldsymbol{l}\times d\boldsymbol{l}']\cdot(\boldsymbol{l}-\boldsymbol{l}')}{|\boldsymbol{l}-\boldsymbol{l}'|^3} \tag{6.24}$$

The expression gives the writhe of a curve C. In contrast to the linking number, the writhe, like the twist, is not a topological invariant. The writhe can be interpreted as the difference in the positive and negative crossings averaged over all projections of the knot or flux tube onto the plane.

Although the twist helicity is compensated by the writhe helicity as the flux tube of the un-knot is deformed out of a planar configuration, the twist creates, by Ampère's circuital law, a current flowing along the flux tube. This in turn generates a poloidal or meridional flux which links the flux tube. It is the associated link helicity that stores at least part of the energy needed to deform the flux tube.

6.5 HELICITY AND THE ASYMPTOTIC HOPF INVARIANT

In the appendix it is shown that the classical Hopf invariant [34] for the mapping of a 3-sphere on a 2-sphere, $\mathbb{S}^3\rightarrow\mathbb{S}^2$, can be expressed as

$$\gamma=\int_{\mathbb{S}^3}\beta\wedge d\beta \tag{6.25}$$

where β is a 1-form. That the classical Hopf invariant is expressible as a volume integral was first shown to be the case by Whitehead [35].

Rañada [36] uses the Hopf invariant to define the concept of electromagnetic knots, which are vacuum radiation fields where $\boldsymbol{E}\cdot\boldsymbol{B}=0$ and any pair of electric or magnetic field lines are linked. He shows that *locally* any radiation field is equivalent to a sum of such fields, except possibly for a set of measure zero. Rañada demonstrates that $d\beta$ can be expressed in terms of a complex scalar field $\phi(\boldsymbol{r})$ as

$$d\beta = \sum_{i<j} \frac{1}{4\pi i} \frac{\partial_i \phi^* \partial_j \phi - \partial_j \phi^* \partial_i \phi}{(1 + \phi^* \phi)^2} dx^i \wedge dx^j = -\frac{1}{2} \varepsilon_{ijk} B_k \, dx^i \wedge dx^j \tag{6.26}$$

where $\phi(\boldsymbol{r})$ must be single valued at infinity. When β is identified with the vector potential, so that $\beta = -A_i dx^i$, the Hopf invariant is the helicity restricted to integer values which correspond to the linking number of the field lines in $\mathbb{R}^3$.

While the classical Hopf invariant is always an integer, the asymptotic Hopf invariant introduced by Arnold can have any real value and, in three dimensions, can be identified with the helicity. The connection between the classical and asymptotic Hopf invariants can be understood by realizing that the Hopf mapping, which is defined in the appendix, may be considered to be the projection of the Hopf bundle—that is, the projection of the locally trivial fiber space whose total space $\mathbb{S}^3$ has a base space $\mathbb{S}^2$ and fiber $\mathbb{S}^1$. The fibers are the great circles of $\mathbb{S}^3$, so that the 3-sphere $\mathbb{S}^3$ is decomposed into a family of great circles with the 2-sphere $\mathbb{S}^2$ as a quotient space. Now the asymptotic Hopf invariant is defined as the mean value of the asymptotic linking number of a pair of field lines. If the field lines correspond to a unitary vector field that is tangent to the Hopf bundle, the asymptotic Hopf invariant and the classical Hopf invariant coincide.

The asymptotic Hopf invariant is defined for a vector field $\boldsymbol{B}$ that is "homologous to zero." This constraint guarantees the existence of a global vector potential. Specifically, the meaning of homologous to zero can be summarized in a 3-dimensional space $\mathbb{M}$ as follows: if β is the 1-form corresponding [37] to a divergence free vector $\boldsymbol{B}$, $*\beta$, where $*$ is the Hodge star or duality operator [38], is a 2-form; $d(*\beta)$ is a 3-form which vanishes since $\boldsymbol{B}$ is divergence free; and since $d(*\beta) = 0$, a 1-form potential α exists *locally* such that $*\beta = d\alpha$; i.e., $*\beta$ is locally exact. For α to exist *globally*, de Rham's theorems [3] require that the closed surface periods of $*\beta$ vanish over any 2-cycle (here, essentially a closed surface). When $*\beta$ corresponds to a magnetic field, this is guaranteed by the non-existence of magnetic monopoles. Given that $\boldsymbol{B}$ is a divergence-free vector field homologous to zero, the asymptotic Hopf invariant has the same form as (6.25) and is defined in terms of the 1-form α as

$$I(B) = \int_{\mathbb{M}} \alpha \wedge d\alpha \tag{6.27}$$

Because $*\beta = d\alpha$ corresponds to a magnetic field, α represents the vector potential and this integral is the helicity given in (6.8). It is therefore equivalent to (6.13). This equivalency will be important in showing that the asymptotic Hopf invariant is the average value of the asymptotic linking number of a pair of trajectories.

As discussed in Section 6.3, non-singular magnetic fields that satisfy the force-free field equations lie on toroidal magnetic surfaces, provided α is not globally constant. This is also true for the more general case of pressure equilibrium given by (6.1) where the field lines lie on surfaces of constant p. It is possible that the field lines do not close, and therefore intersect the surface across the "hole in the donut" an arbitrary number of times. How does one compute the flux through this surface? Or the helicity of the field? The asymptotic Hopf invariant introduced by Arnold supplies a way of answering such questions.

It will be shown that the asymptotic Hopf invariant [39] is the mean value of the asymptotic linking number. This definition will be seen to be equivalent to giving an ergodic interpretation to the total helicity. To define the asymptotic linking number, however, requires a brief excursion into the theory of dynamical systems.

A dynamical system consists of a space $\mathbb{M}$ and a one-parameter mapping $g^t\boldsymbol{x}$ of $\mathbb{M}$ into itself. Let $g^0\boldsymbol{x}$ be the identity, and impose the requirement that $g^t[g^s\boldsymbol{x}] = g^s[g^t\boldsymbol{x}] = g^{t+s}\boldsymbol{x}$. Then g^t defines a semiflow on $\mathbb{M}$. If the inverse $g^{-t}\boldsymbol{x}$ is uniquely defined for all $\boldsymbol{x} \in \mathbb{M}$, then g^t defines a flow, where both past and future behavior is unambiguously determined by the current state of the system. In what follows, $\mathbb{M}$ will be assumed to be a simply connected 3-dimensional manifold. Later, it will be seen that determining the flux through a closed curve effectively doubly connects the space by excising the curve, which itself may be viewed as the retract of a torus. Thus, determining the flux through a closed curve is equivalent to finding the flux through the "hole in the donut."

The purpose of introducing dynamical systems is to make use of Birkhoff's ergodic theorem from statistical mechanics in the derivation of the asymptotic Hopf invariant. This theorem relates the time average of a function to its phase or state space average. Metrical indecomposability of the phase space is a necessary and sufficient condition for the equality (almost everywhere) of the time and phase averages [40]. The meaning of "metric indecomposability" is as follows: the measure used for integration is defined on subsets of the phase space and must be invariant with respect to g^t. A measure μ is said to be ergodic if either $\mu(X) = 0$ or $\mu(X) = 1$, where X is an invariant subspace. A space having such a measure is said to be indecomposable. What this means is that since an invariant set corresponds to a set of complete trajectories, if an invariant set is separated into two sets, then either one of the component parts has measure zero, or both components are non-measurable [41]. "Almost everywhere" means that exceptions exist, but only on sets of measure zero.

If $\boldsymbol{x} \in X$ and X is metrically indecomposable, then Birkhoff's ergodic theorem can be written as

$$\lim_{T\to\infty} \frac{1}{T} \int_0^T f(g^t\boldsymbol{x})\, dt = \int_X f(\boldsymbol{x})\, d\mu(\boldsymbol{x}) \tag{6.28}$$

Birkhoff's theorem will be used to define the mean linking number with a curve, which will be seen to be equivalent to the flux through the surface bounded by the curve, and to relate the asymptotic linking number of a pair of trajectories to the asymptotic Hopf invariant.

Every vector field on a compact manifold is the phase velocity field of a one-parameter group of diffeomorphisms. Thus, for a magnetic field $\boldsymbol{B}$ on a compact space,

$$\boldsymbol{B}(g^t\boldsymbol{x}) = \frac{d}{dt} g^t\boldsymbol{x} \qquad \text{and} \qquad \boldsymbol{B}(g^\tau\boldsymbol{x}) = \left.\frac{d}{dt}\right|_{t=\tau} g^t\boldsymbol{x} \tag{6.29}$$

There is an equivalence between vector fields, systems of ordinary differential equations, and infinitesimal transformations.

Two equivalent definitions of the asymptotic linking number of the orbit $g^t\boldsymbol{x}, \boldsymbol{x} \in \mathbb{M}$, with a closed curve γ bounding a surface σ will be given, and it will be

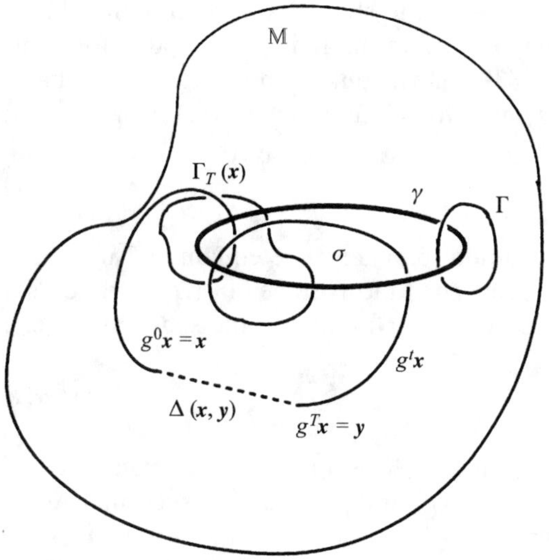

Figure 6.7 To define the asymptotic linking number of a single trajectory with a closed curve γ, short paths $\Delta(\boldsymbol{x}, \boldsymbol{y})$ are introduced to form a closed curve, $\Gamma_T(\boldsymbol{x})$, from the segment of the orbit $g^t\boldsymbol{x}$, $0 \le t \le T$.

shown that they are equal for almost all $\boldsymbol{x}$. To do this, it is necessary to introduce a family of "short paths" $\Delta(\boldsymbol{x}, \boldsymbol{y})$ that can be used to join any pair of points $\boldsymbol{x}, \boldsymbol{y}$ of $\mathbb{M}$. It is assumed that if $\boldsymbol{x}, \boldsymbol{y}$ do not lie on γ, then $\Delta(\boldsymbol{x}, \boldsymbol{y})$ does not intersect γ. As shown in Fig. 6.7, with $g^0\boldsymbol{x} = \boldsymbol{x}$ and $g^T\boldsymbol{x} = \boldsymbol{y}$, a particular $\Delta(\boldsymbol{x}, \boldsymbol{y})$ can be used to form a closed curve $\Gamma_T(\boldsymbol{x})$ from a segment of the orbit $g^t\boldsymbol{x}$ corresponding to $0 \le t \le T$, where T is assumed to be a large number so that the orbit segment is large compared to $\Delta(\boldsymbol{x}, \boldsymbol{y})$.

Now let $N_T(\boldsymbol{x})$ denote the linking number of $\Gamma_T(\boldsymbol{x})$ with γ. Then the asymptotic linking number, $\lambda(\boldsymbol{x})$, with the curve γ is given by

$$\lambda(\boldsymbol{x}) = \lim_{T \to \infty} \frac{1}{T} N_T(\boldsymbol{x}) \tag{6.30}$$

The second definition of the asymptotic linking number is obtained by constructing a closed 1-form α (where $d\alpha = 0$) such that the linking number of every closed curve in $\mathbb{M} - \gamma$ with γ is given by the integral of α over that curve. Thus, $N_T(\boldsymbol{x})$ is replaced by an integral, and for the magnetic field $\boldsymbol{B}$,

$$\hat{\lambda}(\boldsymbol{x}) = \lim_{T \to \infty} \frac{1}{T} \int_0^T \alpha(\boldsymbol{B})\, dt = \lim_{T \to \infty} \frac{1}{T} \int_0^T \alpha\left(\frac{d}{dt} g^t \boldsymbol{x}\right) dt \tag{6.31}$$

where $\alpha(\boldsymbol{B})$ evaluates the 1-form α over the vector field $\boldsymbol{B}$. The limit can be shown to exist, and because $\boldsymbol{B}$ is divergence free, the set of points asymptotic to γ under g^t (those that end on γ as $T \to \infty$) form a set of measure zero. It can also be shown that the limit defining $\hat{\lambda}(\boldsymbol{x})$ is the same for different choices of α, consistent with its definition, except for those points asymptotic to γ. But since this set is of measure zero, $\hat{\lambda}(\boldsymbol{x})$ is independent of the choice of α for almost all $\boldsymbol{x}$. By proving that

$$\lim_{T \to \infty} \frac{1}{T} \int_{\Delta(g^T \boldsymbol{x}, \boldsymbol{x})} \alpha(\boldsymbol{B})\, dt = 0 \tag{6.32}$$

it is possible to show that $\lambda(\boldsymbol{x})$, as defined by (6.30), is independent of the short paths $\Delta(\boldsymbol{x}, \boldsymbol{y})$, and in the limit, $\lambda(\boldsymbol{x}) = \hat{\lambda}(\boldsymbol{x})$ for almost all $\boldsymbol{x}$.

The mean linking number of the phase flow $g^t\boldsymbol{x}$ with γ is the average of the asymptotic linking number $\lambda(\boldsymbol{x})$ over $\mathbb{M}$,

$$\lambda = \int_{\mathbb{M}} \lambda(\boldsymbol{x})\, dV \tag{6.33}$$

Equation (6.31) and the equality of $\lambda(\boldsymbol{x})$ and $\hat{\lambda}(\boldsymbol{x})$ imply that $\lambda(\boldsymbol{x})$ in the integrand of (6.33) is the time average of $\alpha(\boldsymbol{B})$. Birkhoff's theorem can then be used to set the time average of $\alpha(\boldsymbol{B})$ equal to the space average. Thus,

$$\lambda = \int_{\mathbb{M}} \alpha(\boldsymbol{B})\, dV \tag{6.34}$$

Since $\alpha(\boldsymbol{B})$ is a real valued function evaluating the 1-form α over the vector field $\boldsymbol{B}$, there is a unique vector $\boldsymbol{a}$ such that $\alpha(\boldsymbol{B}) = \boldsymbol{a} \cdot \boldsymbol{B}$. The inner product can also be written in terms of differential forms: If β is the 1-form that is the metric dual of the vector $\boldsymbol{B}$, the Hodge star or duality operator allows the inner product between forms [42] to be defined by the relation $(\alpha \cdot \beta)dV = \alpha \wedge *\beta$, where $*\beta$ is a 2-form corresponding to the magnetic field $\boldsymbol{B}$. Thus, in terms of forms, γ may be written as

$$\lambda = \int_{\mathbb{M}} \alpha \wedge *\beta \tag{6.35}$$

$\mathbb{M}$ has been assumed to be simply connected, but the integral above is actually not over $\mathbb{M}$ but over $\mathbb{M} - \gamma$, which means the space is no longer simply connected. The doubly connected nature of the space is critical for obtaining the interesting result that the asymptotic linking number λ is equal to the flux of $\boldsymbol{B}$ through the surface σ bounded by γ. To see this, recall that α is a closed form so that $d\alpha = 0$, and that this guarantees the *local* existence of a function χ such that $\alpha = d\chi$. χ can be made single valued by taking the surface σ to be a cut. Since $d(*\beta)$ is a 3-form, it can be written as the product of the divergence of $\boldsymbol{B}$ and the volume element. But since the divergence of $\boldsymbol{B}$ vanishes,

$$\lambda = \int_{\mathbb{M}-\gamma} d\chi \wedge *\beta = \int_{\mathbb{M}-\gamma} d(\chi * \beta) = \int_{\partial(\mathbb{M}-\gamma)} \chi * \beta + \int_{\sigma} [\chi] * \beta \tag{6.36}$$

where $[\chi]$ is the jump in the function χ in crossing σ. Because $\boldsymbol{B}$ is divergence free, the integral of $*\beta$ over $\partial(\mathbb{M} - \gamma)$, which can be thought of as a tubular neighborhood of γ, vanishes. Since $[\chi]$ is constant on σ, and may be removed from under the integral, the remaining term may be written out explicitly as

$$\lambda = \int_{\Gamma} d\chi \int_{\sigma} *\beta \tag{6.37}$$

The second integral is the flux of $\boldsymbol{B}$ through σ, and the first corresponds to the number of times the path Γ in Fig. 6.7 is traversed, so that the asymptotic linking number, λ, is a multiple of the flux of $\boldsymbol{B}$ through σ.

The discussion thus far has been concerned with the linking of the phase flow $g^t\boldsymbol{x}$ with a closed curve γ. Consider now a pair of points $\boldsymbol{x}_1, \boldsymbol{x}_2 \in \mathbb{M}$ and the phase flows $g^{t_i}\boldsymbol{x}_i$, $i = 1, 2$. The approach taken above of defining the asymptotic linking number in

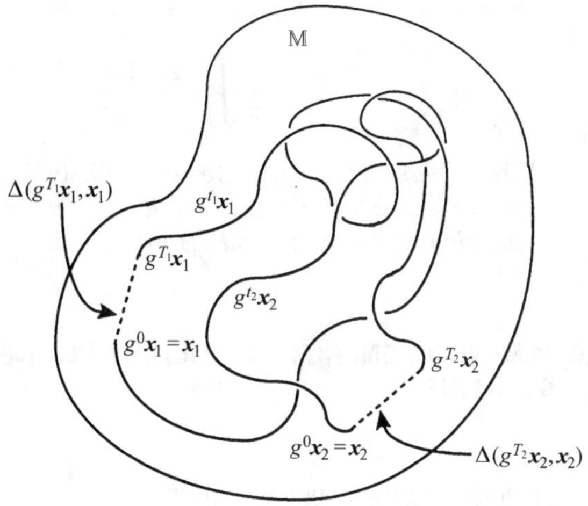

Figure 6.8 To define the asymptotic linking number of a pair of trajectories, closed curves, designated $\Gamma_{T_i}(\boldsymbol{x}_i)$ in the text, are formed by introducing a system of short paths $\Delta(g^{T_i}\boldsymbol{x}_i, \boldsymbol{x}_i)$.

two different ways can be generalized to address the asymptotic linking of the trajectories $g^{t_i}\boldsymbol{x}_i$. Short paths $\Delta(g^{T_i}\boldsymbol{x}_i, \boldsymbol{x}_i)$, are again introduced to close the orbit segments $g^{t_i}\boldsymbol{x}_i$ for $0 \le t_i \le T_i$, where the T_i are chosen to be large numbers. The resulting closed curves $\Gamma_{T_i}(\boldsymbol{x}_i)$ are shown in Fig. 6.8. If $N_{T_1,T_2}(\boldsymbol{x}_1, \boldsymbol{x}_2)$ is the linking number of the closed curves $\Gamma_{T_i}(\boldsymbol{x}_i)$, the asymptotic linking number of the pair of closed trajectories is defined as the limit

$$\lambda(\boldsymbol{x}_1, \boldsymbol{x}_2) = \lim_{T_1, T_2 \to \infty} \frac{N_{T_1,T_2}(\boldsymbol{x}_1, \boldsymbol{x}_2)}{T_1 T_2} \tag{6.38}$$

where it is assumed that the parameters T_i vary so that the closed trajectories $\Gamma_{T_i}(\boldsymbol{x}_i)$ do not intersect.

The form of the Gauss linking number identified in (6.14) suggests that a second form of the asymptotic linking number of the pair of closed trajectories be defined as

$$\hat{\lambda}(\boldsymbol{x}_1, \boldsymbol{x}_2) = \lim_{T_1, T_2 \to \infty} \frac{1}{T_1 T_2} \frac{1}{4\pi} \iint_0^{T_2, T_1} \frac{\boldsymbol{B}_1 \cdot (\boldsymbol{B}_2 \times \{\boldsymbol{x}_1(t_1) - \boldsymbol{x}_2(t_2)\})}{|\boldsymbol{x}_1(t_1) - \boldsymbol{x}_2(t_2)|^3} dt_1 \, dt_2 \tag{6.39}$$

where

$$\boldsymbol{B}_i = \left. \frac{d}{dt_i} \right|_{t_i = 0} g^{t_i}\boldsymbol{x}_i, \qquad i = 1, 2 \tag{6.40}$$

It can again be shown that, for almost all $\boldsymbol{x}_i$, the difference $\lambda(\boldsymbol{x}_1, \boldsymbol{x}_2) - \hat{\lambda}(\boldsymbol{x}_1, \boldsymbol{x}_2)$ vanishes in the limit that T_1, $T_2 \to \infty$, and that the limit is independent of the system of short paths.

The average value of the asymptotic linking number of a pair of trajectories is then defined as

$$\mathbb{\Lambda} = \iint_{\mathbb{M} \times \mathbb{M}} \lambda(\boldsymbol{x}_1, \boldsymbol{x}_2) \, dV_1 \, dV_2 \tag{6.41}$$

Using Birkhoff's ergodic theorem, the time average in (6.39) can be replaced by a space average [43]. Given the equality of $\lambda(\boldsymbol{x}_1,\boldsymbol{x}_2)$ and $\hat{\lambda}(\boldsymbol{x}_1,\boldsymbol{x}_2)$, $\mathbb{\Lambda}$ can be written as

$$\mathbb{\Lambda} = \frac{1}{4\pi}\iint_{\mathbb{M}} \frac{\boldsymbol{B}_1\cdot(\boldsymbol{B}_2\times\{\boldsymbol{x}_1(t_1)-\boldsymbol{x}_2(t_2)\})}{|\boldsymbol{x}_1(t_1)-\boldsymbol{x}_2(t_2)|^3}\,dV_1\,dV_2 \tag{6.42}$$

This is the same as (6.13) for the helicity. *Thus, the helicity can be identified with the average value of the asymptotic linking number of a pair of trajectories and is equivalent to the asymptotic Hopf invariant.*

6.6 MAGNETIC ENERGY IN MULTIPLY CONNECTED DOMAINS

Computing the magnetic energy contained in the uniformly twisted, force-free field of (6.6) and (6.7) in a naive way can further illustrate the importance of topological concepts in electromagnetism. Since the field satisfies the force-free relation, (6.2), as well as $\boldsymbol{A}\cdot\boldsymbol{B}=0$, the energy can be written as

$$E = \frac{1}{2}\int_V \boldsymbol{J}\cdot\boldsymbol{A}\,dV = \frac{1}{2\mu_0}\int_V \boldsymbol{A}\cdot(\nabla\times\boldsymbol{B})\,dV = \frac{1}{2\mu_0}\int_V \alpha\boldsymbol{A}\cdot\boldsymbol{B}\,dV = 0 \tag{6.43}$$

But surely the energy cannot vanish! And indeed it does not, for reasons similar to why the helicity does not vanish, as discussed in Section 6.3.

The second integral on the right-hand side of (6.43) can be expanded as

$$\int_V \boldsymbol{A}\cdot(\nabla\times\boldsymbol{B})\,dV = \int_V \nabla\cdot(\boldsymbol{B}\times\boldsymbol{A})\,dV + \int_V \boldsymbol{B}\cdot(\nabla\times\boldsymbol{A})\,dV = 0 \tag{6.44}$$

And, as a result,

$$\int_V B^2\,dV = \int_S (\boldsymbol{A}\times\boldsymbol{B})\cdot\hat{\boldsymbol{n}}\,dS \tag{6.45}$$

What this equation says is that the energy contained in the bounded force-free field is equal to the surface term which cannot be neglected, as is usually done, by evaluating it at infinity where it vanishes. Since, as was shown in Section 6.3, the surface S cannot be simply connected, one way of computing the value of this term is to again give the solution of (6.6) and (6.7) the topology of a torus by identifying $z = \pm z_0$. As will be shown below, the surface term then has the value $\mu_0 I_\phi \Phi_\Sigma$, where I_ϕ is the net current in the bounded force-free region interior to the torus, and Φ_Σ is the flux through the "hole in the donut."

Returning again to the example given by the physics of a solar prominence, the source of energy for a solar flare is believed to be magnetic in origin, and derived from the prominence's non-potential twisted fields. A knowledge of the field configuration in the force-free region of the chromosphere and above is, however, insufficient to determine the field energy since $\boldsymbol{H}$ in the force-free relation is indeterminate up to an arbitrary multiplicative constant. Nonetheless, as was the case with helicity, it is possible to at least theoretically compute relative energies between different magnetic configurations. To do this one must address the issue of the post-flare magnetic configuration.

If it is assumed that total helicity is conserved, it may well be that a linear force-free field, rather than a potential field, represents the post-flare configuration of the flux tube. The argument that the post-flare configuration should be a linear force-free field is based on the work of Chandrasekhar and Woltjer [44] and has as its basis the Taylor [45] conjecture, which assumes that global helicity is conserved while finite diffusivity effects are invoked to allow the field to relax to a linear one. Unfortunately, Woltjer's result, that a linear force-free field is the lowest energy state that a magnetic field can reach when helicity is conserved, used the approximation of ideal magnetohydrodynamics. But this means that the constant α, or linear, force-free state is topologically inaccessible from most initial configurations. While this situation may not be entirely satisfactory from a theoretical point of view, the Taylor conjecture has been verified by experiment. A recent discussion of the experiments has been given by Taylor [46] in addressing the role of topology in the behavior of turbulent plasmas.

Estimates of the energy available in terms of the topological complexity of the field have also been made by Berger [47]. During the relaxation process, the discussion given in Section 6.4 on the exchange of link, kink, and twist helicity is relevant to the magnetic reconnection problem.

Attempting to compute the energy difference between a twisted arch and a similar arch described by a potential field is, regardless of the true post-flare configuration, another good example of how topological considerations cannot be avoided. Assume again, for the sake of argument, that the magnetic field of a pre-flare solar prominence can be modeled as a flux tube that is uniformly twisted and force-free. Such a field presumably resulted from photospheric motions twisting an initially untwisted uniform field B_0 contained in a cylindrical flux tube of radius R_0. As the tube is twisted, the radius increases to R so as to conserve axial flux. Ignoring for the moment contributions from any surface terms, (6.6) gives the magnetic energy in the tube as

$$\begin{aligned} E &= \frac{1}{\mu_0}\int_V (B_z^2 + B_\phi^2)\,dV = \frac{B_0^2 \pi z}{\mu_0}\int_0^R \frac{2r\,dr}{(1 + r^2/a^2)} \\ &= \frac{B_0^2}{\mu_0}\pi z a^2 \ln(1 + R^2/a^2) \end{aligned} \tag{6.46}$$

On the other hand, the axial flux in the tube is given by

$$\Phi = 2\pi B_0 \int_0^R \frac{r\,dr}{(1 + r^2/a^2)} = B_0 \pi a^2 \ln(1 + R^2/a^2) \tag{6.47}$$

The energy per unit length of the tube, by (6.46), is then $(1/\mu_0)B_0\Phi$. As the twist — as determined by a—varies, the radius R will also vary so as to conserve the axial flux which is equal to the original flux $B_0\pi R_0^2$. *The energy contained in the volume of the flux tube is therefore independent of the twist.* Given the results of Section 6.3, this should not be a surprise. Nor does it mean that energy cannot be stored in uniformly twisted magnetic fields. Twisting the field introduces an axial current density given by $J_z = \alpha H_z$ since the field is force-free. Uniformly twisting a flux tube then results in a current along the tube which creates a magnetic field that links the tube.

The expression for magnetic energy, including the surface term which was ignored in writing (6.46), is

$$E = \frac{1}{2}\int_V \boldsymbol{H}\cdot(\nabla\times\boldsymbol{A})\,dV + \frac{1}{2}\int_{\partial V}(\boldsymbol{H}\times\boldsymbol{A})\cdot d\boldsymbol{S} \tag{6.48}$$

As has been shown above, for a uniformly twisted force-free field, the energy contained in the volume of the flux tube is independent of twist. The first term on the right-hand side of this equation will therefore not contribute when computing the difference in energy between a flux tube containing a potential field and one having a uniform twist. As (6.45) has shown, the surface term can contribute to the energy, and it is the interpretation of this term in finite, multiply connected domains that motivates the following discussion.

The treatment of magnetic field energy in multiply connected domains is again facilitated by using the language of differential forms in 3-dimensional Euclidean space. To that end, the vector potential $\boldsymbol{A}$ is replaced with a 1-form A. Then the magnetic field $\boldsymbol{B}$ is given by the 2-form $B = dA$ (which is equivalent to the local statement $\boldsymbol{B} = \nabla\times\boldsymbol{A}$). Taking the exterior derivative of B shows that B is a closed 2-form; i.e., $dB = 0$ (corresponding to $\nabla\cdot\boldsymbol{B} = 0$). B is also *locally* an exact form since $B = dA$. The question of the existence of A *globally* is the subject of de Rham's theorems. In particular, it is de Rham's first theorem that is most relevant to the topic of magnetic energy in multiply connected domains. The substance of this theorem is as follows: since the magnetic field B is solenoidal ($dB = 0$), a vector potential A exists locally such that $B = dA$. For A to exist globally, the closed surface integrals $\int_z B$, known as periods, must vanish (z being a 2-cycle, essentially a closed surface). The non-existence of magnetic monopoles guarantees, as it did in the discussion preceding (6.27), that this will always be the case. Even though the closed surface periods vanish, it is useful to supplement the solenoidal condition for $\boldsymbol{B}$ with the boundary condition that the normal component of $\boldsymbol{B}$ vanish [48]. While this condition alone would imply that the closed surface periods vanish (even if magnetic monopoles existed), it also allows one to use the concept of relative homology to define *open* surface periods, in terms of the homology modulo the boundary [49]. These open surface periods may then be interpreted as fluxes.

It is also possible to define the homology of open arcs modulo the boundary: Two points are said to be homologous if they bound an arc. If these points are on the boundary of a domain D, then two arcs, C and C' contained in D, are said to be homologous modulo the boundary, $C \sim C'$ [mod ∂D], if their 0-dimensional boundaries are homologous in $\bar{D}$. Thus, when $C \sim C'$ [mod ∂D] it is possible to complete $C - C'$ to a closed, bounding curve in D by adding arcs on the boundary [50].

These concepts are important in what follows because various integrals involving the scalar or vector potential will be evaluated by introducing cuts on open surfaces Σ or along open arcs so as to make either potential single valued.

If the current $\boldsymbol{J}$ is defined as the 1-form J, the 2-form $*J$ can be used to write the differential form of Ampère's circuital law, $\mu_0 {*J} = d(*B)$. The magnetic energy can then be written as

$$\mu_0 E = \frac{1}{2}\int_V A\wedge d(*B) \tag{6.49}$$

Using $d(A \wedge *B) = B \wedge *B - A \wedge d(*B)$ and Stokes's theorem, (6.49) becomes

$$\mu_0 E = \frac{1}{2}\int_V B \wedge *B - \frac{1}{2}\int_{\partial V} A \wedge *B \tag{6.50}$$

Notice that the inner product on p-forms may be defined as

$$\langle \alpha, \beta \rangle = \int_V \alpha \wedge *\beta \tag{6.51}$$

If one then introduces the co-differential operator given in 3-dimensional space by $\delta\omega = (-1)^p * d(*\omega)$, ω being a p-form, and uses the fact that here $** = +1$ applied to 1-forms, the boundary term in (6.50), which will be discussed shortly, can be written as [51]

$$\begin{aligned} \int_{\partial V} A \wedge *dA &= \int_V d(A \wedge *dA) = \int_V B \wedge *B - \int_V A \wedge **d*dA \\ &= \langle B, B \rangle - \langle A, \delta dA \rangle \end{aligned} \tag{6.52}$$

This shows that unless the boundary term vanishes, d and δ will not be adjoint operators as is usually the case for closed (compact) manifolds without boundary. There are some subtleties here in that the inner product defined by (6.51) may not exist if the space V is not compact. Note also that the co-differential operator can be used to write the differential form of Ampère's circuital law as $\mu_0 J = \delta B$.

An example that will bring out the important aspects of computing the magnetic energy in multiply connected domains, and one that has important practical applications, is again given by the torus. The surface or boundary term in the expression for magnetic energy given by (6.50) will not vanish as it does in unbounded 3-dimensional space. Usually, in the case of the latter, this integral is evaluated at infinity, and since $*B$ falls off at least as fast as r^{-2} and A as r^{-1}, the integral does not contribute to the total energy. This will not be the case when the boundary is the surface of a finite torus. What then is the interpretation of the boundary term evaluated on a surface not taken at infinity?

To answer this question, let the domain D be the *exterior* of a toroidal surface S_1, bounded by S_0 and S_1, as shown in Fig. 6.9. $\bar{D}$ then has two components: the doubly connected toroidal space interior to S_1 and the singly connected space exterior to S_0. Assume that the current vanishes in D so that the 1-form $*B$ satisfies $d(*B) = 0$ and the 2-form B satisfies $dB = 0$. Then there exist locally a 0-form (function) ϕ such that $*B = d\phi$ and a 1-form A such that $B = dA$. ϕ can be made single-valued by taking the open surface Σ as a cut.

Consider the integral $\int_D *B \wedge B$. Using the relationship

$$d(\phi B) = d\phi \wedge B + \phi dB \tag{6.53}$$

and the fact that $*B = d\phi$ and $dB = 0$, this integral can be written as

$$\int_D *B \wedge B = \int_D d(\phi B) = \int_{\partial D} \phi B + \int_\Sigma [\phi] B \tag{6.54}$$

where $[\phi] = \int_\Gamma d\phi = \int_\Gamma *B$ is the jump across Σ. This jump is independent of the choice of path Γ provided Γ circles the torus. This can be seen by observing that the two paths

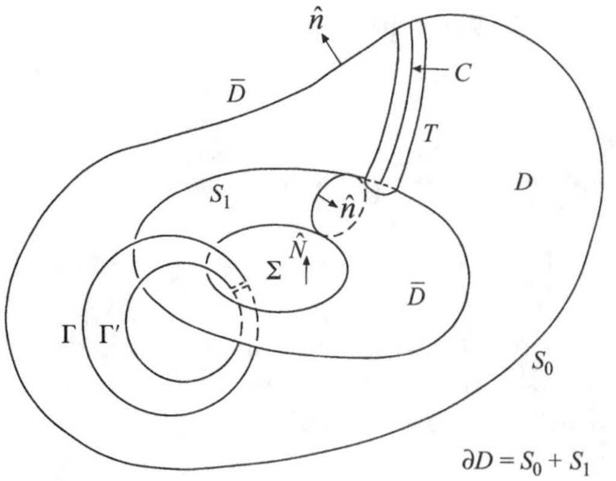

Figure 6.9 The domain D exterior to a toroidal surface S_1 bounded by S_0 and S_1. The open surface Σ and the open arc C are taken as cuts to make ϕ and A, respectively, single-valued.

Γ and Γ' of Fig. 6.9 can be connected along both sides of the open surface cut Σ (as indicated). If one now integrates along the closed path consisting of these connections and Γ and Γ', the contributions of the connections on either side of Σ will cancel. Stokes's theorem then allows the integral over this closed path to be replaced by the integral of $d{*}B$ over the enclosed surface bounded by $\Gamma - \Gamma'$, which vanishes since $d{*}B = 0$. Thus, taking account of the direction of integration along Γ and Γ', the integral over Γ is equal to that over Γ' and $[\phi]$ is independent of the choice of path Γ showing that $[\phi]$ is constant on Σ. This means $[\phi]$ can be removed from under the integral and (6.54) can be written as

$$\int_D {*}B \wedge B = \int_{\partial D} \phi B + \int_\Gamma {*}B \int_\Sigma B \tag{6.55}$$

Alternatively, one can introduce the open arc C (connecting the component of $\bar{D}$ inside the torus with that outside S_0) as a cut so as to make A single-valued. Then instead of (6.53), the relation

$$d(A \wedge {*}B) = dA \wedge {*}B - A \wedge d({*}B) \tag{6.56}$$

and the fact that $d({*}B) = 0$, can be used to write the integral $\int_D {*}B \wedge B$ as

$$\int_D {*}B \wedge B = \int_D d(A \wedge {*}B) \cong \int_{\partial D} A \wedge {*}B + \int_T A \wedge {*}B \tag{6.57}$$

(${*}B \wedge B = B \wedge {*}B$ since ${*}B$ is a 1-form and B a 2-form.)

Here, the arc C has been enclosed in a tube T and the symbol $\cong$ has been used to indicate that small elements of surface, which will vanish as the diameter of the tube is allowed to approach C in the limit, have been ignored. Combining (6.55) and (6.57),

$$\int_{\partial D} A \wedge {*}B \cong \int_{\partial D} \phi B + \int_\Gamma {*}B \int_\Sigma B - \int_T A \wedge {*}B \tag{6.58}$$

Evaluating the last integral of (6.58) is most readily accomplished by transforming to classical notation. The integrand $A \wedge {*}B$ becomes $(\boldsymbol{A} \times \boldsymbol{H}) \cdot \hat{\mathbf{n}}\, \mathrm{d}T$, where $\mathrm{d}\boldsymbol{T}$ is the element of area on the tube oriented along the normal $\hat{\mathbf{n}}$.

Noting that the tube is the topological product of a circle L (with tangent $\boldsymbol{dl}$) and the cut C (with tangent $\boldsymbol{dl}'$), the element of surface area can be written as $\boldsymbol{dT} = \boldsymbol{dl} \times \boldsymbol{dl}'$. The integrand is then $(\boldsymbol{A} \times \boldsymbol{H}) \cdot (\boldsymbol{dl} \times \boldsymbol{dl}') = (\boldsymbol{A} \cdot \boldsymbol{dl})(\boldsymbol{H} \cdot \boldsymbol{dl}') - (\boldsymbol{A} \cdot \boldsymbol{dl}')(\boldsymbol{H} \cdot \boldsymbol{dl})$. The space $\bar{D}$, interior to the toroid, is singly connected by virtue of the cut Σ, while the cut C connects the two components of $\bar{D}$. Then, by Stokes's theorem, the last term can be seen to vanish since $\nabla \times \boldsymbol{H} = 0$ in D. Consider the remaining two terms in the limit that $T \rightarrow C$. Stokes's theorem can again be used to transform the $\boldsymbol{A} \cdot \boldsymbol{dl}$ term; and, because $\boldsymbol{H}$ approaches a value depending only on its limiting position on C, the last integral of (6.58) can be written as the product

$$\int_T (\boldsymbol{A} \times \boldsymbol{H}) \cdot \hat{\mathbf{n}}\, dT = \int_{\partial D} \boldsymbol{B} \cdot \hat{\mathbf{n}}\, dS \int_C \boldsymbol{H} \cdot \boldsymbol{dl}' \tag{6.59}$$

Thus, (6.58) can be written as

$$\int_{\partial D} A \wedge {*B} = \int_{\partial D} \phi B + \int_\Gamma {*B} \int_\Sigma B - \int_{\partial D} B \int_C {*B} \tag{6.60}$$

With the condition that the normal component of $\boldsymbol{B}$ vanish on ∂D, the open surface integral $\int_\Sigma B$ is a period which can be interpreted as a flux, and (6.60) becomes

$$\int_{\partial D} A \wedge {*B} = \int_\Gamma {*B} \int_\Sigma B = \mu_0 I_\phi \Phi \tag{6.61}$$

where I_ϕ is the azimuthal current inside the torus (outside the domain D) and Φ is the flux through the surface Σ. Since the current vanishes in the domain D, Stokes's theorem implies that

$$\int_{\partial D} A \wedge {*B} = \int_D B \wedge {*B} = \langle B, B \rangle_D \tag{6.62}$$

Equation (6.61) can be used to interpret the boundary term of (6.50) in the case that the volume of integration is the *interior* of a torus. Attention, however, must be paid to the orientation of the boundary surface. To arrive at (6.61) and (6.62), the domain D was chosen to be the exterior of the toroidal surface; on the other hand, the volume of integration in (6.50) is now assumed to be the interior of the torus. A surface element on the boundary thus has opposite orientation for the two cases, which will introduce a minus sign when substituting the value of the integral into (6.50).

Thus, as previously maintained, the total energy as expressed by (6.50) is the sum of the energy interior to the torus, where currents may be present, and that in the field exterior to the torus, represented by the boundary term. Since it has been assumed that there are no currents in the domain exterior to the torus, the field in this region must be due to currents within the torus. Currents outside S_0 do not contribute to the magnetic energy in the space bounded by $S_0 + S_1$ because of the boundary conditions imposed and the fact that the space outside S_0 is simply connected. If the boundary configuration were different, so that the space outside S_0 was not simply connected (for example, if S_0 was a second torus so that S_1 was nested within it), then there could be additional contributions to the field energy between the toroids due to currents exterior to S_0.

6.6.1 Gauge Invariance

In a simply connected domain, the integrals $\int_{\partial D} A\wedge *B$ and $\int_D A\wedge *J$ are gauge invariant, as would be expected since the energy must be gauge invariant. Only the second integral, however, need be considered since, under a gauge transformation $A \to A + d\chi$, the term associated with $d\chi$ in the first integral is

$$\int_{\partial D} d\chi\wedge *B = \int_D d(d\chi\wedge *B) = -\int_D d\chi\wedge d(*B) = -\mu_0\int_D d\chi\wedge *J \tag{6.63}$$

where Ampère's circuital law, $\mu_0 *J = d(*B)$, has been used. That the integral $\int_D A\wedge *J$ is gauge invariant in a simply connected domain can be seen as follows: under a gauge transformation, the integral associated with $d\chi$ is

$$\int_D d\chi\wedge *J = \int_D d(\chi *J) = \int_{\partial D} \chi *J = 0 \tag{6.64}$$

where $d*J = 0$ has been used, and it is assumed that the currents occupy a limited region of the domain D so that the normal component of $*J$ vanishes on ∂D.

In a multiply connected domain, such as the interior of a toroid, the terms involving A are not necessarily gauge invariant since the function χ may not be single-valued. Yet, for the energy to be well defined in such domains it must, of course, be gauge invariant. In Section 6.3, which discussed gauge invariance and helicity, it was seen that the gauge term corresponded to the additional helicity due to a linking flux. Similarly, the gauge term in the equations for magnetic energy also has a physical significance.

Consider again a toroidal domain D, as shown in Fig. 6.2. The energy of the configuration can be written in terms of the currents in the domain as

$$E = \frac{1}{2}\int_D A\wedge *J \tag{6.65}$$

Under a gauge transformation, the integral associated with $d\chi$ now has an additional term due to the doubly connected nature of the space. The integral (compare with (6.55)) is given by

$$\int_D d\chi\wedge *J = \int_{\partial D} \chi *J + \int_\Gamma d\chi \int_{\Sigma_1} *J = [\chi]I_\phi = nI_\phi\Phi_{\Sigma_2} \tag{6.66}$$

where the first term on the right-hand side again vanishes and $[\chi] = \int_\Gamma d\chi$ is the jump in the function χ across Σ_1. The latter yields Φ_{Σ_2} for each of n circuits around the torus. In this domain, a gauge transformation corresponds to multiplying the energy in the field by some integer n.

6.7 CONCLUSION

The principal purpose of this chapter has been to show how concepts from topology are needed to properly treat a wide variety of physical phenomena. Whether it is the Aharonov and Bohm effect, magnetic energy in finite, multiply connected domains, such as a Tokamak, or the study of energy storage in solar prominences, topological concepts are crucial, not only as a guide to the intuition, but for computational purposes.

Also properly a part of topology are those elements of knot and braid theory used in Section 6.4. Knot theory is an example of what is known as the placement problem: How many different ways can a circle be embedded in 3-dimensional space? The application of knot and braid theory to electromagnetism has proven to be enormously fruitful.

De Rham's theorems, used in Section 6.6, give the conditions under which a closed 1-form is exact. Since the Poincaré lemma guarantees that every closed form is locally exact, this is an example of a global, as opposed to a differential, property of the space. Local properties are generally different from global ones, and have different implications.

Gauge transformations gave yet another example of the importance of global topology. When the space is not simply connected, it was found that gauge transformations can have physical effects. While a gauge transformation in the doubly connected space of the torus corresponded to simply multiplying the magnetic energy by some integer n, with regard to helicity, understanding the effects of a gauge transformation clarified the underlying physics by showing the role of the corresponding linking flux.

In Section 6.3, it was noted that the Hopf theorem stated that the Klein bottle and the torus are the only smooth, compact, connected surfaces without boundary that can support a nowhere vanishing vector field. The proof of this assertion makes use of the famous Gauss-Bonnet theorem, and is a beautiful example of the importance of global topological concepts. The Gauss-Bonnet theorem relates a differential-geometric quantity, the Gaussian curvature, to a number (the Euler-Poincaré characteristic, or Euler number) which does not depend on the curvature, or even on the existence of a Riemannian metric. The Gauss-Bonnet theorem is

$$\int_{\mathbb{M}} K\,dS = 2\pi\chi(\mathbb{M}) \tag{6.67}$$

where K is the Gaussian curvature of the 2-dimensional manifold $\mathbb{M}$ and $\chi(\mathbb{M})$ is the Euler-Poincaré characteristic of $\mathbb{M}$.

To relate the Gauss-Bonnet theorem to vector fields on $\mathbb{M}$ one uses the Poincaré-Hopf theorem which states that the Euler-Poincaré characteristic, $\chi(\mathbb{M})$, is equal to the sum of the indices [52] of any vector field on $\mathbb{M}$. Now, since the indices of a non-singular vector field vanish, the integral of the Gaussian curvature over the manifold $\mathbb{M}$ must also vanish; that is, $\int_{\mathbb{M}} K\,dS = 0$. The *only* orientable surface satisfying this requirement is the torus. This has some interesting implications for electromagnetism in the large.

The 4-dimensional formulation of Maxwell's equations depends on the existence of a Lorentz metric. But the mere existence of such a metric sets restrictions on the global topology of space-time: any non-compact 4-manifold admits a Lorentz metric, but a compact 4-manifold has a Lorentz metric if and only if its Euler-Poincaré characteristic vanishes [53]; that is, a compact 4-dimensional manifold can have a Lorentz metric defined on it only if it is not simply connected. When this is the case, there exist closed time-like curves, which is generally true for any compact space-time.

Hawking and Ellis [54] have argued that since a compact space-time is equivalent to a non-compact manifold in which points have been identified, on physical grounds one should regard the simply connected covering manifold as representing real space-time. But, nevertheless, multiply connected space-times have been given serious consideration in the literature concerned with cosmological models [55]. To pursue this any further would lead well beyond the limited scope of this chapter.

APPENDIX: THE CLASSICAL HOPF INVARIANT

There are several definitions of the Hopf invariant [56]. Perhaps the simplest, geometrically, is to consider a 3-sphere $\mathbb{S}^3$ in $\mathbb{R}^4$ and define the Hopf map π of $\mathbb{S}^3$ onto the 2-sphere $\mathbb{S}^2$. (The Hopf mapping can be generalized from maps of $\mathbb{S}^3 \to \mathbb{S}^2$ to maps of $\mathbb{S}^{2n-1} \to \mathbb{S}^n$.) If $\mathbb{S}^2$ is regarded as the one-point compactification $\mathbb{C} \cup \{\infty\}$ of the complex numbers, the specific identification of $\mathbb{S}^2$ and $\mathbb{C} \cup \{\infty\}$ can be achieved by use of a stereographic projection. Any point P on a line determined by the two points $P_1(x_1, y_1, z_1)$ and $P_2(x_2, y_2, z_2)$ is given by $P = (1 - \lambda)P_1 + \lambda P_2$, $\lambda \in \mathbb{R}$. With reference to Fig. 6.10, choose $P_1 = (0, 0, 1)$ and $P_2 = (\xi,\ \eta,\ 0)$. Then $x = \lambda\xi$; $y = \lambda\eta$; and $z = (1 - \lambda)$. The constraint given by $x^2 + y^2 + z^2 = 1$ can be used to determine λ as

$$\lambda = \frac{2}{\xi^2 + \eta^2 + 1} = \frac{2}{1 + \zeta\bar{\zeta}} \tag{6.68}$$

where $\zeta = \xi + \mathrm{i}\eta$. The coordinates x, y, and z are then given by

$$x = \frac{2\xi}{1 + \zeta\bar{\zeta}}, \qquad y = \frac{2\eta}{1 + \zeta\bar{\zeta}}, \qquad z = \frac{\zeta\bar{\zeta} - 1}{\zeta\bar{\zeta} + 1}$$
$$x + \mathrm{i}y = \frac{2\zeta}{1 + \zeta\bar{\zeta}} \tag{6.69}$$

The sphere $\mathbb{S}^3$ in $\mathbb{R}^4$ is given by the equation $x_1^2 + x_2^2 + x_3^2 + x_4^2 = 1$. Identify a point on this sphere with two complex numbers (z_1, z_2), where $z_1 = x_1 + \mathrm{i}x_2$ and $z_2 = x_3 + \mathrm{i}x_4$. Then the equation of the sphere can be written as $|z_1|^2 + |z_2|^2 = 1$; i.e., $\mathbb{S}^3 = \{(z_1, z_2) \in \mathbb{C} \times \mathbb{C} \,|\, |z_1|^2 + |z_2|^2 = 1\}$. The Hopf mapping $\pi: \mathbb{S}^3 \to \mathbb{S}^2$ is given by $\pi(z_1, z_2) = z_1/z_2$. To identify the image point on $\mathbb{S}^2$ of the Hopf mapping, set $\zeta = z_1/z_2$ in (6.69). Then

$$x + iy = 2z_1\bar{z}_2, \qquad z = |z_1|^2 - |z_2|^2 \tag{6.70}$$

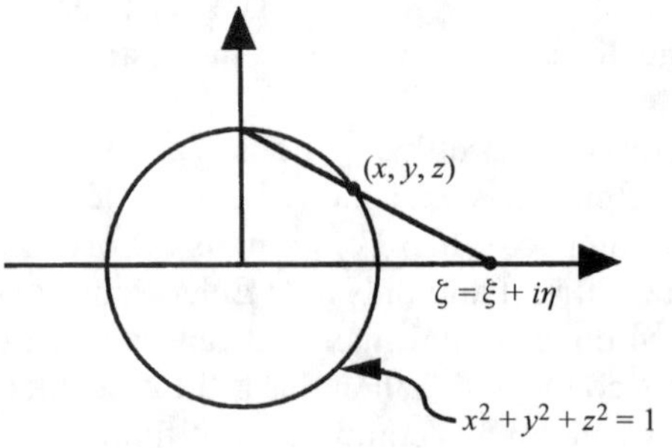

Figure 6.10 Stereographic projection of the complex plane (the one-point compactification of $\mathbb{C} \cup \{\infty\}$).

and

$$\pi(z_1, z_2) \rightarrow [x, y, z] = [2\,\mathrm{Re}(z_1\bar{z}_2), 2\,Im(z_1\bar{z}_2), |z_1|^2 - |z_2|^2] \in \mathbb{S}^2 \tag{6.71}$$

Now if $\lambda = e^{i\theta}, 0 \leq \theta \leq 2\pi$, so that $|\lambda| = 1$, $\pi(\lambda z_1, \lambda z_2) = \pi(z_1, z_2)$. This means that (z_1, z_2) and $(\lambda z_1, \lambda z_2)$ have the same image point in $\mathbb{S}^2$. Therefore, $\pi^{-1}(u) \cong \mathbb{S}^1$ for any $u \in \mathbb{S}^2$. Given any two points $u, v \in \mathbb{S}^2$, the "circumferences" $\pi^{-1}(u)$ and $\pi^{-1}(v)$ are disjoint and linked in $\mathbb{S}^3$. This can be readily seen by choosing, for example, $u = (1, 0, 0)$ and $v = (-1, 0, 0)$ in $\mathbb{S}^2$ and using (6.71) to obtain the additional relations $2z_1\bar{z}_2 = 1$ for u and $2z_1\bar{z}_2 = -1$ for v. These are satisfied, respectively, by

$$z_1 = z_2 = \frac{e^{i\theta}}{\sqrt{2}} \quad \text{and} \quad z_1 = -z_2 = \frac{e^{i\theta}}{\sqrt{2}} \tag{6.72}$$

where $(z_1, z_2) \in \mathbb{S}^3$.

Visualizing the linkage is aided by using the stereographic projection $\Pi: \mathbb{S}^3 \rightarrow \mathbb{R}^3$, with the projection point, which will be chosen here to be $(0, 0, 0, -1) \in \mathbb{S}^3$, identified with the point at infinity. The procedure for doing the projection closely follows the discussion preceding (6.68). Figure 6.10 applies with the complex plane replaced by $\mathbb{R}^3$ and $\mathbb{S}^2$ replaced by $\mathbb{S}^3$. Since it is the inverse transformation that is of interest, choose $P_1 = (0, 0, 0, -1)$ and $P_2 = (x_1, x_2, x_3, x_4)$, both contained in $\mathbb{S}^3$. The relation $P = (1 - \lambda)P_1 + \lambda P_2$, $\lambda \in \mathbb{R}$ then immediately determines λ to be $\lambda = 1/(1 + x_4)$. Thus, if the coordinates in $\mathbb{R}^3$ are given by $(\zeta_1, \zeta_2, \zeta_3)$, since $\zeta_i = \lambda x_i$,

$$\zeta_i = \frac{x_i}{1 + x_4} \tag{6.73}$$

where $x_4 \neq -1$. Using this result and (6.72) results in

$$\begin{aligned} \Pi[\pi^{-1}(u)]&: \zeta_1 = \zeta_3 = \frac{\cos\theta}{\sqrt{2} + \sin\theta}; \qquad \zeta_2 = \frac{\sin\theta}{\sqrt{2} + \sin\theta} \\ \Pi[\pi^{-1}(v)]&: \zeta_1 = -\zeta_3 = \frac{\cos\theta}{\sqrt{2} - \sin\theta}; \qquad \zeta_2 = \frac{\sin\theta}{\sqrt{2} - \sin\theta} \end{aligned} \tag{6.74}$$

If these curves are plotted as a function of θ in the mutually orthogonal planes $\zeta_1 = \pm\zeta_3$, one obtains two disjoint and singly linked ellipses [57]. In general, for π: $\mathbb{S}^{2n-1} \rightarrow \mathbb{S}^n$, $n \geq 2$, if u and v are distinct interior points of $\mathbb{S}^n$, then $\pi^{-1}(u)$ and $\pi^{-1}(v)$ are disjoint $(n - 1)$-manifolds in $\mathbb{S}^{2n-1}$. The classical Hopf invariant is the linking number of these disjoint manifolds.

As mentioned in the text, Whitehead was the first to express the Hopf invariant as a volume integral. Let f be a mapping of $\mathbb{S}^3 \rightarrow \mathbb{S}^2$, and let ω be a 2-form on $\mathbb{S}^2$ such that $\int_{\mathbb{S}^2} \omega = 1$. Then there is a natural induced map f^* that maps 2-forms on $\mathbb{S}^2$ to 2-forms on $\mathbb{S}^3$. The form $f^*\omega$ is a closed 2-form on $\mathbb{S}^3$ because $d(f^*\omega) = f^*d\omega = 0$, where $d\omega = 0$ since there are no 3-forms on $\mathbb{S}^2$. Furthermore, since $\mathbb{S}^3$ has no non-trivial 2-dimensional cycles, there is a 1-form, β, on $\mathbb{S}^3$ such that $f^*\omega = d\beta$. The exterior product $\beta \wedge d\beta$ is then a 3-form on $\mathbb{S}^3$. What Whitehead was able to show is that the Hopf invariant can be written as

$$\gamma = \int_{\mathbb{S}^3} \beta \wedge d\beta \tag{6.75}$$

It is readily shown that γ does not depend on either the choice of β, so long as $f^*\omega = d\beta$, or on the choice of ω [58]. The result, (6.75), can also be generalized to maps of $\mathbb{S}^{2n-1} \to \mathbb{S}^n$.

REFERENCES

[1] Y. Aharonov and D. Bohm, "Significance of Electromagnetic Potentials in the Quantum Theory," *Phys. Rev.*, **115**, p. 485, 1959.

[2] R. G. Chambers, "Shift of an Electron Interference Pattern by Enclosed Magnetic Flux," *Phys. Rev. Lett.*, **5**, p. 3, 1960.

[3] G. E. Marsh, *Force-Free Magnetic Fields: Solutions, Topology and Applications*. Singapore: World Scientific, 1996.

[4] R. Lüst and A. Schlüter, "Kraftfreie Magnet Felder," *Z. Astrophys.*, **34**, p. 263, 1954.

[5] A. D. Jette, "Force-free Magnetic Fields in Resistive Magnetohydrostatics," *J. Math. Annal. Appl.*, **29**, p. 109, 1970.

[6] Chandrasekhar and Kendall, "On Force-free Magnetic Fields," *Astrophys. J.*, **126**, p. 457, 1957.

[7] E. R. Priest, "Basic Magnetic Configuration and Energy Supply Processes for an Interacting Flux Model of Eruptive Solar Flares," in Z. Svestka, B. V. Jackson, and M. E. Machado, eds., *Eruptive Solar Flares*, Springer-Verlag, Berlin, Germany, 1992.

[8] T. Gold and F. Hoyle, "On the Origin of Solar Flares," *Mon. Not. R. Astron. Soc.*, **120**, p. 89, 1960.

[9] P. A. Sturrock, "Solar Flares," in P. A. Sturrock, ed., *Plasma Astrophysics*, Proc. of the Int. School of Phys. "Enrico Fermi," **39**, Academic Press, New York, 1967, p. 168; A. A. Van Ballegooijen and P. C. H. Martens, "Formation and Eruption of Solar Prominences," *Astrophys. J.*, **343**, p. 971, 1989.

[10] E. N. Parker, *Cosmical Magnetic Fields*. Oxford, U.K.: Clarendon Press, 1979, Chapter 9.

[11] V. C. Ferraro and C. Plumpton, *An Introduction to Magneto-Fluid Mechanics*, 2nd ed., Oxford University Press, Oxford, U.K., 1966, p. 47.

[12] L. Woltjer, "A Theorem on Force-free Magnetic Fields," *Proc. Nat. Acad. Sci. USA*, **44**, p. 489, 1958.

[13] H. K. Moffatt, "The Degree of Knottedness of Tangled Vortex Lines," *J. Fluid Mech.*, **35**, p. 117, 1969. For a recent review, see H. K. Moffatt and A. Tsinober, "Helicity in Laminar and Turbulent Flow," *Annu. Rev. Fluid Mech.*, **24**, p. 281, 1992.

[14] H. K. Moffatt and R. L. Ricca, Helicity and the Călugăreanu Invariant," *Proc. R. Soc. London A*, **439**, p. 411, 1992.

[15] V. I. Arnold, "The Asymptotic Hopf Invariant and Its Applications," *Sel. Math. Sov.*, **5**, p. 327, 1986.

[16] M. A. Berger and G. B. Field, "The Topological Properties of Magnetic Helicity," *J. Fluid Mech.*, **147**, p. 133, 1984.

[17] As Ref. 11.

[18] P. Alexandroff and H. Hopf, *Topology*, Berlin, Germany: Springer-Verlag, 1935, p. 552, Satz III. See also C. Godbillon, *Dynamical Systems on Surfaces*, Springer-Verlag, Berlin, Germany, 1980, p. 181.

[19] M. Hénon, "Sur la Topologie des Lignes de Courand dans un Cas Particulier," *C. R. Acad. Sci. (Paris), Ser. A*, **262**, p. 312, 1966.

[20] G. E. Marsh, "Magnetic Energy, Multiply Connected Domains, and Force-free Fields," *Phys. Rev. A*, **46**, p. 2117, 1992.

[21] Two integers are relatively prime if they have no common divisors except ±1.

[22] Ref. 16. This paper also gives a discussion of helicity in terms of the internal structure of a flux tube and the external relation between flux tubes.

[23] G. Călugăreanu, "L'Integral de Gauss et L'Analyse des Nœuds Tridimensionneles," *Rev. Math. Pures Appl.*, **4**, p. 5, 1959.

[24] F. B. Fuller, "The Writhing Number of a Space Curve," *Proc. Nat. Acad. Sci. USA*, **68**, p. 815, 1971; "Decomposition of the Linking Number of a Closed Ribbon: A problem from Molecular Biology," *Proc. Nat. Acad. Sci. USA*, **75**, p. 3557, 1978.

[25] J. H White, "Self-linking and the Gauss Integral in Higher Dimensions," *Am. J. Math.*, **91**, p. 693, 1969; G. Călugăreanu, "Sur les Classes D'Isotopie des Nœuds Tridimensionnels et leurs Invariants," *Czechoslovak Math. J.*, **11**, p. 588, 1961; and Ref. 23.

[26] L. H. Kauffman, *On Knots* (Annals of Mathematics Studies Number 115), Princeton, N.J.: Princeton University Press, 1987, Chapter 2.

[27] H. K. Moffatt, *Magnetic Field Generation in Electrically Conducting Fluids*, Cambridge University Press, Cambridge, U. K., 1978, Section 2.1.

[28] H. K. Moffatt, *Ibid.*, Chapter 2 (See also the comments on the trefoil knot on p. 137 of Ref. 7.); and "The energy spectrum of knots and links," *Nature*, **347**, p. 367, 1990.

[29] More complicated configurations have the kink helicity given by $H_K = \Phi^2(N_+ - N_-)$, where N_+ and N_- are the number of positive and negative crossovers (established by the right-hand rule).

[30] P. Akhmet'ev and A. Ruzmaikin, "Borromeanism and Bordism," in *Topological Aspects of the Dynamics of Fluids and Plasmas*, H. K. Moffatt, et al., eds Dordrecht, Netherlands: Kluwer Academic Publishers, 1992; "A Fourth Order Topological Invariant of Magnetic or Vortex Lines," *J. of Geom. and Phys.*, **15**, p. 95, 1995.

[31] N. W. Evans and M. A. Berger, "Hierarchy of Linking Integrals," in *Topological Aspects of the Dynamics of Fluids and Plasmas*, H. K. Moffatt, et al., eds., Kluwer Academic Publishers, Dordrecht, Netherlands, 1992.

[32] A. Y. K. Chui and H. K. Moffatt, "The Energy and Helicity of Knotted Magnetic Flux Tubes," *Proc. R. Soc. Lond. A*, **451**, p. 609, 1995.

[33] H. Seifert, "Über das Geschlicht von Knoten," *Math. Ann.*, **110**, p. 571, 1934.

[34] See, for example, M. Spivak, *A Comprehensive Introduction to Differential Geometry*. Houston: Publish or Perish, Inc., 1979, Vol. 4; P. J. Hilton, *An Introduction to Homotopy Theory*, Cambridge, U. K.: Cambridge University Press, 1953; Sze-Tsen Hu, *Homotopy Theory*, Academic Press, New York, 1959.

[35] J. H. C. Whitehead, "An Expression of Hopf's Invariant in an Integral," *Proc. Nat. Acad. Sci. USA*, **33**, p. 117, 1947.

[36] A. F. Rañada, "Topological Electromagnetism," *J. Phys A: Math. Gen.*, **25**, p. 1621, 1992; A. F. Rañada and J. L. Trueba, "Electromagnetic Knots," *Phys. Lett. A*, **202**, p. 337, 1995.

[37] The reader is cautioned not to confuse the 1-form β, which is the "metric dual" of the vector $\boldsymbol{B} = B^i \partial_{x^i}$, with $*\beta$, a 2-form which may be identified with a magnetic field.

[38] Here, since the application is to 3-dimensional space, the duality operator maps p-forms onto $(3-p)$-forms. The Hodge star operator has the property that the subspace of $*\omega$ is orthogonal to that of ω, where ω is a p-form. Although the $*$ operator is defined locally, it is independent of local coordinates, but does depend on the existence of an inner product and the orientation of the space.

[39] For reasons of clarity, the exposition given here leaves out some mathematical subtleties that can be found in the literature.

[40] I. E. Farquhar, *Ergodic Theory in Statistical Mechanics*, London, U. K.: Interscience, 1964.

[41] An interesting example of non-measurable sets is the Banach-Tarski Paradox. See, for example, T. J. Jech, "About the Axiom of Choice," contained in J. Barwise, ed., *Handbook of Mathematical Logic*. North-Holland, Amsterdam, Netherlands, 1983.

[42] W. L. Burke, *Applied Differential Geometry*, Cambridge University Press, Cambridge, U. K., 1985, Section 24.

[43] B. A. Khesin and Yu.V. Chekanov, "Invariants of the Euler Equations for Ideal or Barotropic Hydrodynamics and Superconductivity in D Dimensions," *Physica D*, **40**, p. 119, 1989.

[44] S. Chandrasekhar and L. Woltjer, "On Force-free Magnetic Fields," *Proc. Nat. Acad. Sci. USA*, **44**, p. 285, 1958.

[45] J. B. Taylor, "Relaxation of Toroidal Plasma and Generation of Reverse Magnetic Fields," *Phys. Rev. Lett.*, **33**, p. 1139, 1974; "Relaxation and Magnetic Reconnection in Plasmas," *Rev. Mod. Phys.*, **58**, p. 741, 1986.

[46] J. B. Taylor, "Relaxation and Topology in Plasma Experiments;" in, H. K. Moffatt, et al., ed., *Topological Aspects of the Dynamics of Fluids and Plasmas*, Kluwer Academic Publishers, Dordrecht, Netherlands, 1992.

[47] M. A. Berger, "Coronal Heating by Dissipation of Magnetic Structure," *Space Sci. Rev.*, **68**, p. 3, 1994.

[48] In a simply connected domain, where the current vanishes so that $\boldsymbol{B}$ is harmonic ($\nabla^2 \boldsymbol{B} = 0$), if the normal component of $\boldsymbol{B}$ vanishes on the boundary, $\boldsymbol{B}$ will be identically zero.

[49] A. A. Blank, K. O. Friedrichs, and H. Grad, "Theory of Maxwell's Equations Without Displacement Current," Notes on Magneto-Hydrodynamics V, NYO-6486, 1957.

[50] These results are related to the Alexander duality theorem discussed in Ref. 49 and, for example, J. R. Munkres, *Elements of Algebraic Topology*, Addison-Wesley, Inc., New York, 1984.

[51] G. F. D. Duff and D. C. Spencer, "Harmonic Tensors on Riemannian Manifolds with Boundary," *Ann. of Math.*, **56**, p. 128, 1952.

[52] The index of a closed curve is the number of revolutions made by a vector field in traversing the curve once in the counter-clockwise direction. If such a sufficiently small curve circles an isolated singular point of a vector field, the index of the curve is the index of the singular point.

[53] N. Steenrod, *The Topology of Fiber Bundles*, Princeton: Princeton University Press, 1951, p. 207.

[54] S. W. Hawking and G. F. R. Ellis, *The Large Scale Structure of Space-Time*, Cambridge University Press, Cambridge, U. K., 1973, p. 190.

[55] See, for example, M. Lachièze-Rey and J. P. Luminet, "Cosmic topology," *Phys. Rep.*, **254**, p. 135, 1995.

[56] See, for example, Sze-Tsen Hu, *Homotopy Theory*, Academic Press, New York, 1959, p. 334.

[57] G. Morandi, *The Role of Topology in Classical and Quantum Physics*, Lecture notes in Physics, **m7**, Berlin, Germany: Springer-Verlag, 1992, pp. 79–84.

[58] C. Godbillon, *Topologie Algébrique*, Hermann, Paris, France, 1971, pp. 221–222.

Chapter 7

THE ELECTRODYNAMICS OF TORUS KNOTS

Douglas H. Werner

ABSTRACT

The electromagnetic radiation and scattering properties of thin knotted wires are considered in this chapter. A special class of knots, called torus knots, are introduced for the purpose of this investigation. The parameterizations available for torus knots are used in conjunction with Maxwell's equations to formulate useful mathematical representations for the fields radiated by these knots. These representations are then used to derive simple closed-form far-field expressions for electrically small torus knots. The derivation of a new Electric Field Integral Equation (EFIE) suitable for analysis of toroidally knotted wires is also outlined in this chapter. It is also demonstrated that the well-known expressions for the electromagnetic fields radiated by a circular loop antenna (canonical unknot) as well as the linear dipole antenna may be obtained as degenerate forms of the more general torus knot field representations. Finally, a moment method technique is applied to model the backscattering from a threefold rotationally symmetric trefoil knot as a function of frequency, polarization, and incidence angle.

7.1 INTRODUCTION

Over the past 100 years, the study of knots has been primarily confined to the field of mathematics. Knot theory is a subfield of an area of mathematics known as topology, which deals with the properties of geometric objects that are preserved under deformations. A knot is defined as a closed curve in space that does not intersect itself anywhere [1]. In knot theory, there is no distinction placed on the original closed knotted curve and the deformations of that curve through space that do not allow the curve to pass through itself. All of these deformations are considered to be the same knot. In other words, they are topologically equivalent. The simplest of all possible knots is the unknotted circle or loop, which is called the unknot or trivial knot. At first glance, a tangled-up loop of string may appear to be knotted but actually it can be untangled without cutting and gluing. In this case, the tangled-up loop of string degenerates into an unknot.

It has only been over the past few years that knots have been gaining interest outside the field of mathematics in various branches of science and engineering. For instance, applications of knot theory are emerging in such diverse fields as plasma physics, polymer science, and molecular biology. Biochemists have recently discovered that knotting exists in DNA molecules [2], [3]. Since then, synthetic chemists have been studying ways to create knotted molecules in which the type of knot would determine the properties of the molecule [1], [4]. Knots have also been considered in electrostatic and magnetostatic field theory with applications to the study of eruptive solar flares and magnetohydrostatics [5], [6].

The electromagnetic scattering properties of perfectly conducting thin knotted wires were first studied by Manuar and Jaggard [7]. The initial investigations in [7] suggest that the handedness of knots and the degree of knottedness may be deduced from the backscatter signature of electromagnetic waves. The work reported in [7] is based on results obtained for the backscatter differential cross section of the threefold rotationally symmetric trefoil knot and its associated unknot. The trefoil knot is the name given to the simplest nontrivial knot. A trefoil knot can be constructed by tieing a piece of string in a knot and gluing the two ends of the string together. Trefoils were chosen in [7] because they possess a low knottedness and it only takes one crossing switch (cut and reattach) to transform the trefoil into its corresponding unknot. Further discussions of the scattering properties of trefoils and untrefoils may be found in [8] and [9]. The primary objective of this work was to compare the backscatter from a simple knot (the trefoil) to that produced by a simple related unknot (the untrefoil). It was found that a remarkable difference of 25 to 30 orders of magnitude existed between trefoils and untrefoils in the copolarized backscattering of incident circularly polarized plane waves. This difference in the copolarized backscatter was initially attributed to the topological properties of the knots [8]. However, by considering the backscattering from asymmetric trefoils called morphs, Manuar and Jaggard [10] were able to later demonstrate the dominant role that symmetry plays in producing this effect.

One of the major drawbacks for application of knot theory to electromagnetics has been the lack of available parameterizations which can be used to mathematically describe knotted curves. This is because knots have traditionally been studies within a topological context where parameterizations for the curves are not generally required. However, in order to successfully characterize the electromagnetic radiation and scattering properties of knots using Maxwell's equations, it is advantageous to develop parameterizations which can be used to geometrically describe the curves of these knots. This chapter introduces such parameterizations for a family of knots known as (p, q)-torus knots. These knots reside on the surface of a standard torus, thereby making it possible to readily obtain useful parameterizations to describe them. The well-known trefoil is one important example of a (p, q)-torus knot.

The parametric representations for the (p, q)-torus knots are presented in Section 7.2. These parameterizations are then used in combination with Maxwell's equations to derive the vector potential and corresponding electric field expressions which describe radiation from a (p, q)-torus knot. The derivation of an electric field integral equation (EFIE) specifically for (p, q)-torus knots is also outlined in Section 7.2. In Section 7.3, closed-form expressions are derived for the radiation integrals associated with electrically small (p, q)-torus knots. Section 7.3 also shows how the well-known expressions for the electromagnetic fields radiated by a circular loop antenna (canonical unknot) can be obtained as a special case of the more general torus knot field representations derived in Section 7.2. A new set of parametric representations for elliptical (p, q)-torus knots is introduced in Section 7.4.1. It is also demonstrated in Section 7.4.1 how the parameterizations for circular torus knots considered in Sections 7.1–7.3 may be obtained as a special case of the more general elliptical torus knot parameterizations. These parameterizations are then used in Section 7.4.2 to derive expressions for the vector potential and electromagnetic fields of elliptical (p, q)-torus knots made from thin perfectly conducting wire. In Section 7.5, various closed-form expressions are derived for the radiation integrals associated with electrically small elliptical (p, q)-torus knots. Also pointed out in Section 7.5 is the

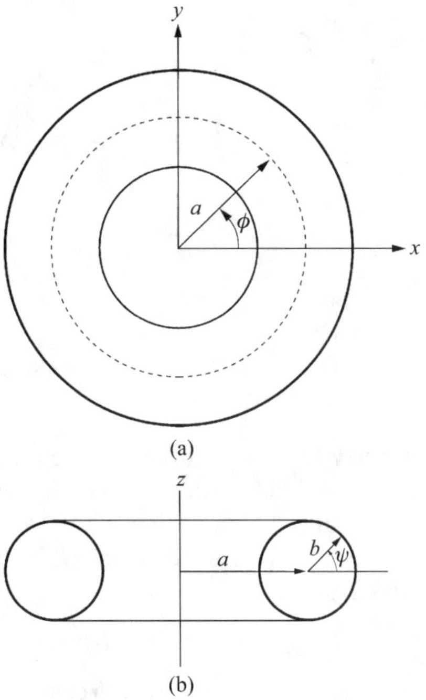

Figure 7.1 Geometry of the solid torus **T**. The top view is shown in (a) and a side view is shown in (b).

interesting fact that the circular loop as well as the linear dipole geometries can both be obtained as degenerate cases of the elliptical torus knot parameterizations. Finally, several examples illustrating the radiation and scattering properties of various (p, q)-torus knots, including the trefoil, are presented in Section 7.6. These results were obtained by performing a method of moments analysis on thin knotted wires, using the parameterizations introduced to define the wire geometry.

7.2 THEORETICAL DEVELOPMENT

7.2.1 Background

Let **T** denote the standard solid torus in $\mathbb{R}^3$ which is depicted in Fig. 7.1. The (x, y, z) coordinates that describe this solid torus **T** are given by

$$x = (a + b\cos\psi)\cos\phi \tag{7.1}$$

$$y = (a + b\cos\psi)\sin\phi \tag{7.2}$$

$$z = b\sin\psi \tag{7.3}$$

where $0 \leq \psi,\ \phi \leq 2\pi$, and $0 \leq b \leq a/2$. There are two types of curves, known as meridian and longitude curves, which are commonly associated with the torus **T**. A meridian curve runs once the short way around the torus, while a longitude curve runs once around the torus the long way. An example of these curves is illustrated in Fig. 7.2.

The so-called (p, q)-torus knots represent a class of knots which live on the surface of the solid torus **T**. These knots are classified by the integers p and q which

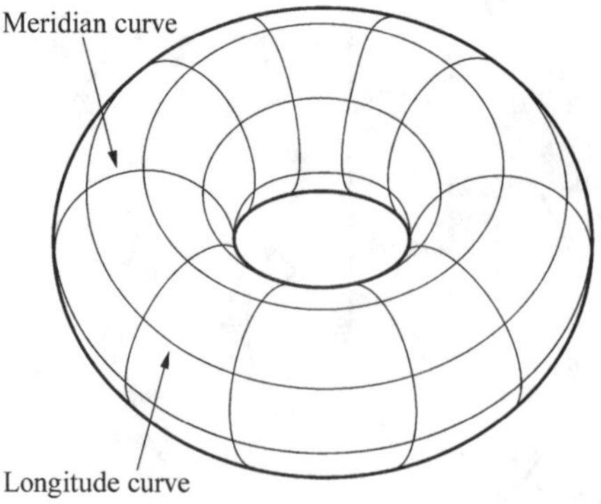

Figure 7.2 Examples of meridian and longitude curves on the surface of a torus **T**.

have the property that they will always be relatively prime (i.e., their greatest common divisor is 1) [1], [11]. The integer p corresponds to how many times the knot wraps around the torus in the longitudinal direction while the integer q indicates the number of times it wraps around the torus in the meridional direction. It can be shown that the curves described by

$$x = (a + b\cos(\psi + qs))\cos(ps) \tag{7.4}$$

$$y = (a + b\cos(\psi + qs))\sin(ps) \tag{7.5}$$

$$z = b\sin(\psi + qs) \tag{7.6}$$

represent the family of (p,q)-torus knots which reside on the surface of **T** provided $0 \le s \le 2\pi$ [5]. Two views of the $(2,3)$-torus knot are shown in Fig. 7.3. It can be seen

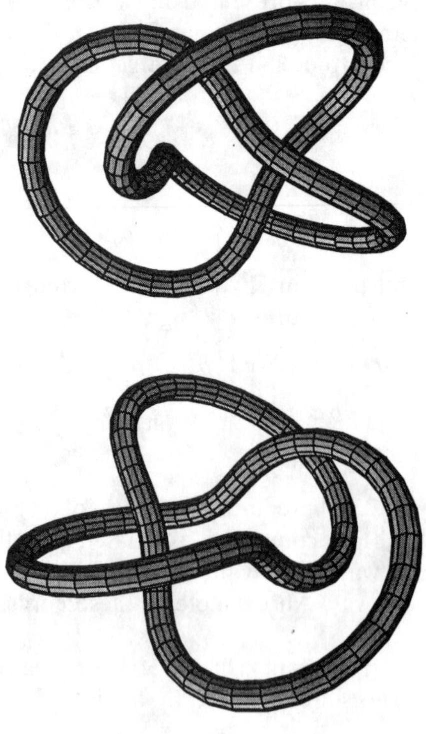

Figure 7.3 Two views of a (2, 3)-torus knot (a trefoil).

Figure 7.4 Two views of a (3, 4)-torus knot.

from Fig. 7.3 that the $(2,3)$-torus knot is a trefoil. The trefoil knot traverses **T** twice longitudinally and three times meridionally. Likewise, Fig. 7.4 shows two views of the higher-order $(3,4)$-torus knot.

Any knot that can be continuously deformed into a circular loop (standard ring) is said to be an unknot or trivial knot. For this reason, the circular loop is regarded as the canonical unknot. Several variations of trivial knots or unknots can be generated as special cases of the (p,q)-torus knots. For instance, the $(1,2)$-torus knot which is shown in Fig. 7.5 represents a trivial knot. The toroidal helix formed by the $(1,4)$-torus knot shown in Fig. 7.6 is also an interesting example of a trivial knot. Finally, we note that the canonical unknot itself may be obtained as a special limiting case of a torus knot. This fact may be demonstrated by setting $b = 0$ and $p = 1$ in (7.4) through (7.6).

Another important property of knots is chirality. A knot is considered chiral if it is topologically distinct from its mirror image [1], [12]. This means that no matter how the knot is deformed, it cannot be superimposed on its mirror image. On the other hand, a knot that can be deformed into its mirror image is called an achiral knot. As pointed out in [7], there is a subtle distinction between the usual geometrical notion of chirality and the topological notion of chirality associated with knots. An object possesses geometrical chirality if it cannot be superimposed on its mirror image through a series of spatial translations and rotations. This definition of chirality is used in relation to inclusion geometries, such as helices, which have been considered for chiral materials [13]. Knots which are topologically chiral, however, have the property that they cannot be superimposed on their mirror image through any kind of a

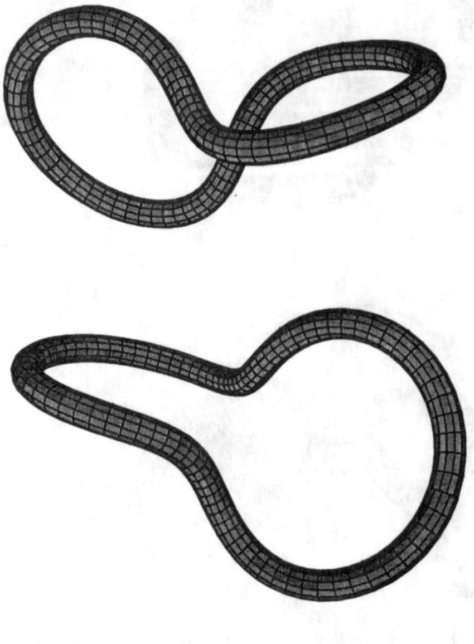

Figure 7.5 Two views of a (1, 2)-torus knot (a trivial knot).

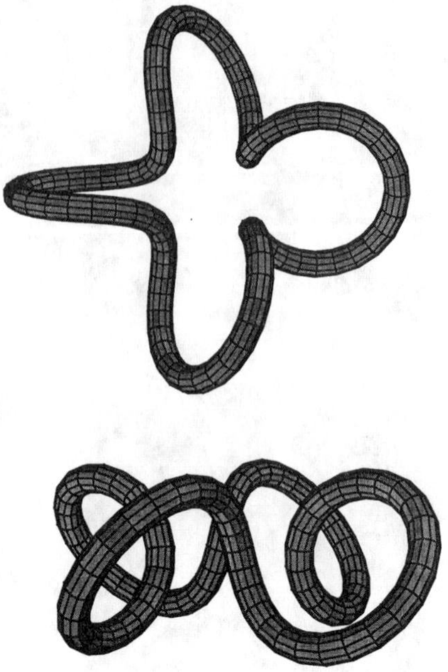

Figure 7.6 Two views of a (1, 4)-torus knot (a toroidal helix).

continuous deformation (i.e., without breaking or cutting the knot). There are some knots, such as the figure-eight knot [12], which are geometrically chiral but topologically achiral. The $(1,2)$-torus knot and the $(1,4)$-torus knot shown in Figs. 7.5 and 7.6, respectively, are also examples of this kind of knot. The trefoil belongs to the class of knots which have both geometrical and topological chirality. In other words, the trefoil is distinct from its mirror image in the geometrical sense as well as the topological sense. This suggests that there are actually two distinct types of trefoil knots rather than just one; namely a right-hand and a left-hand trefoil as shown in Fig. 7.7. This property is also shared by the $(3,4)$-torus knot shown in Fig. 7.4.

7.2.2 Electromagnetic Fields of a Torus Knot

The various properties of the (p,q)-torus knots discussed above make them interesting not only from a topological point of view, but also from an electromagnetics point of view. The remainder of this section will be devoted to developing a theoretical foundation for characterizing the electromagnetic radiation and scattering properties of (p,q)-torus knots. We begin this analysis by considering the vector potential for an arbitrarily shaped wire with an electric current $\vec{I}$, which may be expressed in the general form [14]

$$\vec{A}(x,y,z) = \frac{\mu}{4\pi}\int_C \vec{I}(x',y',z')\frac{e^{-j\beta R}}{R}d\ell' \tag{7.7}$$

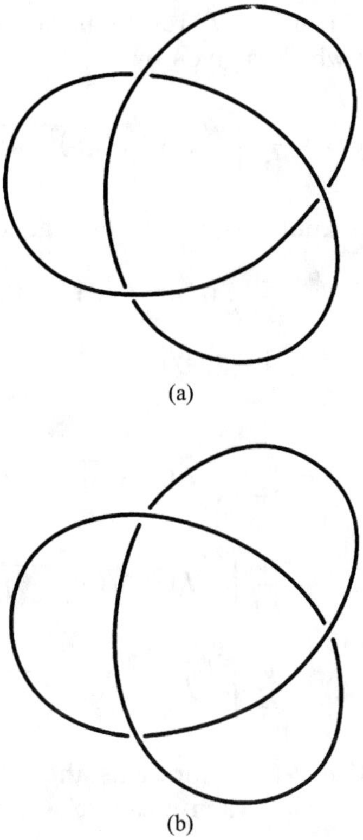

Figure 7.7 Two types of trefoil knots. A left-hand trefoil is shown in (a) while a right-hand trefoil is shown in (b).

where

$$R = \sqrt{(x - x')^2 + (y - y')^2 + (z - z')^2} \tag{7.8}$$

Vector potential expressions which are specialized for the class of (p, q)-torus knots may be derived from (7.7) by making use of the parameterizations introduced in (7.4)–(7.6). By representing these parameterizations in terms of the source coordinates, we have

$$x' = (a + b\cos(\psi + qs'))\cos(ps') \tag{7.9}$$

$$y' = (a + b\cos(\psi + qs'))\sin(ps') \tag{7.10}$$

$$z' = b\sin(\psi + qs') \tag{7.11}$$

It may be shown using (7.9)–(7.11) that the incremental element of length for a (p, q)-torus knot is

$$d\ell' = \sqrt{(qb)^2 + p^2(a + b\cos(\psi + qs'))^2}\, ds' \tag{7.12}$$

which, when integrated over the knot, leads to the following useful formula for arclength L_{pq}:

$$L_{pq} = 2\int_0^{\pi} \sqrt{(qb)^2 + p^2(a + b\cos u)^2}\, du \tag{7.13}$$

Substituting (7.12) into (7.7) results in an expression for the vector potential of a (p, q)-torus knot which is given by

$$\vec{A}(x,y,z) = \frac{\mu}{4\pi}\int_0^{2\pi} \vec{I}(s')\sqrt{(qb)^2 + p^2(a + b\cos(\psi + qs'))^2}\,\frac{e^{-j\beta R}}{R}ds' \qquad (7.14)$$

Furthermore, the knot current $\vec{I}$ has the general form of

$$\vec{I}(s') = I_x(s')\hat{x} + I_y(s')\hat{y} + I_z(s')\hat{z} \qquad (7.15)$$

which implies that (7.14) may be separated into the following three scalar components:

$$A_x(x,y,z) = \frac{\mu}{4\pi}\int_0^{2\pi} I_x(s')\sqrt{(qb)^2 + p^2(a + b\cos(\psi + qs'))^2}\,\frac{e^{-j\beta R}}{R}ds' \qquad (7.16)$$

$$A_y(x,y,z) = \frac{\mu}{4\pi}\int_0^{2\pi} I_y(s')\sqrt{(qb)^2 + p^2(a + b\cos(\psi + qs'))^2}\,\frac{e^{-j\beta R}}{R}ds' \qquad (7.17)$$

$$A_z(x,y,z) = \frac{\mu}{4\pi}\int_0^{2\pi} I_z(s')\sqrt{(qb)^2 + p^2(a + b\cos(\psi + qs'))^2}\,\frac{e^{-j\beta R}}{R}ds' \qquad (7.18)$$

Now suppose that $I_s(s')$ represents the current distribution on the surface of a particular knotted wire. Then it can be shown that

$$I_x(s') = I_s(s')\,T_x(s') \qquad (7.19)$$

$$I_y(s') = I_s(s')\,T_y(s') \qquad (7.20)$$

$$I_z(s') = I_s(s')\,T_z(s') \qquad (7.21)$$

where

$$T_x(s') = \frac{-p(a + b\cos(\psi + qs'))\sin(ps') - qb\sin(\psi + qs')\cos(ps')}{\sqrt{(qb)^2 + p^2(a + b\cos(\psi + qs'))^2}} \qquad (7.22)$$

$$T_y(s') = \frac{p(a + b\cos(\psi + qs'))\cos(ps') - qb\sin(\psi + qs')\sin(ps')}{\sqrt{(qb)^2 + p^2(a + b\cos(\psi + qs'))^2}} \qquad (7.23)$$

$$T_z(s') = \frac{qb\cos(\psi + qs')}{\sqrt{(qb)^2 + p^2(a + b\cos(\psi + qs'))^2}} \qquad (7.24)$$

are the x, y, and z components of the unit vector $\hat{T}$ tangent to the curve of the knot. Substituting (7.19)–(7.21) and (7.22)–(7.24) into (7.16)–(7.18), respectively, yields

$$\begin{aligned} A_x(x,y,z) = &-\frac{\mu}{4\pi}p\int_0^{2\pi} I_s(s')(a + b\cos(\psi + qs'))\sin(ps')\frac{e^{-j\beta R}}{R}ds' \\ &-\frac{\mu}{4\pi}qb\int_0^{2\pi} I_s(s')\sin(\psi + qs')\cos(ps')\frac{e^{-j\beta R}}{R}ds' \end{aligned} \qquad (7.25)$$

$$A_y(x,y,z) = \frac{\mu}{4\pi} p \int_0^{2\pi} I_s(s')(a + b\cos(\psi + qs'))\cos(ps')\frac{e^{-j\beta R}}{R} ds'$$
$$- \frac{\mu}{4\pi} qb \int_0^{2\pi} I_s(s')\sin(\psi + qs')\sin(ps')\frac{e^{-j\beta R}}{R} ds' \tag{7.26}$$

$$A_z(x,y,z) = \frac{\mu}{4\pi} qb \int_0^{2\pi} I_s(s')\cos(\psi + qs')\frac{e^{-j\beta R}}{R} ds' \tag{7.27}$$

Transforming (7.25)–(7.27) to spherical coordinates leads to

$$A_r(r,\theta,\phi) = \frac{\mu\sin\theta}{4\pi}\left\{p\int_0^{2\pi} I_s(s')(a + b\cos(\psi + qs'))\sin(\phi - ps')\frac{e^{-j\beta R}}{R} ds'\right.$$
$$\left. - qb\int_0^{2\pi} I_s(s')\sin(\psi + qs')\cos(\phi - ps')\frac{e^{-j\beta R}}{R} ds'\right\} \tag{7.28}$$
$$+ \frac{\mu\cos\theta}{4\pi}\left\{qb\int_0^{2\pi} I_s(s')\cos(\psi + qs')\frac{e^{-j\beta R}}{R} ds'\right\}$$

$$A_\theta(r,\theta,\phi) = \frac{\mu\cos\theta}{4\pi}\left\{p\int_0^{2\pi} I_s(s')(a + b\cos(\psi + qs'))\sin(\phi - ps')\frac{e^{-j\beta R}}{R} ds'\right.$$
$$\left. - qb\int_0^{2\pi} I_s(s')\sin(\psi + qs')\cos(\phi - ps')\frac{e^{-j\beta R}}{R} ds'\right\} \tag{7.29}$$
$$- \frac{\mu\sin\theta}{4\pi}\left\{qb\int_0^{2\pi} I_s(s')\cos(\psi + qs')\frac{e^{-j\beta R}}{R} ds'\right\}$$

$$A_\phi(r,\theta,\phi) = \frac{\mu}{4\pi} p \int_0^{2\pi} I_s(s')(a + b\cos(\psi + qs'))\cos(\phi - ps')\frac{e^{-j\beta R}}{R} ds'$$
$$+ \frac{\mu}{4\pi} qb \int_0^{2\pi} I_s(s')\sin(\psi + qs')\sin(\phi - ps')\frac{e^{-j\beta R}}{R} ds' \tag{7.30}$$

Finally, general electromagnetic field expressions for the class of (p,q)-torus knots may be obtained directly from the vector potential components given in (7.28)–(7.30) by using the following identities:

$$\vec{E} = \frac{1}{j\omega\mu\varepsilon}\left[\vec{\nabla}(\vec{\nabla}\cdot\vec{A}) + \beta^2\vec{A}\right] \tag{7.31}$$

$$\vec{H} = \frac{1}{\mu}\left[\vec{\nabla}\times\vec{A}\right] \tag{7.32}$$

We next turn our attention to the derivation of appropriate far-zone representations for the electromagnetic fields produced by a (p,q)-torus knot. It is well known that, in the far-zone, simplified relationships exist between the resultant electromagnetic fields and their associated vector potential. These relationships for the fields are given below in component form [14]:

$$E_r \approx 0 \tag{7.33}$$

$$E_\theta \approx -j\omega A_\theta \tag{7.34}$$

$$E_\phi \approx -j\omega A_\phi \tag{7.35}$$

$$H_r \approx 0 \tag{7.36}$$

$$H_\theta \approx j\frac{\omega}{\eta}A_\phi = -\frac{E_\phi}{\eta} \tag{7.37}$$

$$H_\phi \approx -j\frac{\omega}{\eta}A_\theta = \frac{E_\theta}{\eta} \tag{7.38}$$

Hence, in order to complete the derivation, it is only necessary to find far-zone approximations for the vector potential components A_θ and A_ϕ given in (7.29) and (7.30), respectively. In the far zone, it can be shown that

$$\frac{e^{-j\beta R}}{R} \approx \frac{e^{-j\beta r}}{r} e^{j\beta[x' \sin\theta\cos\phi + y' \sin\theta\sin\phi + z' \cos\theta]} \tag{7.39}$$

where

$$r = \sqrt{x^2 + y^2 + z^2} \tag{7.40}$$

At this point in the development, the knot parameterizations introduced in (7.9), (7.10), and (7.11) may be used to transform (7.39) into

$$\frac{e^{-j\beta R}}{R} \approx \frac{e^{-j\beta r}}{r}\Gamma(s') \tag{7.41}$$

where

$$\Gamma(s') = e^{j\beta[(a + b\cos(\psi + qs'))\cos(\phi - ps')\sin\theta + b\sin(\psi + qs')\cos\theta]} \tag{7.42}$$

Substituting (7.41) into (7.29) and (7.30) leads to the required far-zone representations for A_θ and A_ϕ which are given by

$$\begin{aligned} A_\theta(r,\theta,\phi) \approx {} & \frac{\mu\cos\theta}{4\pi}\frac{e^{-j\beta r}}{r}\left\{p\int_0^{2\pi} I_s(s')(a + b\cos(\psi + qs'))\sin(\phi - ps')\Gamma(s')\,ds' \right. \\ & \left. - qb\int_0^{2\pi} I_s(s')\sin(\psi + qs')\cos(\phi - ps')\Gamma(s')\,ds'\right\} \\ & - \frac{\mu\sin\theta}{4\pi}\frac{e^{-j\beta r}}{r}\left\{qb\int_0^{2\pi} I_s(s')\cos(\psi + qs')\Gamma(s')\,ds'\right\} \end{aligned} \tag{7.43}$$

$$\begin{aligned} A_\phi(r,\theta,\phi) \approx {} & \frac{\mu}{4\pi}\frac{e^{-j\beta r}}{r}p\int_0^{2\pi} I_s(s')(a + b\cos(\psi + qs'))\cos(\phi - ps')\Gamma(s')\,ds' \\ & + \frac{\mu}{4\pi}\frac{e^{-j\beta r}}{r}qb\int_0^{2\pi} I_s(s')\sin(\psi + qs')\sin(\phi - ps')\Gamma(s')\,ds' \end{aligned} \tag{7.30}$$

7.2.3 The Torus Knot EFIE

The derivation of an electric field integral equation (EFIE) for perfectly conducting knotted wires is outlined in this section. This derivation is specifically tailored to the development of an EFIE suitable for analyzing closed loops of thin wire bent in the shape of (p,q)-torus knots. A convenient place to start this development is with the general form of the EFIE for an arbitrarily curved wire [15], [16]

$$\int_C I(\ell')\left[\frac{\partial^2}{\partial\ell\,\partial\ell'}G(\ell,\ell') - \beta^2\hat{\ell}\cdot\hat{\ell}'\,G(\ell,\ell')\right]d\ell' = j\omega\varepsilon\hat{\ell}\cdot\vec{E}^i(\ell),\ \ell\in C \tag{7.45}$$

where $\vec{E}^i(\ell)$ represents the incident or impressed electric field and

$$G(\ell,\ell') = \frac{e^{-j\beta|\vec{r}-\vec{r}'|}}{4\pi|\vec{r}-\vec{r}'|} \tag{7.46}$$

is the free-space Green's function. Now suppose we define the source-point coordinates of any (p,q)-torus knot to be

$$x' = (a + b'\cos(\psi + qs'))\cos(ps') \tag{7.47}$$

$$y' = (a + b'\cos(\psi + qs'))\sin(ps') \tag{7.48}$$

$$z' = b'\sin(\psi + qs') \tag{7.49}$$

with the corresponding field-point coordinates defined as

$$x = (a + b\cos(\psi + qs))\cos(ps) \tag{7.50}$$

$$y = (a + b\cos(\psi + qs))\sin(ps) \tag{7.51}$$

$$z = b\sin(\psi + qs) \tag{7.52}$$

The set of knot parameterizations given in (7.47)–(7.49) and (7.50)–(7.52) may be used to obtain expressions for the unit vectors which are tangent to the curve of these knots. A useful form of these expressions was found to be

$$\hat{\ell} = \hat{T} = \frac{\vec{V}}{|\vec{V}|} \tag{7.53}$$

$$\hat{\ell}' = \hat{T}' = \frac{\vec{V}'}{|\vec{V}'|} \tag{7.54}$$

where

$$\vec{V} = \frac{d\vec{r}}{ds} = V_x\hat{x} + V_y\hat{y} + V_z\hat{z} \tag{7.55}$$

$$\vec{V}' = \frac{d\vec{r}'}{ds'} = V'_x\hat{x} + V'_y\hat{y} + V'_z\hat{z} \tag{7.56}$$

$$V_x = \frac{dx}{ds} = -p(a + b\cos(\psi + qs))\sin(ps) - qb\sin(\psi + qs)\cos(ps) \tag{7.57}$$

$$V_y = \frac{dy}{ds} = p(a + b\cos(\psi + qs))\cos(ps) - qb\sin(\psi + qs)\sin(ps) \tag{7.58}$$

$$V_z = \frac{dz}{ds} = qb\cos(\psi + qs) \tag{7.59}$$

$$V'_x = \frac{dx'}{ds'} = -p(a + b'\cos(\psi + qs'))\sin(ps') - qb'\sin(\psi + qs')\cos(ps') \tag{7.60}$$

$$V'_y = \frac{dy'}{ds'} = p(a + b'\cos(\psi + qs'))\cos(ps') - qb'\sin(\psi + qs')\sin(ps') \tag{7.61}$$

$$V'_z = \frac{dz'}{ds'} = qb'\cos(\psi + qs') \tag{7.62}$$

$$|\vec{V}| = \left|\frac{d\vec{r}}{ds}\right| = \sqrt{(qb)^2 + p^2(a + b\cos(\psi + qs))^2} \tag{7.63}$$

$$|\vec{V}'| = \left|\frac{d\vec{r}'}{ds'}\right| = \sqrt{(qb')^2 + p^2(a + b'\cos(\psi + qs'))^2} \tag{7.64}$$

Equations (7.47)–(7.49) and (7.50)–(7.52) may also be used to prove that the following relationship holds for (p, q)-torus knots:

$$\frac{\partial}{\partial \ell}\frac{\partial}{\partial \ell'}G(\ell, \ell') = \frac{\partial s}{\partial \ell}\frac{\partial s'}{\partial \ell'}\frac{\partial}{\partial s}\frac{\partial}{\partial s'}G(s, s') \tag{7.65}$$

where

$$\frac{\partial s}{\partial \ell} = \frac{1}{|\vec{V}|} = \frac{1}{\sqrt{(qb)^2 + p^2(a + b\cos(\psi + qs))^2}} \tag{7.66}$$

$$\frac{\partial s'}{\partial \ell'} = \frac{1}{|\vec{V}'|} = \frac{1}{\sqrt{(qb')^2 + p^2(a + b'\cos(\psi + qs'))^2}} \tag{7.67}$$

$$G(s, s') = \frac{e^{-j\beta R}}{4\pi R} \tag{7.68}$$

$$R = |\vec{R}| = |\vec{r} - \vec{r}'| = \sqrt{(x - x')^2 + (y - y')^2 + (z - z')^2} \tag{7.69}$$

$$\vec{R} = R_x\hat{x} + R_y\hat{y} + R_z\hat{z} \tag{7.70}$$

$$R_x = (a + b\cos(\psi + qs))\cos(ps) - (a + b'\cos(\psi + qs'))\cos(ps') \tag{7.71}$$

$$R_y = (a + b\cos(\psi + qs))\sin(ps) - (a + b'\cos(\psi + qs'))\sin(ps') \tag{7.72}$$

$$R_z = b\sin(\psi + qs) - b'\sin(\psi + qs') \tag{7.73}$$

Substituting (7.53)–(7.54) and (7.65)–(7.68) into (7.45) results in an EFIE of the form

$$\int_0^{2\pi} I_s(s')K(s,s')\,ds' = j\omega\varepsilon\vec{V}(s)\cdot\vec{E}^i(s), \qquad s\in[0,2\pi] \tag{7.74}$$

where $I_s(s')$ represents the current distribution on the surface of a particular (p,q)-torus knot and

$$K(s,s') = \left[\frac{\partial}{\partial s}\frac{\partial}{\partial s'} - \beta^2\vec{V}(s)\cdot\vec{V}'(s')\right]G(s,s') \tag{7.75}$$

is the kernel of the integral equation. It is possible to further simplify the EFIE given in (7.74) by evaluating the derivatives which appear in the expression for the kernel (7.75). This leads to the result

$$\int_0^{2\pi} I_s(s')\tilde{K}(s,s')\,ds' = \frac{j}{\eta}\vec{V}(s)\cdot\vec{E}^i(s), \qquad s\in[0,2\pi] \tag{7.76}$$

where

$$\begin{aligned}\tilde{K}(s,s') = \frac{1}{\beta}K(s,s') &= \Big\{(\beta R)^2\left[(1+j\beta R)-(\beta R)^2\right]F_1(s,s') \\ &\quad -\left[3(1+j\beta R)-(\beta R)^2\right]F_2(s,s')\Big\}\frac{e^{-j\beta R}}{4\pi(\beta R)^5}\end{aligned} \tag{7.77}$$

is a dimensionless form of the kernel in which

$$F_1(s,s') = \beta^2\vec{V}\cdot\vec{V}' = \beta^2\{V_xV_x' + V_yV_y' + V_zV_z'\} \tag{7.78}$$

and

$$\begin{aligned}F_2(s,s') &= (\beta^2\vec{R}\cdot\vec{V})(\beta^2\vec{R}\cdot\vec{V}') \\ &= \left[\beta^2\{R_xV_x + R_yV_y + R_zV_z\}\right] \\ &\quad \times\left[\beta^2\{R_xV_x' + R_yV_y' + R_zV_z'\}\right]\end{aligned} \tag{7.79}$$

The dependence on the knot parameters of (7.78) and (7.79) is implicit. However, explicit representations of these equations may also be found which are given by

$$\begin{aligned}F_1(s,s') &= q^2(\beta b)(\beta b')\cos(\psi+qs)\cos(\psi+qs') \\ &\quad + p^2\left[(\beta a)+(\beta b)\cos(\psi+qs)\right]\left[(\beta a)+(\beta b')\cos(\psi+qs')\right]\cos\left[p(s-s')\right] \\ &\quad + pq(\beta b')\left[(\beta a)+(\beta b)\cos(\psi+qs)\right]\sin(\psi+qs')\sin\left[p(s-s')\right] \\ &\quad - pq(\beta b)\left[(\beta a)+(\beta b')\cos(\psi+qs')\right]\sin(\psi+qs)\sin\left[p(s-s')\right] \\ &\quad + q^2(\beta b)(\beta b')\sin(\psi+qs)\sin(\psi+qs')\cos\left[p(s-s')\right]\end{aligned} \tag{7.80}$$

$$\begin{aligned} F_2(s,s') = &\{p[(\beta a) + (\beta b)\cos(\psi + qs)][(\beta a) + (\beta b')\cos(\psi + qs')]\sin[p(s - s')] \\ &- q(\beta b')[(\beta a) + (\beta b)\cos(\psi + qs)]\sin(\psi + qs')\cos[p(s - s')] \\ &+ q(\beta a)(\beta b')\sin(\psi + qs') + q(\beta b)(\beta b')\sin(\psi + qs)\cos(\psi + qs')\} \\ &\times \{p[(\beta a) + (\beta b)\cos(\psi + qs)][(\beta a) + (\beta b')\cos(\psi + qs')]\sin[p(s - s')] \\ &+ q(\beta b)[(\beta a) + (\beta b')\cos(\psi + qs')]\sin(\psi + qs)\cos[p(s - s')] \\ &- q(\beta a)(\beta b)\sin(\psi + qs) - q(\beta b)(\beta b')\cos(\psi + qs)\sin(\psi + qs')\} \end{aligned} \tag{7.81}$$

Finally, by using (7.69) together with (7.47)–(7.49) and (7.50)–(7.52), the remarkable fact that

$$R|_{s=s'} = |b - b'| \tag{7.82}$$

can easily be shown. This suggests that the knotted wire kernel (7.77) will be nonsingular provided $b \neq b'$. A useful interpretation of the quantity $|b - b'|$ found in (7.82) is that it represents the radius of the wire used to construct the knotted antenna or scatterer. Therefore, numerical solution of the EFIE given in (7.76) should yield accurate results for thin wire torus knots which satisfy the condition $|b - b'| \leq 0.01\lambda$.

The availability of the EFIE derived here is important for several reasons including the fact that it provides the basis for the development of accurate computational electromagnetics modeling techniques for knotted wires. Conventional Method of Moments (MoM) techniques generally approximate curved wires by a series of piecewise-linear wires which can lead to inaccuracies in the modeling results, especially for wires which are highly looped or knotted. This new EFIE provides an alternative to the traditional approaches by allowing a MoM formulation to be developed which is specifically tailored to the analysis of knotted wires. More accurate results could be obtained, for instance, by solving the knotted wire EFIE using either an entire domain or a curved basis function sub-domain MoM formulation.

7.3 SPECIAL CASES

7.3.1 Small Knot Approximation

Simple closed-form expressions are derived in this section for the far-zone electromagnetic fields of electrically small torus knots. Suppose we let $b = \alpha a$ where $0 \leq \alpha \leq \frac{1}{2}$, then we may write (7.42) as

$$\Gamma(s') = e^{j\beta a[(1 + \alpha\cos(\psi + qs'))\cos(\phi - ps')\sin\theta + \alpha\sin(\psi + qs')\cos\theta]} \tag{7.83}$$

For sufficiently small values of α, (7.83) may be approximated by

$$\begin{aligned} \Gamma(s') \approx 1 &+ j\beta a(1 + \alpha\cos(\psi + qs'))\cos(\phi - ps')\sin\theta \\ &+ j\beta a\alpha\sin(\psi + qs')\cos\theta \end{aligned} \tag{7.84}$$

The current distribution on an electrically small torus knot may be assumed to be uniform, i.e., $I_s(s') = I_0$ where I_0 is constant. Hence, under these conditions, the far-zone expressions for A_θ and A_ϕ given in (7.43) and (7.44), respectively, reduce to

$$A_\theta(r,\theta,\phi) \approx \frac{\mu a I_0 \cos\theta}{4\pi}\frac{e^{-j\beta r}}{r}\left\{p\int_0^{2\pi}(1+\alpha\cos(\psi+qs'))\sin(\phi-ps')\Gamma(s')\,ds'\right.$$
$$\left. - q\alpha\int_0^{2\pi}\sin(\psi+qs')\cos(\phi-ps')\Gamma(s')\,ds'\right\} \tag{7.85}$$
$$-\frac{\mu a I_0 \sin\theta}{4\pi}\frac{e^{-j\beta r}}{r}\left\{q\alpha\int_0^{2\pi}\cos(\psi+qs')\Gamma(s')\,ds'\right\}$$

$$A_\phi(r,\theta,\phi) \approx \frac{\mu a I_0}{4\pi}\frac{e^{-j\beta r}}{r}p\int_0^{2\pi}(1+\alpha\cos(\psi+qs'))\cos(\phi-ps')\Gamma(s')\,ds'$$
$$+\frac{\mu a I_0}{4\pi}\frac{e^{-j\beta r}}{r}q\alpha\int_0^{2\pi}\sin(\psi+qs')\sin(\phi-ps')\Gamma(s')\,ds' \tag{7.86}$$

Next, by substituting (7.84) into (7.85)–(7.86) and performing the required integration, we arrive at the following closed-form small-knot approximations:

$$A_\theta \approx 0 \tag{7.87}$$

$$A_\phi \approx \frac{j\mu\beta I_0 \sin\theta}{4}p\left[a^2+\frac{b^2}{2}\right]\frac{e^{-j\beta r}}{r} \tag{7.88}$$

which are valid provided $p \neq q$ and $p \neq 2q$. Finally, the far-field representations for small knots associated with (7.87) and (7.88) may be obtained directly from (7.33)–(7.35). The resulting closed-form expressions are

$$E_r \approx 0 \tag{7.89}$$

$$E_\theta \approx 0 \tag{7.90}$$

$$E_\phi \approx \frac{\eta\beta^2 I_0 \sin\theta}{4}p\left[a^2+\frac{b^2}{2}\right]\frac{e^{-j\beta r}}{r} \tag{7.91}$$

$$H_r \approx 0 \tag{7.92}$$

$$H_\theta \approx -\frac{\beta^2 I_0 \sin\theta}{4}p\left[a^2+\frac{b^2}{2}\right]\frac{e^{-j\beta r}}{r} \tag{7.93}$$

$$H_\phi \approx 0 \tag{7.94}$$

where $p \neq q$ and $p \neq 2q$. It is interesting to note that (7.89)–(7.94) are equivalent to the far fields which would be produced by an electrically small circular loop (canonical unknot) with an effective radius r_e and turns ratio N given by

$$r_e = \sqrt{a^2+\frac{b^2}{2}} = a\sqrt{1+\frac{\alpha^2}{2}} \tag{7.95}$$

$$N = p \tag{7.96}$$

This also suggests that the radiation resistance for a small torus knot will have the form

$$R_r = 20\pi^2 p^2 \left[(\beta a)^2 + \frac{(\beta b)^2}{2} \right]^2 \Omega \tag{7.97}$$

7.3.2 The Canonical Unknot

It was pointed out in Section 7.2.1 that the canonical unknot may be considered as a special limiting case of a torus knot. In particular, the canonical unknot is obtained as a degenerate form of a torus knot when $b = 0$ and $p = 1$. Hence, for this special case, the general vector potential expressions for the (p, q)-torus knots derived in (7.28)–(7.30) will reduce to

$$A_r(r, \theta, \phi) = \frac{\mu a \sin \theta}{4\pi} \int_0^{2\pi} I_s(\phi') \sin(\phi - \phi') \frac{e^{-j\beta R}}{R} d\phi' \tag{7.98}$$

$$A_\theta(r, \theta, \phi) = \frac{\mu a \cos \theta}{4\pi} \int_0^{2\pi} I_s(\phi') \sin(\phi - \phi') \frac{e^{-j\beta R}}{R} d\phi' \tag{7.99}$$

$$A_\phi(r, \theta, \phi) = \frac{\mu a}{4\pi} \int_0^{2\pi} I_s(\phi') \cos(\phi - \phi') \frac{e^{-j\beta R}}{R} d\phi' \tag{7.100}$$

where

$$R = \sqrt{r^2 + a^2 - 2ar \sin \theta \cos(\phi - \phi')} \tag{7.101}$$

These are, as expected, the well-known results for a circular loop antenna of radius a [17].

7.4 ELLIPTICAL TORUS KNOTS

7.4.1 Background

The work originally reported in [18], [19], [23] and reviewed in Sections 7.1.3 of this chapter represents the first attempt to establish a rigorous mathematical foundation from which analysis techniques may be developed and applied toward the study of knot electrodynamics problems. The analysis methodology presented in Section 7.2 was illustrated by considering a special class of knots, known as (p, q)-torus knots, which have interesting topological as well as electromagnetic properties. It was demonstrated that the well-known trefoil originally studied by Manuar and Jaggard [7]–[10] is one important example of a (p, q)-torus knot. A useful set of parametric representations were introduced in Section 7.2 for this particular family of knots by making use of the fact that they may be constructed on the surface of a standard circular torus in $\mathbb{R}^3$. These parameterizations were then used in combination with Maxwell's equations to derive the vector potential and corresponding field expressions which describe the radiation and scattering of electromagnetic waves from

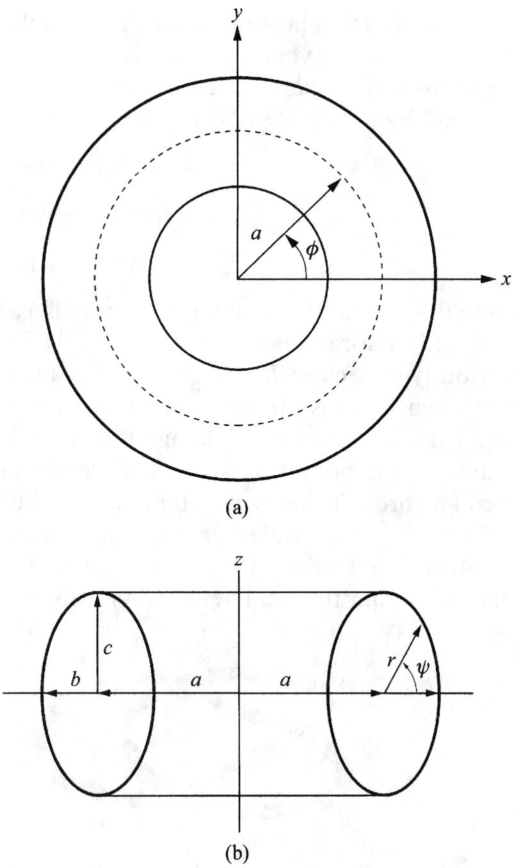

Figure 7.8 Geometrical configuration for a torus with an elliptical cross-section. The top view is shown in (a) and a side view is shown in (b).

circular (p, q)-torus knots. A new electric field integral equation (EFIE) which is well suited for the analysis of thin toroidally knotted wires was also derived in Section 7.2.3.

The constructions considered in the previous sections were restricted to knots which reside on the surface of a torus which has a circular cross section. Here we extend the ideas presented in the previous sections by developing a more general set of parameterizations which are valid not only for circular but also for elliptical (p, q)-torus knots [20]. These knots reside on the surface of a torus which has, in general, an elliptical cross section. The geometry for such an elliptical torus is illustrated in Fig. 7.8. Suppose we let ∂S denote the boundary or surface of the elliptical torus shown in Fig. 7.8, then the Cartesian coordinates for any point $(x, y, z) \in \partial S$ may be represented by

$$x = (a + b\cos\psi)\cos\phi \tag{7.102}$$

$$y = (a + b\cos\psi)\sin\phi \tag{7.103}$$

$$z = c\sin\psi \tag{7.104}$$

where $0 \le \psi,\ \phi \le 2\pi$.

As already discussed in Section 7.2, torus knots are classified by the integers p

and q which are relatively prime. The numbers p and q tell us how many times a particular knot traverses the torus in the longitudinal and meridional directions, respectively. The class of knots which live on the surface of the elliptical torus described by (7.102)–(7.104) can be shown to have parameterizations given by

$$x = (a + b\cos(\psi + qs))\cos(ps) \tag{7.105}$$

$$y = (a + b\cos(\psi + qs))\sin(ps) \tag{7.106}$$

$$z = c\sin(\psi + qs) \tag{7.107}$$

where $0 \leq s \leq 2\pi$. It is of interest to note here that when $c = b$ the parameterizations for elliptical torus knots given in (7.105)–(7.107) will reduce to those considered previously in Section 7.2 for circular torus knots. Hence, from a geometrical point of view, circular torus knots may be considered as a special case of the more general elliptical torus knots, even though topologically speaking they are equivalent. This is a subtle yet important distinction since the introduction of an additional degree of freedom through the parameter c allows the electromagnetic characteristics to be studied for a much wider class of knot geometries. Three different geometries for a (2,3)-torus knot (trefoil) are compared in Fig. 7.9. These three knotted curves were generated from the parameterizations given in (7.105)–(7.107) by choosing $c = b/4$, $c = b$, and $c = 4b$.

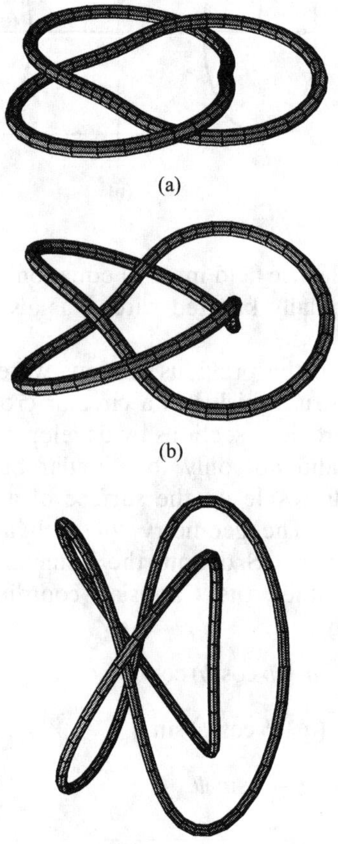

(a)

(b)

(c)

Figure 7.9 A comparison of three different trefoil knot geometries generated by choosing (a) $c = b/4$, (b) $c = b$, and (c) $c = 4b$ for a fixed value of a.

7.4.2 Electromagnetic Fields

Vector potential expressions for the class of elliptical (p, q)-torus knots may be derived directly from (7.7) by employing the parameterizations introduced in (7.105)–(7.107). This approach requires that the knot parameterizations be represented in terms of the source coordinates as

$$x' = (a + b\cos(\psi + qs'))\cos(ps') \tag{7.108}$$

$$y' = (a + b\cos(\psi + qs'))\sin(ps') \tag{7.109}$$

$$z' = c\sin(\psi + qs') \tag{7.110}$$

The next step requires finding a suitable expression for the incremental length element dl' which applies to the particular family of elliptical (p, q)-torus knots. This is accomplished by making use of the knot parameterizations given in (7.108)–(7.110) to show that

$$dl' = \sqrt{p^2(a + b\cos(\psi + qs'))^2 + q^2(b^2\sin^2(\psi + qs') + c^2\cos^2(\psi + qs'))}\,ds' \tag{7.111}$$

This suggests that the arclength L_{pq} of an elliptical torus knot may be obtained directly from (7.111) by integrating over the curve of the knot. In other words, integrating both sides of (7.111) over the appropriate limits leads to a generalized formula for arclength L_{pq} given by

$$L_{pq} = 2\int_0^{\pi} \sqrt{p^2(a + b\cos u)^2 + q^2(b^2\sin^2 u + c^2\cos^2 u)}\,du \tag{7.112}$$

At this stage in the development, we seek to derive a specialized integral representation for the elliptical (p, q)-torus knot vector potential by using (7.7) in conjunction with (7.108)–(7.110) and (7.111). This may be accomplished by following a similar procedure to that outlined in Section 7.2. The resulting expressions for the three components of the vector potential, transformed to spherical coordinates, are given below.

$$\begin{aligned} A_r(r,\theta,\phi) = &\frac{\mu\sin\theta}{4\pi}\left\{p\int_0^{2\pi} I_s(s')(a + b\cos(\psi + qs'))\sin(\phi - ps')\frac{e^{-j\beta R}}{R}ds'\right. \\ &\left. - qb\int_0^{2\pi} I_s(s')\sin(\psi + qs')\cos(\phi - ps')\frac{e^{-j\beta R}}{R}ds'\right\} \\ &+ \frac{\mu\cos\theta}{4\pi}\left\{qc\int_0^{2\pi} I_s(s')\cos(\psi + qs')\frac{e^{-j\beta R}}{R}ds'\right\} \end{aligned} \tag{7.113}$$

$$\begin{aligned} A_\theta(r,\theta,\phi) = &\frac{\mu\cos\theta}{4\pi}\left\{p\int_0^{2\pi} I_s(s')(a + b\cos(\psi + qs'))\sin(\phi - ps')\frac{e^{-j\beta R}}{R}ds'\right. \\ &\left. - qb\int_0^{2\pi} I_s(s')\sin(\psi + qs')\cos(\phi - ps')\frac{e^{-j\beta R}}{R}ds'\right\} \\ &- \frac{\mu\sin\theta}{4\pi}\left\{qc\int_0^{2\pi} I_s(s')\cos(\psi + qs')\frac{e^{-j\beta R}}{R}ds'\right\} \end{aligned} \tag{7.114}$$

$$A_\phi(r,\theta,\phi) = \frac{\mu}{4\pi}\left\{p\int_0^{2\pi} I_s(s')(a+b\cos(\psi+qs'))\cos(\phi-ps')\frac{e^{-j\beta R}}{R}\,ds'\right.$$

$$\left. + qb\int_0^{2\pi} I_s(s')\sin(\psi+qs')\sin(\phi-ps')\frac{e^{-j\beta R}}{R}\,ds'\right\} \tag{7.115}$$

where $I_s(s')$ represents the current distribution on the surface of the knotted wire. These expressions for the vector potential are valid everywhere, including in the near-zone of the knots. However, the approximation

$$\frac{e^{-j\beta R}}{R} \approx \frac{e^{-j\beta r}}{r}\Gamma(s') \tag{7.116}$$

where

$$\Gamma(s') = e^{j\beta[(a+b\cos(\psi+qs'))\cos(\phi-ps')\sin\theta + c\sin(\psi+qs')\cos\theta]} \tag{7.117}$$

may be used to find simplified far-zone vector potential representations. In particular, by substituting (7.116) into (7.114) and (7.115) we obtain the following far-zone approximations for A_θ and A_ϕ:

$$A_\theta(r,\theta,\phi) \approx \frac{\mu\cos\theta}{4\pi}\frac{e^{-j\beta r}}{r}\left\{p\int_0^{2\pi} I_s(s')(a+b\cos(\psi+qs'))\sin(\phi-ps')\Gamma(s')\,ds'\right.$$

$$\left. - qb\int_0^{2\pi} I_s(s')\sin(\psi+qs')\cos(\phi-ps')\Gamma(s')\,ds'\right\} \tag{7.118}$$

$$-\frac{\mu\sin\theta}{4\pi}\frac{e^{-j\beta r}}{r}\left\{qc\int_0^{2\pi} I_s(s')\cos(\psi+qs')\Gamma(s')\,ds'\right\}$$

$$A_\phi(r,\theta,\phi) \approx \frac{\mu}{4\pi}\frac{e^{-j\beta r}}{r}\left\{p\int_0^{2\pi} I_s(s')(a+b\cos(\psi+qs'))\cos(\phi-ps')\Gamma(s')\,ds'\right.$$

$$\left. + qb\int_0^{2\pi} I_s(s')\sin(\psi+qs')\sin(\phi-ps')\Gamma(s')\,ds'\right\} \tag{7.119}$$

Finally, far-zone expressions for the electromagnetic fields produced by elliptical torus knots may be obtained directly from (7.118) and (7.119) by making use of the well-known relationships given in (7.33)–(7.38).

7.5 ADDITIONAL SPECIAL CASES

7.5.1 Circular Torus Knots

The first special case that will be considered involves knots which are formed by winding a piece of wire around the surface of a virtual circular torus. The circular torus knots can be thought of as resulting from a degenerate form of the elliptical torus knots, namely, when $c = b$. In this case, the expressions for the vector potential given in (7.113)–(7.115) will reduce to (7.28)–(7.30), respectively. Likewise, approximate far-zone expressions for the vector potential components associated with circular torus knots may be obtained by setting $c = b$ in (7.118) and (7.119).

7.5.2 Small-Knot Approximation (*b* and *c* are Functions of *a*)

Suppose we assume that b and c are proportional to a such that $b = \alpha a$ and $c = \gamma a$ where $0 \leq \alpha \leq 1/2$. In this case (7.117) may be written as

$$\Gamma(s') = e^{j\beta a[(1+\alpha\cos(\psi+qs'))\cos(\phi-ps')\sin\theta+\gamma\sin(\psi+qs')\cos\theta]} \tag{7.120}$$

When a is small enough, (7.120) can be approximated by

$$\begin{aligned}\Gamma(s') \approx 1 + j\beta a(1+\alpha\cos(\psi+qs'))\cos(\phi-ps')\sin\theta \\ + j\beta a\gamma\sin(\psi+qs')\cos\theta\end{aligned} \tag{7.121}$$

Another approximation which may be made in the case of electrically small torus knots is that the current distribution on these knots will be uniform. Therefore, by setting $I_s(s') = I_0$ where I_0 is a constant, the far-zone vector potential expressions given in (7.118) and (7.119) may be reduced to the form

$$\begin{aligned}A_\theta(r,\theta,\phi) = \frac{\mu a I_0\cos\theta}{4\pi}\frac{e^{-j\beta r}}{r}\left\{p\int_0^{2\pi}(1+\alpha\cos(\psi+qs'))\sin(\phi-ps')\Gamma(s')\,ds'\right. \\ \left. - q\alpha\int_0^{2\pi}\sin(\psi+qs')\cos(\phi-ps')\Gamma(s')\,ds'\right\} \\ -\frac{\mu a I_0\sin\theta}{4\pi}\frac{e^{-j\beta r}}{r}\left\{q\gamma\int_0^{2\pi}\cos(\psi+qs')\Gamma(s')\,ds'\right\}\end{aligned} \tag{7.122}$$

$$\begin{aligned}A_\phi(r,\theta,\phi) = \frac{\mu a I_0}{4\pi}\frac{e^{-j\beta r}}{r}\left\{p\int_0^{2\pi}(1+\alpha\cos(\psi+qs'))\cos(\phi-ps')\Gamma(s')\,ds'\right. \\ \left. + q\alpha\int_0^{2\pi}\sin(\psi+qs')\sin(\phi-ps')\Gamma(s')\,ds'\right\}\end{aligned} \tag{7.123}$$

Finally, simple closed-form representations for the fields produced by electrically small torus knots may be found by substituting (7.121) into (7.122) and (7.123) and evaluating the required integrals. The general case and two special cases ($p = q$ and $p = 2q$) of these small-knot approximations are considered in Sections 7.5.2.1–7.5.2.3.

7.5.2.1 General Case ($p \neq q$ *and* $p \neq 2q$)

The general far-zone expressions presented in this section are valid for all values of p and q, with the exception of the two special cases when $p = q$ and $p = 2q$. For this reason, the general expressions given below hold for most cases of practical interest. It can be shown that the general form of the small-knot far-zone vector potential is

$$A_\theta \approx 0 \tag{7.124}$$

$$A_\phi \approx j\frac{\mu\beta I_0\sin\theta}{4}p\left[a^2+\frac{b^2}{2}\right]\frac{e^{-j\beta r}}{r} \tag{7.125}$$

Using (7.33)–(7.38), the far-field representations for these small knots can then be determined to be

$$E_r \approx 0 \tag{7.126}$$

$$E_\theta \approx 0 \tag{7.127}$$

$$E_\phi \approx \frac{\eta\beta^2 I_0 \sin\theta}{4} p\left[a^2 + \frac{b^2}{2}\right]\frac{e^{-j\beta r}}{r} \tag{7.128}$$

$$H_r \approx 0 \tag{7.129}$$

$$H_\theta \approx -\frac{\beta^2 I_0 \sin\theta}{4} p\left[a^2 + \frac{b^2}{2}\right]\frac{e^{-j\beta r}}{r} \tag{7.130}$$

$$H_\phi \approx 0 \tag{7.131}$$

7.5.2.2 Special Case when p = q

When $p = q$ the following approximations hold for the vector potential:

$$A_\theta \approx -j\frac{\mu I_0}{4}\frac{e^{-j\beta r}}{r} p\beta a^2 \gamma \cos(\psi + \phi) \tag{7.132}$$

$$A_\phi \approx j\frac{\mu I_0}{4}\frac{e^{-j\beta r}}{r} p\beta\left\{\left[a^2 + \frac{b^2}{2}\right]\sin\theta + a^2\gamma\cos\theta\sin(\psi + \phi)\right\} \tag{7.133}$$

The corresponding far-field expressions in this case are

$$E_r \approx 0 \tag{7.134}$$

$$E_\theta \approx -\frac{\eta\beta^2 I_0}{4}\frac{e^{-j\beta r}}{r} pa^2\gamma\cos(\psi + \phi) \tag{7.135}$$

$$E_\phi \approx \frac{\eta\beta^2 I_0}{4}\frac{e^{-j\beta r}}{r} p\left\{\left[a^2 + \frac{b^2}{2}\right]\sin\theta + a^2\gamma\cos\theta\sin(\psi + \phi)\right\} \tag{7.136}$$

$$H_r \approx 0 \tag{7.137}$$

$$H_\theta \approx -\frac{\beta^2 I_0}{4}\frac{e^{-j\beta r}}{r} p\left\{\left[a^2 + \frac{b^2}{2}\right]\sin\theta + a^2\gamma\cos\theta\sin(\psi + \phi)\right\} \tag{7.138}$$

$$H_\phi \approx -\frac{\beta^2 I_0}{4}\frac{e^{-j\beta r}}{r} pa^2\gamma\cos(\psi + \phi) \tag{7.139}$$

7.5.2.3 Special Case when p = 2q

When $p = 2q$ the vector potentials are described by the equations

$$A_\theta \approx -j\frac{\mu I_0}{4}\frac{e^{-j\beta r}}{r}\frac{p}{4}\beta a^2\gamma\alpha\cos(2\psi + \phi) \tag{7.140}$$

$$A_\phi \approx j\frac{\mu I_0}{4}\frac{e^{-j\beta r}}{r} p\beta\left\{\left[a^2 + \frac{b^2}{2}\right]\sin\theta + \frac{1}{4}a^2\gamma\alpha\cos\theta\sin(2\psi + \phi)\right\} \tag{7.141}$$

which give rise to the following far-field representations:

$$E_r \approx 0 \tag{7.142}$$

$$E_\theta \approx -\frac{\eta\beta^2 I_0}{4}\frac{e^{-j\beta r}}{r} pa^2 \gamma\alpha \cos(2\psi + \phi) \tag{7.143}$$

$$E_\phi \approx \frac{\eta\beta^2 I_0}{4}\frac{e^{-j\beta r}}{r} p \left\{ \left[a^2 + \frac{b^2}{2} \right] \sin\theta + \frac{1}{4} a^2 \gamma\alpha \cos\theta \sin(2\psi + \phi) \right\} \tag{7.144}$$

$$H_r \approx 0 \tag{7.145}$$

$$H_\theta \approx -\frac{\beta^2 I_0}{4}\frac{e^{-j\beta r}}{r} p \left\{ \left[a^2 + \frac{b^2}{2} \right] \sin\theta + \frac{1}{4} a^2 \gamma\alpha \cos\theta \sin(2\psi + \phi) \right\} \tag{7.146}$$

$$H_\phi \approx -\frac{\beta^2 I_0 a}{4}\frac{e^{-j\beta r}}{r} pa^2 \gamma\alpha \cos(2\psi + \phi) \tag{7.147}$$

7.5.3 Small-Knot Approximations for Circular Torus Knots (*b* and *c* are Functions of *a*)

We first make note of the fact that the far-zone representations which appear in Section 7.5.2.1 are independent of γ or c. This means that, at least for the general case, the far-zone expressions for circular torus knots will be exactly the same as those derived for the elliptical torus knots in Section 7.5.2.1. However, this does not turn out to be true for the two special cases when $p = q$ and $p = 2q$. These special cases are considered separately in Sections 7.5.3.1 and 7.5.3.2.

7.5.3.1 Special Case when p = q and $\gamma = \alpha$

Far-zone approximations for the two vector potential components can be readily obtained for this case by setting $\gamma = \alpha$ (i.e., $c = b$) in (7.132) and (7.133). The expressions which result are

$$A_\theta \approx -j\frac{\mu I_0}{4}\frac{e^{-j\beta r}}{r} p\beta ab \cos(\psi + \phi) \tag{7.148}$$

$$A_\phi \approx j\frac{\mu I_0}{4}\frac{e^{-j\beta r}}{r} p\beta \left\{ \left[a^2 + \frac{b^2}{2} \right] \sin\theta + ab \cos\theta \sin(\psi + \phi) \right\} \tag{7.149}$$

7.5.3.2 Special Case when p = 2q and $\gamma = \alpha$

To find appropriate expressions for the two components of the vector potential in this case, we set $\gamma = \alpha$ in (7.140) and (7.141). This yields

$$A_\theta \approx -j\frac{\mu I_0}{4}\frac{e^{-j\beta r}}{r}\frac{p}{4} \beta b^2 \cos(2\psi + \phi) \tag{7.150}$$

$$A_\phi \approx j\frac{\mu I_0}{4}\frac{e^{-j\beta r}}{r} p\beta \left\{ \left[a^2 + \frac{b^2}{2} \right] \sin\theta + \frac{b^2}{4} \cos\theta \sin(2\psi + \phi) \right\} \tag{7.151}$$

7.5.4 Small-Knot Approximation (*c* is Independent of *a* and *b*)

The next special case that will be considered is that in which the horizontal radii a and b of the torus are small in comparison to the vertical radius c. In this instance, we let b be proportional to a such that $b = \alpha a$ where $0 \leq \alpha \leq \frac{1}{2}$. We also make the assumption that c is independent of both a and b. Hence, for small enough values of a, the following approximation for (7.117) can be justified:

$$\Gamma(s') \approx [1 + j\beta a(1 + \alpha\cos(\psi + qs'))\cos(\phi - ps')\sin\theta]e^{j\beta c\sin(\psi + qs')\cos\theta} \quad (7.152)$$

Substituting (7.152) into (7.122) and (7.123) and evaluating the required integrals lead to convenient closed-form expressions for A_θ and A_ϕ. The resulting expressions for the general case as well as three special cases are summarized in the following four sections.

7.5.4.1 *General Case* [p/q ≠ 2n, p/q ≠ 2n − 1 *and* p/q ≠ (2n − 1)/2]

The general expressions for the far-zone vector potential and fields presented in this section are valid for the majority of possible torus knot configurations. It can be shown that the general form of the far-zone vector potential components A_θ and A_ϕ are

$$A_\theta \approx 0 \quad (7.153)$$

$$A_\phi \approx j\frac{\mu I_0}{4}\frac{e^{-j\beta r}}{r}p\beta a^2\sin\theta\left\{J_0(x) + \frac{\alpha^2}{2}[J_0(x) + J_2(x)]\right\} \quad (7.154)$$

where

$$x = \beta c\cos\theta \quad (7.155)$$

These expressions hold provided $p/q \neq 2n$, $p/q \neq 2n - 1$ and $p/q \neq (2n-1)/2$ where $n \in \mathbb{N}$. Far-zone representations for the electromagnetic fields in the general case may be obtained directly from (7.153) and (7.154) by making use of (7.33)–(7.38). The resulting far-field expressions are

$$E_r \approx 0 \quad (7.156)$$

$$E_\theta \approx 0 \quad (7.157)$$

$$E_\phi \approx \frac{\eta\beta^2 I_0}{4}\frac{e^{-j\beta r}}{r}pa^2\sin\theta\left\{J_0(x) + \frac{\alpha^2}{2}[J_0(x) + J_2(x)]\right\} \quad (7.158)$$

$$H_r \approx 0 \quad (7.159)$$

$$H_\theta \approx -\frac{\beta^2 I_0}{4}\frac{e^{-j\beta r}}{r}pa^2\sin\theta\left\{J_0(x) + \frac{\alpha^2}{2}[J_0(x) + J_2(x)]\right\} \quad (7.160)$$

$$H_\phi \approx 0 \quad (7.161)$$

7.5.4.2 Special Case when p/q = 2n

The following closed-form expressions for the vector potential can be derived for the special case when $p/q = 2n$ and $n \in \mathbb{N}$:

$$
\begin{aligned}
A_\theta \approx & \frac{\mu a I_0}{4}\frac{e^{-j\beta r}}{r} 2p\cos\theta\sin\left(\phi + \frac{p}{q}\psi\right)J_{p/q}(x) \\
& + j\frac{\mu I_0}{4}\frac{e^{-j\beta r}}{r}\beta a^2\sin\theta\left\{\cos\theta\sin\left(2\phi + \frac{2p}{q}\psi\right)\left[p\left(1+\frac{\alpha^2}{2}\right)J_{2p/q}(x)\right.\right. \\
& + p\frac{\alpha^2}{4}\left(J_{2(p+q)/q}(x) + J_{2(p-q)/q}(x)\right) \\
& \left. + q\frac{\alpha^2}{4}\left(J_{2(p+q)/q}(x) - J_{2(p-q)/q}(x)\right)\right] \\
& \left. - q\gamma\alpha\sin\theta\cos\left(\phi + \frac{p}{q}\psi\right)\left[J_{p/q}(x) + \frac{1}{2}\left(J_{(p+2q)/q}(x) + J_{(p-2q)/q}(x)\right)\right]\right\}
\end{aligned}
\tag{7.162}
$$

$$
\begin{aligned}
A_\phi \approx & \frac{\mu I_0}{4}\frac{e^{-j\beta r}}{r} 2ap\cos\left(\phi + \frac{p}{q}\psi\right)J_{p/q}(x) \\
& + j\frac{\mu I_0}{4}\frac{e^{-j\beta r}}{r}\beta a^2\sin\theta\left\{p\left[J_0(x) + \frac{\alpha^2}{2}(J_0(x) + J_2(x))\right.\right. \\
& + \cos\left(2\phi + \frac{2p}{q}\psi\right)\left(\left(1+\frac{\alpha^2}{2}\right)J_{2p/q}(x)\right. \\
& \left.\left. + \frac{\alpha^2}{4}\left(J_{2(p+q)/q}(x) + J_{2(p-q)/q}(x)\right)\right)\right] \\
& \left. + q\frac{\alpha^2}{4}\cos\left(2\phi + \frac{2p}{q}\psi\right)\left(J_{2(p+q)/q}(x) - J_{2(p-q)/q}(x)\right)\right\}
\end{aligned}
\tag{7.163}
$$

The associated far-field expressions can then be determined from (7.162) and (7.163). This yields

$$E_r \approx 0 \tag{7.164}$$

$$
\begin{aligned}
E_\theta \approx & -j\frac{\eta\beta a I_0}{4}\frac{e^{-j\beta r}}{r} 2p\cos\theta\sin\left(\phi + \frac{p}{q}\psi\right)J_{p/q}(x) \\
& + \frac{\eta\beta^2 I_0}{4}\frac{e^{-j\beta r}}{r} a^2\sin\theta\left\{\cos\theta\sin\left(2\phi + \frac{2p}{q}\psi\right)\left[p\left(1+\frac{\alpha^2}{2}\right)J_{2p/q}(x)\right.\right. \\
& \left. + p\frac{\alpha^2}{4}\left(J_{2(p+q)/q}(x) + J_{2(p-q)/q}(x)\right) + q\frac{\alpha^2}{4}\left(J_{2(p+q)/q}(x) - J_{2(p-q)/q}(x)\right)\right] \\
& \left. - q\gamma\alpha\sin\theta\cos\left(\phi + \frac{p}{q}\psi\right)\left[J_{p/q}(x) + \frac{1}{2}\left(J_{(p+2q)/q}(x) + J_{(p-2q)/q}(x)\right)\right]\right\}
\end{aligned}
\tag{7.165}
$$

$$\begin{aligned}E_\phi \approx & -j\frac{\eta\beta I_0}{4}\frac{e^{-j\beta r}}{r}2ap\cos\left(\phi+\frac{p}{q}\psi\right)J_{p/q}(x)\\ & +\frac{\eta\beta^2 I_0}{4}\frac{e^{-j\beta r}}{r}a^2\sin\theta\left\{p\left[J_0(x)+\frac{\alpha^2}{2}(J_0(x)+J_2(x))\right.\right.\\ & +\cos\left(2\phi+\frac{2p}{q}\psi\right)\left(\left(1+\frac{\alpha^2}{2}\right)J_{2p/q}(x)\right.\\ & \left.\left.+\frac{\alpha^2}{4}\left(J_{2(p+q)/q}(x)+J_{2(p-q)/q}(x)\right)\right)\right]\\ & \left.+q\frac{\alpha^2}{4}\cos\left(2\phi+\frac{2p}{q}\psi\right)\left(J_{2(p+q)/q}(x)-J_{2(p-q)/q}(x)\right)\right\}\end{aligned} \tag{7.166}$$

$$H_r \approx 0 \tag{7.167}$$

$$\begin{aligned}H_\theta \approx & \; j\frac{\beta I_0}{4}\frac{e^{-j\beta r}}{r}2ap\cos\left(\phi+\frac{p}{q}\psi\right)J_{p/q}(x)\\ & -\frac{\beta^2 I_0}{4}\frac{e^{-j\beta r}}{r}a^2\sin\theta\left\{p\left[J_0(x)+\frac{\alpha^2}{2}(J_0(x)+J_2(x))\right.\right.\\ & +\cos\left(2\phi+\frac{2p}{q}\psi\right)\left(\left(1+\frac{\alpha^2}{2}\right)J_{2p/q}(x)\right.\\ & \left.\left.+\frac{\alpha^2}{4}\left(J_{2(p+q)/q}(x)+J_{2(p-q)/q}(x)\right)\right)\right]\\ & \left.+q\frac{\alpha^2}{4}\cos\left(2\phi+\frac{2p}{q}\psi\right)\left(J_{2(p+q)/q}(x)-J_{2(p-q)/q}(x)\right)\right\}\end{aligned} \tag{7.168}$$

$$\begin{aligned}H_\phi \approx & -j\frac{\beta a I_0}{4}\frac{e^{-j\beta r}}{r}2p\cos\theta\sin\left(\phi+\frac{p}{q}\psi\right)J_{p/q}(x)\\ & +\frac{\beta^2 I_0}{4}\frac{e^{-j\beta r}}{r}a^2\sin\theta\left\{\cos\theta\sin\left(2\phi+\frac{2p}{q}\psi\right)\left[p\left(1+\frac{\alpha^2}{2}\right)J_{2p/q}(x)\right.\right.\\ & \left.+p\frac{\alpha^2}{4}\left(J_{2(p+q)/q}(x)+J_{2(p-q)/q}(x)\right)+q\frac{\alpha^2}{4}\left(J_{2(p+q)/q}(x)-J_{2(p-q)/q}(x)\right)\right]\\ & \left.-q\gamma\alpha\sin\theta\cos\left(\phi+\frac{p}{q}\psi\right)\left[J_{p/q}(x)+\frac{1}{2}\left(J_{(p+2q)/q}(x)+J_{(p-2q)/q}(x)\right)\right]\right\}\end{aligned} \tag{7.169}$$

The geometry for the special cases in which $p = 2$ and $q = 1$ ($p/q = 2$) and $p = 4$ and $q = 1$ ($p/q = 4$) is illustrated in Figs. 7.10(a) and 7.10(b), respectively. It is easy to show that these knots are topologically equivalent to the standard circular loop (canonical unknot) and, therefore, both may be classified as trivial knots.

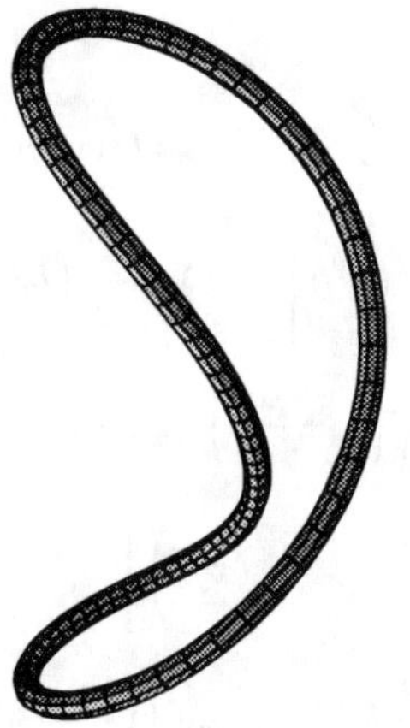

(a)

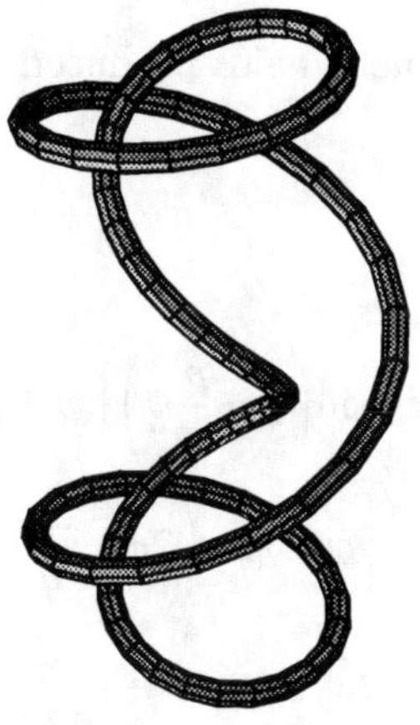

(b)

Figure 7.10 Special elliptical torus knot geometries. A trivial (2, 1)-torus knot is shown in (a) while a trivial (4, 1)-torus knot is shown in (b).

7.5.4.3 Special Case when p/q = 2n − 1

For this special case where $p/q = 2n - 1$ and $n \in \mathbb{N}$, the closed-form expressions for the vector potential which result are given by

$$A_\theta \approx \frac{\mu a I_0}{4} \frac{e^{-j\beta r}}{r} \alpha \cos\theta \sin\left(\phi + \frac{p}{q}\psi\right) \left\{(p+q) J_{(p+q)/q}(x) + (p-q) J_{(p-q)/q}(x)\right\}$$
$$+ j \frac{\mu I_0}{4} \frac{e^{-j\beta r}}{r} \beta a^2 \sin\theta \left\{ \cos\theta \sin\left(2\phi + \frac{2p}{q}\psi\right) \left[p\left(1 + \frac{\alpha^2}{2}\right) J_{2p/q}(x) \right.\right. \tag{7.170}$$
$$\left. + p\frac{\alpha^2}{4}\left(J_{2(p+q)/q}(x) + J_{2(p-q)/q}(x)\right) + q\frac{\alpha^2}{4}\left(J_{2(p+q)/q}(x) - J_{2(p-q)/q}(x)\right) \right]$$
$$\left. - q\gamma \sin\theta \cos\left(\phi + \frac{p}{q}\psi\right)\left(J_{(p+q)/q}(x) + J_{(p-q)/q}(x)\right) \right\}$$

$$
\begin{aligned}
A_\phi \approx & \frac{\mu I_0}{4} \frac{e^{-j\beta r}}{r} a\alpha \cos\left(\phi + \frac{p}{q}\psi\right)\left\{(p+q)J_{(p+q)/q}(x) + (p-q)J_{(p-q)/q}(x)\right\} \\
& + j\frac{\mu I_0}{4} \frac{e^{-j\beta r}}{r} \beta a^2 \sin\theta \left\{ p\left[J_0(x) + \frac{\alpha^2}{2}\left(J_0(x) + J_2(x)\right) \right.\right. \\
& + \cos\left(2\phi + \frac{2p}{q}\psi\right)\left(\left(1 + \frac{\alpha^2}{2}\right)J_{2p/q}(x) \right. \\
& \left.\left. + \frac{\alpha^2}{4}\left(J_{2(p+q)/q}(x) + J_{2(p-q)/q}(x)\right)\right)\right] \\
& \left. + q\frac{\alpha^2}{4}\cos\left(2\phi + \frac{2p}{q}\psi\right)\left(J_{2(p+q)/q}(x) - J_{2(p-q)/q}(x)\right)\right\}
\end{aligned}
\tag{7.171}
$$

The far-zone electromagnetic fields produced in this case were found to be

$$E_r \approx 0 \tag{7.172}$$

$$
\begin{aligned}
E_\theta \approx & -j\frac{\eta\beta a I_0}{4} \frac{e^{-j\beta r}}{r} \alpha \cos\theta \sin\left(\phi + \frac{p}{q}\psi\right)\left\{(p+q)J_{(p+q)/q}(x) + (p-q)J_{(p-q)/q}(x)\right\} \\
& + \frac{\eta\beta^2 I_0}{4} \frac{e^{-j\beta r}}{r} a^2 \sin\theta \left\{ \cos\theta \sin\left(2\phi + \frac{2p}{q}\psi\right)\left[p\left(1 + \frac{\alpha^2}{2}\right)J_{2p/q}(x) \right.\right. \\
& \left. + p\frac{\alpha^2}{4}\left(J_{2(p+q)/q}(x) + J_{2(p-q)/q}(x)\right) + q\frac{\alpha^2}{4}\left(J_{2(p+q)/q}(x) - J_{2(p-q)/q}(x)\right)\right] \\
& \left. - q\gamma \sin\theta \cos\left(\phi + \frac{p}{q}\psi\right)\left(J_{(p+q)/q}(x) + J_{(p-q)/q}(x)\right)\right\}
\end{aligned}
\tag{7.173}
$$

$$
\begin{aligned}
E_\phi \approx & -j\frac{\eta\beta I_0}{4} \frac{e^{-j\beta r}}{r} a\alpha \cos\left(\phi + \frac{p}{q}\psi\right)\left\{(p+q)J_{(p+q)/q}(x) + (p-q)J_{(p-q)/q}(x)\right\} \\
& + \frac{\eta\beta^2 I_0}{4} \frac{e^{-j\beta r}}{r} a^2 \sin\theta \left\{ p\left[J_0(x) + \frac{\alpha^2}{2}\left(J_0(x) + J_2(x)\right) \right.\right. \\
& + \cos\left(2\phi + \frac{2p}{q}\psi\right)\left(\left(1 + \frac{\alpha^2}{2}\right)J_{2p/q}(x) \right. \\
& \left.\left. + \frac{\alpha^2}{4}\left(J_{2(p+q)/q}(x) + J_{2(p-q)/q}(x)\right)\right)\right] \\
& \left. + q\frac{\alpha^2}{4}\cos\left(2\phi + \frac{2p}{q}\psi\right)\left(J_{2(p+q)/q}(x) - J_{2(p-q)/q}(x)\right)\right\}
\end{aligned}
\tag{7.174}
$$

$$H_r \approx 0 \tag{7.175}$$

$$H_\theta \approx j\frac{\beta I_0}{4}\frac{e^{-j\beta r}}{r}a\alpha\cos\left(\phi+\frac{p}{q}\psi\right)\{(p+q)J_{(p+q)/q}(x)+(p-q)J_{(p-q)/q}(x)\}$$
$$-\frac{\beta^2 I_0}{4}\frac{e^{-j\beta r}}{r}a^2\sin\theta\left\{p\left[J_0(x)+\frac{\alpha^2}{2}(J_0(x)+J_2(x))\right.\right.$$
$$+\cos\left(2\phi+\frac{2p}{q}\psi\right)\left(\left(1+\frac{\alpha^2}{2}\right)J_{2p/q}(x)\right. \tag{7.176}$$
$$\left.\left.+\frac{\alpha^2}{4}\left(J_{2(p+q)/q}(x)+J_{2(p-q)/q}(x)\right)\right)\right]$$
$$\left.+q\frac{\alpha^2}{4}\cos\left(2\phi+\frac{2p}{q}\psi\right)\left(J_{2(p+q)/q}(x)-J_{2(p-q)/q}(x)\right)\right\}$$

$$H_\phi \approx -j\frac{\beta a I_0}{4}\frac{e^{-j\beta r}}{r}\alpha\cos\theta\sin\left(\phi+\frac{p}{q}\psi\right)\{(p+q)J_{(p+q)/q}(x)+(p-q)J_{(p-q)/q}(x)\}$$
$$+\frac{\beta^2 I_0}{4}\frac{e^{-j\beta r}}{r}a^2\sin\theta\left\{\cos\theta\sin\left(2\phi+\frac{2p}{q}\psi\right)\left[p\left(1+\frac{\alpha^2}{2}\right)J_{2p/q}(x)\right.\right. \tag{7.177}$$
$$\left.+p\frac{\alpha^2}{4}\left(J_{2(p+q)/q}(x)+J_{2(p-q)/q}(x)\right)+q\frac{\alpha^2}{4}\left(J_{2(p+q)/q}(x)-J_{2(p-q)/q}(x)\right)\right]$$
$$\left.-q\gamma\sin\theta\cos\left(\phi+\frac{p}{q}\psi\right)\left(J_{(p+q)/q}(x)+J_{(p-q)/q}(x)\right)\right\}$$

The geometry for the special cases in which $p = 1$ and $q = 1$ ($p/q = 1$) and $p = 3$ and $q = 1$ ($p/q = 3$) is illustrated in Figs. 7.11(a) and 7.11(b), respectively. These are actually both examples of trivial knots.

7.5.4.4 Special Case when p/q = (2n − 1)/2

The closed-form expressions for the vector potential in this case where $p/q = (2n-1)/2$ and $n \in \mathbb{N}$ are given by

$$A_\theta \approx j\frac{\mu I_0}{4}\frac{e^{-j\beta r}}{r}\beta a^2\frac{\alpha}{2}\sin\theta\cos\theta\sin\left(2\phi+\frac{2p}{q}\psi\right) \tag{7.178}$$
$$\left[(2p+q)J_{(2p+q)/q}(x)+(2p-q)J_{(2p-q)/q}(x)\right]$$

$$A_\phi \approx j\frac{\mu I_0}{4}\frac{e^{-j\beta r}}{r}\beta a^2\sin\theta\left\{p\left[J_0(x)+\frac{\alpha^2}{2}(J_0(x)+J_2(x))\right]\right.$$
$$+\cos\left(2\phi+\frac{2p}{q}\psi\right)\left[p\alpha\left(J_{(2p+q)/q}(x)+J_{(2p-q)/q}(x)\right)\right. \tag{7.179}$$
$$\left.\left.+q\frac{\alpha}{2}\left(J_{(2p+q)/q}(x)-J_{(2p-q)/q}(x)\right)\right]\right\}$$

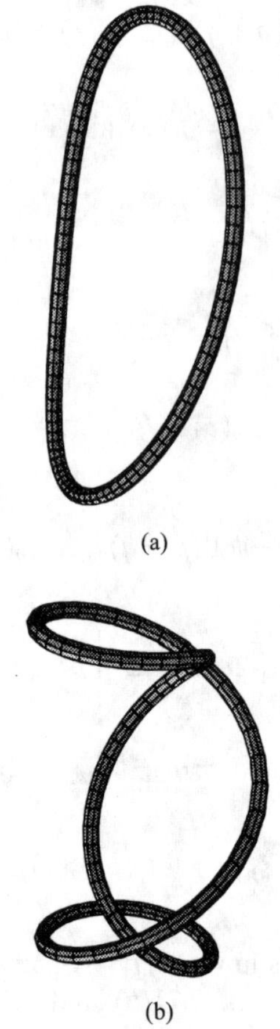

Figure 7.11 Special elliptical torus knot geometries. A trivial (1, 1)-torus knot is shown in (a) while a trivial (3, 1)-torus knot is shown in (b).

and the corresponding far-field representations are

$$E_r \approx 0 \tag{7.180}$$

$$E_\theta \approx \frac{\eta\beta^2 I_0}{4}\frac{e^{-j\beta r}}{r}a^2\frac{\alpha}{2}\sin\theta\cos\theta\sin\left(2\phi+\frac{2p}{q}\psi\right)\left[(2p+q)J_{(2p+q)/q}(x)+(2p-q)J_{(2p-q)/q}(x)\right] \tag{7.181}$$

$$\begin{aligned}E_\phi \approx \frac{\eta\beta^2 \mathrm{I}_0}{4}\frac{e^{-j\beta r}}{r}a^2\sin\theta\Bigg\{p\left[J_0(x)+\frac{\alpha^2}{2}\left(J_0(x)+J_2(x)\right)\right]\\ +\cos\left(2\phi+\frac{2p}{q}\psi\right)\left[p\alpha\left(J_{(2p+q)/q}(x)+J_{(2p-q)/q}(x)\right)\right.\\ \left.+q\frac{\alpha}{2}\left(J_{(2p+q)/q}(x)-J_{(2p-q)/q}(x)\right)\right]\Bigg\}\end{aligned} \tag{7.182}$$

$$H_r \approx 0 \tag{7.183}$$

$$H_\theta \approx \frac{\beta^2 I_0}{4}\frac{e^{-j\beta r}}{r}a^2 \sin\theta \left\{ p\left[J_0(x) + \frac{\alpha^2}{2}\big(J_0(x)+J_2(x)\big)\right] \right.$$
$$+ \cos\left(2\phi + \frac{2p}{q}\psi\right)\left[p\alpha\big(J_{(2p+q)/q}(x) + J_{(2p-q)/q}(x)\big) \right. \tag{7.184}$$
$$\left.\left. + q\frac{\alpha}{2}\big(J_{(2p+q)/q}(x) - J_{(2p-q)/q}(x)\big)\right]\right\}$$

$$H_\phi \approx \frac{\beta^2 I_0}{4}\frac{e^{-j\beta r}}{r}a^2 \frac{\alpha}{2}\sin\theta\cos\theta\sin\left(2\phi + \frac{2p}{q}\psi\right) \tag{7.185}$$
$$\left[(2p+q)J_{(2p+q)/q}(x) + (2p-q)J_{(2p-q)/q}(x)\right]$$

The geometry for the special cases in which $p = 1$ and $q = 2$ ($p/q = 1/2$) and $p = 3$ and $q = 2$ ($p/q = 3/2$) is illustrated in Figs. 7.12(a) and 7.12(b), respectively. The first knot shown in Fig. 7.12(a) is trivial, while the second knot shown in Fig. 7.12(b) is nontrivial.

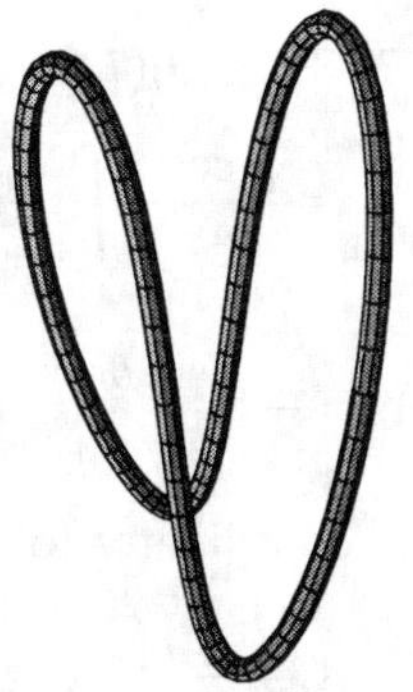

(a)

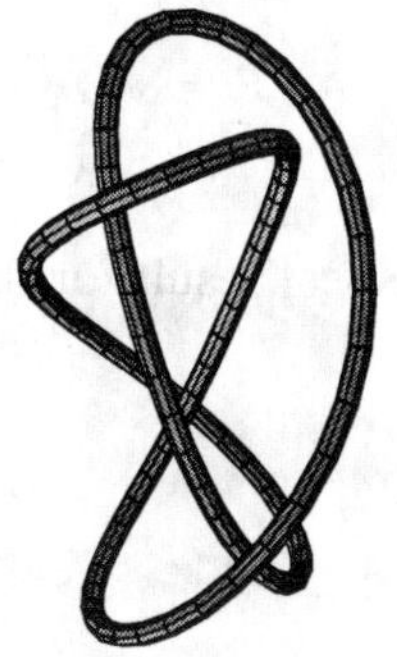

(b)

Figure 7.12 Special elliptical torus knot geometries. A trivial (1, 2)-torus knot is shown in (a) while a nontrivial (3, 2)-torus knot is shown in (b).

7.5.5 Circular Loop and Linear Dipole

For the special case when $b = c = 0$ and $p = 1$, the torus knot parameterizations defined in (7.105)–(7.107) describe a circular loop of radius a. Hence, the general vector potential expressions for the (p, q)-torus knots derived in (7.113)– (7.115) will reduce to the well-known classical results for the circular loop antenna given in (7.98)–(7.100), respectively. In addition to the circular loop, the linear dipole can also be obtained as a degenerate case of the elliptical torus knot parameterizations. To demonstrate this, suppose we let $a = b = 0$, then the expressions for the vector potential components given in (7.113)–(7.115) reduce to

$$A_r = \frac{\mu \cos\theta}{4\pi} qc \int_0^{2\pi} I_s(s') \cos(\psi + qs') \frac{e^{-j\beta R}}{R} ds' \tag{7.186}$$

$$A_\theta = -\frac{\mu \sin\theta}{4\pi} qc \int_0^{2\pi} I_s(s') \cos(\psi + qs') \frac{e^{-j\beta R}}{R} ds' \tag{7.187}$$

$$A_\phi = 0 \tag{7.188}$$

By making the change of variables

$$z' = c \sin(\psi + qs') \tag{7.189}$$

$$dz' = qc \cos(\psi + qs') ds' \tag{7.190}$$

with $q = \frac{1}{2}$ and $\psi = -\pi/2$, (7.186) and (7.187) may be written as

$$A_r = \frac{\mu \cos\theta}{4\pi} \int_{-c}^{c} I_s(z') \frac{e^{-j\beta R}}{R} dz' \tag{7.191}$$

$$A_\theta = -\frac{\mu \sin\theta}{4\pi} \int_{-c}^{c} I_s(z') \frac{e^{-j\beta R}}{R} dz' \tag{7.192}$$

Transforming from spherical coordinates to cylindrical coordinates yields

$$A_z = \frac{\mu}{4\pi} \int_{-c}^{c} I_s(z') \frac{e^{-j\beta R}}{R} dz' \tag{7.193}$$

where

$$R = \sqrt{(z - z')^2 + \rho^2} \tag{7.194}$$

$$\rho = \sqrt{x^2 + y^2} \tag{7.195}$$

This is the well-known classical result for the vector potential of a z-directed linear dipole of length $2c$ [14].

7.6 RESULTS

A plot of the radiation resistance as a function of radius for an electrically small trefoil knot antenna is shown in Fig. 7.13. This plot is based on (7.97) for the special case where $p = 2$ and $b = a/2$. Under these conditions, (7.97) reduces to the simple formula

$$R_r \approx 1000(\beta a)^4\, \Omega \tag{7.196}$$

Figure 7.13 demonstrates that a small trefoil antenna would have a relatively low value of radiation resistance. However, (7.97) suggests that the radiation resistance

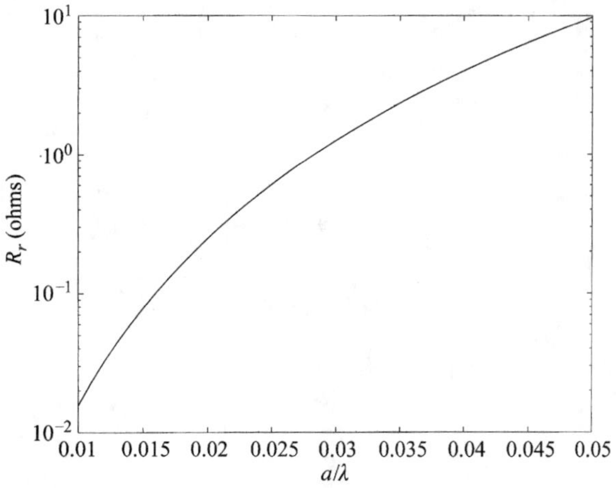

Figure 7.13 The radiation resistance versus radius for an electrically small trefoil knot antenna.

may be increased by considering antennas which possess a higher degree of knottedness.

We next turn our attention to an investigation of the scattering properties of knots. A 3-dimensional view of a trefoil knot is shown in Fig. 7.14. Top and side views

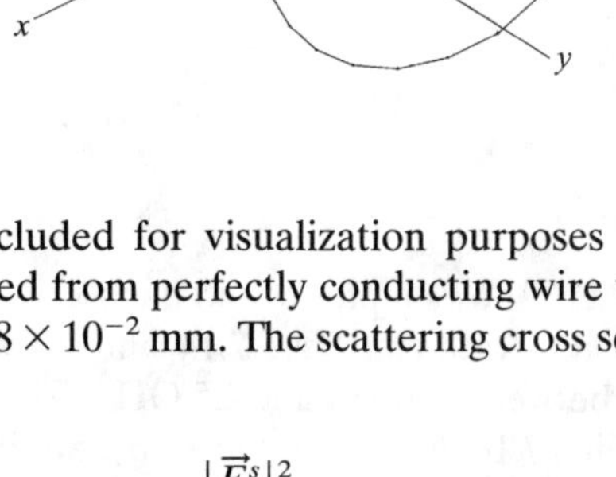

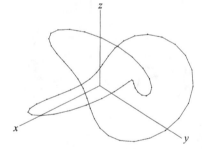

Figure 7.14 A 3-dimensional view of a thin wire model for a (2, 3)-torus knot (i.e., a trefoil).

of this trefoil have also been included for visualization purposes in Fig. 7.15. The trefoil is assumed to be constructed from perfectly conducting wire with an arclength of 41.416 mm and a radius of 5.528×10^{-2} mm. The scattering cross section of the knot may be calculated from

$$\sigma = 4\pi r^2 \frac{|\vec{E}^s|^2}{|\vec{E}^i|^2} \tag{7.197}$$

where $\vec{E}^i$ and $\vec{E}^s$ represent the incident and scattered electric fields, respectively [21]. A linearly polarized plane wave with an intensity of 1 V/m is assumed to be incident on the knot. The corresponding scattered field is determined using a numerical analysis procedure based on the method of moments. This approach is followed in order to calculate the backscatter cross section as a function of frequency for a particular knot using (7.197).

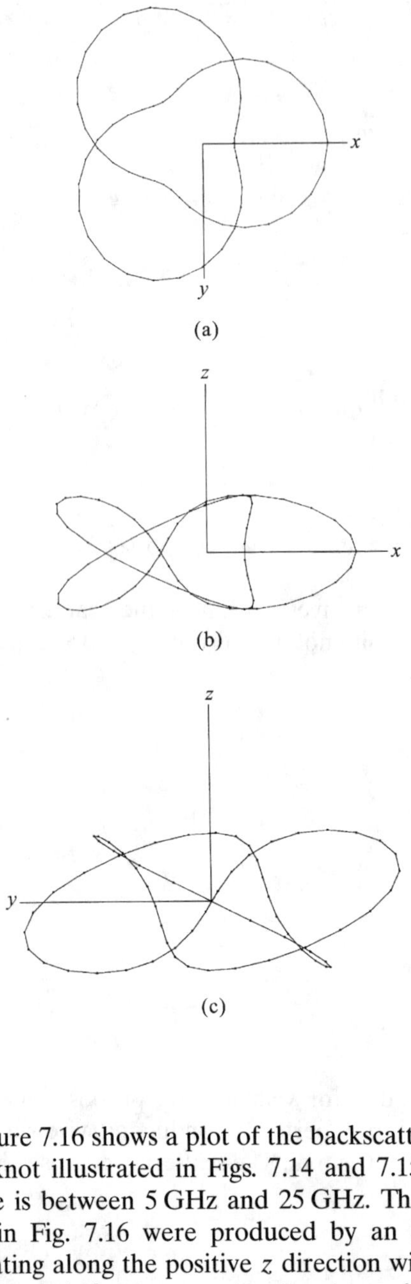

Figure 7.15 Top and side views of a thin wire model for the trefoil knot shown in Fig. 7.14.

Figure 7.16 shows a plot of the backscatter cross section versus frequency for the trefoil knot illustrated in Figs. 7.14 and 7.15. The frequency range chosen for this example is between 5 GHz and 25 GHz. The curves of backscattering cross section shown in Fig. 7.16 were produced by an incident linearly polarized plane wave propagating along the positive z direction with the electric field parallel to the x axis [see Fig. 7.15(a)]. We note that for the trefoil an identical backscattering signature would be obtained if the electric field was parallel to the y axis rather than the x axis. The solid curve shown in Fig. 7.16 represents the total backscatter cross section which is denoted by σ_{total}. The dotted and dashed curves, on the other hand, represent the co-polarized backscatter cross section (σ_{co}) and the cross-polarized backscatter cross section (σ_{cross}), respectively. For the case considered in Fig. 7.16, the total backscatter cross section is essentially the same as the co-polarized backscatter cross section.

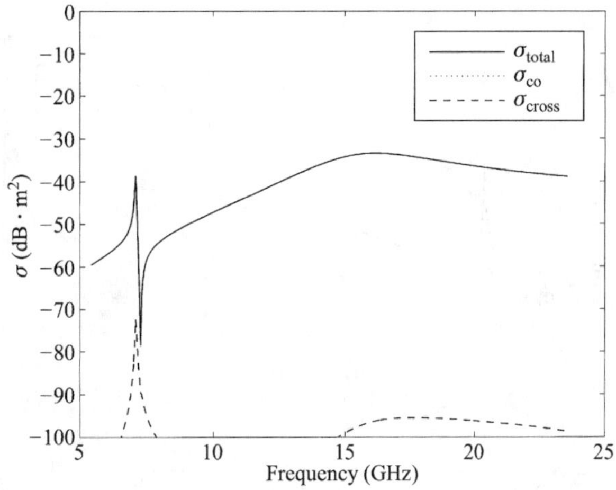

Figure 7.16 Backscatter cross section versus frequency for the trefoil knot illustrated in Fig. 7.15(a). A linearly polarized plane wave is assumed to be incident on the knot traveling in the positive z direction with the electric field parallel to the x axis or the y axis.

The resonance frequencies for a (p, q)-torus knot may be estimated by using the approximate formula

$$f_n = n\left(\frac{c}{L_{pq}}\right) \qquad \text{for } n = 1, 2, \ldots \tag{7.198}$$

where L_{pq} represents the arclength of the knot as defined in (7.112). Note that these are the same resonances that would be associated with a circular loop (canonical unknot) having a circumference equal to L_{pq}. The trefoil knot considered in Figs. 7.14 and 7.15 has an arclength of 41.416 mm which corresponds to one wavelength at 7.24 GHz. This fact is substantiated by (7.198) which suggests that the first resonance should occur at a frequency of approximately $f_1 = 7.24$ GHz. Figure 7.16 demonstrates that a sharp resonance is indeed present at this frequency. However, in this case, all higher-order resonances are suppressed.

The backscatter cross section for a circular loop in the x-y plane with a 41.416 mm circumference ($f_1 = 7.24$ GHz) is shown in Fig. 7.17. A comparison of Fig. 7.16 with

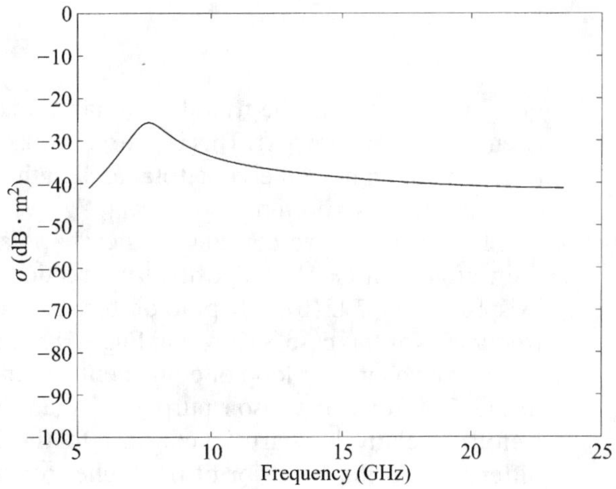

Figure 7.17 Backscatter cross section versus frequency for a circular loop in the x-y plane. A linearly polarized plane wave is assumed to be incident on the loop traveling in the positive z direction with the electric field parallel to the x axis or the y axis.

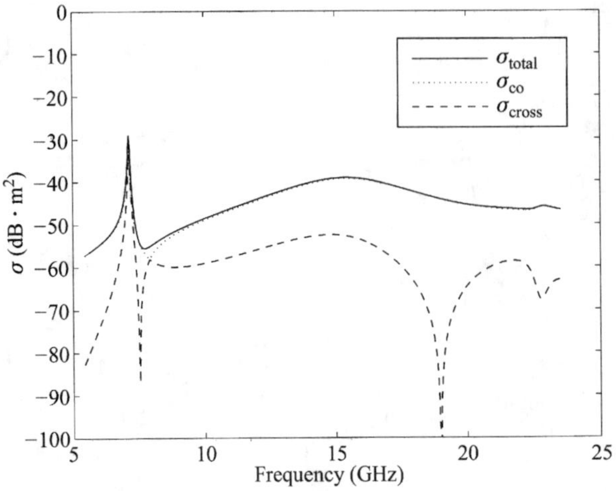

Figure 7.18 Backscatter cross section versus frequency for the trefoil knot illustrated in Fig. 7.15(b). A linearly polarized plane wave is assumed to be incident on the knot traveling in the positive y direction with the electric field parallel to the x axis.

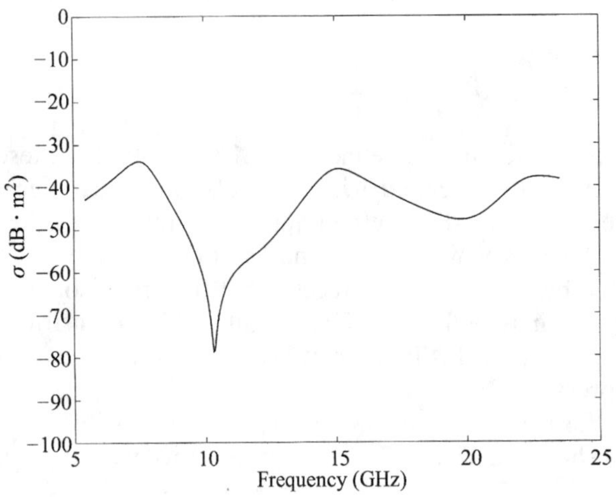

Figure 7.19 Backscatter cross section versus frequency for a circular loop in the x-y plane. A linearly polarized plane wave is assumed to be incident on the loop traveling in the positive y direction with the electric field parallel to the x axis.

Fig. 7.17 reveals that the trefoil knot has a much sharper first resonance than does its circular loop counterpart. This is a direct consequence of the fact that even though the trefoil and loop have the same total arclength, the physical size of the trefoil is smaller since the wire is knotted.

Next suppose we consider a linearly polarized plane wave which is incident on the trefoil knot along the positive y direction with the electric field parallel to the x axis [see Fig. 7.15(b)]. A plot of the backscatter cross section as a function of frequency for this case is shown in Fig. 7.18. The backscatter cross section which would result from a circular loop of equivalent arclength contained in the x-y plane is shown in Fig. 7.19 for comparison purposes. Again, we find that the first resonance of the trefoil is relatively sharp in contrast to the loop. There are also some noticeable differences in the behavior of the higher-order resonance.

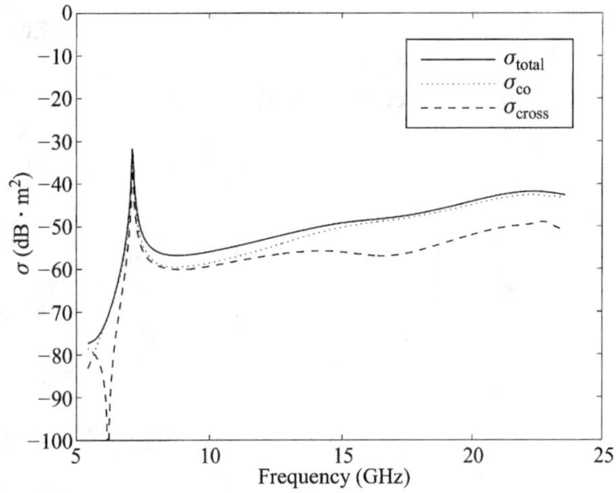

Figure 7.20 Backscatter cross section versus frequency for the trefoil knot illustrated in Fig. 7.15(b). A linearly polarized plane wave is assumed to be incident on the knot traveling in the positive y direction with the electric field parallel to the z axis.

The last case that will be examined for the trefoil involves its response to an incident linearly polarized plane wave propagating along the y direction with the electric field parallel to the z axis [see Fig. 7.15(c)]. Curves of the backscatter cross section for this case are shown in Fig. 7. 20. It is important to point out that, under these circumstances, the incident field would not couple to a circular loop lying in the x-y plane. On the other hand, Fig. 7.20 shows evidence of significant field coupling to the trefoil. This leads to the conclusion that the trefoil knot experiences a strong field coupling for all possible polarizations and angles of incidence, whereas the circular loop does not.

7.7 CONCLUSION

Knot electrodynamics is an emerging area of research which seeks to combine aspects of knot theory with Maxwell's theory of electromagnetism. The primary purpose of this chapter has been to establish a rigorous mathematical foundation from which analysis techniques may be developed and applied toward the study of knot electrodynamics problems. This chapter focuses on the particular class of knots, known as torus knots, which have interesting topological as well as electromagnetic properties. These knots derive their name from the fact that they reside on the surface of a solid torus and, consequently, useful parametric representations for them may be found. A new knotted wire EFIE was derived based on the available parameterizations for (p,q)-torus knots. These parameterizations were also used to derive expressions for the electromagnetic fields radiated by (p,q)-torus knots, including simple closed-form representations for the far-fields of electrically small knots. In addition to this, it was shown that the circular loop as well as the linear dipole geometries may both be obtained as degenerate forms of the parametric representations for elliptical torus knots. Finally, several examples were presented and discussed which illustrate the unique radiation and scattering properties associated with various (p,q)-torus knots.

APPENDIX

In this appendix a methodology is presented for evaluating integrals of the type

$$I(v,x) = \int_{u_1}^{u_2} \cos(vu + x \sin u)\, du \tag{7.199}$$

where

$$v = \frac{n}{q}, \qquad n \in \mathbb{I} \text{ and } q \in \mathbb{N} \tag{7.200}$$

$$x = \beta c \cos\theta \tag{7.201}$$

$$u_1 = 0 \tag{7.202}$$

$$u_2 = 2\pi q \tag{7.203}$$

These integrals are encountered in the process of deriving the far-zone representations for the vector potential components A_θ and A_ϕ considered in Section 7.5.4. The first step toward finding a closed-form solution to (7.199) is to recognize that since v and x are real valued quantities (i.e., $v, x \in \mathbb{R}$), we may write

$$I(v,x) = \Re\left\{\int_{u_1}^{u_2} e^{j[vu + x\sin u]}\, du\right\} \tag{7.204}$$

Next, we make use of the generating function for Bessel functions [22]

$$e^{x(t-1/t)/2} = \sum_{k=-\infty}^{\infty} J_k(x)\, t^k, \qquad t \neq 0 \tag{7.205}$$

Setting $t = e^{ju}$ in (7.205) gives

$$e^{jx\sin u} = \sum_{k=-\infty}^{\infty} J_k(x)\, e^{jku} \tag{7.206}$$

Multiplying both sides of (7.206) by e^{jvu} and integrating with respect to u leads to

$$\int_{u_1}^{u_2} e^{j[vu + x\sin u]}\, du = \sum_{k=-\infty}^{\infty} J_k(x) \int_{u_1}^{u_2} e^{j(v+k)u}\, du \tag{7.207}$$

From (7.204) and (7.207), it follows that

$$\begin{aligned} I(v,x) &= \Re\left\{\sum_{k=-\infty}^{\infty} J_k(x) \int_{u_1}^{u_2} e^{j(v+k)u}\, du\right\} \\ &= \sum_{k=-\infty}^{\infty} J_k(x)\, \Re\left\{\int_{u_1}^{u_2} e^{j(v+k)u}\, du\right\} \\ &= \sum_{k=-\infty}^{\infty} J_k(x) \int_{u_1}^{u_2} \cos\big[(v+k)u\big]\, du \\ &= \sum_{k=-\infty}^{\infty} J_k(x) \left[\frac{\sin\big[(v+k)u_2\big] - \sin\big[(v+k)u_1\big]}{v+k}\right] \end{aligned} \tag{7.208}$$

Finally, by substituting the appropriate values of $u_1 = 0$ and $u_2 = 2\pi q$ into (7.208), we arrive at

$$I(v,x) = \begin{cases} 2\pi q J_{-v}(x), & v \in \mathbb{I} \\ 0, & v \notin \mathbb{I} \end{cases} \tag{7.209}$$

A closed-form solution to the related integrals

$$I(v,-x) = \int_{u_1}^{u_2} \cos(vu - x\sin u)\,du \tag{7.210}$$

may be obtained directly from (7.209) as

$$I(v,-x) = \begin{cases} 2\pi q J_{v}(x), & v \in \mathbb{I} \\ 0, & v \notin \mathbb{I} \end{cases} \tag{7.211}$$

REFERENCES

[1] C. C. Adams, *The Knot Book: An Elementary Introduction to the Mathematical Theory of Knots*, W. H. Freeman and Company, New York, 1994.

[2] S. Wasserman, J. Dungan, and N. Cozzarelli, "Discovery of a Predicted DNA Knot Substantiates a Model for Site-Specific Recombination," *Science*, **229**, pp. 171–174, July 1985.

[3] T. Schlick and W. K. Olson, "Trefoil Knotting Revealed by Molecular Dynamics Simulations of Supercoiled DNA," *Science*, **257**, pp. 1110–1115, 1992.

[4] D. M. Walba, "Topological Stereochemistry," *Tetrahedron*, **41** (16), pp. 3161–3212, 1985.

[5] S. J. Lomonaco, Jr., "The Modern Legacies of Thomson's Atomic Vortex Theory in Classical Electrodynamics," in *The Interface of Knots and Physics*, L. H. Kauffman, ed., Providence, RI: American Mathematical Society, pp. 145–166, 1996.

[6] G. E. Marsh, "Helicity and Electromagnetic Field Topology," in *Advanced Electromagnetism: Foundations, Theory and Applications*, T. W. Barrett and D. M. Grimes, eds., World Scientific, Singapore, pp. 52–76, 1995.

[7] O. Manuar and D. L. Jaggard, "Backscatter Signatures of Knots," *Optics Letters*, **20** (2), pp. 115–117, January 1995.

[8] D. L. Jaggard and O. Manuar, "Can One 'Hear' the Handedness or Topology of a Knot?," *Proceedings of the IEEE Antennas and Propagation Society International Symposium and URSI Radio Science Meeting*, Newport Beach, California, URSI Digest, p. 244, June 18–23, 1995.

[9] O. Manuar and D. L. Jaggard, "Wave Interactions with Trefoils and Untrefoils," *Proceedings of the IEEE Antennas and Propagation Society International Symposium and URSI Radio Science Meeting*, Baltimore, MD, URSI Digest, p. 279, July 21–26, 1996.

[10] O. Manuar and D. L. Jaggard, "Scattering from Knots and Unknots: the Role of Symmetry," *Elec. Lett.*, **33** (4), pp. 278–280, February 1997.

[11] H. Seifert and W. Threlfall, *A Textbook of Topology*, Academic Press, New York, 1980.

[12] L. H. Kauffman, *Knots and Physics*, World Scientific, Singapore, 1991.

[13] D. L. Jaggard, A. R. Mickelson, and C. H. Papas, "On Electromagnetic Waves in Chiral Media," *Appl. Phys.*, **18**, pp. 211–216, 1979.

[14] C. A. Balanis, *Antenna Theory, Analysis and Design*, John Wiley & Sons, New York, 1982.

[15] K. K. Mei, "On the Integral Equations of Thin Wire Antennas," *IEEE Trans. Antennas Propagat.*, **AP-13**, pp. 374– 378, May 1965.

[16] J. J. H. Wang, *Generalized Moment Methods in Electromagnetics: Foundation and Computer Solution of Integral Equations*, John Wiley & Sons, New York, 1991.

[17] D. H. Werner, "An Exact Integration Procedure for Vector Potentials of Thin Circular Loop Antennas," *IEEE Trans. Antennas Propagat.*, **44** (2), pp. 157–165, February 1996.

[18] D. H. Werner, "The Electrodynamics of Torus Knots," *IEEE Antennas and Propagation Society International Symposium Digest*, **2**, pp. 1468–1471, Montreal, Canada, July 1997.

[19] D. H. Werner, "Radiation and Scattering From Thin Toroidally Knotted Wires," to appear in *IEEE Trans. Antennas Propagat.*, August 1999.

[20] D. H. Werner, "Electromagnetic Radiation and Scattering From Elliptical Torus Knots," Proceedings of the *IEEE Antennas and Propagation Society International Symposium*, **2**, pp. 858–861, June 21–26, 1998, Atlanta, GA. See also *Proceedings of the USNC/URSI National Radio Science Meeting*, p. 126, June 21–26, 1998, Atlanta, GA.

[21] J. J. Bowman, T. B. A. Senior, and P. L. E. Uslenghi, *Electromagnetic and Acoustic Scattering by Simple Shapes*, Hemisphere Publishing Co., New York, 1987.

[22] L. C. Andrews, *Special Functions for Engineers and Applied Mathematicians*, Macmillan, New York, 1985.

[23] J. v. Hagen, D. H. Werner, and R. Mittra, "Polarization-Selective Surfaces Composed of Trefoil Knot Elements," *Microwave and Optical Technology Letters*, **21**(3), pp. 170–173, May 5, 1999.

Chapter 8

BIOLOGICAL BEAMFORMING

Randy L. Haupt, Hugh L. Southall, and Teresa H. O'Donnell

8.1 BIOLOGICAL BEAMFORMING

Antenna design usually doesn't invoke biological theories, yet some of the complex processes governing living organisms have application to new and interesting antenna array designs. Superior control of the antenna radiation pattern is the fundamental advantage of an array antenna. Designing the array and controlling the radiation pattern require techniques beyond the elementary adaptive processing and optimization approaches. By modeling such complicated natural processes, such as the brain and natural selection, new and superior antenna arrays can be designed.

Figure 8.1 is a diagram of a linear phased array antenna. Each element samples an incident electromagnetic plane wave and sends the resulting signal to a beamforming network. The beamforming network may use phase shifters, attenuators, power dividers, receivers, transmitters, and other components to coherently combine the signals into one or more antenna patterns. The "beam" in beamforming refers to the main beam of the antenna pattern. A beamformer can be as simple as transmission lines connecting the elements to a power combiner that connects to a receiver/transmitter or as complicated as a digital beamformer, where each element has its own receiver/transmitter. When the signals all add in phase, a maximum or main beam peak occurs. A beamformer may be static or dynamic. Static beamformers can't change how the signals received at each element add together, while dynamic beamformers are able to change the amplitude and/or phase of the signals at each element. Sometimes, beamformers even create multiple main beams.

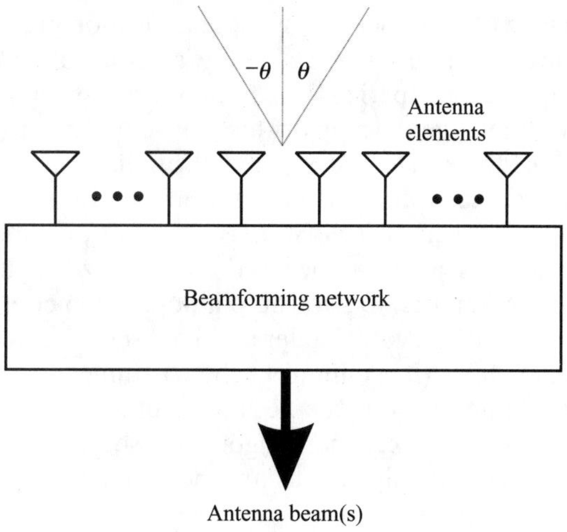

Figure 8.1 A linear array with a beamforming network.

A signal incident upon an array element induces a current given by

$$I_n(t) = A(t)\,e^{[j2\pi(d/\lambda)u]n} \tag{8.1}$$

where

$A(t)$ = signal modulation
n is the half integer with $-(N-1)/2 \leq n \leq (N-1)/2$
N = number of elements (assumed to be even)
$u = \sin(\theta)$
θ = the angle of arrival of a plane wave relative to the array broadside (or normal)
d = spacing between elements
λ = wavelength of incident plane wave

The phase center has been referenced to the array center. The beamformer takes each of these signals and combines them in a way that aids in the detection of a desired signal.

This chapter provides a tutorial for applications of genetic algorithms and neural networks to linear array beamforming networks. The genetic algorithm is used to synthesize optimal low sidelobe arrays via phase tapering. Next, the genetic algorithm places nulls in the sidelobes in the directions of interfering sources. Both applications of the genetic algorithm use traditional array beamforming approaches. Breaking away from traditional beamforming networks, neural networks are applied to direction finding, detection, and beamsteering with a linear array. Examples are presented to demonstrate the superiority of these approaches.

8.2 GENETIC ALGORITHM BEAMFORMING

A genetic algorithm is an optimization process based upon genetics and natural selection. The optimization may be used to design antenna beamforming networks or optimize dynamic beamformer weights to achieve some antenna pattern performance. Genetic algorithms were developed by John Holland [1] over the course of the 1960s and 1970s. Since then, genetic algorithms have been successfully applied to a wide range of extremely complex optimization problems [2]. Genetic algorithms have been successfully used to design low sidelobe antenna arrays in the form of array thinning [3], phase tapering [4], nonuniform spacing [5], and amplitude tapering [6]. In addition, genetic algorithms have been shown to be useful as adaptive nulling algorithms [7]. The successful application of genetic algorithms to antenna array design has spurred significant interest in the antenna community.

The genetic algorithm is a type of evolutionary algorithm that models biological processes and can be used to optimize highly complex cost functions. A genetic algorithm attempts to cause a population composed of many individuals or potential solutions to evolve under specified selection rules to a state which finds the "most fit" individuals (i.e., minimizes the cost function, assuming it has been written so that the minimum value is the desired solution).

What makes these algorithms so good? Some of the advantages they have over conventional algorithms are the genetic algorithm

- Optimizes with continuous or discrete parameters,

- Doesn't require derivative information,
- Simultaneously searches from a wide sampling of the cost surface,
- Deals with a large number of parameters,
- Is well suited for parallel computers,
- Optimizes parameters with extremely complex cost surfaces,
- Provides a list of semi-optimum parameters, not just a single solution,
- May encode the parameters so that the optimization is done with the encoded parameters, and
- Works with numerically generated data, experimental data, or analytical functions.

These advantages are intriguing and often produce stunning results when traditional optimization approaches fail miserably.

Genetic algorithms are a subset of evolutionary computations. Evolutionary computations attempt to use biological processes for the basis of a computer algorithm that optimizes highly complex problems. Figure 8.2 is a flowchart of a simple genetic algorithm. The next few paragraphs explain each block in the flowchart.

A genetic algorithm usually encodes each parameter in a binary sequence called a gene and places the genes in an array known as a chromosome. For instance, $(x, y) = (32, 1)$ can be encoded as

$$\text{chromosome} = (100000000001)$$

Genetic algorithms work well with continuous parameters. Since the examples presented here only deal with digital phase shifters, the binary version of the genetic algorithm is presented. For a more detailed discussion of the continuous parameter genetic algorithm see [8].

The algorithm begins with a population in the form of a random matrix of ones and zeros where each row is a chromosome. Next, the fitness or cost function for each

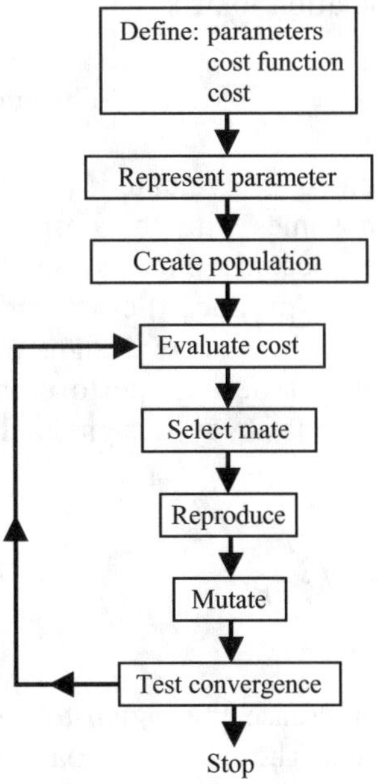

Figure 8.2 Genetic algorithm flowchart.

chromosome is calculated. The chromosomes are ranked from best to worst according to their fitness values. Mating takes place between the most fit chromosomes. Offspring or new chromosomes produced from mating contain parts of the two parents. A simple crossover scheme randomly picks a point in the chromosome. Two offspring result by keeping the binary strings to the left of the crossover point for each parent and swapping the binary strings to the right of the crossover point. Consider two parent chromosomes given by

$$\text{parent\#1} = [1010|1010]$$
$$\text{parent\#2} = [1111|0000]$$

If the random crossover point occurs between bits four and five, the offspring are

$$\text{offspring\#1} = [10100000]$$
$$\text{offspring\#2} = [11111010]$$

The fitness of the new offspring is computed. Mutations change a small percentage of the bits from "1" to "0" or vice versa. The new list is ranked from best to worst and the process repeated. The algorithm stops when an acceptable solution is found or a set number of iterations have lapsed. Many variations and improvements to this process are possible including the use of continuous and binary parameters. Various approaches to the mating, crossover, and mutation operators are possible [2].

The examples in this chapter optimize the phase settings of the phase shifters, so a chromosome looks like

$$\text{chromosome} = [\text{parameter}_1, \text{parameter}_2, \ldots, \text{parameter}_N]$$

This chromosome defines one possible array configuration. A random set of chromosomes are generated and placed in a matrix known as the population matrix.

$$\text{Population matrix} = \begin{bmatrix} \text{chromosome}_1 \\ \text{chromosome}_2 \\ \vdots \\ \text{chromosome}_{xx} \end{bmatrix} \tag{8.2}$$

Each of these vectors is evaluated in an objective function. Those array configurations or chromosomes with the best performance (most closely meet the desired phase tolerance) become parents and the rest are discarded. This set of parents mate and mutate to produce the next generation of potential array designs. After several iterations, the best array designs percolate to the top until a design is found that best matches our desired performance criteria. The robustness of the genetic algorithm to antenna array design is shown in two examples in the two following sections.

8.3 LOW SIDELOBE PHASE TAPERS

An array antenna may increase the signal-to-noise ratio of the received signal by lowering its sidelobe levels. The usual approach to obtaining the low sidelobes is to multiply the signal at each element by a weighting factor then add all the signals together. A less common approach is to multiply the received signals by a phase term [9]. Closed-form solutions exist to find desirable low sidelobe amplitude tapers. Low sidelobe phase tapers, however, must be numerically derived. This example uses a genetic algorithm to find a symmetrical phase taper that reduces the maximum sidelobe level.

Figure 8.3 shows a model of the components of a linear array of point sources with phase-only control of the signals. Mathematically, the array factor is represented as

$$AF(u) = \sum_{n=1}^{N} a_n I_n(t) e^{j\delta_n} \cos\theta \tag{8.3}$$

where

a_n = amplitude weight at element n; in this case $a_n = 1$
$I_n(t)$ is defined in (8.1)
δ_n = phase at element n

The equation in (8.3) is not an objective function, or function to be optimized. Our goal is to minimize the maximum sidelobe level of the array factor by varying δ_n. Thus, the objective function must return the maximum sidelobe level as a function of phase. The objective function is given by

$$f(\delta_n) = \max\left\{\left|A(t)e^{[j2\pi(d/\lambda)u]n}\right|\right\} \quad \text{for} \quad u \geq u_{sll} \tag{8.4}$$

where u_{sll} is the starting-point for checking the sidelobe levels.

Digital phase shifters with B bits have a phase setting given by

$$\delta = \left[b_1 + b_2(0.5) + \cdots + b_B 2^{(B-1)}\right]\pi \tag{8.5}$$

where the b_n coefficients represent the binary sequence associated with a phase

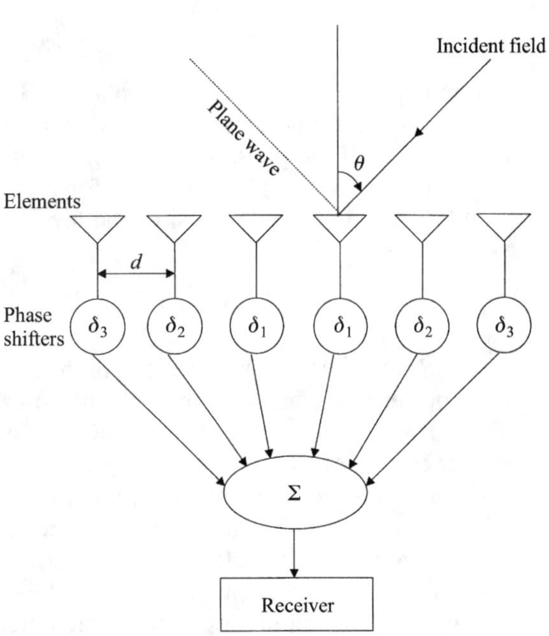

Figure 8.3 Diagram of a linear-phased array antenna.

setting. Thus, a binary encoding of the phase for one element is called a gene. For example, a phase shifter with 3-bit accuracy set at 135° has a gene given by

$$gene = [b_1 b_2 b_3] = [0\,1\,1] \tag{8.6}$$

In turn, the genes are placed together in an array called a chromosome. A four element array with phase settings of 0°, 90°, 270°, and 45° has a chromosome representation of

$$chromosome = [0\,0\,0\,0\,1\,0\,1\,1\,0\,0\,0\,1] \tag{8.7}$$

A chromosome contains the binary encodings for one possible low sidelobe phase taper.

The first optimization example is a 40 element linear array with 3 bit phase shifters, a uniform amplitude taper, and equally spaced elements. Figure 8.4 shows the

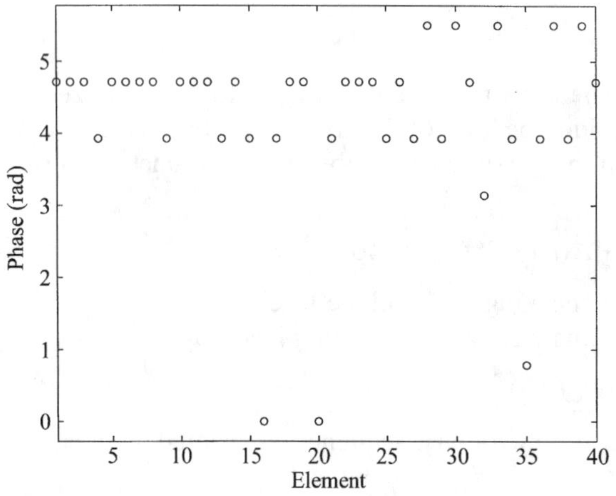

Figure 8.4 Optimized phase taper for a 40-element linear array with 3-bit phase shifters, a uniform amplitude taper, and elements spaced 0.5λ apart.

optimum phase taper that results in a −19.3 dB maximum relative sidelobe level. The corresponding far field pattern is shown in Fig. 8.5. Increasing the number of bits or the accuracy of the phase shifters somewhat improves the solution. For instance, the optimum continuous phase settings for an 80 element array found in [9] result in a −20.1 dB maximum relative sidelobe level. Efficiency can be calculated from

$$\eta = \frac{P}{P_0} \tag{8.8}$$

where P is the peak power of the main beam for the phase tapered array and P_0 is the peak power of the main beam for a uniform array with the same number of elements. The quantized taper has an efficiency of 58%, while the continuous taper had an efficiency of 70%. The quantized optimum solution is not as good as the continuous solution but is more realistic for practical phased arrays.

Practical phased arrays have quantized phase shifters and many elements. Phase quantization restricts sidelobe level performance and array efficiency. A genetic algorithm can find an optimum low sidelobe phase taper for the array and save the expense of an amplitude tapered feed network. This chapter demonstrates the

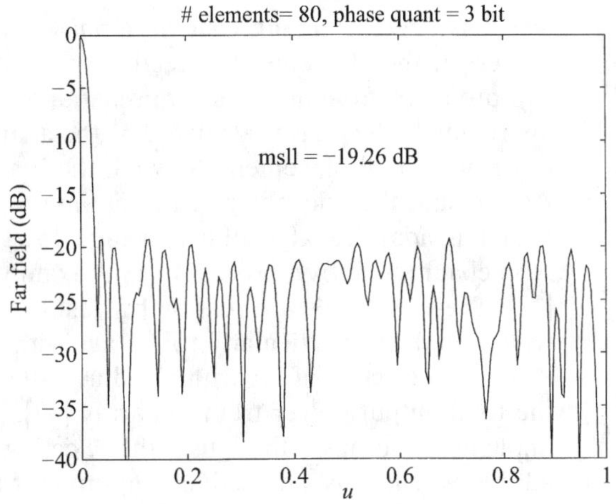

Figure 8.5 Far-field pattern associated with the optimized phase taper.

effective use of a genetic algorithm in finding a quantized low sidelobe phase taper for linear arrays.

8.4 PHASE-ONLY ADAPTIVE NULLING

Low sidelobes don't guarantee adequate reception of a desired signal in the presence of interfering sources. Adaptive nulling complements low sidelobe antennas by placing nulls in some of the low sidelobes to reject the strongest interfering sources. An ideal adaptive algorithm for a phased array antenna has the following desirable characteristics [7]:

- Places multiple deep nulls in the directions of interference,
- Rejects interference over the bandwidth of the antenna,
- Places the nulls very quickly,
- Complements existing phased array technology, and
- Minimizes pattern perturbations.

No adaptive algorithm meets all the characteristics. Selecting the adaptive algorithm, hence the desirable characteristics, depends upon the antenna, the cost, the performance requirements, and the interference environment.

Most adaptive antenna algorithms calculate the adapted weights by multiplying the amplitude and phase weights by the inverse of the sampled covariance matrix to calculate the adapted weights. The resulting complex weights produce nulls in the far-field pattern in the directions of interference. A sampled covariance matrix is formed from the complex signals received at each element in the array. Although mathematically elegant and fast these methods have two impractical hardware requirements on the antenna array. First, the array must have an expensive receiver or correlator at each element. Most arrays have a single receiver at the output of the summer, so the antenna must be designed especially for the algorithm. Not only are multiple receivers expensive, but the receivers require a sophisticated method for calibration [10]. Second, the array must have variable analog amplitude and phase weights at each element. Usually, a phased array has only digital beamsteering phase shifters at the elements. The feed network determines amplitude weights There are two problems from an algorithmic standpoint as well. First, digital phase shifters only approximate the phase calculated by the adaptive algorithms. The weight quantization error limits null placement. Second, these algorithms get stuck in local minima [11]. As a result, they do not find the optimum weights to reject the interference. Some common adaptive algorithms include least mean square algorithm and Howells-Applebaum adaptive processor, and examples can be found in references [11] and [12]. These methods are reasonably fast but the difficulties mentioned prohibit their widespread use, particularly for arrays with more than a handful of elements.

Another class of algorithms adjusts the phase shifter settings in order to reduce the total output power from the array [13], [14], [15]. These algorithms are cheap to implement because they use the existing array architecture without expensive additions, such as adjustable amplitude weights or correlators. Their drawbacks include slow convergence and possibly high pattern distortions.

This class has four approaches. The first approach is the random search algorithm [11]. Random search algorithms randomly sample a small fraction of all possible phase settings in search of the minimum output power. The search space for the current algorithm iteration can be narrowed around the regions of the best weights of the previous iteration. This approach is usually too slow for beamsteering and radar applications. It is less likely to get stuck in a local minimum and does not require an expensive receiver at each element. A second approach forms an approximate numerical gradient and uses a steepest descent algorithm to find the minimum output power [16]. This approach has been experimentally implemented but is slow and gets stuck in local minima. As a result, the best phase settings to achieve appropriate nulls is usually not found. The third approach is a beam space algorithm that assumes the location of the interference is known. This algorithm forms a cancellation beam in the direction of the interference. The height of the cancellation beam is adjusted to cancel the sidelobes and place a null in the interference direction. This approach is fast but requires knowledge of the interference locations and a reasonably accurate estimate of the amplitude and phase weights at each element. Table 8.1 summarizes the advantages and disadvantages of the three phase-only algorithms.

Serious drawbacks to current adaptive algorithms include:

1. Require an expensive receiver at each element—makes array impractical to build.
2. Get stuck in local minima—don't use full potential of the antenna to reject interference.
3. Slowly converge—often not useful for radar or scanning applications.

TABLE 8.1 Advantages and Disadvantages of Three Phase-Only Adaptive Nulling Algorithms

Adaptive Algorithm	Advantages	Disadvantages
Random Search	Hops out of local minima	Very slow
Gradient	Small pattern distortion	Slow—gets stuck in local minima
Beam space	Fast—hops out of local minima small pattern distortions	Must know locations of inferference

4. Can't be implemented on existing antennas—they require adjustable amplitude weights and receivers at every element in addition to beamsteering phase shifters.
5. Cause the main beam to move from its desired pointing direction.
6. Significantly raise the sidelobe levels of a low sidelobe array.

Lowering the sidelobe levels requires an even phase shift about the center of the array [9], while nulling requires an odd phase shift [17]. Since nulling is of importance here, (8.3) simplifies to

$$AF(u) = 2\sin\phi \sum_{n=1}^{N} a_n \cos[(n-1)\Psi + \Delta_n + \delta_n] \tag{8.9}$$

where

$2N$ = number of elements
$\Psi = kdu$
Δ_n = beamsteering phase of element n

Figure 8.6 is a diagram of a phase-only adaptive array with $\Delta_n = 0$.

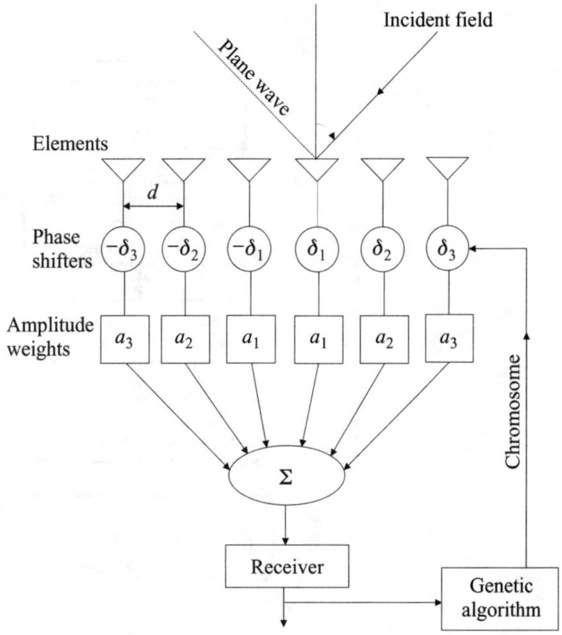

Figure 8.6 Diagram of a phase-only adaptive array.

The digital phase shifters have B bits. B needs to be as small as possible to reduce the cost of the phase shifter but should be large enough to maintain low sidelobes over the scan angles of the array. The assumption here is that only a subset of the phase shifter bits (S) are used to perform the nulling. In order to minimally perturb the array pattern, the adaptive algorithm assumes $S < B$.

8.5 ADAPTIVE ALGORITHM

A phase-only adaptive algorithm modifies the quantized phase weights based on the total output power of the array. If no interference is present, then the algorithm tries to minimize the desired signal. To prevent desired signal degradation, the algorithm should only be turned on when the desired signal becomes swamped by the interference or the nulling phase shifts should be small. This potential problem is discussed in more detail in the next section. This section presents a method that is as fast as the beam space algorithm, doesn't easily get stuck in local minima, and limits pattern distortion.

The adaptive algorithm is based on a genetic algorithm and uses a limited number of bits of the digital phase shifters. A genetic algorithm is a computer program that finds an optimum solution by simulating evolution in nature. In this application the phase shifter settings evolve until the antenna pattern has nulls in the directions of jammers. A genetic algorithm was chosen for this application, because it is an efficient method to perform a search of a very large, discrete space of phase settings for the minimum output power of the array. An adaptive phase-only array has 2^{NB} possible phase settings, many corresponding to local minima in the total power output. Such a large number of phase settings and local minima makes random search and gradient-based algorithms impractical to use.

Figure 8.7 shows a flowchart of the adaptive genetic algorithm. It begins with an initial population consisting of a matrix filled with random ones and zeros. Each row

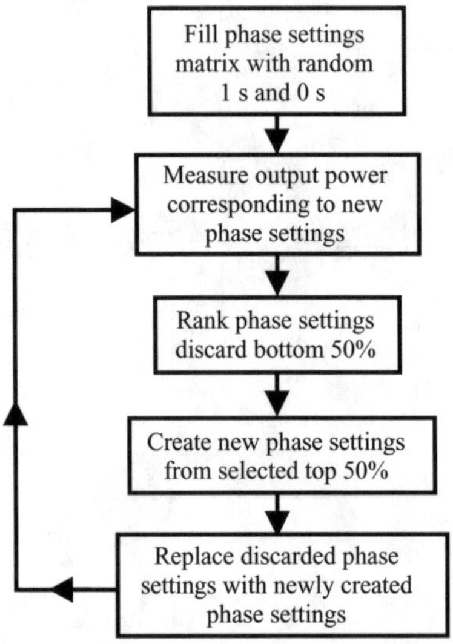

Figure 8.7 Flowchart of an adaptive genetic algorithm.

of the matrix (chromosome) consists of the nulling bits for each element placed side by side. There are NS columns and M rows. The output power corresponding to each chromosome in the matrix is measured and placed in a vector (Fig. 8.8). M must be large enough to adequately search the solution space and help the genetic algorithm arrive at an excellent solution. On the other hand, M needs to be small, so the algorithm is fast. The speed of the algorithm is also a function of N and S. As N and S increase in size, M needs to be larger to keep the algorithm out of local minima, and the number of iterations required for convergence increases. The output power vector and associated chromosomes are ranked with the lowest output power on top and the highest output power on the bottom. The next step discards the bottom X% of the chromosomes, because they have the greatest output power. New nulling

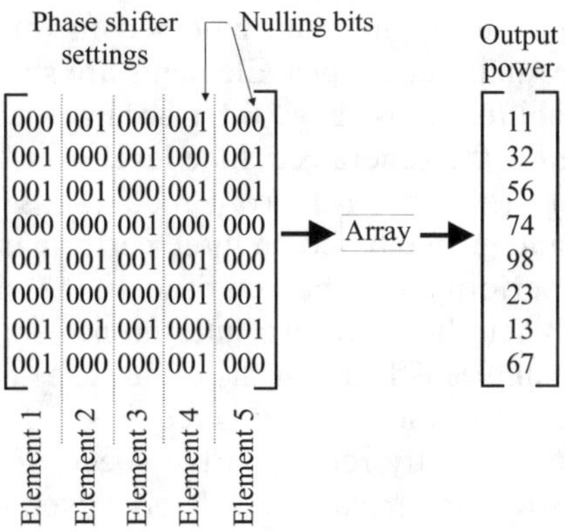

Figure 8.8 A population of phase shifter settings with corresponding output powers.

chromosomes to replace those discarded are created from the chromosomes that were kept (Fig. 8.9). Two chromosomes are selected at random. Chromosomes with lower output power receive a correspondingly higher probability of selection. Next, a random point is selected and bits to the right of the random point are swapped to form two new chromosomes. These new chromosomes are placed in the matrix to replace two settings that were discarded, and their output powers are measured. When enough new chromosomes are created to replace those discarded, the chromosomes are ranked and the process repeated. A small number (less than 1%) of the nulling bits in the matrix can be randomly switched from a one to zero or vice versa. These randomly induced errors allow the algorithm to try new areas of the search space, while it converges on a solution. Usually, the best phase setting is not randomly altered. The next section shows results for determining the best values for S and M.

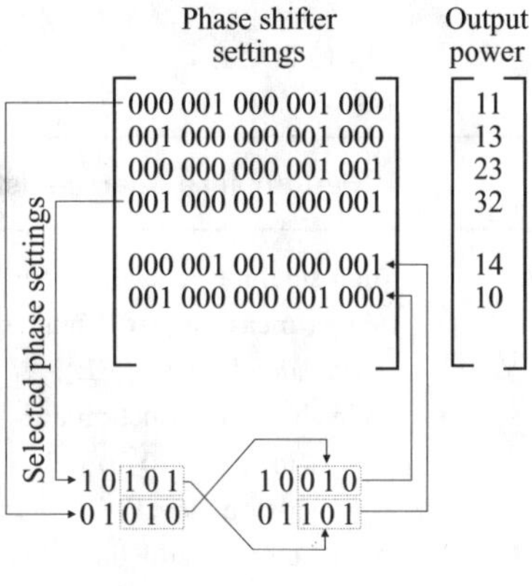

Figure 8.9 Two partners are selected from the mating pool to produce two offspring that are put back into the population to replace those chromosomes that were discarded.

8.6 ADAPTIVE NULLING RESULTS

The genetic algorithm used here begins with 20 chromosomes. Only 10 are kept during the 25 iterations of the simulation. Each iteration, the bottom 6 chromosomes are discarded and replaced by chromosomes generated from the top four chromosomes. These numbers are small enough to keep the algorithm fast but large enough to place the nulls.

The first example array modeled in this paper has 40 elements and a 30 dB Chebychev amplitude taper. Elements are spaced $0.5\lambda_0$ apart at the center frequency f_0. Phase shifters must have 6-bit accuracy in order to keep the quantization lobe slightly below the general sidelobe level over the scan angles of the array ($\pm 30°$ from broadside). The adaptive array model was tested for five different interference scenarios and judged based on three performance criteria. Results appear in Table 8.2. The first performance characteristic is the sidelobe reduction at the interference angle(s). When there are two interference sources, the minimum sidelobe reduction of the two angles is listed in the table. The second performance characteristic is the number of power measurements required for a 10 dB reduction in the sidelobe level. This number directly relates to the speed of the algorithm. The third performance characteristic is the maximum sidelobe level, which is an indication of the amount of pattern distortion. A lower value for any performance characteristic indicates better performance.

Two important parameters are the number of bits used for nulling and the maximum phase shift or the phase of the most significant bit (MSB). A genetic algorithm is often sensitive to the number of bits in a chromosome. Thus, comparing the performance of the algorithm against the number of bits used for nulling is important. The maximum phase shift impacts the main beam and sidelobe distortion

TABLE 8.2 The Genetic Algorithms Adaptive Array was Tested for Five Scenarios. The Nine Columns to the Right List the Number of Nulling Bits and the Phase Value of the Most Significant Bit (MSB). The Three Performance Characteristics Judge the Algorithm for Null Depth, Speed, and Pattern Distortion. A Low Value for a Performance Characteristic is Good.

			NUMBER OF BITS	4	3	2	1	1	1	1
			MSB PHASE	$\frac{\pi}{4}$	$\frac{\pi}{8}$	$\frac{\pi}{16}$	$\frac{\pi}{32}$	$\frac{\pi}{16}$	$\frac{\pi}{8}$	$\frac{\pi}{4}$
Scenario	u_i	u_s	Performance Characteristic							
1	0.62, 0.72	0	min sidelobe reduction at u_i (dB)	−37	−20	−41	−8	−9	−17	−18
			# pwr meas for 10 dB null	68	48	32	–	–	32	26
			max sidelobe level (dB)	−20	−21	−24	−30	−28	−22	−16
2	±0.12	0	min sidelobe reduction at u_i (dB)	−15	−11	0	0	0	0	−14
			# pwr meas for 10 dB null	116	110	–	–	–	–	26
			max sidelobe level (dB)	−22	−22	−28	−33	−28	−32	−18
3	0.61, 0.63	0	min sidelobe reduction at u_i (dB)	−16	−16	−16	−43	−20	−15	−15
			# pwr meas for 10 dB null	26	32	26	26	26	38	26
			max sidelobe level (dB)	−16	−22	−16	−30	−26	−22	−16
4	0	0	min sidelobe reduction at u_i (dB)	−9	−1.5	−0.3	0	−0.1	−0.6	−2.5
			# pwr meas for 10 dB null	–	–	–	–	–	–	–
			max sidelobe level (dB)	–	–	–	–	–	–	–
5	−0.19	0.225	min sidelobe reduction at u_i (dB)	−10	−7	−7	−4	−8	−8	−9
			# pwr meas for 10 dB null	5	–	–	–	–	–	–
			max sidelobe level (dB)	−15	−23	−26	−28	−27	−25	−17

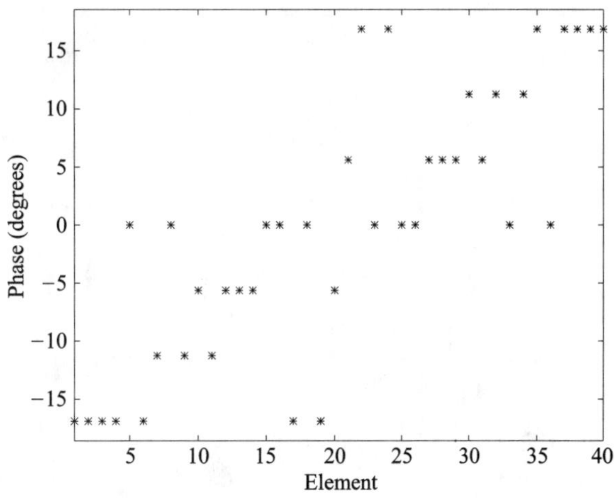

Figure 8.10 Adapted phase weights for the 40-element ($d = 0.5\lambda$) low sidelobe array with 6-bit phase shifters. Two bits were used for nulling with an MSB of $\pi/16$. Interference sources appear at $u = 0.62$ and 0.72.

and the null depth obtainable. Results for MSB phase values above $\pi/4$ produced significant pattern distortions, so they are not listed in Table 8.2.

Scenario 1 is a simple example of two interference sources at $u_i = 0.62$ and 0.72. Sidelobe reduction is best for a middle-sized MSB and S. Convergence is fastest for a smaller S and MSB. As expected, sidelobe distortion increases with larger values of the phase of the MSB. This example suggests that $S = 2$ and MSB phase $= \pi/16$ is the best combination. Figure 8.10 shows the adapted phase weights that produce the nulled pattern in Fig. 8.11. The maximum phase shift to produce the nulling is 16.875°. The maximum sidelobe level increases to -27 dB. Figure 8.12 is a graph of the maximum and average sidelobe reduction of the chromosomes in the population after each iteration. Iteration 0 is the quiescent level of the sidelobe. The average chromosome is of importance, because the antenna receives the desired signal even while it is adapting (measuring the power output for the various chromosomes). If the average chromosome improves the interference rejection, then the likelihood of receiving the desired signal during the adaptation period is improved. The average and best chromosomes are the same at iteration 11. They don't remain the same in later iterations because random mutations are introduced into the population.

Scenario 2 presents the algorithm with the difficult situation where the jammers are at angles symmetric about the main beam. In Fig. 8.11 notice the increase in the sidelobes symmetrically opposed to the nulls placed by the adaptive algorithm. This phenomenon is discussed in [17]. As noted in Table 8.2, symmetric interference sources can only be nulled with large values of phase and many power measurements. Pattern distortions are quite high. Performance improves as the symmetry is broken. When the interference sources are one sidelobe apart, performance similar to Scenario 1 is obtained.

Scenario 3 tests the bandwidth performance of the algorithm by placing two interference sources close together. In this case, the smaller S and MSB phase, the better the algorithm performance. For the most part, null depth is diminished for broadband jammers.

Scenario 4 checks the algorithm when no interference is present, but the desired signal is received by the main beam. The algorithm should not attack the desired signal. Limiting the size of the phase shift keeps the main beam intact. As long as S

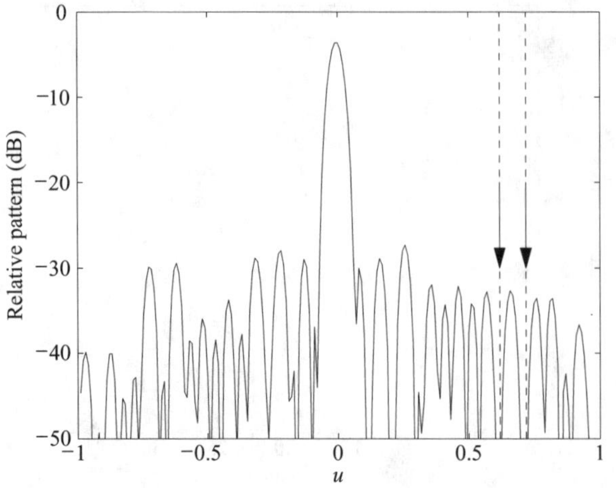

Figure 8.11 Adapted antenna pattern corresponding to the phase shifts in Fig. 8.10.

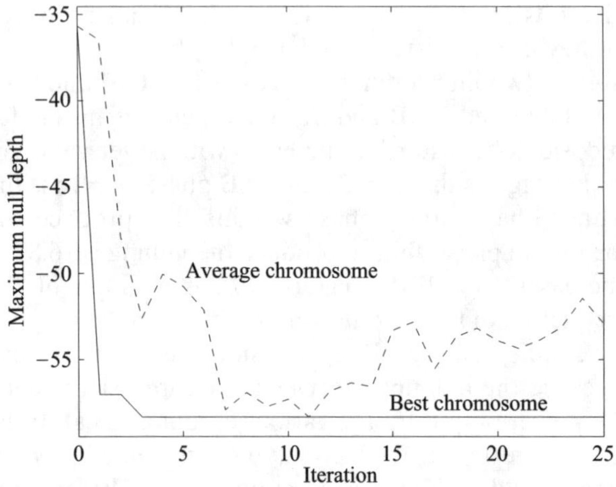

Figure 8.12 Genetic algorithm convergence is graphed over 25 iterations. The solid line represents the output power for the best chromosome, while the dashed line represents the average output power for all the chromosomes in the population.

and MSB are small, then the algorithm has little effect on the main beam. An MSB phase of $\pi/16$ or $\pi/32$ does not require turning the algorithm off when no interference is present, because the algorithm can't attack the main beam.

Scenario 5 has the algorithm attempt to place a null in the quantization sidelobe. Results are marginal for all values of MSB phase. The quantization lobe results from the digital phase shifter quantization.

One last example tests the algorithm for a larger array with high sidelobes. Consider a uniform array of 100 elements spaced 0.5λ apart. Figure 8.13 are the adapted weights for the far-field pattern with interference sources at $u = 0.25$ and $u = 0.35$. Only two nulling bits were used with an MSB of $\pi/16$. The resulting nulled pattern appears in Fig. 8.14. Convergence results are shown in Fig. 8.15. The algorithm takes longer to place deep nulls, because the number of elements is larger.

The algorithm performed well for two jammers that were separated by an angular width of at least one sidelobe and were not symmetric in angular distance from the main beam. Both the 40 element low sidelobe array and the 100 element uniform array showed fast convergence, deep nulling capability, and small pattern distortions.

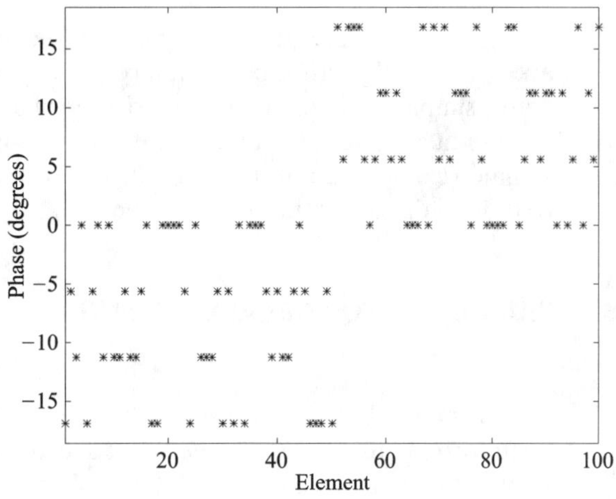

Figure 8.13 Adapted phase weights for the 100-element ($d = 0.5\lambda$) uniform array with 6-bit phase shifters. Two bits were used for nulling with an MSB of $\pi/16$. Interference sources appear at $u = 0.25$ and 0.35.

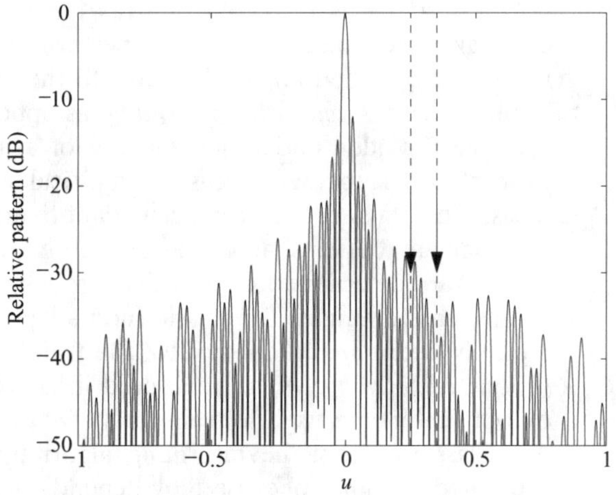

Figure 8.14 Adapted antenna pattern corresponding to the phase shifts in Fig. 8.13.

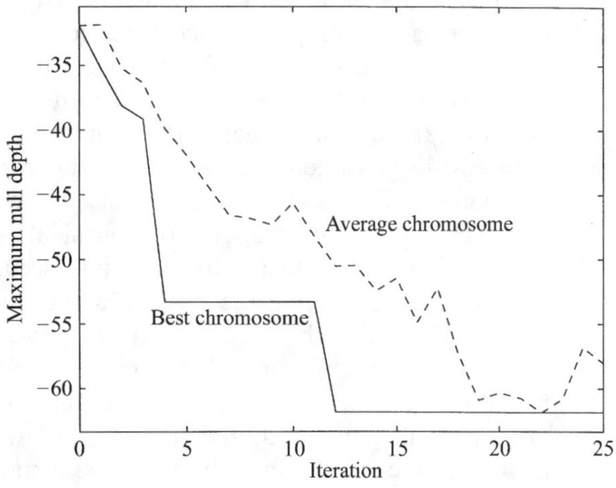

Figure 8.15 Genetic algorithm convergence is graphed over 25 iterations. The solid line represents the output power for the best chromosome, while the dashed line represents the average output power for all the chromosomes in the population.

Using only a couple of least significant bits and small phase values for the MSB are a key to the algorithm performance. This algorithm has the important advantage of being simple to implement on existing phased arrays. Disadvantages include little success at nulling interference entering a quantization sidelobe and interference at symmetric angles about the main beam. Increasing the bandwidth of the interference also degrades algorithm performance.

8.7 NEURAL NETWORK BEAMFORMING

Neural networks can be used with phased array antennas to perform many useful beamforming functions. The network acts not as a true RF beamformer (i.e., a passive network that passes the entire RF signal) but as a nonlinear signal processor [18], [19]. A network might be used to produce an estimate (such as the unknown direction of arrival of a plane wave) from a measurement performed by the array. Another network can be used to control array functionality (such as producing phase shifter commands) to cause the array to perform a specific task, for example to steer the beam to a specified angle. We refer to the application of neural networks to these general areas as *neural beamforming*, as opposed to true RF beamforming.

The key idea underlying the use of a neural network processor as a "beamformer" is that the network is an empirical system which can be trained in situ with measured data and, therefore, can adapt to array non-uniformities, degradations, and environmental effects, and still adequately perform the required function. Network weights are determined empirically through standard neural network learning algorithms, unlike weights for classic (non-adaptive) beamforming algorithms which are typically determined analytically. The training techniques we use are referred to as *supervised learning*, where measured input vectors and their corresponding desired output vectors are used to train the network to learn some unknown function.

Classic antenna beamforming algorithms, such as monopulse, require highly calibrated antennas because they depend on antenna elements having near-identical behavior. These algorithms typically do not adapt well to uncalibrated antennas or antennas with degradations [20], [21]. The requirement for identical elements results from a lack of beamforming algorithms capable of managing the complexities introduced by non-identical elements. Traditional techniques synthesize aperture behavior as a mathematical combination of well-behaved individual element and receiver channel behaviors. Neural beamforming, in contrast, attempts to approximate overall aperture behavior from a finite number of discrete observations of that behavior. If we assume that antenna aperture behavior is a continuous function, it is theoretically possible to model it with a neural network trained at discrete samples of the function. This model can then be used to predict antenna behavior at other points. This is the powerful property of neural networks called *generalization* [22] which allows neural networks to efficiently and accurately approximate complex functions.

We will first define the basic structure and fundamental terminology of neural networks to establish a common ground. We then describe a specific array antenna application where a neural network essentially replaces a piece of microwave hardware (a Butler matrix) to perform the same single-source direction finding (DF) task. Neural network applications in DF will then be discussed in more detail, including a comparison with the classic monopulse DF algorithm and potential applications to multiple-source DF. We then describe the use of neural networks in array control, in particular, the array beamsteering function.

8.8 NEURAL NETWORKS

A neural network is an implementation of a learning algorithm derived from research about the structure of the brain and the way it processes information [22]. It is often referred to as an artificial neural network (ANN) since it is only a theoretical model of biological neurons and the parallel nature of the connections between them via axons and dendrites [23]. An artificial neural network is characterized by: (1) the pattern of interconnections between neurons (called the *architecture*); (2) the method of determining weights (called *training* or *learning*); and (3) the computations performed by the various neurons (called the *activation functions*) [23].

An ANN, which we will hereafter refer to simply as a neural network (NN), can be described on several levels. Mathematically, it can be considered as a collection of vector equations. On another level, it can be considered a computer code which implements the equations. In a hardware implementation, it can be considered as the VLSI circuits which perform the required processing. Perhaps the most visually descriptive level of the network is a layered system of interacting nodes as shown in Fig. 8.16. This network architecture is very common and is similar to the one we use to process signals from phased array antennas. It is referred to as a three-layer, feed forward, fully connected neural network [22].

Each node is a separate processing element which acts on its inputs to produce an output. The nodes are arranged in layers (analogous to the structure in the brain). Nodes in one layer have as inputs the outputs of the preceding layer; these nodes then provide their outputs through weighted connections as inputs to the following layer. A layer is simply a set of nodes connected to common inputs, common outputs, or both [22].

Raw data, which are preprocessed to ease network computational requirements and aid in generalization, enter the network through the input layer, shown on the left in Fig. 8.16. The passive input nodes perform no computations; each simply broadcasts a single data value to the hidden layer nodes. A hidden layer (or layers) has nodes *not* directly connected to the network's input or output. This intermediate layer (or layers) enhances the network's ability to model complex functions. Hidden layer nodes process their inputs via some activation function and broadcast their results to the output layer (or the next hidden layer).

The activation (or transfer) function used in the neural networks described in the following sections is a Gaussian radial basis function (RBF) as indicated in Fig. 8.16. The hidden layer nodes process input data in two steps. First, the radial distance (Euclidean norm, indicated by $\|.\|$ in Fig. 8.16) between the input vector and the node center, m, is calculated for each hidden layer node. The node output is simply evaluated as a Gaussian function of this radial distance. We more precisely define this function in the next section. Output layer nodes perform a linear, weighted sum of the outputs of all the hidden layer nodes, hence the terminology "basis functions."

The selection of the above RBF architecture is based on its capability to approximate unknown continuous functions and on a goal of real-time processing which is independent of antenna array size. This requirement generates two guidelines

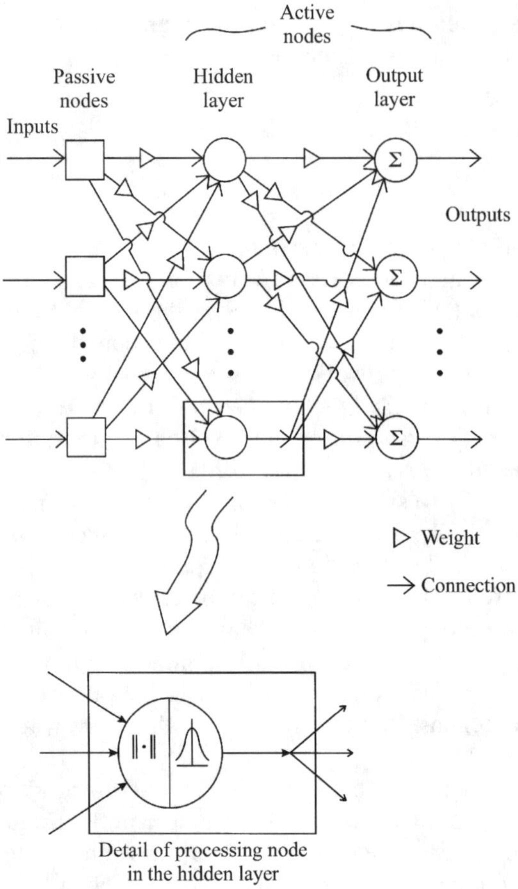

Figure 8.16 Architecture for a three-layer, feedforward radial basis function neural network.

for the NN architecture. First, it is desirable that the network have a processing delay which can be kept constant regardless of the dimension of the inputs. Second, the network should have a minimum number of layers (defined by computations that must be performed sequentially). Thus, as the number of inputs increases (as the dimensions of the antenna increase), the size of the network layers should grow at the same rate and no additional layers should be required. The RBF network architecture reportedly satisfies all of these constraints. The mathematical basis for these networks stems from the fitting (approximation) and regression capabilities of RBFs [24], [25], combined with a feed-forward NN architecture [26], [27].

8.9 DIRECTION FINDING

8.9.1 Analogy Between the Neural Network and the Butler Matrix

Before discussing the general direction finding performance of a NN, it is instructive to describe the analogy between the network and a Butler matrix used as a traditional analog DF beamformer. For this very special case, the NN can be considered as essentially a replacement for the Butler matrix microwave hardware (fixed phase shifters and hybrid couplers arranged in a microwave circuit). Hopefully, this discussion will provide insight into the nature of the beamforming function that the NN synthetically performs. In addition, we use this section to introduce the differences between the network represented in Fig. 8.16 and one used in practice.

An N-element Butler matrix [28] forms N radiated beams, or receives N signals from beams with maxima at angles θ_l. We define the direction cosines, u_l, such that for a 1-dimensional array with element spacing d and wavelength λ:

$$u_l = \sin(\theta_l) = \frac{l\lambda}{Nd} \tag{8.10}$$

for l a half-integer, and N even, i.e.

$$-\frac{N-1}{2} \leq l \leq \frac{N-1}{2} \tag{8.11}$$

The beam directions, θ_l, correspond to the peaks of orthogonal beams in space. The radiated beams have the shape of modified sinc functions, i.e., $\sin(Nx)/N\sin(x)$. Because of orthogonality, the microwave network is lossless and a response due to a wave from one of the "orthogonal beam directions" u_l appears as a signal only at beam port l, corresponding to the lth orthogonal beam. For these incident angles, all other beam ports have zero response.

These element currents (signals) are weighted in the matrix such that for the lth beam, the weights are given by

$$w_{nl} = e^{-j2\pi(d/\lambda)nu_l} \tag{8.12}$$

where u_l is the beam center (in u space) of the lth beam.

The signals are weighted and combined to obtain the following normalized response for the lth beam:

$$S_l(u) = \frac{1}{N} \sum_{n=-(N-1)/2}^{(N-1)/2} I_n(t) e^{-j2\pi(d/\lambda)nu_l} \tag{8.13}$$

We divide by N to normalize the response to $A(t)$ at the beam maxima (peaks) in angle space.

This combines to form the resulting signal at the lth beam port.

$$S_l(u) = A(t) \frac{\sin\left(\frac{N\pi d_x}{\lambda}(u - u_l)\right)}{N\sin\left(\frac{\pi d_x}{\lambda}(u - u_l)\right)} \tag{8.14}$$

This output includes the modulation $A(t)$ weighted by the angular filtering of the antenna. It is obvious from this expression that the u_l are the beam centers (peaks) in u space. For simplicity, hereafter, we assume that the modulation is a pure tone, so that the common factor $e^{+j\omega t}$ can be factored out and disregarded.

We can clearly illustrate the essential concepts and obtain analytic expressions for the neural network weights using an array consisting of only two antenna elements. The circuit diagram for the two-element Butler matrix is shown in Fig. 8.17(a). The element spacing is one-half wavelength at the operating frequency. The weights are implemented as 45° fixed phase shifts.

By noting the relative signal amplitudes at adjacent beam ports, one can estimate the angle of arrival of any incident signal. This is the postprocessing stage, and for the Butler matrix it can be done very simply and accurately. For orthogonal training points, one can show that by comparing the relative amplitudes of the two adjacent output node values (S_k and S_{k+1}) and by approximating $\sin(Nx)/N\sin(x)$ by the sinc function (which is extremely accurate for closely spaced beams), one can obtain a nearly exact interpolation formula:

$$u = u_k + (u_{k+1} - u_k)\frac{S_{k+1}}{S_k + S_{k+1}} \tag{8.15}$$

It can be shown that this expression is valid for $\pi/N \leq 1$. The same interpolation algorithm is used for the NN direction finding array using adjacent output node responses.

The connection diagram for the NN direction finding (DF) array is shown in Fig. 8.17(b) for the simple two element array. This is the network architecture used in implementing the DF and beamsteering functions described in the remainder of the chapter. The only difference between the architecture in Fig. 8.17(b) and the idealized one in Fig. 8.16 is that the weights on the connections between the input layer and the hidden layer nodes are equal to unity. At the equivalent "layer" in the Butler matrix shown in Fig. 8.17(a), the weights can be derived analytically from beam orthogonality considerations [28]. In the NN [Fig. 8.17(b)], the weights on the connections between the hidden layer Gaussian nodes and output summation nodes are determined by training the network with a suitable set of training data, as discussed below.

The NN DF array operates as follows: received signals from the antenna array elements are preprocessed to eliminate extraneous data and convert the available signals into something that the neural network can efficiently use. For example, when the array receives a single incident signal, there is no expected amplitude variation between signals at each element (edge effects neglected). Thus the preprocessor obtains only the phase or phase differences between adjacent elements.

As documented in [29], the use of raw antenna measurements (consisting of either in-phase and quadrature signal components or the amplitudes and phases of the signal) as inputs to the network yielded unacceptable network performance. Our preprocessing method determines the phase difference between signals received at consecutive array elements and uses the sine and cosine of this phase difference as inputs to the network. The phase difference is used because it is a strong function of the angle of arrival (AOA). Taking the sine and cosine of the phase difference overcomes the difficulty of training the network on data which have 2π periodic phase discontinuities (branch cuts) inherent in the phase measurement. Additionally, the branch cuts occur quasi-randomly depending upon arbitrary phase references. As shown in Fig. 8.17(b), the ideal phase difference between the two elements, represented by $x(\theta)$, is given by

$$x(\theta) = \frac{2\pi}{\lambda} d \sin(\theta) = \pi \sin(\theta) = \pi u \tag{8.16}$$

for element separation $d = \lambda/2$ and $u = \sin(\theta)$. The inputs are $\sin(x(\theta))$ at the first node and $\cos(x(\theta))$ at the second node, which we represent as a two-component input vector, $\boldsymbol{I} = [\sin(x(\theta)), \cos(x(\theta))]$.

The input vectors are broadcast to the hidden layer nodes which perform the

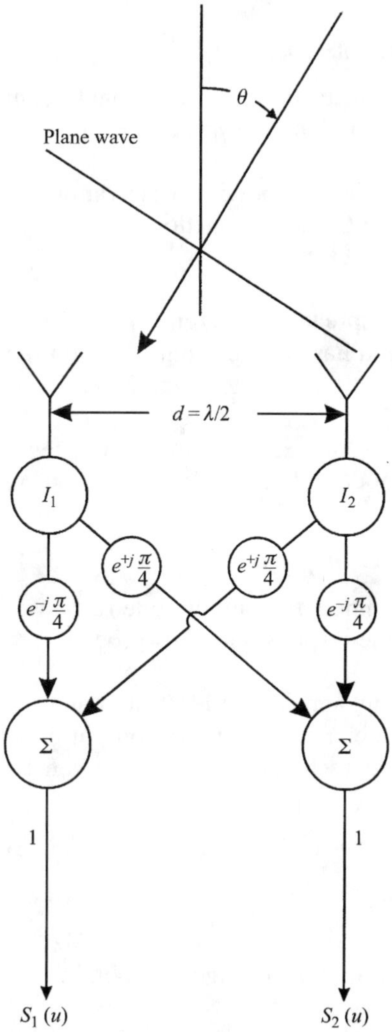

Figure 8.17 (a) Simplified two-element/two-beam Butler matrix for $d = \lambda/2$. For phase reference at the array midpoint, $I_1 = \exp(-j\pi/2u)$ and $I_2 = \exp(+j\pi/2u)$ with $u = \sin(\theta)$. The fixed phase shifts are calculated from (8.12) with orthogonal beam positions $u_l = u_{\pm 1/2} = \pm 0.5$. The ones (1) indicate unit weights.

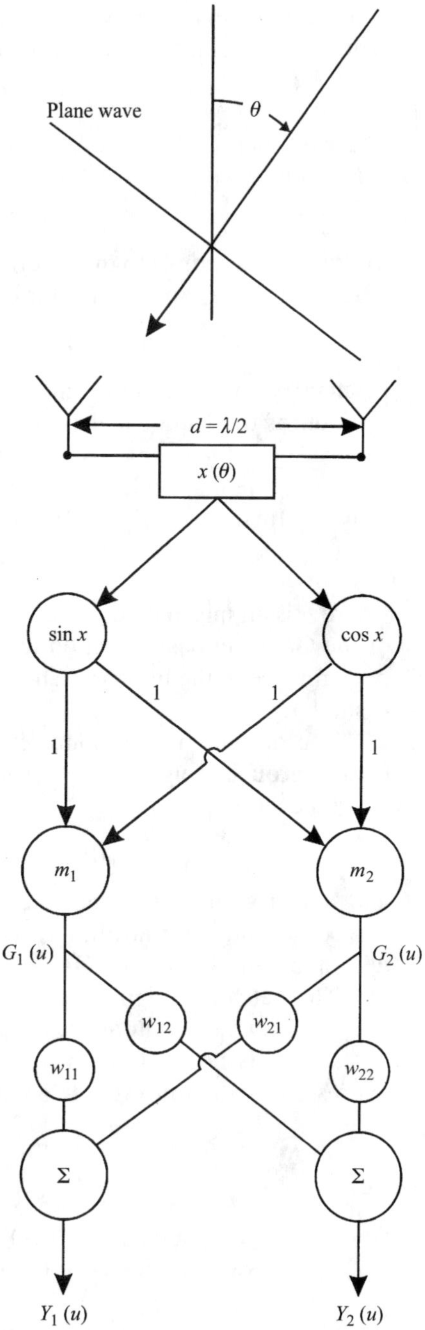

Figure 8.17 (b) Simplified two-element/two-node direction finding array. The phase difference is $\alpha = (2\pi/\lambda)du$, with $u = \sin(\theta)$. The input vector has two components ($\sin\chi$, $\cos\chi$), and the centers have two components, with $m_1 = (\sin\chi_1,\ \cos\chi_1)$ and $m_2 = (\sin\chi_2,\ \cos\chi_2)$ where $\chi_l = \pi d u_l$ and $u_l = \pm 0.5$ are the Butler matrix orthogonal beam locations.

following processing. First, the radial distance, C_i, between the inputs and the node center, m_i, is found. For example, for hidden node 1,

$$C_1 = \|\boldsymbol{I}(u) - m_1\| \tag{8.17}$$

where $u = \sin(\theta)$, $\|.\|$ is the Euclidean norm, and m_1, which is the node center for node 1, is given by

$$m_1 = [\sin(\pi u_1), \cos(\pi u_1)] \tag{8.18}$$

where u_1 corresponds to the first training angle. Therefore, the radial distance between input vector $\boldsymbol{I}(u)$ and the center of node 1 is

$$C_1 = \left[[\sin(x(\theta)) - \sin(x(\theta_1))]^2 + [\cos(x(\theta)) - \cos(x(\theta_1))]^2\right]^{1/2} \tag{8.19}$$

In analogy with the Butler matrix, we make u_1 equal to the first orthogonal beam position of the Butler matrix, i.e., $u_1 = -\frac{1}{2}$. For a two element system with $d = \lambda/2$, $u_{-1/2} = -\frac{1}{2}$ and $u_{+1/2} = +\frac{1}{2}$, corresponding to $\theta_1 = -30°$ and $\theta_2 = +30°$.

Next, the first node output value is calculated from the nonlinear activation (or transfer) function. For our RBF network, we use the Gaussian function

$$G_1(u) = e^{-(C_1^2/2\sigma^2)} \tag{8.20}$$

where σ is an internal network parameter called the spread parameter. It is the width of the Gaussian basis function in the input space and has an important effect on the performance of the network. The calculation of the output value for node 2, i.e., $G_2(u)$ is similar.

The outputs of the hidden layer nodes are fully connected to the output layer nodes through weights. The response at output node 1 is given by the following weighted sum

$$Y_1(u) = w_{11}G_1(u - u_1) + w_{21}G_2(u - u_2) \tag{8.21}$$

and similarly for $Y_2(u)$. Once the weights are determined, the output node responses can be calculated for any input, $x(\theta)$. The output node responses are postprocessed to determine an estimated AOA.

The network weights are determined by training, and can be illustrated analytically for a two element array. Like a Butler matrix beam port response, output node 1 response is normalized to unity at u_1 and forced to be zero at u_2. This can be expressed by the following two equations

$$Y_1(u_1) = w_{11} + w_{21}e^{-(C^2/2\sigma^2)} = 1 \tag{8.22}$$

$$Y_1(u_2) = w_{11}e^{-(C^2/2\sigma^2)} + w_{21} = 0 \tag{8.23}$$

where $C = \|\boldsymbol{I}(u_1) - m_2\| = \|\boldsymbol{I}(u_2) - m_1\| = 2$.

These two equations can be solved for the two unknown weights to obtain

$$w_{11} = \frac{1}{1 - e^{-(C^2/\sigma^2)}} \tag{8.24}$$

$$w_{21} = -w_{11}e^{-(C^2/2\sigma^2)} \tag{8.25}$$

and it can be shown that $w_{22} = w_{11}$ and $w_{12} = w_{21}$. Therefore, we have determined all the network weights and completed network training. The network will output exactly the desired response at the training points, i.e.,

$$Y_j(u_k) = \delta_{jk} \tag{8.26}$$

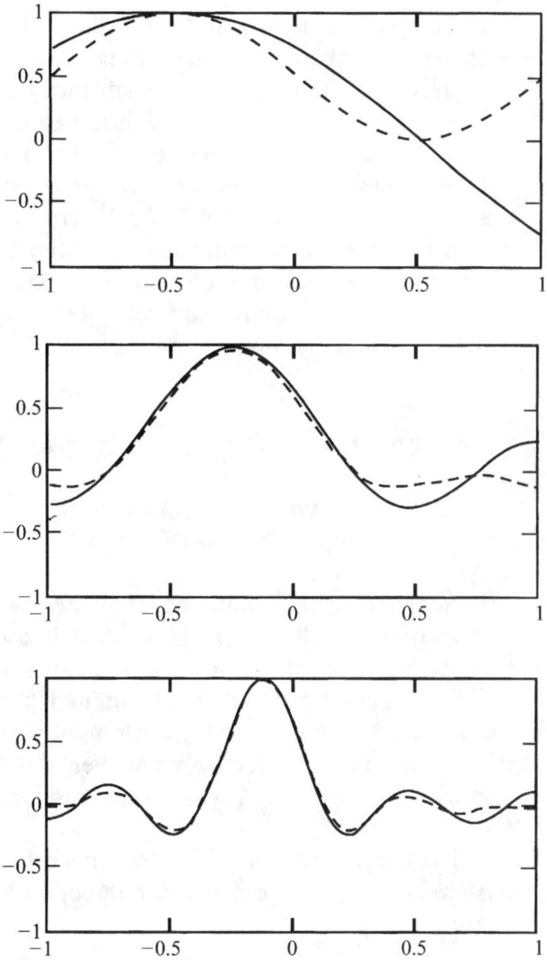

Figure 8.18 Comparison of representative Butler matrix beam port responses and neural network output node responses for two-, four-, and eight-beam systems (top, middle, and bottom, respectively). The Butler matrix beam port response is solid and the neural network output node response is dashed.

where δ_{jk} is the Kronecker delta function. To obtain an estimate of the angle of arrival between training angles, the postprocessing procedure described with (8.15) is used with Y_k and Y_{k+1} representing adjacent output node responses instead of beam port responses S_k and S_{k+1}.

The Butler matrix beam port responses are phase shifted and combined RF element signals which give this beamformer its directional properties. The NN output node responses are not true RF beams, since the network has not passed the entire RF signal; however, directional properties are still present. We can compare the beam port responses to the NN output responses which are weighted sums of the Gaussian processing node outputs. The neural network weights are the full set of weights determined by training the network at certain angles within the field of view of the array. However, to illustrate the comparison with the Butler matrix, we choose training points at the orthogonal beam locations. This forces the output node responses to have the same maximum and zero locations as the Butler matrix. In Fig. 8.18, we show a typical Butler matrix beam port response for two, four and eight beam matrices, i.e., $N = 2$, 4, and 8. For comparison, we show the output response of the corresponding summation (or output) node of a NN DF array. For $N > 4$, the portions of both the Butler matrix and NN responses between the zero crossings on either side of the beam peaks are almost identical. The adaptive network weights, when required to conform to the orthogonal beam location training data, have thus produced an output response that is nearly identical to the Butler matrix beam port response.

Since the postprocessing described above uses only that portion of the beam (or network node) response between the zero crossings on either side of a peak, it follows that the performance of the NN DF array will be very similar to that of the Butler matrix DF system. We emphasize that the performance will be similar only for the ideal systems we have used to derive the system responses. For a degraded system, a neural beamformer trained on data obtained from the degraded system can outperform a conventional beamforming system.

8.9.2 Single-Source DF: Comparison to Monopulse [30]

8.9.2.1 Network Architecture for Single-Source DF

A generalized N antenna element version of the three layer RBF network architecture introduced in Fig. 8.17(b) is used to implement the single source DF function. Each input vector, x, is an element of the n-dimensional input space, $\tilde{X}$. For the experimental eight element antenna used to obtain the data described in this subsection, $N = 8$ and $n = 14$. Each input vector contains the seven cosines and seven sines of the phase differences between the elements. For these n-component input vectors, $x = (x_1, x_2, \ldots, x_n)$, preprocessing ensures such that $x_k \in [-1,1]$, where $k = 1,2,\ldots, n$.

The hidden layer *maps* $\tilde{X}$ into a space $\tilde{\Phi}$, which consists of q-component Gaussian RBF vectors, ϕ, where q is the number of hidden layer nodes. The components of ϕ are

$$\phi_i = e^{-\sum_{k=1}^{n}(x_k - m_{ik})^2/2\sigma^2} \tag{8.27}$$

where ϕ_i is the output of the ith hidden layer node for $i = 1, 2, \ldots, q$. ϕ_i is calculated from the input vector, x, the Gaussian center, m_i, and the spread parameter, σ.

Each output node computes a weighted sum of the outputs generated by the hidden layer nodes,

$$y_j = \sum_{i=1}^{q} \phi_i w_{ij} \tag{8.28}$$

where the w_{ij} values are the output *weights*. Thus, the output layer maps $\tilde{\Phi}$ into $\tilde{Y}$, the space of all possible angular directions to the source, where the output vector $\boldsymbol{y} = (y_1, y_2, \ldots, y_r) \in \tilde{Y}$ for r output nodes. Postprocessing is performed as described in Section 8.9.1 to obtain an estimate of the AOA.

One can avoid some preprocessing and simultaneously reduce the number of input nodes by half, if we replace the Gaussian RBF by a *periodic basis function* (PBF). PBFs have properties similar to RBFs, except that they are periodic in the input variables. See [31] for a discussion.

8.9.2.2 Network Training

In the previous section we presented the forward computational architecture of the RBF network. This computation will only produce correct results if the network has been adequately trained with pairs of inputs and their corresponding outputs. We describe two training techniques for the RBF networks: a standard adaptive RBF training method and a linear algebra technique.

Adaptive RBF training adds Gaussian radial basis functions to the hidden layer as needed. Initially empty, the hidden layer grows as training points are presented to the network. At each training point, either (1) a new Gaussian function is added with its center on the training point and its initial network weights w_{ij} chosen to produce the correct network response for that point, or (2) existing "close" Gaussian functions have their centers moved and weights adjusted to incorporate the new training data. Network weights are adjusted using a modified gradient descent algorithm known as "backpropagation." Additional details regarding the network architecture and training can be found in two articles by Lee [32], [33]. We refer to this implementation of the neural beamformer as ARBF.

Backpropagation of network error, however, does not ensure global optimization and introduces several vague network aspects. Often, the network designer must choose parameters for which the literature contains only general guidelines. Two important parameters include the descent *step size*, and the stopping criteria. The step size dictates the convergence speed and whether the weights will converge or oscillate. One popular approach, which decreases the step size slowly, requires choice of initial step size, decreasing function, and stopping criteria. Even if these parameters are chosen correctly for the application, backpropagation may still stop at local minima.

Unlike ARBF, which adds basis functions as needed, the linear algebra network places Gaussian radial basis functions at all training points and solves for the network weights using linear algebra [34], [21]. We refer to this implementation of the neural beamformer as LINNET. LINNET finds globally optimal weight values, in the least mean squared (LMS) error sense. The weights are only optimal with respect to the training data used to construct the corresponding matrices, not necessarily for any other data.

Consider a specific set of L input vectors from the input space $\tilde{X}$ labeled x_l, where $l = 1, 2, \ldots, L$. For our data sets which measure a source at 1° intervals from $-60°$ to $+60°$, $L = 121$. Each component of the n-dimensional input vector is x_{lk}, where $k = 1, 2, \ldots, n$. Similarly, consider a specific set of L Gaussian RBF vectors $\phi_l \in \tilde{\Phi}$, where $l = 1, 2, \ldots, L$, corresponding to the L input vectors. A subset of t input vectors is selected to train the network. Since we train on measured data at 10° intervals, $t = 13$, and training vectors are chosen where $l = 1, 11, 21, \ldots, 121$. Data not used for training are used to evaluate network performance. We define an $L \times q$ Gaussian function matrix, Φ, where q is the number of hidden nodes. Since LINNET uses one hidden node for each training point, $q = 13$. Φ contains all the Gaussian function evaluations, ϕ_{li}, for every measured input vector, with $l = 1, 2, \ldots, L$ denoting the input vector and i denoting the Gaussian number, $i = 1, \ldots, q$, i.e.,

$$\Phi_{L\times q} = [\phi_{li}] \qquad \text{and} \qquad \phi_{li} = e^{-\sum_{k=1}^{n}(x_{lk}-m_{ik})^2/2\sigma^2} \tag{8.29}$$

To train the network, we select t rows of Φ to form a 13×13 ($t = 13$, $q = 13$) *training Gaussian function matrix*, Φ_t, which is used to solve for the weights.

Output vector $\boldsymbol{y}_l$ has components y_{lj} for $l = 1, \ldots, L$ and $j = 1, \ldots, r$, where r is the number of output nodes. The L rows of Y are the output vectors corresponding to the L input vectors. Each element of the output matrix, Y, is the weighted sum of Gaussian function values

$$Y_{L\times r} = [y_{lj}] = \left[\sum_{i=1}^{q} \phi_{li} w_{ij}\right]$$
$$Y = \Phi W \tag{8.30}$$

Note that since we use the same number of output nodes, r, as training angles, t, and Gaussian nodes, q, then $r = t = q = 13$.

For training, we know the desired outputs at the t training angles (−60°, −50°, . . ., +50°, +60°), corresponding to the rows of Y with $l = 1, 11, \ldots, 111, 121$. We use these rows to form the desired output matrix Y_d. Using Φ_t and Y_d, we can solve for the $q \times t$ weight matrix W from

$$Y_d = \Phi_t W \tag{8.31}$$

where Y_d is a $t \times t$ identity matrix.

Since we chose the number of training vectors, t, equal to the number of Gaussians, q, both Y_d and Φ_t have the same dimension. Equation 8.31 implies that we have a *determined* linear system, i.e. q equations in q unknowns. If, however, the number of Gaussian nodes were reduced so that $t > q$, then the linear system is *over-determined* and LMS can be used to solve for the weights. In either case, the LMS solution is the most general, since the pseudo inverse of the LMS solution becomes a matrix inverse for $q = t$. All calculations in this chapter assume $q = t$.

The steps for computing the optimal W are:

1. Collect input data, i.e., the x_{lk}, for L measurements at L known source angles.
2. Use a subset, q in length, of the L input vectors to form a $q \times q$ matrix Φ_t. These q input vectors are also used as the Gaussian function centers, m_{ik}.
3. Solve for W using the LMS solution [35].

$$W = (\Phi_t^T \Phi_t)^{-1} \Phi_t^T Y_d \tag{8.32}$$

8.9.2.3 Rapid Convergence

One reason we use an RBF network is that they reportedly require orders of magnitude less iterations to converge (solving for weights) than networks based on the sigmoid neuron activation function [35]. Our results support this claim; ARBF achieved near optimum performance after only three weight update iterations. The error surface shown in Fig. 8.19, generated by LINNET, explains the reason for this rapid convergence. We arbitrarily picked two weights from the optimal LINNET weight matrix, W, perturbed them from optimal, and calculated the resulting network performance to obtain the three-dimensional error surface.

In Fig. 8.19, $W(3,3)$ and $W(12,13)$ represent normalized values for weights "connecting" the output of the third Gaussian with the third output vector component, and the output of the twelfth Gaussian with the thirteenth output vector component, respectively. Normalized values for each weight are indicated along the

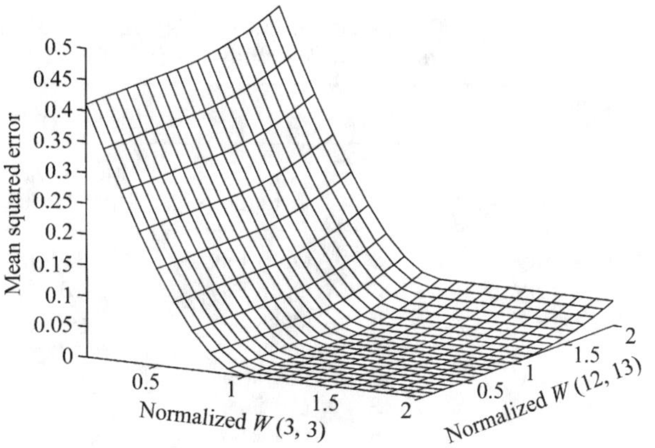

Figure 8.19 LINNET performance vs. weight changes.

axes, with optimal weights occurring at 1, 1. This plot is one of many we investigated; every one had the same steep sides and flat bottom. This held true even for data including severe near-field scattering which resulted in poor DF performance with large rms angle errors (see DATA2 in Table 8.3). Thus, if initial weights lie on a steep side or a flat bottom, the modified gradient descent search should converge quickly to near-optimal performance in only a few iterations for the DF problem.

8.9.2.4 Monopulse Direction Finding

Monopulse is a classic array antenna application for DF which uses high-quality sum and difference beams [37]. We selected aperture distributions developed by Taylor and Bayliss to synthesize sum and difference beams respectively with theoretical −30 dB near-in sidelobes. Of course, beam quality depends upon element match and the received signal-to-noise ratio, *S/N* [37]. We calibrated the array (element amplitude and phase mismatch compensation) by normalizing the ideal Taylor and Bayliss weights with the measured complex signal values obtained at each element for a broadside plane wave. Exact element mismatch compensation occurs at broadside; however, compensation degrades for source angles off array broadside.

We measured received amplitude and phase at each of the eight elements of the array for 121 source azimuth angles between −60° and +60° at 1° steps. The measurements are multiplied by monopulse weights to form beams which supply inputs to the source angle estimation algorithm. Since we know the true source location, we can determine the rms angle error (calculated over all 121 source angles) from this angle estimate. We use rms angle error to compare the performance of both monopulse and neural network DF algorithms.

8.9.2.5 Experimental DF Results

In this section, we describe the experimental array antenna and present monopulse and neural beamformer DF results using measured data from a variety of antenna and environment scenarios.

The antenna was a linear, eight-element array with one half wavelength element spacing at X-band (11.811 GHz). Each "element" was a combined column of eight horizontally polarized, open-ended waveguides (for increased elevation directivity).

TABLE 8.3 Comparison of Monopulse. ARBF, and LINNET RMS Angular Errors (in degrees) under Various Conditions

ANGULAR RMS ERRORS (IN DEGREES)						
Data Set	S/N	Conditions	Monopulse Uncalibrated	Monopulse Calibrated	LINNET	ARBF
TEST97	~35 dB	Open Bay	1.37	0.73	0.41	0.53
TEST92	<10 dB	Open Bay	1.73	1.32	2.01	1.97
TEST87	~35 dB	Open Bay[1]	0.97	1.13	1.46	1.41
TESTSC	~35 dB	Open Bay[2]	5.29	4.30	1.67	1.59
DATA1	~18 dB	Chamber	0.95	1.27	1.43	1.34
DATA2	>30 dB	Chamber[3]	9.45	8.05	14.98	15.06
TEST80	~15 dB	Open Bay	1.90	1.05	0.47	0.54
TEST95	~25 dB	Open Bay	1.37	0.71	1.37	1.29
TEST60	~5 dB	Open Bay	16.59	9.64	1.93	1.95
TEST59	~35 dB	Open Bay[4]	1.53	0.76	0.47	0.56

Note: [1]absorber blocking element 6, [2]metal ground plane screen, [3]coffee cans, [4]repeat of TEST97.

We obtained single-source data for monopulse and NN testing at every degree for azimuth angles from $-60°$ to $+60°$ by stepping a single source on a fixed radius in the far-field of the array. At every angle, we measured and recorded element amplitude and phase by connecting each element, in turn, to a phase/amplitude receiver through a coaxial RF switching matrix.

We collected data for the ten scenarios shown in Table 8.3. The row labels are data set names, where different sets represent changes in either received signal-to-noise ratios, *S/N*, or the physical environment. The location "Chamber" is an anechoic chamber completely encased in microwave absorber, while "Open Bay" designates an open bay style room with the walls partially covered with absorber and an absorber "wall" built into the plane of the array. For TEST87, microwave absorber filled one complete column of open-ended waveguides (one "element"). The fine mesh metal screen used with TESTSC was placed coplanar with and surrounding the face of the array out to many wavelengths. A pair of 2-lb metal coffee cans were stacked on top of one another and centered approximately four wavelengths in front of the array while collecting DATA2. TEST59 was taken 1 month after TEST97 under the same conditions.

Table 8.3 compares the calibrated and uncalibrated monopulse rms error metric for the different measured data sets to the two RBF NNs, ARBF and LINNET. Each simulation uses a single data set, i.e., eight element amplitudes and eight phases at 121 azimuth angles. Note that calibration improves monopulse performance for all data sets except TEST87 and DATA1. The calibration data were obtained from data set TEST97 and used to calibrate the monopulse algorithm for all 10 data sets. Ideally, this removes element mismatch in the array. However, for TEST87 one element was filled with microwave absorber, a condition not accounted for in the calibration data from TEST97. If we calibrate the TEST87 data using the broadside wave data in TEST87, the rms angle error becomes 0.59° instead of the 1.13°. A similar explanation also holds for DATA1. For TEST60, the large rms angle error is due to large errors near $\pm 60°$ where the signal drops below the noise floor. Ignoring these outliers, the error drops to a more reasonable value of 2.5°. The large error for DATA2 is due to the severe near field scattering environment which is not accounted for in the calibration.

In Table 8.3, we also summarize the performance of both RBF neural networks for the same 10 data sets. Each network was trained with input and output vector samples from a single data set at 10° intervals and then tested across that entire data set at 1° intervals. During training, we optimized the Gaussian σ for that particular data set. Note that the performance of LINNET and ARBF is very similar.

Note that neural network training is similar to monopulse calibration, since it uses known external signals to learn something about the array. However, while monopulse calibration attempts to compensate for antenna element mismatch, neural beamformer supervised learning attempts to generalize the entire DF function. The ARBF and LINNET beamformers both approximate the function by locating Gaussian node centers and determining the weight matrix, W. However, LINNET's linear algebra solution for W theoretically yields superior performance (for the same training data) than the modified gradient descent approach used in ARBF.

Table 8.3 shows that, empirically, this is true for training points evenly distributed across the input space. However, for poorly chosen training points (which happens if the antenna or environment conditions degrade), ARBF adapts by moving Gaussian centers to improve performance, while LINNET's Gaussian nodes remain fixed. Therefore, under less than "ideal" conditions (i.e., TEST87 and DATA1) ARBF can outperform LINNET, as shown in Table 8.3.

In most cases, neural beamformer performance is comparable to that of monopulse. For DATA2 data with severe near-field scattering, the monopulse algorithm rms error is half that of the neural net algorithms. However, for TESTSC, which also includes near-field scattering since the large ground screen was not really flat, the neural net algorithms outperformed monopulse by almost a factor of three. The neural net algorithms also significantly outperformed (by more than a factor of two) monopulse at relatively low S/N in a high reflection environment (TEST80).

8.9.3 Multiple-Source Direction Finding

Neural networks have also been used to address the problem of multiple source direction finding; however this is a much more difficult task to accomplish [38]. The difficulty arises because of the breadth of the problem space over which the network must be trained. To date, most success in this area has occurred when individual NNs are used to address subsets of the multisource problem.

For example, consider the possible combinations that may occur when only two sources are present in the signal environment. The signals may be perfectly correlated (coherent), somewhat correlated, or totally uncorrelated. Then, the two signals may be at different strengths (i.e., SNRs) or at the same signal level. The signals can also be amplitude modulated or simple FM signals. Finally, and most importantly, the angles of arrival can be any combination of directions within the antenna field of view. The two-source case alone presents numerous possible signal combinations, and the complexity increases drastically as more sources are introduced.

Despite these difficulties, there are good reasons for considering the use of NNs to address this problem. Traditional multisource DF signal processing approaches like MUSIC and MLE (maximum likelihood estimation), for example, can be hard to implement in real-time hardware. They also are difficult to adapt to antenna arrays with non-linearities or evolving degradations without extensive calibrations and knowledge of the array nonlinearities. A trained NN, on the other hand, has been shown to perform superresolution of several targets significantly faster than MUSIC, with comparable results within the "trained" region of the function space.

The multisource DF problem requires more sophisticated preprocessing techniques than those illustrated here for single-source DF. Simple phase differences between consecutive elements do not contain sufficient information to map the antenna response to an unambiguous set of target directions. Instead, typical multisource DF preprocessing methods involve calculating the spatial covariance matrix (R) of the antenna response over some number of independent samples or "looks." This time-averaged correlation matrix is then used in its entirety or further preprocessed to obtain the network inputs. Reference [38] provides an excellent introduction to an RBF multisource DF network.

8.10 NEURAL NETWORK BEAMSTEERING

The array beamsteering function is the inverse of the direction finding function. In beamsteering, we are given a desired pointing angle and must establish the required phase profile on the array aperture to steer the array main beam to that angle. Therefore, the NN architecture for beamsteering is different than the NN architecture for DF, since the input is a scalar. This scalar is the commanded steering (or scan) angle and can be any value within the antenna field of view.

8.10.1 Network Architecture for Beamsteering

The architecture of the neural network beamsteering system is shown in Fig. 8.20. The Gaussian RBF nodes have centers in the input space of the commanded steering angles (the antenna field of view). Network outputs are phase differences. Postprocessing automatically converts the phase differences to eight digital states (8-bit words) to set the phase shifters to steer the main beam to the commanded angle.

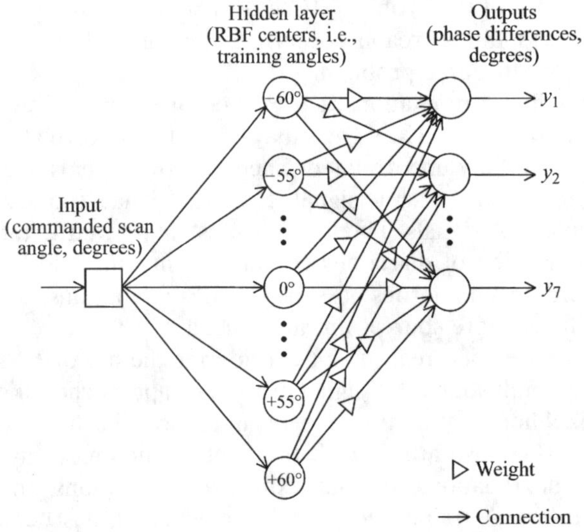

Figure 8.20 RBF neural network architecture used for the beamsteering function.

The role of the network is function approximation [24], [27]. Once again, we train the network with supervised learning by presenting it with beamsteering angles and corresponding "target" or "desired" output vectors. Each target vector consists of the negative of the seven measured phase differences. Negative values are used so we can apply the conjugate phase gradient. The network is "trained" at these angles to output the target vector, and, hopefully, has learned to generalize and accurately steer the beam between training angles.

The RBFs used in the hidden layer nodes (nodes with RBF centers indicated in Fig. 8.20) are Gaussians. The output of the ith Gaussian for steering angle θ is

$$\phi_i(\theta) = e^{-[(\theta-\theta_i)^2/2\sigma^2]} \tag{8.33}$$

for $i = 1, 2, \ldots, t$. There are t training angles in the input space. The θ_i are the Gaussian centers in the input space, and σ is the RBF spread parameter.

A general rule of thumb for determining σ is that it should not be much smaller than the distance between training angles, nor should it be very much larger than this distance. In other words, the Gaussians should overlap sufficiently for generalization between training points, but not be so broad that their responses are the same and the capability for generalization is lost. The jth component of the output vector, y_j, for $j = 1, 2, \ldots, 7$, is computed as a weighted sum of hidden layer node outputs, i.e.,

$$y_j = \sum_{i=1}^{t} \phi_i(\theta) w_{ij} \tag{8.34}$$

To solve for the network weights, w_{ij}, we define an interpolation matrix, Φ, whose elements are given by

$$\phi_{ki} = e^{-[(\theta_k-\theta_i)^2/2\sigma^2]} \tag{8.35}$$

where k and $i = 1, 2, \ldots, t$. Next, define a "desired" output matrix Y_d whose rows are the target output vectors. We can write the relation between Y_d, Φ, and the matrix of network weights W as a linear system.

$$Y_d = \Phi W \tag{8.36}$$

In our case, Φ is square and positive definite, hence invertible. Therefore, we can solve for the weights from

$$W = \Phi^{-1} Y_d \tag{8.37}$$

to complete the training process. The network has then learned the function and we can calculate predicted output vectors for any N commanded scan angles θ_n, for $n = 1, 2, \ldots, N$. We now define a performance matrix $\tilde{\Phi}$ whose elements are calculated from

$$\tilde{\phi}_{ki}(\theta_n) = e^{-(\theta_n-\theta_i)^2/2\sigma^2} \tag{8.38}$$

where $i = 1, 2, \ldots, t$. Note that θ_n must be within the antenna field of view, or, in neural network parlance, the trained region. Output matrix, Y, is calculated from

$$Y = \tilde{\Phi} W, \tag{8.39}$$

where the nth row of Y is the network output vector corresponding to commanded scan angle θ_n.

8.10.2 The Experimental Phased-Array Antenna

The experimental antenna used to test neural beamsteering algorithms was a C-band (7.1 GHz), eight-element, linear array of open-ended waveguides with ferrite phase shifters as shown in Fig. 8.21. Due to mechanical considerations, the elements had to be separated by 0.783λ. Measurements were taken with the array in an

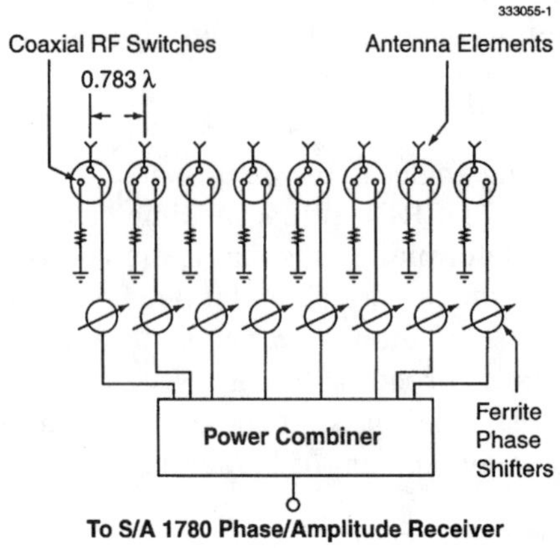

Figure 8.21 Eight element linear array architecture. The coaxial switches switch an element into either a 50 Ω load or into the analog power combiner.

anechoic chamber as shown in Fig. 8.22. The array rotates in front of a stationary source horn located 12 ft away. The linear array is set forward of the center of rotation by 8 in., therefore the array phase center is *not* over the center of rotation.

With all eight coaxial RF switches connected to the antenna elements, the feed is an analog beamformer with a single output port going to a Scientific Atlanta 1780 phase/amplitude receiver. We can then measure array patterns. Individual elements are connected to the receiver as shown in Fig. 8.22 where the leftmost element is connected. We can then measure element patterns. The phase and amplitude response of each element (including the cumulative effects of coaxial feed lines and the power combiner) is measured as the array rotates from −90° to +90°. Measured element phase is the useful information for training our neural network to perform the beam-steering function.

We control only the phase shift in each element channel. The Reggia-Spencer reflective phase shifters are axial-field, ferrite types and, at 7.1 GHz, provide 360° of insertion phase. The amount of phase shift is determined by the current through a magnetic coil. The current magnitude is determined by the voltage output from a digital-to-analog (D/A) converter in the phase shifter driver. An 8-bit word is sent to the D/A from a PC controller. Therefore, we can set 256 states (0–255) corresponding to insertion phases between 0° and 360°.

8.10.3 Experimental Beamsteering Results in a Clean Environment

To train the network, we obtain training vectors by measuring the phase response of each element as described earlier. These data are shown in Fig. 8.23 without the $\pm\pi$ transitions introduced by the receiver. The phase data have been

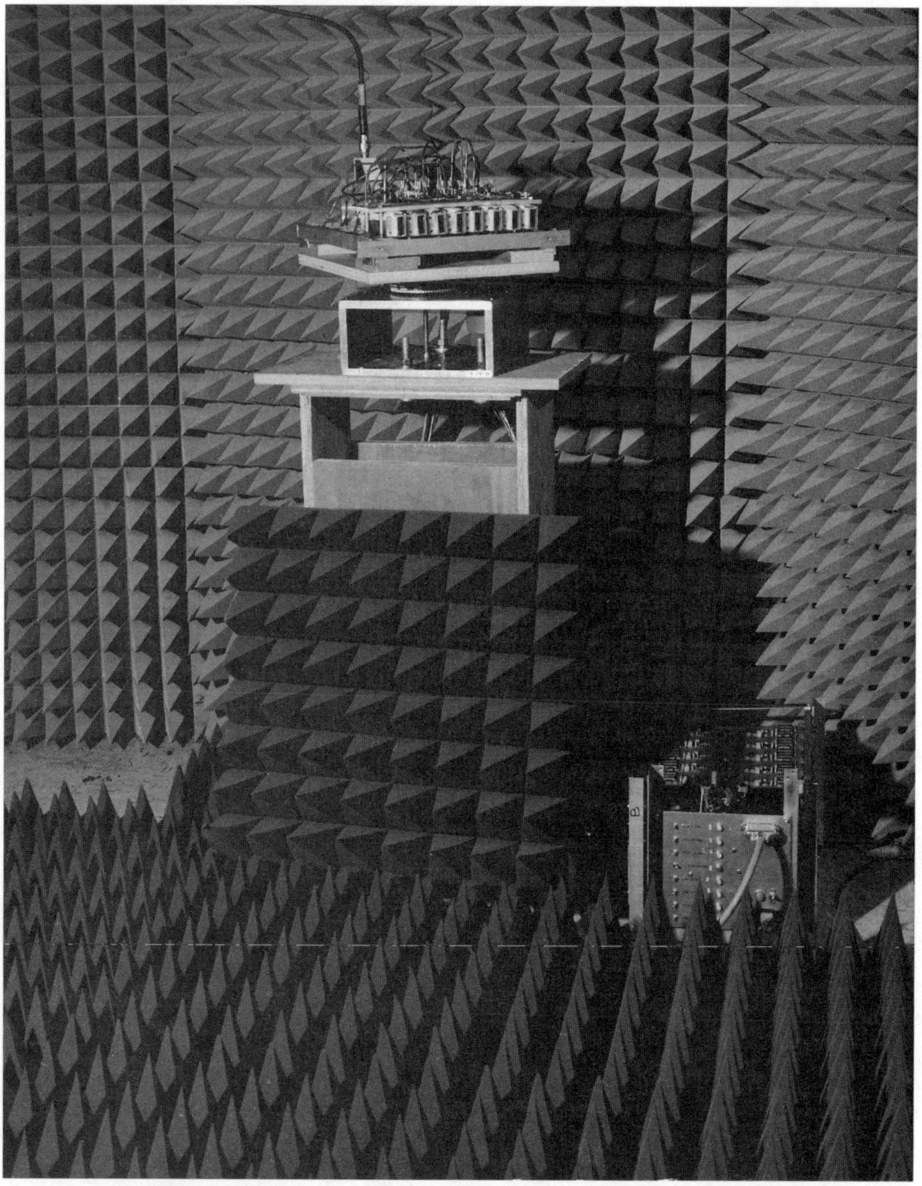

Figure 8.22 Experimental phased array antenna in the anechoic chamber. The phase shifter drivers are in the box in the lower right.

unwrapped using data analysis software, making it easier to see phase differences between elements. For example, to find the seven phase differences at a $-60°$ training angle, just draw a vertical line from $-60°$ and find the difference in phase between successive elements. Note that to the left of 0°, the phase difference between elements 7 and 8 is smallest, while to the right, the phase difference between elements 1 and 2 is smallest. Phase differences are negative to the left of 0° and positive to the right. We define phase difference as the phase of element $m + 1$ minus the phase of element

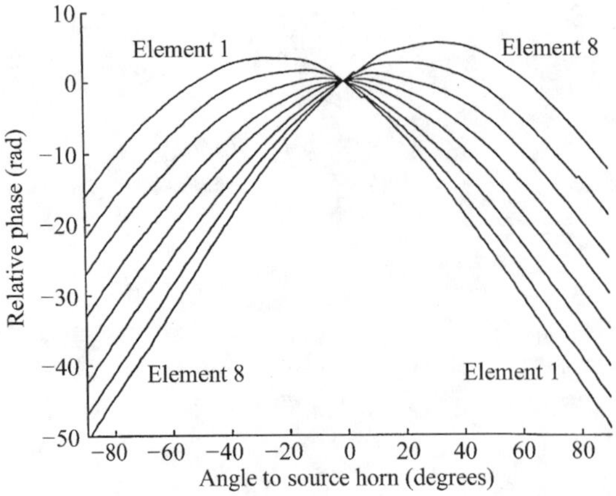

Figure 8.23 Measured element phases in radians (unwrapped). Note that element numbers are sequential 1 to 8 from top to bottom on the left and sequential 8 to 1 from top to bottom on the right.

m. Also, note that the seven phase differences are not the same for a given far-field angle as would be the case for a plane wave hitting an ideal array of identical elements.

The phase differences are zero at 0° since the plane wave from the source horn hits the array broadside. However, this condition occurs only after we remove the measured broadside phases (which we call "offset" or correction phases) due to the insertion phases in the different coaxial line lengths and the power combiner in each of the eight element channels. For this commercial power combiner, typical phase imbalance is ±12°. Phase imbalance due to line lengths are much larger, since no special attention was paid to phase-matching.

After we removed the offset phases with phase shifters, we measured an antenna pattern and obtained the broadside beam shown in Fig. 8.24(a). The higher sidelobes for the measured pattern are due to the amplitude taper induced by setting the phase shifters. The received amplitude as a function of phase state is highly nonlinear and can vary by as much as 8 dB over the 256 states of a single element and by more than this from element to element.

The beam is steered by applying the conjugate phase gradient at the commanded steering angle. To accomplish this, the network outputs the negative of the seven measured phase differences at the training angles. Between training angles, the network produces an approximation based on its generalization capability. In postprocessing, the phase differences are converted to element phases (modulo 2π) by making element one the reference (zero phase). These "beamsteering" element phases are combined with the offset phases on an element-by-element basis to obtain total phase at each element. Care must be taken in combining the two phases. In particular, the two phases must be summed at each element for positive steering angles and subtracted for negative ones.

To obtain the digital state to send to the phase shifter driver, we use a calibration lookup table of digital state (0–255) versus phase shift (0°–360°) for each element. The digital state versus phase shift is a highly nonlinear function for our phase shifters. To illustrate the process, suppose element three requires 57° of phase shift. Postprocessing finds the calibration table for element three and looks up the digital state giving a phase shift closest to 57°, with a typical accuracy of ±2°.

In the remainder of Fig. 8.24, we show scanned antenna patterns for three

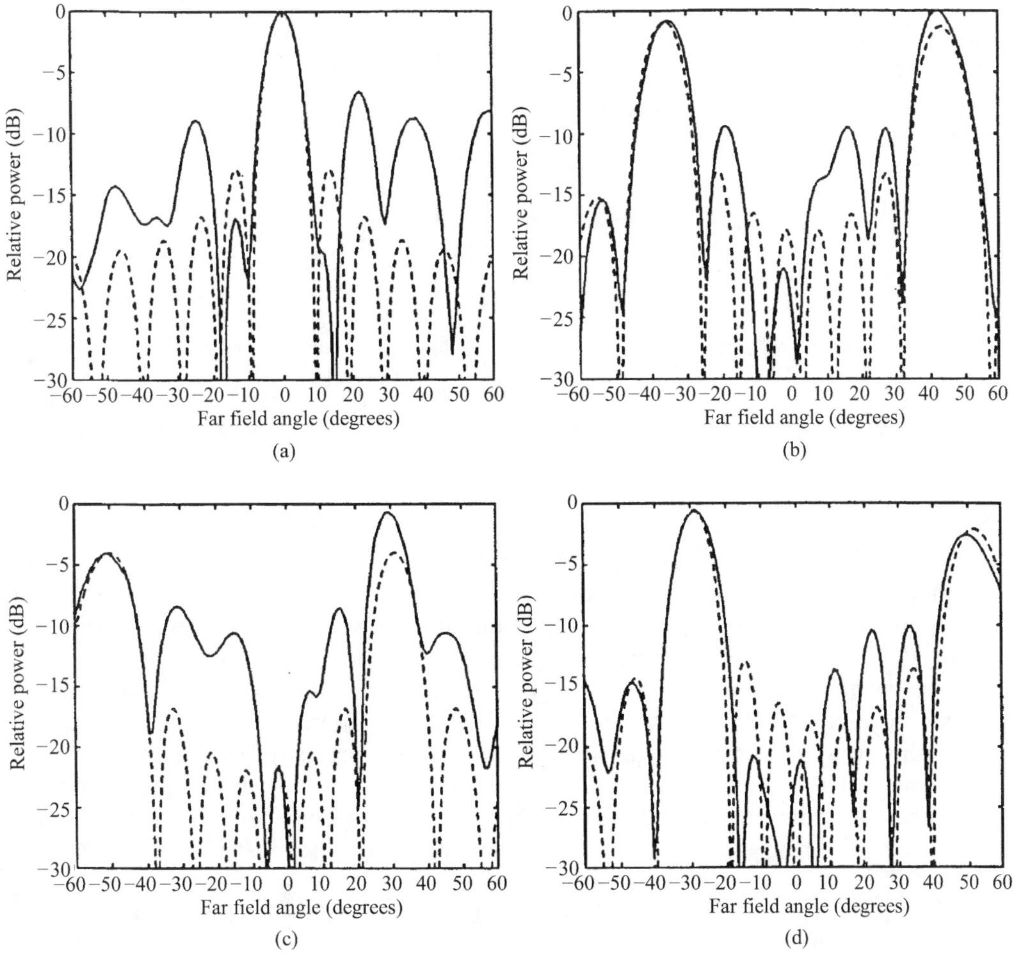

Figure 8.24 Scanned antenna patterns: (a) Broadside with offset phases removed. (b) Training angle −36.37°. (c) Training angle, −50°. (d) Network-predicted angle between training angles, −29.1°. Patterns for an ideal, uniformly illuminated array with linear phase progression for beam-steering are shown as dashed curves. The peak of the ideal main beam has been normalized to be the same as the peak of the measured main beam.

commanded steering angles. The main beam is always to the left of 0°, while the right beam is a grating lobe due to the large element spacing. The scan angles in Fig. 8.24(b) and (c) are for training angles of −36.37° and −50°. Figure 24(d) shows a pattern for a scan angle of −29.1° which is between training angles. Measured steering angles agree with the commanded angles within one degree. Angles used for training were approximately every 5° starting at −59.09° and ending at −0.01°. We trained the network on only half the field of view since we only wanted to demonstrate the concept. The angles are not integers because a stepper motor was used to rotate the array and each step turned out to be slightly less than a degree to obtain equal steps from −90° to +90°.

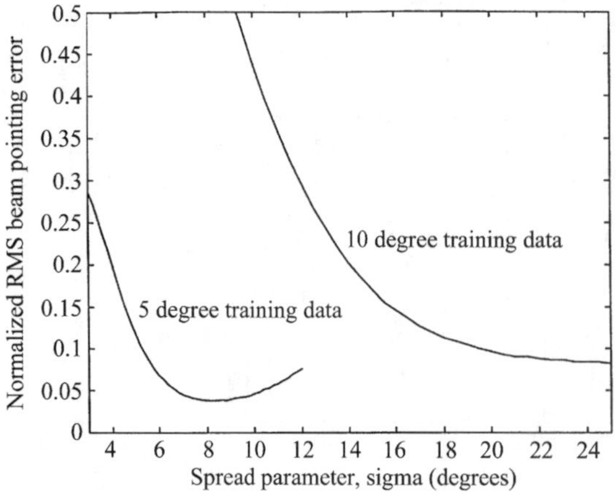

Figure 8.25 RMS beampointing error normalized by θ_{3dB} as a function of σ for two sets of training data. For the 5° data set, there are 14 training angles over the input space and for the 10° data set, there are seven training angles over the input space.

Finally, we investigate network performance as σ and the number of training angles are varied. Once the network architecture has been selected, these are the only parameters left to adjust. In Fig. 8.25, we plot the RMS beamsteering error normalized by the broadside 3 dB beamwidth ($\theta_{3dB} \approx 8.1°$). The RMS error is calculated for steering angles from $-60°$ to 0° in 2° increments. The error is the difference between commanded and network-predicted steering angles. There are too many scan angles in this computation to set phase shifters and measure patterns. Therefore we use element phases predicted by the neural network and mathematically transform to the far-field to obtain predicted steering angles. Based on comparisons with measured patterns, steering angles obtained in this manner should be reasonably accurate. Note that there is an optimum σ for each set of training data. For the 5° set, the optimum value for σ is about 1.6 times the separation between training angles, and slightly more than twice the separation for the 10° set.

8.10.4 Neural Beamsteering in the Presence of a Near-Field Scatterer

We have shown that a neural network can be used to perform the control function to accurately steer the main beam of an array antenna. Phase shifter commands are computed by the network which automatically (through network training) accounts for non-ideal behavior of the array elements. The departure from ideal array behavior [where the phase difference between array elements is a constant equal to $(2\pi/\lambda)d\sin(\theta)$] is relatively small in the discussion and experimental examples above; however, a near field scatterer can be used to produce larger departures from the ideal. In this case, the ideal plane wave front is distorted so the phase differences are no longer constant, but vary across the array face. The near field scatterer becomes a part of a "degraded" antenna system which the neural network can control to produce the desired result, in this case, to accurately steer the array main beam in spite of the disturbing effects of the near field scatter.

This is demonstrated experimentally for two scenarios: a single scatterer fixed relative to the rotating array and located between the array and the transmitting horn;

and, a single scatterer rotating synchronously with the array. The network is trained using phase fronts distorted by the near field scatterer. The network improves beam-pointing accuracy for both scenarios. The improvements are small since the effect of a single near field scatterer on the main beam pointing direction is not large. Therefore, we also use simulated data to more clearly illustrate the network's capability to accurately point the beam in the presence of near field scattering.

8.10.4.1 Neural Network Beamsteering

The NN beamsteering architecture was described in detail earlier in Section 8.10.1. As before, the scalar network input is the commanded beamsteering (or scan) angle and the network outputs are the seven phase differences. Postprocessing converts the phase differences to eight digital states (8-bit words) which are sent to phase shifter drivers. The drivers set the eight phase shifters to steer the array main beam.

The network is trained using supervised learning consisting of beamsteering angles and corresponding "target" or "desired" output vectors. Each target vector consists of the negative of the seven measured phase differences. The network is trained to output the target vectors at these angles. The measured phase differences contain the information in the distorted phase fronts caused by near-field scatterers, therefore the network has "learned about" these distortions and can use this information to accurately point the main beam. The trained NN generalizes from the training data and can produce output steering vectors which accurately steer the beam between training angles.

8.10.4.2 Theoretical Predictions (Simulations)

The electric field at each array element can be calculated assuming that the source horn radiates a spherical wave. For free space propagation, the field at the *i*th element is given by

$$E_i = \frac{e^{jkR_i}}{R_i} \tag{8.40}$$

where $k = 2\pi/\lambda$ and R_i is the distance from the horn to the phase center of the *i*th element. (approximately 136 in.). Since $\lambda = 1.66$ inches and the array length is $D = 10.4$ in., the far-field criterion $(2D^2/\lambda)$ is about 130 in. Therefore, there is a slight curvature to the phase front when it hits the array. However, this effect is small and the quadratic phase error does not squint the beam when ideal (plane wave) steering vectors are used, ignoring any quadratic phase error.

To model the effect of a near field scatterer, we assume that the field incident upon the scatterer is re-radiated isotropically from the center of the scatterer at some fraction of the value of the magnitude of the incident field. We use $\eta = 0.3$ since this resulted in simulated element phases most similar to measured element phases. At the *i*th element, the total field is the sum of the direct field from (8.40) and the field from the scatterer.

$$E_{sum} = \frac{e^{jkR_i}}{R_i} + \eta E_{inc} \frac{e^{jkr_i}}{r_i} \tag{8.41}$$

E_{inc} is the electric field incident upon the scatterer from the horn. A fraction, η, is scattered, and travels a distance r_i from the scatterer to the ith element. This can be modeled based on the geometries described in Section 8.10.4.3 for both rotating and fixed scattering scenarios. The geometries determine the R_i and the r_i for any element at any azimuth angle θ, where θ is the angle between a line connecting the array center of rotation and the array normal.

The presence of a scatterer in the near field of an antenna distorts the arriving wave front so that it is no longer a uniform plane wave. If it were a uniform plane wave, the phase difference between array elements would be constant and equal to

$$\Delta\phi = \frac{2\pi}{\lambda} d \sin\psi \tag{8.42}$$

where ψ is the angle between the incident wave front and the normal to the array and d is the distance between elements (about 0.78λ). For our experiment, the array phase center is offset forward of the array center of rotation, therefore, $\theta \neq \psi$.

As described in Section 8.10.4.1, the NN is trained to output phase differences. These phase differences are used to calculate the phase commands required to steer the main beam. For a uniform plane wave, all seven phase differences are equal to the negative of (8.42). The phase differences are not all equal when we place a scatterer in the near field of the antenna. If we were to apply the ideal phase taper of (8.42), the beam would be steered to the wrong angle, i.e., the observed beam pointing angle would not be the same as the commanded steering angle. We call the difference between observed and commanded steering angles the pointing error. Using a phase taper from the trained NN, we can make this difference smaller than when the ideal phase taper is used.

As a baseline, we calculate the pointing error over the antenna field of view for both neural network and ideal steering vectors for $\eta = 0$, i.e., no scatterer. Note that since θ is not equal to ψ, an error will be incurred even with no scatterer, since we calculate phase differences based on $(2\pi/\lambda)d\sin\theta$, for azimuth angle θ. Of course, we could calibrate for this by calculating the correct phase difference, since we can easily find ψ as a function of the azimuthal angle θ; however, that is not the point. We want to show that the network, trained on raw data, can account for effects which ordinarily require careful calibration. The neural network automatically accounts for the fact

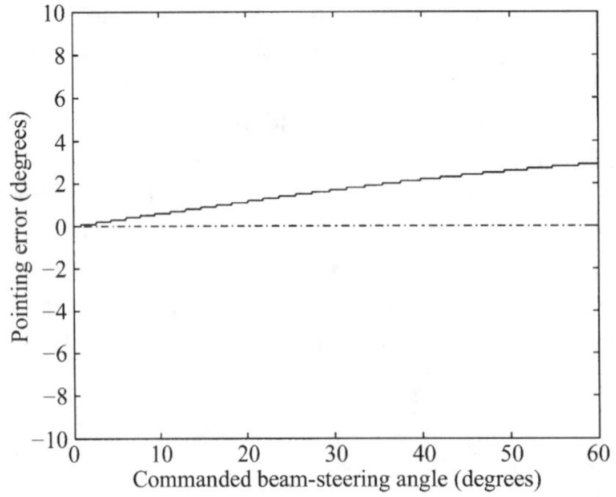

Figure 8.26 Simulated beampointing error for the baseline case: no scatterer. Solid line is for the ideal steering vectors. Dot-dashed line is for network predicted steering vectors. The error for the ideal vectors is due to the fact that $\psi \neq \theta$, since the phase center of the array is not directly over the center of rotation and the source horn is not infinitely far from the array.

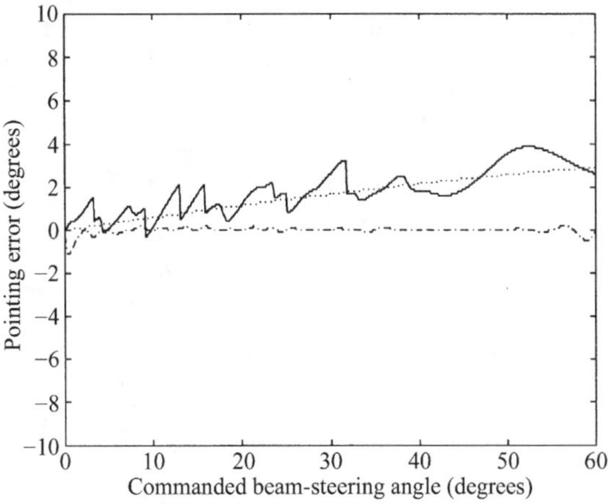

Figure 8.27 Simulated beampointing error for the rotating scatterer. Solid line is for ideal steering vectors. Dot-dashed line is for network-predicted steering vectors. Baseline case is shown dotted for comparison.

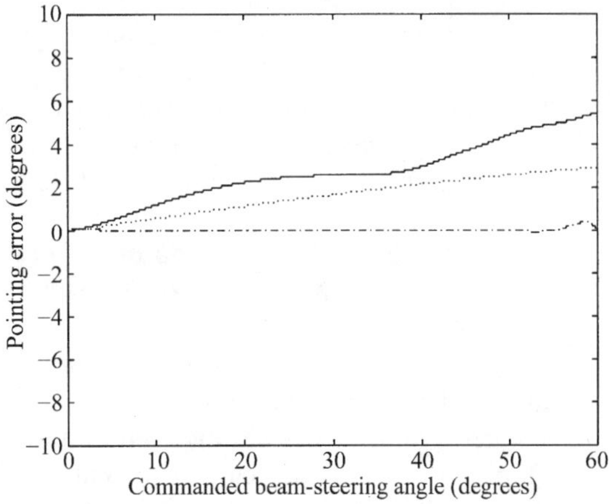

Figure 8.28 Simulated beampointing error for the fixed scatterer. Solid line is of ideal steering vectors. Dot-dashed line is for network-predicted steering vectors. Baseline case is shown dotted for comparison.

that $\theta \neq \psi$, since the network has been trained using measured data which, naturally, include the fact that the array phase center is offset from the center of rotation. The baseline case is shown in Fig. 8.26.

Using simulated data including the presence of the rotating scatterer, with $\eta = 0.3$, we calculated the pointing error as described above. Pointing errors for both ideal and network predicted steering vectors are shown in Fig. 8.27. The simulated pointing error for the fixed scatterer, with $\eta = 0.3$, is shown in Fig. 8.28.

8.10.4.3 Description of the Scattering Experiment

The scatterer rotated synchronously with the C-band array described earlier in Section 8.10.2 and was positioned 14 in. from the right end of the array, at a 45° angle. The array rotated in front of a stationary source horn located 12 ft away. We also used a scenario where the scatterer was fixed relative to the rotating antenna. The fixed scatterer was located 70 in. in front of the array center of rotation, between the array and the source horn. It was positioned 10 in. off to the side of a line connecting the array center of rotation and the source horn.

TABLE 8.4 Pointing Angle RMS Errors for Fixed and Rotating Scatterers. Comparison between Desired, Ideal, and Network Scan Angles

FIXED SCATTERER			ROTATING SCATTERER		
Commanded Scan Angle	Ideal Steering Angle	Network Steering Angle	Commanded Scan Angle	Ideal Steering Angle	Network Steering Angle
0°	0.00°	0.00°	25°	23.19°	23.64°
5°	5.00°	4.54°	30°	28.19°	29.33°
10°	10.23°	10.68°	35°	34.09°	33.64°
15°	13.64°	14.31°	40°	37.05°	38.64°
20°	19.09°	19.09°	45°	42.96°	43.32°
25°	23.64°	23.64°	50°	46.30°	47.28°
30°	28.18°	29.13°			
RMS error	1.064°	0.809°	RMS error	2.37°	1.64°

The antenna patterns used in Section 8.10.4.4 to experimentally determine pointing errors are mechanically scanned patterns, i.e., the main beam was electronically scanned to a fixed azimuthal angle (the commanded beam steering angle) by setting the eight phase shifters; the array was then mechanically rotated on the positioner. The mechanically scanned antenna pattern is the measured response at the output of an eight-way power combiner [39].

8.10.4.4 Experimental Beamsteering Results with a Near-Field Scatterer

Beampointing errors were measured experimentally for the two scattering scenarios discussed in Section 8.10.4.3. We set the phase shifters using element phases calculated from ideal steering vectors and from neural network predictions. Phase shifters were set to the closest available insertion phase from a look-up table of insertion phase versus state, since the insertion phase versus state was highly non-linear for our Reggia-Spencer phase shifters [39]. There were 256 states for approximately 360° of insertion phase; therefore, the phase shifters could be set with an accuracy of about 1°.

Measured pointing errors for the rotating scatterer and the fixed scatterer are shown in Table 8.4. The "ideal steering angle" means the measured beampointing angle obtained by using the ideal phase taper of (8.42). The "network steering angle" means the measured beampointing angle obtained by using a NN generated phase taper. For the rotating scatterer, we used commanded steering angles of 25°, 30°, 35°, 40°, 45°, and 50° to determine the pointing errors. For the fixed scatterer, we used commanded steering angles of 0°, 5°, 10°, 15°, 20°, 25°, and 30°. The experimental beampointing angle was determined by taking the average of the measured beam peak angle and the angle corresponding to the midpoint of the measured 3 dB beamwidth.

REFERENCES

[1] J. H. Holland, "Genetic Algorithms," *Sci. Amer.*, pp. 66–72, July 1992.

[2] D. E. Goldberg, *Genetic Algorithms in Search, Optimization, and Machine Learning*, Addison-Wesley, New York, 1989.

[3] R. L. Haupt, "Thinned arrays using genetic algorithms," *IEEE AP-S Transactions*, **42** (7), July 1994.

[4] R. L. Haupt, "Optimum quantized low sidelobe phase tapers for arrays," *IEE Electronics Letters*, **31** (14), pp. 1117–1118, July 6, 1995.

[5] R. L. Haupt, "An introduction to genetic algorithms for electromagnetics',, *IEEE AP Magazine*, **37** (2), pp. 7–15, April 1995.

[6] R. L. Haupt, "Synthesizing Low Sidelobe Quantized Amplitude and Phase Tapers for Linear Arrays Using Genetic Algorithms," International Conference on Electromagnetics in Advanced Applications, Torino, Italy, September 1995.

[7] R. L. Haupt, "Phase-only Adaptive Nulling with Genetic Algorithms," *IEEE AP-S Trans.*, **45** (5), May 1997.

[8] R. L. Haupt, and S. E. Haupt, *Practical Genetic Algorithms*, John Wiley & Sons, New York, 1998.

[9] J. F. DeFord and O. P. Gandhi, "Phase-only synthesis of minimum peak sidelobe patterns for linear and planar arrays," *IEEE AP-S Trans.*, **36** (2), pp. 191–201, February 1988.

[10] R. Kinsey, "Array Antenna Self-calibration Techniques," AMTA Workshop: Testing Phased Arrays and Diagnostics, San Jose, CA, June 1989.

[11] R. A. Monzingo and T. W. Miller, *Introduction to Adaptive Antennas*, John Wiley and Sons, New York, 1980.

[12] R. T. Compton, *Adaptive Antenna Concepts and Performance*, Prentice Hall, Englewood Cliffs, NJ, 1988.

[13] C. A. Baird and G. G. Rassweiler, "Adaptive sidelobe nulling using digitally controlled phase-shifters," *IEEE AP Trans.*, **24** (5), pp. 638–649, September 1976.

[14] M. K. Leavitt, "A Phase Adapation Algorithm," *IEEE AP-S Trans.*, **24** (5), pp. 754–756, September 1976.

[15] H. Steyskal, "Simple Method for Pattern Nulling by Phase Perturbation," *IEEE AP-S Trans.*, **31**, (1), pp. 163–166, January 1983.

[16] R. L. Haupt, "Adaptive Nulling in Monopulse Antennas," *IEEE AP-S Trans.*, **AP-36** (1), January 1988.

[17] R. A Shore, "A Proof of the Odd-Symmetry of the Phases for Minimum Weight Perturbation Phase-only Null Synthesis," *IEEE AP-S Trans.*, **32** (5), pp. 528–530, May 1984.

[18] R. J. Mailloux, and H. L. Southall, "The Analogy between the Butler Matrix and the Neural Network Direction Finding Array," *IEEE Antennas and Propagation Magazine*, **39**(6), pp. 27–32, December 1997.

[19] H. L. Southall, "Resolution of a Neural Network Direction Finding Array," 1997 IEEE AP-S International Symposium and URSI Radio Science Meeting, Montreal, Canada, 13–18 July 1997.

[20] T. O'Donnell, J. Simmers, and D. J. Jacavanco, "Neural Beamforming for Phased Array Antennas," Proceedings of the 1992 Antenna Applications Symposium (Robert Allerton Park), USAF Rome Laboratory, Griffiss AFB, New York, September 1992.

[21] J. Simmers, H. L. Southall, and T. O'Donnell, "Advances in Neural Beamforming," Proceedings of the 1993 Antenna Applications Symposium (Robert Allerton Park), University of Illinois, USAF Rome Laboratory, Griffiss AFB, NY, September 1993.

[22] D. Hammerstrom, "Working with Neural Networks," *IEEE Spectrum*, **30** (7), July 1963, pp. 46–53.

[23] L. F. Fausett, *Fundamentals of Neural Networks*, Prentice Hall, Englewood Cliffs, NJ, 1994.

[24] M. J. D. Powell, Radial Basis Functions for Multivariable Interpolation: A Review, in

Algorithms for Approximation, J. C. Mason and M. G. Cox, eds., Clarendon Press, Oxford, 1987, pp. 143–167.

[25] D. F. Specht, "A General Regression Neural Network," *IEEE Transactions on Neural Networks*, **2** (6), November 1991, pp. 568–576.

[26] D. S. Broomhead and D. Lowe, "Multivariable Function Interpolation and Adaptive Networks," *Complex Systems*, **2**, 1988, pp. 321–355.

[27] T. Poggio and F. Girosi, "Networks for Approximation and Learning," *Proceedings of the IEEE*, **78** (9), September 1990, pp. 1481–1496.

[28] J. Butler and R. Lowe, "Beamforming Matrix Simplifies Design of Electronically Scanned Antennas," *Electronic Design*, **9**, April 12, 1961, pp. 170–173.

[29] J. A. Simmers and T. O'Donnell, "Adaptive RBF Neural Beamforming," Proceedings of the 1992 IEEE Mohawk Valley Section Command, Control, Communications, and Intelligence (C3I) Technology & Applications Conference, June 1992, pp. 94–98.

[30] H. I. Southall, J. A. Simmers, and T. H. O'Donnell, "Direction Finding in Phased Arrays with a Neural Beamformer," *IEEE Transactions on Antennas and Propagation*, **43**(12), pp. 1369–1374, December 1995.

[31] F. J. Narcowich, "Generalized hermite interpolation and positive definite kernels on a reimannian manifold," *Journal of Mathematical Analysis and Computations*, **190**, 1995, pp. 165–193.

[32] S. Lee, "Supervised Learning with Gaussian Potentials," in *Neural Networks for Signal Processing*, B. Kosko, ed., Prentice Hall, Englewood Cliffs, NJ, 1992, pp. 189–227.

[33] S. Lee and R. M. Kil, "Bidirectional Continuous Associator Based on Gaussian Potential Function Network," Proceedings of the 1989 IEEE IJCNN Vol. I, San Diego, CA, 1989, pp. 45–53.

[34] A. C. Tsoi, "Multilayer Perceptron Trained Using Radial Basis Functions," *Electronics Letters*, **25** (19), September 14, 1989, pp. 1296–1297.

[35] B. D. Carlson and D. Willner, "Antenna Pattern Synthesis Using Weighted Least Squares," *IEE Proceedings* – H, **139** (1), February 1992, pp. 11–16.

[36] B. Kosko, ed., *Neural Networks for Signal Processing*, Prentice Hall, Englewood Cliffs, NJ, 1992.

[37] S. M. Sherman, *Monopulse Principles and Techniques*, Artech House, Norwood, MA, 1984.

[38] T. Lo, H. Leung, and J. Litva, "Radial Basis Function Neural Network for Direction-of-Arrivals Estimation," *IEEE Signal Processing Letters*, **1** (2), February 1994, pp. 45–47.

[39] H. Southall, S. Santarelli, E. Martin, and T. O'Donnell, "Neural Network Beam-Steering for Phased-Array Antennas," Proceedings of the 1995 Antenna Applications Symposium (Robert Allerton Park), University of Illinois, USAF Rome Laboratory, Griffiss AFB, NY, September 1995.

Chapter 9

MODEL-ORDER REDUCTION IN ELECTROMAGNETICS USING MODEL-BASED PARAMETER ESTIMATION

Edmund K. Miller and Tapan K. Sarkar

ABSTRACT

This chapter outlines and demonstrates the use of model-based parameter estimation (MBPE) in electromagnetics. MBPE can be used to circumvent the requirement of obtaining all samples of desired quantities (e.g., impedance, gain, RCS) from a first-principles model (FPM) or from measured data (MD) by instead using a reduced-order, physically based approximation of the sampled data called a fitting model (FM). One application of a FM is interpolating between (pole-series FMs), and/or extrapolating from (exponential-series FMs), samples of FPM or MD observables to reduce the amount of data that is needed. A second is to use a FM in FPM computations by replacing needed mathematical expressions with simpler analytical approximations to reduce the computational cost of the FPM itself. As an added benefit, the FMs can be more suitable for design and optimization purposes than the usual numerical data that comes from a FPM or MD because the FMs can normally be handled analytically rather than via operations on the numerical samples. Attention here is focused on the use of FMs that are described by exponential and pole series, and how data obtained from various kinds of sampling procedures can be used to quantify such models, i.e., to determine numerical values for their coefficients.

9.1 BACKGROUND AND MOTIVATION

Most EM phenomena, whether observed in the time domain (TD) or the frequency domain (FD), or as a function of angle or location, are not of interest at just one or a few discrete times, frequencies, angles, or locations but, instead, require essentially continuous representation over some specified observation interval. Even with increasingly sophisticated instrumentation or computer models, determining EM observables to sufficient resolution as a function of the relevant variable can be expensive as well as potentially error prone. A typical problem now involves determining TD or FD responses over bandwidths that are no longer just a few percent or a few megahertz, but might extend over frequency ranges that are 10:1 or more in relative bandwidth or multiple gigahertz bandwidths in absolute terms. Resolving a response that contains many high-Q, closely spaced resonances requires slower sweeping rates or long observation times experimentally, or an excessive number of frequency samples or time steps computationally.

A computational basis for solving most problems in physics and engineering derives from FPMs of the applicable physics. Depending on the problem parameters and information needed, various physically and mathematically based approximations are almost invariably used subsequently in obtaining numerical results. The approximations can arise from: making appropriate changes in the physical problem (e.g., replacing a solid, metallic surface with a wire mesh); exploiting commonly encountered problem attributes (such as the thin-wire approximation); or taking advantage of asymptotic trends (as in physical and geometrical optics and various diffraction theories). Such approximations can yield acceptable results for a wide variety of problems, and might be regarded as being derived or "second-generation" FPMs, i.e., they originate from a FPM to which they have a well-defined relationship rather than directly and rigorously from Maxwell's equations themselves.

Another category of models might also be recognized which are further removed from the FPM on which they are based, but are instead related to some variable associated with that model and/or to some observable that FPM produces. For example, consider that when wideband, impulsive excitation is used, the EM late-time response can be rigorously represented as an exponential series in time whose exponents are generally complex, problem-dependent, parameters. Such a representation of the late-time response can be directly deduced from a TD FPM [1] and forms the basis for the Singularity Expansion Method (SEM) [2]. Generally, however, the exponential series otherwise retains no specific information about the problem being modeled concerning its physical or electrical characteristics, except to the extent that the parameters of the series might imply them. The value of such a model is that it can substantially reduce the effective rank of the problem description insofar as the complexity of representing its observables is concerned. Thus, rather than needing the complete problem description required to compute the $X_s \times X_s$ interaction coefficients of an IE FPM, this reduced-rank, parametric description can be substituted for the FPM in appropriate circumstances. Such reduced-order models, whose analytic form is known and which are therefore completely specified when their associated parameters are quantitatively determined, provide the basis for MBPE.

Although MBPE is discussed here specifically with respect to some representative EM applications, it should be understood to be a very general procedure that is applicable to essentially any process, physical or otherwise, for which a reduced-order, parametric model can be deduced. Also, it must be noted that MBPE is not "curve fitting" in the sense that term is normally used, which also involves finding the parameters that fit some function to given data. The essential difference between MBPE and curve fitting is that the former uses a FM based on the problem physics, while the latter need not do so, which is why MBPE might be characterized as "smart" curve fitting. Curve fitting that includes the goal of finding the correct FM for the process that generated the given data is also described as "system identification." It's worth emphasizing that MBPE is not limited to physical processes but can be discerned in variously named analytical procedures {e.g., Kummer's method, Richardson extrapolation, and Romberg quadrature [3]} whose purpose is to speed numerical convergence of integrals and infinite sums whose evaluation produces a sequence of results that can be regarded as a generalized "signal." In discussing a non-linear procedure he developed for a similar purpose, Shanks [4] referred to such phenomena as "physical" and "mathematical" transients. In essence, any process that uses or produces a sequence or set of samples is a candidate for MBPE, using as a FM whatever mathematical representation is an appropriate reduced-order description of that process whose FM parameters are determined by whatever computational procedure produces robust-enough results.

This chapter begins by discussing FMs given by exponential and pole series, both of which are widely encountered in electromagnetics and related to each other by the Laplace transform, arising whenever wave-equation problems are encountered. As a paradigm for the general problem we use the time-frequency transform pair to illustrate use of MBPE for time waveforms that are described by sums of complex exponentials. The other half of the transform pair involves using MBPE for spectral estimation, not where time samples provide the starting data as is most often the case, but instead when the original data come from sampling in frequency. It is demonstrated that the parameters of the time-frequency models can be obtained numerically either from samples of the process being modeled, a procedure called "function sampling," or alternatively from samples of the time or frequency derivatives of the process using "derivative sampling." We then demonstrate application of MBPE to various of EM problems where the goal is to minimize the number of FPM samples that are needed to represent EM observables of interest. A discussion of some possibilities for using MBPE to reduce the complexity of FPM computations and thereby reduce their computational cost concludes the chapter, which is excerpted and adapted from three articles by the author that appeared in the IEEE Antennas and Propagation Society Magazine [5], [6], [7].

9.2 WAVEFORM-DOMAIN AND SPECTRAL-DOMAIN MODELING

Although Maxwell's equations are usually first encountered in the TD, traditionally most subsequent analysis has been done in the FD, a situation, however, that is changing with the increasing computer power available. Each of these two domains describes the same physical problem in principle, and either is generally capable of providing the sought-for EM behavior. Therefore, it seems reasonable to seek solutions in that domain and using that formulation best suited to the problem at hand, wherein medium or object non-linearity, dispersivity, inhomogeneity, anisotropy, etc. might be the determining factor in choosing a model. Whatever approach is taken to solve a problem, it must be noted that once the problem's sampled observables have been obtained it is often necessary to examine their behavior in the transform domain as well.

While the TD and FD provide probably the most-used examples of exponential and pole series FMs, respectively, the same transform relationship exists between other observable pairs that are also described by exponential and pole series as listed in Table 9.1. Thus, the terms "waveform domain" (WD) and "spectral domain" (SD) are used for phenomena that are described by exponential series and pole series as a generalization of their more specific and familiar TD and FD forms.

Series comprised of complex exponentials or complex poles can represent the kinds of wave-equation solutions that arise in EM and similar physical phenomena because they occur as solutions to the kinds of differential equations that describe the problem physics. Thus, they might be regarded as providing natural "basis" functions for developing numerical solutions to such problems. One example of this is

TABLE 9.1 Some Waveform- and Spectral-Domain Field Expressions

Domain	Variables		Model	Parameters (both domains)	Comments
Waveform (Exponential)					
	x	g	$f(x)$	Residues	
Spectral (Pole)					
	X		$F(X)$	Poles	
			Model 1a—Time/Frequency		
Time	t	—	$f(t) = \Sigma M_\alpha \exp(s_\alpha t)$	M_α = amplitude of αth mode.	
Complex frequency	s		$F(s) = \Sigma M_\alpha/(s - s_\alpha)$	$s_\alpha = i\omega_\alpha - \sigma_\alpha$ = complex resonance frequency of αth mode.	Provides a way to find SEM poles from transient or spectral data [14]
			Model 1b—Frequency/Time		
Frequency	ω	—	$f(s) = \Sigma M_\alpha \exp(t_\alpha s)$	M_α = amplitude of αth mode	
Time	t		$F(t) = \Sigma M_\alpha/(t - t_\alpha)$	t_α = peak time for a waveform	Superposition of sinusoidal signals
			Model 2a—Frequency/Space		
Frequency	ω	i/c	$f(\omega) = \Sigma S_\alpha \exp(g\omega R_\alpha)$	S_α = amplitude of αth source	
Space	R		$F(R) = \Sigma S_\alpha/[g(R - R_\alpha)]$	R_α = position of αth source along line of view	Related to profiling or inverse imaging along viewing direction [15]
			Model 2b—Space/Frequency		
Space	R	i/c	$f(R) = \Sigma S_\alpha \exp(gR\omega_\alpha)$	S_α = amplitude of αth source	
Frequency			$F(\omega) = \Sigma S_\alpha/[g(\omega - \omega_\alpha)]$	ω_α = frequency of αth source	Observing field of multifrequency sources over a range of spatial locations
			Model 3a—Space/Angle		
Space	x	ik	$f(x) = \Sigma P_\alpha \exp\{gx[\cos(\varphi_\alpha)]\}$	P_α = amplitude of αth incident plane wave	
Angle	φ		$F(\varphi) = \Sigma P_\alpha/\{g[\cos(\varphi) - \cos(\varphi_\alpha)]\}$	φ_α = incidence angle of αth plane wave with respect to line of observation	A basis for direction finding [16]

TABLE 9.1 Some Waveform- and Spectral-Domain Field Expressions (*continued*)

			Model 3b—Angle/Space		
Angle	φ	ik	$f(\varphi) = \Sigma\, S_\alpha \exp[gx_\alpha \cos(\varphi)]$	S_α = amplitude of αth source	
Space	x		$F(x) = \Sigma\, S_\alpha/[g(x - x_\alpha)]$	x_α = position of αth source along linear array	Can be used for pattern synthesis and imaging linear source distributions [17]

demonstrated by SEM [8], [2] which exploits the fact that the EM behavior of conducting objects is described in part by complex-frequency resonances, or poles. While the pole terms alone do not describe the entire response, a summation of poles can be a good approximation to that response and provides a concise way to represent it. For our purpose, which is to reduce the number of samples of FPM observables that are needed in the first place, as well as to develop more efficient FPMs, the FM need not be exact so long as it provides an acceptably accurate and parsimonious representation. This is an important distinction of a FM as compared with what is required of a FPM.

The generic WD and SD FMs can be expressed as

$$f(x) = f_p(x) + f_{np}(x) = \sum R_\alpha \exp(s_\alpha x) + f_{np}(x) \tag{9.1}$$

and

$$F(X) = F_p(X) + F_{np}(X) = \sum R_\alpha/(X - s_\alpha) + F_{np}(X), \qquad \alpha = 1, \ldots, P \tag{9.2}$$

where "x" represents the WD independent variable and "X" is its SD, or transformed, counterpart. For the time-frequency transform pair, x would be the time variable t and X would be the complex frequency s, in general, but which for practical purposes may be limited to radian frequency $i\omega$. While the exponential or pole series contributions, designated respectively above by $f_p(x)$ and $F_p(X)$, represent what we might call the "resonant" response, there is in general a non-pole component as well, denoted by $f_{np}(x)$ and $F_{np}(X)$. The additional non-pole term in the waveform and spectral models is included to account for the fact that the complete response is not entirely described by the exponential or pole terms, except for the TD at "late" times when an object is no longer excited by an incident field, as in the "early" time, the response is not purely resonant. For the FD, there the separation between the driven and source-free behavior is not so clearly demarcated because essentially all frequency components are affected by both time behaviors.

The FM parameters, the complex resonances (or poles), "s_α," and the modal amplitudes (or residues), "R_α" (or their rational-function counterparts which occur as polynomial coefficients), are quantified by fitting samples of the relevant observable to the desired model, $f_p(x)$ or $F_p(X)$. Once these parameters are available from one domain, they can be used to obtain the observable they represent in the transform

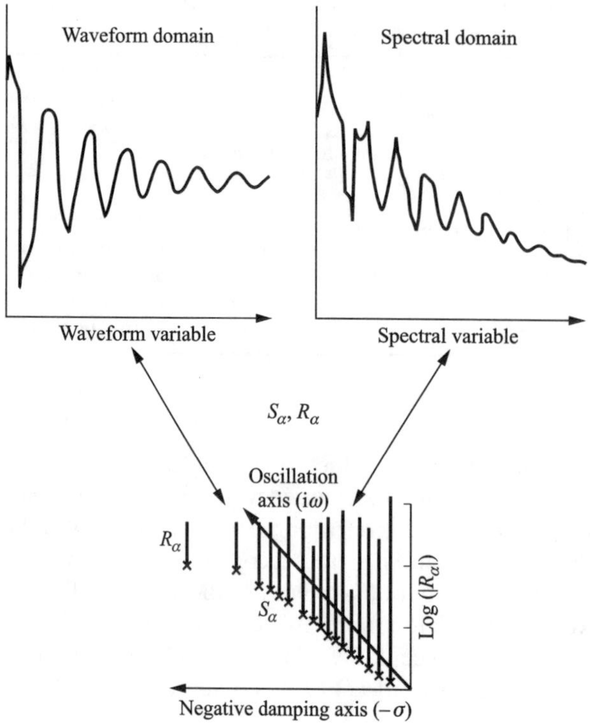

Figure 9.1 Illustrations of the relationship between the waveform-domain (left-hand plot) and spectral-domain (right-hand plot) observables and the two sets of parameters, R_α and s_α, common to them both (the bottom plot). The pole locations are shown in the quadrant II only, since they usually occur in complex conjugate pairs reflected about the damping axis and have only negative real parts.

domain as well. The various FMs and their estimated parameters provide a mathematically concise and physically insightful way to characterize electromagnetic and other wave-equation phenomena. The relationship between the model parameters and observables in the two domains is illustrated conceptually in Fig. 9.1.

Given that exponential and pole series are especially appropriate FMs for EM applications of MBPE, there remain the important questions of how FPMs should be sampled and how the FMs themselves can be most effectively employed. Each of these questions is considered immediately below and in the next two sections. Although exponential- and pole-series models are not the only physics-based FMs that can be used in electromagnetics, because of their ubiquitous presence in first-principles formulations and EM observables they are the ones we choose to emphasize here. Some of the far-field problems suitable for WD FMs are surveyed by Miller [9] and some other kinds of FMs are discussed below in Section 9.7.

9.2.1 Selecting a Fitting Model

A variety of questions arise about using MBPE for EM phenomena, among which are:

What kind of FM should be used, exponential series or pole series?

What order FMs are most suitable, how many data should they span and how much should they overlap?

What kind of GM sampling is available and/or most appropriate, function sampling, derivative sampling, or a combination thereof?

How can the FM parameters be most robustly estimated?

and

Is the data presampled or is it to be sampled as it's modeled?

Note that in the case of presampled data, the problem is to develop the best FM-representation possible, but if the data are undersampled a unique result cannot be obtained, which will be indicated by the mismatch between overlapping FMs. If such data are oversampled, however, then it may be possible to estimate its uncertainty, as is discussed in Section 9.5.4 below. If GM sampling is performed "on the fly," i.e., during the process of its FM computation, then the distinct advantage of being able to sample adaptively becomes possible, as is discussed in Section 9.4.3.

Two of the more useful MBPE applications are those of determining FD transfer functions, for which model 1b of Table 9.1 is applicable, and evaluating radiation and scattering patterns, for which model 3a is appropriate. But if the goal is to model the far-field patterns as a function of frequency and angle, then a hybrid FM would be suitable, where the frequency dependence is modeled using SD FMs while the angle dependence is modeled using WD FMs. It's also interesting to observe that the impedance-matrix coefficients which arise in using a FDIE are WD entities, coming from a Green's function, as are the solution currents, as they are comprised of waves propagating over the surface of the object being modeled.

9.3 SAMPLING FIRST-PRINCIPLE MODELS AND OBSERVABLES IN THE WAVEFORM DOMAIN

As mentioned above, function and derivative sampling of the FPM can be employed in both the WD and SD, a point expanded upon here and in Section 9.4. Furthermore, it will also be shown that a given FM can use essentially any combination of function and derivative samples so long as the sampling satisfies whatever constraints are imposed by the FM. One constraint of a WD FM requires that the sampling be done in uniform steps of the independent variable x. Aside from such constraints, the FM only needs enough information from the FPM or MD to permit the FM parameters to be computed to the needed accuracy. This means that there can be a great deal of latitude available in using MBPE, an advantage on the one hand but one presenting, on the other, an array of options that can sometimes obscure how to obtain the best results from its use.

9.3.1 Waveform-Domain Function Sampling

Our starting-point for function sampling in the WD is provided by Prony's method, a procedure whose presence can be discerned in much of modern signal processing, even though developed originally in the late 18th century [10]. Although other signal-processing approaches may yield better results, especially for noisy signals, e.g., pencil-of-function-based methods [11], [12], [13] (see Appendix 9.2), all begin at the same starting-point using a series of discrete data samples. Prony's method has the advantage of being obviously and directly connected to the physical process being modeled. It begins by assuming the availability of uniformly spaced samples of an exponential series, i.e.,

$$f_i = f(x_i) = f(i\delta x) = \sum R_\alpha \exp(s_\alpha x_i) = \sum R_\alpha \exp(s_\alpha i\delta x)$$
$$\alpha = 1 \text{ to } P, \qquad i = 0, \ldots, D-1 \tag{9.3}$$

where δx is the sampling interval and there are a total of D samples. Rewriting (9.3), we get

$$f_i = \sum R_\alpha (X_\alpha)^i, \qquad i = 0, \ldots, D-1 \tag{9.4}$$

where $X_\alpha = \exp(s_\alpha \delta x)$.

The equations represented by (9.3) and (9.4) above can be written explicitly as

$$f_0 = R_1 + R_2 + \cdots + R_P \tag{9.5a}$$

$$f_1 = R_1 X_1 + R_2 X_2 + \cdots + R_P X_P \tag{9.5b}$$

$$f_2 = R_1 (X_1)^2 + R_2 (X_2)^2 + \cdots + R_P (X_P)^2 \tag{9.5c}$$

$$\vdots$$

$$f_{D-1} R_1 (X_1)^{D-1} + R_2 (X_2)^{D-1} + \cdots + R_P (X_P)^{D-1}, \tag{9.5d}$$

where the pattern of the X_α terms suggests that they might satisfy a polynomial of the form

$$A(X) = a_0 + a_1 X + a_2 X^2 + \cdots + a_P X^P = (X - X_1)(X - X_2)$$
$$\ldots (X - X_P) = \Pi(X - X_\alpha) = 0, \qquad \alpha = 1, \ldots, P \tag{9.6}$$

which is known as the characteristic equation. Upon multiplying the equation for f_0 by a_0, that for f_1 by a_1, etc. through f_P and adding the resulting equations, we obtain

$$f_0 a_0 + f_1 a_1 + \cdots + f_P a_P = 0 \tag{9.7a}$$

where $A(X_1) = A(X_2) = \ldots = A(X_P) = 0$ has been used. Repeating this sequence of operations by multiplying f_1 by a_0, f_2 by a_1, etc. leads to

$$f_1 a_0 + f_2 a_1 + \cdots + f_{P+1} a_P = 0 \tag{9.7b}$$

Continuing these steps to generate a total of $D - P$ equations, we obtain

$$f_2 a_0 + f_3 a_1 + \cdots + f_{P+2} a_P = 0 \tag{9.7c}$$

$$\vdots$$

$$f_{D-P-1} a_0 + f_{D-P} a_1 + \cdots + f_{D-1} a_P = 0 \tag{9.7d}$$

Equation (9.7) forms the basis for finding the coefficients of the characteristic equation. Since (9.7) is homogeneous, some additional information, or a constraint on the characteristic-equation coefficients, is required for the solution to be completed.

Various constraints have been suggested for this purpose, among them being:

$$\sum(a_i)^2 = 1,\ i = 0, 1, \ldots, P; \qquad \text{(energy constraint)} \tag{9.8a}$$

$$a_P = 1; \qquad \text{(linear predictor constraint)} \tag{9.8b}$$

$$a_i = 1; \qquad \text{(non-causal constraint)} \tag{9.8c}$$

Prony's method employs $a_P = 1$, leading to the "linear-predictor" equation

$$f_i a_0 + f_{i+1} a_1 + \cdots + f_{i+P-1} a_{P-1} = -f_{i+P} \tag{9.9a}$$

so-called because knowledge of the P predictor coefficients and the past P samples of the process that generated $f(x)$ yields, or "predicts," the next sample as a weighted sum. If some intermediate coefficient were to be set to unity instead, then (9.9a) would take the form

$$f_i a_0 + f_{i+1} a_1 + \cdots + f_{2i-1} a_{i-1} + f_{2i+1} a_{i+1} + \cdots + f_{i+P} a_P = -f_{2i}, \tag{9.9b}$$

where f_{2i} is given as a weighted sum of "earlier" and "later" samples. The energy constraint is not often employed because it leads to a non-linear system of equations.

The parameters of the exponential-series FM now become accessible by first solving (7) with $a_P = 1$ for the P predictor coefficients and then finding the roots of the characteristic equation from which

$$s_\alpha = (1/\delta)\log(X_\alpha) \tag{9.10}$$

follows. The residues can be subsequently found using various approaches, the most straightforward of which is to solve (9.5) directly. As an alternative, the matrix pencil approach can be used [11], [12], [13], [18], [19], of which Prony's method is a special case.

We note that the number of data samples must be such that $D \geq 2P$ since P R_αs and P s_αs are needed to quantify the exponential model of (9.3). Equivalently, since there are P characteristic-equation coefficients needing solution, $D - P \geq P$ is required in (9.7). Since the value of P is itself often not known, some experimentation is usually needed to determine it, one of which is to examine the singular-value spectrum of the data matrix. While there is a definite lower limit to the number of data samples that are needed, it may happen that many more samples are available than this because a longer data record is available and/or the sampling interval is smaller than the minimum required. Such over sampling represents redundant information that can be used to reduce noise effects. Also, uniform sampling in x of $f(x)$ is required for the estimation procedure outlined above to be realized as non-uniform sampling in x would not produce the integer exponents in (9.5) that permits the polynomial of (9.6) to be utilized.

Many variations of the above procedure can be identified, differing most significantly in how the data samples might be "preprocessed" prior to accomplishing the actual parameter estimation. For example, instead of starting with the data samples themselves, auto-covariance estimates might be used as a means of reducing noise effects. Although the auto-covariance samples might be defined in various ways, for a finite-length data sequence they basically involve replacing the individual f_is by sums of shifted, two-sample products of the data. This is essentially equivalent to solving the data matrix using a pseudo inverse, a process that also replaces data samples by product sums since the matrix then solved is a product of the original matrix by its transpose.

Other approaches for reducing noise effects are also available. If multiple measurements of the same process can be made and if the noise is uncorrelated, averaging will improve the signal-to-noise ratio of the data samples. Alternately, these separate data sets might be used to obtain multiple sets of parameter estimates which can themselves then be averaged. The problem of dealing with noisy, uncertain and incomplete data is an important one that goes beyond the scope of this tutorial discussion, since the primary concern here is where noise is not a limiting factor. However, an FM can also be used to reduce the effects of noise or uncertainty in over-sampled data as is illustrated in Section 9.5.3, and which is a goal of most signal processing.

9.3.2 Waveform-Domain Derivative Sampling

Derivative sampling of WD data proceeds in a fashion analogous to that used for function sampling since the same FM is used for each, but rather than using samples of the process as a function of x we instead have

$$f_n = d^n f(x)/dx^n = \sum R_\alpha (s_\alpha)^n \exp(s_\alpha x), \alpha = 1, \ldots, P; n = 0, \ldots, D-1 \quad (9.11)$$

Making a change of variables like that done above for function sampling, we rewrite (9.11) as

$$f_n = \sum R'_\alpha (s_\alpha)^n, \ \alpha = 1, \ldots, P; n = 0, \ldots, D-1 \quad (9.12)$$

where $R'_\alpha = R_\alpha \exp(s_\alpha x)$, i.e., the variation in f_n due to whatever value is used for the observation variable x is absorbed in the residue.

Upon explicitly writing some of the individual equations implicit in (9.12), we obtain

$$\begin{aligned} f_0 &= R_1 + R_2 + \cdots + R_P, \\ f_1 &= R_1(s_1)^1 + R_2(s_2)^1 + \cdots + R_P(s_P)^1 \\ f_2 &= R_1(s_1)^2 + R_2(s_2)^2 + \cdots + R_P(s_P)^2 \\ &\vdots \\ f_{D-1} &= R_1(s_1)^{D-1} + R_2(s_2)^{D-1} + \cdots + R_P(s_P)^{D-1} \end{aligned} \quad (9.13)$$

Again, a polynomial model is suggested by the terms appearing in (9.11). If we assume that

$$B(s) = b_0 + b_1 s + b_2 s^2 + \ldots b_P s^P = \prod (s - s_\alpha); \ \alpha = 1, \ldots, P \quad (9.14)$$

and multiply the first equation in (9.13) by b_0, the second by b_1, etc., and add the resulting equations together we obtain

$$f_0 b_0 + f_1 b_1 + \cdots + f_P b_P = 0 \tag{9.15a}$$

since $B(s_1) = B(s_2) = \ldots = B(s_P) = 0$. Continuing this process by successively repeating these steps starting with the equation for f_1, f_2, f_3, etc., we get

$$\begin{aligned} & f_1 b_0 + f_2 b_1 + \cdots + f_{P+1} b_P = 0 \\ & f_2 b_0 + f_3 b_1 + \cdots + f_{P+2} b_P = 0 \\ & \vdots \\ & f_{D-P-2} b_0 + f_{D-P-1} b_1 + \cdots + f_{D-2} b_P = 0 \\ & f_{D-P-1} b_0 + f_{D-P} b_1 + \cdots + f_{D-1} b_P = 0 \end{aligned} \tag{9.15b}$$

The set of equations in (9.15) provides a basis for solving for the coefficients of the $B(s)$ polynomial. As was the case for waveform-function sampling, (9.15) is also homogeneous in the $B(s)$ coefficients, and thus requires additional information, or a constraint, to solve it. Lacking any convincing reason to do otherwise, we choose to set b_P to unity, which results in right-hand side entries for the successive equations of (9.15) given by $-f_{P+1}$, $-f_{P+2}$, …, $-f_{D-1}$, respectively. The resulting relationships then defined by (9.15) again take the form of a predictor equation, but where now P lower-order derivatives and the P predictor coefficients yield a derivative one order higher, although other coefficients constraints would lead to other relationships. Again, we need $D \geq 2P$ for a solution.

Thus, either function sampling or derivative sampling can be used to estimate the parameters of an exponential series FM. While function sampling might be achieved analytically, numerically or experimentally, derivative sampling is limited to the former two approaches as there seems to be no practical way of measuring, say, the time derivatives of a transient waveform. Derivative sampling is none-the-less relevant, as significant computational advantages can arise from using derivative information in the SD, as is discussed further in Section 9.8.3.2. Although a similar advantage might also be found for using derivatives in WD modeling, it's mentioned here primarily to emphasize the symmetry that exists between function and derivative sampling and between the WD and SD.

9.3.3 Combining Waveform-Domain Function Sampling and Derivative Sampling

As a generalization of the above discussion, it should be noted that it is also possible to combine function and derivative sampling of WD data. Letting $f_{i,n}$ represent the nth derivative of $f(x)$ at $x = x_i$ leads to

$$f_{i,n} = \sum R_\alpha (s_\alpha)^n (X_\alpha)^i; \; \alpha = 1, \ldots, P; \; i = 0, \ldots, D; \; n = 0, \ldots, F \tag{9.16}$$

where it is assumed that there are F derivatives at each of $D + 1$ sample locations in x and where $n = 0$ is the usual function sample. The same number of derivatives is not needed at each sample point, but this restriction makes the development more straightforward. The following equations are then obtained,

$$
\begin{aligned}
f_{0,0} &= R_1 + R_2 + \cdots + R_P,\\
f_{0,1} &= R_1(s_1)^1 + R_2(s_2)^1 + \cdots + R_P(s_P)^1\\
&\vdots\\
f_{0,F} &= R_1(s_1)^F + R_2(s_2)^F + \cdots + R_P(s_P)^F\\
f_{1,0} &= R_1(X_1)^1 + R_2(X_2)^1 + \cdots + R_P(X_P)^1\\
f_{1,1} &= R_1(s_1)^1(X_1)^1 + R_2(s_2)^1(X_2)^1 + \cdots + R_P(s_P)^1(X_P)^1\\
&\vdots\\
f_{1,F} &= R_1(s_1)^F(X_1)^1 + R_2(s_2)^F(X_2)^1 + \cdots + R_P(s_P)^F(X_1)^1\\
&\vdots\\
f_{D,0} &= R_1(X_1)^D + R_2(X_2)^D + \cdots + R_P(X_P)^D,\\
f_{D,1} &= R_1(s_1)^1(X_1)^D + R_2(s_2)^1(X_2)^D + \cdots + R_P(s_P)^1(X_P)^D\\
&\vdots\\
f_{D,F} &= R_1(s_1)^F(X_1)^D + R_2(s_2)^F(X_2)^D + \cdots + R_P(s_P)^F(X_P)^D
\end{aligned}
\tag{9.17}
$$

The polynomial that might be used to reduce this system of equations is not so obvious as was the case for function sampling or derivative sampling alone. At least two distinct choices might be considered, as

$$P_1(X,s) = A(X)B(s) \tag{9.18a}$$

and

$$P_2(X,s) = C(X,s) \tag{9.18b}$$

the difference being in the number of independent coefficients for given highest-order terms in X and s. The product form of (9.16a), for example, which might be viewed as a separation-of-variables form, yields $\sim D + F$ coefficients whereas that of (9.16b) has instead $\sim DF$. Thus expanding $P_1(X,s)$ in terms of $A(X)$ and $B(s)$, we obtain

$$
\begin{aligned}
P_1(X,s) &= (a_0 + a_1X + a_2X^2 + \cdots + a_{D-1}X^{D-1} + a_DX^D)\\
&\quad\times(b_0 + b_1s + b_2s^2 + \cdots + b_{F-1}s^{F-1} + b_Fs^F)\\
&= a_0b_0 + a_1b_0X + a_0b_1s + a_2b_0X^2\\
&\quad + a_1b_1Xs + a_0b_2s^2 + \cdots + a_Db_FX^Ds^F\\
&= 0
\end{aligned}
\tag{9.19a}
$$

where the cross-term coefficients (in X and s) involve various products and sums of the $A(X)$ and $B(s)$ coefficients for a total of $D + F + 2$ independent coefficients.

If, on the other hand, we use (9.16b), we would have

$$P_2(X,s) = c_{0,0} + c_{1,0}X + c_{0,1}s + c_{2,0}X^2 + c_{1,1}Xs + c_{0,2}s^2 + \cdots + c_{D,F}X^Ds^F = 0 \tag{9.19b}$$

which has a total of $DF+1$ coefficients. Clearly, either approach leads to a linear system of equations for the polynomial coefficients, but without testing it's not obvious which of these, or possibly another, might be better.

Anticipating encountering a similar question in SD sampling where various combinations of function and derivative sampling have been found to be successful, some fundamental questions arise concerning the most effective sampling strategy in the WD. First, a minimum of $2P$ samples are required of a waveform signal having P pole terms whatever combination of function and derivative sampling has been used to generate them. Second, the sample spacing must satisfy the Nyquist rate for the highest-frequency component of the signal, i.e., $\delta \leq 1/(2f_{\max})$ where $f_{\max}$ is the maximum frequency, at least in a generalized sense, noting that this restriction is irrelevant when all samples are taken at a single time. Finally, the data samples must be spaced such as to obtain a polynomial model. A uniform spacing δ is generally used, but any sequence of sample spacings that are related as integer multiples of δ could be used in principle.

A critical question in considering alternate sampling strategies, both those that involve function sampling as well as function and derivative sampling, is "How much information is provided by each additional sample?" Some sampling strategies are clearly less effective. For example, if $2P$ samples were to be used but at an arbitrarily closer spacing $\delta' \ll 1/2f_{\max}$, then the information provided by each sample shifts further to the right in the digits representing that sample's numerical value. This is because a given sample approaches the values of its neighbors in the limit $\delta' \rightarrow 0$ so that all sample values would eventually become identical. The consequence of decreasing the sample spacing is that proportionately higher-accuracy samples would be needed to retain a fixed amount of information when performing function sampling.

It seems likely that derivative sampling might be subject to similar kinds of constraints. One of the first questions to consider might be how the information content of a derivative sample varies as a function of derivative order, and is there a highest "reasonable" order to use? Another question would be how the proper Nyquist rate changes with increasing derivative order. From a computational viewpoint it would also be important to know how function and derivative samples vary with respect to information content and whether there is some combination of both that maximizes the acquired information while minimizing the computer costs of evaluating the FPM.

Proceeding as before, successively multiplying the data-sample equations by the set of coefficients from the chosen polynomial model, equations in the characteristic-equation coefficients can be derived. For example, if we were to use the product-form polynomial in $A(X)$ and $B(s)$ as in (9.18a), equations like the following

$$a_0 b_0 f_{0,0} + a_1 b_0 f_{1,0} + a_0 b_1 f_{0,1} + \cdots + a_{D-1} b_{F-1} f_{D-1,F-1} + f_{D,F} = 0 \tag{9.19c}$$

would be obtained. Thus, the initial solution of the data matrix would provide intermediate results in terms of the coefficient products which could be subsequently solved for the individual coefficients themselves. The approach just described has apparently not been implemented.

9.4 SAMPLING FIRST-PRINCIPLE MODELS AND OBSERVABLES IN THE SPECTRAL DOMAIN

The use of SD data for MBPE is conceptually similar to sampling in the WD, but because the FM for the former is the transform of the latter, significant differences arise, as is now shown where SD function and derivative sampling are respectively described.

9.4.1 Spectral-Domain Function Sampling

Spectral-domain function sampling begins with the FM given by (9.2) and assumes the availability of samples denoted by

$$F_i = F(X_i) = \sum R_\alpha/(X_i - s_\alpha) + F_{np}(X_i),\ \alpha = 1, \ldots, P \tag{9.20}$$

where, in contrast to waveform sampling, there is no requirement that the sampling points X_i be uniformly spaced. However, in contrast to waveform sampling, where the FM can be a purely exponential series at late times, spectral sampling can not avoid the presence of the non-pole term which is generally unknown. Fortunately, an approximation to the F_{np} term can be realized by generalizing the pole series in (9.20) [20]. First note that, as written, the pole series can be developed into a particular rational function where the denominator order exceeds that of the numerator by one. A general rational function has no specified connection between the orders of the polynomials which comprise it, however. The capability of a pole series to model resonances can be retained while changing the numerator-polynomial order relative to that of the denominator. Note that the special case of using rational functions as a model is also known as Padé approximation [21] when the function and derivative sampling is done at a single point although the term has been applied to the more general approach considered here [22].

The possibility of using various numerator and denominator polynomial orders provides a way to approximate the effect of the non-pole term by simply increasing the order of the numerator polynomial. For example, an increase of one in the numerator order has the effect of representing F_{np} by a constant, which, when absorbed into the rational function, results in equal numerator and denominator orders. If F_{np} is represented by a constant and a term linear in X, this has the effect of making the numerator order one greater than the denominator. Thus, by varying the relative orders of the polynomials which comprise the FM, various models of the non-pole contribution are implicitly included. This rather simple way of handling the non-pole contribution does not seem suitable for WD data, unfortunately, since the early-time, or driven, response contains essentially all frequency components whereas approximating the SD non-pole contribution needs only to be effective over a relatively narrow frequency band when using windowed FMs.

Therefore, in general we use a SD FM given by

$$F(X) = N(X)/D(X) \tag{9.21a}$$

$$N(X) = N_0 + N_1 X + N_2 X^2 + \cdots + N_n X^n \tag{9.21b}$$

and

$$D(X) = D_0 + D_1 X + D_2 X^2 + \cdots + D_d X^d \tag{9.21c}$$

The coefficients of the SD FM are also obtained from sampled values of the response as seen by rewriting (9.20)

$$F_i \underline{D}_i = \underline{N}_i,\ i = 0, \ldots, D-1 \tag{9.22a}$$

where

$$F_i = F(X_i) \tag{9.22b}$$

$$\underline{D}_i = D(X_i) = D_0 + D_1 X_i + D_2 (X_i)^2 + \cdots + D_d (X_i)^d \tag{9.22c}$$

and

$$\underline{N}_i = N(X_i) = N_0 + N_1 X_i + N_2 (X_i)^2 + \cdots + N_n (X_i)^n \tag{9.22d}$$

There are $d + n + 2$ unknown coefficients in the two polynomials $D(X)$ and $N(X)$, and as for WD modeling, a constraint or additional condition is needed to make the sampled equations inhomogeneous. Again, there is no unique choice for this constraint, but if we set $D_d = 1$, then the following equations result:

$$\begin{aligned}
&F_0 D_0 + X_0 F_0 D_1 + \cdots + F_0 (X_0)^{d-1} D_{d-1} - N_0 - X_0 N_1 - \cdots - (X_0)^n N_n = -(X_0)^d F_0 \\
&F_1 D_0 + X_1 F_1 D_1 + \cdots + F_1 (X_1)^{d-1} D_{d-1} - N_0 - X_1 N_1 - \cdots - (X_1)^n N_n = -(X_1)^d F_1 \\
&\vdots \\
&F_{D-1} D_0 + X_{D-1} F_{D-1} D_1 + \cdots + F_{D-1} (X_{D-1})^{d-1} D_{d-1} - N_0 \\
&\quad - X_{D-1} N_1 - \cdots - (X_{D-1})^n N_n = -(X_{D-1})^d F_{D-1}
\end{aligned} \tag{9.23a}$$

where $D \geq n + d + 1$ is now required. Note that the matrix coefficients are now comprised of a product of a data sample and the frequency at which the sample is taken raised to a power, in contrast to the time-domain situation where the data samples alone are the data-matrix coefficients. Also observe that the poles in the SD arise directly as the roots of $D(x)$, whereas in the WD, the poles are natural logarithms of the roots of the characteristic equation, (9.6). The exponentiation of the sampling frequencies suggests that large dynamic numerical ranges in the matrix coefficients may result if d and n are very large. One way to avoid this is to scale the frequency, so that, for example, if the sampling range is centered at 1 GHz, a scaling of 10^9 in the frequencies leads to nominal scaled values near unity. It is also possible to center the SD model about a frequency in the interval of interest so that terms like $(X_i - X_{\text{ref}})^n$ result. Combining scaling and translation similarly produces terms like $[(X_i - X_{\text{ref}})/X_{\text{ref}}]^n$.

An oversampled system, i.e., one where $D > n + d + 1$, can be handled in various ways, one of which is to employ a pseudo inverse for the solution. Another approach would be to employ overlapping windows of different data sets, both to reduce the n and d for each FM and to compare their respective results to obtain an error estimate. A third would be to progressively increase the number of data samples while retaining the same number of FM coefficients and comparing the FM spectra to observe their trends.

Derivative sampling can also be performed in the SD, as can various combinations of function and derivative sampling. One result is that the samples can be spaced more widely when derivative information is available. A more important consideration is that in some circumstances a derivative sample can be obtained for a computation operation count that is of order $1/X_i$ of the first function sample alone. Thus, if a derivative sample provides information commensurate with that provided by a function sample, an obvious computation advantage arises from using derivative samples to decrease the number of X_i sample points that are needed. Using derivative sampling in the context of a frequency-domain FPM is discussed more fully in Section 9.8.3.

9.4.2 Spectral-Domain Derivative Sampling

The set of possible sampling strategies concludes with derivative sampling in the SD which was called "Frequency-Derivative Interpolation" (FDI) [23] when first presented. In this case, we rewrite (9.21a) as

$$F(X)D(X) = N(X) \tag{9.21a)'}$$

which upon differentiating with respect to X t times leads to (omitting the explicit X dependence)

$$\begin{aligned} &F'D + FD' = N' \\ &F''D + 2F'D' + FD'' = N'' \\ &F'''D + 3F''D' + 3F'D'' + FD''' = N''' \\ &\vdots \\ &F^{(t)}D + tF^{(t-1)}D^{(1)} + \cdots + C_{t,t-m}F^{(m)}D^{(t-m)} + \cdots + FD^{(t)} = N^{(t)} \end{aligned} \tag{9.23b}$$

where $C_{r,s} = r!/[s!(r-s)!]$ is the binomial coefficient and we require both that the t + 1 samples $F^{(0)}, \ldots, F^{(t)}$ be such that $t + 1 \geq n + d + 1$ and that one of the coefficients is set to unity for a unique solution to be obtained. Clearly these also become a linear system of equations for the polynomial coefficients which can be expressed in the form

$$\begin{aligned} &N_0 = F_0 \\ &N_1 - F_0D_1 = F_1 \\ &N_2 - F_1D_1 - F_0D_2 = F_2 \\ &\vdots \\ &N_n - F_{n-1}D_1 - F_{n-2}D_2 - \cdots - F_{n-d}D_d = F_n \\ &-F_{n+1}D_1 - F_nD_2 - \cdots - F_{n-d+1}D_d = F_{n+1} \\ &\vdots \\ &-F_{D-2}D_1 - F_{D-3}D_2 - \cdots - F_{D-d}D_d = F_{D-1} \end{aligned} \tag{9.24}$$

where the substitution $X' = X - X_s$ has been made, with X_s the sampling frequency, so that all frequency-dependent terms drop out from the final equations and D_0

has been set to unity. The subscripted "F" quantities in the above matrix are $F_m = F^{(m)}/m!$. A technique called "asymptotic waveform expansion" (AWE) was reported for circuit modeling [24] shortly after FDI was presented, and in one implementation leads to equations identical to (9.24) [25]. It's worth noting that AWE was motivated by a need to develop transient responses from SD data, and explicitly determines the FM poles for this purpose using a Laplace transform.

A still more general sampling strategy can be employed where two or more frequencies are used at each of which a variable number of frequency derivatives can be independently specified. This leads to a more complicated system of equations [20], but which is numerically simply still another way to compute the FM parameters, i.e., the FM and the data used for its quantification in the SD can be independently specified. The computational advantages of using SD derivative sampling from a CEM viewpoint are discussed in Section 9.8.3.

9.4.3 Adapting and Optimizing Sampling of the GM

A major advantage of MBPE, and a prime motivating factor for its use, is its potential for minimizing the operation count needed for computing various EM (and other wave-equation phenomena) responses. This is especially the case concerning the time-frequency transform pair, where there is more flexibility with respect to sample placement and model order in frequency than in time, since uniform sampling is required in time but not in frequency. Actually, this requirement is not inconsistent with how most time-domain computations are performed, where equal time steps are most-often used throughout the time interval for which the model is run. However, it also means that if subsequent evaluation of the time response shows that it was under sampled, it's not very practical to add new samples to the original result; instead, the model computation must be entirely repeated with a new time step.

This is not the case for all WD models, however, as is demonstrated by sampling the far fields of a source distribution (model 3b in Table 9.1), where new observation angles can be added to an existing set without repeating any previous computation. This advantage is always true in the SD, where adding one, or several, new function or derivative samples to an existing set can be done while fully retaining the benefit of whatever samples have already been obtained.

Recall that a major goal of using MBPE is minimizing the measurement or computational cost of acquiring and representing observables to a desired accuracy or uncertainty over some specified range of an observation variable. Ideally, this would involve choosing sample locations and derivative orders that maximize the new information provided relative to the cost of obtaining that new sample. If this goal were to be realized for each of the samples ultimately needed for computing the FM parameters while satisfying whatever error criterion is specified, then the overall cost would clearly be a minimum for the chosen error measure. This idea is illustrated conceptually in Fig. 9.2.

There seems to be no obvious or unique approach for achieving this goal, since P in the WD and n and d in the SD, the number of samples D and the data range they span, are all free parameters in developing a FM. This means that some computer experimentation will usually be required to determine suitable numerical values for them. A few of the several sampling strategies that seem plausible and worth

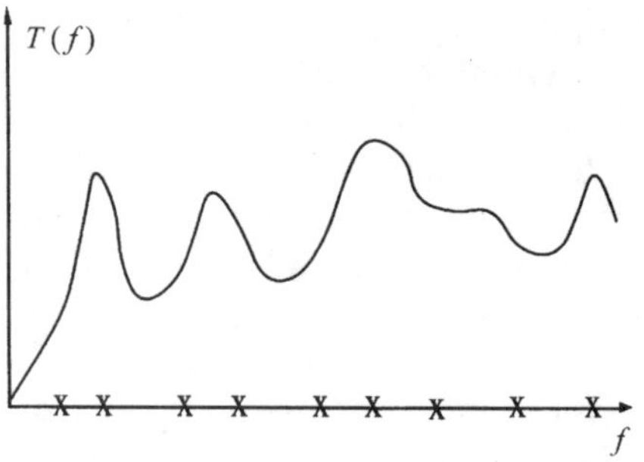

Figure 9.2 Conceptual diagram to illustrate optimal placement of frequency samples in developing a FM representation of a transfer function T(f). Since there is no computational penalty in doing so, we can expect that the samples would generally be nonuniformly spaced as a means of maximally reducing the uncertainty in the FM representation of T(f) with each added GM sample.

considering are discussed immediately below, after which error measures, something that is required by all adaptive procedures, are then considered.

9.4.3.1 Possible Adaptive Sampling Strategies

Some possibilities of adaptively sampling an observable of interest include:

1. A strategy similar to one found useful for adaptive numerical quadrature based on Romberg's method (RM) [26] might be considered for the adaptive-estimation problem. The adaptive RM procedure involves choosing a starting subinterval over which five successive uniformly spaced samples of an integrand are computed upon which RM is applied to the three trapezoidal-rule quadrature values that are obtained (samples 1 and 5, 1, 3, and 5, and all five samples). An error estimate provided by RM indicates whether new samples are required anywhere in this original subinterval. If that is the case, then two new subintervals are formed from each half of the original subinterval and two trapezoidal-rule values are formed from each (using samples 1 and 3 and 1, 2, 3 in the first half, and samples 3 and 5 and 3, 4, 5 in the second). If the RM error estimates show either of these new subintervals to require additional samples, then new samples are added where indicated half way between the original ones and the process is successively repeated. On the other hand, if the initial error is smaller than specified, the process is repeated while doubling the subinterval size, using a minimum of three and maximum of five samples per subinterval. This kind of adaption has been found to work extremely well for numerically integrating sharply peaked integrands, yielding sample spacings that can vary by a factor of 10^6.
2. The first two samples could be taken at the endpoints of the variable range with additional samples subsequently developed by successively subdividing this range into subintervals of evenly spaced samples until the error criterion has been satisfied over all subintervals thus formed.
3. If an error measure is available as a continuous function of the independent variable over the range of interest, then placing a new sample wherever that measure is a maximum seems to be an obvious choice as a way to achieve the greatest amount of additional information from a single additional sample. This is the approach described below.

9.4.3.2 Estimating FM Error or Uncertainty as an Adaptation Strategy

An adaptive process can be only as effective as the error measure used for estimating the degree to which the FM (or its equivalent) differs from the GM (or whatever process whose results the FM is to approximate). This observation is a general one that applies to all manner of numerical processes having as their goal to minimize the number of accurate samples that are needed as a means of reducing the cost of developing a sampled representation of some process over a specified range of independent variable(s). For an SD application, it appears desirable to use lower-order FMs over subintervals of the spectral range to be covered to avoid possible ill-conditioning. It then follows that two or more FMs will be needed to span the spectral range of interest, leading to the situation illustrated in Fig. 9.3. By using

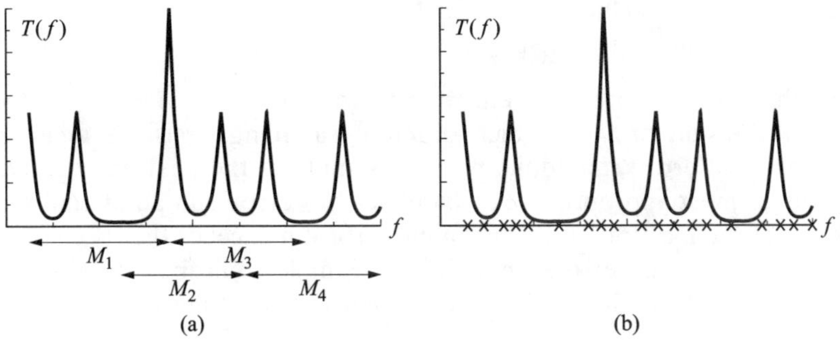

Figure 9.3 The possibility of developing a FM-representation of a transfer function over a wide frequency interval by employing a number of subinterval, lower-order fitting models, M_i, is illustrated here (a). Each FM represents the response over a frequency interval $F_i = F_{i,u} - F_{i,l}$ where the adjacent FM's share some frequency samples and the difference between them, $\Delta ME_{i,j}(f_k) \equiv \{[|M_i(f_k) - M_j(f_k)|]/[|M_i(f_k)| + |M_j(f_k)|]\}$, provides a normalized, mutual-error estimate between FM's i and j. By evaluating $ME_{ijK} = \max_{ijk}\{\Delta ME_{i,j}(f_k)\}$ at $f = f_K$ as a function of frequency, new samples, either function or derivative, can be placed where ME_{ijK} exhibits values that exceed a specified amount and where, therefore, the corresponding FM results are most uncertain. By placing a new GM sample where the FM mismatch exceeds a specified amount, a generally non-uniform placement of the GM samples results as shown conceptually in (b).

overlapping FMs which share common data, their differences, or mismatch errors, can then be used to estimate FM uncertainty as a function of frequency. The mismatch error between FMs i and j, defined as

$$\Delta ME_{i,j}(f_k) \equiv \{[|M_i(f_k) - M_j(f_k)|]/[|M_i(f_k)| + |M_j(f_k)|]\} \tag{9.25}$$

is then computed for each pair of overlapping models at frequencies f_k, $k = 1, \ldots, F$ to find the maximum mismatch error for models i and j, $ME_{ijL} = \max\{\Delta ME_{i,j}(f_k)\}$ at $f = f_L$. Subsequent sample placement and type would then be chosen to maximize the information acquired from a new GM sample by adding it at the frequency f_K, where max{ME_{ijK}}, the overall maximum-mismatch error (MME) for all i,j,k comparisons, occurs. Sampling of the GM would be concluded when the MME is less than a specified value. Also note that, alternatively, an exact error measure results from comparing a FM result with a GM sample $G(f_k)$, using the measure $\Delta GE_{i,k} \equiv [|G(f_i) - M_k(f_i)|]/[|G(f_i) + M_k(f_i)|]$. However, doing this potentially would require more GM samples with a consequent increased computer time, while providing, in addition, only a pointwise error measure in f. Thus, ME_{ijK} requires less computation and yields a global, but approximate, error measure while $\Delta GE_{i,k}$ requires more computation and yields a pointwise, but exact, error measure.

There are at least three kinds of errors that the FM might produce relative to GM results:

1. Non-physical results are produced, e.g., negative conductance or resistance, as illustrated in Fig. 9.4;
2. Amplitude shifts in transfer functions;
3. Resonance shifts between transfer functions;

with the latter two occurring either between two (or more) FMs or between a FM and the GM result, and for which different sampling decisions might be made. For non-physical errors, an additional GM sample at the peak in negative conductance would be most appropriate. For baseline-shift errors, an additional sample could be placed at the location of the maximum error if it exceeds the specified error criterion. For resonance-shift errors, an additional sample could be put between the response peaks, again if the shift exceeds the specified error criterion.

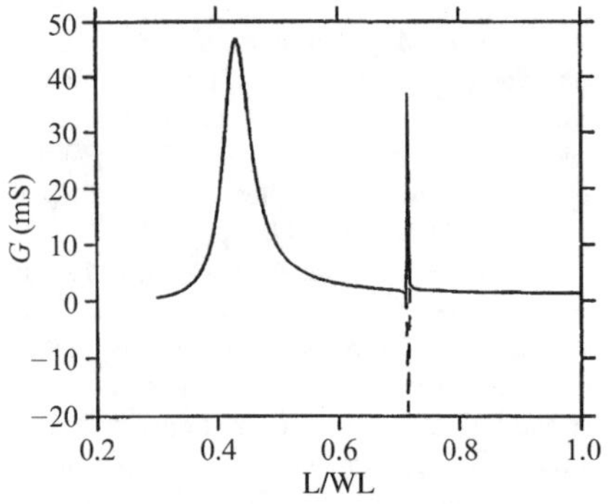

Figure 9.4 Example of an MBPE FM providing a nonphysical result, in this case negative conductance for an antenna [20]. Clearly, the best place to add one additional GM sample would be at the maximum value of negative conductance.

To summarize, for a SD FM having numerator and denominator polynomials of order n and d, respectively, $n + d + 1$ samples of the GM are needed, which can range, on the one hand, from using $n + d + 1$ different frequencies to, on the other hand, using one function sample and $n + d$ derivatives at a single frequency. The "best" approach would be that which minimizes the GM operation count required to achieve some specified accuracy or uncertainty criterion appropriate to the transfer function being estimated, sampling adaptively with the most appropriate mix of function and derivative samples. Further improvement might be realized if better ways to handle the effects of non-pole contributions and poles that lie outside the frequency interval of interest could be found. Finally, use of other FMs is worth exploring as well as considering other signal-processing approaches. An analogous approach is used for a WD observable as described below for far-field pattern estimation.

9.4.4 Initializing and Updating the Fitting Models

Whether using MBPE in the WD or SD, and whether using pre-sampled data or performing adaptive sampling, some initial choices must be made, such as

1. The range of observation variable to be used and initial sampling interval
2. The number of FMs to be used initially and their overlap
3. The mix of function and/or derivative GM samples to use
4. For SD models, the orders of the numerator and denominator polynomials
5. The maximum mismatch error to be specified
6. The trade-off between number of FMs and their order

An FM having P unknown parameters can be quantified from function samples of the GM at P locations on the one hand to one function sample and $P-1$ derivative samples at the same location on the other, to any variation in between. The "best" approach would be that which minimizes the GM operation count required to achieve some specified uncertainty criterion appropriate to the phenomenon being estimated, while sampling adaptively with the most appropriate mix of function and derivative samples. This very general kind of GM sampling strategy can be used in the SD because the function samples are independently obtained, but this is not generally the case for WD models. For example, a transient response is normally computing by "marching on in time," using a fixed time step. In some cases, the time step may be increased one or more times during the computation as the higher-frequency components disappear due to radiation. But it's generally not feasible to add new samples to a time domain GM *a posteriori* without completely repeating the computation. Furthermore, WD FMs quantified using Prony's or a similar method require uniformly spaced GM samples and so for this reason are not suited to adaptive sampling as outlined above. The only feasible adaptation that might be done in this case would be to halve the sampling interval.

There are other WD GMs, however, where new samples can be arbitrarily added. This possibility is demonstrated by sampling the far fields of a source distribution (model 3b in Table 9.1). In this case, new observation angles can be added to an existing set without repeating any previous computation. Of course the uniform-sampling requirement still applies if a Prony-type FM is employed. But a Prony-type FM is not always needed, as is demonstrated in adaptive sampling of radiation and scattering patterns in Section 9.6.2.

9.5 APPLICATION OF MBPE TO SPECTRAL-DOMAIN OBSERVABLES

Considered in turn here is application of MBPE to precomputed or presampled frequency-domain data followed by adaptive MBPE when the GM samples can be obtained as needed.

9.5.1 Non-Adaptive Modeling

The MBPE procedure described above has been tested for a variety of wave-equation phenomena like the examples that follow, where presampled data are used, i.e., no new data samples are available during the process of developing FMs. Plotted in Fig. 9.5 are the input conductance and susceptance of a monopole as a function of antenna size in wavelengths [27]. These results were obtained by sampling a GM (NEC) at 0.15 intervals in L/λ, shown by the open crosses, which are connected by the dashed lines to show the result of straight-line interpolation where the MBPE results are the solid line.

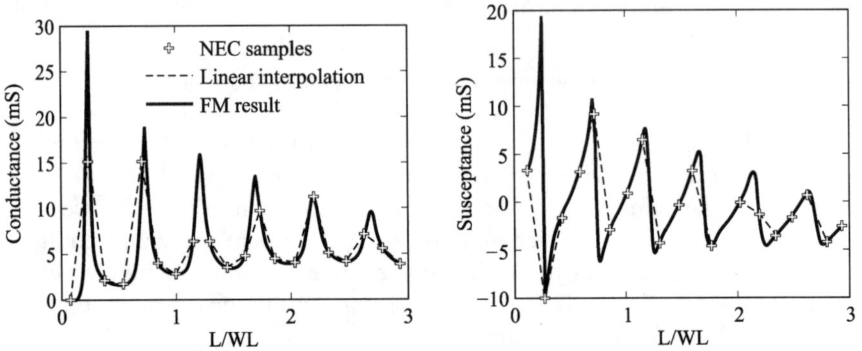

Figure 9.5 Input conductance and susceptance of monopole antenna versus length in wavelengths as obtained from a series of overlapping rational-function fitting models using $d = n = 3$ (the solid line) based on 20 GM samples spaced 0.15 apart in L/wavelength which are shown as open crosses and joined by a dashed line [27]. A comparison of the FM values with GM samples at other frequencies reveals a numerical agreement of 1% or better.

The FM used a $d = n = 3$ rational function, which requires seven data points. The first FM, M_1, was used to develop an estimate out to the fourth data sample, as shown by the solid line. The second model, M_2, was "slid" up in frequency by adding the eighth data sample and dropping the first, and it was used to plot the solid line between the fourth and fifth data samples. This process was continued until model M_{14} was reached, which completed the fitting-model plot from data points 16 to 20. This particular procedure is not unique as other approaches might be used, but it produces results within a few percent of the GM between the data samples actually used to quantify the fitting model. Extensions of this idea include comparing the FMs for mutual consistency in their regions of overlap to estimate the relative uncertainty of the FM result and to determine whether additional GM samples are needed.

A "forked-monopole" antenna, one having a straight center section and a Vee end load whose arms are slightly different in length, has a low-frequency, common-mode resonance with a differential-mode response occurring with increasing frequency, as shown in Fig. 9.6 when driven against a perfect ground [20]. Results obtained directly from NEC are graphically indistinguishable from a FM using $n = 4$

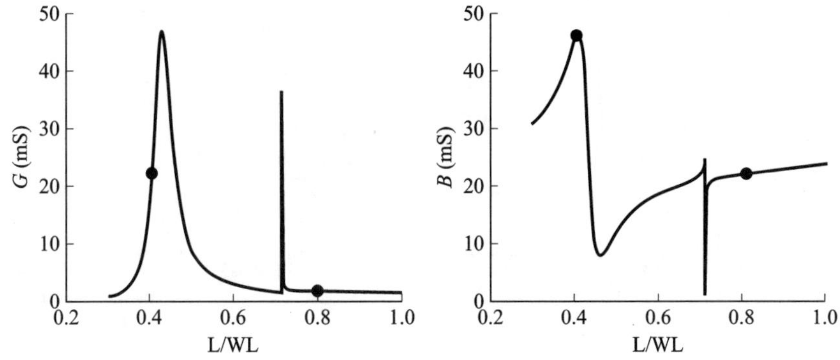

Figure 9.6 Conductance and susceptance of a forked-monopole antenna driven against a ground plane as computed from NEC (solid line) and an MBPE FM (dashed line which here is graphically indistinguishable) using $n = 4$ and $d = 5$ as a function of size in wavelengths (L/WL) [20]. The solid dots show where a function sample and the first four frequency derivatives were obtained from NEC to determine the FM parameters.

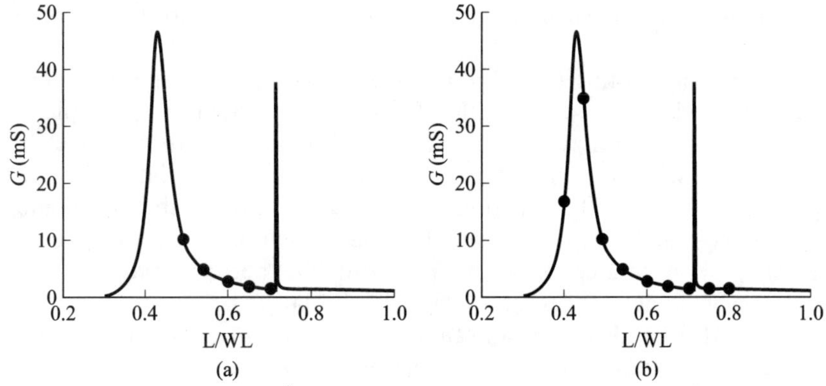

Figure 9.7 Conductance of forked-monopole antenna from NEC (solid line) and MBPE (dashed line which here is graphically indistinguishable) using a function sample and one derivative at five frequencies, and function-sampling only at nine frequencies [20]. For the former, $n = 5$ and $d = 4$ while for the latter $n = 4$ and $d = 4$.

and $d = 5$, based on a function sample and the first four frequency derivatives at the points indicated by the solid dots.

As a demonstration of the flexibility and insensitivity of the FM to the kind of data used to quantify it and the order of the polynomials used for the rational function, some additional conductance results are shown for the forked-monopole antenna in Fig. 9.7 [20]. In Fig. 9.7(a) the result of using a function sample and the first frequency derivative at five frequencies is illustrated, where the FM employs $n = 5$ and $d = 4$. In Fig. 9.7(b), function sampling alone is employed at nine frequencies and using $n = 4$ and $d = 4$. These results demonstrate a high degree of versatility of MBPE, although it should be emphasized that some degree of experimentation may

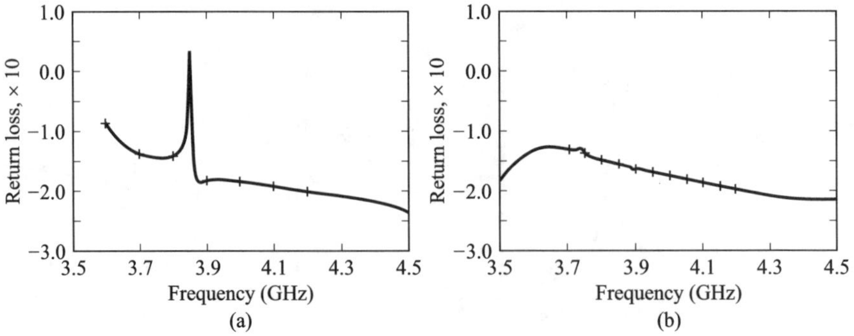

Figure 9.8 Results obtained from synthesizing a corrugated-horn antenna without MBPE (a), and when incorporating a SD FM in the synthesis procedure (b) [28]. The FM revealed a sharp spike in the return loss in the original design, permitting its removal when using MBPE to synthesize the design.

be needed to establish the bandwidth and FM order that produce acceptable results. An adaptive approach may offer the most robust way to do this, as is discussed in Section 9.7.1.

When attempting to design an antenna to achieve a specified performance over some bandwidth, it is often neither feasible nor economical to sample the response so finely as to ensure that important features are not missed. A sparsely sampled set of frequency samples is therefore normally used in the initial design, resulting in the possibility that a final evaluation might discover features that invalidate the design or reduce its effectiveness. An example of such an application was reported by Fermelia et al. [28] in connection with optimizing the performance of a corrugated-horn antenna yielding the results of Fig. 9.8. Initially, the design was developed over a several GHz bandwidth from samples spaced at 0.1 GHz intervals using a minimax procedure as illustrated by the crosses in Fig. 9.8(a). Upon fitting a SD FM to these data, the solid line shown in this figure was obtained, a result confirmed upon subsequent experimental investigation that revealed a spike in the return loss between two of the sampled values. By incorporating MBPE into the design optimization procedure, this spike and its effect were essentially mitigated as shown in Fig. 9.8(b).

An example of estimating the wideband scattered response of a 2-dimensional hollow cylinder with a narrow aperture or slot is presented in Fig. 9.9 [29], [30]. The GM result for the scattered field in (a) is obtained from an analytic solution, sampled at 949 points in ka to produce an essentially continuous curve. In Fig. 9.9(b), a FD integral-equation (FDIE) model is used as the GM to obtain the cylinder current and its first four derivatives at ka values of 2, 3, 4, 5, and 5.3 from which the backscattered field and FM results are computed. It may be observed that the resonances exhibited by the slotted cylinder are extremely sharp, and can be easily missed unless the GM is sampled using appropriately small ka steps, or unless revealed by a FM whose underlying mathematical behavior is capable of estimating the actual solution accurately enough.

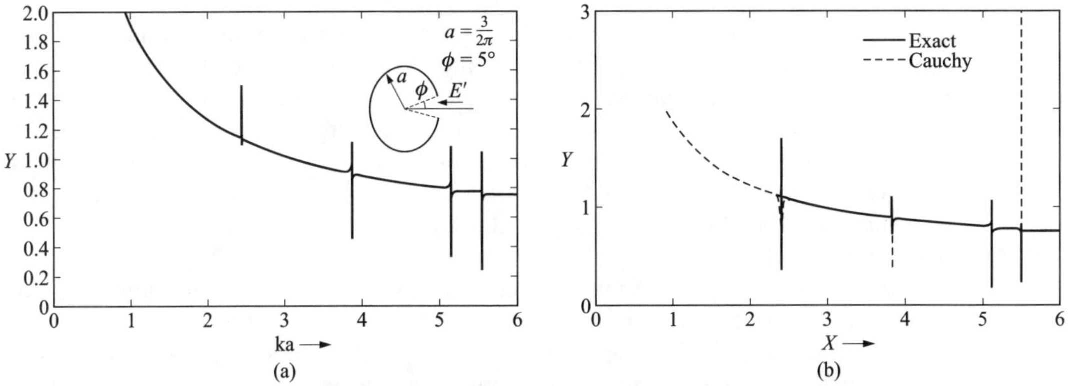

Figure 9.9 Analytical GM backscattered far-field as a function of ka for a slit cylinder for a TM field normally incident on the 20 degrees slit [29], [30] (a). This response is characterized by a series of extremely sharp resonances and thus provides a challenging problem if the transfer function is to be reliably developed using pointwise sampling. Comparison of a FDIE GM (continuous line) and FM (dashed line, denoted as Cauchy) results for the backscattered field from the slit cylinder of Fig. 9.9(a) where the FM is based on values of the scattered fields at its first four derivatives at ka = 2, 3, 4, 5, and 5.3 is shown in (b). The frequencies of the resonances predicted by the FM are within a percent or so of those from the exact solution.

9.5.2 Adaptive Modeling

Implementation of adaptive sampling, for which some earlier results are discussed in [32], [33], is demonstrated in Fig. 9.10. The approach employed is the one illustrated in Fig. 9.3 with the amplitude difference between overlapping FMs being used to determine at what frequency f_L another GM sample is needed until the MME $< \varepsilon$, with $\varepsilon = 0.01$. Results based on this approach as presented in Fig. 9.10 are

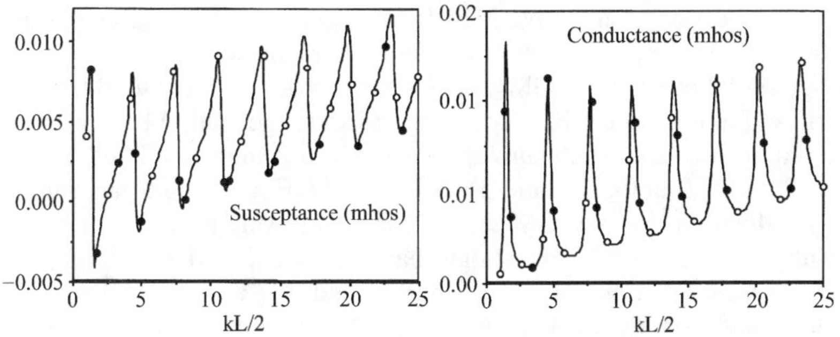

Figure 9.10 Input conductance and susceptance as obtained from an analytical approximation [31] for a center-fed, dipole antenna and modeled using four overlapping FMs and a total of 31 GM samples. The initial FM estimate is based on 16 GM samples (the open circles) located at 1.5 intervals in kL/2. Additional GM samples (the solid circles) were added one at a time at the frequency where the maximum difference was found between the overlapping FMs until a specified mismatch error of ≤1% was obtained between them [66].

obtained using as the GM an analytical approximation to the input admittance of a center-fed dipole antenna [31]. The 12 initial GM samples, spaced at 1.5 intervals over the 1–17.5-kL/2 range covered by the calculations, were used with four overlapping rational-function FMs. Two FMs (numbers 1 and 4) extended from GM samples 1–6 and 7–12 (using $d = 3, n = 2$) and the other two (numbers 2 and 3) extended over the eight GM samples from 1–8 and 5–12 ($d = 4, n = 3$, respectively. This resulted in a minimum of two FMs overlapping at the ends of the frequency range and three overlapping in the center.

Ten subsequent GM samples were successively added, determined by the maximum difference between the overlapping FMs (in units of kL/2), at 9.1, 2.8, 7.9, 12.1, 4.6, 1.6, 10.9, 17.2, 8.2, and 13.9 at which point the convergence criterion had been satisfied at all remaining FM sample frequencies, which were spaced at 0.3 intervals in kL/2. At this point the four FMs had added 5, 7, 7, and 4 additional GM samples, respectively, where the sum is less than 10 because each additional GM sample is shared by two or more FMs. The FMs were changed with each new GM sample by alternately increasing first n, and then d, to maintain $d \leq n \leq d + 1$. Note that the new GM samples are mostly located at extrema of the admittance frequency variation, indicating that the FMs are evidently most sensitive to regions of rapid change in the process being sampled. It's also important to note that differences between the average FM values and additional GM samples used as checks were found to be similar to the differences between the FM themselves, another indication that FM differences are reliable indicators as areas of greatest uncertainty in the FM representation of the GM.

It should be observed that if the number of samples used for a FM in a fixed frequency interval is increased monotonically, the condition number of the FM matrix can increase beyond some acceptable threshold. Consequently, as an alternative to simply increasing the number of GM samples per FM until the maximum mismatch error falls below a specified value, it might be more appropriate, or even necessary, to divide a too-large FM into two smaller ones. Another way to handle the ill-conditioning would be to employ singular-value decomposition as a means of handling a poorly conditioned FM matrix as described in Appendix 9.1.

Application of adaptive MBPE to the current induced on the front side of an infinite circular cylinder by an incident TE plane wave is demonstrated in Fig. 9.11 [66]. A total of 14 initial GM samples are uniformly distributed from ka = 0.1 to 26.1 and used to compute 4 FMs spanning the ka range from 0.1 to 14.1, 0.1 to 18.1, 8.1 to 22.1 and 8.1 to 26.1, with corresponding n and d values of 4 and 3 (numbers 1 and 4) and 5 and 4 (numbers 2 and 3), respectively. Employing a mismatch error of three digits, the 10 additional GM samples are adaptively placed as indicated in Fig. 9.11, resulting in graphically indistinguishable FM and GM results. Thus far, only PEC objects have been modeled. In Figs. 9.12 and 9.13 are shown, respectively, scattered-field results for a dielectric and lossy infinite, circular cylinder using oversampled, precomputed data supplied by Kluskens [63]. The results are for a cylinder of radius 0.159 m and TM plane-wave illumination with a relative permittivity of 16 in Fig. 9.12 and $16 - j16$ in Fig. 9.13 with the original data supplied at 5 MHz steps.

Five initial FMs were used in each case, spanning samples 1–4, 1–5, 3–6, 4–8, 5–8. Additional GM samples, shown by the solid circles, were used from the presampled set until a maximum FM–FM mismatch of 0.01 was achieved. When adding a new GM samples to a given FM resulted in $n + d + 1 > 18$, it was split into two new overlapping

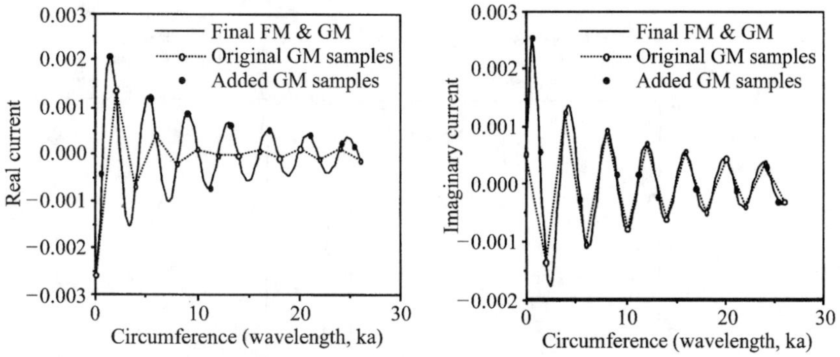

Figure 9.11 Results for adaptive sampling of the frequency-dependent current on the front side of an infinite, circular cylinder excited by a normally incident TE plane wave [66]. Four overlapping FMs are used with 14 initial GM samples and 10 additional GM samples were needed to achieve a mismatch error less than three digits.

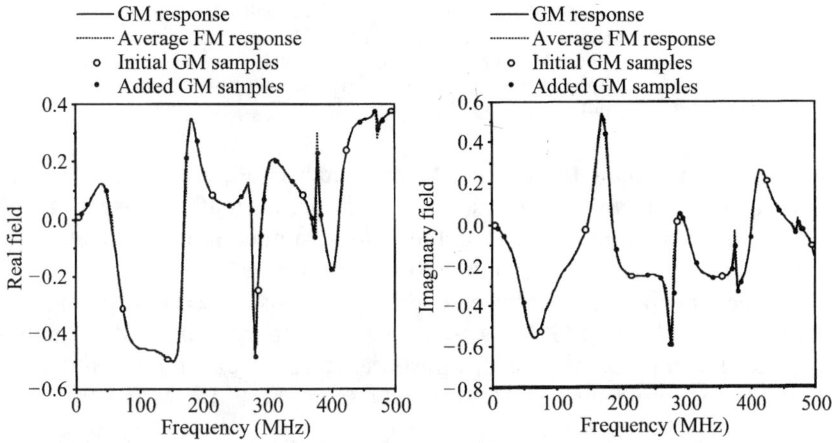

Figure 9.12 Adaptive sampling using precomputed data [63] for the scattered field from an infinite, circular cylinder whose relative permittivity is 16 is illustrated here. The precomputed GM samples, spaced at 5 MHz intervals, are joined by a solid line while the average FM results are shown by the dotted line which here is graphically indistinguishable. The GM samples used for the original FM computation are shown by open circles while the samples added during the adaptation process are indicated by solid circles.

models, resulting in a total of six FMs for Fig. 9.12 and five for Fig. 9.13, in both cases using an estimation error of 0.01. Generally speaking the added GM samples in both cases are located where the response is changing most rapidly. The GM and FM results are generally graphically indistinguishable except in regions of rapid change where the average FM is sampled more finely than the precomputed GM data. It should be realized that the actual mismatch achieved between the final FM estimate and the true GM results, were the latter to be computed more finely in frequency, is not guaranteed to be equal to, or better than, the specified mismatch error. The frequencies at which GM samples are added are limited to those where the FM comparisons are preselected for the adaptation process. It's possible that at fre-

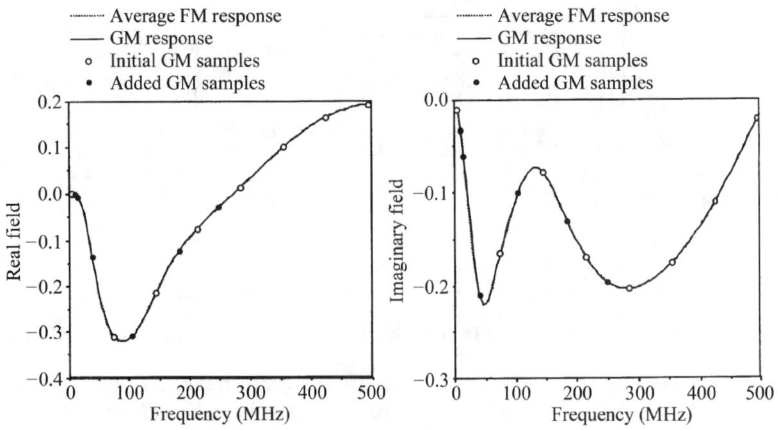

Figure 9.13 Adaptive sampling using precomputed data [63] for the scattered field from an infinite, circular cylinder whose relative permittivity is $16 - j16$ is illustrated here. The precomputed GM samples, spaced at 5 MHz intervals, are joined by a solid line while the average FM results are shown by the dotted line which here is graphically indistinguishable. The GM samples used for the original FM computation are shown by open circles while the samples added during the adaptation process are indicated by solid circles.

quencies other than these the FM mismatch could be larger than those actually computed, and possibly also resulting in a larger mismatch between the FMs and GM. This possibility could be reduced, but not guaranteed to be eliminated, by more finely sampling the FMs when the adaptation is performed.

Since the mismatch between overlapping FMs is used to estimate the actual error between the FMs and GM, it's advisable to compare them to determine how reliable the FM–FM mismatch is for this purpose. Results pertaining to this are presented in Fig. 9.14 for a pole-series GM. As can be seen, the FM–GM match is generally better

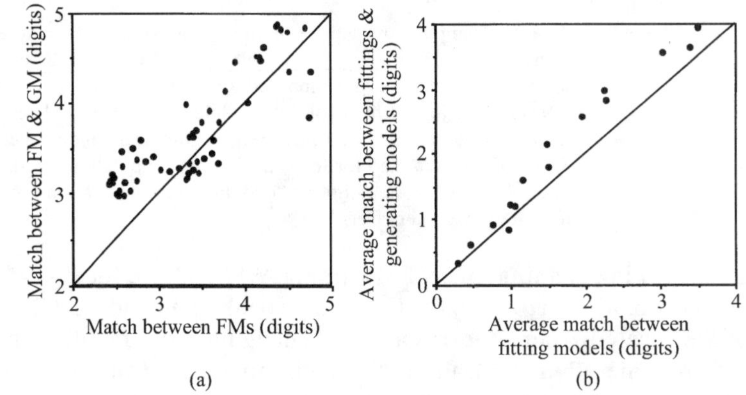

Figure 9.14 The average minimum-match values between the FMs and GM data samples for the 16th model iteration of a pole-series GM (a) and the frequency-averaged values of these quantities for each of the 16 FM sets (b) [33]. In both cases the FM–FM minimum match is seen to be generally less than the FM–GM value, showing that the former is a conservative error estimate of the accuracy achieved by the FMs.

than is the FM–FM match, showing that the latter is, at least for this case, a rather conservative test of the FM accuracy. Similar results are obtained for other GMs, however, indicating the general validity of using the FM–FM mismatch as a measure of the FM accuracy relative to the GM itself.

9.5.3 Filtering Noisy Spectral Data

One application of FMs, whether in the WD or SD, is exploiting redundancy in oversampled data as a way of reducing noise effects. Of course, without any knowledge about how the data of interest have originated, filtering might be done using a simple averaging process, where several sequential data points are averaged over a moving window. However, such a procedure does not exploit knowledge of the process from which the data have been obtained, whether an experimental measurement or from a GM. Using a parameterized FM offers the possibility of more effective noise reduction because the FM itself adds information beyond that available from the data samples themselves.

This possibility has been explored by Lin [34] who applied rational-function FMs to noisy SD data for which one result is presented in Fig. 9.15. It can be observed that the oversampled noisy spectrum is well represented by the four sequential FMs that were used, demonstrating the potential benefit of combining redundant data and an appropriate FM to reduce noise effects in spectral data. It should be realized that successful use of MBPE for the kinds of examples shown here will usually require some experimentation to establish the range of FM parameters that yield the best results. The ill-conditioning of the data matrix that very often occurs may also require the FM computation to be done in higher precision (at least 16 digits) than normally used for the GM, with singular-value decomposition offering an alternative treatment (see Appendix 1).

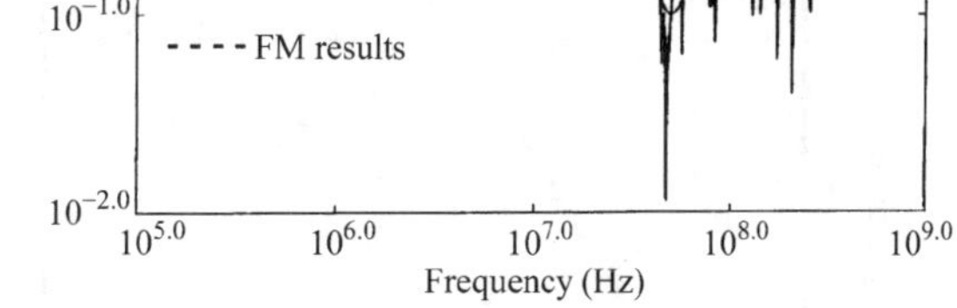

Figure 9.15 Smoothing of noisy spectral data achieved using a rational-function spectral filter (FM) (Fig. 34 from [34]). The original measured data (the solid, more jagged line) has a 20 dB signal-to-noise ratio and 512 total samples of which 12 are randomly deleted. It is modeled using four rational-function FMs in sequence (the dashed, smoother curve) having 125 samples each with $n = d = 12$, 12, 10, and 12, respectively, whose coefficients are computed using a least-squares solution.

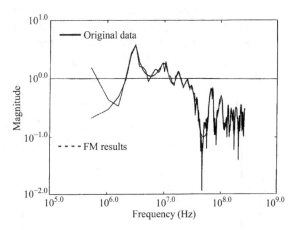

9.5.4 Estimating Data Accuracy

Knowledge of a FM for a physical observable also offers the possibility of estimating the accuracy of data obtained from that observable if a sufficient number of samples is available. The procedure can be compared to linear regression, where

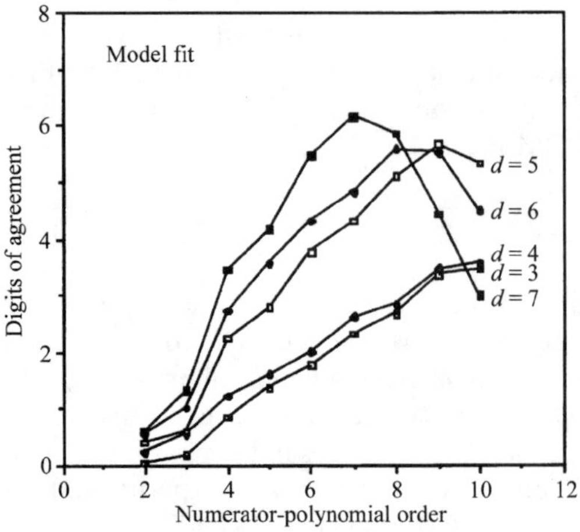

Figure 9.16 This fit between a rational-function FM of variable n and d and a pole-series GM, in digits [35]. The best model fit is about six digits for $N = 7$ and $D = 7$, indicating that the smallest error detectable for this particular example would be about 10^{-6}.

a best fit is determined to given data using a straight-line FM. For EM SD data, various approaches might be used to estimate the data accuracy, one of which is described here.

To illustrate the process, a specified pole spectrum, consisting of 15 poles having $s_i = \sqrt{i/20} + j*i(j = \sqrt{-1})$, is used as the GM between frequency endpoints of 5.5 and 8.5 [35]. Using this fixed-frequency window, FMs of varying n and d are fit to a set of $n + d + 1$ evenly spaced, noise-free data samples. The match, or model fit, between additional GM samples and these FMs is then calculated to determine the best-fit values of n $(= N)$ and d $(= D)$, as is illustrated in Fig. 9.16.

Having determined the best FM for the selected data, the next step is to test this FM against noisy data provided by the same GM. This is illustrated in Fig. 9.17, where uniformly distributed white noise has been added to the GM, with the average FM fit plotted as a function of the average GM accuracy, both given in digits. To obtain these results, $d = 7$ is employed but n is varied from 6 to 9 to further confirm that $n = 7$ yields the best fit. It can be seen that the average FM fit matches the average GM input-data accuracy, showing the potential for estimating the data accuracy using MBPE.

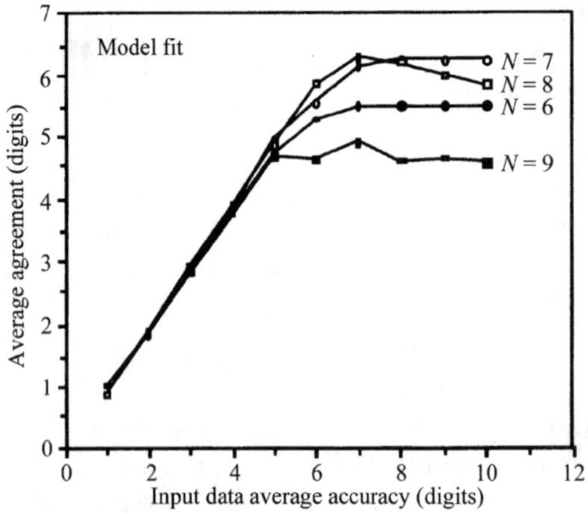

Figure 9.17 The average agreement achieved between a FM and additional samples of a pole-series GM, as a function of the GM sample accuracy [35]. A FM denominator-polynomial order of 7 and a variable numerator-polynomial order were used.

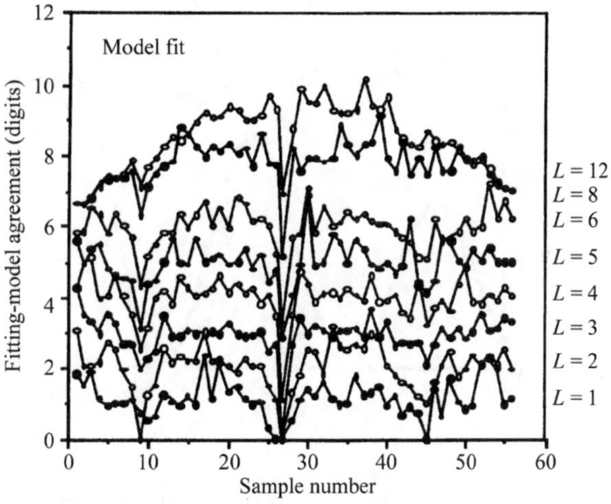

Figure 9.18 FM agreement with additional frequency samples of the current on the front side of an infinite cylinder illuminated by a normally incident TE plane wave (the GM) [35]. Value of L indicates the digit specification of the series that gives the current, i.e., summation is stopped when next term is less than 10^L of the sum to that point.

An example of applying the rational-function accuracy-estimation procedure to an EM transfer function, the current excited by a plane wave on an infinite, circular cylinder as a function of frequency (i.e., the same GM as used for Fig. 9.11), is presented in Fig. 9.18. It can be seen that the FM agreement, in digits, with the additional GM samples is approximately equal to the specified convergence level, L, in digits, for the series that gives the current. For this example, at least, this demonstrates that the approximate accuracy of an EM GM can be estimated from the FM–GM agreement.

Some results of using MBPE for estimating the accuracy of experimental data are presented in Figs. 9.19–9.21 using results for the field scattered from a finite, circular cylinder as obtained from the David Florida Laboratory [64], [35]. Initially, the complex data were employed, yielding the results shown in Fig. 9.19, where rather poor agreement is obtained between the FM and the measured data. Anticipating that

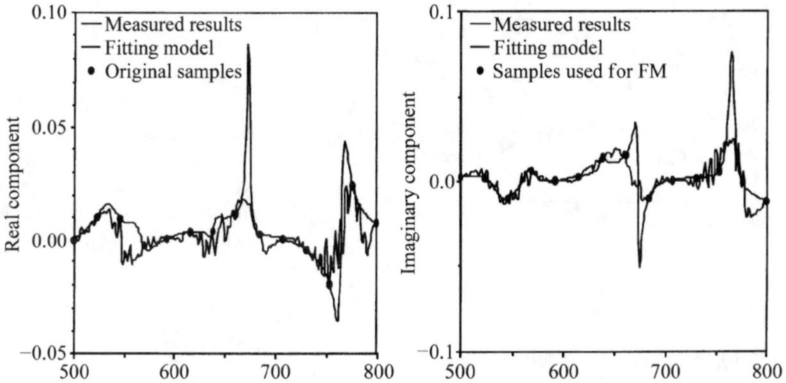

Figure 9.19 Results obtained from a 14-coefficient complex FM with $n = 7$ and $d = 6$ (the dark solid line) using the samples shown by the solid circles, with the narrow line showing the original data [35], [64]. Although the FM seems to average the oscillations seen in the experimental data, the overall agreement is poor.

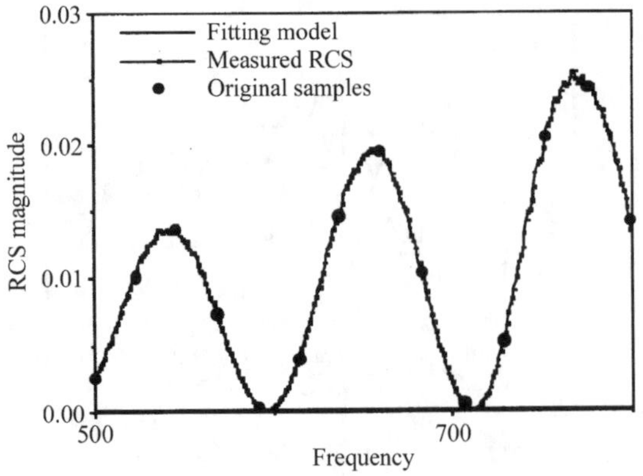

Figure 9.20 Repeats Fig. 9.19 but uses the magnitude of the measured data as the input. The data fit is much better here than when using complex input data.

the measured data might be known more accurately in magnitude than in phase, the FM was recomputed using the magnitude data instead, with the outcome shown in Fig. 9.20. For simplicity, the FM computation continues to use a complex model but since the input data are now pure real, only the real part of the FM is non-zero. Graphically, it's clear that the match obtained between the FM and magnitude data is much better than when using the complex data. This is confirmed by the results of Fig. 9.21, where the mismatch between the original data and these two FMs is shown, indicating that the uncertainty of the measured magnitude data appears to be about 1%. Also implied by this outcome is that the error in the measured phase is substantially greater than that of the magnitude.

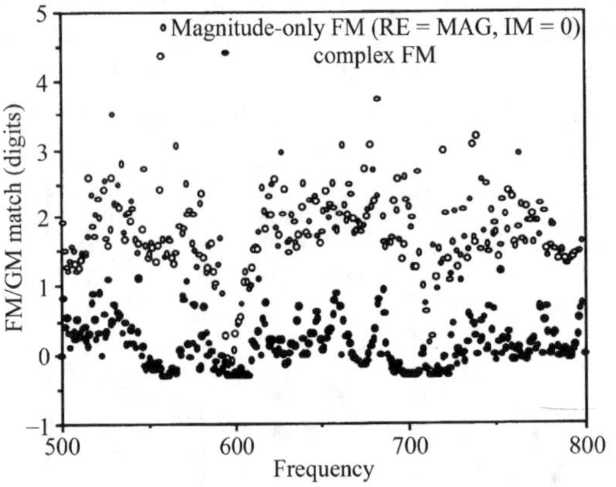

Figure 9.21 The match in digits obtained between the complex FM (solid symbols) and magnitude-only FM (open symbols) for the field scattered from a finite circular cylinder. Average agreement for the magnitude-only FM is about two digits, with minima occurring at minima in the RCS.

9.6 WAVEFORM-DOMAIN MBPE

Waveform-domain MBPE is usually associated with modeling transient data (e.g., [14]). Since numerous examples can be found in the literature, that particular application is not considered here. Instead, WD application to far-field data (radiation and scattering patterns) is demonstrated.

9.6.1 Radiation-Pattern Analysis and Synthesis

One pattern application is that of synthesis where a discrete-source approximation (DSA) required to produce a given far-field pattern is to be determined. The usual procedure is to specify the antenna geometry and to attempt the synthesis by controlling only the amplitude and phase of each source in the array. But a WD FM provides another way of accomplishing the synthesis in which the source locations, as well as their complex strengths, are both determined. An example of using MBPE for pattern synthesis is illustrated in Figs. 9.22 and 9.23 where the Taylor pattern for a continuous aperture width $A_W = 7$ wavelengths [36], [37] is synthesized. The problem in this case is to develop a DSA whose pattern, $P_{\text{dis}}(S;q)$, best matches the specified pattern, $P_{\text{spec}}(q)$, by varying the number, S, of discrete sources in the array. As S increases from a small value relative to A_w, the normalized mismatch error,

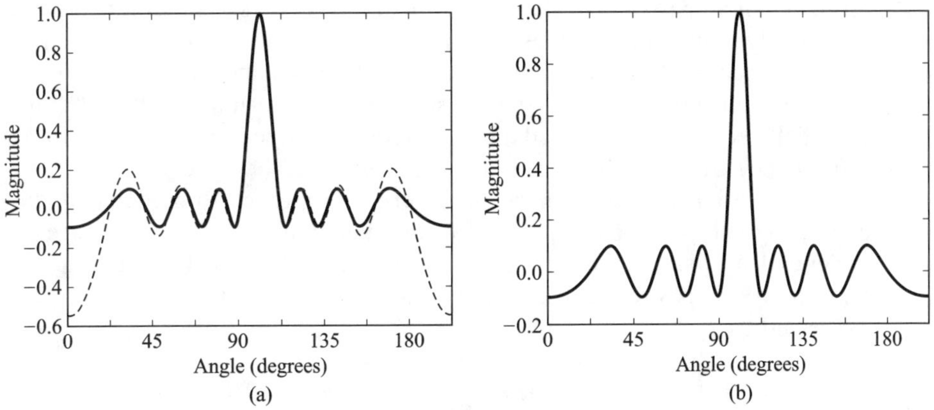

Figure 9.22 Comparison of the pattern produced by a specified continuous Taylor source 7 wavelengths wide [36] and the DSA Prony-synthesized pattern having $S = 8$ (a) and $S = 12$ elements (b) [37]. In the range of $S = 8$ to $S = 14$ elements (see Fig. 9.23), the difference between the desired and synthesized patterns decreases monotonically.

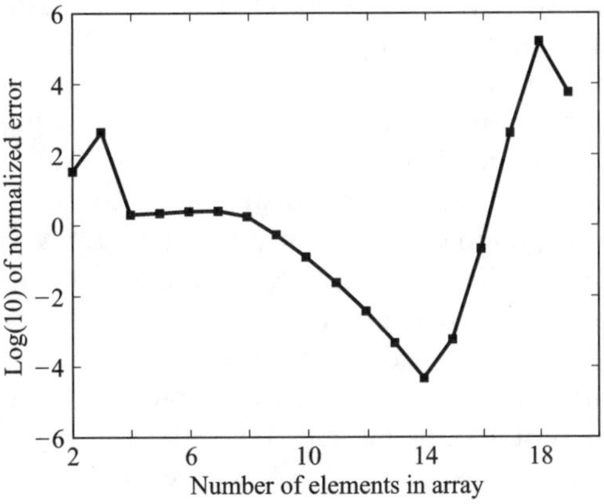

Figure 9.23 The mean-square difference between a specified continuous Taylor pattern 7 wavelengths wide and that provided by a discrete Prony-synthesized array having a variable number of elements [37]. As the number of elements approaches twice the aperture width, the mismatch error becomes a minimum.

$E_n(S;q) = |[P_{\text{dis}}(S;q) - P_{\text{spec}}(q)]/|P_{\text{spec}}(q)||$, initially indicates a poor match with $E_n(S;q) \sim 1$, but for $S > A_w$ it decreases exponentially to a minimum where $S \sim 2A_w$ as shown in Fig. 9.23.

9.6.2 Adaptive Sampling of Far-Field Patterns

A procedure to use MBPE for adaptive far-field pattern estimation called WASPE (Windowed Adaptive Sampling Pattern Estimation), applicable to both radiation and scattering problems, that permits a pattern to be reconstructed to a prescribed uncertainty is next outlined. WASPE begins with sparsely sampled far-field values from a GM at a prespecified set of observation angles which are used to set up an initial sequence of FMs, each of which shares two or more GM samples in their region of overlap in the same fashion as used for adaptive frequency sampling described above. Each new GM-sample angle is then selected where the maximum mismatch of all pairs of overlapping pairs of FMs occurs as determined by computing more finely sampled FM values and then increasing the order of the FMs whose angle windows include that sample. The process of comparing FM values continues until the maximum mismatch falls below some prescribed uncertainty. The approach is essentially equivalent to the adaptive method developed in 9.5.2 above for estimating a frequency transfer function, but for which the FM was a pole series rather than the exponential series used here. An alternate approach to modeling angle dependence using polynomials is reported by Allard et al. [65]. The performance of WASPE is dependent, among other things, on the kind of FM used for its implementation. One approach is provided by Prony's method, where an appropriate FM can be expressed as

$$f(\theta) = \sum_{\alpha=1}^{W} R_\alpha \exp[ikd_\alpha \cos(\theta)] \tag{9.26}$$

with the W point sources of strengths R_α and located at positions d_α along the x axis from which the observation angle θ is measured. By sampling the GM far-field in uniform steps of $\cos\theta$, Prony's method can be used to estimate R_α and d_α and thereby obtain a FM that provides a continuous estimate of the pattern between the GM samples [17].

Strictly speaking, when used in this fashion Prony's method is only applicable to linear arrays. Furthermore, if the number of actual sources, S, and/or their effective aperture size in wavelengths, A, is too large, the matrix needing solution can become very ill-conditioned. Thus, it's logical to window the pattern so that its effective rank is maintained below some threshold. This means that in contrast to the situation where an actual source distribution can be "imaged" from its complete pattern, when using windowed data an equivalent DSA valid over only the window used for its computation will result. Since the goal for the application is not to image the source distribution but rather to compute the pattern more efficiently, not having the actual distribution is not a disadvantage. Also note that were this done for a scattering pattern, whose source distribution depends on the angle of incidence, no single DSA can describe the backscatter pattern in any case. An alternative to Prony's method is to instead assign the DSA source locations and use the GM samples to compute the source amplitudes. This has the advantage that the GM samples may be arbitrarily located in angle, thereby making an adaptive procedure more practical. Also, only half the samples are needed for a given value of W since the source locations are no longer unknowns.

Using this alternative requires some rationale for assigning the source locations, or equivalently, specifying an appropriate FM. Since the highest-angular-frequency component of the pattern is proportional to kA, one possibility is to use as the FM over angle window m

$$f(\theta) = \sum_{n=S}^{N} R_{mn} \exp[ikn\cos(\theta)] + \sum_{n=-S'}^{-N} R'_{mn} \exp[ikn\cos(\theta)] \tag{9.27}$$

where $N = \text{int}(A+1)$ and $2N - S - S' = F$ with F the total number of exponential terms used for the mth FM, int(X) denotes the value of X rounded off to the nearest lower integer and $|S - S'| \leq 1$. Also needed to "initialize" the FM selection is choosing a starting value for F, and the number of FMs, M. Somewhat arbitrarily, we choose a small value for F of 3 or 4, with M determined by the number of anticipated lobes in the pattern being processed and the amount of overlap chosen for the adjacent FMs.

A candidate scattering pattern on which to test the WASPE is the backscattered field of a finite-length circular cylinder for which an angle-dependent approximation is given by [62]

$$E(\theta) \propto \cos(\theta_i)\frac{\sin[2\pi A \sin(\theta_i)]}{2\pi A \sin(\theta_i)} \tag{9.28}$$

If each FM is intended on average to include two lobes of the scattering pattern, then $\sim A$ angular windows would be required to cover this range of incidence angles. Using a total overlap between adjacent FMs of 2/3, then $M \sim 16$ FMs would be needed for $A = 10$. As each new GM sample is added, the FMs that include it are increased in rank by alternately increasing S and S' by one, until a maximum mismatch, $\text{MM}_{i,j}$, between all pairs of overlapping FMs of

$$\text{MM}_{i,j}(\theta) = \max[|\text{FM}_i(\theta) - \text{FM}_j(\theta)|/(|\text{FM}_i(\theta)| + |\text{FM}_j(\theta)|)] \leq 10^{-X} \tag{9.29}$$

is achieved, i.e., an X-digit match is obtained between them. It's also possible to scale X, e.g., so that it depends on the relative magnitude of the FMs being tested relative to the maximum value in the pattern, to thereby allocate the final uncertainty in the estimated pattern according to problem requirements. Note that the FM itself may represent the complex far-field or its magnitude.

The result of using WASPE on the scattered field of a 10-wavelength cylinder is shown in Fig. 9.24. The 16 overlapping FMs used here superimpose to graphical resolution and agree to within two digits of accuracy independent of the actual field level. The converged pattern estimate employs a total of 56 GM samples, or about 2.8 per pattern lobe.

Another candidate pattern on which to test the WASPE is the radiation pattern of a uniform aperture given by [36]:

$$E(\theta) \propto \frac{\sin[\pi A \sin(\theta_i)]}{\pi A \sin(\theta_i)} \tag{9.30}$$

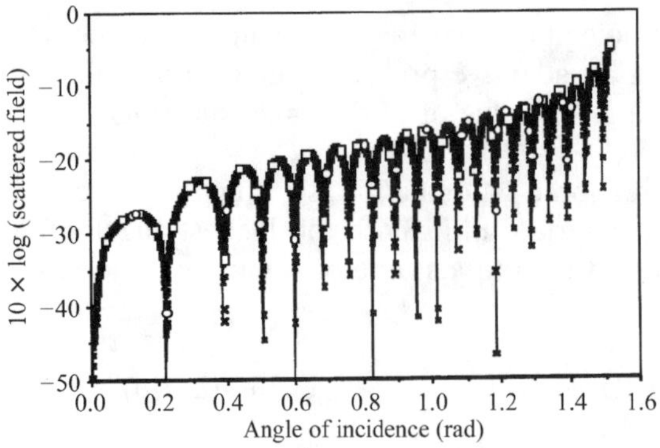

Figure 9.24 Example of using WASPE for estimating the backscatter pattern of a circular cylinder 10 wavelengths long. The 32 initial GM samples are shown by the squares while the 24 adaptively added samples are shown by the circles. The prescribed estimation uncertainty used here was 0.01.

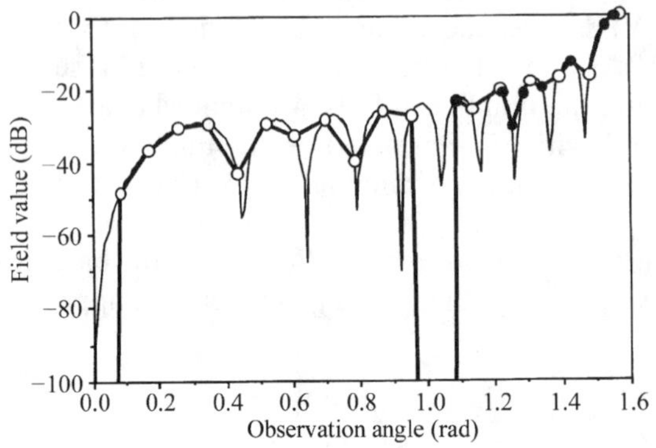

Figure 9.25 Results from a finely sampled GM and those from using WASPE with FMs having initially four terms to estimate the radiation pattern of a uniform-aperture distribution are shown by the thin line. The thicker line joins the GM samples used by WASPE, with the open circles denoting the initial samples and the solid circles showing those added, using an estimation uncertainty of 0.001.

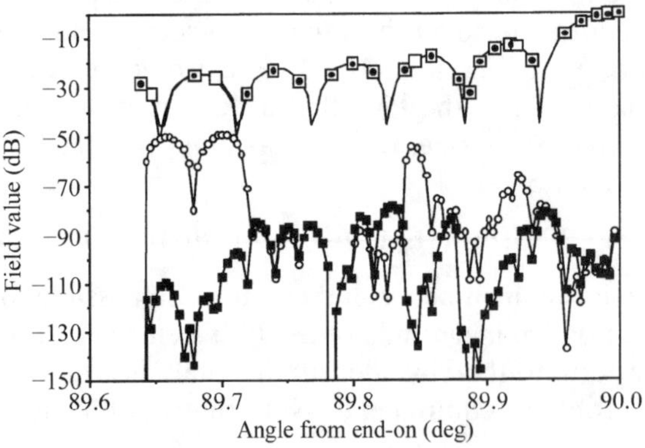

Figure 9.26 Results for estimating a radiation pattern of a constant-amplitude aperture near broadside using WASPE. The upper curve is an overplot of the actual pattern and WASPE estimates obtained using FMs with a convergence specification X of 2.5 and 4 digits, respectively. The solid circles are the GM samples used for $X = 2.5$ and the open squares for $X = 4$. The line with the open circles is the difference between the GM and the FM estimate for $X = 2.5$ and that with the solid squares is the difference for $X = 4$.

and for which some example results are given in Figs. 9.25 and 9.26. In Fig. 9.25, $A = 10$, and the number of initial terms per FM is four with 16 angle windows sliding in angle by one GM sample for a total of 19 original GM samples. Windows 2 and 15 use five samples to overlap FMs 1 and 16 at each end of the angle range. The value used for X is 2. The WASPE-estimated results in Fig. 9.25 are essentially graphically indistinguishable from the actual GM. However, the adaptively added samples are all located between about 1.3 and $\pi/2$ radians, indicating that the initial samples are probably placed too close together between 0 and 1.3.

The effect of modeling an aperture $A = 1000$ wavelengths long near its broadside maximum is illustrated in Fig. 9.26. The WASPE results are obtained using initial FMs having three terms and using $X = 2.5$ and 4, respectively. It can be seen that reducing the estimation error from 2.5 to 4 digits results in adding four more GM samples whose locations are near local maxima in the difference between the actual pattern and that estimated using $X = 2.5$. This is significant as it indicates that basing the adaptation on differences between overlapping FMs is a reliable measure of where the actual estimation error is a maximum. It can also be seen that while the actual FM error generally decreases as new GM samples are added, this is not assured as can be seen in the angle range 89.8 to 89.85°.

9.6.3 Inverse Scattering

When far-field scattered or radiated fields are sampled at a fixed position as the observation frequency is varied, still another kind of WD observable results, of the kind shown in Table 9.1 as model 2, a procedure sometimes called *range profiling*. In this case, a discrete set of point radiators or scatterers produces a superposition of observation phasors whose relative phase depends on source distance for a fixed frequency and on frequency for a fixed source. The separate fields of different sources add with phases that depend on frequency and position as a series of complex exponentials, thus becoming a candidate for WD Prony's method. This kind of data can be modeled using Prony's method for any given observation angle relative to a 3-dimensional (3D) source distribution. However, in order to reconstruct the 3D spatial locations of the sources, various observation angles must be used, requiring either that the changed relative positions of a given source be identifiable from the different reconstructions, or that simultaneous solutions of observations made from multiple viewing angles be obtained, a problem that the basic Prony method is not able to handle [9]. Alternatively, by resorting to a least-squares numerical solution, the WD Prony model can be employed [38], [39], [40] to image a distribution of scattering sources in the observation plane when various viewing angles in that plane are available. If different viewing planes can also be used, then true 3-dimensional image reconstruction can be feasible.

An especially simple, but still relevant, problem of this kind is plane-wave reflection from a layered half-space where, interestingly, both the impulse response and the swept-frequency reflection coefficients can be expressed in WD forms. Thus, a WD FM can be used for the transient response, or a pole series for the SD FM, where the poles are inversely related to the electrical thickness of the individual layers [41]. Alternatively, a WD FM can be used for the swept-frequency response, which also has a pole series SD FM but where now the poles are proportional to the layer electrical thicknesses [15], an observation previously made by Tai [42] in connection with a transmission line having sections of different characteristic impedance.

9.7 OTHER EM FITTING MODELS

Examples of MBPE applications in physics and engineering are bountiful. Perhaps the first encountered by most engineering students is that of estimating physical parameters from laboratory measurements. These can include determining the characteristics of various components such as transformers and power generators by measuring their inputs under varying loads, and estimating the effective permittivity of materials from EM reflection and transmission measurements. All such applications involve use of formulas for the anticipated behavior in which one or more parameters are to be numerically quantified from appropriate measurements.

9.7.1 Antenna Source Modeling Using MBPE

Examples of MBPE more specific to computational electromagnetics (CEM) can be found. It is known that the input admittance of an infinite, circular conducting cylinder excited by a z-directed electric field applied across a finite gap of width δ exhibits a susceptance that goes to infinity as $\delta \to 0$ [43]. Physically, this behavior occurs because the feed region behaves as a circular capacitor whose susceptance is approximately $i\omega C \sim i\omega\pi a^2 \varepsilon_0/\delta$, where a is the radius of the cylinder and ω is the radian frequency.

This effect also extends to antennas whose feed regions are large compared with a, as is the case of a wire antenna modeled using subsectional basis and testing functions with the applied field point-sampled on (or integrated over) the driven segment. If the number of segments used for the moment-method model is systematically increased to test convergence while continuing to use a single segment as the source, then the effects of model convergence and the variable source model can interact, producing results such as those shown in Fig. 9.27 for a 2-wavelength, center-driven dipole. The reactance variation with X_s, the number of unknowns, is observed to be quite extreme, indicating that even with 100 unknowns the results have still not converged. However, the susceptance behavior is much smoother and shows that the conductance has converged for $X_s \leq 10$ while the susceptance varies monotonically out to $X_s = 100$. Actually, the susceptance behavior

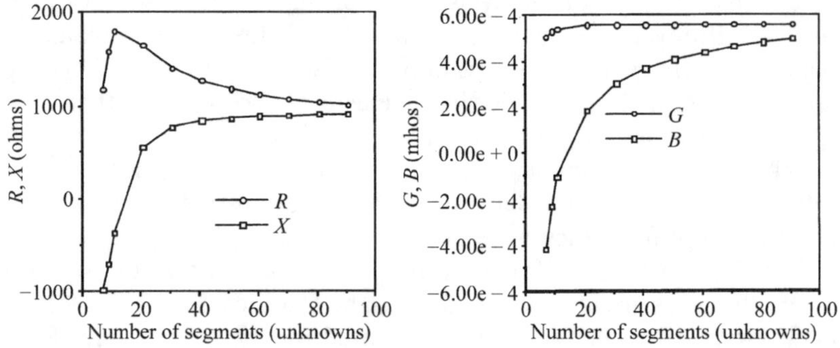

Figure 9.27 Results from an NEC GM for the input impedance and input admittance as a function of the number of unknowns, X_s, used in a model for a center-fed, two-wavelength dipole. The impedance has not satisfactorily converged over the X_s range shown which the admittance results demonstrate is due to a susceptance variation. In this case, the exciting source was a tangential field applied to the center segment of the antenna, whose decreasing length simulates a changing antenna source gap [(9.31)]. The susceptance curve exhibits the kind of variation expected from changing the gap size. In order to avoid this problem, the number of excited segments would have to increase in proportion to X_s to maintain a constant gap width.

is dominated by the source capacitance and is well approximated by the infinite-antenna expression [43]

$$B(X_s) \approx A_1 + A_2/[\text{Ln}(L/X_s) + A_3] \quad (9.31)$$

where L is the antenna length, the A_is are the model parameters and $(L/X_s) = \delta$ is the width of the gap across which an axially directed field is applied to excite the infinite antenna.

9.7.2 MBPE Applied to STEM (STatistical ElectroMagnetics)

Use of statistics in electromagnetics becomes more important when problem size and complexity increase to the point where deterministic answers, even if available, may not be very useful because of problem complexity. Consider, for example, the radar cross section (RCS) of a target such as a B-52 aircraft at a frequency of 10 GHz where the target size exceeds 1000 wavelengths in linear dimension. The aspect-angle variation in the RCS of such a target is extremely "spiky," with changes of 10s of dB possible for incidence-angle changes of a few tenths of a degree. It's reasonable to ask whether, or how such a solution might be used, even were an exact, deterministic solution available for this problem. Of course, in reality the aircraft-to-aircraft variation will be such that an answer for one specific aircraft may not be relevant to another, besides which in-flight measurements are affected by a multitude of additional variations that contribute further uncertainty to what the "correct" answer might be. Thus, at least two reasons for using statistics can be cited, problem complexity and problem uncertainty, where the latter is the usual reason why statistics is used.

As an alternative to deterministic electromagnetics, STatistical ElectroMagnetics (STEM) needs to be considered, wherein various statistical measures of relevant EM observables are used. However, a significant drawback of using statistics arises whenever a probability density function (PDF) and other statistical descriptions of a process are not available from analysis and must instead be inferred from fitting various candidate distribution functions to data. The problem is much simplified if the PDF can be determined analytically instead, which also results in needing substantially fewer data to numerically quantify whatever parameters are contained in the PDF. This possibility is illustrated in Fig. 9.28 where experimental data have been fit

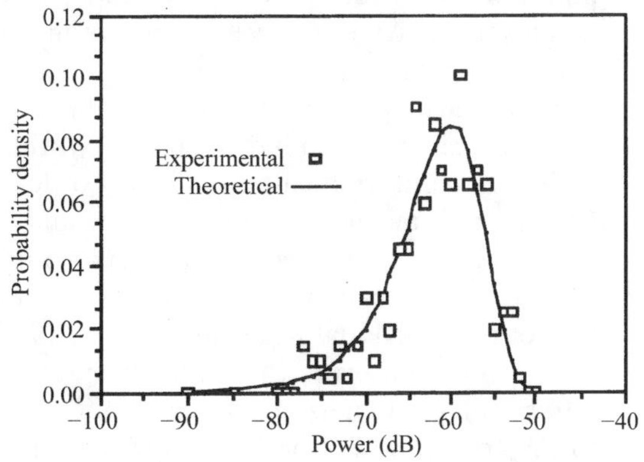

Figure 9.28 Probability densities and cumulative distributions for the power coupled to a shielded cable in a mode-stirred chamber measured by Kaman Sciences [45] compared with Lehman's theory. This result is typical of the comparisons that Lehman's theory has produced with experimental data obtained in such experiments. Note that the only "floating" parameter needed to complete this comparison is knowledge of the experimental power density [44].

to an analytically derived PDF by computing numerical values for the parameters of the normal distribution [44].

9.8 MBPE APPLICATION TO A FREQUENCY-DOMAIN INTEGRAL-EQUATION, FIRST-PRINCIPLES MODELS

Almost all EM boundary-value problems involve finding the fields over some surface or throughout some volume due to sources distributed over that same surface or volume. When using an integral-equation formulation, these source-field relationships are given by a Green's function or some equivalent field propagator, whereas a differential-equation model employs the Maxwell curl equations as the propagator. The spatial behavior of the fields might be viewed for some purposes as a generalized signal, as can angle, time, and frequency dependencies of such fields as well. Such a perspective can suggest alternate ways of representing the fields in signal-processing terms for numerical purposes to simplify whatever computations that must be done, a viewpoint that is explored here.

For moment-method models based on a FDIE (other FPMs can be analyzed in a similar context), the number of arithmetic operations or operation count (OC) required for solution at a single frequency f_i, OC_i, will depend on the number of unknowns, X_{si}, used in the model approximately as

$$OC_i \sim A_x X_{si}^2 + B_x X_{si}^3 \tag{9.32a}$$

or, in terms of frequency, f, as

$$OC_i \sim A_f f_i^4 + B_f f_i^6 \tag{9.32b}$$

for a 3-dimensional object requiring 2-dimensional sampling (i.e., sampling over its surface). When a solution is desired over some frequency interval, as is usually the case, then the total operation count, OC_T, can be estimated as

$$OC_T \sim \sum A_f f_i^4 + B_f f_i^6,\ i = 1, \ldots, F \tag{9.33}$$

where there is a total of F solution frequencies.

The "A" terms here account for filling the "system" matrix [known as the impedance or $\underline{\underline{Z}}$ matrix when an electric-field IE (EFIE) is used], and the "B" terms account for solving the system matrix in factored or inverted form as a "solution" matrix (known as the admittance or $\underline{\underline{Y}}$ matrix when using an EFIE). Use of iterative solution techniques can reduce the above B terms to of order $B'_x X_{si}^2$ and $B'_f f_i^4$ but where B'_f can be much larger than B_f. Various "fast" techniques are being developed with the goal of reducing the highest-order terms in (9.32a) to of order $X_{si}\log(X_{si})$. Clearly, any means of reducing F would also be helpful in decreasing the total computational cost, a point that has also been considered above in Section 9.4.3 in connection with adaptive frequency sampling. First, let's consider how the impact on OC_i of the individual terms in (9.32) or (9.33) can be mitigated.

It can be observed that as the frequency increases from zero for a given problem, OC_i at first grows in proportion to FX_{si}^2 or Ff_i^4, but it eventually becomes proportional to FX_{si}^3 or Ff_i^6 when the higher-order term begins to dominate, assuming a constant

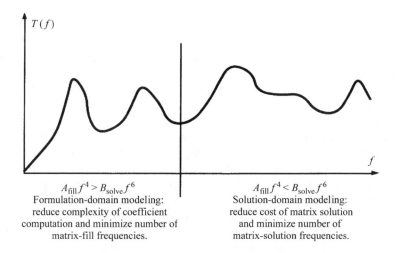

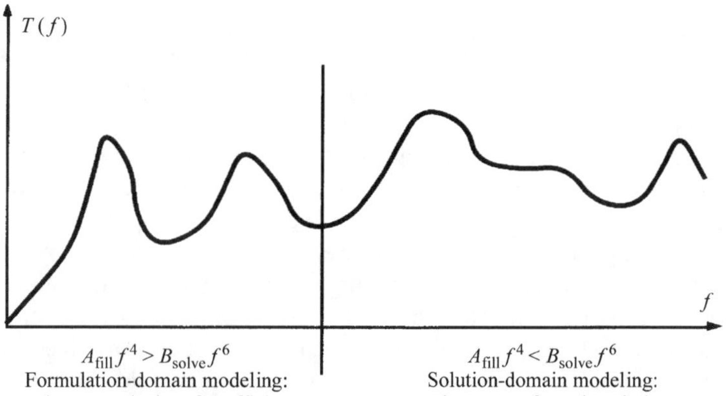

Figure 9.29 Illustration showing where matrix-fill and matrix-solution operation counts dominate for a frequency-domain integral equation solved using L-U decomposition and assuming that a fixed sampling density per wavelength is used as frequency is increased. At lower frequencies a computational benefit can be realized by finding ways to compute interaction coefficients more efficiently or to reduce the number of rigorously computed $\underline{\underline{Z}}$ matrices. At higher frequencies, a computational benefit results from reducing the number of solutions, or $\underline{\underline{Y}}$ matrices.

spatial sampling density per λ^2. By reducing both the number of impedance matrices that need to be computed from the defining formulation when impedance-matrix computation dominates OC_i, and the number of admittance matrices that need to be solved from the impedance matrix when solution time dominates OC_i, it should be possible to significantly reduce the OC_T required to cover a specified bandwidth. This might be done by modeling the frequency behavior of the impedance matrix for smaller problems, and the frequency behavior of the admittance matrix for larger problems, as illustrated in Fig. 9.29. The goal in both cases is to reduce the number of FPM evaluations needed, i.e., to reduce F in (27), and to thus minimize OC_T. It should be noted that the "crossover" point in X_{si} between fill-time and solution-time domination of OC_i, can vary from as few as 200 unknowns to as many as 10,000 unknowns. Thus, the FDIE $\underline{\underline{Z}}$ and $\underline{\underline{Y}}$ matrices, or their interaction coefficients, both become candidates for MBPE, albeit using different kinds of FMs having WD and SD nature, respectively. In addition to modeling the frequency behavior of $\underline{\underline{Z}}$ and $\underline{\underline{Y}}$, it is also worth considering whether the spatial behaviors, which represent the geometrical dependence of field-source interactions, of either of these matrices can be modeled. From a signal-processing perspective, the spatial variation of the field over an object due to a given localized source (e.g., a subdomain basis function) on that object might be regarded as a generalized signal. If this field "signal" can be predicted or estimated using a FM that is computationally simpler than using the defining equations normally employed for calculating the exact interaction coefficients, the magnitude of the A_f term in (9.32) could be commensurately reduced. Alternatively, if a first-principles evaluation of only a subset of the X_s^2 interaction coefficients needs to be done, say CX_s

where $C \ll X_s$, and the rest can be estimated from the rigorously computed coefficients, then the effect would be to change the $A_x X_s^2$ term to of order $CA_x X_s$, if the cost of obtaining the estimated values is much less than that of the rigorously computed coefficients.

A specific example of reducing the A_f term in (9.32) is to compute none of the X_s^2 interaction coefficients in $\underline{\underline{Z}}$ from first principles, but instead to estimate them using some appropriate FM based on a small set of precomputed, first-principles field samples. This approach, summarized in Section 9.8.2.4, was used by Burke and Miller [46] for reducing the cost of evaluating the Sommerfeld integrals that arise when modeling objects near an interface where the FMs are analytical approximations to these integrals. Their approach is an extension of earlier work described by Miller et al. [46] where linear interpolation of sampled Sommerfeld integrals for the matrix coefficients was used as a curve-fitting procedure rather than a full MBPE implementation. Other problems having special Green's functions are also candidates for this procedure, another example of which is provided by modeling sources between infinite parallel planes [48]. In these instances, the reduction in OC_i is reflected in a decrease in the A_f coefficient in (9.32).

Another way of reducing the effect of the A_f term is to exploit the fact that, as the source and observation points become more widely separated, the complexity of their interaction fields is reduced. This is easy to see by considering that the fields in the vicinity of a linear source distribution of a few wavelengths in extent are to be pointwise sampled along a line parallel to, and a distance away from, the source. In order to develop an accurate-enough representation, it would be necessary to sample the field at some minimum density per wavelength. Now consider the situation as the sampling line is moved further and further from the source. The field variation along that line will become less complex as the near-field components decrease with increasing distance and finally only the $1/r$ field remains. Furthermore, with increasing observation distance, the spatial variation of the field along a line of fixed length further decreases in complexity because the angle subtended by this line at the source monotonically decreases. In other words, at a great enough, finite distance away from the source, the field variation becomes a function of subtended angle rather than of linear distance over some fixed observation range.

Thus, the effective rank of the interaction between sources of given size decreases as their separation distance increases, reducing the complexity of that interaction, and, consequently, the amount of computation needed to determine it to some specified accuracy. This idea is exploited in techniques such as the fast multipole method [49], impedance-matrix localization [50], recursive models [51], etc., where rank reduction is explicitly employed, or where more distant interaction coefficients are approximated by simpler expressions [52]. Such approaches effectively employ exact interaction coefficients when the source and observations points are close and controlled approximations as their separation distance increases. This results in an A_f coefficient that becomes smaller as source-observation distance increases, or a reduced matrix-fill time. In addition, because the interaction complexity decreases with increasing distance for fixed source and observations spans, the complexity, or effective number of interaction coefficients also decreases, reducing the OC_i of multiplying the impedance matrix by a source vector from of order $(X_{si})^2$ to of order X_{si}. The OC_i associated with solving a matrix having X_{si} unknowns by iteration thus trends toward order X_{si} or $X_{si} \log X_{si}$ from being of order $(X_{si})^2$.

Whether spatial variations in the solution or admittance matrix might be exploited in a similar fashion is not so clear. Certainly, a valid solution to the original problem must exhibit spatial dependencies consistent with the geometry and excitation involved and be consistent with Maxwell's equations. Graphical examination of the $\underline{\underline{Y}}$ matrix for simple objects like a straight wire ([53] and in Section 9.8.2.6) reveals that it exhibits a standing-wave character, not a surprising result in that the currents on such structures are well known to have such behavior. In other words, the "traveling-wave" nature of the Green's function in the formulation domain, reflected in terms like $\exp(ikR)/R$, is converted to a standing-wave response in the solution domain, where amplitude maxima occur at resonances associated with the object poles. Thus, an appropriate FM for the admittance matrix should apparently be expected to be of wave-domain type as well. These kinds of ideas are now explored from the perspective of MBPE in the following.

9.8.1 The Two Application Domains in Integral-Equation Modeling

We have discussed MBPE in CEM from the perspective of whether the quantities of interest exhibit wavelike or pole-like behavior, referring to their respective occurrences by the designations waveform domain and spectral domain, respectively, determined by the mathematical description that applies to a given quantity. There is another domain pair that is also useful for problem categorization, one describing the domain wherein a boundary-value problem is defined, and the other describing the domain wherein a solution to that problem is presented. We refer to the former as the "formulation" domain, in which a formal mathematical statement originating from Maxwell's equations is developed for a problem, and to the latter as the "solution" domain, in which that original formulation has been mathematically solved. In the formulation domain we begin with known exciting fields to which to-be-found induced sources are required by Maxwell's equations to satisfy the appropriate boundary conditions. Finding these induced sources requires inverting the original source-field relationship. For all but the simplest problems the inversion requires numerical computation.

A potential approach to alleviating the computational requirements that arise in either the formulation domain or solution domain is to exploit the underlying physical and mathematical behavior of EM fields, as is embodied in first-principles analysis of wave-equation problems, through a simplifying, reduced-order, signal-processing formalism. Of course, knowledge of the problem physics is required to begin with, as any solution process, analytical or numerical, cannot be initiated without having the applicable physics captured in appropriate mathematical form, something that in terms of Maxwell's equations might be characterized as a microscopic description. But the physical behavior of greater practical interest is usually macroscopic in nature, as it is not generally the fine details of the current distribution on an antenna or the near fields around a scatterer, but the antenna input impedance and gain, or scattering cross section, that are needed for system design. The macroscopic description is naturally a reduced-order one and provides the context for MBPE.

As previously observed, MBPE involves fitting physically motivated analytical approximations (the model) to accurately computed or measured EM observables from which unknown coefficients (the model parameters) are numerically obtained. These fitting models can then be used in subsequent applications to more efficiently characterize time, frequency, angle, and space responses as well as to provide more insightful access to the underlying physics. In the solution domain, MBPE can be applied directly to the spatial and frequency dependencies of the computed observables themselves, such as currents and fields, or to the solution matrix from which these quantities are obtained, as is discussed here.

Alternatively, we might also employ MBPE in the formulation domain, where it is the behavior of the first-principles analytical representation that is being approximated by reduced-order FMs. In that case, the frequency and space dependence to be represented would be that of the defining source-field relationships as contained in the system matrix. In contrast to working in the solution domain, where resonance effects dominate the EM behavior, growing phase change and geometric attenuation of the fields of increasingly distant sources dominate the behavior when working in the formulation domain. While the most appropriate FM will depend on the particular quantity being modeled, exponential- and pole-series models are widely applicable. For example, the frequency and spatial variations of a FDIE are suitable for exponential, or WD, FMs in the formulation domain, whereas in the solution domain pole-series FMs are suitable for frequency variation and exponential-series FMs for their spatial and angle variations. Formulation-domain approaches are described by Newman [54] and Benthien and Schenck [55], and solution-domain approaches are presented in Burke et al. [56], [23]. Other kinds of FMs can be found useful, e.g., based on the geometrical theory of diffraction, for other analytical formulations and solutions. Some of these possibilities are discussed in the following sections. Each of these two FMs and problem domains are considered in the following.

9.8.2 Formulation-Domain Modeling

9.8.2.1 Waveform-Based MBPE in the Formulation Domain

The question to be considered here is how waveform-based MBPE might be used for improving the numerical solution of a FDIE. First observe that the coefficients that appear in the impedance matrix for an FDIE model can be expressed in the generic form

$$Z_{m,n}(\omega) = \int_{\Delta_n} S_n(\omega) K_{R,m,n}(\omega) K_{\Delta,m,n}(\omega) \, d\Delta_n \tag{9.34a}$$

where m and n denote the observation and source patches, respectively: S_n is the source on path Δ_n; $K_{R,m,n}$ is the patch-to-patch (or P-P) part of the IE kernel; and $K_{\Delta,m,n}$ is the in-patch (or I-P) part of the kernel, where we have assumed a subdomain numerical model is being used. These terms refer, respectively, to that part of the kernel function whose frequency and spatial dependence is driven by an exponential phase change due to the P-P distance as contrasted with those variations due to variable positions within the source (and, possibly, observation) patches. Because increasing the P-P distance and increasing the frequency both increase the phase of $K_{R,m,n}$, changes in one or the other of these variables have similar effects on interaction phase, an effect that is exploited in using scale models in making experimental measurements. Considering frequency variations specifically, except for the patches that are close to each other (with respect to the wavelength), the P-P term would normally represent a faster frequency variation while the I-P would always represent a slower frequency dependence because patches need to be small relative to a wavelength while the interpatch distance can be arbitrarily large. The $X_s \times X_s$ set of interaction coefficients defined by (9.34) provides all the information needed to represent an object's EM characteristics, to the degree permitted by a numerical model based on it, but the source-field, integral-equation relationship represented by $\underline{\underline{Z}} \cdot \underline{I} = \underline{V}$ must be inverted to $\underline{I} = \underline{\underline{Y}} \cdot \underline{V}$ to obtain the desired solution.

The P-P, or fast, term always has the form, for IE-based models,

$$K_{R,m,n}(\omega) = e^{jkr_{m,n}}/r_{m,n} \tag{9.35}$$

where $r_{m,n}$, the separation between the origins of the source- and observation-patch local coordinates, is assumed to be a far-field distance and k is the wavenumber, a form that emphasizes the traveling-wave nature of the impedance-matrix coefficients. Integration over the source patch (and, possibly, the observation patch if using other than delta-function field sampling; though here we explicitly consider only source integration) involves changes in the fast term of order $\exp(jk\Delta r_{mn})$, where $k\Delta r_{mn} \ll 1$ is the distance variation caused by scanning over patch n. The interaction coefficient $Z_{m,n}$ can, thus, be rewritten as

$$Z_{m,n}(\omega) = e^{jkR_{m,n}} \int_{\Delta_n} S_n(\omega) K'_{\Delta,m,n}(\omega)\, d\Delta_n \tag{9.34b}$$

with $R_{m,n} + \Delta r_{mn} = r_{m,n}$ and $K_{\Delta,m,n}$ a modified slow-variation kernel. This form of interaction coefficient suggests that we can estimate $Z_{m,n}(\omega_2)$ at a new frequency ω_2 from an accurately computed value at frequency ω_1 as

$$Z_{m,n}(\omega_2) \approx Z_{m,n}(\omega_1) e^{j(\omega_2 - \omega_1)(R_{m,n}/c)} M_{m,n}(\omega_2 - \omega_1) \tag{9.36}$$

where $M_{m,n}(\omega_2 - \omega_1)$ is an interpolation model that accounts for the slowly varying part of the kernel function whose specific form would depend not only on object geometry but on whatever frequency dependence might have been incorporated into the basis and testing functions that are used. Exploiting a capability for modeling the spatial variation and its decreasing complexity with increasing distance is more involved, but is essentially embodied in "fast" methods which seek to reduce the OC of filling $\underline{\underline{Z}}$ and performing $\underline{\underline{Z}} \cdot \underline{I}$ multiplies from of order X_s^2 to of order $X_s \mathrm{Log}(X_s)$ (e.g., see [49], [50], [51]). The specific problem of modeling the frequency variations in $\underline{\underline{Z}}$ is considered next.

9.8.2.2 Modeling Frequency Variations: Antenna Applications

The model $M_{m,n}$ might be represented in various ways including using low-order polynomials (the model) whose coefficients (the parameters) are computed from samples of $Z_{m,n}$ at selected frequencies [54], [55]. When sinusoidal base functions specifically are used, it may also be advantageous to develop a recursive form for $M_{m,n}$. By accurately computing the impedance matrix at widely spaced frequencies and using estimated values at intervening frequencies, the goal is to obtain acceptably accurate results across a wide bandwidth without the cost of computing the impedance matrix at all sampling frequencies desired. As an example application, the analytical behavior of the impedance-matrix coefficients has been approximated for small variations in frequency about the computation point ω_1 by [54]

$$M_{m,n}(\omega-\omega_1)\big|_{\text{imag}} = A_i + B_i \ln(\omega-\omega_1) + C_i(\omega-\omega_1) \tag{9.37}$$

$$M_{m,n}(\omega-\omega_1)\big|_{\text{real}} = A_r + B_r(\omega-\omega_1) + C_r(\omega-\omega_1)^2 \tag{9.38}$$

A result obtained by Newman is presented in Fig. 9.30 where the input impedance of a center-fed dipole antenna is plotted as a function of frequency over its first two resonance frequencies. Two different curves are shown, one for a GM sample interval of 300 MHz and the other of 600 MHz.

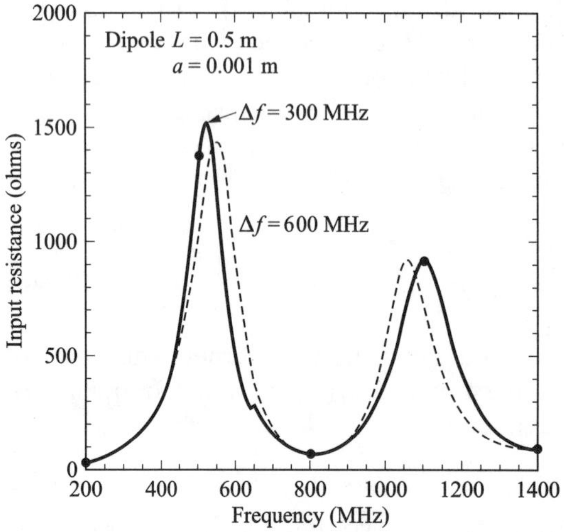

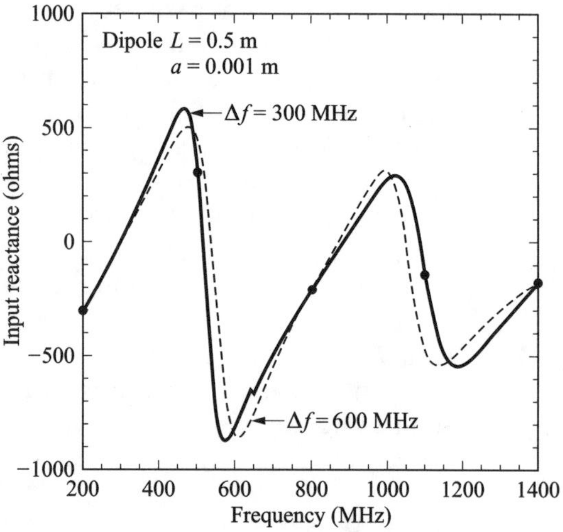

Figure 9.30 Results from using MBPE and two different FMs to represent the interaction coefficients of the $\underline{\underline{Z}}$ matrix of a center-fed, half-wave dipole antenna [54]. The FMs employ the approach of (9.36) to (9.38) and are based on GM samples spaced 300 MHz apart (the solid line) and 600 MHz apart (the dashed line). The discontinuity in the impedance curves occurs at the point where a FM replaces a GM sample at one end of its span with a new one at the other as successive GM samples are employed.

Virga and Rahmat-Samii [57] used $\underline{\underline{Z}}$-matrix frequency modeling for more complex communications antennas, one result for which is shown in Fig. 9.31. There the input impedance of a four-turn helical antenna on an infinite ground plane is plotted versus frequency as obtained by direct evaluation and from two interpolation methods. The interpolated results are obtained using a simple quadratic FM and a second FM that incorporates the singularity of interaction coefficients for segments closer than one-half wavelength and otherwise uses (9.36) to (9.38). The authors report an approximately 30-to-1 computer savings result from modeling the $\underline{\underline{Z}}$ matrix as compared with direct evaluation at a total of 301 frequency samples.

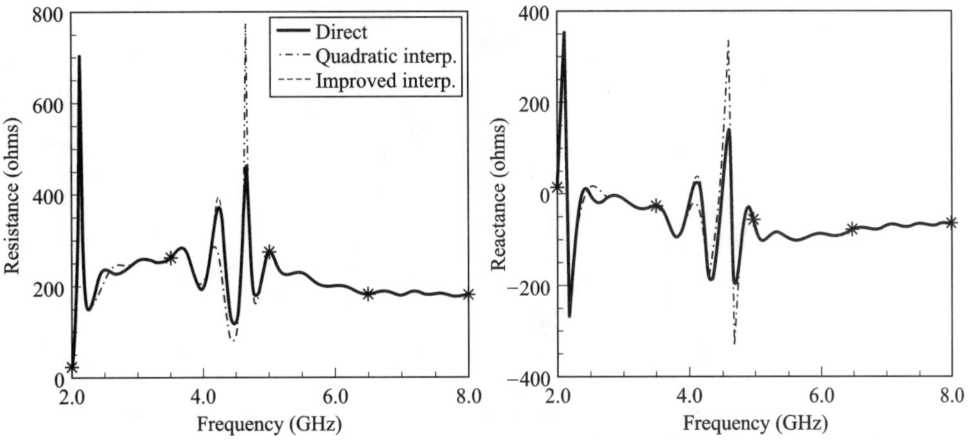

Figure 9.31 Results from using MBPE and two different FMs of the interaction coefficients of the $\underline{\underline{Z}}$ matrix for a helical antenna. Values obtained from straight-line interpolation of 301 GM samples (solid line) are compared with a quadratic FM (dot-dash line) and a FM defined by (9.36) to (9.38) (dashed line) [57]. The GM samples used for the FM results are indicated by the starred points.

9.8.2.3 Modeling Frequency Variations: Elastodynamic Scattering

Benthien and Schenck [55] have used an approach similar to the above for modeling elastodynamic problems whose frequency responses, it should be noted, can be much more complex in structure than typical EM spectra. One interesting aspect of their work is that they are able to span a bandwidth that includes several resonances with only two FPM system-matrix computations at its endpoints while using MBPE for frequencies between them, in contrast to modeling the solution (admittance) matrix where the equivalent of two samples per resonance, or more, is required. The resonance structure is manifested only when the solution has been developed, or the source-field description has been inverted. This difference stems from the fact that, so long as phase changes are handled accurately enough across a given frequency interval over which the phasor between the most widely spaced points on the object being modeled rotates through several (say n) times 2π, then n to $2n$ resonances can be predicted by using only two FPM samples. Examples of such [55] results are included in Figs. 9.32 and 9.33.

9.8.2.4 Modeling Spatial Variations: The Sommerfeld Problem

A computationally demanding problem is that of determining the Sommerfeld fields that result from the interaction of elementary sources with a half space and which become part of the Green's function in an IE model for antennas near ground, microstrip structures, etc. One of the first approaches to this problem to incorporate numerically the rigor of the Sommerfeld integrals while avoiding their computational complexity used a 2-dimensional mesh of precomputed Sommerfeld integrals [46].

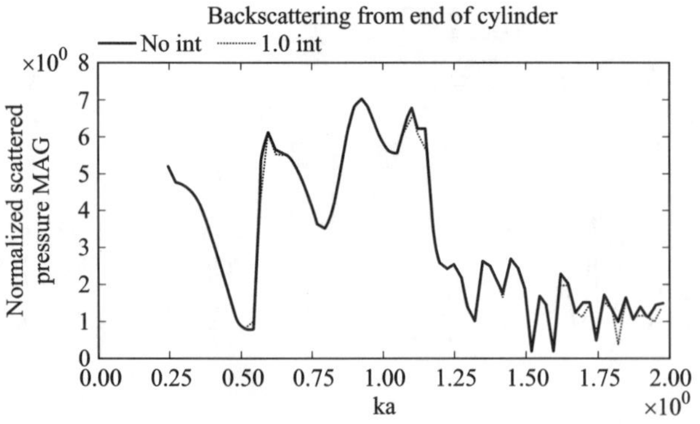

Figure 9.32 Results for acoustic backscattering from the end of a circular cylinder as obtained from FPM samples joined by straight lines (solid line) and using MBPE on $\underline{\underline{Z}}$-matrix samples spaced 1.0 unit apart in ka (dotted line) with the FM evaluated at the same set of frequencies [55]. A linear interpolation model is used for $M_{m,n}(\omega_2 - \omega_1)$ in (9.36). While the samples for both are spaced too far apart to completely resolve the fine structure of the response, a second series of pole-based FMs might be used to rectify this problem, as additional FM samples developed in the solution domain would require much less computation than using the formulation-domain FMs each of which involves solution of a large matrix.

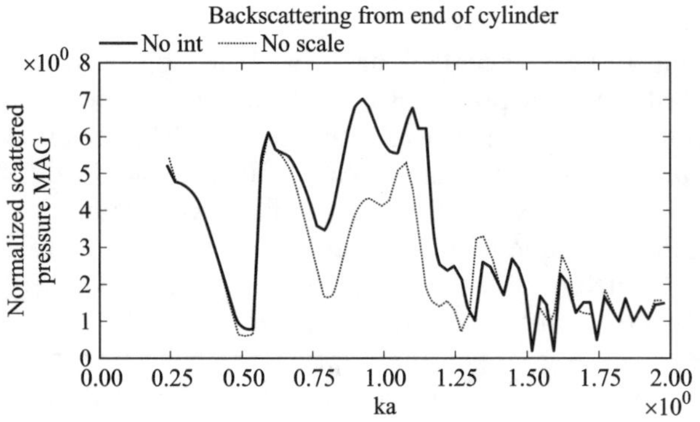

Figure 9.33 Results as for Fig. 9.32 but where the $\underline{\underline{Z}}$-matrix samples used for MBPE are spaced 0.4 units apart in ka (dotted line) and the FM is not scaled. In this case, $Z_{m,n}(\omega_2)$ is represented by $Z_{m,n}(\omega_1)M_{m,n}(\omega_2 - \omega_1)$ with linear interpolation used for the latter [55]. Comparison of Figs. 9.30 and 9.31 emphasizes the importance of using a FM capable of accurately enough representing the variation of the quantity being modeled in the MBPE process.

This permitted evaluating the impedance-matrix coefficients for wire objects located on one side of an interface from simple bivariate interpolation, since the fields are then functions of only two variables, their lateral separation r and the sum of the vertical coordinates, z' and z, of the source and observations points relative to the interface, respectively. This essentially curve-fitting approach, used in NEC-2, reduced

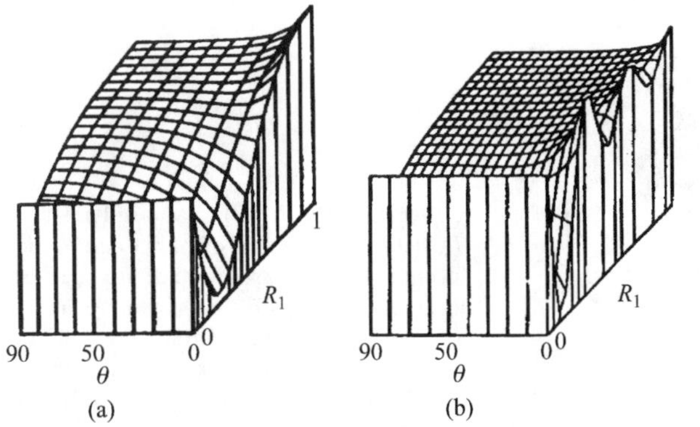

Figure 9.34 These plots exhibit representative electric fields due to a delta-current source located on the air side of an air–earth interface as a function of radial distance R_1 and elevation angle θ for a lossy (a), and a lossless (b) lower medium, respectively [46]. The field is seen to be spatially less variable than the mathematical complexity of the Sommerfeld integral which gives it might imply. An interference between the different-wavelength above-surface and below-surface fields can be observed along the interface (where $\theta = 0$).

the matrix fill time to being little more than what is needed for a perfectly conducting ground where image theory is analytically rigorous. An example of the field variation on the same side of the interface as a fixed source located near an interface is shown as a function of observation position in Fig. 9.34 [46]. Clearly, the field is quite well behaved and can be accurately approximated using linear interpolation.

Later, in extending NEC-2 to model objects interacting across an interface, a problem where 3-dimensional interpolation would have been required since the fields are then functions of lateral separation and the distance from the interface of both the source and observation points, an unattractive prospect, MBPE was then employed. The analytical "models" in this case are extracted from various asymptotic and other approximations that are applicable to the parameter range of interest so that Sommerfeld integrals of the form

$$V_{\pm}^{T} = 2\int_0^{\infty} \frac{e^{-\gamma_{-,+}|z'| - \gamma_{\pm}|z|}}{k_-^2\gamma_+ + k_+^2\gamma_-} J_0(\lambda\rho)\lambda d\lambda \tag{9.39}$$

are replaced by expressions like

$$\tilde{T}_{\rho}^{V} \approx \frac{-i\omega\mu_0 K}{4\pi}\left[\frac{k_+^2}{k_+^2 + k_-^2}\left(\frac{1 - \sin\theta}{\cos\theta}\right) - S\cos\theta\right] R^{-1}$$

where

$$\theta = \tan^{-1}\frac{|z - z'|}{\rho}, \qquad S = \frac{z'}{R}, \qquad \text{and } K = \frac{k_-^2 - k_+^2}{k_-^2 + k_+^2} \tag{9.40}$$

with the details described by Burke and Miller [46]. Finally, the fields needed for the integral-equation model are then approximated by

$$E(r, z, z') \approx \sum A_n f_n(\rho, z, z'); \qquad n = 1, \ldots, N \tag{9.41}$$

where the $f_n(r, z, z')$ functions comprise the model and A_n are the model parameters. This MBPE approach for the interface problem provides an essentially rigorous numerical model for objects interacting across an infinite, planar interface at a cost of increasing the matrix fill time by only 5–10 times over what modeling the same object(s) in free space would require. An example of one electric-field component transmitted to a lower medium from an above-surface source is shown in Fig. 9.35.

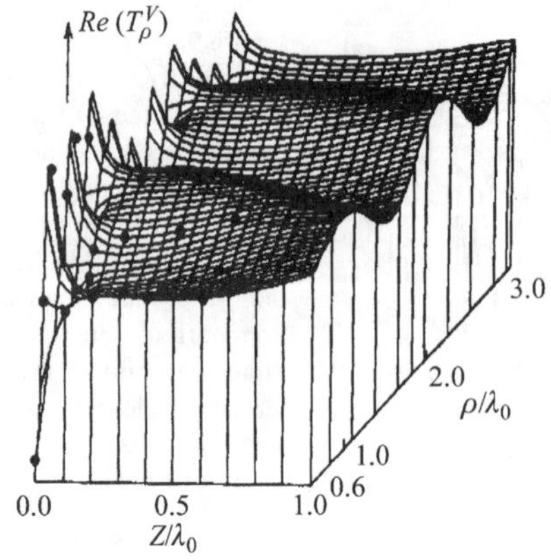

Figure 9.35 The MBPE FM results for the T_ρ^V field as a function of radial and vertical location beneath an interface for a source located above it [46]. Also shown by the dots are the Sommerfeld-integral values used as GMs for the FM approximation. Although not clearly shown because they overlap, the exact and MBPE results are both plotted in this figure.

An alternate approach for the half-space problem based on a series of complex images developed using Prony's method is described by Shubair and Chow [58]. For a vertically oriented antenna, a series of three to five image antennas is found to be adequate for the one-sided problem (antenna on one side of the interface). A similar approach for a horizontal antenna is reported by Fang et al. [59].

9.8.2.5 Modeling Spatial Variations: Waveguide Fields

Other opportunities for Green's-function applications of MBPE arise from separation-of-variables solutions for exterior problems involving cylinders, spheroids, etc. where infinite series of special functions occur and for interior problems where the Green's function can then involve an infinite series of images. An example of the latter application is reported by Demarest et al. [48] for a wire antenna located in the region between infinite, parallel, perfectly conducting planes for which some results are shown in Fig. 9.36. The microscopic source-field description contained in the

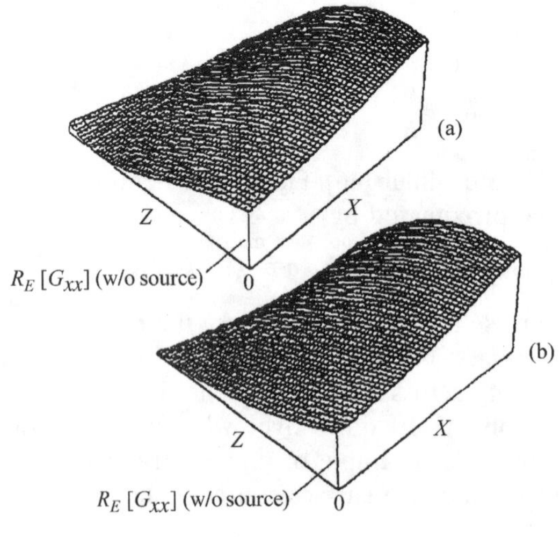

Figure 9.36 Results for the G_{xx} component of the dyadic Green's function (x-directed source perpendicular to waveguide walls, and x-directed field) as obtained from direct evaluation of the defining equation (a), and as evaluated using a FM consisting of two multiplied third-order polynomials in x and z, (b) [48]. Current on a dipole antenna located midway between the waveguide walls obtained from using (b) is within 5% of that resulting from (a).

infinite-series Green's function for this problem is replaced by low-order FMs of the spatial field behavior that provide a much more efficient, yet acceptably accurate, numerical representation of these fields. These FMs substantially increase the efficiency of computing the interaction coefficients in an FDIE model by a factor of about 20 for the example shown. We note that the Green's function for this particular problem has the characteristics of both the WD and SD FMs, since the contribution of each image is pole-like, having a $1/(x - x_n)$ amplitude multiplied by a wave-like phase factor $\exp(-ik(x - x_n))$. The signal fields of this kind of Green's function might be described as having a hybrid character, with the goal of the FM being to replace both with a simpler analytical description.

9.8.2.6 Modeling Spatial Variations: Moment-Method Impedance Matrices

The preceding examples deal with modeling the frequency variation of the coefficients of a FDIE system matrix as a means of reducing the number of needed FPM-matrix evaluations when spanning many resonances across some frequency band to reduce the overall operation count. Alternatively, we might examine the feasibility of reducing the number of FDIE-matrix coefficients that need FPM evaluation at a given frequency as way to reduce the OC at a given frequency. This kind of approach is described by Vecchi et al. [52] and demonstrated by application to a microstrip line with a coupled dipole, an example of which is presented in Fig. 9.37. According to Vecchi et al., there is no appreciable difference between results obtained using the FM and the exact results.

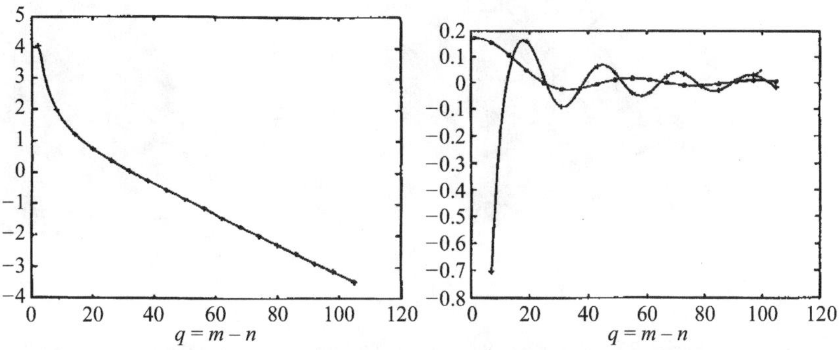

Figure 9.37 Example of modeling the spatial variation of an interaction coefficient for an integral-equation system matrix of a microstrip line with a coupled dipole using a Galerkin subdomain [52]. Left result is $\log_{10}$ of the static, singular part of the self-impedance matrix for the line as a function of the difference between the source and observation subdomain indices. The +s show the points where the interactions are sampled for the FM, the solid line is the exact result, and the dashed line is the FM result. The FM in this case is a polynomial applied to $Z_s[\log(q)]$. Right results are obtained for the frequency-dependent part of the interaction coefficient, where the solid line is the exact result, the dashed line is FM approximation, again using a polynomial, and the ○s and +s indicate the real and imaginary samples used for evaluating the FMs.

Spatial variations of impedance and admittance matrices can be better appreciated by presenting them as surface plots, an example of which is included in Fig. 9.38(a) for a straight wire where the numbering of the subdomains used for this model is sequential from 1 to N. The axes represent the matrix rows and columns and the normalized height of the surface at row $= m$, column $= n$ above the row-column plane represents the magnitude of a given interaction coefficient, $|Z_{m,n}|$. The plotting routine passes a smooth surface through the set of $|Z_{m,n}|$ values, and so may introduce a "fractional" interpolation between the discrete set of row-column indices. It's well known, of course, that the impedance matrix for a straight wire or a strip is of Toeplitz form and so this particular matrix could be fully displayed by a single row or column. The matrix for the same wire bent to form a polygonal approximation of a circular loop would look very similar on this same linear scale except that there would be large components near the corners opposite the main diagonal, assuming the wire segments are numbered in order as well. Were a spatial FM to be used for these impedance matrices, the phase variation would also be needed, or as an alternative the real and imaginary parts instead. The use of magnitude and phase for modeling $\underline{\underline{Z}}$ seems the more appropriate since they are "smoother" [60].

An example for a more interesting structure, an eight-turn helical spiral having a total wire length of 16 wavelengths, is shown in Fig. 9.38(b), also using a linear

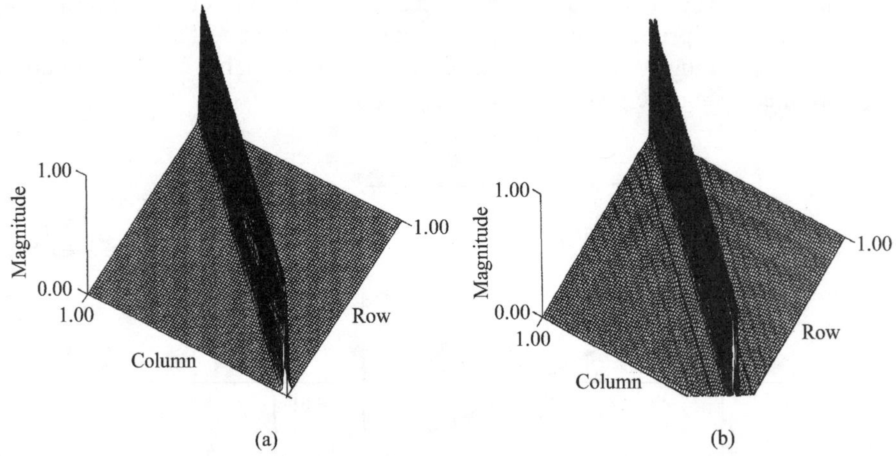

Figure 9.38 Surface plots of the magnitudes of the impedance-matrix coefficients of a straight wire 2 wavelengths long, (a), and an eight-turn helical spiral 16 wavelengths long, (b).

magnitude scale. A "splitting" of the coefficients along the main diagonal may be observed, due to the changing orientation of the neighboring wire segments as they spiral around the helix. This effect can be seen to continue as a ripple in the coefficients further away from the main diagonal.

More information can be conveyed by plotting the log of the matrix coefficients, for which two examples are included in Fig. 9.39. The first, in (a), is for a wire 2 free-space wavelengths in length, located parallel to, and 10^{-4} free-space wavelengths above, the interface between an upper free space and a lower half-space having a

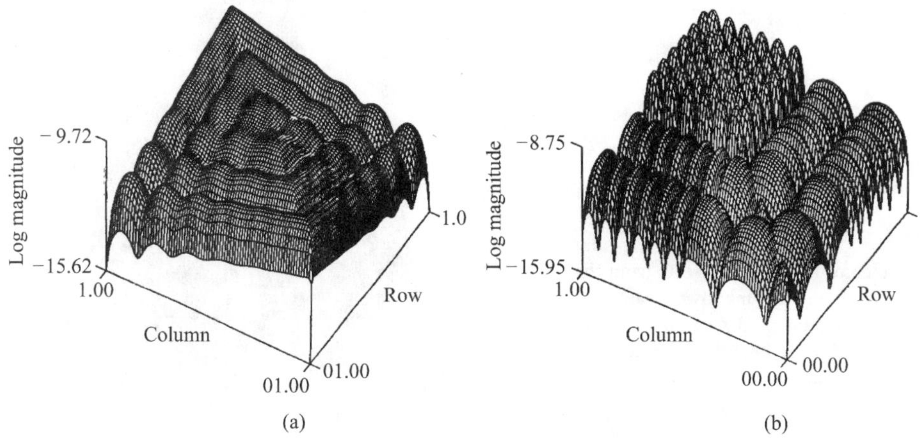

Figure 9.39 Surface plots of the magnitudes of the admittance-matrix coefficients for a two-wavelength wire in free space when parallel to and 10^{-4} free-space wavelengths above a half-space with $\varepsilon_r = 10 - j10$, (a), and for the same wire when perpendicular to a dielectric half-space of $\varepsilon_r = 10$ with its center at the interface, (b). The influence of the lossy lower half space is evident in the attenuation exhibited by the current. The change in dominant current wavelength in the vertical wire is clearly demonstrated in (b) on each half of the wire. Again, an appropriate fitting model for such currents is an exponential series, or waveform-domain, form.

relative dielectric constant of $10 - j10$. There the interference between the waves propagating above and below the interface is seen in a somewhat different way than demonstrated in Figs. 9.34 and 9.35 for the Sommerfeld field alone. Aside from the fact that this matrix is also of Toeplitz form, and therefore more simply filled and solved than an arbitrary matrix, the regular variation of the coefficients in a given row or column indicates the feasibility of using a suitable spatial FM for reducing matrix-fill complexity.

When the same 2-wavelength wire is rotated 90° to penetrate the earth–air interface normally at its midpoint, the impedance matrix shown in Fig. 9.39(b) is obtained. The matrix is now block-Toeplitz but is otherwise nearly as simple spatially as is the case for the same object located in free space.

A much more complicated structure because it has a surface, rather than a linear, geometry is a wire mesh for which an impedance matrix is presented in Fig. 9.40. This plot dramatizes the problem encountered when attempting to visually display source-field relationships over a 2-dimensional surface (or a 3-dimensional volume) in terms of a 2-dimensional matrix of interaction coefficients. The interactions are dependent on the four spatial coordinates that define observation- and source-patch locations Δ_m and Δ_n, respectively, as well on other details of the numerical model, and when projected onto the two-dimensional surface of the impedance matrix their associated spatial relationships are disordered. The field "signal" is no longer simply discerned by observing the behavior of a row or column from $\underline{\underline{Z}}$ but instead requires following a path through the matrix determined by the numbers assigned to the individual unknowns. This does not mean that the spatial variation of the fields can

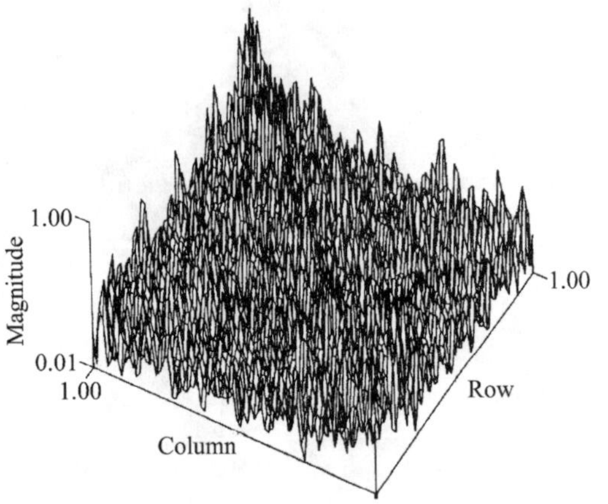

Figure 9.40 Surface plot of the magnitudes of the admittance-matrix coefficients for a 5-wire × 5-wire mesh of wires whose total length is 2 wavelengths. The irregular appearance of this matrix results from the fact that although the unknowns can be numbered in a sequential fashion, their separation is no longer linearly dependent on the respective indices used to construct the system and solution matrices.

no longer be modeled, but that matrix indices can no longer serve as the FM variables.

9.8.3 Using Spectral MBPE in the Solution Domain

In Sections 9.4 and 9.5 of this chapter, the use of FMs to represent the frequency dependence of EM observables was discussed and demonstrated using various examples. Here we consider the more fundamental problem of modeling the frequency dependence of the admittance matrix itself using both function sampling and derivative sampling.

9.8.3.1 Modeling the Admittance Matrix

As previously discussed, the impedance or system matrix arising from an FDIE model contains all the interaction information needed to describe the EM properties of an object being modeled. In a numerical model, this information is represented by the fields produced at observation patches in response to unit-amplitude source patches, both sets of which span the entire object. The relative amplitudes and phase changes associated with the X_s^2 source-field patch pairs convey object size implicitly in these interactions. The inverse of these relationships in the form of the admittance matrix is needed to establish the absolute source amplitudes that satisfy the required boundary conditions. The solution, or admittance, matrix explicitly includes aspects of object size and shape the effects of which are exhibited as periodic body resonances as a function of frequency. Thus, the model appropriate for MBPE representation of the admittance matrix must be capable of handling frequency-dependent resonances.

Since the observables that the solution matrix provides are well approximated by pole series, or more generally rational functions, as demonstrated above in Section 9.5, it follows that the solution matrix itself might also be modeled using rational functions. This conclusion follows by noting that for a single-port antenna its input admittance is defined as the ratio of the feedpoint current to the exciting voltage. For a wire antenna excited at segment j then, having already shown that the admittance can be modeled by a rational function, the $Y_{j,j}$ coefficient of the solution matrix must also have this model. Similarly, the currents on the other wire segments for that excitation, given by $Y_{i,j}$ where $i = 1, \ldots, i-1, i+1, \ldots, X_s$ if there are a total X_s segments, can also be modeled using a rational function. These observations extend to exciting other segments of the wire one at a time, indicating that each coefficient in the solution matrix can be represented by a rational-function FM. Furthermore, since each of these coefficients shares the same resonance structure and thus the same denominator polynomial, the solution matrix can be modeled by a denominator polynomial multiplying a matrix of numerator polynomials, as exhibited by

$$\underline{\underline{Y}}(s) = \frac{1}{D(s)} \begin{bmatrix} X & \cdots & X & n_{1a}(s) & \cdots & n_{1b}(s) & \cdots \\ X & \cdots & X & n_{2a}(s) & \cdots & n_{2b}(s) & \cdots \\ X & \cdots & X & n_{3a}(s) & \cdots & n_{3b}(s) & \cdots \\ \vdots & \vdots & \vdots & \vdots & \vdots & \vdots & \vdots \\ X & \cdots & X & n_{X_s a}(s) & \cdots & n_{X_s b}(s) & \cdots \end{bmatrix} \tag{9.42}$$

where $n_{i,j}(s)$ is the numerator polynomial for coefficient i,j, $D(s)$ is the common denominator and a and b are the indices of those excitation ports whose current responses have been modeled. This form permits direct representation of the wire current for an arbitrary right-hand-side excitation so long as its frequency lies within the valid bandwidth of the solution matrix or of the rational function FMs that comprise its coefficients. Adding a non-pole (or "entire-function") contribution, using $n \geq d$ for various EM observables as mentioned in Section 9.4, offers a similar possibility of generalizing the admittance-matrix model [61].

9.8.3.2 Sampling Admittance-Matrices Derivatives

The FM approaches discussed here for estimating frequency responses require sampled values of the impedance or admittance matrices from which the MBPE parameters can be computed and from which the FM is thus quantified. The sampling can be done either as a function of frequency; as a function of derivative, with respect to frequency at a given frequency; or a combination thereof [20]. Also, since an EM frequency response has complex-conjugate behavior around zero frequency, this knowledge can be employed to provide virtual samples that further improve the MBPE performance, i.e., negative-frequency samples can be employed at essentially no further FPM cost.

For practical reasons of numerical conditioning and accuracy, however, it is advisable not to cover too wide a frequency interval with a single model. The approach that now seems most attractive is to employ a series of frequency windows that slide over the frequency interval to be modeled. These windows can be of lower order to avoid the conditioning problems that can otherwise. As noted in Sections 9.4 and 9.5 above, using sliding, and overlapping, windows also can yield some estimate of the numerical accuracy of the modeled transfer function by comparing the results of two, or more, windows in their region of overlap where they share common samples. Here, we outline specifically the additional computational benefits that arise from derivative sampling.

On writing the moment-method equations that arise from an integral-equation formulation in matrix form, the impedance equation

$$\sum_{i=1}^{X_s} Z_{i,j}(\omega) I_j(\omega) = V_i(\omega) \tag{9.43}$$

is obtained where these various quantities are evaluated at the frequency ω. A solution for the current can then be formally written as an admittance equation

$$I_i(\omega) = \sum_{j=1}^{X_s} Y_{i,j}(\omega) V_j(\omega) \tag{9.44}$$

where $Y_{i,j}$ is the inverse of $Z_{i,j}$. We should note, however, that the approach developed here for the frequency derivatives could be implemented using LU factorization, iteration, or any other solution method.

Upon differentiating the impedance equation with respect to frequency there is obtained

$$\sum_{j=1}^{X_s} [Z_{i,j}(\omega) I_j'(\omega) + Z_{i,j}'(\omega) I_i(\omega)] = V_i'(\omega) \tag{9.45}$$

where the prime denotes a frequency derivative. A solution of the differentiated impedance equation for the differentiated current can then be written

$$I_i'(\omega) = \sum_{j=1}^{X_s} Y_{i,j}(\omega) \left(V_j'(\omega) - \sum_{k=1}^{X_s} Z_{j,k}'(\omega) I_k(\omega) \right) \tag{9.46}$$

where we observe that while the differentiated impedance matrix appears as part of a modified right-hand side of the differentiated admittance equation, I' is given in terms of an undifferentiated admittance matrix. Computing the differentiated current thus requires an additional number of computations beyond those needed for solution of the undifferentiated current proportional to X_s^2 rather than the X_s^3 that would apply to obtain another frequency sample (assuming that LU decomposition is used rather than iteration). Continuing this process, the nth frequency derivative of the current is given by

$$I_i^{(n)}(\omega) = \sum_{j=1}^{X_s} Y_{i,j}(\omega) \left[V_j^{(n)}(\omega) - \sum_{m=1}^{n} C_{nm} \left(\sum_{k=1}^{X_s} Z_{jk}^{(m)}(\omega) I_k^{(n-m)}(\omega) \right) \right] \tag{9.47}$$

where again $C_{n,k}$ is the binomial coefficient and the superscript in parenthesis indicates differentiation with respect to frequency of the order indicated.

It is especially important to observe that information about the nth frequency derivative of the current continues to require an operation count proportional to X_s^2. Expressed in another way, each additional frequency derivative of the solution vector for the current can be computed in a number of operations proportional to $A(n, N_{rhs})/X_s$ where A is a function which depends on the order of the derivative and the number of right-hand sides for which the solution is sought. If the frequency derivatives provide information comparable to that available from the frequency samples themselves, it can be appreciated that there could be a substantial computational advantage to using the solution derivatives in estimating the transfer functions. The problem of implementing the above approach in the NEC code is discussed by Miller and Burke [20], [21].

9.9 OBSERVATIONS AND CONCLUDING COMMENTS

This chapter has considered the rationale and illustrated the application of model-based parameter estimation (MBPE) to achieve reduced-order representations of electromagnetic observables via fitting models (FMs), the "model-based" part of MBPE, that derive from the physics of EM fields. The "parameter-estimation" part of MBPE is the process of obtaining numerical values for the coefficients of the FM by matching or fitting it to sampled values of the EM observable of interest. Although a wider range of FMs is feasible, attention here is focused on what are termed waveform-domain models, comprised of exponential series, and spectral-domain models, comprised of pole series. These kinds of FMs are shown to provide natural "basis functions" for many kinds of EM observables, whether these observables are based on experimental measurement or numerical computation.

It has been long recognized that the role of models in science and engineering is crucial both to demonstrate correct understanding of the applicable physics that the models are intended to represent and to exploit this understanding in developing practical application based on that physical understanding. The models employed can be "first principles" or generating models (GMs) such as Maxwell's equations which can, in principle, provide a numerical solution of arbitrary accuracy to whatever problem is of interest so long as the problem itself can be described to sufficient accuracy. They can also be scale or "real-life" physical models where measurements yield an "experimental solution." In practice, of course, computer resources and other limitations constrain both the accuracy of the numerical solution obtained for the numerical model and the fidelity to which that model replicates the physical problem it is intended to represent. Similar kinds of observations can be made about experimental measurement.

Problem complexity, deriving from the information needed to adequately define its geometrical and electrical properties, and from the information needed to describe its electromagnetic response, further limits what can be done with GM solutions or experimental measurement. Thus, even if there were no limitations to solving first-principles models or conducting measurements, the utility of the results obtained may, paradoxically, be limited by their sheer complexity. Beyond that, practical considerations ensure that relying only on data from either source for design purposes will generally be unacceptable because the costs of those data will generally reduce the coverage of the parameter space that can be explored, leading to uncertainty about whether important behaviors have been missed and whether the performance synthesized numerically will be actually achieved.

Physically based FMs on the other hand complement the GMs on whose samples they are based by offering the possibility of accurately representing the physical observables produced by the GM while doing so with orders-of-magnitude reduced rank or complexity. These FMs, furthermore, provide an analytically continuous quantitative representation of the observable they model while revealing details that a coarsely sampled GM can easily miss. Thus, not only does an appropriate FM greatly decrease the amount of GM sampling required, but it leads to a more useful form for needed observables as analytically useful formulas rather than, for example, tables of numbers. Although a wide variety of FMs relevant to electromagnetic fields can be identified, the more useful seem to be the exponential and pole series to which most attention has been devoted here, since they form a natural basis for many kinds of electromagnetic observables.

The applicability of low-order fitting models (FMs) in computational electromagnetics both to reduce the sampling density of computed observables and to decrease the computational cost of obtaining these observables has been demonstrated here. Both of these possibilities depend on the fact that much EM modeling is redundant, in that source and field variations as a function of time, frequency, angle and space can be accurately described by physically derived FMs that permit equivalent information to be determined from fewer computations. A conclusion to be reached from this observation is that GMs need not be employed in the manner they most often now are used to obtain desired information. Rather, supplementary information is available from our knowledge of EM mathematics and physics, allowing us to employ reduced-order models to represent observables obtained from a GM or to reduce the complexity of the GMs themselves. This substitution offers the possibilities of greatly decreasing the number of evaluations required of FPMs and the cost of their evaluation, with a consequent reduction in the overall computer cost required to obtain the information desired.

In the context of using an FDIE solved using the moment method, modeling the frequency variation of the impedance matrix saves an operation count (OC) proportional to X_s^2 for each frequency sample that can be eliminated. Similarly, modeling the frequency variation of the admittance matrix produces an OC saving proportional to X_s^3 for each sample eliminated. Modeling the spatial variation of the impedance matrix can reduce the OC of a solution toward $X_s \log(X_s)$, which forms the basis for the newer, "fast" techniques. Modeling the spatial variation of the admittance matrix might offer similar kinds of savings, but has not yet been tested. For problems solved using a GM and requiring hours of computer time for each new frequency sample, the savings in computer resources resulting from matrix modeling can be substantial. The models discussed, especially for the admittance matrix, not only provide a more useful representation of the physical behavior continuous in the independent variable, but are valuable for other purposes such as obtaining transient responses.

As a final point, it is worth noting that a FM is well suited to adaptive sampling of the observable it is intended to represent. This means that sampling can be algorithmically tailored to how a given observable changes with the relevant independent variable(s). The usual *ad hoc* sampling approach increases the cost of data acquisition when oversampling occurs, or increases uncertainty about the actual behavior of an observable when undersampling results. On the other hand, adaptive sampling naturally leads to a variable sampling density when the observable behavior permits this, and the total number of samples is controlled by the observable behavior and the uncertainty acceptable in its FM representation.

ACKNOWLEDGMENTS

The authors acknowledge the original contributions of G. J. Burke in developing frequency-derivative interpolation and his implementation of MBPE in NEC and of J. K. Breakall in providing the matrix plots used in this chapter.

SOME ABBREVIATIONS

d	Order of rational-function denominator polynomial
DSA	Discrete-source approximation
EFIE	Electric-field integral equation
EVA	Eigenvalue analysis
FD	Frequency domain
FDI	Frequency-derivative interpolation
FDIE	Frequency-domain integral equation
FM	Fitting model
FPM	First-principles model
GM	Generating model
MBPE	Model-based parameter estimation
MD	Measured data
MME	Maximum mismatch error
n	Order of rational-function numerator polynomial
NEC	Numerical electromagnetics code
OC	Operation count for numerical solution
OC_i	Operation count at frequency f_i
OC_T	Total operation count summed over frequency
PEC	Perfect electric conductor
RCS	Radar cross section
RM	Romberg's method
SD	Spectral domain
SEM	Singularity expansion method
STEM	STatistical ElectroMagnetics
SVD	Singular-value decomposition
TD	Time domain
WASPE	Windowed adaptive sampling pattern estimation
WD	Waveform domain
X_s	Generic number of spatial unknowns
X_{si}	Number of spatial unknowns at frequency f_i

APPENDIX 9.1 ESTIMATING DATA RANK

A requirement for using MBPE in either the WD or SD is determining the effective rank, R, of the data to be modeled. There are two significant difficulties that can arise. First, if R is underestimated, i.e., if a FM order $P < R$ is used, then a unique solution is not possible. If, on the other hand, R becomes too large and/or $P > R$ is used, the system of equations for the FM parameters can become very ill-conditioned, leading to inaccurate or completely erroneous results if simple LU-decomposition is used to solve the data matrix. Since it is rarely a "crisp" number, determining R and choosing P requires some analysis of the data and its effective noise level. An approach that uses eigenvalue analysis (EVA) and singular-value decomposition (SVD) for estimating the effective rank of given data is outlined here. A different approach to establish the order of windowed FMs, motivated by the goal of developing an adaptive sampling strategy, is described in Section 9.4.3. Estimation of data rank is directly related to the problem of data compression while also enabling controllable preservation of the data's information content.

Consider a $D \times 1$ data vector obtained in a process to be modeled, given by

$$\underline{f}(n) = [f_n, f_{n+1}, f_{n+2}, \ldots, f_{n+D-1}]^T \tag{9.48}$$

where T denotes a transpose, and where the data can be complex in general. The goal of rank estimation may be expressed as one of transforming the data into a feature space having the same rank as the original data. As Haykin [66] points out, "However, it would be desirable to design the transformation in such a way that the data vector can be represented by a reduced number of 'effective' features and yet retain most of the intrinsic information content of the input data. In other words, the data vector undergoes a dimensionality reduction." This is, in principle, achievable through the Karhunen–Loeve expansion. To accomplish this, and with H denoting a conjugate transpose and $E[.]$ an expected value of a random variable, form the correlation matrix, $[R]$, of the data

$$[R] = E[\underline{f}(n)\underline{f}^H(n)] \tag{9.49}$$

with a conjugate transpose denoted by the superscript H. The elements $[R]$ can be written $R_{i,j} = r_{i-j}$, $i,j = 0, 1, \ldots, D-1$, and where

$$r_k \equiv E[\underline{f}(n)\underline{f}^*(n-k)] \qquad \text{for } k = 0, \pm1, \pm2, \ldots \tag{9.50}$$

Let the eigenvectors of $[R]$ be given by $\underline{q}_1, \underline{q}_2, \ldots, \underline{q}_M$, with the associated, ordered eigenvalues $\lambda_1 > \lambda_2 > \ldots, \lambda_D$, which together can exactly represent the original data. An approximation $\underline{w}(u)$ for $\underline{f}(n)$, can be developed as

$$\underline{w}(n) = \Sigma c_i(n)\underline{q}_i; \qquad i = 1, \ldots, p < D \tag{9.51}$$

whose rank is lower than that of the original data. An error $\underline{e}(n)$ results, defined by

$$\underline{e}(n) = \underline{f}(n) - \underline{w}(n) = \Sigma c_i(n)\underline{q}_i; \qquad i = p+1, \ldots, D \tag{9.52}$$

of mean-square value

$$e = E[\|\underline{e}(n)\|^2] = \Sigma\Sigma E[c_i^*(n)c_j(n)] = \Sigma(\lambda_i); \qquad i = p+1, \ldots, D \tag{9.53}$$

showing that the accuracy of the approximated vector $\underline{w}(n)$ depends on the "smallness" of the neglected eigenvectors.

If the original data are inexact due to noise or other uncertainty, then we have

$$\underline{y}(n) = \underline{f}(n) + \underline{n}(n) \tag{9.54}$$

where $\underline{n}(n)$ is assumed to be zero-mean noise uncorrelated with the original data and the new data vector is $\underline{y}(n)$. If the noise variance is σ^2, i.e.,

$$E[\underline{n}(n)\underline{n}^H(n)] = \sigma^2[I] \tag{9.55}$$

where $[I]$ is the identity matrix, then the mean-square error is given by

$$\varepsilon_{\underline{y}} = \mathrm{E}[\|\underline{y}(n) - \underline{f}(n)\|^2] = E[\|\underline{n}(n)\|^2] = Ds^2. \tag{9.56}$$

It follows that the error in a reduced-rank approximation $\underline{v}(n)$ of $\underline{f}(n)$ in the presence of noise can be expressed as

$$\varepsilon_{\underline{u}} = E[\|\underline{v}(n) - \underline{f}(n)\|^2] \tag{9.57}$$

and

$$\underline{v}(n) = [\underline{q}_1, \underline{q}_2, \ldots, \underline{q}_m]\underline{c}(n) + [\underline{q}_1, \underline{q}_2, \ldots, \underline{q}_m]\underline{n}(n) \tag{9.58}$$

which gives

$$\varepsilon_{\underline{u}} = \Sigma\lambda_i + p\sigma^2, \qquad i = p+1, \ldots, D \tag{9.59}$$

The first term in the error expression is due, as before, to a reduced-rank description of the original data while the second term is due to the effect of data uncertainty. It can be seen that reduced-rank modeling of noisy data is beneficial provided that

$$\Sigma\lambda_i < (M-p)\sigma^2; \qquad i = p+1, \ldots, D \tag{9.60}$$

i.e., that the smaller eigenvalues should be comparable to the data variance.

Thus, reduced-rank modeling involves an explicit bias-variance trade-off. Rank reduction has the effect of purposely introducing a bias in return for a reduction in the variance due to the effect of data uncertainty. Again to quote Haykin [66], "Indeed, the example clearly illustrates the motivation for using a parsimonious (i.e., simpler) model that may not exactly match the underlying physics responsible for generating the date vector $\underline{u}(n)$, hence the bias; but the model is less susceptible to noise, hence a reduction in variance." Therefore, some types of checks should be built into the computational process so that the user knows whether the data are sufficient and the FM order is appropriate. An available check for this purpose is SVD [67].

Finally, it's worth mentioning that a theoretical choice for choosing exponential- and pole-series FMs is that in the time domain, the complex exponentials form the eigenfunctions of any linear, time-invariant physical system, which in the frequency domain becomes a rational function. Thus, these FMs provide an appropriate, or "natural" basis, for such problems. The trade-off between bias and variance is accomplished through the total least-squares (TLS) implementation of this methodology implemented in SVD as discussed by [68] for the WD and by [69] for the SD.

APPENDIX 9.2 USING THE MATRIX PENCIL TO ESTIMATE WAVEFORM-DOMAIN PARAMETERS

The matrix pencil approach is one of the most computationally efficient and statistically robust ways to fit uniformly sampled noisy data with a sum of complex exponentials. First consider the noise-free case and write the two data matrices [12]

$$Y_{2;i,j} = f_{i+j-1} \text{ and } Y_{1;i,j} = f_{i+j-2};\ i = 1,2, \ldots, D-L+1 \text{ and } j = 1,2, \ldots, L \tag{9.61}$$

where D is the total number of data samples. The matrix pencil, defined as $[Y_2] - I[Y_1]$, can then be written

$$[Y_2] - I[Y_1] = [Z_1][A][Z_0][Z_2] - I[Z_1][A][Z_2] \tag{9.62}$$

where

$$Z_{0;i,j} = \text{diag}[X_1, X_2, \ldots, X_P] \tag{9.63a}$$

$$Z_{1;i,j} = (X_j)^{i-1};\ i = 1, \ldots, D - L + 1 \text{ and } j = 1, \ldots, P \tag{9.63b}$$

$$Z_{2;i,j} = (X_i)^{j-1};\ i = 1, \ldots, P \text{ and } j = 1, \ldots, L \tag{9.63c}$$

$$A_{i,j} = \text{diag}[R_1, R_2, \ldots, R_P] \tag{9.63d}$$

Hence, (23) can be written

$$[Y_2] - \lambda[Y_1] = [Z_1][A]\{[Z_0] - \lambda[I]\}[Z_2] \tag{9.64}$$

where it can be seen that when $\lambda = X_i$, $i = 1, 2, \ldots, P$, the ith row of $[Z_0] - \lambda[I]$ is zero and its rank is then $P - 1$. This means that the X_i can be found as the generalized eigenvalues of the matrix pair $\{[Y_2]; [Y_1]\}$, or equivalently as the ordinary eigenvalues of the matrix $[Y_1]^+[Y_2] = \lambda[I]$ where $[Y_1]^+$ is the Moore–Penrose pseudo inverse of $[Y_1]$, given by $[Y_1]^+ = \{[Y_1]^H[Y_1]\}^{-1}[Y_1]^H$.

If the data are noisy, i.e., if $\underline{y}(n) = \underline{f}(n) + \underline{n}(n)$, the pole locations will exhibit some degree of variance. It's desirable in that case to develop a statistical estimate for the X_i that is not only robust but that comes close as possible to the Cramer–Rao bound which is the least variance that can theoretically be achieved. This can be achieved by performing a TLS, SVD of the composite data matrix given by

$$Y_{i,j} = y_{j-i};\ i = 0, \ldots, L \text{ and } 0 = 1, \ldots, D - L \tag{9.65}$$

with modified $[Y_2]$ and $[Y_1]$ matrices obtained by substituting y_i for f_i and deleting the first and the last columns of $[Y]$, respectively. Now perform a SVD of Y to get

$$[Y] = [U][\Sigma][V]^H \tag{9.66}$$

with $[U]$ and $[V]$ unitary matrices composed of the eigenvectors of $[Y][Y]^H$ and $[Y]^H[Y]$, respectively, and $[\Sigma]$ a rectangular matrix given by

$$\Sigma_{\mathrm{i},j} = \text{diag}[\sigma_1, \sigma_2, \ldots, \sigma_L];\ i = 1, \ldots, D - L + 1 \text{ and } j = 1, \ldots, L \tag{9.67}$$

with σ_i the singular values of $[Y]$. Thus, for $\underline{u}$ and $\underline{v}$ any column of $[U]$ and $[V]$

$$\underline{u}^H[Y] = \sigma\underline{v}^H \text{ and } [Y]\underline{u} = \sigma\underline{u} \tag{9.68}$$

It then follows that the modified matrix pencil becomes

$$[Y_2] - \lambda[Y_1] = [U][\Sigma][V_2^H - \lambda V_1^H] \tag{9.69}$$

where $[V_2]$ and $[V_1]$ are obtained by, respectively, eliminating the first and last rows of $[V]$. The singular values are ordered such that $\sigma_1 \geq \sigma_2 \ldots \sigma_M \geq \sigma_L \geq 0$ and only the first M are retained, so that the dynamic range σ_1/σ_M represents the number of significant bits in the data.

In essence, the noisy data matrix, $[Y]$, is "prefiltered" using SVD, after which the matrix pencil is implemented according to (9.69). The right singular vectors, which comprise the matrix $[V]$, are seen to be sufficient to compute the X_i and the poles. Thus, it can be more efficient to solve the eigenvalue problem

$$[Y]^H[Y][V] = \gamma[V] \tag{9.70}$$

for $[V]$ alone, to accomplish the noise filtering, and so (9.69) is transformed to a generalized eigenvalue problem

$$[V_1][V_2^H] - \lambda[V_1][V_1^H] = 0 \tag{9.71}$$

The generalized eigenvalues from (9.71) will provide the best estimate for the poles in the presence of noise. Also, there is no limitation on the number of eigenvalues that is computationally feasible except as controlled by the data precision and actual data rank. In summary, the SVD of $[Y]$ or the eigenvalue decomposition of $[Y]^H[Y]$ provides a robust, statistical estimate for the data poles. The solution is completed by solving for the residues, one approach being linear least squares given by

$$[A] = \{[Z]^H[Z]\}^{-1}[Z]^H[D] \tag{9.72}$$

with

$$D_i = y_i;\ i = 0, \ldots, N$$

and

$$Z_{i,j} = (X_j)^{i-1};\ i = 0, \ldots, D-1 \text{ and } j = 1, \ldots, M \tag{9.73}$$

REFERENCES

[1] J. T. Cordaro and W. A. Davis, "Time Domain Techniques in the Singularity Expansion Method," *IEEE Transactions on Antennas and Propagation*, **23**, 1980, pp. 358–367.

[2] C. E. Baum, "Emerging Technology for Transient and Broadband Analysis and Synthesis of Antennas and Scatterers," *Proc. IEEE*, **64** (12), 1976, pp. 1598–1616.

[3] A. Ralston, *A First Course in Numerical Analysis*, McGraw-Hill Book Co., New York, 1965.

[4] D. Shanks, "Non-Linear Transformations of Divergent and Slowly Convergent Sequences," *Journal of Mathematics and Physics*, the Technology Press, Massachusetts Institute of Technology, Cambridge, MA, 1955, pp. 1–42.

[5] E. K. Miller, "Model-Based Parameter Estimation in Electromagnetics: I—Background and Theoretical Development," *IEEE Antennas and Propagation Society Magazine*, **40** (1), 1998, February, pp. 42–52.

[6] E. K. Miller, "Model-Based Parameter Estimation in Electromagnetics: II—Applications to EM Observables," *IEEE Antennas and Propagation Society Magazine*, **40** (2), 1998, April, pp. 51–65.

[7] E. K. Miller, "Model-Based Parameter Estimation in Electromagnetics: III—Applications to EM Integral Equations," *IEEE Antennas and Propagation Society Magazine*, **40** (3), 1998, June, pp. 49–66.

[8] S. A. Schelkunoff and H. T. Friis, *Antennas: Theory and Practice*, John Wiley & Sons, Inc., New York, Chapter 9, 1952.

[9] E. K. Miller, "Model-Based Parameter-Estimation Applications in Electromagnetics," in *Electromagnetic Modeling and Measurements for Analysis and Synthesis Problems*, B. de Neumann, ed., Kluwer Academic Publishers, NATO ASI Series E: Applied Sciences, **199**, 1991, pp. 205–256.

[10] R. Prony, "Essai Experimental et Analytique sur les Lois de la Dilatabilité de Fluides Elastiques et sur Celles del la Force Expansive de la Vapeur de l'Alkool, à Differentes Temperatures," *J. l'Ecole Polytech.* (Paris), **1**, 1795, pp. 24–76.

[11] T. K. Sarkar, J. Nebat, D. D. Weiner, and V. K. Jain, "Suboptimal Approximation/Identification of Transient Waveforms from Electromagnetics Systems by Pencil-of-Function Method," *IEEE Trans. on Antennas and Propagation*, **28** (12), 1980, pp. 928–933.

[12] T. K. Sarkar and O. Pereira, "Using the Matrix Pencil Method to Estimate the Parameters

of a Sum of Complex Exponentials," *IEEE Antennas and Propagation Society Magazine*, **37** (1), 1995, pp. 48–55.

[13] Y. Hua and T. K. Sarkar, "Generalized Pencil-of-Functions Method for Extracting the Poles of an Electromagnetic System from its Transient Response," *IEEE Transactions on Antennas and Propagation*, **37** (2), 1989, pp. 229–234.

[14] A. J. Poggio, M. L. Van Blaricum, E. K. Miller, and R. Mittra, "Evaluation of a processing technique for transient data," *IEEE Trans. on Antennas and Propagation*, **26** (1), 1978, pp. 165–173.

[15] E. K. Miller and D. L. Lager, "Inversion of One-Dimensional Scattering Data Using Prony's Method," *Radio Science*, **17**, 1982, pp. 211–217.

[16] C. J. Benning, "Adaptive Array Beamforming and Steering Using Prony's Method," *Proceedings 19th Annual Symposium on USAF Antenna Research and Development Program*, U. of Illinois, Allerton, 1969.

[17] E. K. Miller and D. L. Lager, "Imaging of Linear Source Distributions," *Electromagnetics*, **3**, 1983, pp. 21–40.

[18] R. S. Adve and T. K. Sarkar, "Generation of Accurate Broadband Information from Narrowband Data Using the Cauchy Method," *Microwave and Optical Technology Letters*, **6** (10), 1993, pp. 569–573.

[19] R. S. Adve and T. K. Sarkar, "The Effect of Noise in the Data on the Cauchy Method," *Microwave and Optical Technology Letters*, **7** (5), 1994, pp. 242–247.

[20] E. K. Miller and G. J. Burke, "Using Model-Based Parameter Estimation to Increase the Physical Interpretability and Numerical Efficiency of Computational Electromagnetics," *Computer Physics Communications*, **68**, 1991, pp. 43–75.

[21] W. H. Press, S. A. Teukolsky, W. T. Vettering, and B. P. Flannery, *Numerical Recipes*, 2nd edn, Cambridge University Press, New York, 1992.

[22] V. I. Kukulin, V. M. Krasnopol'sky, and J. Horácek, *Theory of Resonances: Principles and Applications*, Academia, Praha, Czechoslovakia, and Kluwer Academic Publishers, Dordrecht, The Netherlands, 1989.

[23] G. J. Burke and E. K. Miller (1988), "Use of Frequency-Derivative Information to Reconstruct an Electromagnetic Transfer Function," *Fourth Annual ACES Review*, Naval Postgraduate School, Monterey, CA, March 22–24.

[24] L. T. Pillage and R. A. Rohrer, "Asymptotic Waveform Evaluation for Timing Analysis," *IEEE Trans. Computer Aided Design*, 1990, pp. 352–366.

[25] C. J. Reddy, Private communication, 1997.

[26] E. K. Miller, "A Variable Interval Width Quadrature Technique Based on Romberg's Method," *Journal of Computational Physics*, **5**, 1970, pp. 265–279.

[27] G. J. Burke, Private communication, 1992.

[28] L. R. Fermelia, G. Z. Rollins, and P. Ramanujam, "The Use of Model-Based Parameter Estimation (MBPE) for the Design of Corrugated Horns," *Program Digest of US URSI Radio Science Meeting*, University of Michigan, 1993, p. 388.

[29] K. Kottapalli, T. K. Sarkar, R. Adve, Y. Hua, E. K. Miller, and G. J. Burke, "Accurate Computation of Wideband Response of Electromagnetic Systems Utilizing Narrowband Information," *Computer Physics Communications*, **68**, 1991, pp. 126–144.

[30] K. Krishnamoothy, T. K. Sarkar, X. Yang, E. K. Miller, and G. J. Burke, "Use of Frequency-Derivative Information to Reconstruct the Scattered Field of a Conducting Cylinder Over a Wide Frequency Band," *Journal of Electromagnetic Waves and Applications*, **5** (6), pp. 653–663, 1991.

[31] E. K. Miller, "An Approximate Formula for the Admittance of a Long, Thin Antenna,"

IEEE Transactions on Antennas and Propagation, **16**, 1968, pp. 127–128.

[32] G. J. Burke and E. K. Miller, "Developing Optimal and Automatic Frequency Sampling in Moment-Method Solutions," *in Proceedings of the Tenth Annual Review of Progress in Applied Computational Electromagnetics*, Doubletree Hotel, Monterey, CA, March 21–26, 1994, pp. 165–172.

[33] E. K. Miller, "Minimizing the Number of Frequency Samples Needed to Represent a Transfer Function Using Adaptive Sampling," in 12th Annual Review of Progress in Applied Computational Electromagnetics, Naval Postgraduate School, Monterey, CA, 1996, pp. 1132–1139.

[34] S. D. Lin, "*Interpolation Techniques for Noisy Spectral Data*," Ph.D. dissertation, Mississippi State University, 1991.

[35] E. K. Miller, "Using Model-Based Parameter Estimation to Estimate the Accuracy of Numerical Models," in *12th Annual Review of Progress in Applied Computational Electromagnetics*, Naval Postgraduate School, Monterey, CA, 1996, pp. 588–595.

[36] C. A. Balanis, *Antenna Theory*, Harper and Row, New York, 1982, pp. 679–684.

[37] Mathur, Private communication, Ohio University, Columbus, 1995.

[38] M. P. Hurst and R. Mittra, "Scattering Center Analysis Via Prony's Method," *IEEE Transactions on Antennas and Propagation*, **35**, 1987, pp. 986–988.

[39] J. J. Sacchini, "Development of Two-Dimensional Parametric Radar Signal Modeling and Estimation Techniques with Application to Target Identification," Ph.D. dissertation, Ohio State University, Columbus, 1992.

[40] I. J. Gupta, "High-Resolution Radar Imaging Using 2-D Linear Prediction," *IEEE Transactions on Antennas and Propagation*, **42**, 1994, pp. 31–37.

[41] R. J. Lytle and D. L. Lager, "Using the Natural-Frequency Concept in Remote Probing of the Earth," *Radio Science*, **7** (3), 1976, pp. 245–250.

[42] C. R. Tai, "Transients on Lossless Terminated Transmission Lines," *IEEE Transactions on Antennas and Propagation*, **26**, 1978, p. 556.

[43] E. K. Miller, "Admittance Dependence of the Infinite Cylindrical Antenna Upon Exciting Gap Thickness," *Radio Science*, 1967, pp. 1431–1435.

[44] T. H. Lehman, "A Statistical Theory of Electromagnetic Fields in Complex Cavities," Phillips Laboratory, Kirtland AFB, NM, *Interaction Note 494*, 1993.

[45] R. Smith, Private communication, Kaman Sciences, Colorado Springs, CO, 1990.

[46] G. J. Burke and E. K. Miller, "Modeling Antennas Near to and Penetrating a Lossy Interface," *IEEE Transactions on Antennas and Propagation*, **32**, 1984, pp. 1040–1049.

[47] E. K. Miller, J. N. Brittingham, and J. T. Okada, "Bivariate-Interpolation Approach for Efficiently and Accurately Modeling Antennas Near a Half Space," *Electronics Letters*, **13**, 1977, pp. 690–691.

[48] K. R. Demarest, E. K. Miller, K. Kalbasi, and L.-K. Wu, "A Computationally Efficient Method of Evaluating Green's Functions for 1-, 2-, and 3D Enclosures," *IEEE Transactions on Magnetics*, **25** (4), 1989, pp. 2878–2880.

[49] R. Coifman, V. Rokhlin, and S. Wandzura, "The Fast Multipole Method for the Wave Equation: A Pedestrian Prescription," *IEEE Antennas and Propagation Magazine*, **34** (3), 1993, pp. 7–12.

[50] F. X. Canning, "Transformations That Produce a Sparse Moment Matrix," *J. Electromagnetic Waves and Applications*, **4**, 1990, pp. 983–993.

[51] Chew, W. C. (1993), "Fast Algorithms for Wave Scattering Developed at the University of Illinois' Electromagnetics Laboratory," *IEEE Antennas and Propagation Magazine*, **35** (4), pp. 22–32.

[52] G. Vecchi, P. Pirinoli, L. Matekovits, and M. Orefice, "Singular Value Decomposition in Resonant Printed Antenna Analysis," in *Proceedings of Joint 3rd International Conference of Electromagnetics in Aerospace Applications*, 1993, pp. 343–346.

[53] E. K. Miller, F. J. Deadrick, G. J. Burke, and J. A. Landt, "Computer-Graphics Applications in Electromagnetic Computer Modeling," *Electromagnetics*, **1**, 1981, pp. 135–153.

[54] E. H. Newman, "Generation of Wide band Data From the Method of Moments by Interpolating the Impedance Matrix," *IEEE Trans. Antennas Propagat.*, **36** (12), 1988, pp. 1820–1824.

[55] G. W. Benthien and H. A. Schenck, "Structural-Acoustic Coupling," in *Boundary Element Methods in Acoustics*, R. D. Ciskowski and C. A. Brebbia, ed., Computational Mechanics Publications, New York, 1991.

[56] G. J. Burke, E. K. Miller, S. Chakrabarti, and K. Demarest, "Using Model-Based Parameter Estimation to Increase the Efficiency of Computing Electromagnetic Transfer Functions," *IEEE Trans. Magnetics*, **25** (4), 1989, pp. 2807–2809, July.

[57] K. Virga and Y. Rahmat-Samii, "Wide-Band Evaluation of Communications Antennas Using [Z] Matrix Interpolation with the Method of Moments," in *IEEE Antennas and Propagation Society International Symposium*, Newport Beach, CA, 1995, pp. 1262–1265.46.

[58] R. M. Shubair and Y. L. Chow, "A Simple and Accurate Complex Image Interpretation of Vertical Antennas Present in Contiguous Dielectric Half-Spaces," *IEEE Transactions on Antennas and Propagation*, **41**, 1993, pp. 806–812.

[59] D. G. Fang, J. J. Yang, and G. L. Delisle, "Discrete Image Theory for Horizontal Electric Dipoles in a Multilayered Medium," *IEE Proceedings*, **135**, 1988, pp. 297–303.

[60] K. W. Brown and A. Prata, "Efficient Analysis of Large Cylindrical Reflector Antennas Using a Nonlinear Solution of the Electric Field Integral Equation," *IEEE Antennas and Propagation Society International Symposium*, Seattle, WA, 1994, pp. 42–45.

[61] C. E. Baum, "Toward an Engineering Theory of Electromagnetic Scattering: The Singularity and Eigenmode Expansion Methods," in *Electromagnetic Scattering*, P. L. E. Uslenghi, ed., Chapter 15, Academic Press, New York, 1978, pp. 571–651.

[62] E. F. Knott, J. F. Shaeffer, and M. T. Tuley, *RADAR Cross Section*, 2nd edn, Boston: Artech House, 1993.

[63] Kluskens, M., Private communication, 1998.

[64] Shantnu Mishar, David Florida Laboratory, Ottawa, Canada, 1996.

[65] R. J. Allard, D. H. Werner, J. S. Zmyslo, and P. L. Werner, "Model-Based Parameter Estimation of Antenna Radiation Pattern Frequency Spectra," *Proc. IEEE Antennas and Propagation Symposium*, V1, pp. 62–65, 1998.

[66] S. Haykin, *Adaptive Filter Theory*, Prentice Hall, Englewood Cliffs, 1996.

[67] Y. Hua and T. K. Sarkar, "On SVD for Estimating Generalized Eigenvalues of Singular Matrix Pencil in Noise," *IEEE Trans. Acoustics, Speech and Signal Processing*, **39** (4), 1991, pp. 892–900.

[68] Y. Hua and T. K. Sarkar, "On the Total Least Squares Linear Prediction Method for Frequency Estimation," *IEEE Trans. Acoustics, Speech and Signal Processing*, **38** (12), 1990, pp. 1286–2189.

[69] R. S. Adve, T. K. Sarkar, S. M. Rao, E. K. Miller, and D. R. Pflug, "Application of the Cauchy Method for Extrapolating/Interpolating Narrow-Band System Responses," *IEEE Trans. on Microwave Theory*, **45** (5, Pt. II), 1997, pp. 837–845.

Chapter 10

ADAPTIVE DECOMPOSITION IN ELECTROMAGNETICS

Joseph W. Burns and Nikola S. Subotic

ABSTRACT

Definition of a mathematical representation of a signal or physical quantity is a fundamental step in the formulation of forward or inverse scattering problems. Usually, the signal is represented as a linear sum of functions that are chosen to be either physically descriptive or convenient for mathematical analysis. Traditionally, expansions using a single orthonormal basis have been used. These representations produce unique decompositions; however, signals are not always compactly represented using these expansions.

An alternate approach is to develop a compact, or "sparse," representation of signals by decomposing them in terms of a very general "dictionary" of functions that closely represent anticipated signal characteristics. For forward problems, such as prediction of the electromagnetic scattering from an object, sparse representations of the unknown, typically the surface current, can lead to significant reductions in computational complexity. For inverse problems, such as the analysis of a backscattered radar signal, sparse representations can provide more identifiable and interpretable representations of the return.

Recently, several adaptive techniques for generating sparse decompositions in terms of very general, overdetermined dictionaries have appeared in the signal processing literature and have begun to be applied, with promising results, to electromagnetic sensing, prediction and data-processing applications.

We first review the concept of an adaptive decomposition algorithm, both the idea of overdetermined dictionaries and the details of several fitting algorithms. Several applications of these techniques to forward scattering problems, inverse scattering problems, and radar data postprocessing applications are described to illustrate the utility of these techniques.

10.1 INTRODUCTION

A fundamental step in the formulation of forward or inverse scattering problems is the definition of a mathematical representation of a signal or physical quantity. The signal is usually represented as a linear sum of functions that are chosen to be either physically descriptive or convenient for mathematical analysis. Additionally, it is desirable to describe the signal with as few functions as possible, i.e., generate a "sparse" representation. Traditionally, expansions using a single orthonormal basis have been used. These representations produce unique decompositions; however, some signals are not always compactly represented using these expansions. A simple example is a signal with oscillations and spikes being represented in terms of a Fourier (complex exponential) basis. Even though the Fourier decomposition will represent the oscillations sparsely, it takes a large number of Fourier basis elements to describe the spikes.

An alternate approach is to represent signals by decomposing them in terms of a very general "dictionary" of functions that closely represent anticipated signal characteristics. By "tuning" the dictionary to the anticipated signal characteristics, a sparse representation is promoted. In the above example, spikes would be added to the basis set to represent those components of the signal.

For forward-scattering problems, such as prediction of the electromagnetic scattering from an object, sparse representations of the unknown, typically the surface current, can lead to significant reductions in computational complexity. For inverse problems, such as the analysis of a backscattered radar signal, sparse representations can provide more identifiable and interpretable representations of the return. In problems where the signals are derived from some physical interaction such as electromagnetic scattering, the representation should provide some description, either intuitive or directly physical, as to what is going on in the signal.

A frequently cited analogy [1], [2] is to liken these decompositions to processing a speech waveform, which is simply a time-dependent sequence of sounds. The sequence can be projected onto any basis, i.e., Fourier, Gabor, French, German; however, it is preferable to project the speech onto a basis understood by the user, for example English. Thus a specific dictionary of words can be defined to characterize the sounds in a manner that facilitates recognition.

A variety of techniques for obtaining a sparse representation of a signal in terms of an overdetermined dictionary of functions, usually referred to as *adaptive decomposition* techniques, have been described in the signal processing literature over the last decade, usually motivated by speech synthesis and data compression applications. Recently, these techniques have come to the attention of researchers in the electromagnetics field, and are being applied to a variety of electromagnetic sensing and scattering problems.

Note, our interest in overdetermined dictionaries precludes discussions of techniques which, at their heart, use a single basis set. Consequently, more conventional decomposition techniques such as Fourier analysis or the wide variety of modern spectral estimation techniques, all of which are based on decomposing signal in terms of exponentials, will not be discussed in detail, but only cited for comparison with adaptive techniques, as appropriate.

We will first review the concept of an adaptive decomposition algorithm, both the idea of overdetermined dictionaries and the details of several fitting algorithms. Several applications of these techniques to forward-scattering problems, inverse scattering problems, and radar data postprocessing applications will then be described to illustrate the utility of these techniques.

10.2 ADAPTIVE DECOMPOSITION

In many instances, it is desirable to decompose a complex-valued signal, $s(f) \in L_2^c(a, b)$ using a collection of parameterized functions $\{\phi_\gamma(f)\}$, $\phi_\gamma(f) \in L_2^c(a,b)$, $\gamma \in \Gamma$, as

$$s(f) = \sum_\gamma A_\gamma \phi_\gamma(f) \tag{10.1}$$

where A_γ is a complex-valued weight, and $L_2^c(a,b)$ represents the space of all complex-valued, square integrable functions on the interval (a,b)

$$\int_a^b |s(f)|^2 df < \infty \tag{10.2}$$

and Γ is the collection of parameters.

Note, in our general discussion, f and t simply represent parameters in conjugate domains. In our specific discussions of dictionaries and applications, we will generally consider the signal to be measured as a function of frequency f, as is common in electromagnetic applications, and look for representations as a function of time t, or equivalently for radar applications, range r. This does not reflect any limitation on the adaptive techniques discussed since they can just as easily be applied to time domain signals to obtain frequency representations.

Most signal processing techniques decompose a signal onto a single orthonormal basis. For example, Fourier-based techniques use the set of complex exponentials leading to $\gamma = (t)$, decomposing functions $\phi_{(t)}(f) = e^{-j2\pi ft}$, and

$$s(f) = \sum_t A_t e^{-j2\pi ft} \tag{10.3}$$

Decomposition onto a single orthonormal basis can provide a useful decomposition of signals under many conditions. By "useful," we mean that the decomposition requires few basis functions (sparsity) to represent the signal and that the decomposition is easily interpretable. However, when the signal contains characteristics that are not well described by the basis, the sparsity and physical interpretation of the decomposition degrade. For instance, a large number of complex exponentials is required to describe signals with fast transitions such as the step function. Using adaptive decomposition algorithms with a composite set of decomposing functions comprised of various function classes which more closely model the signal characteristics can maintain the usefulness of the decomposition.

Each parameterized function $\phi_\gamma(f)$, in (10.1) can be viewed as an "atom" in the decomposition of the signal $s(f)$. The set of atoms can be viewed as a "dictionary" $\mathscr{D} = \{\phi_\gamma(f) : \gamma \in \Gamma\}$. The dictionary is then populated by atoms which closely model the salient characteristics of the signal. Examples of dictionary elements and their characteristics are: pure tones (Fourier dictionary), bumps (wavelet dictionary), chirps (chirplet dictionary), etc. Each of these dictionaries taken individually can be viewed as a classical orthonormal basis of $L_2^c(a,b)$.

Recently, there has been significant activity in the area of general representations of signals (see, e.g., [1]). Specifically, there has been a movement toward using a broader set of functions to represent a signal rather than using purely a single set of orthogonal functions. The goal is to develop fast, stable signal representations that are sparse and offer separation of disparate phenomena.

We will now allow the dictionary to be comprised of disparate functions which describe different signal characteristics [1]. If a signal exhibits multiple characteristics such as oscillations and steps then a dictionary populated by functions encompassing both characteristics is appropriate. Dictionaries composed of differing sets of orthogonal basis functions can be merged into a composite dictionary, which now can be viewed as an overcomplete basis set for $L_2^c(a,b)$.

The signals and atoms that we use in data-processing applications are sampled. Allow $s = [s(f_1), s(f_2), \ldots, s(f_K)]^\dagger$ and $\phi_\gamma = [\phi_\gamma(f_1), \phi_\gamma(f_2), \ldots, \phi_\gamma(f_K)]^\dagger$ to be vectors comprised of the frequency samples of the signal and atoms, respectively. † represents the vector transpose. The size of the index set Γ is defined as N. If we allow Φ to be a matrix whose columns are the N indexed dictionary element vectors (atoms) as

$$\Phi = [\phi_1 \phi_2 \ldots \phi_N] \tag{10.4}$$

and we let a be a column vector of the coefficients (A_γ), the decomposition problem, (10.1), is that of finding a solution to

$$\Phi a = s \tag{10.5}$$

When the dictionary is overcomplete, that is, when $N > K$, $\Phi a = s$ has more than one solution; the solution is nonunique. Some elements in the dictionary have representations in terms of other elements. The goal of the decomposition method is to find the sparsest representation that best fits the signal in a manner which closely describes the components and/or physical characteristics of the signal.

For many electromagnetic analysis problems, we wish to construct a dictionary populated by atoms which describe the various physical phenomena, such as the scattering mechanisms associated with the backscattering from man-made objects. These atoms and the associated dictionary elements are discussed in Section 10.3. The methodologies of finding the decomposition, given the dictionary and the data, are described in Section 10.4.

10.3 OVERDETERMINED DICTIONARIES

A key new concept of adaptive decomposition is that a signal no longer needs to be represented purely in terms of a single, orthonormal basis set. In the adaptive construction, the set of basis functions is allowed to be overcomplete and be comprised of disparate functions which describe different signal characteristics [1]. This set can be composed of many classes of functions, often referred to as a dictionary, which describe particular signal characteristics. As an example, a composite dictionary may simultaneously contain complex exponentials and wavelets.

As discussed in Section 10.1, an illustrative example of decomposition onto an overdetermined dictionary is the interpretation of speech using a language dictionary. The most descriptive dictionary to use when interpreting a speech sequence is a language dictionary understood by the listener. Further, even in a native language, concepts can be tersely communicated when a large vocabulary of words is available for use in the dictionary.

Note that signal analysis using adaptive decomposition approaches the problem from a fundamentally different view than more traditional techniques, such as Fourier analysis. With a conventional decomposition onto a single basis set, disparate signal components are generally recognized by their usually non-sparse coefficient representations involving several elements of the basis. Identification is achieved by observing how the specific signal component is mapped onto many elements of the single basis. With adaptive decomposition, the objective is to create a dictionary containing several basis sets, each descriptive of anticipated signal characteristics, such that energy corresponding to a specific signal component is mapped onto a single dictionary element.

The constituent functions of the dictionary can be defined via specific analytical models, or classes of functions which closely approximate anticipated signal characteristics. *Physics-based dictionaries* consist of functions defined from first principles analysis of underlying physical processes, while *data-based* dictionaries consist of functions that describe characteristics of the component signals in the data. In addition, the adaptive techniques under consideration can allow the decomposition of the signal onto a very general, *merged* dictionary consisting of both physics-based and data-based functions.

As an example for electromagnetic applications, it is well known that the scattered field produced by a complex target can be modeled as a linear combination of individual wave functions or scattering mechanisms. Further, a measurement of this scattering may be expressed as the sum of target scattering and additive measurement contamination and/or clutter, i.e., for a specific aspect and polarization, the measurement versus frequency can be expressed as

$$s(f) = s^{target}(f) + s^{clutter}(f) \tag{10.6}$$

The ability to decompose the signal onto a general, overdetermined dictionary allows the choice of parameterized functions (the dictionary elements) that best represent the anticipated characteristics of the target scattering and contamination.

A dictionary formulated to decompose radar measurements of target scattering can be formed using functions that closely approximate individual target scattering mechanisms, clutter and interference. Target diffraction and specular scattering can be represented by GTD-based models. Other classes of functions such as wavelets, wavelet packets, and symmlets can characterize both narrowband resonances, signal oscillations, and "bumps" within the signal indicative of interference and clutter. While not tied directly to a specific scattering phenomenon, these functions describe signal characteristics which make them useful as bases for clutter identification and removal.

10.3.1 Physics-Based Dictionaries

The functions that constitute the elements of the dictionary should be chosen to closely model anticipated signal characteristics. Thus, for electromagnetic problems, such as the analysis of radar signals or the computation of the surface current induced on a scatterer, a natural choice for elements of the dictionary are parameterized functional descriptions of wave mechanisms or localized scattering phenomenology that are likely to contribute to the quantity under evaluation. A properly defined, *physics-based* dictionary can provide a close fit to the data and can be used to identify the mechanisms that make up a signal by observing which dictionary elements have significant energy in the decomposition (see Section 10.5.1). Several examples of physical functions that can be used to construct a general dictionary are discussed below. The unique property of adaptive decomposition algorithms is that all of these diverse classes of functions can be combined into a single dictionary for the analysis of complex signals.

The scattered field or induced surface current of a complex target can be modeled as a linear combination of individual mechanisms, which is consistent with the representation given in Section 10.2.1. Accordingly, appropriate dictionary elements ϕ_γ can be defined from standard analytical techniques of electromagnetic analysis. As an example, the direct, or early-time, scattering from a complex target can be modeled as the return from a collection of localized scattering centers. Scattering center information can be used to characterize the target geometry. The location of scattering centers can be used to model the shape and size of the target, and the frequency dependence and/or polarimetric signature of individual scattering centers can be used to characterize local target geometries. Exploitation of scattering center information has been recently investigated by several authors using both more traditional spectral estimation techniques [3], [4], [5], [6], [7], [8], [9] and adaptive techniques [10], [11], [12], [13].

Dictionary elements that describe the behavior of canonical, high-frequency scattering centers in frequency sweep data can be straightforwardly derived using Physical Optics and the Geometrical Theory of Diffraction (GTD) [8]. For a given aspect, a dictionary element to characterize a scattering center backscatter response as a function of frequency f can be modeled as

$$s_n(f) = e_n A_n (f/f_c)^{\alpha_n} e^{\,j4\pi f r_n/c} \tag{10.7}$$

where e_n describes the polarization state, A_n is a complex coefficient, f_c is the sweep center frequency, α_n denotes the frequency dependence of the n the scattering center, r_n is the range from the scatterer to the radar, and c is the speed of light. In the GTD model, the scatterer type is determined by the parameter, which is typically chosen as integer multiples of 1/2. In a dictionary used to fit a discretely sampled frequency sweep, α and r would be discretely parameterized over a range of acceptable values, where the range of α is determined by the number of scattering center types considered and the range of r is determined by the physical area being interrogated.

Note any number of discrete values for α and r could be chosen, but the number of dictionary elements, and thus the computational complexity of the adaptive solution, rises rapidly with the number of parameter values. Consequently, efficient processing motivates the selection of the minimum range and sampling of parameters required to fit the data. Examples of the application of this dictionary to scattering center characterization are discussed in Section 10.5.1.

Other wave objects that have been suggested by McClure and Carin [10] for inclusion in physics-based dictionaries are resonances, which are localized in frequency and can be represented by damped resonant modes [14], [15], [16], and dispersive mechanisms that result from scattering from certain materials [17] or structural geometries [18], [19], [20], [21]. For instance, dispersion has been shown to characterize scattering aircraft engine inlets [22], [23]. Dispersive mechanisms are not localized in time or frequency and are generally modeled by chirp functions with time-dependent instantaneous frequencies. Miller [24] has recently provided a summary of functional forms for a wide variety of scattering mechanisms in both the frequency and time domains.

10.3.2 Data-Based Dictionaries

Analytical, physics-based models for some signal components may not be readily available. Alternately, some signal components may be more compactly described by functions that describe how the signal component is manifest in the data, rather than by functions that are derived from an underlying physical model. In either instance, the functions included in a *data-based* dictionary are chosen to closely model characteristics or signal components in the composite signal being analyzed.

An example of a signal behavior that is difficult to characterize for a wide range of signal sources using physics-based functions is the angular scattering behavior of a general target. Physical models for the angular behavior can be generated using analytical predictions methods; however, they are dependent on a large number of parameters, such as scattering center size, orientation, and visibility. Thus, a very large dictionary of functions is generally required to cover the range of possible angular variations. An alternate approach is to populate the dictionary with a set of functions that model the behavior of the signal components in the data and are parameterized by a small number of data parameters. For the angular scattering example, the nominal aspect and width of the angular scattering lobe would be an appropriate set of parameters.

To characterize angular scattering behavior, Trintinalia et al. [13] have used an adaptive representation, where the angular dependence is represented by Gaussian basis functions with adjustable variances, which characterize the lobes width, i.e., a dictionary element to characterize the scattered field as a function of frequency and angle is given by

$$s_n(f, \theta) = A_n e^{-[(\theta-\theta_n)^2/2\sigma_n^2]} e^{j4\pi fr/c} \tag{10.8}$$

where θ_n locates the center of the lobe and σ_n characterizes the angular width of the lobe.

Other classes of functions such as wavelets, wavelet packets, and symmlets can characterize both narrowband resonances, signal oscillations, and "bumps" within the signal, which are indicative of interference and clutter. An introductory description of these classes functions can be found in [25]. While not tied directly to a specific scattering phenomena, these functions closely describe some characteristics found in radar signals and noise which may make them useful as bases for filtering.

10.4 SOLUTION ALGORITHMS

In the problem of obtaining the atomic decomposition of a signal, $s(f)$, the basic questions are: What are the best sets of atoms to use for the decomposition? What is the best metric and methodology to choose individual atoms from the set that best represent the signal in a sparse manner? Section 10.3 addressed the first question. This section attempts to answer the second question. We will briefly describe a number of algorithms which can be viewed as methods of fitting the dictionary to the data. They, in turn, use various objective function metrics (e.g., L_1 or L_2 norm) and various search strategies to arrive at a "best" solution. These algorithms can be regularized when the signal of interest is in the presence of noise. We refer the reader to, e.g., [26], [27], for references on regularization.

The first technique discussed, the Method of Frames (MOF) [25], does not generally result in a sparse decomposition, but it was one of the first approaches to representing a signal in terms of an overdetermined dictionary. The second technique described, the Best Orthogonal Basis [28], does not decompose a signal onto a general dictionary; however, for signals compactly represented by certain classes of orthogonal functions, it can produce near-optimally sparse representations, which has significant utility for the formulation of forward-scattering problems. We will concentrate on the last three techniques presented: Basis Pursuit (BP) [1], Matching Pursuit (MP) [29], and reweighted minimum norm [30]; all of which achieve the desired result of generating sparse signal decompositions onto general, overdetermined dictionaries.

10.4.1 Method of Frames

One of the first techniques suggested for representing a signal in terms of an overdetermined dictionary was the Method of Frames described by Daubechies [25]. The Method of Frames attempts to solve $\Phi a = s$ by picking a whose coefficients have minimum L_2 norm

$$\min \|a\|_2 \text{ subject to } \Phi a = s \tag{10.9}$$

where $\|a\|_2^2 = \Sigma_{n=1}^{N} |a_n|^2$. The key problem here is that the solution is an average of all possible solutions $\Phi a = s$; so it is typically of very poor sparsity, and also does not super-resolve [1]. This method can also be interpreted as a least squares solution.

10.4.2 Best Orthogonal Basis

Some dictionaries such as wavelet packets and cosine packets have a very special structure, such that certain structured subcollections of the elements amount to orthogonal bases. A method of adaptively picking from among these many bases to provide a single orthogonal basis which is the most sparse has been proposed by Coifman and Wickerhauser [28]. The algorithm in some cases delivers near-optimal sparsity representations. This algorithm has been applied to decomposing classes of operators to sparsify their structure for fast computation. This is of great interest to the forward-scattering community.

The Best Orthogonal Basis method attempts to solve $\Phi a = s$ by picking a whose coefficients have minimum entropy,

$$\min \sum_k a_k \log \frac{1}{a_k} \text{ subject to } \Phi a = s \tag{10.10}$$

The technique begins by constructing a large dictionary, all of whose elements are orthogonal, and structuring it into a tree of subcollections of the basis functions. The inner product of the data with each element of the dictionary is computed, and the entropy of collections of the resulting coefficients is computed. Combinations that minimize entropy are retained. The procedure is repeated until the minimum entropy basis is obtained.

This technique does not work with dictionaries composed of disparate, nonorthogonal functions, such as dictionaries comprised of scattering functions. However, when dictionaries are comprised of elements which qualitatively exhibit the characteristics of signal components and satisfy the structure can be constructed, such as dictionaries composed of wavelets, the Best Orthogonal Basis algorithm can then be applied.

10.4.3 Basis Pursuit

Chen and Donoho [1] have suggested a method of decomposition based on a true global optimization which is at least theoretically feasible, due to recent advances in linear programming. This method is called Basis Pursuit.

Basis pursuit solves $\Phi a = s$ by picking a whose coefficients have minimum L_1 norm, i.e.,

$$\min \|a\|_1 \text{ subject to } \Phi a = s \tag{10.11}$$

where $\|a\|_1 = \Sigma_{n=1}^{N} |a_n|$. Basis Pursuit simply replaces the L_2 norm of the Method of Frames by the L_1 norm. The solution to this problem tends to produce sparse representations. To see this, let

$$s = a_i \phi_i, \qquad s = \sum_k a_k \phi_k \tag{10.12}$$

be two representations of signal s, and $\|\cdot\|$ be the norm for which the dictionary is normalized. We have

$$\|s\| = \|a_i \phi_i\| = |a_i| \, \|\phi_i\| \tag{10.13}$$

$$\|s\| = \| \sum_k a_k \phi_k \| \leq \sum_k \|a_k \phi_k\| = \sum_k |a_k| \, \|\phi_k\| \tag{10.14}$$

Because the dictionary is normalized, we have from (10.13) and (10.14)

$$|a_i| \leq \sum_k |a_k| \tag{10.15}$$

Therefore, the sparse representation has a smaller L_1 norm.

Basis Pursuit starts from an initial decomposition basis (e.g., a basis produced by the Method of Frames) and iteratively improves the basis by replacing atoms in the original basis by other atoms from the dictionary. The flow of the Basis Pursuit algorithm is shown in Figure 10.1. Although this method is also greedy, the theory of linear programming states that solutions based on these methods must converge to a global optimum.

For real signals and dictionaries the problem in (10.11) has an equivalent linear programming formulation [1] and can be solved using well-known methods such as the simplex and interior point methods (see, e.g., Dantzig [31] and Gonzaga [32]). Unfortunately, the above comments do not apply to complex-valued signals and

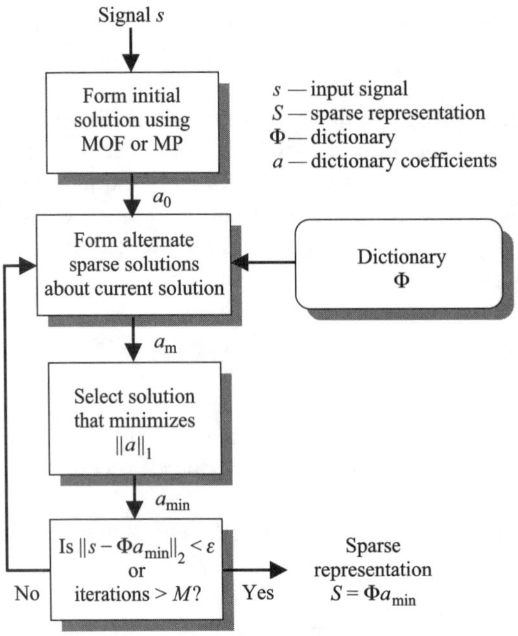

Figure 10.1 Process flow of the Basis Pursuit algorithm showing the use of the L_1 norm to refine the decomposition.

dictionaries. A linear programming formulation is possible with the use of an alternative L_1 norm for which

$$\|a\|_1 = \sum_{n=1}^{N} |\mathscr{R}(a_n)| + |\mathscr{I}(a_n)| \tag{10.16}$$

However, the arguments in (10.13) an (10.14) are not valid in this case, so there is no indication that the solution of the problem in (10.11) will give a sparse decomposition for complex-valued signals.

10.4.3.1 Basis Pursuit Decomposition Example

Figure 10.2 shows an example of decomposition and filtering using the Basis Pursuits algorithm with a merged dictionary. The example was generated using Chen and Donoho's Atomizer software [1]. The figure shows a case where the real-valued input signal is composed of two sinusoids, three resonant responses and additive noise, such that the signal-to-noise ratio (SNR) is 10 dB. The plot in the upper left shows the input signal, and the plot in the lower left shows the decomposition via the Discrete Cosine Transform (DCT) of the signal. The DCT shows the unresolved contributions due to the sinusoids; however, the resonant mechanisms are not evident as localized returns in the range domain.

The noisy input signal was also decomposed onto a merged dictionary consisting of an overly fine cosine basis (to resolve the closely spaced sinusoids) and a Symmlet wavelet packet basis (to resolve the resonant returns) using the Basis Pursuit DeNoising algorithm of Chen and Donoho [1]. The resulting coefficients are shown

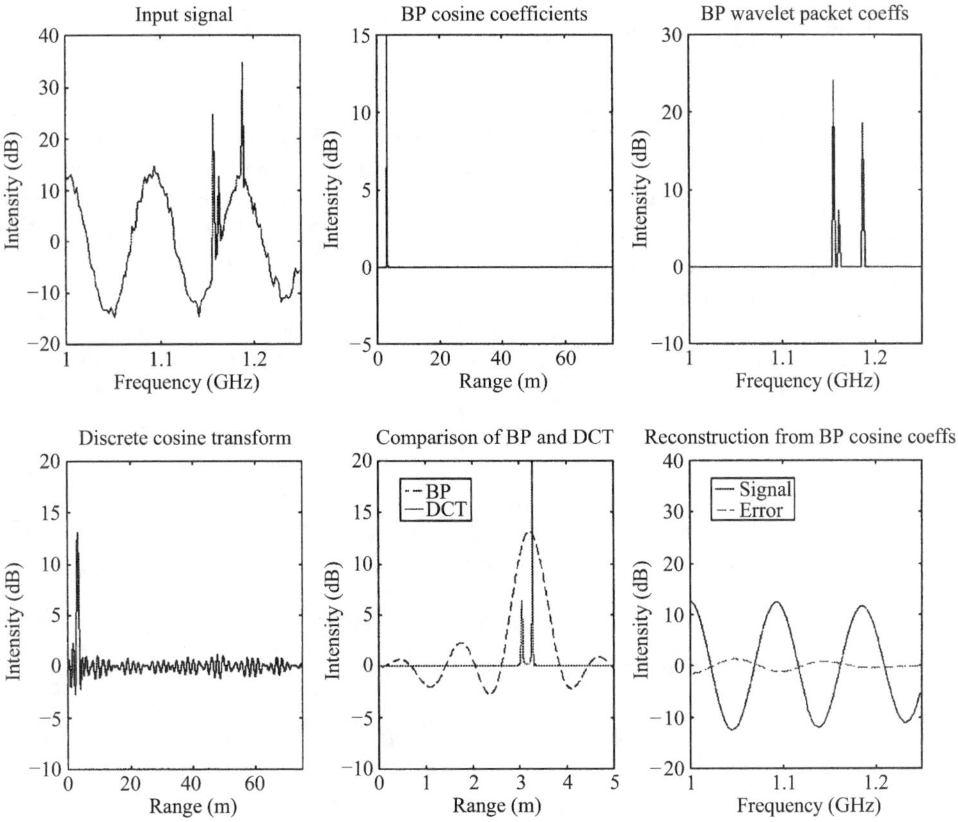

Figure 10.2 Operation of the Basis Pursuits algorithm on real-valued data using a merged dictionary of cosines and wavelet packets.

in the plots to the upper right in the figure, with the cosine coefficients in the center and the wavelet packet coefficients to the right. The cosine coefficients show two isolated peaks corresponding to the two sinusoids, and the wavelet packet coefficients show three strong returns corresponding to the resonances.

The ability of the Basis Pursuit technique to "super-resolve" is demonstrated in the plot in the lower center of the figure where the DCT coefficients and Basis Pursuit cosine coefficients are plotted on an expanded scale. As can be seen, the Basis Pursuit technique allows the sinusoids to be resolved from the narrowband data.

The application of the technique to noise removal is illustrated in the plot to the lower right. The Basis Pursuit DeNoising algorithm inherently removes additive noise in the process of obtaining the expansion coefficients, which uses the "shrinkage" concept of Donoho [33]. If we consider the resonant responses to be undesired "glitches" in the data, we can reconstruct the desired sinusoidal signal from the cosine coefficients. The dotted line shows the very small error in the reconstruction. The examples suggest that the Basis Pursuit decomposition onto appropriately chosen dictionaries could be a technique for automated noise filtering, RFI suppression and scattering mechanism identification.

10.4.4 Matching Pursuit

An algorithm which builds up a sequence of sparse approximations starting from an empty decomposition and builds it in a greedy fashion has been proposed by Mallet and Zhang [29]. This algorithm is called Matching Pursuit. At a given stage p in the construction, that element of the dictionary which best correlates with the residual is added to the approximation. The process is terminated when a desired level of accuracy is achieved. This algorithm has no specific issues with respect to complex-valued signals.

Let $s^{(p)}(f) = \Sigma_{\gamma\in\Gamma^{(p)}} A_\gamma \phi_\gamma(f)$ be the decomposition of $s(f)$ after p steps where $\Gamma^{(p)} \subset \Gamma$ and $\Gamma^{(1)} \subset \Gamma^{(2)} \subset \ldots \subset \Gamma^{(N)} = \Gamma$. In vector form we will describe it as $s^{(p)} = a^{(p)}\Phi^{(p)}$. Also let $R^{(p)}$ be the residual, i.e.,

$$s = s^{(p)} + R^{(p)} \tag{10.17}$$

The algorithm is initialized with $s^{(0)} = 0$ and $R^{(0)} = s$. We choose the dictionary element that best fits $R^{(p)}$ as follows. For a certain element ϕ_n, $n = 1,2, \ldots, N$, of the dictionary we can write the projection of $R^{(p)}$ on ϕ_n as

$$R^{(p)} = \langle R^{(p)},\phi_n \rangle\phi_n + R_n^{(p+1)} \tag{10.18}$$

where $\langle R^{(p)}, \phi_n\rangle$ is the inner product of $R^{(p)}$ and ϕ_n defined for sampled data by

$$\langle R^{(p)}, \phi_n\rangle = \sum_{k=1}^{K} R^{(p)}(k)\phi_n^*(k) \tag{10.19}$$

with * denoting complex conjugation and $R_n^{(p+1)}$ is the residual. Because the projection and the residual are orthogonal we have

$$\begin{aligned}\|R^{(p)}\|^2 &= \|\langle R^{(p)}, \phi_n\rangle \phi_n\|^2 + \|R_n^{(p+1)}\|^2 \\ &= |\langle R^{(p)}, \phi_n\rangle|^2 \|\phi_n\|^2 + \|R_n^{(p+1)}\|^2 \\ &= |\langle R^{(p)}, \phi_n\rangle|^2 + \|R_n^{(p+1)}\|^2\end{aligned} \tag{10.20}$$

where the last equality holds because the dictionary is normalized. Therefore, to minimize $\|R_n^{(p+1)}\|^2$ we need to maximize $|\langle R^{(p)}, \phi_n\rangle|^2$. Let

$$n^* = \arg\max_n |\langle R^{(p)}, \phi_n\rangle|^2 \tag{10.21}$$

Then the best approximation of $R^{(p)}$ is

$$R^{(p)} = \langle R^{(p)}, \phi_{n^*}\rangle\phi_{n^*} + R_{n^*}^{(p+1)} \tag{10.22}$$

so that

$$s = s^{(p+1)} + R^{(p+1)} \tag{10.23}$$

where

$$\begin{aligned}s^{(p+1)} &= s^{(p)} + \langle R^{(p)}, \phi_{n^*}\rangle\phi_{n^*} \\ R^{(p+1)} &= R_{n^*}^{(p+1)}\end{aligned}$$

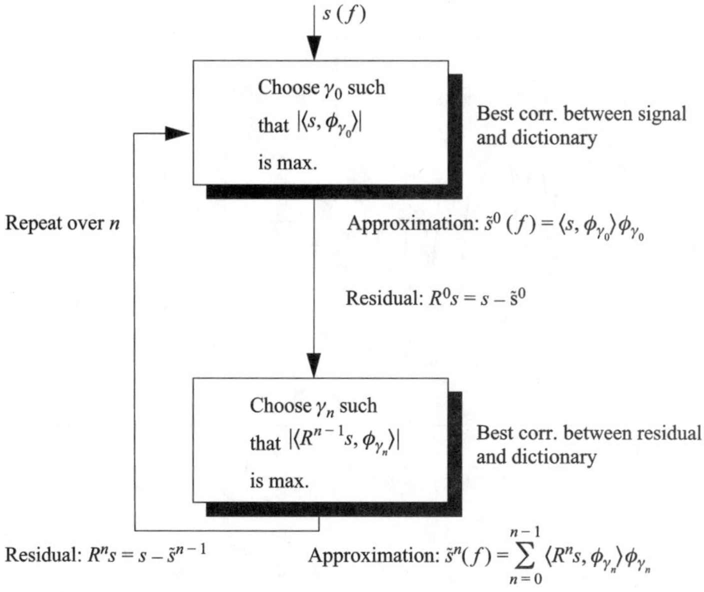

Figure 10.3 Process flow of the Matching Pursuit algorithm showing the successive matched filtering operations of the representation error residual.

For given ε and p^* the algorithm terminates at

$$p = \min(p^*, p_\epsilon) \tag{10.24}$$

where

$$p_\varepsilon = \min\{p : \|R^{(p)}\| \leq \varepsilon \|s\|\} \tag{10.25}$$

A graphical depiction of the algorithm is shown in Fig. 10.3. This algorithm provides a sparse representation by design. However, a potential pitfall occurs because of its greedy nature. The algorithm may spend successive iterations attempting to correct any early mistakes.

10.4.4.1 Matching Pursuit Decomposition Example

Decomposition and identification of diverse scattering mechanisms using Matching Pursuits is demonstrated in the example shown in Fig. 10.4. The intensity of a complex signal comprised of the numerically computed nose-on backscattering from a canonical target, an ogive, added to a resonance-like function of the form $A/(f - f_o)$ is shown to the upper left in the figure. The intensity of the Fourier transform of the signal is shown in the upper right. The resonance-like function is evident in the input signal, but its presence is not obvious in the Fourier transform of the signal. The signal was decomposed using the Matching Pursuit algorithm with a merged dictionary to identify scattering mechanisms using a version of Chen and Donoho's Atomizer software [1], which we modified to include new dictionaries. The merged dictionary consisted of the set of symmlets with eight vanishing moments and a set of GTD-based weighted complex exponentials of the form described in Section 10.3.1.

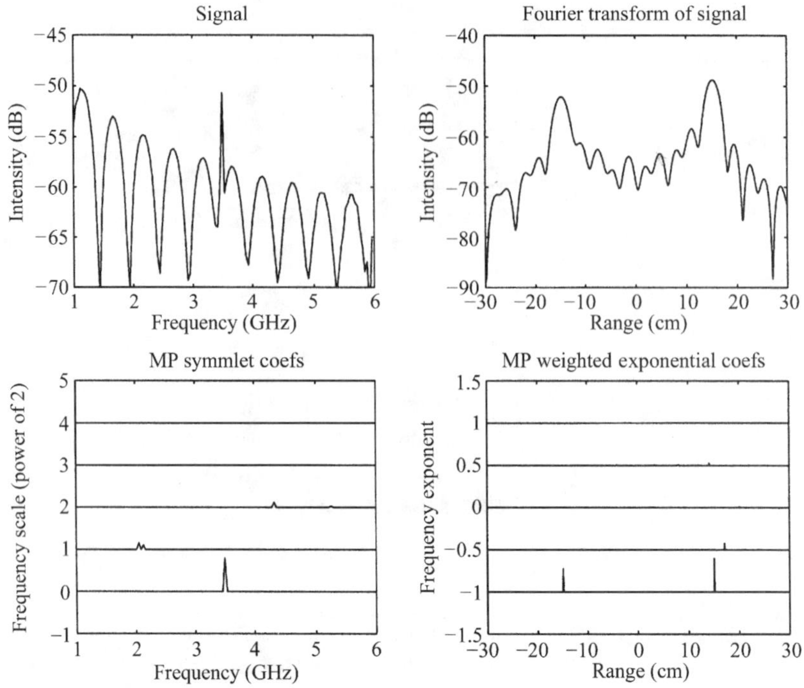

Figure 10.4 Operation of the Matching Pursuits algorithm on complex-valued data using a merged dictionary of symmlets and weighted exponentials.

As seen in the lower two plots in the figure, the symmlet coefficients provide a sparse representation of the resonant behavior and indicate its location in the frequency domain, and the coefficients of the weighted exponentials identify the location in range and intrinsic frequency dependence of localized scattering centers in data. Thus, the Matching Pursuit decomposition provides a sparse representation of relevant scattering mechanisms. Alternately, if we consider the resonant behavior to be measurement contamination, the signal can be filtered to recover the desired target response by reconstructing the signal using only the appropriate portion of the adaptive decomposition, i.e., the weighted exponentials.

10.4.5 Reweighted Minimum Norm

The reweighted minimum norm method of Gorodnitsky and Rao [30], which they refer to as the FOCal Underdetermined System Solver (FOCUSS), has two integral parts: a low-resolution initial estimate of the decomposition (which is equivalent to the Method of Frames result) followed by an iteration process that refines the initial estimate to the final localized energy solution. The iterations are based on weighted norm minimization of the dependent variable with the weights being a function of the preceding iterative solutions. Figure 10.5 shows flowchart of reweighted minimum norm method.

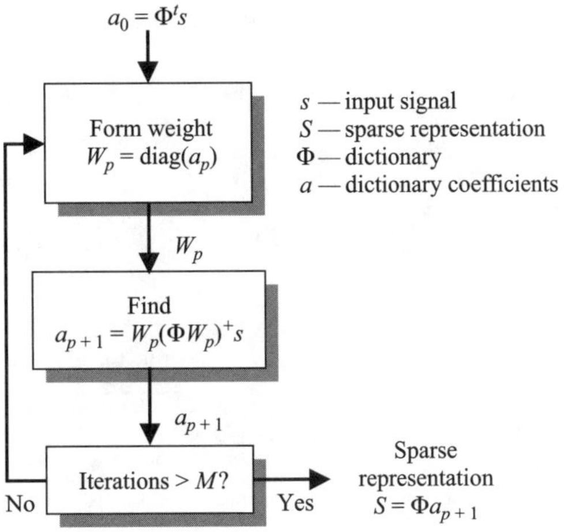

Figure 10.5 Process flow of the reweighted minimum norm algorithm showing the successive refining of the chosen dictionary elements through a refinement of the weighting on the elements.

FOCUSS attempts to solve $\Phi a = s$ by picking a whose coefficients have a minimum weighted L_2 norm.

$$\min \| W^+ a \|_2 \text{ subject to } \Phi a = s \tag{10.26}$$

where $A^+ = A^H (AA^H)^{-1}$ denotes the Moore-Penrose inverse. The solution is given by

$$a = W(\Phi W)^+ s \tag{10.27}$$

The FOCUSS algorithm is iterative and has a three-step form

$$\begin{aligned} &\text{Step 1:} \quad W_{pk} = (\text{diag}(a_{k-1})) \\ &\text{Step 2:} \quad q_k = (\Phi W_{pk}) + s \\ &\text{Step 3:} \quad a_k = W_{pk} q_k \end{aligned}$$

The iterative application of the weighting prunes the solution space (the admitted dictionary elements). To see this, the objective function can be expanded as

$$\| W^+ a \| = \sum_{i=1, w_i \neq 0}^{n} \left(\frac{a_i}{w_i} \right)^2 \tag{10.28}$$

Relatively large entries in W require corresponding large a_i to make a contribution. Larger a_{k-1} are reinforced if the respective columns in Φ are significant in fitting s. Smaller values of the solution are then gradually suppressed.

Gorodnitsky and Rao view the algorithm as a novel optimization method which combines desirable characteristics of both classical optimization and learning-based algorithms. Tikhonov or truncated SVD regularization techniques can be used to improve the solution when the minimum norm problem is ill conditioned or in the presence of significant noise.

10.4.5.1 Reweighted Minimum Norm Decomposition Example

Decomposition and identification of diverse signal components using the reweighted minimum norm (FOCUSS) method is demonstrated in Fig. 10.6. A complex signal was constructed by adding a step function $AU(f - f_o)$ to the

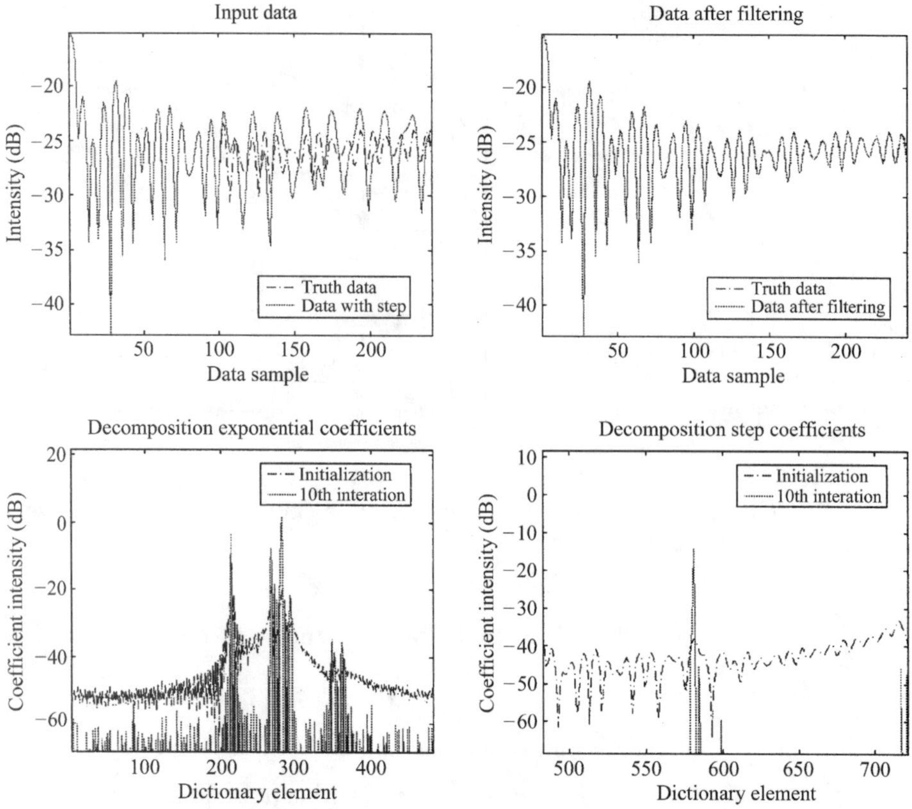

Figure 10.6 Operation of the reweighted minimum norm (FOCUSS) algorithm on complex-valued data using a merged dictionary of overfine exponentials and steps (Walsh functions). The step can be filtered from the data by simply reconstructing the frequency response from the exponential coefficients. The result is shown in the plot in the upper right of the figure, where the intensity of the filtered data is shown to compare very well to that of the truth.

numerically computed backscattering as a function of frequency from a cylinder illuminated end-on. The data simulate a measurement situation where the frequency response of the target over a wide bandwidth has been assembled by combining measurements made independently over two adjacent frequency bands. The step represents an additive bias one measurement band has relative to the other band, which is an error term that needs to be identified and removed.

The intensity of the cylinder backscattering (referred to as the Truth Data) with and without the step is shown in the plot in the upper left of the figure. As can be seen, the step perturbs the signal when it turns on after data sample 100. The data with the step were decomposed using FOCUSS onto a merged dictionary consisting of a set of overfine complex exponentials and a set of Walsh (step) functions. The lower two plots in the figure show the intensity of the dictionary coefficients determined by FOCUSS after the initial minimum norm fit of the data to the dictionary and after 10 iterations of the algorithm. The exponential coefficients are shown to the left, and the step coefficients are shown to the right.

The initial minimum norm fit provides a non-sparse representation of the signal. The initial set of exponential coefficients is roughly equivalent to the result that would be obtained by performing an upsampled Fourier Transform. Energy is concentrated in dictionary elements corresponding to scattering from the ends of the target and multiple interactions between the ends. Target scattering mechanisms can be resolved, but the range profile has significant sidelobes. The initial set of step basis function coefficients show energy spread over all dictionary elements. The initial decomposition could be used to filter the step response, but it provides little information on the strength or location of the step.

The decomposition after 10 iterations provides significantly more information on signal components. Energy in the exponential components is much more localized, improving spatial resolution and identification of target scattering centers. The energy in the step coefficients has now been concentrated into approximately one dictionary element that correctly indicates when the step occurs.

10.5 APPLICATIONS

The novel methods of signal decomposition discussed in the previous sections have numerous potential application in electromagnetics. In this section we will illustrate their use in three such applications by showing example results and providing the reader with references to examine these topics in more detail. The first application will be in the area of inverse scattering, specifically in the interpretation of radar data and synthetic aperture radar images. Here we will examine the ability of the adaptive decompositions to estimate not only the location of the scattering mechanism but also the type of scattering that is present.

The second area is data filtering. Sparse decompositions, by their very nature, attempt to compress "useful" energy into a few isolated, easily identifiable coefficients and spread additive Gaussian noise as low-level energy over the entire set of basis elements. Specific signal components that represent contamination can be isolated and, if appropriate, eliminated from the signal. Noise energy can be reduced by thresholding low-level coefficients in the manner discussed by Donoho [33].

The third application is in the area of forward-scattering, where adaptive decomposition has been recently proposed as a means of improving the computational efficiency of numerical electromagnetic scattering predictions. By looking upon the computation of the forward-scattering from a complicated object as an operator on a signal, methods that sparsify the operator significantly reduce computation time.

A fourth area that has received significant attention is application of adaptive decomposition techniques to the processing of range-Doppler or inverse synthetic aperture radar (ISAR) imagery. Adaptive time-frequency processing has been used to compensate for image blurring due to uncompensated target motion during the data collection. This application will not be discussed in detail due to space limitation, but the interested reader is referred to the recent works of Chen [34], Trinitalia and Ling [35], Chen and Qian [36], and Wang et al. [37].

10.5.1 Scattering Decomposition for Inverse Problems

Adaptive decompositions can be used to extract more information from measured data in inverse scattering problems. As a specific example, adaptive decomposition can be used to estimate the position of and the type of scattering mechanisms found in radar data. Classic range compression of radar data has no ability to directly classify scatterer type; it only estimates range position. Range compression, obtained by taking the Fourier transform of frequency diverse data, is essentially a decomposition of the radar data onto a single basis set, the set of uniform amplitude, complex exponentials, which inherently implies that all target scattering returns can be compactly modeled as point scatterers (e.g., the amplitude of an individual scattering center is constant over the data collection). With this model only the location and relative strength of scattering centers can be determined. In addition, scattering mechanisms that depart significantly from the point scatterer model will result in a non-sparse decomposition, i.e., the corresponding return will be smeared in the compressed data. Adaptive techniques can provide more information by allowing the radar data to be decomposed onto dictionaries that describe frequency-dependent scattering behavior, which allows better isolation and identification of target mechanisms.

We will show the utility of adaptive decomposition for identifying scattering mechanisms by applying these techniques to scalar and polarimetric radar data. The dictionary that we will use is the GTD-based, weighted exponential dictionary described in Section 10.3.1 and its generalization to polarimetric data. The adaptive decomposition algorithms that we will concentrate on are Matching Pursuits and the FOCUSS algorithms. Note parametric spectral estimation algorithms, such as the recent work of Potter et al. [8] can also be used to estimate scatterer type; but the formulation is not easily generalized to include scattering mechanisms, such as resonances, that are not compactly represented by the GTD-based set of functions.

10.5.1.1 Identification of Scattering Centers in Range Profiles

To evaluate the performance of the algorithm on well-characterized, experimentally measured data, the Matching Pursuit algorithm was applied to the backscattering from a simple array of canonical targets, which was collected in ERIM International's compact radar measurement range from 12 to 18 GHz. The resulting data were decomposed using the Matching Pursuit algorithm with the GTD-based dictionary. Table 10.1 shows the different scattering mechanisms that can be determined using frequency dependence.

TABLE 10.1 Types of Scattering Mechanisms which can be Differentiated with Frequency Dependence

α	Scattering Mechanism
1	Trihedral, dihedral, flat plate
1/2	Singly curved surface (e.g., cylinder)
0	Straight edge specular, doubly curved surface (e.g., sphere)
$-1/2$	Curved edge diffraction
-1	Corner diffraction

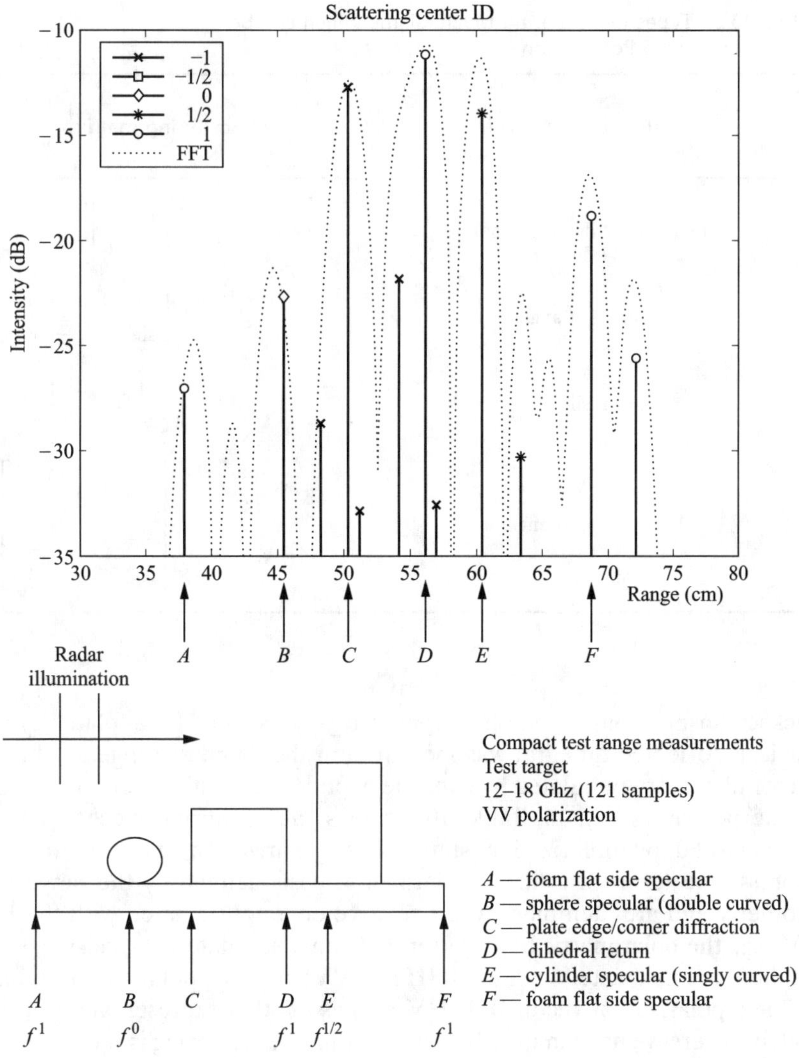

Figure 10.7 Scattering center decomposition of compact range measurements generated using Matching Pursuits with a GTD-based, weighted exponential dictionary.

The results are shown in Fig. 10.7, along with a sketch of the target array. The figure shows the intensity of the Fourier transform of the test signal, along with amplitude, spatial location, and type of the dictionary elements selected by the Matching Pursuit algorithm. The scatterer type is indicated by the data marker. As can be seen, the algorithm accurately locates and predicts the type of the most significant scattering mechanisms. Range compression using Fourier basis elements only provides location information.

To examine the sensitivity of the Matching Pursuit algorithm to the frequency bandwidth, and thus the intrinsic spatial resolution, of the scattering data and to evaluate the utility of using a vector dictionary to decompose polarimetric data, the

TABLE 10.2 Types of Scattering Mechanisms which can be Differentiated with Polarization

Scattering mechanism	Scattering matrix $\begin{bmatrix} HH & HV \\ VH & VV \end{bmatrix}$
Flat plate, trihedral, sphere	$\begin{bmatrix} 1 & 0 \\ 0 & 1 \end{bmatrix}$
Dihedral at angle θ	$\begin{bmatrix} \cos 2\theta & \sin 2\theta \\ \sin 2\theta & -\cos 2\theta \end{bmatrix}$
Helix	$\frac{1}{2}\begin{bmatrix} 1 & \pm i \\ \pm i & -1 \end{bmatrix}$
Linear wire at angle θ	$\begin{bmatrix} \cos^2 \theta & \frac{\sin 2\theta}{2} \\ \frac{\sin 2\theta}{2} & \sin^2 \theta \end{bmatrix}$

backscattering from a simple target array consisting of a plate, cylinder, and a vertically oriented dihedral was measured in the compact range. In this case we fit a vector dictionary populated by overfine exponentials with different polarization states to the polarimetric data. Table 10.2 shows the scattering mechanisms that can be distinguished by polarization state. Note polarization states in these results are expressed in terms of a linear polarization basis defined by the range measurement geometry and are identified by the relative energy associated with the HH, HV, and VV, e.g., the polarimetric backscattering from a flat plate at broadside, a trihedral and a sphere are associated with the [HH HV VV] = [1 0 1] polarization state. Thus, let us define a polarization vector [HH HV VV] as another representation of the scattering matrix, where we have implicitly assumed that the scattering is reciprocal such that the HV return is equal to the VH return.

The canonical objects were offset in range by 10 cm (plate/cylinder) and 20 cm (cylinder/dihedral). Range profiles using a Fourier basis were processed using frequency bandwidths consistent with range resolutions of 2.5 cm, 12 cm, 16 cm, and 24 cm. The Matching Pursuit algorithm was applied to the same data, and the results are summarized in Fig. 10.8. With the Matching Pursuits algorithm, estimates of position and type remain stable out to resolutions of 16 cm where the plate/cylinder combination is unresolved using the Fourier basis.

Our next example in Fig. 10.9 shows the additional information that can be obtained of using a more detailed dictionary. Here we've defined a dictionary with elements that have GTD-based frequency dependencies and polarization states. This combination allows the discrimination of a richer set of scattering mechanisms. The algorithm was applied to a simulation of the backscattering from a sphere, plate, and a dihedral. Note that the sphere and plate, which have the same polarization state, can now be separated when we include the frequency dependence in the dictionary.

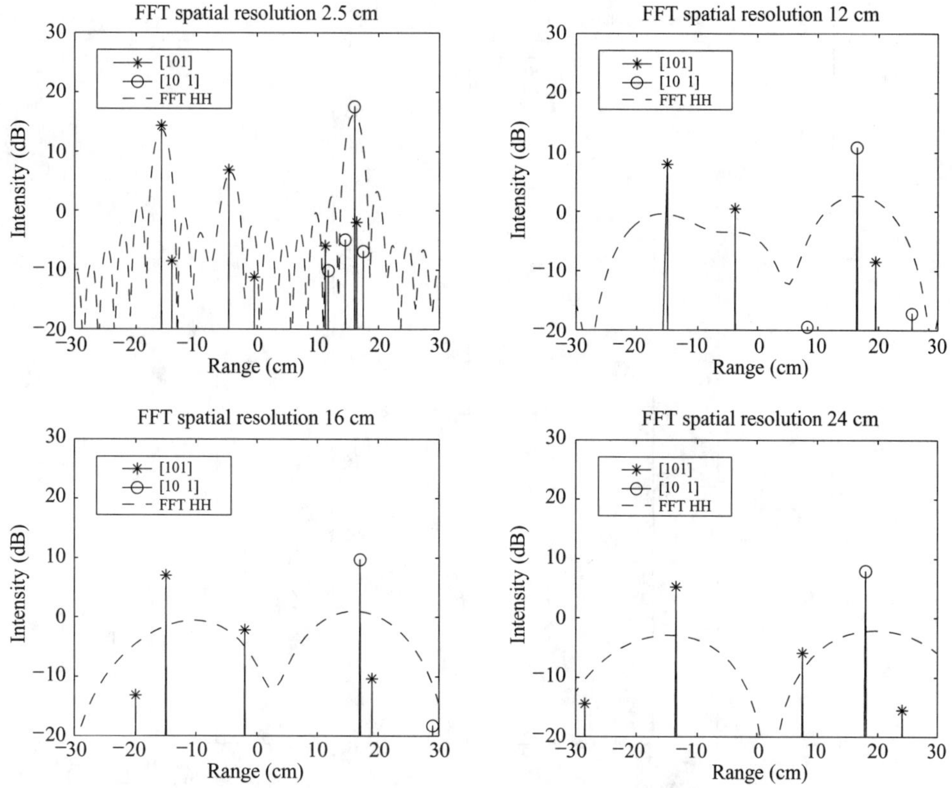

Figure 10.8 Decomposition of compact range data using the Matching Pursuits algorithm and a polarimetric dictionary indexed on resolution.

10.5.1.2 Identification of Scattering Centers in SAR Imagery

The use of adaptive decomposition techniques to identify scatterer types in synthetic aperture radar (SAR) imagery has been investigated. Application of parametric decompositions to SAR data requires some further processing. The reader is referred to Walker [38] or Carrara et al. [39] for a more in-depth discussion of SAR image formation and processing. If phase history data are available, the data are compressed (Fourier transformed) in azimuth only, and the adaptive decomposition applied to each frequency sweep. If complex SAR imagery is available, the data would first be uncompressed in range, and, again, the adaptive decomposition applied to each frequency sweep. Alternatively, a full 2-dimensional formulation of the scattering mechanisms can be formulated (see, e.g., [40]) and used as the dictionary elements. Here the decomposition would then be applied to the data without any azimuth compression.

To show the effect of the adaptive decomposition on SAR imagery, we constructed a simple canonical target consisting of a dihedral and a cylinder separated by approximately 19 cm in range, as shown in Fig. 10.10. The polarimetric scattering from the target was measured in ERIM International's compact range, and the resulting data were processed in the manner described above.

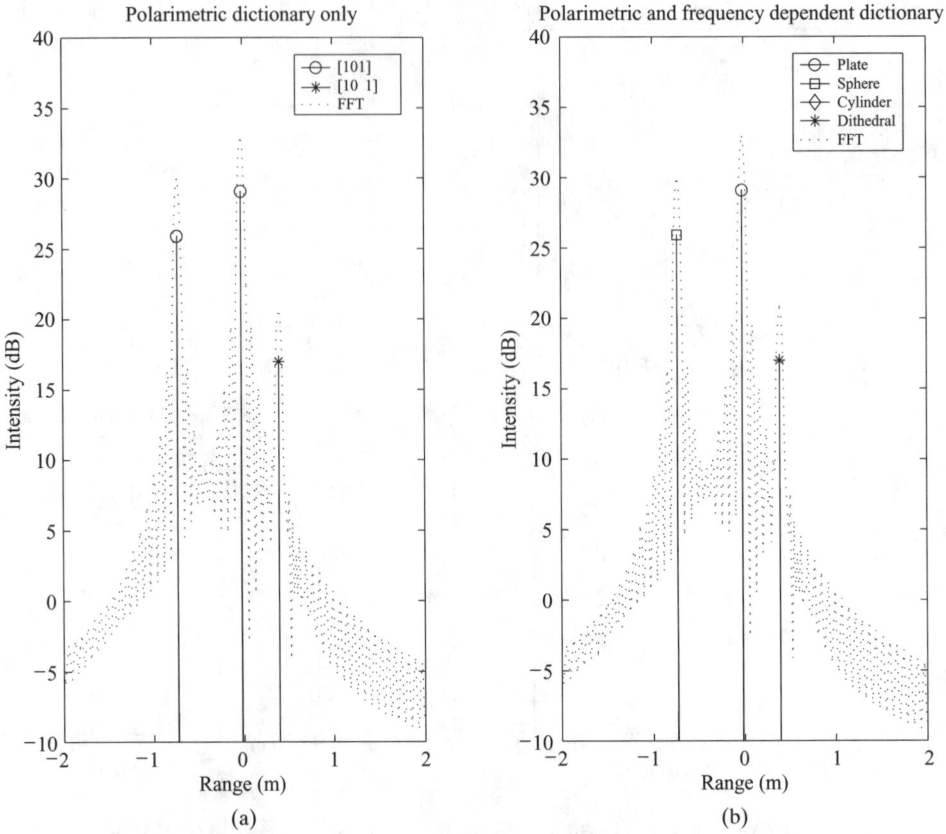

Figure 10.9 Decomposition of simulated data using the Matching Pursuits algorithm and (a) polarimetric dictionary and (b) polarimetric and frequency-dependent dictionary.

Figure 10.10 Object used in SAR image scattering decomposition studies.

Figure 10.11 shows the results of a number of decomposition methods. The data were processed using frequency bandwidth and an angular aperture sufficient to produce a 4-cm resolution in range and cross-range. Figure 10.11(a) shows the result of a Fourier decomposition of each channel of the polarimetric SAR data when the target was oriented such that the cylinder was broadside to the radar illumination. For display, we project the data onto specific polarimetric decompositions. The range direction is vertical in the image, cross-range is horizontal, and the radar illumination

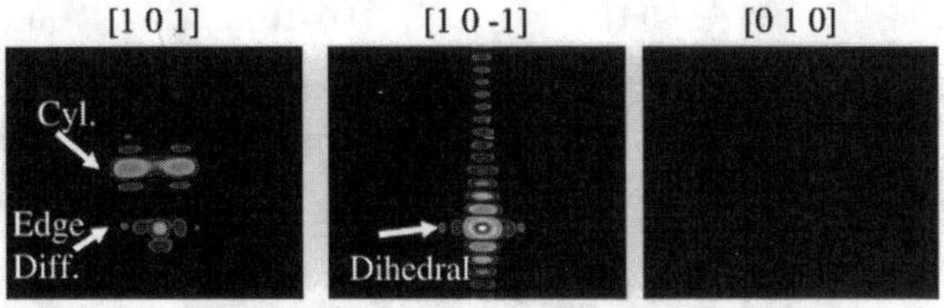

(a) Decomposition based on Fourier transform

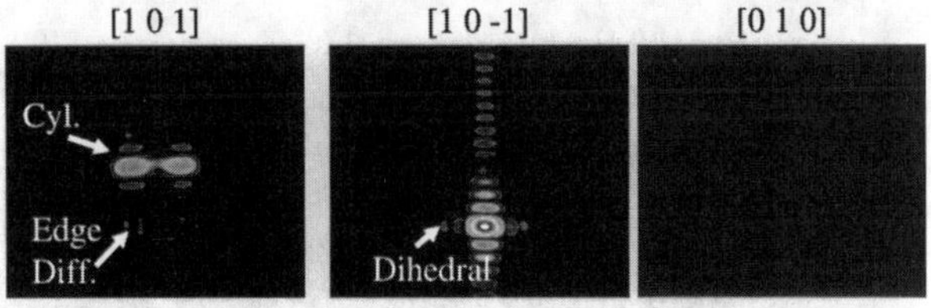

(b) Decomposition based on Maximum Likelihood algorithm

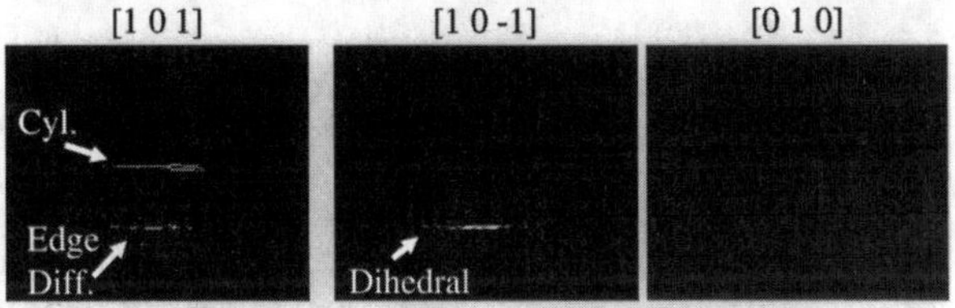

(c) Decomposition based on FOCUSS algorithm

Figure 10.11 Radar imagery with 4-cm-range resolution showing scattering mechanisms using (a) a simple polarimetric Fourier decomposition, (b) a Maximum Likelihood polarimetric decomposition, and (c) an adaptive decomposition algorithm generated using FOCUSS with a polarimetric dictionary.

is from the bottom. The [1 0 1] decomposition shows returns indicative of plates, trihedrals, spheres, cylinders, and edge diffractions. The [1 0 −1] decomposition is indicative of dihedrals whose axis is oriented along either the horizontal or vertical axis of the linear polarization basis. The [0 1 0] channel shows returns indicative of scatterers producing significant cross-polarized scattering, e.g., a dihedral whose axis is oriented at 45° to the polarization basis axes. Scattering from the front surface of the cylinder and the edges of the dihedral corner reflector is evident in the [1 0 1] image. The strong dihedral return from the corner reflector is evident in the [1 0 −1] image. Note no energy is evident in the [0 1 0] channel since there are no strong scattering mechanisms which support this decomposition at this particular target orientation.

Figure 10.11(b) shows the result from a Maximum Likelihood decomposition of the polarimetric signal. The data are projected onto the competing dictionary elements, and the dictionary element with the largest projection is chosen as indicative of the scattering type at that location. We again display the data decomposed into channels indicative of the various scattering mechanisms. Notice that the "winner take all" nature of the Maximum Likelihood algorithm has suppressed the edge diffraction return in the data.

Last, Fig. 10.11(c) shows the result of applying the FOCUSS adaptive decomposition algorithm with a polarimetric dictionary. Each scattering mechanism is clearly

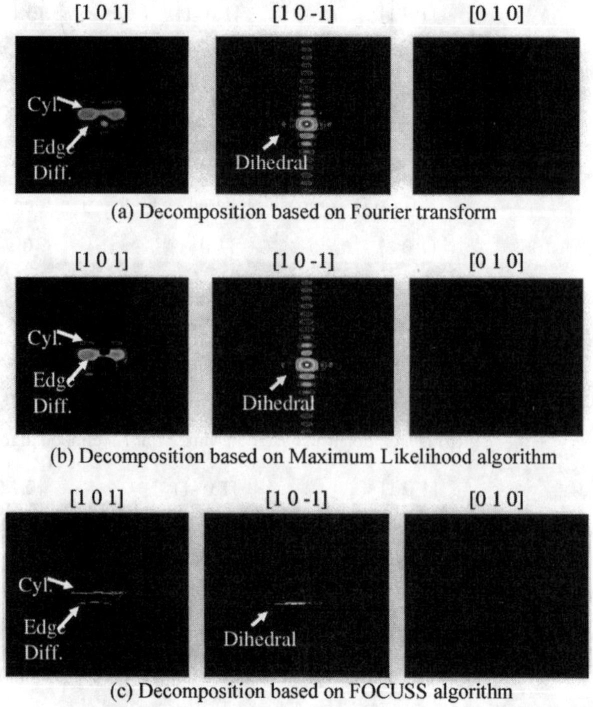

Figure 10.12 Radar imagery with 16-cm-range resolution showing scattering mechanisms using (a) a simple polarimetric Fourier decomposition, (b) a Maximum Likelihood polarimetric decomposition, and (c) an adaptive decomposition algorithm generated using FOCUSS with a polarimetric dictionary.

visible. Also note that the localization of the mechanisms in range is better than that for the Fourier or Maximum likelihood decompositions due to the lack of sidelobes and finer localization of the scattering mechanisms.

Figure 10.12 shows similar decompositions of the same target with a reduced frequency bandwidth. Here the equivalent range resolution is 16 cm. The angular aperture remained the same as previous, so the cross-range resolution was still 4 cm. Note that the edge diffraction and cylinder returns are unresolved in the Fourier and Maximum likelihood decompositions. The FOCUSS algorithm, however, correctly types the various scattering mechanisms and also has the ability to spatially separate the various mechanisms. This is a significant improvement over the previous two methods. The adaptive algorithm provides scatterer-type information and enhanced spatial resolution.

10.5.2 Decompositions for Data Filtering

The data filtering application is distinct from inverse scattering applications in that the objective is to construct a modified version of the signal by modifying the decomposition of the signal. Specifically, the signal is first decomposed onto a dictionary. The decomposition is then modified to affect some change in the data, such as scattering mechanism extraction or removal of measurement contamination, and, finally, the modified decomposition is used to generate the altered version of the signal. With inverse scattering the objective is simply to identify signal components from the decomposition, and the processing stops with the decomposition. Note that the data postprocessing application requires that the decomposition be generated in a manner that allows the signal to be faithfully reconstructed.

Notional examples of mechanism identification and filtering using Matching Pursuits, Basis Pursuits, and FOCUSS were shown previously in Section 10.4. In each instance, the adaptive algorithm was used to produce a sparse decomposition that appropriately allocated signal energy to elements of a diverse dictionary which was chosen to represent desired signal components and contamination. The contamination was effectively removed by performing the reconstruction using only those dictionary elements corresponding to the desired signal. The diverse dictionary allows the identification of disparate mechanisms. In some instances, it is not possible to isolate an unwanted signal component in decompositions generated using only a signal class of functions. A simple example is that a very narrow response may not appear as a localized return in a Fourier transform. Sparse decompositions facilitate filtering by allowing signal components to be more easily isolated and identified.

10.5.2.1 Measurement Contamination Mitigation

As a more practical example, adaptive decomposition techniques have been applied to the filtering of radar data, with the specific objective of removing target support contamination from radar cross-section measurements.

Modern nonlinear spectral estimation techniques rely on damped exponential signal models to increase resolution significantly beyond the levels possible via linear Fourier analysis. When the model is valid, it is well known that parametric variants provide the accurate estimates of the amplitude, phase, and location of scatterers. References [41], [42] developed and demonstrated parametric signal history editing (PSHE) techniques, based on these estimators, for removing additive target support contamination from narrowband measurements. These techniques allow target and support returns to be extracted from frequency sweep data with much greater accuracy and resolution than that afforded by conventional Fourier techniques. These algorithms, however, are formulated to decompose the radar signal with only a single class of functions, namely damped exponentials, which limits the range scattering mechanisms that can be compactly represented. In some measurement situations, the performance of these techniques can degrade because of low signal-to-noise ratios, the presence of data artifacts, and mismatch between the target scattering behavior and the underlying signal model.

To overcome these limitations, the utility of adaptive decomposition techniques to identify, extract, and reconstruct scattering mechanisms from radar measurements was investigated. For the examples considered, an overfine set of weighted complex exponentials were used for the dictionary. The solution algorithm used in the mitigation algorithm was the reweighted minimum norm method (FOCUSS) algorithm described in Section 10.4 since it has the best ability to resolve closely spaced scatterers.

Target Support Contamination of Scattering Measurements. Numerical simulations of scattering from a canonical target, perfectly conducting ogive, on a perfectly conducting, ogival cross-sectional pylon were generated to characterize target-pylon interactions [43] and to test PSHE algorithms. The axis of the pylon was tilted at 60° from horizontal. A flush junction was modeled between the target and pylon. The numerical simulations provide a controlled test of algorithms with data containing realistic scattering mechanisms. As shown in Fig. 10.13, the basic target and pylon scattering mechanisms were identified in range profiles formed from the multifrequency data. A sketch of the target-support geometry is provided, and the spatial locations of significant scattering mechanisms are labeled. The problem under consideration is to remove the contamination associated with the pylon returns from data collected over a much narrower frequency bandwidth where Fourier techniques cannot provide isolation of the individual scattering mechanisms. The mitigation example will concentrate on removing the pylon from the VV polarization data.

At VV polarization, the isolated pylon supports a surface wave across its top. Discontinuities at the leading and trailing edges of the pylon tip cause energy to be scattered back to the radar. The pylon appears to have nearly equal intensity scatterers at each of its top two corners. The addition of a conducting flush-mounted target eliminates the return from the trailing edge of the tip, and the return from the

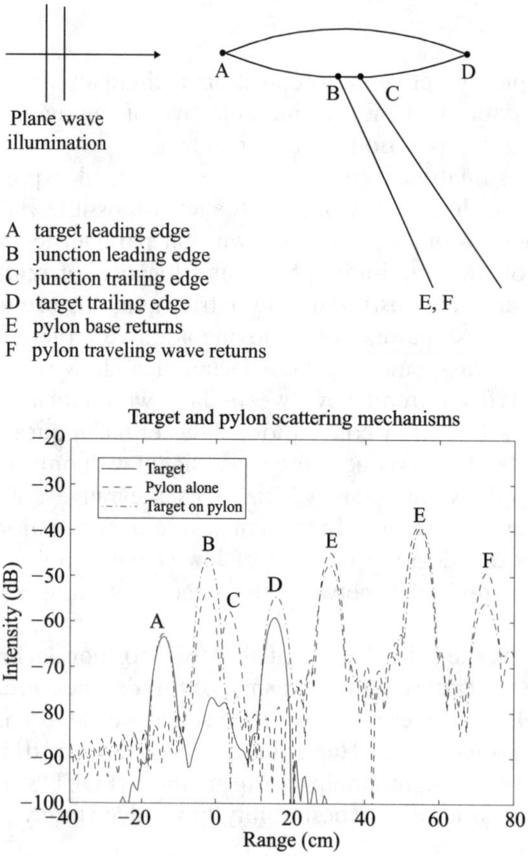

Figure 10.13 Range profiles of ogive termination on pylon illustrating target and support scattering mechanisms.

leading edge of the tip is significantly enhanced due to indirect illumination due to bistatic scattering from the target. The target range profile at VV polarization shows scattering from the leading and trailing edges of ogive. The return from the leading tip of the ogive is not affected by the pylon, but the return from the trailing tip is perturbed by the pylon.

The PSHE algorithms are designed to reduce additive measurement clutter. They do not compensate for perturbative target-support interactions that produce multiplicative modifications of target scattering. To provide a fair evaluation of the algorithm's performance, we need to separate target measurement errors due to additive measurement contamination from those due to perturbative effects.

Before the PSHE algorithms were applied to the VV polarization data, two observations were made. First, the profiles of the target-pylon combination show large returns associated with the bottom corners of the pylon and returns that can be attributed to traveling waves excited on the pylon. The returns from the bottom of the pylon and the traveling wave returns are usually not significant in an actual measurement situation since the bottom of an actual pylon is normally not visible to the radar and the traveling wave returns are usually out of the range gate on a long pylon. Thus, the scattering associated with the bottom of the pylon and the traveling waves were removed from the target-pylon data using coarse resolution range gating prior to processing. The second observation is that the pylon support produces a significant perturbation of the return associated with the rear tip of the target for VV polarization. Consequently, the extracted target response will not match the free-space target response even if all the additive measurement contamination is removed. To evaluate the ability of the PSHE algorithms to eliminate additive clutter contamination, the target response extracted from the full bandwidth target-pylon data will be referred to as the "Truth."

Numerical Demonstration of Editing with Adaptive Decomposition. To test the ability of the adaptive technique is to remove large amounts of additive measurement clutter in narrow bandwidth measurement situations; the algorithm was applied to the VV polarization ogive-on-pylon numerical data over a narrow frequency range. In addition, additive noise was added to the data to simulate a 20 dB SNR. For comparison purposes, our damped exponential total least squares-ESPRIT (TLS-ESPRIT) PSHE algorithm described in [42] was applied to the same data. TLS-ESPRIT is a parametric spectral estimation technique that generates a damped exponential model for the data [44], [45]. With this bandwidth, the target response cannot be reliably isolated and extracted from the target-pylon data using Fourier-based editing techniques.

The narrowband decomposition and editing performance obtained with the adaptive technique are compared to that obtained using the TLS-ESPRIT algorithm in Fig. 10.14. The signal decomposition shown in the upper left shows that FOCUSS produces a range profile that is much more resolved than the FFT and localizes scattering mechanisms at nearly the same locations as the TLS-ESPRIT tones. The target response is extracted using knowledge of the spatial locations of target scattering mechanisms. The target RCS before and after editing is shown to the lower left, where the RCS of the computed target-pylon data (labeled Measurement), the target extracted using FOCUSS (labeled FOCUSS PSHE), the target extracted using damped exponential TLS-ES-PRIT (labeled TLS-ESPRIT PSHE), and the truth

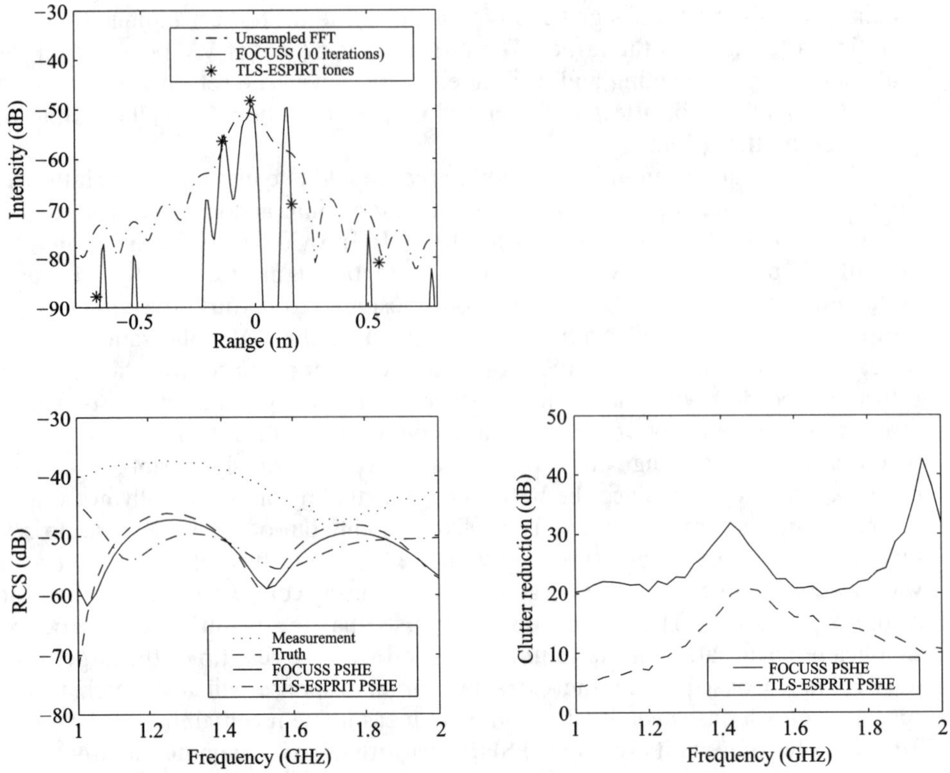

Figure 10.14 Comparison of narrowband PSHE contamination mitigation performance obtained using FOCUSS and TLS-ESPRIT decompositions.

described above are given. The additive clutter reduction for both the algorithms is quantified to the lower right. TLS-ESPRIT was shown in [42] to provide excellent editing results on the same data with high SNR. The significant observation here is that PSHE with FOCUSS provides excellent editing results even at moderate SNR; thus, the adaptive technique provides a potentially more robust extraction method than the traditional spectral techniques.

Experimental Demonstration of Editing with Adaptive Decomposition. To test the ability of the adaptive technique is to remove large amounts of additive measurement contamination in narrow bandwidth experimental data [46], the algorithm was applied to the VV polarization compact range measurements of a canonical target, a metal ogive, terminating a metal, ogival cross-sectional pylon support, as shown in Fig. 10.15. The ogive was supported by dielectric strings and just rested on the pylon tip to simulate a flush target pylon junction. The FOCUSS algorithm with an overfine complex exponential dictionary was applied to nose-on backscattering data over a frequency bandwidth that would provide roughly two Fourier resolution cells on the target, making conventional range gating very difficult.

The range profile generated by taking the Fourier transform of the frequency diverse data is shown in plot to the left in Fig. 10.16, along with the decomposition

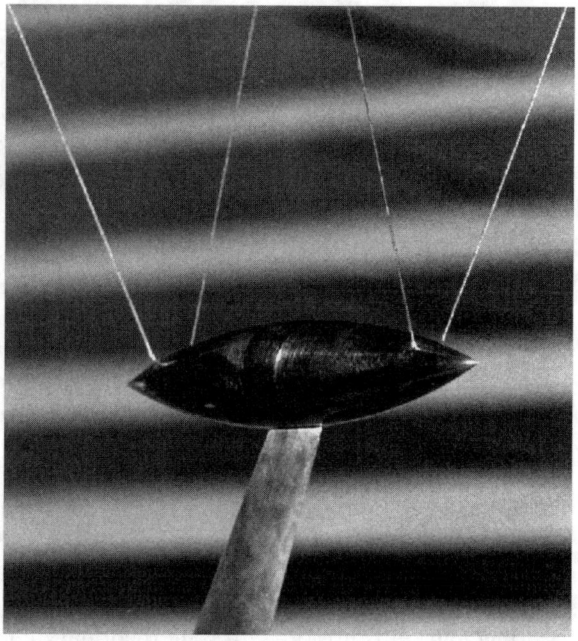

Figure 10.15 Photo of canonical target terminating pylon support in the compact range. Dielectric strings are used to support the target.

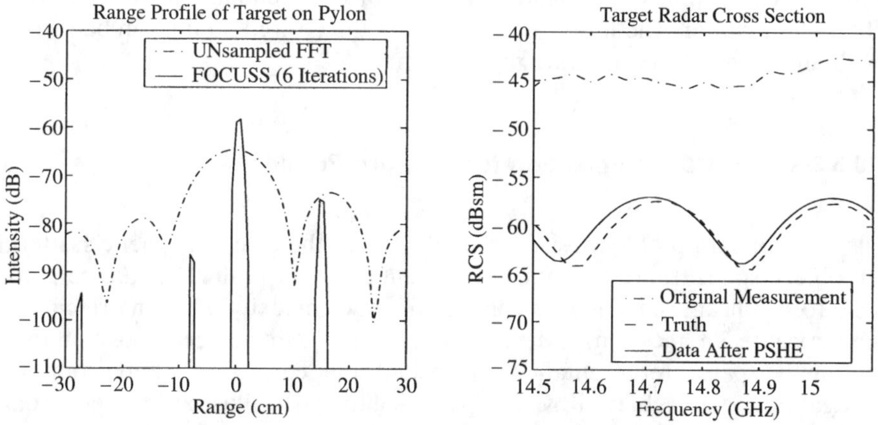

Figure 10.16 Example of narrowband PSHE contamination mitigation performance obtained using FOCUSS on compact range data.

obtained using the adaptive technique. As can be seen, the FOCUSS results are much more isolated and easier to identify. The return near the center is associated with the pylon and removed. The reconstruction of the target response is compared to the original measurement and a numerical prediction of the true free-space target scattering (labeled "truth") in the plot to the right. The agreement between the filtered data and the truth is very good, indicating that the adaptive technique can provide robust filtering of measurement contamination.

Adaptive Editing in Presence of Dissimilar Mechanisms. The examples discussed above emphasized algorithm performance in the presence of noise, but the adaptive PSHE technique can also improve mitigation performance in the presence of data artifacts and model mismatch by performing the signal decomposition with a very general dictionary. This is demonstrated in Fig. 10.17, where a scattering mechanism that was bandlimited in frequency was added to the numerical prediction for the target on the pylon. The input data over a larger frequency bandwidth (1.5–3.5 GHz) is shown in the plot to the upper left. Note the bump near 2.1 GHz that shows the effect of the bandlimited response. The decomposition dictionary was augmented to include a set of exponentials that had limited frequency extent. The Fourier transform is shown to the upper right, and the presence of the bandlimited mechanism is not obvious. The dictionary coefficients obtained using FOCUSS after the first and 20th iterations are shown in the middle plots. The isolation of the target, support, and bandlimited responses is evident. These decompositions can now be used to reconstruct the signal. As above, we remove the pylon support return, and we can either keep or remove the bandlimited response depending on whether we consider it target or clutter. The lower plots show that the desired output can be reconstructed with and without the disparate mechanism.

10.5.3 Current Decomposition for Forward Problems

For forward problems, such as prediction of the electromagnetic scattering from an object, sparse representations of the unknown, typically the surface current, can lead to significant reductions in computational complexity. Thus, in principle, efficient numerical techniques can be developed by using adaptive decomposition to compute optimally sparse representations, provided the cost of decomposition does not mitigate the gain achieved by the sparse solution. To illustrate the potential of this approach, recent applications of adaptive techniques to electromagnetic scattering computations [47], [48] will be summarized in this section.

The scattering from an object can be computed by numerically solving an electromagnetic integral equation for the currents induced on the object by an incident field. The classical method of moments solution [49] discretizes the integral equation by expanding the unknown quantity, the current, in a set of basis functions. The procedure produces a dense linear system with N unknowns, which correspond to the coefficients of the basis functions. Since the basis is typically defined such that the current is spatially sampled at intervals much smaller than the wavelength of the illuminating radiation, the number of unknowns N grows rapidly with the electrical size of the object. Even more significant, direct solvers for dense systems have $O(N^3)$

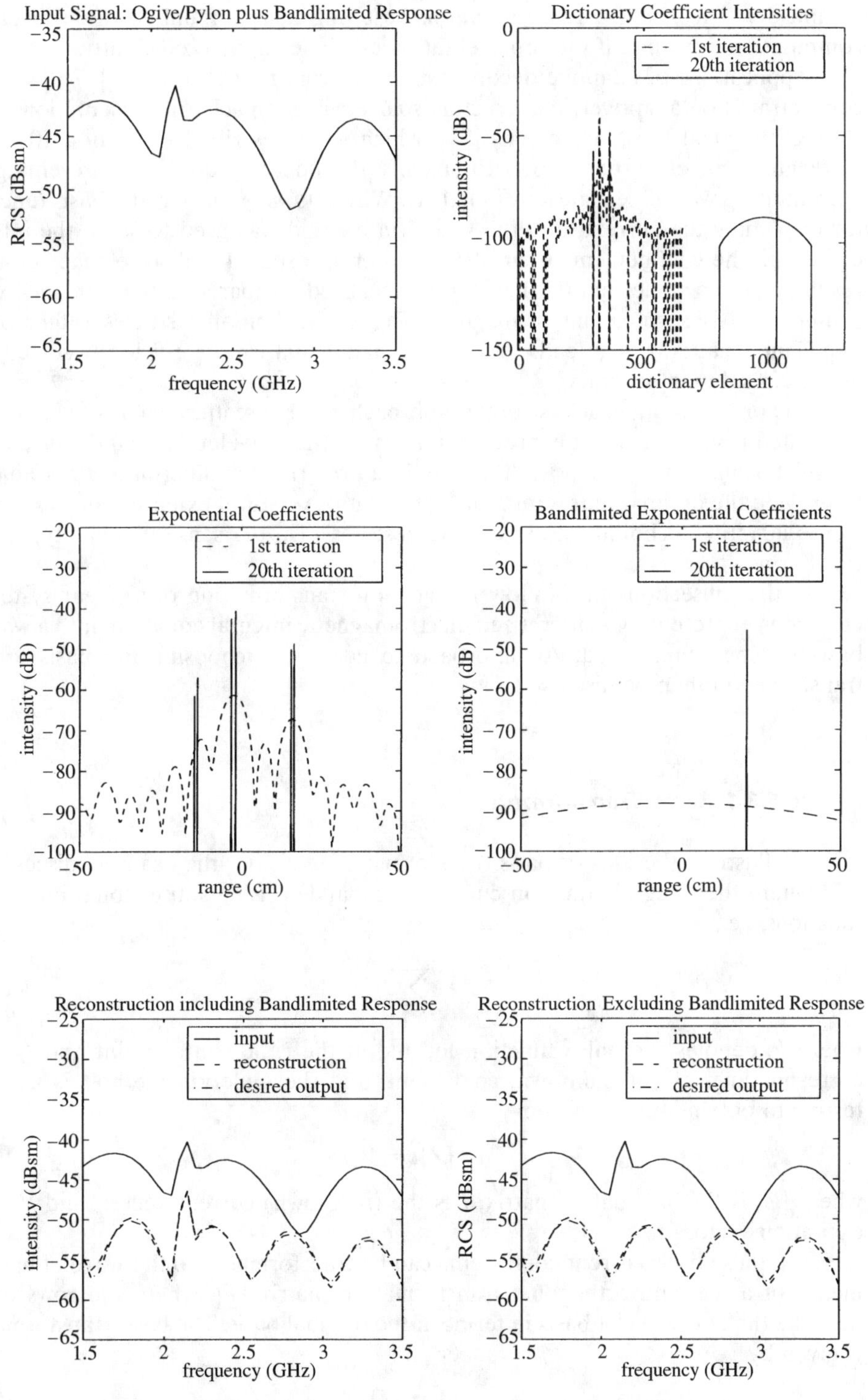

Figure 10.17 Example of support removal in the presence of dissimilar scattering mechanisms using an adaptive technique.

complexity. Thus, sparse representations resulting in fewer unknowns can lead to computational savings, if the representation can be computed efficiently.

Applications of adaptive decomposition to scattering prediction [47], [48] have concentrated on adaptively constructing sparse wavelet packet representations using Best Orthogonal Basis algorithms [28], which were described in Section 10.4. This work grew out of earlier efforts to sparsify the linear system by representing the current using wavelet expansions [50], [51]. Wavelets were used as the basis functions for expanding the current, and the Galerkin method was used to solve the integral equation. The confinement of wavelets in both the spatial and scale-space domains result in an impedance matrix that is more localized compared to that obtained when a simpler basis, such as pulses, are used. The localization allows thresholding of the impedance matrix, which leads to a sparse matrix that can be efficiently solved with no loss in solution accuracy.

The principal drawbacks with this approach are that scattered fields are not easily calculated from the current expressed in terms of the wavelet basis, and the standard Wavelet Transform decomposes the data in a prescribed manner and is not adapted to provide the optimal representation of the data. A natural extension of this work is to use adaptive techniques to select the best basis (in the sense of sparsity) for the transform.

In the subsections that follow the general transformation of a linear system of equations, representing a discretized electromagnetic integral equation, into a wavelet basis will be summarized. An approach to adaptively choosing the basis for the transform will then be discussed.

10.5.3.1 Basis Transformation

To illustrate the idea of basis transformation, consider the example discussed in [47] where the integral equation current is expanded using subsectional pulse basis functions, i.e.,

$$J = \sum_{i=1}^{N} I_i P_i \tag{10.29}$$

where P_i denotes the pulse function located at the i the source point and I_i is the corresponding, as yet unknown, coefficient, and the Galerkin method is used for testing to obtain a linear system

$$[Z]\bar{I} = \bar{V} \tag{10.30}$$

where $[Z]$ is the impedance matrix, $\bar{I}$ is the (unknown) current vector, and $\bar{V}$ is the excitation vector.

The pulse basis current coefficients can be transformed into the wavelet basis by means of a real and orthogonal transformation matrix $[T]$, where the rows of $[T]$ describe the new wavelet basis in terms of the original basis. The basis transformation is given by

$$\bar{A} = [T]\bar{I} \tag{10.31}$$

where the elements of $\bar{A}$ are the new coefficients of the new expansion of J.

The transformation can be applied to the impedance matrix by multiplying from the right by $[\tilde{T}]$, the transpose of $[T]$,

$$[Z_A] = [Z][\tilde{T}] \tag{10.32}$$

Since $[T]$ is orthogonal, the transformed linear system becomes

$$[Z_A]\bar{A} = \bar{V} \tag{10.33}$$

Solving for $\bar{I}$ or $\bar{A}$ will yield the same estimate for the current.

When a Galerkin method is used, the same transformation is used for the testing procedure that was applied to the current basis. Multiplying the excitation vector from the left yields

$$\bar{B} = [\tilde{T}]\bar{V} \tag{10.34}$$

as the excitation vector. Applying the same transformation on the columns of $[Z_A]$ gives the transformed impedance matrix

$$[Z] = [T][Z_A] = [T][Z][\tilde{T}] \tag{10.35}$$

which leads to the transformed system of equations

$$[Z]\bar{A} = \bar{B}. \tag{10.36}$$

Solving this matrix equation is equivalent to solving the original matrix equation; however, the additional sparsity achieved by the basis transformation results in a more computationally efficient solution, albeit at the computational expense of performing the transformation.

Note, as discussed in detail in [47], the transformation can be more efficiently implemented using digital filtering techniques instead of the matrix multiplication.

10.5.3.2 Adaptive Construction of Basis Functions

Many possible wavelet and wavelet packet transformations can be applied to transform the linear system into a new basis that promotes sparsity. The objective of an adaptive construction is to select a basis that best suits the problem at hand.

The work of both Baharav and Leviatan [47] and Golik [48] uses a decomposition of the excitation vector to select the basis. The first step in the procedure is to construct the excitation vector or some variant of the excitation vector, and then in the second step, compute a sparse representation of the excitation using a best basis algorithm. The basis functions selected for the representation define the appropriate transformation. The general assumption is that the physical variation of the excitation vector for a given set of incidence angles and/or parameters is related to the variation of the unknown, the current, in such a manner that a basis function set that can sparsely represent the excitation can also represent the current.

Baharav and Leviatan [47] modify the excitation vector by zeroing testing points in the shadow and provide a smooth transition between the lit and shadow regions.

The appropriate wavelet packet basis is chosen using a "top-down near-best basis algorithm, with an additive cost function," where the cost function used is given by

$$C(a) = -\sum_{i:a_i \neq 0} a_i^2 \ln a_i^2 \tag{10.37}$$

where a is a vector of basis function coefficients associated with the node in the wavelet packet decomposition. Golik [48] used the excitation vector and a different cost function for the basis selection algorithm, i.e.,

$$C(a) = \sum_i |a_i| \tag{10.38}$$

Both studies showed adaptive construction of basis functions, especially when used with wavelet packets, could result in significant reductions in computation burden over standard wavelet transformations for certain illuminations and geometries.

Several issues require further study, such as the general adequacy of using the excitation vector as a metric for determining the basis for the current expansion and the optimality of the basis selection technique. In addition, the computational burden of adaptively constructing the basis needs to be considered when assessing if this prediction technique is competitive with alternate emerging procedures for efficiently solving electrically large problems, such as the fast multipole method [52]. However, the concept of using adaptive decomposition techniques from the signal processing literature to adaptively construct a basis that promotes sparsity, and thus computational efficiency, in the solution of electromagnetic scattering problems appears to have potential and is an interesting area of new research.

10.6 CONCLUSION

Adaptive methods have recently been developed for generating sparse representations of signals in terms of a very general, overdetermined dictionary of functions, which are defined to closely model characteristics of anticipated signal components. These techniques, initially motivated by more traditional signal processing applications, such as speech processing and compression, have a wide variety of potential applications in electromagnetic sensing and prediction. The development of applications using adaptive decomposition, which synthesizes knowledge of signal processing and electromagnetics, is an interesting area of future research.

REFERENCES

[1] S. Chen and D. Donoho, "Basis Pursuits," Tech. Rep., Stanford University, CA, 1995.

[2] M. McClure and L. Carin, "Matching-pursuits Time-frequency Processing of Electromagnetic Scattering Data," in *1996 USNC/URSI Radio Sci. Mtg. Digest*, Baltimore, MD, p. 14, 1996.

[3] R. A. Altes, "Sonar for Generalized Target Description and Its Similarity to Animal Echolocation Systems," *J. Acoust. Soc. Amer.*, **59**, pp. 97–105, January 1976.

[4] M. P. Hurst and R. Mittra, "Scattering Center Analysis via Prony's Method," *IEEE Trans. Antennas Propagat.*, **AP-35**, pp. 986–988, August 1987.

[5] R. Carrierre and R. L. Moses, "High Resolution Parametric Modeling of Canonical Radar Scatterers with Application to Target Identification," *IEEE Trans. Antennas Propagat.*, **40**, pp. 13–18, January 1992.

[6] A. Moghaddar, Y. Ogawa, and E. K. Walton, "Estimating the Time Delay and Frequency Decay Parameter of Scattering Components Using a Modified MUSIC Algorithm," *IEEE Trans. Antennas Propagat.*, **42**, pp. 1412–1418, 1994.

[7] W. M. Steedly and R. I. Moses, "High Resolution Exponential Modeling of Fully Polarized Radar Returns," *IEEE Trans. Aerosp. Electron Syst.*, **27**, pp. 459–469, May 1991.

[8] L. Potter, D.-M. Chiang, R. Carriere, and M. Gerry, "A GTD-based Parametric Model for Radar Scattering," *IEEE Trans. Antennas Propagat.*, **43**, pp. 1058–1067, October 1995.

[9] M. McClure, R. C. Qiu, and L. Carin, "On the Superresolution Identification of Observables from Swept-frequency Scattering Data," *IEEE Trans. Antennas Propagat.*, **45**, pp. 631–640, April 1997.

[10] M. McClure and L. Carin, "Matching Pursuits with a Wave-based Dictionary," *IEEE Trans. Sign. Proc.*, **45**, pp. 2912–2927, December 1997.

[11] N. S. Subotic, J. W. Burns, and D. Pandelis, "Scattering Center Characterization Using Matching Pursuits with a Weighted Exponential Dictionary," in *1997 URSI North American Radio Sci. Mtg. Digest*, Montreal, CA, p. 258, 1997.

[12] N. S. Subotic and J. W. Burns, "Scattering Center Characterization Using Adaptive Decomposition with a Vectorized Dictionary," in *1998 URSI North American Radio Sci. Mtg. Digest*, Atlanta, GA, p. 326, 1998.

[13] L. C. Trinitalia, R. Bhalla, and H. Ling, "Scattering Center Parameterization of Wide Angle Backscattered Data Using Adaptive Gaussian Representation," *IEEE Trans. Antennas Propagat.*, **45**, pp. 1664–1668, November 1997.

[14] E. Heyman and L. B. Felsen, "A Wavefront Interpretation of the Singularity Expansion Method," *IEEE Trans. Antennas Propagat.*, **AP-33**, pp. 706–718, 1985.

[15] H. Shirai and L. B. Felsen, "Modified GTD for Generating Complex Resonances for Flat Strips and Disks," *IEEE Trans. Antennas Propagat.*, **AP-34**, pp. 779–790, 1986.

[16] H. Shirai and L. B. Felsen, "Modified GTD for Generating Complex Resonances for Flat Strips and Disks," *IEEE Trans. Antennas Propagat.*, **AP-34**, pp. 1196–1207, 1986.

[17] K. E. Oughston, J. E. K. Laurens, and C. M. Balictsis, "Asymptotic Description of Electromagnetic Pulse Propagation in a Linear Dispersive Medium," in *Ultra-Wideband Short-Pulse Electromagnetics*, H. L. Bertoni, L. Carin, and L. B. Felsen, eds, pp. 223–240, Plenum Press, New York, 1993.

[18] L. Brillouin, *Wave Propagation and Group Velocity*, Academic Press, New York, 1960.

[19] J. D. Jackson, *Classical Electrodynamics*, John Wiley and Sons, New York, 2nd edn, 1975.

[20] L. Carin and L. B. Felsen, "Time-harmonic and Transient Scattering by Finite Periodic Flat Strip Arrays: Hybrid (Ray)-(Floquet mode)-(MOM) Algorithms and Its GTD Interpretation," *IEEE Trans. Antennas Propagat.*, **41**, pp. 412–421, April 1993.

[21] F. Niu and L. B. Felsen, "Time-domain Leaky Modes on Layer Media: Dispersion Characteristics and Synthesis of Pulsed Radiation," *IEEE Trans. Antennas Propagat.*, **41**, pp. 755–761, June 1993.

[22] A. Moghaddar and E. K. Walton, "Time-frequency Distribution Analysis of Scattering from Waveguide Cavities," *IEEE Trans. Antennas Propagat.*, **41**, pp. 677–680, May 1993.

[23] H. Kim and H. Ling, "Wavelet Analysis of Backscattering Data from an Open-ended Waveguide Cavity," *IEEE Microwave Guided Wave Lett.*, **2**, pp. 140–142, April 1993.

[24] E. K. Miller, "Model-based Parametert Estimation in Electromagnetics: Part I. Background and Theoretical Development," *IEEE Trans. Antennas Propagat. Mag.*, **40**, pp. 42–52, January 1998.

[25] I. Daubechies, "Time-frequency Localization Operators: A Geometric Phase Space Approach," *IEEE Trans. Inf. Th.*, **34**, p. 605, April 1988.

[26] P. C. Hansen and D. P. O'Leary, "The Use of the l-curve in the Regularization of Ill-posed Problems," *SIAM J. Sci. Comput.*, **14**, pp. 1487–1503, November 1993.

[27] A. N. Tikhonov and V. Y. Arsenin, *Solutions of Ill-Posed Problems*, John Wiley and Sons, New York, 1977.

[28] R. R. Coifman and M. V. Wickerhauser, "Entropy-based Algorithms for Best-basis Selection," *IEEE Trans. Inf. Th.*, **38**, p. 713, February 1992.

[29] S. Mallat and Z. Zhang, "Matching Pursuit with Time Frequency Dictionaries," *IEEE Trans. Sign. Proc.*, **41**, pp. 3397–3415, December 1993.

[30] I. F. Gorodnitsky and B. D. Rao, "Sparse Signal Reconstruction from Limited Data Using FOCUSS: A Re-weighted Minimum Norm Algorithm," *IEEE Trans. Sign. Proc.*, **45**, pp. 600–616, March 1997.

[31] G. Dantzig, *Linear Programming and Extensions*, Princeton Univ. Press, Princeton, NJ, 1963.

[32] C. Gonzaga, "Path-following Methods for Linear Programming," *SIAM Review*, **34**, p. 167, February 1992.

[33] D. Donoho, "De-noising by Soft-Thresholding," *IEEE Trans. Inf. Th.*, **41**, p. 613, March 1995.

[34] V. C. Chen, "Applications of Time-frequency Processing to Radar Imaging," *Opt. Eng.*, **36**, pp. 1152–1161, April 1997.

[35] L. C. Trinitalia and H. Ling, "Joint Time-frequency ISAR Using Adaptive Processing," *IEEE Trans. Antennas Propagat.*, **45**, pp. 221–227, 1997.

[36] V. C. Chen and S. Qian, "Joint Time-frequency Transform for Radar Range-dopler Imaging," *IEEE Trans. Aerosp. Electron. Syst.*, **34**, pp. 486–499, February 1998.

[37] H. L. Y. Wang and V. C. Chen, "ISAR Motion Compensation via Adaptive Joint Time-frequency Technique," *IEEE Trans. Aerosp. Electron. Syst.*, **34**, pp. 670–677, February 1998.

[38] J. L. Walker, "Range-doppler Imaging of Rotating Objects," *IEEE Trans. Aerosp. Electron. Syst.*, **AES-16**, pp. 23–52, January 1980

[39] W. G. Carrara, R. S. Goodman, and R. M. Majewski, *Spotlight Synthetic Aperture Radar: Signal Processing Algorithms*, Artech House, Boston, 1995.

[40] L. C. Potter and R. L. Moses, "Attributed scattering centers for SAR ATR," *IEEE Trans. Image Proc.*, **6**, pp. 79–91, January 1997.

[41] S. R. DeGraaf, E. I. LeBaron, K. M. Quinlan, G. G. Fliss, and S. S. LiFliss, "ISAR RCS Editing via Modern Spectral Estimation Methods," in *Proc. 1995 AMTA Symp.*, Williamsburg, VA, pp. 245–250, 1995.

[42] J. W. Burns and S. R. DeGraaf, "Parametric Signal History Editing Techniques for Removal of Additive Support Contamination in Narrowband RCS Measurements," in *Proc. 1996 AMTA Symp.*, Seattle, WA, pp. 314–319, 1996.

[43] J. W. Burns, E. I. LeBaron, and G. G. Fliss, "Characterization of Target-pylon Interactions in RCS Measurements," in *1997 IEEE AP-S Intl. Symp. Digest*, **1**, Montreal, Canada, pp. 144–147, 1997.

[44] R. Roy and T. Kailath, "ESPRIT – Estimation of Signal Parameters via Rotational Invariance Techniques," *IEEE Trans. Acoust. Speech, Sign. Proc.*, **37**, July 1989.

[45] R. Roy and T. Kailath, "ESPRIT and Total Least Squares," in *Conf. Rec of the 21st Asilomar Conf. on Sig. Sys. and Comp.*, Pacific Grove, CA, 1987.

[46] J. W. Burns, "RCS Measurement Contamination Removal Using Adaptive Decomposition Techniques," in *1998 IEEE AP-S Intl. Symp. Digest.*, **3**, Atlanta, GA, pp. 1570–1573, 1998.

[47] Z. Baharav and Y. Leviatan, "Impedance Matrix Compression Using Adaptively Constructed Basis Functions," *IEEE Trans. Antennas Propagat.*, **44**, pp. 1231–1238, September 1996.

[48] W. L. Golik, "Wavelet Packets for Fast Solution of Electromagnetic Integral Equations," *IEEE Trans. Antennas Propagat.*, **46**, pp. 618–624, May 1998.

[49] R. F. Harrington, *Field Computation by Moment Methods*, Robert E. Krieger Publishing Co., Malabar, FL, 1968.

[50] B. Z. Steinberg and Y. Leviatan, "On the Use of Wavelet Expansions in the Method of Moments," *IEEE Trans. Antennas Propagat.*, **41**, pp. 610–619, 1993.

[51] O. P. Franza, R. L. Wagner, and W. C. Chew, "Wavelet-like Basis Functions for Solving Scattering Integral Equations," in *1994 IEEE AP-S Intl. Symp. Digest.*, **1**, Seattle, WA, pp. 3–6, 1994.

[52] V. Rokhlin, "Rapid Solution of Integral Equations of Scattering Theory in Two Dimensions," *J. Comp. Phys.*, **86**, pp. 414–439, February 1990.

Chapter 11

LOMMEL EXPANSIONS IN ELECTROMAGNETICS

Douglas H. Werner

ABSTRACT

Lommel expansions and the important role they play in electromagnetic theory are reviewed in this chapter. These expansions are shown to be useful in developing an exact integration procedure for evaluating electromagnetic field quantities associated with current-carrying wire antennas such as cylindrical dipoles and circular loops. Several exact series representations are introduced in this chapter which are valid not only in the far-field of the source, but also in its near-field region. These series expansions are general and can be shown to contain all of the well-known classical approximations as special limiting cases. Some applications of these recently found exact series expansions are discussed, including their use in the development of efficient yet accurate algorithms for computational modeling of moderately thick cylindrical wire antennas.

11.1 INTRODUCTION

The recent development and application of analysis techniques based on Lommel expansions have led to the solution of several long-standing problems in electromagnetics [1]. This chapter is primarily intended to provide an introduction to these Lommel expansions as well as to acquaint the reader with their utility in the derivation of exact solutions for a number of the more commonly occurring and basic integrals found in antenna theory. Included among these integrals are the cylindrical wire kernel, the uniform current cylindrical dipole vector potential, and various near-field integrals associated with the circular loop antenna. In the past, these integrals were either evaluated by time-consuming numerical techniques or by making certain convenient but restrictive assumptions. The solutions to the integrals presented in this chapter are mathematically exact in the sense that no assumptions or approximations were made in their derivation. These exact expressions may be used to evaluate antenna near fields to any desired degree of accuracy, while computational efficiency is achieved through the use of several recurrence relations. They may also serve as the starting-point for deriving convenient far-field approximations through the application of asymptotic methods.

The solution to many fundamental problems in electromagnetics requires the evaluation of integrals which have the general form

$$\frac{1}{2\pi}\int_0^{2\pi} f(\varphi')G(R')\,d\varphi' \tag{11.1}$$

where

$$G(R') = \frac{e^{-j\beta R'}}{R'} \tag{11.2}$$

and

$$R' = \sqrt{R^2 - 2\rho a \cos(\varphi - \varphi')} \tag{11.3}$$

$$R = \sqrt{\zeta^2 + d^2} \tag{11.4}$$

$$d = \sqrt{\rho^2 + a^2} \tag{11.5}$$

$$\zeta = z - z' \tag{11.6}$$

For instance, the cylindrical wire kernel belongs to this class of integrals [2]. Integrals of this type also appear in the electromagnetic near-field expressions for current-carrying loop antennas [3]. In spite of their basic importance, however, a systematic methodology for finding an analytical solution to these integrals has not been reported in the electromagnetics literature. This is primarily due to the lack of a suitable series expansion for (11.2). Here we outline the derivation for a new series representation for (11.2) which can be used to readily find exact solutions to integrals of the form given in (11.1). Lommel expansions are shown to play a key role in this derivation.

Suppose we consider Lommel's expansions for $(\sqrt{x+y})^{\pm\nu} J_\nu(\sqrt{x+y})$, which are given by [4]

$$(\sqrt{x+y})^{-\nu} J_\nu(\sqrt{x+y}) = \sum_{m=0}^{\infty} \frac{(-y/2)^m}{m!} (\sqrt{x})^{-(\nu+m)} J_{\nu+m}(\sqrt{x}) \tag{11.7}$$

$$(\sqrt{x+y})^{\nu} J_\nu(\sqrt{x+y}) = \sum_{m=0}^{\infty} \frac{(y/2)^m}{m!} (\sqrt{x})^{(\nu-m)} J_{\nu-m}(\sqrt{x}) \tag{11.8}$$

Setting $\nu = -\frac{1}{2}$ in (11.7) and $\nu = \frac{1}{2}$ in (11.8) leads to

$$(\sqrt{x+y})^{1/2} J_{-1/2}(\sqrt{x+y}) = \sum_{m=0}^{\infty} \frac{(-y/2)^m}{m!} (\sqrt{x})^{(1/2-m)} J_{-1/2+m}(\sqrt{x}) \tag{11.9}$$

$$(\sqrt{x+y})^{1/2} J_{1/2}(\sqrt{x+y}) = \sum_{m=0}^{\infty} \frac{(y/2)^m}{m!} (\sqrt{x})^{(1/2-m)} J_{1/2-m}(\sqrt{x}) \tag{11.10}$$

The following relations involving half-integer order Bessel functions

$$J_{1/2}(z) = \sqrt{\frac{2}{\pi z}} \sin(z) \tag{11.11}$$

$$J_{-1/2}(z) = \sqrt{\frac{2}{\pi z}} \cos(z) \tag{11.12}$$

$$j_m(z) = \sqrt{\frac{\pi}{2z}}\, J_{m+1/2}(z) \tag{11.13}$$

may be used to further reduce (11.9) and (11.10) to

$$\cos(\sqrt{x+y}) = \sum_{m=0}^{\infty} \frac{(-y/2)^m}{m!} \frac{j_{m-1}(\sqrt{x})}{(\sqrt{x})^{m-1}} \tag{11.14}$$

$$\sin(\sqrt{x+y}) = \sum_{m=0}^{\infty} \frac{(y/2)^m}{m!} \frac{j_{-m}(\sqrt{x})}{(\sqrt{x})^{m-1}} \tag{11.15}$$

If we let $x = (\beta R)^2$ and $y = -2\beta^2 \rho a \cos(\varphi - \varphi')$, then (11.14) and (11.15) become

$$\cos(\beta R') = \sum_{m=0}^{\infty} \frac{(\beta^2 \rho a)^m}{m!} \frac{j_{m-1}(\beta R)}{(\beta R)^{m-1}} \cos^m(\varphi - \varphi') \tag{11.16}$$

$$\sin(\beta R') = \sum_{m=0}^{\infty} \frac{(\beta^2 \rho a)^m}{m!} \frac{y_{m-1}(\beta R)}{(\beta R)^{m-1}} \cos^m(\varphi - \varphi') \tag{11.17}$$

where $j_{m-1}(\bullet)$ and $y_{m-1}(\bullet)$ are spherical Bessel functions of the first and second kinds, respectively, of order $m-1$. Combining (11.16) and (11.17) using Euler's identity results in an expansion for $e^{-j\beta R'}$ given by

$$e^{-j\beta R'} = e^{-j\beta R} + \sum_{m=1}^{\infty} \frac{(\beta^2 \rho a)^m}{m!} \frac{h_{m-1}^{(2)}(\beta R)}{(\beta R)^{m-1}} \cos^m(\varphi - \varphi') \tag{11.18}$$

Finally, by taking the derivative of both sides of (11.18) with respect to ζ and utilizing a recurrence relation for spherical Hankel functions,

$$\frac{d}{dx}[x^{-m} h_m^{(2)}(x)] = -x^{-m} h_{m+1}^{(2)}(x) \tag{11.19}$$

we arrive at the desired expression for (11.2)

$$\frac{e^{-j\beta R'}}{R'} = \frac{e^{-j\beta R}}{R} - j\beta \sum_{m=1}^{\infty} \frac{(\beta^2 \rho a)^m}{m!} \frac{h_m^{(2)}(\beta R)}{(\beta R)^m} \cos^m(\varphi - \varphi') \tag{11.20}$$

11.2 THE CYLINDRICAL WIRE DIPOLE ANTENNA

The cylindrical wire dipole is one of the simplest and most popular of all antennas. Many antennas are constructed from cylindrical wires arranged in various configurations. Numerically rigorous computational electromagnetic modeling techniques, such as the method of moments, have been developed for the accurate analysis of complex antenna systems composed of cylindrical wires [5–8]. These techniques require the evaluation of various near-zone electric field integrals associated with the cylindrical wire dipole. In this section we introduce analytical methods for evaluating these integrals which are based on the Lommel expansions introduced in Section 11.1.

The geometry for the standard cylindrical wire dipole is illustrated in Fig. 11.1. The length of the antenna is $2h$, and its diameter is $2a$. A circumferentially uniform current $I(z')$ is assumed to be present on the surface of the cylindrical antenna. The

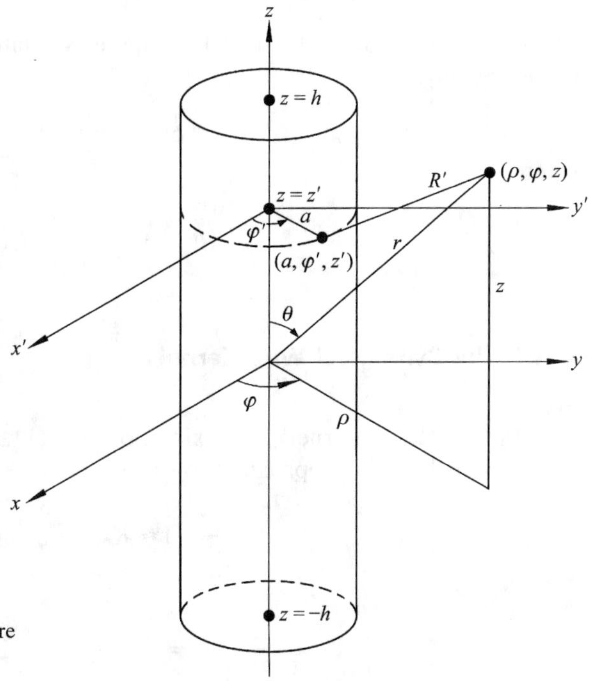

Figure 11.1 Geometry for the cylindrical wire dipole antenna.

cylindrical coordinates of the source point and field point are (a, φ', z') and (ρ, φ, z), respectively. The vector potential associated with this cylindrical antenna may be expressed as [2]

$$\mathbf{A} = A_z(\rho, z)\,\hat{\mathbf{z}} \tag{11.21}$$

where

$$A_z(\rho, z) = \frac{\mu}{4\pi}\int_{-h}^{h} I(z')K(z - z')\,dz' \tag{11.22}$$

and

$$K(z - z') = \frac{1}{2\pi}\int_{0}^{2\pi} \frac{e^{-j\beta R'}}{R'}\,d(\varphi - \varphi') \tag{11.23}$$

is known as the cylindrical wire kernel [2].

It can be shown that the electric field of the dipole antenna shown in Fig. 11.1 may be expressed in terms of the vector potential as

$$\mathbf{E} = E_\rho(\rho, z)\,\hat{\boldsymbol{\rho}} + E_z(\rho, z)\,\hat{\mathbf{z}} \tag{11.24}$$

where

$$E_\rho(\rho, z) = \frac{1}{j\omega\mu\varepsilon}\frac{\partial^2}{\partial\rho\,\partial z}A_z(\rho, z) \tag{11.25}$$

$$E_z(\rho, z) = \frac{1}{j\omega\mu\varepsilon}\left[\frac{\partial^2}{\partial z^2} + \beta^2\right]A_z(\rho, z) \tag{11.26}$$

Likewise, the magnetic field of the dipole is related to the vector potential in the following way:

$$\mathbf{H} = H_\varphi(\rho, z)\,\hat{\varphi} \tag{11.27}$$

where

$$H_\varphi(\rho, z) = -\frac{1}{\mu}\frac{\partial}{\partial\rho}A_z(\rho, z) \tag{11.28}$$

11.2.1 The Cylindrical Wire Kernel

The reduced kernel approximation of (11.23) is often used for thin-wire applications [2]. This approximation is given by

$$K(z - z') \approx K_0(z - z') = \frac{e^{-j\beta r_0}}{r_0} \tag{11.29}$$

where

$$r_0 = \sqrt{(z - z')^2 + \rho^2} \tag{11.30}$$

The reduced kernel approximation is valid for very thin wires; however, it becomes less accurate as the radius of the cylindrical antenna increases relative to its length. Hence, for thicker wires, other more accurate representations of the cylindrical wire kernel defined in (11.23) must be employed.

One such representation may be obtained by integrating both sides of (11.20) with respect to φ' from 0 to 2π and dividing by 2π. This results in an exact series expansion for the kernel which has the form [9], [10]

$$K(z - z') = \frac{e^{-j\beta R}}{R} - j\beta\sum_{n=1}^{\infty}\frac{(2n-1)!!}{(2n)!!}\,\frac{(\beta^2\rho a)^{2n}}{(2n)!}\,\frac{h_{2n}^{(2)}(\beta R)}{(\beta R)^{2n}} \tag{11.31}$$

where use has been made of the fact that

$$\frac{1}{\pi}\int_0^{\pi}\cos^m\varphi'\,d\varphi' = \begin{cases} 0; & m = 2n-1 \quad \text{for} \quad n = 1, 2, \ldots \\ \dfrac{(2n-1)!!}{(2n)!!}; & m = 2n \quad \text{for} \quad n = 1, 2, \ldots \end{cases} \tag{11.32}$$

and where

$$(2n-1)!! = 1\cdot 3 \ldots (2n-1) \tag{11.33}$$

$$(2n)!! = 2\cdot 4 \ldots (2n) \tag{11.34}$$

At this point we recognize that the spherical Hankel functions which appear in (11.31) are closely related to the complex exponential functions. In other words, integer-order spherical Hankel functions of the second kind may be written as [11]

$$h_m^{(2)}(x) = j^{m+1}\frac{e^{-jx}}{x}\sum_{k=0}^{m}\frac{(m+k)!}{k!(m-k)!}(2jx)^{-k} \tag{11.35}$$

Another useful identity is [12]

$$\frac{(2n-1)!!}{(2n)!!} = \binom{1/2}{n}(2n-1)(-1)^{n+1} \tag{11.36}$$

where

$$\binom{1/2}{n} = \begin{cases} 1, & n=0 \\ \dfrac{1/2(1/2-1)(1/2-2)\ldots(1/2-n+1)}{n!}, & n>0 \end{cases} \tag{11.37}$$

Substituting (11.35) and (11.36) into (11.31) results in a convenient expression for the exact kernel given by [10]

$$K(z-z') = -\frac{e^{-j\beta R}}{R}\sum_{n=0}^{\infty}\sum_{k=0}^{2n} A_{n,k}\frac{(\beta^2\rho a)^{2n}}{(\beta R)^{2n+k}} \tag{11.38}$$

where

$$A_{n,k} = \binom{1/2}{n}\frac{(2n-1)}{(2j)^k}\frac{(2n+k)!}{k!(2n)!(2n-k)!} \tag{11.39}$$

Finally, we note that a useful set of recurrence relations may be derived from (11.39) which provide a computationally efficient means for evaluating the coefficients $A_{n,k}$ which appear in (11.38). The resulting recurrence relations are [1]

$$A_{n,k} = \frac{(2n+1-k)(2n+k)}{j(2k)}A_{n,k-1} \qquad \text{for } n>0 \text{ and } k>0 \tag{11.41}$$

$$A_{n,0} = \begin{cases} -1 & \text{for } n=0 \\ -\dfrac{1}{(2n)^2}A_{n-1,0} & \text{for } n>0 \end{cases} \tag{11.42}$$

A more intuitive feel for the behavior of the kernel can be gained from (11.38) since the spherical Hankel functions which were originally present in (11.31) have been replaced by a complex exponential function and terms involving powers of $1/R$. Equation (11.38) has the added advantages that it is much easier to work with from the analytical point of view and more amenable to numerical evaluation. Algorithms based on (11.38) have been successfully developed which can be used for the efficient as well as accurate numerical analysis of moderately thick cylindrical wire antennas [13].

The exact series representation given in (11.38) may be used to derive some new approximations for the cylindrical wire kernel as well as show that several well-known classical results may be obtained as special limiting cases of this series [14]. We first make note of the fact that the commonly used reduced kernel approximation (11.29) may be obtained from either the integral (11.23) or the series (11.38) by allowing the wire radius to become vanishingly thin (i.e., by letting $\beta a \to 0$). Next, we recognize that an extended kernel approximation which is valid for thicker wires may be easily derived from (11.38) by retaining only the first two terms in the series, namely those corresponding to $n = 0$ and $n = 1$. This results in a relatively simple approximation which is given by

$$K(z-z') \approx \frac{e^{-j\beta R}}{R}\left[1-\left(\frac{\beta\rho a}{2R}\right)^2\left\{1-\frac{3j}{\beta R}-\frac{3}{(\beta R)^2}\right\}\right] \tag{11.43}$$

A large argument approximation for the cylindrical wire kernel may also be derived from (11.38). To begin the derivation, suppose we consider the inner summation in (11.38). That is

$$\sum_{k=0}^{2n} A_{n,k}\frac{(\beta^2\rho a)^{2n}}{(\beta R)^{2n+k}} = \binom{1/2}{n}\frac{(2n-1)}{(2n)!}\frac{(\beta^2\rho a)^{2n}}{(\beta R)^{2n}}\left[1+\frac{1}{1!}\frac{(2n+1)!}{(2n-1)!}\frac{1}{(2j\beta R)}\right.$$
$$+\frac{1}{2!}\frac{(2n+2)!}{(2n-2)!}\frac{1}{(2j\beta R)^2}+\frac{1}{3!}\frac{(2n+3)!}{(2n-3)!}\frac{1}{(2j\beta R)^3}+\ldots \tag{11.44}$$
$$\left.\ldots+\frac{1}{(2n)!}\frac{(2n+2n)!}{(2n-2n)!}\frac{1}{(2j\beta R)^{2n}}\right]$$

If we assume that βR is large enough so that it is only necessary to retain the first two terms in this series, then (11.44) may be approximated by

$$\sum_{k=0}^{2n} A_{n,k}\frac{(\beta^2\rho a)^{2n}}{(\beta R)^{2n+k}} \approx A_{n,0}\left(\frac{\beta^2\rho a}{\beta R}\right)^{2n} + A_{n,1}\frac{(\beta^2\rho a)}{(\beta R)^2}\left(\frac{\beta^2\rho a}{\beta R}\right)^{2n-1} \quad \text{for } (\beta R)^2 \gg \frac{1}{4} \tag{11.45}$$

where

$$A_{n,0} = \binom{1/2}{n}\frac{(2n-1)}{(2n)!} = \frac{(-1)^{n+1}}{2^{2n}(n!)^2} \tag{11.46}$$

$$A_{n,1} = \frac{1}{2j}\binom{1/2}{n}\frac{(2n-1)}{(2n)!}\frac{(2n+1)!}{(2n-1)!} = \frac{j(-1)^n}{2^{2n-1}[(n-1)!]^2}+\frac{j(-1)^n}{2^{2n}n!(n-1)!} \tag{11.47}$$

This fact may be used to show that

$$K(z-z') \approx -\frac{e^{-j\beta R}}{R}\left[\sum_{n=0}^{\infty} A_{n,0}\left(\frac{\beta^2\rho a}{\beta R}\right)^{2n} + \frac{(\beta^2\rho a)}{(\beta R)^2}\sum_{n=0}^{\infty} A_{n,1}\left(\frac{\beta^2\rho a}{\beta R}\right)^{2n-1}\right] \tag{11.48}$$

provided $(\beta R)^2 \gg \frac{1}{4}$. Finally, substituting (11.46) and (11.47) into (11.48) leads to a convenient closed-form expression given by

$$K(z-z') \approx \frac{e^{-j\beta R}}{R}\left[\left\{1-\frac{j}{2\beta R}\left(\frac{\beta^2\rho a}{\beta R}\right)^2\right\}J_0\left(\frac{\beta^2\rho a}{\beta R}\right)+\frac{j}{2\beta R}\left(\frac{\beta^2\rho a}{\beta R}\right)J_1\left(\frac{\beta^2\rho a}{\beta R}\right)\right] \tag{11.49}$$

where use has been made of the series representations for zero-order and first-order Bessel functions of the first kind

$$J_0\left(\frac{\beta^2\rho a}{\beta R}\right) = \sum_{n=0}^{\infty}\frac{(-1)^n}{(n!)^2}\left[\frac{(\beta^2\rho a)/(\beta R)}{2}\right]^{2n} \tag{11.50}$$

$$J_1\left(\frac{\beta^2\rho a}{\beta R}\right) = \sum_{n=0}^{\infty}\frac{(-1)^n}{(n!)(n+1)!}\left[\frac{(\beta^2\rho a)/(\beta R)}{2}\right]^{2n+1} \tag{11.51}$$

For sufficiently large values of βR, (11.49) reduces to

$$K(z-z') \sim \frac{e^{-j\beta R}}{R} J_0\left(\frac{\beta^2 \rho a}{\beta R}\right) \qquad \text{as } \beta R \to \infty \tag{11.52}$$

This expression may also be obtained directly from the integral form of the kernel (11.23) by using the standard far-zone approximations

$$R' \approx R - \left(\frac{\rho a}{R}\right) \cos \varphi' \tag{11.53}$$

for phase terms and

$$R' \approx R \tag{11.54}$$

for amplitude terms.

11.2.2 The Uniform Current Vector Potential and Electromagnetic Fields

Any cylindrical wire antenna may be viewed as the superposition of many electrically short segments over which the current distribution remains constant. Consequently, the uniform current short dipole antenna is considered the basic component from which more complex antenna systems can be constructed [15]. An expression for the vector potential of the cylindrical wire dipole shown in Fig. 11.1, which has a uniformly distributed current, may be obtained from (11.22) by assuming that $I(z') = I_0$. This results in

$$A_z(\rho, z) = \frac{\mu I_0}{4\pi} \int_{-h}^{h} K(z-z')\,dz' \tag{11.55}$$

Several approximations are commonly used in an effort to simplify the integrals contained in the uniform current vector potential expression (11.55). These approximations include the "reduced kernel" for electrically thin wire antennas [2], and the far-zone expansion for an "ideal dipole" [16]. A complete exact formulation is presented in this section for the vector potential and corresponding electromagnetic fields of a cylindrical dipole antenna with a uniform current distribution. An exact representation of the vector potential and electromagnetic fields has not previously been available for the uniform current cylindrical wire antenna. The exact expressions given in this section converge rapidly in the near-field region of the antenna. The properties of these exact expressions allow them to be readily used for the efficient as well as accurate computational modeling of cylindrical wire antennas.

Exact expressions for the uniform current vector potential, and ultimately the electromagnetic fields, are derived here using the series representation for the cylindrical wire kernel introduced in the previous section. It will be demonstrated in this section that the uniform current vector potential may be expressed in terms of a series which involves a generalized exponential integral and higher-order associated integrals. A numerically stable three-term forward recurrence relation for the higher-order associated integrals will be presented. Also discussed are efficient methods for the evaluation of the initial three integrals required by the recurrence relation.

Substituting the expansion for the exact kernel (11.38) into the expression for the uniform current vector potential (11.55) and integrating term by term yields

$$A_z(\rho, z) = -\frac{\mu I_0}{4\pi} \sum_{n=0}^{\infty} \sum_{k=0}^{2n} A_{n,k} (\beta^2 \rho a)^{2n} E_n^k(\rho, z) \tag{11.56}$$

where

$$E_n^k(\rho, z) = \int_{-h}^{h} \frac{e^{-j\beta R}}{R} \frac{1}{(\beta R)^{2n+k}} dz' \tag{11.57}$$

and R is defined in (11.4). It is of interest to note that

$$E_0^0(\rho, z) = \int_{-h}^{h} \frac{e^{-j\beta R}}{R} dz' \tag{11.58}$$

is the standard thin-wire generalized exponential integral for a current filament on the wire surface at $\varphi' = 90°$, while $E_n^k(\bullet)$ with $n > 0$ correspond to high-order associated integrals.

A computationally efficient procedure for evaluating the higher-order integrals $E_n^k(\bullet)$ defined in (11.57) is through the use of a numerically stable forward recurrence relation. In order to find this recurrence relation, it is convenient to introduce the notation

$$\Upsilon_m(\rho, z) = \int_{-h}^{h} \frac{e^{-j\beta R}}{R} \frac{1}{(\beta R)^m} dz' \qquad \text{for} \qquad m = 0, 1, 2, \ldots \tag{11.59}$$

such that $\Upsilon_0(\bullet)$ corresponds to the generalized exponential integral $E_0^0(\bullet)$ and $\Upsilon_m(\bullet)$, where $m > 0$, corresponds to higher-order associated integrals when $m = 2n + k$. By introducing the change of variables $\zeta = z - z'$, the integral in (11.59) may be transformed into

$$\Upsilon_m = \int_{\zeta_1}^{\zeta_2} \frac{e^{-j\beta R}}{R} \frac{1}{(\beta R)^m} d\zeta \tag{11.60}$$

where

$$\zeta_1 = z - h \tag{11.61}$$

$$\zeta_2 = z + h \tag{11.62}$$

If we define

$$u = \frac{e^{-j\beta R}}{R} \frac{1}{(\beta R)^m} \tag{11.63}$$

$$dv = d\zeta \tag{11.64}$$

then integration by parts may be used to obtain a recurrence relation from (11.60). This recurrence relation is given by

$$\Upsilon_m = \frac{1}{(m-1)(\beta d)^2}\left[\frac{\beta\zeta_2 e^{-j\beta R_2}}{(\beta R_2)^{m-1}} - \frac{\beta\zeta_1 e^{-j\beta R_1}}{(\beta R_1)^{m-1}} + j\Upsilon_{m-3} + (m-2)\Upsilon_{m-2} - j(\beta d)^2\Upsilon_{m-1}\right], \qquad m \geq 3 \tag{11.65}$$

where

$$R_1 = \sqrt{\zeta_1^2 + d^2} \tag{11.66}$$

$$R_2 = \sqrt{\zeta_2^2 + d^2} \tag{11.67}$$

The recurrence relation given in (11.65) requires a knowledge of the initial three integrals Υ_0, Υ_1, and Υ_2. One approach for evaluating these three integrals involves using a Maclaurin series expansion of the complex exponential function. This approach has been used by Harrington [17] to approximate Υ_0 by retaining the first few terms in the series expansion. Here we present the complete series expansion for Υ_0 as well as for Υ_1 and Υ_2. Before expanding the integrands, however, it is advantageous to express Υ_0, Υ_1, and Υ_2 in the form

$$\Upsilon_0 = e^{-j\beta R_0}\int_{\zeta_1}^{\zeta_2} \frac{e^{-j\beta(R-R_0)}}{R}\,d\zeta \tag{11.68}$$

$$\Upsilon_1 = e^{-j\beta R_0}\frac{1}{\beta}\int_{\zeta_1}^{\zeta_2} \frac{e^{-j\beta(R-R_0)}}{R^2}\,d\zeta \tag{11.69}$$

$$\Upsilon_2 = e^{-j\beta R_0}\frac{1}{\beta^2}\int_{\zeta_1}^{\zeta_2} \frac{e^{-j\beta(R-R_0)}}{R^3}\,d\zeta \tag{11.70}$$

where

$$R_0 = \sqrt{z^2 + d^2} \tag{11.71}$$

such that $|R - R_0| \leq h$. The Maclaurin series expansion for the exponential function is represented by

$$e^{-j\beta(R-R_0)} = \sum_{n=0}^{\infty} \frac{(j\beta)^n}{n!}(R_0 - R)^n \tag{11.72}$$

where we may write

$$(R_0 - R)^n = \sum_{k=0}^{n} \binom{n}{k}(-1)^k R_0^{n-k} R^k \tag{11.73}$$

The series in (11.72) and (11.73) may be combined to produce

$$e^{-j\beta(R-R_0)} = \sum_{n=0}^{\infty}\sum_{k=0}^{n} \frac{(-j\beta)^k (j\beta R_0)^{n-k}}{k!(n-k)!} R^k \tag{11.74}$$

Finally, substituting this expansion into the expressions for Υ_0, Υ_1, and Υ_2 given in

(11.68), (11.69), and (11.70), respectively, yields the following exact representations [1], [10]:

$$\Upsilon_0 = e^{-j\beta R_0} \sum_{n=0}^{\infty} \sum_{k=0}^{n} B_{n,k} F_{k-1} \tag{11.75}$$

$$\Upsilon_1 = e^{-j\beta R_0} \frac{1}{\beta} \sum_{n=0}^{\infty} \sum_{k=0}^{n} B_{n,k} F_{k-2} \tag{11.76}$$

$$\Upsilon_2 = e^{-j\beta R_0} \frac{1}{\beta^2} \sum_{n=0}^{\infty} \sum_{k=0}^{n} B_{n,k} F_{k-3} \tag{11.77}$$

where

$$B_{n,k} = \frac{(-j\beta)^k (j\beta R_0)^{n-k}}{k!(n-k)!} \tag{11.78}$$

$$F_m = \int_{\zeta_1}^{\zeta_2} R^m \, d\zeta \tag{11.79}$$

Computationally useful recurrence relations exist for the coefficients $B_{n,k}$ defined in (11.78), which are given by

$$B_{n,k} = -\frac{1}{R_0}\left(\frac{n+1-k}{k}\right) B_{n,k-1}, \qquad n>0 \text{ and } k>0 \tag{11.80}$$

and

$$B_{n,0} = \begin{cases} 1, & n=0 \\ \dfrac{j\beta R_0}{n} B_{n-1,0}, & n>0 \end{cases} \tag{11.81}$$

Closed-form solutions can be obtained for the integrals F_{-3}, F_{-2}, F_{-1}, and F_0 while a recurrence relation can be used to determine the higher-order integrals F_m for $m \geq 1$. In other words, it can be shown that

$$F_{-3} = \begin{cases} \dfrac{4zh}{R_1 R_2 (\zeta_2 R_1 + \zeta_1 R_2)}, & \zeta_1 > 0 \\ \dfrac{\zeta_2 R_1 - \zeta_1 R_2}{d^2 R_1 R_2}, & \zeta_1 \leq 0 \end{cases} \tag{11.82}$$

$$F_{-2} = \frac{1}{d}\left[\tan^{-1}\left(\frac{\zeta_2}{d}\right) - \tan^{-1}\left(\frac{\zeta_1}{d}\right)\right] \tag{11.83}$$

$$F_{-1} = \begin{cases} \ln\left[\dfrac{R_2 + \zeta_2}{R_1 + \zeta_1}\right], & \zeta_1 > 0 \\ \ln\left[\dfrac{R_1 - \zeta_1}{R_2 - \zeta_2}\right], & \zeta_1 \leq 0 \end{cases} \tag{11.84}$$

$$F_0 = 2h \tag{11.85}$$

$$F_m = \frac{1}{(m+1)}[\zeta_2 R_2^m - \zeta_1 R_1^m + m d^2 F_{m-2}], \qquad m \geq 1 \tag{11.86}$$

where the options for F_{-3} and F_{-1} are intended to avoid loss of numerical precision.

Numerical integration schemes may also be employed to evaluate Υ_0, Υ_1, and Υ_2. However, these integrals are difficult to evaluate directly using numerical techniques because their integrands are sharply peaked. This difficulty may be avoided by extracting the problematic "singularity" from these integrals. Following this procedure results in

$$\Upsilon_0 = F_{-1} + \int_{-h}^{h} \frac{e^{-j\beta R} - 1}{R} dz' \tag{11.87}$$

$$\Upsilon_1 = \frac{1}{\beta} F_{-2} - jF_{-1} + \frac{1}{\beta} \int_{-h}^{h} \frac{e^{-j\beta R} + j\beta R - 1}{R^2} dz' \tag{11.88}$$

$$\Upsilon_2 = \frac{1}{\beta^2} F_{-3} - \frac{j}{\beta} F_{-2} - \frac{1}{2} F_{-1} + \frac{1}{\beta^2} \int_{-h}^{h} \frac{e^{-j\beta R} + (\beta R)^2/2 + j\beta R - 1}{R^3} dz' \tag{11.89}$$

The integrals contained in (11.87), (11.88), and (11.89) may be more efficiently evaluated numerically since their integrands are well behaved.

The exact expressions for the kernel and vector potential may be used to find exact representations for the electric field produced by a cylindrical wire with a uniform current distribution. This may be accomplished by first substituting the integral expression for the uniform current vector potential (11.55) into (11.25) and (11.26). We next make use of some properties of the cylindrical wire kernel [18]

$$\frac{\partial}{\partial z} K(z - z') = -\frac{\partial}{\partial z'} K(z - z') \tag{11.90}$$

$$\frac{\partial^2}{\partial z^2} K(z - z') = \frac{\partial^2}{\partial z'^2} K(z - z') \tag{11.91}$$

in order to show that

$$E_\rho(\rho, z) = E_0 \left[\frac{1}{\beta^2} \frac{\partial}{\partial \rho} K(z - h) - \frac{1}{\beta^2} \frac{\partial}{\partial \rho} K(z + h) \right] \tag{11.92}$$

$$E_z(\rho, z) = E_0 \left[\frac{1}{\beta^2} \frac{\partial}{\partial z} K(z - h) - \frac{1}{\beta^2} \frac{\partial}{\partial z} K(z + h) - \int_{-h}^{h} K(z - z')\, dz' \right] \tag{11.93}$$

where

$$E_0 = \frac{j\eta\beta I_0}{4\pi} \tag{11.94}$$

$$\eta = \sqrt{\frac{\mu}{\varepsilon}} \tag{11.95}$$

Finally, the exact series expansion for the cylindrical wire kernel and for the uniform current vector potential given in (11.38) and (11.56), respectively, may be used to show that

$$\frac{1}{\beta^2}\frac{\partial}{\partial\rho}K(z-z') = \frac{1}{\beta\rho}\frac{e^{-j\beta R}}{(\beta R)^3}\sum_{n=0}^{\infty}\sum_{k=0}^{2n} A_{n,k}(\beta^2\rho a)^{2n} \cdot \left\{\frac{(\beta\rho)^2[(2n+k+1)+j\beta R] - 2n(\beta R)^2}{(\beta R)^{2n+k}}\right\} \tag{11.96}$$

$$\frac{1}{\beta^2}\frac{\partial}{\partial z}K(z-z') = \beta(z-z')\frac{e^{-j\beta R}}{(\beta R)^3}\sum_{n=0}^{\infty}\sum_{k=0}^{2n} A_{n,k}(\beta^2\rho a)^{2n} \cdot \left\{\frac{(2n+k+1)+j\beta R}{(\beta R)^{2n+k}}\right\} \tag{11.97}$$

$$\int_{-h}^{h} K(z-z')\,dz' = -\sum_{n=0}^{\infty}\sum_{k=0}^{2n} A_{n,k}(\beta^2\rho a)^{2n}\Upsilon_{2n+k}(\rho,z) \tag{11.98}$$

A similar procedure may be followed in order to derive an exact expression for the magnetic field produced by a uniform current cylindrical dipole. Hence, using (11.28) in combination with (11.55) leads to

$$H_\varphi(\rho,z) = -H_0\left\{\frac{1}{\beta}\frac{\partial}{\partial\rho}\int_{-h}^{h} K(z-z')\,dz'\right\} \tag{11.99}$$

where

$$H_0 = \frac{\beta I_0}{4\pi} \tag{11.100}$$

Substituting (11.98) into (11.99) and performing the required differentiation with respect to ρ yields

$$H_\varphi(\rho,z) = H_0\sum_{n=0}^{\infty}\sum_{k=0}^{2n} A_{n,k}(\beta^2\rho a)^{2n} B_{2n+k}(\rho,z) \tag{11.101}$$

where

$$B_{2n+k} = \frac{2n}{\beta\rho}\Upsilon_{2n+k} - j\beta\rho\Upsilon_{2n+k+1} - \beta\rho(2n+k+1)\Upsilon_{2n+k+2} \tag{11.102}$$

The terms Υ_{2n+k}, Υ_{2n+k+1}, and Υ_{2n+k+2} appearing in (11.102) may be efficiently computed using the recurrence relation presented in (11.65).

11.3 THE THIN CIRCULAR LOOP ANTENNA

Over the years a substantial body of literature has been devoted to the subject of loop antennas. This is primarily due to the wide variety of practical applications that loop antennas enjoy when operated in the transmitting as well as the receiving mode. One of the most popular configurations for the loop antenna is circular. There are several references available which contain an excellent treatment of the theory and design of circular loop antennas, *viz.*, King et al. [19], Balanis [20], and Smith [21].

Many of the papers which have been published are strictly concerned with the development of far-field expressions for circular loop antennas with various dimensions and assumed current distributions. The far-field intensities for a circular loop carrying uniform current have been calculated by Foster [22] for loops of any size relative to the wavelength. Expressions are obtained in [23] for the distant field intensity in the plane and on the axis of a thin circular loop. This analysis is restricted to circular loops having circumference equal to an integral number of wavelengths with an assumed sinusoidal current distribution. Approximate far-field formulas have been derived by Glinski [24] for the circular loop antenna with an assumed current distribution of the hyperbolic cosine form. These formulas are valid for ratios of loop perimeter to wavelength on the order of 0.5 or less. This restriction was later removed by Lindsay [25], who obtained exact far-field representations which are valid for hyperbolic cosine current loops of all sizes. Starting with the first five terms of a Fourier cosine series representation of the loop current, Rao [26] derived expressions for the associated far-field components. This technique is accurate for circular loops with circumferences of up to two and one-half wavelengths. Still another treatment involves the derivation of the far-zone electric field intensity components for a circular loop with a traveling-wave current distribution which undergoes an arbitrary phase change over one complete revolution [27].

Papers dealing with the derivation of near-field expressions for current-carrying circular loop antennas are much less common than those which only consider the far-field. This may be attributed to the fact that, in the near region of a loop antenna, more complex forms of the vector potential and corresponding field integrals result which have proven to be extremely difficult to evaluate analytically. As a consequence of this, analytical expressions for the near-zone integrals have only been found for the most elementary case of a uniform current circular loop [28], [29].

Because of the lack of available techniques for evaluating vector potential integrals directly for regions near the loop, an alternative approach is sometimes used in which the loop fields are expressed in terms of two scalar potentials, known as the electric and magnetic Debye potentials [30]. These potentials are solutions of the homogeneous scalar Helmholtz wave equation in spherical coordinates which involve spherical Bessel functions and spherical harmonics. This approach is usually used for the solution of boundary value problems in spherical coordinates; however, it may also be successfully applied to the analysis of sources with given current distributions. For example, Smythe [31] employs this method to determine near-field expressions for a uniform current circular loop. Other applications of this technique which have been reported include the analysis of circular loop antennas with spherical insulation [19], [32–33], as well as spherical cores [19], [34–35].

A direct integration procedure for far-zone vector potentials of thin, circular loop antennas has been developed in [36]. This method is general since it leads to simple integrals which have closed-form solutions for most loop current distributions. However, a comparable integration technique has not been available for evaluating the more complicated near-zone vector potential integrals. A systematic approach is introduced in this section, which is based on Lommel expansions, for the exact integration of general near-zone vector potentials associated with current-carrying loop antennas. A particular example is considered where this new integration technique is employed to find exact solutions to the vector potential and electromagnetic field integrals for loops with a Fourier cosine-series expansion of the current.

The observation is made that degenerate forms of these exact representations lead to simplified expressions for the important special cases of a uniform and cosinusoidal current loop. It is shown that the familiar small-loop approximations, as well as the classical far-field expressions, may be obtained as limiting cases of the more general exact series representations for the uniform current loop introduced in this section. Convenient asymptotic far-field expansions will be derived for the loop with a cosinusoidal current distribution. This far-field analysis for the cosinusoidal loop is generalized to loops having an arbitrary current represented by a Fourier cosine series. Finally, exact near-zone and far-zone expansions are derived for the thin circular loop antenna with a traveling-wave current distribution.

The geometry for the standard circular loop antenna of radius a is depicted in Fig. 11.2. The source point and the field point are designated by the spherical coordinates $(r' = a,\ \theta' = 90°,\ \varphi')$ and (r, θ, φ), respectively. Hence, the distance from the source point on the loop to the field point at some arbitrary location in space is

$$R' = \sqrt{R^2 - 2ar\sin\theta\cos(\varphi - \varphi')} \tag{11.103}$$

where

$$R = \sqrt{r^2 + a^2} \tag{11.104}$$

Since the current is flowing in the φ-direction around the loop, i.e., $I(\varphi)$, then the vector potential may be expressed in the general form [20]

$$\mathbf{A} = A_r(r, \theta, \varphi)\,\hat{\mathbf{r}} + A_\theta(r, \theta, \varphi)\,\hat{\boldsymbol{\theta}} + A_\varphi(r, \theta, \varphi)\,\hat{\boldsymbol{\varphi}} \tag{11.105}$$

where

$$A_r(r, \theta, \varphi) = \frac{\mu a \sin\theta}{4\pi}\int_0^{2\pi} I(\varphi')\sin(\varphi - \varphi')\frac{e^{-j\beta R'}}{R'}\,d\varphi' \tag{11.106}$$

$$A_\theta(r, \theta, \varphi) = \frac{\mu a \cos\theta}{4\pi}\int_0^{2\pi} I(\varphi')\sin(\varphi - \varphi')\frac{e^{-j\beta R'}}{R'}\,d\varphi' \tag{11.107}$$

$$A_\varphi(r, \theta, \varphi) = \frac{\mu a}{4\pi}\int_0^{2\pi} I(\varphi')\cos(\varphi - \varphi')\frac{e^{-j\beta R'}}{R'}\,d\varphi' \tag{11.108}$$

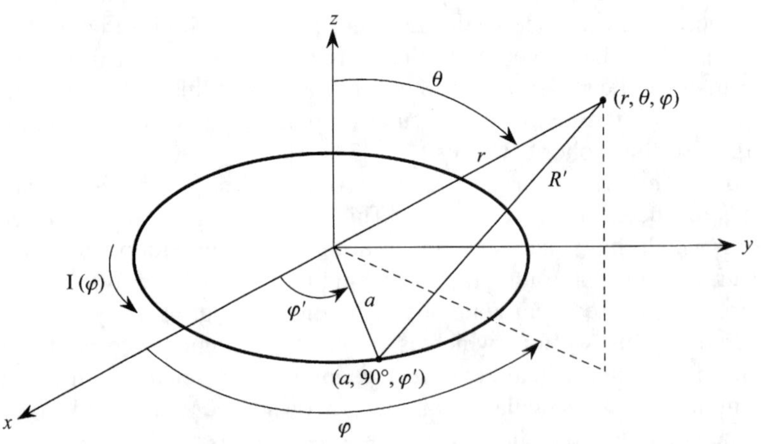

Figure 11.2 Geometry for the circular loop antenna.

Expressions for the electric and magnetic fields of the circular loop may be derived from its vector potential by making use of the relationships

$$\mathbf{H} = \frac{1}{\mu}\nabla \times \mathbf{A} \tag{11.109}$$

$$\mathbf{E} = \frac{1}{j\omega\varepsilon}\nabla \times \mathbf{H} = \frac{1}{j\omega\mu\varepsilon}[\nabla(\nabla \cdot \mathbf{A}) + \beta^2 \mathbf{A}] \tag{11.110}$$

11.3.1 An Exact Integration Procedure for Near-Zone Vector Potentials of Thin Circular Loops

A systematic approach for evaluating the vector potential integrals defined in (11.106)–(11.108) has not been available. In this Section we present a general method for obtaining exact solutions to potential integrals of this type. The resulting series expansions are general in the sense that they are expressed in terms of simple integrals involving the loop current $I(\varphi)$ which, for many cases of practical interest, have closed-form solutions. The derivation of these exact representations begins by recognizing that the vector potential integrals (11.106)–(11.108) may be expressed in the following way:

$$A_r(r, \theta, \varphi) = -\frac{\mu}{j2\beta r}\frac{d}{d\varphi}\Im(r, \theta, \varphi) \tag{11.111}$$

$$A_\theta(r, \theta, \varphi) = -\frac{\mu}{j2\beta r \tan\theta}\frac{d}{d\varphi}\Im(r, \theta, \varphi) \tag{11.112}$$

$$A_\varphi(r, \theta, \varphi) = \frac{\mu}{j2\beta r \cos\theta}\frac{d}{d\theta}\Im(r, \theta, \varphi) \tag{11.113}$$

where the integral defined by

$$\Im(r, \theta, \varphi) = \frac{1}{2\pi}\int_0^{2\pi} I(\varphi')e^{-j\beta R'}\,d\varphi' \tag{11.114}$$

is common to all three components of the vector potential. Next, we make use of the transformations $z = r\cos\theta$, $z' = 0$ and $\rho = r\sin\theta$ in order to show that the series expansion for $e^{-j\beta R'}$ given in (11.18) may be expressed in the form

$$e^{-j\beta R'} = e^{-j\beta R} + \sum_{m=1}^{\infty}\frac{(\beta^2 ar\sin\theta)^m}{m!}\frac{h_{m-1}^{(2)}(\beta R)}{(\beta R)^{m-1}}\cos^m(\varphi - \varphi') \tag{11.115}$$

where R' and R are defined as in (11.103) and (11.104), respectively. Substituting the series (11.115) into (11.114) and integrating term by term leads to

$$\Im(r, \theta, \varphi) = G_0 e^{-j\beta R} + \sum_{m=1}^{\infty} G_m(\varphi)\frac{(\beta^2 ar\sin\theta)^m}{m!}\frac{h_{m-1}^{(2)}(\beta R)}{(\beta R)^{m-1}} \tag{11.116}$$

where

$$G_0 = \frac{1}{2\pi}\int_0^{2\pi} I(\varphi')\,d\varphi' \tag{11.117}$$

and

$$G_m(\varphi) = \frac{1}{2\pi}\int_0^{2\pi} I(\varphi')\cos^m(\varphi - \varphi')\,d\varphi' \tag{11.118}$$

If the current around the loop has the property that $I(\varphi' + 2\pi) = I(\varphi')$, then the integral (11.118) may be expressed in the convenient form

$$G_m(\varphi) = \frac{1}{2\pi}\int_0^{2\pi} I(\varphi + \varphi')\cos^m \varphi'\,d\varphi' \tag{11.119}$$

An exact representation for $\Im(\bullet)$ may be obtained from (11.116), provided closed-form solutions to the integrals (11.117) and (11.118) or (11.119) exist for a particular current distribution. Fortunately, it is possible to evaluate these integrals analytically for the majority of commonly assumed current distributions. Once an exact representation for a specific $\Im(\bullet)$ has been found, then it is a straightforward procedure to determine exact expressions for the vector potential components $A_r(\bullet)$, $A_\theta(\bullet)$, and $A_\varphi(\bullet)$ by substituting the series expansion for $\Im(\bullet)$ into (11.111), (11.112), and (11.113), respectively, and performing the necessary differentiation. Following this procedure leads to a set of exact expansions for the circular loop vector potential components which are given by

$$A_r(r,\theta,\varphi) = -\frac{\mu}{2j\beta r}\sum_{m=1}^{\infty} G'_m(\varphi)\frac{(\beta^2 ar\sin\theta)^m}{m!}\frac{h^{(2)}_{m-1}(\beta R)}{(\beta R)^{m-1}} \tag{11.120}$$

$$A_\theta(r,\theta,\varphi) = -\frac{\mu\cot\theta}{2j\beta r}\sum_{m=1}^{\infty} G'_m(\varphi)\frac{(\beta^2 ar\sin\theta)^m}{m!}\frac{h^{(2)}_{m-1}(\beta R)}{(\beta R)^{m-1}} \tag{11.121}$$

$$A_\varphi(r,\theta,\varphi) = \frac{\mu\beta a}{2j}\sum_{m=1}^{\infty} G_m(\varphi)\frac{(\beta^2 ar\sin\theta)^{m-1}}{(m-1)!}\frac{h^{(2)}_{m-1}(\beta R)}{(\beta R)^{m-1}} \tag{11.122}$$

where

$$G'_m(\varphi) = \frac{d}{d\varphi}G_m(\varphi) \tag{11.123}$$

11.3.2 Examples

11.3.2.1 Fourier Cosine Series Representation of the Loop Current

The procedure introduced in Section 11.3.1 will be followed in this section to derive exact expressions for the vector potential and electromagnetic fields corresponding to loops having an arbitrary current distribution represented by a Fourier decomposition. The representation for the loop current which will be considered is given by the Fourier cosine series

$$I(\varphi) = \sum_{n=0}^{\infty} I_n\cos(n\varphi) \tag{11.124}$$

This form of the current distribution is very general and may be applied to the analysis of the radiation characteristics of loop antennas under a wide variety of circumstances [19], [37–38].

As discussed in the previous section, the key to successfully obtaining exact expressions for the three components of the vector potential lies in the ability to find a closed-form solution to the integral (11.118) or (11.119) for a given current distribution. This may be accomplished in this case by substituting the expression for the loop current (11.124) into (11.119) which then yields

$$G_m(\varphi) = \sum_{n=0}^{\infty} I_n\{A_{mn}\cos(n\varphi) - B_{mn}\sin(n\varphi)\} \tag{11.125}$$

where it may be shown using [39] that

$$A_{mn} = \frac{1}{2\pi}\int_0^{2\pi} \cos^m\varphi'\cos(n\varphi')\,d\varphi' = \frac{1}{\pi}\int_0^{\pi}\cos^m\varphi'\cos(n\varphi')\,d\varphi'$$
$$= \begin{cases} 0, & m<n \\ \dfrac{1}{2^m}\dbinom{m}{k}, & m\geq n \text{ and } m-n=2k \\ 0, & m>n \text{ and } m-n=2k+1 \end{cases} \tag{11.126}$$

and that

$$B_{mn} = \frac{1}{2\pi}\int_0^{2\pi}\cos^m\varphi'\sin(n\varphi')\,d\varphi' = 0 \tag{11.127}$$

The fact that the coefficients $B_{mn} = 0$ for all values of m and n suggests that (11.125) simplifies to

$$G_m(\varphi) = \sum_{n=0}^{\infty} A_{mn}I_n\cos(n\varphi) \tag{11.128}$$

We next use (11.123) in order to show that

$$G'_m(\varphi) = -\sum_{n=1}^{\infty} A_{mn}I_n n\sin(n\varphi) \tag{11.129}$$

Substituting (11.128) and (11.129) into (11.120)–(11.122) leads to the following exact expansions for the vector potential components:

$$A_r(r,\theta,\varphi) = \frac{\mu\beta a\sin\theta}{2j}\sum_{m=1}^{\infty}\sum_{n=1}^{\infty}A_{mn}I_n n\frac{(\beta^2 ar\sin\theta)^{m-1}}{m!}\frac{h_{m-1}^{(2)}(\beta R)}{(\beta R)^{m-1}}\sin(n\varphi) \tag{11.130}$$

$$A_\theta(r,\theta,\varphi) = \frac{\mu\beta a\cos\theta}{2j}\sum_{m=1}^{\infty}\sum_{n=1}^{\infty}A_{mn}I_n n\frac{(\beta^2 ar\sin\theta)^{m-1}}{m!}\frac{h_{m-1}^{(2)}(\beta R)}{(\beta R)^{m-1}}\sin(n\varphi) \tag{11.131}$$

$$A_\phi(r,\theta,\varphi) = \frac{\mu\beta a}{2j}\sum_{m=1}^{\infty}\sum_{n=0}^{\infty}A_{mn}I_n m\frac{(\beta^2 ar\sin\theta)^{m-1}}{m!}\frac{h_{m-1}^{(2)}(\beta R)}{(\beta R)^{m-1}}\cos(n\varphi) \tag{11.132}$$

These expansions may be further reduced to the form

$$A_r(r,\theta,\varphi) = \frac{\mu\beta a\sin\theta}{4j}\sum_{m=1}^{\infty}\sum_{\substack{n=1\\ m-n=2k\\ k=0,1,\ldots}}^{m} I_n n \frac{[(\beta^2 ar\sin\theta)/2]^{m-1}}{[(m-n)/2]!\,[(m+n)/2]!} \times \frac{h_{m-1}^{(2)}(\beta R)}{(\beta R)^{m-1}}\sin(n\varphi) \tag{11.133}$$

$$A_\theta(r,\theta,\varphi) = \frac{\mu\beta a\cos\theta}{4j}\sum_{m=1}^{\infty}\sum_{\substack{n=1\\ m-n=2k\\ k=0,1,\ldots}}^{m} I_n n \frac{[(\beta^2 ar\sin\theta)/2]^{m-1}}{[(m-n)/2]!\,[(m+n)/2]!} \times \frac{h_{m-1}^{(2)}(\beta R)}{(\beta R)^{m-1}}\sin(n\varphi) \tag{11.134}$$

$$A_\varphi(r,\theta,\varphi) = \frac{\mu\beta a}{4j}\sum_{m=1}^{\infty}\sum_{\substack{n=0\\ m-n=2k\\ k=0,1,\ldots}}^{m} I_n m \frac{[(\beta^2 ar\sin\theta)/2]^{m-1}}{[(m-n)/2]!\,[(m+n)/2]!} \times \frac{h_{m-1}^{(2)}(\beta R)}{(\beta R)^{m-1}}\cos(n\varphi) \tag{11.135}$$

by making use of the expression for the coefficients A_{mn} given in (11.126).

Expressions for the electromagnetic field of a circular transmitting loop antenna have been derived in [19, Sec. 9.3] for an assumed Fourier series representation of the current distribution. The final form of these field expressions has been left in terms of various integrals with respect to φ'. The authors of [19, Sec. 9.3] go on to point out that the integrations which appear in these equations have not been carried out analytically, and numerical integration must be used to evaluate them. An alternative approach will be introduced here for evaluating these integrals analytically, thereby avoiding the need to invoke time-consuming numerical integration schemes.

Exact representations for the vector potential components resulting from a general Fourier cosine series expression of the loop current have been successfully derived above. These exact representations will be used here as the basis for the derivation of the associated exact representations for the loop antenna electric and magnetic field components. The availability of these exact expansions is important because they provide an efficient means for evaluation of the near-field quantities and also can lead to a better understanding of the physics involved in the problem.

Expressions for the three components of the magnetic field (which are in terms of the vector potential components) follow directly from (11.109). These expressions are given by

$$H_r(r,\theta,\varphi) = \frac{1}{\mu r\sin\theta}\left[\frac{\partial}{\partial\theta}(\sin\theta A_\varphi) - \frac{\partial}{\partial\varphi}A_\theta\right] \tag{11.136}$$

$$H_\theta(r,\theta,\varphi) = \frac{1}{\mu r}\left[\frac{1}{\sin\theta}\frac{\partial}{\partial\varphi}A_r - \frac{\partial}{\partial r}(rA_\varphi)\right] \tag{11.137}$$

$$H_\varphi(r,\theta,\varphi) = \frac{1}{\mu r}\left[\frac{\partial}{\partial r}(rA_\theta) - \frac{\partial}{\partial\theta}A_r\right] \tag{11.138}$$

Substituting the appropriate expansions from (11.133)–(11.135) into (11.136)–(11.138) and performing the required mathematical operations may be shown to result in the following exact representations for the magnetic field components of the loop

$$H_r(r,\theta,\varphi) = \frac{\beta(\beta a)^2\cos\theta}{8j}\sum_{\substack{m=2\\ m-n=2k\\ k=1,2,\ldots}}^{\infty}\sum_{n=0}^{m-1} C^1_{mn}\cos(n\varphi)[(\beta^2 ar\sin\theta)/2]^{m-2} \times \frac{h^{(2)}_{m-1}(\beta R)}{(\beta R)^{m-1}} \tag{11.139}$$

$$H_\theta(r,\theta,\varphi) = \frac{\beta(\beta r)(\beta a)}{4j}\sum_{\substack{m=1\\ m-n=2k\\ k=0,1,\ldots}}^{\infty}\sum_{n=0}^{m} C^2_{mn}\cos(n\varphi)[(\beta^2 ar\sin\theta)/2]^{m-1} \times \frac{h^{(2)}_{m}(\beta R)}{(\beta R)^{m}} - \frac{\beta(\beta a)^2\sin\theta}{8j}\sum_{\substack{m=2\\ m-n=2k\\ k=1,2,\ldots}}^{\infty}\sum_{n=0}^{m-1} C^1_{mn}\cos(n\varphi) \times [(\beta^2 ar\sin\theta)/2]^{m-2}\frac{h^{(2)}_{m-1}(\beta R)}{(\beta R)^{m-1}} \tag{11.140}$$

$$H_\varphi(r,\theta,\varphi) = -\frac{\beta(\beta r)(\beta a)\cos\theta}{4j}\sum_{\substack{m=1\\ m-n=2k\\ k=0,1,\ldots}}^{\infty}\sum_{n=1}^{m} C^3_{mn}\sin(n\varphi) \times [(\beta^2 ar\sin\theta)/2]^{m-1}\frac{h^{(2)}_{m}(\beta R)}{(\beta R)^{m}} \tag{11.141}$$

where

$$C^1_{mn} = I_n\frac{(m-n)(m+n)}{[(m-n)/2]!\,[(m+n)/2]!} \tag{11.142}$$

$$C^2_{mn} = I_n\frac{m}{[(m-n)/2]!\,[(m+n)/2]!} \tag{11.143}$$

$$C^3_{mn} = I_n\frac{n}{[(m-n)/2]!\,[(m+n)/2]!} \tag{11.144}$$

Decomposing (11.110) into its three scalar components yields

$$E_r(r,\theta,\varphi) = \frac{\eta}{j\beta r\sin\theta}\left[\frac{\partial}{\partial\theta}(\sin\theta H_\varphi) - \frac{\partial}{\partial\varphi}H_\theta\right] \tag{11.145}$$

$$E_\theta(r,\theta,\varphi) = \frac{\eta}{j\beta r}\left[\frac{1}{\sin\theta}\frac{\partial}{\partial\varphi}H_r - \frac{\partial}{\partial r}(rH_\varphi)\right] \tag{11.146}$$

$$E_\varphi(r,\theta,\varphi) = \frac{\eta}{j\beta r}\left[\frac{\partial}{\partial r}(rH_\theta) - \frac{\partial}{\partial\theta}H_r\right] \tag{11.147}$$

Exact representations for the three components of the electric field may be determined from (11.145)–(11.147) by making use of the corresponding exact expansions for the three magnetic field components found in (11.139)–(11.141). This approach leads to the following simplified expansions for the electric field components

$$E_r(r,\theta,\varphi) = \frac{\eta\beta(\beta a)\sin\theta}{4} \times \Bigg[(\beta a)^2 \sum_{\substack{m=1\\ m-n=2k\\ k=0,1,\dots}}^{\infty} \sum_{n=1}^{m} C^3_{mn} \sin(n\varphi)[(\beta^2 ar\sin\theta)/2]^{m-1} \frac{h^{(2)}_{m+1}(\beta R)}{(\beta R)^{m+1}} - \sum_{\substack{m=1\\ m-n=2k\\ k=0,1,\dots}}^{\infty} \sum_{n=1}^{m} C^4_{mn} \sin(n\varphi)[(\beta^2 ar\sin\theta)/2]^{m-1} \frac{h^{(2)}_{m}(\beta R)}{(\beta R)^{m}} \Bigg] \tag{11.148}$$

$$E_\theta(r,\theta,\varphi) = \frac{\eta\beta(\beta a)\cos\theta}{4} \times \Bigg[\sum_{\substack{m=1\\ m-n=2k\\ k=0,1,\dots}}^{\infty} \sum_{n=1}^{m} C^5_{mn} \sin(n\varphi)[(\beta^2 ar\sin\theta)/2]^{m-1} \frac{h^{(2)}_{m}(\beta R)}{(\beta R)^{m}} - \sum_{\substack{m=1\\ m-n=2k\\ k=0,1,\dots}}^{\infty} \sum_{n=1}^{m} C^3_{mn} \sin(n\varphi)[(\beta^2 ar\sin\theta)/2]^{m-1} \frac{h^{(2)}_{m-1}(\beta R)}{(\beta R)^{m-1}} \Bigg] \tag{11.149}$$

$$E_\varphi(r,\theta,\varphi) = \frac{\eta\beta(\beta a)}{4} \times \Bigg[\sum_{\substack{m=1\\ m-n=2k\\ k=0,1,\dots}}^{\infty} \sum_{n=1}^{m} C^6_{mn} \cos(n\varphi)[(\beta^2 ar\sin\theta)/2]^{m-1} \frac{h^{(2)}_{m}(\beta R)}{(\beta R)^{m}} - \sum_{\substack{m=1\\ m-n=2k\\ k=0,1,\dots}}^{\infty} \sum_{n=0}^{m} C^2_{mn} \cos(n\varphi)[(\beta^2 ar\sin\theta)/2]^{m-1} \frac{h^{(2)}_{m-1}(\beta R)}{(\beta R)^{m-1}} \Bigg] \tag{11.150}$$

where

$$C^4_{mn} = (m+1)\,C^3_{mn} = I_n \frac{(m+1)n}{[(m-n)/2]!\,[(m+n)/2]!} \tag{11.151}$$

$$C^5_{mn} = mC^3_{mn} = C^4_{mn} - C^3_{mn} = I_n \frac{mn}{[(m-n)/2]!\,[(m+n)/2]!} \tag{11.152}$$

$$C^6_{mn} = nC^3_{mn} = I_n \frac{n^2}{[(m-n)/2]!\,[(m+n)/2]!} \tag{11.153}$$

11.3.2.2 The Uniform Current Loop Antenna

Suppose that the current distribution on the loop is assumed to be uniform, i.e., $I(\varphi) = I_0$ where I_0 is a constant. The integral form of the vector potential for a uniform current loop is well known and may be expressed as [20]

$$\mathbf{A} = A_\varphi(r, \theta)\,\hat{\boldsymbol{\varphi}} \tag{11.154}$$

where

$$A_\varphi(r, \theta) = \frac{a\mu I_0}{2}\frac{1}{\pi}\int_0^\pi \cos\varphi' \frac{e^{-j\beta R'}}{R'}\,d\varphi' \tag{11.155}$$

and R' from (11.103) reduces to

$$R' = \sqrt{R^2 - 2ar\sin\theta\cos\varphi'} \tag{11.156}$$

The uniform current case may be considered as a degenerate form of the more general Fourier cosine series loop current representation defined in (11.124) for which $I_n = 0$ when $n > 0$. It is easily verified using this interpretation that, for a uniform current loop, the general vector potential expansions given in (11.133)–(11.135) collapse down to

$$A_r = A_\theta = 0 \tag{11.157}$$

$$A_\varphi(r, \theta) = \frac{\beta a\mu I_0}{2j}\sum_{m=1}^{\infty}\frac{[(\beta^2 ar\sin\theta)/2]^{2m-1}}{m!(m-1)!}\frac{h_{2m-1}^{(2)}(\beta R)}{(\beta R)^{2m-1}} \tag{11.158}$$

Hence, we have demonstrated that the integration technique introduced here may be used to derive an exact solution of the vector potential integral for the important case when the loop current is uniform.

In a similar fashion, exact expressions for the magnetic and electric field components for the uniform current circular loop may be obtained as degenerate forms of (11.139)–(11.141) and (11.148)–(11.150). The resulting expansions are

$$H_r(r, \theta) = \frac{\beta(\beta a)^2 I_0\cos\theta}{2j}\sum_{m=1}^{\infty}\frac{[(\beta^2 ar\sin\theta)/2]^{2m-2}}{[(m-1)!]^2}\frac{h_{2m-1}^{(2)}(\beta R)}{(\beta R)^{2m-1}} \tag{11.159}$$

$$H_\theta(r, \theta) = \frac{\beta(\beta a)^2 I_0\sin\theta}{2j}\sum_{m=1}^{\infty}\frac{[(\beta^2 ar\sin\theta)/2]^{2m-2}}{[(m-1)!]^2}\Lambda_m(R) \tag{11.160}$$

$$\mathrm{H}_\varphi = 0 \tag{11.161}$$

$$E_r = E_\theta = 0 \tag{11.162}$$

$$E_\varphi(r, \theta) = -\frac{\eta\beta(\beta a) I_0}{2}\sum_{m=1}^{\infty}\frac{[(\beta^2 ar\sin\theta)/2]^{2m-1}}{m(m-1)!}\frac{h_{2m-1}^{(2)}(\beta R)}{(\beta R)^{2m-1}} \tag{11.163}$$

where

$$\Lambda_m(R) = \frac{(\beta r)^2}{(2m)}\frac{h_{2m}^{(2)}(\beta R)}{(\beta R)^{2m}} - \frac{h_{2m-1}^{(2)}(\beta R)}{(\beta R)^{2m-1}} \tag{11.164}$$

or equivalently

$$\Lambda_m(R) = \left[\frac{4m-1}{2m}\left(\frac{r}{R}\right)^2 - 1\right]\frac{h^{(2)}_{2m-1}(\beta R)}{(\beta R)^{2m-1}} - \frac{1}{2m}\left(\frac{r}{R}\right)^2\frac{h^{(2)}_{2m-2}(\beta R)}{(\beta R)^{2m-2}} \tag{11.165}$$

An alternative series representation for the vector potential given in (11.158) may be found by using the identity (11.35) which provides a relationship between the spherical Hankel and complex exponential functions. The resulting series expansion for the vector potential is

$$A_\varphi(r,\theta) = \frac{\beta a \mu I_0}{2j}e^{-j\beta R}\sum_{m=1}^{\infty}\sum_{k=0}^{2m-1} D^1_{mk}\frac{[(\beta^2 ar\sin\theta)/2]^{2m-1}}{(\beta R)^{2m+k}} \tag{11.166}$$

where

$$D^1_{mk} = \frac{1}{(2j)^k}\frac{(2m+k-1)!}{(2m-k-1)!\,k!}\frac{(-1)^m}{m!(m-1)!} \tag{11.167}$$

An analogous procedure may be followed to obtain an alternative set of expansions for the electromagnetic field components (11.159), (11.160), and (11.163). This may be shown to yield

$$H_r(r,\theta) = \frac{\beta(\beta a)^2 I_0\cos\theta}{2j}e^{-j\beta R}\sum_{m=1}^{\infty}\sum_{k=0}^{2m-1} D^2_{mk}\frac{[(\beta^2 ar\sin\theta)/2]^{2m-2}}{(\beta R)^{2m+k}} \tag{11.168}$$

$$H_\theta(r,\theta) = \frac{\beta(\beta a)^2 I_0\sin\theta}{2j}e^{-j\beta R}\sum_{m=1}^{\infty}\sum_{k=0}^{2m-1} D^2_{mk}\frac{[(\beta^2 ar\sin\theta)/2]^{2m-2}}{(\beta R)^{2m+k}}\Lambda_{mk}(R) \tag{11.169}$$

$$E_\varphi(r,\theta) = -\frac{\eta\beta(\beta a)I_0}{2}e^{-j\beta R}\sum_{m=1}^{\infty}\sum_{k=0}^{2m-1} D^1_{mk}\frac{[(\beta^2 ar\sin\theta)/2]^{2m-1}}{(\beta R)^{2m+k}} \tag{11.170}$$

where

$$\Lambda_{mk}(R) = 1 - \left(\frac{r}{R}\right)^2\frac{(2m+k)+j\beta R}{(2m)} \tag{11.171}$$

$$D^2_{mk} = mD^1_{mk} = \frac{1}{(2j)^k}\frac{(2m+k-1)!}{(2m-k-1)!\,k!}\frac{(-1)^m}{[(m-1)!]^2} \tag{11.172}$$

Computationally efficient recurrence relations for the coefficients D^1_{mk} and D^2_{mk} may be found from (11.167) and (11.172), respectively. These recurrence relations are given by

$$D^{1,2}_{mk} = \frac{(2m-k)(2m+k-1)}{j(2k)}D^{1,2}_{mk-1} \qquad \text{for } m \geq 1 \text{ and } k > 0 \tag{11.173}$$

with

$$D^1_{m0} = \begin{cases} -1, & m = 1 \\ -\dfrac{1}{m(m-1)}D^1_{m-10}, & m > 1 \end{cases} \tag{11.174}$$

and

$$D_{m0}^2 = \begin{cases} -1, & m = 1 \\ -\dfrac{1}{(m-1)^2} D_{m-10}^2, & m > 1 \end{cases} \tag{11.175}$$

The familiar small-loop approximations for the fields [20] may be obtained directly from (11.168)–(11.170) by retaining only the first term ($m = 1$) in each series and recognizing that $R \rightarrow r$ as $a \rightarrow 0$. Under these conditions, (11.168)–(11.170) reduce to the well-known results

$$H_r(r, \theta) \approx \frac{j\beta a^2 I_0 \cos\theta}{2r^2}\left[1 + \frac{1}{j\beta r}\right] e^{-j\beta r} \tag{11.176}$$

$$H_\theta(r, \theta) \approx -\frac{(\beta a)^2 I_0 \sin\theta}{4r}\left[1 + \frac{1}{j\beta r} - \frac{1}{(\beta r)^2}\right] e^{-j\beta r} \tag{11.177}$$

$$E_\varphi(r, \theta) \approx \frac{(\beta a)^2 \eta I_0 \sin\theta}{4r}\left[1 + \frac{1}{j\beta r}\right] e^{-j\beta r} \tag{11.178}$$

In the far zone of the loop, that is, when $\beta r \gg 1$, expressions (11.176)–(11.178) may be used to show that the fields reduce to the simple form given by [20].

$$H_r \approx 0 \tag{11.179}$$

$$H_\theta \approx -\frac{(\beta a)^2 I_0}{4}\frac{e^{-j\beta r}}{r}\sin\theta \tag{11.180}$$

$$H_\varphi = 0 \tag{11.181}$$

$$E_r = E_\theta = 0 \tag{11.182}$$

$$E_\varphi \approx \frac{\eta(\beta a)^2 I_0}{4}\frac{e^{-j\beta r}}{r}\sin\theta \tag{11.183}$$

Hence, we obtain the well-known result that the radiation pattern of an electrically small loop has a $\sin\theta$ variation. The radiation pattern that would be produced by this small loop antenna is illustrated in Fig. 11.3.

Another useful result which may be found from (11.158) is the classical far-field expansion for the uniform current circular loop vector potential. In the far-field region $r \gg a$ and $R \approx r$, which implies that

$$A_\varphi(r, \theta) \approx \frac{\beta a \mu I_0}{2j}\sum_{m=1}^{\infty}\frac{[(\beta a \sin\theta)/2]^{2m-1}}{m!(m-1)!} h_{2m-1}^{(2)}(\beta r) \tag{11.184}$$

Replacing the spherical Hankel functions in (11.184) by their large argument asymptotic expansion

$$h_{2m-1}^{(2)}(\beta r) \sim (-1)^m \frac{e^{-j\beta r}}{(\beta r)} \qquad \text{as} \qquad \beta r \rightarrow \infty \tag{11.185}$$

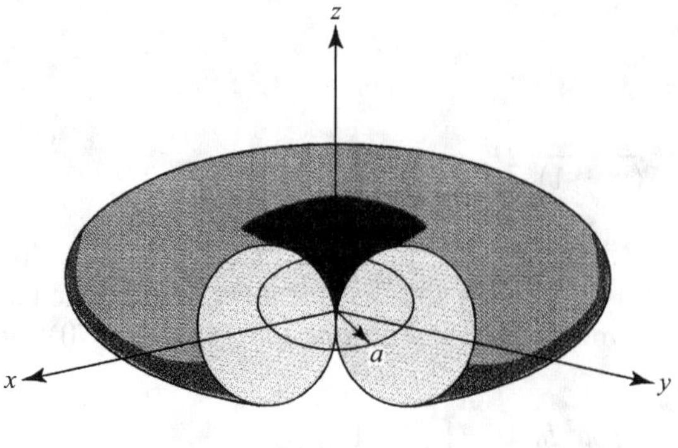

Figure 11.3 Radiation pattern for an electrically small loop antenna.

leads to

$$A_\varphi(r,\theta) \approx \frac{a\mu I_0}{2j}\frac{e^{-j\beta r}}{r}\sum_{m=1}^{\infty}\frac{(-1)^m[(\beta a\sin\theta)/2]^{2m-1}}{m!(m-1)!} \tag{11.186}$$

but

$$J_1(\beta a\, sin\, \theta) = \sum_{m=0}^{\infty}\frac{(-1)^m[(\beta a\sin\theta)/2]^{2m+1}}{m!(m+1)!} \tag{11.187}$$

so that

$$A_\varphi(r,\theta) \approx j\frac{a\mu I_0}{2}\frac{e^{-j\beta r}}{r}J_1(\beta a\sin\theta) \tag{11.188}$$

which is the desired far-field result [20]. The corresponding expressions for the far electric and magnetic field components follow directly as

$$H_r \approx 0 \tag{11.189}$$

$$H_\theta \approx -\frac{(\beta a)I_0}{2}\frac{e^{-j\beta r}}{r}J_1(\beta a\sin\theta) \tag{11.190}$$

$$H_\varphi = 0 \tag{11.191}$$

$$E_r = E_\theta = 0 \tag{11.192}$$

$$E_\varphi \approx \frac{\eta(\beta a)I_0}{2}\frac{e^{-j\beta r}}{r}J_1(\beta a\sin\theta) \tag{11.193}$$

11.3.2.3 The Cosinusoidal Current Loop Antenna

The uniform current analysis which was presented in Section 11.3.2.2 is valid for electrically small loops. However, as the radius of the loop increases, the current is no longer uniform as it begins to vary along the circumference of the loop. One common assumption for the variation of the loop current under these conditions is a cosinusoidal distribution [25]. Just as in the case of a uniformly distributed current, the cosinusoidal current distribution defined by $I(\varphi) = I_p\cos(p\varphi)$ may be considered a degenerate form of the Fourier representation (11.124) for which $I_n = 0$ when $n \neq p$.

This suggests that exact expressions for the three components of the cosinusoidal current vector potential may be obtained as special cases of the more general expansions (11.133)–(11.135). Under these circumstances, it may be shown that (11.133)–(11.135) reduce to

$$A_r(r,\theta,\varphi) = \frac{\mu\beta a I_p \sin(p\varphi)\sin\theta}{4j}\sum_{m=0}^{\infty}\frac{p[(\beta^2 ar\sin\theta)/2]^{2m+p-1}}{m!(m+p)!}\frac{h^{(2)}_{2m+p-1}(\beta R)}{(\beta R)^{2m+p-1}} \tag{11.194}$$

$$A_\theta(r,\theta,\varphi) = \frac{\mu\beta a I_p \sin(p\varphi)\cos\theta}{4j}\sum_{m=0}^{\infty}\frac{p[(\beta^2 ar\sin\theta)/2]^{2m+p-1}}{m!(m+p)!}\frac{h^{(2)}_{2m+p-1}(\beta R)}{(\beta R)^{2m+p-1}} \tag{11.195}$$

$$A_\varphi(r,\theta,\varphi) = \frac{\mu\beta a I_p \cos(p\varphi)}{4j}\sum_{m=0}^{\infty}\frac{(2m+p)[(\beta^2 ar\sin\theta)/2]^{2m+p-1}}{m!(m+p)!} \times\frac{h^{(2)}_{2m+p-1}(\beta R)}{(\beta R)^{2m+p-1}} \tag{11.196}$$

The corresponding degenerate forms of the magnetic field components (11.139)–(1.141) and the electric field components (11.148)–(11.150) are

$$H_r(r,\theta,\varphi) = \frac{\beta(\beta a)^2 I_p\cos(p\varphi)\cos\theta}{2j} \cdot\sum_{m=1}^{\infty}\frac{[(\beta^2 ar\sin\theta)/2]^{2m+p-2}}{(m-1)!(m+p-1)!}\frac{h^{(2)}_{2m+p-1}(\beta R)}{(\beta R)^{2m+p-1}} \tag{11.197}$$

$$\begin{aligned}H_\theta(r,\theta,\varphi) &= \frac{\beta(\beta a)(\beta r) I_p\cos(p\varphi)}{4j} \cdot\sum_{m=0}^{\infty}\frac{(2m+p)[(\beta^2 ar\sin\theta)/2]^{2m+p-1}}{m!(m+p)!}\frac{h^{(2)}_{2m+p}(\beta R)}{(\beta R)^{2m+p}}\\ &\quad -\frac{\beta(\beta a)^2 I_p\cos(p\varphi)\sin\theta}{2j} \cdot\sum_{m=1}^{\infty}\frac{[(\beta^2 ar\sin\theta)/2]^{2m+p-2}}{(m-1)!(m+p-1)!}\frac{h^{(2)}_{2m+p-1}(\beta R)}{(\beta R)^{2m+p-1}}\end{aligned} \tag{11.198}$$

$$H_\varphi(r,\theta,\varphi) = -\frac{\beta(\beta a)(\beta r) I_p\sin(p\varphi)\cos\theta}{4j} \cdot\sum_{m=0}^{\infty}\frac{p[(\beta^2 ar\sin\theta)/2]^{2m+p-1}}{m!(m+p)!}\frac{h^{(2)}_{2m+p}(\beta R)}{(\beta R)^{2m+p}} \tag{11.199}$$

$$\begin{aligned}E_r(r,\theta,\varphi) &= \frac{\eta\beta(\beta a) I_p\sin(p\varphi)\sin\theta}{4}\\ &\quad\cdot\Bigg[(\beta a)^2\sum_{m=0}^{\infty}\frac{p[(\beta^2 ar\sin\theta)/2]^{2m+p-1}}{m!(m+p)!}\frac{h^{(2)}_{2m+p+1}(\beta R)}{(\beta R)^{2m+p+1}}\\ &\quad -\sum_{m=0}^{\infty}\frac{p(2m+p+1)[(\beta^2 ar\sin\theta)/2]^{2m+p-1}}{m!(m+p)!}\frac{h^{(2)}_{2m+p}(\beta R)}{(\beta R)^{2m+p}}\Bigg]\end{aligned} \tag{11.200}$$

$$E_\theta(r,\theta,\varphi) = \frac{\eta\beta(\beta a) I_p \cos(p\varphi)\cos\theta}{4} \cdot \left[\sum_{m=0}^{\infty} \frac{p(2m+p)[(\beta^2 ar\sin\theta)/2]^{2m+p-1}}{m!\,(m+p)!} \frac{h^{(2)}_{2m+p}(\beta R)}{(\beta R)^{2m+p}} - \sum_{m=0}^{\infty} \frac{p[(\beta^2 ar\sin\theta)/2]^{2m+p-1}}{m!\,(m+p)!} \frac{h^{(2)}_{2m+p-1}(\beta R)}{(\beta R)^{2m+p-1}} \right] \tag{11.201}$$

$$E_\varphi(r,\theta,\varphi) = \frac{\eta\beta(\beta a) I_p \cos(p\varphi)}{4} \cdot \left[\sum_{m=0}^{\infty} \frac{p^2[(\beta^2 ar\sin\theta)/2]^{2m+p-1}}{m!\,(m+p)!} \frac{h^{(2)}_{2m+p}(\beta R)}{(\beta R)^{2m+p}} - \sum_{m=0}^{\infty} \frac{(2m+p)[(\beta^2 ar\sin\theta)/2]^{2m+p-1}}{m!\,(m+p)!} \frac{h^{(2)}_{2m+p-1}(\beta R)}{(\beta R)^{2m+p-1}} \right] \tag{11.202}$$

Convenient far-zone expressions for the cosinusoidal current loop vector potential components may be obtained as limiting cases of (11.195) and (11.196). These far-zone representations are

$$A_\theta(r,\theta,\varphi) \approx \frac{\mu(\beta a)p(j)^p I_p \sin(p\varphi)\cos\theta}{2j} \frac{e^{-j\beta r}}{\beta r} \frac{J_p(w)}{w} \tag{11.203}$$

$$A_\varphi(r,\theta,\varphi) \approx \frac{\mu(\beta a)(j)^p I_p \cos(p\varphi)}{2j} \frac{e^{-j\beta r}}{\beta r} J_p'(w) \tag{11.204}$$

where

$$J_p'(w) = \frac{p}{w} J_p(w) - J_{p+1}(w) = \frac{1}{2}[J_{p-1}(w) - J_{p+1}(w)] \tag{11.205}$$

and

$$w = \beta a \sin\theta \tag{11.206}$$

The expressions given in (11.203) and (11.204) were derived from (11.195) and (11.196), respectively, by making use of the fact that in the far-field $R \approx r$ and

$$h^{(2)}_{2m+p-1}(\beta r) \sim (-1)^m (j)^p \frac{e^{-j\beta r}}{\beta r} \quad \text{as} \quad \beta r \to \infty \tag{11.207}$$

Finally, the approximations for the vector potential components given in (11.203) and (11.204) may be used to derive the following far-zone expansions for the fields

$$H_r \approx 0 \tag{11.208}$$

$$H_\theta \approx \frac{(\beta a)(j)^p I_p \cos(p\varphi)}{2} \frac{e^{-j\beta r}}{r} J_p'(w) \tag{11.209}$$

$$H_\varphi \approx -\frac{(\beta a)p(j)^p I_p \sin(p\varphi)\cos\theta}{2} \frac{e^{-j\beta r}}{r} \frac{J_p(w)}{w} \tag{11.210}$$

$$E_r \approx 0 \tag{11.211}$$

$$E_\theta \approx -\frac{\eta(\beta a)p(j)^p I_p \sin(p\varphi)\cos\theta}{2}\frac{e^{-j\beta r}}{r}\frac{J_p(w)}{w} \tag{11.212}$$

$$E_\varphi \approx -\frac{\eta(\beta a)(j)^p I_p \cos(p\varphi)}{2}\frac{e^{-j\beta r}}{r}J_p'(w) \tag{11.213}$$

11.3.2.4 General Far-Field Approximations

The Fourier series representation of the loop current given in (11.124) may be considered a weighted sum of cosinusoidal currents. Far-zone approximations for the vector potential components of a cosinusoidal current loop have been derived previously, in Section 11.3.2.3. Hence, simplified far-zone expressions of A_θ and A_φ for arbitrary loop current distributions represented by Fourier cosine series may be conveniently obtained as linear combinations of (11.203) and (11.204), respectively. The resulting general far-field approximations are

$$A_\theta(r,\theta,\varphi) \approx \frac{\mu a \cos\theta}{2j}\frac{e^{-j\beta r}}{r}\sum_{n=1}^{\infty} n(j)^n I_n \sin(n\varphi)\frac{J_n(w)}{w} \tag{11.214}$$

$$A_\varphi(r,\theta,\varphi) \approx \frac{\mu a}{2j}\frac{e^{-j\beta r}}{r}\sum_{n=0}^{\infty}(j)^n I_n \cos(n\varphi)J_n'(w) \tag{11.215}$$

which may be used to show that

$$H_r \approx 0 \tag{11.216}$$

$$H_\theta \approx \frac{\beta a}{2}\frac{e^{-j\beta r}}{r}\sum_{n=0}^{\infty}(j)^n I_n \cos(n\varphi)J_n'(w) \tag{11.217}$$

$$H_\varphi \approx -\frac{\beta a \cos\theta}{2}\frac{e^{-j\beta r}}{r}\sum_{n=1}^{\infty} n(j)^n I_n \sin(n\varphi)\frac{J_n(w)}{w} \tag{11.218}$$

$$E_r \approx 0 \tag{11.219}$$

$$E_\theta \approx -\frac{\eta\beta a \cos\theta}{2}\frac{e^{-j\beta r}}{r}\sum_{n=1}^{\infty} n(j)^n I_n \sin(n\varphi)\frac{J_n(w)}{w} \tag{11.220}$$

$$E_\varphi \approx -\frac{\eta\beta a}{2}\frac{e^{-j\beta r}}{r}\sum_{n=0}^{\infty}(j)^n I_n \cos(n\varphi)J_n'(w) \tag{11.221}$$

11.3.2.5 The Traveling-Wave Current Loop Antenna

A methodology for evaluating the complicated near-field integrals associated with circular loop antennas has been introduced in Section 11.3.1. This technique is very general by virtue of the fact that it is applicable to loops of any size and with arbitrary current distribution. Thus far, we have only considered loop antennas which have a standing-wave current distribution. In this section, however, we show how this exact integration procedure applies not only to the analysis of standing-wave loop antennas, but also to the analysis of traveling-wave loop antennas.

Previous investigations of traveling-wave loop antennas have focused on the derivation of far-field representations and are only special cases of the analysis presented here. For example, far-field expressions have been found in [23], [40], and [41] for a single-turn traveling-wave loop where the total change in phase over one complete revolution around the loop is an integer multiple of 2π. This analysis was later extended in [27] to include the more general case of a loop with a traveling-wave current distribution of the form

$$I(\varphi) = I_0 e^{-j\gamma\varphi} \tag{11.222}$$

where the parameter γ can in general be complex with a negative imaginary part. In this case, the phase changes over one complete revolution by an amount equal to $2\pi\Re(\gamma)$ radians. A method for producing a traveling-wave current distribution of this type has been demonstrated in [42]. A summary of this technique may also be found in [43].

The first step in the process of obtaining exact near-zone representations for the vector potential components of the traveling-wave loop antenna is to evaluate the integral

$$G_m(\varphi) = \frac{I_0}{2\pi}\int_0^{2\pi} e^{-j\gamma\varphi'} \cos^m(\varphi - \varphi')\,d\varphi' \tag{11.223}$$

which results from substituting (11.222) into (11.118). It may be shown that this integral has the following closed-form solution:

$$\begin{aligned} G_{2n}(\varphi) = I_0 e^{-j\gamma\pi}\,\text{sinc}(\gamma\pi)\Bigg\{&\frac{1}{2^{2n}}\binom{2n}{n} \\ &+ \frac{1}{2^{2n-1}}\sum_{k=1}^{n}\binom{2n}{n-k}\frac{\gamma}{(2k)^2-\gamma^2}[j(2k)\sin(2k)\varphi - \gamma\cos(2k)\varphi]\Bigg\} \end{aligned} \tag{11.224}$$

when m is even and

$$\begin{aligned} G_{2n+1}(\varphi) = I_0 e^{-j\gamma\pi}\,&\text{sinc}(\gamma\pi) \\ &\cdot \frac{1}{2^{2n}}\sum_{k=0}^{n}\binom{2n+1}{n-k}\frac{\gamma}{(2k+1)^2-\gamma^2} \\ &\cdot [j(2k+1)\sin(2k+1)\varphi - \gamma\cos(2k+1)\varphi] \end{aligned} \tag{11.225}$$

when m is odd. The function sinc(●) which appears in (11.224) and (11.225) is defined as

$$\text{sinc}(z) = \frac{\sin(z)}{z} \tag{11.226}$$

Taking the derivative with respect to φ of both sides of (11.224) and (11.225) yields

$$G'_{2n}(\varphi) = I_0 e^{-j\gamma\pi}\,\mathrm{sinc}(\gamma\pi) \cdot \frac{1}{2^{2n-1}} \sum_{k=1}^{n} \binom{2n}{n-k} \frac{\gamma(2k)}{(2k)^2 - \gamma^2} [j(2k)\cos(2k)\varphi + \gamma \sin(2k)\varphi] \tag{11.227}$$

and

$$G'_{2n+1}(\varphi) = I_0 e^{-j\gamma\pi}\,\mathrm{sinc}(\gamma\pi) \cdot \frac{1}{2^{2n}} \sum_{k=0}^{n} \binom{2n+1}{n-k} \frac{\gamma(2k+1)}{(2k+1)^2 - \gamma^2} \cdot [j(2k+1)\cos(2k+1)\varphi + \gamma \sin(2k+1)\varphi] \tag{11.228}$$

At this point in the development, we recognize that (11.120)–(11.122) may be expressed in the convenient form

$$A_r(r,\theta,\varphi) = \frac{j\mu}{2\beta r}\left\{ \sum_{n=0}^{\infty} G'_{2n+1}(\varphi) \frac{(\beta^2 ar\sin\theta)^{2n+1}}{(2n+1)!} \frac{h_{2n}^{(2)}(\beta R)}{(\beta R)^{2n}} + \sum_{n=1}^{\infty} G'_{2n}(\varphi) \frac{(\beta^2 ar\sin\theta)^{2n}}{(2n)!} \frac{h_{2n-1}^{(2)}(\beta R)}{(\beta R)^{2n-1}} \right\} \tag{11.229}$$

$$A_\theta(r,\theta,\varphi) = \cot\theta\, A_r(r,\theta,\varphi) \tag{11.230}$$

$$A_\varphi(r,\theta,\varphi) = \frac{\mu\beta a}{2j}\left\{ \sum_{n=0}^{\infty} G_{2n+1}(\varphi) \frac{(\beta^2 ar\sin\theta)^{2n}}{(2n)!} \frac{h_{2n}^{(2)}(\beta R)}{(\beta R)^{2n}} + \sum_{n=1}^{\infty} G_{2n}(\varphi) \frac{(\beta^2 ar\sin\theta)^{2n-1}}{(2n-1)!} \frac{h_{2n-1}^{(2)}(\beta R)}{(\beta R)^{2n-1}} \right\} \tag{11.231}$$

Substituting (11.227) and (11.228) into (11.229) leads to the following exact series representation for $A_r(\bullet)$:

$$\begin{aligned} A_r(r,\theta,\varphi) = {} & \frac{j\mu I_0}{\beta r} e^{-j\gamma\pi}\,\mathrm{sinc}(\gamma\pi) \\ & \cdot \left\{ \sum_{n=0}^{\infty} \sum_{k=0}^{n} \frac{\gamma(2k+1)}{(2k+1)^2 - \gamma^2} \frac{[(\beta^2 ar\sin\theta)/2]^{2n+1}}{(n-k)!\,(n+k+1)!} \frac{h_{2n}^{(2)}(\beta R)}{(\beta R)^{2n}} \right. \\ & \cdot [j(2k+1)\cos(2k+1)\varphi + \gamma\sin(2k+1)\varphi] \\ & + \sum_{n=1}^{\infty} \sum_{k=1}^{n} \frac{\gamma(2k)}{(2k)^2 - \gamma^2} \frac{[(\beta^2 ar\sin\theta)/2]^{2n}}{(n-k)!\,(n+k)!} \frac{h_{2n-1}^{(2)}(\beta R)}{(\beta R)^{2n-1}} \\ & \left. \cdot [j(2k)\cos(2k)\varphi + \gamma\sin(2k)\varphi] \right\} \end{aligned} \tag{11.232}$$

Likewise, (11.224) and (11.225) may be substituted into (11.231) in order to arrive at a series expansion for $A_\varphi(\bullet)$ given by

$$\begin{aligned} A_\varphi(r,\theta,\varphi) = \frac{\mu\beta a I_0}{2j} e^{-j\gamma\pi}\,\mathrm{sinc}(\gamma\pi)\Bigg\{ &\sum_{n=1}^{\infty} \frac{[(\beta^2 ar\sin\theta)/2]^{2n-1}}{n!(n-1)!} \frac{h_{2n-1}^{(2)}(\beta R)}{(\beta R)^{2n-1}} \\ &+ \sum_{n=0}^{\infty}\sum_{k=0}^{n} \frac{\gamma(2n+1)}{(2k+1)^2-\gamma^2} \frac{[(\beta^2 ar\sin\theta)/2]^{2n}}{(n-k)!(n+k+1)!} \frac{h_{2n}^{(2)}(\beta R)}{(\beta R)^{2n}} \\ &\cdot [j(2k+1)\sin(2k+1)\varphi - \gamma\cos(2k+1)\varphi] \\ &+ \sum_{n=1}^{\infty}\sum_{k=1}^{n} \frac{\gamma(2n)}{(2k)^2-\gamma^2} \frac{[(\beta^2 ar\sin\theta)/2]^{2n-1}}{(n-k)!(n+k)!} \frac{h_{2n-1}^{(2)}(\beta R)}{(\beta R)^{2n-1}} \\ &\cdot [j(2k)\sin(2k)\varphi - \gamma\cos(2k)\varphi] \Bigg\} \end{aligned} \tag{11.233}$$

We next make use of (11.136)–(11.138) in order to derive exact series representations for the near-zone magnetic field components produced by the traveling-wave loop. The resulting magnetic field expansions are

$$\begin{aligned} H_r(r,\theta,\varphi) = \frac{\beta(\beta a)^2 I_0\cos\theta}{4j} e^{-j\gamma\pi}\,\mathrm{sinc}(\gamma\pi)\Bigg\{ &2\sum_{n=1}^{\infty} \frac{[(\beta^2 ar\sin\theta)/2]^{2n-2}}{[(n-1)]!} \frac{h_{2n-1}^{(2)}(\beta R)}{(\beta R)^{2n-1}} \\ &+\gamma \sum_{n=0}^{\infty}\sum_{k=0}^{n} \left[\frac{(2k+1)^2-(2n+1)^2}{(2k+1)^2-\gamma^2}\right] \frac{[(\beta^2 ar\sin\theta)/2]^{2n-1}}{(n-k)!(n+k+1)!} \frac{h_{2n}^{(2)}(\beta R)}{(\beta R)^{2n}} \\ &\cdot [\gamma\cos(2k+1)\varphi - j(2k+1)\sin(2k+1)\varphi] \\ &+\gamma \sum_{n=1}^{\infty}\sum_{k=1}^{n} \left[\frac{(2k)^2-(2n)^2}{(2k)^2-\gamma^2}\right] \frac{[(\beta^2 ar\sin\theta)/2]^{2n-2}}{(n-k)!(n+k)!} \frac{h_{2n-1}^{(2)}(\beta R)}{(\beta R)^{2n-1}} \\ &\cdot [\gamma\cos(2k)\varphi - j(2k)\sin(2k)\varphi] \Bigg\} \end{aligned} \tag{11.234}$$

$$\begin{aligned} H_\theta(r,\theta,\varphi) = \frac{j\beta(\beta a) I_0}{4\beta r} e^{-j\gamma\pi}\,\mathrm{sinc}(\gamma\pi)\Bigg\{ &\gamma^2 \sum_{n=1}^{\infty} \frac{[(\beta^2 ar\sin\theta)/2]^{2n-1}}{(n!)^2} \Lambda_{2n}^{0}(R) \\ &- 2\gamma \sum_{n=0}^{\infty}\sum_{k=0}^{n} \frac{[(\beta^2 ar\sin\theta)/2]^{2n}}{(n-k)!(n+k+1)!} \Lambda_{2n+1}^{2k+1}(R) \\ &\cdot [j(2k+1)\sin(2k+1)\varphi - \gamma\cos(2k+1)\varphi] \\ &- 2\gamma \sum_{n=1}^{\infty}\sum_{k=1}^{n} \frac{[(\beta^2 ar\sin\theta)/2]^{2n-1}}{(n-k)!(n+k)!} \Lambda_{2n}^{2k}(R) \\ &\cdot [j(2k)\sin(2k)\varphi - \gamma\cos(2k)\varphi] \Bigg\} \end{aligned} \tag{11.235}$$

$$H_\varphi(r,\theta,\varphi) = \frac{\beta(\beta a)(\beta r) I_0 \cos\theta}{2j} e^{-j\gamma\pi}\,\mathrm{sinc}(\gamma\pi)$$
$$\cdot\left\{\gamma \sum_{n=0}^{\infty}\sum_{k=0}^{n} \frac{(2k+1)}{(2k+1)^2-\gamma^2}\,\frac{[(\beta^2 ar\sin\theta)/2]^{2n}}{(n-k)!\,(n+k+1)!}\,\frac{h_{2n+1}^{(2)}(\beta R)}{(\beta R)^{2n+1}}\right.$$
$$\cdot[j(2k+1)\cos(2k+1)\varphi + \gamma\sin(2k+1)\varphi] \tag{11.236}$$
$$+\gamma \sum_{n=1}^{\infty}\sum_{k=1}^{n} \frac{(2k)}{(2k)^2-\gamma^2}\,\frac{[(\beta^2 ar\sin\theta)/2]^{2n-1}}{(n-k)!\,(n+k)!}\,\frac{h_{2n}^{(2)}(\beta R)}{(\beta R)^{2n}}$$
$$\left.\cdot[j(2k)\cos(2k)\varphi + \gamma\sin(2k)\varphi]\right\}$$

where

$$\Lambda_n^k(R) = \left[\frac{k^2-n^2}{k^2-\gamma^2}\right]\frac{h_{n-1}^{(2)}(\beta R)}{(\beta R)^{n-1}} + (\beta r)^2\left[\frac{n}{k^2-\gamma^2}\right]\frac{h_n^{(2)}(\beta R)}{(\beta R)^n} \tag{11.237}$$

There are several advantages and useful applications of the mathematically exact near-field expansions for the traveling-wave loop given in (11.234)–(11.236). For instance, they can be used to gain additional insight into the problem physics as well as provide general expressions which contain all of the classical approximations as special cases. These series expansions were implemented on a computer in order to investigate the near-zone magnetic field behavior of a traveling-wave loop antenna. -Figure 11.4 contains plots of the total magnetic field intensity as a function of φ for

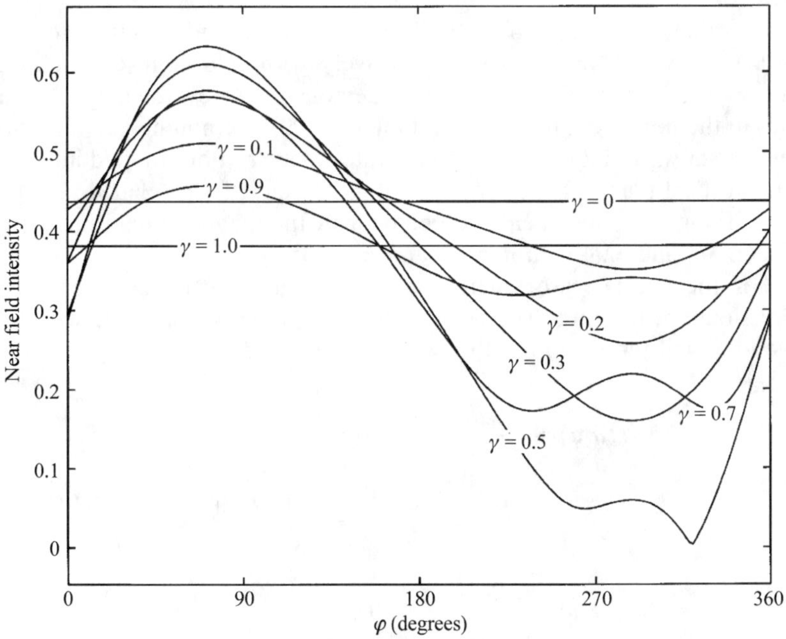

Figure 11.4 Total near-zone magnetic field intensity as a function of φ for several different values of the parameter γ. In this case $I_0 = 1$ A, $a = \lambda/2\pi$ ($\beta a = 1$), $\theta = 90°$, and $r = \lambda/2$ ($\beta r = \pi$).

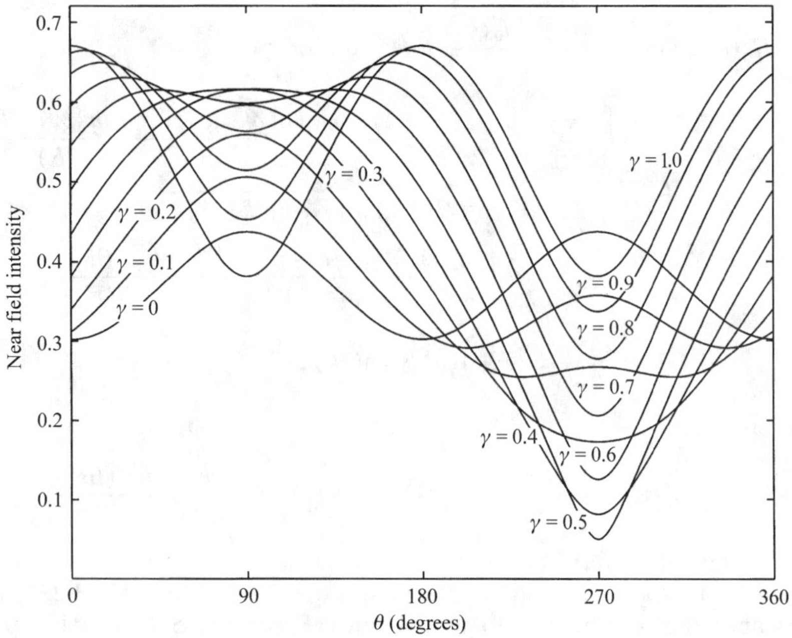

Figure 11.5 Total near-zone magnetic field intensity as a function of θ for several different values of the parameter γ. In this case $I_0 = 1$ A, $a = \lambda/2\pi$ ($\beta a = 1$), $\varphi = 90°$, and $r = \lambda/2$ ($\beta r = \pi$).

various values of the parameter γ. The loop current and radius were assumed to be $I_0 = 1$ A and $a = \lambda/2\pi$ ($\beta a = 1$), respectively, with field point coordinates of $\theta = 90°$ and $r = \lambda/2$ ($\beta r = \pi$). This set of curves demonstrates how the magnetic field intensity varies in the horizontal plane of the loop ($\theta = 90°$) for noninteger values of γ. Similar plots are shown in Fig. 11.5 which illustrates the variations in field intensity produced in the vertical plane of the loop (i.e., $\varphi = 90°$ and $0° \leq \theta \leq 360°$). Finally, the curves shown in Fig. 11.6 document the behavior of the magnetic field intensity as r varies between $\lambda/4$ and $3\lambda/4$ with $\theta = 90°$ and $\varphi = 90°$.

For the special case where γ is an integer, it can be shown that the exact expressions for the three vector potential components associated with the traveling-wave loop will reduce to the following convenient form

$$A_r(r, \theta, \varphi) = \frac{\mu \beta a I_0 \sin\theta}{4}(2p)e^{-j2p\varphi} \cdot \sum_{n=p}^{\infty} \binom{2n}{n-p} \frac{[(\beta^2 ar\sin\theta)/2]^{2n-1}}{(2n)!} \frac{h_{2n-1}^{(2)}(\beta R)}{(\beta R)^{2n-1}} \tag{11.238}$$

$$A_\theta(r, \theta, \varphi) = \frac{\mu \beta a I_0 \cos\theta}{4}(2p)e^{-j2p\varphi} \cdot \sum_{n=p}^{\infty} \binom{2n}{n-p} \frac{[(\beta^2 ar\sin\theta)/2]^{2n-1}}{(2n)!} \frac{h_{2n-1}^{(2)}(\beta R)}{(\beta R)^{2n-1}} \tag{11.239}$$

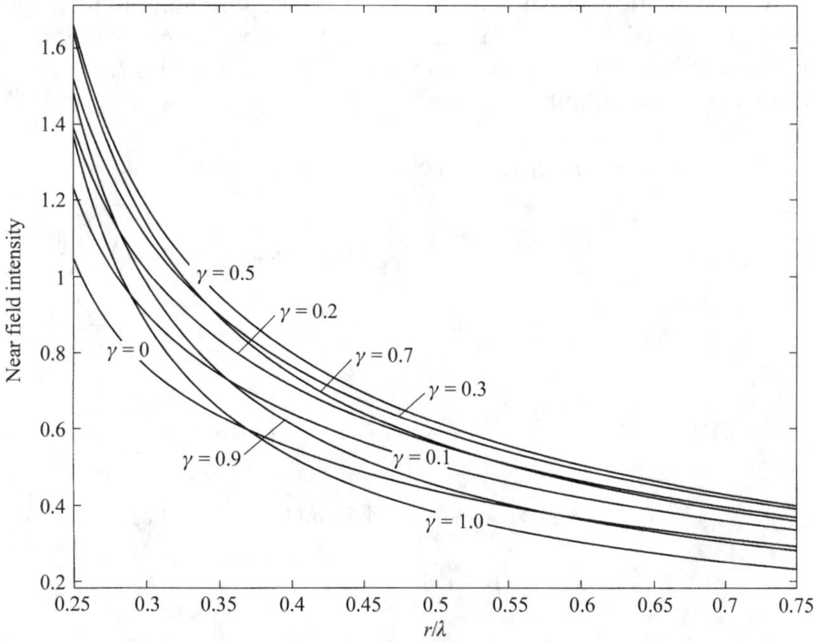

Figure 11.6 Total near-zone magnetic field intensity as a function of r for several different values of the parameter γ. In this case $I_0 = 1$ A, $a = \lambda/2\pi$ ($\beta a = 1$), $\theta = 90°$, and $\varphi = 90°$.

$$A_\varphi(r,\theta,\varphi) = \frac{\mu\beta a I_0}{4j} e^{-j2p\varphi} \cdot \sum_{n=p}^{\infty} \binom{2n+1}{n-p} \frac{[(\beta^2 ar\sin\theta)/2]^{2n-1}}{(2n-1)!} \frac{h_{2n-1}^{(2)}(\beta R)}{(\beta R)^{2n-1}} \tag{11.240}$$

when γ is even (i.e., $\gamma = 2p$ for $p = 0, 1, \ldots$) and

$$A_r(r,\theta,\varphi) = \frac{\mu\beta a I_0 \sin\theta}{4}(2p+1)e^{-j(2p+1)\varphi} \cdot \sum_{n=p}^{\infty} \binom{2n+1}{n-p} \frac{[(\beta^2 ar\sin\theta)/2]^{2n}}{(2n+1)!} \frac{h_{2n}^{(2)}(\beta R)}{(\beta R)^{2n}} \tag{11.241}$$

$$A_\theta(r,\theta,\varphi) = \frac{\mu\beta a I_0 \cos\theta}{4}(2p+1)e^{-j(2p+1)\varphi} \cdot \sum_{n=p}^{\infty} \binom{2n+1}{n-p} \frac{[(\beta^2 ar\sin\theta)/2]^{2n}}{(2n+1)!} \frac{h_{2n}^{(2)}(\beta R)}{(\beta R)^{2n}} \tag{11.242}$$

$$A_\varphi(r,\theta,\varphi) = \frac{\mu\beta a I_0}{4j} e^{-j(2p+1)\varphi} \cdot \sum_{n=p}^{\infty} \binom{2n}{n-p} \frac{[(\beta^2 ar\sin\theta)/2]^{2n}}{(2n)!} \frac{h_{2n}^{(2)}(\beta R)}{(\beta R)^{2n}} \tag{11.243}$$

when γ is odd (i.e., $\gamma = 2p + 1$ for $p = 0, 1, \ldots$).

An asymptotic evaluation of the exact series representations for $A_\theta(\bullet)$ and $A_\varphi(\bullet)$ can be performed in order to derive useful far-zone approximations. Following a similar procedure to that outlined in Sections 11.3.2.2 and 11.3.2.3 leads to the desired set of far-zone expansions

$$A_\theta(r,\theta,\varphi) \sim j\mu(\beta a)\, I_0 \cos\theta\, e^{-j\gamma\pi}\,\mathrm{sinc}(\gamma\pi)\frac{e^{-j\beta r}}{\beta r} \cdot \sum_{k=1}^{\infty}\frac{\gamma k}{k^2-\gamma^2}(j)^k[jk\cos(k\varphi)+\gamma\sin(k\varphi)]\frac{J_k(w)}{w} \quad \text{as} \quad \beta r \to \infty \tag{11.244}$$

and

$$A_\varphi(r,\theta,\varphi) \sim \frac{\mu(\beta a)\, I_0}{2j} e^{-j\gamma\pi}\,\mathrm{sinc}(\gamma\pi)\frac{e^{-j\beta r}}{\beta r}\left\{-J_1(w) + 2\sum_{k=1}^{\infty}\frac{\gamma}{k^2-\gamma^2}(j)^k[jk\sin(k\varphi)-\gamma\cos(k\varphi)]J'_k(w)\right\} \quad \text{as} \quad \beta r \to \infty \tag{11.245}$$

where w has been defined in (11.206). The corresponding far-zone approximations for the electromagnetic fields of the traveling-wave loop may be obtained from (11.244) and (11.245) by making use of the well-known relationships

$$H_r \approx 0 \tag{11.246}$$

$$H_\theta \approx -\frac{1}{\mu r}\frac{\partial}{\partial r}(rA_\varphi) \tag{11.247}$$

$$H_\varphi \approx \frac{1}{\mu r}\frac{\partial}{\partial r}(rA_\theta) \tag{11.248}$$

$$E_r \approx 0 \tag{11.249}$$

$$E_\theta \approx \eta H_\varphi \tag{11.250}$$

$$E_\varphi \approx -\eta H_\theta \tag{11.251}$$

This results in the following far-zone expansions for the nontrivial field components:

$$H_\theta(r,\theta,\varphi) \approx \frac{\beta a I_0}{2} e^{-j\gamma\pi}\,\mathrm{sinc}(\gamma\pi)\frac{e^{-j\beta r}}{r}\left\{-J_1(w) + 2\sum_{k=1}^{\infty}\frac{\gamma}{k^2-\gamma^2}(j)^k[jk\sin(k\varphi)-\gamma\cos(k\varphi)]J'_k(w)\right\} \tag{11.252}$$

$$H_\varphi(r,\theta,\varphi) \approx \beta a I_0 \cos\theta\, e^{-j\gamma\pi}\,\mathrm{sinc}(\gamma\pi)\frac{e^{-j\beta r}}{r} \cdot \sum_{k=1}^{\infty}\frac{\gamma k}{k^2-\gamma^2}(j)^k[jk\cos(k\varphi)+\gamma\sin(k\varphi)]\frac{J_k(w)}{w} \tag{11.253}$$

$$E_\theta(r,\theta,\varphi) \approx \eta\beta a I_0 \cos\theta \, e^{-j\gamma\pi}\,\mathrm{sinc}(\gamma\pi)\frac{e^{-j\beta r}}{r} \cdot \sum_{k=1}^{\infty}\frac{\gamma k}{k^2-\gamma^2}(j)^k[jk\cos(k\varphi)+\gamma\sin(k\varphi)]\frac{J_k(w)}{w} \tag{11.254}$$

$$E_\varphi(r,\theta,\varphi) \approx -\frac{\eta\beta a I_0}{2}e^{-j\gamma\pi}\,\mathrm{sinc}(\gamma\pi)\frac{e^{-j\beta r}}{r}\left\{-J_1(w) + 2\sum_{k=1}^{\infty}\frac{\gamma}{k^2-\gamma^2}(j)^k[jk\sin(k\varphi)-\gamma\cos(k\varphi)]J_k'(w)\right\} \tag{11.255}$$

which are in agreement with the results previously reported in [27].

We next consider the special case when $\gamma = 0$. In this case, the current on the loop is uniform (i.e., $I(\varphi) = I_0$) and the general expansions for the traveling-wave loop given in (11.232), (11.230), and (11.233) reduce to

$$A_r(r,\theta,\varphi) = 0 \tag{11.256}$$

$$A_\theta(r,\theta,\varphi) = 0 \tag{11.257}$$

$$A_\varphi(r,\theta,\varphi) = \frac{\mu\beta a I_0}{2j}\sum_{n=1}^{\infty}\frac{[(\beta^2 ar\sin\theta)/2]^{2n-1}}{n!(n-1)!}\frac{h_{2n-1}^{(2)}(\beta R)}{(\beta R)^{2n-1}} \tag{11.258}$$

These results are in agreement with those reported in Section 11.3.2.2, which provides confirmation that an exact representation for the uniform current vector potential may be obtained as a degenerate form of the traveling-wave loop expansions. Finally, when γ is an integer (i.e., $\gamma = p = 0, 1, \ldots$) it can be shown that the far-field expansions for the traveling-wave current loop given in (11.252)–(11.255) reduce to the simple form

$$H_\theta(r,\theta,\varphi) \approx \frac{\beta a I_0}{2}e^{-jp\varphi}(j)^p\frac{e^{-j\beta r}}{r}J_p'(w) \tag{11.259}$$

$$H_\varphi(r,\theta,\varphi) \approx \frac{p\beta a I_0\cos\theta}{2}e^{-jp\varphi}(j)^{p-1}\frac{e^{-j\beta r}}{r}\frac{J_p(w)}{w} \tag{11.260}$$

$$E_\theta(r,\theta,\varphi) \approx \frac{p\eta\beta a I_0\cos\theta}{2}e^{-jp\varphi}(j)^{p-1}\frac{e^{-j\beta r}}{r}\frac{J_p(w)}{w} \tag{11.261}$$

$$E_\varphi(r,\theta,\varphi) \approx -\frac{\eta\beta a I_0}{2}e^{-jp\varphi}(j)^p\frac{e^{-j\beta r}}{r}J_p'(w) \tag{11.262}$$

11.4 A GENERALIZED SERIES EXPANSION

A family of integrals which frequently arises in antenna theory is considered in this section. The general form of these integrals is defined by [44]

$$I(n) = \frac{1}{2\pi}\int_0^{2\pi}\cos(n\varphi')\frac{e^{-j\beta R'}}{R'}\,d\varphi' \tag{11.263}$$

which are a special case of (11.1) where $f(\varphi') = \cos(n\varphi')$. It is of interest to point out here that the integral in (11.263) reduces to the well-known cylindrical wire kernel when $n = 0$ and, when $n = 1$, it is proportional to the vector potential for a circular loop antenna with a uniformly distributed current. Although the special cases $n = 0$ and $n = 1$ of this integral have been treated earlier in Section 11.2.1 and Section 11.3.2.2, respectively, a general methodology for evaluating integrals of this form has not been developed for arbitrary values of the parameter n. Treating the entire family of integrals given by (11.263) unifies these independent cases as part of a more general framework.

An exact series representation for such integrals will be developed here which is valid for all integer values of the parameter n. The derivation of this series representation is based essentially upon the application of a pair of Lommel expansions, which have been introduced in Section 11.1. It will be shown that this general series representation of (11.263) for arbitrary values of n contains as special cases the previously derived results for the cylindrical wire kernel ($n = 0$) and the uniform current loop antenna ($n = 1$).

We begin this derivation by recognizing that (11.263) may be expressed as

$$I(n) = \frac{j}{(2\pi\beta\zeta)} \frac{d}{d\zeta} \int_0^{2\pi} \cos(n\varphi') e^{-j\beta R'} \, d\varphi' \tag{11.264}$$

which reduces the integrand to a product involving a simple exponential. A procedure has been described in Section 1 that makes use of Lommel expansions in order to derive an exact series representation for the exponential term which appears in the integrand of (11.264). Substituting expression (11.18) for the exponential into (11.264) and integrating term-by-term yields

$$\begin{aligned} I(n) &= \frac{j}{(2\pi\beta\zeta)} \left(\frac{d}{d\zeta} e^{-j\beta R} \right) \int_0^{2\pi} \cos(n\varphi') \, d\varphi' \\ &+ \frac{j}{\beta\zeta} \sum_{m=1}^{\infty} \frac{(\beta^2 \rho a)^m}{m!} \left(\frac{d}{d\zeta} \frac{h_{m-1}^{(2)}(\beta R)}{(\beta R)^{m-1}} \right) \frac{1}{2\pi} \int_0^{2\pi} \cos(n\varphi') \cos^m \varphi' \, d\varphi' \end{aligned} \tag{11.265}$$

The expressions on the right-hand side of (11.265) can be greatly simplified by carrying out the indicated mathematical operations. To evaluate the first term on the right-hand side of (11.265), we observe that

$$\frac{1}{2\pi} \int_0^{2\pi} \cos(n\varphi') \, d\varphi' = \delta_{n0} \tag{11.266}$$

where δ_{n0} represents the Kronecker delta function defined by

$$\delta_{n0} = \begin{cases} 1, & n = 0 \\ 0, & \text{otherwise} \end{cases} \tag{11.267}$$

and also that

$$\frac{d}{d\zeta} e^{-j\beta R} = -\frac{j\beta}{R} \zeta e^{-j\beta R} \tag{11.268}$$

Thus, by applying (11.266), (11.267), and (11.268), it follows that the first term in (11.265) reduces to

$$\frac{j}{(2\pi\beta\zeta)}\left(\frac{d}{d\zeta}e^{-j\beta R}\right)\int_0^{2\pi}\cos(n\varphi')\,d\varphi' = \frac{e^{-j\beta R}}{R}\delta_{n0} \tag{11.269}$$

The integral appearing in the second term of (11.265) can be evaluated by a simple change of variables according to (11.126). Using the recurrence relation (11.19) along with (11.126) we can then express the second term of (11.265) as

$$\begin{aligned}\frac{j}{\beta\zeta}\sum_{m=1}^{\infty}\frac{(\beta^2\rho a)^m}{m!}\left(\frac{d}{d\zeta}\frac{h_{m-1}^{(2)}(\beta R)}{(\beta R)^{m-1}}\right)\frac{1}{2\pi}\int_0^{2\pi}\cos(n\varphi')\cos^m\varphi'\,d\varphi' \\ = -j\beta\sum_{m=1}^{\infty}A_{mn}\frac{(\beta^2\rho a)^m}{m!}\frac{h_m^{(2)}(\beta R)}{(\beta R)^m}\end{aligned} \tag{11.270}$$

Finally, substituting (11.269) and (11.270) into (11.265) leads us to the following exact series representation for the integral $I(n)$:

$$I(n) = \frac{e^{-j\beta R}}{R}\delta_{n0} - j\beta\sum_{m=1}^{\infty}A_{mn}\frac{(\beta^2\rho a)^m}{m!}\frac{h_m^{(2)}(\beta R)}{(\beta R)^m} \tag{11.271}$$

When $n = 0$, the integral (11.263) reduces to the well-known cylindrical wire kernel [10]. In other words,

$$K(\zeta) = I(0) = \frac{1}{2\pi}\int_0^{2\pi}\frac{e^{-j\beta R'}}{R'}d\varphi' \tag{11.272}$$

which, according to (11.271), has the following series representation:

$$K(\zeta) = \frac{e^{-j\beta R}}{R} - j\beta\sum_{m=1}^{\infty}A_{m0}\frac{(\beta^2\rho a)^m}{m!}\frac{h_m^{(2)}(\beta R)}{(\beta R)^m} \tag{11.273}$$

Because A_{m0} vanishes whenever m is odd, we can introduce the new summation index $n = \frac{1}{2}m$, and thereby recast (11.273) into the form

$$K(\zeta) = \frac{e^{-j\beta R}}{R} - j\beta\sum_{n=1}^{\infty}\frac{(\beta^2\rho a/2)^{2n}}{(n!)^2}\frac{h_{2n}^{(2)}(\beta R)}{(\beta R)^{2n}} \tag{11.274}$$

which is equivalent to (11.31).

The vector potential for a uniform current circular loop antenna can be written as [20]

$$A_\varphi(r,\theta) = \frac{a\mu I_0}{2}I(1) = \frac{a\mu I_0}{4\pi}\int_0^{2\pi}\cos\varphi'\frac{e^{-j\beta R'}}{R'}d\varphi' \tag{11.275}$$

It then follows from (11.271) that a series representation for (11.275) may be found that has the form

$$A_\varphi(r,\theta) = \frac{\beta a\mu I_0}{2j}\sum_{m=1}^{\infty}A_{m1}\frac{(\beta^2\rho a)^m}{m!}\frac{h_m^{(2)}(\beta R)}{(\beta R)^m} \tag{11.276}$$

Because the terms of this series are nonzero only if m is odd, we can rewrite (11.276) as

$$A_\varphi(r,\theta) = \frac{\beta a \mu I_0}{2j} \sum_{n=1}^{\infty} \frac{(\beta^2 \rho a/2)^{2n-1}}{n!(n-1)!} \frac{h_{2n-1}^{(2)}(\beta R)}{(\beta R)^{2n-1}} \tag{11.277}$$

which is in agreement with (11.158) when $\rho = r\sin\theta$, $z = r\cos\theta$, and $z' = 0$. Thus, we see from (11.274) and (11.277) that the series representation given in (11.271) provides, as important special cases, exact expressions for the cylindrical wire kernel and the uniform current circular loop vector potential.

Another important application for the series solution (11.271) of the integral (11.263) is in the derivation of an expansion for (11.2). The decomposition of a spherical wave into cylindrical harmonics can be written as [45]

$$\frac{e^{-j\beta R'}}{R'} = \sum_{m=0}^{\infty} \varepsilon_m \cos(m\varphi') \int_0^\infty \lambda J_m(\lambda\rho) J_m(\lambda a) \frac{e^{-|\zeta|\sqrt{\lambda^2-\beta^2}}}{\sqrt{\lambda^2-\beta^2}}\, d\lambda \tag{11.278}$$

where ε_m is Neumann's number, defined by

$$\varepsilon_m = \begin{cases} 1, & m = 0 \\ 2, & m \geq 1 \end{cases} \tag{11.279}$$

It follows immediately upon multiplying both sides of this equation by $\cos(n\varphi')$ and integrating with respect to φ' from 0 to 2π that [45]

$$I(n) = \int_0^\infty \lambda J_n(\lambda\rho) J_n(\lambda a) \frac{e^{-|\zeta|\sqrt{\lambda^2-\beta^2}}}{\sqrt{\lambda^2-\beta^2}}\, d\lambda \tag{11.280}$$

Hence, by substituting (11.280) into (11.278) and making use of (11.271), we arrive at the following exact series expansion for a spherical wave:

$$\begin{aligned} \frac{e^{-j\beta R'}}{R'} &= \sum_{n=0}^{\infty} \varepsilon_n I(n) \cos(n\varphi') \\ &= \frac{e^{-j\beta R}}{R} - j\beta \sum_{n=0}^{\infty} \sum_{m=1}^{\infty} A_{mn} \varepsilon_n \frac{(\beta^2 \rho a)^m}{m!} \frac{h_m^{(2)}(\beta R)}{(\beta R)^m} \cos(n\varphi') \end{aligned} \tag{11.281}$$

This new spherical wave expansion is particularly useful in those situations, which are often encountered in electromagnetics, where integrations of the form given in (11.1) are required. For these cases, term-by-term integration of the series would translate into much simpler integrals of the type

$$\frac{1}{2\pi} \int_0^{2\pi} f(\varphi') \cos(n\varphi')\, d\varphi' \tag{11.282}$$

which, for most cases of practical interest, can be evaluated in closed form. The double series given in (11.281) may be reduced to a single series via the following identity:

$$\cos^m \varphi' = \sum_{n=0}^{\infty} A_{mn} \varepsilon_n \cos(n\varphi') \tag{11.283}$$

The resulting series representation is found to be

$$\frac{e^{-j\beta R'}}{R'} = \frac{e^{-j\beta R}}{R} - j\beta \sum_{m=1}^{\infty} \frac{(\beta^2 \rho a)^m}{m!} \frac{h_m^{(2)}(\beta R)}{(\beta R)^m} \cos^m \varphi' \tag{11.284}$$

which is in agreement with (11.20).

We next consider the far-zone of the source where the following asymptotic expansions are valid:

$$h_m^{(2)}(\beta R) \sim (j)^{m+1} \frac{e^{-j\beta r}}{\beta r} \qquad \text{as} \qquad \beta R \to \infty \tag{11.285}$$

where

$$r = \sqrt{\rho^2 + z^2} \tag{11.286}$$

and

$$\frac{(\beta^2 \rho a)^m}{m!} \frac{1}{(\beta R)^m} \sim \frac{(\beta a \sin\theta)^m}{m!} \qquad \text{as} \qquad \beta R \to \infty \tag{11.287}$$

where θ is the angle from the z axis to the field point vector. Hence, by combining (11.285) and (11.287) with (11.271), we arrive at a far-zone approximation for $I(n)$ which is given by

$$\begin{aligned} I(n) &\approx \frac{e^{-j\beta r}}{r} \delta_{n0} + \frac{e^{-j\beta r}}{r} \sum_{m=1}^{\infty} A_{mn} (j)^m \frac{(\beta a \sin\theta)^m}{m!} \\ &= \frac{e^{-j\beta r}}{r} \sum_{m=0}^{\infty} A_{mn} (j)^m \frac{(\beta a \sin\theta)^m}{m!} \end{aligned} \tag{11.288}$$

At this point in the development, we choose to introduce a new summation index, namely $p = (m - n)/2$, which transforms (11.288) into

$$\begin{aligned} I(n) &\approx \frac{e^{-j\beta r}}{r} \sum_{p=0}^{\infty} \binom{2p+n}{p} \frac{(j)^{2p+n}}{2^{2p+n}} \frac{(\beta a \sin\theta)^{2p+n}}{(2p+n)!} \\ &= \frac{e^{-j\beta r}}{r} (j)^n \sum_{p=0}^{\infty} (-1)^p \frac{(\beta a \sin\theta/2)^{2p+n}}{p!(p+n)!} \\ &= \frac{e^{-j\beta r}}{r} (j)^n J_n(\beta a \sin\theta) \end{aligned} \tag{11.289}$$

where the series representation for Bessel functions of the first kind was used in order to arrive at the final closed-form far-zone representation (11.289). It will be demonstrated below that this general far-zone expansion reduces to known results for the two special cases considered above (i.e., $n = 0$ and $n = 1$).

By setting $n = 0$ we can use (11.272) and (11.289) to obtain the following far-zone approximation for the cylindrical wire kernel:

$$K(\zeta) \approx \frac{e^{-j\beta r}}{r} J_0(\beta a \sin\theta) \tag{11.290}$$

which is equivalent to (11.52) previously derived in Section 11.2. If we now set $n = 1$, we can use (11.275) and (11.289) to obtain a far-zone expression for the uniform current loop vector potential. Combining these equations, we arrive at

$$A_\varphi(r,\theta) \approx j\frac{a\mu \mathrm{I}_0}{2}\frac{e^{-j\beta r}}{r}J_1(\beta a \sin\theta) \tag{11.291}$$

which is in agreement with (11.188).

11.5 APPLICATIONS

There are a wide variety of applications for the electromagnetic field expansions which have been presented and discussed in this chapter. For example, they have been successfully applied in the development of a numerically rigorous moment method formulation which is capable of accurately modeling moderately thick cylindrical wire antennas [13]. New algorithms have been introduced in [13] for the efficient computation of the cylindrical wire kernel and related impedance matrix integrals. These algorithms make use of exact series representations, such as those derived in Section 11.2, as well as efficient numerical procedures and lead to a significant reduction in overall computation time for thicker wires. Another major advantage of this moment method technique is that it is no longer restricted by the segment length-to-radius ratio limitations inherent in past formulations, thereby making it possible to achieve solution convergence for a much wider class of wire antenna structures.

Cylindrical wire method of moments formulations require calculation of the electric field produced by each segment in the problem space, subject to the boundary condition that the tangential component of the total electric field vanishes at the surface of each wire segment. In other words,

$$\hat{\mathbf{s}} \cdot [\mathbf{E}^i + \mathbf{E}^s] = 0 \tag{11.292}$$

where $\mathbf{E}^i$ is the incident electric field, $\mathbf{E}^s$ is the scattered electric field and $\hat{\mathbf{s}}$ is the unit vector in the direction of assumed current. At this point, the presumption is made that the wire current will be flowing axially in a perfectly conducting and infinitely thin current sheath on the wire surface. The scattered field due to a wire segment of length Δ and radius a centered about the z axis with an arbitrary current distribution $I(z')$ may be expressed in the form

$$\mathbf{E}^s = \frac{1}{j\omega\mu\varepsilon}\left[\frac{\partial^2}{\partial\rho\partial z}\hat{\boldsymbol{\rho}} + \left(\beta^2 + \frac{\partial^2}{\partial z^2}\right)\hat{\mathbf{z}}\right]A_z(\rho,z) \tag{11.293}$$

where

$$A_z(\rho,z) = \frac{\mu}{4\pi}\int_{-\Delta/2}^{\Delta/2} I(z')K(z-z')\,dz' \tag{11.294}$$

is the z component of the vector potential and $K(z-z')$ is the corresponding cylindrical wire kernel which has already been defined in (11.23).

The basis functions which have been adopted for use in this application are of the form [46]

$$I_n(z') = A_n + B_n \sin[\beta(z' - z_n)] + C_n \cos[\beta(z' - z_n)] \tag{11.295}$$

with z_n the midpoint of the nth segment such that $|z' - z_n| \leq \Delta_n/2$ where Δ_n represents the segment length. This choice of basis functions is commonly used because it represents very well the actual currents which arise in cylindrical wire applications and therefore leads to rapid convergence [47]. These basis functions along with a point matching technique are employed in conjunction with Pocklington's integral equation in such a way that both current and charge satisfy continuity conditions at segment junctions [48]. This approach allows two of the unknown coefficients in (11.295) to be eliminated, leaving only one constant. Following this procedure leads to a moment method solution of Pocklington's electric field integral equation for arbitrary wire configurations, which may be expressed as

$$\frac{j\omega\mu}{4\pi} \sum_{n=1}^{N} \int_{-\Delta_n/2}^{\Delta_n/2} I_n(z') G(\rho_m, z_m - z')\, dz' = \hat{\mathbf{s}}_m \cdot \mathbf{E}^i(\rho_m, z_m) \tag{11.296}$$

where $m = 1, 2, \ldots, N$ and

$$G(\rho_m, z_m - z') = \left[\left\{ \hat{\mathbf{s}}_m \cdot \hat{\mathbf{z}}_n + \frac{1}{\beta^2} \left(\frac{\partial^2}{\partial\rho\partial z} \hat{\mathbf{s}}_m \cdot \hat{\boldsymbol{\rho}}_n + \frac{\partial^2}{\partial z^2} \hat{\mathbf{s}}_m \cdot \hat{\mathbf{z}}_n \right) \right\} K(z - z') \right]_{\substack{\rho = \rho_m \\ z = z_m}} \tag{11.297}$$

The parameters (ρ_m, z_m) represent the midpoint coordinates of the observation segment, $\hat{\mathbf{s}}_m$ represents the unit vector in the direction of assumed current on the observation segment, and $\hat{\boldsymbol{\rho}}_n$ and $\hat{\mathbf{z}}_n$ are unit vectors for the local coordinate system defined by the source segment.

As discussed above, the method of moments formulation used here assumes a trigonometric current on each segment of the form defined in (11.295). This suggests that the electric field due to each of the three components of the current distribution (i.e., uniform, sine, and cosine) must be computed. It can be shown that the ρ and z components of the electric field due to a segment of length Δ with a uniformly distributed current $I(z') = I_0$ may be expressed as

$$E_\rho^u = E_0 \left[\frac{1}{\beta^2} \frac{\partial}{\partial\rho} K(z - \Delta/2) - \frac{1}{\beta^2} \frac{\partial}{\partial\rho} K(z + \Delta/2) \right] \tag{11.298}$$

$$E_z^u = E_0 \left[\frac{1}{\beta^2} \frac{\partial}{\partial z} K(z - \Delta/2) - \frac{1}{\beta^2} \frac{\partial}{\partial z} K(z + \Delta/2) - \int_{-\Delta/2}^{\Delta/2} K(z - z')\, dz' \right] \tag{11.299}$$

where E_0 is defined in (11.94). The corresponding expressions for the electric field components due to a sinusoidal current distribution of the form $I(z') = I_0 \sin(\beta z')$ are

$$E_z^s = E_0 \left\{ \sin(\beta\Delta/2) \left[\frac{1}{\beta^2} \frac{\partial}{\partial z} K(z - \Delta/2) + \frac{1}{\beta^2} \frac{\partial}{\partial z} K(z + \Delta/2) \right] \right.$$
$$\left. - \frac{1}{\beta} \int_{-\Delta/2}^{\Delta/2} \cos(\beta z') \frac{\partial}{\partial\rho} K(z - z')\, dz' \right\} \tag{11.300}$$

$$E_z^s = E_0\left\{\sin(\beta\Delta/2)\left[\frac{1}{\beta^2}\frac{\partial}{\partial z}K(z-\Delta/2)+\frac{1}{\beta^2}\frac{\partial}{\partial z}K(z+\Delta/2)\right]\right.$$
$$\left.+\cos(\beta\Delta/2)\left[\frac{1}{\beta}K(z-\Delta/2)-\frac{1}{\beta}K(z+\Delta/2)\right]\right\} \quad (11.301)$$

Finally, for an assumed cosinusoidal current distribution of the form $I(z') = I_0\cos(\beta z')$, we have

$$E_\rho^c = E_0\left\{\cos(\beta\Delta/2)\left[\frac{1}{\beta^2}\frac{\partial}{\partial \rho}K(z-\Delta/2)-\frac{1}{\beta^2}\frac{\partial}{\partial \rho}K(z+\Delta/2)\right]\right.$$
$$\left.+\frac{1}{\beta}\int_{-\Delta/2}^{\Delta/2}\sin(\beta z')\frac{\partial}{\partial \rho}K(z-z')dz'\right\}. \quad (11.302)$$

$$E_z^c = E_0\left\{\sin(\beta\Delta/2)\left[\frac{1}{\beta}K(z+\Delta/2)-\frac{1}{\beta}K(z-\Delta/2)\right]\right.$$
$$\left.+\cos(\beta\Delta/2)\left[\frac{1}{\beta^2}\frac{\partial}{\partial z}K(z-\Delta/2)-\frac{1}{\beta^2}\frac{\partial}{\partial z}K(z+\Delta/2)\right]\right\} \quad (11.303)$$

Formulas (11.298) through (11.303) contain a total of six different expressions involving the cylindrical wire kernel $K(z-z')$ which require evaluation. These expressions are summarized as follows:

$$\frac{1}{\beta}K(z-z') = \frac{1}{2\pi}\int_0^{2\pi}\frac{e^{-j\beta R'}}{\beta R'}d\varphi' \quad (11.304)$$

$$D_z K(z-z') = \frac{1}{\beta^2}\frac{\partial}{\partial z}K(z-z') \quad (11.305)$$

$$D_\rho K(z-z') = \frac{1}{\beta^2}\frac{\partial}{\partial \rho}K(z-z') \quad (11.306)$$

$$I(\rho, z; a, \Delta) = \int_{-\Delta/2}^{\Delta/2} K(z-z')dz' \quad (11.307)$$

$$S(\rho, z; a, \Delta) = \frac{1}{\beta}\int_{-\Delta/2}^{\Delta/2}\sin(\beta z')\frac{\partial}{\partial \rho}K(z-z')dz' \quad (11.308)$$

$$C(\rho, z; a, \Delta) = \frac{1}{\beta}\int_{-\Delta/2}^{\Delta/2}\cos(\beta z')\frac{\partial}{\partial \rho}K(z-z')dz' \quad (11.309)$$

It is possible to further simplify the expressions given in (11.305) and (11.306) as well as (11.308) and (11.309). Evaluating the derivatives which appear in (11.305) and (11.306) leads to the following useful integral representations:

$$D_z K(z-z') = -(z-z')\frac{1}{2\pi}\int_0^{2\pi}\frac{(1+j\beta R')}{(\beta R')^2}\frac{e^{-j\beta R'}}{R'}d\varphi' \quad (11.310)$$

$$D_\rho K(z-z') = -\frac{1}{2\pi}\int_0^{2\pi}(\rho - a\cos\varphi')\frac{(1+j\beta R')}{(\beta R')^2}\frac{e^{-j\beta R'}}{R'}d\varphi' \quad (11.311)$$

The double integrals of (11.308) and (11.309) may be reduced to single integrals of the form

$$S(\rho, z; a, \Delta) = \frac{1}{\pi}\int_0^{\pi} \frac{(\rho - a\cos\varphi')}{(\rho^2 + a^2 - 2\rho a\cos\varphi')} \cdot \frac{e^{-j\beta R'}}{\beta R'}\left[(z - z')\sin(\beta z') + jR'\cos(\beta z')\right]\Bigg|_{z'=-\Delta/2}^{z'=\Delta/2} d\varphi' \tag{11.312}$$

$$C(\rho, z; a, \Delta) = \frac{1}{\pi}\int_0^{\pi} \frac{(\rho - a\cos\varphi')}{(\rho^2 + a^2 - 2\rho a\cos\varphi')} \cdot \frac{e^{-j\beta R'}}{\beta R'}\left[(z - z')\cos(\beta z') - jR'\sin(\beta z')\right]\Bigg|_{z'=-\Delta/2}^{z'=\Delta/2} d\varphi' \tag{11.313}$$

by making use of the fact that

$$\int_{-\Delta/2}^{\Delta/2} \frac{\partial}{\partial\rho}\left\{\frac{e^{-j\beta(R' \pm z')}}{\beta R'}\right\} dz' = \pm(\rho - a\cos\varphi')\frac{e^{-j\beta(R' \pm z')}}{\beta R'[R' \pm (z' - z)]}\Bigg|_{z'=-\Delta/2}^{z'=\Delta/2} \tag{11.314}$$

For many thin-wire applications sufficient accuracy can be achieved by invoking the reduced-kernel approximation (11.29) discussed in Section 11.2. The use of this reduced-kernel approximation leads to simplified forms for the expressions which are given in (11.304), (11.307), (11.310), (11.311), (11.312), and (11.313). These simplified expressions are

$$\frac{1}{\beta}K_0(z - z') = \frac{e^{-j\beta r_0}}{\beta r_0} \tag{11.315}$$

$$D_z K_0(z - z') = -(z - z')\frac{(1 + j\beta r_0)}{(\beta r_0)^2}K_0(z - z') \tag{11.316}$$

$$D_\rho K_0(z - z') = -\rho\frac{(1 + j\beta r_0)}{(\beta r_0)^2}K_0(z - z') \tag{11.317}$$

$$I_0(\rho, z; a, \Delta) = \int_{-\Delta/2}^{\Delta/2} K_0(z - z')\,dz' \tag{11.318}$$

$$S_0(\rho, z; a, \Delta) = \frac{1}{\beta\rho}K_0(z - z')\left[(z - z')\sin(\beta z') + jr_0\cos(\beta z')\right]\Bigg|_{z'=-\Delta/2}^{z'=\Delta/2} \tag{11.319}$$

$$C_0(\rho, z; a, \Delta) = \frac{1}{\beta\rho}K_0(z - z')\left[(z - z')\cos(\beta z') - jr_0\sin(\beta z')\right]\Bigg|_{z'=-\Delta/2}^{z'=\Delta/2} \tag{11.320}$$

Hence, in the case of the thin-wire reduced-kernel formulation, (11.304) and (11.310)–(11.313) reduce to simple closed-form expressions, while the double integral of (11.307) reduces to the single integral given in (11.318). Several techniques have been developed for evaluating this integral, which are summarized in [1], [10], [49], [50].

On the other hand, when the full-kernel formulation for thicker wires is used, all six expressions involving the kernel (11.304), (11.307), (11.310)–(11.313) are non-analytic. As a consequence of this, the computational overhead for the moderately thick-wire formulation can be significant if an effort is not made to devise efficient algorithms for evaluating these six expressions. Several techniques have been developed in [13] for the efficient as well as accurate analysis of these integrals. This includes treatments based on the recently found exact series representation for the cylindrical wire kernel and related integrals discussed in Section 11.2 as well as efficient numerical procedures. In particular, efficient algorithms have been developed in [13] for evaluating (11.304), (11.310), and (11.311) which are based on the series representations given in (11.38), (11.96), and (11.97), respectively. Efficient numerical techniques have also been proposed in [13] for evaluating these three integrals.

The integrals given in (11.312) and (11.313) are in a form in which they may be readily integrated numerically. It was found that accurate results may be achieved for the majority of cases by using a very low order numerical integration scheme, such as a three-point Gaussian quadrature. Finally, it should be pointed out that various useful techniques for evaluating (11.307) have been discussed elsewhere in the literature [1], [10], [51–54] and, therefore, will not be considered here.

Either the thin-wire reduced-kernel or the thick-wire full-kernel formulation can be used to calculate the electric fields due to a segment. The effectiveness of these solutions depends on many factors including relative proximity to the source segment and the wire radius under consideration. The reduced-kernel method of moments formulation is only valid for relatively thin wires (i.e., $a \leq 0.01\lambda$). However, it has been demonstrated in [13] that the upper limit on the validity of this moment-method formulation may be extended by at least a factor of 10 to include moderately thick wires with radii in the range $0.01\lambda \leq a \leq 0.1\lambda$. This is accomplished in part by making use of more accurate electric field expressions, which are in terms of the full cylindrical wire kernel (11.23) rather than the reduced kernel (11.29).

Measured input resistance and reactance values for various moderately thick monopoles of length 0.4λ situated over a perfectly conducting ground plane are plotted in Figs. 11.7 and 11.8, respectively [55]. Calculated values of input resistance and reactance have also been included in these figures for comparison purposes. These calculated values were generated using two different moment-method codes, one based on the reduced-kernel thin-wire formulation and the other based on the full-kernel thick-wire formulation. Figures 11.7 and 11.8 graphically illustrate the considerable improvement which can be achieved in input impedance predictions for moderately thick wires when using a full-kernel moment-method code as opposed to a code which is based on the reduced-kernel approximation.

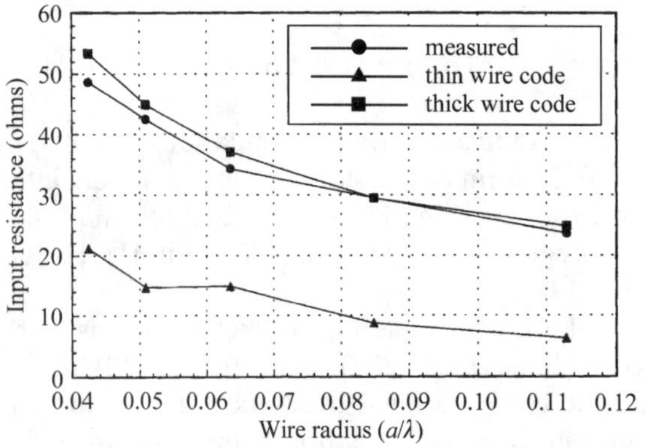

Figure 11.7 Code predictions for the input resistance of a 0.4λ monopole compared to measurements.

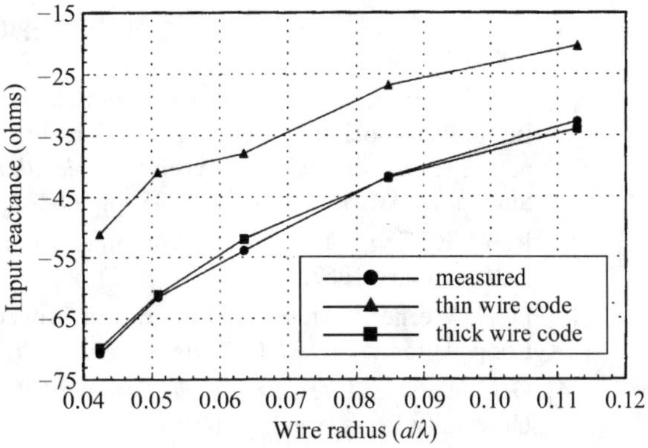

Figure 11.8 Code predictions for the input reactance of a 0.4λ monopole compared to measurements.

11.6 CONCLUSION

Lommel expansions have played an essential role in the recent solution of several long-standing problems in electromagnetic field theory. This chapter was intended to familiarize the reader with the properties of these Lommel expansions as well as to provide a comprehensive treatment of their applications in electromagnetics. The first application which was considered concerns the derivation of an exact series representation for the cylindrical wire kernel. This series expansion for the kernel was then used to obtain a complete exact formulation for the vector potential and corresponding electromagnetic fields of a cylindrical dipole antenna with a uniform current distribution. The usefulness of these exact representations was demonstrated by considering an example in which they are employed to develop a new moment method technique capable of efficiently as well as accurately modeling moderately thick cylindrical wire antennas.

Another important application of Lommel expansions was shown to be in the development of an exact integration procedure for vector potentials of thin circular loop antennas. The technique is very general as well as straightforward to apply, and works for most loop current distributions of practical interest. This new integration technique has been applied in this chapter to find near-field expansions for loop antennas with a wide variety of commonly assumed current distributions. Included among these is the general case of a circular loop with an arbitrary current distribution represented by a Fourier cosine series. Exact vector potential and field expansions were also obtained for the special cases of loops having uniform, cosinusoidal and traveling-wave current distributions. Starting with these exact series representations, asymptotic expansions were used to find convenient far-field approximations.

ACKNOWLEDGMENTS

The author would like to thank Matthew K. Emsley for his assistance with the preparation of Figs. 11.4–6.

REFERENCES

[1] D. H. Werner, "Analytical and Numerical Methods for Evaluating Electromagnetic Field Integrals Associated with Current-Carrying Wire Antennas," in *Advanced Electromagnetism: Foundations, Theory and Applications*, T. W. Barrett and D. M. Grimes, eds., Singapore: World Scientific, 1995, pp. 682–762.

[2] R. W. P. King, "The Linear Antenna—Eighty Years of Progress," *Proc. IEEE*, **55**, pp. 2–26, January 1967.

[3] D. H. Werner, "An Exact Integration Procedure for Vector Potentials of Thin Circular Loop Antennas," *IEEE Trans. Antennas Propagat.*, **44** (2), pp. 157–165, February 1996.

[4] G. N. Watson, *A Treatise on the Theory of Bessel Functions*, Cambridge, U.K.: Cambridge University Press, 1962, p. 140.

[5] R. F. Harrington, *Field Computation by Moment Methods*, MacMillan, New York, 1968.

[6] R. Mittra, ed., *Numerical and Asymptotic Techniques in Electromagnetics,* Springer-Verlag, New York, 1975.

[7] R. Mittra, ed., *Computer Techniques for Electromagnetics*, Hemisphere Publishing Corporation, New York, 1987.

[8] E. K. Miller, L. Medgyesi-Mitschang, and E. H. Newman, eds., *Computational Electromagnetics: Frequency-Domain Method of Moments*, IEEE Press, New York, 1992.

[9] W. Wang, "The Exact Kernel for Cylindrical Antenna," *IEEE Trans. Antennas Propagat.*, **39**, pp. 434–435, April 1991.

[10] D. H. Werner, "An Exact Formulation for the Vector Potential of a Cylindrical Antenna with Uniformly Distributed Current and Arbitrary Radius," *IEEE Trans. Antennas Propagat.*, **41** (8), pp. 1009–1018, August 1993.

[11] M. Abramowitz and I. E. Stegun, *Handbook of Mathematical Functions*, Washington, D.C.: Nat. Bur. Stand., 1972, p. 439, 10.1.17.

[12] C. M. Butler, "Evaluation of Potential Integral at Singularity of Exact Kernel in Thin-Wire Calculations," *IEEE Trans. Antennas Propagat.*, **AP-23**, pp. 293–295, March 1975.

[13] D. H. Werner, "A Method of Moments Approach for the Efficient and Accurate Modeling of Moderately Thick Cylindrical Wire Antennas," *IEEE Trans. Antennas Propagat.* **46** (3), pp. 373–382, March 1998.

[14] P. L. Werner and D. H. Werner, "On Approximations for the Cylindrical Wire Kernel," *Electronics Letters*, **32** (23), pp. 2108–2109, November 1996.

[15] R. F. Harrington, *Time-Harmonic Electromagnetic Fields*, McGraw-Hill, New York, 1961.

[16] W. L. Stutzman and G. A. Thiele, *Antenna Theory and Design*, John Wiley and Sons, New York, 1981.

[17] R. F. Harrington, "Matrix Methods for Field Problems," *Proc. IEEE*, **55**, pp. 136–149, February 1967.

[18] S. A. Schelkunoff and H. T. Friis, *Antenna Theory and Practice*, John Wiley and Sons, New York, 1952, p. 369.

[19] R. W. P. King and G. S. Smith, *Antennas in Matter: Fundamentals, Theory and Applications*, MIT Press, Cambridge, MA, pp. 527–605, 1981.

[20] C. A. Balanis, *Antenna Theory, Analysis and Design*, Harper and Row, New York, pp. 164–203, 1982.

[21] G. S. Smith, "Loop Antennas," in *Antenna Engineering Handbook*, R. C. Johnson and H. Jasik, eds, McGraw-Hill, New York, pp. 5.1–5.24, 1984.

[22] D. Foster, "Loop Antennas With Uniform Current," *Proc. IRE*, **32**, pp. 603–607, October 1944.

[23] J. B. Sherman, "Circular Loop Antennas at Ultra-High Frequences," *Proc. IRE*, **32**, pp. 534–537, September 1944.

[24] G. Glinski, "Note on the Circular Loop Antennas With Nonuniform Current Distribution," *J. Appl. Phys.*, **18**, pp. 638–644, July 1947.

[25] J. E. Lindsay Jr., "A Circular Loop Antenna With Nonuniform Current Distribution," *IRE Trans. Antennas Propagat.*, **AP-8** (4), pp. 439–441, July 1960.

[26] B. R. Rao, "Far Field Patterns of Large Circular Loop Antennas: Theoretical and Experimental Results," *IEEE Trans. Antennas Propagat.*, **AP-16**, pp. 269–270, March 1968.

[27] S. M. Prasad and B. N. Das, "Circular Loop Antenna With Traveling-Wave Current Distribution,"*IEEE Trans. Antennas Propagat.*, **AP-18** (2), pp. 278–280, March 1970.

[28] F. M. Greene, "The Near-Zone Magnetic Field of a Small Circular Loop Antenna," *J. Res. NBS*, **71-C** (4), pp. 319–326, October 1967.

[29] P. L. Overfelt, "Near-Fields of a Constant Current Thin Circular Loop Antenna of Arbitrary Radius," *IEEE Trans. Antennas Propagat.*, **44** (2), pp. 166–171, February 1996.

[30] C. J. Bouwkamp and H. B. G. Casimir, "On Multipole Expansion in the Theory of Electromagnetic Radiation," *Physica*, **20**, pp. 539–554, 1954.

[31] W. R. Smythe, *Static and Dynamic Electricity*, McGraw-Hill, New York, pp. 488–490, 1950.

[32] J. R. Wait, "Insulated Loop Antenna Immersed in a Conducting Medium," *J. Res. NBS*, **59** (2), pp. 133–137, August 1957.

[33] J. R. Wait, "A Note on the Insulated Loop Antenna Immersed in a Conducting Medium," *Radio Sci. J. Res. NBS*, **68-D** (11), pp. 1249–1250, November 1964.

[34] P. P. Pavlov, "Insulated Loop Antenna with Spherical Ferrite Core in a Conducting Medium," *Radio Eng. Electron. Phys. (USSR)*, **8**, pp. 1461–1466, 1963.

[35] R. H. Williams, "Insulated and Loaded Loop Antenna Immersed in a Conducting Medium," *Radio Sci. J. Res. NBS*, **69-D** (2), pp. 287–289, February 1965.

[36] E. J. Martin Jr., "Radiation Fields of Circular Loop Antennas by a Direct Integration Process," *IRE Trans. Antennas Propagat.*, **AP-8** (1), pp. 105–107, January 1960.

[37] R. W. P. King, "The Loop Antenna For Transmission and Reception," in *Antenna Theory: Part I*, R. E. Collin and F. J. Zucker, eds, McGraw-Hill, New York, 1969, pp. 458–482.

[38] R. W. P. King and S. Prasad, *Fundamental Electromagnetic Theory and Applications*, Prentice-Hall, Englewood Cliffs, NJ, 1986, pp. 365–392.

[39] I. S. Gradshteyn and I. M. Ryzhik, *Table of Integrals, Series, and Products*, Academic Press, New York, 1980, pp. 416–417 [Section 3.631, (17)].

[40] H. L. Knudsen, "The Field Radiated by Ring Quasi-Array of a Infinite Number of Tangential or Radial Dipoles," *Proc. IRE*, **41**, pp. 781–789, June 1953.

[41] S. A. Adekola, "On the Excitation of a Circular Loop Antenna by Traveling- and Standing-Wave Current Distributions," *Int. J. Electronics*, **54** (6), pp. 705–732, 1983.

[42] K. M. Chen and R. W. P. King, "A Loop Antenna Coupled to a Four-Wire Line: and its Possible Use as an Element in a Circularly Polarized End-Fire Array," *Proc. Inst. Elect. Engrs.*, **109**, pp. 55–62, 1962.

[43] R. E. Collin and F. J. Zucker, *Antenna Theory – Part 1*, McGraw-Hill, New York, pp. 480–481, 1969.

[44] D. H. Werner and T. W. Colegrove, "On a New Cylindrical Harmonic Representation For Spherical Waves," *IEEE Trans. Antennas Propagat.*, **47**(1), pp. 97–100, January 1999.

[45] C. Monzon, "A Loop Antenna in Front of a Resistive Sheet," *IEEE Trans. Antennas Propagat.*, **44** (3), pp. 405–412, March 1996.

[46] Y. S. Yeh and K. K. Mei, "Theory of Conical Equiangular-Spiral Antennas: Part I – Numerical Technique," *IEEE Trans. Antennas Propagat.*, **AP-15**, pp. 634–639, September 1967.

[47] E. K. Miller and F. J. Deadrick, "Some Computational Aspects of Thin-Wire Modeling," in *Numerical and Asymptotic Techniques in Electromagnetics*, R. Mittra, Ed., Springer-Verlag, New York, 1975, pp. 89–127.

[48] G. J. Burke and A. J. Poggio, "Numerical Electromagnetics Code (NEC) – Method of Moments," *Rep. NOSC TD 116*, Lawrence Livermore Lab., Livermore, CA, January 1981.

[49] D. H. Werner, P. L. Werner, J. A. Huffman, A. J. Ferraro, and J. K. Breakall, "An Exact Solution of the Generalized Exponential Integral and its Application to Moment Method Formulations," *IEEE Trans. Antennas Propagat.*, **41**, pp. 1716–1719, December 1993.

[50] D. H. Werner, "Author's Reply to the Comments of A. E. Gera," *IEEE Trans. Antennas Propagat.*, **42**, pp. 1201–1202, August 1994.

[51] D. H. Werner, J. A. Huffman, and P. L. Werner, "Techniques for Evaluating the Uniform Current Vector Potential at the Isolated Singularity of the Cylindrical Wire Kernel,"*IEEE Trans. Antennas Propagat.*, **42**, pp. 1549–1553, November 1994.

[52] S.-O. Park and C. A. Balanis, "Efficient Kernel Calculation of Cylindrical Antennas," *IEEE Trans. Antennas Propagat.*, **43**, pp. 1328–1331, November 1995.

[53] D. H. Werner and V. Fotopolous, "An Efficient Technique for Self-Term Evaluation of the Uniform Current Vector Potential Integrals," *J. Electromagn. Waves Applicat.*, **10**, pp. 1595–1600, 1996.

[54] M. P. Ramachandran, "On the Bounded Part of the Kernel in the Cylindrical Antenna Integral Equation," *Applied Computational Electromagnetics Society Journal*, **13** (1), pp. 71–77, March 1998.

[55] R. W. P. King, *Table of Antenna Characteristics.* IFI/Plenum, New York, 1971.

Chapter 12

FRACTIONAL PARADIGM IN ELECTROMAGNETIC THEORY

Nader Engheta

ABSTRACT

In this chapter, an overview of some of our recent work in the use of fractional calculus in electromagnetic theory and development of fractional paradigm in electromagnetism is given. The general trend among the cases that we have studied and have reviewed here is fractionalization of some appropriate operators in these problems. The "fractional operator" derived in each case provides us with a tool to obtain "fractional" intermediate situations between the canonical cases in each of these problems. The fractional multipoles, electrostatic fractional image methods, fractional solutions to the conventional Helmholtz equation, and fractional duality in electromagnetism using the fractional curl operator are among the case studies that in recent years we have introduced and investigated, and they are reviewed here. These problems have motivated us toward the development of fractional paradigm in electromagnetic theory. The review that is provided in this chapter highlights some of the main points and major results that we have obtained in these problems.

12.1 INTRODUCTION

In recent years, we have been interested in bringing the tools of fractional derivatives/fractional integrals into the theory of electromagnetism (see, e.g., [1]–[10]), and to develop and study the area that we name *fractional paradigm in electromagnetic theory*. Fractional derivatives/integrals are mathematical operators involving differentiation/integration to arbitrary (non-integer) real or complex orders—operators such as $d^{\nu}f(x)/dx^{\nu}$ where ν can be taken to be a non-integer real or even a complex number (see, e.g., [11]–[15]). These operators have been studied by mathematicians in the field of fractional calculus for many years (see, e.g., [11]–[28]). In a sense, these operators effectively behave as the so-called intermediate cases between the integer-order differentiation and integration. The electromagnetic theory, however, is a classical topic in which the usual differential and integral operators are commonly used in treating various problems and phenomena. So it is interesting to explore what possible mathematical applications and physical justifications one would find if one brought the tools of fractional differentiation/integration, or in general fractionalization of operators, into electromagnetic problems. In the past few years, we have been exploring some possible roles of fractional calculus in electromagnetism. To that end, we have applied the tools of fractional calculus in selected problems in electromagnetics, and obtained some results that highlight certain notable features and potential mathematical applications of these operators in electromagnetic theory [1]–[10]. Some of these results are reviewed in this chapter.

The results of our earlier work in exploring the mathematical applications of fractional calculus in electromagnetics motivated us to look into fractionalization of some other operators commonly used in electromagnetic problems and to search for potential roles and applications and physical meanings of fractionalization of these linear operators in electromagnetic theory. Since fractional derivatives/integrals, as is briefly discussed and reviewed later in this chapter, are effectively the result of fractionalization of differentiation and integration operators, it is a natural next step to inquire whether the fractionalization of some other linear operators may play interesting roles in electromagnetic theory. Searching for such fractionalization has led us to novel solutions, interpretable as "fractional solutions," for certain standard electromagnetic problems.

This chapter provides an overview of some of our ideas and findings in developing the fractional paradigm in electromagnetism. As such it only highlights some of the major issues and main results that we have encountered in our recent work. For more detail, interested readers are referred to our work in [1]–[10].

12.2 WHAT IS MEANT BY FRACTIONAL PARADIGM IN ELECTROMAGNETIC THEORY?

In the mathematical treatment of problems in electromagnetic theory, often for a given general problem we can first solve canonical cases. For example, if the goal is to solve the straightforward problem of electrostatic potential of a static electric charge distribution in free space, which is treated in many standard textbooks in electromagnetics, we can usually start by considering the well-known cases of individual static point multipoles, i.e., point monopole, point dipole, point quadrupole, etc. Let us symbolically show the chart of Fig. 12.1 where the block entitled "Problem" represents the general problem of interest (in this example, the electrostatic potential in free space), and the two blocks "Case 1" and "Case 2" indicate the two canonical situations, which in this example can be taken to be the point monopole and point dipole, respectively. Now, one can ask the following question: **Are there any "intermediate" situations or solutions between these two well-known canonical cases**

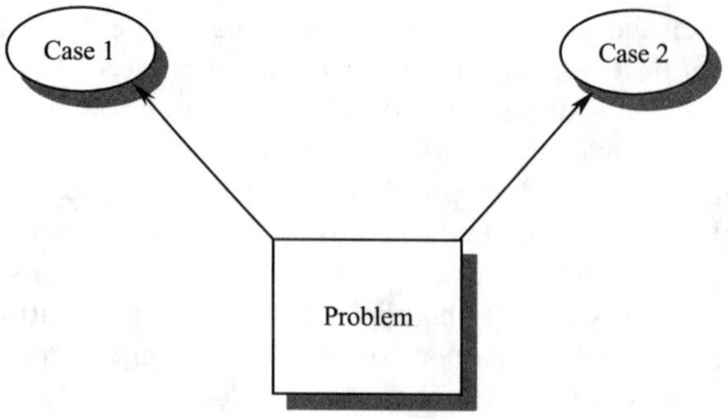

Figure 12.1 A symbolic block diagram representing a generic "Problem" and its two canonical "Cases."

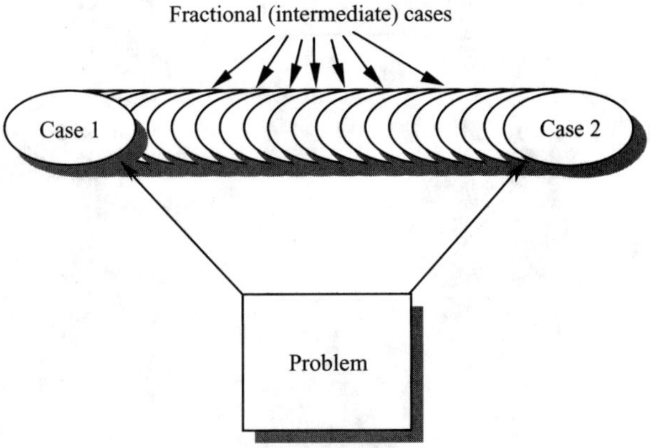

Figure 12.2 The block diagram symbolizing the fractional paradigm. A generic "Problem" and its two canonical "Cases" are shown as blocks along with the "fractional" intermediate situations (or cases) between the two cases 1 and 2.

for this "Problem"? In other words, as symbolically depicted in Fig. 12.2, **Can we have fractional "intermediate" domains or solutions between the two canonical cases shown in Fig. 12.1?** These questions can be rephrased and interpreted in the following way:

> If we consider a certain "quantity" or "entity" (e.g., the electrostatic point monopole or point dipole here) whose properties or identifiers (e.g., potential of these multipoles) depend on a parameter with integer values, can we still consider that quantity when that specific parameter takes a non-integer "fractional" value? In other words, can we conceive an "intermediate" case for that entity?

In order to explain this point more clearly, we follow in more detail the above example of electrostatic potential. The concept of multipoles and multipole expansion in electromagnetic theory has been well known and studied extensively [29]–[33]. As mentioned before, let us take our "Problem" block in Fig. 12.1 as the electrostatic potential distribution, and consider the canonical cases as the "entities" of electric point monopoles, electric point dipoles, electric point quadrupoles, etc. These entities have specific properties. For instance, for the static electric point monopole, the scalar electric potential has the radial dependence of R^{-1} with no angular variation. The scalar potential of electrostatic point dipole, however, varies as R^{-2} with a specific angular variation. Comparing and contrasting the scalar potentials of these two cases, we obviously notice that the R-dependences of these electrostatic multipoles have exponents that are negative *integers*. Now for this special example the above questions can be asked as follows. Can we have an "intermediate" electrostatic multipole whose R-dependence varies as $R^{-\nu}$ where ν takes a non-integer value between 1 and 2? If yes, what would such a "fractional-order" multipole look like? Of course, such an intermediate multipole would not simply be made by a linear combination of *one* point monopole and *one* point dipole, each with appropriate coefficients. Because if that were the case, the scalar potential in the distant region of such a combination would be dominated by the potential of the point monopole only, since the static dipole's scalar potential drops more rapidly than that of the monopole. Therefore, if a *fractional* intermediate multipole is possible, a specific charge distribution should be sought in order to have such an intermediate multipole having a potential distribution with R-dependence of $R^{-\nu}$. As will be shown and reviewed in this chapter later, we

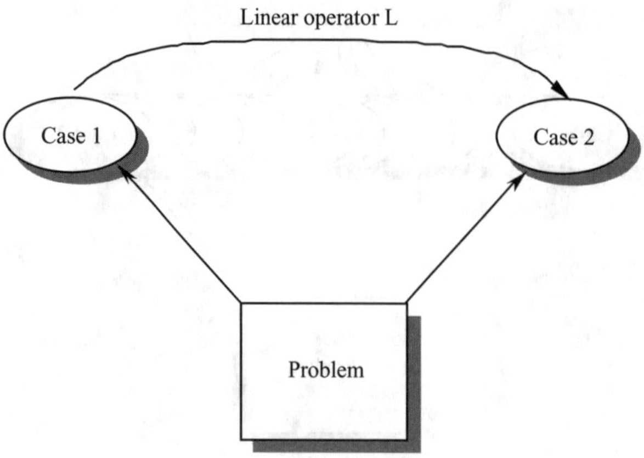

Figure 12.3 A possible linear operator $\boldsymbol{L}$ (or mapping) that maps or "connects" the canonical case 1 to the canonical case 2 of the problem.

have studied this issue and have introduced the idea of fractional-order multipoles [1] using the tools of fractional calculus, and have shown that such fractional multipoles can be used in describing potential distribution in front of perfectly conducting circular cones (for the three-dimensional (3D) case of fractional multipoles) and in front of perfectly conducting two-dimensional (2D) wedges (for the 2D case of fractional multipoles) [2].

What is a recipe to find possible fractional intermediate solutions between the two canonical cases in a given general problem? To answer this question, first we need to search for an operator that "connects" the two cases, i.e., a mapping that takes the case 1 and "brings" it into the case 2. If such an operator exits, we symbolically show this operator as a mapping $\underline{\underline{L}}$ between the case 1 and case 2, as depicted in Fig. 12.3. The range and domain of this operator should also be specified. In the problems that we have studied in recent years we have restricted our attention to situations where $\underline{\underline{L}}$ is a linear operator. We should then "fractionalize" this operator $\underline{\underline{L}}$. The procedure for fractionalization of operators will be addressed shortly. The new fractionalized operator, which we symbolically denote by $\underline{\underline{L}}^{\nu}$ with the fractional parameter ν, can, under certain conditions, be used to obtain the intermediate cases between the original cases 1 and 2 (see, Fig. 12.4). In order for $\underline{\underline{L}}^{\nu}$ to be the fractionalized operator that provides us with the intermediate solutions to our original problem, $\underline{\underline{L}}^{\nu}$ should satisfy the following properties (or desiderata):

- For $\nu = 1$, the fractional operator $\underline{\underline{L}}^{\nu}$ should become the original operator $\underline{\underline{L}}$, which provides us with case 2 when it is applied on case 1;
- For $\nu = 0$, the operator $\underline{\underline{L}}^{\nu}$ should become the identity operator $\underline{\underline{I}}$ and thus case 1 can be mapped onto itself;
- For any two values ν_1 and ν_2 of the fractionalization parameter ν, we should have the additivity property in ν for the operator $\underline{\underline{L}}^{\nu}$, i.e., $\underline{\underline{L}}^{\nu_1} \circ \underline{\underline{L}}^{\nu_2} = \underline{\underline{L}}^{\nu_1 + \nu_2}$; and
- The operator $\underline{\underline{L}}^{\nu}$ should commute with the operators involved in the mathematical description of the original problem. For example, if the problem of interest is to find the electrostatic potential for a given charge distribution in free space, the mathematical description of this problem is the Poisson

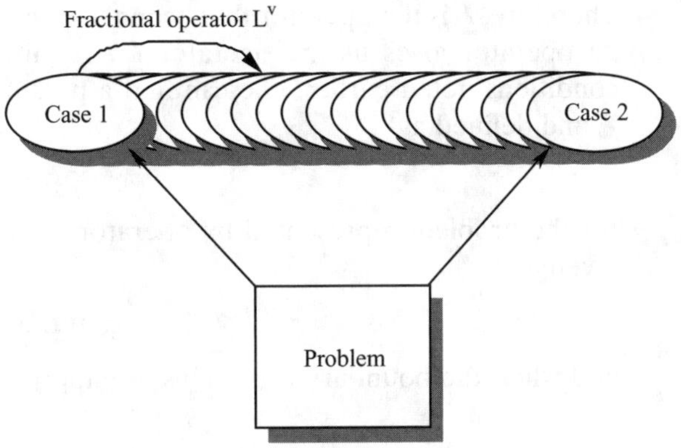

Figure 12.4 A possible "fractional" operator $\underline{\underline{L}}^{\nu}$ that may be used to find the "fractional" intermediate cases between the cases 1 and 2.

equation, $\nabla^2\psi = -\rho/\varepsilon_o$. So the fractional operator $\underline{\underline{L}}^{\nu}$ should commute with the Laplacian operator ∇^2.

If these four conditions are satisfied, the fractional operator $\underline{\underline{L}}^{\nu}$ may be used to obtain the intermediate solutions for the problem of interest. In general, when the problem at hand is described by the following general mathematical relation

$$\underline{\underline{A}}f = g \tag{12.1}$$

where $\underline{\underline{A}}$ is the operator that represents the problem, f is the solution or quantity of interest in the problem (subject to appropriate boundary conditions),[1] and g is the given "source" term, by applying the operator $\underline{\underline{L}}^{\nu}$ on both sides of (12.1) one obtains

$$\underline{\underline{L}}^{\nu} \circ \underline{\underline{A}}f = \underline{\underline{L}}^{\nu}g \tag{12.2}$$

If $\underline{\underline{L}}^{\nu}$ commutes with $\underline{\underline{A}}$ (condition 4), we will have $\underline{\underline{L}}^{\nu} \circ \underline{\underline{A}} = \underline{\underline{A}} \circ \underline{\underline{L}}^{\nu}$, thus we will get

$$\underline{\underline{A}} \circ \underline{\underline{L}}^{\nu}f = \underline{\underline{L}}^{\nu}g \tag{12.3}$$

which shows that for the problem of interest represented by the operator $\underline{\underline{A}}$ but now with the new source term $\underline{\underline{L}}^{\nu}g$, a possible new solution is $\underline{\underline{L}}^{\nu}f$ subject to appropriate (and, if needed, modified) boundary conditions.[2] This solution is obtained by applying the fractional operator $\underline{\underline{L}}^{\nu}$ on the original solution f.

Considering the symbolic block diagram given in Figs. 12.3 and 12.4, let us denote the source terms of the problem by g_1 and g_2 for cases 1 and 2, respectively, and their corresponding solutions by f_1 and f_2. Both solutions satisfy their relevant boundary conditions. If the operator $\underline{\underline{L}}$ can be found as an operator that satisfies the relation

$$\underline{\underline{L}}g_1 = g_2 \tag{12.4}$$

and if $\underline{\underline{L}}$ and $\underline{\underline{A}}$ commute, we can then have a possible solution f_2 as[2]

$$f_2 = \underline{\underline{L}}f_1 \tag{12.5}$$

[1] The term "boundary conditions" is used here in the general sense, embracing boundary, initial, and radiation conditions.

[2] In dealing with the solutions $\underline{\underline{L}}^{\nu}f$, $\underline{\underline{L}}f_1$, or $\underline{\underline{L}}^{\nu}f_1$ care must be taken in considering the possible role of any homogenous part of the solution, f_h, which satisfies $\underline{\underline{A}}f_h = 0$.

Therefore, $\underline{\underline{L}}$ is the operator that "connects" case 1 to case 2. Fractionalization of such an operator gives us the operator $\underline{\underline{L}}^{\nu}$. If this fractional operator satisfies the four conditions noted earlier, one can find a possible "fractionalized" solution shown by f_ν and defined as[3]

$$f_\nu \equiv \underline{\underline{L}}^{\nu} f_1 \tag{12.6}$$

for the problem represented by operator $\underline{\underline{A}}$ when the "intermediate" source term is given as

$$g_\nu \equiv \underline{\underline{L}}^{\nu} g_1 \tag{12.7}$$

and when the boundary conditions are appropriately modified if necessary.

12.2.1 A Recipe for Fractionalization of a Linear Operator $\underline{\underline{L}}$

We consider the class of linear operators (or mappings) in which the domain and range of any operator from this class are similar spaces of the same dimensions. So any linear operator of this class should map an element from, say, the space C^n, onto generally another element in the same space C^n. That is, $\underline{\underline{L}}:C^n \rightarrow C^n$ where C^n is an n-dimensional vector space over the field of complex numbers. The fractionalized operator $\underline{\underline{L}}^{\nu}$, which should satisfy the four conditions, is also an operator with the domain and range in the space C^n. So $\underline{\underline{L}}^{\nu}:C^n \rightarrow C^n$. A recipe for constructing the fractional operator $\underline{\underline{L}}^{\nu}$ is described in the following steps

1. One first solves the eigenvalue problem for the operator $\underline{\underline{L}}$ and obtains the eigenfunctions ("eigenvectors") and eigenvalues of this operator in the space C^n. So one solves

$$\underline{\underline{L}} u_m = a_m u_m \tag{12.8}$$

where u_m and a_m for $m = 1, 2, 3, ..., n$ are the eigenfunctions and eigenvalues of the operator $\underline{\underline{L}}$ in the space C^n, respectively;

2. When u_m's form a complete set of basis functions in the space C^n, any function (vector) in this space can be expressed as a linear combination of u_m's. So, an arbitrary vector h in space C^n can be written as $h = \sum_{m=1}^{n} c_m u_m$ where c_m's are the coefficients of expansion of h in terms of u_m's.
3. With the knowledge of the eigenfunctions and eigenvalues of the operator $\underline{\underline{L}}$, the fractional operator $\underline{\underline{L}}^{\nu}$ with parameter ν can be defined to be an operator that has the same eigenfunctions u_m's, but with the eigenvalues of a_m^{ν}. Thus, the fractional operator $\underline{\underline{L}}^{\nu}$ satisfies the eigenvalue problem $\underline{\underline{L}}^{\nu} u_m = a_m^{\nu} u_m$.[3] Applying this fractional operator $\underline{\underline{L}}^{\nu}$ on an arbitrary vector function h in the space C^n, we get

$$\underline{\underline{L}}^{\nu} h = \underline{\underline{L}}^{\nu} \sum_{m=1}^{n} c_m u_m = \sum_{m=1}^{n} c_m \underline{\underline{L}}^{\nu} u_m = \sum_{m=1}^{n} c_m a_m^{\nu} u_m \tag{12.9}$$

[3] Since the terms a_m^{ν} may be multivalued, care must be exercised in using the proper branch cuts for these functions when we deal with physical conditions.

subject to the possibility of interchanging the order of $\underline{\underline{L}}^{\nu}$ and $\Sigma_{m=1}^{n}$ with appropriate consideration for convergence. The above equation effectively defines the fractional operator $\underline{\underline{L}}^{\nu}$ from the knowledge of operator $\underline{\underline{L}}$ and its eigenfunctions and eigenvalues. Using the above definition, one can easily show that the conditions 1, 2, and 3, listed earlier, are satisfied by operator $\underline{\underline{L}}^{\nu}$. That is,

$$\underline{\underline{L}}^{1} = \underline{\underline{L}}, \qquad \underline{\underline{L}}^{0} = \underline{\underline{I}}, \qquad \underline{\underline{L}}^{\nu_1} \circ \underline{\underline{L}}^{\nu_2} = \underline{\underline{L}}^{\nu_1 + \nu_2} \tag{12.10a}$$

The satisfaction of the 4th condition, however, should only be examined when the specific operator describing the problem of interest (i.e., $\underline{\underline{A}}$) and the operator $\underline{\underline{L}}$ are chosen.

It must be pointed out that there has been much work conducted on fractionalization of the Fourier transform operator by several researchers such as Condon [34], Bargmann [35], Namias [36], McBride and Kerr [37], Lohmann (see, e.g., [38], [39]), Mendlovic and Ozaktas and their coworkers (see, e.g., [39]–[41]), Bernardo and Soares [42], Almeida [43], Pellat-Finet and Bonnet [44], and Shamir and Cohen [45], to name a few. Moreover, some other fractional operators such as fractional Gabor transform [46], fractional Wigner distribution function [47], fractional Hilbert transform [48], fractional Talbot effects [49], [50], and fractional Legendre Transforms [51] have also been investigated. These studies show the richness and power of fractionalization of operators.

Having described the concept of fractional paradigm and fractionalization of linear operators, we now consider the roles of fractional operators and the issues of such paradigm in some of the specific electromagnetic problems that we have studied in recent years. Although our original motivation behind some of these case studies was to explore the role of fractional derivatives and fractional integrals in certain electromagnetic problems and to find out what can be gained by bringing the fractional calculus into electromagnetic theory, in the following sections in this chapter when we provide an overview of our recent work we will present and review the results of our study in terms of the notion of fractional paradigm and fractionalization of operators. This approach hopefully demonstrates the point that these specific case studies indeed represent some samples of what fractional paradigm in electromagnetic theory may embrace.

12.3 FRACTIONAL PARADIGM AND ELECTROMAGNETIC MULTIPOLES

In order to explore the notion of fractional paradigm in the specific topic of multipole expansion of sources, potentials, and fields in electromagnetic theory, we return to the first example of electrostatic point monopole and point dipole and their corresponding potential distributions described earlier in this chapter. The governing equation for the electrostatic potential distribution in free space is the Poisson equation $\nabla^2 \psi = -\rho/\varepsilon_o$ where ρ, a scalar function of position, is the volume charge density of the source. It is common knowledge that point multipoles such as monopoles, dipoles, quadrupoles, octopoles, etc., are well-defined sources with specific potential and field distributions. The spatial distributions of these point multipoles can be expressed in terms of the Dirac delta function and its spatial derivatives (see, e.g., [30, Chapter 2], [31]–[33]). The volume charge distribution of a point monopole can be written as $\rho_o = q\delta(\boldsymbol{r})$, and that of a point dipole can be expressed as $\rho_1 = -\boldsymbol{p}\cdot\nabla\delta(\boldsymbol{r})$ where q is the strength of the point charge, $\boldsymbol{p}$ represents the dipole vector with the intensity and direction, and $\boldsymbol{r}$ is the position vector. It can be seen that for a point monopole, the Dirac delta function is used while for a point dipole the first spatial derivatives of the Dirac delta function are involved.[4] Without loss of generality in our discussion here, let us assume that the dipole vector is oriented along the z axis, i.e., $\boldsymbol{p} = p\hat{\boldsymbol{z}}$, and thus $\rho_1 = -p(\partial\delta(\boldsymbol{r})/\partial z)$. Following what we described for the block diagrams shown in Figs. 12.1–12.4, we take the case of the point monopole as "Case 1," and the point dipole as "Case 2." We can then identify the operator $\underline{\underline{\boldsymbol{L}}}$ that relates the source term in case 1, ρ_o, with the source in case 2, ρ_1. That is $\underline{\underline{\boldsymbol{L}}}\rho_o = \rho_1$. By comparing the expressions of the two sources, the operator $\underline{\underline{\boldsymbol{L}}}$ can be written as $\underline{\underline{\boldsymbol{L}}} = -(p/q)(\partial/\partial z)$. The multiplicative constant p/q, which has the physical dimension of *length*, directly depends on the strength of the two sources, and thus it is insignificant in our discussion here. Therefore, we can write the operator $\underline{\underline{\boldsymbol{L}}}$ as

$$\underline{\underline{\boldsymbol{L}}} = l\frac{\partial}{\partial z} \tag{12.10b}$$

where l represents a constant with dimension *length*. The next step is to consider the fractional operator $\underline{\underline{\boldsymbol{L}}}^{\nu}$ where parameter ν can be a non-integer number. This operator can be symbolically expressed as

$$\underline{\underline{\boldsymbol{L}}}^{\nu} = l^{\nu}\frac{\partial^{\nu}}{\partial z^{\nu}} \tag{12.11}$$

where the symbol $\partial^{\nu}/\partial z^{\nu}$ represents the *fractional*-order differentiation operator.

Fractional derivatives, fractional integrals, and their properties are the subject of study in the field of fractional calculus, and have been studied by mathematicians over many years (see, e.g., [11]–[28]).[5] Fractional derivatives and integrals, which are also called "fractional differintegrals" by Oldham and Spanier [11], can be shown by the symbol ${}_aD_x^{\nu}f(x)$ (following Davis [14]) or $d^{\nu}f(x)/d(x-a)^{\nu}$ (following Oldham and Spanier [11]) where ν is an arbitrary (in general non-integer) order of the operator, and a is the lower limit of the integrals used to define these operators. There are several definitions given by mathematicians for the fractional derivatives and fractional integrals (see, e.g., [11, Chapter 3]). Among these definitions, one that has been often used in the literature on fractional calculus is the one known as the Riemann-Liouville fractional integral [11, p. 49]. It is written as

$$ {}_aD_x^{\nu}f(x) \equiv \frac{1}{\Gamma(-\nu)}\int_a^x (x-u)^{-\nu-1}f(u)\,du \qquad \text{for } \nu<0, \text{ and } x>a \tag{12.12}$$

[4] Strictly speaking, Dirac delta function should be treated as a *distribution* [52] or a *generalized* function [25, Vol. 1].

[5] Although the name of *fractional* calculus has been used for this field in mathematics, it must be noted that orders of integration and differentiation may in general be also irrational numbers, or even complex numbers (see, e.g., [20]).

where ${}_aD_x^\nu$ denotes the (fractional) νth-order integration of function $f(x)$ with the lower limit of integration being a, and $\Gamma(\cdot)$ is the Gamma function. This definition can be interpreted as a generalization of Cauchy's repeated integration formula [11, p. 38]. That is for $\nu = -n$, the above definition provides us with

$$ {}_aD_x^{-n} f(x) = \frac{1}{(n-1)!} \int_a^x (x-u)^{n-1} f(u)\, du $$

which, according to Cauchy's repeated integration formula, is equivalent to

$$ \frac{1}{(n-1)!} \int_a^x (x-u)^{n-1} f(u)\, du = \int_a^x dx_{n-1} \int_a^{x_{n-1}} dx_{n-2} \cdots \int_a^{x_1} f(x_0)\, dx_0 \qquad (12.13) $$

The Riemann-Liouville fractional integral is for $\nu < 0$. However, for fractional derivatives where $\nu > 0$, the above Riemann-Liouville fractional integration definition can still be applied if it is used in conjunction with the following additional step

$$ {}_aD_x^\nu f(x) = \frac{d^m}{dx^m} {}_aD_x^{\nu-m} f(x) $$

for $\nu > 0$, where m is a positive integer large enough so that $(\nu - m) < 0$ and thus the Riemann–Liouville integration can be utilized for ${}_aD_x^{\nu-m}f(x)$ [11, p. 50]. The operator d^m/dx^m is the ordinary mth-order differential operator [11, p. 50]. Several other definitions for fractional derivatives and integrals can be found in various literatures on fractional calculus (see, e.g., [11, Chapter 3], [12], [13]). For easy reference, some of these definitions are briefly mentioned in the Appendix.

Applications of fractional calculus can be found in several topics in mathematics such as differential equations, generalized functions, Mellin transforms, and complex analysis to name a few (see, e.g., [11]–[28]). One of the earliest applications of fractional calculus appears to be the one given by Abel in 1823 in his study of the tautochrone problem [28]. The historical bibliography, prepared by Ross and reprinted in pp. 3–15 of the book by Oldham and Spanier [11], provides an excellent historical account of development of the fractional calculus. Moreover, Ross in his article [23] gives an informative concise history of this field. Samko et al. also include a brief historical outline of the fractional calculus in their book on this topic [12, pp. xxvii–xxxvi]. It is also interesting to note that Lützen in his article [19] has mentioned that Liouville, who was one of the major pioneers in the development of fractional calculus, was probably inspired by the problem of fundamental force law in Ampère's electrodynamics and treated a fractional differential equation in that problem. Heaviside, who was the pioneer in the development of his operational calculus, used fractional derivatives in his work [53, Vol. 2, Chapters 6–8]. There have been many other renowned mathematicians who worked in the area of fractional calculus (see, e.g., the historical outlines in [11], [12], [23]). In recent years, researchers interested in the field of fractals and related subjects have considered using the concept of fractional calculus in some of their investigations. Among those, one should mention the work of LeMéhauté, Héliodore, and their co-workers [54], [55], Nigmatulin [56], [57], and Kolwankar and Gangal [58], [59]. Furthermore, researchers in some fields of applied science have used some aspects of fractional derivatives/integrals in their work (see, e.g., [60]–[65]). Some optical implementations of fractional-order differentiation have also been investigated [66]– [68].

Returning to the fractional operator $\underline{\underline{L}}^{\nu}$ given in (12.11) and following the idea presented in (12.3), (12.6), and (12.7), we find the "intermediate source" ρ_{ν}, by applying the operator $\underline{\underline{L}}^{\nu}$ on the source of case 1, i.e., on the charge density ρ_o of the point monopole. (In applying the fractional derivative involved in the fractional operator $\underline{\underline{L}}^{\nu}$, we choose to take the lower limit a to be $-\infty$.) So we have

$$\begin{aligned}\rho_{\nu} &= \underline{\underline{L}}^{\nu}\rho_o \\ &= l^{\nu}\,{}_{-\infty}D_z^{\nu}[q\delta(x)\delta(y)\delta(z)]\end{aligned} \tag{12.14}$$

where $\delta(\boldsymbol{r})$ is substituted by its equivalent in the Cartesian coordinate system. If ν is a non-integer real number between zero and unity, what form of charge distribution do we have for ρ_{ν}? In our previous work [1], we have shown that if we start with a charge distribution of a simple point monopole $\rho_o = q\delta(\boldsymbol{r})$ and we take the fractional νth-order derivative of this function with respect to the z coordinate, we will obtain the following charge distribution

$$\begin{aligned}\rho_{\nu}(\boldsymbol{r}) \equiv ql^{\nu}\,{}_{-\infty}D_z^{\nu}\delta(\boldsymbol{r}) &= ql^{\nu}\frac{\partial^2}{\partial z^2}\begin{bmatrix}\delta(x)\,\delta(y)\dfrac{1}{\Gamma(2-\nu)}z^{1-\nu} & \text{for } z>0 \\ 0 & \text{for } z<0\end{bmatrix} \\ &= ql^{\nu}\frac{\partial^2}{\partial z^2}\left[\delta(x)\,\delta(y)\,U(z)\frac{1}{\Gamma(2-\nu)}z^{1-\nu}\right]\end{aligned} \tag{12.15}$$

Here $U(z)$ is the unit step function, the subscript z in ${}_{-\infty}D_z^{\nu}$ denotes fractional differintegration with respect to z, and the lower limit $a = -\infty$. At $z = 0$, the value of $\rho_{\nu}(\boldsymbol{r})$ has a singularity, and strictly speaking this should be treated as a *generalized* function.. For $0<\nu<1$, as we mentioned earlier in the definition of fractional derivatives, the operator ${}_{-\infty}D_z^{\nu}\delta(z)$ should be treated as ${}_{-\infty}D_z^{\nu}\delta(z) = (\partial^m/\partial z^m)\,{}_{-\infty}D_z^{\nu-m}\,\delta(z)$ where $(\nu - m)<0$. In the above equation, $m = 2$. Then, when evaluating $\partial^m/\partial z^m$, care must be taken at $z = 0$. Figure 12.5 provides a sketch of the charge distribution given in (12.15). To gain some insights, the two limiting cases of $\nu = 0$, for which ρ_{ν} becomes ρ_o of the point monopole, and $\nu = 1$, which gives us ρ_1 of the point dipole, are also sketched. When $0<\nu<1$, the volume charge density ρ_{ν} represents a line charge along the positive z axis with volume charge density given in (12.15). The volume charge densities of these charge distributions are also sketched in Fig. 12.5 as a function of z. This charge distribution, which can be regarded as a "fractional-order" multipole, is *effectively* an "intermediate" charge distribution between the cases of the point monopole and the point dipole, and it approaches these two limiting cases when $\nu \rightarrow 0$ or $\nu \rightarrow 1$, respectively. Some of the interesting features of this fractional-order multipole are described in our previous work reported in [1].

Before analyzing the electrostatic scalar potential of such a fractional-order multipole, we should examine whether the above operator $\underline{\underline{L}}^{\nu}$ commutes with ∇^2, which is the operator in the Poisson equation for the electrostatic problem under consideration. In general, attention must be paid in dealing with commutativity of fractional derivative/integral operators among themselves and with other operators.

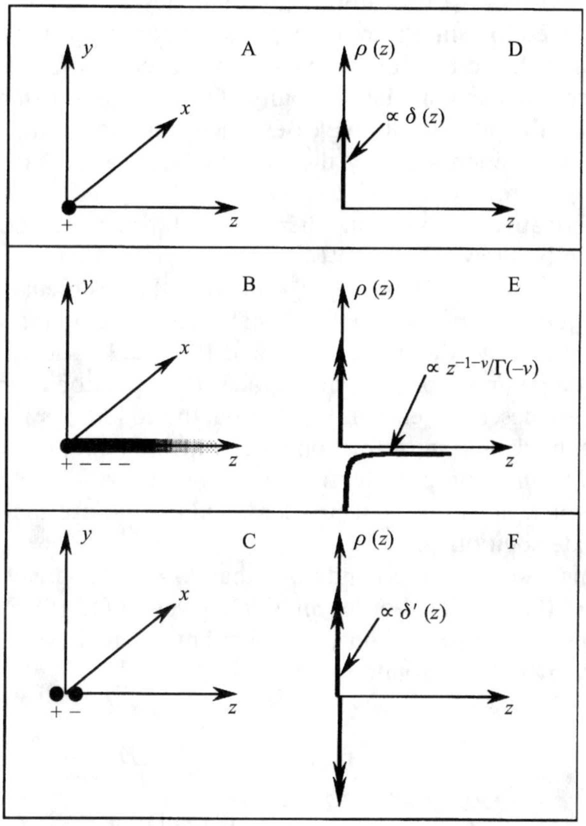

Figure 12.5 (From [1].) Fractional 2^ν-pole charge distributions (with $0 \le \nu \le 1$) are sketched in Panels A, B, and C. In A and C an electric point monopole ($\nu = 0$) and an electric point-dipole ($\nu = 1$) are shown, respectively. In B, a case of a fractional-order 2^ν-pole is shown with $0 < \nu < 1$. (The νth-order fractional derivative is taken with respect to z.) In Panels D, E, and F, the forms of the volume charge densities for the charge distributions shown in A, B, and C are illustrated, respectively. These volume charge densities have the form of $\rho(z)\delta(x)\delta(y)$, where the form of the function $\rho(z)$ as a function of z is only shown. In D, $\delta(z)\delta(x)\delta(y)$ corresponds to the point monopole (in A). In F, $\delta'(z)\delta(x)\delta(y)$ corresponds to the point dipole (in C). Panel E shows the form of $\rho(z)$ for the fractional 2^ν-pole (shown in B) for a fractional value for ν. In E at $z = 0$, the expression for (12.15) has singularity in the form resulting from $\partial^2/\partial z^2[U(z)z^{1-\nu}/\Gamma(2-\nu)]$ at $z = 0$. [*This figure was originally published in N. Engheta, "On Fractional Calculus and Fractional Multipoles in Electromagnetism,"* IEEE Trans. Antennas and Propagation, *vol. 44, no. 4, pp. 554–566, April 1996, Copyright © 1996, IEEE.*]

Considering the Cartesian coordinates for the operators in the present case, the fractional z-derivative operator ${}_{-\infty}D_z^{\nu}$ commutes with $\partial^2/\partial x^2$ and $\partial^2/\partial y^2$ parts of the Laplacian operator, since they are independent by the fact that they apply to different variables. For the commutativity condition between $\partial^2/\partial z^2$ and ${}_{-\infty}D_z^{\nu}$, however, one in general needs to examine the suitability of function ψ, which here satisfies the Poisson equation for the point multipole cases. Such conditions can be interpreted and taken from the composition rules for the fractional derivatives/integrals given in ([23], [11, pp. 82–87]). Here in our case, ψ is zero at $z = -\infty$, and so are all its positive-integer-order z-derivatives at $z = -\infty$. Therefore, following the composition rules in ([23], [11]), it can be shown that ${}_{-\infty}D_z^{\nu}(\partial^2/\partial z^2)\,\psi = {}_{-\infty}D_z^{\nu+2}\,\psi$ and $(\partial^2/\partial z^2)\,{}_{-\infty}D_z^{\nu}\psi = {}_{-\infty}D_z^{2+\nu}\,\psi$. Thus ${}_{-\infty}D_z^{\nu}(\partial^2/\partial z^2)\,\psi = (\partial^2/\partial z^2)\,{}_{-\infty}D_z^{\nu}\psi$ for ψ in this problem. Furthermore, due to the same composition rules for the fractional derivatives/integrals it can also be seen that $\underline{\underline{L}}^{\nu_1}(\underline{\underline{L}}^{\nu_2}\psi) = \underline{\underline{L}}^{\nu_1+\nu_2}\psi$ where $0 \le \nu_1 \le 1$ and $0 \le \nu_2 \le 1$. Therefore, for the problem of electrostatic potential of point multipoles, the operator $\underline{\underline{L}}^{\nu}$ satisfies the conditions 3 and 4 that we described earlier. In addition, the conditions 1 and 2 are clearly satisfied in light of the definition of fractional derivative/integral that is used for $\underline{\underline{L}}^{\nu}$. Thus, the operator $\underline{\underline{L}}^{\nu}$ can be used as the fractional operator resulting from fractionalization of the operator $\underline{\underline{L}}$ in the present problem, and can be utilized to obtain the fractional intermediate solutions here.

Having examined the conditions that $\underline{\underline{L}}^{\nu}$ should satisfy, the electrostatic scalar potential of the fractional-order multipole given in (12.15) can be analyzed. We have studied this "intermediate" potential distribution and have reported the results in [1]. This scalar potential is found to be

$$\begin{aligned}\psi_\nu(x,y,z) &= \underline{\underline{L}}^{\nu}\psi_o = l^{\nu}\,{}_{-\infty}D_z^{\nu}\frac{q}{4\pi\varepsilon R}\\ &= \frac{ql^{\nu}\Gamma(\nu+1)}{4\pi\varepsilon_o R^{1+\nu}}P_\nu(-\cos\theta)\end{aligned} \tag{12.17}$$

where $P_\nu(-\cos\theta)$ is the Legendre function of the first kind and the *non-integer* degree ν and $R = \sqrt{x^2+y^2+z^2}$. In Fig. 12.6, we present a series of contour plots, which we originally reported in [1], for this potential in the x–z plane, in the region $-2 \le z \le 2$ and $0 < x \le 2$, for several values of ν between zero and unity. From (12.16), it can be easily seen that this potential drops as $R^{-1-\nu}$ (with $0 < \nu \le 1$) as R increases. Therefore this is a potential distribution that can be *effectively* regarded as the "intermediate" case between the potential of a monopole (when $\nu = 0$), which drops as R^{-1}, and the potential of a point dipole (when $\nu = 1$), which drops as R^{-2}. So the charge distribution $\underline{\underline{L}}^{\nu}\rho_o$, given explicitly in (12.15), can be considered as a fractional-order multipole, an intermediate case between a point monopole and a point dipole. It must also be noted that although the electrostatic scalar potential of such a fractional-order multipole drops as $R^{-1-\nu}$, Gauss' law is certainly satisfied as we have shown in [1]. Some of the notable features of this potential distribution and fractional-order multipole are reported in [1].

We have also extended the concept of fractional multipoles to the two-dimensional cases of line multipoles. The "intermediate" fractional line multipoles and their corresponding electrostatic scalar potential distributions in the two-dimensional case have also been analyzed and reported in [1].

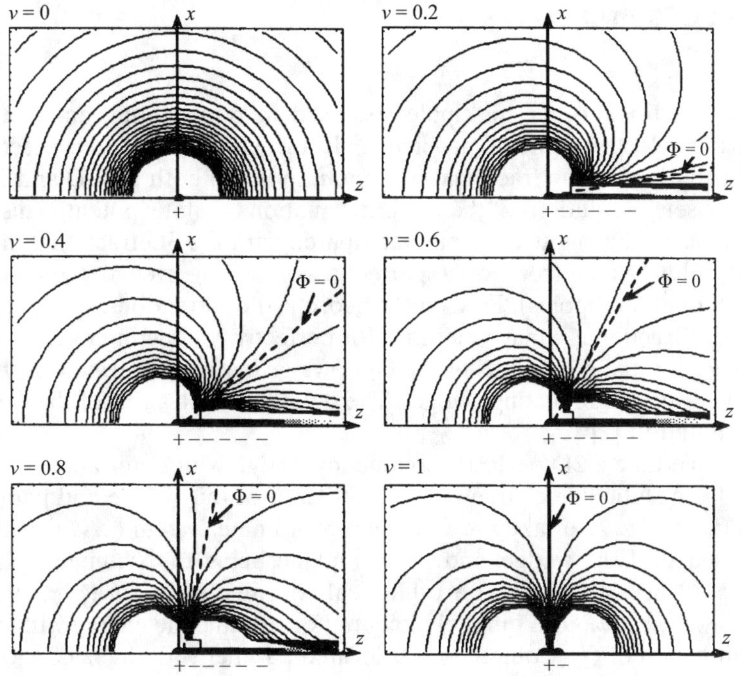

Figure 12.6 (From [1].) Contour plots for the electrostatic scalar potential of the fractional 2^{ν}-pole charge distribution of (12.15) for $0 \le \nu \le 1$. The expression for the potential is given in (12.16). This scalar potential is azimuthally symmetric, and therefore the intersection of equipotential surfaces with the x–z plane is only shown here. The contours are shown for the region $-2 \le z \le +2$ and $0 < x \le 2$, and for six different values of ν. For $\nu = 0$ (top left panel), the source is a single electric point monopole (shown as +) located at $x = y = z = 0$, and the contours are independent of θ. For $\nu = 1$ (bottom right panel), the source is an electric point dipole (shown as +−) located at the origin, and the angular dependence of the potential is $-\cos\theta$. For fractional values of ν between zero and unity, we can see from the above contour plots that the scalar potentials of fractional 2^{ν}-poles are "intermediate" cases between those of the point monopole and point dipole, and *effectively* they "evolve" from one case to the other. The dashed lines, which are added later on top of the contour plots, show the approximate location of zero potential (root of $P_{\nu}(-\cos\theta) = 0$.) In the region of the x–z plane shown above, to the left of this dashed line, the potential is positive and to the right it is negative. Along the positive z axis, the charge distribution is also sketched. The plus and minus signs below the z axis indicate the sign of the charge distribution along the z axis. [*This figure was originally published in N. Engheta, "On Fractional Calculus and Fractional Multipoles in Electromagnetism,"* IEEE Trans. Antennas and Propagation, *vol. 44, no. 4, pp. 554–566, April 1996, Copyright © 1996, IEEE.*]

12.4 FRACTIONAL PARADIGM AND ELECTROSTATIC IMAGE METHODS FOR PERFECTLY CONDUCTING WEDGES AND CONES

Can fractional-order multipoles be used in representing an alternative model for certain electromagnetic problems? It turned out that the electrostatic potential distributions of the fractional-order multipoles in both the 3D and 2D cases resemble some sets of solutions of the Laplace equation for static potential distributions in front of a perfectly conducting cone (compared with the 3D fractional multipoles) and the 2D solutions in front of a perfectly conducting wedge (compared with the 2D fractional multipoles). From such theoretical observations, we developed the electrostatic "fractional" image methods for perfectly conducting wedges and cones, and we have reported our results in [2]. Here we give a brief overview of this method for the 2D perfectly conducting wedges. The detail for both cases of the wedge and cone can be found in [2].

We take a 2D perfectly conducting wedge with outer angle β. An infinitely long uniform static line charge is placed in front of this wedge and parallel with the edge of the wedge. We take the Cartesian coordinate system (x, y, z) with the z axis along the edge of the wedge, and the x–z plane being the symmetry plane of the wedge (Fig. 12.7a). We also use a cylindrical coordinate system (ρ, φ, z) where $x = \rho \cos \varphi$ and $y = \rho \sin \varphi$.[6] The infinitely long uniform static line charge with charge density per unit length of Q_l (Coulomb/m) is located parallel with the z axis at an arbitrary point with coordinates (ρ_o, φ_o). This problem is a 2D problem, and the electrostatic potential $\psi(\rho, \varphi)$ is independent of the z coordinate. It is known that if the outer angle β takes the specific value of π/s where s is a positive integer $s = 1, 2, 3, \ldots$, the conventional method of images gives us a set of $2s - 1$ discrete image charges [69, p. 70]. As we asked in [2], what would happen if the outer angle β were not π/s? What kind of image charges, if any, would one expect to obtain when $\beta \neq \pi/s$? One method of solving this problem is obtained by Nikoskinen and Lindell who have done an excellent analysis for the problem of image solutions for the Poisson equation for the dielectric wedge geometry [70]. In their solutions, they found that for the perfectly conducting wedge with an arbitrary angle, their image solutions have a distributed portion in the imaginary angular domain [70]. We have introduced another way to approach this problem using the idea of fractional-order multipoles [2]. In our method, we have shown that for the general case of $\beta \neq \pi/s$ we can have equivalent charge distributions or "images" that effectively behave as distributed "intermediate" cases between the cases of discrete images obtained for the wedge with specific wedge angles $\beta = \pi/s$. We have also shown that the fractional orders of these equivalent charges, which we have called "fractional" images, depend on β, the wedge's angle.

The electrostatic potential $\psi(\rho, \varphi)$ due to the line charge in front of the

[6] To be consistent with the notations we used in [2], here in this section the symbol ρ is used for the radial distance in the cylindrical coordinate systems (ρ, φ, z). This should not be confused with ρ as a volume charge density discussed in Section 12.3.

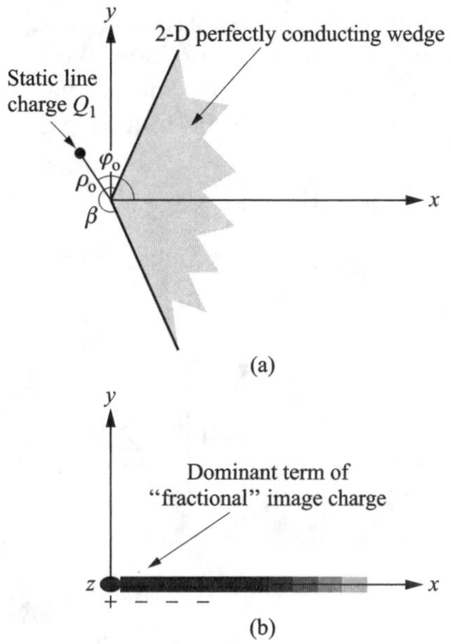

Figure 12.7 (From [2].) Panel (a) shows a two-dimensional perfectly conducting wedge with the outer angle β. An infinitely long static line charge with uniform line charge density of Q_l (Coulomb/m) is located at (ρ_o, φ_o). The Cartesian (x, y, z) and cylindrical (ρ, φ, z) coordinate systems are used. The geometry of the problem is independent of z. For the electrostatic potential in the distant region (where $\rho \gg \rho_o$) and in $\pi - \beta/2 < \varphi < \pi + \beta/2$, the wedge and the original line charge can be replaced with the dominant term of the equivalent charge given in (12.20), which is proportional to the fractional π/β-derivative of the line monopole. This fractional $2^{\pi/\beta}$-pole, sketched in Panel (b), gives rise to a potential that is similar to the potential in front of the wedge in Panel (a) for observers in the far region (where $\rho \gg \rho_o$) in $\pi - \beta/2 < \varphi < \pi + \beta/2$. [*This figure was originally published in N. Engheta, "Electrostatic 'Fractional' Image Methods for Perfectly Conducting Wedges and Cones,"* IEEE Trans. Antennas & Propagation, *vol. 44, no. 12, pp. 1565–1574, December 1996, Copyright © 1996, IEEE.*]

perfectly conducting wedge (i.e., in the angular range of $\pi - (\beta/2) \le \varphi \le \pi + (\beta/2)$) is expressed as

$$\psi(\rho, \varphi) = \sum_{m=1}^{\infty} \frac{Q_l}{m\pi\varepsilon}\left(\frac{\rho_<}{\rho_>}\right)^{m\pi/\beta} \sin\left[\frac{m\pi}{2} - \frac{m\pi}{\beta}(\pi - \varphi)\right] \sin\left[\frac{m\pi}{2} - \frac{m\pi}{\beta}(\pi - \varphi_o)\right] \quad (12.18)$$

where $\rho_<(\rho_>)$ is the smaller (larger) of ρ_o and ρ, respectively, and ε is the permittivity of the homogeneous isotropic medium in front of the wedge [71, p. 76]. Let us first consider the region where $\rho > \rho_o$ and $\pi - (\beta/2) \le \varphi \le \pi + (\beta/2)$. From this equation, it can be seen that in this region the ρ-dependences of various terms of the potential involve the generally non-integer powers $-m\pi/\beta$. We know that for any 2D electrostatic charge density with finite-region cross section in the x–y plane (not the case under study here) the conventional multipole expansion of the electrostatic potential involves terms whose ρ-dependences have negative integer powers (except the zeroth-order line-monopole whose potential has logarithmic dependence on ρ). So from the fact that $-m\pi/\beta$ is, in general, not an integer when $\beta \neq \pi/s$, we infer that for the wedge problem with $\beta \neq \pi/s$ an equivalent representation of charge distribution, which would provide a potential distribution similar to that of the wedge in the region $\rho > \rho_o$ and $\pi - (\beta/2) \le \varphi \le \pi + (\beta/2)$ is not limited in a certain bounded region in the two-dimensional x–y plane (i.e., it would not be localized charge distribution in the x–y plane). To demonstrate this point, we take the dominant term in (12.18) for $\rho \gg \rho_o$. The ρ-dependence of this term has the form $\rho^{-\pi/\beta}$. As $\rho \to \infty$, this term drops more rapidly than $-\ln \rho$ of a line monopole, and drops less rapidly than ρ^{-n} of a line 2^n-pole where n is the first positive integer larger than π/β. Therefore, for a 2D wedge with $\beta > \pi$, if we remove the wedge (and the original line charge) we should substitute them with an equivalent charge distribution that produces a potential with $\rho^{-\pi/\beta}$-dependence in the region $\rho \gg \rho_o$. This equivalent charge should have a form that is neither a line monopole nor a line dipole at the origin. Instead it should effectively be an "intermediate" case between these two line multipoles. Thus, a 2D fractional-order multipole can be of use here. We have reported the mathematical detail of this analysis in [2]. For $\rho > \rho_o$ our analysis has shown that if the wedge and its original line charge are replaced with the following equivalent charge distribution,

$$Q_{equiv.}(x,y) = \sum_{n=0}^{\infty} \left\{ \frac{2\pi\rho_o^{(2n+1)\pi/\beta} \cos\left[\dfrac{(2n+1)\pi}{\beta}(\pi-\varphi_o)\right]}{\beta\Gamma\left[\dfrac{(2n+1)\pi}{\beta}+1\right]} \times {}_{-\infty}D_x^{(2n+1)\pi/\beta}[Q_l\delta(x)\delta(y)] \right\} + \sum_{n=0}^{\infty} \left\{ \frac{2\pi\rho_o^{(2n+2)\pi/\beta} \sin\left[\dfrac{(2n+2)\pi}{\beta}(\pi-\varphi_o)\right]}{\beta\Gamma\left[\dfrac{(2n+2)\pi}{\beta}+1\right]} \times {}_{-\infty}D_x^{((2n+2)\pi/\beta)-1}\left[-Q_l\delta(x)\delta^{(1)}(y)\right] \right\} \quad (12.19)$$

the same electrostatic potential in the region $\rho > \rho_o$ and $\pi - (\beta/2) \le \varphi \le \pi + (\beta/2)$ is obtained. In [2] we have discussed in detail some of the interesting properties of this equivalent charge distribution, which involves fractional derivatives of the line charge $Q_l\delta(x)\delta(y)$ and of the line dipole $-Q_l\delta(x)\delta^{(1)}(y)$.

The dominant term in the above expression is

$$Q_{equiv.}(x,y) \approx \frac{2\pi\rho_o^{\pi/\beta}\cos\left[\dfrac{\pi}{\beta}(\pi-\varphi_o)\right]}{\beta\Gamma\left[\dfrac{\pi}{\beta}+1\right]} {}_{-\infty}D_x^{\pi/\beta}\left[Q_l\delta(x)\delta(y)\right]$$
$$= \frac{2\pi Q_l\rho_o^{\pi/\beta}\cos\left[\dfrac{\pi}{\beta}(\pi-\varphi_o)\right]}{\beta\Gamma\left[\dfrac{\pi}{\beta}+1\right]} \frac{\partial^p}{\partial x^p} \begin{bmatrix} \delta(y)\dfrac{1}{\Gamma\left[p-\dfrac{\pi}{\beta}\right]}x^{-(\pi/\beta)+p-1} & \text{for } x>0 \\ 0 & \text{for } x<0 \end{bmatrix} \quad (12.20)$$

If we remove the wedge and the line charge in front of it, and replace them with the above equivalent charge, for the region $\rho \gg \rho_o$ and $\pi - (\beta/2) \le \varphi \le \pi + (\beta/2)$ we will get the same potential as the dominant term in (12.18). The ρ-dependence of this potential has the form $\rho^{-\pi/\beta}$. In the above equation, the integer p is chosen such that $(\pi/\beta - p) < 0$, and thus the Riemann-Liouville fractional integral for the definition of fractional integral can be used. For non-integer values of π/β, this two-dimensional charge density distribution is a (non-localized) surface charge distribution on the x–z plane for $x > 0$, with volume charge density given in the above equation. Figure 12.7b shows the sketch of this charge distribution that replaces the wedge and its neighboring line charge.

As we have shown in [2], when π/β is an integer, the orders of derivatives involved in the definition of equivalent charge distribution become integer numbers, thus resulting in conventional differentiation of the Dirac delta function. For example, if $\beta = \pi$ the wedge becomes a two-dimensional perfectly conducting half space with flat interface on the y–z plane. From (12.18), one gets

$$\psi(\rho, \varphi)_{\beta=\pi} \approx Q_l \rho_o \cos(\varphi_o) \frac{\cos(\varphi)}{\pi\varepsilon\rho} \quad \text{for } \rho \gg \rho_o \quad \text{and} \quad \frac{\pi}{2} < \varphi < \frac{3\pi}{2}, \quad \frac{\pi}{2} < \varphi_o < \frac{3\pi}{2} \tag{12.21}$$

which is effectively the potential of a 2D line dipole with lines parallel with the z axis, located at the origin, and with the vector of the dipole, $\boldsymbol{p} = 2Q_l\rho_o\cos(\varphi_o)\hat{\boldsymbol{x}}$ ($\hat{\boldsymbol{x}}$ being the unit vector along the x axis). Equation (12.20) for this case ($\beta = \pi$) is written as

$$Q_{equiv.}(x, y)_{\beta=\pi} \approx -2\rho_o \cos(\varphi_o)\ {}_{-\infty}D_x^1\left[Q_l\delta(x)\delta(y)\right] = 2Q_l\rho_o\cos(\varphi_o)\left[-\delta^{(1)}(x)\delta(y)\right] \quad \text{for} \quad \rho \gg \rho_o \quad \text{and} \quad \frac{\pi}{2} \le \varphi_o \le \frac{3\pi}{2} \tag{12.22}$$

This is in fact the volume charge density of a line dipole located at the origin with the same dipole vector $\boldsymbol{p} = 2Q_l\rho_o\cos(\varphi_o)\hat{\boldsymbol{x}}$ as mentioned above. This is expected, because when the original line charge at (ρ_o, φ_o) is parallel with the flat perfectly conducting interface at the y–z plane, there is an image line charge with charge density per unit length $-Q_l$ located at $(\rho_o, \pi - \varphi_o)$ and parallel with the original line charge. This image line charge ($-Q_l$) and the original line charge Q_l together form a line-dipole that provides the dominant term of the potential in the far region as given in (12.21). Similar steps can be shown for the case of $\beta = \pi/2$ that leads to an equivalent line quadrupole [2]. Thus, when π/β is an integer number, the conventional *integer-order* derivatives of the Dirac delta function in (12.19) lead to a charge distribution that effectively behaves as discrete (localized) image line charges. When π/β is not an integer, then *fractional* derivatives of Dirac delta function are used, which yield distributed (non-localized) charge densities for the equivalent charges. So, effectively as the angle of the wedge varies, the equivalent charge (12.19) or (12.20) "evolves" between discrete (localized) and distributed (non-localized) charge densities, continually "filling the gap" between the discrete cases of localized image line charges as line dipole, line quadrupole, line octopole, etc. We have named such "intermediate" charge distributions the "fractional images."

We have mentioned in [2] that when $\beta > \pi$ for observation points close to the edge of the wedge where $\rho \ll \rho_o$, the equivalent charge distribution is (12.20) but with π/β replaced with $-\pi/\beta$. We have also extended the idea of fractional image method described above to the case of an electrostatic problem for a 2D dielectric wedge, and presented the results in a recent symposium [72]. Furthermore, we have analyzed the electrostatic fractional image method for the perfectly conducting circular cone with arbitrary cone angle. In this case, the 3D fractional-order multipoles that we have introduced in [1] and reviewed in Section 12.3 here were used. The detail can be found in [2].

12.5 FRACTIONAL PARADIGM IN WAVE PROPAGATION

In this section, we return to the charts shown in Figs. 12.1–12.4, and specify another problem for our "Problem" block and two of its canonical cases for the cases 1 and 2. The problem of interest in this section is the monochromatic scalar wave propagation. We know for such a problem the scalar Helmholtz equation $\nabla^2 f + k^2 f = g$ is the governing differential equation where f is the scalar quantity of interest, the constant k is the wave number, and g is the source term. In solving the scalar Helmholtz equation, the standard canonical solutions for the one-, two-, and three-dimensional cases can be identified as plane, cylindrical, and spherical waves, respectively. The corresponding sources of these canonical solutions can be one-, two-, and three-dimensional Dirac delta functions, respectively. Now for case 1 in the block diagrams of Figs. 12.1–12.4 we choose the canonical case of plane wave propagation with its one-dimensional Dirac delta function source (i.e., uniform sheet source), and for case 2 we pick the azimuthally symmetric cylindrical wave and its corresponding two-dimensional Dirac delta function source (i.e., uniform line source). Following the similar line of questions as in the previous sections, we can ask whether there are intermediate "fractional" solutions for the same standard Helmholtz equation—solutions that are "intermediate" between the cases of a plane wave and a cylindrical wave. If there are solutions of this sort, what are their corresponding sources that would effectively behave as the intermediate sources between the integer-dimensional Dirac delta functions, i.e., between the one- and two-dimensional Dirac delta functions? One way to find the answers to these questions is to search for the corresponding operator $\underline{\underline{L}}$ in this case. To that end, we take a Cartesian coordinate system (x, y, z) in a three-dimensional physical space. The Green function $G(\boldsymbol{r}, \boldsymbol{r}_o; k)$ for the scalar Helmholtz equation should satisfy

$$\nabla^2 G(\boldsymbol{r}, \boldsymbol{r}_o; k) + k^2 G(\boldsymbol{r}, \boldsymbol{r}_o; k) = -\delta(\boldsymbol{r} - \boldsymbol{r}_o) \tag{12.23}$$

where $\boldsymbol{r} = x\hat{\boldsymbol{x}} + y\hat{\boldsymbol{y}} + z\hat{\boldsymbol{z}}$ and $\boldsymbol{r}_o = x_o\hat{\boldsymbol{x}} + y_o\hat{\boldsymbol{y}} + z_o\hat{\boldsymbol{z}}$ are the position vectors for the observation and source points, respectively, and $\hat{\boldsymbol{x}}$, $\hat{\boldsymbol{y}}$, and $\hat{\boldsymbol{z}}$ are the unit vectors in the coordinate system. For the one-dimensional Dirac delta function source $\delta_1(\boldsymbol{r}) \equiv \delta(x)$ located at the y–z plane, the solution to the scalar Helmholtz equation is a plane wave [73, Vol. 1, p. 811] that is written as

$$G_1(\boldsymbol{r}, \boldsymbol{r}_o; k) = G_1(|x|; k) = i\frac{e^{ik|x|}}{2k} \tag{12.24}$$

The time dependence of $e^{-i\omega t}$ is assumed hereafter. When the source is the two-dimensional Dirac delta function $\delta_2(\boldsymbol{r}) \equiv \delta(x)\delta(y)$ (i.e., a uniform line source along the z axis), the solution is an azimuthally symmetric cylindrical wave [73, vol. 1, p. 811] as

$$G_2(\boldsymbol{r}, \boldsymbol{r}_o; k) = G_2(\sqrt{x^2 + y^2}; k) = \frac{i}{4} H_o^{(1)}(k\sqrt{x^2 + y^2}) \tag{12.25}$$

where $H_o^{(1)}(\cdot)$ is the zeroth-order Hankel function of the first kind [74, Chapter 9]. Comparing the two sources $g_1 = \delta(x)$ and $g_2 = \delta(x)\delta(y)$, one can readily see from the mathematical point of view that if g_2 is integrated over the y coordinate from $y = -\infty$ to some arbitrary point on the y axis, one obtains[7]

$$\int_{-\infty}^{y} g_2(x,u)\,du = \int_{-\infty}^{y} \delta(x)\delta(u)\,du = \delta(x)U(y) \tag{12.26}$$

where u is a dummy variable of integration, and $U(y)$ is the unit step function with the value of zero for $y < 0$ and the value of unity for $y > 0$. If we take this result and write its *even* (symmetric) portion, we will get

$$\frac{1}{2}\big[\delta(x)U(y) + \delta(x)U(-y)\big] = \frac{1}{2}\delta(x) \tag{12.27}$$

which is proportional with the one-dimensional Dirac delta function source. So aside from the factor (1/2), the source g_1 here can be obtained from g_2 by the two-step process of first integrating g_2 over the y coordinate from $y = -\infty$ to some arbitrary point on the y axis, and then finding the even portion of the result. For the sake of simplicity, here for the corresponding operator $\underline{\underline{L}}$ we choose only the integration operator over the y coordinate from $y = -\infty$ to some arbitrary point on the y axis. Finding the even (symmetric) part of the results can be done at the end when the final result of this operation (or its fractionalized operation) is obtained. Furthermore, to keep units and physical dimensions of the sources consistent, if the y coordinate has the dimension of *length* we will multiply the integration operator with the insignificant multiplicative constant l^{-1} where l has the dimension of *length*. Therefore, we choose $\underline{\underline{L}}$ as

$$\underline{\underline{L}} = l^{-1}\int_{-\infty}^{y} \cdots du = l^{-1}\,{}_{-\infty}D_y^{-1} \tag{12.28}$$

When we fractionalize this operator $\underline{\underline{L}}$ to get the fractional operator $\underline{\underline{L}}^{\nu}$, the result is the *fractional* integration operator, which can be expressed as the Riemann-Liouville fractional integral given in (12.12). Thus, we have[8]

$$\underline{\underline{L}}^{\nu} = l^{-\nu}\,{}_{-\infty}D_y^{-\nu} = \frac{l^{-\nu}}{\Gamma(\nu)}\int_{-\infty}^{y} (y-u)^{\nu-1} \cdots du \qquad \text{for } \nu > 0 \tag{12.29}$$

It can be easily shown that this fractional integration operator readily satisfies the first two conditions of fractional operators given earlier in this chapter. Recognizing that the operator representing the problem here is $\nabla^2 + k^2$, for the conditions 3 and 4, when $0 < \nu_1 + \nu_2 < 1$ with $\nu_1 > 0$ and $\nu_2 > 0$ we invoke again the composition rules of the fractional integration ([23], [11, pp. 82–87]) and the condition on function f that at $y = -\infty$ the function f and all its positive-integer-order derivatives with respect to y vanish. Thus, in these cases conditions 3 and 4 can also be satisfied, and therefore the operator $\underline{\underline{L}}^{\nu}$ can be used as a fractional operator that can provide us with the fractional solutions to the problem of interest here.

We apply this fractional operator $\underline{\underline{L}}^{\nu}$ on the source $g_2 = \delta(x)\delta(y)$ and we obtain,

[7]At this point, we are not concerned with the units and physical dimensions of these sources. This point will be addressed shortly.

[8]Here we use $-\nu$ instead of ν in the Riemann-Liouville fractional integral. Thus, we should have $\nu > 0$.

after finding the even portion of the result, the following expression for the symmetric (even) fractional intermediate source

$$ {}_eS_\nu(x,y) = \frac{l^{-\nu}}{2}\left[{}_{-\infty}D_y^{-\nu}\delta(x)\delta(y) + {}_{-\infty}D_{-y}^{-\nu}\delta(x)\delta(y)\right] = \frac{l^{-\nu}\delta(x)|y|^{-1+\nu}}{2\Gamma(\nu)} \quad \text{for } 0<\nu<1 \tag{12.30} $$

The pre-subscript "e" signifies the "even" symmetric nature of this source with respect to the y coordinate. This source, which involves fractional integrals of the two-dimensional Dirac delta function $\delta(x)\delta(y)$, is a distributed source in the y–z plane. When $\nu\to 0$, the above source approaches the two-dimensional Dirac delta function $\delta(x)\delta(y)$ located at $x = y = 0$ (i.e., a line at the z axis), and when $\nu\to 1$ the above source becomes $\frac{1}{2}l^{-1}\delta(x)$, which is a one-dimensional source located at the plane (i.e., a sheet source in the y–z plane). When $0<\nu<1$ the above source can effectively be considered as an "intermediate" source between the two cases of one- and two-dimensional Dirac delta function sources [4].

For the above "intermediate" source as the source term in the right-hand side of the inhomogeneous Helmholtz equation, we have the solution to this equation for the observation points on the symmetry plane (i.e., on the x–z plane) and for $x>0$ [4]. That is,

$$ f_\nu(x, y=0;k) = \frac{il^{-\nu}\Gamma[(1-\nu)/2]\cos(\nu\pi/2)}{4\sqrt{\pi}}\left(\frac{x}{2k}\right)^{\nu/2} H^{(1)}_{-\nu/2}(kx) \quad \text{for} \quad 0<\nu<1, \quad \text{and} \quad \begin{matrix} x>0 \\ y=0 \end{matrix} \tag{12.31} $$

where $H^{(1)}_{-\nu/2}$ is the Hankel function of order $-\nu/2$ and of the first kind [74, p. 358]. For observation points outside the x–z plane, we have found the asymptotic solution suitable for the far-zone region [4]. Denoting $\rho \equiv \sqrt{x^2+y^2}$ and $\varphi \equiv \tan^{-1}(y/x)$ in the cylindrical coordinate system (ρ, φ, z), for $k\rho \gg 1$ and $|\varphi|>0$ but not too small, the far-zone solution is given as

$$ {}_ef_\nu(x,y;k) \underset{k\rho\gg 1}{\cong} \frac{il^{-\nu}}{4\pi}\cos\left(\frac{\nu\pi}{2}\right)(k\sin|\varphi|)^{-\nu}\sqrt{\frac{2\pi}{k\rho}}e^{ik\rho-i\pi/4} + \frac{il^{-\nu}}{4k^\nu\Gamma(\nu)}\frac{e^{ik|x|}}{(k|y|)^{1-\nu}} \quad \text{for } 0<\nu<1, \quad \text{and} \quad |\varphi|>0 \quad \text{but not too small} \tag{12.32} $$

This far-zone solution to the Helmholtz equation with the "intermediate" source ${}_eS_\nu(x,y)$ in (12.30) has two parts: a cylindrical wave which drops as $\rho^{-1/2}$ in the far zone, and a non-uniform plane wave that propagates in the x direction but whose amplitude drops with y as $|y|^{\nu-1}$ for $0<\nu<1$. It can be easily shown that when ν approaches one of the limits of 0 or 1, the above solution becomes the asymptotic form for the two- or one-dimensional case. Specifically, when $\nu\to 0$, the source ${}_eS_\nu(x,y)$ becomes a two-dimensional Dirac delta function at $x = y = 0$. In this case, the second term in (12.32) that represents a "plane-wave" portion of the solution disappears. The first term becomes

$$ {}_ef_{\nu=0}(x,y;k) \underset{k\rho\gg 1}{\cong} \frac{i}{4}\sqrt{\frac{2}{\pi k\rho}}e^{ik\rho-i\pi/4} $$

which is the asymptotic form for $(i/4)H_o^{(1)}(k\rho)$, as expected. For $\nu = 1$, ${}_eS_\nu(x,y)$ gets the form $\frac{1}{2}\delta(x)$. In this case, the cylindrical term disappears, and the plane-wave portion becomes $(i/4kl)\exp(ik|x|)$ which is half of the value of $G_1(x;k)$ for the source is $l^{-1}\delta(x)$. So indeed the above solution effectively behaves as an "intermediate wave" or "fractional wave" between the cases of plane and cylindrical waves. On the x–z plane for the far-zone observation points, the magnitude of the exact solution given in (12.31) has been analyzed for $kx \gg 1$ [4]. This result can be expressed as

$$\left|f_\nu(x,y=0;k)\right|_{kx\gg 1} = \frac{l^{-\nu}\Gamma[(1-\nu)/2]\left|\cos(\nu\pi/2)\right|}{4\pi 2^{(\nu-1)/2}k^{(1+\nu)/2}}|x|^{(\nu-1)/2} \quad \text{for } 0<\nu<1 \tag{12.33}$$

Therefore, on the x–z plane, the magnitude of the solution drops as $|x|^{(\nu-1)/2}$ in the far zone. The rate of drop of this magnitude along the x axis depends on the fractional order ν. When $\nu\to 0$, this magnitude drops as $|x|^{-1/2}$, and this is expected since in this case we have the cylindrical wave as the solution to the Helmholtz equation. When $\nu\to 1$, the magnitude does not drop with distance x because in this case the solution is the plane wave $(i/4kl)\exp(ik|x|)$. So *effectively*, the two- and one-dimensional Green functions of (12.23) have been "smoothly connected" by varying the order of fractional integration involved in the source ${}_eS_\nu(x,y)$. For observation points outside the x-z plane (and not too close to this plane), the far-zone solution given in (12.32) has both cylindrical- and plane-wave portions, and in each of the limits as $\nu\to 0$ and $\nu\to 1$, one of these parts disappears and the other represents the far-zone solution for a limiting case. We have also addressed the extension of the above analysis to the cases that are intermediate between n- and $(n-1)$-dimensional Dirac delta functions for the scalar Helmholtz equation in n-dimensional space (n being a positive integer). This can be found in [4]. In this case, starting from the Green function for the scalar Helmholtz operator with the n-dimensional delta function source we can "smoothly collapse" to the $(n-1)$ dimensional case using sources described as νth-order fractional integrals of n-dimensional Dirac delta functions, where $0<\nu<1$ is a real non-integer number. One can then continue this process from $(n-1)$ to $(n-2)$, and from $(n-2)$ to $(n-3)$, and so on until one reaches the Green function for the one-dimensional delta function source.

12.6 FRACTIONALIZATION OF THE DUALITY PRINCIPLE IN ELECTROMAGNETISM

As the final case reviewed in this chapter, we examine the possibility of "fractionalization" of the duality principle in electromagnetic theory. It is well known that in a given electromagnetic problem for a set of electromagnetic fields satisfying the Maxwell equations, there is another set of fields (and sources) named *dual* fields that also satisfy the Maxwell equations for the dual problem. The dual problem can be obtained by appropriate transformation of fields and sources (see, e.g., [29, p. 6], [75, p. 21]). One such transformation is $\boldsymbol{E}\to -\eta\boldsymbol{H}_d$, $\boldsymbol{H}\to \eta^{-1}\boldsymbol{E}_d$, $\boldsymbol{J}_e\to -\eta^{-1}\boldsymbol{J}_{md}$, $\boldsymbol{J}_m\to \eta\boldsymbol{J}_{ed}$, $\rho_e\to -\eta^{-1}\rho_{md}$, and $\rho_m\to \eta\rho_{e_d}$ where $\eta\equiv\sqrt{\mu/\varepsilon}$. Owing to the symmetry of the Maxwell equations, it is well known that if the fields $\boldsymbol{E}$ and $\boldsymbol{H}$ together with the sources $\boldsymbol{J}_e$, ρ_e, $\boldsymbol{J}_m$, and ρ_m satisfy the Maxwell equations, the dual fields $\boldsymbol{E}_d$ and $\boldsymbol{H}_d$ also satisfy the

Maxwell equations when the dual sources $\boldsymbol{J}_{ed}$, ρ_{ed}, $\boldsymbol{J}_{md}$, and ρ_{md} are used.[9] Using the terminology of Figs. 12.1–12.4, in any given electromagnetic problem assigned as our "Problem" box, the original problem and its corresponding solutions can be taken to be our "Case 1" box, and the *dual* problem with its fields can be assigned as our "Case 2" box. Then we can repeat our questions about intermediate situations again. Can we have an intermediate situation that would be regarded as the "fractional" situation between the original problem (Case 1) and its dual problem (Case 2), and that would satisfy the Maxwell equations with appropriate boundary conditions? If so, can the duality principle be "fractionalized"? As we have mentioned in the previous sections, in order to find answers to these questions first we need to search for an operator $\underline{\underline{L}}$ that "connects" case 1 to case 2. For the sake of simplicity here, let us consider the cases where the time-harmonic source-free Maxwell equations are only needed. (The time dependence is $e^{-i\omega t}$.) Furthermore, we assume that we have a homogeneous simple material medium with parameters ε and μ. Therefore we can write

$$\begin{aligned} \nabla \times \boldsymbol{E} &= i\omega\mu\boldsymbol{H} \qquad & \nabla \times \boldsymbol{H} &= -i\omega\varepsilon\boldsymbol{E} \\ \nabla \cdot \boldsymbol{E} &= 0 \qquad & \nabla \cdot \boldsymbol{H} &= 0 \end{aligned} \tag{12.34}$$

These equations can be easily rearranged and normalized as follows:

$$\begin{aligned} \overline{curl}\,\hat{\boldsymbol{E}} &= \hat{\boldsymbol{H}} \qquad & \overline{curl}\,\hat{\boldsymbol{H}} &= -\hat{\boldsymbol{E}} \\ \overline{div}\hat{\boldsymbol{E}} &= 0 \qquad & \overline{div}\,\hat{\boldsymbol{H}} &= 0 \end{aligned} \tag{12.35}$$

where $\overline{curl}$ and $\overline{div}$ are shorthand for $(1/ik)\nabla\times$ and $(1/ik)\nabla\cdot$ with $k \equiv \omega\sqrt{\mu\varepsilon}$, respectively, $\hat{\boldsymbol{E}} \equiv \boldsymbol{E}$, and $\hat{\boldsymbol{H}} \equiv \eta\boldsymbol{H}$ with $\eta = \sqrt{\mu/\varepsilon}$. Considering the above source-free time-harmonic normalized Maxwell equations and the duality transformation of $\hat{\boldsymbol{E}} \rightarrow -\hat{\boldsymbol{H}}_d$ and $\hat{\boldsymbol{H}} \rightarrow \hat{\boldsymbol{E}}_d$, we can easily see that the normalized dual fields $\hat{\boldsymbol{E}}_d$ and $\hat{\boldsymbol{H}}_d$ can also be written as

$$\begin{aligned} \hat{\boldsymbol{E}}_d &= \overline{curl}\,\hat{\boldsymbol{E}} \\ \hat{\boldsymbol{H}}_d &= \overline{curl}\,\hat{\boldsymbol{H}} \end{aligned} \tag{12.36}$$

The above equations reveal that the operator $\overline{curl}$, which is the link between the original fields and the dual fields, is indeed the operator $\underline{\underline{L}}$ that we have been searching for in this problem. That is

$$\underline{\underline{L}} = \overline{curl} = \frac{1}{ik}\nabla\times \tag{12.37}$$

The next step is to explore the possibility of fractionalization of this operator. This involves fractionalizing the curl operator. We have introduced elsewhere the idea of fractional curl operator ([6], [8]). The fractional curl operator, as we addressed in [6], is an operator shown symbolically as $curl^{\nu}$ with parameter ν being in general non-integer. When $\nu = 1$, we get the conventional *curl* operator. When $\nu = 0$, we should obtain the identity operator. For values of ν other than zero and unity, one

[9] In this transformation, the material parameters ε and μ are not interchanged and thus the medium is the same for both the original and dual problems. However, as is well known there is another way of finding the dual fields in which the material parameters also have to be interchanged (see, e.g., [75, p. 22]).

then gets operator $curl^{\nu}$. With appropriate mathematical conditions, this operator, when applied repeatedly, is additive in the index parameter ν, and also commutes, i.e., $curl^{\nu_1} curl^{\nu_2} = curl^{\nu_2} curl^{\nu_1} = curl^{\nu_1+\nu_2}$. In [6], we have presented a method for fractionalization of the curl operator. In the course of this method, we also needed to fractionalize the cross-product operator. For the sake of brevity and saving space here, these mathematical steps are not repeated here and the interested readers are referred to [6] for detail. There, as an illustrative example, we explicitly wrote the expression of $curl^{\nu}$ for the special case where three-dimensional vector $\boldsymbol{F}$ is a function of z coordinate only. That is $\boldsymbol{F} = \boldsymbol{F}(z) = F_x(z)\hat{\boldsymbol{x}} + F_y(z)\hat{\boldsymbol{y}} + F_z(z)\hat{\boldsymbol{z}}$. The expression for this special case of $curl^{\nu}\boldsymbol{F}(z)$ can then be written as

$$
\begin{aligned}
curl^{\nu}\boldsymbol{F}(z) = &\left[\cos\left(\frac{\nu\pi}{2}\right) {}_{-\infty}D_z^{\nu}F_x(z) - \sin\left(\frac{\nu\pi}{2}\right) {}_{-\infty}D_z^{\nu}F_y(z)\right]\hat{\boldsymbol{x}} \\
&+ \left[\sin\left(\frac{\nu\pi}{2}\right) {}_{-\infty}D_z^{\nu}F_x(z) + \cos\left(\frac{\nu\pi}{2}\right) {}_{-\infty}D_z^{\nu}F_y(z)\right]\hat{\boldsymbol{y}} \\
&+ \delta_{\nu 0}\, {}_{-\infty}D_z^{\nu}F_z(z)\hat{\boldsymbol{z}}
\end{aligned}
\tag{12.38}
$$

where $\delta_{\nu 0}$ is the Kronecker delta with value of unity when $\nu = 0$, and zero when $\nu \neq 0$. By examining the above expression for $curl^{\nu}\boldsymbol{F}(z)$, it can be easily seen that when $\nu = 1$ we have $curl^{1}\boldsymbol{F}(z) = \nabla \times \boldsymbol{F}(z)$, i.e., the conventional curl operator for $\boldsymbol{F} = \boldsymbol{F}(z)$ is obtained. Furthermore, when $\nu = 0$ we get $curl^{0}\boldsymbol{F}(z) = \underline{\underline{\boldsymbol{I}}} \cdot \boldsymbol{F}(z) = \boldsymbol{F}(z)$, as expected. We also notice that the above equation provides $curl^{\nu_1} curl^{\nu_2}\boldsymbol{F}(z) = curl^{\nu_2} curl^{\nu_1}\boldsymbol{F}(z) = curl^{\nu_1+\nu_2}\boldsymbol{F}(z)$ where ν_1 and ν_2 can be non-integer numbers. Although for the cases of interest here the parameter ν can be a non-integer real number between zero and unity, it is interesting to note that for ν being integer numbers greater than unity we also get the appropriate expressions for $curl^{\nu}$. Specifically when $\nu = 2$, we obtain $\nabla \times \nabla \times \boldsymbol{F}(z)$ for the vector $\boldsymbol{F} = \boldsymbol{F}(z)$, and for $\nu = 3$, we get $\nabla \times \nabla \times \nabla \times \boldsymbol{F}(z)$, and so forth.

With the use of the fractional curl operator, we now express the fractional operator $\underline{\underline{\boldsymbol{L}}}^{\nu}$ as $\underline{\underline{\boldsymbol{L}}}^{\nu} \equiv \overline{curl}^{\nu}$ and then utilize this in defining the following vector fields

$$
\begin{aligned}
\hat{\boldsymbol{E}}_{fd} &= \overline{curl}^{\nu}\hat{\boldsymbol{E}} \\
\hat{\boldsymbol{H}}_{fd} &= \overline{curl}^{\nu}\hat{\boldsymbol{H}}
\end{aligned}
\qquad 0 < \nu < 1
\tag{12.39}
$$

where $\overline{curl}^{\nu} \equiv [1/(ik)^{\nu}]curl^{\nu}$ and ν is in general a non-integer number between zero and unity. We have shown in [6] that these vector fields satisfy the time-harmonic source- free normalized Maxwell equations in (12.35). Therefore the vector fields $\hat{\boldsymbol{E}}_{fd}$ and $\hat{\boldsymbol{H}}_{fd}$ are normalized electromagnetic fields in the source-free situations. For $\nu = 0$, we find $\hat{\boldsymbol{E}}_{fd} \equiv \hat{\boldsymbol{E}}$ and $\hat{\boldsymbol{H}}_{fd} \equiv \hat{\boldsymbol{H}}$ which is the original set of fields in the problem. When $\nu = 1$, we obtain

$$
\begin{aligned}
\hat{\boldsymbol{E}}_{fd} &= \overline{curl}\,\hat{\boldsymbol{E}} = \hat{\boldsymbol{H}} \\
\hat{\boldsymbol{H}}_{fd} &= \overline{curl}\,\hat{\boldsymbol{H}} = -\hat{\boldsymbol{E}}
\end{aligned}
\tag{12.40}
$$

which are the normalized dual fields that satisfy the source-free Maxwell equations. Therefore, for all values of ν between zero and unity the fields $\hat{\boldsymbol{E}}_{fd}$ and $\hat{\boldsymbol{H}}_{fd}$ defined by (12.39) provide us with the new set of solutions to the source-free Maxwell equations. These fields can *effectively* be regarded as the intermediate solutions between the original fields and the dual fields. We have therefore suggested the name *fractional duality* and have used the subscript "*fd*" for the fields introduced in (12.39).

To offer physical insights into the concept of fractional duality, let us consider the following example.

We consider a standing wave that results from two transverse electromagnetic uniform plane waves propagating in opposite directions along the z axis. For this well-known case, the electric and magnetic fields can be easily written as

$$\begin{aligned} \boldsymbol{E} &= \hat{\boldsymbol{x}}[E_o \exp(ikz) - E_o \exp(-ikz)] = \hat{\boldsymbol{x}} 2iE_o \sin(kz) \\ \eta \boldsymbol{H} &= \hat{\boldsymbol{y}} 2E_o \cos(kz) \end{aligned} \tag{12.41}$$

At $z = 0$, the transverse impedance is $Z \equiv E_x/H_y = i\eta \tan(kz)|_{z=0} = 0$. Therefore, if one puts a perfect electric conducting (PEC) plane of infinite extent at the x–y plane where $z = 0$, the standing wave in the region $z \le 0$ will be unchanged. Now we find the fractional dual fields $\hat{\boldsymbol{E}}_{fd}$ and $\hat{\boldsymbol{H}}_{fd}$ by applying the fractional operator $\overline{curl^\nu}$ on the original fields of this standing wave. We obtain

$$\begin{aligned} \hat{\boldsymbol{E}}_{fd} = \boldsymbol{E}_{fd} &= 2i \exp\left(-i\frac{\nu\pi}{2}\right) E_o \sin\left(kz + \frac{\nu\pi}{2}\right)\left[\cos\left(\frac{\nu\pi}{2}\right)\hat{\boldsymbol{x}} + \sin\left(\frac{\nu\pi}{2}\right)\hat{\boldsymbol{y}}\right] \\ \hat{\boldsymbol{H}}_{fd} = \eta \boldsymbol{H}_{fd} &= 2 \exp\left(-i\frac{\nu\pi}{2}\right) E_o \cos\left(kz + \frac{\nu\pi}{2}\right)\left[-\sin\left(\frac{\nu\pi}{2}\right)\hat{\boldsymbol{x}} + \cos\left(\frac{\nu\pi}{2}\right)\hat{\boldsymbol{y}}\right] \end{aligned} \tag{12.42}$$

These dual fields do also satisfy the Maxwell equations, and represent another set of standing waves. However, these transverse electric and magnetic fields have been rotated in the x–y plane by an angle $\nu\pi/2$. When $\nu = 0$, these fractional dual fields become the fields of the original standing wave given in (12.41). For $\nu = 1$, we obtain, from (12.42), the conventional dual fields. Considering the transverse impedance of this fractional dual standing wave at $z = 0$, we write

$$Z_{fd} \equiv \frac{E_{fd_x}}{H_{fd_y}} = -\frac{E_{fd_y}}{H_{fd_x}} = i\eta \tan\left(kz + \frac{\nu\pi}{2}\right)\Big|_{z=0} = i\eta \tan\left(\frac{\nu\pi}{2}\right) \tag{12.43}$$

If we now put at the x–y plane a two-dimensional plane of infinite extent with the surface impedance of Z_{fd}, the fractional dual fields $\hat{\boldsymbol{E}}_{fd}$ and $\hat{\boldsymbol{H}}_{fd}$ in the region $z \le 0$ will not change. So effectively (12.49) presents the solutions of a normally incidence standing plane wave on an infinitely extent flat surface with the surface impedance Z_{fd}. It should be noted that when $\nu = 0$, we get $Z_{fd} = 0$ indicating that the flat surface is a PEC surface. For $\nu = 1$, this flat surface has the surface impedance of $Z_{fd} = i\infty$, thus denoting a perfectly magnetic conducting (PMC) surface. As mentioned before, in this case, our fractional dual fields become the conventional dual fields. However, when $0 < \nu < 1$, one gets flat surfaces with surface impedance Z_{fd} that is effectively "intermediate" cases between the PEC and PMC surfaces, and the fractional dual fields also represent the "intermediate" case between the cases of original fields and the conventional dual fields. This example suggests that the concept of fractional duality may perhaps have interesting applications in treating certain scattering problems in electromagnetism.

When the sources are included the fractional dual field solutions to the time-harmonic Maxwell equation can still be obtained by applying the fractional operator $\overline{curl^{\nu}}$ on the original fields $\hat{\boldsymbol{E}}$ and $\hat{\boldsymbol{H}}$. However, the current sources for such fractional situations have terms such as

$$\boldsymbol{J}_{e,fd} = \overline{curl^{\nu}}\boldsymbol{J}_e \qquad \text{and} \qquad \boldsymbol{J}_{m,fd} = \overline{curl^{\nu}}\boldsymbol{J}_m \tag{12.44}$$

and the corresponding electric and magnetic charges may be expressed as

$$\rho_{e,fd} = \frac{1}{i\omega} div\, \boldsymbol{J}_{e,fd} \qquad \text{and} \qquad \rho_{m,fd} = \frac{1}{i\omega} div\, \boldsymbol{J}_{m,fd} \tag{12.45}$$

12.7 SUMMARY

We have provided a brief overview of some of the ideas and case studies that in recent years we have investigated in the development of fractional paradigm in electromagnetic theory. These specific problems represent only samples of what the notion of fractional paradigm may embrace, and they highlight some of the roles of fractionalization of operators in electromagnetics. Needless to say, more work needs to be done to explore further the other possible roles of fractional domains in electromagnetism. We speculate that this will open doors to novel physical, mathematical, and engineering problems in electromagnetic theory. If the intermediate situations can be discovered and treated with the elegant mathematical tools of fractional calculus, or in general the tool of fractionalization of operators, the so-called well-known canonical problems may be generalized into the fractional domain. This notion may break the barrier of conventional ways of thinking about the canonical problems in classical electrodynamics and may bring about the paradigm in which continuous fractional parameters (either real or complex parameters) can provide novel physical and mathematical situations in electromagnetic theory.

ACKNOWLEDGMENTS

Portions of the work reviewed in this chapter were supported in part by the U.S. National Science Foundation under Grant No. ECS-96-12634.

APPENDIX

In the literature on fractional calculus one may find several definitions given for fractional differentiation/integration. Interested readers may consult some of these references (see, e.g., [11]–[16]). In this appendix, we briefly review some of the definitions provided for these operators.

We have already mentioned about the Riemann-Liouville fractional integration. More detail can be found in ([11]–[13]).

Another definition for fractional differintegration was given by Liouville for functions that can be expanded in a series of exponentials. When a function $f(x)$ can be expressed as $f(x) = \Sigma_{i=0}^{\infty} c_i e^{u_i x}$, according to Liouville ([17], [11, p. 53]) the νth-order fractional differentiation/integration (with lower limit $a = -\infty$) can be given as

$$\frac{d^\nu}{dx^\nu} f(x) \equiv \sum_{i=0}^{\infty} c_i u_i^\nu e^{u_i x} \tag{12.46}$$

Another definition is given for the power series with fractional powers. If one considers a function of one variable, $g(x)$, defined over a closed interval $[a,b]$, and represented as the following power series $g(x) = \Sigma_{i=0}^{\infty} A_i (x-a)^{\gamma + i/n}$ for $\gamma > -1$, and $a \le x \le b$, where n is a positive integer [11, p. 46]. The condition $\gamma > -1$ is needed in order to have $g(x)$ in the class of functions that can be differintegrated [11, p. 46]. Each term in this series may have a non-integer power. Now if one considers one of these terms, e.g., $(x-a)^r$ with the non-integer exponent denoted by r where $r > -1$, according to the definition given by Riemann ([11, p. 53], [18]) a fractional νth-order derivative for such a term can be written as

$$_aD_x^\nu (x-a)^r = \frac{\Gamma(r+1)}{\Gamma(r-\nu+1)}(x-a)^{r-\nu} \qquad \text{for } x > a \tag{12.47}$$

where ν may be a non-integer number ([18], [11]). (Riemann considered the case of $a = 0$ ([11, p. 53], [18])). It can be easily observed that for ν being a positive integer m, one gets the conventional definition for mth-order derivative of $(x-a)^r$.

It can be shown that, with proper mathematical care, the Riemann-Liouville fractional integration shown in (12.12) provides the definition given by Liouville when $a = -\infty$ in (12.12), and it leads to the definition by Riemann when $a = 0$, hence the name Riemann-Liouville fractional integral [23].

As mentioned in [11, p. 53], another definition is given by Civin. In this definition, if a function $f(x)$ is expressed as

$$f(x) \equiv \int_{-\pi}^{+\pi} e^{ixt}\, ds(t) \tag{12.48}$$

the νth-order fractional differintegration of $f(x)$ can be defined as

$$\frac{d^\nu}{dx^\nu} f(x) \equiv \int_{-\pi}^{+\pi} i^\nu t^\nu e^{ixt}\, ds(t) \tag{12.49}$$

According to [11, Chapter 3], there are several other definitions for fractional derivatives/integrals put forward by mathematicians such as Grunwald, Weyl, Nekrassov, and Erdelyi to name a few.

Fractional differentiation and integration have interesting features. These can be found in various references (see, e.g., [11]–[16]). Some of these features were also briefly reviewed in [1].

REFERENCES

[1] N. Engheta, "On Fractional Calculus and Fractional Multipoles in Electromagnetism," *IEEE Trans. Antennas & Propagation*, **44** (4), pp. 554–566, April 1996. Erratum, **44** (9), 1307, September 1996.

[2] N. Engheta, "Electrostatic 'Fractional' Image Methods for Perfectly Conducting Wedges and Cones," *IEEE Trans. Antennas & Propagation*, **44** (12), pp. 1565–1574, December 1996.

[3] N. Engheta, "A Note on Fractional Calculus and the Image Method for Dielectric Spheres," *J. of Electromagnetic Waves and Applications*, **9** (9), pp. 1179–1188, September 1995.

[4] N. Engheta, "Use of Fractional Integration to Propose Some 'Fractional' Solutions for the Scalar Helmholtz Equation," a chapter in *Progress in Electromagnetics Research (PIER), monograph Series, Vol. 12*, Jin A. Kong, ed., EMW Pub., Cambridge, MA, pp. 107– 132, Chapter 5, 1996.

[5] N. Engheta, "On the Role of Fractional Calculus in Electromagnetic Theory," in *IEEE Antennas and Propagation Magazine*, **39** (4), pp. 35–46, August 1997.

[6] N. Engheta, "Fractional Curl Operator in Electromagnetics," *Microwave and Optical Technology Letters*, **17** (2), pp. 86–91, February 5, 1998.

[7] N. Engheta, "Fractional Duality in Electromagnetic Theory," a talk presented in *The 1998 URSI International Symposium on EM Theory*, Thessaloniki, Greece, May 25–28, 1998. The summary appeared in Vol. 2, pp. 814–816 of the proceedings.

[8] N. Engheta, "Fractionalization of the Curl Operator and Its Electromagnetic Application," a talk presented in *The 1997 IEEE Antennas and Propagation International Symposium/North America Radio Science (URSI) Meeting*, Montreal, Québec, Canada, July 13–18, 1997. The summary appeared in the Digest of the IEEE AP-S Symposium, **2**, pp. 1480–1483.

[9] N. Engheta, "Complex-Order Fractional Curl in Electromagnetism," a talk presented in *the 1998 USNC/URSI National Radio Science Meeting, Atlanta, Georgia*, June 21–26, 1998. The one-page abstract is in p. 163 of the digest.

[10] N. Engheta, "Fractional Kernels and Intermediate Zones in Electromagnetism: Planar Geometries," a talk presented in *The 1998 IEEE Antennas and Propagation International Symposium/USNC/URSI National Radio Science Meeting, Atlanta, Georgia*, June 21–26, 1998. The summary appeared in the Digest of the IEEE AP-S Symposium, **2**, pp. 879–881.

[11] K. B. Oldham and J. Spanier, *The Fractional Calculus*, Academic Press, New York, 1974.

[12] S. G. Samko, A. A. Kilbas, and O. I. Marichev, *Fractional Integrals and Derivatives, Theory and Applications*, Gordon and Breach Science Publishers, Langhorne, PA, 1993. (Originally published in Russian by Nauka i. Tekhnika, Minsk, 1987.)

[13] K. S. Miller and B. Ross, *An Introduction to the Fractional Calculus and Fractional Differential Equations*, New York, John Wiley & Sons, 1993.

[14] H. T. Davis, *The Theory of Linear Operators*, The Principia Press, Bloomington, IN, 1936, pp. 64–77, and pp. 276–292.

[15] A. C. McBride and G. F. Roach, eds, *Fractional Calculus*, Research Notes in Mathematics, Pitman Advanced Publishing Program, Boston, 1985.

[16] V. Kiryakova, *Generalized Fractional Calculus and Its Applications*, Pitman Research Notes in Mathematics Series 301, Longman Scientific & Technical, Longman Group, Essex, U.K., 1994.

[17] J. Liouville, "Mémoire: Sur le Calcul des Différentielles à Indices Quelconques," *J. Ecole Polytechn.*, **13**, pp. 71–162, 1832.

[18] B. Riemann, "Versuch einer allgemeinen Auffasung der Integration und Differentiation," in *The Collected Works of Bernhard Riemann*, H. Weber, edn, 2nd ed., Dover, New York, Chapter 19, pp. 353–366, 1953.

[19] J. Lützen, "Liouville's Differential Calculus of Arbitrary Order and Its Electrodynamical Origin," in *Proceedings of the Nineteenth Nordic Congress of Mathematicians, Reykjavik, 1984*, published by the Icelandic Mathematical Society, 1985, pp. 149–160.

[20] E. R. Love, "Fractional Derivatives of Imaginary Order," *Journal of the London Mathematical Society*, **3**, Part 2, Second Series, pp. 241–259, February 1971.

[21] A. Erdélyi and I. N. Sneddon, "Fractional Integration and Dual Integral Equations," *Canadian Journal of Mathematics*, **XIV** (4), pp. 685–693, 1962.

[22] G. K. Kalisch, "On Fractional Integrals of Pure Imaginary Order in L_p," *Proceedings of the American Mathematical Society*, **18**, pp. 136–139, 1967.

[23] B. Ross, "Fractional Calculus – An historical apologia for the development of a calculus using differentiation and antidifferentiation of non-integral orders," *Mathematics Magazine*, **50** (3), pp. 115–122, May 1977.

[24] T. J. Osler, "The Fractional Derivative of a Composite Function," *SIAM Journal on Mathematical Analysis*, **1** (2), pp. 288–293, May 1970.

[25] I. M. Gel'fand and G. E. Shilov, *Generalized Functions*, Academic Press, New York, five volumes, 1964–68.

[26] J. A. de Durán and S. L. Kalla, "An Application of Fractional Calculus to the Solution of $(n+1)$th-order Ordinary and Partial Differential Equations," *Journal of Fractional Calculus*, **2**, pp. 67–76, November 1992.

[27] B. Rubin, "Fractional Integrals and Weakly Singular Integral Equations of the First Kind in the n-Dimensional Ball," *Journal d'Analyse Mathématique*, **63**, pp. 55–102, 1994.

[28] N. H. Abel, "Solution de Quelques Problèmes à E l'Aide d'Intégrales Définies," in *Œuvres Complètes de Niels Henrik Abel*, vol. 1, L. Sylow and S. Lie, eds., Grondahl, Christiana, Norway, Chapter 2, pp. 11–18, 1881. (As mentioned in this reference, the original article was published in *Magazin for Naturvidenskaberne*, Aargang 1, Bind 2, 1823.)

[29] C. H. Papas, *Theory of Electromagnetic Wave Propagation*, Dover Publications, New York, 1988. (Unabridged republication of work first published by McGraw-Hill, New York, 1965.)

[30] J. Van Bladel, *Singular Electromagnetic Fields and Sources*, Oxford Science Publications, Oxford University Press, New York, 1991.

[31] J. Van Bladel, "The Multipole Expansion Revisited," *Archiv für Elektronik und Übertragungstechnik*, **31**, pp. 407–411, 1977.

[32] V. Namias, "Application of the Dirac delta function to Electric Charge and Multipole Distributions," *American Journal of Physics*, **45** (7), pp. 624–630, July 1977.

[33] C. A. Kocher, "Point-Multipole Expansion for Charge and Current Distributions," *American Journal of Physics*, **46** (5), pp. 578–579, May 1978.

[34] E. U. Condon, "Immersion of the Fourier Transform in a Continuous Group of Functional Transformations," *Proceedings of the National Academy of Science, U.S.A.*, **23**, 158–164, 1937.

[35] V. Bargmann, "On a Hilbert Space of Analytic Functions and an Associated Integral Transform, Part I," *Commun. Pure Appl. Math.*, **14**, pp. 187–214, 1961.

[36] V. Namias, "The Fractional Order Fourier Transform and its Application to Quantum Mechanics," *J. Inst. Maths Applics*, **25**, pp. 241–265, 1980.

[37] A. C. McBride and F. H. Kerr, "On Namias's Fractional Fourier Transforms," *IMA Journal of Applied Mathematics*, **39**, pp. 159–175, 1987.

[38] A. W. Lohmann, "Image Rotation, Wigner Rotation, and the Fractional Fourier Transform," *J. Opt. Soc. Am. A.*, **10** (10), pp. 2181–2186, 1993.

[39] D. Mendlovic, H. M. Ozaktas, and A. W. Lohmann, "Graded-Index Fibers, Wigner-

Distribution Functions, and the Fractional Fourier Transform," *Applied Optics*, **33** (26), pp. 6188–6193, September 10, 1994.

[40] D. Mendlovic and H. M. Ozaktas, "Fractional Fourier Transform and Their Optical Implementation: I," *J. Opt. Soc. Am. A.*, **10** (9), pp. 1875–1881, 1993.

[41] H. M. Ozaktas, B. Barshan, D. Mendlovic, and L. Onural, "Convolution, Filtering, and Multiplexing in Fractional Fourier Domains and Their Relation to Chirp and Wavelet Transforms," *Journal of the Optical Society of America, A.*, **11** (2), pp. 547–559, February 1994.

[42] L. M. Bernardo and O. D. D. Soares, "Fractional Fourier Transforms and Optical Systems," *Optics Communications*, **110**, pp. 517–522, September 1, 1994.

[43] L. B. Almeida, "The Fractional Fourier Transform and Time-Frequency Representations," *IEEE Transactions on Signal Processing*, **42** (11), pp. 3084–3091, November 1994.

[44] P. Pellat-Finet and G. Bonnet, "Fractional Order Fourier Transform and Fourier Optics," *Optics Communications*, **111**, pp. 141–154, September 15, 1994.

[45] J. Shamir and N. Cohen, "Root and Power Transformations in Optics," *J. Opt. Soc. Am. A.*, **12** (11), pp. 2415–2425, November 1995.

[46] Y. Zhang, B.-Y. Gu, B.-Z. Dong, G.-Z. Yang, H. Ren, X. Zhang, and S. Liu, "Fractional Gabor Transform," *Optics Letters*, **22** (21), pp. 1583–1585, November 1, 1997.

[47] D. Dragoman, "Fractional Wigner Distribution Function," *Journal of the Optical Society of America, A.*, **13** (3), pp. 474–478, March 1996.

[48] A. W. Lohmann, D. Mendlovic, and Z. Zalevsky, "Fractional Hilbert Transform," *Optics Letters*, **21** (4), pp. 281–283, February 15, 1996.

[49] H. Hamam and J. L. de Bougrenet de la Tocnaye, "Programmable Joint Fractional Talbot Computer-Generated Holograms," *Journal of the Optical Society of America, A.*, **12** (2), pp. 314–324, February 1995.

[50] M. Testorf and J. Ojeda-Castaneda, "Fractional Talbot Effect: Analysis in Phase Space," *Journal of the Optical Society of America, A.*, **13** (1), pp. 119–125, January 1996.

[51] M. A. Alonso and G. W. Forbes, "Fractional Legendre Transformation," *J. Phys. A: Math. Gen.*, **28**, pp. 5509–5527, 1995.

[52] L. Schwartz, *Theorie des Distributions*, Hermann, Paris, France, 1966.

[53] O. Heaviside, *Electromagnetic Theory*, Vol. II, second printing, Ernest Benn Pub., London, 1925.

[54] A. Le Méhauté, F. Héliodore, and V. Dionnet, "Overview of Electrical Processes in Fractal Geometry: From Electrodynamic Relaxation to Superconductivity," *Proceedings of the IEEE*, **81** (10), pp. 1500–1510, 1993.

[55] A. Lé Méhauté, F. Héliodore, D. Cottevieille, and F. Latreille, "Introduction to Wave Phenomena and Uncertainty in a Fractal Space-I," *Chaos, Solitons & Fractals*, **4** (3), pp. 389–402, 1994.

[56] R. R. Nigmatulin, "On the Theory of Relaxation for Systems with 'Remnant' Memory," *Physica Status Solidi* (b), **124** (1), pp. 389–393, July 1984.

[57] R. R. Nigmatulin, "The Temporal Fractional Integral and Its Physical Sense," in Ecole d'ÉTÉ Internationale sur *Geometrie Fractale et Hyperbolique, Derivation Fractionnaire et Fractal, Applications dans les Sciences de l'Ingenieur et en Economie*, Bordeaux, France, July 3–8, 1994 organized by Alain Le Méhauté from Institut Superieur des Materiaux du Mans (ISMANS), Le Mans, France and A. Oustaloup from Equipe CRONE, Laboratoire d'Automatique et de Productique. ENSERB Université Bordeaux I, Bordeaux, France.

[58] K. M. Kolwankar and A. D. Gangal, "Fractional Differentiability of Nowhere Differentiable Functions and Dimensions," *Chaos*, **6** (4), pp. 505–513, December 1996.

[59] K. M. Kolwankar and A. D. Gangal, "Local Fractional Fokker-Planck Equation," *Physical Review Letters*, **80** (2), pp. 214–217, 1998.

[60] M. Ochmann and S. Makarov, "Representation of the Absorption of Nonlinear Waves by Fractional Derivatives," *Journal of Acoustical Society of America*, **94** (6), 3392–3399, December 1993.

[61] W. R. Schneider and W. Wyss, "Fractional Diffusion and Wave Equation," *J. Math. Phys.*, **30** (1), pp. 134–144, January 1989.

[62] M. Caputo and F. Mainardi, "Linear Models of Dissipation in Anelastic Solids," *La Rivista del Nuovo Cimento* (della Società Italiana di Fisica), **1**, series 2 (2), 161–198, April–June 1971.

[63] F. Mainardi and P. Pironi, "The Fractional Langevin Equation: Brownian Motion Revisited," *Extracta Mathematicae*, **11** (1), pp. 140–154, 1996.

[64] C. F. Chen, Y. T. Tsay, and T. T. Wu, "Walsh Operational Matrices for Fractional Calculus and Their Application to Distributed Systems," *Journal of the Franklin Institute*, **303** (3), pp. 267–284, March 1977.

[65] Z. Altman, D. Renaud, and H. Baudrand, "On the Use of Differential Equations of Nonentire Order to Generate Entire Domain Basis Functions with Edge Singularity," *IEEE Transactions on Microwave Theory and Techniques*, **42** (10), pp. 1966–1972, October 1994.

[66] H. Kasprzak, "Differentiation of a Noninteger Order and Its Optical Implementation," *Applied Optics*, **21** (18), pp. 3287–3291, September 15, 1982.

[67] T. Szoplik, V. Climent, E. Tajahuerce, J. Lancis, and M. Fernandez-Alonso, "Phase-Change Visualization in Two-Dimensional Phase Objects with a Semiderivative Real Filter," *Applied Optics*, **37** (23), pp. 5472–5478, August 10, 1998.

[68] J. Lancis, T. Szoplik, E. Tajahuerce, V. Climent, and M. Fernandez-Alonso, "Fractional Derivative Fourier Plane Filter for Phase-Change Visualization," *Applied Optics*, **36** (29), pp. 7461–7464, October 10, 1997.

[69] W. R. Smythe, *Static and Dynamic Electricity*, McGraw-Hill, New York, 3rd edn, 1968.

[70] K. I. Nikoskinen and I. V. Lindell, "Image Solution for Poisson's Equation in Wedge Geometry," *IEEE Trans. on Antennas & Propagation*, **43** (2), pp. 179–187, 1995.

[71] J. D. Jackson, *Classical Electrodynamics*, New York, John Wiley and Sons, 2nd edn, 1975.

[72] N. Engheta, "Extension of the Electrostatic Fractional Image Method to the Case of Electrostatic Dielectric Wedge," a talk presented at the *1997 North American Radio Science Meeting, Montreal, Canada*, July 13–18, 1997. The one-page abstract is on p. 729 of the digest of abstracts.

[73] P. M. Morse and H. Feshbach, *Methods of Theoretical Physics*, New York, McGraw-Hill, Two volumes, 1953.

[74] M. Abramowitz and I. A. Stegun, *Handbook of Mathematical Functions*, Dover, New York, 1970.

[75] A. Ishimaru, *Electromagnetic Wave Propagation, Radiation, and Scattering*, Prentice-Hall, Englewood, NJ, 1991.

Chapter 13

SPHERICAL-MULTIPOLE ANALYSIS IN ELECTROMAGNETICS

Siegfried Blume and Ludger Klinkenbusch

ABSTRACT

The vector spherical-multipole analysis is a systematic technique to represent any solenoidal electromagnetic field within a linear and partially homogeneous medium by means of a compact, orthogonal, and physically interpretable series expansion. By utilizing sphero-conal coordinates a large variety of interesting scattering and diffraction problems can be solved mostly analytically. Since sphero-conal coordinates are generalized spherical coordinates, the theory is valid for ordinary spherical coordinates as well.

The vector spherical-multipole functions can be reduced to appropriately chosen elementary solutions of the scalar homogeneous Helmholtz equation in sphero-conal coordinates, i.e., products of spherical cylinder functions and Lamé products. The latter ones are found as eigensolutions of a two-parametric eigenvalue problem with two coupled Lamé differential equations. Bilinear forms of scalar and dyadic Green's functions are derived as well as electromagnetic plane wave expansions.

Three examples for the application in electrical engineering are given. First the scattering by a perfectly conducting semi-infinite elliptic cone is investigated. Since the obtained series do not converge in the usual sense, suitable sequence transformations (e.g., Euler's summation techniques) are applied to achieve the physically correct values. Then the scattering by a finite elliptic cone is treated analytically as a classical two-domain approach. Finally the solution for a canonical EMC-problem, i.e., a loaded spherical shell with an elliptic aperture, is found via an Electric Field Integral Equation (EFIE) with a special integral kernel. The EFIE is solved mostly analytically using a complete set of entire domain functions and Galerkin's method.

13.1 INTRODUCTION

The spherical-multipole analysis is one of the classical methods in electromagnetics; however there is still a scientific interest in this subject. The multipole analysis is based on a series expansion with suitable elementary solutions of the concerning differential equation, e.g., the Helmholtz equation. Hence multipole techniques count to the analytical methods, but they play a considerable role also in computational electromagnetics. Appropriate examples are the Generalized Multipole Technique (GMT) [21], the Fast Multipole Method [18], the hybridization of the Finite Element Method [32], and the hybridization of the Integral Equation Technique with a problem-adapted integral kernel. The spherical-multipole expansion is also a tool to efficiently perform a near-field-to-far-field transformation in antenna measurement [31].

This chapter deals with the "classical" spherical-multipole analysis, however, in sphero-conal coordinates. These are to be understood as the most general spherical coordinates, in which the Helmholtz equation is rigorously separable. The large variety of coordinate surfaces allows us to investigate mostly analytically the scattering and diffraction by some interesting geometries.

Section 13.2 provides the reader with the properties of the sphero-conal coordinate system, as far as necessary for the following sections. In Section 13.3 the spherical-multipole analysis of scalar, e.g., acoustical fields is carried out. First the elementary solutions (scalar spherical-multipole functions) of the scalar homogeneous Helmholtz equation in sphero-conal coordinates are deduced. As in ordinary spherical coordinates the radial dependence of the spherical-multipole functions is described by means of spherical cylinder (Bessel) functions. The angular (ϑ- and φ-) depending part satisfies the eigenvalue equation of the so-called Lamé products, since it can be represented by products of periodic Lamé functions in φ and non-periodic Lamé functions in ϑ. The corresponding Lamé differential equations are coupled by two independent parameters, which are to be determined by enforcing the relevant boundary and/or continuity conditions. It is shown how this two-parametric eigenvalue problem with two Lamé differential equations can be solved by utilizing suitable analytical and numerical techniques. This section is completed by some orthogonality relations and by the bilinear representation of the Green's function, where all necessary proofs are outlined.

Section 13.4 deals with the spherical-multipole analysis of electromagnetic fields. The spherical-multipole expansion is built up by means of the two vector spherical-multipole functions, which are the solenoidal solutions of the vector Helmholtz equation and which are directly related to the scalar spherical-multipole functions derived in Section 13.3. The deduced multipole expansions are valid for the free space as well as for a domain outside a perfectly conducting elliptic cone. This section also contains the relevant vector orthogonalities and dyadic completeness relations, where again all proofs will be given. One main part is the derivation of the dyadic Green's function for the outer domain of the semi-infinite elliptic cone and for the free space. In addition the dyadic Green's function will be utilized to deduce the multipole expansion of an arbitrary plane wave incident upon the semi-infinite elliptic cone as well as in the free space.

In Section 13.5 three different boundary value problems will be solved by the spherical-multipole technique. First the scattering of plane waves by a perfectly conducting semi-infinite elliptic cone will be investigated. This is motivated by the role which this canonical geometry plays for the improvement of asymptotic high-frequency techniques like GTD and UTD. These ray optical techniques represent extensions of classical geometrical optics by introducing diffracted rays produced when incident rays hit edges, corners, tips, or impinge tangentially on smoothly curved surfaces of scattering objects. The initial value of the field on a diffracted ray is determined from the incident field at the point of diffraction with the aid of an appropriate scalar or dyadic diffraction coefficient. These diffraction coefficients are to be determined from the solutions of boundary value problems for objects with simple shapes (canonical structures) capable of approaching the local geometry at the point of diffraction. In the classical monograph about electromagnetic and acoustic scattering by simple shapes edited by Bowman, Senior, and Uslenghi [15], a large variety of canonical boundary value problems is treated by the method of separation of variables in problem-adapted orthogonal coordinates. The solutions given in the form of infinite series using modal techniques are—if possible—transformed by analytic means into others that are better convergent, or are converted into integrals that can be evaluated asymptotically for far-field observations. Shapes excluded from that book are, among others, the semi-infinite elliptic cone and the plane angular sector. The treatment of scattering by these structures will be contributed to in this section.

The investigations concerning the elliptic cone and its degenerations, the circular cone and the plane angular sector, have a long history. A review concerning this history is carried out in [4].

By using the spherical-multipole expansion of a plane wave incident upon the semi-infinite elliptic cone, the rigorous spherical multipole representations of the nose-on radar cross sections (RCS) for different states of the incident plane wave's polarization are quickly derived. In case of a circular cone these expressions coincide and give exactly the expression for the nose-on RCS of the circular cone originally derived by Hansen and Schiff [24]. However, the modal series representing the field scattered by a semi-infinite elliptic cone are slowly convergent. It appears to be nearly impossible to obtain numerical results from these analytically exact solutions if both source and field points are far from the tip. In the attempt to obtain better convergent series suitable summation techniques will be utilized. Such techniques have been *analytically* applied by Felsen [19] and by Siegel, Chrispin, and Schensted [49] in transforming the eigenfunction expansions and in evaluating them asymptotically for wide-angle and narrow-angle circular cones. According to our two parameter eigenvalue problem with two coupled Lamé differential equations the application of this mathematical method appears to be hopeless. However with the aid of modern computers it is possible to apply Euler's summation technique *numerically* in transforming the relevant series for arbitrary opening angles of elliptic cones into better convergent series.

Scattering and diffraction by finite circular cones have often been discussed in the literature. By using the spherical-multipole expansion in sphero-conal coordinate, scattering by a finite elliptic cone which is bounded by a spherical cap can be treated as well. The pertinent two-domain problem is solved by the classical mode-matching technique, where the coupling-integrals can be solved analytically and determine the convergence properties of the resulting system of linear equations. Since this finite geometry possesses a tip and edges, it is also of interest as a reference for numerically obtained results.

Finally the electromagnetic behavior of a canonical shielding structure will be analyzed. A perfectly conducting spherical shell with an elliptical aperture includes a concentrically arranged spherical load (homogeneous sphere) and is illuminated by the interfering field. Since the ellipticity parameters of the aperture can be chosen arbitrarily, also a model for a shield with a slit aperture is obtained. The boundary value problem is solved via an Electric Field Integral Equation (EFIE) approach, where the kernel of the EFIE is the bilinear form of the homogeneous sphere's dyadic Green's function. The multipole solution allows a fast computation of exact reference results and it is the base for further analysis of the obtained formulas, e.g., to deduce appropriate definitions for the characterization of the shielding efficiency.

13.2 SPHERO-CONAL COORDINATES

The sphero-conal (or conical) coordinate system belongs to those famous 11 coordinate systems, in which the Helmholtz equation is separable. They are denoted by r, ϑ, φ and are related to the Cartesian coordinates x, y, z by

$$\begin{aligned} x &= r \sin\vartheta \cos\varphi & 0 &\leq r \\ y &= r\sqrt{1 - k^2\cos^2\vartheta}\,\sin\varphi & 0 &\leq \vartheta \leq \pi \\ z &= r\cos\vartheta\sqrt{1 - k'^2\sin^2\varphi} & 0 &\leq \varphi \leq 2\pi \end{aligned} \tag{13.1}$$

The parameters k and k' are to be chosen such that

$$0 \leq k, k' \leq 1; \qquad k^2 + k'^2 = 1 \tag{13.2}$$

Sphero-conal coordinates are generalized spherical coordinates. In case of $k = 1$ the sphero-conal coordinates turn into ordinary spherical coordinates with the z axis as the polar axis.

Figure 13.1 shows coordinate surfaces. $r = r_0$ describes a sphere with radius r_0, concentrically arranged around the origin. $\vartheta = \vartheta_0$ is a semi-infinite elliptic cone, symmetrically arranged around the positive z axis ($\vartheta_0 < \pi/2$), or around the negative z axis ($\vartheta_0 > \pi/2$). The half opening angles (Fig. 13.1) are found by

$$\begin{aligned} \vartheta_x &= \vartheta_0 \\ \vartheta_y &= \arccos(k\cos\vartheta_0) \end{aligned} \tag{13.3}$$

The coordinate surface $\varphi = \varphi_0$ is one half of a semi-infinite elliptic cone around the y axis.

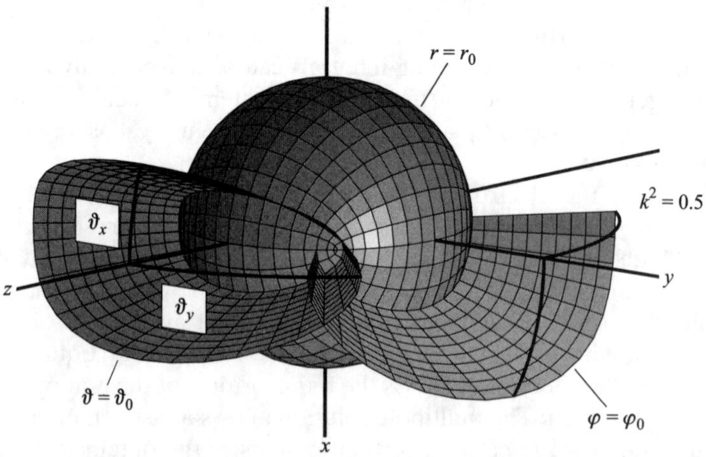

Figure 13.1 Sphero-conal coordinate surfaces.

There are a lot of interesting degenerations of the coordinate surfaces:

$k = 1; \vartheta = \vartheta_0$:	circular cone with half opening angle ϑ_0,
$\vartheta = \pi/2$:	x-y plane,
$\vartheta = 0(\pi)$:	plane angular sector in the y-z plane around the positive (negative) z axis and with half opening angle $\vartheta_y^s = \arccos(k)$ (see Fig. 13.2),
$k = 0; \vartheta = 0\,(\pi)$:	half plane in the y-z plane with the y axis as the edge and with $z \geq 0$ ($z \leq 0$),
$\varphi = \pi/2\,(3\pi/2)$:	plane angular sector in the y-z plane around the positive (negative) y axis; for any given k these sectors are the complements of the sectors $\vartheta = 0$ and $\vartheta = \pi$ to the y-z plane (see Fig. 13.2),
$k = 1; \varphi = \varphi_0$:	half plane with the z axis as the edge.

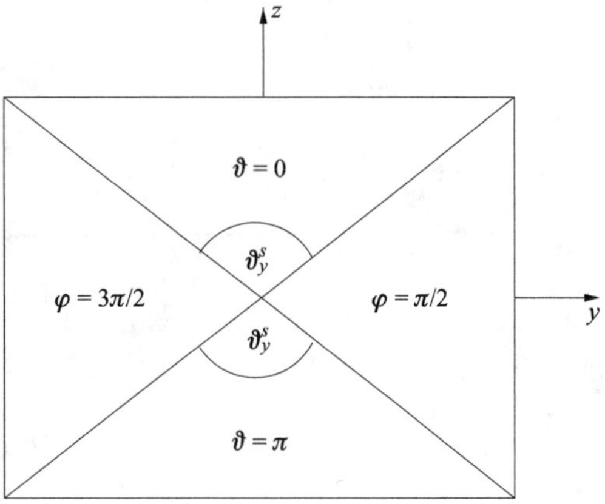

Figure 13.2 The y-z plane in sphero-conal coordinates.

The (metric) scaling coefficients are found to be

$$\begin{aligned} h_r &= \left|\frac{\partial \vec{r}}{\partial r}\right| = 1 \\ h_\vartheta &= \left|\frac{\partial \vec{r}}{\partial \vartheta}\right| = r\sqrt{\frac{k^2\sin^2\vartheta + k'^2\cos^2\varphi}{1 - k^2\cos^2\vartheta}} = rs_\vartheta \\ h_\varphi &= \left|\frac{\partial \vec{r}}{\partial \varphi}\right| = r\sqrt{\frac{k^2\sin^2\vartheta + k'^2\cos^2\varphi}{1 - k'^2\sin^2\varphi}} = rs_\varphi \end{aligned} \tag{13.4}$$

where s_ϑ and s_φ will be referred to as the normalized scaling coefficients. Singular points of the coordinate system, i.e. points, where the functional determinant vanishes, are the origin ($r = 0$) and the sector edges $\{(\vartheta, \varphi)|\vartheta = 0, \pi; \varphi = \pi/2, 3\pi/2\}$. Special roles are played by the sector-like coordinate surfaces $\vartheta = 0$ and $\vartheta = \pi$. For example, a fixed point $(r, 0, \varphi)$ on one side of the sector $\vartheta = 0$ is described by $(r, 0, \pi - \varphi)$ on the other side. This fact has to be considered if a function Ψ (r, ϑ, φ) and its spatial

TABLE 13.1 Differential Operations in Sphero-Conal Coordinates

	$f(\vec{r}) = f(r, \vartheta, \varphi); \qquad \vec{F}(\vec{r}) = F_r\hat{r} + F_\vartheta\hat{\vartheta} + F_\varphi\hat{\varphi}$
	$s_\vartheta = \sqrt{\dfrac{k^2\sin^2\vartheta + k'^2\cos^2\varphi}{1 - k^2\cos^2\vartheta}}; \qquad s_\varphi = \sqrt{\dfrac{k^2\sin^2\vartheta + k'^2\cos^2\varphi}{1 - k'^2\sin^2\varphi}}$
$\vec{ds}$	$\hat{r}dr + \hat{\vartheta}rs_\vartheta d\vartheta + \hat{\varphi}rs_\varphi d\varphi$
$\vec{\nabla}f$	$\hat{r}\dfrac{\partial f}{\partial r} + \hat{\vartheta}\dfrac{1}{rs_\vartheta}\dfrac{\partial f}{\partial\vartheta} + \hat{\varphi}\dfrac{1}{rs_\varphi}\dfrac{\partial f}{\partial\varphi}$
$\vec{\nabla}^2 f$	$\dfrac{1}{r^2}\dfrac{\partial}{\partial r}\left(r^2\dfrac{\partial f}{\partial r}\right) + \dfrac{1}{r^2}(\vec{r}\times\vec{\nabla})^2 f$
$(\vec{r}\times\vec{\nabla})^2 f$	$\dfrac{1}{s_\vartheta s_\varphi}\left\{\dfrac{\partial}{\partial\vartheta}\left(\dfrac{s_\varphi}{s_\vartheta}\dfrac{\partial f}{\partial\vartheta}\right) + \dfrac{\partial}{\partial\varphi}\left(\dfrac{s_\vartheta}{s_\varphi}\dfrac{\partial f}{\partial\varphi}\right)\right\}$
$\vec{\nabla}\cdot\vec{F}$	$\dfrac{1}{r^2}\dfrac{\partial}{\partial r}(r^2F_r) + \dfrac{1}{r^2 s_\vartheta s_\varphi}\left\{\dfrac{\partial}{\partial\vartheta}(rs_\varphi F_\vartheta) + \dfrac{\partial}{\partial\varphi}(rs_\vartheta F_\varphi)\right\}$
$\vec{\nabla}\times\vec{F}$	$\hat{r}\dfrac{1}{r^2 s_\vartheta s_\varphi}\left\{\dfrac{\partial}{\partial\vartheta}(rs_\varphi F_\varphi) - \dfrac{\partial}{\partial\varphi}(rs_\vartheta F_\vartheta)\right\}$
	$+\hat{\vartheta}\dfrac{1}{rs_\varphi}\left\{\dfrac{\partial}{\partial\varphi}F_r - \dfrac{\partial}{\partial r}(rs_\varphi F_\varphi)\right\} + \hat{\varphi}\dfrac{1}{rs_\vartheta}\left\{-\dfrac{\partial}{\partial\vartheta}Fr + \dfrac{\partial}{\partial r}(rs_\vartheta F_\vartheta)\right\}$

derivatives must yield the same values at both sides of this sector-like coordinate surface, e.g., in the free space. It can be shown that the following conditions hold:

$$\begin{aligned} \left.\frac{\partial\Psi(\vartheta)}{\partial\vartheta}\right|_{\vartheta=0} &= 0 \qquad \text{if} \qquad \Psi(r, \vartheta, \varphi) = \Psi(r, \vartheta, \pi - \varphi) \\ \Psi(\vartheta)\left.\right|_{\vartheta=0} &= 0 \qquad \text{if} \qquad \Psi(r, \vartheta, \varphi) = -\Psi(r, \vartheta, \pi - \varphi) \end{aligned} \tag{13.5}$$

The same relations hold for $\vartheta = \pi$. The relations (13.5) will be referred to as the *geometrical boundary conditions*. These will be of particular interest for the creation of suitable functions to describe the solutions of the scalar Helmholtz equation in sphero-conal coordinates. Finally, Tables 13.1 and 13.2 give an overview about the most relevant equations and relations in sphero-conal coordinates.

13.3 SPHERICAL-MULTIPOLE ANALYSIS OF SCALAR FIELDS

13.3.1 Scalar Spherical-Multipole Expansion in Sphero-Conal Coordinates

The scalar homogeneous Helmholtz equation

$$\vec{\nabla}^2\Psi(\vec{r}) + \kappa^2\Psi(\vec{r}) = 0 \tag{13.6}$$

TABLE 13.2 Relations between Cartesian and Sphero-Conal Coordinates and Components

$$x = r\sin\vartheta\cos\varphi \qquad 0\le r$$
$$y = r\sqrt{1-k^2\cos^2\vartheta}\sin\varphi \; ; \qquad 0\le\vartheta\le\pi \; ; \qquad 0\le k, k'\le 1$$
$$z = r\cos\vartheta\sqrt{1-k'^2\sin^2\varphi} \qquad 0\le\varphi\le 2\pi \qquad k^2+k'^2=1$$

$$r = \sqrt{x^2+y^2+z^2}$$

$$\vartheta = \begin{cases} \arcsin c_\vartheta & (z\ge 0) \\ \pi - \arcsin c_\vartheta & (z<0) \end{cases} \; ; \qquad c_\vartheta = \left[p + \left[p^2 + \frac{k'^2x^2}{k^2r^2}\right]^{1/2}\right]^{1/2}$$
$$p = \frac{k^2r^2 - k'^2x^2 - z^2}{2k^2r^2} (k \ne 0)$$

$$\varphi = \begin{cases} \arccos c_\varphi & (x\ge 0,\, y\ge 0) \\ \pi - \arccos c_\varphi & (x<0,\, y\ge 0) \\ \pi + \arccos c_\varphi & (x<0,\, y<0) \\ 2\pi - \arccos c_\varphi & (x\ge 0,\, y<0) \end{cases} \; ; \qquad c_\varphi = \left[q + \left[q^2 + \frac{k^2x^2}{k'^2r^2}\right]^{1/2}\right]^{1/2}$$
$$q = \frac{k'^2r^2 - k^2x^2 - y^2}{2k'^2r^2} (k' \ne 0)$$

$$F_x = F_r\sin\vartheta\cos\varphi + F_\vartheta\cos\vartheta\cos\varphi\frac{1}{s_\vartheta} - F_\varphi\sin\vartheta\sin\varphi\frac{1}{s_\varphi}$$

$$F_y = F_r\sqrt{1-k^2\cos^2\vartheta}\sin\varphi + F_\vartheta\frac{k^2\sin\vartheta\cos\vartheta\sin\varphi}{\sqrt{1-k^2\cos^2\vartheta}}\frac{1}{s_\vartheta} + F_\varphi\cos\sqrt{1-k^2\cos^2\vartheta}\frac{1}{s_\varphi}$$

$$F_z = F_r\cos\vartheta\sqrt{1-k'^2\sin^2\varphi} - F_\vartheta\sin\vartheta\sqrt{1-k'^2\sin^2\varphi}\frac{1}{s_\vartheta} - F_\varphi\frac{k'^2\cos\vartheta\cos\varphi\sin\varphi}{\sqrt{1-k'^2\sin^2\varphi}}\frac{1}{s_\varphi}$$

$$F_r = F_x\sin\vartheta\cos\varphi + F_y\sqrt{1-k^2\cos^2\vartheta}\sin\varphi + F_z\cos\vartheta\sqrt{1-k'^2\sin^2\varphi}$$

$$F_\vartheta = F_x\cos\vartheta\cos\varphi\frac{1}{s_\vartheta} + F_y\frac{k^2\cos\vartheta\sin\vartheta\sin\varphi}{\sqrt{1-k^2\cos^2\vartheta}}\frac{1}{s_\vartheta} - F_z\sin\vartheta\sqrt{1-k'^2\sin^2\varphi}\frac{1}{s_\vartheta}$$

$$F_\varphi = -F_x\sin\vartheta\sin\varphi\frac{1}{s_\varphi} + F_y\cos\varphi\sqrt{1-k^2\cos^2\vartheta}\frac{1}{s_\varphi} - F_z\frac{k'^2\cos\vartheta\cos\varphi\sin\varphi}{\sqrt{1-k'^2\sin^2\varphi}}\frac{1}{s_\varphi}$$

reads in sphero-conal coordinates (Table 13.1):

$$\frac{1}{r^2}\frac{\partial}{\partial r}\left(r^2\frac{\partial\Psi}{\partial r}\right) + \frac{1}{r^2}(\vec{r}\times\vec{\nabla})^2\Psi + \kappa^2\Psi = 0 \tag{13.7}$$

A separation expression

$$\Psi(r,\vartheta,\varphi) = z(r)Y(\vartheta,\varphi) \tag{13.8}$$

leads with the separation constant $\nu(\nu+1)$ to the differential equation of the spherical cylinder functions

$$\frac{d}{dr}\left[r^2\frac{d}{dr}z_\nu(r)\right] + [(\kappa r)^2 - \nu(\nu+1)]z_\nu(r) = 0 \tag{13.9}$$

and to the eigenvalue equation

$$-(\vec{r}\times\vec{\nabla})^2 Y_\nu(\vartheta,\varphi) = \nu(\nu+1)Y_\nu(\vartheta,\varphi) \tag{13.10}$$

Among the spherical cylinder functions $z_\nu(\kappa r)$, which are related to the (ordinary) cylinder functions $Z_\nu(\kappa r)$ by

$$z_\nu(\kappa r) = \sqrt{\frac{\pi}{2\kappa r}}\, Z_{\nu+1/2}(\kappa r)$$

we will especially need the spherical Bessel functions $j_\nu(\kappa r)$, spherical Neumann functions $n_\nu(\kappa r)$, and spherical Hankel functions of the first and second kind, $h_\nu^{(1)}(\kappa r)$ and $h_\nu^{(2)}(\kappa r)$, respectively. The eigensolutions of (13.10) in sphero-conal coordinates will be denoted as the *Lamé products* $Y_\nu(\vartheta,\varphi)$. The eigenvalues $\nu(\nu+1)$ and the corresponding Lamé products $Y_\nu(\vartheta,\varphi)$ are to be chosen so that specific conditions, e.g., boundary conditions, continuity conditions, regularity conditions, are satisfied. In the context of the examples of use later in this chapter we will especially consider three cases:

1. Dirichlet—case: Single-valued, real and regular functions $Y_\sigma(\vartheta,\varphi)$ on $0\le\vartheta\le\vartheta_0\le\pi$, $0\le\varphi\le 2\pi$, satisfying the geometric boundary conditions (13.5) at $\vartheta=0$ and the Dirichlet-condition at $\vartheta=\vartheta_0$:
$$Y_\sigma(\vartheta,\varphi)|_{\vartheta=\vartheta_0} = 0 \tag{13.11}$$
2. Neumann—case: Single-valued, real, and regular functions $Y_\tau(\vartheta,\varphi)$ on $0\le\vartheta\le\vartheta_0\le\pi$, $0\le\varphi\le 2\pi$, satisfying the geometric boundary conditions (13.5) at $\vartheta=0$ and the Neumann-condition at $\vartheta=\vartheta_0$:
$$\left.\frac{\partial Y_\tau(\vartheta,\varphi)}{\partial\vartheta}\right|_{\vartheta=\vartheta_0} = 0 \tag{13.12}$$
3. Free space—case: Single-valued, real, and regular functions $Y_{n,m}(\vartheta,\varphi)$ on the complete spherical surface $0\le\vartheta\le\pi$, $0\le\varphi\le 2\pi$, satisfying the geometric boundary conditions (13.5) at $\vartheta=0$ and at $\vartheta=\pi$. This eigenvalue problem has often been discussed in the literature using ordinary spherical coordinates [55], and it has been shown that the corresponding eigenvalues have to be integral ($n=0,1,2,3,\ldots$) and that to every eigenvalue n there exist $2n+1$ linearly independent eigenfunctions $Y_{n,m}(\vartheta,\varphi)$ with $-n\le m\le +n$.

For the Dirichlet- and Neumann-cases the eigenvalues σ_i and τ_i are real and distinct such that:

$$0<\sigma_1<\sigma_2<\sigma_3<\ldots \qquad \text{and} \qquad 0\le\tau_0<\tau_1<\tau_2<\ldots \tag{13.13}$$

respectively. As shown in the next subsection, these Lamé products form an orthonormal set of eigenfunctions. For the determination of the eigenfunctions and of the eigenvalues, we now further investigate the eigenvalue equation (13.10) in sphero-conal coordinates:

$$\begin{aligned}&\sqrt{1-k^2\cos^2\vartheta}\,\frac{\partial}{\partial\vartheta}\left(\sqrt{1-k^2\cos^2\vartheta}\,\frac{\partial Y_\nu}{\partial\vartheta}\right) + \sqrt{1-k'^2\sin^2\varphi}\\ &\quad\times\frac{\partial}{\partial\varphi}\left(\sqrt{1-k'^2\sin^2\varphi}\,\frac{\partial Y_\nu}{\partial\varphi}\right) + (k^2\sin^2\vartheta + k'^2\cos^2\varphi)\nu(\nu+1)Y_\nu = 0\end{aligned} \tag{13.14}$$

TABLE 13.3 Periodic Lamé Functions

Type	Series Representation	Behavior at $\varphi = 0$	Behavior at $\varphi = \pi/2$	Period
1	$\Phi_\nu^{(1)} = \sum_{i=0}^{\infty} A_{2i} \cos 2i\varphi$	$\left.\frac{d\Phi_\nu^{(1)}(\varphi)}{d\varphi}\right\vert_{\varphi=0} = 0$	$\left.\frac{d\Phi_\nu^{(1)}(\varphi)}{d\varphi}\right\vert_{\varphi=\pi/2} = 0$	π
2	$\Phi_\nu^{(2)} = \sum_{i=0}^{\infty} A_{2i+1} \cos (2i+1)\varphi$	$\left.\frac{d\Phi_\nu^{(2)}(\varphi)}{d\varphi}\right\vert_{\varphi=0} = 0$	$\Phi_\nu^{(2)}(\pi/2) = 0$	2π
3	$\Phi_\nu^{(3)} = \sum_{i=0}^{\infty} B_{2i+2} \sin(2i+2)\varphi$	$\Phi_\nu^{(3)}(0) = 0$	$\Phi_\nu^{(3)}(\pi/2) = 0$	π
4	$\Phi_\nu^{(4)} = \sum_{i=0}^{\infty} B_{2i+1} \sin(2i+1)\varphi$	$\Phi_\nu^{(4)}(0) = 0$	$\left.\frac{d\Phi_\nu^{(4)}(\varphi)}{d\varphi}\right\vert_{\varphi=\pi/2} = 0$	2π

A second separation expression

$$Y_\nu(\vartheta, \varphi) = \Theta_\nu(\vartheta)\Phi_\nu(\varphi) \tag{13.15}$$

leads after some algebraic manipulation to the following *Lamé differential equations*

$$\sqrt{1 - k'^2 \sin^2 \varphi}\,\frac{d}{d\varphi}\left(\sqrt{1 - k'^2 \sin^2 \varphi}\,\frac{d\Phi_\nu}{d\varphi}\right) + [\lambda - \nu(\nu+1)k'^2 \sin^2 \varphi]\Phi_\nu = 0 \tag{13.16}$$

$$\sqrt{1 - k^2 \cos^2 \vartheta}\,\frac{d}{d\vartheta}\left(\sqrt{1 - k^2 \cos^2 \vartheta}\,\frac{d\Theta_\nu}{d\vartheta}\right) + [\nu(\nu+1)(1 - k^2 \cos^2 \vartheta) - \lambda]\Theta_\nu = 0 \tag{13.17}$$

which are coupled by the two constants ν and λ. For the cases under consideration we expect in φ periodic solutions with the largest period 2π, hence (13.16) is denoted as the differential equation of the periodic Lamé functions. In general the solutions in ϑ will be non-periodic, so (13.17) is referred to as the differential equation of the non-periodic Lamé functions.

It is noted that subject to the considered boundary conditions both the periodic and the non-periodic Lamé differential equations can be understood as Sturm-Liouville-Problems, hence all the relations known for the eigenfunctions and eigenvalues of Sturm-Liouville problems are valid. Since the values for ν and λ have to be the same in the differential equations (13.16) and (13.17), it is clear that the Sturm-Liouville problems concerned are coupled.

For the solution of the differential equation of the periodic Lamé functions (13.16) we will use four different types of Fourier series, which differ with respect to their symmetry and periodicity properties and which are represented in Table 13.3.

Inserting these four "function types" $\Phi_\nu^{(p)}(\varphi)$ $(p = 1, 2, 3, 4)$ into the differential equation (13.16) leads to four different three-term recurrence relations. For example, we derive for the coefficients A_{2i} of the function type 1:

$$\begin{aligned} (a_0 - \lambda)A_0 + c_0 A_2 &= 0 \\ b_{2i-2}A_{2i-2} + (a_{2i} - \lambda)A_{2i} + c_{2i}A_{2i+2} &= 0 \qquad (i = 1, 2, 3, \ldots) \end{aligned} \tag{13.18}$$

where the constants a_i, b_i and c_i only depend on the ellipticity parameter k' and on $\nu(\nu+1)$. The three-term recurrence relations (13.18) may be written in the form of an algebraic eigenvalue equation

$$\mathcal{M}\vec{A} = \lambda\vec{A} \tag{13.19}$$

with the eigenvalue λ, the eigenvector $\vec{A} = (A_0, A_2, A_4, A_6, \ldots)$ and with a tridiagonal matrix $\mathcal{M}$. For the other function types other three-term recurrence relations and other algebraic eigenvalue equations are obtained [30], [13], [20].

For given values of k' and $\nu(\nu+1)$ the eigenvalue equation (13.19) can be solved, which may be done efficiently by using appropriate numerical techniques suitable to the tridiagonal structure. In general the dimension of the matrix in (13.19) will be infinite, but it can be shown that a specific finite dimension leads to a solution with a sufficient accuracy of the resultant periodic Lamé functions' representation.

To any of the four function types of the periodic Lamé functions there can be defined a corresponding function type of the non-periodic Lamé functions $\Theta_\nu^{(p)}(\vartheta)$ $(p = 1, 2, 3, 4)$. The series representations shown in Table 13.4 were first introduced by Jansen and Boersma [30] and have one particular advantage: inserting of the series representations for the non-periodic Lamé functions $\Theta_\nu^{(p)}(\vartheta)$ (Table 13.4) into the

TABLE 13.4 Non-periodic Lamé Functions

Type	Series Representation	Behavior at $\vartheta = 0$
1	$\Theta_\nu^{(1)}(\vartheta) = \sum_{i=0}^{\infty} A_{2i}T(2i)P_\nu^{2i}(\cos\vartheta)$	$\left.\frac{d\Theta_\nu^{(1)}(\vartheta)}{d\vartheta}\right\|_{\vartheta=0} = 0$
2	$\Theta_\nu^{(2)}(\vartheta) = \sum_{i=0}^{\infty} A_{2i+1}T(2i+1)P_\nu^{2i+1}(\cos\vartheta)$	$\Theta_\nu^{(2)}(0) = 0$
3	$\Theta_\nu^{(3)}(\vartheta) = \frac{\sqrt{1-k^2\cos^2\vartheta}}{\sin\vartheta}\sum_{i=0}^{\infty} B_{2i+2}(2i+2)T(2i+2)P_\nu^{2i+2}(\cos\vartheta)$	$\Theta_\nu^{(3)}(0) = 0$
4	$\Theta_\nu^{(4)}(\vartheta) = \frac{\sqrt{1-k^2\cos^2\vartheta}}{\sin\vartheta}\sum_{i=0}^{\infty} B_{2i+1}(2i+1)T(2i+1)P_\nu^{2i+1}(\cos\vartheta)$	$\left.\frac{d\Theta_\nu^{(4)}(\vartheta)}{d\vartheta}\right\|_{\vartheta=0} = 0$

$T(i) = -(\nu-1)(\nu+i+1)T(i+2); \qquad T(0) = T(1) = 1$

differential equation (13.17) leads exactly to the same three-term recurrence relations and to the same algebraic eigenvalue equations as obtained for the corresponding periodic Lamé functions representation $\Phi_\nu^{(p)}(\varphi)$. Moreover, the four types of "Lamé products"

$$Y_\nu^{(p)}(\vartheta,\varphi) = \Theta_\nu^{(p)}(\vartheta)\Phi_\nu^{(p)}(\varphi) \qquad p = 1;2;3;4 \tag{13.20}$$

automatically satisfy the geometrical boundary conditions (13.5) at $\vartheta = 0$. In order to find the values of ν and the corresponding values of λ, which satisfy the Dirichlet- or Neumann-boundary condition at $\vartheta = \vartheta_0$, a numerical procedure is utilized. First, it is clear that for a given (finite) dimension of the algebraic eigenvalue equation (13.19) there will be an infinite number of eigenvalues λ and corresponding eigenvectors in general. However, from the properties of the differential equation of the non-periodic Lamé functions (13.17) with the considered boundary-conditions it follows that only the eigenvalues with

$$\lambda \leq \nu(\nu+1) \tag{13.21}$$

lead to a meaningful Lamé product. We solve the eigenvalue equation (13.19) numerically and obtain relations between the λ_l $(l = 0,1,2,\ldots)$ and ν, which are graphically represented in Fig. 13.3 and which are denoted as "eigenvalue curves". Due to the Sturm-Liouville properties the periodic Lamé function belonging to an eigenvalue on the lth eigenvalue curve $(l = 0,1,2,\ldots)$ has exactly l zeros in the interval $0 \leq \varphi < \pi$. Using a numerical search algorithm we then will find the discrete pairs of eigenvalues ν_i, λ_i lying on the eigenvalue curves and satisfying either the Dirichlet- or the Neumann-condition. The Dirichlet- and the Neumann-eigenvalues rigorously interchange, which is due to the properties of the underlying Sturm-Liouville problems.

It should be remarked that the resulting series representations for the periodic Lamé functions (Table 13.3) are uniformly convergent in the whole domain $(0 \leq \varphi \leq 2\pi)$, while the series for the non-periodic Lamé functions (Table 13.4) are uniformly convergent only on the interval $0 \leq \vartheta < \arccos(-k)$, i.e., for $k \neq 0$ at least

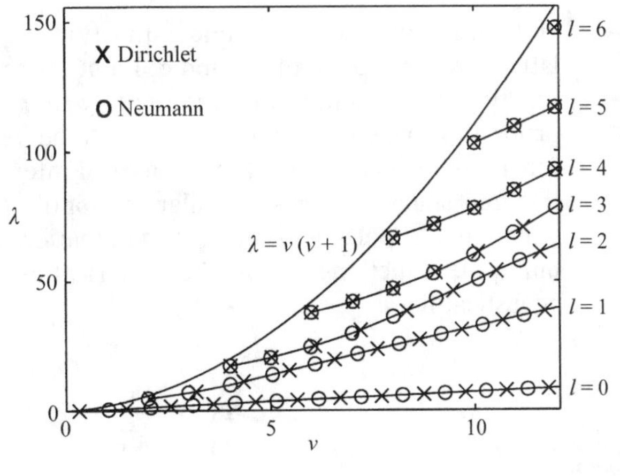

Figure 13.3 Eigenvalue curves and eigenvalues for the first function type with $\vartheta_0 = 175°$ and $k^2 = 0.5$.

on the interval $0 \le \vartheta \le \pi/2$. For $\vartheta > \pi/2$ we integrate the differential equation (13.17) numerically, starting from the values at $\vartheta = \pi/2$ obtained with the series representation.

In the free-space case ($0 \le \vartheta \le \pi; 0 \le \varphi \le 2\pi$) the eigenvalues are integral $\nu = n$ and it can be shown that the Lamé functions series in Tables 13.3 and 13.4 are finite, hence we obtain four types of "Lamé polynomials," which are denoted by:

$$Y^{(p)}_{n,m}(\vartheta, \varphi) = \Theta^{(p)}_{n,m}(\vartheta)\,\Phi^{(p)}_{n,m}(\varphi) \qquad p = 1; 2; 3; 4 \tag{13.22}$$

The $2n+1$ linearly independent eigenfunctions belonging to the eigenvalue n (index m) are distributed among the four function types.

It should be noted that in case of $k^2 = 1$, i.e., in case of ordinary spherical coordinates, the normalized Lamé products for the free-space case [normalization see (13.29)] exactly turn into the well-known representation of the normalized spherical surface harmonics:

$$Y_{n,m}(\vartheta, \varphi) = \sqrt{\frac{2n+1}{(1+\varepsilon_m)\pi}\frac{(n-m)!}{(n+m)!}}\,P_n^m(\cos\vartheta)\begin{matrix}\cos m\varphi \\ \sin m\varphi\end{matrix} \tag{13.23}$$

with $0 \le n$ and $0 \le m \le n$ and where the Neumann number is defined by

$$\varepsilon_m = \begin{cases} 2 & \text{if } n = 0 \\ 1 & \text{if } n > 0 \end{cases} \tag{13.24}$$

Summing up, it may be said that we have found a base to construct any solution of the eigenvalue equation (13.10) in sphero-conal coordinates for the Dirichlet-case, for the Neumann-case, and for the free-space case. The obtained elementary solutions of the scalar homogeneous Helmholtz equation are given by

$$\Psi_{\nu,\kappa}(r, \vartheta, \varphi) = z_\nu(\kappa r) Y_\nu(\vartheta, \varphi) \tag{13.25}$$

and consist of spherical cylinder functions $z_\nu(\kappa r)$ and Lamé products $Y_\nu(\vartheta, \varphi)$. The latter are products of periodic Lamé functions $\Phi_\nu(\varphi)$ and non-periodic Lamé functions $\Theta_\nu(\vartheta)$ and are represented by four function types with different symmetry properties. The functions $\Psi_{\nu,\kappa}(\vec{r})$ will be termed the "scalar spherical-multipole functions," which relates to the physical interpretation of the terms in (13.25) within the free space ($\nu = n = 0$: scalar monopole field; $n = 1$: scalar dipole field; $n = 2$: scalar quadrupole field, etc.). A summation over all the relevant scalar spherical-multipole functions yields the spherical-multipole expansion of the scalar, e.g., acoustical field:

$$\Psi(\vec{r}) = \sum_\nu \alpha_\nu z_\nu(\kappa r) Y_\nu(\vartheta, \varphi) \tag{13.26}$$

13.3.2 Scalar Orthogonality Relations

13.3.2.1 Orthogonality of Lamé Products

Subject to an appropriate normalization the Lamé products $Y_\nu(\vartheta, \varphi)$ satisfy the following orthogonality relations:

$$\int_0^{\vartheta_0}\int_0^{2\pi} Y_\sigma(\vartheta,\varphi)Y_{\sigma'}(\vartheta,\varphi)s_\vartheta s_\varphi\, d\vartheta\, d\varphi = \delta_{\sigma\sigma'}$$
$$\int_0^{\vartheta_0}\int_0^{2\pi} Y_\tau(\vartheta,\varphi)Y_{\tau'}(\vartheta,\varphi)s_\vartheta s_\varphi\, d\vartheta\, d\varphi = \delta_{\tau\tau'} \tag{13.27}$$

where σ, σ' are eigenvalues belonging to the Dirichlet condition (13.11) and τ, τ' are eigenvalues belonging to the Neumann condition (13.12) The Kronecker symbol is defined in the usual way as

$$\delta_{ij} = \begin{cases} 1 & \text{if } i = j \\ 0 & \text{if } i \neq j \end{cases} \tag{13.28}$$

In the free-space case we have

$$\int_0^{\pi}\int_0^{2\pi} Y_{n,m}(\vartheta,\varphi)Y_{n',m'}(\vartheta,\varphi)s_\vartheta s_\varphi\, d\vartheta\, d\varphi = \delta_{nn'}\delta_{mm'} \tag{13.29}$$

For the proof, remember that the Lamé products are eigenfunctions of the operator $(\vec{r}\times\vec{\nabla})^2$ subject to the considered boundary conditions at $\vartheta = 0$ and at $\vartheta = \vartheta_0$. We show that this operator is symmetric (self-adjoint), i.e., that

$$\int_0^{\vartheta_0}\int_0^{2\pi}\left[(\vec{r}\times\vec{\nabla})^2 Y_\nu\right]Y_{\nu'}s_\vartheta s_\varphi\, d\vartheta\, d\varphi - \int_0^{\vartheta_0}\int_0^{2\pi} Y_\nu\left[(\vec{r}\times\vec{\nabla})^2 Y_{\nu'}\right]s_\vartheta s_\varphi\, d\vartheta\, d\varphi \tag{13.30}$$

vanishes ($\nu = \sigma$ or $\nu = \tau$), since it can be shown that eigenfunctions of symmetric operators are orthogonal [25]. With the representation of the operator $(\vec{r}\times\vec{\nabla})^2$ in sphero-conal coordinates (Table 13.1), by utilizing the method of partial integration, and by considering the φ-periodicity as well as the behavior at $\vartheta = 0$ [geometric boundary conditions, (13.5)] and at $\vartheta = \vartheta_0$ [Dirichlet or Neumann condition, (13.11) and (13.12)], we obtain from (13.30):

$$\begin{aligned}&\int_0^{\vartheta_0}\int_0^{2\pi}\left\{\left[\frac{\partial}{\partial\vartheta}\left(\frac{s_\varphi}{s_\vartheta}\frac{\partial Y_\nu}{\partial\vartheta}\right)+\frac{\partial}{\partial\varphi}\left(\frac{s_\vartheta}{s_\varphi}\frac{\partial Y_\nu}{\partial\varphi}\right)\right]Y_{\nu'} - Y_\nu\left[\frac{\partial}{\partial\vartheta}\left(\frac{s_\varphi}{s_\vartheta}\frac{\partial Y_{\nu'}}{\partial\vartheta}\right)+\frac{\partial}{\partial\varphi}\left(\frac{s_\vartheta}{s_\varphi}\frac{\partial Y_{\nu'}}{\partial\varphi}\right)\right]\right\}d\vartheta\, d\varphi \\ &= \underbrace{\int_0^{2\pi}\frac{s_\varphi}{s_\vartheta}\left[\frac{\partial Y_{\nu'}}{\partial\vartheta}Y_\nu - \frac{\partial Y_\nu}{\partial\vartheta}Y_{\nu'}\right]\Bigg|_0^{\vartheta_0} d\varphi}_{=0} + \underbrace{\int_0^{\vartheta_0}\frac{s_\vartheta}{s_\varphi}\left[\frac{\partial Y_{\nu'}}{\partial\varphi}Y_\nu - \frac{\partial Y_\nu}{\partial\varphi}Y_{\nu'}\right]\Bigg|_0^{2\pi} d\vartheta}_{=0} = 0\end{aligned} \tag{13.31}$$

The orthogonality relation (13.29) can be proved in a similar way by taking into consideration the geometrical boundary conditions at $\vartheta = \pi$ as well as at $\vartheta = 0$. The completeness relation belonging to the orthonormal systems of Lamé products can be built up in the usual way

$$\frac{\delta(\vartheta-\vartheta')\delta(\varphi-\varphi')}{s_\vartheta s_\varphi}=\sum_\nu Y_\nu(\vartheta,\varphi)Y_\nu(\vartheta',\varphi') \tag{13.32}$$

and is to be understood as the equality between two distributions.

13.3.2.2 Orthogonality of Scalar Multipole Functions

The scalar multipole functions, which are built up by means of spherical Bessel functions

$$\Psi^I_{\nu,\kappa}(\vec{r})=j_\nu(\kappa r)Y_\nu(\vartheta,\varphi) \tag{13.33}$$

where $\nu=\sigma$ are eigenvalues belonging to the Dirichlet condition (13.11) or $\nu=\tau$ are eigenvalues belonging to the Neumann condition (13.12), or $\nu=n,m$ (free-space eigenvalues), satisfy the following orthogonality relation:

$$\iiint_V \Psi^I_{\nu,\kappa}(\vec{r})\Psi^I_{\nu',\kappa'}(\vec{r})\,dv=\frac{\pi}{2}\frac{\delta(\kappa-\kappa')}{\kappa^2}\delta_{\nu\nu'} \tag{13.34}$$

The volume integral is to be taken over the domain outside the elliptic cone (Dirichlet case and Neumann case) or over the complete free space. The proof of (13.34) is found quickly by utilizing the orthogonalities of the Lamé products (13.27), (13.29) and by using the orthogonality of the spherical Bessel functions on a semi-infinite domain [56]:

$$\int_0^\infty j_\nu(\kappa r)j_\nu(\kappa' r)r^2\,dr=\frac{\pi}{2}\frac{\delta(\kappa-\kappa')}{\kappa^2} \tag{13.35}$$

For the derivation of the corresponding completeness relation consider an arbitrary function which can be described completely by the expansion

$$f(\vec{r})=\int_0^\infty\sum_\nu a_{\nu,\kappa}\Psi^I_{\nu,\kappa}(\vec{r})\,d\kappa \tag{13.36}$$

The coefficients $a_{\nu,\kappa}$ are found by using the orthogonality relation (13.34) and by employing the filter property of the δ-function as

$$a_{\nu,\kappa}=\frac{2}{\pi}\kappa^2\iiint_V f(\vec{r})\Psi^I_{\nu,\kappa}(\vec{r})\,dv \tag{13.37}$$

We insert (13.37) into (13.36) and obtain

$$f(\vec{r})=\iiint f(\vec{r}\,')\frac{2}{\pi}\int_0^\infty\kappa^2\sum_\nu\Psi^I_{\nu,\kappa}(\vec{r})\Psi^I_{\nu,\kappa}(\vec{r}\,')\,d\kappa\,dv' \tag{13.38}$$

Comparison of (13.38) with the relation

$$f(\vec{r})=\iiint f(\vec{r}\,')\delta(\vec{r}-\vec{r}\,')\,dv' \tag{13.39}$$

finally yields the desired completeness relation

$$\delta(\vec{r}-\vec{r}\,') = \frac{2}{\pi}\int_0^\infty \kappa^2 \sum_\nu \Psi^I_{\nu,\kappa}(\vec{r})\Psi^I_{\nu,\kappa}(\vec{r}\,')\,d\kappa \tag{13.40}$$

13.3.3 Scalar Green's Functions in Sphero-Conal Coordinates

We are now interested in solutions of the scalar inhomogeneous Helmholtz equation in sphero-conal coordinates

$$\Delta\Psi(r,\vartheta,\varphi) + \kappa^2\Psi(r,\vartheta,\varphi) = -q(r,\vartheta,\varphi) \tag{13.41}$$

in the domain outside of an acoustically soft elliptic cone ($\Psi = 0$ at $\vartheta = \vartheta_0$), an acoustically hard elliptic cone ($\partial\Psi/\partial\vartheta = 0$ at $\vartheta = \vartheta_0$) or within the free space.

Moreover Ψ has to satisfy the Sommerfeld radiation condition at infinity and the condition of finite energy within any finite region. The solutions are found via the (scalar) Green's functions $G(\vec{r},\vec{r}\,')$ associated to the scalar Helmholtz equation and defined by the differential equation

$$\Delta G(r,\vartheta,\varphi;r',\vartheta',\varphi') + \kappa^2 G(r,\vartheta,\varphi;r',\vartheta',\varphi') = -\frac{\delta(r-r')}{r^2}\frac{\delta(\vartheta-\vartheta')\delta(\varphi-\varphi')}{s_\vartheta s_\varphi} \tag{13.42}$$

Once an appropriate Green's function has been found, the solution of the inhomogeneous Helmholtz equation (13.41) subject to the considered boundary conditions is given by

$$\Psi(\vec{r}) = \iiint_V G(\vec{r},\vec{r}\,')q(\vec{r}\,')\,dv' \tag{13.43}$$

which may be shown by utilizing Green's second identity.

For the determination of the Green's function we first replace a part of the distributional term on the right side of (13.42) by the completeness relation (13.32) and describe the Green's function by the complete bilinear approach

$$G(\vec{r},\vec{r}\,') = \sum_\nu F_\nu j_\nu(\kappa r_<)h_\nu^{(2)}(\kappa r_>)Y_\nu(\vartheta,\varphi)Y_\nu(\vartheta',\varphi') \tag{13.44}$$

For the acoustically soft cone we use eigenvalues and eigenfunctions of the Dirichlet type ($\nu = \sigma$), for the acoustically hard cone eigenvalues and functions of the Neumann type ($\nu = \tau$), and for the free-space case integral eigenvalues and corresponding eigenfunctions $\nu = n, m$. We have introduced the abbreviations

$$r_< = \begin{cases} r & \text{if } r \le r' \\ r' & \text{if } r > r' \end{cases}; \qquad r_> = \begin{cases} r' & \text{if } r \le r' \\ r & \text{if } r > r' \end{cases} \tag{13.45}$$

Spherical Bessel functions j_ν are to be chosen to satisfy the energy condition for $r_< = 0$ as well, and spherical Hankel functions of the second kind $h_\nu^{(2)}$ guarantee the right radiation behavior at $r_> \to \infty$ for the chosen time factor $e^{+j\omega t}$. Insertion of (13.44) into (13.42), utilizing the eigenvalue equation of the Lamé products, and comparing both sides of (13.42) leads to the differential equation:

$$F_\nu\left\{\frac{d}{dr}\left(r^2\frac{d}{dr}\right)-[\nu(\nu+1)-(\kappa r)^2]\right\}j_\nu(\kappa r_<)h_\nu^{(2)}(\kappa r_>)=\delta(r-r') \tag{13.46}$$

Integrating (13.46) within an infinitesimal domain around $r = r'$ and making use of the Wronskian determinant $\mathscr{W}(\kappa r'; j_\nu, h_\nu^{(2)})$ yields the factor F_ν and finally we derive

$$G(\vec{r},\vec{r}')=-j\kappa\sum_\nu \mathrm{j}_\nu(\kappa r_<)h_\nu^{(2)}(\kappa r_>)Y_\nu(\vartheta,\varphi)Y_\nu(\vartheta',\varphi') \tag{13.47}$$

13.4 SPHERICAL-MULTIPOLE ANALYSIS OF ELECTROMAGNETIC FIELDS

13.4.1 Vector Spherical-Multipole Expansion of Solenoidal Electromagnetic Fields

We start with the symmetric form of Maxwell's equations within a linear homogeneous and isotropic medium and consider a harmonic time dependence (time factor $e^{+j\omega t}$):

$$\begin{aligned}\vec{\nabla}\times\vec{H}(\vec{r}) &= +\vec{J}_e^{imp}(\vec{r})+\mathrm{j}\omega\varepsilon\vec{E}(\vec{r})\\ \vec{\nabla}\times\vec{E}(\vec{r}) &= -\vec{J}_m^{imp}(\vec{r})-\mathrm{j}\omega\mu\vec{H}(\vec{r})\end{aligned} \tag{13.48}$$

$\vec{E}$ and $\vec{H}$ are the electric and magnetic field intensities, $\vec{J}_e^{imp}$ and $\vec{J}_m^{imp}$ are the impressed electric and (fictitious) magnetic current densities, and ε and μ are the effective permittivity and permeability, respectively. Decoupling of the field intensities leads to the system of second-order differential equations:

$$\begin{aligned}\vec{\nabla}\times\vec{\nabla}\times\vec{E}(\vec{r})-\kappa^2\vec{E}(\vec{r}) &= -\mathrm{j}\omega\mu\underline{\vec{J}}_e^{imp}(\vec{r})\\ \vec{\nabla}\times\vec{\nabla}\times\vec{H}(\vec{r})-\kappa^2\vec{H}(\vec{r}) &= -\mathrm{j}\omega\varepsilon\underline{\vec{J}}_m^{imp}(\vec{r})\end{aligned} \tag{13.49}$$

where $\kappa = \omega\sqrt{\varepsilon\mu}$ is the wave number and where the effective impressed electric and magnetic current densities are given by

$$\begin{aligned}\underline{\vec{J}}_e^{imp}(\vec{r}) &= \vec{J}_e^{imp}(\vec{r})+\frac{1}{\mathrm{j}\omega\mu}\vec{\nabla}\times\vec{J}_m^{imp}(\vec{r})\\ \underline{\vec{J}}_m^{imp}(\vec{r}) &= \vec{J}_m^{imp}(\vec{r})-\frac{1}{\mathrm{j}\omega\varepsilon}\vec{\nabla}\times\vec{J}_e^{imp}(\vec{r})\end{aligned} \tag{13.50}$$

respectively. To begin with, we are interested in the electromagnetic fields outside the regions containing the impressed current densities, i.e., in the fundamental solutions of the homogeneous equation

$$\vec{\nabla}\times\vec{\nabla}\times\vec{F}(\vec{r})-\kappa^2\vec{F}(\vec{r})=0 \tag{13.51}$$

Provided that $\kappa \neq 0$ these solutions are solenoidal and identical to the solenoidal solutions of the homogeneous vector Helmholtz equation, which are found by (Appendix 13.1)

$$\vec{M}(\vec{r}) = (\vec{r} \times \vec{\nabla})\Psi(\vec{r})$$
$$\vec{N}(\vec{r}) = \left[\frac{1}{\kappa}\vec{\nabla} \times (\vec{r} \times \vec{\nabla})\right]\Psi(\vec{r}) \tag{13.52}$$

where $\Psi(\vec{r})$ satisfies the homogeneous scalar Helmholtz equation

$$\Delta\Psi(\vec{r}) + \kappa^2\Psi(\vec{r}) = 0 \tag{13.53}$$

As shown in Section 13.3, the elementary solutions of (13.53) in sphero-conal coordinates (i.e., the scalar spherical-multipole functions) are found as products of spherical cylinder functions and Lamé products $\Psi_{\nu,\kappa}(\vec{r}) = z_\nu(\kappa r)Y_\nu(\vartheta, \varphi)$. Insertion into (13.52) yields the *vector spherical-multipole functions*:

$$\vec{M}_{\nu,\kappa}(\vec{r}) = z_\nu(\kappa r)\left[-\frac{1}{s_\varphi}\frac{\partial Y_\nu(\vartheta, \varphi)}{\partial \varphi}\hat{\vartheta} + \frac{1}{s_\vartheta}\frac{\partial Y_\nu(\vartheta, \varphi)}{\partial \vartheta}\hat{\varphi}\right]$$

$$\vec{N}_{\nu,\kappa}(\vec{r}) = -\frac{z_\nu(\kappa r)}{\kappa r}\nu(\nu+1)Y_\nu(\vartheta, \varphi)\hat{r} \tag{13.54}$$
$$-\frac{1}{\kappa r}\frac{d}{dr}[rz_\nu(\kappa r)]\left[\frac{1}{s_\vartheta}\frac{\partial Y_\nu(\vartheta, \varphi)}{\partial \vartheta}\hat{\vartheta} + \frac{1}{s_\varphi}\frac{\partial Y_\nu(\vartheta, \varphi)}{\partial \varphi}\hat{\varphi}\right]$$

The terms in brackets in (13.54) are referred to as the *transverse vector functions* and given by

$$\vec{m}_\nu(\vartheta, \varphi) = -\frac{1}{s_\varphi}\frac{\partial Y_\nu(\vartheta, \varphi)}{\partial \varphi}\hat{\vartheta} + \frac{1}{s_\vartheta}\frac{\partial Y_\nu(\vartheta, \varphi)}{\partial \vartheta}\hat{\varphi}$$
$$\vec{n}_\nu(\vartheta, \varphi) = +\frac{1}{s_\vartheta}\frac{\partial Y_\nu(\vartheta, \varphi)}{\partial \vartheta}\hat{\vartheta} + \frac{1}{s_\varphi}\frac{\partial Y_\nu(\vartheta, \varphi)}{\partial \varphi}\hat{\varphi} \tag{13.55}$$

The summation over all relevant $\vec{N}_{\nu,\kappa}(\vec{r})$- and $\vec{M}_{\nu,\kappa}(\vec{r})$-functions is called the *spherical-multipole expansion of the electromagnetic field*

$$\vec{E}(\vec{r}) = \sum_\nu a_\nu \vec{N}_{\nu,\kappa}(\vec{r}) + \frac{Z}{\mathrm{j}}\sum_\eta b_\eta \vec{M}_{\eta,\kappa}(\vec{r})$$
$$\vec{H}(\vec{r}) = \frac{\mathrm{j}}{Z}\sum_\nu a_\nu \vec{M}_{\nu,\kappa}(\vec{r}) + \sum_\eta b_\eta \vec{N}_{\eta,\kappa}(\vec{r}) \tag{13.56}$$

In (13.56) we have introduced the intrinsic wave impedance of the medium by $Z = \sqrt{\mu/\varepsilon}$. The coefficients of the spherical-multipole expansion are the *electric multipole amplitudes* a_ν, measured in volts per meter (as the electric field intensity) and the *magnetic multipole amplitudes* b_η, measured in amperes per meter (as the magnetic field intensity). The eigenvalues ν and η are to be chosen so that the physically required boundary and/or continuity conditions of the electromagnetic field are fulfilled. We are especially interested in two cases:

- The electromagnetic field in the domain bounded by a perfectly electrically conducting (PEC) semi-infinite elliptic cone $\vartheta = \vartheta_0$. The tangential electric field components (i.e., E_r and E_φ) have to vanish at the cone's surface, hence from the multipole expansion (13.56) and from the representation of the vector spherical-multipole functions (13.54) it follows that the eigenvalues ν have to be of the Dirichlet type $\nu = \sigma$, defined by (13.11) and that the eigenvalues η have to be of the Neumann type $\eta = \tau$, as defined by (13.12).
- The electromagnetic field in the free space. In this case both the eigenvalues ν and η have to be integral, and for every eigenvalue there exist $2n+1$ linearly independent $\vec{N}_{n,m,\kappa}$-functions and $2n+1$ linearly independent $\vec{M}_{n,m,\kappa}$-functions.

The spherical-multipole expansion (13.56) can be understood as a modal expansion. We remark that the $\vec{M}_{\nu,\kappa}$-functions have no radial component in contrast to the $\vec{N}_{\nu,\kappa}$-functions, which may be seen from their definitions (13.54). Hence the magnetic field of that part in (13.56), which belongs to the electric multipole amplitudes a_ν, is purely transverse with respect to the radial direction, i.e., this is the TM-part of the electromagnetic field. By analogy the part belonging to the magnetic multipole amplitudes is the TE part of the electromagnetic field. Table 13.5 gives an overview this modal decomposition.

The name "spherical-multipole analysis" is related to the physical interpretation of each term of the multipole expansion in the free space. Consider for example the lowest terms of the free space TM-modes, which is $n = 1$, because $n = 0$ yields a zero

TABLE 13.5 Field Decomposition in Spherical TM-Modes and TE-Modes

	TM-Modes	TE-Modes
E_r	$-a_\nu \nu(\nu+1)\dfrac{z_\nu(\kappa r)}{\kappa r}Y_\nu(\vartheta,\varphi)$	0
E_ϑ	$-a_\nu \dfrac{1}{\kappa r}\dfrac{d[rz_\nu(\kappa r)]}{dr}\dfrac{1}{s_\vartheta}\dfrac{\partial Y_\nu(\vartheta,\varphi)}{\partial\vartheta}$	$-\dfrac{Z}{\mathrm{j}}b_\eta z_\eta(\kappa r)\dfrac{1}{s_\varphi}\dfrac{\partial Y_\eta(\vartheta,\varphi)}{\partial\varphi}$
E_φ	$-a_\nu \dfrac{1}{\kappa r}\dfrac{d[rz_\nu(\kappa r)]}{dr}\dfrac{1}{s_\varphi}\dfrac{\partial Y_\nu(\vartheta,\varphi)}{\partial\varphi}$	$\dfrac{Z}{\mathrm{j}}b_\eta z_\eta(\kappa r)\dfrac{1}{s_\vartheta}\dfrac{\partial Y_\eta(\vartheta,\varphi)}{\partial\vartheta}$
H_r	0	$-b_\eta \eta(\eta+1)\dfrac{z_\eta(\kappa r)}{\kappa r}Y_\eta(\vartheta,\varphi)$
H_ϑ	$-\dfrac{\mathrm{j}}{Z}a_\nu z_\nu(\kappa r)\dfrac{1}{s_\varphi}\dfrac{\partial Y_\nu(\vartheta,\varphi)}{\partial\varphi}$	$-b_\eta \dfrac{1}{\kappa r}\dfrac{d[rz_\eta(\kappa r)]}{dr}\dfrac{1}{s_\vartheta}\dfrac{\partial Y_\eta(\vartheta,\varphi)}{\partial\vartheta}$
H_φ	$\dfrac{\mathrm{j}}{Z}a_\nu z_\nu(\kappa r)\dfrac{1}{s_\vartheta}\dfrac{\partial Y_\nu(\vartheta,\varphi)}{\partial\vartheta}$	$-b_\eta \dfrac{1}{\kappa r}\dfrac{d[rz_\eta(\kappa r)]}{dr}\dfrac{1}{s_\varphi}\dfrac{\partial Y_\eta(\vartheta,\varphi)}{\partial\varphi}$

electromagnetic field. Provided that spherical Hankel functions of the second kind are used, the electromagnetic field corresponding to the electric multipole amplitudes $a_{1,m}$ can be understood as produced by an elementary electric (Hertzian) dipole located at the origin. The three possible values of m stand for the three orthogonal dipole's polarizations. Consequently, the fields corresponding to the five TM terms with $n = 2$ and to the seven with $n = 3$ can be interpreted as produced by electric quadrupoles and by electric octupoles, respectively, each located at the center $r = 0$. By duality, the terms in the TE part correspond to fields produced by elementary magnetic (Fitzgerald) dipoles ($n = 1$), magnetic quadrupoles ($n = 2$), magnetic octupoles ($n = 3$), etc. Hence the spherical-multipole expansion is an analysis (decomposition) of the electromagnetic field into physically interpretable elementary (canonical) fields.

The duality between electric and magnetic quantities can be symbolized by means of the *Fitzgerald transform*, which for example can be helpful while checking analytical results or investigating problems with complementary structures. Table 13.6 gives the basic rules for this transform, where in addition the multipole amplitudes are included.

TABLE 13.6 Fitzgerald Transform

$\vec{E} \rightarrow +\vec{H}$	$\vec{H} \rightarrow -\vec{E}$
$\vec{J}_e^{imp} \rightarrow +\vec{J}_m^{imp}$	$\vec{J}_m^{imp} \rightarrow -\vec{J}_e^{imp}$
$a_\nu \rightarrow +b_\eta$	$b_\eta \rightarrow -a_\nu$
$\varepsilon \rightarrow +\mu$	$\mu \rightarrow +\varepsilon$
$\nu \rightarrow +\eta$	$\eta \rightarrow +\nu$

Application of the Fitzgerald transform to the spherical-multipole expansion (13.56) yields the multipole expansion itself, as is expected for a complete representation of the electromagnetic field.

13.4.2 Vector Orthogonality Relations

13.4.2.1 Orthogonality of the Transverse Vector Functions

The transverse vector functions $\vec{m}_\nu(\vartheta, \varphi)$ and $\vec{n}_\eta(\vartheta, \varphi)$ are orthogonal according to

$$\begin{aligned}
&\int_0^{\vartheta_0}\int_0^{2\pi} \vec{m}_\nu(\vartheta, \varphi)\cdot\vec{m}_{\nu'}(\vartheta, \varphi)s_\vartheta s_\varphi \, d\vartheta \, d\varphi = \nu(\nu+1)\delta_{\nu\nu'} \\
&\int_0^{\vartheta_0}\int_0^{2\pi} \vec{n}_\eta(\vartheta, \varphi)\cdot\vec{n}_{\eta'}(\vartheta, \varphi)s_\vartheta s_\varphi \, d\vartheta \, d\varphi = \eta(\eta+1)\delta_{\eta\eta'} \\
&\int_0^{\vartheta_0}\int_0^{2\pi} \vec{m}_\nu(\vartheta, \varphi)\cdot\vec{n}_\eta(\vartheta, \varphi)s_\vartheta s_\varphi \, d\vartheta \, d\varphi = 0
\end{aligned} \tag{13.57}$$

provided that

- ν, ν' are eigenvalues σ belonging to the Dirichlet condition (13.11) and η,η' are eigenvalues τ belonging to the Neumann condition (13.12), or:
- ν, ν' are eigenvalues τ belonging to the Neumann condition (13.12) and η,η' are eigenvalues σ belonging to the Dirichlet condition (13.11).

In the free-space case we have

$$\begin{aligned}
&\int_0^{\pi}\int_0^{2\pi} \vec{m}_{n,m}(\vartheta,\varphi)\cdot\vec{m}_{n',m'}(\vartheta,\varphi) s_\vartheta s_\varphi \, d\vartheta \, d\varphi = n(n+1)\delta_{nn'}\,\delta_{mm'}\\
&\int_0^{\pi}\int_0^{2\pi} \vec{n}_{n,m}(\vartheta,\varphi)\cdot\vec{n}_{n',m'}(\vartheta,\varphi) s_\vartheta s_\varphi \, d\vartheta \, d\varphi = n(n+1)\delta_{nn'}\,\delta_{mm'}\\
&\int_0^{\pi}\int_0^{2\pi} \vec{m}_{n,m}(\vartheta,\varphi)\cdot\vec{n}_{n',m'}(\vartheta,\varphi) s_\vartheta s_\varphi \, d\vartheta \, d\varphi = 0
\end{aligned} \tag{13.58}$$

To prove the first two equations in (13.57) we insert (13.55) into (13.57), apply the method of integration by parts, consider the 2π-periodicity of the integrands and utilize the eigenvalue equation of the Lamé products:

$$\begin{aligned}
&\int_0^{\vartheta_0}\int_0^{2\pi} \vec{m}_\nu\cdot\vec{m}_{\nu'} s_\vartheta s_\varphi \, d\vartheta\, d\varphi = \int_0^{\vartheta_0}\int_0^{2\pi} \vec{n}_\nu\cdot\vec{n}_{\nu'} s_\vartheta s_\varphi \, d\vartheta\, d\varphi\\
&= \int_0^{\vartheta_0}\int_0^{2\pi}\left[\frac{\partial Y_\nu}{\partial\varphi}\frac{s_\vartheta}{s_\varphi}\frac{\partial Y_{\nu'}}{\partial\varphi} + \frac{\partial Y_\nu}{\partial\vartheta}\frac{s_\varphi}{s_\vartheta}\frac{\partial Y_{\nu'}}{\partial\vartheta}\right] d\vartheta\, d\varphi\\
&= \underbrace{\int_0^{\vartheta_0} Y_\nu \frac{s_\vartheta}{s_\varphi}\frac{\partial Y_{\nu'}}{\partial\varphi}\bigg|_0^{2\pi} d\vartheta}_{=0} + \int_0^{2\pi} Y_\nu \frac{s_\varphi}{s_\vartheta}\frac{\partial Y_{\nu'}}{\partial\vartheta}\bigg|_0^{\vartheta_0} d\varphi\\
&\quad - \int_0^{\vartheta_0}\int_0^{2\pi} Y_\nu \underbrace{\left\{\frac{\partial}{\partial\vartheta}\left[\frac{s_\varphi}{s_\vartheta}\frac{\partial Y_{\nu'}}{\partial\vartheta}\right] + \frac{\partial}{\partial\varphi}\left[\frac{s_\vartheta}{s_\varphi}\frac{\partial Y_{\nu'}}{\partial\varphi}\right]\right\}}_{= s_\vartheta s_\varphi (\vec{r}\times\vec{\nabla})^2 Y_{\nu'}} d\vartheta\, d\varphi\\
&= \int_0^{2\pi} Y_\nu \frac{s_\varphi}{s_\vartheta}\frac{\partial Y_{\nu'}}{\partial\vartheta}\bigg|_0^{\vartheta_0} d\varphi + \nu(\nu+1)\int_0^{\vartheta_0}\int_0^{2\pi} Y_\nu Y_{\nu'} s_\vartheta s_\varphi \, d\vartheta\, d\varphi
\end{aligned} \tag{13.59}$$

Because of the considered boundary conditions the integrand of the first integral in (13.59) vanishes at $\vartheta = \vartheta_0$. The integrand vanishes at $\vartheta = 0$ because of the behavior of the non-periodic Lamé functions at $\vartheta = \vartheta_0$, if ν and ν' belong to eigenfunctions of the same function type. If ν and ν' belong to eigenfunctions of different function types the integral vanishes. The last integral in (13.59) is identical to the orthogonality relation of the Lamé products (13.27); hence, we derive

$$\int_0^{\vartheta_0}\int_0^{2\pi} \vec{m}_\nu\cdot\vec{m}_{\nu'} s_\vartheta s_\varphi \, d\vartheta\, d\varphi = \int_0^{\vartheta_0}\int_0^{2\pi} \vec{n}_\nu\cdot\vec{n}_{\nu'} s_\vartheta s_\varphi \, d\vartheta\, d\varphi = \nu(\nu+1)\delta_{\nu\nu'} \tag{13.60}$$

In proof of the last equation in (13.57) we apply integration by parts and consider the boundary conditions at ϑ_0 (different for Y_ν and Y_η), the behavior of the non-periodic Lamé functions at $\vartheta = 0$, and the periodicity properties of the Lamé products:

$$\int_0^{\vartheta_0}\int_0^{2\pi} \vec{m}_\nu \cdot \vec{n}_\eta s_\vartheta s_\varphi \, d\vartheta \, d\varphi = \int_0^{\vartheta_0}\int_0^{2\pi}\left[\frac{\partial Y_\nu}{\partial \vartheta}\frac{\partial Y_\eta}{\partial \varphi} - \frac{\partial Y_\nu}{\partial \varphi}\frac{\partial Y_\eta}{\partial \vartheta}\right] d\vartheta \, d\varphi$$
$$= \underbrace{\int_0^{2\pi} Y_\nu \frac{\partial Y_\eta}{\partial \varphi}\bigg|_0^{\vartheta_0} d\varphi}_{=0} - \int_0^{\vartheta_0} \underbrace{Y_\nu \frac{\partial Y_\eta}{\partial \vartheta}\bigg|_0^{2\pi}}_{=0} d\vartheta \tag{13.61}$$
$$= 0$$

The orthogonality relations for the free-space case (13.58) can be proved in a similar way by taking into consideration the geometrical boundary conditions at $\vartheta = \pi$ as well as at $\vartheta = 0$.

The corresponding dyadic completeness relations for the transverse vector functions can be stated by

$$(\hat{\vartheta}\hat{\vartheta} + \hat{\varphi}\hat{\varphi})\frac{\delta(\vartheta - \vartheta')\delta(\varphi - \varphi')}{s_\vartheta s_\varphi} = \sum_\nu \frac{\vec{n}_\nu(\vartheta, \varphi)\vec{n}_\nu(\vartheta', \varphi')}{\nu(\nu+1)} + \sum_\eta \frac{\vec{m}_\eta(\vartheta, \varphi)\vec{m}_\eta(\vartheta', \varphi')}{\eta(\eta+1)} \tag{13.62}$$

where ν and η are eigenvalues belonging to different boundary conditions at $\vartheta = \vartheta_0$ ($\nu = \sigma$ (Dirichlet) and $\eta = \tau$ (Neumann) or vice versa). In the free-space case we have

$$(\hat{\vartheta}\hat{\vartheta} + \hat{\varphi}\hat{\varphi})\frac{\delta(\vartheta - \vartheta')\delta(\varphi - \varphi')}{s_\vartheta s_\varphi} = \sum_{n,m} \frac{\vec{n}_{n,m}(\vartheta, \varphi)\vec{n}_{n,m}(\vartheta', \varphi')}{n(n+1)} + \sum_{n,m} \frac{\vec{m}_{n,m}(\vartheta, \varphi)\vec{m}_{n,m}(\vartheta', \varphi')}{n(n+1)} \tag{13.63}$$

13.4.2.2 Orthogonality of the Vector Spherical-Multipole Functions

The vector spherical-multipole functions, which are built up by means of spherical Bessel functions and which are related to the scalar spherical-multipole functions $\Psi^I_{\nu,\kappa}(\vec{r}) = j_\nu(\kappa r)Y_\nu(\vartheta, \varphi)$ by

$$\vec{L}^I_{\nu,\kappa}(\vec{r}) = \frac{1}{\kappa}\vec{\nabla}\Psi^I_{\nu,\kappa}(\vec{r}) = \frac{dj_\nu(\kappa r)}{d\kappa r} Y_\nu(\vartheta, \varphi)\hat{r} + \frac{j_\nu(\kappa r)}{\kappa r}\vec{n}_\nu(\vartheta, \varphi)$$
$$\vec{M}^I_{\eta,\kappa}(\vec{r}) = (\vec{r} \times \vec{\nabla})\Psi^I_{\eta,\kappa}(\vec{r}) = j_\eta(\kappa r)\vec{m}_\eta(\vartheta, \varphi) \tag{13.64}$$
$$\vec{N}^I_{\nu,\kappa}(\vec{r}) = \left[\frac{1}{\kappa}\vec{\nabla} \times (\vec{r} \times \vec{\nabla})\right]\Psi^I_{\nu,\kappa}(\vec{r}) = -\frac{j_\nu(\kappa r)}{\kappa r}\nu(\nu+1)Y_\nu(\vartheta, \varphi)\hat{r} - \frac{1}{\kappa r}\frac{d}{dr}[r j_\nu(\kappa r)]\vec{n}_\nu(\vartheta, \varphi)$$

satisfy the following orthogonality relations:

$$\iiint_V \vec{L}^I_{\nu,\kappa}(\vec{r}) \cdot \vec{L}^I_{\nu',\kappa'}(\vec{r}) dv = \frac{\pi}{2} \frac{\delta(\kappa - \kappa')}{\kappa^2} \delta_{\nu\nu'} \tag{13.65a}$$

$$\iiint_V \vec{M}^I_{\eta,\kappa}(\vec{r}) \cdot \vec{M}^I_{\eta',\kappa'}(\vec{r}) dv = \frac{\pi}{2} \frac{\delta(\kappa - \kappa')}{\kappa^2} \eta(\eta + 1) \delta_{\eta\eta'} \tag{13.65b}$$

$$\iiint_V \vec{N}^I_{\nu,\kappa}(\vec{r}) \cdot \vec{N}^I_{\nu',\kappa'}(\vec{r}) dv = \frac{\pi}{2} \frac{\delta(\kappa - \kappa')}{\kappa^2} \nu(\nu + 1) \delta_{\nu\nu'} \tag{13.65c}$$

$$\iiint_V \vec{L}^I_{\nu,\kappa}(\vec{r}) \cdot \vec{M}^I_{\eta,\kappa'}(\vec{r}) dv = 0 \tag{13.65d}$$

$$\iiint_V \vec{L}^I_{\nu,\kappa}(\vec{r}) \cdot \vec{N}^I_{\nu',\kappa'}(\vec{r}) dv = 0 \tag{13.65e}$$

$$\iiint_V \vec{N}^I_{\nu,\kappa}(\vec{r}) \cdot \vec{M}^I_{\eta,\kappa'}(\vec{r}) dv = 0 \tag{13.65f}$$

provided that the volume integrals are taken about the domain outside the semi-infinite elliptic cone and

- ν, ν' are eigenvalues σ belonging to the Dirichlet condition (13.11) and η,η' are eigenvalues τ belonging to the Neumann condition (13.12), or
- ν, ν' are eigenvalues τ belonging to the Neumann condition (13.12) and η,η' are eigenvalues σ belonging to the Dirichlet condition (13.11).

By analogy in the free-space case the relations (13.65) are valid for the functions $\vec{L}^I_{n,m,\kappa}(\vec{r})$, $\vec{M}^I_{n,m,\kappa}(\vec{r})$ and $\vec{N}^I_{n,m,\kappa}(\vec{r})$, if the eigenvalues ν and η both are integral eigenvalues n and the volume of integration is the complete free space.

To begin with the proof of (13.65a) we have

$$\begin{aligned} \iiint_V \vec{L}^I_{\nu,\kappa}(\vec{r}) \cdot \vec{L}^I_{\nu',\kappa'}(\vec{r}) dv &= \frac{1}{\kappa\kappa'} \int_0^\infty \frac{d}{dr}[j_\nu(\kappa r)] \frac{d}{dr}[j_{\nu'}(\kappa' r)] r^2 \, dr \int_0^{\vartheta_0} \int_0^{2\pi} Y_\nu Y_{\nu'} s_\vartheta s_\varphi d\vartheta d\varphi \\ &\quad + \int_0^\infty \frac{j_\nu(\kappa r)}{\kappa r} \frac{j_{\nu'}(\kappa' r)}{\kappa' r} r^2 \, dr \int_0^{\vartheta_0} \int_0^{2\pi} \vec{n}_\nu \cdot \vec{n}_{\nu'} s_\vartheta s_\varphi \, d\vartheta \, d\varphi \\ &= \int_0^\infty \left[\frac{d}{d(\kappa r)} [j_\nu(\kappa r)] \frac{d}{d(\kappa' r)} [j_\nu(\kappa' r)] r^2 \right. \\ &\quad \left. + \nu(\nu+1) \frac{j_\nu(\kappa r) j_\nu(\kappa' r)}{\kappa\kappa'} \right] dr \, \delta_{\nu\nu'} \end{aligned} \tag{13.66}$$

where we have utilized the orthogonality relations (13.27) and (13.57). With the recurrence relations of the spherical Bessel functions

$$j_\nu(x) = \frac{x}{2\nu + 1}[j_{\nu-1}(x) + j_{\nu+1}(x)] \tag{13.67a}$$

$$\frac{d}{dx} j_\nu(x) = \frac{1}{2\nu + 1}[\nu j_{\nu-1}(x) - (\nu + 1) j_{\nu+1}(x)] \tag{13.67b}$$

and with the orthogonality relation of the spherical Bessel functions (13.35) we obtain from (13.66):

$$\iiint_V \vec{L}^I_{\nu,\kappa}(\vec{r})\cdot\vec{L}^I_{\nu',\kappa'}(\vec{r})dv = \frac{1}{2\nu+1}\left\{\nu\int_0^\infty j_{\nu-1}(\kappa r)j_{\nu-1}(\kappa' r)r^2\,dr\right.$$

$$\left.+(\nu+1)\int_0^\infty j_{\nu+1}(\kappa r)j_{\nu+1}(\kappa' r)r^2\,dr\right\}\delta_{\nu\nu'} \quad (13.68)$$

$$= \frac{\pi}{2}\frac{\delta(\kappa-\kappa')}{\kappa^2}\delta_{\nu\nu'}$$

The proof of (13.65) is simply a combination of the orthogonality relations of the $\vec{m}_\eta$-functions (13.57) and of the spherical Bessel functions (13.35).

By analogy, (13.65c) can be shown using the orthogonality relations of the Lamé products Y_ν (13.27) and of the transverse vector functions $\vec{n}_\nu$ (13.57) as well as the orthogonality relation (13.35) and the recurrence relations (13.67a, b) of the spherical Bessel functions.

The evidence of (13.65d) is shown by

$$\iiint_V \vec{L}^I_{\nu,\kappa}(\vec{r})\cdot\vec{M}^I_{\eta,\kappa'}(\vec{r})\,dv = \frac{1}{\kappa}\int_0^\infty j_\nu(\kappa r)j_\eta(\kappa' r)r^2\int_0^{\vartheta_0}\int_0^{2\pi}\vec{\nabla}Y_\nu\cdot\vec{m}_\eta s_\vartheta s_\varphi\,d\vartheta\,d\varphi\,dr$$

$$= \frac{1}{\kappa}\int_0^\infty j_\nu(\kappa r)j_\eta(\kappa' r)r\,dr\underbrace{\int_0^{\vartheta_0}\int_0^{2\pi}\vec{n}_\nu\cdot\vec{m}_\eta s_\vartheta s_\varphi\,d\vartheta\,d\varphi}_{=0\ [\text{eqn. (13.57)}]} \quad (13.69)$$

$$= 0$$

and (13.65e) can be proved as follows:

$$\iiint_V \vec{L}^I_{\nu,\kappa}(\vec{r})\cdot\vec{N}^I_{\nu',\kappa'}(\vec{r})dv = -\nu(\nu+1)\frac{1}{\kappa\kappa'}\int_0^\infty \frac{d}{dr}[j_\nu(\kappa r)]\frac{j_{\nu'}(\kappa' r)}{r}r^2\,dr$$

$$\times\int_0^{\vartheta_0}\int_0^{2\pi}Y_\nu Y_{\nu'}s_\vartheta s_\varphi\,d\vartheta\,d\varphi$$

$$-\frac{1}{\kappa\kappa'}\int_0^\infty j_\nu(\kappa r)\frac{1}{r}\frac{d}{dr}[rj_{\nu'}(\kappa' r)]r$$

$$\times\int_0^{\vartheta_0}\int_0^{2\pi}\vec{n}_\nu\cdot\vec{n}_{\nu'}s_\vartheta s_\varphi\,d\vartheta\,d\varphi\,dr \quad (13.70)$$

$$= -\frac{\nu(\nu+1)}{\kappa\kappa'}\delta_{\nu\nu'}\left\{\int_0^\infty rj_\nu(\kappa' r)\frac{d}{dr}[j_\nu(\kappa r)]dr\right.$$

$$\left.+\int_0^\infty j_\nu(\kappa r)\frac{d}{dr}[rj_\nu(\kappa' r)]dr\right\}$$

Integrating one of the integrals in (13.70) by parts yields for any finite value of ν

$$\iiint_V \vec{L}^I_{\nu,\kappa}(\vec{r}) \cdot \vec{N}^I_{\nu',\kappa'}(\vec{r})dv = rj_\nu(\kappa' r)j_\nu(\kappa r)\Big|_{r=0}^{r\to\infty} = 0 \tag{13.71}$$

Finally, (13.65f) can be shown easily with the aid of the orthogonality relation (13.57). Following the procedure described in Section 13.3 [(13.36)–(13.40)], the corresponding completeness relation is found as

$$\begin{aligned}\tilde{I}\delta(\vec{r}-\vec{r}\,') = \frac{2}{\pi}\int_0^\infty \kappa^2 \Bigg[\sum_\nu \frac{\vec{N}^I_{\nu,\kappa}(\vec{r})\vec{N}^I_{\nu,\kappa}(\vec{r}\,')}{\nu(\nu+1)} + \sum_\eta \frac{\vec{M}^I_{\eta,\kappa}(\vec{r})\vec{M}^I_{\eta,\kappa}(\vec{r}\,')}{\eta(\eta+1)} \\ + \sum_\nu \vec{L}^I_{\nu,\kappa}(\vec{r})\vec{L}^I_{\nu,\kappa}(\vec{r}\,') \Bigg] d\kappa\end{aligned} \tag{13.72}$$

where $\tilde{I}$ is the unity dyad and where ν and η are eigenvalues belonging to different boundary conditions at $\vartheta = \vartheta_0$ [$\nu = \sigma$ (Dirichlet) and $\eta = \tau$ (Neumann) or vice versa]. In the free-space case we have:

$$\begin{aligned}\tilde{I}\delta(\vec{r}-\vec{r}\,') = \frac{2}{\pi}\int_0^\infty \kappa^2 \Bigg[\sum_{n,m} \frac{\vec{N}^I_{n,m,\kappa}(\vec{r})\vec{N}^I_{n,m,\kappa}(\vec{r}\,')}{n(n+1)} + \sum_{n,m} \frac{\vec{M}^I_{n,m,\kappa}(\vec{r})\vec{M}^I_{n,m,\kappa}(\vec{r}\,')}{n(n+1)} \\ + \sum_{n,m} \vec{L}^I_{n,m,\kappa}(\vec{r})\vec{L}^I_{n,m,\kappa}(\vec{r}\,') \Bigg] d\kappa\end{aligned} \tag{13.73}$$

13.4.3 Dyadic Green's Functions in Sphero-Conal Coordinates

We are now interested in the solutions of the inhomogeneous differential equation (13.49) for the electric field in vacuum ($Z = Z_0 = \sqrt{\mu_0/\varepsilon_0}$, $\kappa = \kappa_0 = \omega\sqrt{\varepsilon_0\mu_0} = 2\pi/\Lambda$; Λ is the wavelength)

- within the domain bounded by a PEC semi-infinite elliptic cone

$$\vec{\nabla} \times \vec{\nabla} \times \vec{E}(\vec{r}) - \kappa_0^2 \vec{E}(\vec{r}) = -j\omega\mu_0 \underline{\vec{J}}_e^{imp}(\vec{r}) \tag{13.74}$$

- within the free space.

$$\hat{\vartheta} \times \vec{E}(\vec{r})\Big|_{\vartheta=\vartheta_0} = 0 \tag{13.75}$$

In both cases, besides the regularity condition, which says that the field energy must be finite in any finite domain, the electric field has to satisfy the radiation condition [52]

$$\lim_{r\to\infty} r\left\{\vec{\nabla} \times \vec{E}(\vec{r}) + j\kappa_0 \hat{r} \times \vec{E}(\vec{r})\right\} = 0 \tag{13.76}$$

provided the field sources are localized within a finite domain around the origin $r = 0$.

The solution of this vector problem is found via the corresponding dyadic Green's function $\tilde{\Gamma}(\vec{r},\vec{r}\,')$, which is defined by the dyadic differential equation

$$\vec{\nabla}\times\vec{\nabla}\times\tilde{\Gamma}(\vec{r},\vec{r}\,')-\kappa_0^2\tilde{\Gamma}(\vec{r},\vec{r}\,') = -\tilde{I}\delta(\vec{r}-\vec{r}\,') \tag{13.77}$$

Employing Green's second theorem (in vector form)

$$\begin{aligned}\iiint_V \left[\vec{A}\cdot\left(\vec{\nabla}\times\vec{\nabla}\times\vec{B}\right)-\vec{B}\cdot\left(\vec{\nabla}\times\vec{\nabla}\times\vec{A}\right)\right]dv \\ = \oiint_{S(V)} \left[\left(\vec{\nabla}\times\vec{A}\right)\cdot\left(\hat{n}\times\vec{B}\right)-\left(\vec{\nabla}\times\vec{B}\right)\cdot\left(\hat{n}\times\vec{A}\right)\right]ds\end{aligned} \tag{13.78}$$

it can be shown that any solution of (13.74)/(13.76) can be represented by

$$\vec{E}(\vec{r}) = \mathrm{j}\omega\mu_0 \iiint_V \tilde{\Gamma}(\vec{r},\vec{r}\,')\cdot\underline{\vec{J}}_e^{imp}(\vec{r}\,')\,dv' \tag{13.79}$$

under the assumptions that the dyadic Green's function satisfies the "radiation" condition

$$\lim_{r\to\infty} \vec{r}\times\left\{\vec{\nabla}\times\tilde{\Gamma}(\vec{r},\vec{r}\,')+\mathrm{j}\kappa_0\hat{r}\times\tilde{\Gamma}(\vec{r},\vec{r}\,')\right\}=0 \tag{13.80}$$

and, in case of the PEC semi-infinite elliptic cone, that the dyadic Green's function possesses the same boundary condition (13.75) as the electric field

$$\hat{\vartheta}\times\tilde{\Gamma}(\vec{r},\vec{r}\,')\Big|_{\vartheta=\vartheta_0}=0 \tag{13.81}$$

Moreover, in the cases under consideration the dyadic Green's functions satisfy the symmetry relation

$$\tilde{\Gamma}(\vec{r},\vec{r}\,')=\tilde{\Gamma}^T(\vec{r}\,',\vec{r}) \tag{13.82}$$

in which $\tilde{\Gamma}^T$ is the transpose of the Green's dyadic $\tilde{\Gamma}$.

For the derivation we describe the dyadic Green's function of the PEC semi-infinite elliptic cone by a bilinear, symmetric approach which consists of vector spherical-multipole functions and which is complete for $\vec{r}=\vec{r}\,'$ as well:

$$\begin{aligned}\tilde{\Gamma}(\vec{r},\vec{r}\,') = \frac{2}{\pi}\int_0^\infty \kappa^2\Bigg[\sum_\sigma A_{\sigma,\kappa}\frac{\vec{N}^I_{\sigma,\kappa}(\vec{r})\vec{N}^I_{\sigma,\kappa}(\vec{r}\,')}{\sigma(\sigma+1)}+\sum_\tau B_{\tau,\kappa}\frac{\vec{M}^I_{\tau,\kappa}(\vec{r})\vec{M}^I_{\tau,\kappa}(\vec{r}\,')}{\tau(\tau+1)} \\ +\sum_\sigma C_{\sigma,\kappa}\vec{L}^I_{\sigma,\kappa}(\vec{r})\vec{L}^I_{\sigma,\kappa}(\vec{r}\,')\Bigg]d\kappa\end{aligned} \tag{13.83}$$

The upper index ^{I} stands for the use of spherical Bessel functions $j_\nu(\kappa r)$ to guarantee regularity everywhere. The choice of the eigenvalues σ [Dirichlet condition, see (13.11)] and τ [Neumann condition, see (13.12)] automatically satisfies (13.81), while the continuous eigenvalue spectrum κ extends over the complete domain from 0 to ∞. For the determination of the coefficients $A_{\sigma,\kappa}$, $B_{\tau,\kappa}$ and $C_{\sigma,\kappa}$ we insert (13.83) into (13.77) and replace the right part in (13.77) by the completeness relation (13.72), in which we have set $\nu=\sigma$ and $\eta=\tau$. The operation $\vec{\nabla}\times\vec{\nabla}\times$ reproduces the solenoidal vector spherical multipole functions $\vec{N}^I_{\sigma,\kappa}$ and $\vec{M}^I_{\tau,\kappa}$ except for the factor κ^2, while the $\vec{L}^I_{\sigma,\kappa}$ functions vanish. Now comparing both sides of the resulting orthogonal expansions, we have

$$A_{\sigma,\kappa} = B_{\tau,\kappa} = -\frac{1}{(\kappa^2 - \kappa_0^2)}$$
$$C_{\sigma,\kappa} = \frac{1}{\kappa_0^2} \tag{13.84}$$

and obtain the *eigenfunction expansion* of the dyadic Green's function for the PEC semi-infinite elliptic cone

$$\tilde{\Gamma}(\vec{r},\vec{r}\,') = \frac{2}{\pi}\int_0^\infty \left\{\frac{\kappa^2}{\kappa_0^2}\sum_\sigma \vec{L}^I_{\sigma,\kappa}(\vec{r})\vec{L}^I_{\sigma,\kappa}(\vec{r}\,') - \frac{\kappa^2}{\kappa^2-\kappa_0^2}\left[\sum_\sigma \frac{\vec{N}^I_{\sigma,\kappa}(\vec{r})\vec{N}^I_{\sigma,\kappa}(\vec{r}\,')}{\sigma(\sigma+1)} + \sum_\tau \frac{\vec{M}^I_{\tau,\kappa}(\vec{r})\vec{M}^I_{\tau,\kappa}(\vec{r}\,')}{\tau(\tau+1)}\right]\right\} d\kappa \tag{13.85}$$

We will now evaluate the integrals in (13.85) in the complex κ-plane by utilizing appropriate distributional and functional analysis. We write (13.85) as

$$\begin{aligned}
\tilde{\Gamma}(\vec{r},\vec{r}\,') = &-\frac{2}{\pi}\sum_\tau \int_0^\infty \frac{\kappa^2}{\kappa^2-\kappa_0^2} j_\tau(\kappa r) j_\tau(\kappa r')\,d\kappa \frac{\vec{m}_\tau(\vartheta,\varphi)\vec{m}_\tau(\vartheta',\varphi')}{\tau(\tau+1)} \\
&-\frac{2}{\pi}\sum_\sigma \Bigg\{ \int_0^\infty \Bigg\{\sigma(\sigma+1)\frac{1}{\kappa^2-\kappa_0^2}\frac{j_\sigma(\kappa r)}{r}\frac{j_\sigma(\kappa r')}{r'} \\
&-\frac{1}{\kappa_0^2}\frac{d[j_\sigma(\kappa r)]}{dr}\frac{d[j_\sigma(\kappa r')]}{dr'}\Bigg\} d\kappa\, Y_\sigma(\vartheta,\varphi)\hat{r} Y_\sigma(\vartheta',\varphi')\hat{r} \\
&+\int_0^\infty \Bigg\{\frac{1}{\kappa^2-\kappa_0^2}\frac{1}{r}\frac{d[r j_\sigma(\kappa r)]}{dr}\frac{1}{r'}\frac{d[r' j_\sigma(\kappa r')]}{dr'} \\
&-\frac{1}{\kappa_0^2}\sigma(\sigma+1)\frac{j_\sigma(\kappa r)}{r}\frac{j_\sigma(\kappa r')}{r'}\Bigg\} d\kappa \frac{\vec{n}_\sigma(\vartheta,\varphi)\vec{n}_\sigma(\vartheta',\varphi')}{\sigma(\sigma+1)} \\
&+\int_0^\infty \Bigg\{\frac{1}{\kappa^2-\kappa_0^2}\frac{j_\sigma(\kappa r)}{r}\frac{1}{r'}\frac{d[r' j_\sigma(\kappa r')]}{dr'} \\
&-\frac{1}{\kappa_0^2}\frac{d j_\sigma(\kappa r)}{dr}\frac{j_\sigma(\kappa r')}{r'}\Bigg\} d\kappa\, Y_\sigma(\vartheta,\varphi)\hat{r}\vec{n}_\sigma(\vartheta',\varphi') \\
&+\int_0^\infty \Bigg\{\frac{1}{\kappa^2-\kappa_0^2}\frac{1}{r}\frac{d[r j_\sigma(\kappa r)]}{dr}\frac{j_\sigma(\kappa r')}{r'} \\
&-\frac{1}{\kappa_0^2}\frac{j_\sigma(\kappa r)}{r}\frac{d[j_\sigma(\kappa r')]}{dr'}\Bigg\} d\kappa\, \vec{n}_\sigma(\vartheta,\varphi) Y_\sigma(\vartheta',\varphi')\hat{r}\Bigg\}
\end{aligned} \tag{13.86}$$

Special attention must be paid to the second integral in (13.86). Using the addition theorems (13.67a, b) and the orthogonality relation of the spherical Bessel functions (13.35), in which r and κ are exchanged, it is seen that

$$\begin{aligned}
&\int_0^\infty \left\{\sigma(\sigma+1)\frac{1}{\kappa^2-\kappa_0^2}\frac{j_\sigma(\kappa r)}{r}\frac{j_\sigma(\kappa r')}{r'} - \frac{1}{\kappa_0^2}\frac{d[j_\sigma(\kappa r)]}{dr}\frac{d[j_\sigma(\kappa r')]}{dr'}\right\}d\kappa \\
&= \int_0^\infty \left\{\sigma(\sigma+1)\left[\frac{1}{\kappa^2-\kappa_0^2}-\frac{1}{\kappa_0^2}\right]\frac{j_\sigma(\kappa r)}{r}\frac{j_\sigma(\kappa r')}{r'}\right. \\
&\quad \left. -\frac{\kappa^2}{\kappa_0}\frac{1}{2\sigma+1}[\sigma j_{\sigma-1}(\kappa r)j_{\sigma-1}(\kappa r')+(\sigma+1)j_{\sigma+1}(\kappa r)j_{\sigma+1}(\kappa r')]\right\}d\kappa \\
&= \int_0^\infty \sigma(\sigma+1)\left[\frac{1}{\kappa^2-\kappa_0^2}-\frac{1}{\kappa_0^2}\right]\frac{j_\sigma(\kappa r)}{r}\frac{j_\sigma(\kappa r')}{r'}d\kappa - \frac{\pi}{2}\frac{1}{\kappa_0^2 r^2}\delta(r-r')
\end{aligned} \tag{13.87}$$

The other integrals in (13.86) do not yield distributional parts. By using the relations

$$j_\nu(\kappa r) = \tfrac{1}{2}[h_\nu^{(1)}(\kappa r)+h_\nu^{(2)}(\kappa r)] \tag{13.88}$$

$$\begin{aligned}
h_\nu^{(1)}(-\kappa r) &= -e^{\mathrm{j}(\nu+1/2)\pi}h_\nu^{(2)}(\kappa r) \\
j_\nu(-\kappa r) &= e^{-\mathrm{j}(\nu+1/2)\pi}j_\nu(\kappa r)
\end{aligned} \tag{13.89}$$

the remaining integrals in (13.86) and (13.87) can be modified as

$$\begin{aligned}
\tilde{\Gamma}(\vec{r},\vec{r}') &= -\frac{1}{\pi}\sum_\tau \int_{-\infty}^\infty \frac{\kappa^2}{\kappa^2-\kappa_0^2}j_\tau(\kappa r_<)h_\tau^{(2)}(\kappa r_>)d\kappa \frac{\vec{m}_\tau(\vartheta,\varphi)\vec{m}_\tau(\vartheta',\varphi')}{\tau(\tau+1)} \\
&\quad -\frac{1}{\pi}\sum_\sigma \left\{\left[\int_{-\infty}^\infty \sigma(\sigma+1)\left[\frac{1}{\kappa^2-\kappa_0^2}-\frac{1}{\kappa_0^2}\right]\frac{j_\sigma(\kappa r_<)}{r_<}\frac{h_\sigma^{(2)}(\kappa r_>)}{r_>}d\kappa\right.\right. \\
&\quad \times Y_\sigma(\vartheta,\varphi)\hat{r}Y_\sigma(\vartheta',\varphi')\hat{r} - \frac{\pi}{\kappa_0^2 r^2}\delta(r-r')Y_\sigma(\vartheta,\varphi)\hat{r}Y_\sigma(\vartheta',\varphi')\hat{r} \\
&\quad + \int_{-\infty}^\infty \left\{\frac{1}{\kappa^2-\kappa_0^2}\frac{1}{r_<}\frac{d[r_< j_\sigma(\kappa r_<)]}{dr_<}\frac{1}{r_>}\frac{d[r_> h_\sigma^{(2)}(\kappa r_>)]}{dr_>}\right. \\
&\quad \left. -\frac{1}{\kappa_0^2}\sigma(\sigma+1)\frac{j_\sigma(\kappa r_<)}{r_<}\frac{h_\sigma^{(2)}(\kappa r_>)}{r_>}\right\}d\kappa \frac{\vec{n}_\sigma(\vartheta,\varphi)\vec{n}_\sigma(\vartheta',\varphi')}{\sigma(\sigma+1)} \\
&\quad + \int_{-\infty}^\infty \left\{\frac{1}{\kappa^2-\kappa_0^2}\frac{j_\sigma(\kappa r_<)}{r_<}\frac{1}{r_>}\frac{d[r_> h_\sigma^{(2)}(\kappa r_>)]}{dr_>}\right. \\
&\quad \left. -\frac{1}{\kappa_0^2}\frac{dj_\sigma(\kappa r_<)}{dr_<}\frac{h_\sigma^{(2)}(\kappa r_>)}{r_>}\right\}d\kappa\, Y_\sigma(\vartheta,\varphi)\hat{r}\vec{n}_\sigma(\vartheta',\varphi') \\
&\quad + \int_{-\infty}^\infty \left\{\frac{1}{\kappa^2-\kappa_0^2}\frac{1}{r_<}\frac{d[r_< j_\sigma(\kappa r_<)]}{dr_<}\frac{h_\sigma^{(2)}(\kappa r_>)}{r_>}\right. \\
&\quad \left.\left. -\frac{1}{\kappa_0^2}\frac{j_\sigma(\kappa r_<)}{r_<}\frac{dh_\sigma^{(2)}(\kappa r_>)}{dr_>}\right\}d\kappa\, \vec{n}_\sigma(\vartheta,\varphi)Y_\sigma(\vartheta',\varphi')\hat{r}\right\}
\end{aligned} \tag{13.90}$$

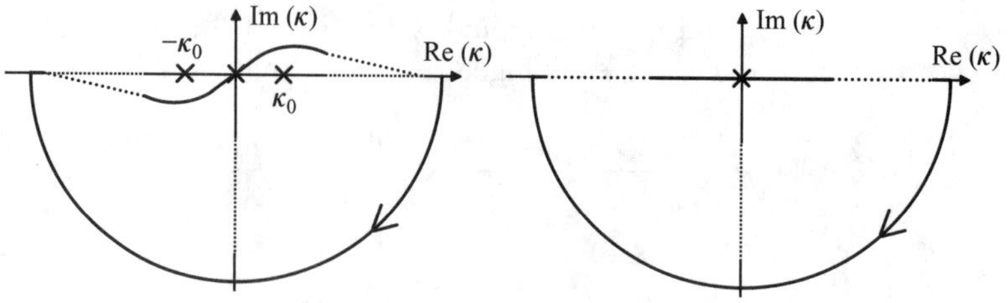

Figure 13.4 Paths of integration corresponding to the radiation condition (13.80) for the solenoidal ($\vec{N}$ and $\vec{M}$) terms (a) and for the divergence ($\vec{L}$) terms (b).

where we have used the abbreviation (13.45) in the indicated way, to be sure that the integrands vanish at infinity in the lower part of the complex κ-plane. Hence the integrals in (13.90) may be solved with the aid of residuum calculus. As shown in Appendix 13.2, the paths of integration follow explicitly from the "radiation" condition (13.80) of the Green's dyadic. The terms in the Green's dyadic derived from the parts with $\vec{N}$ and $\vec{M}$ functions, i.e., the parts proportional to the factor $1/(\kappa^2 - \kappa_0^2)$ in (13.90), have to be integrated along the contour represented in Fig. 13.4a, while the terms derived from parts with $\vec{L}$-functions, i.e., the parts proportional to the factor $1/\kappa_0^2$ in (13.90), have to be integrated along the contour shown in Fig. 13.4b. The crosses in the Figs. 13.4a, b are possible single poles of the integrands, which may be seen for the poles at $\kappa = \pm\kappa_0$ directly from (13.90) and for the poles at $\kappa = 0$ from the asymptotic behavior of the product of spherical Bessel and Hankel functions:

$$r_<, r_>, \nu \text{ fixed and } |\kappa| \to 0: \quad j_\nu(\kappa r_<) h_\nu^{(2)}(\kappa r_>) \approx \frac{\mathrm{j}}{\kappa} \frac{1}{(2\nu+1)} \frac{r_<^\nu}{r_>^{\nu+1}} \tag{13.91}$$

Now we evaluate the integrals in (13.90) by utilizing usual residue calculus and remark that the poles at $\kappa = 0$ are located directly on the path of integration, hence, in these cases one half of the complete residuum has to be considered. However, it is seen that the parts in (13.90) due to the poles at $\kappa = 0$ which arise from the integrals with paths as shown in Fig. 13.4a are completely eliminated by those parts, which arise from the integrals with paths as shown in Fig. 13.4b, and the following result is obtained:

$$\begin{aligned}
\tilde{\Gamma}(\vec{r}, \vec{r}') = {} & \mathrm{j}\kappa_0 \sum_\tau j_\tau(\kappa_0 r_<) h_\tau^{(2)}(\kappa_0 r_>) \frac{\vec{m}_\tau(\vartheta, \varphi) \vec{m}_\tau(\vartheta', \varphi')}{\tau(\tau+1)} \\
& + \mathrm{j}\kappa_0 \sum_\sigma \left\{ \sigma(\sigma+1) \frac{j_\sigma(\kappa_0 r_<)}{\kappa_0 r_<} \frac{h_\sigma^{(2)}(\kappa_0 r_>)}{\kappa_0 r_>} Y_\sigma(\vartheta, \varphi) \hat{r} Y_\sigma(\vartheta', \varphi') \hat{r} \right. \\
& + \frac{1}{\kappa_0 r_<} \frac{d[r_< j_\sigma(\kappa_0 r_<)]}{dr_<} \frac{1}{\kappa_0 r_>} \frac{d[r_> h_\sigma^{(2)}(\kappa_0 r_>)]}{dr_>} \frac{\vec{n}_\sigma(\vartheta, \varphi) \vec{n}_\sigma(\vartheta', \varphi')}{\sigma(\sigma+1)} \\
& + \frac{j_\sigma(\kappa_0 r_<)}{\kappa_0 r_<} \frac{1}{\kappa_0 r_>} \frac{d[r_> h_\sigma^{(2)}(\kappa_0 r_>)]}{dr_>} Y_\sigma(\vartheta, \varphi) \hat{r} \vec{n}_\sigma(\vartheta', \varphi')
\end{aligned} \tag{13.92}$$

$$+\frac{1}{\kappa_0 r_<}\frac{d[r_< j_\sigma(\kappa_0 r_<)]}{dr_<}\frac{h_\sigma^{(2)}(\kappa_0 r_>)}{\kappa_0 r_>}\vec{n}_\sigma(\vartheta,\varphi)Y_\sigma(\vartheta',\varphi')\hat{r}\Bigg\}$$

$$+\frac{1}{\kappa_0^2 r^2}\delta(r-r')\sum_\sigma Y_\sigma(\vartheta,\varphi)\hat{r}Y_\sigma(\vartheta',\varphi')\hat{r}$$

With the completeness relation of the Lamé products (13.32) the *modal expansion* of the dyadic Green's function for the PEC semi-infinite elliptic cone is found as

$$\tilde{\Gamma}(\vec{r},\vec{r}') = \mathrm{j}\kappa_0\left\{\sum_\sigma \frac{\vec{N}^{I(II)}_{\sigma,\kappa_0}(\vec{r})\vec{N}^{II(I)}_{\sigma,\kappa_0}(\vec{r}')}{\sigma(\sigma+1)} + \sum_\tau \frac{\vec{M}^{I(II)}_{\tau,\kappa_0}(\vec{r})\vec{M}^{II(I)}_{\tau,\kappa_0}(\vec{r}')}{\tau(\tau+1)}\right\} + \hat{r}\hat{r}\frac{1}{\kappa_0^2}\delta(\vec{r}-\vec{r}') \qquad r\leq r'(r>r') \tag{13.93}$$

where the upper index I stands for the use of spherical Bessel functions $j_\nu(\kappa_0 r_<)$ and the upper index II for spherical Hankel functions of the second kind $h_\nu^{(2)}(\kappa_0 r_>)$, respectively.

By analogy, the modal expansion for the free-space dyadic Green's function is derived as

$$\tilde{\Gamma}(\vec{r},\vec{r}') = \mathrm{j}\kappa_0\left\{\sum_{n,m} \frac{\vec{N}^{I(II)}_{n,m,\kappa_0}(\vec{r})\vec{N}^{II(I)}_{n,m,\kappa_0}(\vec{r}')}{n(n+1)} + \sum_{n,m} \frac{\vec{M}^{I(II)}_{n,m,\kappa_0}(\vec{r})\vec{M}^{II(I)}_{n,m,\kappa_0}(\vec{r}')}{n(n+1)}\right\} + \hat{r}\hat{r}\frac{1}{\kappa_0^2}\delta(\vec{r}-\vec{r}') \qquad r\leq r'(r>r') \tag{13.94}$$

13.4.4 Plane Electromagnetic Waves in Sphero-Conal Coordinates

We are searching for the multipole amplitudes a_σ and b_τ of the multipole expansion

$$\begin{aligned}\vec{E}(\vec{r}) &= \sum_\sigma a_\sigma \vec{N}^I_{\sigma,\kappa_0}(\vec{r}) + \frac{Z_0}{\mathrm{j}}\sum_\tau b_\tau \vec{M}^I_{\tau,\kappa 0}(\vec{r}) \\ \vec{H}(\vec{r}) &= \frac{\mathrm{j}}{Z_0}\sum_\sigma a_\sigma \vec{M}^I_{\sigma,\kappa_0}(\vec{r}) + \sum_\tau b_\tau \vec{N}^I_{\tau,\kappa_0}(\vec{r})\end{aligned} \tag{13.95}$$

which describes the field caused by a plane electromagnetic wave with amplitude E_0, (electrically) polarized in the direction $\hat{\xi}$ and incident from the direction ϑ_i, φ_i upon a PEC semi-infinite elliptic cone.

To come to the multipole amplitudes a_σ and b_τ we will locate a short electric (Hertzian) dipole at infinity and then multiply the spherical-multipole expansion of the corresponding electromagnetic field by a factor to transform the amplitude and phase dependences of the dipole field to those of a plane wave field. This factor is found by inserting the closed form of the free-space dyadic Green's function [56]:

$$\tilde{\Gamma}(\vec{r},\vec{r}') = -\frac{\kappa_0}{4\pi}\left[\tilde{I} + \frac{1}{\kappa_0^2}\nabla\nabla\right]\frac{e^{-\mathrm{j}\kappa_0|\vec{r}-\vec{r}'|}}{\kappa_0|\vec{r}-\vec{r}'|} \tag{13.96}$$

and the current distribution of a Hertzian dipole located at $\vec{r}_i$ with the electric current moment $\vec{c}_{el} = c_{el}\hat{\xi}$ $(\hat{\xi}\cdot\hat{r}_i = 0)$:

$$\underline{\vec{J}}_e^{imp}(\vec{r}) = \vec{c}_{el}\,\delta(\vec{r} - \vec{r}_i) \tag{13.97}$$

into (13.79). We utilize the filter property of the δ-distribution, let $r_i \to \infty$ and obtain with the usual Taylor expansion of the spherical-wave term:

$$\vec{E}(\vec{r}) = -\frac{\mathrm{j}\kappa_0\,\omega\mu_0}{4\pi}\frac{e^{-\mathrm{j}\kappa_0 r_i}}{\kappa_0 r_i}\left[\tilde{I} + \frac{1}{\kappa_0^2}\nabla\nabla\right]e^{\mathrm{j}\kappa_0\hat{r}_i\cdot\vec{r}}\cdot\vec{c}_{el} \tag{13.98}$$

Now performing the gradients in (13.98) and with

$$\kappa_0\,\omega\mu_0\,c_{el} = E_0 \tag{13.99}$$

we have

$$\vec{E}(\vec{r}) = -\frac{\mathrm{j}}{4\pi}\frac{e^{-\mathrm{j}\kappa_0 r_i}}{\kappa_0 r_i}E_0\,\hat{\xi}e^{\mathrm{j}\kappa_0\hat{r}_i\cdot\vec{r}} \tag{13.100}$$

Obviously the desired factor, which transforms this dipole field into the field of a plane electromagnetic wave, polarized in the direction $\hat{\xi}$ and incident from ϑ_i, φ_i, is given by

$$\Upsilon = \mathrm{j}4\pi\kappa_0 r_i e^{+\mathrm{j}\kappa_0 r_i} \tag{13.101}$$

Now we insert the modal expansion of the dyadic Green's function for a PEC semi-infinite elliptic cone (13.93) and the dipole's current distribution (13.97) into (13.79), let $r_i \to \infty$, utilize the asymptotic representation of the spherical Hankel functions of the second kind for large values of its argument and multiply the result by the factor Υ in (13.101). With the abbreviation (13.99) the result can be written as a spherical-multipole expansion (13.95) with the multipole amplitudes:

$$\begin{aligned} a_\sigma &= 4\pi E_0\frac{e^{\mathrm{j}(\sigma+1)\pi/2}}{\sigma(\sigma+1)}[\vec{n}_\sigma(\vartheta_i, \varphi_i)\cdot\hat{\xi}] \\ b_\tau &= 4\pi\frac{E_0}{Z_0}\frac{e^{\mathrm{j}(\tau+1)\pi/2}}{\tau(\tau+1)}[\vec{m}_\tau(\vartheta_i, \varphi_i)\cdot\hat{\xi}] \end{aligned} \tag{13.102}$$

The resulting field is the *diffracted* field caused by a $\hat{\xi}$-polarized plane electromagnetic wave incident from the direction $\hat{r}_i$ upon a PEC semi-infinite elliptic cone.

While starting from the free-space dyadic Green's function (13.94) the plane electromagnetic wave expansion for the free space is found in a similar way as

$$\begin{aligned} \vec{E}(\vec{r}) &= \sum_{n,m} a_{n,m}\vec{N}^I_{n,m,\kappa_0}(\vec{r}) + \frac{Z_0}{\mathrm{j}}\sum_{n,m} b_{n,m}\vec{M}^I_{n,m,\kappa_0}(\vec{r}) \\ \vec{H}(\vec{r}) &= \frac{\mathrm{j}}{Z_0}\sum_{n,m} a_{n,m}\vec{M}^I_{n,m,\kappa_0}(\vec{r}) + \sum_{n,m} b_{n,m}\vec{N}^I_{n,m,\kappa_0}(\vec{r}) \end{aligned} \tag{13.103}$$

with the multipole amplitudes

$$a_{n,m} = 4\pi E_0 \frac{\mathrm{j}^{n+1}}{n(n+1)} [\vec{n}_{n,m}(\vartheta_i, \varphi_i) \cdot \hat{\xi}]$$
$$b_{n,m} = 4\pi \frac{E_0}{Z_0} \frac{\mathrm{j}^{n+1}}{n(n+1)} [\vec{m}_{n,m}(\vartheta_i, \varphi_i) \cdot \hat{\xi}] \tag{13.104}$$

For the problems discussed later in this chapter it is useful to evaluate the asymptotic behavior of the free-space plane electromagnetic wave (13.103)/(13.104) in the far-field. For this purpose we use the asymptotic expansions of the spherical Bessel functions

$$r \to \infty; \quad n \ll \kappa_0 r: \quad \begin{aligned} j_n(\kappa_0 r) &= \frac{1}{2}\left[(-\mathrm{j})^{n+1}\frac{e^{+\mathrm{j}\kappa_0 r}}{\kappa_0 r} + (\mathrm{j})^{n+1}\frac{e^{-\mathrm{j}\kappa_0 r}}{\kappa_0 r}\right] \\ \frac{1}{\kappa_0 r}\frac{d}{dr}[r j_n(\kappa_0 r)] &= \frac{1}{2}\left[(-\mathrm{j})^{n}\frac{e^{+\mathrm{j}\kappa_0 r}}{\kappa_0 r} + (\mathrm{j})^{n}\frac{e^{-\mathrm{j}\kappa_0 r}}{\kappa_0 r}\right] \end{aligned} \tag{13.105}$$

which is valid if $n \ll \kappa_0 r$. Since we have an infinite series, n tends to ∞ and the use of the asymptotic representations (13.105) is not allowed in the sense of a rigorous mathematical analysis. However, as will be shown, the numerical evaluation of the multipole expansions in the context of the semi-infinite elliptic cone will be performed by taking into account only a finite number of the first series terms while utilizing suitable series transformation techniques. Hence from this point of view the condition $n \ll \kappa_0 r$ will not be violated. By inserting (13.104) and (13.105) into (13.103) we now obtain the far-field expression for the electric field:

$$\begin{aligned} \vec{E}_\infty(\vec{r}) = -E_0 2\pi\mathrm{j}\Bigg\{ &\frac{e^{+\mathrm{j}\kappa_0 r}}{\kappa_0 r} \sum_{n,m} \left[\frac{\vec{n}_{nm}(\vartheta, \varphi)\vec{n}_{n,m}(\vartheta_i, \varphi_i)}{n(n+1)} + \frac{\vec{m}_{n,m}(\vartheta, \varphi)\vec{m}_{n,m}(\vartheta_i, \varphi_i)}{n(n+1)}\right] \cdot \hat{\xi} \\ &+ \frac{e^{-\mathrm{j}\kappa_0 r}}{\kappa_0 r} \sum_{n,m} (-1)^n \left[\frac{\vec{n}_{n,m}(\vartheta, \varphi)\vec{n}_{n,m}(\vartheta_i, \varphi_i)}{n(n+1)} \right. \\ &\left. - \frac{\vec{m}_{n,m}(\vartheta, \varphi)\vec{m}_{n,m}(\vartheta_i, \varphi_i)}{n(n+1)}\right] \cdot \hat{\xi} \Bigg\} \end{aligned} \tag{13.106}$$

Because of the properties of the transverse vector functions ($\pi - \vartheta, \varphi + \pi$ is the opposite point of ϑ, φ on a sphere)

$$\begin{aligned} \vec{m}_{n,m}(\vartheta, \varphi) &= (-1)^n \vec{m}_{n,m}(\pi - \vartheta, \varphi + \pi) \\ \vec{n}_{n,m}(\vartheta, \varphi) &= (-1)^{n+1} \vec{n}_{n,m}(\pi - \vartheta, \varphi + \pi) \end{aligned} \tag{13.107}$$

(13.106) can be written as

$$\begin{aligned} \vec{E}_\infty(\vec{r}) = -E_0 2\pi\mathrm{j}\Bigg\{ &\frac{e^{+\mathrm{j}\kappa_0 r}}{\kappa_0 r} \sum_{n,m} \left[\frac{\vec{n}_{n,m}(\vartheta, \varphi)\vec{n}_{n,m}(\vartheta_i, \varphi_i)}{n(n+1)} + \frac{\vec{m}_{n,m}(\vartheta, \varphi)\vec{m}_{n,m}(\vartheta_i, \varphi_i)}{n(n+1)}\right] \cdot \hat{\xi} \\ &+ \frac{e^{-\mathrm{j}\kappa_0 r}}{\kappa_0 r} \sum_{n,m} - \left[\frac{\vec{n}_{n,m}(\vartheta, \varphi)\vec{n}_{n,m}(\pi - \vartheta_i, \varphi_i + \pi)}{n(n+1)} \right. \\ &\left. + \frac{\vec{m}_{n,m}(\vartheta, \varphi)\vec{m}_{n,m}(\pi - \vartheta_i, \varphi_i + \pi)}{n(n+1)}\right] \cdot \hat{\xi} \Bigg\} \end{aligned} \tag{13.108}$$

In (13.108) there appear expressions similar to the completeness relation of the transverse vector functions (13.63). However, as mentioned earlier, we can use only a finite number of series terms, and a finite number of series terms does not lead to the exact δ-functional on the left side of (13.63), but to a δ-convergent sequence δ_N, which can be shown by employing a suitable transformation technique to the first N terms of the completeness relation (13.63). With these assumptions we may finally write (13.108) as

$$\vec{E}_\infty(\vec{r}) = -E_0 2\pi \mathrm{j}\hat{\xi}\left\{\frac{\delta_N(\vartheta-\vartheta_i)\,\delta_N(\varphi-\varphi_i)}{s_\vartheta s_\varphi}\frac{e^{+\mathrm{j}\kappa_0 r}}{\kappa_0 r} - \frac{\delta_N[\vartheta-(\pi-\vartheta_i)]\,\delta_N[\varphi-(\varphi_i+\pi)]}{s_\vartheta s_\varphi}\frac{e^{-\mathrm{j}\kappa_0 r}}{\kappa_0 r}\right\} \tag{13.109}$$

Choosing the origin as the point of view, the first term in this far-field expression asymptotically represents the incident part, while the second term is an asymptotically valid expression of the transmitted electric field of the plane electromagnetic wave.

13.5 APPLICATIONS IN ELECTRICAL ENGINEERING

13.5.1 Electromagnetic Scattering by a PEC Semi-Infinite Elliptic Cone

Basically, the solution of the boundary value problem for a PEC semi-infinite elliptic cone, which is described by the sphero-conal coordinate surface $\vartheta = \vartheta_0$ and by the parameter k and which is illuminated by a plane electromagnetic wave, has been deduced in the previous section and is given by the spherical-multipole expansion (13.95) and the corresponding multipole amplitudes (13.102). However, since we are interested in the scattered far-field and (13.95)/(13.102) describes the total field, our first job is to separate the scattered from the total field. For this purpose we insert the asymptotic expansions of the spherical Bessel functions

$$r \to \infty; \quad \nu \ll \kappa_0 r: \quad \begin{aligned} j_\nu(\kappa_0 r) &= \frac{1}{2}\left[e^{-\mathrm{j}\pi(\nu+1)/2}\frac{e^{+\mathrm{j}\kappa_0 r}}{\kappa_0 r} + e^{\mathrm{j}\pi(\nu+1)/2}\frac{e^{-\mathrm{j}\kappa_0 r}}{\kappa_0 r}\right] \\ \frac{1}{\kappa_0 r}\frac{d}{dr}[r j_\nu(\kappa_0 r)] &= \frac{1}{2}\left[e^{-\mathrm{j}\pi\nu/2}\frac{e^{+\mathrm{j}\kappa_0 r}}{\kappa_0 r} + e^{\mathrm{j}\pi\nu/2}\frac{e^{-\mathrm{j}\kappa_0 r}}{\kappa_0 r}\right] \end{aligned}$$

into (13.95) and obtain with (13.102) the *total* far-field of a plane electromagnetic wave with amplitude E_0 and electrical polarization in the direction $\hat{\xi}$, which is incident upon the PEC semi-infinite elliptic cone as

$$\begin{aligned} \vec{E}_\infty(\vec{r}) = -E_0 2\pi \mathrm{j}\Bigg\{&\frac{e^{+\mathrm{j}\kappa_0 r}}{\kappa_0 r}\left[\sum_\sigma \frac{\vec{n}_\sigma(\vartheta,\varphi)\vec{n}_\sigma(\vartheta_i,\varphi_i)}{\sigma(\sigma+1)} + \sum_\tau \frac{\vec{m}_\tau(\vartheta,\varphi)\vec{m}_\tau(\vartheta_i,\varphi_i)}{\tau(\tau+1)}\right]\cdot\hat{\xi} \\ &+ \frac{e^{-\mathrm{j}\kappa_0 r}}{\kappa_0 r}\left[\sum_\sigma e^{\mathrm{j}\sigma\pi}\frac{\vec{n}_\sigma(\vartheta,\varphi)\vec{n}_\sigma(\vartheta_i,\varphi_i)}{\sigma(\sigma+1)} - \sum_\tau e^{\mathrm{j}\tau\pi}\frac{\vec{m}_\tau(\vartheta,\varphi)\vec{m}_\tau(\vartheta_i,\varphi_i)}{\tau(\tau+1)}\right]\cdot\hat{\xi}\Bigg\} \end{aligned} \tag{13.110}$$

With the completeness relation (13.62) for a δ-convergent sequence δ_N, we can write (13.10) as

$$\begin{aligned}\vec{E}_\infty(\vec{r}) = &-E_0 2\pi \mathrm{j}\hat{\xi}\frac{e^{+\mathrm{j}\kappa_0 r}}{\kappa_0 r}\frac{\delta_N(\vartheta-\vartheta_i)\delta_N(\varphi-\varphi_i)}{s_\vartheta s_\varphi} \\ &- E_0 2\pi \mathrm{j}\frac{e^{-\mathrm{j}\kappa_0 r}}{\kappa_0 r}\left[\sum_\sigma e^{\mathrm{j}\sigma\pi}\frac{\vec{n}_\sigma(\vartheta,\varphi)\vec{n}_\sigma(\vartheta_i,\varphi_i)}{\sigma(\sigma+1)}\right. \\ &\left.-\sum_\tau e^{\mathrm{j}\tau\pi}\frac{\vec{m}_\tau(\vartheta,\varphi)\vec{m}_\tau(\vartheta_i,\varphi_i)}{\tau(\tau+1)}\right]\cdot\hat{\xi}\end{aligned} \tag{13.111}$$

where we assume that only the first N series terms are used. Now the first part in (13.111) asymptotically represents the incident plane wave, which has been shown in the previous section, see (13.109). Hence the second part contains the far-field approximations of the transmitted plane wave as well as of the scattered field $\vec{E}^{sc}(\vec{r})$. Since in the far-field the transmitted plane wave is viewable only at $\pi-\vartheta_i, \varphi_i+\pi$ (which is denoted as the transition direction), the evaluation of the second part in (13.111) must yield the scattered field at each direction different from the transition direction:

$$\begin{aligned}\vec{E}^{sc}(\vec{r}) = &-E_0 2\pi \mathrm{j}\frac{e^{-\mathrm{j}\kappa_0 r}}{\kappa_0 r}\left[\sum_\sigma e^{\mathrm{j}\sigma\pi}\frac{\vec{n}_\sigma(\vartheta,\varphi)\vec{n}_\sigma(\vartheta_i,\varphi_i)}{\sigma(\sigma+1)}\right. \\ &\left.-\sum_\tau e^{\mathrm{j}\tau\pi}\frac{\vec{m}_\tau(\vartheta,\varphi)\vec{m}_\tau(\vartheta_i,\varphi_i)}{\tau(\tau+1)}\right]\cdot\hat{\xi} \qquad (\vartheta,\varphi)\neq(\pi-\vartheta_i,\ \varphi_i+\pi)\end{aligned} \tag{13.112}$$

In general this result can be used to calculate the bistatic radar cross section (RCS) $\sigma^{sc}(\vartheta,\varphi)$ of the PEC semi-infinite elliptic cone for plane wave incidence by inserting (13.112) into the pertinent definition [15]:

$$\begin{aligned}\sigma^{sc}(\vartheta,\varphi) &= \lim_{r\to\infty} 4\pi r^2\frac{|\vec{E}^{sc}(\vec{r})|^2}{|E_0|^2} \\ &= \frac{\Lambda^2}{\pi}\left|2\pi\sum_\sigma e^{\mathrm{j}\sigma\pi}\frac{\vec{n}_\sigma(\vartheta,\varphi)\vec{n}_\sigma(\vartheta_i,\varphi_i)\cdot\hat{\xi}}{\sigma(\sigma+1)}\right. \\ &\quad\left.-2\pi\sum_\tau e^{\mathrm{j}\tau\pi}\frac{\vec{m}_\tau(\vartheta,\varphi)\vec{m}_\tau(\vartheta_i,\varphi_i)\cdot\hat{\xi}}{\tau(\tau+1)}\right|^2 \\ &\quad(\vartheta,\varphi)\neq(\pi-\vartheta_i,\ \varphi_i+\pi)\end{aligned} \tag{13.113}$$

However, the attempt to simply sum up the series terms in (13.113) fails because it does not yield a limiting value. One reason for this behavior is rooted in the fact that we try to evaluate a multipole expansion of a Green's function for a semi-infinite scattering object with the source point $\vec{r}_i$ as well as the observation point $\vec{r}$ lying at infinity. Hence in this expansion a strong convergence-forcing factor, i.e., the spherical Bessel function with a finite argument, is missing.

However, obviously there should exist a meaningful limiting value of this series, which becomes apparent while moving an observation point from a finite $\vec{r}$ (with a convergent series) to infinity ($r\to\infty$). To obtain this meaningful and physically correct limiting value we utilize a suitable technique to transform the non-convergent series into a convergent one. As an example for a powerful transformation technique in Appendix 13.3 the Euler summation technique is summarized.

In order to numerically evaluate the result (13.113) we will now specialize it first to obtain the backscattering RCS, which can be calculated for the co-polarization case by the expression

$$\begin{aligned}\sigma^{sc}(\vartheta_i,\varphi_i) &= \lim_{r\to\infty} 4\pi r^2 \frac{|\vec{E}^{sc}(\vec{r}_i)\cdot\hat{\xi}|^2}{|E_0|^2} \\ &= \frac{\Lambda^2}{\pi}\left| 2\pi\sum_\sigma e^{j\sigma\pi}\frac{\vec{n}_\sigma(\vartheta_i,\varphi_i)\cdot\hat{\xi}\,\vec{n}_\sigma(\vartheta_i,\varphi_i)\cdot\hat{\xi}}{\sigma(\sigma+1)} \right. \\ &\quad \left. - 2\pi\sum_\tau e^{j\tau\pi}\frac{\vec{m}_\tau(\vartheta_i,\varphi_i)\cdot\hat{\xi}\,\vec{m}_\tau(\vartheta_i,\varphi_i)\cdot\hat{\xi}}{\tau(\tau+1)}\right|^2\end{aligned} \tag{13.114}$$

An important special case is that of the "nose-on" incidence of the plane wave $\vartheta_i = \varphi_i = 0$. Furthermore we consider two orthogonal polarizations $\hat{\xi} = \hat{x}$ and $\hat{\xi} = \hat{y}$ and denote the pertinent cross sections by σ_x^{sc} and σ_y^{sc}, respectively. Using the definitions of the transverse vector functions and the properties of the Lamé products in case of $\vartheta_i = \varphi_i = 0$, we obtain

$$\begin{aligned}\sigma_x^{sc}(0,0) &= \frac{\Lambda^2}{\pi}\left| 2\pi\sum_\sigma \frac{e^{j\sigma\pi}}{\sigma(\sigma+1)}\left[\frac{\partial Y_\sigma^{(2)}}{\partial\vartheta}\right]^2_{\substack{\vartheta_i=0\\ \varphi_i=0}} - 2\pi\sum_\tau \frac{e^{j\tau\pi}}{\tau(\tau+1)}\left[\frac{1}{k'}\frac{\partial Y_\tau^{(4)}}{\partial\varphi}\right]^2_{\substack{\vartheta_i=0\\ \varphi_i=0}}\right|^2 \\ \sigma_y^{sc}(0,0) &= \frac{\Lambda^2}{\pi}\left| 2\pi\sum_\sigma \frac{e^{j\sigma\pi}}{\sigma(\sigma+1)}\left[\frac{1}{k'}\frac{\partial Y_\sigma^{(4)}}{\partial\varphi}\right]^2_{\substack{\vartheta_i=0\\ \varphi_i=0}} - 2\pi\sum_\tau \frac{e^{j\tau\pi}}{\tau(\tau+1)}\left[\frac{\partial Y_\tau^{(2)}}{\partial\vartheta}\right]^2_{\substack{\vartheta_i=0\\ \varphi_i=0}}\right|^2\end{aligned} \tag{13.115}$$

where the upper index of the Lamé products stands for the function types, as explained in Section 13.3. As has been shown [11], in the case of a circular cone ($k = 1$) the results in (13.115) turn into

$$\begin{aligned}\sigma_x^{sc}(0,0) = \sigma_y^{sc}(0,0) &= \frac{\Lambda^2}{\pi}\left|\sum_\sigma \frac{\sigma(\sigma+1)e^{j\sigma\pi}}{2\int_0^{\vartheta_0}[P_\sigma^1(\cos\vartheta)]^2\sin\vartheta\, d\vartheta}\right. \\ &\quad \left. - \sum_\tau \frac{\tau(\tau+1)e^{j\tau\pi}}{2\int_0^{\vartheta_0}[P_\tau^1(\cos\vartheta)]^2\sin\vartheta\, d\vartheta}\right|^2\end{aligned} \tag{13.116}$$

This nose-on backscattering RCS of a circular cone, described by the spherical coordinate surface $\vartheta = \vartheta_0$, is identical to the expression probably first found by Hansen and Schiff [24].

The Physical Optics approximation of the nose-on backscattering RCS of the semi-infinite elliptic cone has been deduced by Blume and Kahl [6] as

$$\sigma_{PO}^{sc}(0,0) = \frac{\Lambda^2}{16\pi}\frac{1}{k^2}\frac{(1-k^2\cos^2\vartheta_0)\sin^2\vartheta_0}{\cos^4\vartheta_0} \tag{13.117}$$

which obviously possesses an error, since it is identical for the x- and for the y-polarization cases. For the circular cone ($k = 1$) we obtain from (13.117):

$$\sigma_{PO}^{sc}(0,0) = \frac{\Lambda^2}{16\pi} \tan^4 \vartheta_0 \tag{13.118}$$

also found by Spencer in 1951 [54].

The nose-on backscattering RCS of the semi-infinite elliptic cone (13.115) and of the circular cone (13.116) have been numerically evaluated by utilizing the complex Euler's summation technique described in Appendix 13.3. The results have been compared to those obtained by the corresponding Physical Optics approximations (13.117) and (13.118).

Figure 13.5(a) shows the normalized backscattering RCS of a semi-infinite circular cone and of a plane angular sector. The crosses represent values computed from the multipole expansions (13.115) and (13.116) while utilizing the complex Euler summation technique. The Physical Optics approximation for the circular cone (13.118) is represented in Fig. 13.5a) as the solid curve. Obviously, this is at least an excellent approximation to the multipole result. It has often been conjectured in the literature that the nose-on PO result is the exact RCS for any symmetrically illuminated semi-infinite body of revolution; it has been proved analytically only for the circular paraboloid by Schensted [47]. The RCS of the angular sector is computed exclusively by the evaluation of the multipole expansion (13.115) for y polarization, since the PO approximation (13.117) would yield an obviously wrong result $\sigma_{PO}^{sc} = 0$. Moreover, it has been shown analytically, that the RCS of the sector in the x-polarization case vanishes identically [11].

In Fig. 13.5(b) we see the normalized backscattering RCS of semi-infinite elliptic cones for some values of the half cone opening angle α_x as a function of the other half cone opening angle α_y. It is seen that the PO approximations of the RCS (solid curves) always are arranged between the multipole results for the x-pol. (dotted curves) and the y-pol. (dashed curves) cases.

13.5.2 Electromagnetic Scattering by a PEC Finite Elliptic Cone

Consider the PEC finite elliptic cone as shown in Fig. 13.6, the flank of which is described in terms of sphero-conal coordinates by $r = r_0$, $\vartheta = \vartheta_0$, while the spherical cap is given by $r = r_0$, $\vartheta_0 \leq \vartheta \leq \pi$. Since the finite cone is lying around the negative z axis, the cone's half opening angles are found according to (13.3) by

$$\begin{aligned} \alpha_x &= \pi - \vartheta_x = \pi - \vartheta_0 \\ \alpha_y &= \pi - \vartheta_y = \pi - \arccos(k \cos \vartheta_0) \end{aligned} \tag{13.119}$$

for a given ellipticity parameter k. The corresponding electromagnetic boundary value problem is formulated as a classical two-domain problem by defining a *complementary region A*, which is the complement of the elliptic cone to the sphere $r = r_0$, centered at the cone's tip, and by defining a *radiation region B*, which is the complete space outside the sphere $r = r_0$. It is supposed that the source of the electromagnetic field is localized within the radiation region B.

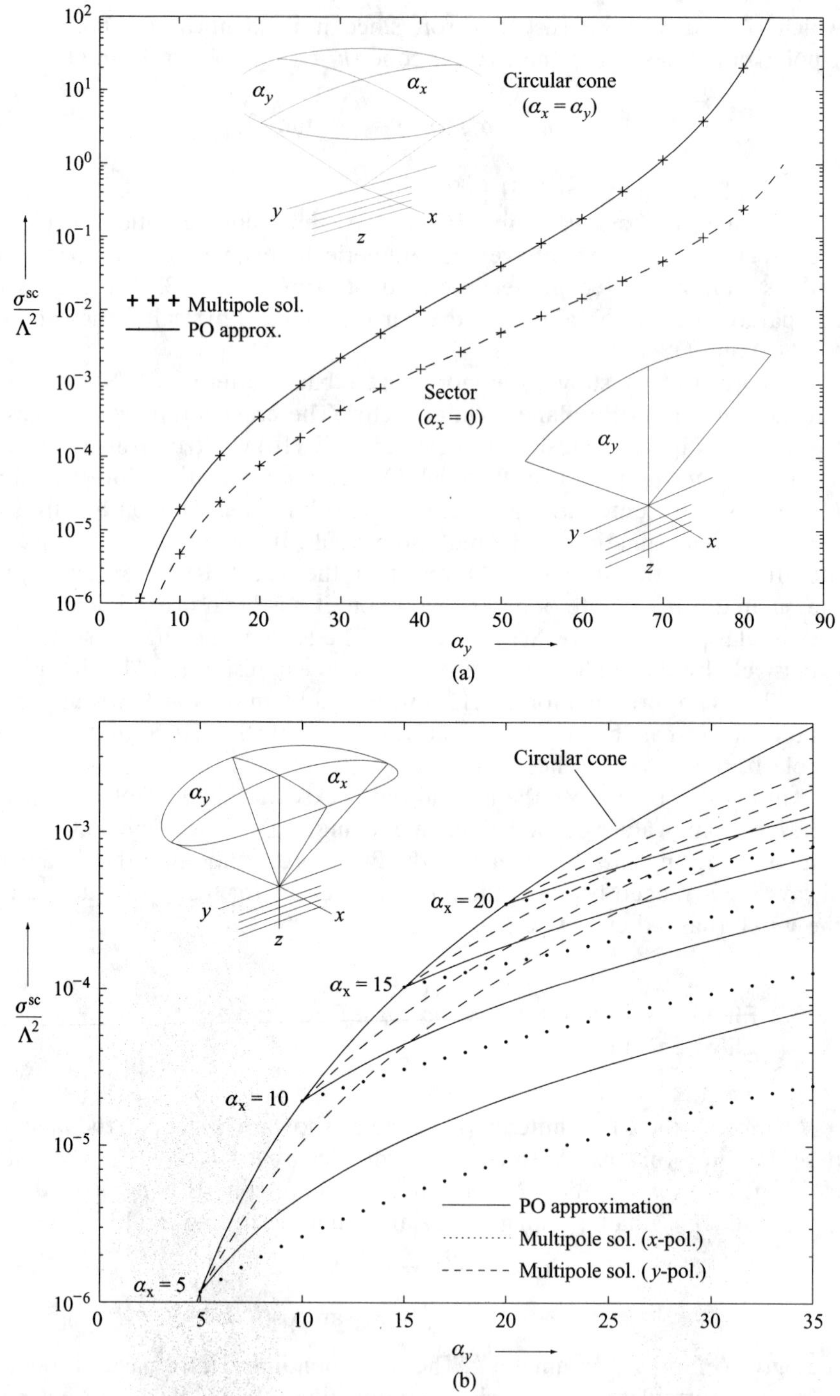

Figure 13.5 Normalized (to the squared wavelength Λ^2) nose-on backscattering RCS of semi-infinite circular cones, of plane angular sectors and of semi-infinite elliptic cones with different values of the half cone opening angle α_x, all as functions of α_y.

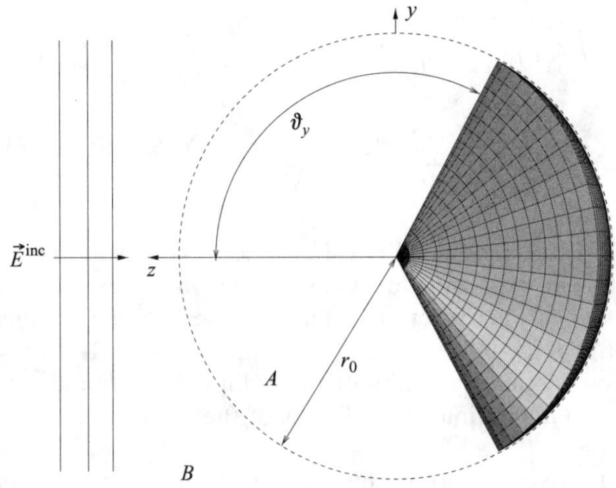

Figure 13.6 Plane wave incident upon a finite elliptic cone.

For the solution of the boundary value problem the electromagnetic field is described by spherical-multipole expansions in each domain. Within the complementary region A we use the spherical-multipole expansion

$$\begin{aligned}\vec{E}^A(\vec{r}) &= \sum_{\sigma} a_\sigma \vec{N}^I_{\sigma,\kappa_0}(\vec{r}) + \frac{Z_0}{\mathrm{j}} \sum_{\tau} b_\tau \vec{M}^I_{\tau,\kappa_0}(\vec{r}) \\ \vec{H}^A(\vec{r}) &= \frac{\mathrm{j}}{Z_0} \sum_{\sigma} a_\sigma \vec{M}^I_{\sigma,\kappa_0}(\vec{r}) + \sum_{\tau} b_\tau \vec{N}^I_{\tau,\kappa_0}(\vec{r})\end{aligned} \tag{13.120}$$

with eigenvalues σ and τ to the Dirichlet condition and to the Neumann condition as defined by (13.11) and (13.12), respectively, to guarantee the vanishing of the tangential electric field components at the cone's flank $\vartheta = \vartheta_0$. The upper index I of the vector spherical-multipole functions stands for the use of spherical Bessel functions $j_\nu(\kappa_0 r)$ to satisfy the energy condition for $r = 0$ as well. Z_0 and κ_0 are the free-space wave impedance and wave number, respectively. Within the radiation region B, we first split the total field into a known incident part $\vec{E}^{inc}, \vec{H}^{inc}$, which would be the total field in the absence of the finite elliptic cone, and into an unknown scattered part $\vec{E}^{sc}, \vec{H}^{sc}$

$$\begin{aligned}\vec{E}^B(\vec{r}) &= \vec{E}^{inc}(\vec{r}) + \vec{E}^{sc}(\vec{r}) \\ \vec{H}^B(\vec{r}) &= \vec{H}^{inc}(\vec{r}) + \vec{H}^{sc}(\vec{r})\end{aligned} \tag{13.121}$$

and describe each of these field parts by a spherical-multipole expansion of the free-space type:

$$\begin{aligned}\vec{E}^{inc}(\vec{r}) &= \sum_{n,m} a^{inc}_{n,m} \vec{N}_{n,m,\kappa_0}(\vec{r}) + \frac{Z_0}{\mathrm{j}} \sum_{n,m} b^{inc}_{n,m} \vec{M}_{n,m,\kappa_0}(\vec{r}) \\ \vec{H}^{inc}(\vec{r}) &= \frac{\mathrm{j}}{Z_0} \sum_{n,m} a^{inc}_{n,m} \vec{M}_{n,m,\kappa_0}(\vec{r}) + \sum_{n,m} b^{inc}_{n,m} \vec{N}_{n,m,\kappa_0}(\vec{r})\end{aligned} \tag{13.122}$$

$$\vec{E}^{sc}(\vec{r}) = \sum_{n,m} a^{sc}_{n,m}\vec{N}^{II}_{n,m,\kappa_0}(\vec{r}) + \frac{Z_0}{\mathrm{j}}\sum_{n,m} b^{sc}_{n,m}\vec{M}^{II}_{n,m,\kappa_0}(\vec{r})$$
$$\vec{H}^{sc}(\vec{r}) = \frac{\mathrm{j}}{Z_0}\sum_{n,m} a^{sc}_{n,m}\vec{M}^{II}_{n,m,\kappa_0}(\vec{r}) + \sum_{n,m} b^{sc}_{n,m}\vec{N}^{II}_{n,m,\kappa_0}(\vec{r}) \tag{13.123}$$

While the choice of special spherical cylinder functions in the multipole functions of the incident field expansion depends on the locations of the desired source and observation points, the scattered field must be described by means of spherical Hankel functions of the second kind $h_n^{(2)}(\kappa_0 r)$ (upper index II), to satisfy the radiation condition. For the determination of the unknown multipole amplitudes we consider the boundary- and continuity-conditions of the electromagnetic field:

1. The transverse part of the electric field has to be continuous across the aperture:
$$\hat{r}\times\vec{E}^{A}(\vec{r}) = \hat{r}\times[\vec{E}^{sc}(\vec{r}) + \vec{E}^{inc}(\vec{r})]$$
$$r = r_0; 0\le\vartheta<\vartheta_0; 0\le\varphi\le 2\pi \tag{13.124}$$
2. The transverse part of the electric field has to vanish on the spherical cap of the perfectly conducting elliptic cone:
$$\hat{r}\times[\vec{E}^{sc}(\vec{r}) + \vec{E}^{inc}(\vec{r})] = 0 \qquad r = r_0;\ \vartheta_0\le\vartheta\le\pi; 0\le\varphi\le 2\pi \tag{13.125}$$
3. The transverse part of the magnetic field has to be continuous across the aperture:
$$\hat{r}\times\vec{H}^{A}(\vec{r}) = \hat{r}\times[\vec{H}^{sc}(\vec{r}) + \vec{H}^{inc}(\vec{r})]$$
$$r = r_0; 0\le\vartheta<\vartheta_0; 0\le\varphi\le 2\pi \tag{13.126}$$

We insert (13.120), (13.122), (13.123) into (13.124) and (13.125) and obtain

$$\sum_{n,m}\left[a^{inc}_{n,m}\left.\frac{[rj_n(\kappa_0 r)]'}{\kappa_0 r}\right|_{r_0} + a^{sc}_{n,m}\left.\frac{[rh_n^{(2)}(\kappa_0 r)]'}{\kappa_0 r}\right|_{r_0}\right]\vec{m}_{n,m}$$
$$+\frac{Z_0}{j}\sum_{n,m}\left[b^{inc}_{n,m}j_n(\kappa_0 r_0) + b^{sc}_{n,m}h_n^{(2)}(\kappa_0 r_0)\right]\vec{n}_{n,m}$$
$$=\begin{cases}\displaystyle\sum_{\sigma} a_{\sigma}\left.\frac{[rj_{\sigma}(\kappa_0 r)]'}{\kappa_0 r}\right|_{r_0}\vec{m}_{\sigma} + \frac{Z_0}{j}\sum_{\tau} b_{\tau}j_{\tau}(\kappa_0 r_0)\vec{n}_{\tau} & 0\le\vartheta<\vartheta_0\\ 0 & \vartheta_0\le\vartheta\le\pi\end{cases} \tag{13.127}$$

where the prime indicates the differentiation with respect to r. With the orthogonality relations for the functions $\vec{n}_{nm}, \vec{m}_{nm}$ we have from (13.127)

$$a^{inc}_{n,m}\left.\frac{[rj_n(\kappa_0 r)]'}{\kappa_0 r}\right|_{r_0} n(n+1) + a^{sc}_{n,m}\left.\frac{[rh_n^{(2)}(\kappa_0 r)]'}{\kappa_0 r}\right|_{r_0} n(n+1)$$
$$=\sum_{\sigma} a_{\sigma}\left.\frac{[rj_{\sigma}(\kappa_0 r)]'}{\kappa_0 r}\right|_{r_0}\langle\vec{m}_{\sigma},\vec{m}_{n,m}\rangle + \frac{Z_0}{j}\sum_{\tau} b_{\tau}j_{\tau}(\kappa_0 r_0)\langle\vec{n}_{\tau},\vec{m}_{n,m}\rangle \tag{13.128a}$$

and

$$\frac{Z_0}{j}\left[b_{n,m}^{inc} j_n(\kappa_0 r_0) n(n+1) + b_{n,m}^{sc} h_n^{(2)}(\kappa_0 r_0) n(n+1)\right]$$
$$= \sum_\sigma a_\sigma \left.\frac{[r j_\sigma(\kappa_0 r)]'}{\kappa_0 r}\right|_{r_0} \langle \vec{m}_\sigma, \vec{n}_{nm}\rangle + \frac{Z_0}{j}\sum_\tau b_\tau j_\tau(\kappa_0 r_0)\langle \vec{n}_\tau, \vec{n}_{nm}\rangle \tag{13.128b}$$

In (13.128a, b) we have introduced the "coupling integrals" as

$$\langle \vec{a}_\nu, \vec{b}_\eta\rangle = \int_0^{\vartheta_0} \vec{a}_\nu(\vartheta,\varphi)\cdot\vec{b}_\eta(\vartheta,\varphi) s_\vartheta s_\varphi \, d\vartheta \, d\varphi \tag{13.129}$$

which can be evaluated completely analytically [33].

Now we apply the operation $\hat{r}\times$ to (13.126) and obtain

$$-\frac{j}{Z_0}\sum_\sigma a_\sigma j_\sigma(\kappa_0 r_0)\vec{m}_\sigma + \sum_\tau b_\tau \left.\frac{[r j_\tau(\kappa_0 r)]'}{\kappa_0 r}\right|_{r_0} \vec{n}_\tau =$$
$$-\frac{j}{Z_0}\sum_{n,m}\left[a_{n,m}^{inc} j_n(\kappa_0 r_0) + a_{n,m}^{sc} h_n^{(2)}(\kappa_0 r_0)\right]\vec{m}_{n,m} \tag{13.130}$$
$$+\sum_{n,m}\left[b_{n,m}^{inc}\left.\frac{[r j_n(\kappa_0 r)]'}{\kappa_0 r}\right|_{r_0} + b_{n,m}^{sc}\left.\frac{[r h_n^{(2)}(\kappa_0 r)]'}{\kappa_0 r}\right|_{r_0}\right]\vec{n}_{n,m}$$

With the orthogonality relations of the $\vec{n}_\sigma$ and $\vec{m}_\tau$ functions we derive from (13.130)

$$a_\sigma = \frac{1}{j_\sigma(\kappa_0 r_0)\sigma(\sigma+1)}\left\{\sum_{n,m}\left[a_{n,m}^{inc} j_n(\kappa_0 r_0) + a_{n,m}^{sc} h_n^{(2)}(\kappa_0 r_0)\right]\langle \vec{m}_\sigma, \vec{m}_{n,m}\rangle\right.$$
$$\left. - \frac{Z_0}{j}\sum_{n,m}\left[b_{n,m}^{inc}\left.\frac{[r j_n(\kappa_0 r)]'}{\kappa_0 r}\right|_{r_0} + b_{n,m}^{sc}\left.\frac{[r h_n^{(2)}(\kappa_0 r)]'}{\kappa_0 r}\right|_{r_0}\right]\langle \vec{m}_\sigma, \vec{n}_{n,m}\rangle\right\} \tag{13.131a}$$

and

$$b_\tau = \frac{1}{\left.\frac{[r j_\tau(\kappa_0 r)]'}{\kappa_0 r}\right|_{r_0}\tau(\tau+1)}\left\{-\frac{j}{Z_0}\sum_{n,m}\left[a_{n,m}^{inc} j_n(\kappa_0 r_0) + a_{n,m}^{sc} h_n^{(2)}(\kappa_0 r_0)\right]\langle \vec{n}_\tau, \vec{m}_{n,m}\rangle\right.$$
$$\left. + \sum_{n,m}\left[b_{n,m}^{inc}\left.\frac{[r j_n(\kappa_0 r)]'}{\kappa_0 r}\right|_{r_0} + b_{n,m}^{sc}\left.\frac{[r h_n^{(2)}(\kappa_0 r)]'}{\kappa_0 r}\right|_{r_0}\right]\langle \vec{n}_\tau, \vec{n}_{n,m}\rangle\right\} \tag{13.131b}$$

Inserting (13.131a, b) into (13.128a, b) yields a system of linear equations, which can be written as

$$\begin{pmatrix} \vdots \\ c_{n,m}^{inc} \\ \vdots \\ \vdots \\ d_{n,m}^{inc} \\ \vdots \end{pmatrix} = \left(\begin{array}{c|c} \vdots & \vdots \\ \cdots\ M_1 & M_2\ \cdots \\ \vdots & \vdots \\ \hline \vdots & \vdots \\ \cdots\ M_3 & M_4\ \cdots \\ \vdots & \vdots \end{array}\right) \begin{pmatrix} \vdots \\ -h_l^{(2)}(\kappa_0 r_0) a_{l,k}^{sc} \\ \vdots \\ \vdots \\ \left.\frac{[r h_l^{(2)}(\kappa_0 r)]'}{\kappa r}\right|_{r_0} \frac{Z_0}{j} b_{l,k}^{sc} \\ \vdots \end{pmatrix} \tag{13.132}$$

where the amplitudes $c_{n,m}^{inc}$ and $d_{n,m}^{inc}$ can be calculated uniquely from the incident field amplitudes $a_{n,m}^{inc}$ and $b_{n,m}^{inc}$ as

$$c_{n,m}^{inc} = a_{n,m}^{inc} \left. \frac{[rj_n(\kappa_0 r)]'}{\kappa_0 r} \right|_{r_0} n(n+1)$$
$$- \sum_{\sigma} \frac{\left. \frac{[rj_\sigma(\kappa_0 r)]'}{\kappa_0 r} \right|_{r_0}}{j_\sigma(\kappa_0 r_0)} \frac{1}{\sigma(\sigma+1)} \langle \vec{m}_\sigma, \vec{m}_{n,m} \rangle$$
$$\left\{ \sum_{l,k} a_{l,k}^{inc} j_l(\kappa_0 r_0) \langle \vec{m}_\sigma, \vec{m}_{l,k} \rangle - \frac{Z_0}{j} \sum_{l,k} b_{l,k}^{inc} \left. \frac{[rj_l(\kappa_0 r)]'}{\kappa_0 r} \right|_{r_0} \langle \vec{m}_\sigma, \vec{n}_{l,k} \rangle \right\} \tag{13.133a}$$
$$- \sum_{\tau} \frac{j_\tau(\kappa_0 r_0)}{\left. \frac{[rj_\tau(\kappa_0 r)]'}{\kappa_0 r} \right|_{r_0}} \frac{1}{\tau(\tau+1)} \langle \vec{n}_\tau, \vec{m}_{n,m} \rangle$$
$$\left\{ - \sum_{l,k} a_{l,k}^{inc} j_l(\kappa_0 r_0) \langle \vec{n}_\tau, \vec{m}_{l,k} \rangle + \frac{Z_0}{j} \sum_{l,k} b_{l,k}^{inc} \left. \frac{[rj_l(\kappa_0 r)]'}{\kappa_0 r} \right|_{r_0} \langle \vec{n}_\tau, \vec{n}_{l,k} \rangle \right\}$$

and

$$d_{n,m}^{inc} = \frac{Z_0}{j} b_{n,m}^{inc} j_n(\kappa_0 r_0) n(n+1)$$
$$- \sum_{\sigma} \frac{\left. \frac{[rj_\sigma(\kappa_0 r)]'}{\kappa_0 r} \right|_{r_0}}{j_\sigma(\kappa_0 r_0)} \frac{1}{\sigma(\sigma+1)} \langle \vec{m}_\sigma, \vec{n}_{n,m} \rangle$$
$$\left\{ \sum_{l,k} a_{l,k}^{inc} j_l(\kappa_0 r_0) \langle \vec{m}_\sigma, \vec{m}_{l,k} \rangle - \frac{Z_0}{j} \sum_{l,k} b_{l,k}^{inc} \left. \frac{[rj_l(\kappa_0 r)]'}{\kappa_0 r} \right|_{r_0} \langle \vec{m}_\sigma, \vec{n}_{l,k} \rangle \right\} \tag{13.133b}$$
$$- \sum_{\tau} \frac{j_\tau(\kappa_0 r_0)}{\left. \frac{[rj_\tau(\kappa_0 r)]'}{\kappa_0 r} \right|_{r_0}} \frac{1}{\tau(\tau+1)} \langle \vec{n}_\tau, \vec{n}_{n,m} \rangle$$
$$\left\{ - \sum_{l,k} a_{l,k}^{inc} j_l(\kappa_0 r_0) \langle \vec{n}_\tau, \vec{m}_{l,k} \rangle + \frac{Z_0}{j} \sum_{l,k} b_{l,k}^{inc} \left. \frac{[rj_l(\kappa_0 r)]'}{\kappa_0 r} \right|_{r_0} \langle \vec{n}_\tau, \vec{n}_{l,k} \rangle \right\}$$

The matrix elements of the symmetric matrix (13.132) are determined by

$$M_1 = - \sum_{\sigma} \frac{\left. \frac{[rj_\sigma(\kappa_0 r)]'}{\kappa_0 r} \right|_{r_0}}{j_\sigma(\kappa_0 r_0)} \frac{1}{\sigma(\sigma+1)} \langle \vec{m}_\sigma, \vec{m}_{n,m} \rangle \langle \vec{m}_\sigma, \vec{m}_{l,k} \rangle \tag{13.134a}$$
$$+ \sum_{\tau} \frac{j_\tau(\kappa r_0)}{\left. \frac{[rj_\tau(\kappa_0 r)]'}{\kappa_0 r} \right|_{r_0}} \frac{1}{\tau(\tau+1)} \langle \vec{n}_\tau, \vec{m}_{n,m} \rangle \langle \vec{n}_\tau, \vec{m}_{l,k} \rangle + \frac{\left. \frac{[rh_n^{(2)}(\kappa_0 r)]'}{\kappa_0 r} \right|_{r_0}}{h_n^{(2)}(\kappa_0 r_0)} n(n+1) \delta_{nl} \delta_{mk}$$

$$M_2 = -\sum_{\sigma} \frac{\left.\frac{[rj_\sigma(\kappa_0 r)]'}{\kappa_0 r}\right|_{r_0}}{j_\sigma(\kappa_0 r_0)} \frac{1}{\sigma(\sigma+1)} \langle \vec{m}_\sigma, \vec{m}_{n,m}\rangle \langle \vec{m}_\sigma, \vec{n}_{l,k}\rangle$$

$$+ \sum_{\tau} \frac{j_\tau(\kappa_0 r_0)}{\left.\frac{[rj_\tau(\kappa_0 r)]'}{\kappa_0 r}\right|_{r_0}} \frac{1}{\tau(\tau+1)} \langle \vec{n}_\tau, \vec{m}_{n,m}\rangle \langle \vec{n}_\tau, \vec{n}_{l,k}\rangle \tag{13.134b}$$

$$M_3 = -\sum_{\sigma} \frac{\left.\frac{[rj_\sigma(\kappa_0 r)]'}{\kappa_0 r}\right|_{r_0}}{j_\sigma(\kappa_0 r_0)} \frac{1}{\sigma(\sigma+1)} \langle \vec{m}_\sigma, \vec{n}_{n,m}\rangle \langle \vec{m}_\sigma, \vec{m}_{l,k}\rangle$$

$$+ \sum_{\tau} \frac{j_\tau(\kappa_0 r_0)}{\left.\frac{[rj_\tau(\kappa_0 r)]'}{\kappa_0 r}\right|_{r_0}} \frac{1}{\tau(\tau+1)} \langle \vec{n}_\tau, \vec{n}_{n,m}\rangle \langle \vec{n}_\tau, \vec{m}_{l,k}\rangle \tag{13.134c}$$

$$M_4 = -\sum_{\sigma} \frac{\left.\frac{[rj_\sigma(\kappa_0 r)]'}{\kappa_0 r}\right|_{r_0}}{j_\sigma(\kappa_0 r_0)} \frac{1}{\sigma(\sigma+1)} \langle \vec{m}_\sigma, \vec{n}_{n,m}\rangle \langle \vec{m}_\sigma, \vec{n}_{l,k}\rangle$$

$$+ \sum_{\tau} \frac{j_\tau(\kappa_0 r_0)}{\left.\frac{[rj_\tau(\kappa_0 r)]'}{\kappa_0 r}\right|_{r_0}} \frac{1}{\tau(\tau+1)} \langle \vec{n}_\tau, \vec{n}_{n,m}\rangle \langle \vec{n}_\tau, \vec{n}_{l,k}\rangle - \frac{h_n^{(2)}(\kappa_0 r_0)}{\left.\frac{[rh_n^{(2)}(\kappa_0 r)]'}{\kappa_0 r}\right|_{r_0}} n(n+1)\delta_{nl}\delta_{mk} \tag{13.134d}$$

It has been shown analytically, that in the free-space case the solution of (13.132) yields a vanishing scattered field, that in the limiting case of a PEC sphere the classical Mie-series solution is obtained, and that if the finite elliptic cone degenerates into a finite sector in the y-z plane, an incident wave which is polarized in the x direction is not disturbed [35].

The convergence properties of the obtained system of equations are strongly related to the correct choice of the upper limits of all sums. The rules for choosing these upper limits are found from the behavior of the spherical Bessel functions $j_\nu(\kappa_0 r_0)$ with respect to the order ν and from the properties of the used coupling integrals $\langle \vec{f}_\nu, \vec{g}_\eta \rangle$, where the maximal coupling which is obtained if $\nu \approx \eta$ (equal eigenvalues), has to be considered.

As an example for the numerical evaluation Fig. 13.7a shows the lines of constant magnetic field intensity in the x-z plane of a PEC finite elliptic cone, which is illuminated by a nose-on traveling plane wave, electrically polarized in the x direction. It is seen that the tip and the edges of the finite cone are centers for the scattered field; however the structure of the plane wave in the backscattering direction is nearly undisturbed at this frequency. Finally, Fig. 13.7b shows the nose-on backscattering RCS for finite elliptic cones with different half opening angles α_x and α_y as a function

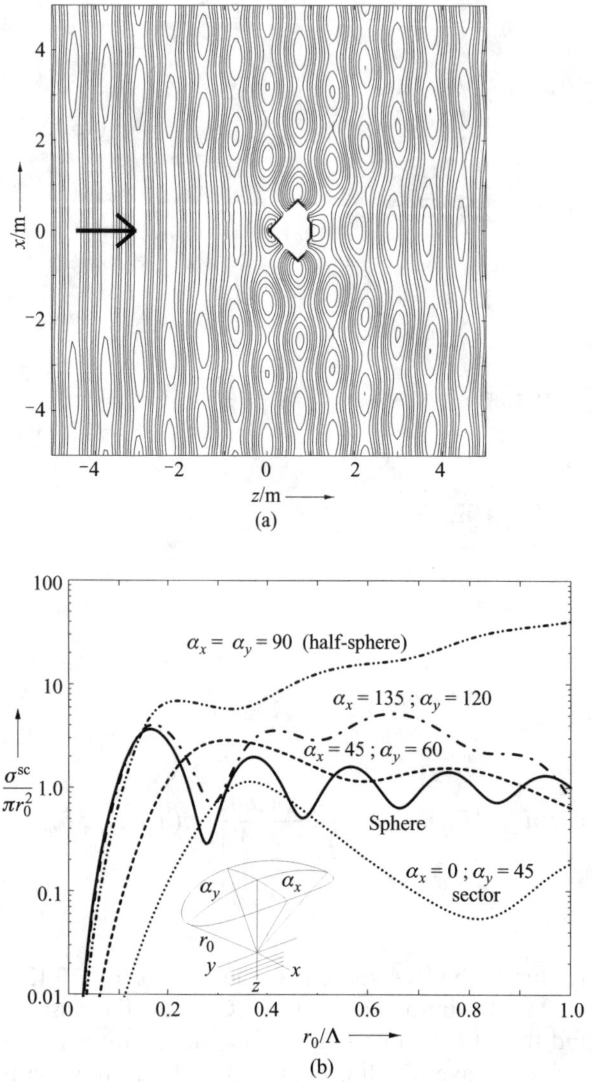

Figure 13.7 Snap-shot of the lines of constant (total) magnetic field intensity H_y in the x-z plane of a PEC finite elliptic cone with flank-length $r_0 = 1$ m and half opening angles $\alpha_x = 45°$ and $\alpha_y = 60°$ for a nose-on incident plane wave (wavelength $\Lambda = 1.5$ m) electrically polarized in the x direction (a) and normalized backscattering RCS of a PEC finite elliptic cone for different half cone-opening angles α_x and α_y as a function of the normalized cone's flank-length for a nose-on incident plane wave electrically polarized in the y direction (b).

of the ratio cone-flank length to wavelength, while the incident plane wave is polarized in the y direction.

13.5.3 Shielding Properties of a Loaded Spherical Shell with an Elliptic Aperture

Consider an infinitesimally thin PEC spherical shell with an elliptical aperture as seen in Fig. 13.8a, which is described in sphero-conal coordinates by $r = r_0, 0 \le \vartheta \le \vartheta_0$, $0 \le \varphi \le 2\pi$. The shell encloses concentrically a homogeneous sphere with radius r_A ($r_A < r_0$), which consists of a material with the wavenumber κ_A and the intrinsic wave

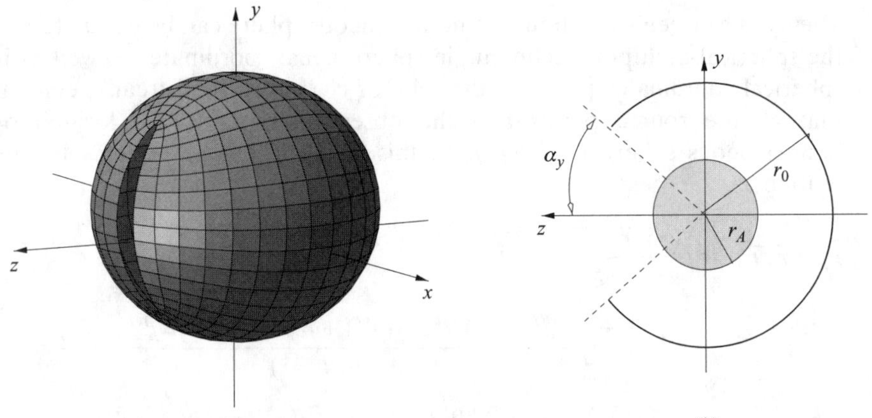

(a) (b)

Figure 13.8 Loaded spherical shell with an elliptical aperture.

impedance Z_A (Fig. 13.8b). The source of the electromagnetic field is given by an impressed electric current distribution $\vec{J}_e^{imp}(\vec{r}'')$, located outside the homogeneous sphere, i.e., $r'' > r_0$.

The electromagnetic boundary value problem is solved mostly analytically via an Electric Field Integral Equation (EFIE) approach. For this purpose we first split the total electric field into a known incident part $\vec{E}^{inc}(\vec{r})$ and into an unknown scattered part $\vec{E}^{sc}(\vec{r})$:

$$\vec{E}(\vec{r}) = \vec{E}^{inc}(\vec{r}) + \vec{E}^{sc}(\vec{r}) \tag{13.135}$$

The incident field is defined as that one which appears in the absence of the spherical shell, while the homogeneous sphere is still present. Hence, the incident field can be described utilizing the dyadic Green's function of the homogeneous sphere $\tilde{G}_{h.s.}(\vec{r},\vec{r}'')$ by

$$\vec{E}^{inc}(\vec{r}) = \mathrm{j}\omega\mu_0 \iiint_V \tilde{G}_{h.s.}(\vec{r},\vec{r}'')\cdot\vec{J}_e^{imp}(\vec{r}'')\,dv'' \tag{13.136}$$

Consequently, the scattered field is caused only by the electric surface current $\vec{J}_{S,e}(\vec{r}')$ on the PEC spherical shell S and is calculated by

$$\vec{E}^{sc}(\vec{r}) = \mathrm{j}\omega\mu_0 \iint_S \tilde{G}_{h.s.}(\vec{r},\vec{r}')\cdot\vec{J}_{S,e}(\vec{r}')\,ds' \tag{13.137}$$

since this surface current radiates in the presence of the homogeneous sphere. By enforcing the boundary condition for the tangential electric field at the PEC spherical shell the following EFIE is obtained:

$$\left.\iint_S \hat{r}\times\tilde{G}_{h.s.}(\vec{r},\vec{r}')\cdot\vec{J}_{S,e}(\vec{r}')\,ds'\right|_{\vec{r}\in S} = -\left.\iiint_V \hat{r}\times\tilde{G}_{h.s.}(\vec{r},\vec{r}'')\cdot\vec{J}_e^{imp}(\vec{r}'')\,dv''\right|_{\vec{r}\in S} \tag{13.138}$$

The dyadic Green's function of a homogeneous sphere can be calculated by means of the spherical-multipole technique in sphero-conal coordinates as well as in ordinary spherical coordinates [56]. For the sake of clarity we shall treat exemplary the case that all the source points and the observation points are located outside the homogeneous sphere $(r, r' > r_A)$. In this case the dyadic Green's function can be written as

$$\begin{aligned}\tilde{\Gamma}_{h.s.}(\vec{r},\vec{r}\,') = {}& \hat{r}\hat{r}'\frac{\delta(\vec{r}-\vec{r}\,')}{\kappa_0^2} \\ & + \mathrm{j}\kappa_0\left[\sum_{n,m}\frac{\vec{N}^{I(II)}_{n,m,\kappa_0}(\vec{r})\vec{N}^{II(I)}_{n,m,\kappa_0}(\vec{r}\,') + \alpha_n\vec{N}^{II}_{n,m,\kappa_0}(\vec{r})\vec{N}^{II}_{n,m,\kappa_0}(\vec{r}\,')}{n(n+1)}\right. \\ & \left. + \sum_{n,m}\frac{\vec{M}^{I(II)}_{n,m,\kappa_0}(\vec{r})\vec{M}^{II(I)}_{n,m,\kappa_0}(\vec{r}\,') + \beta_n\vec{M}^{II}_{n,m,\kappa_0}(\vec{r})\vec{M}^{II}_{n,m,\kappa_0}(\vec{r}\,')}{n(n+1)}\right]; \\ & r \le r'(r > r')\end{aligned} \tag{13.139}$$

where the "scattering coefficients" α_n and β_n depend on the properties of the homogeneous sphere:

$$\begin{aligned}\alpha_n &= -\frac{\dfrac{\kappa_A}{\kappa_0}[rj_n(\kappa_0 r)]'_{r_A} j_n(\kappa_A r_A) - \dfrac{Z_A}{Z_0}[rj_n(\kappa_A r)]'_{r_A} j_n(\kappa_0 r_A)}{\dfrac{\kappa_A}{\kappa_0}[rh_n^{(2)}(\kappa_0 r)]'_{r_A} j_n(\kappa_A r_A) - \dfrac{Z_A}{Z_0}[rj_n(\kappa_A r)]'_{r_A} h_n^{(2)}(\kappa_0 r_A)} \\ \beta_n &= -\frac{\dfrac{\kappa_A}{\kappa_0}[rj_n(\kappa_0 r)]'_{r_A} j_n(\kappa_A r_A) - \dfrac{Z_0}{Z_A}[rj_n(\kappa_A r)]'_{r_A} j_n(\kappa_0 r_A)}{\dfrac{\kappa_A}{\kappa_0}[rh_n^{(2)}(\kappa_0 r)]'_{r_A} j_n(\kappa_A r_A) - \dfrac{Z_0}{Z_A}[rj_n(\kappa_A r)]'_{r_A} h_n^{(2)}(\kappa_0 r_A)}\end{aligned} \tag{13.140}$$

Note that the upper indices I and II in (13.139) symbolize the use of spherical Bessel functions $j_n(\kappa_0 r)$ and spherical Hankel functions of the second kind $h_n^{(2)}(\kappa_0 r)$, respectively.

The surface-current on the PEC spherical shell is described by means of a complete orthogonal expansion with transverse vector functions:

$$\vec{J}_{S,e}(\vartheta,\varphi) = \begin{cases}\sum_{\sigma} K_\sigma \vec{m}_\sigma(\vartheta,\varphi) + \sum_{\tau} K_\tau \vec{n}_\tau(\vartheta,\varphi) & r = r_0, 0 \le \vartheta \le \vartheta_0, 0 \le \varphi \le 2\pi \\ 0 & r = r_0, \vartheta_0 < \vartheta \le \pi, 0 \le \varphi \le 2\pi\end{cases} \tag{13.141}$$

The choice of the eigenvalues σ to the Dirichlet-type and of τ to the Neumann type [see (13.11) and (13.12)] automatically guarantees the vanishing of the normal surface current component at the edge of the spherical shell $\vartheta = \vartheta_0$, which is a physical boundary condition. We now insert (13.139)–(13.141) into (13.138) and remark, first, that the δ-functional term in the dyadic Green's function does not play a role while solving this special boundary value problem, since the surface current on the PEC shell has no radial component. However, it is noted that for diffraction geometries which possess a radial current the δ-function must be considered, since the case $\vec{r} = \vec{r}\,'$ occurs in the left side of the EFIE (13.138).

As remarked earlier, it is supposed that all the source points $\vec{r}\,''$ are located within the domain outside the PEC spherical shell ($r'' > r_0$). Hence, for the dyadic Green's function in the right side of the EFIE (13.138) only the case $r < r''$ must be considered, while in the left side of the EFIE only the case $r = r'$ occurs.

Now we employ the Galerkin method to transform the EFIE into a system of linear equations, i.e., we choose weighting functions $\vec{m}_{\sigma'}(\vartheta, \varphi)$ and $\vec{n}_{\tau'}(\vartheta, \varphi)$. Utilizing the orthogonality relations of the transverse vector functions we are led to a system of linear equations, which can be written as

$$(\kappa r_0)^2 Z_0 \left(\begin{array}{c|c} \mathcal{M}_{\sigma,\sigma'} & \mathcal{M}_{\tau,\sigma'} \\ \hline \mathcal{M}_{\sigma,\tau'} & \mathcal{M}_{\tau,\tau'} \end{array} \right) \begin{pmatrix} \vdots \\ K_\sigma \\ \vdots \\ \vdots \\ K_\tau \\ \vdots \end{pmatrix} = \begin{pmatrix} \vdots \\ S_{\sigma'} \\ \vdots \\ \vdots \\ S_{\tau'} \\ \vdots \end{pmatrix} \tag{13.142}$$

where the matrix elements are given by

$$\mathcal{M}_{\sigma,\sigma'} = \sum_{n,m} \frac{1}{n(n+1)} \left[\frac{[rj_n(\kappa_0 r)]'_{r_0}}{\kappa_0 r_0} - \alpha_n \frac{[rh_n^{(2)}(\kappa_0 r)]'_{r_0}}{\kappa_0 r_0} \right] \frac{[rh_n^{(2)}(\kappa_0 r)]'_{r_0}}{\kappa_0 r_0} \times \langle \vec{m}_\sigma, \vec{n}_{n,m} \rangle \langle \vec{m}_{\sigma'}, \vec{n}_{n,m} \rangle \tag{13.143a}$$
$$+ \sum_{n,m} \frac{1}{n(n+1)} [j_n(\kappa_0 r_0) - \beta_n h_n^{(2)}(\kappa_0 r_0)] h_n^{(2)}(\kappa_0 r_0) \langle \vec{m}_\sigma, \vec{m}_{n,m} \rangle \langle \vec{m}_{\sigma'}, \vec{m}_{n,m} \rangle$$

$$\mathcal{M}_{\tau,\sigma'} = \sum_{n,m} \frac{1}{n(n+1)} \left[\frac{[rj_n(\kappa_0 r)]'_{r_0}}{\kappa_0 r_0} - \alpha_n \frac{[rh_n^{(2)}(\kappa_0 r)]'_{r_0}}{\kappa_0 r_0} \right] \frac{[rh_n^{(2)}(\kappa_0 r)]'_{r_0}}{\kappa_0 r_0} \times \langle \vec{n}_\tau, \vec{n}_{n,m} \rangle \langle \vec{m}_{\sigma'}, \vec{n}_{n,m} \rangle \tag{13.143b}$$
$$+ \sum_{n,m} \frac{1}{n(n+1)} [j_n(\kappa_0 r_0) - \beta_n h_n^{(2)}(\kappa_0 r_0)] h_n^{(2)}(\kappa_0 r_0) \langle \vec{n}_\tau, \vec{m}_{n,m} \rangle \langle \vec{m}_{\sigma'}, \vec{m}_{n,m} \rangle$$

$$\mathcal{M}_{\sigma,\tau'} = \sum_{n,m} \frac{1}{n(n+1)} \left[\frac{[rj_n(\kappa_0 r)]'_{r_0}}{\kappa_0 r_0} - \alpha_n \frac{[rh_n^{(2)}(\kappa_0 r)]'_{r_0}}{\kappa_0 r_0} \right] \frac{[rh_n^{(2)}(\kappa_0 r)]'_{r_0}}{\kappa_0 r_0} \times \langle \vec{m}_\sigma, \vec{n}_{n,m} \rangle \langle \vec{n}_{\tau'}, \vec{n}_{n,m} \rangle \tag{13.143c}$$
$$+ \sum_{n,m} \frac{1}{n(n+1)} [j_n(\kappa_0 r_0) - \beta_n h_n^{(2)}(\kappa_0 r_0)] h_n^{(2)}(\kappa_0 r_0) \langle \vec{m}_\sigma, \vec{m}_{n,m} \rangle \langle \vec{n}_{\tau'}, \vec{m}_{n,m} \rangle$$

$$\mathcal{M}_{\tau,\tau'} = \sum_{n,m} \frac{1}{n(n+1)} \left[\frac{[rj_n(\kappa_0 r)]'_{r_0}}{\kappa_0 r_0} - \alpha_n \frac{[rh_n^{(2)}(\kappa_0 r)]'_{r_0}}{\kappa_0 r_0} \right] \frac{[rh_n^{(2)}(\kappa_0 r)]'_{r_0}}{\kappa_0 r_0} \times \langle \vec{n}_\tau, \vec{n}_{n,m} \rangle \langle \vec{n}_{\tau'}, \vec{n}_{n,m} \rangle$$

$$+ \sum_{n,m} \frac{1}{n(n+1)} [j_n(\kappa_0 r_0) - \beta_n h_n^{(2)}(\kappa_0 r_0)] h_n^{(2)}(\kappa_0 r_0) \langle \vec{n}_\tau, \vec{m}_{n,m} \rangle \langle \vec{n}_{\tau'}, \vec{m}_{n,m} \rangle \quad (13.143d)$$

The coupling integrals introduced in (13.143a–d) and defined as

$$\langle \vec{a}_\nu, \vec{b}_\eta \rangle = \int_0^{\vartheta_0} \vec{a}_\nu(\vartheta, \varphi) \cdot \vec{b}_\eta(\vartheta, \varphi) s_\vartheta s_\varphi \, d\vartheta \, d\varphi \quad (13.144)$$

can be solved completely analytically [33]. The elements of the right part of (13.142) are calculated by

$$S_{\sigma'} = -\sum_{n,m} a^i_{n,m} \left[\frac{[rj_n(\kappa_0 r)]'_{r_0}}{\kappa_0 r_0} - \alpha_n \frac{[rh_n^{(2)}(\kappa_0 r)]'_{r_0}}{\kappa_0 r_0} \right] \langle \vec{m}_{\sigma'}, \vec{n}_{n,m} \rangle + \frac{Z_0}{\mathrm{j}} \sum_{n,m} b^i_{n,m} [j_n(\kappa_0 r_0) - \beta_n h_n^{(2)}(\kappa_0 r_0)] \langle \vec{m}_{\sigma'}, \vec{m}_{n,m} \rangle \quad (13.145a)$$

$$S_{\tau'} = -\sum_{n,m} a^i_{n,m} \left[\frac{[rj_n(\kappa_0 r)]'_{r_0}}{\kappa_0 r_0} - \alpha_n \frac{[rh_n^{(2)}(\kappa_0 r)]'_{r_0}}{\kappa_0 r_0} \right] \langle \vec{n}_{\tau'}, \vec{n}_{n,m} \rangle + \frac{Z_0}{\mathrm{j}} \sum_{n,m} b^i_{n,m} [j_n(\kappa_0 r_0) - \beta_n h_n^{(2)}(\kappa_0 r_0)] \langle \vec{n}_{\tau'}, \vec{m}_{n,m} \rangle \quad (13.145b)$$

where the multipole amplitudes $a^i_{n,m}$ and $b^i_{n,m}$ belong to the electromagnetic field radiated by the impressed electric current density in the free space:

$$a^i_{nm} = -\kappa_0^2 Z_0 \frac{1}{n(n+1)} \iiint_V \vec{N}^{II}_{n,m,\kappa_0}(\vec{r}\,'') \cdot \vec{J}^{imp}_e(\vec{r}\,'') \, dv'' \quad (13.146a)$$

$$b^i_{nm} = -\mathrm{j}\kappa_0^2 \frac{1}{n(n+1)} \iiint_V \vec{M}^{II}_{n,m,\kappa_0}(\vec{r}\,'') \cdot \vec{J}^{imp}_e(\vec{r}\,'') \, dv'' \quad (13.146b)$$

Note that the multipole amplitudes $a^i_{n,m}$ and $b^i_{n,m}$ of an incident plane electromagnetic wave are given by (13.104). The convergence properties of the method mainly depend on the choice of the upper limits of the used expansions, which can be found by taking into account the properties of the coupling integrals as well as the behavior of the spherical Bessel functions $j_n(\kappa_0 r_0)$ for increasing n.

It has been shown that in the limiting case of a PEC sphere exactly the well-known Mie-series results are obtained [33]. Of course, also dyadic Green's function for suitable geometries other than that one of the homogeneous sphere can be inserted into the EFIE. For example, by using the dyadic Green's function of a homogeneous spherical shell we have also investigated the problem of a PEC spherical shell with an elliptical aperture, which is interiorly covered by an absorbing layer.

Obviously, the loaded spherical shell with an elliptical aperture can serve as a canonical model for a partially filled electromagnetic shield. Commonly used definitions to characterize the properties of an *empty* shield are the electric shielding effectiveness a_e and the magnetic shielding effectiveness a_m, which are defined as

$$a_e = 20\log_{10}\frac{|\vec{E}^{unsh.}(\vec{r})|_{r=0}}{|\vec{E}^{sh.}(\vec{r})|_{r=0}} \quad \text{and} \quad a_m = 20\log_{10}\frac{|\vec{H}^{unsh.}(\vec{r})|_{r=0}}{|\vec{H}^{sh.}(\vec{r})|_{r=0}} \tag{13.147}$$

where $\vec{E}^{unsh.}, \vec{H}^{unsh.}$ are the field intensities which would be present without a shield, and $\vec{E}^{sh.}, \vec{H}^{sh.}$ are the field intensities with a shield, each measured at the shield's center ($r = 0$). Because these definitions do not take into consideration the effect of a load, we introduce the *special shielding effectiveness* a_P as the ratio of the time-average power absorbed by the unshielded load and the time-average power absorbed by the shielded load:

$$a_P = 10\log_{10}\frac{\bar{P}^{unsh.}}{\bar{P}^{sh.}} \tag{13.148}$$

It has been shown analytically that in the limiting case of a vanishing spherical load with radius r_A the special shielding effectiveness can be expressed by [34]

$$a_{em} = \lim_{r_A \to 0} a_P = 10\log_{10}\frac{2}{10^{-a_e/10} + 10^{-a_m/10}} \tag{13.149}$$

i.e., by a simple combination of the commonly used (and easily measurable) electric and magnetic shielding effectiveness (13.147).

Finally, two numerical results are represented. Figure 13.9a shows the time-domain electric field intensity at the center of a non-loaded spherical shell with an elliptical aperture, which is illuminated by a Gaussian-type electromagnetic plane wave. The shell is optionally interiorly lined by a 5-cm-thick homogeneous absorber. The effect of the absorber can be observed in a significant reduction of the electric field's amplitude as well as of its oscillation time. This time-domain result has been obtained by performing a discrete Fourier transform from 2000 frequency-domain runs for each curve.

In Fig. 13.9b we see the special shielding effectiveness a_P for a concentrically arranged spherical load with different values of the load's radius r_A. It is remarkable that for frequencies lower than the first resonance frequency the special shielding effectiveness for non-empty shields is significantly smaller than in case of an empty shield ($r_A = 0$).

APPENDIX 13.1 SOLUTIONS OF THE VECTOR HELMHOLTZ EQUATION

The vector homogeneous Helmholtz equation

$$\Delta\vec{F}(\vec{r}) + \kappa^2\vec{F}(\vec{r}) = 0 \tag{13.150}$$

($\kappa \neq 0$) is satisfied by the vector functions

$$\vec{L}(\vec{r}) = \frac{1}{\kappa}\vec{\nabla}\Psi(\vec{r}) \tag{13.151a}$$

$$\vec{M}(\vec{r}) = (\vec{r} \times \vec{\nabla})\Psi(\vec{r}) \tag{13.151b}$$

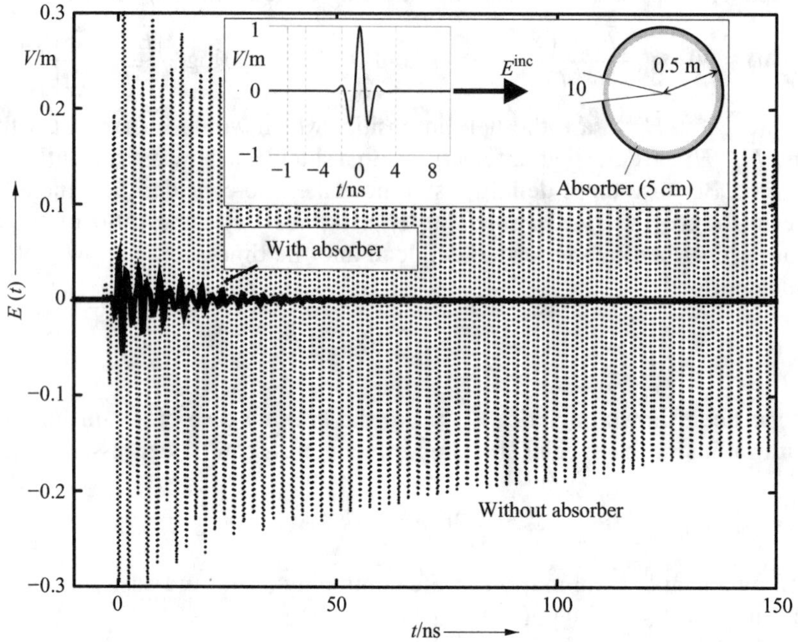
0.3
V/m
0.2
0.1
0
−0.1
−0.2
−0.3
E (t)
1
V/m
0
−1
−1 −4 0 4 8
t/ns
E^inc
0.5 m
10
Absorber (5 cm)
With absorber
Without absorber
0
50
100
150
t/ns

(a)

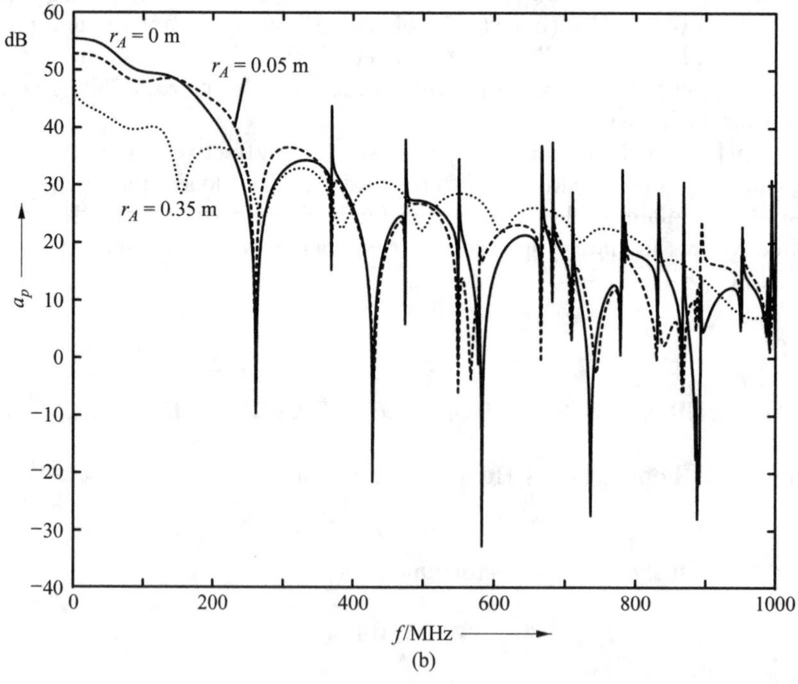
60
dB
50
40
30
20
10
0
−10
−20
−30
−40
a_p
r_A = 0 m
r_A = 0.05 m
r_A = 0.35 m
0
200
400
600
800
1000
f/MHz

(b)

$$\vec{N}(\vec{r}) = \left[\frac{1}{\kappa}\vec{\nabla}\times(\vec{r}\times\vec{\nabla})\right]\Psi(\vec{r}) \tag{13.151c}$$

provided that $\Psi(\vec{r})$ is a solution of the scalar homogeneous Helmholtz equation:

$$\Delta\Psi(\vec{r}) + \kappa^2\Psi(\vec{r}) = 0 \tag{13.152}$$

(a) For the proof of (13.151a) we use the vector identity

$$\Delta\vec{A} = \vec{\nabla}\vec{\nabla}\cdot\vec{A} - \vec{\nabla}\times\vec{\nabla}\times\vec{A} \tag{13.153}$$

and utilize that $\vec{\nabla}\times\vec{\nabla}\Psi \equiv 0$:

$$\Delta\vec{L} + \kappa^2\vec{L} = \frac{1}{\kappa}\vec{\nabla}[\vec{\nabla}\cdot\vec{\nabla}\Psi + \kappa^2\Psi] = \frac{1}{\kappa}\vec{\nabla}[\Delta\Psi + \kappa^2\Psi] = 0 \tag{13.154}$$

(b) The key to show the evidence of (13.151b) is the vector identity [3]:

$$\begin{aligned}\vec{\nabla}\times\left(\vec{A}\times\vec{B}\right) &= -\vec{\nabla}\left(\vec{A}\cdot\vec{B}\right) + 2\left(\vec{B}\cdot\vec{\nabla}\right)\vec{A} + \vec{A}\left(\vec{\nabla}\cdot\vec{B}\right) - \vec{B}\left(\vec{\nabla}\cdot\vec{A}\right)\\ &\quad+\vec{A}\times\left(\vec{\nabla}\times\vec{B}\right) + \vec{B}\times\left(\vec{\nabla}\times\vec{A}\right)\end{aligned} \tag{13.155}$$

With (13.153) and because $\vec{M} = \vec{r}\times\vec{\nabla}\Psi = -\vec{\nabla}\times(\vec{r}\,\Psi)$, we obtain from (13.151b)

$$\begin{aligned}\Delta\vec{M} + \kappa^2\vec{M} &= -\vec{\nabla}\times\vec{\nabla}\times\vec{M} + \kappa^2\vec{M} =\\ &\quad-\vec{\nabla}\times\vec{\nabla}\times(\vec{r}\times\vec{\nabla}\Psi) - \kappa^2\vec{\nabla}\times(\vec{r}\,\Psi)\end{aligned} \tag{13.156}$$

Identification of $\vec{A} = \vec{r}$ and $\vec{B} = \vec{\nabla}\Psi$ in (13.155) yields:

$$\begin{aligned}\vec{\nabla}\times\left(\vec{r}\times\vec{\nabla}\Psi\right) &= -\vec{\nabla}\left(\vec{r}\cdot\vec{\nabla}\Psi\right) + \underbrace{2\left(\vec{\nabla}\Psi\cdot\vec{\nabla}\right)\vec{r}}_{=\vec{\nabla}\Psi} + \vec{r}\underbrace{\left(\vec{\nabla}\cdot\vec{\nabla}\Psi\right)}_{=\Delta\Psi} - \vec{\nabla}\Psi\underbrace{\left(\vec{\nabla}\cdot\vec{r}\right)}_{=3}\\ &\quad+\vec{r}\times\underbrace{\left(\vec{\nabla}\times\vec{\nabla}\Psi\right)}_{=0} + \vec{\nabla}\Psi\times\underbrace{\left(\vec{\nabla}\times\vec{r}\right)}_{=0}\end{aligned} \tag{13.157}$$

We insert (13.157) into (13.156), utilize that $\vec{\nabla}\times\vec{\nabla}\Psi \equiv 0$ and obtain

$$\Delta\vec{M} + \kappa^2\vec{M} = -\vec{\nabla}\times(\vec{r}\Delta\Psi) - \kappa^2\vec{\nabla}\times(\vec{r}\,\Psi) = -\vec{\nabla}\times\vec{r}\,[\Delta\Psi + \kappa^2\Psi] = 0 \tag{13.158}$$

Figure 13.9 Spherical shell (radius $r_0 = 0.5$ m) with an elliptic aperture, illuminated by a plane electromagnetic wave, traveling in z direction (toward the aperture) and polarized in the y direction. The half opening angles of the aperture are $\alpha_x = 5°$, $\alpha_y = 45.2°$. (a) Electric field at $r = 0$ for a Gaussian-type impulse (see sub-figure) incident on an empty shield without (dotted) and with (solid) an absorbing layer (layer thickness 5 cm; layer material: $\varepsilon_r = \mu_r = 5$, $\sigma_e = 0.015$ 1/(Ωm), $\sigma_m = 0.015$ Ω/m). (b) Special shielding effectiveness a_P in dB as a function of the frequency for a concentrically arranged spherical load with different values of its radius r_A ($\varepsilon_A/\varepsilon_0 = 1$; $\mu_A/\mu_0 = 10$; $\sigma_{e,A} = 10$ 1/(Ωm)).

(c) With (13.153) and because $\vec{N} = (1/\kappa)\vec{\nabla}\times\vec{M}$ it is seen that

$$\Delta\vec{N} + \kappa^2\vec{N} = -\vec{\nabla}\times\vec{\nabla}\times\vec{N} + \kappa^2\vec{N} = -\vec{\nabla}\times\frac{1}{\kappa}[\vec{\nabla}\times\vec{\nabla}\times\vec{M} - \kappa^2\vec{M}] = 0 \quad (13.159)$$

which completes the proof of (13.151c).

APPENDIX 13.2 PATHS OF INTEGRATION FOR THE EIGENFUNCTION EXPANSION OF THE DYADIC GREEN'S FUNCTION

We are searching for the paths of integration in the complex κ-plane for the eigenfunction expansion of the dyadic Green's function (13.85):

$$\tilde{\Gamma}(\vec{r},\vec{r}\,') = \frac{2}{\pi}\int_0^\infty \left\{\frac{\kappa^2}{\kappa_0^2}\sum_\sigma \vec{L}^I_{\sigma,\kappa}(\vec{r})\vec{L}^I_{\sigma,\kappa}(\vec{r}\,') - \frac{\kappa^2}{\kappa^2-\kappa_0^2}\left[\sum_\sigma \frac{\vec{N}^I_{\sigma,\kappa}(\vec{r})\vec{N}^I_{\sigma,\kappa}(\vec{r}\,')}{\sigma(\sigma+1)} + \sum_\tau \frac{\vec{M}^I_{\tau,\kappa}(\vec{r})\vec{M}^I_{\tau,\kappa}(\vec{r}\,')}{\tau(\tau+1)}\right]\right\} d\kappa \quad (13.160)$$

which corresponds to the "radiation" condition (13.80):

$$\lim_{r\to\infty} \vec{r}\times\left\{\vec{\nabla}\times\tilde{\Gamma}(\vec{r},\vec{r}\,') + j\kappa_0\hat{r}\times\tilde{\Gamma}(\vec{r},\vec{r}\,')\right\} = 0 \quad (13.161)$$

We insert (13.160) into (13.161), utilize the properties and orthogonality relations of the vector spherical-multipole functions, and obtain first

$$\lim_{r\to\infty} \vec{r}\times\int_0^\infty d\kappa \frac{\delta(\kappa-\kappa')}{\kappa^2-\kappa_0^2}\left[\kappa\vec{M}^I_{\sigma,\kappa}(\vec{r}) + j\kappa_0\hat{r}\times\vec{N}^I_{\sigma,\kappa}(\vec{r})\right] = 0 \quad (13.162a)$$

$$\lim_{r\to\infty} \vec{r}\times\int_0^\infty d\kappa \frac{\delta(\kappa-\kappa')}{\kappa^2-\kappa_0^2}\left[\kappa\vec{N}^I_{\tau,\kappa}(\vec{r}) + j\kappa_0\hat{r}\times\vec{M}^I_{\tau,\kappa}(\vec{r})\right] = 0 \quad (13.162b)$$

$$\lim_{r\to\infty} \vec{r}\times\int_0^\infty d\kappa\,\delta(\kappa-\kappa')j\kappa_0\hat{r}\times\vec{L}^I_{\sigma,\kappa}(\vec{r}) = 0 \quad (13.162c)$$

and derive then with the aid of the filter-property of the δ-functional:

$$\lim_{r\to\infty} \vec{r}\times\left[\kappa\vec{M}^I_{\sigma,\kappa}(\vec{r}) + j\kappa_0\hat{r}\times\vec{N}^I_{\sigma,\kappa}(\vec{r})\right] = 0 \quad (13.163a)$$

$$\lim_{r\to\infty} \vec{r}\times\left[\kappa\vec{N}^I_{\tau,\kappa}(\vec{r}) + j\kappa_0\hat{r}\times\vec{M}^I_{\tau,\kappa}(\vec{r})\right] = 0 \quad (13.163b)$$

$$\lim_{r\to\infty} \vec{r}\times\hat{r}\times\vec{L}^I_{\sigma,\kappa}(\vec{r}) = 0 \quad (13.163c)$$

With the relations $\hat{r}\times\vec{n}_\nu = \vec{m}_\nu$, $\hat{r}\times\vec{m}_\nu = -\vec{n}_\nu$ and subject to the fact that (13.163a–c) must be valid for any ϑ, φ it is seen that

$$\lim_{r\to\infty}\left\{\kappa r j_\sigma(\kappa r) - j\frac{\kappa_0}{\kappa}\frac{d}{dr}[r j_\sigma(\kappa r)]\right\} = 0 \quad (13.164a)$$

$$\lim_{r\to\infty}\left\{\frac{d}{dr}[rj_\tau(\kappa r)] + \mathrm{j}\kappa_0 r j_\tau(\kappa r)\right\} = 0 \tag{13.164b}$$

$$\lim_{r\to\infty} j_n(\kappa r) = 0 \tag{13.164c}$$

With the asymptotic representation of the spherical Bessel functions for large values of its argument

$$\nu \text{ fixed}, |\kappa r|\to\infty \qquad \begin{aligned} j_\nu(\kappa r) &\approx \frac{1}{\kappa r}\cos\left[\kappa r - \frac{\pi}{2}(\nu+1)\right] \\ \frac{d}{dr}[rj_\nu(\kappa r)] &\approx -\sin\left[\kappa r - \frac{\pi}{2}(\nu+1)\right] \end{aligned} \tag{13.165}$$

we derive from (13.164a–c):

$$\tan\left[\kappa r - \frac{\pi}{2}(\sigma+1)\right] = \mathrm{j}\frac{\kappa}{\kappa_0} \tag{13.166a}$$

$$\sigma, \tau \text{ fixed}, r\to\infty \qquad \tan\left[\kappa r - \frac{\pi}{2}(\tau+1)\right] = \mathrm{j}\frac{\kappa_0}{\kappa} \tag{13.166b}$$

$$\frac{1}{\kappa r}\cos\left[\kappa r - \frac{\pi}{2}(\sigma+1)\right] = 0 \tag{13.166c}$$

Note that (13.166a), (13.166b), and (13.166c) are transcendental equations for finding the paths of integration in the complex κ-plane for the parts due to the $\vec{N}^I_{\sigma,\kappa}$ functions, the $\vec{M}^I_{\tau,\kappa}$ functions, and the $\vec{L}^I_{\sigma,\kappa}$ functions, respectively, in the Green's dyadic (13.160).

For fixed and real σ and $r\to\infty$ obviously (13.166c) can be satisfied only if κ lies on the real axis, as shown in Fig. (13.10a). This is the path of integration for the terms with $\vec{L}$-functions in the Green's dyadic. With Euler's formula we can write (13.166a, b) as

$$e^{2\mathrm{j}[\kappa r - \pi/2(\nu+1)]} = \frac{\kappa_0 - \kappa}{\kappa_0 + \kappa} \tag{13.167}$$

where $\nu = \sigma$ or $\nu = \tau$. If $|\kappa| \ll \kappa_0$, we have as a first-order approximation [51]

$$e^{2\mathrm{j}[\kappa r - \pi/2(\nu+1)]} = 1 \tag{13.168}$$

with the solutions

$$\kappa = \frac{\pi}{r}\left(l + \frac{\nu+1}{2}\right), \qquad l = 0, 1, 2, 3, \ldots \tag{13.169}$$

A second-order approximation is derived from (13.169) as

$$\kappa \approx \frac{\pi}{r}\left(l + \frac{\nu+1}{2} + \varepsilon\right) \qquad l = 0, 1, 2, 3, \ldots; \qquad \varepsilon \ll l + \frac{\nu+1}{2}$$

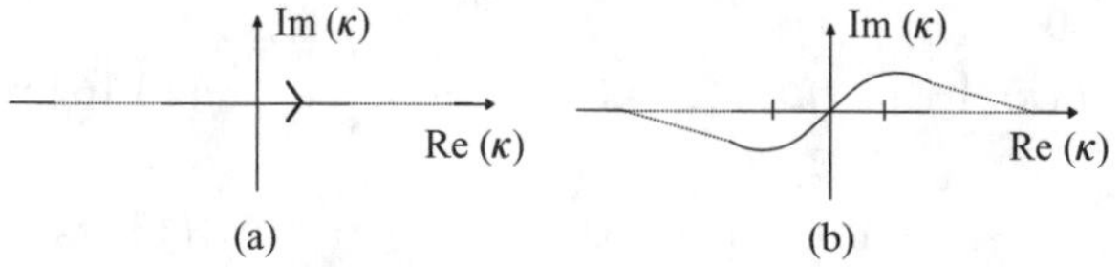

Figure 13.10 Path of integration for the terms with $\vec{L}$ functions (a) and principle path of integration for the terms with $\vec{N}$ and $\vec{M}$ functions (b) in the Green's dyadic.

and is inserted into (13.167). Using the first two terms of the exponential function's Taylor series approximation, we derive as a second-order approximation for the case $\kappa \ll \kappa_0$

$$\kappa \approx \frac{\pi}{r}\left(l + \frac{\nu+1}{2}\right)\left(1 + \frac{\mathrm{j}}{\kappa_0 r}\right) \qquad l = 0, 1, 2, 3, \ldots \tag{13.170}$$

A similar procedure for the case $\kappa \gg \kappa_0$ yields the second-order approximation

$$\kappa \approx \frac{\pi}{r}\left(l + \frac{\nu+1}{2}\right)\left(1 + \mathrm{j}\frac{\kappa_0 r}{\pi^2}\frac{1}{l^2}\right), \qquad l \to \infty \tag{13.171}$$

The results in (13.170) and (13.171) are represented graphically in Fig. 13.10b), which shows the principal path of integration for the terms with $\vec{N}$ and $\vec{M}$ functions in the Green's dyadic. Note that the path cannot have any further intersections with the real axis, which may be seen by setting $\mathrm{Im}(\overline{\kappa}) = 0$ in (13.167).

APPENDIX 13.3 THE EULER SUMMATION TECHNIQUE

Consider a series with terms v_n, which converges slowly or even seems to diverge, although it should yield a finite limiting value. In order to obtain a new and better convergent series which yields the same finite limiting value, introduce a parameter y whose choice will determine the rapidity of convergence of the transformed series

$$\begin{aligned}\sum_{n=1}^{\infty} v_n &= \sum_{n=1}^{\infty} \frac{1-y}{1-y} v_n \\ &= \frac{1}{1-y}\left[v_1 + \sum_{n=1}^{\infty}(v_{n+1} - y v_n)\right]\end{aligned} \tag{13.172}$$

If the transformed series converges slowly also, then apply Euler's transformation a second time, a third time (leaving out the first term each time) ... until the newest series is of sufficient convergence. It can be shown that the rapidity of convergence is maximized if the parameter y is chosen as

$$y = \lim_{n\to\infty} \frac{v_{n+1}}{v_n}, \qquad y \neq 1 \tag{13.173}$$

As a first example consider an alternating series

$$\sum_{n=1}^{\infty} v_n = +1 - 1 + 1 - 1 + \ldots \tag{13.174}$$

Choosing $y = -1$ and applying Euler's technique once, the result is $+1/2$. This is a

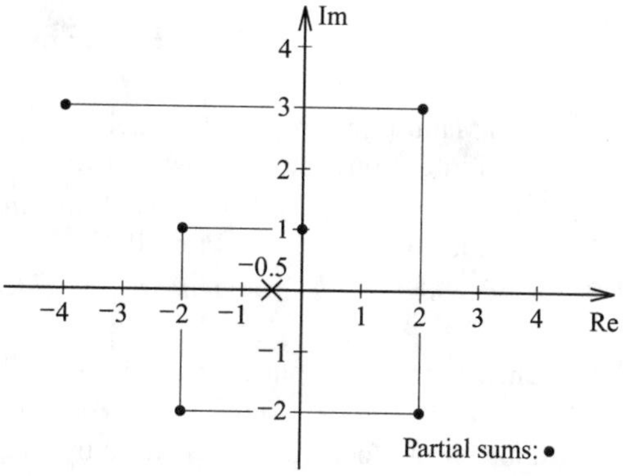

Figure 13.11 Graphical representation of complex-valued partial sums of the series (13.176).

plausible result, since it represents the mean value of 0 and 1. As a further example consider the following series with complex valued terms

$$\sum_{n=1}^{\infty} v_{\rm n} = +1e^{j\pi/2} + 2e^{j\pi} + 3e^{j3\pi/2} + 4e^{j2\pi} + \cdots \tag{13.175}$$

Regarding the amplitudes of the complex terms we will call it a "quasi-divergent" series. Choosing

$$y = \lim_{n\to\infty} \frac{v_{n+1}}{v_n} = e^{j\pi/2} = j \tag{13.176}$$

and applying Euler's summation technique two times in sequence, the result is $-1/2$. This is a plausible result too, which can be demonstrated by plotting the partial sums of the series (13.176) in the complex plane (Fig. 13.11). It is apparent that the partial sums are circulating asymptotically around the point $-1/2$, i.e., around the value attributed to the series by Euler's summation technique automatically.

A general algorithm for an N-fold Euler transformation is given by

$$\sum_{n=1}^{\infty} v_n = \sum_{m=0}^{N-1} \left(\frac{y}{1-y}\right)^{m+1} \Delta^m \left(\frac{v_n}{y^n}\right)_{n=1} + \left(\frac{y}{1-y}\right)^{N} \sum_{n=1}^{\infty} y^n \Delta^N \left(\frac{v_n}{y^n}\right) \tag{13.177}$$

where we have introduced

$$\Delta^0\left(\frac{v_n}{y^n}\right) = \frac{v_n}{y^n} \quad \text{and} \quad \Delta^m\left(\frac{v_n}{y^n}\right) = \Delta^{m-1}\left(\frac{v_{n+1}}{y^{n+1}}\right) - \Delta^{m-1}\left(\frac{v_n}{y^n}\right), \quad m \geq 1 \tag{13.178}$$

Finally, it is noted that the series (13.115) representing the backscattering RCS of a semi-infinite elliptic cone are of this "quasi-divergent" type, at least asymptotically. To evaluate these series, we take into account that the difference between successive eigenvalues σ or τ becomes π/ϑ_0 asymptotically. On behalf of the phase factors $e^{j\sigma\pi}$ or $e^{j\tau\pi}$ in the series (13.115), the rapidity of convergence of the RCS series is maximized if the parameter y is properly chosen to be

$$y = e^{j\pi^2/\vartheta_0} \tag{13.179}$$

REFERENCES

[1] Bateman Manuscript Project, A. Erdélyi (ed.), *Higher Transcendental Functions (III)*, New York, Toronto, London, McGraw-Hill Book Co., 1955.

[2] R. M. Bevensee, *Handbook of Conical Antennas and Scatterers*, New York, London, Paris, Gordon and Breach Science Pub., 1973.

[3] S. Blume, *Theorie Elektromagnetischer Felder*, 4th edn, Heidelberg, Germany, Hüthig-Verlag, 1994.

[4] S. Blume, "Spherical Multipole Analysis of Electromagnetic and Acoustical Scattering by a Semi-infinite Elliptic Cone," *IEEE Antennas and Propagat. Magazine*, **38**, 33–44, 1996.

[5] S. Blume and J. Boersma, "Diffraction by an Elliptic Cone and the Solution of the Wedge Diffraction Problem," *Arch. Elektrotech.*, **75**, pp. 403–409, 1992.

[6] S. Blume and G. Kahl, "The Physical Optics Radar Cross Section of an Elliptic Cone," *IEEE Trans. Antennas Propagat.*, **AP-35**, pp. 457–460, 1987.

[7] S. Blume and J. Schulte, "An Investigation of the Diffraction of Waves by Complementary Infinite Double Sectors," *Electrical Engineering*, **81**, 1–10, 1998.

[8] S. Blume and Ch. Czeslik, "Diffraction and Scattering of Electromagnetic Waves by Spheres with Elliptic Slit Aperture and Interior Loading," *Electrical Engineering*, **80**, 145–154, 1997.

[9] S. Blume and V. Krebs, "Numerical Evaluation of Dyadic Diffraction Coefficients and Bistatic Radar Cross Sections for a Perfectly Conducting Semi-infinite Elliptic Cone," *IEEE Trans. Antennas and Propagat.*, **46**, 414–424, 1998.

[10] S. Blume and L. Klinkenbusch, "Scattering of a Plane Electromagnetic Wave by a Spherical Shell with an Elliptical Aperture," *IEEE Trans. Electromagn. Compat.*, **34**, 308–314, 1992.

[11] S. Blume, L. Klinkenbusch, and U. Uschkerat, "The Radar Cross Section of the Semi-infinite Elliptic Cone," *Wave Motion*, **17**, 365–389, 1993.

[12] S. Blume and U. Uschkerat, "The Radar Cross Section of the Semi-infinite Elliptic Cone: Numerical Evaluation," *Wave Motion*, **22**, 311–324, 1995.

[13] J. Boersma and J. K. M Jansen, Electromagnetic Field Singularities at the Tip of an Elliptic Cone, *EUT Report*, **90-WSK-01**, Department of Mathematics and Computing Science, Eindhoven University of Technology, Eindhoven, 1990.

[14] C. J. Bouwkamp and H. B. G. Casimir, "On Multipole Expansions in the Theory of Electromagnetic Radiation." *Physica*, **XX**, 539–554, 1954.

[15] J. J. Bowman, T. B. A. Senior, and P. L. E. Uslenghi, *Electromagnetic and Acoustic Scattering by Simple Shapes*, revised printing, Hemisphere Publishing Corporation, New York, 1987.

[16] R. E. Collin, *Field Theory of Guided Waves* (2nd ed.), N.J., Piscataway, IEEE Press, 1991.

[17] P. Debye, "Der Lichtdruck auf Kugeln von Beliebigem Material," *Ann. Physik* (4. Folge), **30**, 57–136, 1909.

[18] N. Engheta, W. D. Murphy, and V. Rokhlin, "The Fast Multipole Method for Electromagnetics," *IEEE Trans. Antennas and Propagat.*, **40**, 634–641, 1992.

[19] L. B. Felsen, "Backscattering from Wide-angle and Narrow-angle Cones," *J. Appl. Phys.*, **26**, 138–151, 1955.

[20] B. Grafmüller, *Kegelleitungen und Kegelhörner Elliptischen Querschnitts*, Dissertation, Ruhr-Universität, Bochum, Germany, 1985.

[21] Ch. Hafner and L. Bomholt, *The 3D Electromagnetic Wave Simulator*, John Wiley-

Interscience, New York, Chichester, 1993.

[22] T. B. Hansen, "Corner Diffraction Coefficients for the Quarter Plane," *IEEE Trans. Antennas Propagat.*, **AP-39**, 976–984, 1991.

[23] W. W. Hansen, "A New Type of Expansion in Radiation Problems," *Physical Review*, **47**, 139–143, 1935.

[24] W. W. Hansen and L. I. Schiff, *Theoretical Study of Electromagnetic Waves Scattered from Shaped Metallic Surfaces*, Stanford University Microwave Laboratory Quarterly Reports 1 to 4, Contract W28-099-333, 1948.

[25] G. Hellwig, *Differentialoperatoren der Mathematischen Physik*, Springer Verlag, Berlin, Göttingen, Heidelberg, 1964.

[26] K. C. Hill, *A UTD Solution to the EM Scattering by the Vertex of a Perfectly Conducting Plane Angular Sector*, Ph.D. Thesis, The Ohio State University, Columbus, 1990.

[27] E. L. Ince, "The Periodic Lamé Functions," *Proc. Roy. Soc. Edinburgh*, **60**, 47–63, 1939/40.

[28] E. L. Ince, "Further Investigations into the Periodic Lamé Functions," *Proc. Roy. Soc. Edinburgh*, **60**, 83–99, 1939/40.

[29] J. D. Jackson, *Classical Electrodynamics*, 2nd edn, John Wiley and Sons, Inc., New York, 1975.

[30] J. K. M. Jansen, *Simple-periodic and Non-periodic Lamé Functions and their Application in the Theory of Conical Waveguides*, Dissertation, Techn. Hogeschool, Eindhoven, 1976.

[31] J. E. Jensen, ed., "Spherical Near-field Antenna Measurements," *IEE electromagnetic wave series*, **26**, Peter Peregrinus Ltd., London, 1988.

[32] J. Jin, *The Finite Element Method in Electromagnetics*, John Wiley and Sons, Chichester and New York, 1993.

[33] L. Klinkenbusch, *Sphärische Multipolanalyse der Beugung und Streuung an elliptisch berandeten Kugelschalen*, Dissertation, Ruhr-Universität, Bochum, Germany, 1991.

[34] L. Klinkenbusch, *Theorie der Sphärischen Absorberkammer und des Mehrschaligen Kugelschirmes*, Habilitation Thesis, Ruhr-Universität, Bochum, Germany, 1996.

[35] L. Klinkenbusch, "Scattering of an Arbitrary Plane Electromagnetic Wave by a Finite Elliptic Cone." *Archiv für Elektrotechnik*, **76**, 181–193, 1993.

[36] L. Kraus, *Diffraction by a Plane Angular Sector*. Dissertation, New York University, 1955.

[37] L. Kraus and L. M. Levine, "Diffraction by an Elliptic Cone," Comm., *Pure Appl. Math.*, **14**, 49–68, 1961.

[38] W. Magnus, F. Oberhettinger, and R. P. Soni, *Formulas and Theorems for Special Functions of Mathematical Physics*, Springer Verlag, New York, 1966.

[39] J. R. Mentzer, *Scattering and Diffraction of Radio Waves*, London, New York, Pergamon Press, 1955.

[40] G. Mie, "Beiträge zur Optik trüber Medien, Speziell Kolloidaler Metallösungen," *Ann. Physik* (4. Folge), **25**, 379–445, 1908.

[41] F. H. Northover, "The Diffraction of Electromagnetic Waves Around a Finite, Perfectly Conducting Cone," *Quart. Journ. Mech. Appl. Math.*, **XV**, Pt. 1, 1–9, 1962.

[42] P. H. Pathak, "On the Eigenfunction Expansions of Electromagnetic Dyadic Green's Functions," *IEEE Trans. Ant. Propag.*, **AP-31**, 837–846, 1983.

[43] J. Radlow, "Note on the Diffraction at a Corner," *Arch. Rational Mech. Anal.*, **19**, 62–70, 1965.

[44] C. C. Rogers, J. R. Schindler, and F. K. Schultz, "The Scattering of a Plane Electromagnetic

Wave by a Finite Cone." In *Electromagnetic Theory and Antennas*, E. C. Jordan, ed., Pergamon Press, Oxford, 1963.

[45] J. N. Sahalos and G. A. Thiele, "Periodic Lamé Functions and Applications to the Diffraction by a Plane Angular Sector," *Canadian Journal of Physics*, **61**, 1583–1591, 1983.

[46] R. S. Satterwhite, *Diffraction by a Plane Angular Sector*, Dissertation, The Ohio State University, Columbus, 1969.

[47] C. E. Schensted, "Electromagnetic and Acoustic Scattering by a Semi-infinite Body of Revolution," *J. Appl. Phys.*, **26**, 306–308, 1955.

[48] T. B. A. Senior, A Note on Hansen's Vector Wave Functions," *Can. J. Phys.*, **38**, 1702–1705, 1960.

[49] K. M. Siegel, J. W. Crispin Jr., and C. E. Schensted, "Electromagnetic and Acoustical Scattering from a Semi-infinite Cone," *J. Appl. Phys.*, **26**, 309–313, 1955.

[50] V. P. Smyshlyaev, "The High-frequency Diffraction of Electromagnetic Waves by Cones of Arbitrary Cross Sections," *SIAM J. Appl. Math.*, **53**, 670–688, 1993.

[51] A. Sommerfeld, *Partial Differential Equations*, Academic Press, New York, 1949.

[52] A. Sommerfeld, *Optics*, Academic Press, New York, 1954.

[53] A. Sommerfeld, "Die Greensche Funktion der Schwingungsgleichung," *Jahresbericht der Deutschen Mathem.-Vereinigung*, **21**, 309–353, 1912.

[54] R. C. Spencer, *Back Scattering from Conducting Surfaces*, Air Force Cambridge Research Center Report No. E5070, 1951.

[55] J. A. Stratton, *Electromagnetic Theory*, McGraw-Hill Book Co., New York, London, 1941.

[56] C. T. Tai, *Dyadic Green Functions in Electromagnetic Theory*, 2nd edn, IEEE Press, Piscataway, NJ, 1991.

[57] E. Vafiadis and J. N. Sahalos, "The Electromagnetic Field of a Slotted Elliptic Cone," *IEEE Trans. Antennas Propagat.*, **AP-38**, 1894–1898, 1990.

[58] D.-S. Wang and L. N. Medgyesi-Mitschang, "Electromagnetic Scattering from Finite Circular and Elliptic Cones," *IEEE Trans. Antennas Propagat.*, **AP33**, 488–497, 1985.

[59] R. W. Ziolkowski and W. A. Johnson, "Electromagnetic Scattering of an Arbitrary Plane Wave from a Spherical Shell with a Circular Aperture," *J. Math. Phys.*, **28**, 1293–1314, 1987.

Chapter 14

A SYSTEMATIC STUDY OF PERFECTLY MATCHED ABSORBERS

Mustafa Kuzuoglu and Raj Mittra

ABSTRACT

In the numerical modeling of electromagnetic radiation and/or scattering problems by finite methods, the computational domain is usually truncated by an artificial boundary on which suitable absorbing boundary conditions (ABCs) are imposed. An alternative approach to mesh truncation, which was introduced by Berenger for finite difference time domain (FDTD) implementation, employs a region which is designed to absorb plane waves whose frequency and incident angle are arbitrary. Since the plane waves are transmitted into the region without any reflection, the region is called a perfectly matched layer (PML). In this chapter, our main objective is to systematically derive the partial differential equations satisfied within the PML media. It is demonstrated that both Maxwellian PMLs (the anisotropic and the bianisotropic PML media) as well as non-Maxwellian PMLs (Berenger PML) can be realized by assuming a field decay behavior within the PML. Causality and reciprocity issues, and their implications in the proper design and operation of perfectly matched absorbers, are also discussed. It is shown that if the constitutive parameters of the PML medium satisfy the Kramers-Kronig relations, the medium is causal and does not exhibit a singular behavior at lower frequencies. This, in turn, enables us to extend the PML concept to the static case. The reciprocity concept is also important in the design of PMLs, since the PML medium occupies a bounded domain in mesh truncation applications. The medium must be reciprocal and, as a consequence, the decay behavior of the waves traveling in the longitudinal direction must be identical to that of the waves reflected from the terminating boundary and traveling in the opposite direction. Some examples of causal/non-causal and reciprocal/non-reciprocal PMLs are given in the work to illustrate these behaviors.

14.1 INTRODUCTION

During the past several years, a variety of techniques have been developed for the numerical modeling and solution of electromagnetic scattering and radiation problems involving objects with arbitrary geometry and constitutive parameters. These computational methods are either based on the direct discretization of the governing partial differential equations, or on the discretization of the integral equations that are essentially reformulations of the boundary value problems associated with the differential equations. The well-known *Method of Moments* (*MoM*) has been successfully applied to the solution of the integral equations arising in radiation and scattering problems of electromagnetics [1]. Recently, however, there has been a trend toward solution of differential equations because of the versatility of the approach in handling inhomogeneous media. A popular approach for the direct discretization of Maxwell's equations is the *Finite Difference Time Domain* (*FDTD*) method, wherein the partial derivatives with respect to space and time variables are discretized by using finite difference expressions, and the solution is obtained iteratively by marching-on-in-time [2].

Another technique, which is also characterized as a differential equation-based approach, is the finite element method (FEM) [3]. In FEM, one begins with a weak variational form that embodies the original boundary value problem and then discretizes it by approximating the unknown function in terms of a finite number of trial (or shape) functions, each of which is non-zero over a small portion of the computational domain. This, in turn, results in a sparse matrix equation for the unknown degrees of freedom. The sparsity of the matrix equation is a major advantage of FEM over MoM. This is because, in contrast to the integral operators used in the MoM that yield dense matrices after discretization, the local character of the partial differential equation formulation is preserved in the FEM.

In electromagnetic scattering and/or radiation problems involving open regions, the physical domain extends to infinity. For this case, the existence and uniqueness of the solution of Maxwell's equations can be guaranteed by imposing the Sommerfeld radiation condition, which is governed by the behavior of the electromagnetic field at points that are sufficiently far away from the sources. This asymptotic behavior is built into the kernels of the integral operators, and, hence, no additional effort is necessary to enforce it. In contrast, auxiliary constraints must be imposed to account for the proper extension of the fields to infinity in the finite methods.

In order to employ finite methods (FDTD or FEM) to the solution of radiation and/or scattering problems involving spatially unbounded domains, the physical domain is truncated by an artificial boundary to obtain a bounded computational domain. On this boundary, the unknown function (either a scalar or a vector field) must satisfy an operator equation, so that the computational domain is effectively coupled to the free space region that extends to infinity. This coupling is achieved by designing the boundary operator in a manner such that the outgoing waves that arrive at the boundary from the interior domain are absorbed, with negligible reflection.

Basically, there are two different types of boundary operators, each of which has its own advantages as well as disadvantages. The first of these operators has a non-local character, i.e., it relates the value of the function at each point on the boundary to all other points on the same boundary. This non-local boundary operator equation can be defined either as an infinite modal expansion with unknown coefficients [4], or as a boundary integral equation [5]. In either case, the formulation based on the coupling of the interior partial differential equation and the boundary operator equation is exact. In the modal expansion approach, the FEM equations are coupled to the truncated modal equation, and the coefficients of the truncated series appear as additional unknown degrees of freedom. In the integral equation approach, the integral equation is discretized by the boundary element method (BEM) and the resulting linear system of equations is solved simultaneously with those arising from the FEM implementation [6]. Both of these approaches have a major disadvantage however, since the linear system of equations resulting from the discretization of the non-local boundary operator is a dense matrix equation, which, when coupled to the sparse FEM matrix, spoils its sparsity and, hence, increases the computational cost.

The second type of boundary operator is local in nature, and is a partial differential operator designed to absorb the outgoing waves. The boundary partial differential equations, which are basically annihilators of outgoing waves, are called absorbing boundary conditions (ABCs). The expressions of these ABCs not only depend on the geometry of the artificial boundary, but also on the partial differential equation governing the problem. ABCs have been derived for planar [7], cylindrical [8], and spherical boundaries [9], for the Helmholtz as well as for the vector wave equation. ABCs are usually derived by designing partial differential operators which annihilate the first few dominant terms of the asymptotic expansions (with respect to distance measured from the sources) of the unknown field [8]. However, ABCs are not very accurate for evanescent plane waves or for plane waves with large angle of incidence with respect to the normal to the boundary. This, in turn, dictates that the artificial boundary must be removed sufficiently far away from the scatterer or antenna. One consequence of this is that the FEM mesh size may increase dramatically, especially at lower frequencies.

In addition to the analytically defined ABCs, numerical absorbing boundary conditions (NABCs) have been developed, as a linear system of equations that relate the values of the function at the boundary nodes to those at the neighboring points. The equations themselves are derived by requiring that the outgoing waves be effectively absorbed by the boundary [10]–[13]. The relationship between ABCs and NABCs has been demonstrated by Stupfel and Mittra [14]–[15], who have shown that suitably defined NABCs tend to analytical ABCs on locally planar, cylindrical, or spherical boundaries, in the limit of mesh size tending to zero.

An alternative to using the local or non-local ABCs, is to combine the finite elements with elements of infinite extent. The concept of an infinite element was introduced by Bettess [16], for problems governed by Laplace's equation in exterior domains. This method has also been used in electromagnetic wave-scattering problems [17]. However, it has been observed that there are difficulties in the implementation of numerical integration that must be performed over the infinite element, especially in wave propagation problems [18].

The most recent approach to the problem of mesh truncation has been introduced by Berenger [19]. This method, which is investigated in detail in this chapter, is basically the truncation of the computational domain by a layer which absorbs impinging plane waves, without any reflection, irrespective of their frequency and angle of incidence. In view of the reflectionless character of the layer, it is referred to as a perfectly matched layer (PML). Following the verification of the superiority of Berenger's PML [20], [21], the PML concept has been applied extensively in FDTD simulations [22]–[32]. Although Berenger's PML has been introduced in FDTD applications, the PML concept has also been used in FEM, especially after the realization of a perfectly matched absorber as an anisotropic medium, by Sacks et al. [33], who have interpreted the PML as a medium with suitably defined complex permittivity and permeability tensors. The performance of the anisotropic PML has been investigated in time and frequency domain applications in conjunction with the FDTD scheme [34]–[38], and FEM [39]–[47], respectively. The anisotropic PML, which has been originally introduced in Cartesian coordinates, has been extended to cylindrical and spherical coordinates in time and frequency domains [48]–[54]. There are also studies where PML media are interpreted in terms of complex coordinate stretching, and it has been shown that the anisotropic PML can be constructed by applying a complex coordinate transformation in the direction normal to the PML-free space interface [55]–[56]. The other aspects of PMLs that appear in the literature can be summarized as: performance analysis of PMLs for evanescent and/or low-frequency waves, [57]–[62], Maxwellian absorbers [63]–[64], and implementation of PMLs in TLM analysis [66]–[69]. A recent research area is the design of material-independent PMLs [70]–[73], where the PML medium is designed to match the interior domain with arbitrary constitutive parameters.

In this chapter, we discuss some aspects of perfectly matched absorbers in a systematic way. In Section 14.2, we demonstrate how different kinds of PMLs (Maxwellian or non-Maxwellian) can be realized, by postulating the field variation within the absorbing medium, and deriving the partial differential equations whose solutions exhibit this desired behavior. Section 14.3 is mainly about the discussion of causality in PMLs, and the importance of this concept in extending the PML action to the static case. In Section 14.4, we discuss reciprocity in PMLs, and give examples of reciprocal and non-reciprocal perfectly matched absorbers. Our main emphasis in this chapter is on the conceptual aspects of the PML method, rather that its utilization as a mesh truncation approach in numerical simulations.

14.2 SYSTEMATIC DERIVATION OF THE EQUATIONS GOVERNING PERFECTLY MATCHED ABSORBERS

In this section, we derive the partial differential equations satisfied by Maxwellian (or non-Maxwellian) fields within perfectly matched absorbers. For this purpose, we decompose the three-dimensional space $\mathcal{R}^3$ into two components Ω_{FS} and Ω_{PML}, such that

$$\Re^3 = \Omega_{FS} \cup \Omega_{PML} \tag{14.1}$$

where Ω_{FS} represents the region occupied by free space, and Ω_{PML} is the region occupied by the perfectly matched absorber. As a first step, Ω_{FS} and Ω_{PML} are assumed to be half-spaces, and their planar interface, denoted by Γ, is taken as a constant Cartesian coordinate surface for simplicity. The basic hypothesis in the construction of a PML region Ω_{PML}, is that, a plane wave in Ω_{FS} with arbitrary frequency and angle of incidence must be transmitted into Ω_{PML} without any reflection, and the transmitted field must decay in the longitudinal direction perpendicular to Γ. The partial differential equations satisfied by the fields within Ω_{PML} are derived according to the basic PML hypothesis given above, and the steps given below are followed in the systematic realization of a PML medium:

Step 1 Postulate the nature of the field variation within the absorbing medium.

Step 2 Derive the second-order partial differential equations whose solutions exhibit the desired decay behavior.

Step 3 Rewrite these partial differential equations in the format of Maxwell's equations.

Step 4 Deduce the constitutive relations in Ω_{PML} to identify the absorber as a material with appropriate constitutive parameters.

It will be demonstrated in this section that the most crucial step in the derivation is Step 3, since it is usually possible to deduce different types of Maxwell's equations from the second-order partial differential equation obtained in Step 2. It may even be possible to obtain equations that are non-Maxwellian (as in the case of Berenger's PML). This implies that it is possible to obtain the same PML action by means of different Maxwellian or non-Maxwellian partial differential equations. These equations are obviously related to different media, and, we name each equation set (and the corresponding medium) as a different *realization*. This not only enables us to derive the PMLs that are well known in the literature, such as the Berenger's PML and the anisotropic PML, but to also develop new types of PMLs. The details of the derivations will be given for a TM plane wave in the following subsection. Since this assumption reduces the problem to a two-dimensional one, it becomes relatively easy to follow the steps of the derivations.

14.2.1 Different PML Realizations for a TM Model Problem

Let Ω_{FS} and Ω_{PML} be defined as

$$\Omega_{FS} = \{(x, y)|x < 0\} \tag{14.2}$$

$$\Omega_{PML} = \{(x, y)|x > 0\} \tag{14.3}$$

and the interface Γ is the y axis, as shown in Fig. 14.1. In Ω_{FS}, the plane wave $E_z(x, y)$ is given by (a suppressed time dependence $e^{j\omega t}$ is assumed in this chapter):

$$E_z(x, y) = e^{-jk[\cos\theta x + \sin\theta y]} \tag{14.4}$$

where k is the wave number and θ is the angle between the direction of propagation of the plane wave and the x axis. According to the PML hypothesis, this plane wave must be transmitted into the region Ω_{PML} without any reflection, and must be attenuated in the $+x$ direction. To obtain the partial differential equation satisfied by $E_z(x, y)$ in Ω_{PML}, we follow the steps given above.

As a first step, we stipulate that in Ω_{PML}, $E_z(x, y)$ be given by

$$E_z(x, y) = f(x)e^{-jk[\cos\theta x + \sin\theta y]} \tag{14.5}$$

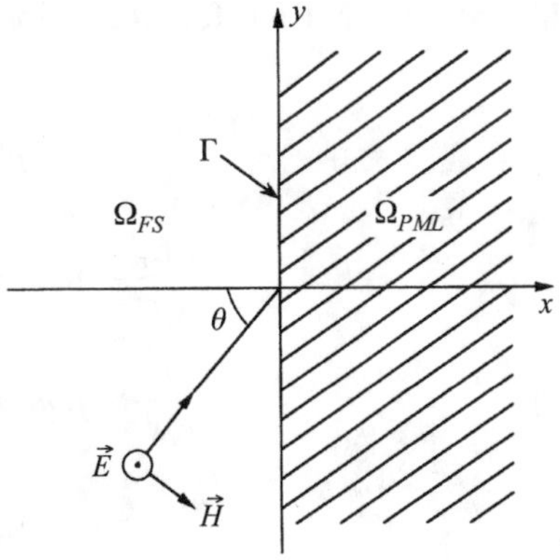

Figure 14.1 A TM wave with arbitrary frequency and direction of propagation incident on the free space–PML interface Γ, where both of the regions are half spaces.

where $f(x)$ is a function satisfying the following two properties:

1. $f(0) = 1$.
2. $f(x)$ decreases monotonically for $x > 0$.

For instance, we can choose $f(x)$ as

$$f(x) = e^{-\alpha \cos \theta x} \tag{14.6}$$

where $\alpha > 0$ is a constant. An important aspect of (14.6) is the dependence of the decay characteristics on the incidence angle θ. This dependence has important consequences, as will be explained in Section 14.4 on reciprocity in PMLs. Now, let the parameter a be defined as

$$a = 1 + \frac{\alpha}{jk} \tag{14.7}$$

Then, (14.5) becomes

$$E_z(x, y) = e^{-jk[a \cos \theta x + \sin \theta y]} \tag{14.8}$$

By direct differentiation, we can readily see that $E_z(x,y)$ satisfies the partial differential equation

$$\frac{1}{a^2}\frac{\partial^2 E_z}{\partial x^2} + \frac{\partial^2 E_z}{\partial y^2} + k^2 E_z = 0 \tag{14.9}$$

Hence, we have obtained the second-order partial differential equation as implied by the second step of our derivation scheme. The next step will be the extraction of Maxwell's equations (or, Maxwell-like equations) from (14.9). Below, we show that it is possible to obtain three different realizations, namely, the split-field formulation of Berenger, the anisotropic PML and the bianisotropic PML.

14.2.1.1 The Split-Field Realization (Berenger's PML)

In this realization, E_z is decomposed as [19]

$$E_z = E_{zx} + E_{zy} \tag{14.10}$$

and from (14.9), we can extract four equations for the unknowns, H_x, H_y, E_{zx}, and E_{zy} as

$$\frac{\partial E_z}{\partial y} = -j\omega\mu_0 H_x \tag{14.11}$$

$$\frac{\partial E_z}{\partial x} = j\omega\mu_0 a H_y \tag{14.12}$$

$$\frac{\partial H_y}{\partial x} = j\omega\varepsilon_0 a E_{zx} \tag{14.13}$$

$$\frac{\partial H_x}{\partial y} = -j\omega\varepsilon_0 E_{zy} \tag{14.14}$$

It is evident from (14.12) and (14.13) that the basic advantage of this realization is the appearance of the parameter a as a simple multiplier. By using the fact that $j\omega a = j\omega[1 + \alpha c/j\omega]$ we can verify that the factor $j\omega a$ in the frequency domain corresponds to the operator $(\partial/\partial t + \alpha c)$ in the time domain. However, due to the splitting of the E_z field, the equations are non-Maxwellian. In the time domain (14.11)–(14.14) become

$$\frac{\partial E_z}{\partial y} = -\mu_0 \frac{\partial H_x}{\partial t} \tag{14.15}$$

$$\frac{\partial E_z}{\partial x} = \mu_0 \frac{\partial H_y}{\partial t} + \alpha c \mu_0 H_y \tag{14.16}$$

$$\frac{\partial H_y}{\partial x} = \varepsilon_0 \frac{\partial E_{zx}}{\partial t} + \alpha c \varepsilon_0 E_{zx} \tag{14.17}$$

$$\frac{\partial H_x}{\partial y} = -\varepsilon_0 \frac{\partial E_{zy}}{\partial t} \tag{14.18}$$

Implementation of FDTD to (14.15)–(14.18) is straightforward, owing to the split-field derivation and this serves to explain the popularity of Berenger's PML in the FDTD community [20]–[32].

14.2.1.2 The Anisotropic Realization

For this case we extract three equations for the unknowns H_x, H_y, and E_z from (14.9), as follows:

$$\frac{\partial E_z}{\partial y} = -j\omega\mu_0 \frac{1}{a} H_x \tag{14.19}$$

$$\frac{\partial E_z}{\partial x} = j\omega\mu_0 a H_y \tag{14.20}$$

$$\frac{\partial H_y}{\partial x} - \frac{\partial H_x}{\partial y} = j\omega\varepsilon_0 a E_z \tag{14.21}$$

The basic difference of this realization from the previous one is the appearance of the factor $j\omega/a$ in (14.19), which corresponds to a convolution operation in the time domain. In other words, the versatility of the Berenger PML is not present in this case in time domain applications. However, the equations (14.19)–(14.21) are Maxwellian, and we can obtain the following constitutive relations [33]:

$$\vec{B} = \mu_0[\Lambda]\vec{H} \tag{14.22}$$

$$\vec{D} = \varepsilon_0[\Lambda]\vec{E} \tag{14.23}$$

where $[\Lambda]$ is the diagonal tensor given by

$$[\Lambda] = \begin{bmatrix} 1/a & 0 & 0 \\ 0 & a & 0 \\ 0 & 0 & a \end{bmatrix} \tag{14.24}$$

Since the permittivity and the permeability tensors are diagonal, we have a uniaxial anisotropic PML medium in this case. This result immediately raises the possibility of realizing an anisotropic PML as a material medium. However, this is not possible in terms of passive inclusions, because negative electric (or magnetic) conductivity terms appear in (14.22) and (14.23), which implies that this medium is active. The anisotropic PML has been successfully used as a mesh truncation method [36]–[38], and the time domain equations can be obtained easily from (14.19)–(14.21) by noting the fact that convolution operation can be represented by an equivalent differential equation:

$$\frac{\partial E_z}{\partial y} = -\frac{\partial B_x}{\partial t} \tag{14.25}$$

$$\frac{\partial E_z}{\partial x} = \frac{\partial B_y}{\partial t} \tag{14.26}$$

$$\frac{\partial H_y}{\partial x} - \frac{\partial H_x}{\partial y} = \frac{\partial D_z}{\partial t} \tag{14.27}$$

$$\mu_0 \frac{\partial H_x}{\partial t} = \frac{\partial B_x}{\partial t} + \alpha c B_x \tag{14.28}$$

$$\mu_0 \frac{\partial H_y}{\partial t} + \alpha c \mu_0 H_y = \frac{\partial B_y}{\partial t} \tag{14.29}$$

$$\varepsilon_0 \frac{\partial E_z}{\partial t} + \alpha c \varepsilon_0 E_z = \frac{\partial D_z}{\partial t} \tag{14.30}$$

14.2.1.3 The Bianisotropic Realization

From (14.9), once more we can extract three following equations for the unknowns H_x, H_y, and E_z:

$$\frac{\partial E_z}{\partial y} = -j\omega\mu_0 H_x \tag{14.31}$$

$$\frac{\partial E_z}{\partial x} = j\omega\mu_0 H_y + \frac{\alpha}{jk}\frac{1}{a}\frac{\partial E_z}{\partial x} \tag{14.32}$$

$$\frac{\partial H_y}{\partial x} - \frac{\partial H_x}{\partial y} = j\omega\varepsilon_0 E_z + \frac{\alpha}{jk}\frac{1}{a}\frac{\partial H_y}{\partial x} \tag{14.33}$$

The equations given above can be obtained from (14.19)–(14.21) by some field transformations. Such transformations have been used by Werner and Mittra [74] to demonstrate the relationship between Berenger's PML and the anisotropic PML. A different derivation, having important implications, will be given in Section 14.4.

Next, we note that the factor $1/a$ appears in (14.32) and (14.33), and its presence implies convolutions in the time domain. However, it is also possible to interpret this realization as a material medium. From (14.31)–(14.33), we can extract the constitutive relations of a bianisotropic medium with constitutive dyadics, $\bar{\bar{\varepsilon}}$, $\bar{\bar{\xi}}$, $\bar{\bar{\zeta}}$, and $\bar{\bar{\mu}}$ as follows:

$$\vec{D} = \bar{\bar{\varepsilon}} \cdot \vec{E} + \bar{\bar{\xi}} \cdot \vec{H} \tag{14.34}$$

$$\vec{B} = \bar{\bar{\zeta}} \cdot \vec{E} + \bar{\bar{\mu}} \cdot \vec{H} \tag{14.35}$$

In (14.34)–(14.35), $\bar{\bar{\varepsilon}} = \varepsilon_0 \bar{\bar{I}}$, $\bar{\bar{\mu}} = \mu_0 \bar{\bar{I}}$, where $\bar{\bar{I}}$ is the identity dyadic. All entries of $\bar{\bar{\xi}}$ and $\bar{\bar{\zeta}}$ are zero, except for ξ_{zy} and ζ_{yz}, and those are given by $\xi_{zy} = \zeta_{yz} = -(\alpha/ck^2)(1/a)(\partial/\partial x)$. Since $\bar{\bar{\xi}}$ and $\bar{\bar{\zeta}}$ are functions of k, the medium is time-dispersive. In addition, the medium is space-dispersive (or non-local) [75], because of the appearance of the space derivative $\partial/\partial x$ in the constitutive parameters.

The time domain equations read

$$\frac{\partial E_z}{\partial y} = -\frac{\partial B_x}{\partial t} \tag{14.36}$$

$$\frac{\partial E_z}{\partial x} = \frac{\partial B_y}{\partial t} \tag{14.37}$$

$$\frac{\partial H_y}{\partial x} - \frac{\partial H_x}{\partial y} = \frac{\partial D_z}{\partial t} \tag{14.38}$$

$$B_x = \mu_0 H_x \tag{14.39}$$

$$\frac{\partial B_y}{\partial t} + \alpha c B_y = \mu_0 \frac{\partial H_y}{\partial t} + \mu_0 \alpha c H_y + \alpha c E_z \tag{14.40}$$

$$\frac{\partial D_z}{\partial t} + \alpha c D_y = \varepsilon_0 \frac{\partial E_y}{\partial t} + \varepsilon_0 \alpha c E_z + \alpha c H_y \tag{14.41}$$

14.2.2 Cartesian Mesh Truncations and Corner Regions

Our discussions in the previous section were based on the assumption that Ω_{PML} is a half space. In fact, in mesh truncation applications, Ω_{PML} must be a bounded region enclosing all sources and inhomogeneities, as shown in Fig. 14.2. It is evident from Fig. 14.2 that the boundary of Ω_{PML} must be composed of constant Cartesian coordinate planes in order to make use of the theory we have developed in the previous section. The inner and outer boundaries of Ω_{PML} are defined as Γ_i and Γ_o. A successful PML design will imply that the waves incident to Γ_i from the inner domain Ω_{FS} will be transmitted into Ω_{PML} without any reflection and they will be attenuated as they traverse Ω_{PML} toward the outer boundary Γ_o, which is either a perfect electric conductor (PEC) or a perfect magnetic conductor (PMC). The PML is designed such that the field reflected from Γ_o is negligible when it reaches Γ_i and the field variation in Ω_{FS} is not affected by the fields in Ω_{PML}. If the Ω_{PML} is thick enough for the fields to decay sufficiently in the normal direction, the PML design successfully truncates the mesh, and the fields in Ω_{FS} behave as though the domain extends to infinity.

Another important aspect of realistic Cartesian PML is the determination of its parameters in a corner region. Again, we consider an idealized case where the corner

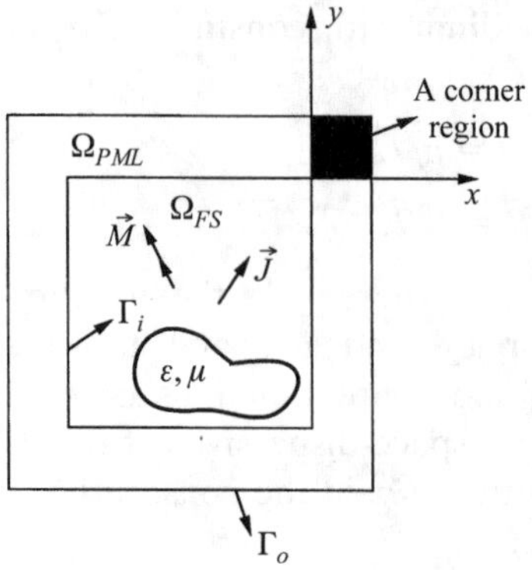

Figure 14.2 A Cartesian PML region enclosing scatterers and radiators. Such a domain necessarily has corner regions as shown in the figure.

region extends to infinity, as shown in Fig. 14.3. In this case, as well, the partial differential equations in the PML region can be derived by using the systematic approach developed in the previous section. We define the sets Ω_{FS} and Ω_{PML} as

$$\Omega_{FS} = \{(x, y)|x<0, y<0\} \tag{14.42}$$

$$\Omega_{PML} = \Omega_{PML,x} \cup \Omega_{PML,y} \cup \Omega_{PML,xy} \tag{14.43}$$

where $\Omega_{PML,x} = \{(x,y)|x>0, y<0\}$, $\Omega_{PML,y} = \{(x,y)|x<0, y>0\}$ and $\Omega_{PML,xy} = \{(x,y)|x>0, y>0\}$. These sets are shown in Fig. 14.3 where $\Omega_{PML,xy}$ corresponds to the corner set extending to infinity.

In a similar fashion, we start with a plane wave solution in Ω_{FS}:

$$E_z(x, y) = e^{-jk[\cos\theta x+\sin\theta y]} \tag{14.44}$$

We postulate the following fields in $\Omega_{PML,x}$, $\Omega_{PML,y}$, and $\Omega_{PML,xy}$:

$$E_z(x, y) = e^{-\alpha_1 \cos\theta x} e^{-jk[\cos\theta x+\sin\theta y]}, \ (x, y) \in \Omega_{PML,x} \tag{14.45}$$

$$E_z(x, y) = e^{-\alpha_2 \sin\theta y} e^{-jk[\cos\theta x+\sin\theta y]}, \ (x, y) \in \Omega_{PML,y} \tag{14.46}$$

$$E_z(x, y) = e^{-\alpha_1 \cos\theta x} e^{-\alpha_2 \sin\theta y} e^{-jk[\cos\theta x+\sin\theta y]}, \ (x, y) \in \Omega_{PML,xy} \tag{14.47}$$

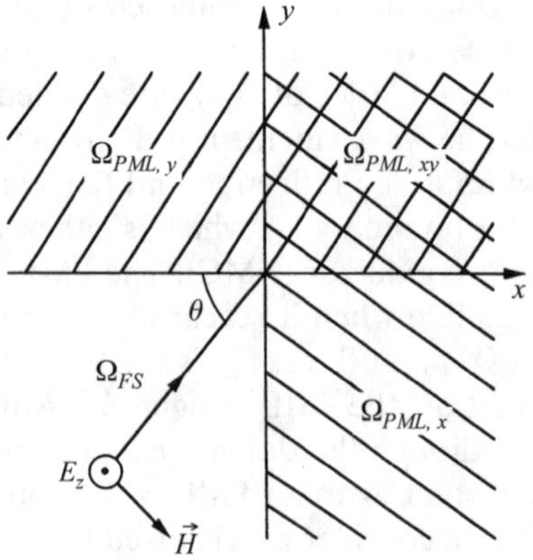

Figure 14.3 An idealized corner region, where two Cartesian half space PMLs intersect.

Letting $a = 1 + (\alpha_1/jk)$ and $b = 1 + (\alpha_2/jk)$, $E_z(x, y)$ satisfies the following partial differential equations:

$$\frac{1}{a^2}\frac{\partial^2 E_z}{\partial x^2} + \frac{\partial^2 E_z}{\partial y^2} + k^2 E_z = 0, \qquad (x, y) \in \Omega_{PML,x} \tag{14.48}$$

$$\frac{\partial^2 E_z}{\partial x^2} + \frac{1}{b^2}\frac{\partial^2 E_z}{\partial y^2} + k^2 E_z = 0, \qquad (x, y) \in \Omega_{PML,y} \tag{14.49}$$

$$\frac{1}{a^2}\frac{\partial^2 E_z}{\partial x^2} + \frac{1}{b^2}\frac{\partial^2 E_z}{\partial y^2} + k^2 E_z = 0, \qquad (x, y) \in \Omega_{PML,xy} \tag{14.50}$$

It is evident that Maxwell's (or Maxwell-type) equations can be extracted from (14.48)–(14.50). An interesting situation arises in the anisotropic realization of a corner region, where the dyadic $[\Lambda]$ appears in both $\bar{\bar{\varepsilon}}$ and $\bar{\bar{\mu}}$. For this case,

$$[\Lambda] = [\Lambda]_x = \begin{bmatrix} 1/a & 0 & 0 \\ 0 & a & 0 \\ 0 & 0 & a \end{bmatrix}, \qquad (x, y) \in \Omega_{PML,x} \tag{14.51}$$

$$[\Lambda] = [\Lambda]_y = \begin{bmatrix} b & 0 & 0 \\ 0 & 1/b & 0 \\ 0 & 0 & b \end{bmatrix}, \qquad (x, y) \in \Omega_{PML,y} \tag{14.52}$$

and

$$[\Lambda] = [\Lambda]_{xy} = \begin{bmatrix} b/a & 0 & 0 \\ 0 & a/b & 0 \\ 0 & 0 & ab \end{bmatrix}, \qquad (x, y) \in \Omega_{PML,xy} \tag{14.53}$$

From (14.51)–(14.53) we note that the dyadic $[\Lambda]$ of the corner region of a PML medium can be obtained by multiplying the dyadics belonging to those half-spaces intersecting at that corner [37].

14.2.3 Example of FEM Implementation of the Cartesian PML

As an example of how the cartesian PML can be implemented as a mesh truncation method in FEM, consider the problem of constructing the two-dimensional Green's function where the partial differential equation is the well-known Helmholtz equation, as given below:

$$\nabla^2 u + k^2 u = -\delta(\vec{r}) \tag{14.54}$$

and the solution is given by $u(\vec{r}) = (j/4) H_0^{(2)}(k|\vec{r}|)$ [76]. Let Ω be the computational domain shown in Fig. 14.4. By using the symmetry in the problem, only one-fourth of the real computational domain is used. In Ω_{FS}, u satisfies (14.54), and in Ω_{PML}, the partial differential equation is given by

$$\nabla \cdot ([\Lambda] \nabla u) + k^2 \gamma u = 0 \tag{14.55}$$

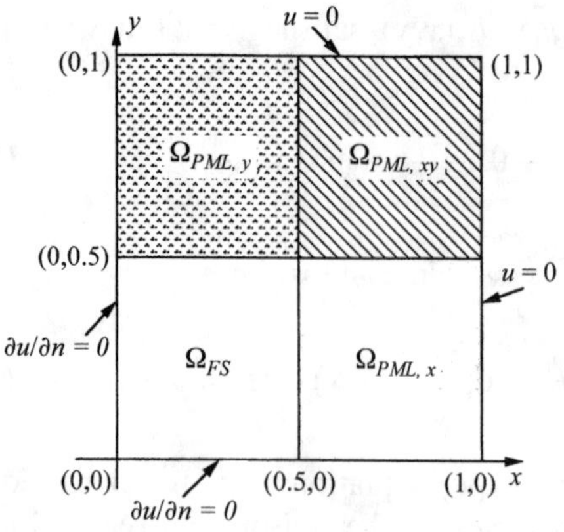

Figure 14.4 A quarter of the computational domain for the FEM evaluation of two-dimensional free-space Green's function for the Helmholtz equation.

where $[\Lambda]$ is a 2×2 diagonal matrix with nonzero entries Λ_{11} and Λ_{22}. In $\Omega_{PML,x}$, $\Omega_{PML,y}$ and $\Omega_{PML,xy}$, the parameters Λ_{11}, Λ_{22}, and γ become

$$\Lambda_{11} = 1/a,\ \Lambda_{22} = a,\ \gamma = a \qquad (x, y) \in \Omega_{PML,x} \tag{14.56}$$

$$\Lambda_{11} = b,\ \Lambda_{22} = 1/b,\ \gamma = b \qquad (x, y) \in \Omega_{PML,y} \tag{14.57}$$

$$\Lambda_{11} = b/a,\ \Lambda_{22} = a/b,\ \gamma = ab \qquad (x, y) \in \Omega_{PML,xy} \tag{14.58}$$

where $a = 1 + (\alpha_1/jk)$ and $b = 1 + (\alpha_2/jk)$.

The weak forms of (14.54) and (14.55) can be obtained as [54]

$$\int_{\Omega_{PS}} \nabla u \cdot \nabla \psi d\Omega - k^2 \int_{\Omega_{FS}} u\psi d\Omega = \int_{\Omega_{FS}} \delta(\vec{r})\, \psi d\Omega \tag{14.59}$$

$$\int_{\Omega_{PML}} [\Lambda] \nabla u \cdot \nabla \psi d\Omega - k^2 \int_{\Omega_{PML}} \gamma u \psi d\Omega = 0 \tag{14.60}$$

where ψ is a suitable weighting function.

In Figs. 14.5a and 14.5b, we present the magnitude and phase plots of u versus the y-coordinate variation for $0 \leq y \leq 1$, where $k = 2\pi$. In this case, an FEM mesh composed of eight-noded isoparametric quadrilaterals has been generated to cover the computational region $\Omega = \{(x,y)|0 \leq x \leq 1, 0 \leq y \leq 1\}$, where $\Omega_{FS} = \{(x,y)|0 \leq x \leq 1/2, 0 \leq y \leq 1/2\}$. In Ω_{PML}, the parameters a and b are chosen as $a = b = 1 + 10/jk$.

14.2.4 Interpretation of the Cartesian PML in Terms of Complex Coordinate Stretching

If we reconsider the model problem discussed in Section 14.2.1, where Ω_{PML} occupies the half space $x > 0$, we note that

$$E_z(x, y) = e^{-jk_x ax} e^{-jk_y y}, \qquad (x, y) \in \Omega_{PML} \tag{14.61}$$

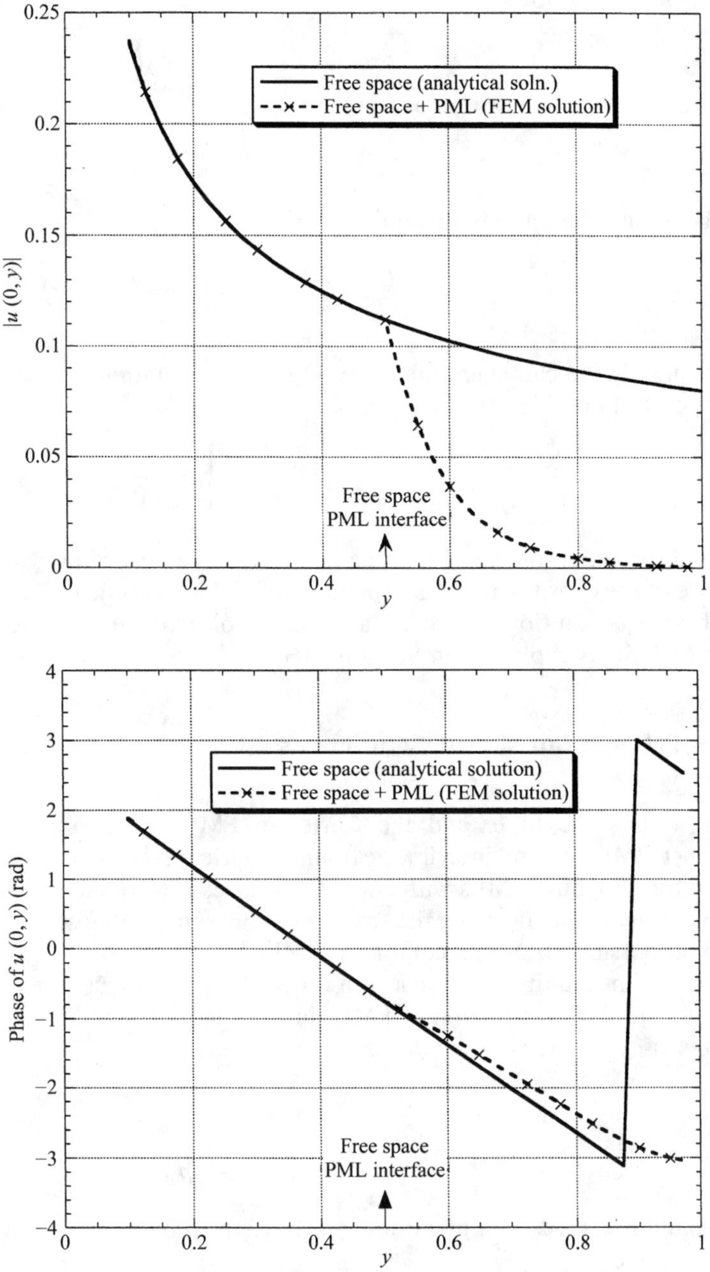

Figure 14.5 Magnitude plot of the two-dimensional free-space Green's function for the Helmholtz equation along the y axis (top). Phase plot of the two-dimensional free-space Green's function for the Helmholtz equation along the y axis (bottom).

where $k_x = k\cos\theta$, $k_y = k\sin\theta$, and $a = 1 + (\alpha/jk)$. Next, we define a new (complex) coordinate $\tilde{x}$ as follows:

$$\tilde{x} = ax \tag{14.62}$$

Then, $E_z(x, y)$ can be written as

$$E_z(\tilde{x}, y) = e^{-jk_x\tilde{x}}e^{-jk_y y} \tag{14.63}$$

which satisfies the Helmholtz equation

$$\frac{\partial^2 E_z}{\partial \tilde{x}^2} + \frac{\partial^2 E_z}{\partial y^2} + k^2 E_z = 0 \tag{14.64}$$

Consequently, we can claim that $E_z(x, y)$ satisfies Helmholtz equation (14.64) everywhere on $\Re^2$, provided that

$$\tilde{x} = \begin{cases} x & x < 0 \\ ax & x > 0 \end{cases} \tag{14.65}$$

This interpretation of the PML is named as complex coordinate stretching [55], [56] because of the coordinate transformation defined in (14.65). The idea is attractive, but one has to be cautious, since an arbitrary choice of the parameter a may violate causality [65], as explained in Section 14.3.

14.2.5 PMLs in Curvilinear Coordinates

It is possible to extend the Cartesian PML to curvilinear coordinates and construct PML regions in cylindrical and spherical coordinates [48]–[54]. The main reason for designing PMLs with curved interfaces is to reduce the white space in the computational domain, when the scatterer or antenna is almost a sphere or a cylinder. It is also possible to design conformal PMLs [48] where the absorber surrounds the object, minimizing the white space. In this section, we discuss the cylindrical PML, to demonstrate the construction of PMLs in curvilinear coordinates.

Let us define Ω_{FS} and Ω_{PML} as

$$\Omega_{FS} = \{(\rho, \varphi) \,|\, \rho < \rho_0\} \tag{14.66}$$

$$\Omega_{PML} = \{(\rho, \varphi) \,|\, \rho > \rho_0\} \tag{14.67}$$

The solution $E_z(\rho, \phi)$ is a plane wave in Ω_{FS}, given by

$$E_z(\rho, \varphi) = e^{-jk\rho\cos(\varphi - \tilde{\varphi})} \tag{14.68}$$

Next, we postulate a decay behavior in Ω_{PML} and derive the partial differential equation satisfied by $E_z(\rho, \varphi)$ in this region. The approach in this derivation is identical to the one we used in Section 14.2.1. In Ω_{PML}, we assume that

$$E_z(\rho, \varphi) = e^{-\alpha(\rho - \rho_0)\cos(\varphi - \tilde{\varphi})}e^{-jk\rho\cos(\varphi - \tilde{\varphi})} \tag{14.69}$$

where α is a constant. Defining $a = 1 + [\alpha(\rho - \rho_0)/jk\rho]$, (14.69) reduces to

$$E_z(\rho, \varphi) = e^{-jka\rho\cos(\varphi - \tilde{\varphi})} \tag{14.70}$$

Then, by direct differentiation,

$$\frac{1}{b}\frac{\partial}{\partial\rho}\left(\frac{1}{b}\frac{\partial E_z}{\partial\rho}\right) = -k^2\cos^2(\varphi - \tilde{\varphi})\,E_z \tag{14.71}$$

where $b = (d/d\rho)[a\rho]$. Similarly,

$$\frac{\partial^2 E_z}{\partial\varphi^2} = -\frac{\rho a}{b}\frac{\partial E_z}{\partial\rho} - k^2(\rho a)^2\sin^2(\varphi - \tilde{\varphi})\,E_z \tag{14.72}$$

Dividing (14.72) by $(\rho a)^2$ and adding it to (14.71), we obtain

$$\frac{1}{b}\frac{\partial}{\partial\rho}\left(\frac{1}{b}\frac{\partial E_z}{\partial\rho}\right) + \frac{1}{(\rho a)^2}\frac{\partial^2 E_z}{\partial\varphi^2} + \frac{1}{\rho ab}\frac{\partial E_z}{\partial\rho} + k^2 E_z = 0 \tag{14.73}$$

The next step is to extract Maxwell's equations from (14.73), that enables us to obtain the anisotropic and bianisotropic PMLs in cylindrical coordinates. Non-Maxwellian split-field PML equations can also be obtained from (14.73). These realizations are summarized below.

14.2.5.1 Split-Field (Non-Maxwellian) Realization

$$\frac{1}{\rho}\frac{\partial E_z}{\partial\varphi} = -j\omega\mu_0 a H_\rho \tag{14.74}$$

$$\frac{\partial E_z}{\partial\rho} = j\omega\mu_0 b H_\varphi \tag{14.75}$$

$$\frac{H_\varphi}{\rho} - \frac{1}{\rho}\frac{\partial H_\rho}{\partial\varphi} = -j\omega\varepsilon_0 a E_{z\rho} \tag{14.76}$$

$$\frac{\partial H_\varphi}{\partial\rho} = j\omega\varepsilon_0 b E_{z\varphi} \tag{14.77}$$

where $E_z = E_{z\rho} + E_{z\varphi}$.

14.2.5.2 Anisotropic Realization

$$\frac{1}{\rho}\frac{\partial E_z}{\partial\varphi} = -j\omega\mu_0\frac{a}{b}H_\rho \tag{14.78}$$

$$\frac{\partial E_z}{\partial\rho} = j\omega\mu_0\frac{b}{a}H_\varphi \tag{14.79}$$

$$\frac{1}{\rho}\left[\frac{\partial}{\partial\rho}(\rho H_\varphi) - \frac{\partial H_\rho}{\partial\varphi}\right] = j\omega\varepsilon_0 ab E_z \tag{14.80}$$

14.2.5.3 Bianisotropic Realization

$$\frac{1}{\rho}\frac{\partial E_z}{\partial \varphi} = -j\omega\mu_0 H_\rho \tag{14.81}$$

$$\frac{\partial E_z}{\partial \rho} = j\omega\mu_0 H_\varphi + \frac{\alpha}{jk}\frac{1}{b}\frac{\partial E_z}{\partial \rho} \tag{14.82}$$

$$\frac{1}{\rho}\left[\frac{\partial}{\partial \rho}(\rho H_\varphi) - \frac{\partial H_\rho}{\partial \varphi}\right] = j\omega\varepsilon_0 E_z + \frac{\alpha}{jk}\frac{1}{b}\frac{\partial H_\varphi}{\partial \rho} \tag{14.83}$$

14.3 CAUSALITY AND STATIC PMLs

In the previous section, we have obtained the realization of an anisotropic PML where the tensor $[\Lambda]$ appears in both the permittivity and permeability tensors as [33], i.e.,

$$\bar{\bar{\varepsilon}} = \varepsilon_0[\Lambda] \quad \bar{\bar{\mu}} = \mu_0[\Lambda] \tag{14.84}$$

Also in the previous section, if the longitudinal direction within Ω_{PML} is taken as $+x$ direction, $[\Lambda]$ has been obtained as

$$[\Lambda] = \begin{bmatrix} 1/a & 0 & 0 \\ 0 & a & 0 \\ 0 & 0 & a \end{bmatrix} \tag{14.85}$$

where

$$a = 1 + \frac{\alpha}{jk} \tag{14.86}$$

It is evident that if the free-space–PML interface is not a constant coordinate surface, but any arbitrary plane in $\Re^3$, our longitudinal direction vector perpendicular to this plane will no longer be along a Cartesian coordinate. In this case, it is still possible to find a local coordinate system (ξ, η, ν) such that ξ is the direction normal to the (η, ν) plane, which now represents the free-space PML interface. The representation of $[\Lambda]$ in this local coordinate system will have the form

$$[\Lambda]_{(\xi,\eta,\nu)}(\omega) = \begin{bmatrix} 1/a(\omega) & 0 & 0 \\ 0 & a(\omega) & 0 \\ 0 & 0 & a(\omega) \end{bmatrix} \tag{14.87}$$

The parameter a, appearing in (14.87), and hence in the constitutive parameters $\bar{\bar{\varepsilon}}$ and $\bar{\bar{\mu}}$, plays an important role in the characterization of the frequency dependence of the PML. In FEM applications, the solutions are derived for a single frequency at a time; however, the variation of $a(\omega)$ (and $1/a(\omega)$) over the entire spectrum cannot be defined arbitrarily. In fact, if $a(\omega)$ is given by

$$a(\omega) = 1 + \frac{\alpha c}{j\omega} \tag{14.88}$$

where c is the speed of light in vacuum, magnitude of $a(\omega)$ increases, without any bound, as ω tends to zero. Therefore, the frequency dependence given by (14.88) cannot possibly belong to a material medium. As will be shown below, the real and imaginary parts of $a(\omega)$ are not independent and they can be determined from each other in a unique way.

14.3.1 Constitutive Relations of a Causal PML

Let us consider the constitutive relationship $\vec{D}(\omega) = \bar{\bar{\varepsilon}}(\omega)\vec{E}(\omega)$. This relationship can be interpreted as one that relates the input and output of a linear, time-invariant system, whose transfer function matrix is $\varepsilon_0[\Lambda](\omega)$. The matrix $[\Lambda](\omega)$ (which is a diagonal matrix in a local coordinate system), must satisfy [77]

$$\text{(i)}\ \ [\Lambda](\omega) = [\Lambda]^*(-\omega) \tag{14.89}$$

where * denotes complex conjugation.

$$\text{(ii)}\ \ \mathrm{Re}([\Lambda](\omega) - I) = \frac{2}{\pi}P\int_0^\infty \frac{x}{x^2 - \omega^2}\mathrm{Im}([\Lambda](x))\,dx \tag{14.90}$$

and

$$\mathrm{Im}([\Lambda](\omega)) = -\frac{2\omega}{\pi}P\int_0^\infty \frac{1}{x^2 - \omega^2}\mathrm{Re}([\Lambda](x) - I)\,dx \tag{14.91}$$

where I is the identity matrix and P denotes the Cauchy principal value.

The condition (i) arises from the fact that both $\vec{E}(t)$ and $\vec{D}(t)$, that are inverse Fourier transforms of $\vec{E}(\omega)$ and $\vec{D}(\omega)$, respectively, are real signals. Equations (14.90) and (14.91) are known as Kramers-Kronig relations, and they must be satisfied if the system is to be causal. Specifically, this means that $\vec{D}(t)$ depends only on $\vec{E}(t')$ if the time t' precedes t. Another important consequence of (14.90) and (14.91) is that the real and imaginary parts of the tensor $[\Lambda](\omega)$ are related to each other, i.e., they are not independent. Since the frequency dependence of $[\Lambda](\omega)$ is determined by $a(\omega)$ (and $1/a(\omega)$), the conditions (i) and (ii) must be satisfied by these functions.

Let $a(\omega) = a_r(\omega) + ja_i(\omega)$. Condition (i) implies that $a_r(\omega)$ and $a_i(\omega)$ are even and odd functions of ω, respectively. Thus for $a(\omega)$, the Kramers-Kronig relations can be written as

$$a_r(\omega) - 1 = \frac{2}{\pi}P\int_0^\infty \frac{xa_i(x)}{x^2 - \omega^2}dx \tag{14.92}$$

$$a_i(\omega) = -\frac{2\omega}{\pi}P\int_0^\infty \frac{a_r(x) - 1}{x^2 - \omega^2}dx \tag{14.93}$$

An important consequence of (14.93) is that if $a_r(\omega) = 1$ for $\omega \in [0, \infty)$, then $a_i(\omega)$ must vanish for $\omega \in [0, \infty)$, which is consistent with the fact that for a dispersionless medium there can be no absorption. It can be seen from (14.92) and (14.93) that there can be dispersion only if the medium has some absorption. Hence, it is impossible to realize a causal PML medium by choosing $\mathrm{Re}(a(\omega))$ as unity over the entire frequency spectrum.

Let us now discuss how $a(\omega)$ can be defined properly, such that conditions (14.92) and (14.93) are satisfied. We postulate the frequency dependence of $a(\omega)$ to be [65]

$$a(\omega) = 1 + \frac{\beta}{1 + j\alpha\omega} \tag{14.94}$$

where α and β are constants. It is straightforward to verify that (14.94) satisfies (14.92) and (14.93), owing to the special form of the ω dependence. Now, let us consider the term $1/a(\omega)$, which can be written as

$$\frac{1}{a(\omega)} = 1 - \frac{\beta}{(1+\beta) + j\alpha\omega} \tag{14.95}$$

It is also possible to show that (14.95) satisfies the Kramers-Kronig relations. Hence, the tensor $[\Lambda](\omega)$, which is defined in terms of $a(\omega)$ and $1/a(\omega)$, can be used to realize a causal PML medium. Since $[\Lambda](\omega)$ appears in the expressions of both $\bar{\bar{\varepsilon}}(\omega)$ and $\bar{\bar{\mu}}(\omega)$, the electromagnetic wave propagation within the PML obeys the law of causality if the expression of $a(\omega)$ is given by (14.94).

Finally, let us consider the low- and high-frequency limits of $a(\omega)$. Equation (14.94) can be rewritten as

$$a(\omega) = \frac{1 + \alpha^2\omega^2 + \beta}{1 + \alpha^2\omega^2} - j\frac{\alpha\omega\beta}{1 + \alpha^2\omega^2} \tag{14.96}$$

(i) Low-frequency limit:

If $\alpha\omega \ll 1$, (14.96) becomes

$$a(\omega) \approx 1 + \beta \tag{14.97}$$

(ii) High-frequency limit:

If $\alpha\omega \gg 1$, (14.96) becomes

$$a(\omega) \approx 1 - j\frac{\beta}{\alpha\omega} \tag{14.98}$$

It is interesting to note that $a(\omega)$, given by (14.98), has been used in PML applications in FEM or FDTD formulations. Therefore, it is not surprising to find that there are difficulties [61] associated with the performance of the PML chosen as above as the frequency of operation becomes very low.

In order to obtain the time domain equations that are going to be used in FDTD applications, let us assume that the longitudinal direction of decay along the PML is $+z$ direction. Then $[\Lambda]$ becomes:

$$[\Lambda] = \begin{bmatrix} a & 0 & 0 \\ 0 & a & 0 \\ 0 & 0 & 1/a \end{bmatrix} \tag{14.99}$$

where $a = 1 + \beta/(1 + j\omega\alpha)$ and $1/a = 1 - \beta/((1+\beta) + j\omega\alpha)$. Let us consider the constitutive relation $\vec{D} = \varepsilon_0[\Lambda]\vec{E}$. The components of this equation can be written as

$$D_x = \varepsilon_0 E_x + \frac{\varepsilon_0 \beta E_x}{1 + j\omega\alpha} \tag{14.100}$$

$$D_y = \varepsilon_0 E_y + \frac{\varepsilon_0 \beta E_y}{1 + j\omega\alpha} \tag{14.101}$$

$$D_z = \varepsilon_0 E_z - \frac{\varepsilon_0 \beta E_z}{(1+\beta) + j\omega\alpha} \tag{14.102}$$

Remembering that multiplication by $j\omega$ in the frequency domain corresponds to a differentiation with respect to time in the time domain, (14.100)–(14.102) can be reduced to the following differential equations in time domain:

$$(D_x - \varepsilon_0 E_x) + \alpha \frac{\partial}{\partial t}(D_x - \varepsilon_0 E_x) = \varepsilon_0 \beta E_x \tag{14.103}$$

$$(D_y - \varepsilon_0 E_y) + \alpha \frac{\partial}{\partial t}(D_y - \varepsilon_0 E_y) = \varepsilon_0 \beta E_y \tag{14.104}$$

$$(1+\beta)(D_z - \varepsilon_0 E_z) + \alpha \frac{\partial}{\partial t}(D_z - \varepsilon_0 E_z) = -\varepsilon_0 \beta E_z \tag{14.105}$$

In a similar fashion the frequency domain constitutive relation $\vec{B} = \mu_0[\Lambda]\vec{H}$ reduces to the following expressions in time domain:

$$(B_x - \mu_0 H_x) + \alpha \frac{\partial}{\partial t}(B_x - \mu_0 H_x) = \mu_0 \beta H_x \tag{14.106}$$

$$(B_y - \mu_0 H_y) + \alpha \frac{\partial}{\partial t}(B_y - \mu_0 H_y) = \mu_0 \beta H_y \tag{14.107}$$

$$(1+\beta)(B_z - \mu_0 H_z) + \alpha \frac{\partial}{\partial t}(B_z - \mu_0 H_z) = -\mu_0 \beta H_z \tag{14.108}$$

The discretized version of (14.103)–(14.108) must be coupled to the discretized curl equations $\nabla \times \vec{E} = -(\partial\vec{B}/\partial t)$ and $\nabla \times \vec{H} = (\partial\vec{D}/\partial t)$ in the FDTD algorithm.

14.3.2 Non-Causal PML Media (an Example)

In Section 14.3.1, we have demonstrated that the parameter a plays an important role in the design of PML media, and the frequency dependence of this parameter cannot be chosen arbitrarily since this choice may violate causality. To demonstrate this, we construct a specific example where $[\Lambda]$ is given as

$$[\Lambda] = \begin{bmatrix} a & 0 & 0 \\ 0 & a & 0 \\ 0 & 0 & 1/a \end{bmatrix} \tag{14.109}$$

but we no longer insist that the parameter a represents a material property. In other words, $[\Lambda]$ is now an operator acting on $\vec{H}$ and $\vec{E}$ such that we get a PML action (i.e., the PML hypothesis given in Section 14.2 is satisfied by the medium). Now, let a be given by

$$a(\omega) = 1 - j\,\text{sgn}(\omega)\,\alpha \tag{14.110}$$

Note that the main difference between the parameter a given by (14.110) and the previous definition (14.88) is the modified form of the imaginary part of $a(\omega)$ in (14.110). In the previous section, we have deduced that the imaginary part of a must be an odd function with respect to frequency so that the time domain field expressions are real. Equation (14.110) satisfies this condition, since using the definition of $\text{sgn}(\omega)$, it can be expressed as

$$a(\omega) = \begin{cases} 1 - j\alpha & \omega > 0 \\ 1 + j\alpha & \omega < 0 \end{cases} \tag{14.111}$$

In frequency domain, the constitutive relation $\vec{D} = \varepsilon_0[\Lambda]\vec{E}$, becomes

$$D_x = \varepsilon_0(1 - j\,\text{sgn}(\omega)\,\alpha)\,E_x \tag{14.112}$$

$$D_y = \varepsilon_0(1 - j\,\text{sgn}(\omega)\,\alpha)\,E_y \tag{14.113}$$

$$D_z = \frac{\varepsilon_0}{1 - j\,\text{sgn}(\omega)\,\alpha}E_z \tag{14.114}$$

In order to obtain the time domain expressions, we must find the inverse Fourier transform of the term $j\,\text{sgn}(\omega)$ appearing in (14.110). In fact [78],

$$F\left\{\frac{1}{\pi t}\right\} = j\,\text{sgn}(\omega) \tag{14.115}$$

Hence, the time domain expressions become

$$D_x = \varepsilon_0 E_x - \alpha\frac{1}{\pi t} * E_x \tag{14.116}$$

$$D_y = \varepsilon_0 E_y - \alpha\frac{1}{\pi t} * E_y \tag{14.117}$$

$$D_z - \alpha\frac{1}{\pi t} * D_z = \varepsilon_0 E_z \tag{14.118}$$

It is known that the convolution of a function $f(t)$ with $1/\pi t$ is defined as the Hilbert Transform of $f(t)$ [78], given by

$$H(f)(t) = \int_{-\infty}^{\infty} \frac{1}{\pi\tau} f(t-\tau)\,d\tau \tag{14.119}$$

which is obviously a non-causal transformation owing to the fact that $H(f)(t)$ depends on the future values of the function $f(t)$. As a result, the time domain expressions (14.116)–(14.118) are effectively non-causal and, hence, they cannot be employed in iterative time-marching schemes like the FDTD algorithm. However, in FEM, since we are concerned with obtaining a solution at a single frequency, this PML can be used effectively.

14.3.3 Static PMLs

The main superiority of the causal PML developed in Section 14.3.1 is its well-defined behavior at low frequencies. In fact, the constitutive parameters can be extended to the static case, where $\omega = 0$. In this case, the parameter a becomes

$$a = 1 + \beta \tag{14.120}$$

where $\beta > 0$. In other words, the parameter a, which is complex when $\omega \neq 0$, becomes a real quantity for $\omega = 0$. The static PML has been employed in the finite difference calculation of capacitance matrices of interconnect structures by Veremey and Mittra [79]. Our aim in this section is to discuss the behavior of the static PML by considering the two examples given below.

Example 1: As shown in Fig. 14.6, suppose that $\Re^3 = \Omega_{FS} \cup \Omega_{PML}$, where Ω_{FS} and Ω_{PML} are the free space and PML regions, respectively, according to the notation of Section 14.2. These regions are expressed as

$$\Omega_{FS} = \{(x, y, z) | z < z_0\} \tag{14.121}$$

$$\Omega_{PML} = \{(x, y, z) | z > z_0\} \tag{14.122}$$

where $z_0 > 0$. The permittivity tensor $\bar{\bar{\varepsilon}}$ of the PML region is given as

$$\bar{\bar{\varepsilon}} = \varepsilon_0 [\Lambda] = \varepsilon_0 \begin{bmatrix} a & 0 & 0 \\ 0 & a & 0 \\ 0 & 0 & 1/a \end{bmatrix} \tag{14.123}$$

where $a > 1$ is a real constant given by (14.120).

Now, assume that a point charge Q resides at the origin $(0,0,0)$. The partial differential equations satisfied by the electrostatic potential φ in Ω_{FS} and Ω_{PML} are given by

$$\varepsilon_0 \nabla \cdot (\nabla \varphi) = -Q\delta(\vec{r}) \qquad (x, y, z) \in \Omega_{FS} \tag{14.124}$$

$$\varepsilon_0 \nabla \cdot ([\Lambda] \nabla \varphi) = 0 \qquad (x, y, z) \in \Omega_{PML} \tag{14.125}$$

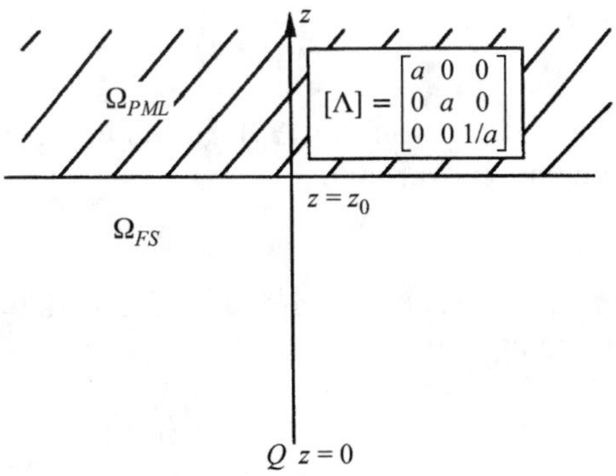

Figure 14.6 A point charge Q located at a distance z_0 to the planar surface of a static PML which occupies a half space.

The solutions of (14.124) and (14.125) can be obtained by using separation of variables in cylindrical coordinates and the Fourier-Bessel transform [80], yielding

$$\varphi_{FS}(\rho, z) = \frac{Q}{4\pi\varepsilon_0}\int_0^\infty e^{-kz} J_0(k\rho)\,dk + \int_0^\infty f(k) e^{kz} J_0(k\rho)\,dk \tag{14.126}$$

$$\varphi_{PML}(\rho, z) = \int_0^\infty h(k) e^{-kaz} J_0(k\rho)\,dk \tag{14.127}$$

Using the identity

$$\int_0^\infty e^{-kz} J_0(k\rho)\,dk = \frac{1}{\sqrt{\rho^2 + z^2}} \tag{14.128}$$

the first term of (14.126) becomes $Q/(4\pi\varepsilon_0\sqrt{\rho^2 + z^2})$, which is the electrostatic potential of the point charge Q in free space, in the absence of the PML. To find the unknowns $f(k)$ and $h(k)$, we employ the interface conditions

$$\varphi_{FS}(\rho, z_0) = \varphi_{PML}(\rho, z_0) \tag{14.129}$$

$$\frac{\partial \varphi_{FS}}{\partial z}(\rho, z_0) = \frac{1}{a}\frac{\partial \varphi_{PML}}{\partial z}(\rho, z_0) \tag{14.130}$$

which yield

$$f(k) = 0 \tag{14.131}$$

$$h(k) = \frac{Q}{4\pi\varepsilon_0} e^{-k(1-a)z_0} \tag{14.132}$$

Equation (14.131) implies that the half space Ω_{PML} is perfectly matched to Ω_{FS}, and $\varphi_{FS}\ (\rho, z)$ becomes

$$\varphi_{FS}(\rho, z) = \frac{Q}{4\pi\varepsilon_0\sqrt{\rho^2 + z^2}} \tag{14.133}$$

which is in fact the potential field of a point charge in free space. The potential $\varphi_{PML}(\rho, z)$ can be obtained by substituting (14.132) in (14.127), which yields

$$\varphi_{PML}(\rho, z) = \frac{Q}{4\pi\varepsilon_0}\int_0^\infty e^{-k[z_0 + a(z - z_0)]} J_0(k\rho)\,dk \tag{14.134}$$

The above can be simplified to

$$\varphi_{PML}(\rho, z) = \frac{Q}{4\pi\varepsilon_0}\frac{1}{\sqrt{\rho^2 + [z_0 + a(z - z_0)]^2}} \tag{14.135}$$

The field expression (14.135) can be interpreted in terms of coordinate stretching by defining a new variable $\tilde{z}$ as

$$\tilde{z} = z_0 + a(z - z_0) \tag{14.136}$$

In terms of this stretched coordinate variable, we may obtain the potential function of the point charge as

$$\varphi_{PML}(\rho, \tilde{z}) = \frac{Q}{4\pi\varepsilon_0}\frac{1}{\sqrt{\rho^2 + \tilde{z}^2}} \tag{14.137}$$

Example 2: In the previous section, we have mentioned that PML regions occupying half-spaces are only idealizations, and in real mesh truncation problems Ω_{PML} must be bounded. Therefore, also in static PML applications, a more realistic problem is the one with the geometry shown in Fig. 14.7. In this case, Ω_{FS} and Ω_{PML} are defined as

$$\Omega_{FS} = \{(x, y, z) | z < z_1\} \tag{14.138}$$

$$\Omega_{PML} = \{(x, y, z) | z_1 < z < z_2\} \tag{14.139}$$

Again, by using separation of variables in cylindrical coordinates, we obtain

$$\varphi_{FS}(\rho, z) = \frac{Q}{4\pi\varepsilon_0} \int_0^\infty e^{-kz} J_0(k\rho)\, dk + \int_0^\infty f(k) e^{kz} J_0(k\rho)\, dk \tag{14.140}$$

$$\varphi_{PML}(\rho, z) = \int_0^\infty g(k) e^{-kaz} J_0(k\rho)\, dk + \int_0^\infty h(k) e^{kaz} J_0(k\rho)\, dk \tag{14.141}$$

After applying the interface conditions at $z = z_1$ and the boundary condition at $z = z_2$, we obtain

$$g(k) = \frac{Q}{4\pi\varepsilon_0} e^{-k(1-a)z_1} \tag{14.142}$$

$$h(k) = -\frac{Q}{4\pi\varepsilon_0} e^{-k(1-a)z_1} e^{-2kaz_2} \tag{14.143}$$

$$f(k) = -\frac{Q}{4\pi\varepsilon_0} e^{k[2z_1 + 2a(z_2 - z_1)]} \tag{14.144}$$

By substituting (14.144) into (14.140), we obtain the expression for the potential in free space as

$$\varphi_{FS}(\rho, z) = \frac{Q}{4\pi\varepsilon_0\sqrt{\rho^2 + z^2}} - \frac{Q}{4\pi\varepsilon_0\sqrt{\rho^2 + (z - \hat{z})^2}} \tag{14.145}$$

where $\hat{z} = 2z_1 + 2a(z_2 - z_1)$. If $a = 1$ (i.e., the permittivity of the PML layer is ε_0), $\hat{z}$ reduces to $\hat{z} = 2z_2$, which is the z coordinate of the image of the point charge Q with respect to the PEC plane at $z = z_2$. For $a > 1$, $\hat{z} = 2z_1 + 2a(z_2 - z_1)$, still gives us the

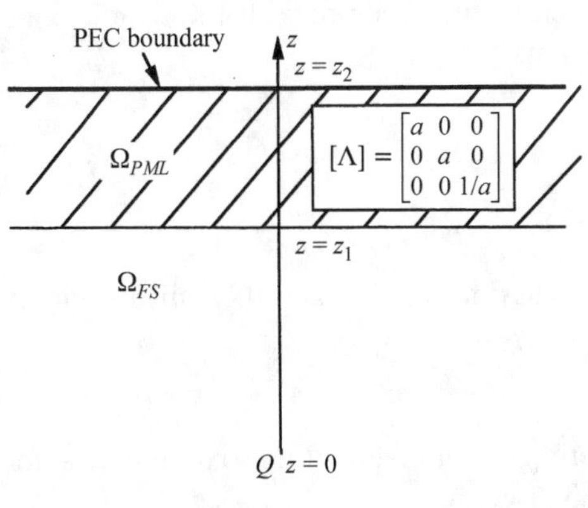

Figure 14.7 A point charge Q located at a distance z_1 to the planar surface of a static PML of thickness $z_2 - z_1$ which is terminated by a PEC boundary.

z coordinate of the image charge, but in this case the image charge moves farther away along the z axis, as the value of a increases.

14.4 RECIPROCITY IN PERFECTLY MATCHED ABSORBERS

In Section 14.2, we have derived the equations governing three different PML media, namely, Berenger's PML medium (a non-Maxwellian realization with a split-field formulation), and two Maxwellian realizations: the anisotropic and bianisotropic PML media. In the last two realizations, the governing equations are Maxwell's equations and for each realization we have suitably defined constitutive relations. Our principal concern in this section is to discuss the reciprocity issue in these material media. This issue turns out to be very relevant for perfectly matched absorbers because the medium occupies a finite, bounded region $\Omega_{PML} \subset \Re^3$, with PEC or PMC terminations at the outer boundary. Therefore, in a successful PML design, we expect the material to be reciprocal, such that the decay characteristics of the wave that reach the inner boundary, after being reflected from the outer boundary should be the same as that of the wave that traverses through the PML and decays in the longitudinal direction. The equivalence of the decay characteristics in the opposing directions can be guaranteed by choosing a reciprocal medium. In this section, we prove that the anisotropic and bianisotropic PMLs derived in Section 14.2 are reciprocal media. Furthermore, we provide an example of a non-reciprocal medium, namely the transparent absorbing boundary (TAB) medium [81].

14.4.1 Verification of Reciprocity in the Anisotropic and Bianisotropic Realizations

Consider the constitutive relations of a linear bianisotropic medium which is not space-dispersive (i.e., the constitutive relations do not contain any partial differential operators with respect to the space variables):

$$\vec{D} = \bar{\bar{\varepsilon}} \cdot \vec{E} + \bar{\bar{\xi}} \cdot \vec{H} \tag{14.146}$$

$$\vec{B} = \bar{\bar{\zeta}} \cdot \vec{E} + \bar{\bar{\mu}} \cdot \vec{H} \tag{14.147}$$

The medium is reciprocal if the dyadic constitutive parameters satisfy the following relations [82]:

$$\bar{\bar{\varepsilon}} = \bar{\bar{\varepsilon}}^T \tag{14.148}$$

$$\bar{\bar{\mu}} = \bar{\bar{\mu}}^T \tag{14.149}$$

$$\bar{\bar{\xi}} = -\bar{\bar{\zeta}}^T \tag{14.150}$$

It is easy to verify that the anisotropic PML is reciprocal because $\bar{\bar{\varepsilon}}$ and $\bar{\bar{\mu}}$ are given by

$$\bar{\bar{\varepsilon}} = \varepsilon_0[\Lambda] \qquad \bar{\bar{\mu}} = \mu_0[\Lambda] \tag{14.151}$$

where $[\Lambda]$ is a diagonal matrix. Hence, $\bar{\bar{\varepsilon}}$ and $\bar{\bar{\mu}}$ satisfy (14.148) and (14.149).

In order to prove reciprocity in the bianisotropic PML, we must first derive Maxwell's equations for this particular medium [83]. Although this medium has been introduced in Section 14.2, the derivation of the governing equations has been postponed to this section, since the reciprocity proof is closely related to the derivation. To obtain the governing equations, we follow the steps given in Section 14.2. We start by defining the free space and PML regions as

$$\Omega_{FS} = \{(x, y)|x<0\} \tag{14.152}$$

$$\Omega_{PML} = \{(x, y)|x>0\} \tag{14.153}$$

We assume that, in Ω_{FS}, we have a TM plane wave given by

$$E_z(x, y) = e^{-jk[\cos\theta x+\sin\theta y]} \tag{14.154}$$

Next, we stipulate that in the region Ω_{PML}, $E_z(x, y)$ be given by

$$E_z(x, y) = f(x)e^{-jk[\cos\theta x+\sin\theta y]} \tag{14.155}$$

where the function $f(x)$ satisfies the following conditions:

1. $f(0) = 1$
2. $f(x)$ is a monotonically decreasing function of x (i.e., $f'(x) \leq 0$), for $x \geq 0$.
3. $f(x) \neq 0$ for $x \geq 0$.

In Section 14.2, we have chosen $f(x)$ as $f(x) = e^{-\alpha\cos\theta x}$, but now, we leave $f(x)$ as an arbitrary function as long as the three conditions given above are satisfied. In this section, our main objective is to demonstrate the implications of such an arbitrary choice of the decay characteristics on the reciprocity of the medium.

The next step in our derivation is to determine the partial differential equation whose solution yields $E_z(x,y)$ given by (14.155). For this purpose, we divide both sides of (14.155) by $f(x)$ and operate $(\nabla^2 + k^2)$ on both sides. Noting that the right-hand side vanishes, we obtain [83]

$$\frac{\partial}{\partial x}\left[\frac{\partial E_x}{\partial x} - \frac{f'(x)}{f(x)}E_z\right] + \frac{\partial^2 E_z}{\partial y^2} + k^2 E_z - \frac{f'(x)}{f(x)}\frac{\partial E_z}{\partial x} + \left[\frac{f'(x)}{f(x)}\right]^2 E_z = 0 \tag{14.156}$$

From (14.156), we derive Maxwell's curl equations as

$$-\frac{\partial E_z}{\partial y}\hat{a}_x + \left(\frac{\partial E_z}{\partial x} - \frac{f'(x)}{f(x)}E_z\right)\hat{a}_y = j\omega\mu_0[H_x\hat{a}_x + H_y\hat{a}_y] \tag{14.157}$$

$$\frac{\partial H_y}{\partial x} - \frac{\partial H_x}{\partial y} - \frac{f'(x)}{f(x)}H_y = j\omega\varepsilon_0 E_z \tag{14.158}$$

Next, we express (14.157) and (14.158) in a more compact form as

$$-\nabla \times \vec{E} = j\omega\mu_0\vec{H} - \frac{f'(x)}{f(x)}\hat{a}_x \times \vec{E} \tag{14.159}$$

$$\nabla \times \vec{H} = j\omega\varepsilon_0\vec{E} + \frac{f'(x)}{f(x)}\hat{a}_x \times \vec{H} \tag{14.160}$$

Although (14.159) and (14.160) are derived for the TM case, they are valid for the general case as well. The constitutive relations can be expressed as

$$\vec{D} = \varepsilon_0 \vec{E} + \frac{1}{j\omega} \frac{f'(x)}{f(x)} \hat{a}_x \times \vec{H} \tag{14.161}$$

$$\vec{B} = \mu_0 \vec{H} - \frac{1}{j\omega} \frac{f'(x)}{f(x)} \hat{a}_x \times \vec{E} \tag{14.162}$$

A comparison of (14.161) and (14.162) with the constitutive relations of a bianisotropic medium, given in (14.146) and (14.147), leads to the following results:

$$\overline{\overline{\varepsilon}} = \varepsilon_0 \overline{\overline{I}} \tag{14.163}$$

$$\overline{\overline{\mu}} = \mu_0 \overline{\overline{I}} \tag{14.164}$$

where $\overline{\overline{I}}$ is the identity matrix. The other two constitutive dyadics are given by

$$\overline{\overline{\xi}} = \overline{\overline{\zeta}}^T = \frac{1}{j\omega} \frac{f'(x)}{f(x)} \begin{bmatrix} 0 & 0 & 0 \\ 0 & 0 & -1 \\ 0 & 1 & 0 \end{bmatrix} \tag{14.165}$$

In order to show that the bianisotropic PML realization obtained in Section 14.2 is reciprocal, we must first obtain the factor $f'(x)/f(x)$ for the decay characteristics $f(x) = e^{-\alpha \cos \theta x}$, which was given in Section 14.2.3.1. For this particular choice, it follows that

$$\frac{f'(x)}{f(x)} u(x, y) = -\alpha \cos \theta u(x, y) \tag{14.166}$$

where u is any one of the field components. By direct differentiation, we have

$$\frac{\partial u(x, y)}{\partial x} = \frac{f'(x)}{f(x)} u(x, y) - jk \cos \theta u(x, y) \tag{14.167}$$

Using (14.166), (14.167) can be reduced to

$$\frac{\partial u(x, y)}{\partial x} = \frac{f'(x)}{f(x)} u(x, y) + \frac{jk}{\alpha} \frac{f'(x)}{f(x)} u(x, y) \tag{14.168}$$

Consequently,

$$\frac{f'(x)}{f(x)} u(x, y) = \frac{1}{1 + \dfrac{jk}{\alpha}} \frac{\partial u(x, y)}{\partial x} \tag{14.169}$$

A more convenient form of (14.169) can be obtained as

$$\frac{f'(x)}{f(x)} u(x, y) = \frac{\alpha}{jk} \frac{1}{a} \frac{\partial u(x, y)}{\partial x} \tag{14.170}$$

where a is the parameter defined in the previous sections as $a = 1 + \alpha/jk$. From (14.170), we conclude that $f'(x)/f(x)$ is a differential operator defined by

$$\frac{f'(x)}{f(x)} = \frac{\alpha}{jk} \frac{1}{a} \frac{\partial}{\partial x} \tag{14.171}$$

An interesting result can be obtained by substituting (14.171) in the partial differential equation (14.156). This leads to the partial differential equation given by

$$\frac{1}{a^2}\frac{\partial^2 E_z}{\partial x^2}+\frac{\partial^2 E_z}{\partial y^2}+k^2 E_z=0 \tag{14.172}$$

which is identical to the equation that was obtained for the anisotropic realization. This result explains the equivalence of the anisotropic and bianisotropic realizations discussed in Section 14.2.

The constitutive dyadics $\bar{\bar{\xi}}$ and $\bar{\bar{\zeta}}$ become differential operators if (14.171) is substituted in (14.165):

$$\bar{\bar{\xi}}=\bar{\bar{\zeta}}^T=\frac{1}{j\omega}\frac{\alpha}{c\alpha+j\omega}\frac{\partial}{\partial x}\begin{bmatrix}0 & 0 & 0\\ 0 & 0 & -1\\ 0 & 1 & 0\end{bmatrix} \tag{14.173}$$

The derivative term $\partial/\partial x$ in (14.173) implies that this bianisotropic realization is non-local (or, space-dispersive). As mentioned at the beginning of this section, the reciprocity relations given by (14.148)–(14.150) are valid for media that are not space-dispersive. Since $\bar{\bar{\xi}}$ and $\bar{\bar{\zeta}}$ are now matrix differential operators, matrix transpose operation in (14.150) must be replaced by a new operation, denoted by $\bar{\bar{\xi}}^*$, and the formal adjoint of the differential operator $\partial/\partial x$ must be taken into account. Since the formal adjoint of the differential operator $\partial/\partial x$ is $(-\partial/\partial x)$ [84], the relationship between $\bar{\bar{\xi}}^*$ and $\bar{\bar{\zeta}}$ is given by

$$\bar{\bar{\xi}}^*=\frac{1}{j\omega}\frac{\alpha}{c\alpha+j\omega}\left(-\frac{\partial}{\partial x}\right)\begin{bmatrix}0 & 0 & 0\\ 0 & 0 & 1\\ 0 & -1 & 0\end{bmatrix}=-\bar{\bar{\zeta}} \tag{14.174}$$

which implies that the medium is reciprocal. This result can be verified by using the definition of reciprocity directly, as given below.

Let $(\vec{E}_a,\vec{H}_a)$ and $(\vec{E}_b,\vec{H}_b)$ be a pair of electromagnetic fields in this medium. Then we have [see 82]

$$\langle a,b\rangle-\langle b,a\rangle=\iiint_{\Re^3}(\vec{E}_b\cdot\vec{D}_a-\vec{E}_a\cdot\vec{D}_b+\vec{H}_a\cdot\vec{B}_b-\vec{H}_b\cdot\vec{B}_a)\,dx\,dy\,dz \tag{14.175}$$

where the symmetric product forms, $\langle .,. \rangle$, are the reactions.

If the medium is reciprocal, (14.175) must vanish [82]. Substituting the constitutive relations (14.146)–(14.147) into (14.175), we obtain

$$\begin{aligned}\langle a,b\rangle-\langle b,a\rangle=\iiint_{\Re^3}&\{\vec{E}_b\cdot(\bar{\bar{\varepsilon}}-\bar{\bar{\varepsilon}}^T)\cdot\vec{E}_a+\vec{H}_a\cdot(\bar{\bar{\mu}}-\bar{\bar{\mu}}^T)\cdot\vec{H}_b\\ &+\vec{E}_b\cdot\bar{\bar{\xi}}\cdot\vec{H}_a+\vec{H}_a\cdot\bar{\bar{\zeta}}\cdot\vec{E}_b-\vec{E}_a\cdot\bar{\bar{\xi}}\cdot\vec{H}_b-\vec{H}_b\cdot\bar{\bar{\zeta}}\cdot\vec{E}_a\}\,dx\,dy\,dz\end{aligned} \tag{14.176}$$

The first two terms in (14.176) vanish, since $\bar{\bar{\varepsilon}}$ and $\bar{\bar{\mu}}$ are diagonal tensors given by (14.163) and (14.164). The crucial terms are those containing $\bar{\bar{\xi}}$ and $\bar{\bar{\zeta}}$. Consider the integral

$$I=\iiint_{\Re^3}\{E_b\cdot\bar{\bar{\xi}}\cdot\vec{H}_a+\vec{H}_a\cdot\bar{\bar{\zeta}}\cdot\vec{E}_b\}\,dx\,dy\,dz \tag{14.177}$$

which is part of the integral in (14.176). Using the expressions of $\bar{\bar{\xi}}$ and $\bar{\bar{\zeta}}$, (14.177) reduces to

$$I = \frac{1}{j\omega}\frac{\alpha}{c\alpha + j\omega}\iiint_{\Re^3}\left\{ E_b \cdot \begin{bmatrix} 0 & 0 & 0 \\ 0 & 0 & -\frac{\partial}{\partial x} \\ 0 & \frac{\partial}{\partial x} & 0 \end{bmatrix} \cdot \vec{H}_a + \vec{H}_a \cdot \begin{bmatrix} 0 & 0 & 0 \\ 0 & 0 & \frac{\partial}{\partial x} \\ 0 & -\frac{\partial}{\partial x} & 0 \end{bmatrix} \cdot \vec{E}_b \right\} dx\,dy\,dz \tag{14.178}$$

which may be written as

$$I = \frac{1}{j\omega}\frac{\alpha}{c\alpha + j\omega}\iiint_{\Re^3}\left\{ E_{by}\left(-\frac{\partial H_{az}}{\partial x}\right) + E_{bz}\frac{\partial H_{ay}}{\partial x} + H_{ay}\frac{\partial E_{bz}}{\partial x} + H_{az}\left(-\frac{\partial E_{by}}{\partial x}\right)\right\} dx\,dy\,dz \tag{14.179}$$

It is evident that (14.179) vanishes when we integrate the third and fourth terms by parts. In a similar manner, it can be shown that the last remaining pair of terms also vanish in (14.176), which implies that

$$\langle a, b\rangle - \langle b, a\rangle = 0 \tag{14.180}$$

which proves that the medium is reciprocal.

It may be verified that this result is closely connected to the special choice for the decay function $f(x)$ as $f(x) = e^{-\alpha \cos\theta x}$. The presence of $\cos\theta$ in the exponent leads to a reciprocal non-local medium [85].

14.4.2 Example of a Non-Reciprocal PML

It is evident from (14.165) that if the medium is not space-dispersive, it can never be a reciprocal medium. Let $f(x)$ be an arbitrary function satisfying the three conditions mentioned in the beginning of this section. Then,

$$\bar{\bar{\xi}} = \bar{\bar{\zeta}}^T = \frac{1}{j\omega}\frac{f'(x)}{f(x)}\begin{bmatrix} 0 & 0 & 0 \\ 0 & 0 & -1 \\ 0 & 1 & 0 \end{bmatrix} \tag{14.181}$$

which, obviously, does not satisfy (14.150). This medium, which is a realization of the transparent absorbing boundary approach of Peng and Balanis [81], fails to satisfy reciprocity. For instance, let $f(x) = e^{-\alpha x}$, where $\alpha > 0$. Then, (14.181) becomes [86]

$$\overline{\overline{\xi}} = \overline{\overline{\zeta}}^T = \frac{-\alpha}{j\omega}\begin{bmatrix} 0 & 0 & 0 \\ 0 & 0 & -1 \\ 0 & 1 & 0 \end{bmatrix} \tag{14.182}$$

which yields a non-reciprocal medium. The implications of this result can be demonstrated for a TM plane wave, $E_z = e^{-jkx}$, which is normally incident to the free-space PML interface. In Ω_{FS}, we have

$$\frac{\partial E_z}{\partial x} = j\omega\mu_0 H_y \tag{14.183}$$

$$\frac{\partial H_y}{\partial x} = j\omega\varepsilon_0 E_z \tag{14.184}$$

In Ω_{PML}, the relevant equations are:

$$\frac{\partial E_z}{\partial x} + \alpha E_z = j\omega\mu_0 H_y \tag{14.185}$$

$$\frac{\partial H_y}{\partial x} + \alpha H_y = j\omega\varepsilon_0 E_z \tag{14.186}$$

From (14.183)–(14.186), we obtain the solutions as

$$E_z = e^{-jkx}, \qquad H_y = -\eta_0 e^{-jkx} \qquad \text{in } \Omega_{FS} \tag{14.187}$$

$$E_z = e^{-\alpha x} e^{-jkx}, \qquad H_y = -\eta_0 e^{-\alpha x} e^{-jkx} \qquad \text{in } \Omega_{PML} \tag{14.188}$$

where η_0 is the intrinsic impedance of free space. From (14.188), we conclude that the second medium acts as a PML as long as it extends to infinity (i.e., there are no reflected waves traveling in the $-x$ direction.

However, if the PML region is terminated by a PEC at $x = a$, then the solutions of (14.183)–(14.186) may be written as

$$E_z = e^{-jkx} + be^{jkx} \qquad \text{in } \Omega_{FS} \tag{14.189}$$

$$E_z = ce^{-\alpha x} e^{-jkx} + de^{-\alpha x} e^{jkx} \text{ in } \Omega_{PML} \tag{14.190}$$

The non-reciprocal nature of the medium is now evident from (14.190) where the decay factor $e^{-\alpha x}$ appears as a multiplier of waves going in both $+x$ and $-x$ directions. The constants b, c, and d can be evaluated by using the interface conditions at $x = 0$, and the boundary condition at $x = a$. As a result, we find $c = 1$, which means that the field is totally transmitted. However, $b = d = -e^{-jk2a}$, which implies that the reflection from the outer boundary is intensively observed in Ω_{FS}, which is of course not desirable in a PML. Finally, the solutions for H_y can also be calculated as

$$H_y = -\eta_0 e^{-jkx} + b\eta_0 e^{jkx} \qquad \text{in } \Omega_{FS} \tag{14.191}$$

$$H_y = -c\eta_0 e^{-\alpha x} e^{-jkx} + d\eta_0 e^{-\alpha x} e^{jkx} \qquad \text{in } \Omega_{PML} \tag{14.192}$$

14.5 CONCLUSION

In this chapter, we have discussed three important aspects of perfectly matched absorbers:

1. Systematic derivation of the PML partial differential equations, and the extraction of Maxwellian or non-Maxwellian realizations from the governing partial differential equations.
2. Investigation of the conditions under which the PML medium is causal, and the extension of PML to lower frequencies and even to the static case.
3. Restrictions on the constitutive parameters of the PML such that it is a reciprocal medium.

In the systematic derivation of the partial differential equations of the PML media, one first postulates the behavior of the field in the medium, and then deduces the partial differential equations whose solution yields this particular field variation. This approach is more powerful than the conventional approaches in the literature, where the starting-point is either the medium itself, or the governing equations. An important consequence of the systematic approach is the possibility of obtaining more than one realization yielding the identical field behavior in the medium. In Section 14.2, we have shown that it is possible to generate the same field behavior by Berenger's PML (which is non-Maxwellian), the anisotropic PML and the bianisotropic PML. The systematic approach is applicable to Cartesian as well as curvilinear PMLs effectively.

Causality is obviously an important factor in the design of PMLs, especially in time domain applications. The PML parameter $a = 1 + (\alpha/jk)$ is not valid over the whole frequency interval, because of its singular behavior as $k \to 0$. However, the parameter a can be modified to the form $a = 1 + \beta/(1 + jk\alpha)$ to get around this difficulty. It is also fortuitous that, after this modified value of a is substituted in the constitutive relations, they satisfy the well-known Kramers-Kronig relations. In Section 14.3, we have also shown that arbitrary choices of this parameter may yield non-causal media.

Satisfaction of reciprocity is yet another important issue in bounded PML media. When PMLs are used in mesh truncation, they always occupy a finite region in space, and they have an outer boundary (usually a PEC termination), over which suitable boundary conditions are imposed. Although the fields start to decay as soon as they enter into the PML medium, their magnitude is still non-zero when they arrive at the outer boundary. After they are reflected by this outer boundary, they travel in the PML, now in the opposite direction, toward the free-space PML interface. Reciprocity plays a very important role in this case, since the decay characteristics of the waves traveling toward and away from the outer boundary must be identical. This condition is satisfied if the medium is reciprocal. In Section 14.4, we have discussed the conditions under which the PML medium is reciprocal, and gave an example of a reciprocal space-dispersive (non-local) bianisotropic PML.

Although the PML concept has originally been introduced as a mesh truncation approach in FDTD simulations, its interpretation as an anisotropic (or, more generally, bianisotropic) material medium engender new research areas in this field, for instance the physical realization of perfectly matched absorbers, which is a very challenging problem indeed.

REFERENCES

[1] R. F. Harrington, *Field Computation by Moment Methods*, Macmillan, New York, 1968.

[2] K. S. Kunz and R. J. Luebbers, *The Finite Difference Time Domain Method for Electromagnetics*, CRC Press, Boca Raton, FL, 1993.

[3] O. C. Zienkiewicz and K. Morgan, *Finite Elements and Approximation*, John Wiley and Sons, New York, 1983.

[4] K. K. Mei, M. A. Morgan, and S.-K. Chang, "Finite Methods in Electromagnetic Scattering," in *Electromagnetic Scattering*, P. L. E. Uslenghi, ed., Academic Press, New York, pp. 359–392, 1978.

[5] X. Yuan, "Three-dimensional Electromagnetic Scattering from Inhomogeneous Objects by the Hybrid Moment and Finite Element Method," *IEEE Trans. Microwave Theory Tech.*, **MTT-38**, pp. 1053–1058, 1990.

[6] O. C. Zienkiewicz, D. W. Kelly, and P. Bettess, "Marriage a la Mode—The Best of Two Worlds (finite elements and boundary integrals)," in *Energy Methods in Finite Element Analysis*, R. Glowinski, E. Y. Rodin, and O. C. Zienkiewicz, eds., John Wiley Interscience, New York, 1979.

[7] R. L. Higdon, "Absorbing Boundary Conditions for Difference Approximations to the Multi-dimensional Wave Equation," *Math. Comput.*, **47**, pp. 437–459, 1986.

[8] A. Bayliss, M. Gunzburger, and E. Turkel, "Boundary Conditions for the Numerical Solution of Elliptic Equations in Exterior Domains," *SIAM J. Appl. Math.*, **42**, pp. 430–451, 1982.

[9] B. Stupfel, "Absorbing Boundary Conditions on Arbitrary Boundaries for the Scalar and Vector Wave Equations," *IEEE Trans. Antennas Propagat.*, **42**, pp. 773–780, 1994.

[10] R. Gordon, R. Mittra, A. Glisson, and E. Michielssen, "Finite Element Analysis of Electromagnetic Scattering by Complex Bodies Using an Efficient Numerical Boundary Condition for Mesh Truncation," *Electron. Lett.*, **29**, pp. 1102–1103, 1993.

[11] R. Mittra, A. Boag, and A. Boag, "A New Look at the Absorbing Boundary Condition Issue for FEM Mesh Truncation in Open Region Problems," in Second International Workshop on Finite Element Methods Electromagnetic Wave Problems, COMPEL, Siena, Italy, May 1994, **13**, Supplement A, pp. 211–216.

[12] A. Boag, A. Boag, R. Mittra, and Y. Leviatan, "A Numerical Absorbing Boundary Condition for Finite-difference and Finite-element Analysis of Open Structures," *Microwave Opt. Tech. Lett.*, **7**, pp. 395–398, 1994.

[13] A. Boag and R. Mittra, "A Numerical Absorbing Boundary Condition for 3D Edge-based Finite-element Analysis of Very Low-frequency Fields," *Microwave Opt. Tech. Lett.*, **9**, pp. 22–27, 1995.

[14] B. Stupfel and R. Mittra, "A Theoretical Study of Numerical Absorbing Boundary Conditions," *IEEE Trans. Antennas Propagat.*, **43**, pp. 478–487, 1995.

[15] B. Stupfel and R. Mittra, "Numerical Absorbing Boundary Conditions for the Scalar and Vector Wave Equations," *IEEE Trans. Antennas Propagat.*, **44**, pp. 1015–1022, 1996.

[16] P. Bettess, "Infinite Elements," *Int. J. Numer. Methods Eng.*, **11**, pp. 53–64, 1977.

[17] M. S. Towers, A. McCowan, and J. A. R. Macnab, "Electromagnetic Scattering from an Arbitrary, Inhomogeneous 2-D Object—A Finite and Infinite Element Solution," *IEEE*

Trans. Antennas Propagat., **41**, pp. 770–777, 1993.

[18] P. Bettess, "More on Infinite Elements," *Int. J. Numer. Methods Eng.*, **15**, pp. 1613–1626, 1980.

[19] J. P. Berenger, "A Perfectly Matched Layer for the Absorption of Electromagnetic Waves," *J. Comput. Physics*, **114**, pp. 185–200, 1994.

[20] D. S. Katz, E. T. Thiele, and A. Taflove, "Validation and Extension to Three Dimensions of the Berenger PML Absorbing Boundary Condition for FD-TD Meshes," *IEEE Microwave Guided Wave Lett.*, **4**, pp. 268–270, 1994.

[21] C. E. Reuter, R. M. Joseph, E. T. Thiele, D. S. Katz, and A. Taflove, "Ultrawideband Absorbing Boundary Condition for Termination of Waveguiding Structures in FD-TD Simulations," *IEEE Microwave Guided Wave Lett.*, **4**, pp. 344–346, 1994.

[22] D. T. Presscott and N. V. Shuley, "Reflection Analysis of FDTD Boundary Conditions – Part II: Berenger's PML Absorbing Layers," *IEEE Trans. on Microwave Theory and Techniques*, **45**, pp. 1171–1178, 1997.

[23] M. Chen, B. Houshmand, and T. Itoh, "FDTD Analysis of Metal-strip-loaded Dielectric Leaky Wave Antenna," *IEEE Trans. Antennas and Propagat.*, **45**, pp. 1294–1301, 1997.

[24] C. Furse, C. Yan, G. Lazzi, and O. Gandhi, "Use of PML Boundary Conditions for Wireless Telephone Simulations," *Microwave and Optical Tech. Lett.*, **15**, pp. 95–98, 1997.

[25] J.-P. Berenger, "Improved PML for the FDTD Solution of Wave-structure Interaction Problems," *IEEE Trans. Antennas and Propagat.*, **45**, pp. 466–473.

[26] A. R. Roberts and J. Joubert, "PML Absorbing Boundary Condition for Higher-order FDTD Schemes," *Electronics Letters*, **33**, pp. 32–34, 1997.

[27] E. Michielssen, W. C. Chew, and D. S. Weile, "Genetic Algorithm Optimized Perfectly Matched Layers for Finite Difference Frequency Domain Applications," *1996 AP-S International Symposium Proceedings*, **3**, pp. 2106–2109, 1996.

[28] J. Fang and Z. Wu, "Closed-form Expression of Numerical Reflection Coefficient at PML Interfaces and Optimization of PML Performance,"*IEEE Microwave and Guided Wave Letters*, **6**, pp. 332–334, 1996.

[29] M. F. Pasik, G. Aguirre, and A. C. Cangellaris, "Application of the PML Boundary Condition to the Analysis of Patch Antennas," *Electromagnetics*, **16**, pp. 435–449.

[30] J. C. Weihl and R. Mittra, "Efficient Implementation of Berenger's Perfectly Matched Layer (PML) for Finite-difference-time-domain Mesh Truncation," *IEEE Microwave and Guided Wave Letters*, **6**, pp. 94–96, 1996.

[31] Z. Wu and J. Fang, "Numerical Implementation and Performance of Perfectly Matched Layer Boundary Condition for Waveguide Structures," *IEEE Trans. Microwave Theory and Techniques*, **43**, pp. 2676–2683, 1995.

[32] J. B. Verdu, R. Gillard, K. Moustadir, and J. Citerne, "Extension of the PML Technique to the FDTD Analysis of Multilayer Planar Circuits and Antennas," *Microwave and Optical Technology Letters*, **10**, pp. 323–327, 1995.

[33] Z. S. Sacks, D. M. Kingsland, R. Lee, and J.-F. Lee, "A Perfectly Matched Anisotropic Absorber for Use as an Absorbing Boundary Condition," *IEEE Trans. Antennas Propagat.*, **43**, pp. 1460–1463, 1995.

[34] D. M. Sullivan, "Unsplit Step 3-D PML for Use with the FDTD Method," *IEEE Microwave and Guided Wave Letters*, **7**, pp. 184–186, 1997.

[35] S. Abardanel and D. Gottlieb, "A Mathematical Analysis of the PML Method," *Journal of Computational Physics*, **134**, p. 357, 1997.

[36] A. J. Roden and S. D. Gedney, "Efficient Implementation of the Uniaxial-based PML

Media in Three-dimensional Nonorthogonal Coordinates with the use of FDTD Technique," *Microwave and Optical Technology Letters*, **14**, pp. 71–75, 1997.

[37] S. D. Gedney, "Anisotropic Perfectly Matched Layer-Absorbing Medium for the Truncation of FDTD Lattices," *IEEE Trans. Antennas Propagat.*, **44**, pp. 1630–1639, 1996.

[38] S. D. Gedney, "Anisotropic PML Absorbing Media for the FDTD Simulation of Fields in Lossy and Dispersive Media," *Electromagnetics*, **16**, pp. 399–415, 1996.

[39] J. Tang, K. D. Paulsen, and S. A. Haider, "Perfectly Matched Layer Mesh Terminations for Nodal-based Finite Element Methods in Electromagnetic Scattering," *IEEE Trans. Antennas Propagat.*, **46**, pp. 507–516, 1998.

[40] S. Hyun, J. Hwang, Y. Lee, and S. Kim, "Computation of Resonant Modes of Open Resonators using the FEM and the Anisotropic Perfectly Matched Layer Boundary Condition," *Microwave and Optical Technology Letters*, **16**, pp. 352–356, 1997.

[41] J.-Y. Wu, D. M. Kingsland, J.-F. Lee, and R. Lee, "Comparison of Anisotropic PML to Berenger's PML and its Application to the Finite-element Method for EM Scattering," *IEEE Trans. Antennas Propagat.*, **45**, pp. 40–50, 1997.

[42] A. C. Polycarpou, M. R. Lyons, and C. Balanis, "Two-dimensional Finite Element Formulation of the Perfectly Matched Layer," *IEEE Microwave and Guided Wave Letters*, **6**, pp. 338–340, 1996.

[43] K. Rada, M. Aubourg, and P. Guillon, "Perfectly Matched Layer Formulation for Simulating Microwave Axisymmetric Structures Using Finite Elements," *Electronics Letters*, **32**, pp. 1456–1457, 1996.

[44] J.-M. Jin and W. C. Chew, "Combining PML and ABC for Finite-element Analysis of Scattering Problems," *Microwave and Optical Technology Letters*, **12**, pp. 192–197, 1996.

[45] U. Pekel and R. Mittra, "Application of the Perfectly Matched Layer (PML) Concept to the Finite Element Method Frequency Domain Analysis of Scattering Problems," *IEEE Microwave and Guided Wave Letters*, **5**, pp. 258–260, 1995.

[46] U. Pekel and R. Mittra, "Mesh Truncation in the Finite-element-frequency-domain Method with a Perfectly Matched Layer (PML) Applied in Conjunction with Analytic and Numerical Absorbing Boundary Conditions," *Microwave and Optical Technology Letters*, **9**, pp. 176–180, 1995.

[47] U. Pekel and R. Mittra, "Finite-element-method Frequency-domain Application of the Perfectly Matched Layer (PML) Concept," *Microwave and Optical Technology Letters*, **9**, pp. 117–122, 1995.

[48] M. Kuzuoglu and R. Mittra, "Mesh Truncation by Perfectly Matched Anisotropic Absorbers in the Finite Element Method," *Microwave Opt. Technol. Lett.*, **12**, pp. 136–140, 1996.

[49] F. L. Teixeira and W. C. Chew, "Analytical Derivation of a Conformal Perfectly Matched Absorber for Electromagnetic Waves," *Microwave and Optical Technology Letters*, **17**, pp. 231–236, 1998.

[50] F. L. Teixeira and W. C. Chew, "Systematic Derivation of Anisotropic PML Absorbing Media in Cylindrical and Spherical Coordinates," *IEEE Microwave and Guided Wave Letters*, **7**, pp. 371–373, 1997.

[51] M. Kuzuoglu and R. Mittra, "Review of Some Recent Advances in Perfectly-matched Absorbers for Mesh Truncation in FEM," *AP-S International Symposium Proceedings*, **2**, pp. 1302–1305, 1997.

[52] F. L. Teixeira and W. C. Chew, "Perfectly Matched Layer in Cylindrical Coordinates," *AP-S International Symposium Proceedings*, **3**, pp. 1908–1911, 1997.

[53] F. L. Teixeira and W. C. Chew, "PML-FDTD in Cylindrical and Spherical Grids," *IEEE Microwave and Guided Wave Letters*, **7**, pp. 285–287, 1997.

[54] M. Kuzuoglu and R. Mittra, "Investigation of Nonplanar Perfectly Matched Absorbers for Finite Element Mesh Truncation," *IEEE Trans. Antennas Propagat.*, **45**, pp. 474–486, 1997.

[55] W. C. Chew, J. M. Jin, and E. Michielssen, "Complex Coordinate System as a Generalized Absorbing Boundary Condition," *AP-S International Symposium Proceedings*, **3**, pp. 2060–2063, 1997.

[56] W. C. Chew, J. M. Jin, and E. Michielssen, "Complex Coordinate Stretching as a Generalized Absorbing Boundary Condition," *Microwave and Optical Technology Letters*, **15**, pp. 363–369, 1997.

[57] J.-P. Berenger, "Effective PML for the Absorption of Evanescent Waves in Waveguides," *IEEE Microwave and Guided Wave Letters*, **8**, pp. 188–190, 1998.

[58] M. Okoniewski, M. A. Stuchly, M. Mrozowski, and J. DeMoerloose, "Modal PML," *IEEE Microwave and Guided Wave Letters*, **7**, pp. 33–35, 1997.

[59] J. Fang and Z. Wu, "Generalized Perfectly Matched Layer for the Absorption of Propagating and Evanescent Waves in Lossless and Lossy Media," *IEEE Trans. Microwave Theory and Techniques*, **44**, pp. 2216–2222, 1996.

[60] Y. C. Lau, M. S. Leong, and P. S. Kooi, "Extension of Berenger's PML Boundary Condition in Matching Lossy Medium and Evanescent Waves," *Electronics Letters*, **32**, pp. 974–976, 1996.

[61] J. DeMoerloose and A. M. Stuchly, "Reflection Analysis of PML ABC's for Low Frequency Applications," *IEEE Microwave and Guided Wave Letters*, **6**, pp. 177–179, 1996.

[62] J. DeMoerloose and A. M. Stuchly, "Behavior of Berenger's ABC for Evanescent Waves," *IEEE Microwave and Guided Wave Letters*, **5**, pp. 344–346, 1995.

[63] R. W. Ziolkowski, "Design of Maxwellian Absorbers for Numerical Boundary Conditions and for Practical Applications Using Engineered Artificial Materials," *IEEE Trans. Antennas Propagat.*, **45**, pp. 656–671, 1997.

[64] R. W. Ziolkowski, "Time-derivative Lorentz Material Model-based Absorbing Boundary Condition," *IEEE Trans. Antennas Propagat.*, **45**, pp. 1530–1535, 1997.

[65] M. Kuzuoglu and R. Mittra, "Frequency Dependence of the Constitutive Parameters of Causal Perfectly Matched Anisotropic Absorbers," *IEEE Microwave and Guided Wave Letters*, **6**, pp. 447–449, 1996.

[66] N. Pena and M. M. Ney, "Absorbing-boundary Conditions Using Perfectly Matched Layer (PML) Technique for Three-dimensional TLM Simulations," *IEEE Trans. Microwave Theory and Techniques*, **45**, pp. 1749–1755, 1997.

[67] J. Xu, Z. Chen, and J. Chuang, "Numerical Implementation of PML Boundary Conditions in the TLM-based SCN FDTD Grid," *IEEE Trans. Microwave Theory and Techniques*, **45**, pp. 1263–1267, 1997.

[68] Z. Chen and J. Xu, "Generalized TLM-based FDTD-summary of Recent Progress," *IEEE Microwave and Guided Wave Letters*, **7**, pp. 12–14, 1997.

[69] C. Eswarappa and W. J. R. Hoefer, "Absorbing Boundary Conditions for Time-domain TLM and FDTD Analysis of Electromagnetic Structures," *Electromagnetics*, **16**, pp. 489–519, 1996.

[70] F. L. Teixeira and W. C. Chew, "General Closed-form PML Constitutive Tensors to Match Arbitrary Bianisotropic and Dispersive Linear Media," *IEEE Microwave and Guided Wave Letters*, **8**, pp. 223–225, 1998.

[71] A. P. Zhao, "Application of the Material-independent PML Absorbers to the FDTD Analysis of Electromagnetic Waves in Nonlinear Media," *Microwave and Optical Technology Letters*, **17**, pp. 164–168, 1998.

[72] A. P. Zhao, J. Juntunen, and A. V. Raisanen, "Generalized Material-independent PML Absorbers for the FDTD Simulation of Electromagnetic Waves in Arbitrary Anisotropic Dielectric and Magnetic Media," *IEEE Microwave and Guided Wave Letters*, **8**, pp. 52–54, 1998.

[73] A. P. Zhao and A. V. Raisanen, "Extension of Berenger's PML Absorbing Boundary Conditions to Arbitrary Anisotropic Magnetic Media," *IEEE Microwave and Guided Wave Letters*, **8**, pp. 15–17, 1998.

[74] D. H. Werner and R. Mittra, "New Field Scaling Interpretation of Berenger's PML and its Comparison to other PML Formulations," *Microwave and Optical Technology Letters*, **16**, pp. 103–106, 1997.

[75] I. V. Lindell, *Methods for Electromagnetic Field Analysis*, Clarendon Press, Oxford, 1992.

[76] D. Colton, *Partial Differential Equations – An Introduction*, Random House, New York, 1988.

[77] R. H. Good and T. J. Nelson, *Classical Theory of Electric and Magnetic Fields*, Academic Press, New York, 1971.

[78] A. Papoulis, *The Fourier Integral and its Applications*, McGraw Hill, New York, 1962.

[79] V. Veremey and R. Mittra, "A Technique for Fast Calculation of Capacitance Matrices of Interconnect Structures," to appear in *IEEE Transactions on Components, Packaging, and Manufacturing Technology-Part B*.

[80] D. G. Duffy, *Transform Methods for Solving Partial Differential Equations*, CRC Press, New York, 1994.

[81] J. Peng and C. A. Balanis, "A Generalized Reflection-Free Domain-truncation Method: Transparent Absorbing Boundary," *IEEE Trans. Antennas and Propagat.*, **46**, pp. 1015–1022, 1998.

[82] J. A. Kong, *Electromagnetic Wave Theory*, John Wiley & Sons, New York, 1986.

[83] M. Kuzuoglu and R. Mittra, "Anisotropic and Bianisotropic Realizations of Perfectly Matched Absorbers," to appear.

[84] I. Stakgold, *Boundary Value Problems of Mathematical Physics*, Vol. 1, Macmillan, 1967.

[85] M. Kuzuoglu and R. Mittra, "Reciprocity in Perfectly Matched Absorbers," to appear.

[86] M. Kuzuoglu and R. Mittra, "A Systematic Approach to the Derivation of the Constitutive Parameters of a Perfectly Matched Absorber," to appear in *IEEE Microwave and Guided Wave Lett.*

Chapter 15

FAST CALCULATION OF INTERCONNECT CAPACITANCES USING THE FINITE DIFFERENCE METHOD APPLIED IN CONJUNCTION WITH THE PERFECTLY MATCHED LAYER (PML) APPROACH FOR MESH TRUNCATION

Vladimir Veremey and Raj Mittra

ABSTRACT

A finite difference (FD) method for rapid and accurate evaluation of capacitance matrices of interconnect configurations is described in this work. The method utilizes the newly developed *Perfectly Matching Layer* technique for mesh truncation, specially adapted to the static case in conjunction with a mixed boundary condition, referred to herein as the α-technique. Applications of the proposed approach to the modeling of complex structures, comprising multiple metal layers, crossovers, vias and bends embedded in a layered dielectric medium, are illustrated in this chapter. The technique has been combined with the domain decomposition approach to compute the capacitance matrices of complex interconnect configurations in an efficient manner. Both the overlapping and nonoverlapping domain types of approaches are employed to model the interconnects, and convergence and efficiency issues of the proposed algorithms are examined. A novel approach for efficient truncation of a class of large interconnects called the wraparound scheme, is introduced in this work. Several numerical examples that illustrate the efficiency and flexibility of the approaches, described above, are included.

15.1 INTRODUCTION

Currently, there exists a great demand for efficient modeling and simulation of electronic packages, which, in turn, requires rapid parameter extraction of general interconnect configurations. This is especially true in the deep submicron range, where the line widths are 0.25μ or smaller. In this range the interconnect may account for 60 to 70% of the total delay, and an accurate estimate of this on-chip delay plays a crucial role in the design of the chip.

A wide variety of approaches has been employed in the literature for interconnect modeling. These include the *Boundary Element Method* (*BEM*) and its variations [1], e.g., the *Fast Multipole Method* (*FMM*) [2] and the *Closed Form Green's* function

(*CFG*) approach [3]. the *Finite Element Method* (*FEM*) [4]; the *Finite Difference Methods* (*FDMs*) that include the *Measured Equation of Invariance* (*MEI*) technique [5] and the *Finite Difference Time Domain* (*FDTD*) method [6].

As a general rule, the boundary element methods are numerically more efficient than the other techniques provided the environment of the interconnect is homogeneous, which, however, is seldom ever the case in real-world interconnect structures. A considerable improvement in the computational efficiency of the BEM has been achieved by using the FMM technique; however, there is a time penalty to be paid in using this method when the interconnect is embedded in a multilayer dielectric. Though the latter problem is very conveniently and efficiently handled by using the CFG, it is not well suited for analyzing non-planar or conformal dielectrics. Although a totally general interconnect geometry can be analyzed by using the quasi-static or full wave FEM, as well as by the full wave Maxwell solver FDTD, these are typically computationally intensive. Recently, the Finite Difference/MEI combination has been shown to be almost an order of magnitude faster than the FMM-based algorithm FASTCAP [2] for a class of interconnect configurations, and it appears to be an attractive approach to computing the capacitance matrix of a general interconnect configuration. However, it only appears to be able to deal with "truncated" structures that have finite dimensions—unless it employs a large computation domain to reduce the truncation effects—and this, in turn, places a heavy burden on the CPU memory and time.

The purpose of this chapter is threefold. First, we introduce some new concepts for mesh truncation in the finite difference method that are based on combinations of the newly discovered [7] perfectly matched layer (PML) and a mixed absorbing boundary condition. In common with the technique in [5], the use of these truncation techniques in the FDM also makes it faster than FASTCAP for complex inhomogeneous geometries. Second, we utilize a two-step procedure that first solves the problem in a coarse grid and then refines it by deriving an accurate solution of the problem with fine geometrical details, without paying the price that such a procedure would exact if a one-step solution were used over a fine grid throughout.

Next, we present a wraparound technique for efficiently and accurately truncating structures with large (even infinite) dimensions. The technique enables us to reduce the computation time for "excess capacitance" calculations substantially, without the loss of accuracy experienced in the conventional approaches because of subtraction of large numbers representing the capacitances of the two truncated segments of slightly different dimensions.

Computer programs for evaluating the self and mutual capacitances of general off-chip and on-chip interconnect configurations play a very important role in high-speed digital design. The complexity of these structures often makes these computations highly demanding, both in terms of the CPU memory and time. It is not uncommon to find that the above CPU requirements are impractically large for many realistic interconnect structures, especially in the deep submicron range, where it becomes necessary to deal with a large number of metal layers embedded in multilayer dielectrics. This prompts us to seek alternative techniques, e.g., domain decomposition schemes, that not only render the problem manageable on a computer with a single CPU, but are also well suited for parallel processing. Classical domain decomposition algorithms, such as Schwartz or Schur complement methods [9], [10], have been developed for discrete approximations in the context of finite element or finite difference methods. While, in principle, the domain decomposition scheme can be applied in conjunction with any number of different solvers for the Poisson's equation with associated boundary conditions, the FD method, applied in the various subdomains, appears to be the preferred choice when used in conjunction with iteration algorithms that play a key role in the domain decomposition schemes utilizing overlap regions or common boundaries. The application of the FD method enables us to compute the values of the electric potential at all of the mesh points in the subdomain, including in the overlap region. The solution to the original problem is then generated via an appropriate iteration scheme in the context of domain decomposition.

Finally, in Section 15.7 we extend the above method to apply to larger structures by taking advantage of the characteristics of the impedance boundary condition, which is particularly useful for truncating the subdomain boundaries that are penetrated by conducting traces.

15.2 FINITE DIFFERENCE MESH TRUNCATION BY MEANS OF ANISOTROPIC DIELECTRIC LAYERS

In this section we present a novel technique for mesh truncation in the *Finite Difference* technique for solving electrostatic problems. The technique is based on a generalization of the *Berenger Perfectly Matched Layer* (*PML*) concept [7] to the static case. The method employs dielectric layers with anisotropic properties to reduce the influence of mesh truncation on the computation of the capacitance matrix of interconnect configurations.

One of the key issues arising in the analysis of interconnect structures via the method of *Finite Difference* (*FD*) is that of mesh truncation. The problem is to estimate the size of the buffer region—or the white space—that would provide the desired accuracy of the solution, and to determine what type of condition should be imposed at the outermost boundary to truncate the FD region. The answer is obvious if the original structure is bounded by a *Perfect Electric Conductor* (PEC). However, if the geometry of interest is surrounded by free space, the mesh truncation problem can be rather difficult, especially when the accuracy desired is high. One obvious approach to addressing this problem is to enlarge the FD volume and to place the terminating walls far from the structure. Unfortunately, the matrix size associated with the system of linear algebraic equations that has to be solved can increase considerably if this approach is followed, and this can lead to a dramatic increase in the CPU memory and time requirements. An alternative to employing an absorbing boundary condition (ABC) for mesh truncation is to use the *Perfectly Matching Layers* (*PMLs*), a relatively new technique that has found widespread use in Finite Element and FD solution of Maxwell's equations. Although, unlike in the frequency domain PML, we cannot consider lossy dielectric layers in the static case, the idea of using matching layers with anisotropic properties turns out to be useful in the static case as well. It has been shown that the application of such anisotropic layered structure can stretch the distances between the mesh nodes and thus decrease the influence of the boundaries on the results for the capacitance.

15.2.1 Perfectly Matched Layers for Mesh Truncation in Electrostatics

Let us consider a three-dimensional interconnect geometry described in the Cartesian system of coordinates, and let the walls that terminate the FD region be covered by dielectric layers described by a diagonal tensor

$$\hat{\varepsilon} = \begin{pmatrix} \varepsilon_x & 0 & 0 \\ 0 & \varepsilon_y & 0 \\ 0 & 0 & \varepsilon_z \end{pmatrix}$$

In the layers that are parallel to the x-y plane, $\varepsilon_x = \varepsilon_y$, while in those that are parallel to the y-z plane $\varepsilon_y = \varepsilon_z$, etc. The most important feature of these layers is that the corresponding tensor element of the $\hat{\varepsilon}$ is less then 1.0. Such values of the tensor elements introduce a "stretching" of the distance between the mesh nodes in the corresponding planes. For example, the layers that are parallel to the x-y plane have the following dielectric permittivity tensor

$$\hat{\varepsilon}_s = \begin{pmatrix} a_s & 0 & 0 \\ 0 & a_s & 0 \\ 0 & 0 & \tilde{a}_s \end{pmatrix}$$

with $\tilde{a}_s = 1/a_s$ and $a_s > 1$. This choice tapers the PML layer widths, which play the role of stretching factors, when the tensor elements for the layers parallel to the boundary are increased gradually. The dielectric tensors in the corners of the FD domain are diagonal too, and the tensor elements are determined by the properties of the layers that intersect in the corner region. The truncation efficiency and the reflection properties of the PML layers depend on the number of layers and on the values of the parameters a_s.

A computer program based on the PML technique has been developed and applied to a number of representative geometries arising in the interconnect modeling problem, viz., bends, crossovers and vias. The capacitance matrices of these structures have been calculated by using the FD method. The particular case of a 2×2 line crossover above a PEC plane is presented in Fig. 15.1a. Ten PML layers are used to terminate the FD domain, with their stretching factors gradually increasing as explained earlier.

The distribution of the electric potential in the vicinity of the crossover is presented in Fig. 15.1b. The plot shows the potential distribution in the cross-section that is parallel to the y-z plane, and is located flush with the left facet of line 1, whose potential is set at 1 V while all other lines are grounded. The potential function decreases swiftly inside the PML region, as shown in Figs. 15.1c and d. Furthermore, the PML with properties described above makes the results of the calculation independent of the boundary conditions imposed on the outermost boundary of the FD domain, This is because the potential function in the PML region decreases in the same manner as it would if the thicknesses of these layers were increased progressively, and the outer boundary was placed at a large distance away.

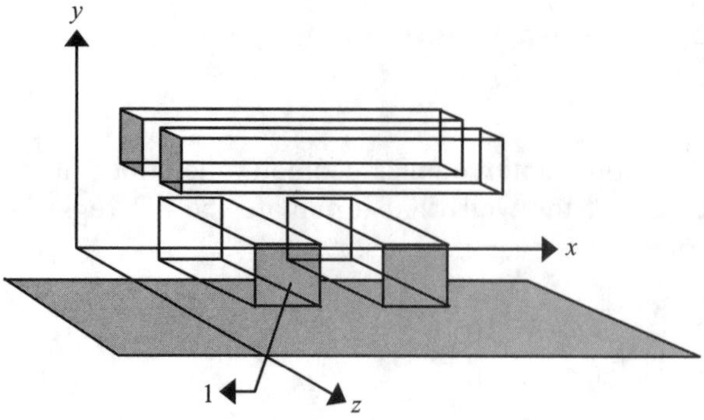

Figure 15.1a Crossover above a conducting plane.

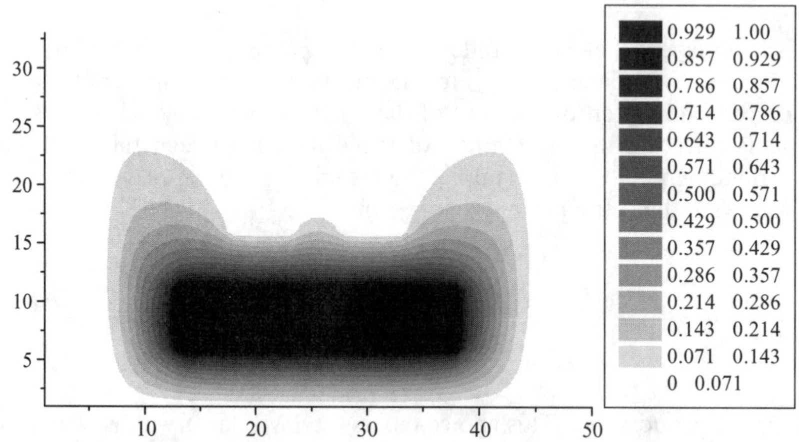

Figure 15.1b Electric potential distribution across the cross section parallel to the y-z plane. The x coordinate of the cross section and the left facet of line 1 are identical.

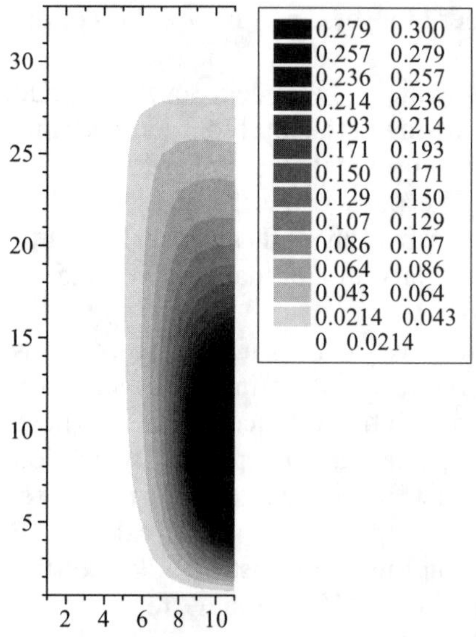

Figure 15.1c Enlarged view of the potential inside the PML region. The cross section is the same as in Fig. 15.1b.

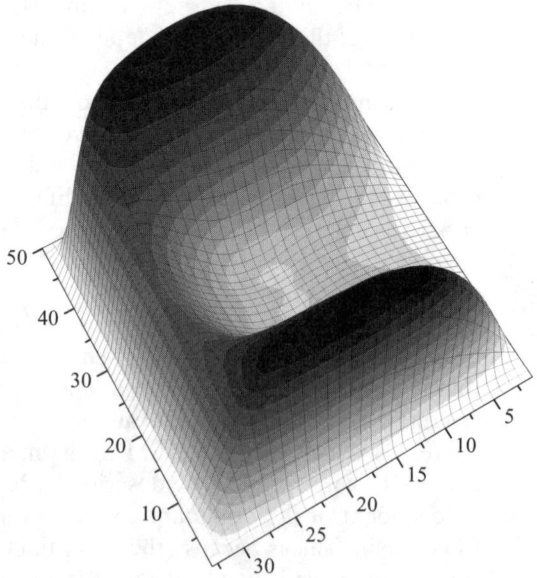

Figure 15.1d Potential distribution across the cross section parallel to the y-z plane, located between the right side of the lines and the PML region.

15.3 α-TECHNIQUE FOR FD MESH TRUNCATION

The FD/PML technique described above offers reliable and accurate means for decreasing the distance between the structure and the boundary of the computational domain. About 20 or 30 PML are sufficient to calculate the mutual capacitance matrices with an accuracy of better than 1%, albeit at the expense of CPU time that obviously increases with the number of PML layers. Although the above solution is still useful as a reference, for many applications it is highly desirable to reduce the computation time without an undue sacrifice in the accuracy of the results. Toward this end, we propose a method that requires only a few PML layers, typically three to six, and is therefore considerably faster than the one requiring a large number of layers. The method will be referred to herein as the α-technique.

For an open region problem, the truncation of a FD grid with a PEC boundary always introduces an error, which can be minimized either by placing this boundary far away from the structure, or by employing a sufficient number of PML layers to achieve the same effect. To reduce the number of PML layers we use an impedance boundary condition, also referred to herein as the α-technique, as described below. The method is based on the premise, which is easily verifiable, that the potential decreases in an exponential fashion in the asymptotic region as we recede away from the structure in the direction of the boundary. At points near the x boundary, say for the point E in Fig. 15.2, we can then replace the conventional FD equations [8], viz.,

$$[\Phi_A(i+1,j) + \Phi_B(i-1,j) + \Phi_C(i,j+1) + \Phi_D(i,j-1)] = 4\Phi_E(i,j) \tag{15.1}$$

with the following:

$$[\Phi_A(i+1,j) + \Phi_C(i,j+1) + \Phi_D(i,j-1)] = 3\Phi_E(i,j) + \frac{\alpha}{2}\Phi_E(i,j) \tag{15.2}$$

where we assume that the potential varies as $e^{\alpha x}$, the mesh is locally uniform at all points A, B, C, D, and E, and that the medium is also locally homogeneous. However, if desired, these equations can be generalized for the case of a non-uniform mesh with inhomogeneous dielectrics. Equation (15.2) serves as the truncation condition at the

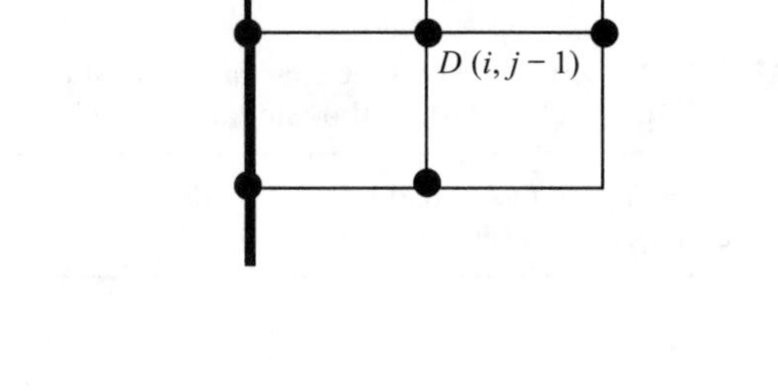

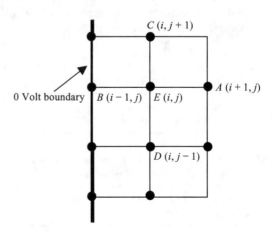

Figure 15.2 Mesh points near the boundary.

Figure 15.3 Two conducting lines located over a ground plane.

boundary, in lieu of the PEC condition, which corresponds to the limit of a large negative α. The key question that we now need to answer is how to determine the parameter α.

For the case of a single line, a single, global α-parameter can be found, and the self capacitance can be computed with an accuracy of better than 5% by using such an α-technique. The computational domain can be relatively small when the global α is used and, consequently, the computation time also becomes small, say only a few seconds even for a moderately large problem. However, for the case of a structure formed by two parallel lines, the global α-technique can only be used to accurately compute either the self or the mutual capacitance, but not both of them simultaneously with the same level of accuracy. This is because the required values of the global α-parameter are different for these two calculations.

To illustrate this point we will now present some numerical results for the two-line geometry shown in Fig. 15.3. The lines have identical dimensions, and their widths, heights, and lengths are 0.6, 0.6, and 20.0μ, respectively. We set the potential of the bottom line to 0 V and locate it at a distance of 1.55μ above the ground plane, while setting the potential of the top line to 1 V and placing it 4.35μ above the ground plane. The dielectric permittivity of the medium is set at $\varepsilon = 3.9$.

Using the FD/PML technique, and with a large number of layers that ensure the convergence of the result, we obtain the results given in Table 15.1.

If we use a single α over the entire boundary for the calculation of both the self and mutual capacitances, the accuracy of calculation decreases. For instance, we find the following results in Table 15.2.

TABLE 15.1 Self and Mutual Capacitances of the Two-Line Problem (see Fig. 15.3) Calculated by the FD/PML Technique

Self capacitance of the second line: 1.6869 fF
Mutual capacitance of the first line: 0.5059 fF

TABLE 15.2 Self and Mutual Capacitances of the Two-Line Problem (see Fig. 15.3) Calculated by the Global α-Technique

Self capacitance of the second line: 1.9276 fF
Mutual capacitance of the first line: 0.3301 fF

Thus, we conclude from the above example that the global α-technique is not suitable for calculating, simultaneously, both the self and mutual capacitances of crossovers with good accuracy.

In light of this, we refine the α-technique with a view of rectifying the above situation in a manner explained below. The clue to modifying the global α-technique comes from an examination of the behavior of the potential near the boundary. When the outer boundary is quite far away, say 30 or more cells, the potential function exhibits a decaying asymptotic behavior (near the boundary), as one recedes from the structure. However, this is no longer true in general, especially when the truncation boundary is brought close to the structure. In fact, in certain cases the potential can show a locally increasing behavior near the close-in truncation boundary, especially in the vicinity of the conductors with zero potential. This leads us to conclude that a locally varying α is needed to correctly represent the behavior of the potential in the general situation.

Now we address the problem of numerically determining the α-parameter as a function of the local coordinates (i, j) of the points near the boundary, and present a two-step procedure to accomplish this goal. As a first step, we employ a global α-parameter to compute the derivatives of the potential function at points near the boundary. We then use these local α parameters, thus determined, in the boundary conditions in the second step to re-solve the potential problem. The use of such a two-step approach enables us to simultaneously compute the self and mutual capacitance values with comparable accuracy.

To illustrate the improvement achieved by using the two-step α-technique we revisit the two-line structure shown in Fig. 15.3. The results we now obtain with the local α-technique are presented in Table 15.3.

TABLE 15.3 Self and Mutual Capacitances of the Two-Line Problem (see Fig. 15.3) Calculated by the Two-Step α-Technique

Self capacitance of the second line: 1.6984 fF
Mutual capacitance of the first line: 0.4968 fF

Comparing the results in Table 15.3 with the reference values in Table 15.1, we observe that both the self and mutual capacitance values are accurate to within 2%. It is evident, therefore, that when applied to this configuration, the two-step global technique remedies the accuracy problem encountered in the one-step global α procedure.

Calculated results for the electric potential in the vicinity of a structure formed by a via and two parallel lines are presented in Fig. 15.4 (see the full-color insert). The potential distribution is computed on a plane that is parallel to the ground plane and is located at the same level as the T-shaped upper conductor of the via. Shown in the figure are the outermost boundary Γ, and the boundary Γ_α used in the algorithm for the recalculation of the coefficient α.

15.4 WRAPAROUND TECHNIQUE FOR MESH TRUNCATION

It is not uncommon in interconnect modeling to come across a situation where the transmission lines forming the interconnect are long and, hence, penetrate through the truncation boundary of the FD mesh. In this section we present a technique for efficiently and accurately handling geometries of this type. The technique utilizes the solution of the corresponding two-dimensional problem in a domain that is formed by unwrapping the corresponding side facets of the three-dimensional FD region. To illustrate this technique we again consider a 2×2 crossover configuration, located between two conducting planes as shown in Fig. 15.5a, though either of these planes may be removed, if desired.

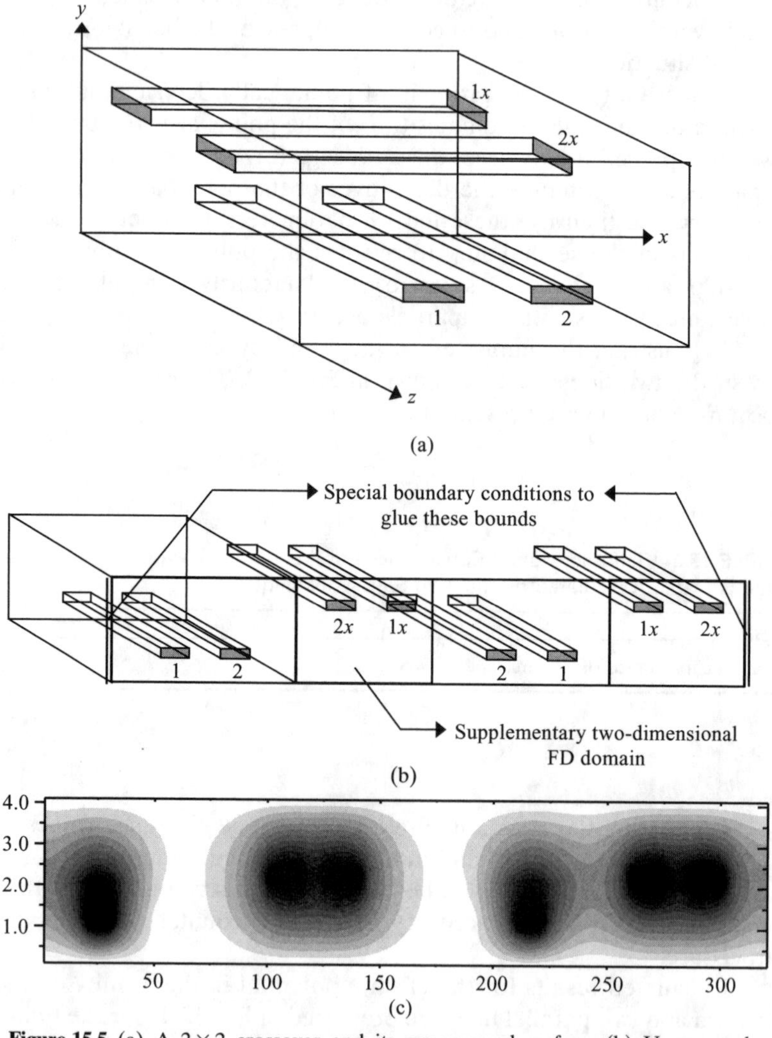

Figure 15.5 (a) A 2×2 crossover and its wraparound surface. (b) Unwrapped geometry and supplementary two-dimensional domains. (c) Potential distribution across a cross-section in the unwrapped problem when the conductors 1, $1x$, and $2x$ are set to 1 volt and conductor 2 is grounded.

Our objective is to compute the excess capacitance when the two lines are essentially infinite in extent, along the x and y axes, respectively. The problem of the mesh truncation becomes trivial if one knows the values of the electric potential distribution over the four planes that surround the crossover. Our approach is to determine the potential distributions on the boundary surfaces by solving a related two-dimensional problem in a region, which is developed by unwrapping the four side walls and placing them on a flat surface. It is assumed that the four walls are sufficiently far away from the crossover region, and that the normal derivative of the potential distribution on these surfaces is very small. It is also to be mentioned that the y borders of the two-dimensional region are glued together and this provides additional conditions for the border points. Finally, the solution to the original three-dimensional problem is derived by utilizing the values obtained for the potential function on the wraparound surface obtained for the two-dimensional problem (see Figs. 15.5b and c).

15.5 TWO-STEP CALCULATION METHOD

To further enhance the efficiency of the capacitance computation using the FD/PML approach, we introduce a two-step mesh refinement procedure. As a first step, we compute the electric potential at the nodes of a coarse grid, used to discretize the computational domain that includes the PMLs. Next, we recalculate the electric potential at the nodes of the refined grid, whose cell size is smaller by a factor of 2 (or more) as compared to that of the coarse grid. The computational domain for the refined calculation includes only the points in the close vicinity of the structure. The initial values of the potential at the grid points on the FD boundary are derived by using a spline interpolation of the results obtained in the first step.

Illustrative examples of the coarse and fine grids with a 2:1 refinement are presented in Fig. 15.6 for the case of a crossover configuration formed by six and three lines that are parallel to the z and x axes, respectively.

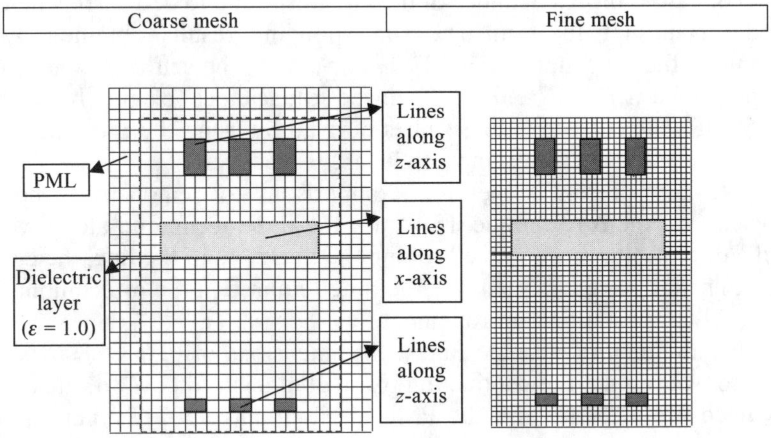

Figure 15.6 Illustrative examples of coarse and fine grids for capacitance computation of a crossover comprising six and three lines parallel to the z and x axes, respectively.

15.6 NUMERICAL RESULTS

In this section, we present some numerical results to illustrate the application of the FD/PML approach described in Section 15.2. A variety of geometries has been analyzed by using this method to illustrate its versatility. We demonstrate that, compared to the PML-only mesh truncation method, the computation time can be significantly reduced (by an order of magnitude or more) via the use of the α-technique, with little loss of accuracy.

15.6.1 Microstrip Line Over a Conducting Plane

The first geometry considered is the simple case of a finite length microstrip line over either an infinite (case 1) ground plane, or a finite-size (case 2) conducting ground plane, whose dimensions are $(15.0\mu \times 15.0\mu)$. Both of these are shown in Fig. 15.7. The dielectric constant of the substrate is $\alpha = 3.9$. The width of the line is 5μ, its height and length are 1.0μ and 10.0μ, respectively. This geometry was chosen because the result for the capacitance was readily available from other sources for convenient comparison.

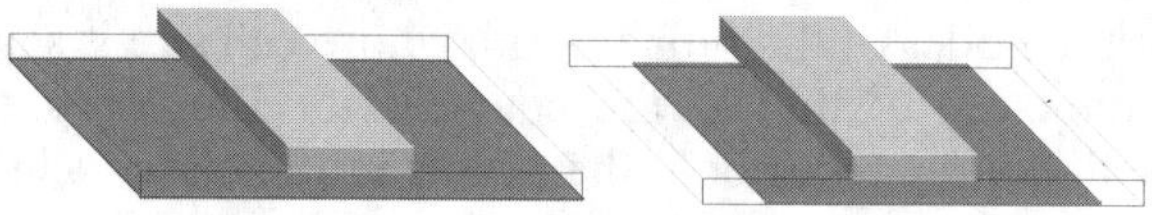

Figure 15.7 Microstrip line over an infinite ground plane, and over a finite-size conducting ground plane.

For case 1, the capacitance of the line was found to be 1.70 fF when 30 PMLs were used for mesh truncation. The corresponding results obtained by using the α-technique the capacitance is 1.688 fF. For case 2, the self capacitance of the line was found to be 1.711 fF and the induced capacitance of the background line was -1.596 fF. The corresponding values derived by using the fast calculation technique were found to be 1.728 fF and -1.564 fF, respectively.

We now present some time comparisons to demonstrate the computational efficiency of the α-technique. If we use the PML technique along with 20–30 layers and a discretization of four cells along the thickness of the line, the computation time is 20 min. and 38 sec. In contrast, the time required in the α-technique is only 1 sec — obviously a substantial improvement.

For the case of a single line with dimensions of $(0.6 \times 0.806 \times 15.0\mu)$, which is located over an infinite conducting plane at a height of 3.913μ, the capacitance value obtained was 1.297 fF with the PML method, and 1.307 fF when the α-technique is used.

The distribution of the electric potential across the cross section that is parallel to the z-y plane is presented in Fig. 15.8 where the cross section is also shown.

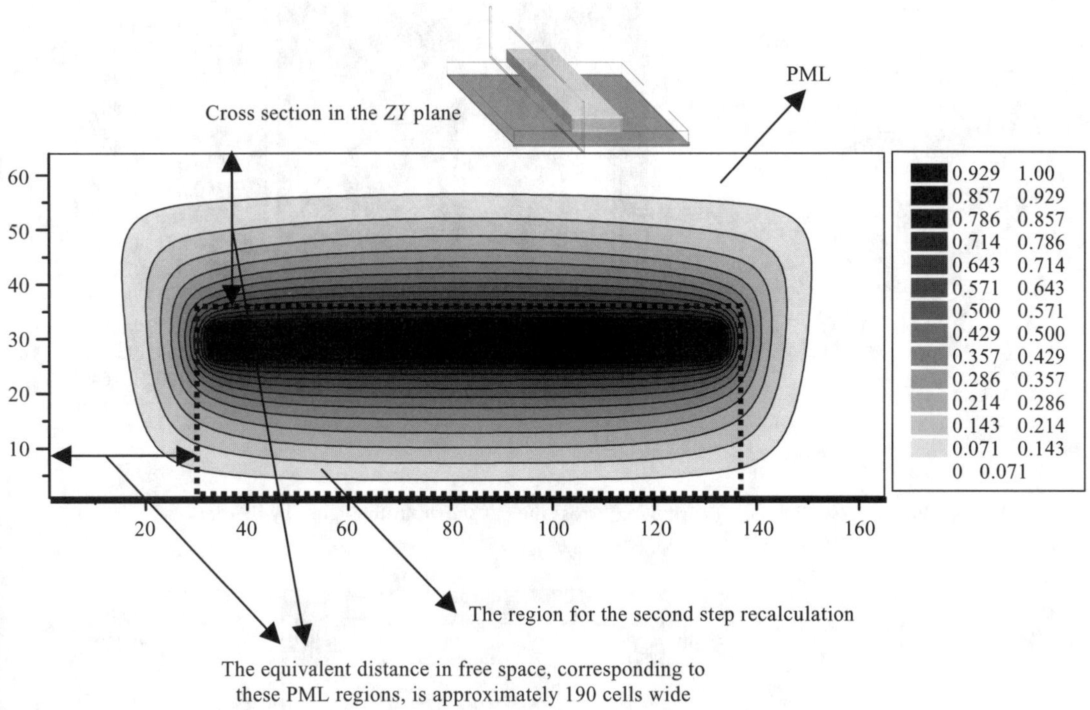

Figure 15.8 Potential distribution across the cut that is parallel to the *y-z* plane.

15.6.2 Coupled Microstrip Bends Over a Conducting Plane

We now consider the geometry formed by two bends that are stacked vertically above a conducting plane. The dimensions of the bends are identical and the widths of the lines that form the bends are 2.0 and 3.0μ, respectively. The line thicknesses are 5.0μ, and the length of each of the two arms of the bends is 15.0μ. The heights of the bend above the ground plane are 5.0μ and 15.0μ, respectively. The structure is embedded in a two-layer dielectric with permittivities of 2.0 and 3.9, and thicknesses of 5.0μ and 15.0μ. If both the bends are set at a potential of 1 V, we obtain the capacitance values of 2.286 fF and 0.8493 fF for the lower and upper bends, respectively. In addition to computing the capacitances, we have also studied the behavior of the electric potential in the vicinity of this structure. Some of these potential distributions are presented in the Figs. 15.9a, b, and c.

15.6.3 Crossover

Next, we consider a crossover structure which is frequently encountered in interconnects. The 3×3 crossover is shown in Fig. 15.10. All of the lines in the crossover have identical dimensions, and it is located above a ground plane covered

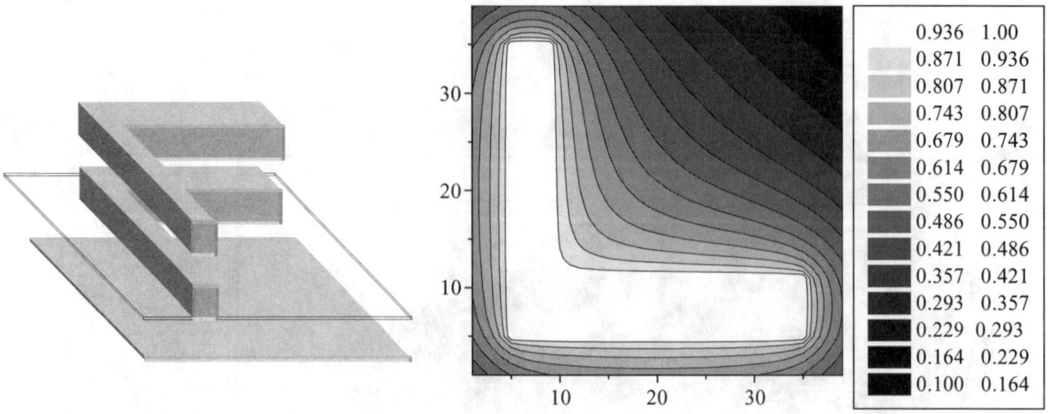

(a)

Figure 15.9a Potential distribution across the cut shown in the figure by the dark rectangle.

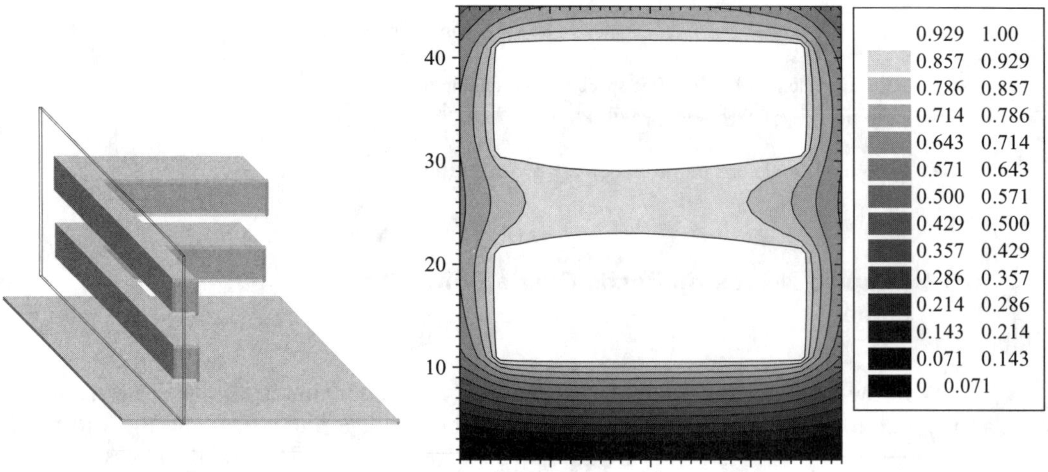

(b)

Figure 15.9b Potential distribution across the cut shown in the figure by the dark rectangle.

by a six-layer dielectric substrate. The widths and the thicknesses of the lines are 0.5 and 1.0μ, respectively, and their lengths are 4.5μ. The height of the lower surfaces of the bottom lines is 2.1μ, and the separation between the lines is 0.5μ. The lines parallel to the x axis are located on the top of those that are parallel to the z axis. The distance from the lower surfaces of the x lines and the ground plane is 3.8μ, and the distance between the x lines is also 0.5μ. The dielectric layers have the following permittivities: 3.7, 4.1, 3.7, 4.3, 3.7, and 4.3, and their heights above the ground plane are: 0.5, 0.6, 1.9, 2.1, 3.6, and 3.8μ, respectively. All of the lines are embedded in a

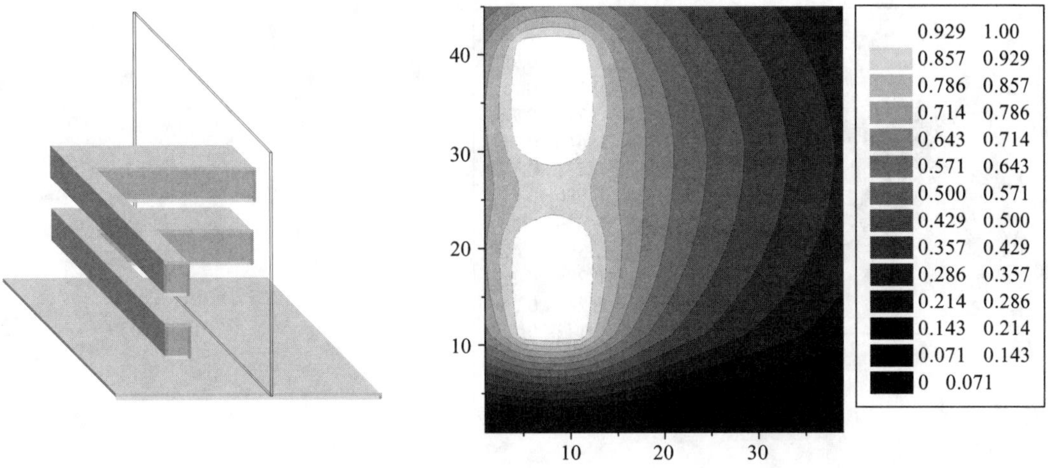

(c)

Figure 15.9c Potential distribution across the cut shown in the figure by the dark rectangle.

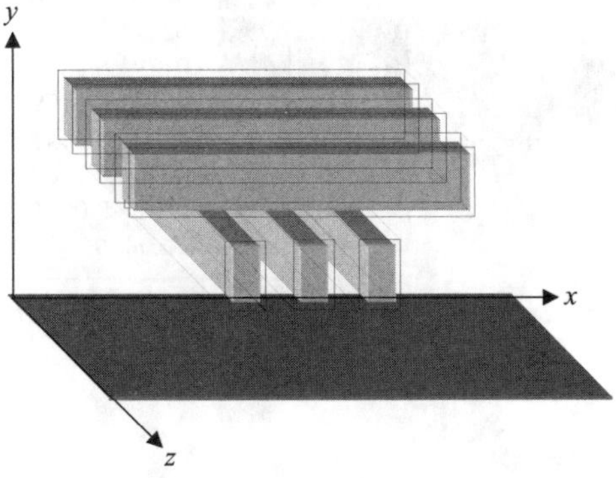

Figure 15.10 Crossover above the ground plane.

dielectric that wraps around their cross-section. The thickness of the dielectric is 0.1μ and its $\varepsilon = 4.0$. The permittivity of the medium above the lines is 3.7. We set the potential of the first line in the lower layer equal to 1, while the rest are set equal to zero. When 30 PMLs are used for mesh truncation, capacitances of the lines are found to be the following:

C11	1.0129 fF
C12	0.45633 fF
C13	0.021564 fF
C14	0.095532 fF
C15	0.068588 fF
C16	0.095532 fF

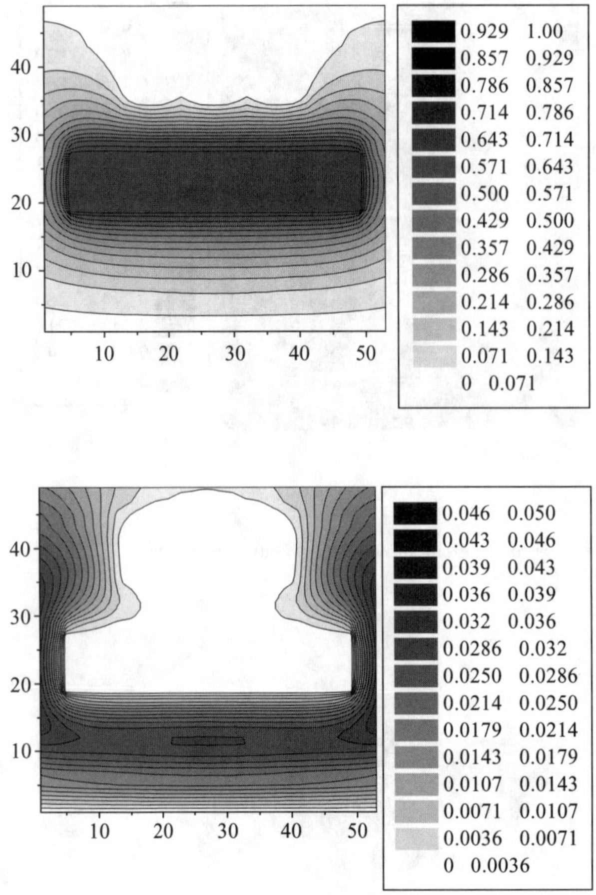

Figure 15.11a Potential distribution across the cross sections that are parallel to x-y plane and juxtaposed with the left sides of (a) the first line; (b) the third line.

The number of cells along the width of the lines in the refined calculation was six and the computation time was 40 min. and 19 sec. The corresponding results obtained by using the α-technique are

C11	1.01277 fF
C12	0.46258 fF
C13	0.021403 fF
C14	0.094819 fF
C15	0.067806 fF
C16	0.094819 fF

and the computation time was 2 min. and 1 sec. This time can be further reduced by relaxing the accuracy criterion, which was quite stringent in the above example. Some representative results for the electric potential in the vicinity of a structure are presented in Figs. 15.11a, b, and c.

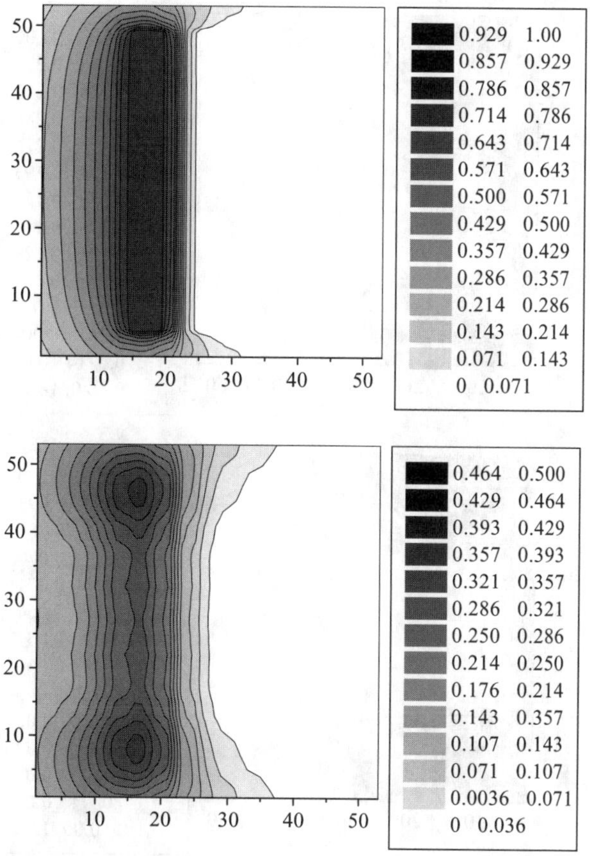

Figure 15.11b Potential distribution across the cross sections that are parallel to x-z plane (a) juxtaposed with the lower sides of the z line; (b) located between the z and x lines.

15.6.4 Combinations of Bends and Crossovers Above a Conducting Plane

In this example, we consider a complex structure formed by a combination of a bend and a crossover. The latter is formed by two lines parallel to the z axis, and one that is parallel to the x axis. The dimensions of the bend are as follows: the widths of the line that form the bend are 2.0μ, their thicknesses are 1.0μ and the lengths of the bend arms are 7.0μ. The bend is located at 3.0μ above the conducting plane. The widths of the lines that are parallel to the z and x axes are 2.0μ and their thicknesses are 1.0μ. The lengths of the straight lines are identical and are equal 11.0μ. The lines parallel to the z and x axes are located 6.0μ and 9.0μ, respectively, above the ground plane. The structure is embedded in a two-layer dielectric region with permittivities of 2.0 and 3.9, and thicknesses of 3.0 and 9.0μ, respectively. By setting the potential of the first line to 1 V and the rest equal to 0, we compute the following values for the capacitances:

−0.5344 fF	for the bend
1.778 fF	for the first line (that is parallel to the z axis)
−0.4568 fF	for the second line (that is parallel to the z axis)
−0.3240 fF	for the third line (that is parallel to the x axis)

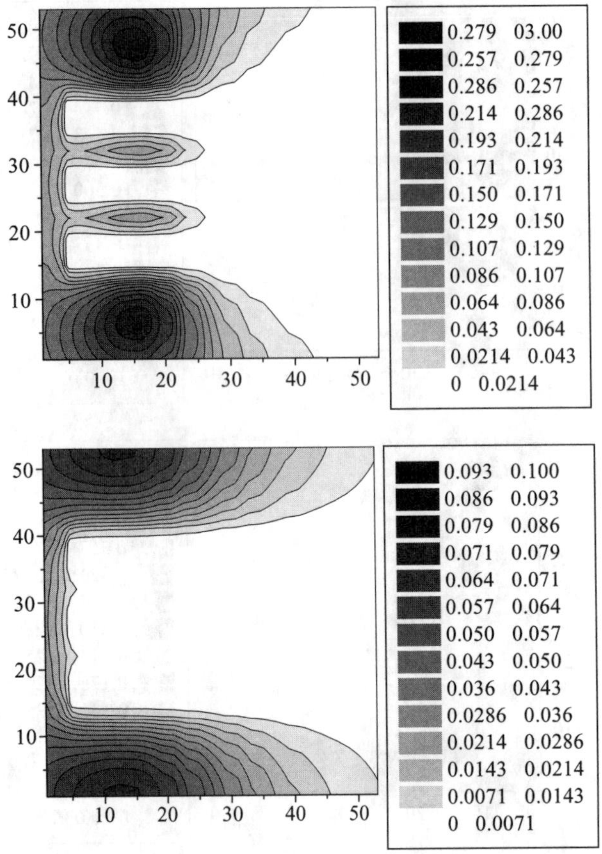

Figure 15.11c Potential distribution across cross-sections that are parallel to *x*-*z* plane juxtaposed with the (a) lower and (b) upper sides of the *x* lines.

If the lines parallel to the *z* axis are coated by dielectric with permittivity 4.0 and thickness 0.1μ (see Fig. 15.12), the capacitances are found to be

−0.54616 fF	for the bend
1.85966 fF	for the first line (that is parallel to the *z* axis)
−0.49167 fF	for the second line (that is parallel to the *z* axis)
−0.33077 fF	for the third line (that is parallel to the *x* axis)

Typical potential distributions are presented in Figs. 15.3a, b, and c.

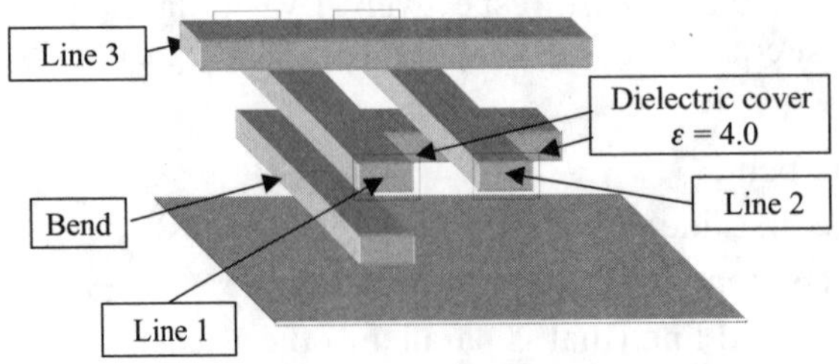

Figure 15.12 Geometry view of the structure comprising a bend, two lines that are parallel to *z* axis, and one line that is parallel to *x* axis.

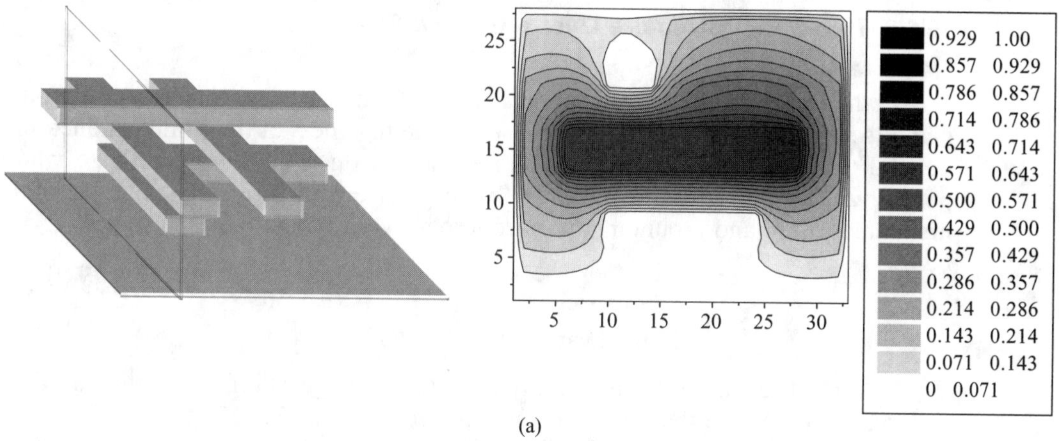

(a)

Figure 15.13a Potential distribution across cross-section that is parallel to the *y-z* plane.

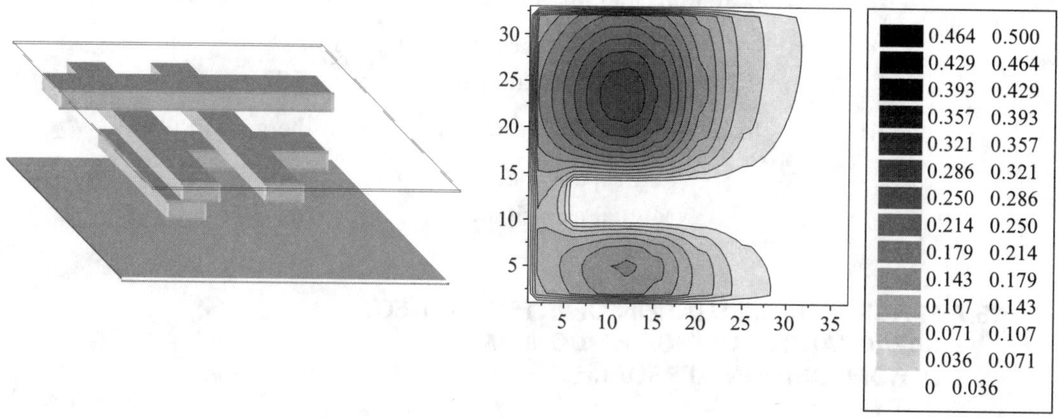

(b)

Figure 15.13b Potential distribution across cross-section that is parallel to the *x-z* plane.

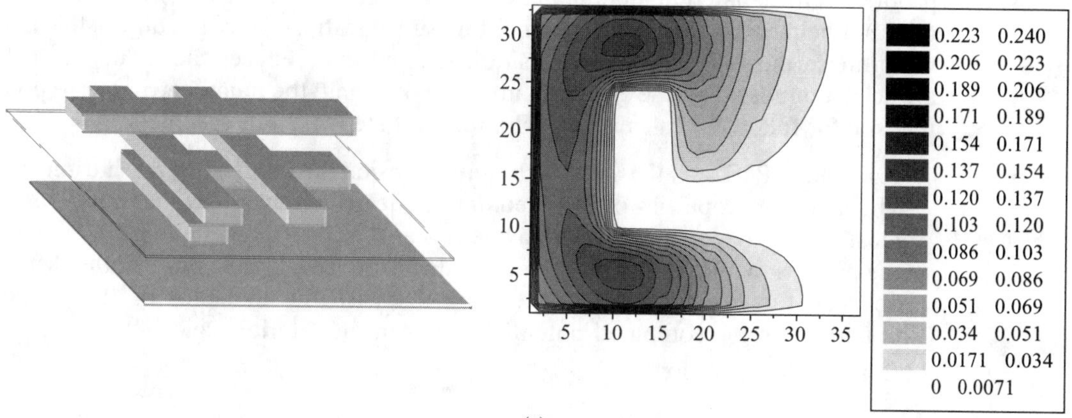

(c)

Figure 15.13c Potential distributions across the cross section that is parallel to the *x-z* plane.

15.6.5 Two-Comb Structure Over a Ground Plane

In the last example, we consider a structure formed by two combs over a ground plane (see Fig. 15.14 in the full-color insert). The thickness of the comb conductors is 2μ and their width is 3μ. The combs are located on a dielectric substrate whose permittivity and thickness are 3.9 and 5.0μ, respectively. By setting the potential of the first comb to 1 V and grounding the second one, we compute the capacitances shown below:

Self capacitance	$C_1 = 19.3236$
Mutual capacitance	$C_2 = -7.449$

The electric potential distributions of this configuration, computed at three different levels above the ground plane, are presented in Fig. 15.14.

If the combs are located parallel to each other at different distances from the ground plane the capacitance values are given below:

(i) Combs located 5μ and 6μ respectively above the ground plane.

Self capacitance	$C_1 = 18.7429$
Mutual capacitance	$C_2 = -5.537$

(ii) Combs located 5μ and 8μ, respectively, above the ground plane.

Self capacitance	$C_1 = 18.189$
Mutual capacitance	$C_2 = -4.1271$

15.7 EFFICIENT COMPUTATION OF INTERCONNECT CAPACITANCES USING THE DOMAIN DECOMPOSITION APPROACH

The interconnect structures comprise metal layers located in layered dielectrics, with or without a ground plane. We point out that since the original FD boundary value problem is formulated in an unbounded domain, the decomposition into subdomains does not help circumvent the problem of mesh truncation. Thus, to address this issue, we propose the following decomposition scheme that reduces the influence of the artificial boundary of the subdomain while keeping the number of mesh points realistic for follow-up calculations. The relevant steps for this scheme are

1. Place static PMLs on the artificial boundary inside the domain for mesh truncation. The material properties of this medium are described by a diagonal tensor $\hat{\varepsilon}$. The most important feature of these layers is that the corresponding tensor element of the $\hat{\varepsilon}$ is less than 1.0. Such tensor elements introduce a "stretching" of the distance between the mesh nodes, and thereby decrease the influence of the artificial boundary on the computed potential values in the interior region.
2. Enforce the condition

$$\left\{ \frac{\partial}{\partial n} U(\vec{p}) + \alpha(\vec{p}) U(\vec{p}) \right\} \Bigg|_{\vec{p} \in \Gamma} = 0$$

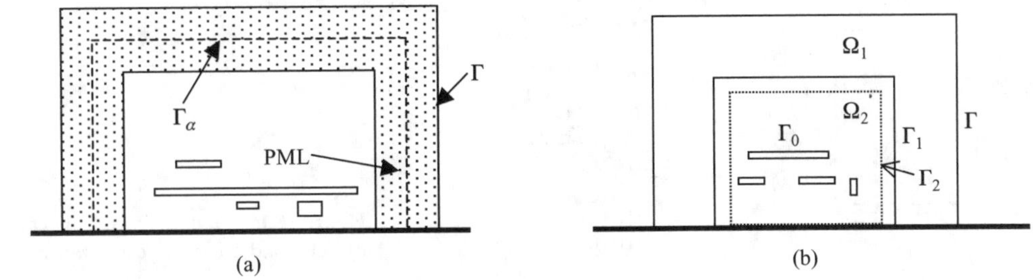

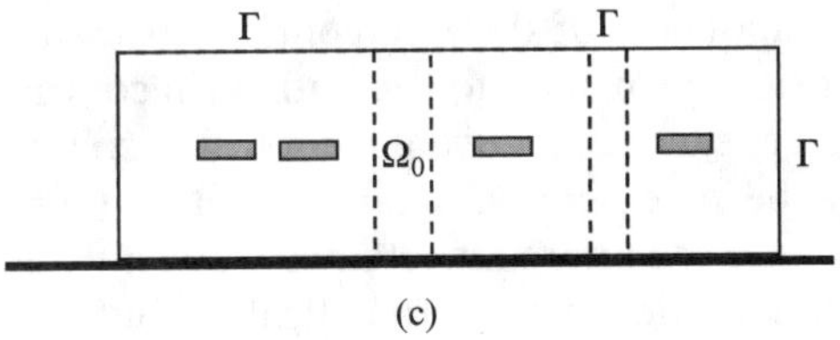

Figure 15.15 Domain decomposition schemes.

on the potential function U at the outermost boundary Γ (see Fig. 15.15). The coefficient α is a function of the coordinates and this function is determined by iterative recalculation, assuming that the normal derivative of U on the boundary Γ_α, which is just inside of Γ (see Fig. 15.15a), is equal to that on Γ. Using this boundary condition in conjunction with the PMLs allows us to obtain accurate results with only a single iterative refinement of α.

Various domain decomposition schemes, which utilize the technique described above for truncating open subdomains with artificial boundaries, have been devised, and have been applied to the problem of computing interconnect capacitances. The geometries of the subdomains are defined with or without overlap regions by taking into account the geometry of the interconnect structure in a systematic manner, as described below via some illustrative examples.

Consider, for instance, the geometry shown in Fig. 15.15b, consisting of a collection of conducting traces whose boundaries are defined by Γ_0. First we truncate the original unbounded region by using the boundary Γ and then divide it into two subdomains, viz., Ω_1 and Ω_2, as shown in Fig. 15.15b, with an overlapping region Ω_d, which is bounded by Γ_2 and Γ_1. Next, we solve the problem in Ω_1 using a coarse grid and then in Ω_2 with a discrete approximation on a refined grid, whose cell size is smaller by a factor of 2 (or more) relative to that in the coarse grid. The computational domain Ω_2 for the refined calculation with the fine grid is a relatively small region, which is confined to the near vicinity of the etches in the interconnect.

For more general and complex interconnect structures, it becomes necessary to divide the computational domain into a multitude of subdomains, again with or without overlap regions, and to use the PML- and α-techniques to reduce the truncation errors introduced by the artificial boundary Γ (see Fig. 15.15c). An alternating algorithm using the overlap regions (Ω in Fig. 15.15c), has been employed and its convergence has been demonstrated numerically.

We have found that the introduction of the overlap regions between the sub-domains and the application of the Schwartz alternating method becomes

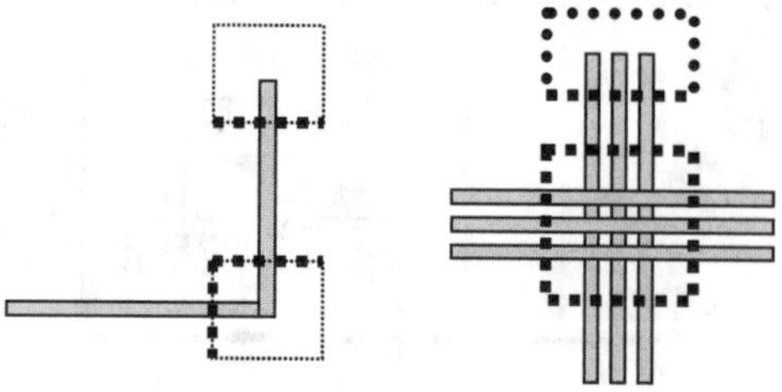

Figure 15.16 Subdomains for the calculation of bend and crossover capacitances.

necessary only when the constituent components of the interconnect structure, viz., crossovers, bends, or vias, are located in close proximity to each other. In contrast, for the geometries shown in Fig. 15.16, we can take advantage of the solution of two-dimensional problems and employ the reflecting boundary condition ($\alpha = 0$) at the boundaries of the subdomains that cross the interconnect lines (shown by dashed lines in Fig. 15.16) to reduce the computation time dramatically. The *Domain Decomposition* approach is efficient for calculating the self and mutual capacitances of complex interconnect structures. An example of such a structure is shown in Fig. 15.17a. It consists of three conducting lines and includes a crossover, a via, and a bend embedded in a layered dielectric medium with three different dielectric permittivities. We set the potential of net-1 to 1 V and ground nets 2 and 3. To apply the Domain Decomposition approach we define the four subdomains as shown in Fig. 15.17b.

The self capacitance of net-1 is determined as a sum of the fringe capacitances of the subdomains 1 and 4, crossover capacitances of the subdomains 2 and 3, the capacitance of the via in the subdomain 3, and the capacitance of the uniform regions between the subdomains 1 and 2, 2 and 3, 3 and 4. The capacitance value of the net-1,

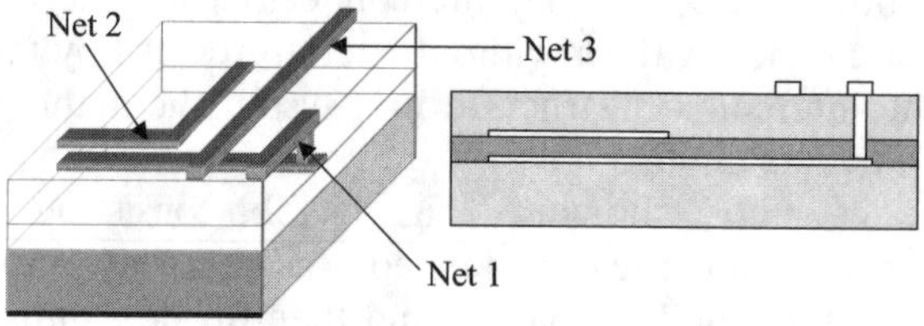

(a)

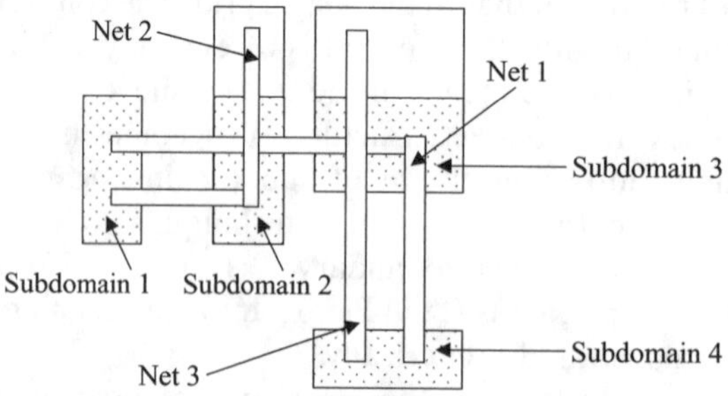

(b)

Figure 15.17 Domain decomposition for capacitance calculation of the structure formed by a straight line, a bend, and a via.

calculated by using the conventional approach, is $C_{net1} = 9.19$ fF, while the result derived via the decomposition scheme described above is 9.33 fF.

Calculated results for the electric potential in the subdomain 3 are presented in Fig. 15.18 (see the full-color insert), at different levels with respect to the ground plane.

15.8 CONCLUSION

This chapter has introduced an efficient Finite Difference (FD) technique for capacitance computations of interconnect structures. Several novel embellishments have been added to the conventional Finite Difference approach to enhance its numerical efficiency. These include: (1) quasi-static version of Perfectly Matched Layer for mesh truncation in conjunction with a mixed boundary condition at the outer layer (α-technique); (2) wraparound technique that constructs the solution to a three-dimensional problem by making efficient use of the solution of a related two-dimensional structure; and (3) two-step calculation for systematic refinement of the solution. Extensive numerical results, that illustrate the versatility of the approach and its applicability to a variety of practical interconnect configurations, have been included along with sample field distributions and computation times for representative interconnect structures.

REFERENCES

[1] V. Pan, G. F. Wang, and B. K. Gilbert, "Edge effect enforced boundary element analysis of multilayered transmission lines," *IEEE Trans. Circuits Syst.* I, **39**, pp. 955–963, November 1992.

[2] K. Nabors and J. White, "Multipole-accelerated capacitance extraction algorithm for 3-D structures with multiple dielectrics," *IEEE Trans. Circuits Syst.* I, **39**, pp. 946–954, November 1992.

[3] S. Oh, V. Jandhyala, J. E. Schutt-Aine, E. Michielssen, and R. Mittra, "Equivalent capacitance of a strip crossover in a multilayered dielectric medium," *IEEE Trans. Microwave Theory Tech.*, **MTT-44**, pp. 347–349, February 1996.

[4] I. Costache, "Finite element method applied to skin-effect problems in strip transmission lines," *IEEE Trans. Microwave Theory Tech.*, **MTT-35**, pp. 1009–1013, November 1987.

[5] W. Sun, W. Wei-Ming Dai, and W. Hong, "Fast parameter extraction of general interconnects using geometry independent Measured Equation of Invariance," *IEEE Trans. Microwave Theory Tech.*, **MTT-45**, pp. 827–836, May 1997.

[6] K. S. Kunz and R. J. Luebbers, *Finite Difference Time Domain Method for Electromagnetics*, CRC Press, Boca Raton, FL, 1993.

[7] P. Berenger, "A perfectly matched layer for the absorption of electromagnetic waves," *J. Comput. Physics*, **114**, pp. 185–200, 1994.

[8] M. Sadiku, *Numerical Techniques in Electromagnetics*, CRC Press, Boca Raton, FL, 1992.

[9] S. Barry, P. Bjrstad, and W. Gropp, *Domain Decomposition*, Cambridge University Press, New York, 1996.

[10] R. Glowinski, J. Periaux, Z.-C. Shi, and O. Widlund, "Domain decomposition methods in sciences and engineering," 8th International Conference, Beijing, P.R. China, John Wiley & Sons, New York, 1997.

Chapter 16

FINITE-DIFFERENCE TIME-DOMAIN METHODOLOGIES FOR ELECTROMAGNETIC WAVE PROPAGATION IN COMPLEX MEDIA

Jeffrey L. Young

16.1 INTRODUCTION

The finite-difference, time-domain method (FDTD) has emerged as a leading numerical tool for the solution of the temporally dependent Maxwell's equations [1], [2]. Originally the scheme was created to discretize Maxwell's equations, with the assumption that the medium that hosted the electromagnetic wave was isotropic, non-dispersive, linear, and time- invariant. In many applications that arise in the study of electromagnetic waves, such assumptions are indeed valid.

As the method grew in popularity and proved its worth in the prediction of scattering, diffraction, and propagation events, many researchers postulated and then devised ways that the algorithm could be applied to problems for which the medium was no longer simple in its composition. Such media types include the ionosphere, biological tissues, crystalline structures, ferrites, optical fibers, and radar absorbing materials, to name a few. Whatever the case may be, however, these media are subsets of what is generally termed herein as *complex* media—media that are described by one or many of the following descriptors: dispersive, anisotropic, non-linear, and time-variant [3].

This chapter examines the most recent and popular advances in FDTD algorithm development as applied to electromagnetic wave propagation in complex media. In some cases, several methodologies are presented and comparisons are made in order to demonstrate the strengths and weaknesses of the various approaches. Such is the case with respect to wave propagation in an isotropic cold plasma, Debye dielectric and the Lorentz dielectric. This chapter presents several different direct integration techniques and two convolution-based techniques. In other cases, only one technique is provided and the reader is referred to the literature for other approaches.

Where appropriate or insightful, error analyses are also provided. Since the discretized equations are approximations of Maxwell's equations, the data obtained from the FDTD algorithm will be corrupted by errors of the dissipative or dispersive kind. These errors linearly accumulate after each time step, and hence, the data will be attenuated and out of phase with the data obtained from the exact solution. To quantify these errors, numerical dispersion relationships are derived. These same dispersion relationships also provide information on the stability of the scheme; this information is couched in terms of a Courant–Friedrichs–Lewy (CFL) number [4]. Thus, for those schemes for which the stability properties are known, that information is also provided.

This chapter is subdivided into 12 sections. The first two sections provide an overview of Maxwell's equations (with respect to complex media) and the FDTD method. Subsequent sections consider the following media types: nondispersive anisotropic media, isotropic cold plasma, magnetoionic media, isotropic warm plasma, Debye dielectric, Lorentz dielectric, ferrites, and non-linear optical media. The chapter closes with a short summary of the information provided herein.

One topic that has been purposefully omitted is a discussion on domain truncation using perfectly matched layers [5]. Since the perfectly matched layer by construction is a complex medium, a discussion of its design could be included. However, given the breadth of this topic, we refer the reader to Chapter 14 and to [6].

16.2 MAXWELL'S EQUATIONS AND COMPLEX MEDIA

According to Maxwell, the electromagnetic field is comprised of an electric field $\mathbf{E}(\mathbf{x}, t)$ and a magnetic field $\mathbf{B}(\mathbf{x}, t)$ that satisfy, at a point, two curl equations and two divergence equations [7]:

$$\nabla \times \mathbf{E} = -\frac{\partial \mathbf{B}}{\partial t} - \mathbf{M} \tag{16.1}$$

$$\nabla \times \mathbf{H} = \frac{\partial \mathbf{D}}{\partial t} + \mathbf{J} \tag{16.2}$$

$$\nabla \cdot \mathbf{D} = 0 \tag{16.3}$$

and

$$\nabla \cdot \mathbf{B} = 0 \tag{16.4}$$

Here $\mathbf{D}$ is the electric flux density and $\mathbf{H}$ is the magnetic intensity. The terms $\mathbf{J}$ and $\mathbf{M}$ are often regarded as source terms, and in that context they are the electric and magnetic current densities, respectively. For purposes of numerical computation, the two curl equations are discretized and advanced in time; the two divergence equations are additional constraints imposed upon the fields. To bring closure to this system of equations, a set of constitutive relationships are imposed that quantify the effects of a material medium on an applied electromagnetic field. In general, we write

$$\mathbf{D} = \mathbf{D}(\mathbf{E}, \mathbf{H}) \qquad \mathbf{B} = \mathbf{B}(\mathbf{E}, \mathbf{H}) \tag{16.5}$$

In addition, the currents may also be coupled to the fields, in which case

$$\mathbf{J} = \mathbf{J}(\mathbf{E}, \mathbf{H}) \qquad \mathbf{M} = \mathbf{M}(\mathbf{E}, \mathbf{H}) \tag{16.6}$$

It is in this latter context that we will invoke the notion of current. That is, rather than the current being typified as *impressed*, it will be considered *induced*—a current generated by a field. For example, in the study of wave propagation in a cold plasma,

$\mathbf{J}$ is replaced with $\mathbf{J}_p$, the polarization current generated in the plasma by an applied field. By definition, the polarization current is viewed as the time rate of change of the polarization vector $\mathbf{P}$ (i.e., $\mathbf{J}_p = \partial\mathbf{P}/\partial t$). The inclusion of $\mathbf{P}$ into the field description proves useful when studying propagation in Lorentz or Debye-type materials.

If the material is *simple* and source free, the previous relationships reduce to $\mathbf{D} = \varepsilon\mathbf{E}$, $\mathbf{B} = \mu\mathbf{H}$, $\mathbf{J} = \sigma\mathbf{E}$, and $\mathbf{M} = 0$, where the constants ε, μ, and σ are the electric permittivity, magnetic permeability, and electric conductivity, respectively. Written as such, $\mathbf{J}$ is now interpreted as the induced conduction current. For *complex* materials, the constitutive relationships do not reduce to such simple forms. Instead, the relationships may be cast in terms of convolution integrals, matrices, or derivatives. For these situations, the material may be characterized as dispersive, anisotropic, or non-linear, to name a few possibilities.[1]

For example, consider the earth's ionosphere, which from a macroscopic point of view, is comprised of a continuum of free electrons and positive ions (i.e., an inviscid fluid of charge) that is trapped within the atmosphere and is polarized by the earth's magnetic field; a zero temperature model is assumed [8]. When an electromagnetic disturbance of angular frequency ω is excited within the ionosphere, a force of $-n_o q(\mathbf{E} + \mathbf{u} \times \mathbf{B}_o)$ is exerted on the electrons; $\mathbf{B}_o = B_o\mathbf{b}_o$ is the earth's static magnetic field, qn_o is the charge density and $\mathbf{u}$ is the average electron velocity. As a result of this force, the electron accelerates and a polarization current of $-n_o q\mathbf{u}$ is established. For source-free conditions, Ampère's law is then given by

$$\varepsilon_o \frac{\partial \mathbf{E}}{\partial t} = \nabla \times \mathbf{H} + n_o q\mathbf{u} \tag{16.7}$$

where ε_o is the free-space permittivity. Additionally, conservation of momentum requires that

$$\frac{\partial \mathbf{u}}{\partial t} = -\nu\mathbf{u} - (q/m)\mathbf{E} + \omega_c \mathbf{b}_o \times \mathbf{u} \tag{16.8}$$

where ν is the collision frequency, $\omega_c = qB_o/m$, which is the cyclotron frequency, and $n_o m$ is the mass density. Together with Faraday's law, (16.7) and (16.8) constitute the governing equations for the case at hand. Under the assumption that the fields are time-harmonic, it is a straightforward process to show that the momentum equation can be embedded into Ampère's law, provided that $\hat{\mathbf{D}}(\omega) = \bar{\bar{\varepsilon}}(\omega) \cdot \hat{\mathbf{E}}(\omega)$, where $\bar{\bar{\varepsilon}}$ is now understood to be a dyad, as defined in the frequency domain; $\hat{\mathbf{D}}$ and $\hat{\mathbf{E}}$ are the phasor representations of $\mathbf{D}$ and $\mathbf{E}$, respectively. For the situation that $\mathbf{b}_o$ points in the z direction, the matrix definition of $\bar{\bar{\varepsilon}}$ is

$$\bar{\bar{\varepsilon}}(\omega) = \begin{bmatrix} \hat{\varepsilon}_1(\omega) & -j\hat{\varepsilon}_2(\omega) & 0 \\ j\hat{\varepsilon}_2(\omega) & \hat{\varepsilon}_1(\omega) & 0 \\ 0 & 0 & \hat{\varepsilon}_z \end{bmatrix} \tag{16.29}$$

[1] Inhomogeneous media can also be classified as complex. However, from a computational point of view, an inhomogeneous medium does not impose fundamental changes to the FDTD methodology. Hence, further discussion of inhomogeneous media is not provided.

where

$$\frac{\hat{\varepsilon}_1}{\varepsilon_o} = 1 + \frac{\omega_p^2(j\omega + \nu)}{j\omega[(j\omega + \nu)^2 + \omega_c^2]} \tag{16.10}$$

$$\frac{\hat{\varepsilon}_2}{\varepsilon_o} = \frac{\omega_p^2 \omega_c}{\omega[(j\omega + \nu)^2 + \omega_c^2]} \tag{16.11}$$

and

$$\frac{\hat{\varepsilon}_z}{\varepsilon_o} = 1 + \frac{\omega_p^2}{j\omega(j\omega + \nu)} \tag{16.12}$$

Given that $\overline{\overline{\varepsilon}}$ is both frequency dependent and non-scalar, the ionosphere is readily seen to be both dispersive and anisotropic. Moreover, if the power level is increased, nonlinearities will occur due to plasma heating. Finally, it is well known that the parameters of the ionosphere change with sunspot cycle, season, and the hour of the day, thus giving it a time-variant nature. All of these properties of the ionosphere make it a truly complex medium.

16.3 FDTD METHOD

Consider a medium that is simple and source-free, but lossy. The FDTD method is based upon the application of central differences for both temporal and spatial discretization [9] which, for the sake of future developments, will be considered separately. With respect to the former, the temporal discretization of Maxwell's curl equations yields

$$\mu\left(\frac{\mathbf{H}^{n+1/2} - \mathbf{H}^{n-1/2}}{\delta_t}\right) = -\nabla \times \mathbf{E}^n \tag{16.13}$$

and

$$\varepsilon\left(\frac{\mathbf{E}^{n+1} - \mathbf{E}^n}{\delta_t}\right) = \nabla \times \mathbf{H}^{n+1/2} - \sigma\left(\frac{\mathbf{E}^{n+1} + \mathbf{E}^n}{2}\right) \tag{16.14}$$

where we have employed central averages for the loss term in Ampère's law. Here δ_t is the time increment and $\mathbf{E}^n$ denotes the value of $\mathbf{E}$ at time $n\delta_t$. Upon the rearrangement of terms,

$$\mu\mathbf{H}^{n+1/2} = \mu\mathbf{H}^{n-1/2} - \delta_t\nabla \times \mathbf{E}^n \tag{16.15}$$

and

$$\varepsilon\mathbf{E}^{n+1} = \varepsilon\left(\frac{1 - \sigma\delta_t/2}{1 + \sigma\delta_t/2}\right)\mathbf{E}^n + \left(\frac{\delta_t}{1 + \sigma\delta_t/2}\right)\nabla \times \mathbf{H}^{n+1/2} \tag{16.16}$$

whence, the update equation for the field constituents is readily observed. Given that $\mathbf{E}$ and $\mathbf{H}$ are synchronized, but syncopated by a half-time step, this type of integration scheme is commonly referred to as leapfrog integration. The previous equations constitute a second-order temporal method, as deduced from the truncation term of the central difference and average approximations.

To accomplish the spatial discretization, the Yee grid is invoked, as shown in Fig. 16.1. The components of the Yee grid are arranged in accordance with the natural

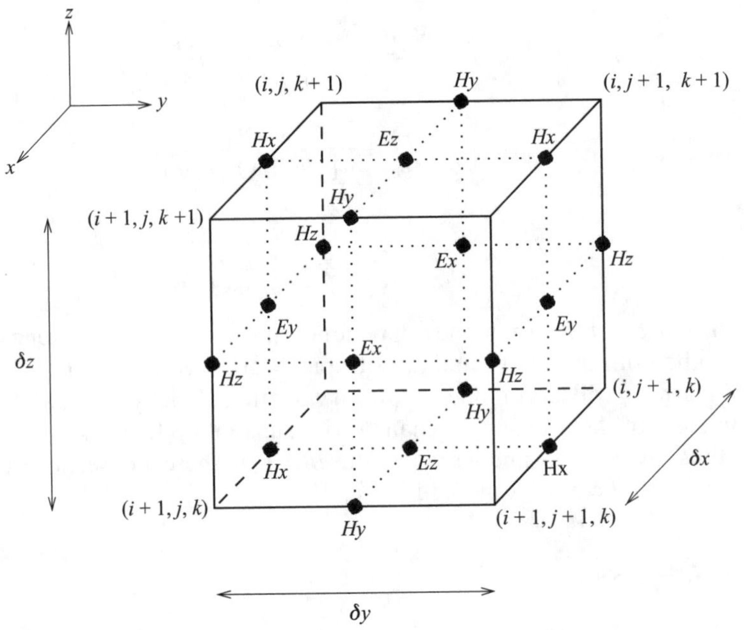

Figure 16.1 The Yee cell.

rotational structure of the curl operator acting upon **H**, which is also spatially proportional to **E** (or vice versa). For example, upon the spatial discretization of the z component of Ampère's law, we obtain,

$$\mathbf{a}_z \cdot \nabla \times \mathbf{H}\big|_{i+1/2,j+1/2,k} = \left[\frac{H_y\big|_{i+1} - H_y\big|_i}{\delta_x}\right]_{j+1/2,k} - \left[\frac{H_x\big|_{j+1} - H_x\big|_j}{\delta_y}\right]_{i+1/2,k} \tag{16.17}$$

The indices (i, j, k) denote the spatial coordinates $(i\delta_x, j\delta_y, k\delta_z)$ with δ_x, δ_y, and δ_z being the length of the cell in the x, y, z directions, respectively. The remaining five components of electromagnetic field are derived in a similar manner.

To quantify the FDTD method in terms of its dissipation and dispersion errors, consider a plane wave propagating in the **k** direction and in an homogeneous open domain. That is, let $\mathbf{E} = \mathbf{E}_o e^{-j\mathbf{k}\cdot\mathbf{x}} e^{j\omega t}$ and $\mathbf{H} = \mathbf{H}_o e^{-j\mathbf{k}\cdot\mathbf{x}} e^{j\omega t}$; $\mathbf{E}_o$ and $\mathbf{H}_o$ are constant vectors. If **E** and **H** are to be solutions of the discretized form of Maxwell's equations, then

$$-j\mathbf{K} \times \mathbf{E}_o = -j\Omega\mu\mathbf{H}_o \tag{16.18}$$

and

$$-j\mathbf{K} \times \mathbf{H}_o = j\Omega\varepsilon\mathbf{E}_o + \Lambda\sigma\mathbf{E}_o \tag{16.19}$$

moreover, $\mathbf{K} \cdot \mathbf{E}_o = 0$ and $\mathbf{K} \cdot \mathbf{H}_o = 0$. Here

$$\mathbf{K} = \left(\frac{2}{\delta_x}\right) \sin\left(\frac{k_x \delta_x}{2}\right)\mathbf{a}_x + \left(\frac{2}{\delta_y}\right) \sin\left(\frac{k_y \delta_y}{2}\right)\mathbf{a}_y + \left(\frac{2}{\delta_z}\right) \sin\left(\frac{k_z \delta_z}{2}\right)\mathbf{a}_z \tag{16.20}$$

$$\Omega = \left(\frac{2}{\delta_t}\right) \sin\left(\frac{\omega\delta_t}{2}\right) \tag{16.21}$$

and

$$\Lambda = \cos\left(\frac{\omega\delta_t}{2}\right) \tag{16.22}$$

Obviously, $\mathbf{K}\rightarrow k$, $\Omega\rightarrow\omega$ and $\Lambda\rightarrow 1$ as $\delta_x, \delta_y, \delta_z, \delta_t \rightarrow 0$, as required for a consistent scheme. Further manipulation of (16.18) and (16.19) leads to the equation

$$\mathbf{K}\cdot\mathbf{K} = -j\Omega\mu(j\Omega\varepsilon + \Lambda\sigma) = \Omega^2\mu\hat{\varepsilon} \tag{16.23}$$

which is the numerical dispersion equation and is the counterpart of the analytical dispersion equation $\mathbf{k}\cdot\mathbf{k} = -j\omega\mu(j\omega\varepsilon + \sigma)$; here $\hat{\varepsilon} = [\varepsilon + \Lambda\sigma/(j\Omega)]$. In many of the subsequent sections this same equation will be referenced, but different permittivity relationships will be supplied. Whatever the case, (16.23) provides the necessary information for understanding the scheme's dispersion, dissipation, anisotropicity, and stability properties [10], [11].

Stability (i.e., non-exponential growth in time) is insured if $\mathrm{Im}\{\omega\delta_t\} > 0$. From (16.23) and for $\sigma = 0$, this constraint is satisfied for all directions of propagation provided that [2]

$$\delta_t < \frac{1}{v_p}\left[\left(\frac{1}{\delta_x}\right)^2 + \left(\frac{1}{\delta_y}\right)^2 + \left(\frac{1}{\delta_z}\right)^2\right]^{-1/2} \tag{16.24}$$

Here $v_p = 1/\sqrt{\mu\varepsilon}$, which is the phase velocity (if $\varepsilon = \varepsilon_o$ and $\mu = \mu_o$ then $v_p = c$, the speed of light). Note: If (16.24) is satisfied, $\mathrm{Im}\{\omega\delta_t\} = 0$, which implies a dissipationless numerical scheme. Finally, given that the FDTD method is both consistent and stable, Lax's theorem states that the data will converge to the exact solution as the cell and time-step become infinitesimally small [12].

In the ensuing sections, the FDTD method will be modified or extended to model wave propagation in complex media. However, in all cases considered, the Yee grid of Fig. 16.1 and the discretization of the curl operators will be retained.

16.4 NON-DISPERSIVE, ANISOTROPIC MEDIA

Certain natural materials, such as crystals, and man-made materials, such as fibers, exhibit anisotropic behaviors. Within these materials, the coupling between components of $\mathbf{D}$ and the components of $\mathbf{E}$ is no longer one to one. To account for anisotropic effects within the scope of FDTD analysis, we consider in this section a methodology proposed by Schneider et al. [13].

The FDTD grid is designed to capture naturally the transverse (or rotational) nature of the electromagnetic field in an isotropic medium. By uncollocating the field components and using central differences, Yee developed an efficient discretization procedure for the curl operator in Maxwell's equations. When, however, the medium is anisotropic, the Yee grid complicates the discretization procedure since the component, say, D_x, is coupled to E_x as well as E_y and E_z. As observed from the grid of Fig. 16.1, these latter two components are not spatially registered with D_x. For this reason, the FDTD methodology must be modified.

Consider an arbitrary homogeneous, non-magnetic, non-dispersive, lossy anisotropic medium which is defined by a permittivity dyad $\bar{\bar{\varepsilon}}$ and a free-space permeability scalar μ_o. The constitutive relationship $\mathbf{D} = \bar{\bar{\varepsilon}} \cdot \mathbf{E}$ is then interpreted as a set of three scalar equations which, in matrix form, is given by

$$\begin{bmatrix} D_x \\ D_y \\ D_z \end{bmatrix} = \begin{bmatrix} \varepsilon_{xx} & \varepsilon_{xy} & \varepsilon_{xz} \\ \varepsilon_{yx} & \varepsilon_{yy} & \varepsilon_{yz} \\ \varepsilon_{zx} & \varepsilon_{zy} & \varepsilon_{zz} \end{bmatrix} \begin{bmatrix} E_x \\ E_y \\ E_z \end{bmatrix} \tag{16.25}$$

Furthermore, to account for loss, we write for Ohm's law $\mathbf{J} = \bar{\bar{\sigma}} \cdot \mathbf{E}$, where the conductivity dyad $\bar{\bar{\sigma}}$ has a similar matrix representation as $\bar{\bar{\varepsilon}}$. Hence, the modified form of (16.2) is

$$\nabla \times \mathbf{H} = \bar{\bar{\varepsilon}} \cdot \frac{\partial \mathbf{E}}{\partial t} + \bar{\bar{\sigma}} \cdot \mathbf{E} \tag{16.26}$$

First consider the temporal discretization of the previous set of equations at time $(n + 1/2)\delta_t$. As before, central difference and average approximations are invoked for the temporal derivative and loss terms, respectively. Doing so, and rearranging the terms, we obtain the following difference equation for $\mathbf{E}$:

$$\mathbf{E}^{n+1} = \left(\frac{\bar{\bar{\varepsilon}}}{\delta_t} + \frac{\bar{\bar{\sigma}}}{2} \right)^{-1} \cdot \left[\nabla \times \mathbf{H}^{n+1/2} + \left(\frac{\bar{\bar{\varepsilon}}}{\delta_t} - \frac{\bar{\bar{\sigma}}}{2} \right) \cdot \mathbf{E}^n \right] \tag{16.27}$$

Inverting the required matrices and executing the appropriate multiplications, we obtain the update equations for each of the components of $\mathbf{E}$. For example [13],

$$\begin{aligned} E_z^{n+1} = \frac{1}{\mathscr{D}} \Bigg\{ & \left[\left(\frac{\varepsilon_{xy}}{\delta t} + \frac{\sigma_{xy}}{2} \right) \left(\frac{\varepsilon_{zx}}{\delta t} + \frac{\sigma_{zx}}{2} \right) - \left(\frac{\varepsilon_{xx}}{\delta t} + \frac{\sigma_{xx}}{2} \right) \left(\frac{\varepsilon_{zy}}{\delta t} + \frac{\sigma_{zy}}{2} \right) \right] \\ & \left[\left(\frac{\varepsilon_{yx}}{\delta t} - \frac{\sigma_{yx}}{2} \right) E_x^n + \left(\frac{\varepsilon_{yy}}{\delta t} - \frac{\sigma_{yy}}{2} \right) E_y^n + \left(\frac{\varepsilon_{yz}}{\delta t} - \frac{\sigma_{yz}}{2} \right) E_z^n + \frac{\partial H_x^{n+1/2}}{\partial z} - \frac{\partial H_z^{n+1/2}}{\partial x} \right] \\ & + \left[\left(\frac{\varepsilon_{yx}}{\delta t} + \frac{\sigma_{yx}}{2} \right) \left(\frac{\varepsilon_{zy}}{\delta t} + \frac{\sigma_{zy}}{2} \right) - \left(\frac{\varepsilon_{yy}}{\delta t} + \frac{\sigma_{yy}}{2} \right) \left(\frac{\varepsilon_{zx}}{\delta t} + \frac{\sigma_{zx}}{2} \right) \right] \\ & \left[\left(\frac{\varepsilon_{xx}}{\delta t} - \frac{\sigma_{xx}}{2} \right) E_x^n + \left(\frac{\varepsilon_{xy}}{\delta t} - \frac{\sigma_{xy}}{2} \right) E_y^n + \left(\frac{\varepsilon_{xz}}{\delta t} - \frac{\sigma_{xz}}{2} \right) E_z^n + \frac{\partial H_y^{n+1/2}}{\partial z} - \frac{\partial H_z^{n+1/2}}{\partial y} \right] \\ & + \left[\left(\frac{\varepsilon_{xx}}{\delta t} + \frac{\sigma_{xx}}{2} \right) \left(\frac{\varepsilon_{yy}}{\delta t} + \frac{\sigma_{yy}}{2} \right) - \left(\frac{\varepsilon_{xy}}{\delta t} + \frac{\sigma_{xy}}{2} \right) \left(\frac{\varepsilon_{yx}}{\delta t} + \frac{\sigma_{yx}}{2} \right) \right] \\ & \left[\left(\frac{\varepsilon_{zx}}{\delta t} - \frac{\sigma_{zx}}{2} \right) E_x^n + \left(\frac{\varepsilon_{zy}}{\delta t} - \frac{\sigma_{zy}}{2} \right) E_y^n + \left(\frac{\varepsilon_{zz}}{\delta t} - \frac{\sigma_{zz}}{2} \right) E_z^n + \frac{\partial H_y^{n+1/2}}{\partial x} - \frac{\partial H_x^{n+1/2}}{\partial y} \right] \Bigg\} \end{aligned} \tag{16.28}$$

where $\mathscr{D}$ is the determinant of the matrix $(\bar{\bar{\varepsilon}}/\delta_t + \bar{\bar{\sigma}}/2)$.

To accomplish the spatial discretization, central differences are used for the spatial derivatives and central averages are used to interpolate the value of a field component from its nearest neighbors. The use of central averages is necessary since each component of **E** and **H** resides in a different location on the Yee cell. For example, to compute the value of E_x and E_y at $i+1/2, j+1/2, k$, we use

$$
\begin{aligned}
E_x|^n_{i+1/2,j+1/2,k} = \tfrac{1}{4}(E_x|^n_{i+1,j+1/2,k+1/2} \\
+ E_x|^n_{i,j+1/2,k+1/2} + E_x|^n_{i+1,j+1/2,k-1/2} + E_x|^n_{i,j+1/2,k-1/2})
\end{aligned}
\tag{16.29}
$$

and

$$
\begin{aligned}
E_y|^n_{i+1/2,j+1/2,k} = \tfrac{1}{4}(E_y|^n_{i+1/2,j+1,k+1/2} \\
+ E_y|^n_{i+1/2,j,k+1/2} + E_y|^n_{i+1/2,j+1,k-1/2} + E_y|^n_{i+1/2,j,k-1/2})
\end{aligned}
\tag{16.30}
$$

As for some of the spatial derivatives that appear in (16.28), we use

$$
\begin{aligned}
\left.\frac{\partial H_x^{n+1/2}}{\partial z}\right|_{i+1/2,j+1/2,k} &= \left(\frac{H_x^{n+1/2}|_{i+1/2,j,k+1} + H_x^{n+1/2}|_{i+1/2,j+1,k+1}}{4\delta_z}\right) \\
&- \left(\frac{H_x^{n+1/2}|_{i+1/2,j,k-1} + H_x^{n+1/2}|_{i+1/2,j+1,k-1}}{4\delta_z}\right)
\end{aligned}
\tag{16.31}
$$

and

$$
\begin{aligned}
\left.\frac{\partial H_z^{n+1/2}}{\partial x}\right|_{i+1/2,j+1/2,k} &= \left(\frac{H_z^{n+1/2}|_{i+1,j+1,k+1/2} - H_z^{n+1/2}|_{i,j+1,k+1/2}}{4\delta_x}\right) \\
&+ \left(\frac{H_z^{n+1/2}|_{i+1,j,k+1/2} - H_z^{n+1/2}|_{i,j,k+1/2}}{4\delta_x}\right) \\
&+ \left(\frac{H_z^{n+1/2}|_{i+1,j+1,k-1/2} - H_z^{n+1/2}|_{i,j+1,k-1/2}}{4\delta_x}\right) \\
&+ \left(\frac{H_z^{n+1/2}|_{i+1,j,k-1/2} - H_z^{n+1/2}|_{i,j,k-1/2}}{4\delta_x}\right)
\end{aligned}
\tag{16.32}
$$

The other derivatives are evaluated in a similar fashion.

Complete error and stability analyses are not provided due to the complexity of the final equations. However, we observe that the invocation of both spatial and temporal averages will introduce a certain amount of numerical dissipation. Additionally, in (16.31), the derivative is estimated over two cells. As a result, dispersive errors will accumulate more rapidly with respect to the standard FDTD scheme.

To validate the aforementioned procedure, consider the case of scattering from a layered medium that consists of three slabs [13]. Consult Table 16.1 for the specifics of the materials. For this situation, $\delta_t = 156.25\,\text{fs}$ and $\delta_x = 93.75\ \mu\text{m}$; a gaussian pulse polarized in the z direction and propagating in the x-direction is used as the excitation signal. The total simulation time is $32{,}768\delta_t$. Thus, given the anisotropicity of the slab and the polarization of the incident field, the scattered field will have both E_z and E_y components.

Consider Fig. 16.2, which shows the frequency-domain reflection coefficients for

TABLE 16.1 Material Parameters. The Permeability and Permittivity of the Sample are Isotropic while the Conductivity is Anisotropic

Layer	Thickness	Permeability	Permittivity	Conductivity (S/m)
#1	3.75 mm	μ_0	$43\varepsilon_0$	$\begin{pmatrix} 0.0 & 0.0 & 0.0 \\ 0.0 & 12.0 & 0.0 \\ 0.0 & 0.0 & 0.0 \end{pmatrix}$
#2	3.75 mm	μ_0	$43\varepsilon_0$	$\begin{pmatrix} 0.0 & 0.0 & 0.0 \\ 0.0 & 8.5 & 8.5 \\ 0.0 & 8.5 & 8.5 \end{pmatrix}$
#3	3.75 mm	μ_0	$43\varepsilon_0$	$\begin{pmatrix} 0.0 & 0.0 & 0.0 \\ 0.0 & 0.0 & 0.0 \\ 0.0 & 0.0 & 12.0 \end{pmatrix}$

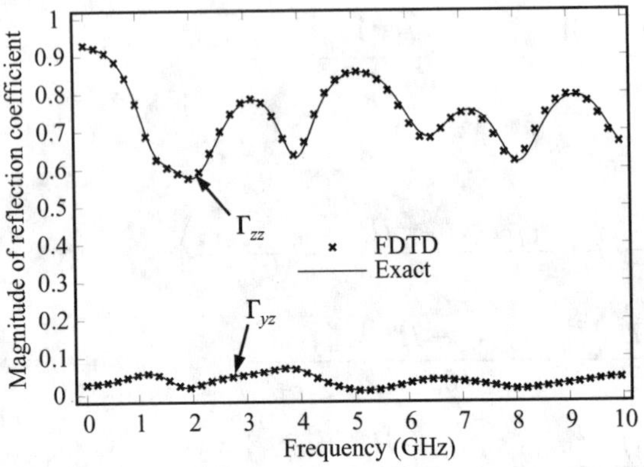

Figure 16.2 Reflection coefficient data as a function of frequency. (a) Γ_{zz}. (b) Γ_{yz}. (See [13]. Used by permission of IEEE.)

this problem; both computed data and data obtained from the exact solution are plotted. In this figure

$$\Gamma_{zz} = \frac{\hat{E}_z^{scat}}{\hat{E}_z^{inc}} \qquad \Gamma_{yz} = \frac{\hat{E}_y^{scat}}{\hat{E}_z^{inc}} \tag{16.33}$$

A casual inspection of this figure reveals that the agreement between data sets is very good.

16.5 COLD PLASMA

To demonstrate the wide variety of techniques available to model numerically wave propagation in linearly dispersive media, the case of wave propagation in an isotropic, lossy, cold plasma is considered next. For this scenario, Ampère's law and the momentum equation are [8]

$$\varepsilon_o \frac{\partial \mathbf{E}}{\partial t} = \nabla \times \mathbf{H} - \mathbf{J}_p \tag{16.34}$$

and

$$\frac{\partial \mathbf{J}_p}{\partial t} = -\nu \mathbf{J}_p + \varepsilon_o \omega_p^2 \mathbf{E} \tag{16.35}$$

where $\mathbf{J}_p$ is the polarization current, ν is the collision frequency and ω_p is the plasma frequency; specifically $\mathbf{J}_p = -n_o q \mathbf{u}$ and $\omega_p = \sqrt{n_o q^2/(m\varepsilon_o)}$. For time-harmonic signals of dependence $e^{j\omega t}$, it is a simple exercise to show that the momentum equation can be embedded into Ampère's law, from which a scalar frequency-domain permittivity $\hat{\varepsilon}$ can be defined. Suppressing that analysis, one finds that

$$\hat{\varepsilon} = \varepsilon_o\left(1 + \frac{\omega_p^2}{\omega(j\nu - \omega)}\right) = \varepsilon_o(1 + \hat{\chi}(\omega)) \tag{16.36}$$

where $\hat{\chi}$ is the scalar frequency-domain susceptibility. Since $\hat{\mathbf{D}} = \hat{\varepsilon}\hat{\mathbf{E}}$, the corresponding time-domain constitutive relationship, cast in terms of a convolution integral, is deducible from Fourier theory [14]:

$$\mathbf{D}(\mathbf{x}, t) = \varepsilon_o \mathbf{E}(\mathbf{x}, t) + \varepsilon_o \int_{-\infty}^{t} \chi(\tau)\mathbf{E}(\mathbf{x}, t - \tau)d\tau \tag{16.37}$$

where

$$\chi(\tau) = \frac{\omega_p^2}{\nu}[1 - e^{-\nu\tau}]U(\tau) \tag{16.38}$$

Here U is the unit step function.

Each equation in the previous development gives a different insight into the phenomenology of wave propagation in a cold plasma. And, for each of those insights, there exists a corresponding FDTD discretization methodology. In this section, we explore several such methodologies and classify those methodologies into two categories: (1) direct integration methods and (2) recursive convolution methods. The direct integration methods are either based on the fundamental state equations of (16.34) and (16.35) or on the constitutive relation $\hat{\mathbf{D}} = \hat{\varepsilon}\hat{\mathbf{E}}$. The recursive methods incorporate the convolution integral of (16.37) into the FDTD paradigm.[2] Although each discretization procedure is consistent with the governing equations, their accuracy order and memory requirements are different. Such similarities and differences are documented in the last subsection.

In the following subsections, (16.15) is the temporally discretized form of Faraday's law. Additionally, the components of $\mathbf{J}_p$, $\mathbf{D}$, and $\mathbf{E}$ are assumed to occupy the same locations in the Yee grid.

16.5.1 Direct Integration Method One: CP-DIM1

Consider first a state–space integration technique. For this situation, (16.34) and (16.35) are discretized using central differences for the derivatives and central averages for the collision term. To maintain synchronization with the FDTD leapfrog

[2] Since the Z-transform method is a subset of the recursive method, it will not be documented herein. For more information on using Z-transforms in conjunction with the FDTD method consult [15]–[17].

integrator, the momentum equation and Faraday's law are advanced simultaneously and collectively leaped with Ampère's law. For source-free media, this procedure results in the following system of equations [18]:

$$\frac{\mathbf{J}_p^{n+1/2} - \mathbf{J}_p^{n-1/2}}{\delta_t} = -\nu\left(\frac{\mathbf{J}_p^{n+1/2} + \mathbf{J}_p^{n-1/2}}{2}\right) + \varepsilon_o\,\omega_p^2\,\mathbf{E}^n \tag{16.39}$$

and

$$\varepsilon_o\left(\frac{\mathbf{E}^{n+1} - \mathbf{E}^n}{\delta_t}\right) = \nabla\times\mathrm{H}^{n+1/2} - \mathbf{J}_p^{n+1/2} \tag{16.40}$$

Given the simplicity of the algebraic manipulations that solve for $\mathbf{J}_p^{n+1/2}$ in (16.39), the final expression is not given. This will be the case for many of the other direct integration methods as well.

16.5.2 Direct Integration Method Two: CP-DIM2

Instead of averaging the loss term, as in CP-DIM1, averaging the current term in Ampère's law is equally possible. For this case, the momentum equation and Ampère's law are synchronized and collectively leaped with Faraday's law [19]. That is,

$$\frac{\mathbf{J}_p^{n+1} - \mathbf{J}_p^{n-1}}{2\delta_t} = -\nu\mathbf{J}_p^n + \varepsilon_o\,\omega_p^2\,\mathbf{E}^n \tag{16.41}$$

and

$$\varepsilon_o\left(\frac{\mathbf{E}^{n+1} - \mathbf{E}^n}{\delta_t}\right) = \nabla\times\mathbf{H}^{n+1/2} - \left(\frac{\mathbf{J}_p^{n+1} + \mathbf{J}_p^n}{2}\right) \tag{16.42}$$

16.5.3 Direct Integration Method Three: CP-DIM3

Next consider the case in which the polarization current in Ampère's law, the electric field term in the momentum equation, and the dissipation term in the momentum equation are all time averaged. The synchronization of the momentum equation and Ampère's law is maintained [20]. The resulting scheme is

$$\frac{\mathbf{J}_p^{n+1} - \mathbf{J}_p^n}{\delta_t} = -\nu\frac{\mathbf{J}_p^{n+1} + \mathbf{J}_p^n}{2} + \varepsilon_0\,\omega_p^2\left(\frac{\mathbf{E}^{n+1} + \mathbf{E}^n}{2}\right) \tag{16.43}$$

and

$$\varepsilon_o\left(\frac{\mathbf{E}^{n+1} - \mathbf{E}^n}{\delta_t}\right) = \nabla\times\mathbf{H}^{n+1/2} - \left(\frac{\mathbf{J}_p^{n+1} + \mathbf{J}_p^n}{2}\right) \tag{16.44}$$

16.5.4 Direct Integration Method Four: CP-DIM4

Observing that the momentum equation has solutions that decay exponentially like $e^{-\nu t}$, Cummer suggests that exponential fitting is necessary, particularly when the dissipation term is large [20]. To that end, the exponentially fitted equation is

$$\mathbf{J}_p^{n+1} = e^{-\nu\delta_t}\mathbf{J}_p^n + \frac{\varepsilon_o\,\omega_p^2}{\nu^2\,\delta_t}(\nu\delta_t + e^{-\nu\delta_t} - 1)\mathbf{E}^{n+1} + \frac{\varepsilon_o\,\omega_p^2}{\nu^2\,\delta_t}(1 - e^{-\nu\delta_t} - \nu\delta_t e^{-\nu\delta_t})\mathbf{E}^n \quad (16.45)$$

Ampère's law still requires the averaging of the polarization current:

$$\varepsilon_o\left(\frac{\mathbf{E}^{n+1} - \mathbf{E}^n}{\delta_t}\right) = \nabla\times\mathbf{H}^{n+1/2} - \left(\frac{\mathbf{J}_p^{n+1} + \mathbf{J}_p^n}{2}\right) \quad (16.46)$$

16.5.5 Direct Integration Method Five: CP-DIM5

The next direct integration technique considers the frequency-domain constitutive relationship $\hat{\mathbf{D}} = \hat{\varepsilon}\hat{\mathbf{E}}$ and (16.36) as the basis of temporal discretization [21]. Combining these two equations, we obtain

$$(\omega^2 - j\omega\nu)\hat{\mathbf{D}} = \varepsilon_o(\omega^2 - j\omega\nu - \omega_p^2)\hat{\mathbf{E}} \quad (16.47)$$

Since multiplications of $j\omega$ in the frequency domain are equivalent to time derivatives in the time domain, (16.47) is equivalent to

$$\frac{\partial^2\mathbf{D}}{\partial t^2} + \nu\frac{\partial\mathbf{D}}{\partial t} = \varepsilon_o\frac{\partial^2\mathbf{E}}{\partial t^2} + \varepsilon_o\,\nu\frac{\partial\mathbf{E}}{\partial t} + \varepsilon_o\,\omega_p^2\,\mathbf{E} \quad (16.98)$$

Discretization of (16.48) with second-order central differences, along with (16.2), yields the following system of semidiscrete numerical equations:

$$\mathbf{D}^{n+1} = \mathbf{D}^n + \delta_t\nabla\times\mathbf{H}^{n+1/2} \quad (16.49)$$

and

$$\varepsilon_o(2 + \nu\delta_t)\mathbf{E}^{n+1} = (2 + \nu\delta_t)\mathbf{D}^{n+1} - 4\mathbf{D}^n + (2 - \nu\delta_t)\mathbf{D}^{n-1}$$
$$- 2\varepsilon_o(\omega_p^2\delta_t^2 - 2)\mathbf{E}^n - \varepsilon_o(2 - \nu\delta_t)\mathbf{E}^{n-1} \quad (16.50)$$

16.5.6 Recursive Convolution Method One: CP-RCM1

The recursive convolution method is based upon the discretization of the convolution integral of (16.37) [22]. At time $n\delta_t$,

$$\mathbf{D}(\mathbf{x}, n\delta_t) = \varepsilon_o\mathbf{E}(\mathbf{x}, n\delta_t) + \varepsilon_o\int_0^{n\delta_t}\mathbf{E}(\mathbf{x}, n\delta_t - \tau)\chi(\tau)d\tau \quad (16.51)$$

Under the assumption that **E** is piecewise constant,

$$\mathbf{D}^n = \varepsilon_o \mathbf{E}^n + \varepsilon_o \sum_{m=0}^{n-1} \mathbf{E}^{n-m} \int_{m\delta_t}^{(m+1)\delta_t} \chi(\tau) d\tau \tag{16.52}$$

Likewise,

$$\mathbf{D}^{n+1} = \varepsilon_o \mathbf{E}^{n+1} + \varepsilon_o \sum_{m=0}^{n} \mathbf{E}^{n-m+1} \int_{m\delta_t}^{(m+1)\delta_t} \chi(\tau) d\tau \tag{16.53}$$

The discretization of Ampère's law, as given by (16.2), requires us to take the difference of the two preceding equations. Doing so and replacing that difference into Ampère's law, we obtain

$$\mathbf{E}^{n+1} = \alpha \mathbf{E}^n + \alpha \Psi^n + \frac{\alpha \delta_t}{\varepsilon_o} \nabla \times \mathbf{H}^{n+1/2} \tag{16.54}$$

where

$$\Psi^n = \sum_{m=0}^{n-1} \mathbf{E}^{n-m} \Delta\chi^m \tag{16.55}$$

and

$$\alpha = \frac{1}{1 + \chi^0} \tag{16.56}$$

Here

$$\Delta\chi^m = \chi^m - \chi^{m+1} \tag{16.57}$$

and

$$\chi^m = \int_{m\delta_t}^{(m+1)\delta_t} \chi(\tau) d\tau \tag{16.58}$$

Performing the calculation for χ^0 and $\Delta\chi^m$ in conjunction with (16.38), we obtain

$$\chi^0 = \frac{\omega_p^2 \delta_t}{\nu} - \left(\frac{\omega_p}{\nu}\right)^2 (1 - e^{-\nu\delta_t}) \tag{16.59}$$

and

$$\Delta\chi^m = -\left(\frac{\omega_p}{\nu}\right)^2 e^{-m\nu\delta_t}(1 - e^{-\nu\delta_t})^2 = \Delta\chi^0 e^{-m\nu\delta_t} \tag{16.60}$$

Although Maxwell's equations are temporally discretized, they are not in a form suitable for computation. For we note that to compute (16.55), one would have to store the complete time history of **E**. However, given the exponential form of $\Delta\chi^m$,

$$\Psi^n = \Delta\chi^0 \sum_{m=0}^{n-1} \mathbf{E}^{n-m} e^{-m\nu\delta_t} \tag{16.61}$$

hence,

$$\Psi^n = \Delta\chi^0 \mathbf{E}^n + e^{-\nu\delta_t} \Psi^{n-1} \tag{16.62}$$

Thus, the equation for Ψ reduces to a simple recursive relationship.

16.5.7 Recursive Convolution Method Two: CP-RCM2

Numerical accuracy of CP-RCM1 can be increased by assuming piecewise linear line segments in the integration of the convolution integral [23]. For this situation, one approximates **E** over the interval $[n\delta_t,(n+1)\delta_t]$ with

$$\mathbf{E}(\mathbf{x},t) = \mathbf{E}(\mathbf{x},n\delta_t) + (t-n\delta_t)\left(\frac{\mathbf{E}(\mathbf{x},(n+1)\delta_t) - \mathbf{E}(\mathbf{x},n\delta_t)}{\delta_t}\right) \tag{16.63}$$

Substitution of this approximation into (16.51) yields the following result:

$$\mathbf{D}^n = \varepsilon_o\mathbf{E}^n + \varepsilon_o\sum_{m=0}^{n-1}\mathbf{E}^{n-m}\chi^m + \varepsilon_o\sum_{m=0}^{n-1}(\mathbf{E}^{n-m-1} - \mathbf{E}^{n-m})\xi^m \tag{16.64}$$

where χ^m is given by (16.58) and

$$\xi^m = \frac{1}{\delta_t}\int_{m\delta_t}^{(m+1)\delta_t}(\tau - m\delta_t)\chi(\tau)\,d\tau \tag{16.65}$$

As before, the time derivative of **D** at time $(n+1/2)\delta_t$ is approximated to second order by $(\mathbf{D}^{n+1} - \mathbf{D}^n)/\delta_t$. From (16.2) and (16.64),

$$\mathbf{E}^{n+1} = \alpha(1-\xi^0)\mathbf{E}^n + \frac{\alpha\delta_t}{\varepsilon_o}\nabla\times\mathbf{H}^{n+1/2} + \alpha\Psi^n \tag{16.66}$$

Here

$$\Psi^n = \sum_{m=0}^{n-1}\mathbf{E}^{n-m}\Delta\chi^m + (\mathbf{E}^{n-m-1} - \mathbf{E}^{n-m})\Delta\xi^m \tag{16.67}$$

$$\alpha = \frac{1}{(1+\chi^0-\xi^0)} \tag{16.68}$$

and

$$\Delta\xi^m = \xi^m - \xi^{m+1} \tag{16.69}$$

The expression for $\Delta\chi^m$ is given by (16.60).

For the time-domain susceptibility function given by (16.38), it is a simple matter to show that

$$\xi^m = \frac{\omega_p^2}{\delta_t\nu}\left[\frac{\delta_t^2}{2} + \frac{e^{-\nu m\delta_t}}{\nu}\left(\delta_t e^{-\nu\delta_t} + \frac{e^{-\nu\delta_t}}{\nu} - \frac{1}{\nu}\right)\right] \tag{16.70}$$

and

$$\Delta\xi^m = \frac{\omega_P^2}{\delta_t\nu^2}\left(\delta_t e^{-\nu\delta_t} + \frac{e^{-\nu\delta_t}}{\nu} - \frac{1}{\nu}\right)(1-e^{-\nu\delta_t})e^{-m\nu\delta_t} = \Delta\xi^0 e^{-m\nu\delta_t} \tag{16.71}$$

Finally, the summation over all time in (16.67) is replaced with a recursive factor—to wit,

$$\Psi^n = \Delta\chi^0\mathbf{E}^n + (\mathbf{E}^{n-1} - \mathbf{E}^n)\Delta\xi^0 + e^{-\nu\delta_t}\Psi^{n-1} \tag{16.72}$$

The expression for $\Delta\chi^0$ is given by (16.60) when $m = 0$.

Although the previous equation appears to require an additional storage location for $\mathbf{E}^{n-1}$, such is not the case. As pointed out by Kelley et al. [23], a low storage implementation is possible by making judicious use of a temporary variable.

16.5.8 Comparative Analysis

The numerical errors of each of the aforementioned schemes are deduced in a similar way as in Section 16.3. Assuming again plane wave propagation and an unbounded domain, the numerical dispersion relationship is similar to that given by (16.23). Of course, each scheme produces a different $\hat{\varepsilon}$, as seen in Table 16.2. Note the equivalence between CP-DIM1 and CP-DIM5. Even though both of these procedures are derived from different equations, the final numerical permittivities are identical. However, as seen in Table 16.4, CP-DIM5 requires more memory. Next, as pointed out by Cummer [20], there exists an equivalence between exponential differencing (i.e., CP-DIM4) and the recursive convolution methods; this observation is confirmed in Table 16.2, where it is seen that the numerical permittivities bear some semblance to one another. Finally, if $\xi^0 = \Delta\xi^0 = 0$, CP-RCM1 and CP-RCM2 are one and the same, as expected.

To appreciate the subtle differences between each of the schemes, Table 16.3 is provided. In this table, each entry corresponds to the leading term of the Taylor series of the function $\Omega^2\hat{\varepsilon}/\varepsilon_o$. As seen from (16.23), the function $\Omega^2\hat{\varepsilon}$ contains the complete temporal information of the numerical scheme. Except for CP-RCM1, these numerical permittivity relationships reduce to the exact analytical permittivity relationship to the order of δ_t^2; CP-RCM1 is temporally accurate to first order. Of course, for certain

TABLE 16.2 Numerical Permittivities for Various Schemes Associated with Propagation in an Isotropic Cold Plasma

Method	Numerical Permittivity: $\hat{\varepsilon}$
CP-DIM1	$\hat{\varepsilon} = \varepsilon_o\left(1 + \dfrac{\omega_p^2}{\Omega(j\nu\Lambda - \Omega)}\right)$
CP-DIM2	$\hat{\varepsilon} = \varepsilon_o\left(1 + \dfrac{\omega_p^2\Lambda}{\Omega(j\nu - \Omega\Lambda)}\right)$
CP-DIM3	$\hat{\varepsilon} = \varepsilon_o\left(1 + \dfrac{\omega_p^2\Lambda^2}{\Omega(j\nu\Lambda - \Omega)}\right)$
CP-DIM4	$\hat{\varepsilon} = \varepsilon_o\left[1 + \dfrac{\Lambda\omega_p^2\{\nu\sin[(\omega - j\nu)\delta_t/2] - \Omega\sinh(\nu\delta_t/2)\}}{j\Omega\nu^2\sin[(\omega - j\nu)\delta_t/2]}\right]$
CP-DIM5	$\hat{\varepsilon} = \varepsilon_o\left(1 + \dfrac{\omega_p^2}{\Omega(j\nu\Lambda - \Omega)}\right)$
CP-RCM1	$\hat{\varepsilon} = \varepsilon_o\left(1 + \dfrac{\chi^0 e^{j\omega\delta_t/2}}{j\Omega\delta_t} + \dfrac{\Delta\chi^0 e^{\nu\delta_t/2}}{2\Omega\delta_t\sin[(\omega - j\nu)\delta_t/2]}\right)$
CP-RCM2	$\hat{\varepsilon} = \varepsilon_o\left[1 - \xi^0 + \dfrac{\chi^0 e^{j\omega\delta_t/2}}{j\Omega\delta_t} + \left(\dfrac{\Delta\chi^0 - j\Omega\delta_t\Delta\xi^0 e^{-j\omega\delta_t/2}}{2\Omega\delta_t\sin[(\omega - j\nu)\delta_t/2]}\right)e^{\nu\delta_t/2}\right]$

TABLE 16.3 Truncation Term from the Taylor Analysis of $\Omega^2 \hat{\varepsilon}/\varepsilon_o$: Cold Plasma

Method	Truncation Term	Truncation Term: $\nu = 0$
CP-DIM1	$\dfrac{(\nu^2\omega^4 + 2j\nu\omega^5 - \omega^6 + j\nu\omega^3\omega_p^2)\delta_t^2}{12(\omega - j\nu)^2}$	$\dfrac{-\omega^4\delta_t^2}{12}$
CP-DIM2	$\dfrac{(\nu^2\omega^4 + 2j\nu\omega^5 - \omega^6 - 2j\nu\omega^3\omega_p^2)\delta_t^2}{12(\omega - j\nu)^2}$	$\dfrac{-\omega^4\delta_t^2}{12}$
CP-DIM3	$\left(-\omega^4 + \dfrac{2j\omega^3\omega_p^2}{\nu + j\omega} - \dfrac{\omega^4\omega_p^2}{(\nu + j\omega)^2}\right)\left(\dfrac{\delta_t^2}{12}\right)$	$\dfrac{\omega^2(-\omega^2 + 3\omega_p^2)\delta_t^2}{12}$
CP-DIM4	$\left(-\omega^4 + \dfrac{3j\omega^3\omega_p^2}{\nu + j\omega}\right)\left(\dfrac{\delta_t^2}{12}\right)$	$\dfrac{\omega^2(-\omega^2 + 3\omega_p^2)\delta_t^2}{12}$
CP-DIM5	$\dfrac{(\nu^2\omega^4 + 2j\nu\omega^5 - \omega^6 + j\nu\omega^3\omega_p^2)\delta_t^2}{12(\omega - j\nu)^2}$	$\dfrac{-\omega^4\delta_t^2}{12}$
CP-RCM1	$\dfrac{\omega_p^2\omega^2\delta_t}{2(\nu - j\omega)}$	$\dfrac{\omega_p^2\omega\delta_t}{2j}$
CP-RCM2	$\left(-\omega^4 + \dfrac{2j\omega^3\omega_p^2}{\nu + j\omega}\right)\left(\dfrac{\delta_t^2}{12}\right)$	$\dfrac{\omega^2(-\omega^2 + 2\omega_p^2)\delta_t^2}{12}$

choices of ω, ω_p, $\nu\delta_t$, the scheme with lowest truncation term is deemed temporally best, from an accuracy point of view.

A special case of interest is when $\nu = 0$. From Table 16.3, it is readily seen that CP-DIM1, CP-DIM2, and CP-DIM5 all have the same leading truncation term, which is independent ω_p. CP-DIM3 and CP-DIM4 also have the same truncation term; the truncation term for CP-RCM2 is close in value to that of CP-DIM3 and CP-DIM4. Since the truncation term for CP-DIM3, CP-DIM4, and CP-RCM2 are dependent on ω_p^2, the truncation terms are smaller than those associated with CP-DIM1, CP-DIM2, and CP-DIM5. Moreover, there exists a frequency for which the truncation terms of CP-DIM3, CP-DIM4, and CP-RCM2 vanish. Finally, if $\omega_p^2\delta_t^2 \ll 1$, then all of the second-order schemes produce a truncation term of $-\omega^4\delta_t^2/12$, which is the truncation term of the standard FDTD algorithm.

Other properties of these algorithms, including memory and one-dimensional CFL equations, are shown in Table 16.4 [20]. As expected from the previous discussion on the truncation terms, CP-DIM1, CP-DIM2, and CP-DIM5 have the same stability equations, which are dependent on the plasma frequency, when $\nu = 0$. When $\nu \neq 0$, CP-DIM2 is unconditionally unstable. CP-DIM3 and CP-DIM4 have a CFL of unity. This is to be expected since the leading truncation term of $\Omega^2\hat{\varepsilon}/\varepsilon_o$ for CP-DIM3 and CP-DIM4 contains the factor $(\omega_p\delta_t/2)^2$, which is also the same factor that appears in the CFL equation for CP-DIM1, CP-DIM2, and CP-DIM5. As for CP-RCM1 and CP-RCM2, the CFL is found experimentally to be near unity [20]. From a memory point of view, CP-DIM2 and CP-DIM5 require additional memory (compared to the other schemes) for the back storage of a vector.

TABLE 16.4 Comparison Table of Various Schemes in Terms of Memory per Cell and Stability Equations: Cold Plasma

Method	3D Memory/Cell	1D Stability
CP-DIM1	9	$\delta_t < (\delta_x/c)\sqrt{1-(\omega_p\delta_t/2)^2};\ \nu = 0$
CP-DIM2	12	$\delta_t < (\delta_x/c)\sqrt{1-(\omega_p\delta_t/2)^2};\ \nu = 0$
CP-DIM3	9	$\delta_t < \delta_x/c$
CP-DIM4	9	$\delta_t < \delta_x/c$
CP-DIM5	12	$\delta_t < (\delta_x/c)\sqrt{1-(\omega_p\delta_t/2)^2};\ \nu = 0$
CP-RCM1	9	$\delta_t^{max} \approx \delta_x/c$
CP-RCM2	9	$\delta_t^{max} \approx \delta_x/c$

16.6 MAGNETOIONIC MEDIA

When the cold plasma is anisotropic due to an applied static magnetic field $\mathbf{B}_o$, the FDTD procedure for the momentum equation must be modified. Consider the dyadic form of the momentum conservation equation given by (16.8):

$$\frac{\partial \mathbf{u}}{\partial t} = \bar{\mathbf{R}} \cdot \mathbf{u} - (q/m)\mathbf{E} \tag{16.73}$$

where $\bar{\mathbf{R}}$ is a dyad; in matrix form,

$$\bar{\mathbf{R}} = \begin{bmatrix} -\nu & -\omega_c b_z & \omega_c b_y \\ \omega_c b_z & -\nu & -\omega_c b_x \\ -\omega_c b_y & \omega_c b_x & -\nu \end{bmatrix} \tag{16.74}$$

with ν representing the effective collision frequency and ω_c denoting the gyro frequency, the latter being equivalent to $q|\mathbf{B}_o|/m$; also, $\mathbf{B}_o = |\mathbf{B}_o|\mathbf{b}_o$ with $\mathbf{b}_o = b_x\mathbf{a}_x + b_y\mathbf{a}_y + b_z\mathbf{a}_z$. (In this discussion we choose to work with the velocity $\mathbf{u}$ rather than the polarization current $\mathbf{J}_p$ in order to match notation with that available in the literature [24].)

For the temporal discretization, second-order central differences and averages are again employed [18]. Thus, the discretized version of (16.73) is

$$\mathbf{u}^{n+1/2} = \bar{\mathbf{T}} \cdot \mathbf{u}^{n-1/2} - (q\delta_t/m)\bar{\mathbf{S}} \cdot \mathbf{E}^n \tag{16.75}$$

Here $\bar{\mathbf{S}}$ and $\bar{\mathbf{T}}$ are also dyads and are given by $\bar{\mathbf{S}} = [\mathbf{1} - \delta_t\bar{\mathbf{R}}/2]^{-1}$ and $\bar{\mathbf{T}} = \bar{\mathbf{S}} \cdot [\mathbf{1} + \delta_t\bar{\mathbf{R}}/2]$, with $\mathbf{1}$ representing the idemfactor.

As in the case of the isotropic plasma, the components of $\mathbf{u}$ are conjoined with the components of $\mathbf{E}$ on the Yee grid. To accomplish the quantization of the spatial operators, averaging is required—this time, however, with respect to nodal nearest neighbors. This is borne out of the fact that the presence of $\mathbf{B}_o$ couples the three components of $\mathbf{u}$, which by construction are not nodally collocated. For example, the components u_x and u_y, evaluated at $(i + 1/2, j + 1/2, k)$, are given by

$$\begin{aligned} 4u_x\big|_{i+1/2,j+1/2,k}^{n-1/2} = {} & u_x\big|_{i+1,j+1/2,k+1/2}^{n-1/2} + u_x\big|_{i,j+1/2,k+1/2}^{n-1/2} + u_x\big|_{i+1,j+1/2,k-1/2}^{n-1/2} \\ & + u_x\big|_{i,j+1/2,k-1/2}^{n-1/2} \end{aligned} \tag{16.76}$$

and

$$4u_y\big|_{i+1/2,j+1/2,k}^{n-1/2} = u_x\big|_{i+1/2,j+1,k+1/2}^{n-1/2} + u_x\big|_{i+1/2,j,k+1/2}^{n-1/2} + u_x\big|_{i+1/2,j+1,k-1/2}^{n-1/2} + u_x\big|_{i+1/2,j,k-1/2}^{n-1/2} \qquad (16.77)$$

The average values for E_x and E_y at $(i + 1/2,\ j + 1/2,\ k)$ are given by (16.29) and (16.30), respectively. The discretized form for Ampère's law is given by (16.40) with $\mathbf{J}_p$ being replaced with $-n_o q\mathbf{u}$.

Instead of quantifying the errors for the full three-dimensional problem, consider the one-dimensional case where only the components E_x, H_x, u_x, E_y, H_y, and u_y are assumed to exist. For this situation, one can spatially register the components H_x and H_y with each other; similarly, E_x, u_x, E_y, and u_y are spatially registered. Thus, for this special case, spatial averaging is not necessary. Carrying out the required steps for the numerical dispersion relation and assuming that the static magnetic field is z directed, we obtain the following numerical permittivities associated with the dyad of (16.9):

$$\frac{\hat{\varepsilon}_1}{\varepsilon_o} = 1 + \frac{\omega_p^2(j\Omega + \nu\Lambda)}{j\Omega[(j\Omega + \nu\Lambda)^2 + \omega_c^2\Lambda^2]} \qquad (16.78)$$

and

$$\frac{\hat{\varepsilon}_2}{\varepsilon_o} = \frac{\omega_p^2\omega_c\Lambda}{\Omega[(j\Omega + \nu\Lambda)^2 + \omega_c^2\Lambda^2]} \qquad (16.79)$$

As expected, as δ_t becomes infinitesimally small, the numerical permittivities reduce to their analytical counterparts: (16.10) and (16.11). In this limit, the accuracy is seen to be of order δ_t^2. The effects of anisotropicity on stability have not been investigated. However, when $\omega_c = \nu = 0$, the previous scheme reduces to CP-DIM1 of Section 16.5. Hence from Table 16.4, $\delta_t < (\delta_x/c)\sqrt{1 - (\omega_p\delta_t/2)^2}$.

Recursive methods and other direct integration methods may also be employed for augmenting the FDTD procedure associated with wave propagation in an anisotropic cold plasma. For more information consult [25], [26].

16.7 ISOTROPIC, COLLISIONLESS WARM PLASMA

To expand upon the previous section, a one-component, fluid model for radio wave propagation in a warm, isotropic, collisionless plasma now is assumed [3]. This being the case, effects due to plasma compression are no longer negligible, but must be included in the governing equations [27]. The state variable description of the source-free governing equations is given by

$$\mu_o\frac{\partial \mathbf{H}}{\partial t} = -\nabla \times \mathbf{E} \qquad (16.80)$$

$$\varepsilon_o\frac{\partial \mathbf{E}}{\partial t} = \nabla \times \mathbf{H} + n_o q\mathbf{u} \qquad (16.81)$$

$$\frac{1}{\gamma p_o}\frac{\partial p}{\partial t} = -\nabla \cdot \mathbf{u} \qquad (16.82)$$

and

$$n_o m \frac{\partial \mathbf{u}}{\partial t} = -n_o q \mathbf{E} - \nabla p \tag{16.83}$$

where p is the pressure, p_o is the background pressure and γ is the ratio of specific heats; all other quantities are defined in previous sections. In essence, the latter two equations constitute Euler's linearized acoustic equations. Maxwell's equations are coupled to Euler's equations via the polarization current $-n_o q\mathbf{u}$ and the Lorentzian force $n_o q\mathbf{E}$. Thus, the electromagnetic wave can impart acoustical compression and acoustical compression can excite the electromagnetic field. As a result, both spatial and temporal dispersion exist (up to now, we have only considered temporally dispersive mediums). This interaction between acoustical waves, which are inherently longitudinal, and electromagnetic waves, which are inherently transverse, is most noticeable in the near-field of an antenna that is immersed in the plasma.

The previous system of equations is amenable to a full, finite-difference time-domain methodology. (To facilitate the presentation, the temporal and spatial discretization schemes are presented separately.) First, to accomplish the spatial discretization, Yee's cell is augmented to incorporate the dependent variables $\mathbf{u}$ and p [18]. This is shown in Fig. 16.3, which shows that $\mathbf{u}$ and $\mathbf{E}$ are nodally conjoined; p is located at the cell center. This particular arrangement of components is compatible with the central difference approximations of the curl, divergence, and gradient operators that appear in (16.80)–(16.83).[3] Moreover, by letting the components of $\mathbf{E}$

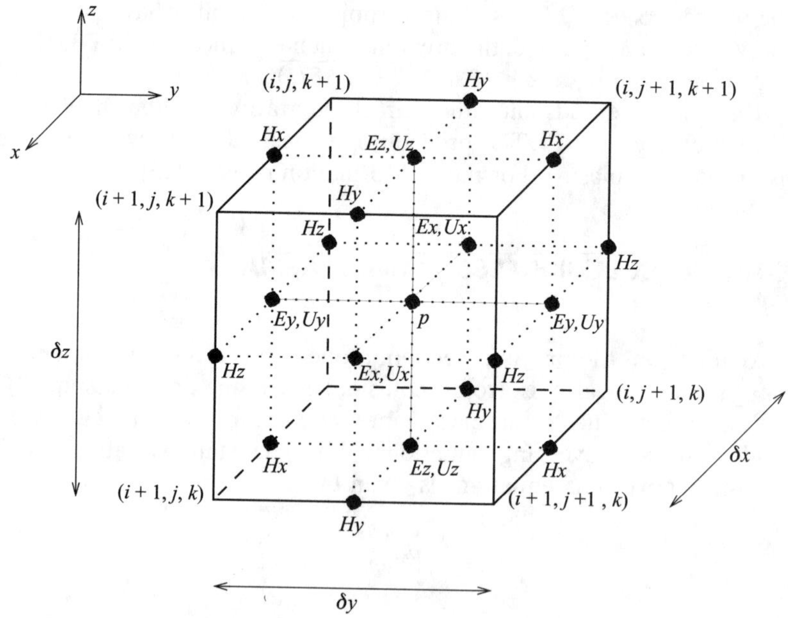

Figure 16.3 The modified Yee cell for both Maxwell's and Euler's equations.

[3] If Euler's and Maxwell's equations are decoupled by setting q to zero, then the grid for $\mathbf{u}$ and p is the equivalent Yee grid for the linearized acoustic equations.

and **u** share the same nodes, we effect a natural coupling between the equations of Euler and Maxwell. Upon the invocation of the central difference approximation for the spatial derivatives, the right-hand side of (16.82)–(16.83) becomes,

$$\frac{1}{\gamma p_o}\frac{\partial}{\partial t}p\,|_{i+1/2,j+1/2,k+1/2} =$$

$$-\left[\frac{u_x|_{i+1} + u_x|_i}{\delta_x}\right]_{j+1/2,k+1/2} - \left[\frac{u_y|_{j+1} - u_y|_j}{\delta_y}\right]_{i+1/2,k+1/2} - \left[\frac{u_z|_{k+1} - u_z|_k}{\delta_z}\right]_{i+1/2,j+1/2} \tag{16.84}$$

and

$$n_o m\frac{\partial}{\partial t}u_z|_{i+1/2,j+1/2,k} = -n_o qE_z|_{i+1/2,j+1/2,k} - \left[\frac{p|_{k+1/2} - p|_{k-1/2}}{\delta_z}\right]_{i+1/2,j+1/2} \tag{16.85}$$

The representation for the x and y components of **u** has a similar form and will be omitted. A representative equation for the spatial derivatives in Maxwell's equations is given by (16.17).

To form a fully explicit integration scheme, the leapfrog concept in conjunction with the central difference approximation for the temporal derivative is also invoked:

$$\mu_o\left(\frac{\mathbf{H}^{n+1/2} - \mathbf{H}^{n-1/2}}{\delta_t}\right) = -\nabla \times \mathbf{E}^n \tag{16.86}$$

$$n_o m\left(\frac{\mathbf{u}^{n+1/2} - \mathbf{u}^{n-1/2}}{\delta_t}\right) = -n_o q\mathbf{E}^n - \nabla p^n \tag{16.87}$$

$$\frac{1}{\gamma p_o}\left(\frac{p^{n+1} - p^n}{\delta_t}\right) = -\nabla \cdot \mathbf{u}^{n+1/2} \tag{16.88}$$

and

$$\varepsilon_o\left(\frac{\mathbf{E}^{n+1} - \mathbf{E}^n}{\delta_t}\right) = \nabla \times \mathbf{H}^{n+1/2} + n_o q\mathbf{u}^{n+1/2} \tag{16.89}$$

The natural interaction between field constituents is again exhibited in the leapfrog integrator. That is, once the electromagnetic disturbance is initiated, we observe that **H** is driven by **E**, **u** is driven by **E** and p, p is driven by **u**, and **E** is driven by **u** and **H**.

A dispersion analysis of this scheme is affected by assuming the usual plane wave space–time dependence: $e^{-j\mathbf{k}\cdot\mathbf{x}}e^{j\omega t}$. Doing so and omitting the intermediate steps, we find that the analysis emulates the well-known theoretical analysis one for one. If

$$\omega_p^2 = \frac{n_o q^2}{m\varepsilon_o} \qquad c^2 = \frac{1}{\mu_o \varepsilon_o} \qquad a^2 = \frac{p_o \gamma}{n_o m} \tag{16.90}$$

(c is the speed of light, a is the acoustical wave speed, and ω_p is the plasma frequency), then the longitudinal waves (or acoustical) satisfy

$$\Omega^2 = \omega_p^2 + a^2\mathbf{K}\cdot\mathbf{K} \tag{16.91}$$

whereas, for the transverse (or optical) waves,

$$\Omega^2 = \omega_p^2 + c^2\mathbf{K}\cdot\mathbf{K} \tag{16.92}$$

Naturally, as the cell size and time step become infinitesimally small, (16.91) and (16.92) reduce to $\omega^2 = \omega_p^2 + a^2\mathbf{k}\cdot\mathbf{k}$ and $\omega^2 = \omega_p^2 + c^2\mathbf{k}\cdot\mathbf{k}$, respectively. It is a simple exercise to show from (16.91) and (16.92) that the maximum δ_t associated with one-dimensional simulations is given by $\delta_t = (\delta_x/c)\sqrt{1-(\omega_p\delta_t/2)^2}$ for optical waves and $\delta_t = (\delta_x/a)\sqrt{1-(\omega_p\delta_t/2)^2}$ for acoustical waves. Since $c > a$, the former equation is more restrictive.

In the above discussion, a wave species, say $\mathbf{E}$, is considered longitudinal if $\mathbf{k}\times\mathbf{E} = 0$ and transverse if $\mathbf{k}\cdot\mathbf{E} = 0$. For a warm, isotropic plasma, $\mathbf{H}$ is entirely transverse, ∇_p is entirely longitudinal, and $\mathbf{u}$ and $\mathbf{E}$ are a linear combination of both types. As expected, when $\omega_p = 0$, the previous dispersion relations describe the separate numerical propagation characteristics of acoustic and electromagnetic waves.

Although the previous numerical algorithm is consistent with the governing equations, there are limitations to its overall utility. Specifically, practical issues arise when physical parameters are used in the equations. Since a/c is small (i.e., 10^{-3}), the ratio of the acoustical wavenumber to the electromagnetic wavenumber is quite large. Thus, for practical problems, the computational domain will span over many acoustical wavelengths, but only over a fraction of the optical wavelength. To handle the wide disparity between optical and acoustical wavelengths, the scheme must be modified. One such modification is the decomposition of the governing equations into a longitudinal set and a transverse set. Associated with these two sets are the optical and acoustical grids and the corresponding CFL equations.

16.8 DEBYE DIELECTRIC

The Debye dielectric has received considerable attention in the literature due to the fast-growing interest in electromagnetic wave interactions with biological and water-based substances [28]–[30]. Given the water content of these substances, they are well modeled by a combination of single-pole models, which account for permanent dipole moments established within the water [3]. Moreover, since the Debye model is consistent with the requirements associated with the Kramers–Kronig relationships, the model inherently satisfies causality requirements. For this reason, an M-pole model is often used to construct an analytical model from empirical data. Since in the frequency domain the permittivity of the Debye dielectric is given by

$$\hat{\varepsilon} = \varepsilon_o\varepsilon_\infty + \varepsilon_o\sum_{k=1}^{M}\frac{\varepsilon_{sk}-\varepsilon_\infty}{1+j\omega\tau_k} = \varepsilon_o(1+\hat{\chi}(\omega)) \tag{16.93}$$

the modeling task is to deduce values for τ_k, ε_∞, and ε_{sk} that best fit $\hat{\varepsilon}$ to the data [31]. Here τ_k is the kth relaxation time, $\varepsilon_o\varepsilon_\infty$ is the value of $\hat{\varepsilon}$ at infinite frequency and $\varepsilon_{sk}\varepsilon_o$ is the zero frequency permittivity of the kth relaxation. Note: In the time domain,

$$\chi(t) = \sum_{k=1}^{M}\left(\frac{\varepsilon_{sk}-\varepsilon_\infty}{\tau_k}\right)e^{-t/\tau_k}U(t) \tag{16.94}$$

The Debye dielectric may also be characterized in state variable form (excluding Faraday's law) in terms of the polarization vector **P** [32]:

$$\varepsilon_o \varepsilon_\infty \frac{\partial \mathbf{E}}{\partial t} = \nabla \times \mathbf{H} - \sum_{k=1}^{M} \frac{1}{\tau_k} [(\varepsilon_{sk} - \varepsilon_\infty)\varepsilon_o \mathbf{E} - \mathbf{P}_k] \tag{16.95}$$

and

$$\frac{\partial \mathbf{P}_k}{\partial t} = \frac{1}{\tau_k} [(\varepsilon_{sk} - \varepsilon_\infty)\varepsilon_o \mathbf{E} - \mathbf{P}_k] \tag{16.96}$$

where $k = 1, 2, \ldots, M$ in the last equation. (Note: The subscript k denotes the pole number, not the spatial index of z.) These latter two equations form the basis for two of the direct integration schemes.

16.8.1 Direct Integration Method One: D-DIM1

For the first direct integration technique, consider initially a single-pole model. Then

$$\varepsilon_o \varepsilon_\infty \frac{\partial \mathbf{E}}{\partial t} = \nabla \times \mathbf{H} - \frac{1}{\tau} [(\varepsilon_s - \varepsilon_\infty)\varepsilon_o \mathbf{E} - \mathbf{P}] \tag{16.97}$$

and

$$\frac{\partial \mathbf{P}}{\partial t} = \frac{1}{\tau} [(\varepsilon_s - \varepsilon_\infty)\varepsilon_o \mathbf{E} - \mathbf{P}] \tag{16.98}$$

The components of **P** and **E** are conjoined on the Yee grid and the discretization of the curl operators is accomplished in the usual manner. Regarding the temporal discretization, we invoke the notion of the central difference and average approximations and the leapfrog time advancement [33]. To this end,

$$\frac{\mathbf{P}^{n+1/2} - \mathbf{P}^{n-1/2}}{\delta_t} = \frac{1}{\tau}\left[(\varepsilon_s - \varepsilon_\infty)\varepsilon_o \mathbf{E}^n - \left(\frac{\mathbf{P}^{n+1/2} + \mathbf{P}^{n-1/2}}{2}\right)\right] \tag{16.99}$$

and

$$\varepsilon_o \varepsilon_\infty \left(\frac{\mathbf{E}^{n+1} - \mathbf{E}^n}{\delta_t}\right) = \nabla \times \mathbf{H}^{n+1/2} - \frac{1}{\tau}\left[(\varepsilon_s - \varepsilon_\infty)\varepsilon_o \left(\frac{\mathbf{E}^{n+1} + \mathbf{E}^n}{2}\right) - \mathbf{P}^{n+1/2}\right] \tag{16.100}$$

The preceding equations can be algebraically arranged so that only $\mathbf{P}^{n+1/2}$ and $\mathbf{E}^{n+1}$ appear on the left-hand side of the equations; see [33]. Note that in this implementation, the polarization vector **P** and the magnetic field vector **H** are advanced simultaneously and leaped with **E**.

The basic strategy for M-poles is essentially the same as before, with the exception that all of the polarization equations are advanced simultaneously with

Faraday's law [33]. When temporal averaging is invoked, the discretized temporal form becomes

$$\frac{\mathbf{P}_k^{n+1/2} - \mathbf{P}_k^{n-1/2}}{\delta_t} = \frac{1}{\tau_k}\left[(\varepsilon_{sk} - \varepsilon_\infty)\varepsilon_o \mathbf{E}^n - \left(\frac{\mathbf{P}_k^{n+1/2} + \mathbf{P}_k^{n-1/2}}{2}\right)\right] \tag{16.101}$$

and

$$\varepsilon_o \varepsilon_\infty \left(\frac{\mathbf{E}^{n+1} - \mathbf{E}^n}{\delta_t}\right) = \nabla \times \mathbf{H}^{n+1/2} - \sum_{k=1}^{M} \frac{1}{\tau_k}\left[(\varepsilon_{sk} - \varepsilon_\infty)\varepsilon_o \left(\frac{\mathbf{E}^{n+1} + \mathbf{E}^n}{2}\right) - \mathbf{P}_k^{n+1/2}\right] \tag{16.102}$$

Further manipulation of (16.101) and (16.102) leads to the following update equations for **P** and **E**:

$$\mathbf{P}_k^{n+1/2} = \alpha_k \mathbf{P}_k^{n-1/2} + \beta_k \delta_t \tau_k^{-1}(\varepsilon_{sk} - \varepsilon_\infty)\varepsilon_o \mathbf{E}^n \tag{16.103}$$

where

$$\alpha_k = \frac{2\tau_k - \delta_t}{2\tau_k + \delta_t} \tag{16.104}$$

and

$$\beta_k = \frac{2\tau_k}{2\tau_k + \delta_t} \tag{16.105}$$

$$\mathbf{E}^{n+1} = c_1 \mathbf{E}^n + c_2 \delta_t \nabla \times \mathbf{H}^{n+1/2} + c_2 \delta_t \sum_{k=1}^{M} \tau_k^{-1} \mathbf{P}_k^{n+1/2} \tag{16.106}$$

where

$$c_1 = \left[\varepsilon_\infty - \frac{\delta_t}{2}\sum_{k=1}^{M} \frac{\varepsilon_{sk} - \varepsilon_\infty}{\tau_k}\right]\left[\varepsilon_\infty + \frac{\delta_t}{2}\sum_{k=1}^{M} \frac{\varepsilon_{sk} - \varepsilon_\infty}{\tau_k}\right]^{-1} \tag{16.107}$$

and

$$c_2 = \left[\varepsilon_\infty \varepsilon_o + \frac{\delta_t}{2}\sum_{k=1}^{M} \frac{(\varepsilon_{sk} - \varepsilon_\infty)\varepsilon_o}{\tau_k}\right]^{-1} \tag{16.108}$$

16.8.2 Direct Integration Method Two: D-DIM2

To have greater control over the stability properties of the scheme, semi-implicit schemes, such as those proposed by Kashiwa [32] and Petropolous [34], must be considered. (The reader may also consult [35] for further information on the semi-implicit approach.) For example, for Mth-order media,

$$\varepsilon_o \varepsilon_\infty \left(\frac{\mathbf{E}^{n+1} - \mathbf{E}^n}{\delta_t}\right) + \sum_{k=1}^{M} \left(\frac{\mathbf{P}_k^{n+1} - \mathbf{P}_k^n}{\delta_t}\right) = \nabla \times \mathbf{H}^{n+1/2} \tag{16.109}$$

and

$$\left(\frac{\mathbf{P}_k^{n+1} - \mathbf{P}_k^n}{\delta_t}\right) = \frac{1}{\tau_k}\left[(\varepsilon_{sk} - \varepsilon_\infty)\varepsilon_o \left(\frac{\mathbf{E}^{n+1} + \mathbf{E}^n}{2}\right) - \left(\frac{\mathbf{P}_k^{n+1} + \mathbf{P}_k^n}{2}\right)\right] \tag{16.110}$$

here $k = 1, 2, \ldots, M$.

16.8.3 Direct Integration Method Three: D-DIM3

Before leaving this section on direct integration techniques, we wish to consider an integration technique based upon the frequency-domain permittivity relationship: $\hat{\mathbf{D}} = \hat{\varepsilon}\hat{\mathbf{E}}$ [21]. As an example, consider the technique for a one-pole Debye dielectric. For this situation, the constitutive relationship can be written as

$$\hat{\mathbf{D}} = \varepsilon_o \left[\varepsilon_\infty + \frac{\varepsilon_s - \varepsilon_\infty}{1 + j\omega\tau} \right] \hat{\mathbf{E}} \tag{16.111}$$

Or, in the time domain,

$$\mathbf{D} + \tau \frac{\partial \mathbf{D}}{\partial t} = \varepsilon_s \varepsilon_o \mathbf{E} + \varepsilon_\infty \varepsilon_o \tau \frac{\partial \mathbf{E}}{\partial t} \tag{16.112}$$

As before, central differences and averages are applied to the previous ordinary differential equation, whereby an update equation for $\mathbf{E}$ is obtained. Using this update equation and the corresponding update equation for Ampère's law, we obtain

$$\mathbf{D}^{n+1} = \mathbf{D}^n + \delta_t \nabla \times \mathbf{H}^{n+1/2} \tag{16.113}$$

and

$$\mathbf{E}^{n+1} = \left(\frac{\delta_t + 2\tau}{\varepsilon_o(2\tau\varepsilon_\infty + \varepsilon_s \delta_t)} \right) \mathbf{D}^{n+1} + \left(\frac{\delta_t - 2\tau}{\varepsilon_o(2\tau\varepsilon_\infty + \varepsilon_s \delta_t)} \right) \mathbf{D}^n + \left(\frac{2\tau\varepsilon_\infty - \varepsilon_s \delta_t}{2\tau\varepsilon_\infty + \varepsilon_s \delta_t} \right) \mathbf{E}^n \tag{16.114}$$

Note: The update equation for E^{n+1} must be rederived for each additional pole considered in the dielectric model.

16.8.4 Recursive Convolution Method One: D-RCM1

The formulation of the recursive convolution method follows one for one with that of CP-RCM1. The only difference in that formulation and the one presented below is the equation for the accumulator, since the time-domain susceptibility function is different. To account for ε_∞, the first-order update equation is rewritten as [36]

$$\mathbf{E}^{n+1} = \varepsilon_\infty \alpha \mathbf{E}^n + \alpha \Psi^n + \frac{\alpha \delta_t}{\varepsilon_o} \nabla \times \mathbf{H}^{n+1/2} \tag{16.115}$$

where

$$\alpha = \frac{1}{\varepsilon_\infty + \chi^0} \tag{16.116}$$

here

$$\chi^0 = (\varepsilon_s - \varepsilon_\infty)(1 - e^{-\delta_t/\tau}) \tag{16.117}$$

With respect to the accumulator,

$$\Psi^n = \Delta\chi^0 \mathbf{E}^n + e^{-\delta_t/\tau} \Psi^{n-1} \tag{16.118}$$

where

$$\Delta\chi^0 = \chi^0(1 - e^{-\delta_t/\tau}) \tag{16.119}$$

16.8.5 Recursive Convolution Method Two: D-RCM2

To improve the temporal accuracy, a piecewise linear model may be employed in the integrand of the convolution integral [23]. (See also CP-RCM2 for the cold plasma.) The final equations are

$$\mathbf{E}^{n+1} = \alpha(\varepsilon_\infty - \xi^0)\mathbf{E}^n + \frac{\alpha\delta_t}{\varepsilon_o}\nabla\times\mathbf{H}^{n+1/2} + \alpha\Psi^n \tag{16.120}$$

where

$$\alpha = \frac{1}{(\varepsilon_\infty + \chi^0 - \xi^0)} \tag{16.121}$$

and

$$\Psi^n = \Delta\chi^0\mathbf{E}^n + (\mathbf{E}^{n-1} - \mathbf{E}^n)\Delta\xi^0 + e^{-\delta_t/\tau}\Psi^{n-1} \tag{16.122}$$

Here χ^0 and $\Delta\chi^0$ are given by (16.117) and (16.119), respectively. As for ξ^0 and $\Delta\xi^0$,

$$\xi^0 = -(\varepsilon_s - \varepsilon_\infty)\left(\frac{\tau}{\delta_t}\right)\left[\left(\frac{\delta_t}{\tau} + 1\right)e^{-\delta_t/\tau} - 1\right] \tag{16.123}$$

and

$$\Delta\xi^0 = \xi^0(1 - e^{-\delta_t/\tau}) \tag{16.124}$$

16.8.6 Comparative Analysis

As with the cold plasma, Table 16.5 documents the numerical permittivities for the previously reported schemes. From this table, the equivalence between D-DIM2 and D-DIM3 is seen. However, since D-DIM3 is constructed from the frequency-domain constitutive equation, D-DIM3 requires more memory than D-DIM2. Finally, D-RCM1 is first-order accurate in time; all other schemes are temporally second-order accurate.

To appreciate many of the subtle features of each of the aforementioned schemes, Tables 16.6 and 16.7 are given. These tables show the leading truncation term of the function $\Omega^2\hat{\varepsilon}/\varepsilon_o$. From a temporal accuracy point of view, the scheme with the lowest truncation term for a given ω, τ, ε_s, ε_∞, and δ_t is deemed best. Consider Table 16.7, which shows the leading truncation term for the two cases $\omega\tau \ll 1$ and $\omega\tau \gg 1$. Except for D-RCM1, all other schemes have identical high-frequency truncation terms. These terms are the same truncation terms that the standard FDTD algorithm would produce in a dielectric of permittivity $\varepsilon_\infty\varepsilon_o$. In the low-frequency limit only D-DIM2 and D-DIM3 predict the anticipated truncation term. Thus, it can be argued that D-DIM2 and D-DIM3 are the most compatible with the standard FDTD algorithm.

A summary of the stability properties and memory requirements of the various

TABLE 16.5 Numerical Permittivities for Various Schemes Associated with Propagation in Single-Pole Debye Dielectric

Method	Numerical Permittivity: $\hat{\varepsilon}$
D-DIM1	$\hat{\varepsilon} = \varepsilon_\infty \varepsilon_o + \varepsilon_o(\varepsilon_s - \varepsilon_\infty)\left[\dfrac{\Lambda + j(\delta_t/2)^2(\Omega/\tau)}{\Lambda + j\Omega\tau}\right]$
D-DIM2	$\hat{\varepsilon} = \varepsilon_\infty \varepsilon_o + (\varepsilon_s - \varepsilon_\infty)\varepsilon_o\left[\dfrac{\Lambda}{\Lambda + j\Omega\tau}\right]$
D-DIM3	$\hat{\varepsilon} = \varepsilon_\infty \varepsilon_o + (\varepsilon_s - \varepsilon_\infty)\varepsilon_o\left[\dfrac{\Lambda}{\Lambda + j\Omega\tau}\right]$
D-RCM1	$\hat{\varepsilon} = \varepsilon_o\left[\varepsilon_\infty + \dfrac{\chi^0 e^{j\omega\delta_t/2}}{j\Omega\delta_t} - \dfrac{\Delta\chi^0 e^{\delta_t/(2\tau)}}{2j\Omega\delta_t \sinh[(\delta_t/\tau)(1 + j\omega\tau)/2]}\right]$
D-RCM2	$\hat{\varepsilon} = \varepsilon_o\left[\varepsilon_\infty - \xi^0 + \dfrac{\chi^0 e^{j\omega\delta_t/2}}{j\Omega\delta_t} - \left(\dfrac{\Delta\chi^0 - j\Omega\delta_t\Delta\xi^0 e^{-j\omega\delta_t/2}}{2j\Omega\delta_t \sinh[(\delta_t/\tau)(1 + j\omega\tau)/2]}\right)e^{\delta_t/(2\tau)}\right]$

TABLE 16.6 Truncation Term from the Taylor Analysis of $\Omega^2\hat{\varepsilon}/\varepsilon_o$: Debye Dielectric

Method	Truncation Term
D-DIM1	$\dfrac{\omega^4\delta_t^2}{12}\left[(\varepsilon_s - \varepsilon_\infty)\left(\dfrac{3j}{\omega\tau} + \dfrac{j}{j - \omega\tau} - \dfrac{1}{(j - \omega\tau)^2}\right) - \varepsilon_\infty\right]$
D-DIM2	$\dfrac{\omega^4\delta_t^2}{12}\left[(\varepsilon_\infty - \varepsilon_s)\left(\dfrac{2j}{j - \omega\tau} + \dfrac{1}{(j - \omega\tau)^2}\right) - \varepsilon_\infty\right]$
D-DIM3	$\dfrac{\omega^4\delta_t^2}{12}\left[(\varepsilon_\infty - \varepsilon_s)\left(\dfrac{2j}{j - \omega\tau} + \dfrac{1}{(j - \omega\tau)^2}\right) - \varepsilon_\infty\right]$
D-RCM1	$\dfrac{j\omega^3\delta_t}{2}\left[\dfrac{\varepsilon_s - \varepsilon_\infty}{1 + j\omega\tau}\right]$
D-RCM2	$\dfrac{\omega^4\delta_t^2}{12}\left[\dfrac{j(\varepsilon_\infty - 2\varepsilon_s) + \varepsilon_\infty\omega\tau}{j - \omega\tau}\right]$

methodologies is provided in Table 16.8. Here we note that D-DIM2 is the most memory intensive. With respect to the stability equations, the entries for D-DIM1, D-RCM1, and D-RCM2 have not been rigorously established. The entry for D-DIM3 follows directly from the semi-implicit nature of the scheme [37]. Finally, since D-DIM2 and D-DIM3 are equivalent to one another, they also have the same stability equation.

To close this section, a comment on multipole Debye dielectrics is in order. Each of the schemes reviewed herein is capable of modeling the multipole dielectric. Some schemes like D-DIM1, D-RCM1, and D-RCM2 can be directly extended without any special treatment. Other schemes like D-DIM3 require a new derivation for each new

TABLE 16.7 Truncation Term from the Taylor Analysis of $\Omega^2 \hat{\varepsilon}/\varepsilon_o$ for Two Special Cases: Debye Dielectric

Method	Truncation Term: $\omega\tau \ll 1$	Truncation Term: $\omega\tau \gg 1$
D-DIM1	$\left(\frac{\omega^4 \delta_t^2}{12}\right)\left[(\varepsilon_s - \varepsilon_\infty)\left(2 + \frac{3j}{\omega\tau}\right) - \varepsilon_\infty\right]$	$\frac{-\omega^4 \delta_t^2 \varepsilon_\infty}{12}$
D-DIM2	$\frac{-\omega^4 \delta_t^2 \varepsilon_s}{12}$	$\frac{-\omega^4 \delta_t^2 \varepsilon_\infty}{12}$
D-DIM3	$\frac{-\omega^4 \delta_t^2 \varepsilon_s}{12}$	$\frac{-\omega^4 \delta_t^2 \varepsilon_\infty}{12}$
D-RCM1	$\left(\frac{j\omega^3 \delta_t}{2}\right)(\varepsilon_s - \varepsilon_\infty)$	$\left(\frac{\omega^2 \delta_t}{2\tau}\right)(\varepsilon_s - \varepsilon_\infty)$
D-RCM2	$\left(\frac{\omega^4 \delta_t^2}{12}\right)(\varepsilon_\infty - 2\varepsilon_s)$	$\frac{-\omega^4 \delta_t^2 \varepsilon_\infty}{12}$

TABLE 16.8 Comparison Table of Various Schemes in Terms of Memory per Cell and Stability Equations: Single-Pole Debye Dielectric

Method	3D Memory/Cell	1D Stability
D-DIM1	9	$\delta_t < \delta_x / v_p^{max}$
D-DIM2	9	$\delta_t < \sqrt{\varepsilon_\infty}\, \delta_x / c$
D-DIM3	12	$\delta_t < \sqrt{\varepsilon_\infty}\, \delta_x / c$
D-RCM1	9	$\delta_t < \delta_x / v_p^{max}$
D-RCM2	9	$\delta_t < \delta_x / v_p^{max}$

pole added to the model. As for D-DIM2, the semi-implicit nature requires the loading and the inverting of a new matrix (each time a pole is added) to find the necessary coefficients of the algorithm.

16.8.7 Parameter Selection

To set the parameters δ_t and δ_x for the simulation, consider the example in which muscle is the medium. For a two-pole model, $\varepsilon_\infty = 40$, $\varepsilon_{s1} = 3948$, $\varepsilon_{s2} = 59.09$, $\tau_1 = 46.25$ ns and $\tau_2 = 0.0907$ ns [30]. One possible way to deduce a value for δ_t is to make it a fraction of the smallest relaxation; for this example, let $\delta_t = \tau_2/10 = 9.07$ ps. To minimize dispersion errors, the CFL number should be maximized. Thus, from the definition of CFL let $\delta_x = \delta_t v_p^{max} = 0.430$ mm, where, for this example, $v_p^{max} = 4.74 \times 10^7$ m/s. A second way is to set δ_x in accordance with the minimum wavelength of interest. Let $\delta_x = \lambda^{min}/20$, where λ^{min} is the minimum wavelength at some angular frequency; for this example let $\omega = 10/\tau_2$, in which case $\lambda^{min} = 2.69$ mm and $\delta_x = 0.135$ mm. Finally, from the stability relationship, the maximum time step is $\delta_t = \delta_x / v_p^{max} = 2.84$ ps.

Based upon the previous example, the more conservative choice for the values of δ_x and δ_t is based upon the minimum wavelength argument. Unfortunately, this argument produces a time-step that over-resolves the time-domain driving signal. Moreover, by doing so, we incur the cost of increasing the simulation time by about a factor of 3.2 and the computational memory by a factor of 32.8 (three-dimensional domain), as compared with the other method. Had we chosen the first method, the high-frequency information of the medium would not be captured and hence, the early-time information of the output data would be in error. One possible way to resolve these contrary specifications is to increase the spatial accuracy order of the scheme [34]. For example, if a fourth order in space and a second order in time scheme is employed, the sampling requirements on the minimum wavelength can be relaxed and the resulting time-step will be commensurate with the smallest relaxation.

16.9 LORENTZ DIELECTRIC

The Lorentz dielectric is one in which the atomic structure of the material is modeled as a damped vibrating system. From an electromagnetic point of view, the atom consists of two charged concentric surfaces; the innermost surface is positively charged and is highly massive; the outermost surface is negatively charged and is lightly massive. Atomic forces exist that couple these two surfaces together via a tension coefficient. When exposed to an electric field, the outermost shell is free to move in accordance with inertial laws, whereas the innermost shell is assumed to be stationary. Hence, upon the invocation of classical Newtonian laws, the displacement of the outermost shell, relative to its quiescent location, is found to be a solution to a second-order, ordinary differential equation (ODE). Further manipulation of this ODE, and the realization that the polarization vector is proportional to the shell displacement and the polarization current is proportional to velocity, allows one to write [38]:

$$\frac{\partial \mathbf{J}_p}{\partial t} = -\nu_1 \mathbf{J}_p + (\varepsilon_s - \varepsilon_\infty)\varepsilon_o \omega_1^2 \mathbf{E} - {\omega_1}^2 \mathbf{P} \tag{16.125}$$

and

$$\frac{\partial \mathbf{P}}{\partial t} = \mathbf{J}_p \tag{16.126}$$

where $\varepsilon_s \varepsilon_o$ and $\varepsilon_\infty \varepsilon_o$ are the static and infinite permittivities, respectively; furthermore, ω_1 is the resonant frequency and ν_1 is the damping frequency. As for Ampère's law,

$$\varepsilon_o \varepsilon_\infty \frac{\partial \mathbf{E}}{\partial t} = \nabla \times \mathbf{H} - \mathbf{J}_p \tag{16.127}$$

If the dependent variables are cast in terms of their time-harmonic counterparts, the mechanical properties of the atom can be embedded into the macroscopic permittivity [3]. To this end,

$$\hat{\varepsilon} = \varepsilon_\infty \varepsilon_o + \frac{(\varepsilon_s - \varepsilon_\infty)\varepsilon_o \omega_1^2}{\omega_1^2 + j\omega\nu_1 - \omega^2} = \varepsilon_o(1 + \hat{\chi}(\omega)) \quad (16.128)$$

With respect to the time-domain susceptibility function,

$$\chi(t) = \gamma e^{-\zeta t} \sin(\beta t) U(t) \quad (16.129)$$

which is the inverse Fourier transform of $\hat{\chi}(\omega)$. Here $\zeta = \nu_1/2$, $\beta = \sqrt{\omega_1^2 - \zeta^2}$ and $\gamma = \omega_1^2(\varepsilon_s - \varepsilon_\infty)/\beta$.

16.9.1 Direct Integration Method One: L-DIM1

The first discretization procedure is based upon the governing equations of (16.125)–(16.127). The temporal discretization employs central differences and averages. Assuming that the components of $\mathbf{E}$, $\mathbf{J}_p$, and $\mathbf{P}$ are spatially collocated, we write for the temporally discretized equations the following [33]:

$$\left(\frac{\mathbf{J}_p^{n+1/2} - \mathbf{J}_p^{n-1/2}}{\delta_t}\right) = -\nu_1\left(\frac{\mathbf{J}_p^{n+1/2} + \mathbf{J}_p^{n-1/2}}{2}\right) + (\varepsilon_s - \varepsilon_\infty)\varepsilon_o \omega_1^2 \mathbf{E}^n - \omega_1^2 \mathbf{P}^n \quad (16.130)$$

$$\varepsilon_o \varepsilon_\infty \left(\frac{\mathbf{E}^{n+1} - \mathbf{E}^n}{\delta_t}\right) = \nabla \times \mathbf{H}^{n+1/2} - \mathbf{J}_p^{n+1/2} \quad (16.131)$$

and

$$\left(\frac{\mathbf{P}^{n+1} - \mathbf{P}^n}{\delta_t}\right) = \mathbf{J}_p^{n+1/2} \quad (16.132)$$

As with the standard FDTD method, the leapfrog scheme is naturally compatible with the physics. That is, one set of field constituents are effective sources for the other set of constituents, and vice versa. This same relationship is also observed in the isotropic cold plasma; see CP-DIM1. Moreover, for loss-free media, the scheme exhibits a perfect leapfrog structure.

Although the previous model assumes a single resonant frequency, multiple resonant frequencies are equally possible. Hence, when the media are modeled by N resonances, the pertinent equations become

$$\varepsilon_o \varepsilon_\infty \frac{\partial \mathbf{E}}{\partial t} = \nabla \times \mathbf{H} - \sum_{n=1}^{N} \mathbf{J}_{pn} \quad (16.133)$$

$$\frac{\partial \mathbf{J}_{p1}}{\partial t} = -\nu_1 \mathbf{J}_{p1} + G_1(\varepsilon_s - \varepsilon_\infty)\varepsilon_o \omega_1^2 \mathbf{E} - \omega_1^2 \mathbf{P}_1 \quad (16.134)$$

$$\frac{\partial \mathbf{P}_1}{\partial t} = \mathbf{J}_{p1} \quad (16.135)$$

$$\vdots = \vdots$$

$$\frac{\partial \mathbf{J}_{pN}}{\partial t} = -\nu_N \mathbf{J}_{pN} + G_N(\varepsilon_s - \varepsilon_\infty)\varepsilon_o \omega_N^2 \mathbf{E} - \omega_N^2 \mathbf{P}_N \quad (16.136)$$

and

$$\frac{\partial \mathbf{P}_N}{\partial t} = \mathbf{J}_{pN} \tag{16.137}$$

Here $\Sigma_{n=1}^{N} G_n = 1$. As before, the components of $\mathbf{E}$, $\mathbf{P}_1$, ..., $\mathbf{P}_N$ and $\mathbf{J}_{p1}$ ... $\mathbf{J}_{pN}$ are nodally collocated. The leapfrog methodology remains essentially the same as in the previous discussion, except that now all of the polarization equations are advanced simultaneously with Ampère's law; similarly, all force equations are advanced simultaneously with Faraday's law; see [33] for further details.

16.9.2 Direct Integration Method Two: L-DIM2

To decouple the CFL relationship from the parameters of the medium, temporal, implicit methods are required [37]. For example, the employment of trapezoidal integration yields [38]:

$$\left(\frac{\mathbf{J}_p^{n+1} - \mathbf{J}_p^n}{\delta_t}\right) = -\nu_1\left(\frac{\mathbf{J}_p^{n+1} + \mathbf{J}_p^n}{2}\right) + (\varepsilon_s - \varepsilon_\infty)\varepsilon_o \omega_1^2\left(\frac{\mathbf{E}^{n+1} + \mathbf{E}^n}{2}\right) - \omega_1^2\left(\frac{\mathbf{P}^{n+1} + \mathbf{P}^n}{2}\right) \tag{16.138}$$

$$\varepsilon_o \varepsilon_\infty\left(\frac{\mathbf{E}^{n+1} - \mathbf{E}^n}{\delta_t}\right) = \nabla \times \mathbf{H}^{n+1/2} - \left(\frac{\mathbf{P}^{n+1} - \mathbf{P}^n}{\delta_t}\right) \tag{16.139}$$

and

$$\left(\frac{\mathbf{P}^{n+1} - \mathbf{P}^n}{\delta_t}\right) = \left(\frac{\mathbf{J}_p^{n+1} + \mathbf{J}_p^n}{2}\right) \tag{16.140}$$

16.9.3 Direct Integration Method Three: L-DIM3

To conclude the discussion on direct integration techniques, the technique based upon the frequency-domain permittivity is provided next [21]. Utilizing the constitutive relationship and unpacking (16.128), we deduce that

$$\omega_1^2 \mathbf{D} + \nu_1 \frac{\partial \mathbf{D}}{\partial t} + \frac{\partial^2 \mathbf{D}}{\partial t^2} = \varepsilon_s \varepsilon_o \omega_1^2 \mathbf{E} + \varepsilon_\infty \varepsilon_o \nu_1 \frac{\partial \mathbf{E}}{\partial t} + \varepsilon_\infty \varepsilon_o \frac{\partial^2 \mathbf{E}}{\partial t^2} \tag{16.141}$$

Application of central differences and averages to the various terms in the above equation and discretization of Ampère's law yields

$$\mathbf{D}^{n+1} = \mathbf{D}^n + \delta_t \nabla \times \mathbf{H}^{n+1/2} \tag{16.142}$$

and

$$\begin{aligned} \mathscr{D}\mathbf{E}^{n+1} &= (\omega_1^2 \delta_t^2 + \nu_1 \delta_t + 2)\mathbf{D}^{n+1} - 4\mathbf{D}^n + (\omega_1^2 \delta_t^2 - \nu_1 \delta_t + 2)\mathbf{D}^{n-1} \\ &\quad + 4\varepsilon_\infty \varepsilon_o \mathbf{E}^n - (\omega_1^2 \delta_t^2 \varepsilon_s \varepsilon_o - \nu_1 \delta_t \varepsilon_\infty \varepsilon_o + 2\varepsilon_\infty \varepsilon_o)\mathbf{E}^{n-1} \end{aligned} \tag{16.143}$$

where $\mathscr{D} = (\omega_1^2 \delta_t^2 \varepsilon_s \varepsilon_o + \nu_1 \delta_t \varepsilon_\infty \varepsilon_o + 2\varepsilon_\infty \varepsilon_o)$.

16.9.4 Recursive Convolution Method One: L-RCM1

For the Lorentz dielectric, the recursive convolution method developed for the cold plasma (i.e., CP-RCM1) and the Debye dielectric (i.e., D-RCM1) applies as well [39], [40]. From that discussion,

$$\mathbf{E}^{n+1} = \varepsilon_\infty \alpha \mathbf{E}^n + \alpha \Psi^n + \frac{\alpha \delta_t}{\varepsilon_o} \nabla \times \mathbf{H}^{n+1/2} \tag{16.144}$$

where

$$\alpha = \frac{1}{\varepsilon_\infty + \chi^0} \tag{16.145}$$

To compute χ^0, complex variables are invoked by letting $\chi^0 = \mathrm{Re}\{\tilde{\chi}^0\}$[4] and by letting

$$\tilde{\chi}^m = \int_{m\delta_t}^{(m+1)\delta_t} \tilde{\chi}(t)\, dt \tag{16.146}$$

Considering the time-domain susceptibility function defined by (16.129), we write instead, $\chi(t) = \mathrm{Re}\{\tilde{\chi}(t)\}$, where

$$\tilde{\chi}(t) = -j\gamma e^{(-\zeta + j\beta)t} U(t) \tag{16.147}$$

Thus, upon the integration of (16.146), the setting of m to a value of zero yields

$$\tilde{\chi}^0 = \frac{-j\gamma}{\zeta - j\beta} [1 - e^{(-\zeta + j\beta)\delta_t}] \tag{16.148}$$

With respect to the accumulator, it is a simple matter to show that $\Psi^n = \mathrm{Re}\{\tilde{\Psi}^n\}$, where

$$\tilde{\Psi}^n = \sum_{m=0}^{n-1} \mathbf{E}^{n-m} \Delta\tilde{\chi}^m = \Delta\tilde{\chi}^0 \mathbf{E}^n + e^{(-\zeta + j\beta)\delta_t} \tilde{\Psi}^{n-1} \tag{16.149}$$

Here

$$\Delta\tilde{\chi}^0 = \tilde{\chi}^0 (1 - e^{(-\zeta + j\beta)\delta_t}) \tag{16.150}$$

The recursive convolution method requires two additional vectors since the accumulator sums vectors of the complex kind.

16.9.5 Recursive Convolution Method Two: L-RCM2

To increase the accuracy, a piecewise approximation may be invoked; see [23]. The update equation is still given by (16.120) with α defined by (16.121). That is,

$$\mathbf{E}^{n+1} = \alpha(\varepsilon_\infty - \xi^0)\mathbf{E}^n + \frac{\alpha \delta_t}{\varepsilon_o} \nabla \times \mathbf{H}^{n+1/2} + \alpha \Psi^n \tag{16.151}$$

[4]The tilde denotes a complex time-domain quantity.

where

$$\alpha = \frac{1}{(\varepsilon_\infty + \chi^0 - \xi^0)} \tag{16.152}$$

here $\chi^0 = \text{Re}\{\tilde{\chi}^0\}$ with $\tilde{\chi}^0$ given by (16.148). As for ξ^0, $\xi^0 = \text{Re}\{\tilde{\xi}^0\}$, where

$$\tilde{\xi}^0 = \left(\frac{j\gamma}{\delta_t(\zeta - j\beta)^2}\right)\left([(\zeta - j\beta)\delta_t + 1]e^{(-\zeta + j\beta)\delta_t} - 1\right) \tag{16.153}$$

Consider the accumulator and let $\Psi = \text{Re}\{\tilde{\Psi}\}$; then

$$\tilde{\Psi}^n = \Delta\tilde{\chi}^0 \mathbf{E}^n + (\mathbf{E}^{n-1} - \mathbf{E}^n)\Delta\tilde{\xi}^0 + e^{(-\zeta + j\beta)\delta_t}\tilde{\Psi}^{n-1} \tag{16.154}$$

Here $\Delta\tilde{\chi}^0$ is given by (16.150) and

$$\Delta\tilde{\xi}^0 = \tilde{\xi}^0(1 - e^{(-\zeta + j\beta)\delta_t}) \tag{16.155}$$

16.9.6 Comparative Analysis

Consider Table 16.9, which documents the numerical permittivities for the previously reported schemes. Note: The following notation is used in the entries for L-RCM1 and L-RCM2:

$$S_1 = \frac{\Delta\tilde{\chi}^0 e^{(\zeta - j\beta)\delta_t/2}}{4j\sin[(\beta - \omega + j\zeta)\delta_t/2]} \tag{16.156}$$

and

$$S_2 = \frac{(\Delta\tilde{\chi}^0)^* e^{(\zeta + j\beta)\delta_t/2}}{4j\sin[(-\beta - \omega + j\zeta)\delta_t/2]} \tag{16.157}$$

$$S_3 = \frac{(\Delta\tilde{\chi}^0 - j\Omega\delta_t\Delta\tilde{\xi}^0 e^{-j\omega\delta_t/2})e^{(\zeta - j\beta)\delta_t/2}}{4j\sin[(\beta - \omega + j\zeta)\delta_t/2]} \tag{16.158}$$

and

$$S_4 = \frac{\left((\Delta\tilde{\chi}^0)^* - j\Omega\delta_t(\Delta\tilde{\xi}^0)^* e^{-j\omega\delta_t/2}\right)e^{(\zeta + j\beta)\delta_t/2}}{4j\sin[(-\beta - \omega + j\zeta)\delta_t/2]} \tag{16.159}$$

Unlike the schemes for the cold plasma and the Debye dielectric, there exists no equivalence between the various methodologies. Although not apparent from this table, all methods except for L-RCM1 are second-order accurate in time; L-RCM1 is first-order accurate. To deduce the complete temporal error mechanisms of each scheme, the function $\Omega^2\hat{\varepsilon}$ should be computed, as well as the corresponding truncation term. (This analysis is not provided due to the complexity of the final expressions.) As one might expect, all schemes except for L-RCM1 are second-order accurate; L-RCM1 is first-order accurate.

The CFL equations for each scheme are complicated functions of the dielectric's parameters; for that reason closed-form expressions are difficult to deduce. However, since L-DIM2 is semi-implicit, the CFL is medium independent and hence $\delta_t < \sqrt{\varepsilon_\infty}\,\delta_x/c$ in a one-dimensional space. With respect to L-DIM1, the CFL equation for the special case when $\nu = 0$ can be found in [41].

TABLE 16.9 Numerical Permittivities for Various Schemes Associated with Propagation in an Isotropic Lorentz Dielectric

Method	Numerical Permittivity: $\hat{\varepsilon}$
L-DIM1	$\hat{\varepsilon} = \varepsilon_\infty \varepsilon_o + \dfrac{(\varepsilon_s - \varepsilon_\infty)\varepsilon_o \omega_1^2}{\omega_1^2 + j\Omega\nu_1\Lambda - \Omega^2}$
L-DIM2	$\hat{\varepsilon} = \varepsilon_\infty \varepsilon_o + \dfrac{(\varepsilon_s - \varepsilon_\infty)\varepsilon_o \omega_1^2 \Lambda^2}{\omega_1^2 \Lambda^2 + j\Omega\nu_1\Lambda - \Omega^2}$
L-DIM3	$\hat{\varepsilon} = \varepsilon_\infty \varepsilon_o + \dfrac{(\varepsilon_s - \varepsilon_\infty)\varepsilon_o \omega_1^2 \cos(\omega\delta_t)}{\omega_1^2 \cos(\omega\delta_t) + j\nu_1\Omega\Lambda - \Omega^2}$
L-RCM1	$\hat{\varepsilon} = \varepsilon_\infty \varepsilon_o + \varepsilon_o \dfrac{\chi^0 e^{j\omega\delta_t/2} + (S_1 + S_2)}{j\Omega\delta_t}$
L-RCM2	$\hat{\varepsilon} = (\varepsilon_\infty - \xi^0)\varepsilon_o + \varepsilon_o \dfrac{\chi^0 e^{j\omega\delta_t/2} + (S_3 + S_4)}{j\Omega\delta_t}$

The memory requirement for each scheme, except L-DIM3, is 12 unknowns per cell. Due to the back storage of both **D** and **E**, L-DIM3 requires 15 unknowns per cell.

As with the Debye dielectric, the problem of setting the time step and the spatial cell size is not straightforward. Again, two choices exist: 1) set δ_t to be a fraction of ω_1 or ν_1 and use the stability relationship to set δ_x or 2) set δ_x to correspond to the minimum wavelength of interest and use the stability relationship to deduce δ_t. Note: Unlike the Debye dielectric, all algorithms in this section will produce inaccurate data for frequencies near the resonance frequency and for loss-free medium. In this case when $\omega \rightarrow \omega_1$, $\varepsilon_r \rightarrow \infty$ and the associated wavelength tends toward zero. Thus, there exists no δ_x for which this frequency component can be spatially resolved.

Finally, the extension to Nth-order Lorentzian media is straightforward for each algorithm, except L-DIM3. For L-DIM3, the update equation must be rederived each time a set of poles is added to the model.

16.9.7 Numerical Results

To demonstrate typical results, the one-dimensional slab problem is considered [33] in conjunction with the first method presented in this section. Let the computational domain consist of 1000 cells with the Lorentzian dielectric slab occupying the region from the 500th cell to the 750th cell. We will assume an initial field on the left-hand side of the slab; the incident field is a Gaussian pulse of the form

$$E_y(x, t) = e^{-w^2(t-(x-x_o)/c)^2} \tag{16.160}$$

where $w = 1.66 \times 10^{11}$ s¯1. From Fourier analysis, it is found that the pulse contains a usable frequency content up to 80 GHz (i.e., $\omega_{max} = 2\pi 80 \times 10^9$ r/s). If we require that the pulse be sampled at 10π points per free space wavelength when $\omega = \omega_{max}$ and set $\Delta_t = 0.5\Delta_x/c$, then $\Delta_x = 0.119$ mm and $\Delta_t = 0.199$ ps. For the Lorentzian dielectric,

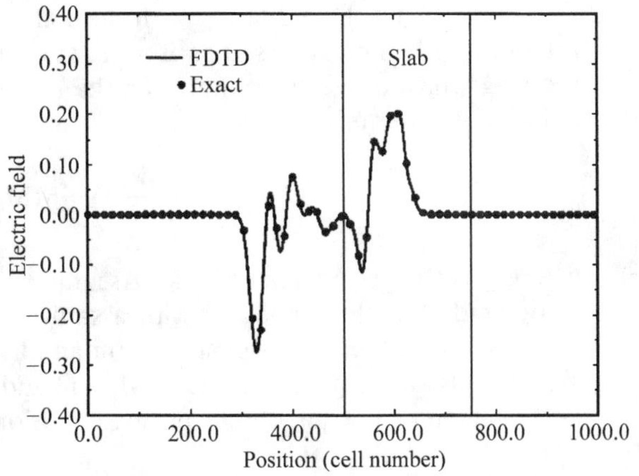

Figure 16.4 The exact and computed time-domain solution for E_y at $t = 750\Delta_t$; four-pole Lorentz dielectric.

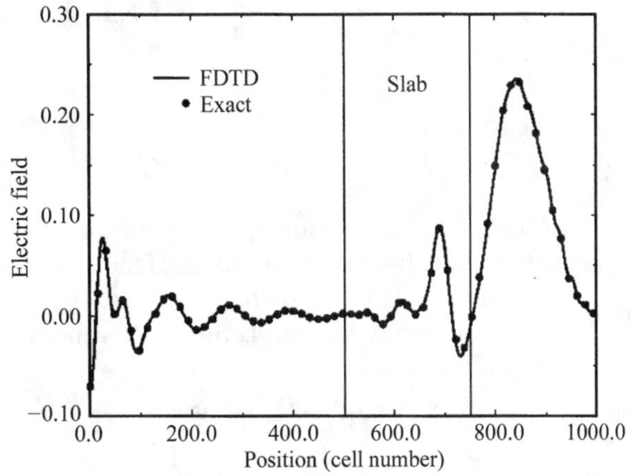

Figure 16.5 The exact and computed time-domain solution for E_y at $t = 1500\Delta_t$: four-pole Lorentz dielectric.

a four-pole model is assumed: $\varepsilon_s = 3.0$, $\varepsilon_\infty = 1.5$, $\omega_1 = 40\pi \times 10^9$ r/s, $\nu_1 = 0.2\omega_1$, $G_1 = 0.4$, $\omega_2 = 100\pi \times 10^9$ r/s, $\nu_2 = 0.2\omega_2$, and $G_2 = 0.6$. Spatial plots of the pulse are provided in Figs. 16.4 and 16.5 when $t = 750\Delta_t$ and $t = 1500\Delta_t$. As seen from the plots, the correlation between the analytical data and the FDTD data is consistent with numerical theory. Clearly, one can argue that the dispersive nature of the medium significantly corrupts the original shape of the pulse.

16.10 MAGNETIC FERRITES

The use of the FDTD algorithm to model electromagnetic wave propagation in magnetic ferrites has grown in recent years. With the ferrite being a key element in many microwave circuits such as the circulator, isolator, and tunable filter, there exists a clear need to model such circuits numerically. As with most schemes, there are many approaches to accomplish this end. In this section, however, we will consider a direct integration technique [42]–[46].

Consider (16.1) and (16.2) with the source terms set to zero and the constitutive relationship being expressed as $\mathbf{B} = \mu_o(\mathbf{H}+\mathbf{M})$. In this context, $\mathbf{M}$ is interpreted as the magnetization vector rather than the magnetic current as in (16.1). The auxiliary equation for the ferrite is

$$\frac{\partial \mathbf{M}}{\partial t} = -\gamma(\mathbf{M}\times\mathbf{H}) \tag{16.161}$$

where γ is the gyromagnetic ratio. Assume that the ferrite is biased with an applied static field $\mathbf{H}_i$ and is endowed with a saturation magnetization $\mathbf{M}_s$. Added to these static fields are the small signal fields $\mathbf{m}$ and $\mathbf{h}$. Superposition of the biased and small signal fields allows us to write $\mathbf{M} = \mathbf{M}_s + \mathbf{m}$ and $\mathbf{H} = \mathbf{H}_i + \mathbf{h}$. With the assumption that $|\mathbf{M}_s| \gg |\mathbf{m}|$ and $|\mathbf{H}_i| \gg |\mathbf{h}|$, the governing equations to be considered are

$$\varepsilon_o \frac{\partial \mathbf{E}}{\partial t} = \nabla\times\mathbf{h} \tag{16.162}$$

$$\mu_o \frac{\partial \mathbf{h}}{\partial t} = -\nabla\times\mathbf{E} + \mu_o\gamma(\mathbf{m}\times\mathbf{H}_i + \mathbf{M}_s\times\mathbf{h}) \tag{16.163}$$

and

$$\frac{\partial \mathbf{m}}{\partial t} = -\gamma(\mathbf{m}\times\mathbf{H}_i + \mathbf{M}_s\times\mathbf{h}) \tag{16.164}$$

Except for the anisotropic nature of the ferrite, the ferrite is in many ways mathematically analogous to the Debye material. For this reason, the discretization procedure is equally analogous.

In the usual manner, central differences are applied to the time derivatives in the following fashion:

$$\mu_o\left(\frac{\mathbf{h}^{n+1/2}-\mathbf{h}^{n-1/2}}{\delta_t}\right) = -\nabla\times\mathbf{E}^n + \mu_o\gamma(\mathbf{m}^n\times\mathbf{H}_i + \mathbf{M}_s\times\mathbf{h}^n) \tag{16.165}$$

$$\frac{\mathbf{m}^{n+1/2}-\mathbf{m}^{n-1/2}}{\delta_t} = -\gamma(\mathbf{m}^n\times\mathbf{H}_i + \mathbf{M}_s\times\mathbf{h}^n) \tag{16.166}$$

and

$$\varepsilon_o\left(\frac{\mathbf{E}^{n+1}-\mathbf{E}^n}{\delta_t}\right) = \nabla\times\mathbf{h}^{n+1/2} \tag{16.167}$$

Unfortunately, neither $\mathbf{m}$ nor $\mathbf{h}$ is known at time $n\delta_t$. Hence, central averages need to be employed for these terms. Doing so, we convert (16.165) and (16.166) into

$$\mu_o\left(\frac{\mathbf{h}^{n+1/2}-\mathbf{h}^{n-1/2}}{\delta t}\right) = -\nabla\times\mathbf{E}^n + \frac{\mu_o\gamma}{2}[(\mathbf{m}^{n+1/2}+\mathbf{m}^{n-1/2})\times\mathbf{H}_i + \mathbf{M}_s\times(\mathbf{h}^{n+1/2}+\mathbf{h}^{n-1/2})] \tag{16.168}$$

and

$$\frac{\mathbf{m}^{n+1/2}-\mathbf{m}^{n-1/2}}{\delta_t} = -\frac{\gamma}{2}[(\mathbf{m}^{n+1/2}+\mathbf{m}^{n-1/2})\times\mathbf{H}_i + \mathbf{M}_s\times(\mathbf{h}^{n+1/2}+\mathbf{h}^{n-1/2})] \tag{16.169}$$

In dyadic form, (16.168) and (16.169) are equivalent to

$$\bar{\mathbf{A}}_1 \cdot \mathbf{h}^{n+1/2} + \bar{\mathbf{A}}_2 \cdot \mathbf{m}^{n+1/2} = \bar{\mathbf{A}}_3 \cdot \mathbf{h}^{n-1/2} + \bar{\mathbf{A}}_4 \cdot \mathbf{m}^{n-1/2} - \nabla \times \mathbf{E}^n \equiv \mathbf{b}_1 \tag{16.170}$$

and

$$\bar{\mathbf{A}}_5 \cdot \mathbf{h}^{n+1/2} + \bar{\mathbf{A}}_6 \cdot \mathbf{m}^{n+1/2} = \bar{\mathbf{A}}_7 \cdot \mathbf{h}^{n-1/2} + \bar{\mathbf{A}}_8 \cdot \mathbf{m}^{n-1/2} \equiv \mathbf{b}_2 \tag{16.171}$$

where, for example,

$$\bar{\mathbf{A}}_1 = \mu_o \begin{bmatrix} 1/\delta_t & \gamma M_3/2 & -\gamma M_2/2 \\ -\gamma M_3/2 & 1/\delta_t & \gamma M_1/2 \\ \gamma M_2/2 & -\gamma M_1/2 & 1/\delta_t \end{bmatrix} \tag{16.172}$$

here $\mathbf{M}_s = M_1 \mathbf{a}_x + M_2 \mathbf{a}_y + M_3 \mathbf{a}_z$. Equations (16.170) and (16.171) represent a system of six equations with six unknowns, which can be solved for $\mathbf{m}^{n+1/2}$ and $\mathbf{h}^{n+1/2}$ in a straightforward manner using a symbolic manipulator, such as *Mathematica* [47], or numerically, for a given set of parameters. In symbolic form,

$$\mathbf{h}^{n+1/2} = (\bar{\mathbf{A}}_1^{-1} \cdot \bar{\mathbf{A}}_2 - \bar{\mathbf{A}}_5^{-1} \cdot \bar{\mathbf{A}}_6) \cdot (\bar{\mathbf{A}}_2^{-1} \cdot \mathbf{b}_1 - \bar{\mathbf{A}}_6^{-1} \cdot \mathbf{b}_2) \tag{16.173}$$

and

$$\mathbf{m}^{n+1/2} = (\bar{\mathbf{A}}_2^{-1} \cdot \bar{\mathbf{A}}_1 - \bar{\mathbf{A}}_6^{-1} \cdot \bar{\mathbf{A}}_5) \cdot (\bar{\mathbf{A}}_1^{-1} \cdot \mathbf{b}_1 - \bar{\mathbf{A}}_5^{-1} \cdot \mathbf{b}_2) \tag{16.174}$$

To simplify results, consider the special case when $\mathbf{H}_i$ is zero and $\mathbf{M}_s = M_2 \mathbf{a}_y$. For this situation, (16.170) reduces to [42],

$$h_x^{n+1/2} - \alpha h_z^{n+1/2} = h_x^{n-1/2} + \alpha h_z^{n-1/2} - \frac{\delta_t}{\mu_o} \mathbf{a}_x \cdot \nabla \times \mathbf{E}^n \equiv \beta_1 \tag{16.175}$$

and

$$\alpha h_x^{n+1/2} + h_z^{n+1/2} = h_z^{n-1/2} - \alpha h_x^{n-1/2} - \frac{\delta_t}{\mu_o} \mathbf{a}_z \cdot \nabla \times \mathbf{E}^n \equiv \beta_2 \tag{16.176}$$

where $\alpha = \delta_t \gamma M_2/2$. Hence,

$$h_x^{n+1/2} = \frac{\beta_1 + \alpha\beta_2}{1 + \alpha^2} \tag{16.177}$$

and

$$h_z^{n+1/2} = \frac{\beta_2 - \alpha\beta_1}{1 + \alpha^2} \tag{16.178}$$

The equation for $h_y^{n+1/2}$ is the standard FDTD equation. With respect to (16.171),

$$m_x^{n+1/2} = m_x^{n-1/2} - \alpha(h_z^{n+1/2} + h_z^{n-1/2}) \tag{16.179}$$

and

$$m_z^{n+1/2} = m_z^{n-1/2} + \alpha(h_x^{n+1/2} + h_x^{n-1/2}) \tag{16.180}$$

Or, to eliminate the need for the back-storage of $h_z^{n-1/2}$ and $h_x^{n-1/2}$,

$$m_x^{n+1/2} = m_x^{n-1/2} - \alpha\left(\frac{\beta_2 - \alpha\beta_1}{1 + \alpha^2} + h_z^{n-1/2}\right) \tag{16.181}$$

and

$$m_z^{n+1/2} = m_z^{n-1/2} + \alpha\left(\frac{\beta_1 - \alpha\beta_2}{1+\alpha^2} + h_x^{n-1/2}\right) \tag{16.182}$$

Next consider the spatial discretization. The computation of the curl operators is accomplished in the usual manner. However, due to the anisotropic nature of the ferrite, the Yee grid is no longer optimal for this scheme. To circumvent this problem, spatial averaging of the components is necessary. This has been explained in Sections 16.4 and 16.6.

An error analysis can be affected by assuming a time-harmonic plane wave as a solution to the discretized equations. Due to the mathematical complexity of the task, the derivation of the dispersion relationship is not provided. However, given that the techniques employed herein were found to provide acceptable results when applied to other media types, we surmise that the aforementioned procedure will in fact produce acceptable data as well.

16.11 NONLINEAR DISPERSIVE MEDIA

In the strictest sense, all media are both dispersive and non-linear. To limit the scope of this section, we consider non-linearities associated with optical devices that exhibit the Kerr effect and have a finite response time. To model the finite response time, Debye or Lorentz models may be employed, depending on the physics of the situation. These models, as in the case of linear dispersive media, appear in Maxwell's equations as a polarization current. This polarization vector is further described by a set of auxiliary equations. However, unlike the examples associated with linear media, the auxiliary equations have non-linear source terms that alter the susceptibility of the medium in proportion with power level. In essence then, the major challenge of devising a suitable discretization procedure is not with the dispersive effects but with the non-linear source term.

The investigation of non-linear effects using a modified FDTD algorithm has received considerable attention in the last few years. In this section, we wish to demonstrate one such algorithm developed by Ziolkowski [48]. For other approaches and applications, consult [49]–[51].

A Kerr–Raman–Lorentz model of a non-linear medium is described by the following system of equations:

$$\varepsilon_L \frac{\partial \mathbf{E}}{\partial t} = \nabla \times \mathbf{H} - \frac{\partial \mathbf{P}}{\partial t} \tag{16.183}$$

$$\frac{\partial^2 \mathbf{P}_L}{\partial t^2} = -\Gamma_L \frac{\partial \mathbf{P}_L}{\partial t} - \omega_L^2 \mathbf{P}_L + \varepsilon_o \chi_o \omega_L^2 \mathbf{E} \tag{16.184}$$

and

$$\frac{\partial^2 \chi_{NL}}{\partial t^2} = -\Gamma_R \frac{\partial \chi_{NL}}{\partial t} - \omega_R^2 \chi_{NL} + \varepsilon_R \omega_R^2 |\mathbf{E}|^2 \tag{16.185}$$

where $\mathbf{P} = \mathbf{P}_L + \mathbf{P}_{NL}$ (i.e., the total polarization equals the sum of the linear and nonlinear polarizations), $\mathbf{P}_{NL} = \varepsilon_o \chi_{NL}\mathbf{E}$ and χ_{NL} is the non-linear susceptibility. In addition, ε_L, ε_R, Γ_L, Γ_R, ω_L, ω_R are parameters that describe the combined Lorentz–Raman dispersive medium and its associated frequency-domain response.

From our knowledge of the linear problem, it can be shown that the previous equations are equivalent to

$$\varepsilon_{eff}\frac{\partial \mathbf{E}}{\partial t} = \nabla \times \mathbf{H} - \mathbf{J}_L - \varepsilon_o \Omega_{NL}\mathbf{E} \tag{16.186}$$

$$\frac{\partial \mathbf{J}_L}{\partial t} = -\Gamma_L \mathbf{J}_L - \omega_L^2 \mathbf{P}_L + \varepsilon_o \chi_o \omega_L^2 \mathbf{E} \tag{16.187}$$

$$\frac{\partial \mathbf{P}_L}{\partial t} = \mathbf{J}_L \tag{16.188}$$

$$\frac{\partial \Omega_{NL}}{\partial t} = -\Gamma_R \Omega_{NL} - \omega_R^2 \chi_{NL} + \varepsilon_R \omega_R^2 |\mathbf{E}|^2 \tag{16.189}$$

and

$$\frac{\partial \chi_{NL}}{\partial t} = \Omega_{NL} \tag{16.190}$$

where $\varepsilon_{eff} = \varepsilon_L + \varepsilon_o \chi_{NL}$.

With the equations written in state equation form, it is apparent that leapfrog concepts postulated for the linear Lorentzian case still apply. That is, both central difference and average approximations are invoked, where necessary, and (16.187) and (16.189) are advanced simultaneously and leaped with the simultaneous advancement of (16.186), (16.188), and (16.190): The first set of equations are

$$\frac{\mathbf{J}_L^{n+1/2} - \mathbf{J}_L^{n-1/2}}{\delta_t} = -\Gamma_L\left(\frac{\mathbf{J}_L^{n+1/2} + \mathbf{J}_L^{n-1/2}}{2}\right) - \omega_L^2 \mathbf{P}_L^n + \varepsilon_o \chi_o \omega_L^2 \mathbf{E}^n \tag{16.191}$$

and

$$\frac{\Omega_{NL}^{n+1/2} - \Omega_{NL}^{n-1/2}}{\delta_t} = -\Gamma_R\left(\frac{\Omega_{NL}^{n+1/2} + \Omega_{NL}^{n-1/2}}{2}\right) - \omega_R^2 \chi_{NL}^n + \varepsilon_R \omega_R^2 |\mathbf{E}^n|^2 \tag{16.192}$$

As for the next set of equations, we first calculate the next time-step for χ_{NL}

$$\frac{\chi_{NL}^{n+1} - \chi_{NL}^n}{\delta_t} = \Omega_{NL}^{n+1/2} \tag{16.193}$$

Since χ_{NL} is known at both the previous and current time-steps and ε_{eff} is a function of χ_{NL},

$$\varepsilon_{eff}^{n+1/2} = \varepsilon_L + \varepsilon_o(\chi_{NL}^{n+1} + \chi_{NL}^n)/2 \tag{16.194}$$

Hence, the second set of equations are

$$\varepsilon_{eff}^{n+1/2}\left(\frac{\mathbf{E}^{n+1} - \mathbf{E}^n}{\delta_t}\right) = \nabla \times \mathbf{H}^{n+1/2} - \mathbf{J}_L^{n+1/2} - \varepsilon_o \Omega_{NL}^{n+1/2}\left(\frac{\mathbf{E}^{n+1} + \mathbf{E}^n}{2}\right) \tag{16.195}$$

and

$$\frac{\mathbf{P}^{n+1} - \mathbf{P}^{n}}{\delta_t} = \mathbf{J}_L^{n+1/2} \tag{16.196}$$

With respect to the spatial discretization, the components of **E**, **J**, and **P** are conjoined on the Yee cell; **H** occupies its usual place. As for the scalars χ_{NL} and Ω_{NL}, these are positioned at the center of the cell. By doing so, the calculation for the energy term $|\mathbf{E}|^2$ in (16.192) is easily accomplished by central averages:

$$\begin{aligned} 4\left|\mathbf{E}\right|\Big|^2_{i+1/2,j+1/2,k+1/2} &= \left[E_x\Big|_{i+1,j+1/2,k+1/2} + E_x\Big|_{i,j+1/2,k+1/2}\right]^2 \\ &+ \left[E_y\Big|_{i+1/2,j+1,k+1/2} + E_y\Big|_{i+1/2,j,k+1/2}\right]^2 \\ &+ \left[E_z\Big|_{i+1/2,j+1/2,k+1} + E_z\Big|_{i+1/2,j+1/2,k}\right]^2 \end{aligned} \tag{16.197}$$

The aforementioned scheme is entirely second order in its implementation. Given the non-linear nature of the equations, a numerical permittivity is not deducible. However, for *small* non-linear effects, we may use the numerical permittivity relationship of L-DIM1 (see Table 16.9) as a basis for an approximate dispersion analysis. And, from that analysis, the parameters δ_t and δ_x may be set. Consider as an example the choice to sample the electromagnetic wave at 15 points per wavelength and consider a domain that is 10 wavelengths long. That is, in one direction, the domain spans 150 cells. Now with respect to memory, it is observed that the scheme is memory intensive—15 data points per cell in a full three-dimensional simulation. This extreme use of memory is not, however, due to the scheme, for we note that the physics requires 14 dependent variables to describe the phenomenon and the algorithm requires one back-storage of χ_{NL}. Thus, for a domain of $150 \times 150 \times 150$ cells, there exist roughly 50 million unknowns to advance per time step. Certainly 50 million unknowns per time step is quite excessive for desk-top machines. However, in the age of the supercomputer and the advent of the parallel architectures, such a number is within scope of these machines. Hence, the ability to model a non-linear optical device in terms of the full Maxwellian vectors system is certainly possible.

16.12 SUMMARY

A survey of some of the more popular schemes associated with FDTD analysis of electromagnetic wave propagation in complex media has been given. These schemes have been characterized in terms of their numerical permittivities in conjunction with a numerical dispersion relationship. In addition, the CFL number of each scheme is given or postulated when such information has been provided in the literature. In some cases, discussions on memory requirements have also been provided.

It is not the intention of this chapter to rank such schemes in terms of *best* or *worst*. Such terms are subjective, since best or worst designators require one to quantify both the tangible (e.g., accuracy) and intangible (e.g., simplicity) features of a scheme into a single number. Instead, each scheme should be analyzed with a particular application in mind and, from that analysis, an informed selection can be made.

ACKNOWLEDGMENTS

Gratitude is offered to Professor J. Schneider for his technical assistance with the section on anisotropic media, to Mr. R. Nelsen for his technical assistance with the numerical permittivity expressions, to Mrs. E. Young for the proofing of the manuscript, and to Dr. J. Shang for his hospitality at the Wright Patterson Air Force Base, where the manuscript was prepared.

REFERENCES

[1] K. S. Yee, "Numerical Solution of Initial Boundary Value Problems Involving Maxwell's Equations in Isotropic Media," *IEEE Trans. Ant. Propagat.*, **14** (3), pp. 302–307, 1966.

[2] A. Taflove and M. E. Brodwin, "Numerical Solution of Steady-state Electromagnetic Scattering Problems Using the Time-dependent Maxwell's Equations," *IEEE Trans. Microwave Theory Tech.*, **23** (8), pp. 623–630, 1975.

[3] A. Ishimaru, *Electromagnetic Wave Propagation, Radiation, and Scattering*, Prentice-Hall, Englewood Cliffs, NJ, 1991.

[4] R. Richtmyer and K. Morton, *Difference Methods for Initial-Value Problems*, Wiley, New York, 1967.

[5] J.-P. Berenger, "A Perfectly Matched Layer for the Absorption of Electromagnetic Waves," *J. Comput. Phys.*, **114** (1), pp. 185–200, 1994.

[6] R. W. Ziolkowski, "The design of Maxwellian absorbers for numerical boundary conditions and for practical applications using engineered artificial materials," *IEEE Trans. Ant. Propagat.*, **45** (4), pp. 656–671, 1997.

[7] R. F. Harrington, *Time-Harmonic Electromagnetic Fields*, McGraw-Hill, New York, 1961.

[8] J. R. Wait, *Electromagnetic Waves in Stratified Media*, Pergamon Press, Oxford, U.K., 1970.

[9] K. S. Kunz and R. J. Luebbers, *The Finite Difference Time Domain Method for Electromagnetics*, CRC Press, Boca Raton, FL, 1993.

[10] Y. Liu, "Fourier Analysis of Numerical Algorithms for the Maxwell Equations," *31st Aerospace Sciences Meeting & Exhibit*, **AIAA-93-0368**, Reno, NV, 1993.

[11] A. Taflove and K. R. Umashankar, "The Finite-difference Time-domain Method for Numerical Modeling of Electromagnetic Wave Interactions," *Electromagnetics*, **10** (1/2), pp. 105–126, 1990.

[12] D. A. Anderson, J. C. Tannehill, and R. H. Pletcher, *Computational Fluid and Mechanics and Heat Transfer*, Taylor & Francis, Bristol, PA, 1984.

[13] J. B. Schneider and S. Hudson, "The Finite-difference Time-domain Method Applied to Anisotropic Material," *IEEE Trans. Ant. Propagat.*, **41** (7), pp. 994–999, 1993.

[14] J. D. Jackson, *Classical Electrodynamics*, John Wiley and Sons, New York, 1975.

[15] D. M. Sullivan, "Frequency-dependent FDTD Methods Using T transforms," *IEEE Trans. Ant. Propagat.*, **40** (10), pp. 1223–1230, 1992.

[16] D. M. Sullivan, "Z-transform Theory and the FDTD Method," *IEEE Trans. Ant. Propagat.*, **44** (1), pp. 28–34, 1996.

[17] W. H. Weedon and C. M. Rappaport, "A General Method for FDTD Modeling of Wave Propagation in Arbitrary Frequency-dependent Media," *IEEE Trans. Ant. Propagat.*, **45** (3), pp. 401–410, 1997.

[18] J. L. Young, "A Full Finite Difference Time Domain Implementation for Radio Wave Propagation in a Plasma," *Radio Sci.*, **29** (6), pp. 1513–1522, 1994.

[19] L. J. Nickisch and P. M. Franke, "Finite-difference Time-domain Solution of Maxwell's Equations for the Dispersive Ionosphere," *IEEE Ant. Propagat. Mag.*, **34** (5), pp. 33–39, 1992.

[20] S. A. Cummer, "An Analysis of New and Existing FDTD Methods for Isotropic Cold Plasma and a Method for Improving Their Accuracy," *IEEE Trans. Ant. Propagat.*, **45** (3), pp. 392–400, 1997.

[21] R. M. Joseph, S. C. Hagness, and A. Taflove, "Direct Time Integration of Maxwell's Equations in Linear Dispersive Media with Absorption for Scattering and Propagation of Femtosecond Electromagnetic Pulses," *Opt. Lett.*, **16** (18), pp. 1412–1414, 1991.

[22] R. J. Luebbers, F. Hunsberger, and K. S. Kunz, "A Frequency-dependent Finite-difference Time-domain Formulation for Transient Propagation in Plasma," *IEEE Trans. Ant. Propagat.*, **39** (1), pp. 29–34, 1991.

[23] D. F. Kelley and R. J. Luebbers, "Piecewise Linear Recursive Convolution for Dispersive Media Using FDTD," *IEEE Trans. Ant. Propagat.*, **44** (6), pp. 792–797, 1996.

[24] L. B. Felsen and N. Marcuvitz, *Radiation and Scattering of Waves*, Prentice-Hall, Englewood Cliffs, NJ, 1973.

[25] F. Hunsberger, R. Luebbers, and K. Kunz, "Finite- difference Time-domain Analysis of Gyrotropic Media—I: Magnetized Plasma," *IEEE Trans. Ant. Propagat.*, **40** (12), pp. 1489–1495, 1992.

[26] T. Kashiwa, N. Yoshida, and I. Fukai, "Transient Analysis of a Magnetized Plasma in Three-dimensional Space," *IEEE Trans. Ant. Propagat.*, **36** (8), pp. 1096–1105, 1988.

[27] J. R. Wait, "On the Theory of Reflection of Electromagnetic Waves from the Interface Between a Compressible Magnetoplasma and a Dielectric," *Radio Sci.*, **68D** (11), pp. 1187–1191, 1964.

[28] D. M. Sullivan, "A Frequency-dependent FDTD Method for Biological Applications," *IEEE Trans. Microwave Theory Tech.*, **40** (3), pp. 532–539, 1992.

[29] O. P. Gandhi and C. M. Furse, "Currents Induced in the Human Body for Exposure to Ultrawideband Electromagnetic Pulses," *IEEE Trans. Electromagn. Compat.*, **39** (2), pp. 174–180, 1997.

[30] C. M. Furse, J.-Y. Chen, and O. P. Gandhi, "The Use of Frequency-dependent Finite-difference Time-domain Method for Induced Current and SAR Calculations for a Heterogeneous Model of the Human Body," *IEEE Trans. Electromagn. Compat.*, **36** (2), pp. 128–133, 1994.

[31] W. D. Hurt, "Multiterm Debye Dispersion Relations for Permittivity of Muscle," *IEEE Trans. Biomed. Engr.*, **32** (1), pp. 60–64, 1985.

[32] T. Kashiwa, N. Yoshida, and I. Fukai, "A Treatment by the Finite-difference Time-domain Method of the Dispersive Characteristics Associated with Orientation Polarization," *IEICE Transactions*, **E73** (8), pp. 1326–1328, 1990.

[33] J. L. Young, "Propagation in Linear Dispersive Media: Finite Difference Time-domain Methodologies," *IEEE Trans. Ant. Propagat.*, **43** (4), pp. 422–426, 1995.

[34] P. G. Petropoulos, "The Computation of Linear Dispersive Electromagnetic Waves," *Applied Computational Electromagnetics Society Journal*, **11** (1), pp. 8–16, 1996.

[35] M. Okoniewski, M. Mrozowski, and M. A. Stuchly, "Simple Treatment of Multi-term Dispersion in FDTD," *IEEE Microwave Guided Wave Lett.*, **7** (5), pp. 121–123, 1997.

[36] R. Luebbers, F. P. Hunsberger, K. S. Kunz, R. B. Standler, and M. Schneider, "A Frequency-dependent Finite-difference Time-domain Formulation for Dispersive

Materials," *IEEE Trans. Electromagn. Compat.*, **32** (3), pp. 222–227, 1990.

[37] P. G. Petropoulos, "Stability and Phase Error Analysis of FD-TD in Dispersive Dielectrics," *IEEE Trans. Ant. Propagat.*, **42** (1), pp. 62–69, 1994.

[38] T. Kashiwa and I. Fukai, "A Treatment by the FD-TD Method of the Dispersive Characteristics Associated with Electronic Polarization," *Microwave Optic. Tech. Lett.*, **3** (6), pp. 203–205, 1990.

[39] R. J. Luebbers and F. Hunsberger, "FDTD for *N*th-order Dispersive Media," *IEEE Trans. Ant. Propagat.*, **40** (11), pp. 1297–1301, 1992.

[40] R. Luebbers, D. Steich, and K. Kunz, "FDTD Calculation of Scattering from Frequency-dependent Materials," *IEEE Trans. Ant. Propagat.*, **41** (9), pp. 1249–1257, 1993.

[41] J. L. Young, A. Kittichartphayak, Y. M. Kwok, and D. Sullivan, "On the Dispersion Errors Related to $(FD)^2$-TD Type Schemes," *IEEE Trans. Microwave Theory Tech.*, **43** (8), pp. 1902–1910, 1995.

[42] M. Okoniewski and E. Okoniewski, "FDTD Analysis of Magnetized Ferrites: A More Efficient Algorithm," *IEEE Microwave Guided Wave Lett.*, **4** (6), pp. 169–171, 1994.

[43] J. A. Pereda, L. A. Vielva, A. Vegas, and A. Prieto, "A Treatment of Magnetized Ferrites Using the FDTD Method," *IEEE Microwave Guided Wave Lett.*, **3** (5), pp. 136–138, 1993.

[44] J. A. Pereda, L. A. Vielva, A. Vegas, and A. Prieto, "FDTD Analysis of Magnetized Ferrites: An Approach Based on the Rotated Richtmyer Difference Scheme," *IEEE Microwave Guided Wave Lett.*, **3** (9), pp. 322–324, 1993.

[45] J. W. Schuster and R. J. Luebbers, "Finite Difference Time Domain Analysis of Arbitrarily Biased Magnetized Ferrites," *Radio Sci.*, **31** (4), pp. 923–929, 1996.

[46] A. Reinex, T. Monediere, and F. Jecko, "Ferrite Analysis Using the Finite-difference Time-domain (FDTD) Method," *Microwave and Opt. Tech. Lett.*, **5** (13), pp. 685–686, 1993.

[47] S. Wolfram, *Mathematica*, Addison-Wesley, Redwood City, CA, 1992.

[48] R. W. Ziolkowski and J. B. Judkins, "Applications of the Nonlinear Finite Difference Time Domain (NL-FDTD) Method to Pulse Propagation in Nonlinear Media: Self-focusing and Linear–Nonlinear Interfaces, *Radio Sci.*, **28** (5), 901–911, 1993.

[49] P. M. Goorjian, A. Taflove, R. M. Joseph, and S. C. Hagness, "Computational Modeling of Femtosecond Optical Solitons from Maxwell's Equations," *IEEE J. Quantum Electronics*, **28** (10), pp. 2416–2422, 1992.

[50] D. M. Sullivan, "Nonlinear FDTD Formulation Using *Z* Transforms," *IEEE Trans. Microwave Theory Tech.*, **43** (3), pp. 676–682, 1995.

[51] R. M. Joseph, P. M. Goorjian, and A. Taflove, "Direct Time Integration of Maxwell's Equations in Two-dimensional Dielectric Waveguides for Propagation and Scattering of Femtosecond Electromagnetic Solitons," *Opt. Lett.*, **18** (7), pp. 491–493, 1993.

Chapter
17 A NEW COMPUTATIONAL ELECTROMAGNETICS METHOD BASED ON DISCRETE MECHANICS

Rodolfo E. Diaz, Franco DeFlaviis, Massimo Noro, and Nicolaos G. Alexopoulos

ABSTRACT

It has been known since the late nineteenth century that electromagnetic phenomena can be simulated by the dynamics of a mechanical system, as long as the Hamiltonian of the mechanical and electromagnetic systems coincide. George Francis FitzGerald's 1885 model of electromagnetic propagation meets this requirement. When implemented numerically it leads to a new computational method for the simulation of the interaction between electromagnetic fields and physically realistic media, that gives a principal role to the vector potential. In this model, the properties of free space and the constitutive properties of material media reside in physically distinct entities. Based on Discrete Mechanics, the simulation propagates the fields in time according to Newton's laws, as strains and stresses due to the interacting forces in the medium. There is no inherent limitation to linear phenomena: dispersive as well as non-linear materials are just as easily accommodated. Furthermore, a limitation to purely electromagnetic entities need not be assumed.

After derivation of the model for propagation in the presence of linear dispersive magnetodielectrics, its generality is demonstrated through the time-domain simulation of a non-linear transmission line for the generation of solitons. No assumption is made about a small signal limit. Finally, the ability of the model to simulate the interaction between matter and waves is illustrated by considering collisions between ponderable media and electromagnetic pulses. The Doppler shift of the pulse reflected from a moving object is correctly obtained, as well as a classical Compton Effect for the case of a very light object. Through collisions between an electromagnetic pulse and a penetrable dielectric obstacle, it is shown that conservation of total momentum arises naturally out of the simulation, without the need to impose any special moving boundary conditions on the field. The method is ideal for the simulation of environments where electromagnetic forces are on par with, and even may modify, the mechanical forces of the devices involved. Applications to electromagnetic propulsion, microelectromechanics, and nanomachines are envisioned.

17.1 INTRODUCTION

Why a new method? The reader would be well justified in asking the question. After all, with the proliferation of inexpensive computational power has come a proliferation of computational electromagnetic methods, methods applicable in the time domain, the frequency domain, hybrid time-frequency domains, based on integral equations, differential equations, minimized error functionals, discretized in finite differences, finite elements, finite volumes, and so on. Indeed, we could wonder with Shankar if the need for a new method is not comparable to the need for a publication of the Latest Table of Integers [1]. However, there is a reason why a new way of looking at things seems to be in order at this time: we live in the age of the engineered complex material. "Stealth technology" requires aerospace composite structures to be tailored for mechanical as well as electromagnetic properties. Atomic Force Microscopy is based on miniaturized electromechanical devices whose structures are, by design, deformable in the presence of the very fields they are measuring. Nanomachines are envisioned that would work inside biological environments where the chemical forces might be comparable to the electrical forces. By definition, the real engineering problems of our day are multidisciplinary. And the common denominator is the underlying material: changeable, deformable, and, in general, non-linear.

Under these conditions, what requirements should be levied on a new computational approach? The general nature of the non-linearities that could be involved in real problems, together with the requirement to preserve causality, suggests a time-domain method. However, since many applications employ frequency as a fundamental variable, the computational code must be able to generate such information efficiently. Therefore, dispersive material properties must be included to allow for single-pulse solutions of the entire spectrum, where appropriate. In engineering design problems it is often desirable to examine a feature of the system in isolation before addressing the entire system. This is best accomplished with a "nearest-neighbor" propagation approach (such as finite difference PDE codes), as long as efficient absorbing boundary conditions can be implemented. Finally, the computational approach must allow the incorporation of material media parameters beyond the electromagnetic constitutives, such as thermal and mechanical properties. Their interactions with the electromagnetic fields must preserve causality and energy conservation with a minimum of assumptions and approximations. In summary, the computational approach must provide a faithful and automatic simulation of reality.

In the analysis of non-electromagnetic phenomena such an approach has existed for some time and is known as Discrete Mechanics [2]. It differs from other PDE methods in the way the "Particle in a Cell" method of modeling hydrodynamic problems differs from the Euler or Lagrange grid approaches [3]. The system is not assumed to be subject to a particular differential equation in which the various material regions represent different boundary conditions. Instead the whole system is recognized to consist of material objects composed of "quasi-molecules" that interact with each other strictly through Newtonian Forces. The elastic behavior of a metal bar automatically results from the Lennard–Jones-like potential that keeps the molecules at a preferred distance from each other. The conduction of heat through the same metal bar is a result of the transmission of vibrational kinetic energy through that same lattice of molecules. Thus, one computer code can simulate the vibrational eigenmodes of an object, the transmission of heat through its length, and even its dissociation (melting) under extreme conditions [2], (pp. 73–86). Since Discrete Mechanics already contains the multidisciplinary aspect of reality that we seek, and since it places its emphasis on the materials involved, it is the natural framework for the development of the new Computational Electromagnetics approach to be presented in this chapter.

In the next section, the two-dimensional mechanical model of electromagnetism conceived by George Francis FitzGerald in 1885 [4], is used to derive the Discrete Mechanics version of Maxwell's equations. It is shown that these equations correspond to a finite difference leapfrog scheme in which the two principal variables are the electric field and the vector potential. In contrast with conventional FDTD, all computed field quantities exist at the same point in space, and their propagation in time exhibits less grid dispersion.

The next step in the development is the inclusion of general dispersive media. This is done in Section 17.3. The intuitive mechanical analogues used clearly display the separation between the energy stored in free space and the energy stored in the material. It is shown that this method of simulation corresponds to a finite difference discretization in which the polarization vector is the principal variable needed to describe the material media. Simulations of the interaction between electromagnetic waves and dispersive materials in canonical configurations are used to prove the validity of the formulation.

Finally, in Section 17.4, the method is extended to non-linear and non-electromagnetic phenomena. Using the expression for the capacitance of a typical varactor as a function of applied voltage, a non-linear transmission line for the generation of solitons is constructed and compared to the work of Jager [5]. Since the material properties reside on entities separate from free space, it follows that the movement of objects through space must consist of the displacement of these properties across the free-space grid. When the material properties are moved they must carry with them their corresponding instantaneous momenta. And since these momenta are coupled to free space through "friction" terms, it follows automatically that the movement of dielectrics through an electromagnetic field creates new (induced) electromagnetic fields in free space. These motional fields occur automatically as a result of the simulation. They do not need to be added to the equations.

The intuitive power of the model and its pedagogical value can be seen in these final examples. Given the physical entities used to represent the fields in the medium there is only one possible definition for the force exerted on ponderable media by electromagnetic fields. It becomes clear that "magnetic pressure" [6] does not exist. Similarly, despite the often cited differences of opinion about what constitutes the electromagnetic momentum inside a material body [7] the simulation proves that there is only one definition that conserves total momentum in the interaction between waves and general material media.

17.2 THE FITZGERALD MECHANICAL MODEL

Consider an array of pulleys of moment of inertia $I_{i,j}$ connected to each other through rubber bands of elasticity $k_{i+1/2,j+1/2}$. An angular velocity pulse applied to the central pulley (i,j) propagates outwards by action and reaction to all pulleys of the system. In fact, the kinetic energy of the central pulley is transferred as potential elastic energy to the neighboring rubber bands. The total force due to tension and compression of the rubber bands exerts on the pulleys an angular acceleration α according to Newton's second law: their angular velocity ω increases by αdt after a time-step dt (see Fig. 17.1) [8].

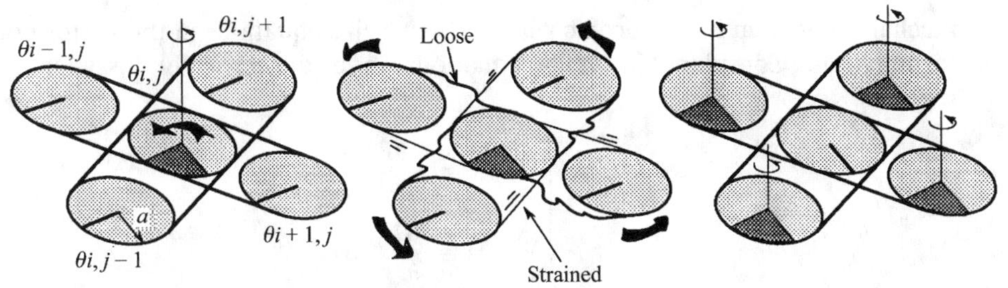

Figure 17.1 Action–reaction mechanism of propagation of motion; the dark gray represents the spanned angle due to the initial action and consequent reaction.

Application of Greenspan's approach [2] to this mechanical system yields a set of finite difference equations. We shall derive here the equation of motion for the simple one-dimensional case; the extension to two dimensions is trivial. The overall torque acting on the ith pulley is caused by the combination of the four elastic restoring forces:

$$T_i = (F_i^{(1)} + F_{i-1}^{(1)})a - (F_i^{(2)} + F_{i-1}^{(2)})a = k_i a^2(\theta_{i+1} - \theta_i) + k_{i-1}a^2(\theta_{i-1} - \theta_i) \tag{17.1}$$

where k is the elastic constant of the rubber band, a the radius of the pulley, and θ the displacement angle. We now introduce discrete mechanics by writing the equation for the angular velocity of the ith pulley after a time-step n:

$$\left(\frac{\Delta\omega}{\Delta t}\right)_i^n = \frac{k_i a^2}{I_i}(\theta_{i+1}^n - \theta_i^n) + \frac{k_{i-1}a^2}{I_i}(\theta_{i-1}^n - \theta_i^n) \tag{17.2}$$

If we now identify the diameter of the pulley with the grid size, $\Delta x = a$, the previous expression becomes

$$\left(\frac{\Delta\omega}{\Delta t}\right)_i^n = \frac{(\Delta x)^4}{I_i}\left\{\frac{k_i\dfrac{\theta_{i+1}^n - \theta_i^n}{\Delta x} - k_{i-1}\dfrac{\theta_i^n - \theta_{i-1}^n}{\Delta x}}{\Delta x}\right\} \tag{17.3}$$

It can be shown that moment of inertia itself is a quantity proportional to the fourth power of the grid size, therefore we are allowed to take the limit for $\Delta t \to 0$ and $\Delta x \to 0$, and recover the continuum differential equation which we write below together with the definition of angular velocity:

$$\begin{cases} \dfrac{\partial\omega}{\partial t} = \dfrac{a^2}{I}\dfrac{\partial}{\partial x}\left(ka^2\dfrac{\partial\theta}{\partial x}\right) \\ \dfrac{\partial\theta}{\partial t} = \omega \end{cases} \tag{17.4}$$

FitzGerald showed that the two-step mechanical model mimics exactly the generation and propagation of electromagnetic waves in Maxwell's curl equations. In this spirit, the inertia of the pulley I represents the medium permittivity (ε), and the elastic constant k represents the medium inverse permeability ($1/\mu$). In fact, we show that the mechanical system is isomorphic with the Maxwell's equations in the vector potential formulation. Consider Maxwell's equations in a dielectrically lossy source-free region:

$$\begin{cases} \nabla \times \mathbf{E} = -\dfrac{\partial \mathbf{B}}{\partial t} & (17.5) \\ \nabla \times \mathbf{H} = \dfrac{\partial \mathbf{D}}{\partial t} + \sigma \mathbf{E} & (17.6) \\ \nabla \cdot \mathbf{D} = 0 & (17.7) \\ \nabla \cdot \mathbf{B} = 0 & (17.8) \end{cases}$$

From (17.8)

$$\mathbf{B} = \nabla \times \mathbf{A} \tag{17.9}$$

and combining (17.5) and (17.9)

$$\mathbf{E} = -\frac{\partial \mathbf{A}}{\partial t} - \nabla \phi \tag{17.10}$$

It is well known that the two variables ϕ and $\mathbf{A}$ can be decoupled with the use of the Lorentz-Gauge condition,

$$\nabla \cdot \mathbf{A} = -\varepsilon\mu \frac{\partial \phi}{\partial t} \tag{17.11}$$

If we restrict our analysis to the case above, taking the curl of both sides of (17.9) and combining with (17.6), we obtain

$$\frac{\partial(\varepsilon \mathbf{E})}{\partial t} = \nabla \times \frac{1}{\mu} \nabla \times \mathbf{A} - \sigma \mathbf{E} \tag{17.12}$$

Therefore, summarizing (17.10), (17.11), and (17.12) we obtain

$$\begin{cases} \dfrac{\partial(\varepsilon \mathbf{E})}{\partial t} = \nabla \times \dfrac{1}{\mu} \nabla \times \mathbf{A} - \sigma \mathbf{E} \\ \dfrac{\partial \mathbf{A}}{\partial t} = -\mathbf{E} - \nabla \phi \\ \nabla \cdot \mathbf{A} = -\varepsilon\mu \dfrac{\partial \phi}{\partial t} \end{cases} \tag{17.13}$$

For the one-dimensional case, all the space derivative in the y and z direction disappears, so that we obtain

$$\begin{cases} \dfrac{\partial E_z}{\partial t} = \dfrac{1}{\varepsilon}\left[\dfrac{\partial}{\partial x}\left(\dfrac{1}{\mu}\dfrac{\partial A_z}{\partial x}\right)\right] - \dfrac{\sigma}{\varepsilon} E_z \\ \dfrac{\partial A_z}{\partial t} = -E_z \end{cases} \tag{17.14}$$

Which is essentially the same as the mechanical equation (17.4) once we identify the correspondence between the mechanical and electromagnetic quantities.

In the mechanical model the presence of a dielectric medium ($\varepsilon \neq \varepsilon_0$) is taken

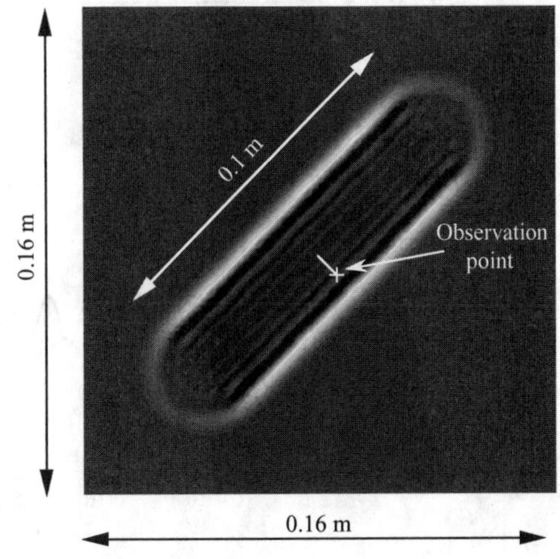

Figure 17.2 Field distribution and experimental geometry for the grid dispersion numerical experiment.

into account by increasing the moment of inertia of the pulley; while the presence of a magnetically permeable medium ($\mu \neq \mu_0$) is modeled by modifying the elasticity of the rubber (the elastic constant of the spring).

In the following numerical experiment we evaluate the cutoff frequency due to grid dispersion in a computer experiment and we compare it with the same quantity evaluated in the analogous FDTD experiment. A locally plane wave tilted 45° with respect to the grid main axes is generated as shown in Fig. 17.2. The geometry was chosen so that the wave is almost plane when it leaves the observation point. The mesh size is coarse enough such that grid dispersion will occur. As shown in [9], for a wave propagating at 45 degrees along the FDTD grid, the grid dispersion effects are expected for $\Delta s \approx 0.5\lambda$, where λ is the wavelength of the electromagnetic wave examined. The corresponding cutoff frequency for our particular choice of Δs is $f_{\max} = 150$ GHz.

Due to grid dispersion, the original pulse is altered and develops a ringing tail as shown in Fig. 17.3. DFT analysis gives the spectrum content which is shown in Fig. 17.4, where we plot the magnitude of the two spectra, and the corresponding phase versus frequency. Notice the sudden drop of the magnitude at the cutoff frequency and the corresponding loss of phase linearity. The observed FDTD cutoff frequency agrees with the predicted value of 150 GHz, while our technique exhibits the cutoff at 170 GHz, corresponding to a 13% improvement in bandwidth. The same bandwidth (170 GHz) can be achieved by FDTD if the mesh size is reduced to $\Delta s = 8.5 \cdot 10^{-4}\, m$ corresponding to a memory increase of 38% and execution time increase of 38%.

17.3 EXTENSION TO DEBYE MATERIALS

In this section we extend the Fitzgerald mechanical model to represent frequency-dependent complex materials, where the dispersive behavior can be modeled by expressing both the displacement and the magnetization with a sum of Debye terms

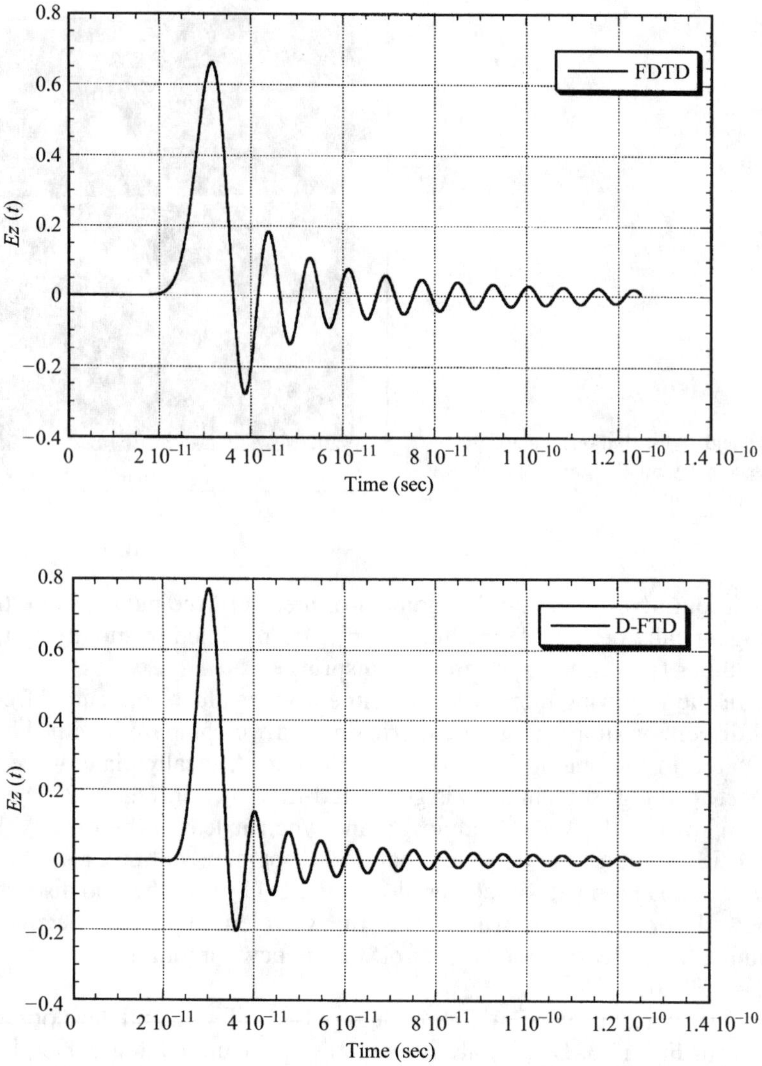

Figure 17.3 Time response of FDTD due to locally plane wave (top). Same quantity for D-FTD formulation (bottom).

[10]. It is key to realize that the dispersive effect in a dielectric Debye material can be ascribed to the polarization:

$$\mathbf{D} = \varepsilon_\infty \mathbf{E} + \mathbf{P} \tag{17.15}$$

where the polarization changes in time in the following way:

$$\frac{\partial \mathbf{P}}{\partial t} = \frac{\varepsilon_s - \varepsilon_\infty}{\tau}\mathbf{E} - \frac{1}{\tau}\mathbf{P} \tag{17.16}$$

In (17.16) ε_s and ε_∞ are the dielectric constants of zero (static) and "infinite" frequency, and τ is the relaxation time constant. The mechanical analogue of the

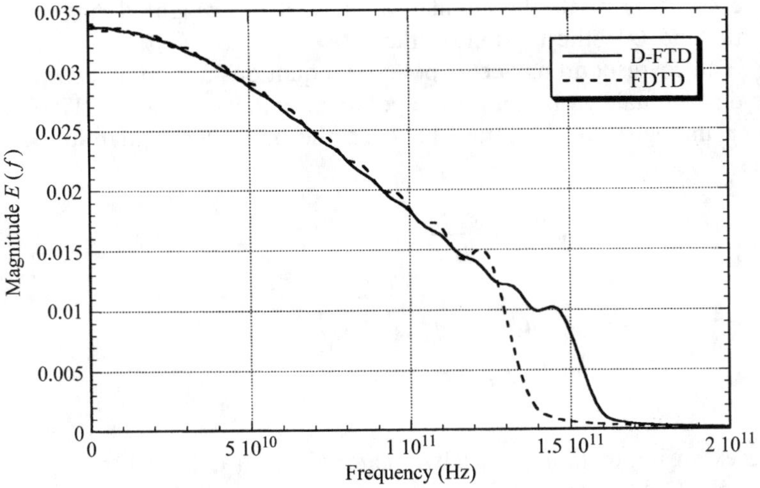

Figure 17.4 Spectrum content of the signal observed at the same observation point for FDTD and D-FTD.

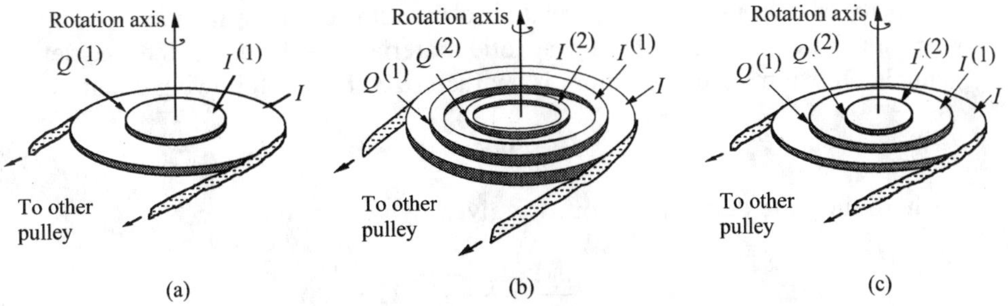

Figure 17.5 Mechanical model for single and multiple electrical Debye materials.

polarization is the angular velocity of an additional weighted ring (of moment of inertia $I^{(1)}$) resting on the pulley and connected to it through a coefficient of friction ($Q^{(1)}$) as shown in Fig. 17.5a. The time constant of the Debye pole τ corresponds to the ratio $I^{(1)}/Q^{(1)}$. Action and reaction of the top pulley then simulate the storage and dissipation of polarization. The new set of equations for this system can be derived from the original equations in the one-dimensional case, by adding an inertial reaction due to the top pulley which is coupled to the bottom one through a friction coefficient [11]

$$\begin{cases} \dfrac{\partial \omega_i^{(1)}}{\partial t} = \dfrac{a^2 Q_i^{(1)}}{I_i^{(1)}}(\omega_i - \omega_i^{(1)}) \\ \dfrac{\partial \omega_i}{\partial t} = \dfrac{a^2}{I_i}\dfrac{\partial}{\partial x}\left(ka^2\dfrac{\partial \theta}{\partial x}\right) - \dfrac{a^2 Q_i^{(1)}}{I_i}(\omega_i - \omega_i^{(1)}) \\ \dfrac{\partial \theta_i}{\partial t} = \omega_i \end{cases} \tag{17.17}$$

Here $\omega_i^{(1)}$ represents the angular velocity of the weighted ring, $I_i^{(1)}$ the moment of inertia, and $Q_i^{(1)}$ the friction coefficient.

The corresponding vector potential equations are obtained upon introduction of the time-domain equation for the polarization; by comparing (17.17) to (17.4) several other analogies can be found between the mechanical quantities and the electrical ones

$$\begin{cases} \dfrac{\partial P_z}{\partial t} = \dfrac{\varepsilon_s - \varepsilon_\infty}{\tau} E_z - \dfrac{1}{\tau} P_z \\ \dfrac{\partial E_z}{\partial t} = \dfrac{1}{\varepsilon_\infty} \dfrac{\partial}{\partial x} \left(\dfrac{1}{\mu} \dfrac{\partial A_z}{\partial x} \right) + \dfrac{\varepsilon_s - \varepsilon_\infty}{\tau \varepsilon_\infty} \left(\dfrac{P_z}{\varepsilon_s - \varepsilon_\infty} - E_z \right) \\ \dfrac{\partial A_z}{\partial t} = -E_z \end{cases} \tag{17.18}$$

The extension to multiple Debye materials is straightforward, in fact we can mimic multiple Debye terms by adding rings surmounting the original pulley arranged concentrically as visible in Fig. 17.5b. An alternative, and completely equivalent, mechanical arrangement for modeling multiple Debye materials consists of a stack of pulleys resting on the bottom one as in Fig. 17.5c. Each pulley is coupled to the one right below and above it through an appropriate friction coefficient. Again the angular velocity of the pulleys can be related to the different polarizations.

In the case of dispersive magnetic materials, we focus on the magnetization. A single Debye magnetic medium is characterized by the following equation:

$$\mathbf{B} = \mu_\infty \mathbf{H} + \mathbf{M} \tag{17.19}$$

where the time dependence of $\mathbf{M}$ is given by

$$\frac{\partial \mathbf{M}}{\partial t} = \frac{\mu_s - \mu_\infty}{\tau} \mathbf{H} - \frac{1}{\tau} \mathbf{M} \tag{17.20}$$

Here μ_s and μ_∞ are the permeability at zero (static) and "infinite" frequency and τ is the relaxation time constants. In order to include this material property in the mechanical model, we modify the rubber bands' elastic property: an additional spring is connected in series with the original one. The new spring is connected with a device immersed in a viscous fluid, so that the oscillation of the second spring is damped according to the viscous coefficient of the fluid γ, see Fig. 17.6a. We introduce the time constant of the magnetic Debye material τ as corresponding to the ratio γ/k_1.

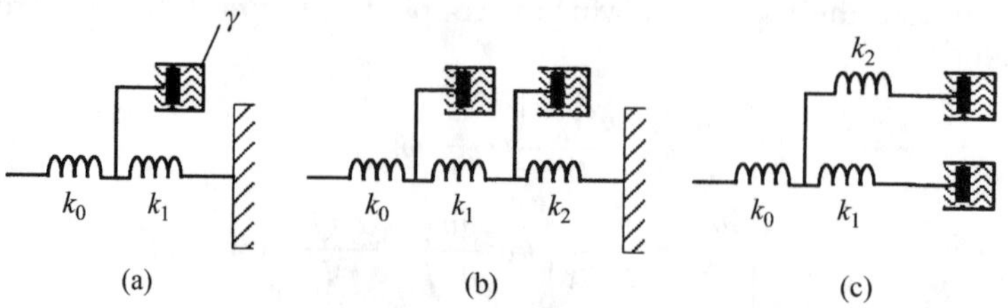

Figure 17.6 Mechanical model for single and multiple magnetic Debye materials.

When the spring system connecting two neighboring pulleys is elongated by

$$x_{eq} = a\Delta\theta = a(\theta_{i+1} - \theta_i) \tag{17.21}$$

the elastic restoring force obeys the following differential equation:

$$\frac{\partial F}{\partial t} = -k_0 \frac{\partial x_{eq}}{\partial t} - \frac{k_0 k_1}{\gamma} x_{eq} - \frac{k_0 + k_1}{\gamma} F \tag{17.22}$$

In the continuum limit, the new set of equations for the system becomes

$$\begin{cases} \dfrac{\partial F}{\partial t} = -a^2 k_0 \dfrac{\partial}{\partial t}\left(\dfrac{\partial \theta}{\partial x}\right) - \dfrac{a^2 k_0 k_1}{\gamma}\dfrac{\partial \theta}{\partial x} - \dfrac{k_0 + k_1}{\gamma} F \\ \dfrac{\partial \omega_i}{\partial t} = -\dfrac{a^2}{I_i}\dfrac{\partial F}{\partial x} \\ \dfrac{\partial \theta_i}{\partial t} = \omega_i \end{cases} \tag{17.23}$$

In (17.23) F represents the restoring force, k_0 and k_1 the elastic constant of the springs, and γ the damping coefficient. We can derive the analogous vector potential equations in the electromagnetic language. In fact in the time domain a magnetic Debye permeability can be written as follows:

$$\mu H_y = \mu_\infty H_y + \frac{\mu_s - \mu_\infty}{1 + \tau\partial/\partial t} H_y = \mu_\infty H_y + \frac{\mu_M}{1 + \tau\partial/\partial t} H_y \tag{17.24}$$

Upon introduction of the vector potential

$$\hat{B}_y = -\frac{\partial \hat{A}_z}{\partial x} \tag{17.25}$$

Equations (17.24) and (17.25) are combined, to yield a differential equation for H_y:

$$\frac{\partial H_y}{\partial t} = -\frac{1}{\mu_\infty}\frac{\partial}{\partial t}\frac{\partial A_z}{\partial x} - \frac{1}{\mu_\infty \tau}\frac{\partial A_z}{\partial x} - \frac{\mu_s}{\mu_\infty \tau} H_y \tag{17.26}$$

In one dimension $(1/\mu)(\partial A_z/\partial x) = -H_y$; therefore, the time-evolution equation for the electric field

$$\frac{\partial E_z}{\partial t} = -\frac{1}{\varepsilon}\frac{\partial}{\partial x}\left(\frac{1}{\mu}\frac{\partial A_z}{\partial x}\right) \tag{17.27}$$

can be replaced by the equations

$$\begin{cases} \dfrac{1}{\mu}\dfrac{\partial A_z}{\partial x} = -H_y \\ \dfrac{\partial E_z}{\partial t} = \dfrac{1}{\varepsilon}\dfrac{\partial}{\partial x}(H_y) \end{cases} \tag{17.28}$$

where it is clear that the top equation is a differential equation for H_y. Combining (17.28) with (17.26) we obtain

$$\begin{cases} \dfrac{\partial H_y}{\partial t} = -\dfrac{1}{\mu_\infty}\dfrac{\partial}{\partial t}\left(\dfrac{\partial A_z}{\partial x}\right) - \dfrac{1}{\mu_\infty \tau}\dfrac{\partial A_z}{\partial x} - \dfrac{\mu_s}{\mu_\infty \tau} H_y \\ \dfrac{\partial E_z}{\partial t} = \dfrac{1}{\varepsilon}\dfrac{\partial H_y}{\partial x} \\ \dfrac{\partial A_z}{\partial t} = -E_z \end{cases} \tag{17.29}$$

Direct comparison of (17.23) to (17.29) shows several other analogies between the mechanical quantities and the electrical ones. Multiple magnetic Debye relaxations can be modeled in two symmetric arrangements. One possibility is to add several damped oscillator devices to the ideal spring representing the elasticity of the original rubber band as in Fig. 17.6b. Alternatively, one could arrange the additional damping devices connected in parallel with the original ideal spring as in Fig. 17.6c. The two approaches are different, but lead to the same result; therefore one is free to choose the model that suits best the material one wants to model. Nothing prevents the electric and magnetic Debye properties from appearing simultaneously in the same material. In fact these are extremely interesting from both a theoretical and practical point of view because of the possible application in the modeling of particular absorbing materials.

The mechanical model is built combining the features needed to model the electric and magnetic frequency-dependent materials. Each pulley is surmounted by one (possibly many) weighted rings coupled to the bottom supporting pulley though a friction coefficient. The action and reaction of the top rings simulate the energy storage of the polarization vector, and ensures the electrical Debye character. On the other hand the frequency-dependent magnetic character is ensured by the peculiar elastic properties of the connecting rubber bands, which can be schematically represented as an ideal spring connected in series with one (possibly many) damped oscillators. By using the many analogies we have derived before, we do not need to re-derive the equation of motion for the quite complex system, but merely add to the basic equation of motion all the necessary ingredients in a heuristic fashion:

$$\begin{cases} \dfrac{\partial \omega_i^{(1)}}{\partial t} = \dfrac{a^2 Q_i}{I_i^{(1)}}(\omega_i - \omega_i^{(1)}) \\ \dfrac{\partial F}{\partial t} = -a^2 k_0 \dfrac{\partial}{\partial t}\dfrac{\partial \theta}{\partial x} - \dfrac{a^2 k_0 k_1}{\gamma}\dfrac{\partial \theta}{\partial x} - \dfrac{k_0 + k_1}{\gamma} F \\ \dfrac{\partial \omega_i}{\partial t} = -\dfrac{a^2}{I_i}\dfrac{\partial F}{\partial x} - \dfrac{a^2 Q_i}{I_i}(\omega_i - \omega_i^{(1)}) \\ \dfrac{\partial \theta_i}{\partial t} = \omega_i \end{cases} \tag{17.30}$$

By proceeding along the same lines, it is possible to build a generic complex material, characterized by multiple electric and magnetic Debye poles, by simply adding the proper number of rings and damped oscillators to the simple mechanical model.

As an application we present here the result of an echo experiment, where a plane wave is backscattered from the air–medium interphase. The reflection coefficient as a function of frequency is therefore calculated as the ratio of the reflected and incoming spectra. The calculated reflection coefficient is compared to the corresponding analytical quantity obtained in the frequency domain from the following relation:

$$|R(\omega)| = \left| \frac{\eta_1 - \eta_0}{\eta_1 + \eta_0} \right| \tag{17.31}$$

where η_0 and η_1 are the characteristic impedance of free space and the complex medium, respectively, and are given by

$$\eta_0 = \sqrt{\frac{\mu_0}{\varepsilon_0}} \tag{17.32a}$$

$$\eta_1 = \sqrt{\frac{\mu_1(\omega)}{\varepsilon_1(\omega)}} \tag{17.32b}$$

As an example we have modeled the air–water interphase, where the complex permittivity of water can be approximated by a single-order Debye relaxation. We have used $\varepsilon_s = 81.0$, $\varepsilon_\infty = 1.8$, and $\tau_0 = 9.4 \cdot 10^{-12}$ sec [10]. In Fig. 17.7 the calculated

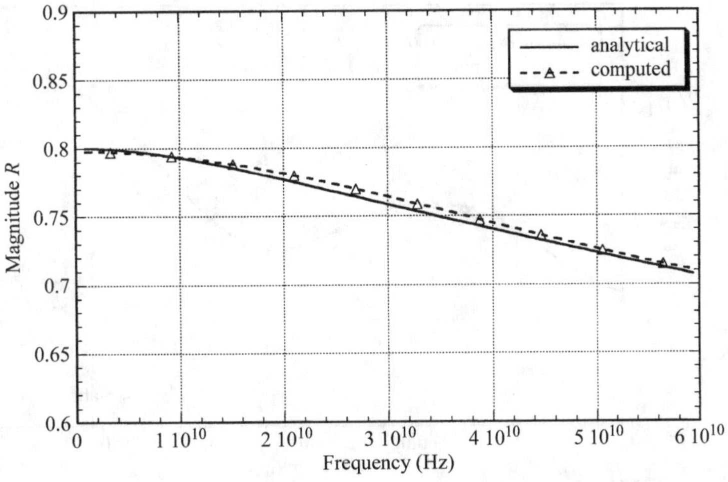

Figure 17.7 Reflection coefficient for single electric Debye interface.

reflection coefficient is compared to the analytical one. On the other hand, lossy ferrites can be properly modeled as magnetic Debye materials with at least two decay times. For this echo experiment we have chosen $\mu_s = 100.0$, $\mu_\infty = 4.0$, $\tau_1 = 10^{-11}$ sec, $\tau_2 = 5.3 \cdot 10^{-11}$ sec, and we plot the reflection coefficient calculated from the simulation together with the analytical one in the same Fig. 17.8.

Dielectric lossy materials can be viewed as the limiting case of dielectric Debye materials where the moments of inertia tend to infinity, i.e., the additional rings (pulleys) on top of the original one are essentially fixed in space. We present here numerical experiment results for the internal electric field of a uniform, circular

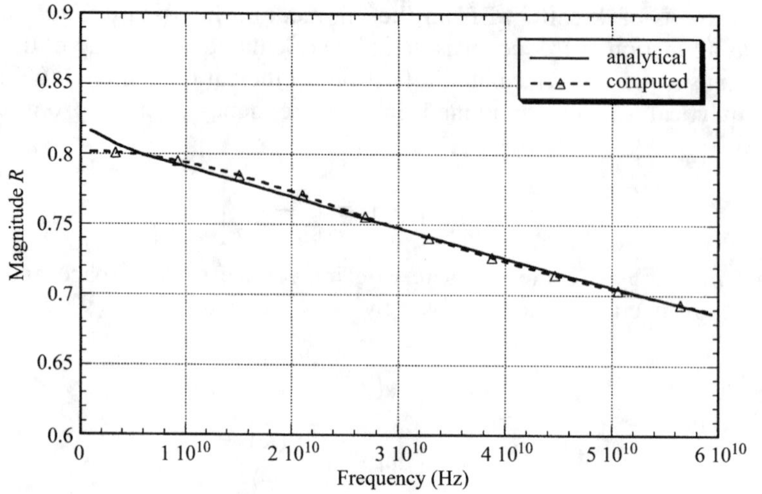

Figure 17.8 Reflection coefficient for double magnetic Debye interface.

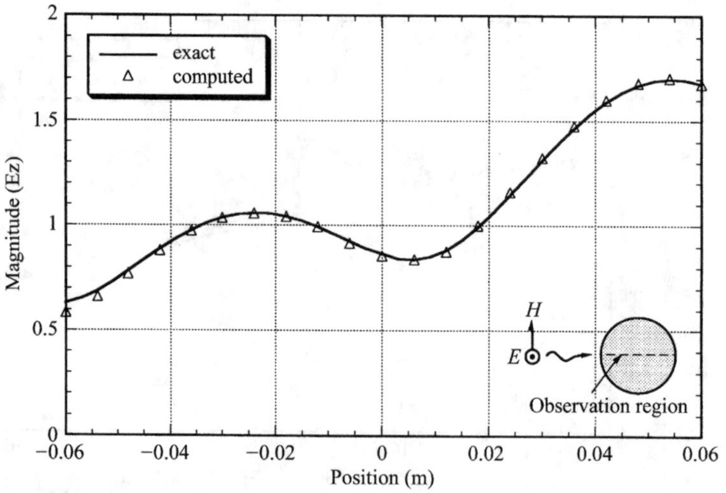

Figure 17.9 Comparison between computed and exact solution of the inner field for a lossless dielectric cylinder.

dielectric cylindrical scatterer. The incident radiation is a wave TM with respect to the cylinder symmetry axis. The plane wave source obtained as series of Gaussian pulses for the z component of the field invests the cylinder and is scattered from it. In Fig. 17.9 we plot the intensity of the scattered electric field along the cylinder diameter in a direction orthogonal to the incident plane wave, and compare it to the analytical solution calculated using the summed series technique as in Jones [12]. We show the results for a lossy cylinder in Fig. 17.10.

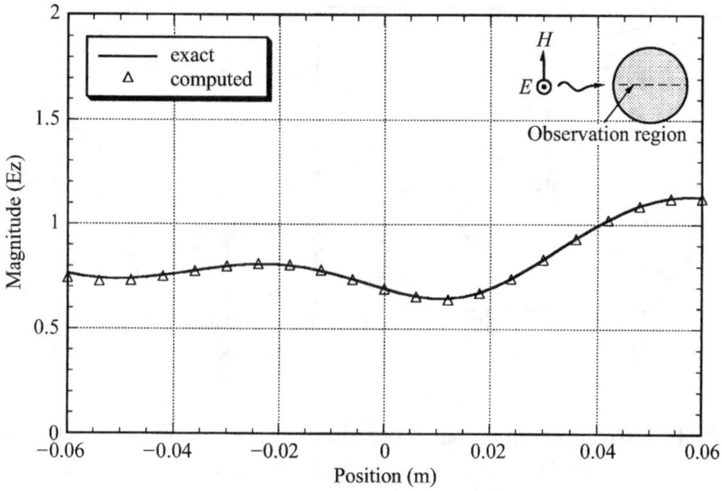

Figure 17.10 Comparison between computed and exact solution of the inner field for a lossy dielectric cylinder.

17.4 THE SIMULATION OF GENERAL PONDERABLE MEDIA

17.4.1 Non-Linear Dielectrics

In the search for faster and more compact communication schemes for circuits and systems, the use of ultrashort time-domain pulses seems inevitable. The generation of such pulses through non-linear transformation methods has been an area of active research, both analytic and experimental, for many years. In [5] Jager reports the successful modeling of the generation of ultrashort soliton pulses by a transmission line periodically loaded with non-linear capacitors. His basic assumption is that in the limit of a small signal and weak dispersion the capacitance can be approximated by

$$C(V) = C_0(1 - 0.1V) \tag{17.33}$$

In practice, the voltage-dependent capacitance required can be obtained using Schottky or varactor diodes [13]. Over a small range of voltages the capacitance of such a device obeys (17.33). However its response over the complete range of voltages is illustrated in Fig. 17.11, and is best modeled by an equation of the form,

$$C = C_0\left\{1.0 + \frac{1}{1 + \frac{\beta}{\pi^2}\left[\frac{\pi}{2} + \arctan(\alpha V - V_0)\right]^2}\right\} \tag{17.34}$$

Limiting this investigation to the case of a one-dimensional transmission line, the periodic loading of the line is simulated by periodically loading the free-space lattice with a non-linear material obeying equation (17.34), with $C_0 = 1.0$, $\beta = 5$, $\alpha = 2$, and

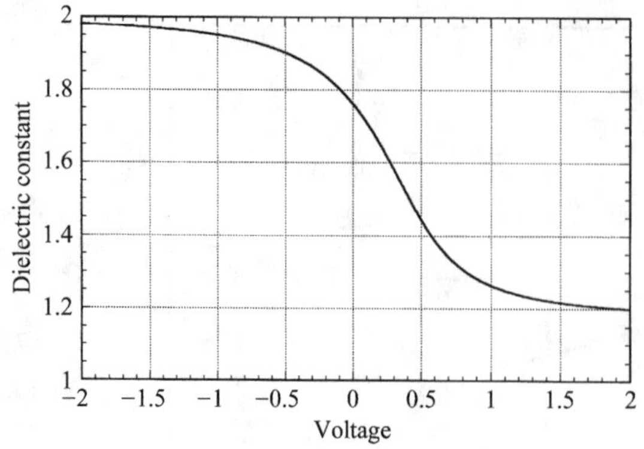

Figure 17.11 Non-linear dependence of dielectric constant versus bias voltage.

$V_0 = 1.0$. In this case we choose the dual definition in which the spring constant models the elastance (inverse permittivity) of the medium. Then, ignoring dissipation to follow [5], the mechanical equations used are

$$k_1 = 1 + \frac{5}{\pi^2}\left[\frac{\pi}{2} + \arctan(2x_0 - 1)\right]^2 \tag{17.35a}$$

$$F_{spring} = a(\theta_i - \theta_{i-1})\frac{k^{(0)}k^{(1)}}{k^{(0)} + k^{(1)}} \tag{17.35b}$$

As before, k_1 represents the spring constant of the material spring and x_0 the elongation of the free-space spring. Thus, for very negative voltages across the device ($x_1 \rightarrow -\infty$) the permittivity tends to the maximum value $\varepsilon = 2$. Whereas for very high voltages it drops to 1.2. The periodic loading is accomplished by placing the material in two adjacent cells of every five, as suggested in Fig. 17.12. Figures 17.13a–d show the results of the simulation under an input trapezoidal voltage pulse. At low voltages, the transmission line just behaves as a linear free-space transmission line periodically loaded with obstacles of nearly constant 1.76 relative permittivity. The wave sees an average effective dielectric constant of 1.3. As the voltage is increased we see the

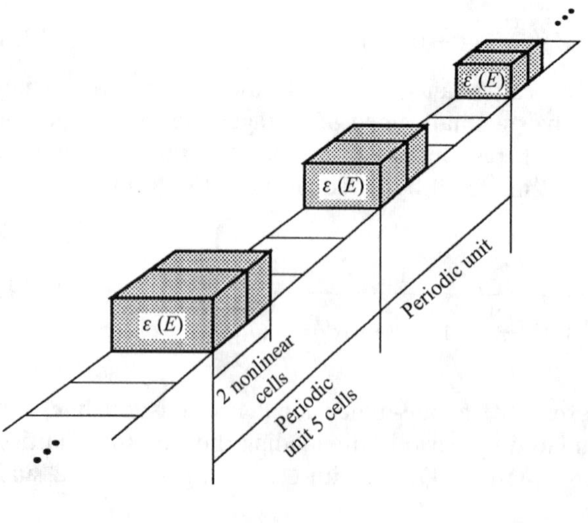

Figure 17.12 Non-linear medium model for one-dimensional transmission line for soliton generation.

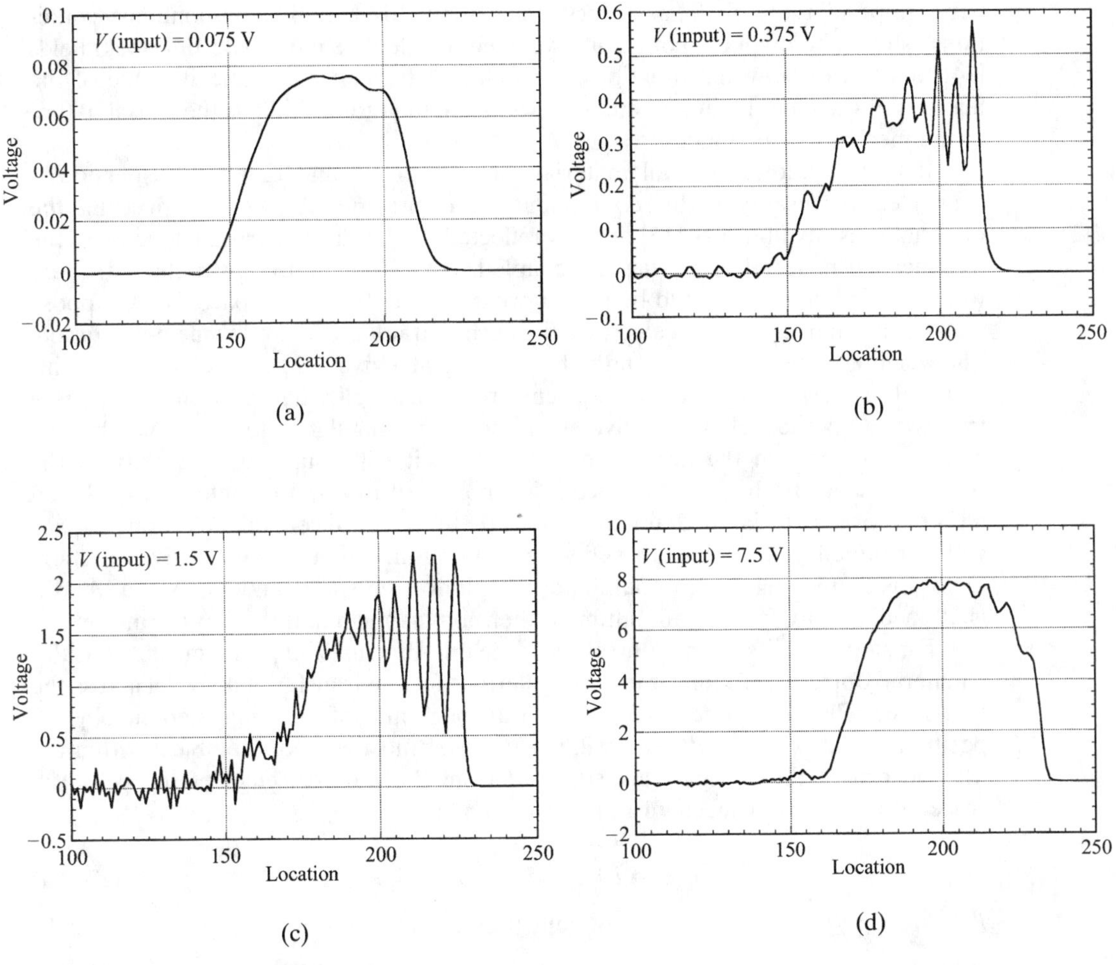

Figure 17.13 Snap shot of the transmitted pulse in the non-linear medium as input voltage is varied.

onset of soliton formation. At voltages that elicit rapid changes in capacitance, we see very strong soliton generation. But then, at very high voltages, well into the saturation region of the devices, the soliton generation slows down and eventually ceases.

17.4.2 How Should Moving Ponderable Media be Modeled?

Since the weighted rings of Fig. 17.5 represent the material medium, the movement of that medium through space must be represented by a sliding of those rings across the free-space grid. That movement must transfer both the moment of inertia and the angular momentum contained in the material, incrementally, from one free-space cell to another. The correct simulation of this transfer is all that should be required in a Discrete Mechanics approach to obtain a valid model of the interaction between ponderable media and electromagnetic fields. No "moving boundary conditions" should be invoked, nor should we need to calculate motional Minkowski fields [14] to add to Maxwell's equations. It should all happen automatically. One of the features of the model that guarantees it will happen automatically is the causal nature of the mechanical analogues.

It is well known that all materials are dispersive and that this dispersion is connected to causality by the requirement that over some region of the spectrum, the material must exhibit loss [15]. This is reflected in the mechanical analogues by the fact that the material properties are coupled to free space through friction. If there were no friction, there would be no materials distinct from free space. Now, suppose a material is immersed in an electric field, such that it has acquired some polarization. The weighted ring I_1 will be found to be spinning at a given angular velocity ω_1. If this material is now moved to an adjacent free-space cell, that portion of the ring transferred to the cell will arrive with the same angular velocity it had before, whereupon, through the accompanying friction, it will couple that velocity to the free-space cell. By the inverse process, if a region of free space contains a field, and a portion of a material medium moves into that free-space cell, the field in free space will be coupled to the newly arrived weighted ring through the newly arrived friction. Fields inside moving media, continually couple to free space, and vice versa. And it is all mediated by friction, the ultimate thermodynamic manifestation of causality.

For simplicity, if we limit ourselves to the one-dimensional problem, and consider a material object x cells long, then the equations for the transfer of material across the free-space cells are encoded as follows: At the beginning of each time-step the object's position is updated by $v*dt$. This will place the leftmost edge of the object within the ith grid cell of the space at a distance d from the leftmost edge of the cell. The susceptibility transfer mechanism is then defined by

$$L = i + 1 - d, \qquad R = 1 - L \tag{17.36a}$$

$$I^{(1)} = I^{(1)}_{object},\ Q = Q_{object} \qquad \text{for cells fully occupied} \tag{17.36b}$$

$$I^{(1)} = I^{(1)}_{object} L,\ Q = Q_{object} L \qquad \text{for the partially occupied leftmost cell} \tag{17.36c}$$

$$I^{(1)} = I^{(1)}_{object} R,\ Q = Q_{object} R \qquad \text{for the partially occupied rightmost cell} \tag{17.36d}$$

and all other cells are reset to free space. The spin transfer mechanism is similarly defined. Given that the object is moving at velocity v, the following quantity is evaluated for all cells fully occupied:

$$u_i = \omega_i^{(1)}(1 - |v|) + |v|\,\omega^{(1)}_{i-sign(v)} \tag{17.36e}$$

then all the corresponding $\omega_i^{(1)}$ are reset by

$$\omega_i^{(1)} = u_i \tag{17.36f}$$

The $sign(v)$ function in the subscript in (17.36e) ensures that the transfer occurs in the direction of motion. The end cells cannot take any $\omega^{(1)}$ spin from or give any $\omega^{(1)}$ spin to free space. Throughout, careful book keeping must be performed, to reset all values properly whenever the object crosses integer cell boundaries.

Since we know electromagnetic fields exert forces on material objects, we must now formulate the force equation in terms of the variables available in the model. It is sometimes stated in textbooks [6] that an imbalance in magnetic pressure across the face of a metallic object exerts the moving force. This is unacceptable on two counts. In a realistic complex environment, an integration over all space of the magnetic fields on either side of the boundary cannot possibly give a local force. But worse yet, the supposed imbalance in tangential magnetic pressure cannot exist across any real boundary because tangential magnetic fields are always continuous. This would seem to leave as the only viable force expression, the Lorentz force:

$$\mathbf{F} = \mathbf{I} \times \mathbf{B} \cdot \mathbf{dl} \tag{17.37}$$

However, if the object is not a metal, what is the meaning of $\boldsymbol{I}$? The only unambiguous answer that the mechanical system will admit is that the force must be due to the interaction between the material body's stored angular momentum and the field in free space. That is, the current in equation (17.37) stands for Maxwell's displacement current. More precisely, it can only stand for the polarization current $\partial \mathbf{P}/\partial t = [\partial(\varepsilon - 1)\mathbf{E}]/\partial t = [\partial(\mathbf{I}^{(1)}\omega)]/\partial t$. And so, the generalized Lorentz force exerted on an arbitrary magnetodielectric by an electromagnetic field must be

$$\mathbf{F} = \frac{\partial(\varepsilon - 1)\mathbf{E}}{\partial t} \times \mathbf{B}_0 - \frac{\partial(\mu - 1)\mathbf{H}}{\partial t} \times \mathbf{D}_0 \tag{17.38}$$

where the subscript "0" in the field densities indicates the field densities in free space: $\mathbf{B}_0 = \mu_0 \mathbf{H}$, $\mathbf{D}_0 = \varepsilon_0 \mathbf{E}$. This convention is consistent with reported experimental results that electrons inside permeable metals do not experience an enhanced Lorentz force due to the permeability of the metal in which they move. Heaviside anticipated this result in 1905 [16].

Given that this expression has been known for almost a century, it is surprising that there is still discussion about the proper form for the electromagnetic momentum inside material objects [7]. That the above expression is self consistent can be ascertained by considering the problem of a plane electromagnetic wave incident on a semi-infinite, low-loss, dielectric half space. Since in free space and inside the dielectric the electric and magnetic fields are always in phase, the time derivatives of (17.38) represent quantities that are in phase quadrature with the other field. As a result, no net work can be performed by the passage of the wave according to the force law of (17.38). Therefore, the transmitted and reflected waves obtained from the Fresnel coefficients, taken together with the incident wave, must conserve both total energy and total momentum during the interaction. It is easy to show that the only expression for the instantaneous electromagnetic momentum per unit volume that can do this is

$$\mathbf{p} = \frac{1}{2}(\mathbf{D} \times \mathbf{B}_0 + \mathbf{D}_0 \times \mathbf{B}) \tag{17.39}$$

In general, the only time an electromagnetic wave exerts a force on a low-loss dieletric is when it is trying to exit the object, since then, the reflected wave interferes with the incident wave inside the medium, creating a standing wave in which the force of (17.38) does not vanish on the average. Whenever the standing wave reaches a peak, the force is at a maximum. Whenever the standing wave vanishes, the force is zero. In the simulations reported here, the resulting pulsating nature of the exerted force is clearly seen. The reason metallic objects are so efficiently moved by electromagnetic fields lies in the fact that the electric and magnetic fields inside metallic objects are always 90° out of phase with each other, making the force of (17.38) nonzero at all times. (Heaviside does point out that in the case of the moving medium, a net force should arise, even inside a lossless dielectric, of the form $v\mathbf{B}_0(\partial\mathbf{D}/\partial t)$ [16, p. 193].)

17.4.3 Collisions Between Pulses and Objects

Consider first a conducting object of small mass. Upon being hit by an electromagnetic pulse in the simulation it acquired a velocity of 0.03c. The Compton effect would require the reflected wave to suffer a red shift of 0.971. Figure 17.14a shows the incident spectrum (solid line) and the reflected spectrum (dashed line). The net computed shift is 0.975. If the same object now moving at 0.03c is hit by an identical pulse coming from the other direction, it would be expected to stop the object and Doppler shift blue by 1.03. Indeed, the final computed velocity was 0.0 and, as Fig. 17.14b shows, a shift of 1.02 was obtained. If the mass of the object is halved, and it is moving at the same velocity, 0.03c, then a collision with an identical pulse should reverse the velocity of the object to −0.03c. Since the object's kinetic energy is unchanged, this would be an elastic collision and the net Doppler shift should be zero. This is what happened in Fig. 17.14c.

In the above examples the object was an impenetrable metal obstacle which suffered acceleration during the entire interaction with an incident Gaussian pulse. A more general and realistic problem is the interaction of a penetrable dielectric obstacle with a pulse that carries several cycles of a "center" frequency. This case is particularly interesting when the frequency in the pulse is either a resonance or an antiresonance of the obstacle. Again, for simplicity, we limit ourselves to one dimension, so that the penetrable object is a dielectric slab. The first antiresonance of such an object occurs when its electrical thickness is one quarter of the wavelength. At that frequency we find a maximum of the reflection coefficient and thus expect the object to interact strongly with the pulse, and to acquire a significant amount of momentum from the wave. On the other hand, at the first resonance, which occurs when the object is one-half wavelength thick, the reflection coefficient vanishes. The wave should be mostly transmitted through the object and should therefore impart little momentum to it. Note that the above expectations are based on the frequency-domain, steady-state, equations for a dielectric slab. In the simulation, a time-domain interaction will take place, governed by (17.38). After the collision takes place and the reflected and transmitted pulses recede from the object, the results should agree with these expectations, with the one caution that the finite extent of the incident pulse implies that the slab interacts with a finite spectrum of frequencies and not just the center frequency.

To first order, when the kinetic energy acquired by the obstacle is small compared to the energy in the wave (non-relativistic limit), conservation of energy demands that the field transmission and reflection coefficients obey

$$T^2 + R^2 = 1.0 \tag{17.40}$$

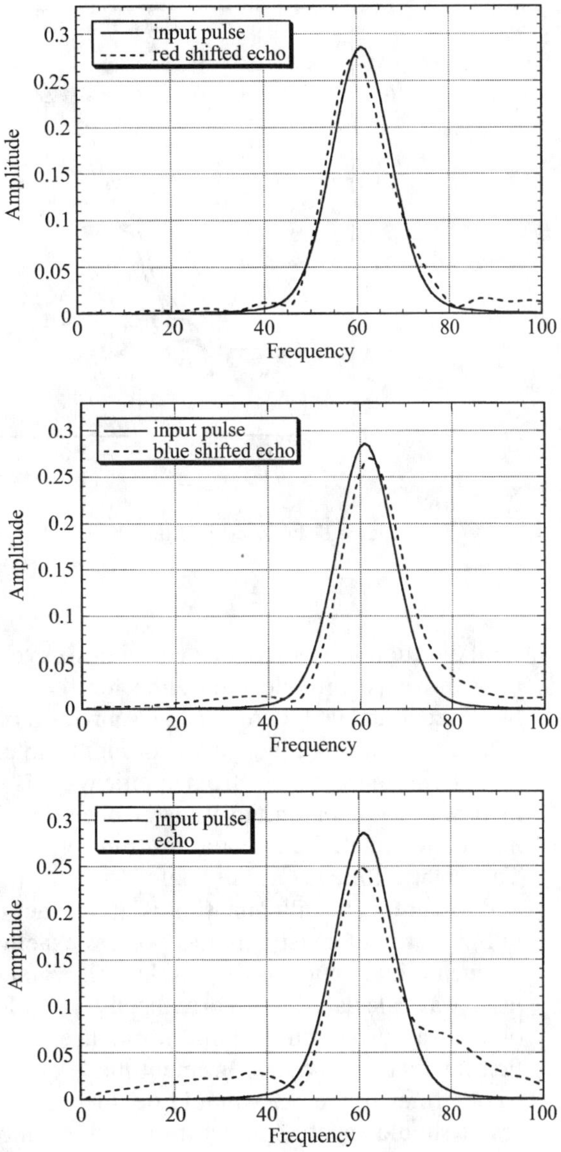

Figure 17.14 Incident spectrum (solid line) and reflected spectrum (dashed line) (top) Red shifted echo (center) blue shifted echo (bottom) zero shifted echo.

assuming unity incident power; whereas conservation of momentum demands that

$$T^2 - R^2 + m\nu = 1.0 \tag{17.41}$$

normalizing to unity incident momentum. Therefore,

$$m\nu = 2R^2 \tag{17.42}$$

The mechanical momentum acquired by the obstacle must be directly proportional to the momentum carried by the reflected wave.

In the simulation, the obstacle had a fixed length. Two incident pulses of equal

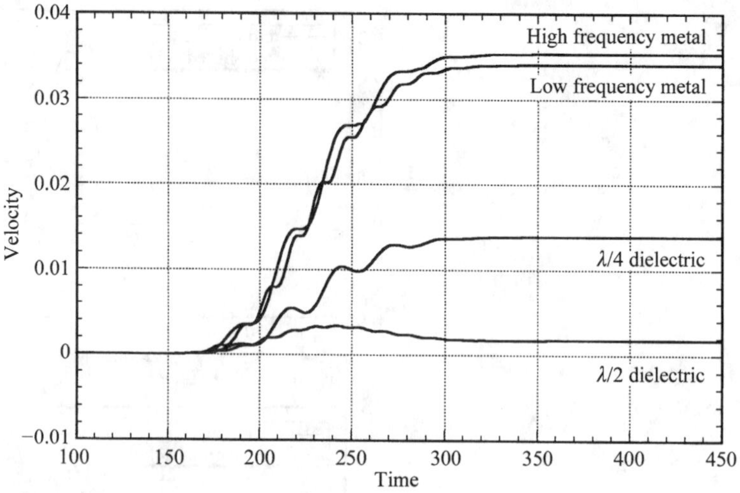

Figure 17.15 Velocity of the obstacle during the collision plotted versus time.

total duration were used, one carrying six cycles of its center frequency and the other only three. To provide a baseline for the experiment, the obstacle was first made highly conducting. Figure (17.15) is a record of the velocity of the obstacle during the collisions. The two upper curves of Fig. 17.15 correspond to the metallic obstacle case. Because a sinusoidal electromagnetic wave is made up of a train of peaks and valleys of power, the momentum transfer occurs in spurts. Since (17.38) is always positive inside the metal, the momentum transfer, that is the velocity increase, is monotonic. The slight difference in the final velocity of the metallic object under high-frequency and low-frequency illumination is an artifact of the simulation, because the incident pulses were not constructed to possess exactly the same total energy. Using the high frequency interaction as the baseline, the total momentum in the reflected electromagnetic was calculated and normalized to 1.0. The total mechanical momentum of the object was also calculated and normalized to 1.0. In other words, we considered this first interaction as a measurement made on the obstacle to determine its mass. Once so determined, the mass should be invariant and all momenta normalized to this first result should satisfy conservation of total momentum as expressed in (17.42). The low-frequency interaction with the metal object gave a final mechanical momentum of 1.037. The total electromagnetic momentum in that reflected wave was 1.029. The agreement with (17.42) is within 1% of the momentum contained in the incident pulse.

The obstacle was then given a dielectric constant of 4.0, making it one-quarter wave thick at the low frequency and one-half wavelength thick at the high. Figure (17.15) shows that when the slab is a quarter wave thick it acquires a significant amount of net momentum. It does so in spurts as before, but note that it is not monotonic. There are times during the interaction that the slab loses momentum to the wave. The net result is a gain of normalized mechanical momentum equal to 0.407. The total electromagnetic momentum in the reflected wave is 0.399. The proportionality of (17.42) is again satisfied to within 1% of the incident momentum.

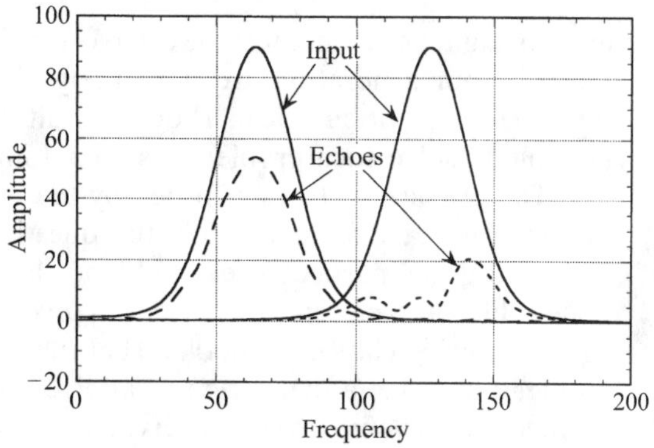

Figure 17.16 Spectrum of the input (solid line) and echoes (dashed lines).

Finally, the case when the slab is a half wave thick is, as expected, the case when the obstacle acquires the least momentum. The non-zero amount of momentum acquired is the result of the finiteness of the pulse and therefore the finite linewidth of the contained spectrum. In the time domain, the obstacle acquires momentum during the first half of the interaction and then tries to give it back during the second half. The total mechanical momentum acquired is 0.052. The total electromagnetic momentum in the reflected wave is 0.042. Again the result is correct within 1% of the total incident momentum. In this case, because we are working close to a "null" of the problem, the exchanged momenta in question are only a few percent of the incident momentum, and numerical effects such as grid dispersion may become contributors to the overall noise. In Fig. 17.16 we plot the spectrum of the input (solid line) and echoes (dashed line) corresponding to the $\lambda/2$ and $\lambda/4$ slab thickness.

That total momentum is conserved in these simulations proves that no special boundary conditions or motional fields need to be added to the computational electromagnetics code to correctly model the physics of wave-matter interactions. All that is required in Discrete Mechanics is that the correct energy-conserving formulation of the forces be implemented and that the motion of matter be correctly modeled. If this is done, all known physical effects should follow automatically.

17.5 CONCLUSION

A Computational Electromagnetics tool based on the principles of Discrete Mechanics has been presented. Although formally it is a PDE approach, much like FDTD, it possesses certain differences that commend it as a new and useful method. The fully condensed nature of its field components, the reduced dispersion apparent in Fig. 17.4, and even the ease with which non-linearities are incorporated explicitly, could be cited as significant advantages. But many of these features, with the correct translation, could also be incorporated into FDTD. What we feel is the greatest contribution of this method to electromagnetic engineering is that it points out a new way of thinking about these problems. It invokes the mechanical intuition natural to the engineer as an essential tool of discovery in, potentially, all fields of engineering. By demonstrating the correct and simultaneous coupling of mechanical and electromagnetic phenomena, we have emphasized that the engineering challenges faced by our generation are not purely electromagnetic, or purely mechanical, or purely thermal, or chemical. They are multidisciplinary. And we have shown that it is possible to solve such problems as an automatic simulation of the total reality.

The physical pictures afforded by the mechanical constructs clarify the meaning of curl. They answer concretely the question, Where does the energy in the wave reside? In the particular example of the ponderomotive interaction between a penetrable obstacle and a wave, they even help in ascertaining the correctness of formulations in classic textbooks. That the correct expression for the electromagnetic momentum inside a dielectric is that given by (17.39) is proven by the fact that total momentum so defined was conserved in the simulation, even though only (17.38) was enforced. Yet these equations were selected not based on arguments of symmetry or covariance but on the logical basis that if the force exerted on matter is due to the fact that matter differs from free space, it must be local and due to the weighted rings and friction that model the reality of matter. The seemingly counterintuitive result that a wave exerts no force upon entering a dielectric slab despite the fact that there is a reflected wave is also borne out by the calculations. The momentum transfer all occurs as a result of the reflection at the far wall. Just as the success of these calculations proves that (17.38) and (17.39) are the correct equations for modeling ponderomotive phenomena in penetrable objects, similar calculations could settle many of the other classical questions of electromagnetism and matter, such as the question of the effective velocity of light inside a moving medium. Thus, in addition to its pedagogical benefits for the visualization of electromagnetic phenomena, this method's natural suitability for the solution of multidisciplinary problems opens a new door for discovery of fundamental physical truth, because it allows the automatic simulation of complex reality. Realistic simulations of nanomachines, electromagnetic railguns, and even the scattering of waves by moveable, resonant objects (as occurs in Quantum Mechanics), are all candidates for the application of this method.

REFERENCES

[1] Ramamurti Shankar, *Principles of Quantum Mechanics*, Plenum Press, New York, 1980.

[2] D. Greenspan, *Discrete Models*, Addison-Wesley, London, 1973.

[3] A. Fernback and Rotenberg, *Methods in Computational Physics*, Academic Press, New York, 1964.

[4] J. Bruce and Hunt, *The Maxwellians*, Cornell University Press, Ithaca, 1991.

[5] D. Jager, "Pulse Generation and Compression on Nonlinear Transmission Lines," in *IEEE MTT-S International Microwave Symposium*, Atlanta, Georgia, June 14, 1993.

[6] P. Lorrain and D. Corson, *Electromagnetic Fields and Waves*, 2nd edn, W. H. Freeman and Company, San Francisco, 1962.

[7] J. D. Jackson, *Classical Electrodynamics*, John Wiley and Sons, New York, 1975.

[8] F. De Flaviis, M. Noro, R.E. Diaz, and N. G. Alexopoulos, "Diaz-Fitzgerald Time Domain (D-FTD) Technique Applied to Electromagnetic Problems," in *IEEE MTT-S Int. Microwave Symp.*, San Francisco pp. 1047–1050, June 1996.

[9] A. Taflove, *Computational Electromagnetics – The Finite Difference Time Domain Method*, Boston, 1995.

[10] K. Luebbers, *The Finite Difference Time Domain Method for Electromagnetics*, CRC Press, Boca Raton, FL, 1993.

[11] F. De Flaviis, M. Noro, N. G. Alexopoulos, R. E. Diaz, and G. Franceschetti, "Extensions to Complex Materials of the Fitzgerald Model for the Solution of Electromagnetic Problems," *Electromagnetics*, **18**, pp. 35–65, 1998.

[12] D. S. Jones, *The Theory of Electromagnetics*, Macmillan, New York, 1964.

[13] S. M. Sze, *Physics of Semiconductor Devices*, John Wiley and Sons, New York, 1981.

[14] O. D. Jefimenko, *Electricity and Magnetism*, Appleton Century Crofts, New York, 1966.

[15] J. S. Toll, "Causality and the Dispersion Relations: Logical Foundations," *Phys. Rev.*, **104**, pp. 1760–1770, 1956.

[16] O. Heaviside, *Electromagnetic Theory*, Chelsea Publishing Co., New York, 1971.

GLOSSARY

Debye materials. Complex material whose permittivity e/o permeability depends on frequency according to specific law.

D-FTD. Variation of FDTD based on Fitzgerald mechanical model.

Doppler shift. Frequency shift of a wave caused by reflection from a moving obstacle

FDTD. Numerical method for the solution of Maxwell equations based on Finite Difference Time Domain.

Grid dispersion. Low pass characteristic behavior of a computational method ascribed to the mesh size.

Hamiltonian. Operator corresponding to sum of the total energy of the system.

Ponderable medium. Medium which possesses inertial mass.

Soliton. A wave whose amplitude increases while propagating through a non-linear medium.

TE mode. A plane wave whose electric field is transverse with respect to the direction of propagation.

TM mode. A plane wave whose magnetic field is transverse with respect to the direction of propagation.

Chapter 18

ARTIFICIAL BIANISOTROPIC COMPOSITES

Frédéric Mariotte, Bruno Sauviac, and Sergei A. Tretyakov

18.1 INTRODUCTION

There are many beautiful colors in Nature, which we can see in the visible spectrum. But for the microwave engineers most of the natural materials look dull gray, as they work in the microwave region of the electromagnetic spectrum. That is because the characteristic molecule size is indeed very small compared to the wavelength at microwave frequencies, and electron transition resonance frequencies usually lay much higher than the working range of radars and other microwave devices. On the other hand, material scientists are in a good position to create new microwave materials: at the millimeter scale it is much easier to shape inclusions (instead of shaping molecules) in their composite materials. A good example is artificial chiral composites, whose inclusions are mirror-asymmetric. Chiral effects in optics are always very small, just because the molecules are small, but in the microwave region the corresponding effects can be very significant, since an engineer can freely change the inclusion shape, size, and concentration. Generally speaking, to provide necessary materials for microwave applications we have to work with artificial materials, and we hope to be able to shape their properties for our needs. This chapter is devoted to electromagnetic properties of composite media with complex-shaped inclusions whose size is comparable to the wavelength.

When the wavelength of electromagnetic waves traveling in a medium and characteristic size of the constituents of the medium (molecules in a natural medium or inclusions in a composite) become comparable, electromagnetic response of the medium becomes quite different from that conventionally described by the effective permittivity and permeability. This phenomenon is called "spatial dispersion." If the spatial dispersion is weak (ka is very small compared to unity, where k is the wave number, and a is the inclusion size), so that only the first-order effects in ka are significant, the spatial dispersion phenomenon is manifested as a magnetoelectric coupling, and the medium can be described in terms of bianisotropic constitutive relations

$$\begin{cases} \mathbf{D} = \bar{\bar{\varepsilon}}\cdot\mathbf{E} + \bar{\bar{\alpha}}\cdot\mathbf{H} \\ \mathbf{B} = \bar{\bar{\mu}}\cdot\mathbf{H} + \bar{\bar{\beta}}\cdot\mathbf{E} \end{cases} \tag{18.1}$$

These relations with their additional field-coupling members describe weak spatial dispersion (the first-order effects in term of ka). There exists an alternative model of the same effects based on the constitutive relations with the first-order spatial derivatives of the electric field included in the relation for the displacement vector (the magnetic induction in that case is identical with the magnetic field), see [1]. That description, however, demands the use of more complicated boundary conditions of media interfaces, although the model (18.1) is compatible with the standard electromagnetic boundary conditions. The model used in this chapter is based on the bianisotropic relations. Here we consider reciprocal media (no external bias fields, no non-electromagnetic interactions of the constituents), thus the medium parameters are subject to the following restrictions [2]:

$$\overline{\overline{\varepsilon}} = \overline{\overline{\varepsilon}}^T,\ \overline{\overline{\mu}} = \overline{\overline{\mu}}^T,\ \overline{\overline{\beta}} = -\overline{\overline{\alpha}}^T \tag{18.2}$$

where T denotes the transpose operation. Since the two coupling dyadics of reciprocal media are connected, we will also use a more convenient notation

$$\begin{cases} \mathbf{D} = \overline{\overline{\varepsilon}} \cdot \mathbf{E} + j\sqrt{\varepsilon_0 \mu_0}\,\overline{\overline{\chi}}^T \cdot \mathbf{H} \\ \mathbf{B} = \overline{\overline{\mu}} \cdot \mathbf{H} - j\sqrt{\varepsilon_0 \mu_0}\,\overline{\overline{\chi}} \cdot \mathbf{E} \end{cases} \tag{18.3}$$

In this chapter, properties of spatially dispersive artificial media (or *bianisotropic composites*) will be reviewed, with the emphasis on isotropic chiral materials and uniaxial omega composites.

Historically, the concept of electromagnetic chirality originated from the optical activity phenomenon discovered in the beginning of the nineteenth century. In 1811, Arago found that the plane of polarization of linearly polarized light traveling through a crystal of quartz was rotated by this crystal when the direction of propagation was along the crystal axis. Then, the experiments of Biot, in 1812, showed that the optical activity was dependent on the thickness of the plate of quartz and on the wavelength of exciting light. In 1822, Fresnel explained these physical properties using the assumption that two different modes were propagating inside the crystal. Next, in 1848, Pasteur first postulated that optically active molecules are three-dimensional chiral figures and that the handedness of these molecules produces the optical activity. By definition, a three dimensional object is chiral if it cannot be superimposed on its mirror image by translations and rotations. Until the beginning of the twentieth century, research on the subject went on and, with the works of Drude, Born, Oseen, Gray, etc., progress in the field was made with many useful applications in chemistry and biology. In 1920, Lindman showed that optical activity could also be observed for lower frequencies using metallic helices of some centimeters length. The optical activity was to become electromagnetic chirality. After that, contributions in the field were essentially done in the modeling of the constitutive equations for these materials (here we should mention important contributions of Fedorov [3]). During those years electromagnetic chirality was lying dormant but in the 1970s renewed interest in the field appeared with the works of Bohren, Engheta, Jaggard, Lakhtakia, and the Varadan family. Toward the end of the 1980s, Jaggard and Engheta [4], [8] and Varadans and Lakhtakia [5] proposed a new application for electromagnetic chiral materials: the invisible medium and the radar stealth covering. Numerous studies on the subject have certainly originated from the idea of this promising application. So far, several samples of artificial chiral materials have been fabricated in different laboratories by embedding miniature helices in a non-chiral host medium [6]. Considerable progress has been made in the theoretical modeling of chiral media and in the theoretical predictions of their properties [7]. Some novel theoretical developments in the field are summarized in the recent book [9].

However, it has been recognized that macroscopic chirality of the medium does not actually influence the reflection coefficient from planar structures at the normal incidence [9], [24]. Its manifestation is rather in polarization transformation in the transmitted field and also in reflection at oblique incidence [10]. This does not mean that chiral inclusions like helices are of no practical interest in absorber design; in fact, helices are very compact resonant structures, whose use can help in reducing the weight and thickness of absorbers. Later on, some progress was made in understanding more complex bianisotropic structures. In particular, it was shown that, in contrast to the role of the chirality parameter, the omega coupling coefficient in uniaxial composites can be chosen so that the reflection coefficient from a bianisotropic slab vanishes for the normal incidence. Indeed, the omega coefficient is the extra parameter which can be in principle used to tune the reflective properties of bianisotropic absorbing layers, which is not possible to do varying the chirality parameter. Also, planar arrangements of omega-shaped inclusions offer a possibility of a still better "packaging" of scatterers in very thin sheets.

Advances in electromagnetics of bianisotropic media are reported at a series of conferences [11] initiated in 1993 [12].

18.2 CHIRAL MEDIA AND OMEGA MEDIA

18.2.1 Classification of Bianisotropic Composites

We start from the classification of bianisotropic composites based on the structure of coupling dyadics in their constitutive relations [13]. For this purpose we first represent the coupling dyadic in (18.3) as the sum of a trace-free dyadic $\overline{\overline{M}}$ and its isotropic part. Namely, we write $\overline{\overline{\chi}} = \chi\overline{\overline{I}} + \overline{\overline{M}}$, where $\overline{\overline{I}}$ is the unit dyadic, and the trace of $\overline{\overline{M}}$ is zero. Here the pseudoscalar (this number changes its sign with the coordinate inversion) complex parameter $\chi = \frac{1}{3}trace(\overline{\overline{\chi}})$ is called the *chirality parameter*. The reason for this name is that in bianisotropic media with mirror-symmetrical microstructures $\chi = 0$ [14]. Thus media with non-zero χ are called *chiral* media, since this is the name of mirror-asymmetric structures. Effects measured by the χ-parameter are isotropic, and media with non-zero chirality parameter can be manufactured as composites with randomly distributed chiral inclusions, such as helices.

In contrast to the permittivity and permeability which are always symmetric dyadics in reciprocal media, the coupling dyadic can be of the general type: indeed, there is only one reciprocity restriction (18.2) imposed on two coupling coefficients. Thus, we further decompose the trace-free dyadic $\overline{\overline{M}}$ into symmetric and antisymmetric parts: $\overline{\overline{M}} = \overline{\overline{N}} + \overline{\overline{J}}$, where $\overline{\overline{N}} = (\overline{\overline{M}} + \overline{\overline{M}}^T)/2$ is a symmetric, and $\overline{\overline{J}} = (\overline{\overline{M}} - \overline{\overline{M}}^T)/2$ is an antisymmetric dyadic. The symmetric part $\overline{\overline{N}}$ can be diagonalized: it is always possible to express $\overline{\overline{N}} = \Sigma\chi_i\mathbf{a}_i\mathbf{a}_i$ where χ_i are complex numbers and $\mathbf{a}_i$ are real vectors. In the general complex case there are more than three terms in this decomposition (no more than nine components appear in the sum in the most general case). In the expansion $\overline{\overline{N}} = \Sigma\chi_i\mathbf{a}_i\mathbf{a}_i$, the coefficients satisfy $\Sigma\chi_i = 0$. This dyadic describes effects similar to that measured by the chirality parameter: indeed, that part can be also written as $\chi\overline{\overline{I}} = \Sigma\chi\mathbf{b}_i\mathbf{b}_i$ where unit vectors $\mathbf{b}_i$ form a basis. It becomes clear that the magnetoelectric coupling modeled by $\overline{\overline{N}}$ can be described as that provided by virtual uniaxial chiral inclusions (thin uniaxial helices), oriented along the axes $\mathbf{a}_i$. There must be at least two sets of virtual inclusions of opposite handedness, because the trace of $\overline{\overline{N}}$ is zero. Electromagnetic phenomena in these media resemble those known in chiral media (e.g., electric field along the $\mathbf{a}_1$ direction causes magnetic polarization in the same direction, which means that for specific directions of propagation there can exist *electromagnetic activity* [15]. However, the microstructure of these composites does not have to be chiral. For these reasons we name composites with $\chi = 0$ but $\overline{\overline{N}} \neq 0$ *pseudochiral*. (This name was originally suggested by Saadoun and Engheta for omega media in [16].)

The antisymmetric dyadic $\overline{\overline{J}}$ in the decomposition of the coupling dyadic can be always presented as

$$\overline{\overline{J}} = \Omega_1 \mathbf{v}_1 \times \overline{\overline{I}} + j\Omega_2 \mathbf{v}_2 \times \overline{\overline{I}}$$

where $\mathbf{v}_{1,2}$ are real unit vectors and $\Omega_{1,2}$ are real numbers. If both the real and imaginary parts of $\mathbf{v}_{1,2}$ point along the same direction (we will further see that this is the case of artificial omega composite materials, for example), one can normalize that vector and write $\overline{\overline{J}} = \Omega \mathbf{v}_0 \times \overline{\overline{I}}$, where Ω is a scalar complex material parameter. In this case, Ω is the field coupling parameter of an effective uniaxial omega medium [17]. Materials with non-zero coefficients $\Omega_{1,2}$ can be called *omega* media although, of course, they do not necessarily contain Ω-shaped inclusions. In the general case we can model "antisymmetric" coupling effects by two sets of uniaxial omega elements ("hats") oriented along vectors $\mathbf{v}_{1,2}$.

This simple classification reveals a very important fact: whatever complex geometrical structure a composite might have, the first-order spatial effects are the same as in certain collections of idealized uniaxial helices and uniaxial pairs of omega inclusions (hats). In other words, these two basic elements are sufficient for describing arbitrary composite structures.

In this chapter we will consider the most important electromagnetic properties of chiral inclusions and chiral media as well as that of omega inclusions and omega media.

18.2.2 Constitutive Equations and Electromagnetic Properties of Chiral Media

Chirality is a geometrical property. An object is called chiral if it cannot be superimposed with its mirror image by translations and rotations. As examples of chiral objects (Fig. 18.1), one can cite helices of one or several turns and, of course, a hand. The very name "chiral" originates from the Greek *kheir* which means "hand." One can define right- (dextrorotatory) and left-handed (levorotatory) chiral objects. Mirror-symmetric objects are called achiral or non-chiral. It is however interesting to note that the sense of rotation of a plane wave propagating in a given chiral medium changes with the frequency, so the same medium is dextrorotatory or levorotatory at different frequencies. In fact, the integral of the real part of the chirality parameter over the frequency axis is exactly zero.

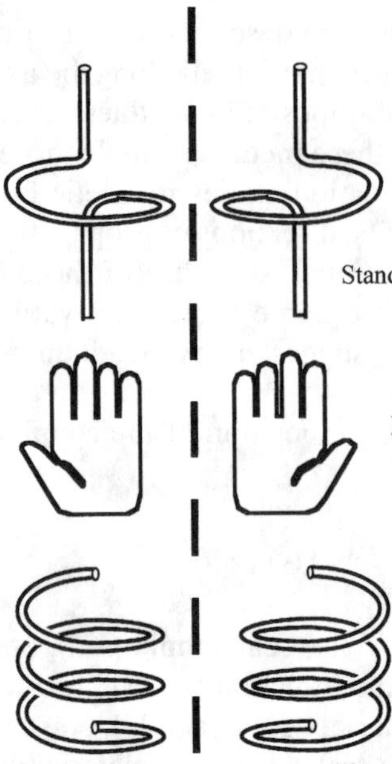

Figure 18.1 Chiral shapes and objects.

18.2.2.1 The Three General Formulations

Isotropic chiral media form a special subgroup of bianisotropic media where the constitutive parameters are no longer tensors, but scalars and pseudoscalars.[1] In the frequency domain, such media can be described by three complex parameters in the general lossy case. Several constitutive formalisms are currently being used to describe the electromagnetic coupling due to chirality that exists in these materials.

18.2.2.1.1 Lindell-Sihvola formalism. One possibility is to express the **D** and **B** vectors in the frequency domain as functions of the **E** and **H** vectors through [18]:

$$\begin{cases} \mathbf{D} = \varepsilon_{LS}\mathbf{E} + j\chi_{LS}\mathbf{H} \\ \mathbf{B} = \mu_{LS}\mathbf{H} - j\chi_{LS}\mathbf{E} \end{cases} \tag{18.4}$$

Here ε_{LS} and μ_{LS} represent the usual permittivity and permeability, and χ_{LS} is the chirality parameter which is an electromagnetic measure of the handedness of the medium. The imaginary unit is j (in this notation, all the material parameters are real for lossless media).

[1]The chirality parameter changes sign with an inversion of the spatial coordinates and is then a pseudoscalar.

18.2.2.1.2 Jaggard-Post-Kong equations. A second way to represent the chiral medium properties has been proposed in [19]. Now **D** and **H** vectors are functions of **B** and **E**:

$$\begin{cases} \mathbf{D} = \varepsilon_J \mathbf{E} + j\xi_J \mathbf{B} \\ \mathbf{H} = j\xi_J \mathbf{E} + \dfrac{\mathbf{B}}{\mu_J} \end{cases} \tag{18.5}$$

18.2.2.1.3 Drude-Born-Fedorov notations. The third kind of equations mentioned originates from the results of [20]. **D** vector is expressed only as a function of the electric field **E** and its first derivatives, and **B** is obtained from the **H** field:

$$\begin{cases} \mathbf{D} = \varepsilon_D \mathbf{E} + \varepsilon_D \beta_D \nabla \times \mathbf{E} \\ \mathbf{B} = \mu_D \mathbf{H} + \mu_D \beta_D \nabla \times \mathbf{H} \end{cases} \tag{18.6}$$

This form stresses the fact that chirality of media is a spatial dispersion effect.

18.2.2.1.4 From one representation to another one. Of course, all the above representations are equivalent for frequency-domain fields. Actually, there does not yet exist a standard notation to describe the physical properties of these materials. So research in the field goes on using several formalisms. Simple equations allow us to translate formulations between formalisms. These relations are very useful in studies and comparisons of the results of different research groups. Here, the $e^{-j\omega t}$ time dependence is assumed. Table 18.1 gives the connections between the different sets of parameters; it is extracted from [21].

TABLE 18.1 Relations between the Parameters in the Different Sets of Constitutive Equations

Translation →	$(\varepsilon_{LS}, \mu_{LS}, \chi_{LS})$	$(\varepsilon_J, \mu_J, \xi_J)$	$(\varepsilon_D, \mu_D, \beta_D)$
$(\varepsilon_{LS}, \mu_{LS}, \chi_{LS})$		$\varepsilon_J = \varepsilon_{LS} - \dfrac{\chi_{LS}^2}{\mu_{LS}}$ $\mu_J = \mu_{LS}$ $\xi_J = \dfrac{\chi_{LS}}{\mu_{LS}}$	$\varepsilon_D = \varepsilon_{LS}\left(1 - \dfrac{\chi_{LS}^2}{\varepsilon_{LS}\mu_{LS}}\right)$ $\mu_D = \mu_{LS}\left(1 - \dfrac{\chi_{LS}^2}{\varepsilon_{LS}\mu_{LS}}\right)$ $\beta_D = \dfrac{\chi_{LS}}{\omega(\varepsilon_{LS}\mu_{LS} - \chi_{LS}^2)}$
$(\varepsilon_J, \mu_J, \xi_J)$	$\varepsilon_{LS} = \varepsilon_J - \mu_J \xi_J^2$ $\mu_{LS} = \mu_J$ $\chi_{LS} = \mu_J \xi_J$		$\varepsilon_D = \varepsilon_J$ $\mu_D = \dfrac{\mu_J}{1 + \mu_J(\xi_J^2/\varepsilon_J)}$ $\beta_D = \dfrac{\xi}{\omega\varepsilon_J}$
$(\varepsilon_D, \mu_D, \beta_D)$	$\varepsilon_{LS} = \dfrac{\varepsilon_D}{1 - \omega^2 \varepsilon_D \mu_D \beta_D^2}$ $\mu_{LS} = \dfrac{\mu_D}{1 - \omega^2 \varepsilon_D \mu_D \beta_D^2}$ $\chi_{LS} = \dfrac{\omega\varepsilon_D \mu_D \beta_D}{1 - \omega^2 \varepsilon_D \mu_D \beta_D^2}$	$\varepsilon_J = \varepsilon_D$ $\mu_J = \dfrac{\mu_D}{1 - \omega^2 \varepsilon_D \mu_D \beta_D^2}$ $\xi_J = \omega\varepsilon_D \beta_D$	

18.2.2.2 Energy Considerations for Material Parameters

Power and energy considerations give certain restrictions on the allowed values of the constitutive parameters [22]. Indeed, the medium under consideration is a passive one. This means that the time average power flux density must be a positive quantity. For Lindell formalism (18.4), the parameters should satisfy the following inequalities:

$$\begin{cases} \Im m(\varepsilon_{LS}) \geq 0 \\ \Im m(\mu_{LS}) \geq 0 \\ \Im m(\varepsilon_{LS}) \Im m(\mu_{LS}) \geq \Im m(\gamma_{LS}) \end{cases} \tag{18.7}$$

Using Jaggard equations (18.5), similar conditions can be obtained:

$$\begin{cases} \Im m(\varepsilon_J + \mu_J \xi_J^2) \geq 0 \\ \Im m(\mu_J) \geq 0 \\ \Im m(\varepsilon_J + \mu_J \xi_J^2) \Im m(\mu_J) \geq [\Im m(\mu_J \xi_J)]^2 \end{cases} \tag{18.8}$$

And finally Drude notations (18.6) lead to the following requirements:

$$\begin{cases} \Im m\left(\dfrac{\varepsilon_D}{1-\omega^2 \varepsilon_D \mu_D \beta_D^2}\right) \geq 0 \\ \Im m\left(\dfrac{\mu_D}{1-\omega^2 \varepsilon_D \mu_D \beta_D^2}\right) \geq 0 \\ \Im m\left(\dfrac{\varepsilon_D}{1-\omega^2 \varepsilon_D \mu_D \beta_D^2}\right) \Im m\left(\dfrac{\mu_D}{1-\omega^2 \varepsilon_D \mu_D \beta_D^2}\right) \geq \left[\Im m\left(\dfrac{\omega \varepsilon_D \mu_D \beta_D}{1-\omega^2 \varepsilon_D \mu_D \beta_D^2}\right)\right]^2 \end{cases} \tag{18.9}$$

These relations are useful to verify that the results of modeling or measurement are physically sound. Considering these energetic requirements, it can be noted that the meaning of ε_{LS} and μ_{LS} is the same as for the usual permittivity and permeability of isotropic dielectric or magnetic materials. This remark has motivated our personal choice for the Lindell-Sihvola formalism. In the following, a chiral medium will be always represented by relation (18.4).

18.2.3 Wave Propagation in Chiral Materials

When the electromagnetic properties of chiral composites are known, general problems, like plane-wave propagation, can be treated. The chiral medium is electromagnetically represented by its permittivity and permeability. Another parameter must be added to completely describe the behavior of this material. As a consequence, this third variable gives modifications to the propagating modes compared to the classical dielectric medium. Macroscopic equations in use are the same as for formulas (18.4). The time convention $e^{-j\omega t}$ is assumed. Maxwell's equations are modified as follows:

$$\nabla \times \mathbf{E} = -\frac{\partial \mathbf{B}}{\partial t} = +j\omega\mu\mathbf{H} + \omega\chi\mathbf{E} \tag{18.10a}$$

$$\nabla \times \mathbf{H} = +\frac{\partial \mathbf{D}}{\partial t} = -j\omega\varepsilon\mathbf{E} + \omega\chi\mathbf{H} \tag{18.10b}$$

To obtain a representative equation of the propagation, the classic scheme is used, taking double curl of the previous equations ($\nabla \times \nabla \times \cdot = \nabla(\nabla \cdot) - \nabla^2 \cdot$):

$$\nabla^2\mathbf{E} + \omega^2(\varepsilon\mu - \chi^2)\mathbf{E} + 2\omega\chi\nabla \times \mathbf{E} = 0 \tag{18.11a}$$

$$\nabla^2\mathbf{H} + \omega^2(\varepsilon\mu - \chi^2)\mathbf{H} + 2\omega\chi\nabla \times \mathbf{H} = 0 \tag{18.11b}$$

Let us focus on the electric field equation (18.11a). The proposed solution is supposed to propagate along the (Oz) direction. It can be written as

$$\mathbf{E} = (E_{ox}\mathbf{e_x} + E_{oy}\mathbf{e_y} + E_{oz}\mathbf{e_z})e^{+jkz} \tag{18.12}$$

With $\nabla \cdot \mathbf{E} = 0$ and $\partial_x = \partial_y = 0$ it comes that $\nabla \cdot \mathbf{E} = jkE_{oz} = 0$, so $E_{oz} = 0$. Equation (18.11a) can be reduced to the following:

$$\begin{cases} [\omega^2(\varepsilon\mu - \chi^2) - k^2]E_{ox} - j2k\omega\chi E_{oy} = 0 \\ +j2k\omega\chi E_{ox} + [\omega^2(\varepsilon\mu - \chi^2) - k^2]E_{oy} = 0 \end{cases} \tag{18.13}$$

We are looking for a non-trivial solution, whence the determinant of (18.13) must vanish. Searching for k values that satisfy the following equation will give the propagation numbers:

$$[k^2 + 2\omega\chi k - \omega^2(\varepsilon\mu - \chi^2)][k^2 - 2\omega\chi k - \omega^2(\varepsilon\mu - \chi^2)] = 0 \tag{18.14}$$

Solving (18.14) gives four solutions, two are for progressive waves (propagation toward $z > 0$):

$$k_1 = \omega(\sqrt{\varepsilon\mu} + \chi) \text{ and } k_2 = \omega(\sqrt{\varepsilon\mu} - \chi) \tag{18.15}$$

The two other solutions are propagating toward $z < 0$:

$$k_3 = -\omega(\sqrt{\varepsilon\mu} + \chi) = -k_1 \text{ and } k_4 = -\omega(\sqrt{\varepsilon\mu} - \chi) = -k_2 \tag{18.16}$$

Using (18.15) and (18.16) with (18.13) we get relations between E_{ox} and E_{oy} components for each solution:

- *Forward waves*:
  - If $k = k_1$, the electric field can be written as: $\mathbf{E}_1 = E_{o1}\,(\mathbf{e_x} + j\mathbf{e_y})e^{+jk_1z}$. The propagating wave is Left Circularly Polarized (LCP) according to the IEEE convention.
  - If $k = k_2$, the electric field could be written as: $\mathbf{E}_2 = E_{o2}\,(\mathbf{e_x} - j\mathbf{e_y})e^{+jk_2z}$. The propagating wave is Right Circularly Polarized (RCP).
- *Backward waves:*
  - If $k = k_3$, the electric field can be written as: $\mathbf{E}_3 = E_{o3}\,(\mathbf{e_x} - j\mathbf{e_y})e^{-jk_1z}$. According to the IEEE convention, the wave is Left Circularly Polarized (LCP).

- If $k = k_4$, the electric field could be written as: $\mathbf{E}_4 = E_{o4}\,(\mathbf{e_x} + j\mathbf{e_y})e^{-jk_2 z}$. The propagating wave is Right Circularly Polarized (RCP).

Associated magnetic fields for each wave solution are derived from Maxwell's equation (18.10). Hence, we find

$$\mathbf{H}_1 = -j\sqrt{\varepsilon/\mu}\,\mathbf{E}_1 \qquad \mathbf{H}_2 = +j\sqrt{\varepsilon/\mu}\,\mathbf{E}_2$$

$$\mathbf{H}_3 = -j\sqrt{\varepsilon/\mu}\,\mathbf{E}_3 \qquad \mathbf{H}_4 = +j\sqrt{\varepsilon/\mu}\,\mathbf{E}_4$$

One can define the wave impedance as the constant that relates electric and magnetic fields:

$$Z(\mathbf{n} \times \mathbf{H}) = \mathbf{E} \tag{18.18}$$

Then, we find that the wave impedance is the same for each propagating wave in the medium, and has the same expression as for the classical dielectric medium:

$$Z = \sqrt{\varepsilon/\mu} \tag{18.19}$$

To summarize, wave propagation inside the isotropic chiral medium is realized with a *right circularly polarized* (*RCP*) wave and a *left circularly polarized wave* (*LCP*). The two propagating modes have the same wave impedance which does not depend on the chirality parameter. At this point, complete electric and magnetic fields vectors can be expressed. As a convention, we mark by (+) waves traveling in the positive z direction, and by (−) the opposite-bound waves.

The total field vector at a point with the coordinate z is given by

$$\mathbf{E}(z) = \mathbf{E}^+(z) + \mathbf{E}^-(z) = \mathbf{E}^+_{\mathbf{LCP}}(z) + \mathbf{E}^+_{\mathbf{RCP}}(z) + \mathbf{E}^-_{\mathbf{LCP}}(z) + \mathbf{E}^-_{\mathbf{RCP}}(z) \tag{18.20a}$$

and

$$\mathbf{H}(z) = \mathbf{H}^+(z) + \mathbf{H}^-(z) = \mathbf{H}^+_{\mathbf{LCP}}(z) + \mathbf{H}^+_{\mathbf{RCP}}(z) + \mathbf{H}^-_{\mathbf{LCP}}(z) + \mathbf{H}^-_{\mathbf{RCP}}(z) \tag{18.20b}$$

For waves traveling toward $z > 0$, we have the following expressions:

$$\mathbf{E}^+_{\mathrm{LCP}} = \begin{bmatrix} +1 \\ +j \\ 0 \end{bmatrix} E^+_L e^{jk_L z} \quad \mathbf{H}^+_{\mathrm{LCP}} = \begin{bmatrix} -j/Z \\ +1/Z \\ 0 \end{bmatrix} E^+_L e^{jk_L z} \quad \text{with} \quad k_L = \omega(\sqrt{\varepsilon\mu} + \chi) \tag{18.21}$$

$$\mathbf{E}^+_{\mathrm{RCP}} = \begin{bmatrix} +1 \\ -j \\ 0 \end{bmatrix} E^+_R e^{jk_R z} \quad \mathbf{H}^+_{\mathrm{RCP}} = \begin{bmatrix} +j/Z \\ +1/Z \\ 0 \end{bmatrix} E^+_R e^{jk_R z} \quad \text{with} \quad k_R = \omega(\sqrt{\varepsilon\mu} - \chi) \tag{18.22}$$

On the other hand, for waves traveling toward $z < 0$, we have the following expressions:

$$\mathbf{E}^-_{\mathrm{LCP}} = \begin{bmatrix} +1 \\ -j \\ 0 \end{bmatrix} E^-_L e^{-jk_L z} \quad \mathbf{H}^-_{\mathrm{LCP}} = \begin{bmatrix} -j/Z \\ -1/Z \\ 0 \end{bmatrix} E^-_L e^{-jk_L z} \quad \text{with} \quad k_L = \omega(\sqrt{\varepsilon\mu} + \chi) \tag{18.23}$$

$$\mathbf{E}^-_{\mathrm{RCP}} = \begin{bmatrix} +1 \\ +j \\ 0 \end{bmatrix} E^-_R e^{-jk_R z} \quad \mathbf{H}^-_{\mathrm{RCP}} = \begin{bmatrix} +j/Z \\ -1/Z \\ 0 \end{bmatrix} E^-_R e^{-jk_R z} \quad \text{with} \quad k_R = \omega(\sqrt{\varepsilon\mu} - \chi) \tag{18.24}$$

18.2.4 Field Equations for Uniaxial Omega Regions

Consider non-chiral reciprocal bianisotropic media with antisymmetric coupling dyadics. In Section 18.2.1, these media were identified as omega media, because the corresponding effects can be modeled by Ω-shaped inclusions, arranged in orthogonal pairs. Constitutive equations for uniaxial composite materials with two orthogonal sets of Ω-shaped elements arranged in "hats" we write for the harmonic time dependence $e^{-j\omega t}$ in the general form (18.3):

$$\begin{cases} \mathbf{D} = \overline{\overline{\varepsilon}} \cdot \mathbf{E} + j\sqrt{\varepsilon_0 \mu_0}\, \overline{\overline{\chi}}^T \cdot \mathbf{H} \\ \mathbf{B} = \overline{\overline{\mu}} \cdot \mathbf{H} - j\sqrt{\varepsilon_0 \mu_0}\, \overline{\overline{\chi}} \cdot \mathbf{E} \end{cases}$$

This general form holds when the two orthogonal sets of omega particles may be different (in size or in volume concentrations). This corresponds to two different scalar coupling coefficients χ_x and χ_y: $\overline{\overline{\chi}} = \chi_x \mathbf{y}_0 \mathbf{x}_0 - \chi_y \mathbf{x}_0 \mathbf{y}_0$. Such a medium can be considered as a combination of a uniaxial omega medium and a pseudo-chiral composite of uniaxial helices. Here we will concentrate on the role of the omega coupling and consider the case when the omega coupling has the uniaxial symmetry (the two sets of helices are identical, $\chi_x = \chi_y = \Omega$). Thus, both the coupling dyadics are antisymmetric and proportional to the 90 degree rotator $\overline{\overline{J}} = \mathbf{y}_0 \mathbf{x}_0 - \mathbf{x}_0 \mathbf{y}_0$ in the $x - y$ plane:

$$\begin{cases} \mathbf{D} = \overline{\overline{\varepsilon}} \cdot \mathbf{E} - j\Omega\sqrt{\varepsilon_0 \mu_0}\, \overline{\overline{J}} \cdot \mathbf{H} \\ \mathbf{B} = \overline{\overline{\mu}} \cdot \mathbf{H} - j\Omega\sqrt{\varepsilon_0 \mu_0}\, \overline{\overline{J}} \cdot \mathbf{E} \end{cases} \tag{18.25}$$

In this case $\overline{\overline{\varepsilon}}$ and $\overline{\overline{\mu}}$ are uniaxial dyadics with transverse (t) and normal (n) components:

$$\overline{\overline{\varepsilon}} = \varepsilon_0(\varepsilon_t \overline{\overline{I}}_t + \varepsilon_n \mathbf{z}_0 \mathbf{z}_0), \qquad \overline{\overline{\mu}} = \mu_0(\mu_t \overline{\overline{I}}_t + \mu_n \mathbf{z}_0 \mathbf{z}_0) \tag{18.26}$$

where $\mathbf{z}_0$ stands for the unit vector along the z axis (the geometrical axis of the structure). $\overline{\overline{I}}_t = \mathbf{x}_0 \mathbf{x}_0 + \mathbf{y}_0 \mathbf{y}_0$ is the transverse unit dyadic and the rotator $\overline{\overline{J}}$ can be expressed also as $\overline{\overline{J}} = \mathbf{z}_0 \times \overline{\overline{I}}_0$. The coupling provided by the omega particles is proportional to the dimensionless parameter Ω. In lossless media all the material parameters are real quantities.

18.2.5 Plane Eigenwaves, Propagation Factors and Wave Impedances of Omega Media

To study electromagnetic waves in uniaxial omega media we Fourier transform Maxwell's equations in the transverse plane and eliminate the longitudinal field components. Splitting the fields into normal and transverse parts

$$\mathbf{E} = E_n \mathbf{z}_0 + \mathbf{E}_t, \qquad \mathbf{H} = H_n \mathbf{z}_0 + \mathbf{H}_t \tag{18.27}$$

Maxwell's equations take the form

$$j\mathbf{k}_t \times \mathbf{E} + \frac{\partial}{\partial z} \mathbf{z}_0 \times \mathbf{E}_t = j\omega(\overline{\overline{\mu}} \cdot \mathbf{H} - j\Omega\sqrt{\varepsilon_0 \mu_0}\, \mathbf{z}_0 \times \mathbf{E}_t) \tag{18.28}$$

$$j\mathbf{k}_t \times \mathbf{H} + \frac{\partial}{\partial z} \mathbf{z}_0 \times \mathbf{H}_t = -j\omega(\overline{\overline{\varepsilon}} \cdot \mathbf{E} - j\Omega\sqrt{\varepsilon_0 \mu_0}\, \mathbf{z}_0 \times \mathbf{H}_t) \tag{18.29}$$

Here, $\mathbf{k}_t$ is the two-dimensional Fourier variable. The normal field components can be expressed in terms of the transverse fields, just like in simple isotropic medium:

$$E_n \mathbf{z}_0 = -\frac{1}{\omega\varepsilon_0\varepsilon_n}\mathbf{k}_t \times \mathbf{H}_t, \qquad H_n \mathbf{z}_0 = \frac{1}{\omega\mu_0\mu_n}\mathbf{k}_t \times \mathbf{E}_t \tag{18.30}$$

The equations for the transverse field components read

$$\left(\frac{\partial}{\partial z} - k_0\Omega\right)\mathbf{E}_t = -\left(j\omega\mu_0\mu_t\overline{\overline{I}}_t - \frac{j}{\omega\varepsilon_0\varepsilon_n}\mathbf{k}_t\mathbf{k}_t\right)\cdot \mathbf{z}_0 \times \mathbf{H}_t \tag{18.31}$$

$$\left(\frac{\partial}{\partial z} + k_0\Omega\right)\cdot \mathbf{z}_0 \times \mathbf{H}_t = -\left(j\omega\varepsilon_0\varepsilon_t\overline{\overline{I}}_t - \frac{j}{\omega\mu_0\mu_n}\mathbf{z}_0 \times \mathbf{k}_t\mathbf{z}_0 \times \mathbf{k}_t\right)\cdot \mathbf{E}_t \tag{18.32}$$

where we use the conventional notation for the free-space wave number $k_0 = \omega\sqrt{\varepsilon_0\mu_0}$. From these equations, one of the fields can be eliminated, which results in the second-order wave equation:

$$\frac{\partial^2}{\partial z^2}\mathbf{E}_t + \left(\beta_{TM}^2\frac{\mathbf{k}_t\mathbf{k}_t}{k_t^2} + \beta_{TE}^2\frac{\mathbf{z}_0 \times \mathbf{k}_t\mathbf{z}_0 \times \mathbf{k}_t}{k_t^2}\right)\cdot \mathbf{E}_t = 0 \tag{18.33}$$

Because the dyadic in the last equation is diagonal, the eigensolutions of the wave equation are two linearly polarized vectors: one is proportional to $\mathbf{k}_t$ and the other one to $\mathbf{z}_0 \times \mathbf{k}_t$. The first solution corresponds to a TM-polarized wave, with the magnetic field orthogonal to vector $\mathbf{k}_t$, and the second one is a TE-wave. β_{TM} and β_{TE} are the longitudinal components of the propagation factors for the TM- and TE-polarized eigenwaves, respectively:

$$\beta_{TM}^2 = \frac{\varepsilon_t}{\varepsilon_n}(k_0^2\varepsilon_n\mu_t - k_t^2) - k_t^2\Omega^2, \qquad \beta_{TE}^2 = \frac{\mu_t}{\mu_n}(k_0^2\varepsilon_t\mu_n - k_t^2) - k_t^2\Omega^2 \tag{18.34}$$

The relations between the transverse fields in these eigenwaves we write in terms of dyadic wave admittances: $\mathbf{z}_0 \times \mathbf{H}_t = \mp\overline{\overline{Y}}_\pm \cdot \mathbf{E}_t$. The dyadic admittances are diagonal, with the scalar coefficients corresponding to the two eigenpolarizations:

$$\overline{\overline{Y}} = Y_\pm^{TM}\frac{\mathbf{k}_t\mathbf{k}_t}{k_t^2} + Y_\pm^{TE}\frac{\mathbf{z}_0 \times \mathbf{k}_t\mathbf{z}_0 \times \mathbf{k}_t}{k_t^2} \tag{18.35}$$

Indices $\pm$ correspond to the waves propagating in the positive and the negative directions of the z axis, respectively. Unlike simple media, these impedances depend on the direction of propagation, although the medium is reciprocal. This reflects the fact that the two directions along the geometrical axis are physically different. In simple uniaxial media (e.g., with elongated inclusions oriented along a given direction), there is a preferred axis. In uniaxial omega medium, one has a preferred direction along the preferred axis. The characteristic admittances can be found from the field equations after the propagation factors are known. The result reads

$$Y_\pm^{TM} = \sqrt{\frac{\varepsilon_0\varepsilon_t}{\mu_0\mu_t}}\frac{1}{1 - \dfrac{k_t^2}{k_0^2\varepsilon_n\mu_t}}\left(\sqrt{1 - \frac{k_t^2}{k_0^2\varepsilon_n\mu_t} - \frac{\Omega^2}{\varepsilon_0\mu_0}} \pm j\frac{\Omega}{\sqrt{\varepsilon_0\mu_0}}\right) \tag{18.36}$$

$$Y_\pm^{TE} = \sqrt{\frac{\varepsilon_0\varepsilon_t}{\mu_0\mu_t}}\left(\sqrt{1 - \frac{k_t^2}{k_0^2\varepsilon_t\mu_n} - \frac{\Omega^2}{\varepsilon_0\mu_0}} \pm j\frac{\Omega}{\sqrt{\varepsilon_0\mu_0}}\right) \tag{18.37}$$

If the omega parameter vanishes, the above results reduce to the known solutions for simple uniaxial media, with symmetric wave admittances. With the knowledge of the eigensolution, reflective and transmitting properties of slabs can be studied, which we will do in one of the following sections. It will be shown that the omega parameter can be chosen so as to match an air-composite interface.

18.3 ELECTROMAGNETIC SCATTERING BY CHIRAL OBJECTS AND MEDIUM MODELING

18.3.1 Baseline to Model Bianisotropic Composites

It appears to be necessary to develop modeling tools to predict the performance of chiral and omega composites taking into account the inhomogeneity of these materials and the particular morphology of the inclusions. Complex composites are realized by embedding helices or more general shapes in a simple host medium. Basically, the composite modeling can be divided in two steps, which are the following:

Step 1 To describe the interactions between an electromagnetic (EM) wave and a single inclusion. It consists in computing the induced current on the inclusion and the scattered field. In some numerically intensive models, we may directly study the interaction of an EM wave with a collection of interacting scatterers.

Step 2 In the second step, these results are used to estimate the average electromagnetic properties of the composite.

Thus, the study of the scattering of a single inclusion in a composite (a chiral object or an omega-shaped scatterer) is the first step toward the final composite modeling. Different ways and methods developed for this purpose are presented.

18.3.2 Analytical Integral Equation Method for a Standard Helix

In this section, the electromagnetic response of some chiral and non-chiral scatterers embedded in host materials is studied without any restriction on the inclusion size with respect to the wavelength. Numerical or analytical calculations are performed in order to compute the field scattered by these shapes and then to design numerical and analytical models of chiral and some non-chiral magnetoelectric composite materials.

Gogny and Mariotte [23], [24] have introduced this method in 1992. The purpose is to calculate analytically the current induced by an electromagnetic wave illuminating a thin wire perfectly conducting helix (standard helix as presented on Fig. 18.1a). In order to calculate the current, we start from the integral equation as provided by

$$\mathbf{E}^t_{tot}(r = r_s) = \mathbf{E}^t_{inc}(r = r_s) + j\omega\mathbf{A} + \frac{j}{\omega\varepsilon\mu}\nabla(\nabla \cdot \mathbf{A}) \tag{18.38}$$

Here $\mathbf{E}^t_{tot}(r = r_s)$ is the tangential component of the total electric field at the surface of the helix, $\mathbf{E}^t_{inc}(r = r_s)$ is the tangential component of the incident electric field, and $\mathbf{A}$ is the potential vector given by

$$\mathbf{A}(r) = \frac{\mu}{4\pi} \int_S \frac{e^{jk|r-r'|}}{|r-r'|} J_S(r')dr' \tag{18.39}$$

$J_S(r')$ is the current density induced at the surface of the helix and $k = \omega\sqrt{\varepsilon\mu}$. The wire diameter is assumed to be very small (compared to the wavelength and the other helix dimensions). Consequently, the thin wire approximation is used, which means the following hypothesis:

- The current $J_S(r')$ is concentrated on the axis of the thin wire.
- The boundary condition $\mathbf{E}^t_{tot}(r = r_s) = 0$ is taken at the surface of the wire of the helix whose radius is equal to a.

Three integral equations have to be solved for each part of the helix: one for each of the two wires of length l and one for the loop of radius R. Using cylindrical coordinates $(\mathbf{e}_\rho, \mathbf{e}_\theta, \mathbf{e}_z)$ to define the currents on the linear portions, we can show that the current densities J_ρ and J_θ on the straight wires are negligible as compared to J_z. For the circular loop, which is actually a torus of radius R and of cross-sectional area πa^2, an axis tangential to the torus $(\mathbf{e}_{\rho_1}, \mathbf{e}_{\theta_1}, \mathbf{e}_\varphi)$ is defined. We can also show that the current densities J_{ρ_1} and J_{θ_1} on the loop are negligible as compared to J_φ. Using methods similar to that developed by Hallen and Einarsson [25], [26], the integral equations for the linear parts are solved. For the circular loop, we expand the integral in function of a/R (which is very small as compared to unity). By imposing boundary conditions (zero value for the currents at each end of the linear portions) and current continuity between wires and loop, the current along the whole helix is solved.

By integrating the currents on the thin wire helix, the scattered field in the far zone is calculated. The electric and magnetic dipole moments $\mathbf{p}$ and $\mathbf{m}$ are calculated using the current distribution on the wire helix. The advantage of such analytical methods is that they give a possibility for parameter studies. In fact, it is quite easy to adjust the dimensions of the helices in order to provide desirable properties for any direction or polarization of the incident wave.

The limitation of the method comes from the thin wire approximation, which restricts the quantities a/R and l/R to be very small as compared to unity. Another limitation could be dictated by the computational power available. In fact, to relax this limitation it is better to work with helices comparable or smaller than the wavelength of the incident wave.

18.3.3 Numerical Integral Equation Method Using the Thin Wire Approximation

To study objects with more complicated shapes (as in Fig. 18.1c), we need a more general method. We have chosen to use an integral equation method that allows us to study arbitrary-shaped three-dimensional thin-wire objects [27]. As generally used chiral objects are made of thin metallic wires, the general method can be simplified using the thin wire approximation. That essentially means that only one-dimensional finite elements can be used to mesh 3D chiral objects.

The purpose is to study interaction phenomena between an electromagnetic wave and a metallic thin-wire body. Suppose the electric incident field $\mathbf{E}_{inc}$ is scattered by a metal obstacle of conductivity σ. This body is a closed volume Ω bounded by its surface Γ. The total field $(\mathbf{E}, \mathbf{H})$ must satisfy Maxwell's equations everywhere, and it must satisfy the far-field Sommerfeld conditions at infinity. As Ω is not a perfect conductor and σ has a finite value, $(\mathbf{E}, \mathbf{H})$ exist inside Ω and the tangential components of $\mathbf{E}$ on Γ can be written as

$$\mathbf{E} - \mathbf{n}(\mathbf{n} \cdot \mathbf{E}) = Z_C(\mathbf{n} \times \mathbf{H}) \tag{18.40}$$

where Z_C is the surface impedance of the conducting body and $\mathbf{n}$ is the outward unit normal vector. The incident field induces a current density $\mathbf{J}_s$ on the wire surface and this current radiates scattered field $\mathbf{E}_d$ that is the solution to the following equation:

$$-\Delta \mathbf{E}_d - k^2 \mathbf{E}_d = j\omega\mu \mathbf{J}_s + j\frac{1}{\omega\varepsilon}\nabla(\nabla \cdot \mathbf{J}_s) \tag{18.41}$$

Here $k = \omega\sqrt{\varepsilon\mu}$, ω is the frequency and ε, μ are the electromagnetic parameters of the surrounding medium.

The thin-wire approximation can be used if the object has the following properties:

- two of its dimensions are very small compared to the third dimension
- the curvature radius of the biggest dimension is large enough

The usual case of thin-wire approximation is the problem of a circular cylinder with radius a which is very small compared to its length L ($a \ll L$).

The cylindrical wire Ω considered here (Fig. 18.2) has a large bend radius, the length is denoted by L, and the radius is equal to a. Considering the geometry, it is obvious that the problem should be expressed in cylindrical coordinates (ρ, θ, s).

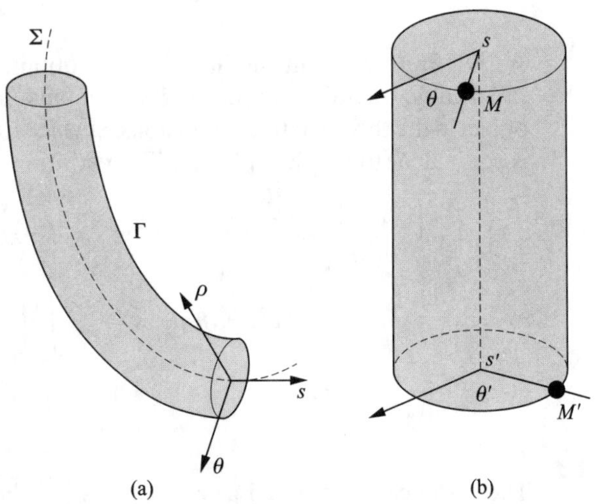

Figure 18.2 Thin wire geometry problem (a) Cylindrical coordinates system. Σ represents the axis of the cylinder. Γ is the surface of the shape. (b) Case for two points M and M' located close to each other.

Thin-wire approximation comes from the observation that the current density is directed along the axis because the angular and radial currents can be neglected. As a consequence, the current density $\mathbf{J}_s(s, \theta)$ does not depend on coordinate θ and remains parallel to the tangential direction (to the axis Σ of the cylinder). Using the curvilinear coordinate s, we obtain the relation

$$\mathbf{J}_s(s, \theta) = |\mathbf{J}_s(s)|\,\boldsymbol{\tau}(s) = J_s(s)\,\boldsymbol{\tau}(s) \tag{18.42}$$

where $\boldsymbol{\tau}(s)$ is the tangential unit vector on Σ. The solution to (18.41) is written as follows:

$$G(R) = \frac{\exp(jkR)}{4\pi R} \tag{18.43}$$

The scattered field is obtained as a convolution of G and the second part of (18.41):

$$\mathbf{E}_d = j\omega\mu \int_\Gamma G(|r - r'|)\mathbf{J}_S(r')dr' + \frac{j}{\omega\varepsilon}\nabla\cdot \int_\Gamma G(|r - r'|)(\nabla\cdot\mathbf{J}_S(r'))\,dr' \tag{18.44}$$

where r' is located on Γ, and r is a point somewhere in the surround medium. Finally, the total electric field $\mathbf{E}$ can be known as

$$\mathbf{E}(r) = \mathbf{E}_{inc}(r) + \mathbf{E}_d(r) \tag{18.45}$$

The current density is supposed to be decomposed using the vector basis functions $\mathbf{J}'$. These vectors are directed along the axis of the wire. Equation (18.40) can be now rewritten as

$$\int_\Gamma \mathbf{E}(r)\cdot\mathbf{J}'(r)\,dr = \int_\Gamma Z_c\mathbf{J}_S(r)\cdot\mathbf{J}'(r)\,dr \tag{18.46}$$

Substituting (18.44) and (18.45) in (18.46) we get finally the variational formula (after integrating by parts):

$$\begin{aligned}\int_\Gamma G(|r - r'|)\left\{j\omega\mu\mathbf{J}_S(r')\cdot\mathbf{J}'(r) - \frac{j}{\omega\varepsilon}(\nabla\cdot\mathbf{J}'(r))(\nabla\cdot\mathbf{J}_S(r'))\right\}dr\,dr' \\ = -\int_\Gamma \mathbf{E}_{inc}(r)\cdot\mathbf{J}'(r)dr + \int_\Gamma Z_c\mathbf{J}_S(r')\cdot\mathbf{J}'(r)dr\end{aligned} \tag{18.47}$$

With a finite element method one can obtain the value of $\mathbf{J}_S$ anywhere on the wire. Practically, one-dimensional finite elements are used to represent geometrically the object in the three spatial dimensions. Surface Γ is described with s and θ coordinates and is equivalent to $\Gamma = \Sigma \times [0, 2\pi]_\theta$. The variational equation (18.47) now becomes

$$\begin{aligned}\int_\Sigma\int_\Sigma\int_0^{2\pi}\int_0^{2\pi} \frac{e^{jkr}}{4\pi r}\left\{j\omega\mu J_S(s')\cdot J'(s)(\boldsymbol{\tau}(s')\cdot\boldsymbol{\tau}'(s)) - \frac{j}{\omega\varepsilon}\frac{dJ_S(s')}{ds'}\frac{dJ'(s)}{ds}\right\} \\ \times a^2\,d\theta\,d\theta'\,ds\,ds' = -\int_\Sigma\int_0^{2\pi} \mathbf{E}_{inc}(s)\cdot\boldsymbol{\tau}'(s)J'(s)a\,ds\,d\theta \\ + \int_\Sigma\int_0^{2\pi} Z_cJ_s(s)\cdot J'(s)\,a\,d\theta\end{aligned} \tag{18.48}$$

The distance r is defined as $r = \sqrt{|s - s'|^2 + 4a^2\sin^2[(\theta - \theta')/2]}$.

Considering that r is a periodic function (2π) of θ, the Green kernel can be simplified to

$$G(u) = \int_0^{2\pi}\int_0^{2\pi} \frac{e^{jkr(u,(\theta-\theta')/2)}}{4\pi r(u,(\theta-\theta')/2)} d\theta d\theta' = 2\int_0^{2\pi} \frac{e^{jkr(u,v)}}{4\pi r(u,v)} dv \tag{18.49}$$

with $u = |s - s'| \in [0, L]$ and $v = (\theta - \theta')/2$.

The Green kernel can be separated into a singular part and a regular part as follows:

$$G_s(u) = \frac{e^{jku}}{a} \ln\left(\frac{1}{u}\right) \tag{18.50}$$

$$G_R(u) = 2\int_0^{2\pi} \frac{e^{jkr(u,v)} - \cos(v)\, e^{jku}}{r(u,v)} dv + \frac{e^{jku}}{a} \ln(2a + \sqrt{u^2 + 4a^2}) \tag{18.51}$$

Inserting the results of the angular integration into (18.48) we obtain

$$\begin{aligned} &\int_\Sigma\int_\Sigma G(|s-s'|)\left\{ j\omega\mu J_S(s')\cdot J'(s)(\boldsymbol{\tau}(s')\cdot\boldsymbol{\tau}'(s)) - \frac{j}{\omega\varepsilon}\frac{dJ_S(s')}{ds'}\frac{dJ'(s)}{ds}\right\} \\ &\times a'\,ds\,ds' = -\int_\Sigma\int_0^{2\pi} \mathbf{E}_{inc}(s)\cdot\boldsymbol{\tau}'(s) J'(s)\,ds\,d\theta + \int_\Sigma\int_0^{2\pi} Z_c J_s(s)\cdot J'(s)\,ds\,d\theta \end{aligned} \tag{18.52}$$

with the Green function expressed as $G = G_R + G_S$.

The solution for the current is constructed as the following decomposition:

$$J_s(s) = \sum_{j=1}^{N} J_j \Phi_j(s) \tag{18.53}$$

Basis functions Φ_j are polynomial functions of the first order; they are also used as the test functions. The problem now becomes a matrix equation [A] [J] = [B]. The matrix coefficients are defined as follows:

$$\begin{aligned} a_{ij} = &\int_\Sigma\int_\Sigma G(|s-s'|)\left\{ j\omega\mu\Phi_j(s')\cdot\Phi_i(s)\cos(\boldsymbol{\tau}(s'),\boldsymbol{\tau}'(s)) - \frac{j}{\omega\varepsilon}\frac{d\Phi_j(s')}{ds'}\frac{d\Phi_i(s)}{ds}\right\} \\ &\times a\,ds\,ds' - \int_\Sigma\int_0^{2\pi} Z_c\Phi_j(s)\cdot\Phi_i(s)\,ds\,d\theta \end{aligned} \tag{18.54}$$

$$b_i = -\int_\Sigma\int_0^{2\pi} [\mathbf{E}_{inc}(s)\cdot\boldsymbol{\tau}'(s)]\Phi_i(s)\,ds\,d\theta \tag{18.55}$$

with

$$[A] = \{a_{ij}\},\ [B] = \{b_i\},\ \text{and } [J] = \{J_i\} \tag{18.56}$$

Solution of the system gives J_i values and with (18.53) we find the current density J_s at any point on the wire.

The presented method is focused on metallic objects. Very recent mathematical developments have been done to study a dielectric or magnetic object whose shape is still a thin-wire shape [27]. Now, a thin-wire approximation is also available for non-metallic wires.

18.3.4 Dipole Representation and Equivalent Polarizabilities for Chiral Scatterers

The next step toward the complete composite modeling is the representation of the scatterers as equivalent dipoles or multipoles. Here we discuss the possibility to represent a chiral scatterer only by its electric and magnetic dipole moments. Different ways to obtain the polarizability tensors are presented.

18.3.4.1 Calculation of Dipole Moments

At this point, surface currents induced on the chiral object under study are available. So it becomes possible to calculate rigorously the scattered fields. Assume that the object is surrounded by a medium with the electromagnetic parameters (ε, μ) and is excited by a plane electromagnetic wave. Incident fields, interacting with the chiral scatterer, induce electric and magnetic multipole moments which radiate energy in all directions. The scattered field can be described as a sum of the fields radiated by all these elementary multipole moments. As the particle is electrically small compared to the wavelength, the modeling can be simplified. Hence, only electric $\mathbf{p}$ and magnetic $\mathbf{m}$ dipole moments will have a significant influence on scattering. The scattered field by the object at a distance r in the direction of the unit vector $\mathbf{n}$ is given by

$$\mathbf{E}_{sc}(r) = \frac{k^2}{4\pi\varepsilon}\frac{e^{jkr}}{r}\left[(\mathbf{n}\times\mathbf{p})\times\mathbf{n} - \mathbf{n}\times\frac{\mathbf{m}}{c}\right] \tag{18.57}$$

Dipole moments $\mathbf{p}$ and $\mathbf{m}$ are defined by the following integrals:

$$\mathbf{p} = -\frac{1}{j\omega}\int \mathbf{J}_s(s')\,ds' \tag{18.58}$$

$$\mathbf{m} = \frac{1}{2}\int \mathbf{r}\times\mathbf{J}_s(s')\,ds' \tag{18.59}$$

$\mathbf{r}$ represents the $\mathbf{GM}$ distance, where G is the geometrical barycenter of the chiral object, and M is a point located at the surface of the wire; ω is the frequency of the incident plane wave, and $\mathbf{J}_s(s')$ is the induced surface current on the wire.

In doing so the electromagnetic scattering of a small chiral particle can be described using its $(\mathbf{p}, \mathbf{m})$ values. A verification of the validity of this approximation has been done. First we have computed the RCS of a single helix using a full-wave complete numerical method (that uses thin-wire meshing or surface meshing and the finite element method) and we have calculated the RCS using the scattered field of (18.57) and the following definition:

$$RCS(\text{dB.m}^2) = \lim_{r\to\infty}\left[10\log_{10}\left(4\pi r^2\frac{\mathbf{E}_{sc}(r)\cdot\mathbf{E}^*_{sc}(r)}{\mathbf{E}_{inc}(r)\cdot\mathbf{E}^*_{inc}(r)}\right)\right] \tag{18.60}$$

Obtained results are exactly the same in a wide frequency range. Another comparison has been made with experimental results. Results are in very good agreement from 1 to 18 GHz for a standard helix (Fig. 18.1a), which is 24 mm in height and 6 mm in width. Examining the results, it can be noticed that resonance phenomena appear at frequencies such as $L = n\,\lambda/2$, where L is the stretched length of the helix, λ is the wavelength of the incident wave, and n is an integer.

This quite simple case of a classical helix (Fig. 18.1a) is a good first step to understanding rather complicated physical phenomena involved in such problems. Consider the case when the electric field component directed along the axial direction of the helix creates surface current i_{E} in the straight wires of the helix. As the loop and these wires are connected, current i_{E} flows in the loop part and as a consequence induces magnetic field. Hence, the electric field **E** induces electric dipole moment **p** by coupling with the straight wires and also contributes to the magnetic dipole moment **m** by the effects acting in the loop. At the same time, the magnetic field component **H** directed along the axis of the helix gives a magnetic flux variation in the plane of the loop that induces current i_{H} in the loop. This current circulates in the straight wire portions and creates a potential difference between the ends of the straight wires. So, we can see that the incident magnetic field **H** has influence on the magnetic dipole moment **m** and also gives rise to an electric dipole moment **p**. So it is now clear that a chiral object provides coupling between the electric field **E** and magnetic field **H**.

Small particles of complex shape can be characterized by dyadic electric and magnetic polarizabilities, which define relations between induced electric and magnetic dipole moments **p**, **m** and external electric and magnetic fields **E** and **H**:

$$\mathbf{p} = \varepsilon(\bar{\bar{a}}_{ee}\mathbf{E} + \bar{\bar{a}}_{em}\,\eta\mathbf{H}) \tag{18.61}$$

$$\mathbf{m} = \frac{\bar{\bar{a}}_{me}}{\eta}\mathbf{E} + \bar{\bar{a}}_{mm}\mathbf{H} \tag{18.62}$$

where $\bar{\bar{a}}_{ee}$, $\bar{\bar{a}}_{mm}$, $\bar{\bar{a}}_{em}$, and $\bar{\bar{a}}_{me}$ are polarizability dyadics, and $\eta = \sqrt{\mu/\varepsilon}$ is the wave impedance of the background medium with the parameters (ε, μ). These relations clearly show the coupling between electric and magnetic field induced by the chiral shape of the object. The determination of the components of these dyadics depends on the method used to study the scattering behavior of the considered chiral or other complicated shape.

18.3.4.2 Polarizabilities Calculation

For the analytical integral method (Section 18.3.2), since the geometry under study is very special, calculation can be made using two incident waves that leads to a 2×2 system of equations. For the Numerical Integral Method (Section 18.3.3), the calculation is slightly more complicated. The study under six different incidences is needed to obtain all the components of the four polarizability tensors (see Fig. 18.3). For each configuration, currents are solved, then **p** and **m** vectors are evaluated using integrals (18.58 and 18.59). Considering all the equations for all the components of dipole moments, we get 36 equations and we need to determine 36 terms of dyadics polarizabilities. The particular choice of the incidence gives a very easy way to solve

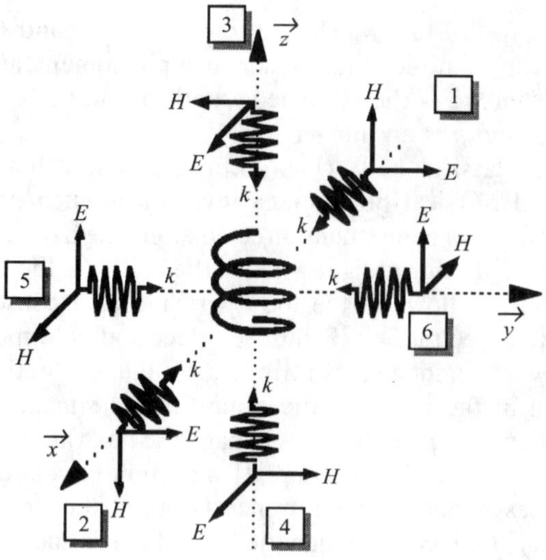

Figure 18.3 Incident fields dedicated to calculate all components of polarizability tensors.

the system of equations. For example, using incidences 1 and 2 leads to the following:

$$\begin{bmatrix}(p_{1x}+p_{2x})/2\\(p_{1y}+p_{2y})/2\\(p_{1z}+p_{2z})/2\end{bmatrix}=\varepsilon\begin{bmatrix}* & a_{ee12} & *\\ * & a_{ee22} & *\\ * & a_{ee32} & *\end{bmatrix}\begin{bmatrix}0\\E_y\\0\end{bmatrix}\qquad\begin{bmatrix}(m_{1x}-m_{2x})/2\\(m_{1y}-m_{2y})/2\\(m_{1z}-m_{2z})/2\end{bmatrix}=\begin{bmatrix}* & * & a_{mm13}\\ * & * & a_{mm23}\\ * & * & a_{mm33}\end{bmatrix}\begin{bmatrix}0\\0\\-H_z\end{bmatrix}$$

$$\begin{bmatrix}(p_{1x}-p_{2x})/2\\(p_{1y}-p_{2y})/2\\(p_{1z}-p_{2z})/2\end{bmatrix}=\varepsilon\eta\begin{bmatrix}* & * & a_{em13}\\ * & * & a_{em23}\\ * & * & a_{em33}\end{bmatrix}\begin{bmatrix}0\\0\\H_z\end{bmatrix}\qquad\begin{bmatrix}(m_{1x}+m_{2x})/2\\(m_{1y}+m_{2y})/2\\(m_{1z}+m_{2z})/2\end{bmatrix}=\frac{1}{\eta}\begin{bmatrix}* & a_{me12} & *\\ * & a_{me22} & *\\ * & a_{me32} & *\end{bmatrix}\begin{bmatrix}0\\E_y\\0\end{bmatrix}$$

The physical requirements, especially reciprocal properties, imposed on the components of the dyadics, give

$$\overline{\overline{a}}_{ee}=\overline{\overline{a}}_{ee}^{T}\qquad\overline{\overline{a}}_{mm}=\overline{\overline{a}}_{mm}^{T}\qquad\overline{\overline{a}}_{me}=-\overline{\overline{a}}_{em}^{T}\tag{18.63}$$

Here T denotes the transpose operation. Generally, these properties are verified in numerical results. To improve the numerical results, and obtain perfect numerical reciprocal values, an easy modification of the scheme can be done. It consists in doing again the same calculation with the dual polarization for each incident wave. As an example for incidence 2 (see Fig. 18.3), we can take an incident field with E_z instead of E_y. Averaging the obtained components of the dyadics from the two calculations will give perfectly symmetrical numerical values.

18.3.5 Analytical Antenna Model for Canonical Chiral Objects and Omega Scatterers

This method employs the antenna theory. The standard helix (Fig. 18.1a) or an omega particle is considered as a system of two connected antennas. These objects are combinations of straight wire portions (which is a dipole antenna) and a looped wire (considered as a loop antenna). The difference between the two shapes is in the mutual position of the two parts: they either belong to the same plane or are located in two orthogonal planes.

Small particles of complex shape can be characterized by dyadic electric and magnetic polarizabilities which define the bianisotropic relations between induced electric and magnetic dipole moments $\mathbf{p}, \mathbf{m}$ and external electric and magnetic fields $\mathbf{E}, \mathbf{H}$:

$$\mathbf{p} = \bar{\bar{a}}_{ee} \cdot \mathbf{E} + \bar{\bar{a}}_{em} \cdot \mathbf{H}, \qquad \mathbf{m} = \bar{\bar{a}}_{me} \cdot \mathbf{E} + \bar{\bar{a}}_{mm} \cdot \mathbf{H} \tag{18.64}$$

The method allows us to express the components of the polarizability dyadics in terms of the parameters of the two antenna constituents.

18.3.5.1 Antenna Representation for the Chiral Scatterer—Polarizability Dyadic

In this analytical model, one-turn helices are modeled by a wire loop connected to two straight-wire elements. The electromagnetic analysis is based on replacing the particles by two connected antennas, representing the wire and loop portions. Fortunately, for the present analysis we do not need to handle antennas of arbitrary sizes since in the practical design of composite materials comparatively small particles are used. In the case of this study, we can assume that the electrical sizes are moderate, so that $|k|L < 0.3$ and $|k|R < 1$, where $k = \omega\sqrt{\varepsilon\mu}$ is the wave number in the background medium. In such case, we can use analytical expressions for the antenna parameters. Corresponding expressions can be found in the literature, and several alternative models are available. For the wire antenna we apply the analytical approach by Wu and King [28]. If terms of the order $(\beta L)^5$ and higher-order terms are neglected, the input admittance of the linear wire antenna can be expressed as [29], [30]:

$$Y_w = 2j\pi \frac{\beta L}{\eta \Psi_{dr}} \left[1 + \beta^2 L^2 \frac{F}{3} - j\beta^3 L^3 \frac{1}{3(\Omega - 3)} \right] \tag{18.65}$$

with the following parameters:

$$F = 1 + \frac{1.08}{2\log(2L/a) - 3}, \qquad \Omega = 2\log(2L/a),$$

$$\Psi_{dr} = 2\log(L/a) - 2 \qquad \text{and} \qquad \eta = \sqrt{\mu/\varepsilon} \tag{18.66}$$

Here, the current distribution function has been approximated by a second-order polynomial. Circular loop antennas are analyzed using the Fourier series expansion of the current distribution function. An analytical model of a circular loop antenna of a moderate size can be constructed by keeping only the most important terms in the expansion, which results in the following approximate formula for the input admittance [30], [31]:

$$Y_L = \frac{-j}{\eta\pi} \left[\frac{1}{A_0} + \frac{2}{A_1} + \frac{2}{A_2} \right] \tag{18.67}$$

with the first three Fourier coefficients

$$A_0 = \frac{\beta R}{\pi}[\log(8R/a) - 2] + \frac{1}{\pi}[0.667(\beta R)^3 - 0.267(\beta R)^5] - j[0.167(\beta R)^4 - 0.033(\beta R)^6]$$

$$A_1 = \left(\beta R - \frac{1}{\beta R}\right)\frac{1}{\pi}[\log(8R/a) - 2]$$

$$+\frac{1}{\pi}[0.667(\beta R)^3 - 0.207(\beta R)^5] - j[0.333(\beta R)^2 - 0.133(\beta R)^4 - 0.026(\beta R)^6]$$

$$A_2 = \left(\beta R - \frac{4}{\beta R}\right)\frac{1}{\pi}[\log(8R/a) - 2.667]$$

$$+\frac{1}{\pi}[-0.4(\beta R)^1 + 0.21(\beta R)^3 - 0.086(\beta R)^5] - j[0.05(\beta R)^4 - 0.012(\beta R)^6] \tag{18.68}$$

This expression corresponds to the approximation of the loop current distribution in the form

$$I(\phi) = \text{const}\left[\frac{1}{A_0} + 2\frac{\cos(\phi)}{A_1} + 2\frac{\cos(2\phi)}{A_2}\right] \tag{18.69}$$

Let us first consider the case when the particle is excited by external high-frequency electric field polarized along the straight wire. In that case the wire antenna operates as a receptor, which supplies power to the radiating loop. Scattering response of the particle can be found as the sum of the power radiated by the currents both in the loop and the wire portions.

Electromotive force V generated in the short dipole antenna by the incident electric field E_z at the point where the two antennas are connected can be expressed as $V = E_z l_{eff}$. The effective length l_{eff} of short dipole antennas is $l_{eff} = l$, while the total geometrical length of the antenna is $2l$. This corresponds to an approximation of the current distribution function by a linear function.

The current in the center of the particle is then $I = V/(Z_l + Z_w)$. Knowing the current, we can find the induced dipole moment: $p = Il_{eff}/(j\omega)$, and the corresponding polarizability component follows as

$$a_{ee}^{zz} = \frac{l_{eff}^2}{j\omega}\frac{1}{Z_l + Z_w} \tag{18.70}$$

The same electromotive force V excites the loop also. The uniform part of the loop current sets a magnetic dipole along the z axis, and the corresponding cross-polarizability component reads

$$a_{me}^{zz} = \mp j\sqrt{\varepsilon\mu}\frac{a^2 l_{eff}}{A_0}\frac{Z_w}{Z_l + Z_w} \tag{18.71}$$

The second Fourier component of the current radiates as an electric dipole directed along the y axis. The electric dipole moment is determined by integration of the $\cos(\phi)$ current component, which gives the polarizability coefficient

$$a_{ee}^{yz} = \mp\frac{2al_{eff}}{\omega\eta}\frac{1}{A_1}\frac{Y_w}{Y_l + Y_w} \tag{18.72}$$

In the above equations $\mp$ corresponds to the right- and left-handed helices, respectively. The other polarizability components can be found in a similar manner, see [30].

The radar cross section of the object can be expressed in the following form:

$$\mathrm{RCS(dB.m^2)} = 10\log_{10}\left[\frac{\eta^2}{4\pi}\frac{(kL^2)^2}{|(Z_l+Z_w)^2|}\right] \tag{18.73}$$

or found in terms of the particle polarizabilities, see [30]. Analytically estimated RCSs have been compared with that computed numerically and experimentally measured. Good agreement is observed at frequencies below and around the main particle resonance.

18.3.5.2 Antenna Representation for the Omega Scatterer

Omega inclusions can be also represented as two connected antennas, only positioned in the same coordinate plane. That means that if the direct field coupling between the loop and dipole antennas can be neglected, the polarizabilities of the omega inclusion can be found if the polarisabilities of the chiral configuration (of the same size) are known. Simple geometrical consideration for the shapes shown in Fig. 18.4 gives the following relations:

$$\begin{aligned} a_{ee}^{zz}\big|_{omega} &= a_{ee}^{zz} + a_{ee}^{yy} + 2a_{ee}^{yz}\big|_{chiral}, & a_{ee}^{xx}\big|_{omega} &= a_{ee}^{xx}\big|_{chiral} \\ a_{mm}^{yy}\big|_{omega} &= a_{mm}^{zz}\big|_{chiral}, & a_{me}^{yz}\big|_{omega} &= a_{me}^{zz} + a_{me}^{zy}\big|_{chiral} \end{aligned} \tag{18.74}$$

Here the polarizability coefficients of the chiral particle are given by formulae (13), (14), (17), (19), (21), (23), (25), (27), (28) from [30].

An important issue is whether the direct interaction (mutual inductance) between the loop and dipole portion is essential and how to calculate that. This question was considered in detail in [32]. Fortunately, in most practical cases this mutual inductance can be neglected.

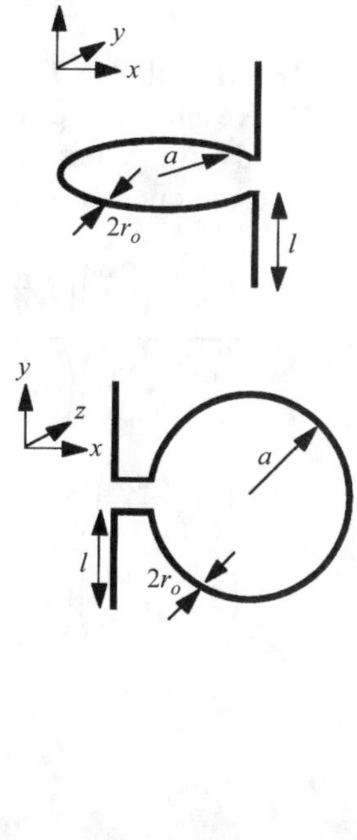

Figure 18.4 Geometry of omega scatterers.

18.3.6 Composite Modeling: Effective Medium Parameters

18.3.6.1 Isotropic Chiral Composites

After studying the scattering behavior of a single inclusion, the final step of the composite modeling is the averaging procedure that results in the effective material parameters. To study a chiral material with N inclusions in a unit volume, we use the generalized Maxwell Garnett mixing rule.

In this approach, macroscopic properties of an isotropic medium are described by its polarization **P** and magnetization **M** vectors

$$\mathbf{D} = \varepsilon\mathbf{E} + \mathbf{P} \tag{18.75}$$

$$\mathbf{B} = \mu(\mathbf{H} + \mathbf{M}) \tag{18.76}$$

where ε and μ are permittivity and permeability of the host medium. **P** and **M** vectors are obtained by addition of all dipole moments per unit volume. For an isotropic chiral medium with randomly dispersed and oriented helices, we can write

$$\mathbf{P} = N\mathbf{p} = N\varepsilon(a_{ee}\mathbf{E}_{loc} + a_{em}\,\eta\mathbf{H}_{loc}) \tag{18.77}$$

$$\mathbf{M} = N\mathbf{m} = N\left(\frac{a_{em}}{\eta}\mathbf{E}_{loc} + a_{mm}\mathbf{H}_{loc}\right) \tag{18.78}$$

where a_{ee}, a_{me}, and $a_{em} = -a_{me}$ are scalar polarizabilities that are obtained by averaging diagonal terms of polarizabilities tensors $\bar{\bar{a}}_{ee}$, $\bar{\bar{a}}_{mm}$, $\bar{\bar{a}}_{em}$, and $\bar{\bar{a}}_{me}$. For example, $a_{em} = (a_{em11} + a_{em22} + a_{em33})/3$. In (18.77) and (18.78), one can notice that **E** and **H** vectors have been replaced by $\mathbf{E}_{loc}$ and $\mathbf{H}_{loc}$ which represent local field to the inclusion. These local fields result in the addition of the external applied field and the average contribution of other surrounding chiral inclusions. Considering Lorentzian fields as for spherical particles we can write

$$\mathbf{E}_{loc} = \mathbf{E} + \mathbf{P}/3\varepsilon \tag{18.79}$$

$$\mathbf{H}_{loc} = \mathbf{H} + \mathbf{M}/3 \tag{18.80}$$

Combining (18.79) and (18.80) with (18.77) and (18.78), **P** and **M** are expressed as functions of **E** and **H**. Results are put in (18.75) and (18.76), and the obtained formulas are compared with general constitutive (18.4). Then expressions of electromagnetic effective parameters can be derived:

$$\varepsilon_{LS} = \varepsilon\frac{\left(1 + 2N\dfrac{\alpha_{ee}}{3}\right)\left(1 - N\dfrac{\alpha_{mm}}{3}\right) - 2N^2\dfrac{\alpha_{em}^2}{9}}{\left(1 - N\dfrac{\alpha_{ee}}{3}\right)\left(1 - N\dfrac{\alpha_{mm}}{3}\right) + N^2\dfrac{\alpha_{em}^2}{9}} \tag{18.81}$$

$$\mu_{LS} = \mu\frac{\left(1 - N\dfrac{a_{ee}}{3}\right)\left(1 + 2N\dfrac{a_{mm}}{3}\right) - 2N^2\dfrac{a_{em}^2}{9}}{\left(1 - N\dfrac{a_{ee}}{3}\right)\left(1 - N\dfrac{a_{mm}}{3}\right) + N^2\dfrac{a_{em}^2}{9}} \tag{18.82}$$

$$j\chi_{LS} = \frac{-N\alpha_{em}\sqrt{\varepsilon\mu}}{\left(1 - N\frac{a_{ee}}{3}\right)\left(1 - N\frac{a_{mm}}{3}\right) + N^2\frac{\alpha_{em}^2}{9}} \tag{18.83}$$

These parameters can now describe the chiral medium under study. This formulation is valid for isotropic chiral medium. It is limited to low concentrations, because for dense media, (18.79) and (18.80) are no longer valid. Nevertheless, this modeling is powerful for most of the artificial chiral composites that can be realized.

18.3.6.2 Bianisotropic Composites

Similar physical ideas can be applied to the most general bianisotropic composite when all the medium parameters are dyadic quantities. Fields within a small spherical shell cut out of the bulk of a uniform bianisotropic composite can be solved from the quasistatic field equations, which gives (18.79 and 18.80). Polarizations can then be written as (18.77 and 18.78) but with dyadic coefficients. Elimination of the local fields from the above relations eventually leads to constitutive relations for the composite. This elimination can be done in different fashions which give equivalent, but formally different final expressions. In the present case the difference in the mathematical form is quite important, since some of the involved dyadics are incomplete. It appears then of advantage to choose such a form which does not involve any inverses of these dyadics.

One such procedure is to substitute first the local field expressions in the expressions for the polarizabilities. Then, eliminating vector **M** we get the expression for the effective permittivity and one of the coupling dyadics in the form

$$\bar{\bar{\varepsilon}}_{eff} = \varepsilon\varepsilon_0\bar{\bar{I}} + \bar{\bar{K}}_1^{-1}\cdot\left[N\bar{\bar{a}}_{ee} + N^2\frac{\bar{\bar{a}}_{em}}{3\mu\mu_0}\cdot\left(\bar{\bar{I}} - N\frac{\bar{\bar{a}}_{mm}}{3\mu\mu_0}\right)^{-1}\cdot\bar{\bar{a}}_{me}\right] \tag{18.84}$$

$$\bar{\bar{\alpha}}_{eff} = \bar{\bar{K}}_1^{-1}\cdot N\bar{\bar{a}}_{em}\left[\bar{\bar{I}} + \left(\bar{\bar{I}} - N\frac{\bar{\bar{a}}_{mm}}{3\mu\mu_0}\right)^{-1}\cdot N\frac{\bar{\bar{a}}_{mm}}{3\mu\mu_0}\right] \tag{18.85}$$

where $\bar{\bar{K}}_1 = \bar{\bar{I}} - N\frac{\bar{\bar{a}}_{ee}}{3\varepsilon\varepsilon_0} - N^2\frac{\bar{\bar{a}}_{em}}{3\mu\mu_0}\cdot\left(\bar{\bar{I}} - N\frac{\bar{\bar{a}}_{mm}}{3\mu\mu_0}\right)^{-1}\cdot\frac{\bar{\bar{a}}_{me}}{3\varepsilon\varepsilon_0}$

In the same way, eliminating first vector **P**, we obtain the permeability dyadic and the other cross-coupling coefficient:

$$\bar{\bar{\mu}}_{eff} = \mu\mu_0\bar{\bar{I}} + \bar{\bar{K}}_2^{-1}\cdot\left[N\bar{\bar{a}}_{mm} + N^2\frac{\bar{\bar{a}}_{me}}{3\varepsilon\varepsilon_0}\cdot\left(\bar{\bar{I}} - N\frac{\bar{\bar{a}}_{ee}}{3\varepsilon\varepsilon_0}\right)^{-1}\cdot\bar{\bar{a}}_{em}\right] \tag{18.86}$$

$$\bar{\bar{\beta}}_{eff} = \bar{\bar{K}}_2^{-1}\cdot N\bar{\bar{a}}_{me}\left[\bar{\bar{I}} + \left(\bar{\bar{I}} - N\frac{\bar{\bar{a}}_{ee}}{3\varepsilon\varepsilon_0}\right)^{-1}\cdot N\frac{\bar{\bar{a}}_{ee}}{3\varepsilon\varepsilon_0}\right] \tag{18.87}$$

Here $\bar{\bar{K}}_2 = \bar{\bar{I}} - N\frac{\bar{\bar{a}}_{mm}}{3\mu\mu_0} - N^2\frac{\bar{\bar{a}}_{me}}{3\varepsilon\varepsilon_0}\cdot\left(\bar{\bar{I}} - N\frac{\bar{\bar{a}}_{ee}}{3\varepsilon\varepsilon_0}\right)^{-1}\cdot\frac{\bar{\bar{a}}_{em}}{3\mu\mu_0}$. Since no inverses of incomplete dyadics are involved here, the parameters can be easily evaluated upon substitution of expressions for the individual polarizabilities. In the general analysis of mixing rules for bianisotropic media, six-dyadic formalism [33] is very convenient.

That allows us to obtain the general results in rather compact and physically clear form.

18.3.6.3 A Relation Between the Polarizabilities

Using a simple quasistatic model of symmetrical wire-and-loop particles [9], it was shown that the polarizabilities of such particles are related as $a_{ee}a_{mm} = a_{em}a_{me}$. In that study, the particles were assumed to be uniaxial objects, so that the polarizabilities along all the directions but the particle axis could be neglected. Also in more general cases of bianisotropic scatterers the relation

$$\overline{\overline{a}}_{ee}^{T} \cdot \overline{\overline{a}}_{mm} = \overline{\overline{a}}_{em}^{T} \cdot \overline{\overline{a}}_{me} \tag{18.88}$$

holds [34]. This is not a universal relation since it reflects a certain symmetry of particles. An example when the relation does not hold is given in [34]. However, it holds in most practical cases, at least approximately. First of all, the existence of such relation clearly shows that the effective medium parameters of chiral and other bianisotropic mixtures are not independent. The permittivity (determined basically by the electric polarizability of inclusions $\overline{\overline{a}}_{ee}$), permeability (dependent on $\overline{\overline{a}}_{mm}$), and the effective chirality parameter (in low-density mixtures that is proportional to the average of $\overline{\overline{a}}_{em}$, see the above mixture rules) are all connected, so that all three parameters have resonances at common resonance frequencies. Moreover, one can not increase the chirality parameter without affecting the permittivity and permeability. Physically that is a consequence of the fact that it is the same current excited in an inclusion which determines all the polarization phenomena.

Second, the relation between the polarizabilities is important in analyzing mixtures of complex bianisotropic inclusions, since exactly the difference between the right- and left-hand sides of (18.88) comes into the Maxwell Garnett averaging relations for chiral mixtures (18.81)–(18.83). This simplifies the calculations and also leads to the conclusion that the second-order terms (with respect to the concentration) in the Maxwell Garnett relation vanish, so that the effective permittivity and permeability in that approximation do not depend on the position-averaged chirality parameter of a single inclusion.

18.4 REFLECTION AND TRANSMISSION IN CHIRAL AND OMEGA SLABS: APPLICATIONS

The general problem of reflection and transmission for chiral and uniaxial omega slabs, placed in vacuum or metal-backed, is treated. The problem is studied for an arbitrary incidence of a plane wave.

18.4.1 Continuity Problems with a Chiral Medium

The first problem to study is the field continuity along a dielectric/chiral interface between two semi-infinite media. The problem under study is presented in Fig. 18.5.

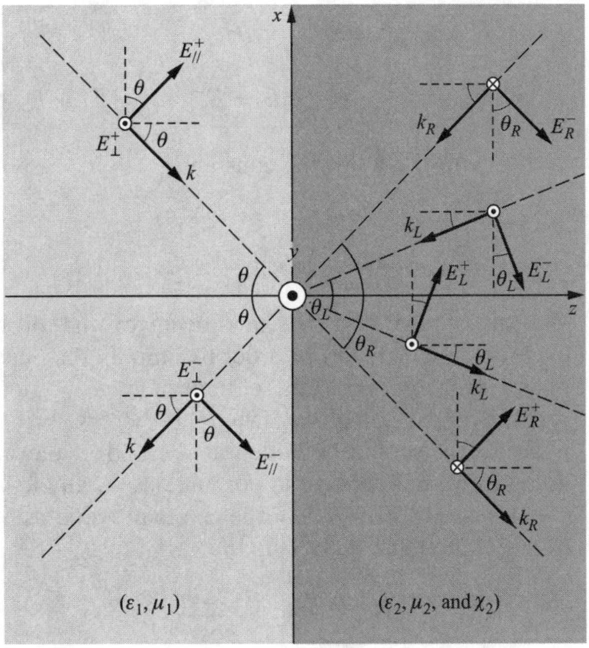

Figure 18.5 Dielectric/chiral interface.

We assume that the incoming wave is a plane wave. The discontinuity plane is located at a general z coordinate. On the left the propagation is a classic plane wave propagation in an isotropic medium. On the right, propagation is divided into RCP and LCP modes as presented previously, and electromagnetic fields are given by (18.21) to (18.24).

The use of Snell-Descartes's law allows us to obtain a relation between the angles of incidence, reflection, and transmission for each wave. It can be expressed by

$$k \sin\theta = k_L \sin\theta_L = k_R \sin\theta_R \tag{18.89}$$

where k is the propagation constant in the left medium: $k = \sqrt{\varepsilon_1 \mu_1}$. The next step is to apply the continuity condition for the tangential components of electromagnetic fields at the interface:

- E_x *component*: $E_{x1} = E_{x2}$

$$E_{\parallel}^{+}\cos\theta - E_{\parallel}^{-}\cos\theta = E_L^{+}\cos\theta_L + E_R^{+}\cos\theta_R - E_L^{-}\cos\theta_L - E_R^{-}\cos\theta_R \tag{18.90}$$

- E_y *component*: $E_{y1} = E_{y2}$

$$E_{\perp}^{+} + E_{\perp}^{-} = j(E_L^{+} - E_R^{+} + E_L^{-} - E_R^{-}) \tag{18.91}$$

- H_x *component*: $H_{x1} = H_{x2}$

$$\frac{1}{Z_1}(-E_{\perp}^{+}\cos\theta + E_{\perp}^{-}\cos\theta) =$$

$$\frac{j}{Z_2}(-E_L^{+}\cos\theta_L + E_R^{+}\cos\theta_R + E_L^{-}\cos\theta_L - E_R^{-}\cos\theta_R) \tag{18.92}$$

- H_y *component*: $H_{y1} = H_{y2}$

$$\frac{1}{Z_1}(E_{\parallel}^{+} + E_{\parallel}^{-}) = \frac{1}{Z_2}(E_L^{+} + E_R^{+} + E_L^{-} + E_R^{-}) \tag{18.93}$$

with the following constants:

$$Z_1 = \sqrt{\mu_1/\varepsilon_1} \qquad Z_2 = \sqrt{\mu_2/\varepsilon_2}$$

$$\text{and} \qquad k_R = \omega(\sqrt{\varepsilon_2\mu_2} - \chi_2) \qquad k_L = \omega(\sqrt{\varepsilon_2\mu_2} + \chi_2) \tag{18.94}$$

We can notice that the unknowns of the problem are the amplitudes of the electromagnetic fields. In order to simplify the equations, we have chosen to omit the phase term in each of these amplitudes to be determined. So, we have written E_L^{+} instead of the complete term $E_L^{+}e^{-jk_L x \sin\theta_L}\, e^{+jk_L z \cos\theta_L}$. This detail is important: the phase terms should be added at the end of the development, because the phase is going to have a decisive importance especially for slabs with finite thickness.

Using (18.90)–(18.93) one can express fields of medium 1 as a function of fields of chiral medium. Assuming that

$$\beta_L = \frac{\cos\theta_L}{\cos\theta} \qquad \text{and} \qquad \beta_R = \frac{\cos\theta_R}{\cos\theta} \tag{18.95}$$

$$\alpha = \frac{Z_1}{Z_2} \tag{18.96}$$

we finally get a matrix system that represents the electromagnetic phenomena at the discontinuity surface:

$$\begin{bmatrix} E_{\parallel}^{+} \\ E_{\perp}^{+} \\ E_{\parallel}^{-} \\ E_{\perp}^{-} \end{bmatrix} = \frac{1}{2}\begin{bmatrix} (\alpha+\beta_L) & (\alpha+\beta_R) & (\alpha-\beta_L) & (\alpha-\beta_R) \\ j(1+\alpha\beta_L) & -j(1+\alpha\beta_R) & j(1-\alpha\beta_L) & -j(1-\alpha\beta_R) \\ (\alpha-\beta_L) & (\alpha-\beta_R) & (\alpha+\beta_L) & (\alpha+\beta_R) \\ j(1-\alpha\beta_L) & -j(1-\alpha\beta_R) & j(1+\alpha\beta_L) & -j(1+\alpha\beta_R) \end{bmatrix}\begin{bmatrix} E_L^{+} \\ E_R^{+} \\ E_L^{-} \\ E_R^{-} \end{bmatrix} \tag{18.97}$$

To study a chiral/non-chiral interface, imagine that the incident wave is already coming from the left but the left medium is now a chiral one. As a consequence the right medium is a dielectric or magnetic material. Using the same method or just inverting the matrix (18.97) will give the result:

$$\begin{bmatrix} E_L^{+} \\ E_R^{+} \\ E_L^{-} \\ E_R^{-} \end{bmatrix} = \frac{1}{4}\begin{bmatrix} \left(\frac{1}{\alpha}+\frac{1}{\beta_L}\right) & -j\left(1+\frac{1}{\alpha\beta_L}\right) & \left(\frac{1}{\alpha}-\frac{1}{\beta_L}\right) & -j\left(1-\frac{1}{\alpha\beta_L}\right) \\ \left(\frac{1}{\alpha}+\frac{1}{\beta_R}\right) & +j\left(1+\frac{1}{\alpha\beta_R}\right) & \left(\frac{1}{\alpha}-\frac{1}{\beta_R}\right) & +j\left(1-\frac{1}{\alpha\beta_R}\right) \\ \left(\frac{1}{\alpha}-\frac{1}{\beta_L}\right) & -j\left(1-\frac{1}{\alpha\beta_L}\right) & \left(\frac{1}{\alpha}+\frac{1}{\beta_L}\right) & -j\left(1+\frac{1}{\alpha\beta_L}\right) \\ \left(\frac{1}{\alpha}-\frac{1}{\beta_R}\right) & +j\left(1-\frac{1}{\alpha\beta_R}\right) & \left(\frac{1}{\alpha}+\frac{1}{\beta_R}\right) & +j\left(1+\frac{1}{\alpha\beta_R}\right) \end{bmatrix}\begin{bmatrix} E_{\parallel}^{+} \\ E_{\perp}^{+} \\ E_{\parallel}^{-} \\ E_{\perp}^{-} \end{bmatrix} \tag{18.98}$$

As a second problem, we are now going to study the interaction of a chiral medium with a perfect metallic conductor plane as shown in Fig. 18.6.

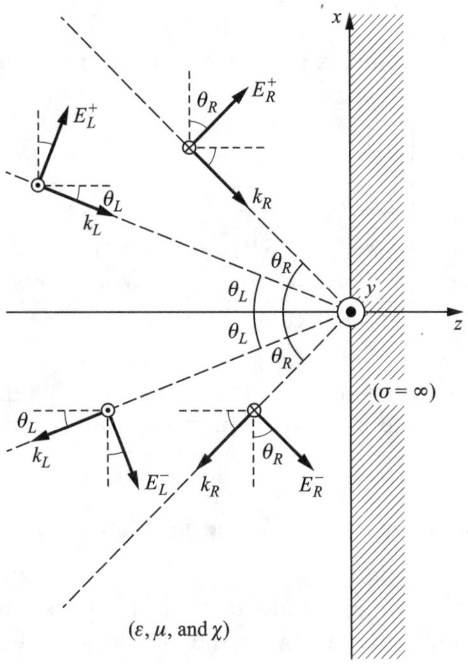

Figure 18.6 Continuity problem between a chiral medium and a metallic plane.

The mathematical development is done in the same way as previously described. As the fields vanish in the ideally conducting medium, we are going to obtain relations between the incident and reflected waves.

- *Boundary condition on E_x component*: $E_x = 0$

$$E_L^+ \cos\theta_L + E_R^+ \cos\theta_R - E_L^- \cos\theta_L - E_R^- \cos\theta_R = 0 \tag{18.99}$$

- *Boundary condition on E_y component*: $E_y = 0$

$$E_L^+ - E_R^+ + E_L^- - E_R^- = 0 \tag{18.100}$$

By combination of these two equations we get the matrix relation

$$\begin{bmatrix} E_L^- \\ E_R^- \end{bmatrix} = \begin{bmatrix} \dfrac{\cos\theta_L - \cos\theta_R}{\cos\theta_L + \cos\theta_R} & \dfrac{2\cos\theta_R}{\cos\theta_L + \cos\theta_R} \\ \dfrac{2\cos\theta_L}{\cos\theta_L + \cos\theta_R} & \dfrac{\cos\theta_R - \cos\theta_L}{\cos\theta_L + \cos\theta_R} \end{bmatrix} \begin{bmatrix} E_L^+ \\ E_R^+ \end{bmatrix} \tag{18.101}$$

As the problem is supposed to be invariant along the x direction, the phase term in the unknown does not depend on the x-coordinate. Assuming that, we write

$$e^{\varphi_L} = e^{+jk_L z\cos\theta_L} \tag{18.102}$$

$$e^{\varphi_R} = e^{+jk_R z\cos\theta_R} \tag{18.103}$$

Expanding the matrix to the 4×4 size, gives a more general matrix that will be useful for more complex problems with several interfaces. So, relation (18.101) becomes

$$\begin{bmatrix} E_L^+ \\ E_R^+ \\ E_L^- \\ E_R^- \end{bmatrix} = \begin{bmatrix} 1 & 0 & 0 & 0 \\ 0 & 1 & 0 & 0 \\ \dfrac{\cos\theta_L - \cos\theta_R}{\cos\theta_L + \cos\theta_R} e^{2\varphi_L} & \dfrac{2\cos\theta_R}{\cos\theta_L + \cos\theta_R} e^{\varphi_L \varphi_R} & 0 & 0 \\ \dfrac{2\cos\theta_L}{\cos\theta_L + \cos\theta_R} e^{\varphi_L \varphi_R} & \dfrac{\cos\theta_R - \cos\theta_L}{\cos\theta_L + \cos\theta_R} e^{2\varphi_R} & 0 & 0 \end{bmatrix} \begin{bmatrix} E_L^+ \\ E_R^+ \\ E_L^- \\ E_R^- \end{bmatrix} \quad (18.104)$$

In this equation E_L^+, E_R^+, E_L^-, and E_R^- only represent the amplitude of the fields since the phase terms have been included in the matrix.

18.4.2 Properties of a Single Slab

Now we can study a practical application. A chiral slab is placed in the air (see Fig. 18.7). Referring to the previous study, now the structure has two separate interfaces where the fields continuity must be enforced. First, the fields must be expressed in each medium ($z < 0$; $0 < z < d$ and $z > d$). Fields in the chiral medium are given by (18.21) to (18.24). For each interface, continuity conditions must be applied as we have done previously. So, electromagnetic interactions along the plane $z = 0$ are described by the matrix equation (18.97). At $z = d$, the problem can be represented by (18.98). With the phase term included, it leads to

$$\begin{bmatrix} E_L^+ \\ E_R^+ \\ E_L^- \\ E_R^- \end{bmatrix} = \begin{bmatrix} e^{-j\varphi_L} & 0 & 0 & 0 \\ 0 & e^{-j\varphi_R} & 0 & 0 \\ 0 & 0 & e^{+j\varphi_L} & 0 \\ 0 & 0 & 0 & e^{+j\varphi_R} \end{bmatrix}$$

$$\cdot \frac{1}{4} \begin{bmatrix} \left(\dfrac{1}{\alpha} + \dfrac{1}{\beta_L}\right) & -j\left(1 + \dfrac{1}{\alpha\beta_L}\right) & \left(\dfrac{1}{\alpha} - \dfrac{1}{\beta_L}\right) & -j\left(1 - \dfrac{1}{\alpha\beta_L}\right) \\ \left(\dfrac{1}{\alpha} + \dfrac{1}{\beta_R}\right) & +j\left(1 + \dfrac{1}{\alpha\beta_R}\right) & \left(\dfrac{1}{\alpha} - \dfrac{1}{\beta_R}\right) & +j\left(1 - \dfrac{1}{\alpha\beta_R}\right) \\ \left(\dfrac{1}{\alpha} - \dfrac{1}{\beta_L}\right) & -j\left(1 - \dfrac{1}{\alpha\beta_L}\right) & \left(\dfrac{1}{\alpha} + \dfrac{1}{\beta_L}\right) & -j\left(1 + \dfrac{1}{\alpha\beta_L}\right) \\ \left(\dfrac{1}{\alpha} - \dfrac{1}{\beta_R}\right) & +j\left(1 - \dfrac{1}{\alpha\beta_R}\right) & \left(\dfrac{1}{\alpha} + \dfrac{1}{\beta_R}\right) & +j\left(1 + \dfrac{1}{\alpha\beta_R}\right) \end{bmatrix} \begin{bmatrix} E_{\|3}^+ \\ E_{\perp 3}^+ \\ E_{\|3}^- \\ E_{\perp 3}^- \end{bmatrix}$$

with

$$\varphi_L = k_L d \cos\theta_L \qquad \varphi_R = k_R d \cos\theta_R \qquad \text{and} \qquad \varphi_3 = k_o d \cos\theta \quad (18.105)$$

The purpose is now to express the fields in medium 1 (incident and reflected waves)

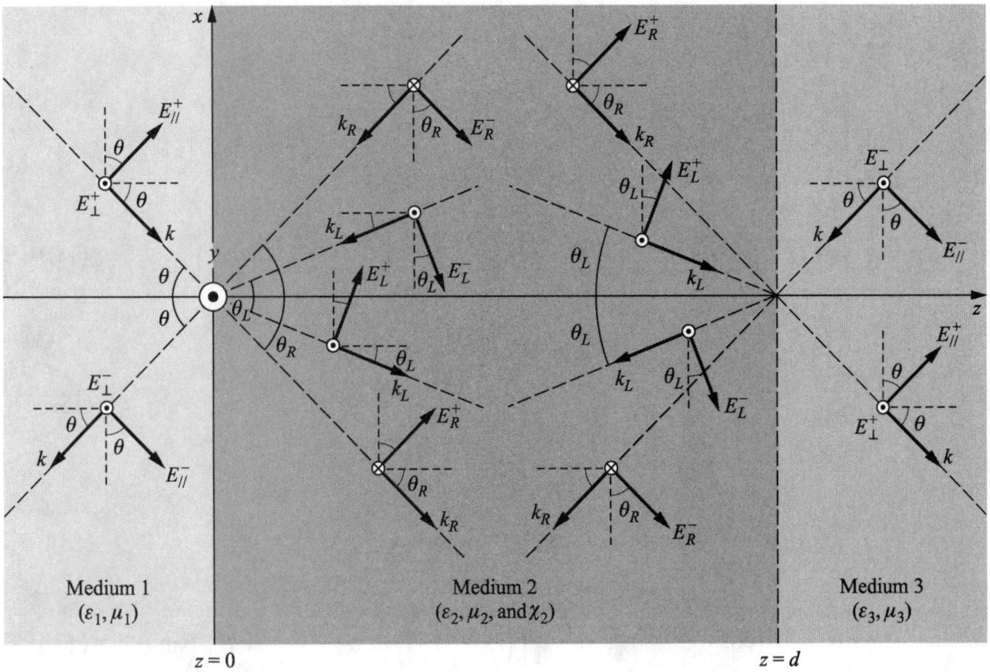

Figure 18.7 Continuity problems of a chiral slab. General structure of a chiral slab in the air.

as a function of fields in medium 3 (transmitted waves). Using (18.105) and (18.97) we get

$$\begin{bmatrix} E^+_{\|1} \\ E^+_{\perp 1} \\ E^-_{\|1} \\ E^-_{\perp 1} \end{bmatrix} = \begin{bmatrix} a_{11} & -a_{21} & -a_{31} & +a_{41} \\ a_{21} & a_{22} & +a_{32} & -a_{42} \\ a_{31} & a_{32} & a_{33} & -a_{43} \\ a_{41} & a_{42} & a_{43} & a_{44} \end{bmatrix} \begin{bmatrix} E^+_{\|3} \\ E^+_{\perp 3} \\ E^-_{\|3} \\ E^-_{\perp 3} \end{bmatrix} \tag{18.106}$$

where the components of the matrix are

$$a_{11} = \frac{1}{2}(\cos\varphi_L + \cos\varphi_R) - \frac{j}{4}\left[\left(\frac{\alpha}{\beta_L} + \frac{\beta_L}{\alpha}\right)\sin\varphi_L + \left(\frac{\alpha}{\beta_R} + \frac{\beta_R}{\alpha}\right)\sin\varphi_R\right] \tag{18.107a}$$

$$a_{21} = \frac{j}{4}\left(\alpha + \frac{1}{\alpha}\right)(\cos\varphi_L - \cos\varphi_R) + \frac{1}{4}\left[\left(\frac{1}{\beta_L} + \beta_L\right)\sin\varphi_L - \left(\frac{1}{\beta_R} + \beta_R\right)\sin\varphi_R\right] \tag{18.107b}$$

$$a_{22} = \frac{1}{2}(\cos\varphi_L + \cos\varphi_R) - \frac{j}{4}\left[\left(\alpha\beta_L + \frac{1}{\alpha\beta_L}\right)\sin\varphi_L + \left(\alpha\beta_R + \frac{1}{\alpha\beta_R}\right)\sin\varphi_R\right] \tag{18.107c}$$

$$a_{31} = \frac{j}{4}\left(\frac{\beta_L}{\alpha} - \frac{\alpha}{\beta_L}\right)\sin\varphi_L + \frac{j}{4}\left(\frac{\beta_R}{\alpha} - \frac{\alpha}{\beta_R}\right)\sin\varphi_R \tag{18.107d}$$

$$a_{32} = -\frac{j}{4}\left(\alpha - \frac{1}{\alpha}\right)(\cos\varphi_L - \cos\varphi_R) + \frac{1}{4}\left[\left(\beta_L - \frac{1}{\beta_L}\right)\sin\varphi_L - \left(\beta_R - \frac{1}{\beta_R}\right)\sin\varphi_R\right] \tag{18.107e}$$

$$a_{33} = \frac{1}{2}(\cos\varphi_L + \cos\varphi_R) + \frac{j}{4}\left[\left(\frac{\alpha}{\beta_L} + \frac{\beta_L}{\alpha}\right)\sin\varphi_L + \left(\frac{\alpha}{\beta_R} + \frac{\beta_R}{\alpha}\right)\sin\varphi_R\right] \tag{18.107f}$$

$$a_{41} = -\frac{j}{4}\left(\alpha - \frac{1}{\alpha}\right)(\cos\varphi_L - \cos\varphi_R) + \frac{1}{4}\left[\left(\frac{1}{\beta_L} - \beta_L\right)\sin\varphi_L - \left(\frac{1}{\beta_R} - \beta_R\right)\sin\varphi_R\right] \tag{18.107g}$$

$$a_{42} = \frac{j}{4}\left[\left(\alpha\beta_L - \frac{1}{\alpha\beta_L}\right)\sin\varphi_L + \left(\alpha\beta_R - \frac{1}{\alpha\beta_R}\right)\sin\varphi_R\right] \tag{18.107h}$$

$$a_{43} = \frac{j}{4}\left(\alpha + \frac{1}{\alpha}\right)(\cos\varphi_L - \cos\varphi_R) - \frac{1}{4}\left[\left(\frac{1}{\beta_L} + \beta_L\right)\sin\varphi_L - \left(\frac{1}{\beta_R} + \beta_R\right)\sin\varphi_R\right] \tag{18.107i}$$

$$a_{44} = \frac{1}{2}(\cos\varphi_L + \cos\varphi_R) + \frac{j}{4}\left[\left(\alpha\beta_L + \frac{1}{\alpha\beta_L}\right)\sin\varphi_L + \left(\alpha\beta_R + \frac{1}{\alpha\beta_R}\right)\sin\varphi_R\right] \tag{18.107j}$$

with α, β_L, and β_R given by relations (18.95) and (18.96).

If the medium number 3 is a semi-infinite medium, it is obvious that $E_{\parallel 3}^-$ and $E_{\perp 3}^-$ do not exist. So $E_{\parallel 3}^- = E_{\perp 3}^- = 0$, and (18.106) becomes

$$\begin{bmatrix} E_{\parallel 1}^+ \\ E_{\perp 1}^+ \end{bmatrix} = \begin{bmatrix} a_{11} & -a_{21} \\ a_{21} & a_{22} \end{bmatrix}\begin{bmatrix} E_{\parallel 3}^+ \\ E_{\perp 3}^+ \end{bmatrix} \quad \text{and} \quad \begin{bmatrix} E_{\parallel 1}^- \\ E_{\perp 1}^- \end{bmatrix} = \begin{bmatrix} a_{31} & a_{32} \\ a_{41} & a_{42} \end{bmatrix}\begin{bmatrix} E_{\parallel 3}^+ \\ E_{\perp 3}^+ \end{bmatrix} \tag{18.108}$$

To obtain the transmission matrix, we must obtain a relation as follows:

$$\begin{bmatrix} E_{\parallel 3}^+ \\ E_{\perp 3}^+ \end{bmatrix} = [T]\begin{bmatrix} E_{\parallel 1}^+ \\ E_{\perp 1}^+ \end{bmatrix} \tag{18.109}$$

It can be easily derived from (18.108):

$$\begin{bmatrix} E_{\parallel 3}^+ \\ E_{\perp 3}^+ \end{bmatrix} = \begin{bmatrix} a_{11} & -a_{21} \\ a_{21} & a_{22} \end{bmatrix}^{-1}\begin{bmatrix} E_{\parallel 1}^+ \\ E_{\perp 1}^+ \end{bmatrix} = \frac{1}{a_{11}a_{22} + a_{21}^2}\begin{bmatrix} a_{22} & +a_{21} \\ -a_{21} & a_{11} \end{bmatrix} \tag{18.110}$$

On the other side of the structure under study, the reflection matrix relates the reflected fields to the incident wave in the following way:

$$\begin{bmatrix} E_{\parallel 1}^- \\ E_{\perp 1}^- \end{bmatrix} = [R]\begin{bmatrix} E_{\parallel 1}^+ \\ E_{\perp 1}^+ \end{bmatrix} \tag{18.111}$$

The solution comes from (18.108) and (18.110):

$$\begin{bmatrix} E_{\|1}^{-} \\ E_{\perp 1}^{-} \end{bmatrix} = \begin{bmatrix} a_{31} & a_{32} \\ a_{41} & a_{42} \end{bmatrix} [T] \begin{bmatrix} E_{\|1}^{+} \\ E_{\perp 1}^{+} \end{bmatrix} = \frac{1}{a_{11}a_{22} + a_{21}^2} \begin{bmatrix} a_{31}a_{22} - a_{32}a_{21} & a_{31}a_{21} + a_{32}a_{11} \\ a_{41}a_{22} - a_{42}a_{21} & a_{41}a_{21} - a_{42}a_{11} \end{bmatrix} \tag{18.112}$$

If the wave is coming upon the slab at the normal incidence, it leads to a lot of simplifications. Indeed, if $\theta = 0$ then $\beta_L = \beta_R = 1$. So, some relations between a_{ij} terms of the general matrix follow. In particular, $a_{11} = a_{22}$, $a_{41} = a_{32}$ and $a_{42} = -a_{31}$. Transmission and reflection matrices for the normal incidence have the form:

$$\begin{bmatrix} E_{\|3}^{+} \\ E_{\perp 3}^{+} \end{bmatrix} = [T] \begin{bmatrix} E_{\|1}^{+} \\ E_{\perp 1}^{+} \end{bmatrix} = \begin{bmatrix} S_{21co} & -S_{21cross} \\ S_{21cross} & S_{21co} \end{bmatrix} \begin{bmatrix} E_{\|1}^{+} \\ E_{\perp 1}^{+} \end{bmatrix} \tag{18.113}$$

$$\begin{bmatrix} E_{\|1}^{-} \\ E_{\perp 1}^{-} \end{bmatrix} = [R] \begin{bmatrix} E_{\|1}^{+} \\ E_{\perp 1}^{+} \end{bmatrix} = \begin{bmatrix} S_{11} & 0 \\ 0 & -S_{11} \end{bmatrix} \begin{bmatrix} E_{\|1}^{+} \\ E_{\perp 1}^{+} \end{bmatrix} \tag{18.114}$$

with

$$S_{21co} = \frac{2Z_0 Z \cos(\omega\chi d)}{2Z_0 Z \cos(kd) - j(Z_0^2 + Z^2)\sin(kd)} \tag{18.115}$$

$$S_{21cross} = \frac{2Z_0 Z \sin(\omega\chi d)}{2Z_0 Z \cos(kd) - j(Z_0^2 + Z^2)\sin(kd)} \tag{18.116}$$

$$S_{11} = \frac{-j(Z_0^2 - Z^2) Z_0 Z \sin(kd)}{2Z_0 Z \cos(kd) - j(Z_0^2 + Z^2)\sin(kd)} \tag{18.117}$$

and $\qquad k = \omega\sqrt{\varepsilon\mu} \qquad Z = \sqrt{\mu/\varepsilon} \qquad$ and $\qquad Z_0 = \sqrt{\mu_0/\varepsilon_0}$

We can notice that the opposite sign that appears in these matrices just comes from an arbitrarily chosen directions convention employed as is seen in Fig. 18.7. It means that actually the fields components are directed in the opposite sense compared to that in the picture.

Physically, we can remark that the chirality coefficient χ does not influence the reflection coefficient (S_{11}) at the normal incidence. Now, if we assume that the chirality parameter χ is purely real, the wave transmitted to the output medium (medium 3) is linearly polarized. However, the transmitted electric field has both a copolarized component (parallel to incident electric field) and a cross-polarized component (perpendicular to the incident electric field). So it appears that the polarization plane has been rotated at an angle of $\omega\chi$d (since the ratio between the parallel and the perpendicular components of transmitted field is equal to tan($\omega\chi$d)). This phenomenon is well known as the optical activity.

In case the chiral coefficient is complex, the wave suffers an additional phenomenon called *circular dichroism*. The two propagation modes (RCP and LCP) in the chiral medium will have different attenuation constants due to the imaginary part of χ. Their combination will give no longer a linear polarization but an elliptical polarization. For small chirality, rotation and ellipticity can be described in the following way (we denote $\chi = \chi' + j\chi''$):

$$\textbf{Rotation: } \theta = \omega\chi' d \qquad \text{and} \qquad \textbf{Ellipticity: } \varphi = \omega\chi'' d \tag{18.118}$$

The phenomenon can be also expressed using the Stokes parameters:

$$\begin{aligned} S_0 &= \|S_{21co}\|^2 + \|S_{21cross}\|^2 \\ S_1 &= \|S_{21co}\|^2 - \|S_{21cross}\|^2 \\ S_2 &= -2\Re e(S_{21co} \cdot S^*_{21cross}) \\ S_3 &= +2\Im m(S_{21co} \cdot S^*_{21cross}) \end{aligned} \tag{18.119}$$

The angles can be obtained from the following relations:

$$\begin{aligned} &\textbf{Rotation:} \qquad \theta = \frac{1}{2}\arctan\left(\frac{S_2}{S_1}\right) \pm n\pi \\ &\textbf{Ellipticity:} \qquad \varphi = \frac{1}{2}\arctan\left(\frac{S_3}{S_0}\right) \pm n\pi \end{aligned} \tag{18.120}$$

18.4.3 Properties of a Chiral Dällenbach Screen

The general chiral structure called "Dällenbach screen" can be depicted using Fig. 18.7. It can be described as a metal-backed chiral slab, as to say that medium 3 (in Fig. 18.7) is now replaced by a metallic medium. The study of this problem can be done using matrix (18.97) for the interface at $z = 0$ and matrix (18.104) for the plane located at $z = d$. Combining these two matrices leads to the following system of equations:

$$\begin{bmatrix} E_{\|}^{+} \\ E_{\perp}^{+} \end{bmatrix} = \begin{bmatrix} A \end{bmatrix} \begin{bmatrix} E_L^{+} \\ E_R^{+} \end{bmatrix} \tag{18.121}$$

$$\begin{bmatrix} E_{\|}^{-} \\ E_{\perp}^{-} \end{bmatrix} = \begin{bmatrix} B \end{bmatrix} \begin{bmatrix} E_L^{+} \\ E_R^{+} \end{bmatrix} \tag{18.122}$$

with

$$A_{11} = (\alpha + \beta_L) + \alpha_{11}(\alpha - \beta_L) + \alpha_{21}(\alpha - \beta_R) \tag{18.123a}$$

$$A_{12} = (\alpha + \beta_R) + \alpha_{12}(\alpha - \beta_L) + \alpha_{22}(\alpha - \beta_R) \tag{18.123b}$$

$$A_{21} = j(1 + \alpha\beta_L) + j\alpha_{11}(1 - \alpha\beta_L) - j\alpha_{21}(1 - \alpha\beta_R) \tag{18.123c}$$

$$A_{22} = -j(1 + \alpha\beta_R) + j\alpha_{12}(1 - \alpha\beta_L) - j\alpha_{22}(1 - \alpha\beta_R) \tag{18.123d}$$

$$B_{11} = (\alpha - \beta_L) + \alpha_{11}(\alpha + \beta_L) + \alpha_{21}(\alpha + \beta_R) \tag{18.123e}$$

$$B_{12} = (\alpha - \beta_R) + \alpha_{12}(\alpha + \beta_L) + \alpha_{22}(\alpha + \beta_R) \tag{18.123f}$$

$$B_{21} = j(1 - \alpha\beta_L) + j\alpha_{11}(1 + \alpha\beta_L) - j\alpha_{21}(1 + \alpha\beta_R) \tag{18.123g}$$

$$B_{22} = -j(1 - \alpha\beta_R) + j\alpha_{12}(1 + \alpha\beta_L) - j\alpha_{22}(1 + \alpha\beta_R) \tag{18.123h}$$

$$\alpha_{11} = \frac{\beta_L - \beta_R}{\beta_L + \beta_R} e^{jk_L 2d\cos\theta_L} \tag{18.123i}$$

$$\alpha_{12} = \frac{2\beta_L}{\beta_L + \beta_R} e^{jk_L d\cos\theta_L} e^{jk_R d\cos\theta_R} \tag{18.123j}$$

$$\alpha_{21} = \frac{2\beta_R}{\beta_L + \beta_R} e^{jk_L d\cos\theta_L} e^{jk_R d\cos\theta_R} \tag{18.123k}$$

$$\alpha_{22} = \frac{\beta_R - \beta_L}{\beta_L + \beta_R} e^{jk_R 2d\cos\theta_R} \tag{18.123l}$$

Solving the problem gives the following equation that relates the incident and reflected waves:

$$\begin{bmatrix} E_\parallel^- \\ E_\perp^- \end{bmatrix} = \begin{bmatrix} B \end{bmatrix} \begin{bmatrix} A \end{bmatrix}^{-1} \begin{bmatrix} E_\parallel^+ \\ E_\perp^+ \end{bmatrix} = \begin{bmatrix} R_{cc} \end{bmatrix} \begin{bmatrix} E_\parallel^+ \\ E_\perp^+ \end{bmatrix} \tag{18.124}$$

Generally R_{cc} is a complete matrix, which confirms the depolarizing properties of the chiral medium. For the special case of the normal incidence, equations are quite simple. The matrix R_{cc} becomes a scalar and can be expressed by

$$R_{cc} = \frac{jZ_0 \tan(\omega\sqrt{\varepsilon\mu}d) + Z}{jZ_0 \tan(\omega\sqrt{\varepsilon\mu}d) - Z} \tag{18.125}$$

where $Z_0 = \sqrt{\mu_0/\varepsilon_0}$ and $Z = \sqrt{\mu/\varepsilon}$.

On this final result, we can notice that the reflection under the normal incidence on a chiral Dällenbach screen does not depend on the chirality coefficient of the slab material.

18.4.4 Reflection and Transmission in Uniaxial Omega Slabs

In this section we will find the reflection and transmission coefficients for plane layers filled with uniaxial omega materials. We consider most interesting for applications the case when the whole structure is uniaxial, i.e., the axis of the medium is orthogonal to the interfaces, so only the one specific direction exists. As has been shown above, the eigenwaves in uniaxial omega media are TE and TM polarized plane waves, which are of course also eigensolutions in simple isotropic media. That observation makes the solution for the reflection and transmission coefficients quite simple. The only complication comes from the fact that the wave admittances in omega media depend on the propagation direction. Thus, formulating the equivalent transmission-line equations one should take this into account (in fact, the transmission-line theory has been extended even for more general propagation properties, see [35]). In particular, the general solution for the transverse field components in the slab should be written as

$$\mathbf{E}_t = \mathbf{A}e^{-j\beta z} + \mathbf{B}e^{j\beta z}, \quad -\mathbf{z}_0 \times \mathbf{H}_t = Y_+\mathbf{A}e^{-j\beta z} + Y_-\mathbf{B}e^{j\beta z} \tag{18.126}$$

where $Y_\pm$ are the admittances for the two propagation directions given by (18.69)–(18.71), and $\mathbf{A}$ and $\mathbf{B}$ are amplitude vectors. Using this representation and the usual solution for simple media at both sides of the slab, the reflection and transmission coefficients can be solved. The result reads

$$R = \frac{(Y_1 - Y_+)(Y_2 + Y_-) - (Y_1 + Y_-)(Y_2 - Y_+)e^{-2j\beta d}}{(Y_1 + Y_+)(Y_2 + Y_-) - (Y_1 - Y_-)(Y_2 - Y_+)e^{-2j\beta d}} \tag{18.127}$$

$$T = \frac{2(Y_- + Y_+)\,Y_1 e^{-j\beta d}}{(Y_1 + Y_+)(Y_2 + Y_-) - (Y_1 - Y_-)(Y_2 - Y_+)e^{-2j\beta d}} \tag{18.128}$$

Here Y_1 and Y_2 are wave admittances in two semi-infinite isotropic spaces which surround the slab (the incident wave comes from medium 1). β is the propagation factor in the slab given by (18.68), and d is the slab thickness.

18.4.5 Zero-Reflection Condition. Omega Slabs on Metal Surface

For an omega slab on an ideally conducting surface the reflection coefficient can be found as a special case of the general formula. The result is

$$R = \frac{Y_1 - Y_+ - (Y_1 + Y_-)e^{-2j\beta d}}{Y_1 + Y_+ - (Y_1 - Y_-)e^{-2j\beta d}} \tag{18.129}$$

Considering plane-wave reflection from a uniaxial omega slab in air ($Y_1 = Y_2$) one can easily see that the reflection coefficient equals zero when the wave admittance Y_+ equals the free-space admittance Y_1. This means that the characteristic admittance for the waves traveling in the slab in the direction of the incident plane wave matches the impedance of the waves in free space. Obviously, in this case there is no reflection on both the interfaces. For the TM-polarized fields the matching requirement is satisfied when

$$\Omega = \frac{j}{2\cos\phi}\left(\mu_t - \varepsilon_t \cos^2\phi - \frac{1}{\varepsilon_n}\sin^2\phi\right) \tag{18.130}$$

and for the TE-polarized fields the no-reflection condition reads

$$\Omega = \frac{j}{2\cos\phi}\left(\mu_t \cos^2\phi - \varepsilon_t + \frac{1}{\mu_n}\sin^2\phi\right) \tag{18.131}$$

Here, ϕ is the incidence angle. For the normal incidence $\phi = 0$ and both these requirements simplify to

$$\Omega = \frac{j}{2}(\mu_t - \varepsilon_t) \tag{18.132}$$

The transmission coefficient for a slab with the parameters satisfying (18.130) or (18.131) is an exponential function $T = e^{-j\beta d}$. We can conclude that for a slab with given arbitrary permittivity parameters and any permeabilities there exists a specific value of the coupling parameter Ω, such that the reflection coefficient is zero for any given polarization and an arbitrary incidence angle. The effect is specific for materials modeled by non-symmetric transmission-line equations. Indeed, any symmetric admittances reduce to the known formulas for simple media with $\Omega = 0$.

The matching condition can be satisfied for one propagation direction only. If a wave is incident on the opposite side of the slab, the reflection coefficient is not zero, so the reflective properties depend on which side of the slab is illuminated. Since the filling composite is reciprocal, the transmission coefficients do not depend on that. This means that the losses are different in the two cases. The waves which come from the side where the interface is matched suffer much higher attenuation in the slab compared to those which come from the opposite side.

For the slab on a metal surface the reflection cannot be made exactly zero because some energy will be eventually reflected from the metal boundary. However, if the air–material interface is matched, the absorption increases because all the energy is transmitted inside the lossy region. In that case the absorption simply exponentially increases with the increase in the slab thickness or in the loss factor.

18.5 FUTURE DEVELOPMENTS AND APPLICATIONS

As has been shown in this chapter, the first-order spatial dispersion in non-local media shows many new and interesting phenomena which can be useful for applications. In the most general case, the first-order spatial dispersion can be modeled by the bianisotropic constitutive relations whose four constitutive dyadics are only restricted by such general laws as the reciprocity conditions, Onsager-Casimir and Kramers-Krönig relations. The classification of reciprocal bianisotropic media reveals that such general media can be constructed using just two basic inclusion shapes (just two "brick forms"): idealized uniaxial helices and "hats" formed of two omega particles.

With the first type of inclusions we can make chiral and pseudochiral media. In these materials, the eigenwaves are circularly (or, in anisotropic media, elliptically) polarized. The main electromagnetic effect is the polarization plane rotation in transmitted field (called optical or electromagnetic activity). In contrast to chiral media, in uniaxial omega composites the eigenwaves are linearly polarized. The most interesting peculiarity of omega media is that the wave admittances and impedances are different for plane waves traveling in the opposite directions. Using this property, it is possible to choose the additional parameter so as the air/omega slab interface is matched for the waves coming from one side of the slab at the normal incidence. Such slab acts as a good absorber of electromagnetic energy.

Considerable progress has been made in practical realizations and measurements of isotropic chiral samples. Such samples have been manufactured in several laboratories. Special methods have been developed to allow extraction of the effective material parameters from the experimental data. Both free-space [36] and waveguide [37] measurement techniques have been applied.

Regarding potential applications of chiral materials as absorbers, it is possible to conclude that the same or very similar effect can be reached using racemic (i.e., non-chiral in the average) mixtures of helices [10]. In other words, the macroscopic chirality of a composite does not influence the normal-incidence reflection (however, one should remember that the permittivity and permeability significantly depend on the inclusion shape). The macroscopic chirality influences reflection at oblique incidence angles, and some polarization transformation takes place in this case. This fact suggests that chirality may possibly significantly change scattering properties of wedge-shaped bodies, but this problem is still not properly understood (see some results in [38]). The other possible application of chiral materials is in antenna techniques for polarization control of radiation. For example, polarization correction lenses can be made using chiral materials. Or indeed, polarization transformers which are able to transform any given polarization state into any other state can be designed.

The effect of the interface matching for omega slabs has been experimentally observed in [39]. It has been shown that the matching condition can be realized at a certain frequency close to the resonant frequency of omega inclusions.

Recently it has been shown that the use of media with still stronger spatial dispersion gives a possibility to physically realize slabs perfectly matched with free space for arbitrary incidence angles (as in the perfectly matched layers used in finite difference numerical techniques) [40]. These highly dispersive materials are non-local composites whose inclusion size is still larger as compared to that in bianisotropic composites. The analysis of such complicated materials needs more development (see, e.g., [41], [1]) before any practical conclusion can be drawn. These studies are also relevant to the problems of understanding the properties of composites with stick-shaped inclusions, since the size of such inclusions can be significant, and spatial dispersion determines the composite properties.

REFERENCES

[1] V. M. Agranovich and V. L. Ginzburg, *Kristal Optics with Spatial Dispersion, and Exitons*, 2nd edn, Springer-Verlag, Berlin, 1984.

[2] J. A. Kong, *Electromagnetic Wave Theory*, John Wiley and Sons, New York, 1986.

[3] F. I. Fedorov, *Theory of Gyrotropy*, Nauka i Tekhnika, Minsk, 1976 (in Russian).

[4] D. L. Jaggard and N. Engheta, "Chirosorb™ as an invisible medium," *Electronics Lett.*, **25**, 173–174, 1989; see also *Electronics Lett.*, **25**, 1060–1061, 1989.

[5] V. K. Varadan, V.V. Varadan, and A. Lakhtakia, "On the Possibility of Designing Anti-reflection Coating Using Chiral Composites," *J. Wave-Mater. Interact.*, **2**, 71–81, 1987.

[6] F. Mariotte and J. P. Parneix (eds.), *Proceedings of 3rd International Workshop on Chiral, Bi-isotropic and Bi-anisotropic Media, Chiral'94*, Périgueux, May 1994.

[7] F. Mariotte, S. A. Tretyakov, and B. Sauviac, "Modeling Effective Properties of Chiral Composites," invited paper in *IEEE Antennas and Propagation Magazine*, **39** (n°2), pp. 22–32, April 1996.

[8] N. Engheta and D. L. Jaggard, "Electromagnetic Chirality and Its Applications," *IEEE Antennas and Propagation Society Newsletter*, October, 6–12, 1988.

[9] I. V. Lindell, A. H. Sihvola, S. A. Tretyakov, and A. J. Viitanen, *Electromagnetic Waves in Chiral and Bi-isotropic Media*, Artech House, Boston and London, 1994.

[10] S. A. Tretyakov, A. A. Sochava, and C. R. Simovski, "Influence of Chiral Shapes of Individual Inclusions on the Absorption in Chiral Composite Coatings," *Electromagnetics*, **16**, (2), pp. 113–127, 1996.

[11] A. Priou, A. Sihvola, S. Tretyakov, and A. Vinogradov, eds., *Advances in Complex Electromagnetic Materials*, Kluwer, 1997, *Proceedings of Bianisotropics'97*, University of Glasgow, U.K., 1997; *Bianisotropics'98, Seventh International Conference on Complex Media*, Braunschweig, Germany, 1998.

[12] Proceedings of Bi-isotropics'93, Workshop on Novel Microwave Materials, *Helsinki University of Technology, Electromagnetics Laboratory Report Series*, No. 137, February 1993; Proceedings of Bianisotropics'93, International Seminar on the Electrodynamics of Chiral and Bianisotropic Media, *Helsinki University of Technology, Electromagnetics Laboratory Report Series*, No. 159, December 1993.

[13] S. A. Tretyakov, A. H. Sihvola, A. A. Sochava, and C. R. Simovski, "Magnetoelectric Interactions in Bi-anisotropic Media," *J. Electromagn. Waves Applic.*, **12**, pp. 481–497, 1998.

[14] L. D. Landau and E. M. Lifshitz, *Electrodynamics of Continuous Media*, 2nd edn, Pergamon Press, Oxford, 1984.

[15] A. A. Sochava, C. R. Simovski, and S. A. Tretyakov, "Chiral Effects and Eigenwaves in Bi-Anisotropic Omega Structures," in *Advances in Complex Electromagnetic Materials*, A. Priou, A. Sihvola, S. Tretyakov, and A. Vinogradov, eds, Kluwer, 1997, pp. 85–102.

[16] M. Saadoun and N. Engheta, "A Reciprocal Phase Shifter Using Novel Pseudochiral or Ω Medium," *Microw. and Optical Technol. Lett.*, **5**, pp. 184–188, 1992.

[17] S. A. Tretyakov and A. A. Sochava, "Proposed Composite Material for Nonreflecting Shields and Antenna Radomes," *Electron. Lett.*, **29**, pp. 1048–1049, 1993.

[18] A. H. Sihvola and I. V. Lindell, "Bi-isotropic constitutive relations," *Microw. and Optical Technol. Lett.*, **4** (n°3), pp. 391–396, 1991.

[19] E. J. Post, *Formal Structure of Electromagnetics*, North Holland, Amsterdam, 1962.

[20] F. I. Fedorov, "On the Theory of Optical Activity in Crystals—I: The Law of Conservation of Energy and the Optical Activity Tensors," *Optics and Spectroscopy*, **6** (n°1), pp. 49–53, 1959.

[21] F. Guérin, "Contribution à l'étude théorique et expérimentale des matériaux composites chiraux et bianisotropes dans le domaine microonde," Doctoral Thesis Dissertation, University of Limoges, France, 1995.

[22] F. Guérin, "Energy Dissipation and Absorption in Bi-isotropic Media Described by Different Formalisms," *Progress in Electromagnetics Research (PIER 9): Bianisotropic and Bi-isotropic Media and Application*, A. Priou, ed., EMW Publishing, Cambridge, MA, 1994, pp. 31–44.

[23] F. Mariotte, D. Gogny, A. Bourgeade, and F. Farail, "Backscattering of the thin wire helix: analytical model, numerical study and free space measurements: Application to chiral composite modelling," *Proceedings of Bianisotropics'93*, Gomel, October 1993.

[24] F. Mariotte, "Heterogeneous chiral materials: modeling and applications," Doctoral Thesis Dissertation, University of Bordeaux I, France, 1994.

[25] O. Einarsson, "Electromagnetic Scattering by a Thin Finite Wire," *Acta Polytechnica Scandinavia*, **N°23**, Stockholm, 1969.

[26] E. Hallén, "Exact Solution of the Antenna Equation," *Trans. Roy. Inst. Tech.*, **183**, Stockholm, 1961; and *Electromagnetic Theory*, John Wiley and Sons Inc., New York, 1962.

[27] J. P. Héliot, "Prise en compte des problèmes d'antennes dans un code d'équations intégrales," Doctoral Thesis Dissertation in Applied Mathematics, **n°1495b**, University of Bordeaux I, France, 1996.

[28] W. P. King and G. S. Smith with M. Owens and T. T. Wu, *Antennas in Matter: Fundamentals, Theory and Applications*, The MIT Press, Cambridge, MA and London, England, 1981.

[29] P. King, and C. W. Harrison, *Antennas and Waves: A Modern Approach*, The MIT Press, Cambridge, MA, and London, England, 1969.

[30] S. A. Tretyakov, F. Mariotte, C. Simovski, T. Kharina, and J. Ph. Héliot, "Analytical Model

for Chiral Scatterers: Comparison with Numerical and Experimental Data," *IEEE Transactions on Antennas and Propagation*, **AP-44** (7), pp. 1006–1015, July 1996.

[31] *Antenna Handbook: Theory, Applications and Design*, Chapter 7, Y. T. Lo and S. W. Lee, eds, Van Nostrand Reinhold Co., New York, 1988.

[32] C. R. Simovski, S. A. Tretyakov, A. A. Sochava, B. Sauviac, and F. Mariotte, "Antenna Model for Conductive Omega Particles," *J. Electomagn. Waves Applic*, **11** (11), pp. 1509–1530, 1997.

[33] I. V. Lindell, A. H. Sihvola, and K. Suchy, "Six-vector Formalism in Electromagnetics of Bi-anisotropic Media," *J. Electromagn.Waves Applic.*, **9** (7/8), pp. 887–903, 1995.

[34] S. A. Tretyakov, C. R. Simovski, and A. A. Sochava, "The Relation between Co- and Cross-polarizabilities of Small Conductive Bi-anisotropic Particles," in *Advances in Complex Electromagnetic Materials*, A. Priou, A. Sihvola, S. Tretyakov, and A. Vinogradov, eds, Kluwer, 1997, pp. 271–280.

[35] I. V. Lindell, S. A. Tretyakov, and M. I. Oksanen, "Vector Transmission-line and Circuit Theory for Bi-isotropic Layered Structures," *J. Electromagn. Waves Applic.*, **7** (1), pp. 147–173, 1993.

[36] T. V. Guire, V. V. Varadan, and V. K. Varadan, "Influence of Chirality on Reflection of EM Waves by Planar Dielectric Slabs," *IEEE Trans. Electromagn. Compat.*, **32** (4), pp. 300–303, 1990; A. G. Smith, "An Experimental Study of Artificial Isotropic Chiral Media at Microwave Frequencies," PhD Dissertation, University of Stellenbosch, South Africa, 1994; K. W. Whites and C. Y. Chung, "Composite uniaxial bianisotropic chiral materials characterization: comparison of predicted and measured scattering," *J. Electromagn. Waves Applic.*, **11** (3), pp. 371–394, 1997; E. Chung, B. Sauviac, V. Vigneras-Lefebvre, and J.-P. Parneix, "Measurements of Heterogeneous Chiral Slabs in C, X, and Ku Bands," *Bianisotropics'98, Seventh International Conference on Complex Media*, Braunschweig, Germany, pp. 309–312, 1998.

[37] G. Busse and A. Jacob, "Generalized constitutive parameter extraction of chiral material from waveguide measurements," *Bianisotropics'98*, 7th International Conference on Complex Media, Braunschweig, Germany, pp. 305–308, 1998; I. P. Theron, "The Circular Birefrigence of Artificial Chiral Crystals at Microwave Frequencies," PhD Dissertation, University of Stellenbosch, South Africa, 1995.

[38] V. V. Fisanov and S. G. Vashtalov, "Diffraction by a Conducting Wedge in a Chiroplasma," *Bianisotropics'98, Seventh International Conference on Complex Media*, Braunschweig, Germany, pp. 229–232, 1998.

[39] T. G. Kharina, S. A. Tretyakov, A. A. Sochava, C. R. Simovski, and S. Bolioli, "Experimental Studies of Artificial Omega Media," *Electromagnetics*, **18**, (4), pp. 423–437, 1998.

[40] J.-P. Berenger, "A Perfectly Matched Layer for the Absorption of Electromagnetic Waves," *J. Comput. Physics*, **114**, 185–200, 1994.

[41] S. I. Maslovski, C. R. Simovski, and S. A. Tretyakov, "Constitutive Equations for Media with Second-order Spatial Dispersion," *Bianisotropics'98, Seventh International Conference on Complex Media*, Braunschweig, Germany, pp. 197–200, 1998.

INDEX

D

E

F

G

T

ABOUT THE EDITORS

Douglas H. Werner is an associate professor in the Pennsylvania State University Department of Electrical Engineering. He is a member of the Communications and Space Sciences Laboratory (CSSL) and is affiliated with the Electromagnetic Communication Research Laboratory. He is also a senior research associate in the Intelligence and Information Operations Department of the Applied Research Laboratory at Penn State.

Dr. Werner received the B.S., M.S., and Ph.D. degrees in electrical engineering from the Pennsylvania State University in 1983, 1985, and 1989 respectively. He also received the M.A. degree in mathematics there in 1986. Dr. Werner was presented with the 1993 Applied Computational Electromagnetics Society (ACES) Best Paper Award and was the recipient of a 1993 International Union of Radio Science (URSI) Young Scientist Award. In 1994 Dr. Werner received the Pennsylvania State University Applied Research Laboratory Outstanding Publication Award. He has also received several Letters of Commendation from the Pennsylvania State University Department of Electrical Engineering for outstanding teaching and research.

Dr. Werner is an associate editor of *Radio Science*, a senior member of the Institute of Electrical and Electronics Engineers (IEEE), a member of the American

Geophysical Union (AGU), URSI Commissions B and G, the Applied Computational Electromagnetics Society (ACES), Eta Kappa Nu, Tau Beta Pi, and Sigma Xi. He has published numerous technical papers and proceedings articles and is the author of six book chapters. His research interests include theoretical and computational electromagnetics with applications to antenna theory and design, microwaves, wireless and personal communication systems, electromagnetic wave interactions with complex materials, fractal and knot electrodynamics, and genetic algorithms.

Raj Mittra is professor in the Electrical Engineering Department and a senior research scientist at the Applied Research Laboratory of the Pennsylvania State University. He is also the director of the Electromagnetic Communication Research Laboratory, which is affiliated with the Communication and Space Sciences Laboratory of the Electrical Engineering Department. Prior to joining Penn State, he was a professor in the Electrical and Computer Engineering Department at the University of Illinois in Urbana-Champaign.

Dr. Mittra is a Life Fellow of the IEEE, a past-president of the IEEE Antennas and Propagation Society, and he has served as the editor of the *Transactions of the Antennas and Propagation Society*. He won the Guggenheim Fellowship Award in 1965 and the IEEE Centennial Medal in 1994. He has been a visiting professor at Oxford University, Oxford, England, and at the Technical University of Denmark, Lyngby, Denmark. Dr. Mittra is the president of RM Associates, which is a consulting organization that provides services to industrial and governmental organizations, both in the U.S. and abroad. His professional interests include the areas of computational electromagnetics, electromagnetic modeling and simulation of electronic packages, communication antenna design including GPS, broadband antennas, EMC analysis, radar scattering, frequency selective surfaces, microwave and millimeter wave integrated circuits, and satellite antennas.

Currently, Dr. Mittra serves as the North American editor of the journal *Archiv f r Elektronik und bertragungstechnik (AE)*. He has published about 500 journal papers and 25 books or book chapters on various topics related to electromagnetics, antennas, microwaves, and electronic packaging. For the past 15 years, he has directed as well as lectured in numerous short courses on Electronic Packaging and Computational Electromagnetics, both nationally and internationally.